# Units

| SYMBOL | NAME | MEASURE OF | CONVERSION FACTORS |
|---|---|---|---|
| A | ampere | electrical current | 1 C/sec |
| Å | Ångström | length | $10^{-10}$ m, 0.1 nm |
| Bq | becquerel | radioactivity | 1 disintegration/sec, 60 dpm |
| C | coulomb | electrical charge | 1 A sec |
| °C | centigrade degree | temperature | K – 273 |
| Ci | curie | radioactivity | $3.7 \times 10^{10}$ Bq, $2.22 \times 10^{12}$ dpm |
| cm | centimeter | length | $10^{-2}$ m, $10^{7}$ nm |
| cpm | counts/min | radioactivity | dpm × counting efficiency[a] |
| d | dalton | molecular mass | $1.66 \times 10^{-24}$ g |
| | | | ($^{1}/_{12}$ mass of a carbon atom) |
| dpm | disintegrations/min | radioactivity | 0.016 Bq, cpm/counting efficiency[a] |
| g | gram | mass | $6.02 \times 10^{23}$ daltons |
| J | joule | energy | 1 kg $m^2/sec^2$, $10^{7}$ ergs, 0.239 cal |
| K | Kelvin | temperature | °C + 273 |
| kb | kilobase | nucleotides | 1000 bases or base pairs |
| kcal | kilocalorie | energy | 4.18 kilojoules |
| kd | kilodalton | molecular mass | 1000 d |
| kJ | kilojoule | energy | 0.24 kilocalories |
| L | liter | volume | 1000 mL |
| m | meter | length | 100 cm, $10^{9}$ nm |
| M | molar | concentration | moles solute per liter of solution |
| µg | microgram | mass | $10^{-6}$ g |
| min | minute | time | 60 sec |
| mL | milliliter | volume | 1 $cm^3$ |
| mole | mole | number | $6.02 \times 10^{23}$ molecules |
| mV | millivolt | electrical potential | $10^{-3}$ volts |
| N | newton | force | 1 kg $m/sec^2$, 1 J/m, $10^{5}$ dynes |
| nm | nanometer | length | $10^{-9}$ m, 10 Å |
| Pa | pascal | pressure | 1 $N/m^2$, $9.87 \times 10^{-6}$ atm |
| S | siemens | electrical conductance | 1 A/V |
| sec | second | time | 3600 sec/hour; 86,400 sec/day |
| V | volt | electrical potential | 1 W/A, 1 J/C, 1000 mV |
| W | watt | power | 1 J/sec, 1 V A |

[a]See table of radioactive isotopes (inside back cover) for efficiency of counting of specific isotopes.

Molecular Biology of

# THE CELL

Fifth Edition

# The Problems Book

# Molecular Biology of THE CELL

Fifth Edition

# The Problems Book

JOHN WILSON and TIM HUNT

*Garland Science*
Vice President: Denise Schanck
Assistant Editor: Sigrid Masson
Production Editor and Layout: Emma Jeffcock
Master Reviewer: Alastair Ewing, The Open University
Copy Editor: Bruce Goatly
Illustrator: Nigel Orme
Cover Designer: Matthew McClements, Blink Studio, Ltd.
Indexer: Merrall-Ross International, Ltd.
Permissions Coordinator: Mary Dispenza



**John Wilson** received his Ph.D. from the California Institute of Technology and pursued his postdoctoral work at Stanford University. He is currently Professor of Biochemistry and Molecular Biology at Baylor College of Medicine in Houston. His research interests include genetic recombination, genome stability, and gene therapy. He has taught medical and graduate students for many years, co-authored books on immunology, molecular biology, and biochemistry, and received numerous teaching honors, including the Distinguished Faculty Award and Robertson Presidential Award for excellence in education.

**Tim Hunt** received his Ph.D. from the University of Cambridge where he taught biochemistry and cell biology for more than 20 years. In the late 1970s and early 1980s he spent his summers teaching cell and molecular biology at the Marine Biological Laboratory, Woods Hole, Massachusetts. He left Cambridge in 1990 and moved to the Cancer Research UK Clare Hall Laboratories, just outside London, where he works on the control of the cell cycle. He shared the 2001 Nobel Prize in Physiology or Medicine with Lee Hartwell and Paul Nurse.



**Library of Congress Cataloging-in-Publication Data**
Wilson, John H., 1944-
Molecular biology of the cell, 5th edition. A problems approach/John Wilson & Tim Hunt. -- 5th ed.
p. cm.
ISBN 978-0-8153-4110-9 (softcover)
1. Cytology--Problems, exercises, etc. 2. Molecular biology--Problems, exercises, etc. I. Hunt, Tim, 1943- II. Title.
QH581.2.W555 2008
571.6'076--dc22

2007005477

Published by Garland Science, Taylor & Francis Group, LLC, an informa business, 270 Madison Avenue, New York NY 10016, USA, and 2 Park Square, Milton Park, Abingdon, OX14 4RN, UK.

Printed in the United States of America

15 14 13 12 11 10 9 8 7 6 5 4 3 2 1

For Lynda and Sofia
Mary, Celia, and Agnes

# Preface

*"You know, the proper method for inquiring after the properties of things is to deduce them from experiments"*

Isaac Newton, 1672

*The Problems Book* aims to provide a running commentary for *Molecular Biology of the Cell, Fifth Edition* by Alberts et al. As we wrote in earlier prefaces, we would like to stimulate our readers to ask questions as well as to accept, digest, and learn the stories that 'the big book' tells. In real life, however, knowledge and understanding come from research, which entails curiosity, puzzlement, doubt, criticism, and debate. Groping one's way through the fog of uncertainty during a research project is a slow and often discouraging process; eureka moments (even if one is lucky) are few and far between. Nevertheless, those moments catch the essence of the drama, and we have tended to focus on them, where we have been able to cast them in the form of a problem. In this way, for student and teacher alike, we hope to encourage a questioning attitude to biology. Without curiosity there would be neither science nor scientists.

We have been making up problems together for more than twenty years, and the revision leading to this new edition of *The Problems Book* has taken us more than three years. There are several new things about this edition. We are proud to say that its 20 chapters now match the first twenty chapters of *Molecular Biology of the Cell,* which means that there are three entirely new chapters: on microscopy (Chapter 9), on the extracellular matrix and cell–cell interactions (Chapter 19) and on cancer (Chapter 20). Elsewhere, the organization of each chapter has undergone major revision besides minor modifications and additions to existing problems. As before, sections start by listing the terms in bold from *MBoC.* As a simple test of memory and comprehension, we have added a new type of problem, which we call "Definitions," where we ask the reader to identify these terms from a one-sentence description of their meaning. The following "True/False" section consists of a set of simple statements, whose truth the reader must judge and justify. Next come short questions we call "Thought Problems," modeled on the kinds of problems presented in *Essential Cell Biology* also by Alberts et al. Some of these are more challenging than others, some are playful, some are serious, but all are designed to make the reader think. After this comes a section called "Calculations," which is designed to help deal with quantitative aspects of cell biology. The calculations in this book are mostly very straightforward, usually involving no more than the interconversion of units, yet they provide a solid framework for thinking about the cell. Are cell-surface receptors sparse in the plasma membrane, or jam-packed? Do molecules diffuse across a cell slowly, or in the blink of an eye? Does chromatin occupy most of the nuclear volume, or just a tiny fraction? Numerical analysis of such questions is very important if one is to gain a feel for the molecular basis of cell biology. Last but not least, the "Data Handling" section contains research-based problems, which arguably form the most important part of the book. Our original brief was to compose problems based on experiments so as to allow readers to get a better feel for the way in which biological knowledge is obtained. It is tremendously important to keep asking, "How do we know that? What's the evidence?" or to wonder how one might go about finding something out. Often it's not at all obvious, often the initial breakthrough was a lucky chance observation, made while investigating some completely different business. In fact, it takes most of us years of research experience to grasp the idea of how one simple fact "can illuminate a distant area, hitherto dark" (Boveri, 1902). Seeing how these tiny shards of evidence give rise to the big picture often involves considerable imagination, as well as a certain discipline, to know how much weight the evidence will bear. We hope we have sometimes, at least, been able to capture the essence of how experiments lead to understanding. To do justice to the authors of the experiments we use in these problems, however, we strongly recommend recourse to the original papers, whose references we always provide.

We hope that the organization and classification of problems will help both student and teacher to find what they are looking for. As far as possible, the order of questions closely follows *Molecular Biology of the Cell.*

Another big change in this edition pleases us very much. For this edition, we have chosen to include the answers to every problem on the CD that comes with this book. We think this is a thoroughly good thing for readers. Many of these problems are difficult to answer, and are not really intended to be set as tests. Rather, we hope that readers will be intrigued (as we were) by the questions we ask, and after thinking a bit will want to see what the answer is, what form the discussion takes, how to get at thinking about this particular kind of a problem. Having used these problems ourselves, we know that even a problem with an answer can serve as the basis for a stimulating discussion in class. And if students are told in advance that a few problems from a larger set will be on an exam, they will be motivated to grapple with the reasoning behind *all* the answers—a lot of learning.

Another departure from previous editions is that a selection of questions from *The Problems Book* now appears in *Molecular Biology of the Cell* at the end of each chapter. We picked these problems in consultation with the authors of *MBoC* to highlight important issues in the text and to cover the range of problem styles. We are pleased with the final selections; they include some of our all-time favorites. The solutions to all these problems are printed in a separate section at the end of *The Problems Book.* We hope that many more readers of the main textbook will try working problems as a result of this change.

As always, we want to hear from our readers, for despite our best efforts, we do not always get things right. Please email John Wilson at jwilson@bcm.edu or Tim Hunt at tim.hunt@cancer.org.uk with your comments or queries, and we'll do our best to answer them.

# Acknowledgments

We are slightly bemused to find that our rate of production, averaged over the history of this project, stands at about one chapter per year. But even this glacial progress would not have been possible without a tremendous amount of help from friends and colleagues whose names are recorded in previous editions of The Problems Book, which appeared in 1989, 1994 and 2002. The three new chapters in this fourth edition had special help from Ralph Baeirlein, who explained in terms we could understand why light slows down in glass, from Richard Peto, who explained how Richard Doll and his colleagues first found the connection between smoking and lung cancer, and from Robin Weiss for help with the cellular transmission of tumors in dogs and Tasmanian devils. Doug Sipp at the Center for Developmental Biology in Kobe provided reprints that were unavailable online. We are grateful to Martin Rees for explaining how to estimate the size of the known universe, to Russ Doolittle, who showed us how to construct simple phylogenetic trees based on protein sequences, and to Niles Eldridge and David Kohn, who helped us to understand Charles Darwin's first (theoretical) sketch of the tree of life. As has been the case from the very first edition, we owe a huge debt to Alastair Ewing of the Open University, who worked through almost all the problems, new and old, discovering embarrassing mistakes and finding better, clearer and more graceful ways of putting things. Denise Schanck has been a tower of strength, as always, and Emma Jeffcock a brilliant designer, coordinator and friend throughout this edition. Mike Morales helped to set up an instant home-away-from-home during meetings in California, and his cheerful humor was much appreciated. Adam Sendroff, who took care of publicity and gave us useful audience feedback, was unfailingly supportive. We are especially grateful to all the authors of *Molecular Biology of the Cell,* who have been extremely helpful in the selection and refinement of problems that appear in the main text. We thank them most warmly for their suggestions. Once again, Nigel Orme has been a great help with the illustrations, particularly in adding color to the selection of images that appear in *Molecular Biology of the Cell,* Fifth Edition.

# A Couple of Things to Know

## Avogadro's Number ($6.02 \times 10^{23}$ molecules/mole)

Avogadro's number ($N$) is perhaps the most important constant in molecular sciences, and it appears again and again in this book. Do you know how it was determined? We didn't, or had forgotten if we ever knew. How can one measure the number of molecules in a mole? And who did it first? You will not find this information in modern biology books, partly because it is ancient history, and partly because it was the business of physicists; some pretty good physicists too, as we shall see.

Amadeo Avogadro had no idea how many molecules there were in 22.4 L of a gas. His hypothesis, presented in 1811, was simply that equal volumes of all gases contained the same number of molecules, irrespective of their size or density. Not until much later, when the reality of molecules was more widely accepted and the microscopic basis for the properties of gases was being worked out, were the first estimates attempted. An Austrian high school teacher called Josef Loschmidt used James Clerk Maxwell's recently developed kinetic theory of gases to estimate how many molecules there were in a cubic centimeter of air. Maxwell had derived an expression for the viscosity of a gas, which is proportional to the density of the gas, to the mean velocity of the molecules, and to their mean free path. The latter could be estimated if one knew the size and number of the molecules. Loschmidt simply made the assumption that when a gas was condensed into a liquid, its molecules were packed as closely as they could be, like oranges in a display on a fruit stand, and from this he was able to get a pretty accurate value for Avogadro's number. Not surprisingly, in Austria they often refer to $N$ as 'Loschmidt's number.' In fact, it wasn't until 1909 that the term 'Avogadro's number' was suggested by Jean Perrin, who won the 1926 Nobel prize for physics (his lecture is available on the Nobel web site, and his book on Atoms [Les Atomes, 1913, translated from the original French by D. Ll. Hammick, reprinted in 1990 by Ox Bow Press] is highly recommended—and accessible—reading. It has been called the finest book on physics of the 20th century).

You may be surprised to discover, as we were, that estimating Avogadro's number was an important topic of Albert Einstein's Ph.D. thesis. Abraham Pais's wonderful biography of Einstein, *Subtle is the Lord* (subtitled *The Science and the Life of Albert Einstein*, 1982 Oxford University Press) devotes Chapter 5, 'The Reality of Molecules,' to this period of the great physicist's life and work. Einstein found three independent ways to estimate $N$: from the viscosity of dilute sucrose solutions, from his analysis of Brownian motion, and from light scattering by gases near the critical point, including the blueness of the sky. Because the sky is five million times less bright than direct sunlight, Avogadro's number is $6 \times 10^{23}$. Isn't that romantic?

But Einstein's was not the last word on the subject. Indeed, according to Pais, he made an "elementary but nontrivial mistake" in his thesis that was later corrected, and it was really Perrin who brought the whole field together with his experiments on Brownian motion. The Nobel presentation speech contains this line:

> "His [Perrin's] measurements on the Brownian movement showed that Einstein's theory was in perfect agreement with reality. Through these measurements a new determination of Avogadro's number was obtained."

For most methods of counting molecules, neither the physics nor the math is easy to follow, but two are simple to understand. The first comes from radioactive decay, and another Nobel prize-winning physicist, Ernest Rutherford. When radium decays, it emits alpha particles, which are helium nuclei. If you can count the radioactive decay events with a Geiger counter and measure the volume of helium emitted, you can estimate Avogadro's number. The second way is much more modern. You can see large proteins and nucleic acids with the aid of an electron microscope.

## Calculations and Unit Analysis

Many of the problems in this book involve calculations. Where the calculations are based on an equation (for example, the Nernst equation or the equation for volume of a sphere), we provide the equation along with a brief explanation of symbols, and often their values. Many calculations, however, involve the conversion of information from one form into another, equivalent form. For example, if the concentration of a protein is $10^{-9}$ M, how many molecules of it would be present in a mammalian nucleus with a volume of 500 μm$^3$? Here, a concentration is given as M (moles/L), whereas the desired answer is molecules/nucleus; both values are expressed as 'number/volume' and the problem is to convert one into the other.

Both kinds of calculation use constants and conversion factors that may or may not be included in the problem. The Nernst equation, for example, uses the gas constant $R$ ($2.0 \times 10^{-3}$ kcal/°K mole) and the Faraday constant $F$ (23 kcal/V mole). And conversion of moles/L to molecules/nucleus requires Avogadro's number $N$ ($6.0 \times 10^{23}$ molecules/mole). All of the constants, symbols, and conversion factors that are used in this book are listed inside the book covers (along with the standard genetic code, the one-letter amino acid code, useful geometric formulas, and data on common radioisotopes used in biology).

For each type of calculation, we strongly recommend the powerful general strategy known as unit analysis (or dimensional analysis). If units (for example, moles/L) are included along with the numbers in the calculations, they provide an internal check on whether the numbers have been combined correctly. If you've made a mistake in your math, the units will not help, but if you've divided where you should have multiplied, for example, the units of the answer will be nonsensical: they will shout 'error.' Consider the conversion of $10^{-9}$ M (moles/L) to molecules/nucleus. In the conversion of moles to molecules, do you multiply $10^{-9}$ by $6 \times 10^{23}$ (Avogadro's number) or do you divide by it? If units are included, the answer is clear.

$$\frac{10^{-9}\,\text{moles}}{\text{L}} \times \frac{6 \times 10^{23}\,\text{molecules}}{\text{mole}} = \frac{6 \times 10^{14}\,\text{molecules}}{\text{L}} \quad \textbf{YES}$$

$$\frac{10^{-9}\,\text{moles}}{\text{L}} \times \frac{\text{mole}}{6 \times 10^{23}\,\text{molecules}} = \frac{1.7 \times 10^{-33}\,\text{mole}^2}{\text{molecules L}} \quad \textbf{NO}$$

Similarly, in the conversion of liters to nuclei, the goal is to organize the conversion factors to transform the units to the desired form.

$$\frac{6 \times 10^{14}\,\text{molecules}}{\text{L}} \times \frac{1\,\text{L}}{1000\,\text{mL}} \times \frac{\text{mL}}{\text{cm}^3} \times \frac{\text{cm}^3}{(10^4\,\mu\text{m})^3} \times \frac{500\,\mu\text{m}^3}{\text{nucleus}} = \frac{300\,\text{molecules}}{\text{nucleus}}$$

If you do this calculation with pure numbers, you must worry at each step whether to divide or multiply. If you attach the units, however, the decision is obvious. It is important to realize that any set of (correct) conversion factors will give the same answer. If you are more comfortable converting liters to ounces, that's fine, so long as you know a string of conversion factors that will ultimately transform ounces to μm$^3$.

There are a few simple rules for handling units in calculations.

1. Quantities with different units cannot be added or subtracted. (You cannot subtract 3 meters from 10 kcal.)
2. Quantities with different units can be multiplied or divided; just multiply or divide the units along with the numbers. (You can multiply 3 meters times 10 kcal; the answer is 30 kcal meters.)
3. All exponents are unitless. (You can't use $10^{6\,\text{mL}}$.)
4. You cannot take the logarithm of a quantity with units.

Throughout this book, we have included the units for each element in every calculation. If the units are arranged so that they cancel to give the correct units for the answer, the numbers will take care of themselves.

# Contents


## Chapter 1 Cells and Genomes — 1

THE UNIVERSAL FEATURES OF CELLS ON EARTH — 1
THE DIVERSITY OF GENOMES AND THE TREE OF LIFE — 4
GENETIC INFORMATION IN EUCARYOTES — 6

## Chapter 2 Cell Chemistry and Biosynthesis — 11

THE CHEMICAL COMPONENTS OF A CELL — 11
CATALYSIS AND THE USE OF ENERGY BY CELLS — 23
HOW CELLS OBTAIN ENERGY FROM FOOD — 30

## Chapter 3 Proteins — 39

THE SHAPE AND STRUCTURE OF PROTEINS — 39
PROTEIN FUNCTION — 47

## Chapter 4 DNA, Chromosomes, and Genomes — 63

THE STRUCTURE AND FUNCTION OF DNA — 63
CHROMOSOMAL DNA AND ITS PACKAGING IN THE CHROMATIN FIBER — 65
THE REGULATION OF CHROMATIN STRUCTURE — 72
THE GLOBAL STRUCTURE OF CHROMOSOMES — 76
HOW GENOMES EVOLVE — 81

## Chapter 5 DNA Replication, Repair, and Recombination — 87

THE MAINTENANCE OF DNA SEQUENCES — 87
DNA REPLICATION MECHANISMS — 88
THE INITIATION AND COMPLETION OF DNA REPLICATION IN CHROMOSOMES — 96
DNA REPAIR — 103
HOMOLOGOUS RECOMBINATION — 111
TRANSPOSITION AND CONSERVATIVE SITE-SPECIFIC RECOMBINATION — 114

## Chapter 6 How Cells Read the Genome: From DNA to Protein — 119

FROM DNA TO RNA — 119
FROM RNA TO PROTEIN — 134
THE RNA WORLD AND THE ORIGINS OF LIFE — 147

## Chapter 7 Control of Gene Expression 151

AN OVERVIEW OF GENE CONTROL 151
DNA-BINDING MOTIFS IN GENE REGULATORY PROTEINS 153
HOW GENETIC SWITCHES WORK 160
THE MOLECULAR GENETIC MECHANISMS THAT CREATE SPECIALIZED CELL TYPES 172
POST-TRANSCRIPTIONAL CONTROLS 182

## Chapter 8 Manipulating Proteins, DNA, and RNA 191

ISOLATING CELLS AND GROWING THEM IN CULTURE 191
PURIFYING PROTEINS 192
ANALYZING PROTEINS 195
ANALYZING AND MANIPULATING DNA 203
STUDYING GENE EXPRESSION AND FUNCTION 214

## Chapter 9 Visualizing Cells 221

LOOKING AT CELLS IN THE LIGHT MICROSCOPE 221
LOOKING AT CELLS AND MOLECULES IN THE ELECTRON MICROSCOPE 228

## Chapter 10 Membrane Structure 231

THE LIPID BILAYER 231
MEMBRANE PROTEINS 238

## Chapter 11 Membrane Transport of Small Molecules and the Electrical Properties of Membranes 243

PRINCIPLES OF MEMBRANE TRANSPORT 243
TRANSPORTERS AND ACTIVE MEMBRANE TRANSPORT 246
ION CHANNELS AND THE ELECTRICAL PROPERTIES OF MEMBRANES 252

## Chapter 12 Intracellular Compartments and Protein Sorting 263

THE COMPARTMENTALIZATION OF CELLS 263
THE TRANSPORT OF MOLECULES BETWEEN THE NUCLEUS AND THE CYTOSOL 266
THE TRANSPORT OF PROTEINS INTO MITOCHONDRIA AND CHLOROPLASTS 274
PEROXISOMES 279
THE ENDOPLASMIC RETICULUM 282

## Chapter 13 Intracellular Vesicular Traffic 289

THE MOLECULAR MECHANISMS OF MEMBRANE TRANSPORT AND THE MAINTENANCE OF COMPARTMENT DIVERSITY 289
TRANSPORT FROM THE ER THROUGH THE GOLGI APPARATUS 296
TRANSPORT FROM THE *TRANS* GOLGI NETWORK TO LYSOSOMES 302
TRANSPORT INTO THE CELL FROM THE PLASMA MEMBRANE: ENDOCYTOSIS 305
TRANSPORT FROM THE *TRANS* GOLGI NETWORK TO THE CELL EXTERIOR: EXOCYTOSIS 310

## Chapter 14 Energy Conversion: Mitochondria and Chloroplasts 315

THE MITOCHONDRION 315

ELECTRON-TRANSPORT CHAINS AND THEIR PROTON PUMPS 322

CHLOROPLASTS AND PHOTOSYNTHESIS 329

THE GENETIC SYSTEMS OF MITOCHONDRIA AND PLASTIDS 336

## Chapter 15 Mechanisms of Cell Communication 343

GENERAL PRINCIPLES OF CELL COMMUNICATION 343

SIGNALING THROUGH G-PROTEIN-COUPLED CELL-SURFACE RECEPTORS (GPCRS) AND SMALL INTRACELLULAR MEDIATORS 352

SIGNALING THROUGH ENZYME-COUPLED CELL-SURFACE RECEPTORS 361

SIGNALING PATHWAYS DEPENDENT ON REGULATED PROTEOLYSIS OF LATENT GENE REGULATORY PROTEINS 367

SIGNALING IN PLANTS 370

## Chapter 16 The Cytoskeleton 373

THE SELF-ASSEMBLY AND DYNAMIC STRUCTURE OF CYTOSKELETAL FILAMENTS 373

HOW CELLS REGULATE THEIR CYTOSKELETAL FILAMENTS 382

MOLECULAR MOTORS 388

THE CYTOSKELETON AND CELL BEHAVIOR 391

## Chapter 17 The Cell Cycle 403

OVERVIEW OF THE CELL CYCLE 403

THE CELL-CYCLE CONTROL SYSTEM 407

S PHASE 410

MITOSIS 413

CYTOKINESIS 427

CONTROL OF CELL DIVISION AND CELL GROWTH 430

## Chapter 18 Apoptosis 439

## Chapter 19 Cell Junctions, Cell Adhesion, and the Extracellular Matrix 445

CADHERINS AND CELL–CELL ADHESION 445

TIGHT JUNCTIONS AND THE ORGANIZATION OF EPITHELIA 449

PASSAGEWAYS FROM CELL TO CELL: GAP JUNCTIONS AND PLASMODESMATA 455

THE BASAL LAMINA 457

INTEGRINS AND CELL-MATRIX ADHESION 459

THE EXTRACELLULAR MATRIX OF ANIMAL CONNECTIVE TISSUES 461

THE PLANT CELL WALL 465

**Chapter 20 Cancer** 469

CANCER AS A MICROEVOLUTIONARY PROCESS 469

THE PREVENTABLE CAUSES OF CANCER 472

FINDING THE CANCER-CRITICAL GENES 473

THE MOLECULAR BASIS OF CANCER-CELL BEHAVIOR 477

CANCER TREATMENT: PRESENT AND FUTURE 481

**Answers to Problems in *Molecular Biology of the Cell*, Fifth Edition** 485

**Credits** 545

**References** 549

**Index** 563

# Problems

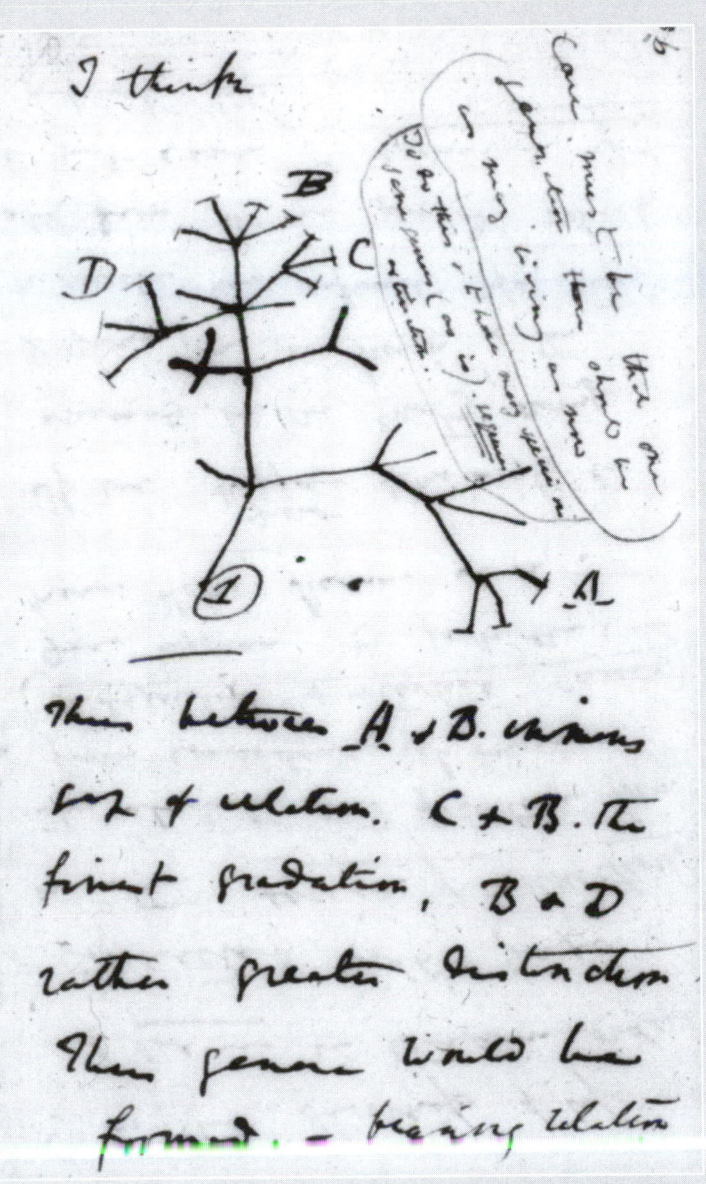


**Darwin's first known (July 1837) sketch of the tree of life.** The writing reads:
"I think"
"Thus between A & B enormous gap of relation. C + B. The finest gradation, B + D rather greater distinction Thus genera would have formed. - bearing relation"

In the bubbles, added later (probably in 1839):
"Case must be that one generation then should be as many living as now"
"To do this & to have many species in same genus (as is) requires extinction."

See Charles Darwin's Notebooks (1836–1844): Geology, Transmutation of Species, Metaphysical Enquiries. Edited with Paul Barrett, Peter Gautrey, Sandra Herbert and Sydney Smith. Ithaca: British Museum (Natural History), Cornell University Press, and Cambridge University Press, 1987. Thanks to David Kohn for the image and helpful comments on its significance.

# Cells and Genomes

# 1

## In This Chapter

THE UNIVERSAL FEATURES OF CELLS ON EARTH 1

THE DIVERSITY OF GENOMES AND THE TREE OF LIFE 4

GENETIC INFORMATION IN EUCARYOTES 6

## THE UNIVERSAL FEATURES OF CELLS ON EARTH

| TERMS TO LEARN | | |
|---|---|---|
| amino acid | messenger RNA (mRNA) | ribonucleic acid (RNA) |
| DNA replication | nucleotide | ribosomal RNA (rRNA) |
| enzyme | plasma membrane | transcription |
| gene | polypeptide | transfer RNA (tRNA) |
| genome | protein | translation |

### DEFINITIONS

Match each definition below with its term from the list above.

**1–1** The selective barrier composed of a lipid bilayer and embedded proteins that surrounds a living cell.

**1–2** A protein that catalyzes a specific chemical reaction.

**1–3** Copying of one strand of DNA into a complementary RNA sequence by the enzyme RNA polymerase.

**1–4** Process by which the sequence of nucleotides in an mRNA molecule directs the incorporation of amino acids into protein.

**1–5** Region of DNA that controls a discrete hereditary characteristic of an organism, usually corresponding to a single protein or RNA.

**1–6** RNA molecule that specifies the amino acid sequence of a protein.

**1–7** Organic molecule containing both an amino group and a carboxyl group. Those in which the amino and carboxyl groups are linked to the same carbon atom serve as the building blocks of proteins.

**1–8** The total genetic information carried by a cell or an organism (or the DNA molecules that carry this information).

### TRUE/FALSE

Decide whether each of these statements is true or false, and then explain why.

**1–9** Genes and their encoded proteins are co-linear; that is, the order of amino acids in proteins is the same as the order of the codons in the RNA and DNA.

**1–10** DNA and RNA use the same four-letter alphabet.

### THOUGHT PROBLEMS

**1–11** 'Life' is easy to recognize but difficult to define. The dictionary defines life as "The state or quality that distinguishes living beings or organisms from dead ones and from inorganic matter, characterized chiefly by metabolism, growth, and the ability to reproduce and respond to stimuli." Biology textbooks usually elaborate slightly; for example, according to a popular text, living things

1. Are highly organized compared with natural inanimate objects.
2. Display homeostasis, maintaining a relatively constant internal environment.
3. Reproduce themselves.
4. Grow and develop from simple beginnings.
5. Take energy and matter from the environment and transform it.
6. Respond to stimuli.
7. Show adaptation to their environment.

Score a car, a cactus, and yourself with respect to these seven characteristics.

**1–12** NASA has asked you to design a module that will identify signs of life on Mars. What will your module look for?

**1–13** You have embarked on an ambitious research project: to create life in a test tube. You boil up a rich mixture of yeast extract and amino acids in a flask along with a sprinkling of the inorganic salts known to be essential for life. You seal the flask and allow it to cool. After several months, the liquid is as clear as ever, and there are no signs of life. A friend suggests that excluding air was a mistake, since most life as we know it requires oxygen. You repeat the experiment, but this time you leave the flask open to the atmosphere. To your great delight, the liquid becomes cloudy after a few days and under the microscope you see beautiful small cells that are clearly growing and dividing. Does this experiment prove that you managed to generate a novel life form? How might you redesign your experiment to allow air into the flask, yet eliminate the possibility that contamination is the explanation for the results?

**1–14** The genetic code (see inside back cover) specifies the entire set of codons that relate the nucleotide sequence of mRNA to the amino acid sequence of encoded proteins. Ever since the code was deciphered nearly four decades ago, some have claimed that it must be a frozen accident, while others have argued that it was shaped by natural selection.

A striking feature of the genetic code is its inherent resistance to the effects of mutation. For example, a change in the third position of a codon often specifies the same amino acid or one with similar chemical properties. But is the natural code more resistant to mutation (less susceptible to error) than other possible versions? The answer is an emphatic "Yes," as illustrated in Figure 1–1. Only one in a million computer-generated 'random' codes is more error-resistant than the natural genetic code.

Does the extraordinary mutation resistance of the genetic code argue in favor of its origin as a frozen accident or as a result of natural selection? Explain your reasoning.

**1–15** You have begun to characterize a sample obtained from the depths of the oceans on Europa, one of Jupiter's moons. Much to your surprise, the sample contains a life-form that grows well in a rich broth. Your preliminary analysis shows that it is cellular and contains DNA, RNA, and protein. When

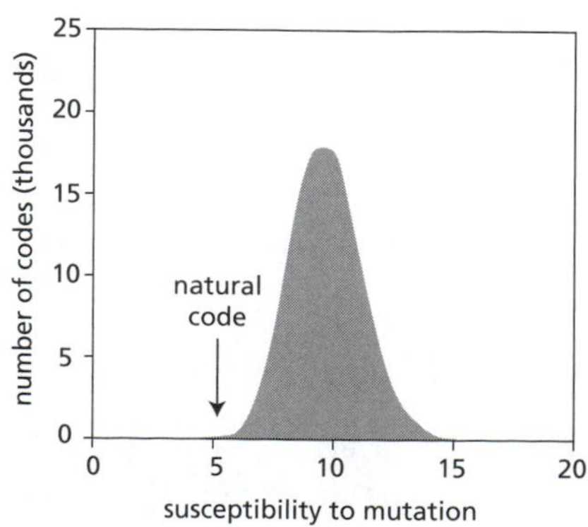


**Figure 1–1** Susceptibility to mutation of the natural code shown relative to that of millions of other computer-generated codes (Problem 1–14). Susceptibility measures the average change in amino acid properties caused by random mutations in a genetic code. A small value indicates that mutations tend to cause only minor changes.

you show your results to a colleague, she suggests that your sample was contaminated with an organism from Earth. What approaches might you try to distinguish between contamination and a novel cellular life-form based on DNA, RNA, and protein?

**1–16** In the 1940s Erwin Chargaff made the remarkable observation that in samples of DNA from a wide range of organisms the mole percent of G [G/(A+T+C+G)] was equal to the mole percent of C, and the mole percents of A and T were equal. This was an essential clue to the structure of DNA. Nevertheless, Chargaff's 'rules' were not universal. For example, in DNA from the virus ϕX174, which has a single-stranded genome, the mole percents are A = 24, C = 22, G = 23, and T = 31. What is the structural basis for Chargaff's rules, and how is it that DNA from ϕX174 doesn't obey the rules?

**1–17** In 1944, at the beginning of his book *What is Life*, the great physicist Erwin Schrödinger (of cat fame) asked the following question: "How can the events *in time and space* which take place within the spatial boundary of a living organism be accounted for by physics and chemistry?" What would be your answer today? Do you think there are peculiar properties of living systems that disobey the laws of physics and chemistry?

**1–18** Which of the following correctly describe the coding relationships (template → product) for replication, transcription, and translation?

A. DNA → DNA
B. DNA → RNA
C. DNA → protein
D. RNA → DNA
E. RNA → RNA
F. RNA → protein
G. Protein → DNA
H. Protein → RNA
I. Protein → protein

## CALCULATIONS

**1–19** An adult human is composed of about $10^{13}$ cells, all of which are derived by cell division from a single fertilized egg.

A. Assuming that all cells continue to divide (like bacteria in rich media), how many generations of cell divisions would be required to produce $10^{13}$ cells?
B. Human cells in culture divide about once per day. Assuming that all cells continue to divide at this rate during development, how long would it take to generate an adult organism?
C. Why is it, do you think, that adult humans take longer to develop than these calculations might suggest?

**1–20** There are about 25,000 genes in the human genome. If you wanted to use a stretch of the DNA of each gene as a unique identification tag, roughly what minimum length of DNA sequence would you need? To be unique, the length of DNA in nucleotides would have to have a diversity (the number of different possible sequences) equivalent to at least 25,000 and would have to be present once in the haploid human genome ($3.2 \times 10^9$ nucleotides)? (Assume that A, T, C, and G are present in equal amounts in the human genome.)

**1–21** Cell growth depends on nutrient uptake and waste disposal. You might imagine, therefore, that the rate of movement of nutrients and waste products across the cell membrane would be an important determinant of the rate of cell growth. Is there a correlation between a cell's growth rate and its surface-to-volume ratio? Assuming that the cells are spheres, compare a bacterium (radius 1 μm), which divides every 20 minutes, with a human cell (radius 10 μm), which divides every 24 hours. Is there a match between the surface-to-volume ratios and the doubling times for these cells? [The surface area of a sphere = $4\pi r^2$; the volume = $(4/3)\pi r^3$.]

# THE DIVERSITY OF GENOMES AND THE TREE OF LIFE

TERMS TO LEARN

| | | |
|---|---|---|
| archaea | model organism | procaryote |
| bacteria | mutation | ortholog |
| eucaryote | paralog | virus |
| homolog | | |

## DEFINITIONS

Match each definition below with its term from the list above.

**1–22** A small packet of genetic material that has evolved as a parasite on the reproductive and biosynthetic machinery of host cells.

**1–23** Organism selected for intensive study as a representative of a large group of species.

**1–24** One of the two divisions of procaryotes, typically found in hostile environments such as hot springs or concentrated brine.

**1–25** A homologous chromosome or, more generally, a macromolecule that has a close evolutionary relationship to another.

**1–26** Living organism composed of one or more cells with a distinct nucleus and cytoplasm.

**1–27** Major category of living cells distinguished by the absence of a nucleus.

## TRUE/FALSE

Decide whether each of these statements is true or false, and then explain why.

**1–28** The vast majority of $CO_2$ fixation into the organic compounds needed for further biosynthesis is carried out by phototrophs.

**1–29** The human hemoglobin genes, which are arranged in two clusters on two chromosomes, provide a good example of an orthologous set of genes.

## THOUGHT PROBLEMS

**1–30** It is not so difficult to imagine what it means to feed on the organic molecules that living things produce. That is, after all, what we do. But what does it mean to 'feed' on sunlight, as phototrophs do? Or, even stranger, to 'feed' on rocks, as lithotrophs do? Where is the 'food,' for example, in the mixture of chemicals ($H_2S$, $H_2$, CO, $Mn^+$, $Fe^{2+}$, $Ni^{2+}$, $CH_4$, and $NH_4^+$) spewed forth from a hydrothermal vent?

**1–31** At the bottom of the seas where hydrothermal vents pour their chemicals into the ocean, there is no light and little oxygen, yet giant (2-meter long) tube worms live there happily. These remarkable creatures have no mouth and no anus, living instead off the excretory products and dead cells of their symbiotic lithotrophic bacteria. These tube worms are bright red because they contain large amounts of hemoglobin, which is critical to the survival of their symbiotic bacteria, and, hence, the worms. This specialized hemoglobin carries $O_2$ and $H_2S$. In addition to providing $O_2$ for its own oxidative metabolism, what role might this specialized hemoglobin play in the symbiotic relationship that is crucial for life in this hostile environment?

**1–32** The overall reaction for the production of glucose ($C_6H_{12}O_6$) by oxygenic (oxygen-generating) photosynthesis,

$$6\ CO_2 + 6\ H_2O + \text{light} \rightarrow C_6H_{12}O_6 + 6\ O_2 \qquad \text{(Equation 1)}$$

was widely interpreted as meaning that light split $CO_2$ to generate $O_2$, and that the carbon was joined with water to generate glucose. In the 1930s a graduate student at Stanford University, C.B. van Neil, showed that the stoichiometry for photosynthesis by purple sulfur bacteria was

$$6\ CO_2 + 12\ H_2S + \text{light} \rightarrow C_6H_{12}O_6 + 6\ H_2O + 12\ S \qquad \text{(Equation 2)}$$

On the basis of this stoichiometry, he suggested that the oxygen generated during oxygenic photosynthesis derived from water, not $CO_2$. His hypothesis was confirmed two decades later using isotopically labeled water. Yet how is it that the 6 $H_2O$ in Equation 1 can give rise to 6 $O_2$? Can you suggest how Equation 1 might be modified to clarify exactly how the products are derived from the reactants?

**1–33** How many possible different trees (branching patterns) can be drawn for eubacteria, archaea, and eucaryotes, assuming that they all arose from a common ancestor?

**1–34** The genes for ribosomal RNA are highly conserved (relatively few sequence changes) in all organisms on Earth; thus, they have evolved very slowly over time. Were such genes 'born' perfect?

**1–35** Several procaryotic genomes have been completely sequenced and their genes have been counted. But how do you suppose one recognizes a gene in a string of Ts, As, Cs, and Gs?

**1–36** Which one of the processes listed below is NOT thought to contribute significantly to the evolution of new genes? Why not?

A. Duplication of genes to create extra copies that can acquire new functions.
B. Formation of new genes *de novo* from noncoding DNA in the genome.
C. Horizontal transfer of DNA between cells of different species.
D. Mutation of existing genes to create new functions.
E. Shuffling of domains of genes by gene rearrangement.

**1–37** Genes participating in informational processes such as replication, transcription, and translation are transferred between species much less often than are genes involved in metabolism. The basis for this inequality is unclear at present, but one suggestion is that it relates to the underlying complexity of the two types of processes. Informational processes tend to involve large aggregates of different gene products, whereas metabolic reactions are usually catalyzed by enzymes composed of a single protein.

A. Archaea are more closely related to eubacteria in their metabolic genes, but are more similar to eucaryotes in the genes involved in informational processes. In terms of evolutionary descent, do you think archaea separated more recently from eubacteria or eucaryotes?
B. Why would the complexity of the underlying process—informational or metabolic—have any effect on the rate of horizontal gene transfer?

**1–38** Why do you suppose that horizontal gene transfer is more prevalent in single-celled organisms than in multicellular organisms?

**1–39** You are interested in finding out the function of a particular gene in the mouse genome. You have sequenced the gene, defined the portion that codes for its protein product, and searched the appropriate databases; however, neither the gene nor the encoded protein resembles anything seen before. What types of information about the gene or the encoded protein would you like to know in order to narrow down the possible functions, and why? Focus on the information you want, rather than on the techniques you might use to get that information.

## CALCULATIONS

**1–40** Natural selection is such a powerful force in evolution because cells with even a small growth advantage quickly outgrow their competitors. To illustrate this process, consider a cell culture that initially contains $10^6$ bacterial cells, which divide every 20 minutes. A single cell in this culture acquires a mutation that allows it to divide with a generation time of only 15 minutes. Assuming that there is an unlimited food supply and no cell death, how long would it take before the progeny of the mutated cell became predominant in the culture? The number of cells $N$ in the culture at time $t$ is described by the equation $N = N_0 \times 2^{t/G}$, where $N_0$ is the number of cells at zero time and $G$ is the generation time. (Before you go through the calculation, make a guess: do you think it would take about a day, a week, a month, or a year?)

# GENETIC INFORMATION IN EUCARYOTES

TERM TO LEARN

genetic redundancy

## TRUE/FALSE

Decide whether each of these statements is true or false, and then explain why.

**1–41** Eucaryotic cells contain either mitochondria or chloroplasts, but not both.

**1–42** Most of the DNA sequences in a bacterial genome code for proteins, whereas most of the sequences in the human genome do not.

**1–43** The only horizontal gene transfer that has occurred in animals is from the mitochondrial genome to the nuclear genome.

## THOUGHT PROBLEMS

**1–44** Animal cells have neither cell walls nor chloroplasts, whereas plant cells have both. Fungal cells are somewhere in between; they have cell walls but lack chloroplasts. Are fungal cells more likely to be animal cells that gained the ability to make cell walls, or plant cells that lost their chloroplasts? This question represented a difficult issue for early investigators who sought to assign evolutionary relationships based solely on cell characteristics and morphology. How do you suppose that this question was eventually decided?

**1–45** Giardiasis is an acute form of gastroenteritis caused by the protozoan parasite *Giardia lamblia*. *Giardia* is a fascinating eucaryote; it contains a nucleus but no mitochondria and no recognizable endoplasmic reticulum or Golgi apparatus—one of the very rare examples of such a cellular organization among eucaryotes. This organization might arise because *Giardia* is an ancient lineage that separated from the rest of eucaryotes before mitochondria were acquired and internal membranes were developed. Or it might be a stripped-down version of a more standard eucaryote that has lost these structures because they are not necessary in the parasitic lifestyle it has adopted. How might you use nucleotide sequence comparisons to distinguish between these alternatives?

**1–46** Rates of evolution appear to vary in different lineages. For example, the rate of evolution in the rat lineage is significantly higher than in the human lineage. These rate differences are apparent whether one looks at changes in nucleotide sequences that encode proteins and are subject to selective pressure or at changes in noncoding nucleotide sequences, which are not under obvious selection pressure. Can you offer one or more possible explanations for the slower rate of evolutionary change in the human lineage versus the rat lineage?

## DATA HANDLING

**1–47** It is difficult to obtain information about the process of gene transfer from the mitochondrial to the nuclear genome in animals because there are few differences among their mitochondrial genomes. The same set of 13 (or occasionally 12) protein genes is encoded in all the numerous animal mitochondrial genomes that have been sequenced. In plants, however, the situation is different, with quite a bit more variability in the sets of proteins encoded in mitochondrial genomes. Analysis of plants can thus provide valuable information on the process of gene transfer.

The respiratory gene *Cox2*, which encodes subunit 2 of cytochrome oxidase, was functionally transferred to the nucleus during flowering plant evolution. Extensive analyses of plant genera have pinpointed the time of appearance of the nuclear form of the gene and identified several likely intermediates in the ultimate loss from the mitochondrial genome. A summary of *Cox2* gene distributions between mitochondria and nuclei, along with data on their transcription, is shown in a phylogenetic context in Figure 1–2.

A. Assuming that transfer of the mitochondrial gene to the nucleus occurred only once (an assumption supported by the structures of the nuclear genes), indicate the point in the phylogenetic tree where the transfer occurred.

B. Are there any examples of genera in which the transferred gene and the mitochondrial gene both appear functional? Indicate them.

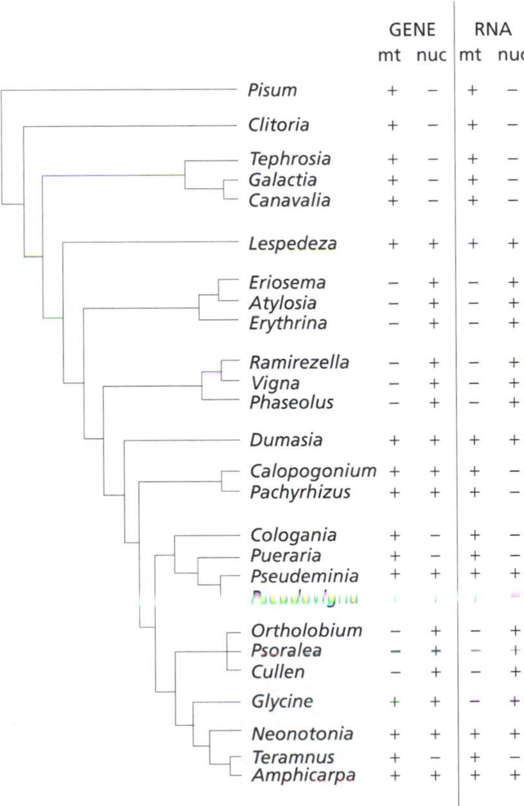

GENE RNA
mt nuc mt nuc
Pisum + – + –
Clitoria + – + –
Tephrosia + – + –
Galactia + – + –
Canavalia + – + –
Lespedeza + + + +
Eriosema – + – +
Atylosia – + – +
Erythrina – + – +
Ramirezella – + – +
Vigna – + – +
Phaseolus – + – +
Dumasia + + + +
Calopogonium + + + –
Pachyrhizus + + + –
Cologania + – + –
Pueraria + – + –
Pseudeminia + + + +
Ortholobium – + – +
Psoralea – + – +
Cullen – + – +
Glycine + + – +
Neonotonia + + + +
Teramnus + – + –
Amphicarpa + + + +


**Figure 1–2** Summary of *Cox2* gene distribution and transcript data in a phylogenetic context (Problem 1–47). The presence of the intact gene or a functional transcript is indicated by (+); the absence of the intact gene or a functional transcript is indicated by (–). mt, mitochondria; nuc, nuclei.

C. What is the minimal number of times that the mitochondrial gene has been inactivated or lost? Indicate those events on the phylogenetic tree.
D. What is the minimal number of times that the nuclear gene has been inactivated or lost? Indicate those events on the phylogenetic tree.
E. Based on this information, propose a general scheme for transfer of mitochondrial genes to the nuclear genome.

**1–48** Although stages in the process of mitochondrial gene transfer can be deduced from studies such as the one in the previous question, there is much less information on the mechanism by which the gene is transferred from mitochondria to the nucleus. Does a fragment of DNA escape the mitochondria and enter the nucleus? Or does the transfer somehow involve an RNA transcript of the gene as the intermediary? The *Cox2* gene provides a unique window on this question. In some species it is found in the mitochondrial genome, in others, in the nuclear genome. The initial transcript of the mitochondrial *Cox2* gene is modified by RNA editing, a process that changes several specific cytosines to uracils. How might this observation allow you to decide whether the informational intermediary in transfer was DNA or RNA? What do you think the answer is?

**1–49** Some genes evolve rapidly, whereas others are highly conserved. But how can we tell whether a gene has evolved rapidly or has simply had a long time to diverge from its relatives? The most reliable approach is to compare several genes from the same two species, as shown for rat and human in Table 1–1. Two measures of rates of nucleotide substitution are indicated in the table. Nonsynonymous changes refer to single nucleotide changes in the DNA sequence that alter the encoded amino acid (ATC → TTC, which is I → F, for example). Synonymous changes refer to those that do not alter the encoded amino acid (ATC → ATT, which is I → I, for example). (As is apparent in the genetic code, inside back cover, individual amino acids are typically encoded by multiple codons.)
A. Why are there such large differences between the synonymous and nonsynonymous rates of nucleotide substitution?
B. Considering that the rates of synonymous changes are about the same for all three genes, how is it possible for the histone H3 gene to resist so effectively those nucleotide changes that alter the amino acid sequence?
C. In principle, a gene might be highly conserved because it exists in a 'privileged' site in the genome that is subject to very low mutation rates. What feature of the data in Table 1–1 argues against this possibility for the histone H3 gene?

**1–50** Plant hemoglobins were found initially in legumes, where they function in root nodules to lower the oxygen concentration so that the resident bacteria can fix nitrogen. These hemoglobins impart a characteristic pink color to the root nodules. When these genes were first discovered, it was so surprising to find a gene typical of animal blood that it was hypothesized that the plant gene arose by horizontal transfer from some animal. Many more hemoglobin genes have now been sequenced, and a phylogenetic tree based on some of these sequences is shown in Figure 1–3.

**Table 1–1 Rates of nucleotide substitutions in three genes from rat and human** (Problem 1–49).

| | | RATES OF CHANGE | |
|---|---|---|---|
| GENE | AMINO ACIDS | NONSYNONYMOUS | SYNONYMOUS |
| Histone H3 | 135 | 0.0 | 4.5 |
| Hemoglobin α | 141 | 0.6 | 4.4 |
| Interferon γ | 136 | 3.1 | 5.5 |

Rates are expressed as nucleotide changes per site per $10^9$ years. The average rate of nonsynonymous changes for several dozen rat and human genes is about 0.8.

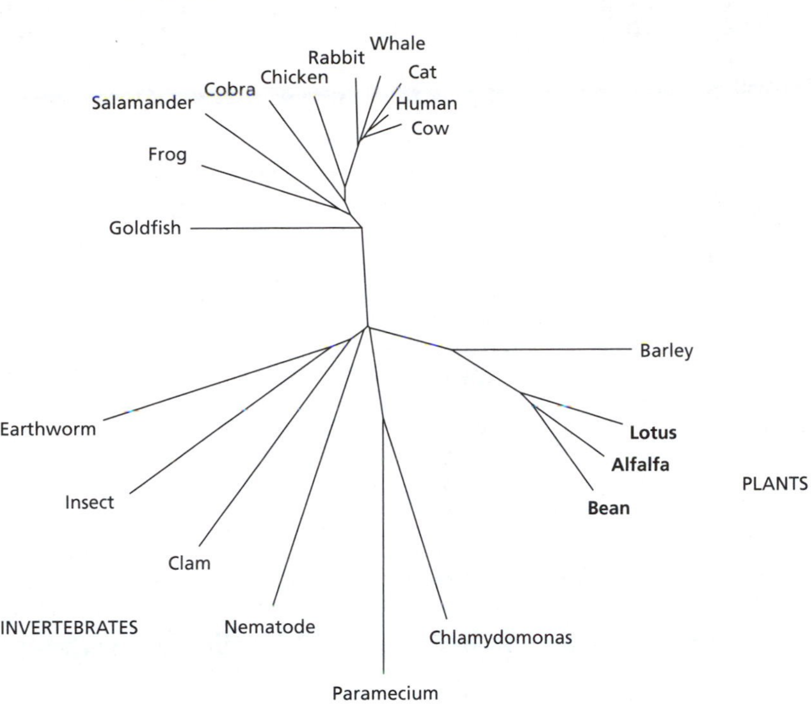


**Figure 1–3** Phylogenetic tree for hemoglobin genes from a variety of species (Problem 1–50). The legumes are shown in *bold*.

A. Does this tree support or refute the hypothesis that the plant hemoglobins arose by horizontal gene transfer?
B. Supposing that the plant hemoglobin genes were originally derived from a parasitic nematode, for example, what would you expect the phylogenetic tree to look like?

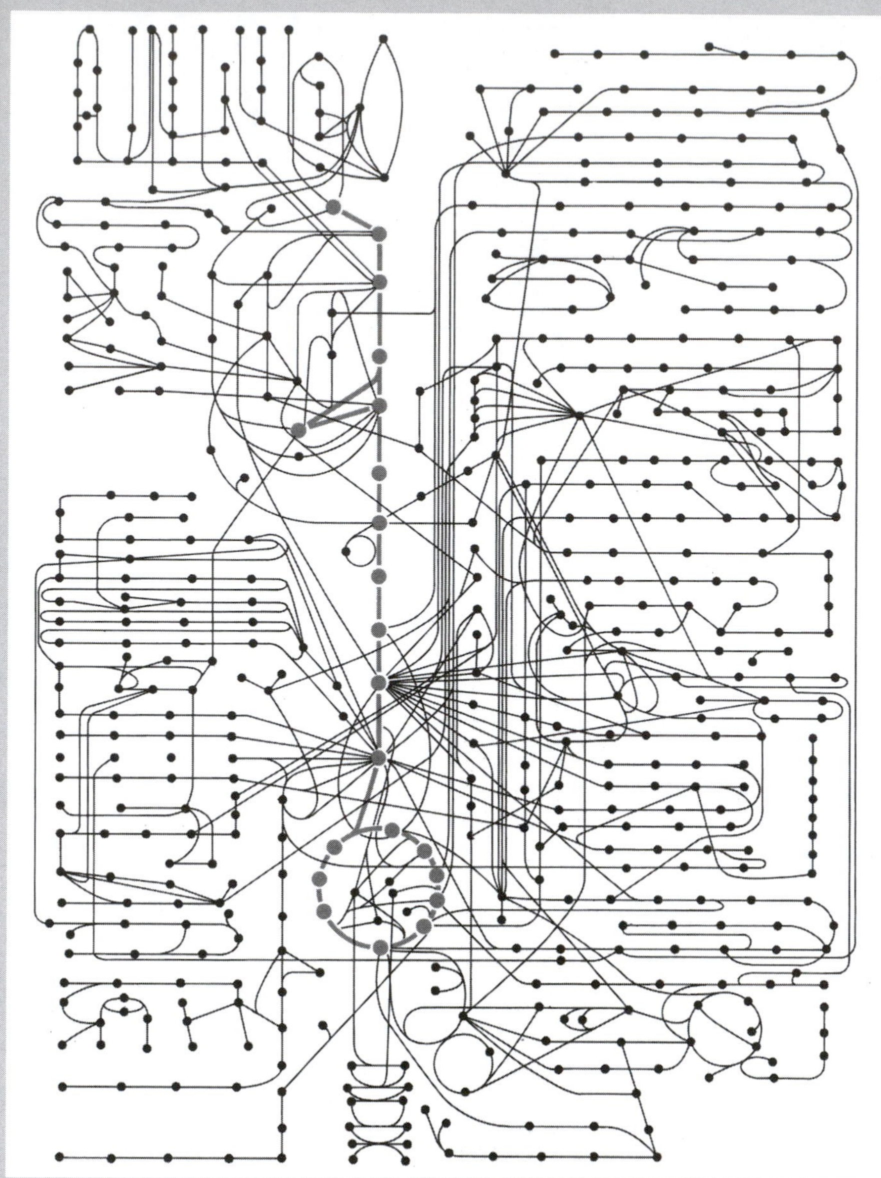

**A map of metabolism.** The aim of classical biochemistry was to explain how food and drink was transformed into flesh and blood. The transformations of simple precursors into more complex molecules, and the breakdown of complex molecules to simpler ones occupied thousands of researchers for at least 50 years. The fruits of this prodigious labor, now augmented by genome sequencing projects, are summarized by diagrams such as these, based on the Biochemical Pathways Wallcharts published by the firm of Boehringer Mannheim (now Roche Biochemicals). These maps, edited by Dr. Gerhard Michal, have a long tradition on the walls of life science laboratories.

# Cells Chemistry and Biosynthesis

# 2

## In This Chapter

THE CHEMICAL COMPONENTS OF A CELL 11

CATALYSIS AND THE USE OF ENERGY BY CELLS 23

HOW CELLS OBTAIN ENERGY FROM FOOD 30

## THE CHEMICAL COMPONENTS OF A CELL

TERMS TO LEARN

acid
adenosine triphosphate (ATP)
atomic weight
Avogadro's number
base
chemical bond
chemical group
condensation reaction
covalent bond
deoxyribonucleic acid (DNA)
electrostatic attraction
hydrogen bond
hydrolysis
hydronium ion ($H_3O^+$)
hydrophilic
hydrophobic
hydrophobic force
lipid
lipid bilayer
mole
molecular weight
molecule
nucleotide
peptide bond
pH scale
phosphodiester bond
polar
protein
proton ($H^+$)
ribonucleic acid (RNA)
sugar
van der Waals attraction

### DEFINITIONS

Match each definition below with its term from the list above.

**2–1** The number of atoms in 1 gram of hydrogen ($6 \times 10^{23}$), and thus in the atomic or molecular weight equivalent in grams of any element or molecule.

**2–2** Force exerted by the hydrogen-bonded network of water molecules that brings two nonpolar surfaces together by excluding water between them.

**2–3** Group of atoms joined together by covalent bonds.

**2–4** Mass of an atom relative to the mass of a hydrogen atom. Essentially equal to the number of protons plus neutrons.

**2–5** Noncovalent bond in which an electropositive hydrogen atom is partially shared by two electronegative atoms.

**2–6** Substance that releases protons when dissolved in water, forming a hydronium ion ($H_3O^+$).

**2–7** Type of (individually weak) noncovalent bond that is formed at close range between nonpolar atoms.

### TRUE/FALSE

Decide whether each of these statements is true or false, and then explain why.

**2–8** Of the original radioactivity in a sample, only about 1/1000 will remain after 10 half-lives.

**2–9** A $10^{-8}$ M solution of HCl has a pH of 8.

**2–10** Strong acids bind protons strongly.

**2–11** Most of the interactions between macromolecules could be mediated just as well by covalent bonds as by noncovalent bonds.

## THOUGHT PROBLEMS

**2–12** The organic chemistry of living cells is said to be special for two reasons: it occurs in an aqueous environment and it accomplishes some very complex reactions. But do you suppose it's really all that much different from the organic chemistry carried out in the top laboratories in the world? Why or why not?

**2–13** The mass of a hydrogen atom—and thus of a proton—is almost exactly 1 dalton. If protons and neutrons have virtually identical masses, and the mass of an electron is negligible, shouldn't all elements have atomic weights that are nearly integers? A perusal of the periodic table shows that this simple expectation is not true. Chlorine, for example, has an atomic weight of 35.5. How is it that elements can have atomic weights that are not integers?

**2–14** A carbon atom contains six protons and six neutrons.

A. What are its atomic number and atomic weight?

B. How many electrons does it have?

C. How many additional electrons must it add to fill its outermost shell? How does this affect carbon's chemical behavior?

D. Carbon with an atomic weight of 14 is radioactive. How does it differ in structure from nonradioactive carbon? How does this difference affect its chemical behavior?

**2–15** A few of the radioactive isotopes that are commonly used in biological experiments are listed in Table 2–1, along with some of their properties.

A. How does each of these unstable isotopes differ in atomic structure from the most common isotope for that element; that is, $^{12}C$, $^{1}H$, $^{32}S$, and $^{31}P$?

B. $^{32}P$ decays to a stable structure by emitting a β particle—an electron—according to the equation: $^{32}P \rightarrow {}^{32}S + e^-$. The product sulfur atom has the same atomic weight as the radioactive phosphorus atom. What has happened?

C. $^{14}C$, $^{3}H$, and $^{35}S$ also decay by emitting an electron. (The electron can be readily detected, which is one reason why these isotopes are so useful in biology.) Write the decay equations for each of these radioactive isotopes and indicate whether the product atom is the most common isotope of the element generated by the decay.

D. Would you expect the product atom in each of these reactions to be charged or uncharged? Explain your answer.

**2–16** Imagine that a $^{32}P$-phosphate has been incorporated into the backbone of DNA. When the atom decays, would you expect the DNA backbone to remain intact? Why or why not?

**2–17** As indicated in Table 2–1, the time it takes for half of a population of radioactive atoms to decay (their half-life) ranges from about two weeks for $^{32}P$ to more than 5000 years for $^{14}C$. Imagine that you created two atoms of $^{32}P$ today. When would you expect the first atom to decay? When would the second atom decay?

**2–18** Specific activity refers to the amount of radioactivity per unit amount of substance, most commonly in biology expressed on a molar basis, for example,

**Table 2–1 Radioactive isotopes and some of their properties** (Problem 2–15).

| RADIOACTIVE ISOTOPE | EMISSION | HALF-LIFE | MAXIMUM SPECIFIC ACTIVITY (Ci/mmol) |
|---|---|---|---|
| $^{14}C$ | β particle | 5730 years | 0.062 |
| $^{3}H$ | β particle | 12.3 years | 29 |
| $^{35}S$ | β particle | 87.4 days | 1490 |
| $^{32}P$ | β particle | 14.3 days | 9120 |

as Ci/mmol. [One curie (Ci), which is the standard unit of radioactive decay, corresponds to $2.22 \times 10^{12}$ disintegrations per minute (dpm).] If you examine Table 2–1, you will see that there seems to be an inverse relationship between maximum specific activity and half-life. Do you suppose this is just a coincidence or is there an underlying reason? Explain your answer.

**2–19** C, H, and O account for 95% of the elements in living organisms (Figure 2–1). These atoms are present in the ratio C:2H:O, which is equivalent to the general formula for carbohydrates ($CH_2O$). Does this mean living organisms are mostly sugar? Why or why not?

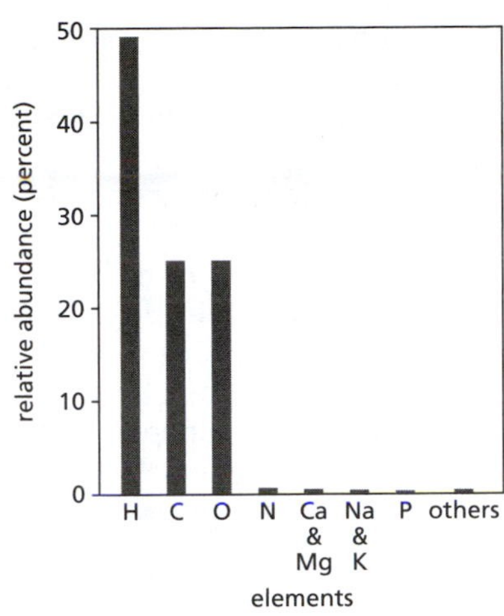


**Figure 2–1** Abundance of elements in living organisms (Problem 2–19).

**2–20** The chemical properties of elements are determined by the behavior of electrons in their outer shell.
A. How many electrons can be accommodated in the first, second, and third electron shells of an atom?
B. How many electrons would atoms of H, C, N, O, P, and S preferentially gain or lose to obtain a completely filled outer electron shell?
C. What are the valences of H, C, N, O, P, and S?

**2–21** Order the following list of processes in terms of their energy content from smallest to largest.
A. ATP hydrolysis in cells.
B. Average thermal motions.
C. C–C bond.
D. Complete oxidation of glucose.
E. Noncovalent bond in water.

**2–22** Why are polar covalent bonds and the resulting permanent dipoles so important in biology?

**2–23** Hydrogen bonds and van der Waals attractions are important in the interactions between molecules in biology.
A. Describe the differences and similarities between van der Waals attractions and hydrogen bonds.
B. Which of the two types of interactions would form (1) between two hydrogens bound to carbon atoms, (2) between a nitrogen atom and a hydrogen bound to a carbon atom, and (3) between a nitrogen atom and a hydrogen bound to an oxygen atom?

**2–24** Oxygen and sulfur have similar chemical properties because both elements have six electrons in their outermost electron shells. Indeed, both oxygen and sulfur form molecules with two hydrogen atoms: water ($H_2O$) and hydrogen sulfide ($H_2S$) (Figure 2–2). Surprisingly, water is a liquid, yet $H_2S$ is a gas, even though sulfur is much larger and heavier than oxygen. Propose an explanation for this striking difference.

**2–25** What do you think the ‘p’ in pH stands for?

**2–26** Imagine that you put some crystals of sodium chloride, potassium acetate, and ammonium chloride into separate beakers of water. Predict whether the pH values of the resulting solutions would be acidic, neutral, or basic. Explain your reasoning.

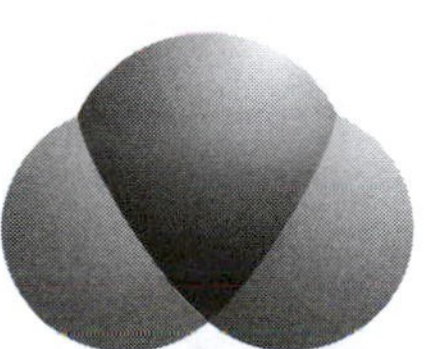

water ($H_2O$)

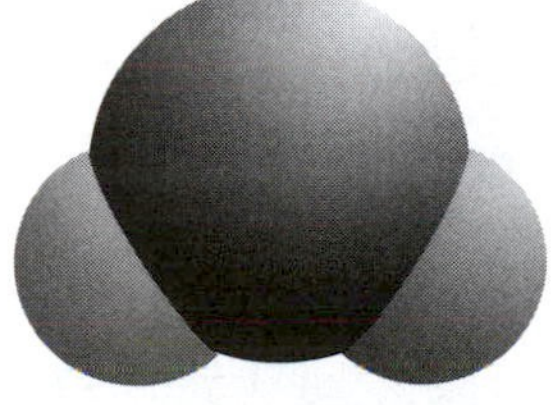

hydrogen sulfide ($H_2S$)

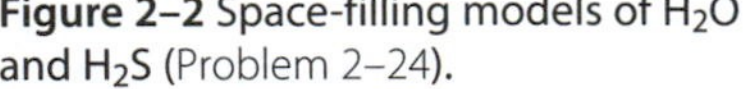

**Figure 2–2** Space-filling models of $H_2O$ and $H_2S$ (Problem 2–24).

**2–27** The amino acid glycine ($H_2NCH_2COOH$) has two ionizable groups: the carboxylic acid group (–COOH) and the basic amine group ($-NH_3^+$). Adding NaOH to a solution of glycine at pH = 1 gives the titration curve shown in Figure 2–3.

A. Write the expressions ($HA \rightleftharpoons H^+ + A^-$) for dissociation of the carboxylic acid (–COOH) and amine groups ($-NH_3^+$).

B. Recall that p*K* is the pH at which exactly half of the carboxylic acid or amine groups are charged. Estimate the p*K* values for the carboxylate and amine groups of glycine.

C. Indicate the predominant ionic species of glycine at each point shown on the curve in Figure 2–3.

D. The isoelectric point of a solute is the pH at which it carries no net charge. Estimate the isoelectric point for glycine from the curve in Figure 2–3.

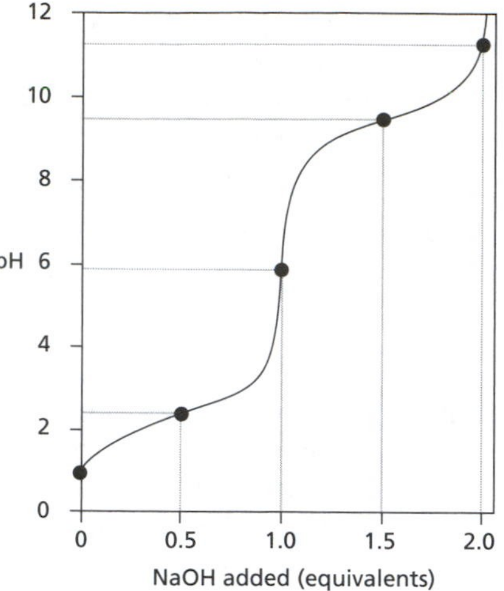


**Figure 2–3** Titration of a solution of glycine (Problem 2–27). One equivalent of $OH^-$ is the amount required to completely neutralize one acidic group.

**2–28** If you want to order glycine from a chemical supplier, you have three choices: glycine, glycine sodium salt, and glycine hydrochloride. Write the structures of these three compounds.

**2–29** From the p*K* values listed in Table 2–2, decide which amino acids were used in the titration curves shown in Figure 2–4.

**2–30** Suggest a rank order for the p*K* values (from lowest to highest) for the carboxyl group on the aspartate side chain in the following environments in a protein. Explain your ranking.

1. An aspartate side chain on the surface of a protein with no other ionizable groups nearby.
2. An aspartate side chain buried in a hydrophobic pocket on the surface of a protein.
3. An aspartate side chain in a hydrophobic pocket adjacent to a glutamate side chain.
4. An aspartate side chain in a hydrophobic pocket adjacent to a lysine side chain.

**2–31** During an all-out sprint, muscles metabolize glucose anaerobically, producing a high concentration of lactic acid, which lowers the pH of the blood and

**Table 2–2 Values for the ionizable groups of several amino acids** (Problem 2–29).

| | p*K* VALUES | | |
|---|---|---|---|
| AMINO ACID | –COOH | $-NH_3^+$ | R GROUP |
| Leucine | 2.4 | 9.6 | |
| Proline | 2.0 | 10.6 | |
| Glutamate | 2.2 | 9.7 | 4.3 (carboxyl) |
| Histidine | 1.8 | 9.2 | 6.0 (imidazole) |
| Cysteine | 1.8 | 10.8 | 8.3 (sulfhydryl) |
| Arginine | 1.8 | 9.0 | 12.5 (guanidino) |
| Lysine | 2.2 | 9.2 | 10.8 (amino) |

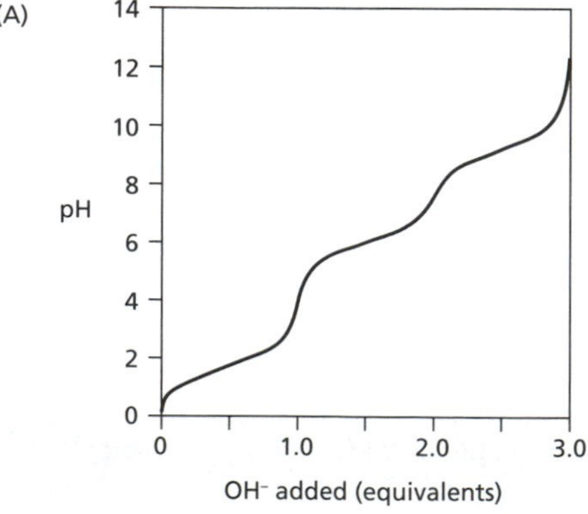


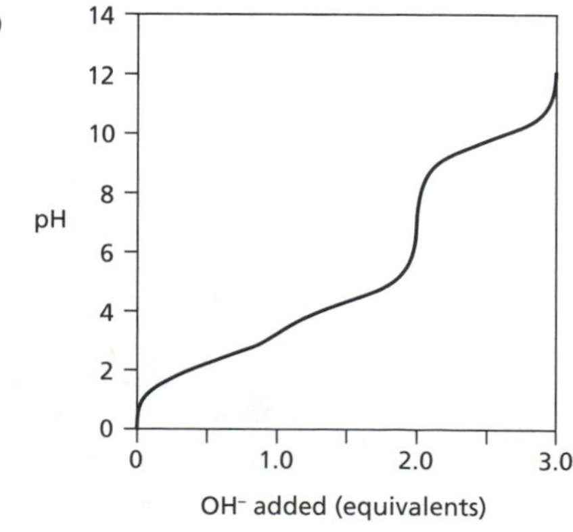


**Figure 2–4** Titration curves for two amino acids (Problem 2–29).

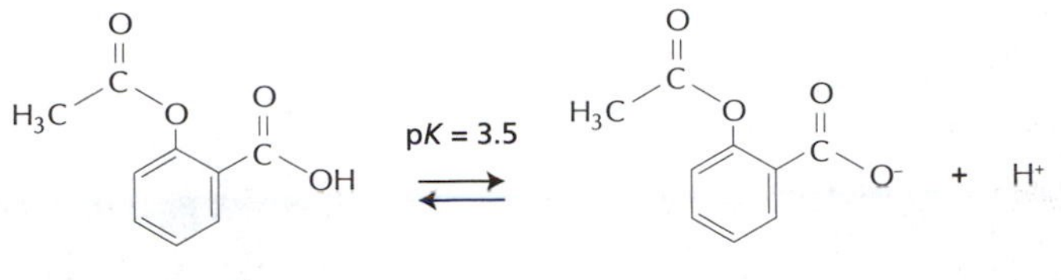

$$\text{uncharged form} \underset{}{\overset{pK = 3.5}{\rightleftharpoons}} \text{charged form} + H^+$$

uncharged form

charged form

**Figure 2–5** Aspirin (Problem 2–32).

of the cytosol. The lower pH inside the cell decreases the efficiency of certain glycolytic enzymes, which reduces the rate of ATP production and contributes to the fatigue that sprinters experience well before their fuel reserves are exhausted. The main blood buffer against pH changes is the bicarbonate/$CO_2$ system.

$$\underset{\text{(gas)}}{CO_2} \rightleftharpoons \underset{\text{(dissolved)}}{CO_2} \overset{pK_1 = 2.3}{\rightleftharpoons} H_2CO_3 \overset{pK_2 = 3.8}{\rightleftharpoons} H^+ + HCO_3^- \overset{pK_3 = 10.3}{\rightleftharpoons} H^+ + CO_3^{2-}$$

To improve their performance, would you advise sprinters to hold their breath or to breathe rapidly for a minute immediately before the race? Explain your answer.

**2–32** Aspirin is a weak acid (Figure 2–5) that is taken up into the bloodstream by diffusion through cells lining the stomach and the small intestine. Aspirin crosses the plasma membrane of a cell most effectively in its uncharged form; in its charged form it cannot cross the hydrophobic lipid bilayer of the membrane. The pH of the stomach is about 1.5 and that of the lumen of the small intestine is about 6.0. Is the majority of the aspirin absorbed in the stomach or in the intestine? Explain your reasoning.

**2–33** What, if anything, is wrong with the following statement: "When NaCl is dissolved in water, the water molecules closest to the ions will tend to orient themselves so that their oxygen atoms point toward the sodium ions and away from the chloride ions." Explain your answer.

**2–34** If noncovalent interactions are so weak in a water environment, how can they possibly be important for holding molecules together in cells?

**2–35** The three molecules in Figure 2–6 contain the seven most common reactive groups in biology. Most molecules in the cell are built from these functional groups. Indicate and name the functional groups in these molecules.

**2–36** Ball-and-stick and space-filling models of glucose are shown in Figure 2–7. In both illustrations there are two different sizes of hydrogen atoms (identified by arrows). Is this accurate or a mistake? Explain your answer.

**2–37** In solution, linear D-glucose forms a ring by reaction of the hydroxyl oxygen on carbon 5 with the carbon of the aldehyde at position 1 (Figure 2–8A). Depending on which side of carbon 1 is attacked, the resulting hydroxyl

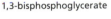

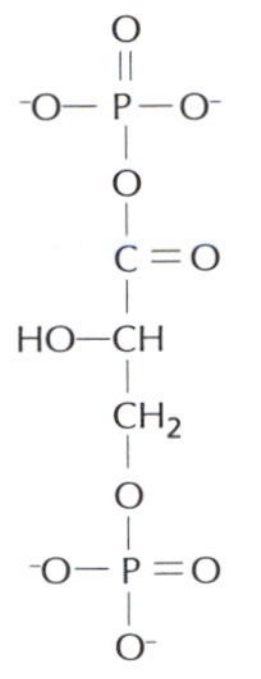

1,3-bisphosphoglycerate

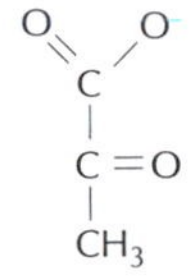

pyruvate

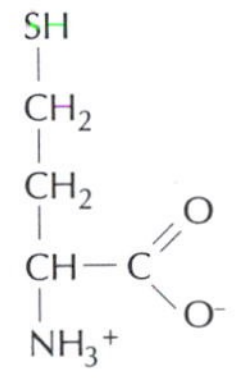

cysteine

**Figure 2–6** Three molecules that illustrate the seven most common functional groups in biology (Problem 2–35). 1,3-Bisphosphoglycerate and pyruvate are intermediates in glycolysis and cysteine is an amino acid.

**Problems 3–30, 3–31,** and **3–33** look at the noncovalent interactions that hold proteins together.

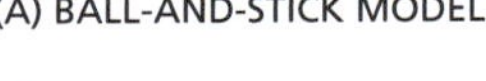


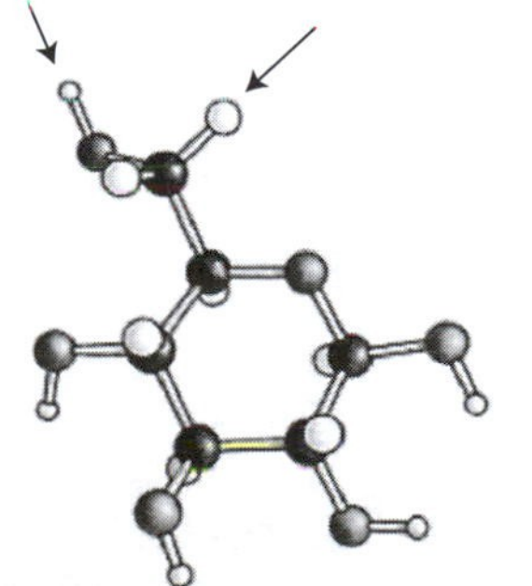

(B) SPACE-FILLING MODEL

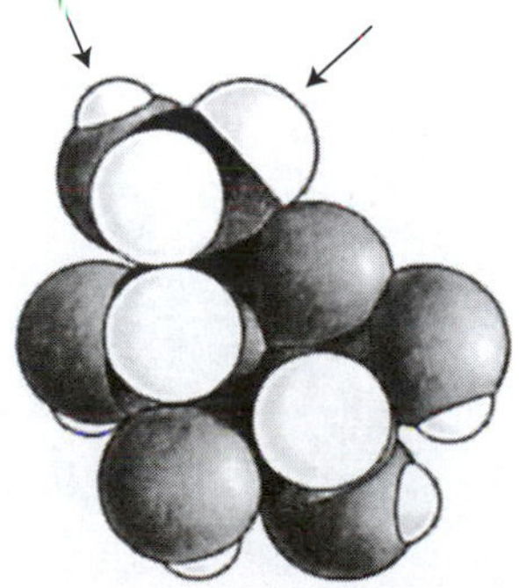

**Figure 2–7** Ball-and-stick and space-filling models of a glucose molecule (Problem 2–36). Arrows identify a 'small' and a 'large' hydrogen atom in each model.

(A) D-GLUCOSE

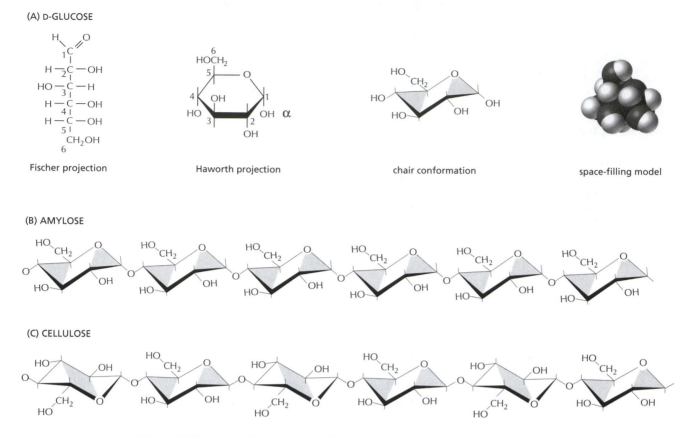

Fischer projection

Haworth projection

chair conformation

space-filling model

(B) AMYLOSE

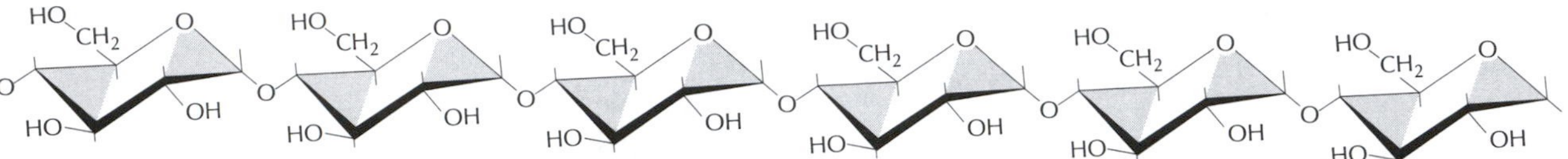


(C) CELLULOSE

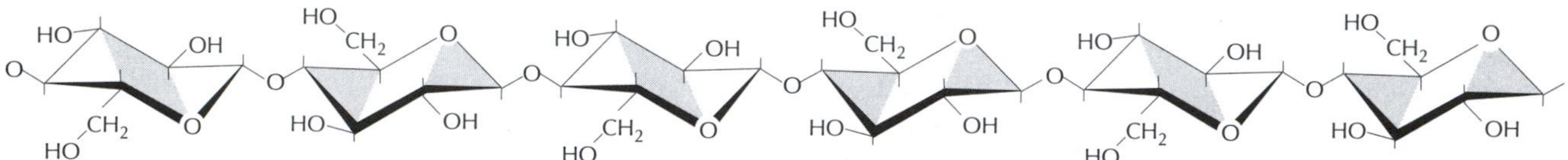


**Figure 2–8** Structures of carbohydrates (Problem 2–37). (A) Several structural representations of D-glucose. (B) Amylose. (C) Cellulose. The glycosidic bonds that connect the hexose monomers in the polysaccharide chains are all 1 → 4, which indicates the numbers of the two carbons that are linked.

group can be located either above the ring (β) or below the ring (α). The α form is represented in Figure 2–8A. Shown along with the Fischer projection of linear D-glucose and the Haworth projection of circular α-D-glucose are two more realistic representations of the energetically favorable chair conformation of α-D-glucose.

Polysaccharides are formed by linkage of sugar monomers via glycosidic bonds. Amylose and cellulose, for example, are polysaccharides composed entirely of hexoses (six-carbon sugars) that are linked together by glycosidic bonds between the number 1 and 4 carbons of adjacent monomers. Structures of amylose and cellulose with the hexoses in the chair conformation are shown in Figure 2–8B and C. Which, if either, of these polysaccharides is composed entirely of D-glucose? What are the similarities and differences between the structures of amylose and cellulose?

**2–38** The drug thalidomide was once prescribed as a sedative to help with nausea during the early stages of pregnancy. One of its optical isomers, (*R*)-thalidomide (Figure 2–9), is the active agent responsible for its sedative effects.

(A) THALIDOMIDE CHEMICAL FORMULA

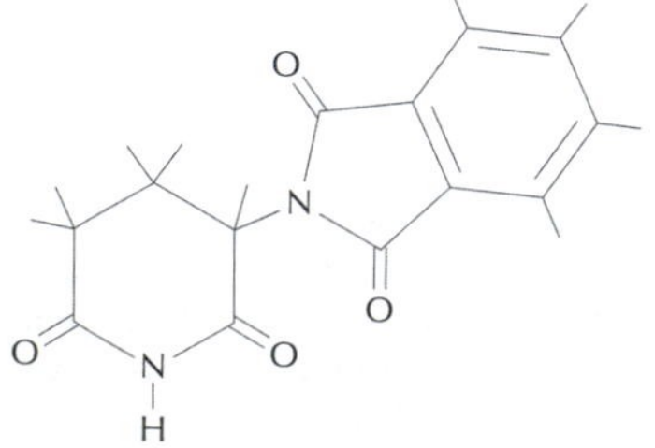


(B) THALIDOMIDE SPACE-FILLING MODEL

**Figure 2–9** The structure of the sedative (*R*)-thalidomide (Problem 2–38).

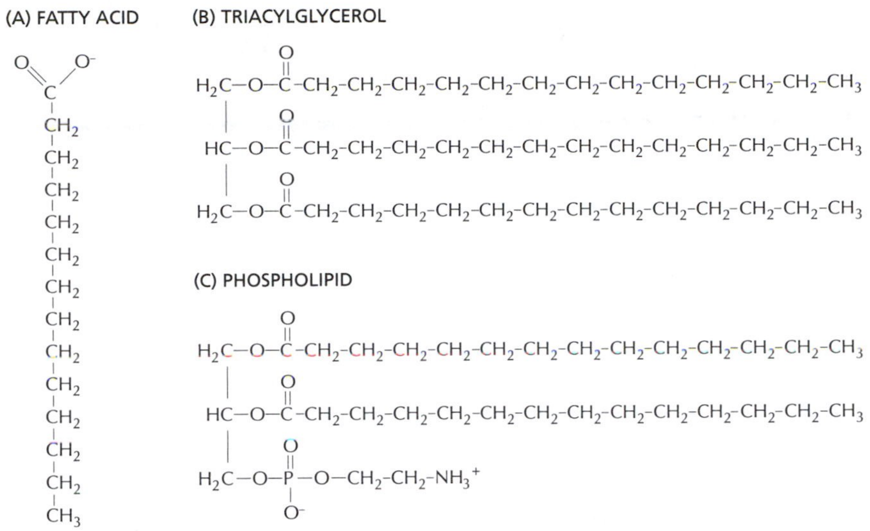

(A) FATTY ACID

$O^-$ / O=C–$(CH_2)_{12}$–$CH_3$

(B) TRIACYLGLYCEROL

$H_2C–O–C(=O)–CH_2–CH_2–CH_2–CH_2–CH_2–CH_2–CH_2–CH_2–CH_2–CH_2–CH_2–CH_2–CH_3$

$HC–O–C(=O)–CH_2–CH_2–CH_2–CH_2–CH_2–CH_2–CH_2–CH_2–CH_2–CH_2–CH_2–CH_2–CH_3$

$H_2C–O–C(=O)–CH_2–CH_2–CH_2–CH_2–CH_2–CH_2–CH_2–CH_2–CH_2–CH_2–CH_2–CH_2–CH_3$

(C) PHOSPHOLIPID

$H_2C–O–C(=O)–CH_2–CH_2–CH_2–CH_2–CH_2–CH_2–CH_2–CH_2–CH_2–CH_2–CH_2–CH_2–CH_3$

$HC–O–C(=O)–CH_2–CH_2–CH_2–CH_2–CH_2–CH_2–CH_2–CH_2–CH_2–CH_2–CH_2–CH_2–CH_3$

$H_2C–O–P(=O)(O^-)–O–CH_2–CH_2–NH_3^+$

**Figure 2–10** A fatty acid, a triacylglycerol, and a phospholipid (Problem 2–39).

It was synthesized, however, as a mixture of both optical isomers—a not uncommon practice that usually causes no problems. Unfortunately, the other optical isomer is a teratogen that led to a horrific series of birth defects characterized by malformed or absent limbs. On the structural formula in Figure 2–9A identify the carbon that is responsible for its optical activity (its chiral center) and sketch the structure of the teratogenic form of thalidomide.

**Problems 3–18** and **3–20** deal with amphiphilic α helices and β sheets.

**2–39** What does the term 'amphiphilic' mean? Figure 2–10 shows a fatty acid, a triacylglycerol, and a phospholipid. Indicate which of these molecules are amphiphilic and illustrate why. How do their amphiphilic characteristics account for the typical structures that these molecules form in cells?

**2–40** A short polypeptide is shown in Figure 2–11. Identify the N-terminus, the C-terminus, the α carbons, and the side chains. Mark the atoms involved in the peptide bonds and those that form each amino acid.

**Problem 3–72** explores the specificity of an enzyme composed entirely of D-amino acids.

**2–41** Why do you suppose that only L-amino acids and not a random mixture of L- and D-amino acids are used to make proteins?

**2–42** There are many different, chemically diverse ways in which small molecules can be linked to form polymers. For example, ethene ($CH_2=CH_2$) is used commercially to make the plastic polymer polyethylene ($…–CH_2–CH_2–CH_2–CH_2–…$). The individual subunits of the three major classes of biological macromolecules, however, are all linked by similar reaction mechanisms, namely, by condensation reactions that eliminate water. Can you think of any benefits that this chemistry offers and why it might have been selected in evolution?

**Problem 3–38** asks how many different proteins of a given size can be made with 20 amino acids.

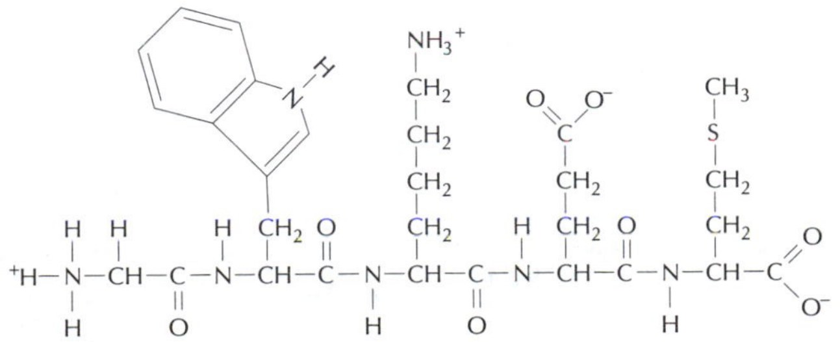

$^+H–N(H)(H)–CH(H)–C(=O)–N(H)–CH(CH_2\text{-indolyl})–C(=O)–N(H)–CH(CH_2CH_2CH_2CH_2NH_3^+)–C(=O)–N(H)–CH(CH_2COO^-)–C(=O)–N(H)–CH(CH_2CH_2SCH_3)–COO^-$

**Figure 2–11** A polypeptide (Problem 2–40).

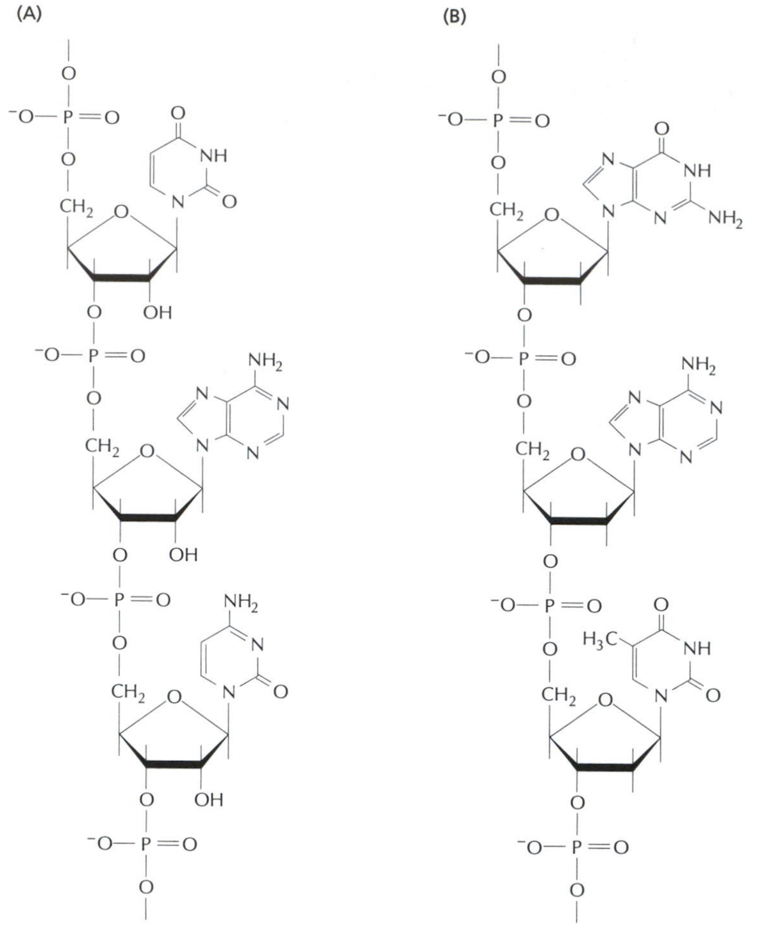

Figure 2–12 Two oligonucleotides (Problem 2–43).

2–43 Two short chains of nucleic acids are shown in Figure 2–12. Identify the 5′ and 3′ ends of each oligonucleotide, name the component bases and sugars, and indicate the components that make up a nucleoside and a nucleotide. Which one is RNA and which is DNA? How can you tell?

## CALCULATIONS

2–44 To gain a better feeling for atomic dimensions, assume that the page on which this question is printed is made entirely of the polysaccharide cellulose (Figure 2–13). Cellulose is described by the formula $(C_6H_{12}O_6)_n$, where $n$ is a large number that varies from one molecule to another. The atomic weights of carbon, hydrogen, and oxygen are 12, 1, and 16, respectively, and this page weighs 5 grams.

A. How many carbon atoms are there in this page?

B. In paper made of pure cellulose, how many carbon atoms would be stacked on top of each other to span the thickness of this page (the page is 21 cm × 27.5 cm × 0.07 mm)? (Rather than solving three simultaneous equations for the carbon atoms in each dimension, you might try a shortcut. Determine the linear density of carbon atoms by calculating the number of carbon atoms on the edge of a cube with the same volume as this page, and then adjust that number to the thickness of the page.)

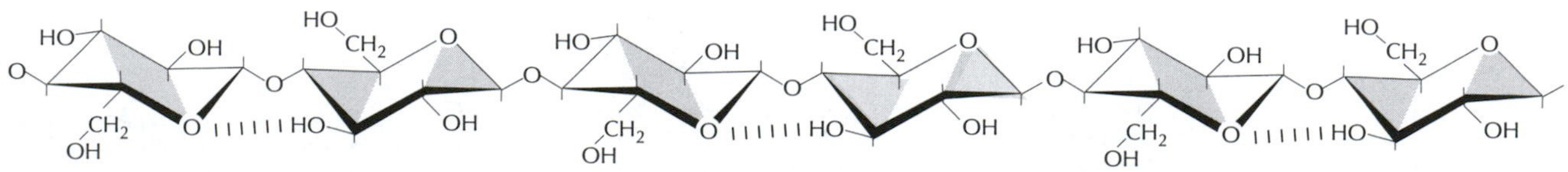

Figure 2–13 Structure of the polysaccharide cellulose (Problem 2–44).

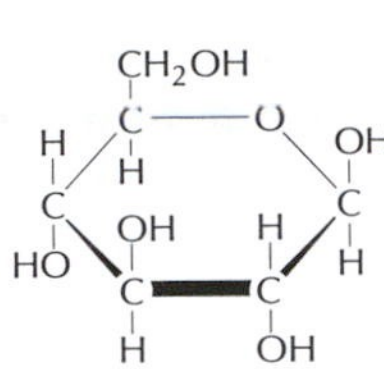


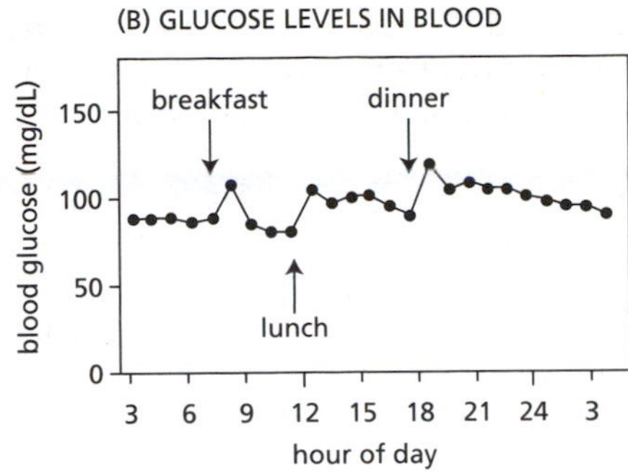


**Figure 2–14** Circulating blood glucose (Problem 2–45). (A) Structure of glucose. (B) Typical variation in blood glucose over the course of a day.

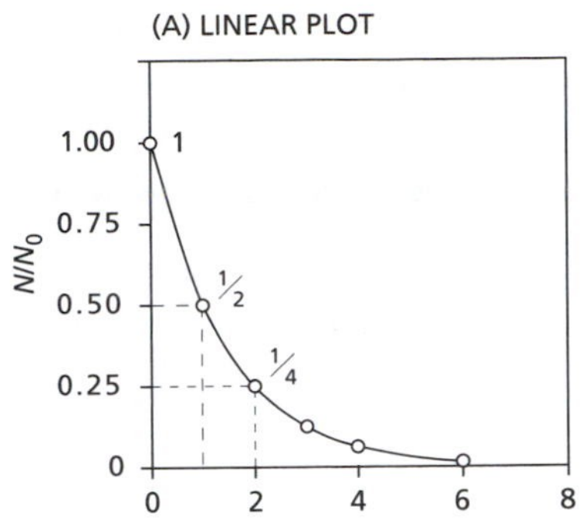


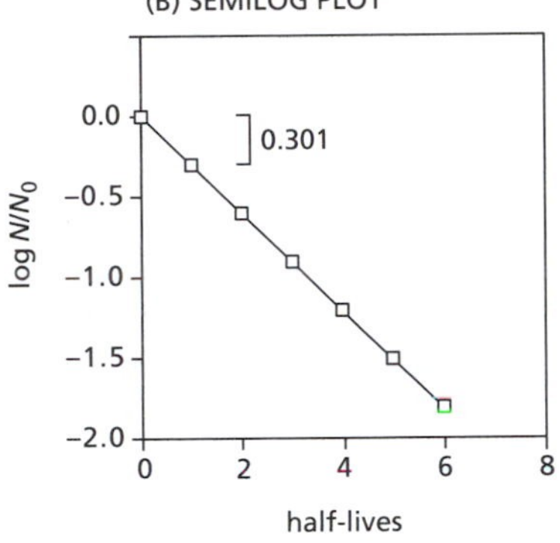


**Figure 2–15** First order decay of a radioactive isotope (Problem 2–46). (A) A linear plot. (B) A semilog plot. The log of 2 is 0.301.

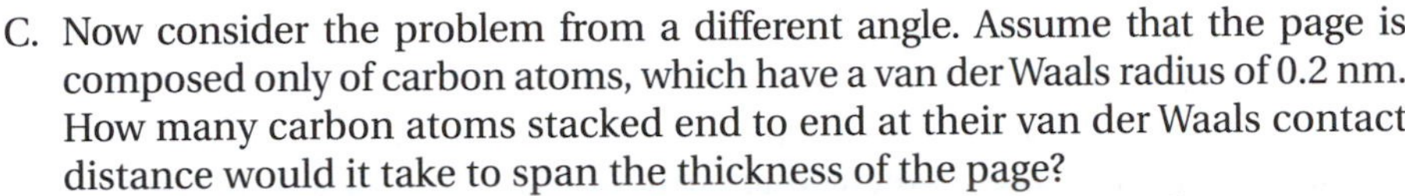

C. Now consider the problem from a different angle. Assume that the page is composed only of carbon atoms, which have a van der Waals radius of 0.2 nm. How many carbon atoms stacked end to end at their van der Waals contact distance would it take to span the thickness of the page?

D. Compare your answers from parts B and C and explain any differences.

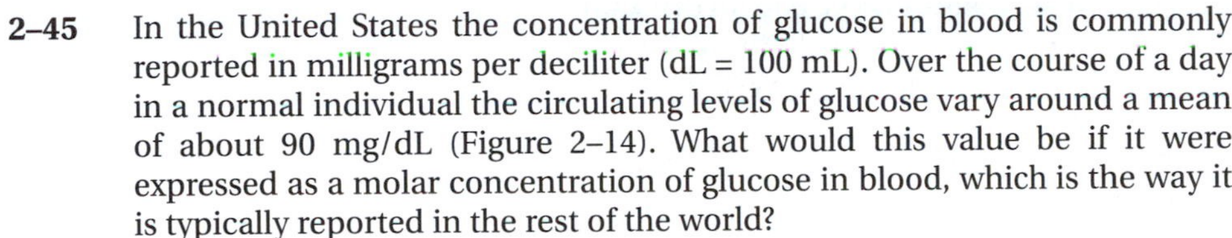

**2–45** In the United States the concentration of glucose in blood is commonly reported in milligrams per deciliter (dL = 100 mL). Over the course of a day in a normal individual the circulating levels of glucose vary around a mean of about 90 mg/dL (Figure 2–14). What would this value be if it were expressed as a molar concentration of glucose in blood, which is the way it is typically reported in the rest of the world?

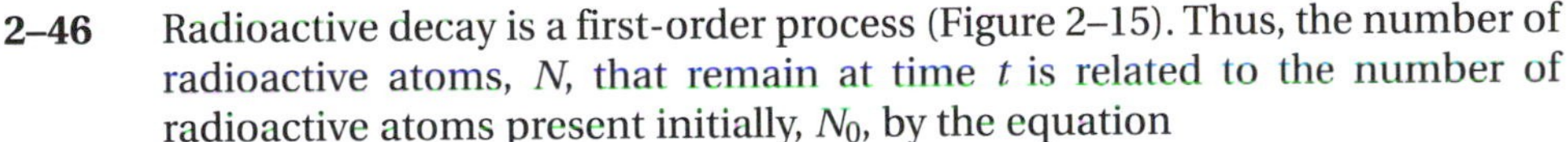

**2–46** Radioactive decay is a first-order process (Figure 2–15). Thus, the number of radioactive atoms, $N$, that remain at time $t$ is related to the number of radioactive atoms present initially, $N_0$, by the equation

$$N = N_0\, e^{-\lambda t}$$

or rearranging and taking the log of both sides,

$$2.303 \log \frac{N}{N_0} = -\lambda t$$

where $\lambda$ is a decay constant, which is different for each radioactive isotope.

The half-life ($t_{1/2}$) of a radioactive isotope is the time required for half the original number of atoms to decay; that is, when $N/N_0$ equals 0.5. For each of the isotopes in Table 2–1 (p. 12), calculate the decay constant $\lambda$ in the same units as the half-life and then in minutes.

**2–47** Neutrons of cosmic radiation constantly bombard Earth's upper atmosphere, converting a fairly constant fraction of $^{14}N_7$ to $^{14}C_6$, as shown in Figure 2–16. This $^{14}C$ enters the biosphere as $CO_2$, being incorporated first into plants via photosynthesis and then via the food chain into animals. As a result, all living plants and animals contain the same fraction of $^{14}C$ as in

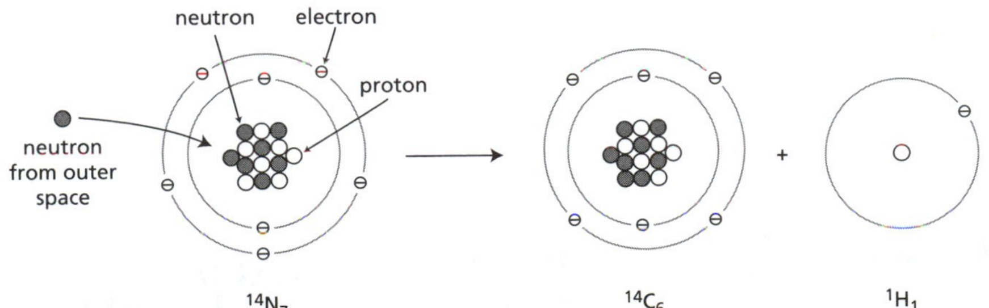


**Figure 2–16** Formation of $^{14}C_6$ from $^{14}N_7$ in the upper atmosphere (Problem 2–47).

atmospheric $CO_2$, which is sufficient to yield 15.3 disintegrations per minute (dpm) per gram of carbon. The number of dpm equals the number of radioactive atoms, *N*, times the decay constant λ, expressed in $min^{-1}$ (see Problem 2–46).

A. What fraction of the carbon atoms in living organisms does $^{14}C$ constitute?

B. The curie (Ci) corresponds to $2.22 \times 10^{12}$ dpm. How many curies of $^{14}C$ are in a 70 kg person? Humans are about 18.5% carbon by weight.

C. When an organism dies, it no longer incorporates carbon from the environment so that the quantity of $^{14}C$ decays with a half-life of 5730 years. What is the age of a biological sample that contains 3.0 dpm/g of carbon?

**2–48** If both carbon atoms in every molecule of glycine were $^{14}C$, what would its specific activity be in Ci/mmol? What proportion of glycine molecules is labeled in a preparation that has a specific activity of 200 μCi/mmol?

**2–49** The molecular weight of ethanol ($CH_3CH_2OH$) is 46 and its density is 0.789 g/cm³.

A. What is the molarity of ethanol in beer that is 5% ethanol by volume? [Alcohol content of beer varies from about 4% (lite beer) to 8% (stout beer).]

B. The legal limit for a driver's blood alcohol content varies, but 80 mg of ethanol per 100 mL of blood (usually referred to as a blood alcohol level of 0.08) is typical. What is the molarity of ethanol in a person at this legal limit?

C. How many 12-oz (355-mL) bottles of 5% beer could a 70-kg person drink and remain under the legal limit? A 70-kg person contains about 40 liters of water. Ignore the metabolism of ethanol, and assume that the water content of the person remains constant.

D. Ethanol is metabolized at a constant rate of about 120 mg per hour per kg body weight, regardless of its concentration. If a 70-kg person were at twice the legal limit (160 mg/100 mL), how long would it take for their blood alcohol level to fall below the legal limit?

**2–50** Imagine that you have a beaker of pure water at neutral pH (pH 7.0).

A. What is the concentration of $H_3O^+$ ions and how were they formed?

B. What is the molarity of pure water? (Hint: 1 liter of water weighs 1 kg.)

C. What is the ratio of $H_3O^+$ ions to $H_2O$ molecules?

**2–51** By a convenient coincidence the ion product of water, $K_w = [H^+][OH^-]$, is a nice round number: $1.0 \times 10^{-14}$ M².

A. Why is a solution at pH 7.0 said to be neutral?

B. What is the $H^+$ concentration and pH of a 1 mM solution of NaOH?

C. If the pH of a solution is 5.0, what is the concentration of $OH^-$ ions?

**2–52** Solutions containing 500 mL of 0.1 M HCl or 500 mL of 0.1 M acetic acid were titrated by addition of increasing amounts of 1 M KOH as shown in Figure 2–17.

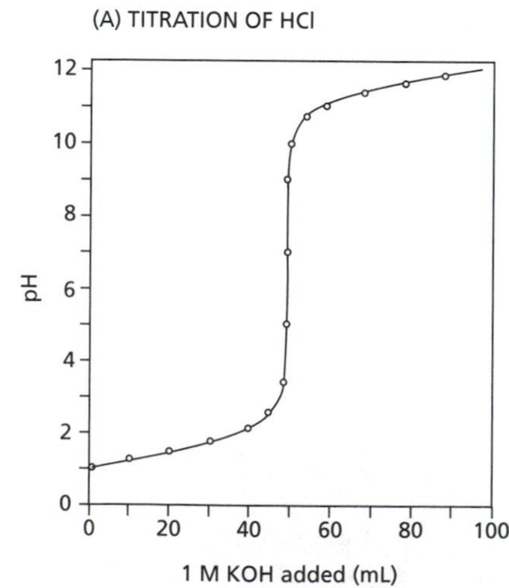


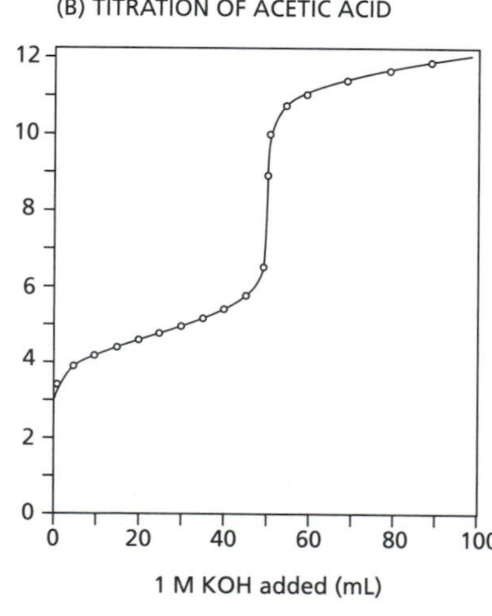


**Figure 2–17** Titration curves for solutions of HCl and acetic acid (Problem 2–52).

**Table 2–3 Dissociation of a weak acid at pH values above and below the p*K*** (Problem 2–53).

| pH | $\log \frac{[A^-]}{[HA]}$ | $\frac{[A^-]}{[HA]}$ | % DISSOCIATION |
|---|---|---|---|
| p*K* +4 | | | |
| p*K* +3 | | | |
| p*K* +2 | | | |
| p*K* +1 | | | |
| p*K* | | | |
| p*K* –1 | | | |
| p*K* –2 | | | |
| p*K* –3 | | | |
| p*K* –4 | | | |

A. How many mL of KOH were added to neutralize all of the protons derived from HCl or acetic acid? The point at which neutralization occurs is termed the equivalence point.
B. For each solution estimate the pH at the equivalence point.
C. Estimate the p*K* for acetic acid.

**2–53** The Henderson–Hasselbalch equation

$$pH = pK + \log \frac{[A^-]}{[HA]}$$

is a useful transformation of the equation for dissociation of a weak acid, HA:

$$K = \frac{[H^+][A^-]}{[HA]}$$

A. It is instructive to use the Henderson–Hasselbalch equation to determine the extent of dissociation of an acid at pH values above and below the p*K*. For the pH values listed in Table 2–3, fill in the values for log $[A^-]/[HA]$ and $[A^-]/[HA]$, and indicate the percentage of the acid that has dissociated.
B. Using the graph in Figure 2–18, sketch the relationship between pH of the solution and the fractional dissociation of a weak acid. Will the shape of this curve be the same for all weak acids?

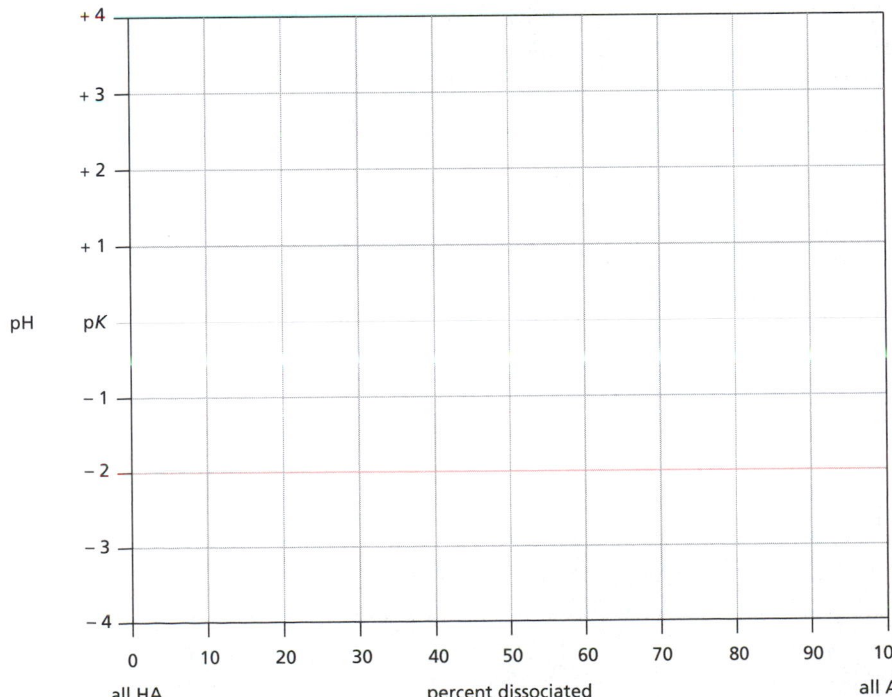


**Figure 2–18** Graph for plotting values from Table 2–3 (Problem 2–53).

**Problem 11–42** considers the use of fluorescent probes to measure intracellular pH.

**2–54** Cells maintain their cytosolic pH in a narrow range around pH 7.0 by using a variety of weak acids to buffer against changes in pH. This is essential because a large number of processes in cells generate or consume $H^+$ ions. Weak acids resist changes in pH—the definition of a buffer—most effectively within about one pH unit on either side of their p*K* values, as can be seen, for example, from the titration curve for acetic acid (see Figure 2–17B). Ionization of phosphoric acid provides an important buffering system in cells. Phosphoric acid has three ionizable protons, each with a unique p*K*.

$$H_3PO_4 \underset{}{\overset{pK = 2.1}{\rightleftharpoons}} H^+ + H_2PO_4^- \overset{pK = 6.9}{\rightleftharpoons} H^+ + HPO_4^{2-} \overset{pK = 12.4}{\rightleftharpoons} H^+ + PO_4^{3-}$$

A. Using the values derived in Problem 2–53, estimate how much of each of the four forms of phosphate ($H_3PO_4$, $H_2PO_4^-$, $HPO_4^{2-}$, and $PO_4^{3-}$), as percentage of the total, are present in the cytosol of cells at pH 7. (No calculators permitted.)

B. What is the ratio of [$HPO_4^{2-}$] to [$H_2PO_4^-$] ([$A^-$]/[HA]) in the cytosol at pH 7? If the cytosol is 1 mM phosphate (sum of all forms), what are the concentrations of $H_2PO_4^-$ and $HPO_4^{2-}$ in the cytosol? (Calculators permitted.)

**2–55** Describe in a general way how you might prepare 1 liter of 1 mM phosphate buffer at pH 6.9. You have available 0.1 M solutions of $H_3PO_4$, $KH_2PO_4$, $K_2HPO_4$, and $K_3PO_4$, 1.0 M solutions of HCl and KOH, and a pH meter. There are several possible (correct) answers. Be creative. Do all methods lead to the same final solution?

**2–56** Inside cells the two most important buffer systems are provided by phosphate and proteins. The quantitative aspects of a buffer system pertain to both the effective buffering range (how near the pH is to the p*K* for the buffer) and the overall concentration of the buffering species (which determines the number of protons that can be handled). As discussed in Problem 2–54, $H_2PO_4^- \rightleftharpoons HPO_4^{2-}$ has a p*K* of 6.9 with an overall intracellular phosphate concentration of about 1 mM. In red blood cells the concentration of globin chains (molecular weight = 15,000) is about 100 mg/mL and each has 10 histidines, with p*K* values between 6.5 and 7.0. Which of these two buffering systems do you think is quantitatively the more important in red blood cells, and why do you think so?

**2–57** The most important buffer in the bloodstream is the bicarbonate/$CO_2$ system. It is much more important than might be expected from its p*K* because it is an open system in which the $CO_2$ is maintained at a relatively constant value by exchange with the atmosphere. (By contrast, the buffering systems described in Problem 2–56 are closed systems with no exchange.) The equilibria involved in the bicarbonate/$CO_2$ buffering system are

$$CO_2(\text{gas}) \rightleftharpoons CO_2(\text{dis}) \overset{pK_1 = 2.3}{\rightleftharpoons} H_2CO_3 \overset{pK_2 = 3.8}{\rightleftharpoons} H^+ + HCO_3^- \overset{pK_3 = 10.3}{\rightleftharpoons} H^+ + CO_3^{2-}$$

$pK_3$ ($HCO_3^- \rightleftharpoons H^+ + CO_3^{2-}$) is so high that it never comes into play in biological systems. $pK_2$ ($H_2CO_3 \rightleftharpoons H^+ + HCO_3^-$) seems much too low to be useful, but it is influenced by the dissolved $CO_2$, which is directly proportional to the partial pressure of $CO_2$ in the gas phase. The dissolved $CO_2$ in turn is kept in equilibrium with $H_2CO_3$ by the enzyme carbonic anhydrase.

$$K_1 = \frac{[H_2CO_3]}{[CO_2(\text{dis})]} = 5 \times 10^{-3}, \text{ or } pK_1 = 2.3$$

The equilibrium for hydration of $CO_2$(dis) can be combined with the equilibrium for dissociation of $H_2CO_3$ to give a $pK'$ for $CO_2(\text{dis}) \rightleftharpoons H^+ + HCO_3^-$

$$K' = \frac{[H^+][HCO_3^-]}{[CO_2(\text{dis})]} = K_1 \times K_2$$

$$pK' = pK_1 + pK_2 = 2.3 + 3.8 = 6.1$$

| CHEMICAL FORMULAS | | pK VALUES $-COO^-$ | pK VALUES $^+H_3N-$ |
|---|---|---|---|
| Ala | $^+H_3N-CH(CH_3)-COO^-$ | 2.34 | 9.69 |
| $Ala_2$ | $^+H_3N-CH(CH_3)-C(=O)-NH-CH(CH_3)-COO^-$ | 3.12 | 8.30 |
| $Ala_3$ | $^+H_3N-CH(CH_3)-C(=O)-NH-CH(CH_3)-C(=O)-NH-CH(CH_3)-COO^-$ | 3.39 | 8.03 |
| $Ala_4$ | $^+H_3N-CH(CH_3)-C(=O)-NH-CH(CH_3)-C(=O)-NH-CH(CH_3)-C(=O)-NH-CH(CH_3)-COO^-$ | 3.42 | 7.94 |

**Figure 2–19** p$K$ values for the carboxyl and amino groups in oligomers of alanine (Problem 2–59).

Even the p$K'$ of 6.1 seems too low to maintain the blood pH around 7.4, yet this open system is very effective, as can be illustrated by a few calculations. The total concentration of carbonate in its various forms, but almost entirely $CO_2$(dis) and $HCO_3^-$, is about 25 mM.

A. Using the Henderson–Hasselbalch equation, calculate the ratio of $HCO_3^-$ to $CO_2$(dis) at pH 7.4. What are the concentrations of $HCO_3^-$ and $CO_2$(dis)?

B. What would the pH be if 5 mM $H^+$ were added under conditions where $CO_2$ was not allowed to leave the system; that is, if the concentration of $CO_2$(dis) was not maintained at a constant value?

C. What would the pH be if 5 mM $H^+$ were added under conditions where $CO_2$ was permitted to leave the system; that is, if the concentration of $CO_2$(dis) was maintained at a constant value?

**2–58** The proteins in a mammalian cell account for 18% of its net weight. If the density of a typical mammalian cell is about 1.1 g/mL and the volume of the cell is $4 \times 10^{-9}$ mL, what is the concentration of protein in mg/mL?

## DATA HANDLING

**2–59** The ionizable groups in amino acids can influence one another, as shown by the p$K$ values for the carboxyl and amino groups of alanine and various oligomers of alanine (Figure 2–19). Suggest an explanation for why the p$K$ of the carboxyl group increases with oligomer size, while that of the amino group decreases.

**Problem 3–74** explores the role of ionizable side chains in the active site of lysozyme.

**2–60** A histidine side chain is known to have an important role in the catalytic mechanism of an enzyme; however, it is not clear whether histidine is required in its protonated (charged) or unprotonated (uncharged) state. To answer this question you measure enzyme activity over a range of pH, with the results shown in Figure 2–20. Which form of histidine is required for enzyme activity?

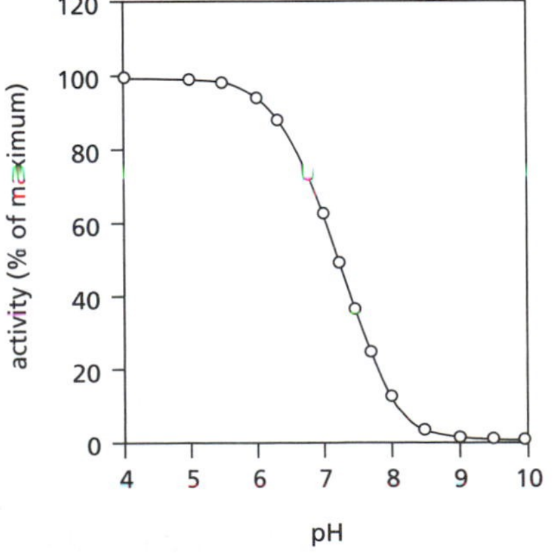


**Figure 2–20** Enzyme activity as a function of pH (Problem 2–60).

# CATALYSIS AND THE USE OF ENERGY BY CELLS

TERMS TO LEARN

| | | |
|---|---|---|
| acetyl CoA | entropy | $NADP^+$/NADPH |
| activated carrier | enzyme | oxidation |
| activation energy | equilibrium | photosynthesis |
| ADP | equilibrium constant ($K$) | reduction |
| ATP | free energy ($G$) | respiration |
| catalyst | free-energy change ($\Delta G$) | standard free-energy change ($\Delta G°$) |
| coupled reaction | metabolism | substrate |
| diffusion | $NAD^+$/NADH | |

## DEFINITIONS

Match each definition below with its term from the list above.

**2–61** Extra energy that must be possessed by atoms or molecules in addition to their ground-state energy in order to undergo a particular chemical reaction.

**2–62** Free-energy change of two reacting molecules at standard temperature and pressure when all components are present at a concentration of 1 mole per liter.

**2–63** Loss of electrons from an atom, as occurs during the addition of oxygen to a molecule or when a hydrogen is removed.

**2–64** Molecule on which an enzyme acts.

**2–65** Net drift of molecules in the direction of lower concentration due to random thermal movement.

**2–66** Protein that catalyzes a specific chemical reaction.

**2–67** Linked pair of chemical reactions in which the free energy released by one of the reactions serves to drive the other.

**2–68** State at which there is no net change in a system. In a chemical reaction, this state is reached when the forward and reverse rates are equal.

**2–69** The energy that can be extracted from a system to drive reactions. Takes into account changes in both energy and entropy.

## TRUE/FALSE

Decide whether each of these statements is true or false, and then explain why.

**2–70** Animals and plants use oxidation to extract energy from food molecules.

**2–71** If an oxidation occurs in a reaction, it must be accompanied by a reduction.

**2–72** Linking the energetically unfavorable reaction $A \rightarrow B$ to a second, favorable reaction $B \rightarrow C$ will shift the equilibrium constant for the first reaction.

## THOUGHT PROBLEMS

**2–73** Distinguish between catabolic and anabolic pathways of metabolism, and indicate in a general way how such pathways are linked to one another in cells.

**2–74** The second law of thermodynamics states that systems will change spontaneously toward arrangements with greater entropy (disorder). Living systems are so intricately ordered, however, it seems they must surely violate the second law. Explain briefly—and in a way your parents could understand—how life is fully compatible with the laws of thermodynamics.

**2–75** The equation for photosynthesis in green plants is

$$\text{light energy} + CO_2 + H_2O \rightarrow \text{sugars} + O_2 + \text{heat energy}$$

Would you expect this reaction to be carried out by a single enzyme? Why is heat energy also produced along with sugars and $O_2$?

**2–76** In the reaction $2\ Na + Cl_2 \rightarrow 2\ Na^+ + 2\ Cl^-$, what is being oxidized and what is being reduced? How can you tell?

**2–77** If a cell in mitosis is cooled to 0°C, the microtubules in the spindle depolymerize into tubulin subunits. The same is true for microtubules made from pure tubulin in a test tube; they assemble readily at 37°C, but disassemble at low temperature. In fact, many protein assemblies that are held together by

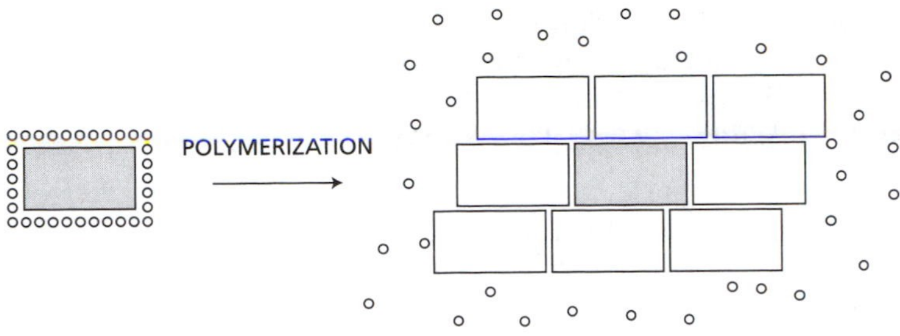


**Figure 2–21** Polymerization of tubulin subunits into a microtubule (Problem 2–77). The fates of one subunit *(shaded)* and its associated water molecules *(small spheres)* are shown.

noncovalent bonds show the same behavior: they disassemble when cooled. This behavior is governed by the basic thermodynamic equation

$$\Delta G = \Delta H - T\Delta S$$

where $\Delta H$ is the change in enthalpy (chemical bond energy), $\Delta S$ is the change in entropy (disorder of the system), and $T$ is the absolute temperature.

A. The change in free energy ($\Delta G$) must be negative for the reaction (tubulin subunits → microtubules) to proceed at high temperature. At low temperature, $\Delta G$ must be positive to permit disassembly; that is, to favor the reverse reaction. Decide what the signs (positive or negative) of $\Delta H$ and $\Delta S$ must be, and show how your choices account for polymerization of tubulin at high temperature and its depolymerization at low temperature. (Assume that the $\Delta H$ and $\Delta S$ values themselves do not change with temperature.)

B. Polymerization of tubulin subunits into microtubules at body temperature clearly occurs with an increase in the orderliness of the subunits (Figure 2–21). Yet tubulin polymerization occurs with an increase in entropy (decrease in order). How can that be?

**2–78** The enzyme carbonic anhydrase is one of the speediest enzymes known. It catalyzes the hydration of $CO_2$ to $H_2CO_3$, which then rapidly dissociates as described in Problem 2–57. Carbonic anhydrase accelerates the reaction $10^7$-fold, hydrating $10^5$ $CO_2$ molecules per second at its maximal speed.

A. What factors might limit the speed of the enzyme?

B. The curve in Figure 2–22 shows the distribution of energy of molecules in solution, and illustrates schematically the proportions that have sufficient energy to undergo an uncatalyzed reaction (to the right of threshold B, Figure 2–22) or a catalyzed reaction (to the right of threshold A, Figure 2–22). On such a curve, to what does the $10^7$-fold rate enhancement by carbonic anhydrase correspond?

**2–79** Discuss the statement: "The criterion for whether a reaction proceeds spontaneously is $\Delta G$ not $\Delta G°$, because $\Delta G$ takes into account the concentrations of the substrates and products."

**2–80** At a particular concentration of substrates and products the reaction below has a negative $\Delta G$.

$$A + B \rightarrow C + D \qquad \Delta G = -4.5 \text{ kcal/mole}$$

At the same concentrations, what is $\Delta G$ for the reverse reaction?

$$C + D \rightarrow A + B$$

**2–81** The values for $\Delta G°$ and for $\Delta G$ in cells have been determined for many different metabolic reactions. What information do these values provide about the rates of these reactions?

**2–82** Thermodynamically, it is perfectly valid to consider the cellular phosphorylation of glucose as the sum of two reactions.

$$\begin{array}{lll} (1) & \text{glucose} + P_i \rightarrow \text{G6P} + H_2O & \Delta G° = 3.3 \text{ kcal/mole} \\ (2) & \underline{ATP + H_2O \rightarrow ADP + P_i} & \Delta G° = -7.3 \text{ kcal/mole} \\ \text{NET:} & \text{glucose} + \text{ATP} \rightarrow \text{G6P} + \text{ADP} & \end{array}$$

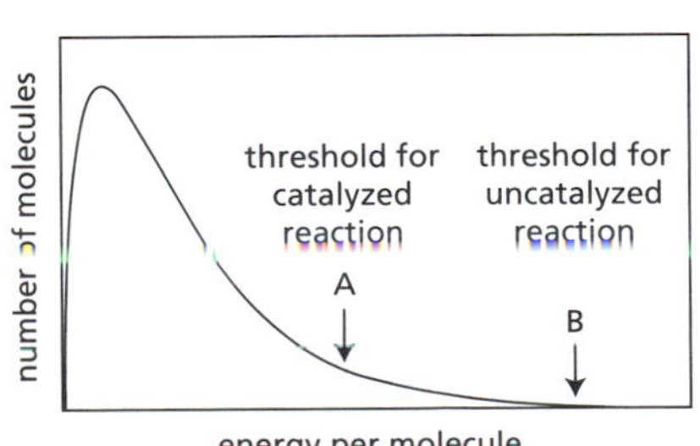


**Figure 2–22** Energy distribution of molecules in solution (Problem 2–78). Molecules with sufficient energy to undergo a catalyzed reaction have energies above (to the *right* of) threshold A. Molecules with sufficient energy to undergo an uncatalyzed reaction have energies above (to the *right* of) threshold B.

But biologically it makes no sense at all. Hydrolysis of ATP (reaction 2) in one part of the cell can have no effect on phosphorylation of glucose (reaction 1) elsewhere in the cell, given that [ATP], [ADP], and [$P_i$] are maintained within narrow limits. How does the cell manage to link these two reactions to achieve the phosphorylation of glucose?

**2–83** Successive steps in a metabolic pathway such as glycolysis, which converts glucose to pyruvate in 10 steps, are connected by metabolic intermediates. The product of the first reaction in the pathway provides a substrate for the second reaction, which in turn generates a product that is a substrate for the third step, and so on. Such pathways are not at equilibrium. Instead, intermediates flow along pathways so that net conversion, for example, of glucose to pyruvate can occur. Flow along such pathways is what distinguishes living cells from dead ones (which are at equilibrium).

The flow of metabolites through a metabolic pathway can occur *only* when the $\Delta G$ value for *each* step is negative. This is a true statement. Convince yourself of its validity by considering a segment of a pathway D → E → F, which has an overall (D → F) $\Delta G$ that is negative under cellular conditions. When there is a flow through the pathway, D is constantly added and F is constantly removed. Consider what would happen if the concentrations of the intermediates were such that the $\Delta G$ for D → E was momentarily positive and the $\Delta G$ for E → F was negative.

**2–84** Each phosphoanhydride bond between the phosphate groups in ATP is a high-energy linkage with a $\Delta G°$ value of –7.3 kcal/mole. Hydrolysis of this bond in cells normally liberates usable energy in the range of 11 to 13 kcal/mole. Why do you think a range of values for released energy is given for $\Delta G$, rather than a precise number, as for $\Delta G°$?

**2–85** Consider the effects of two enzymes. Enzyme A catalyzes the reaction

$$ATP + GDP \rightleftharpoons ADP + GTP$$

whereas enzyme B catalyzes the reaction

$$NADH + NADP^+ \rightleftharpoons NAD^+ + NADPH$$

Discuss whether the enzymes would be beneficial or detrimental to cells.

**2–86** Match the activated carriers below with the group carried in high-energy linkage.

A. Acetyl CoA
B. *S*-Adenosylmethionine
C. ATP
D. Carboxylated biotin
E. NADH, NADPH, $FADH_2$
F. Uridine diphosphate glucose

1. acetyl group
2. carboxyl group
3. electrons and hydrogens
4. glucose
5. methyl group
6. phosphate

**2–87** Which of the following reactions will occur only if coupled to a second, energetically favorable reaction?

A. glucose + $O_2 \rightarrow CO_2 + H_2O$
B. $CO_2 + H_2O \rightarrow$ glucose + $O_2$
C. nucleoside triphosphate + $DNA_n \rightarrow DNA_{n+1} + 2\ P_i$
D. nucleosides → nucleoside triphosphates
E. $ADP + P_i \rightarrow ATP$

## CALCULATIONS

**2–88** It is not easy to follow oxidations and reductions (redox reactions) in organic molecules. In biological redox reactions involving organic molecules in solution, it is the carbon atoms that are usually oxidized or reduced. (That's not always the case, but it's a good place to start.) To follow what is happening,

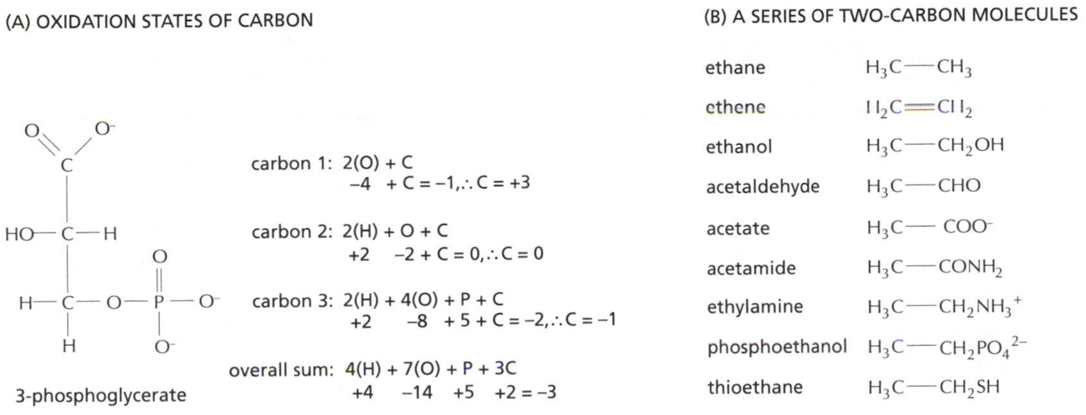

(A) OXIDATION STATES OF CARBON

3-phosphoglycerate

carbon 1: 2(O) + C
−4 + C = −1, ∴ C = +3

carbon 2: 2(H) + O + C
+2 −2 + C = 0, ∴ C = 0

carbon 3: 2(H) + 4(O) + P + C
+2 −8 + 5 + C = −2, ∴ C = −1

overall sum: 4(H) + 7(O) + P + 3C
+4 −14 +5 +2 = −3

(B) A SERIES OF TWO-CARBON MOLECULES

| | |
|---|---|
| ethane | $H_3C—CH_3$ |
| ethene | $H_2C{=}CH_2$ |
| ethanol | $H_3C—CH_2OH$ |
| acetaldehyde | $H_3C—CHO$ |
| acetate | $H_3C—COO^-$ |
| acetamide | $H_3C—CONH_2$ |
| ethylamine | $H_3C—CH_2NH_3^+$ |
| phosphoethanol | $H_3C—CH_2PO_4^{2-}$ |
| thioethane | $H_3C—CH_2SH$ |

**Figure 2–23** Oxidation states of carbon atoms in molecules (Problem 2–88). (A) Method for assigning oxidation states to individual carbon atoms in a molecule. The sum of the oxidation states of the attached atoms plus the carbon atom equals the charge, if any, on those atoms. (B) A list of two-carbon molecules on which to try the method.

it is necessary to ascertain the oxidation states of the carbon atoms in the molecules being studied. This can be done by following these three rules.

1. The oxidation states of all the atoms in the molecule—except carbon—are designated as follows: H is +1, O is –2, N is –3, S is –2, and P is +5.
2. The oxidation state of each carbon atom is calculated as the overall charge (if any) on the attached atoms minus the sum of the oxidation states of all the attached atoms. For this calculation attached carbon atoms—and all atoms connected through such carbon atoms—are ignored.
3. As a check, the sum of the oxidation states on all the atoms should equal the overall charge on the molecule.

An example of this method of assessing the oxidation states of carbon atoms in 3-phosphoglycerate is shown in Figure 2–23A.

A. Use this method to assign oxidation states to each of the carbon atoms in the molecules listed in Figure 2–23B.
B. Using the sum of the oxidation states of the carbon atoms as a measure of the overall oxidation state of a molecule, order the molecules in Figure 2–23B from most reduced to most oxidized.
C. Are the differences in oxidation state between the molecules equal to one electron or two electrons? Do you think there is any particular significance to your answer?
D. Which molecules are at the same overall oxidation state? If you can remember which functional groups are at the same oxidation state, then you can usually tell at a glance whether a redox reaction has occurred—without going through this method of assigning oxidation states.

**2–89** By comparing the oxidation states before and after a reaction, one can decide whether a redox reaction has occurred. A short segment of the pathway of reactions that occur in the citric acid cycle is shown in Figure 2–24.

A. Which of the reactions are redox reactions? Have the molecules involved in the redox reactions been oxidized or have they been reduced? How can you tell? Which carbons have lost electrons and which have gained them?
B. The redox reactions involve electron carriers that are not shown. Have the electron carriers been reduced or have they been oxidized?

**2–90** If the uncatalyzed reaction occurred at the rate of 1 event per century, and if an enzyme speeded up the rate by a factor of $10^{14}$, how many seconds would it take the enzyme to catalyze one event?

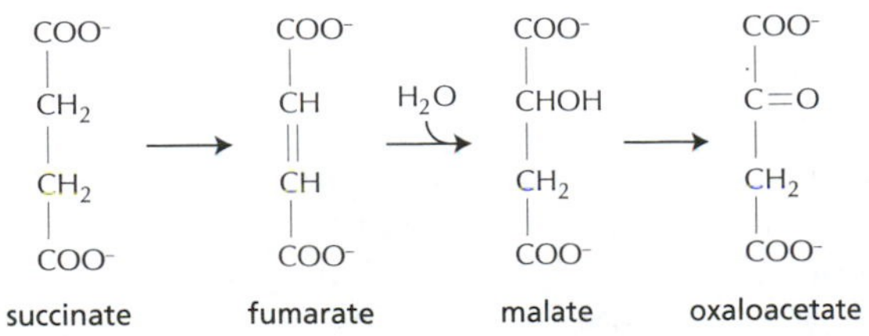

succinate ($COO^-$—$CH_2$—$CH_2$—$COO^-$) → fumarate ($COO^-$—CH=CH—$COO^-$) —$H_2O$→ malate ($COO^-$—CHOH—$CH_2$—$COO^-$) → oxaloacetate ($COO^-$—C=O—$CH_2$—$COO^-$)

**Figure 2–24** A segment of the citric acid cycle (Problem 2–89).

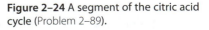

**2–91** 'Diffusion' sounds slow—and over everyday distances it is—but on the scale of a cell it is very fast. The average instantaneous velocity of a particle in solution, that is, the velocity between collisions, is

$$v = (kT/m)^{½}$$

where $k = 1.38 \times 10^{-16}$ g cm$^2$/K sec$^2$, $T$ = temperature in K (37°C is 310 K), $m$ = mass in g/molecule.

Calculate the instantaneous velocity of a water molecule (molecular mass = 18 daltons), a glucose molecule (molecular mass = 180 daltons), and a myoglobin molecule (molecular mass = 15,000 daltons) at 37°C. Just for fun, convert these numbers into kilometers/hour. Before you do any calculations, you might try to guess whether the molecules are moving at a slow crawl (<1 km/hr), an easy walk (5 km/hr), or a record-setting sprint (40 km/hr).

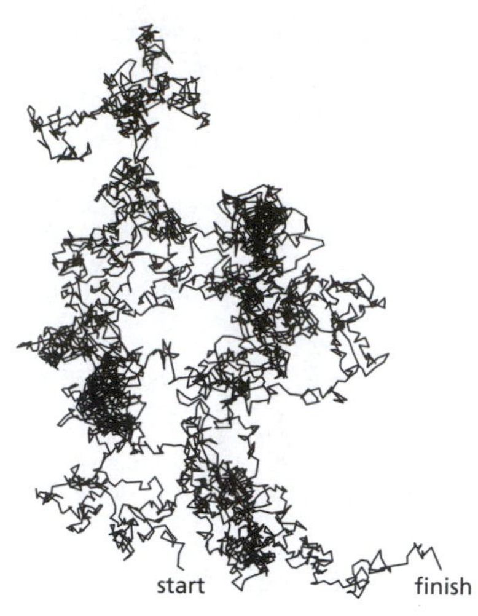


**Figure 2–25** A two-dimensional, simulated random walk of a molecule in solution (Problem 2–92).

**2–92** The instantaneous velocity tells you little about the time it takes for a molecule to move cellular distances because its trajectory is constantly altered by collisions with other molecules in solution (Figure 2–25). The average time it takes for a molecule to travel $x$ cm by diffusion in three dimensions is

$$t = x^2/6D$$

where $t$ is the time in seconds and $D$ is the diffusion coefficient, which is a constant that depends on the size and shape of the particle. Glucose and myoglobin, for example, have diffusion coefficients of about $5 \times 10^{-6}$ cm$^2$/sec and $5 \times 10^{-7}$ cm$^2$/sec, respectively. Calculate the average time it would take for glucose and myoglobin to diffuse a distance of 20 μm, which is approximately the width of a mammalian cell.

**Problems 16–30** and **16–102** deal with diffusion as a means for distributing cellular constituents.

**2–93** Phosphoglucose isomerase catalyzes the interconversion of glucose 6-phosphate (G6P) and fructose 6-phosphate (F6P):

$$\text{G6P} \rightleftharpoons \text{F6P}$$

The $\Delta G$ for this reaction is given by the equation

$$\Delta G = \Delta G° + 2.3\ RT \log \frac{[\text{F6P}]}{[\text{G6P}]}$$

where $R = 1.98 \times 10^{-3}$ kcal/K mole and $T$ = 310 K. A useful number to remember is that 2.3 $RT$ = 1.41 kcal/mole at 37°C, which is body temperature.

**Problem 14–28** examines the relationship between concentration and $\Delta G$ for ATP hydrolysis.

A. At equilibrium, [F6P]/[G6P] is equal to the equilibrium constant ($K$) for the reaction. Rewrite the above equation for the reaction at equilibrium.
B. At equilibrium, the ratio of [F6P] to [G6P] is observed to be 0.5. At this equilibrium ratio, what are the values of $\Delta G$ and $\Delta G°$?
C. Inside a cell the value of $\Delta G$ for this reaction is –0.6 kcal/mole. What is the ratio of [F6P] to [G6P]? What is $\Delta G°$?

**2–94** Phosphorylation of glucose (GLC) by ATP to produce glucose 6-phosphate (G6P) and ADP is the first step in glucose metabolism after entry into cells. It can be written as the sum of two reactions

| | | |
|---|---|---|
| (1) | GLC + $P_i \rightarrow$ G6P + $H_2O$ | $\Delta G°$ = 3.3 kcal/mole |
| (2) | ATP + $H_2O \rightarrow$ ADP + $P_i$ | $\Delta G°$ = –7.3 kcal/mole |
| NET: | GLC + ATP $\rightarrow$ G6P + ADP | |

A. What is the value of the equilibrium constant $K$ for reaction (1)?
B. In a liver cell [GLC] and [$P_i$] are both maintained at about 5 mM. What would the equilibrium concentration of G6P be under these conditions if reaction (1) were the sole source of G6P? Does this concentration of G6P seem reasonable for the initial step in glucose metabolism in cells? Why or why not?
C. In cells the phosphorylation of glucose is accomplished by addition of phosphate from ATP; that is, by the sum of reactions (1) and (2). What is $\Delta G°$ for the net reaction? What is the equilibrium constant $K$ for the net reaction?

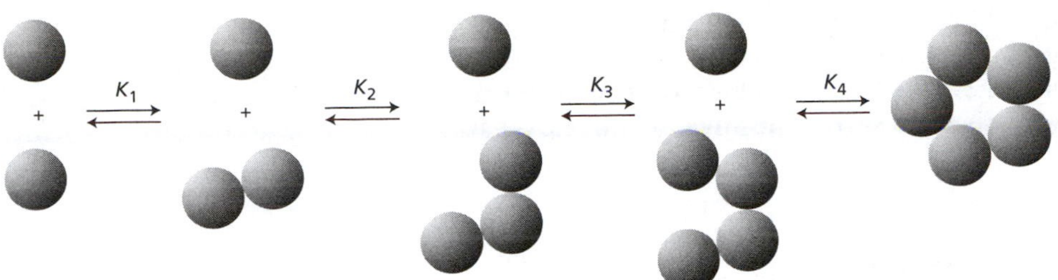


**Figure 2–26** Polymerization of subunits into a pentameric ring (Problem 2–96).

D. In liver cells [ATP] is maintained at about 3 mM and [ADP] at about 1 mM. Under these conditions what would be the equilibrium concentration of G6P? Does this concentration of G6P seem reasonable for the initial step in glucose metabolism in cells? Why or why not?

E. The concentration of G6P in liver cells is around 200 µM. Why is this concentration different than the one you calculated in part D? What is $\Delta G$ for the phosphorylation of glucose as it occurs under cellular conditions?

**2–95** A 70-kg adult human (154 lb) could meet his or her entire energy needs for one day by eating 3 moles of glucose (540 g). (We don't recommend this.) Each molecule of glucose generates 30 ATP when it is oxidized to $CO_2$. The concentration of ATP is maintained in cells at about 2 mM, and a 70-kg adult has about 25 L of intracellular fluid. Given that the ATP concentration remains constant in cells, calculate how many times per day, on average, each ATP molecule in the body is hydrolyzed and resynthesized.

## DATA HANDLING

**2–96** The polymerization of subunits into a pentameric ring is shown in Figure 2–26. The equilibrium constants for association of a subunit at each step in the assembly of the tetramer (that is, $K_1$, $K_2$, and $K_3$) are approximately equal at $10^6$ $M^{-1}$. The equilibrium constant for association of the final subunit in the ring ($K_4$), however, is $>10^{12}$ $M^{-1}$. Why is association of the final subunit so much more highly favored than association of the initial subunits? Why do you suppose the equilibrium constant for the association of the final subunit is approximately the square of the equilibrium constants for the earlier steps?

**2–97** Red blood cells obtain energy in the form of ATP by converting glucose to pyruvate via the glycolytic pathway (Figure 2–27). The values of $\Delta G°$ and $\Delta G$ have been calculated for each of the steps in glycolysis in red blood cells that are actively metabolizing glucose, as summarized in Table 2–4. The $\Delta G°$ values are based on the known equilibrium constants for the reactions; the $\Delta G$

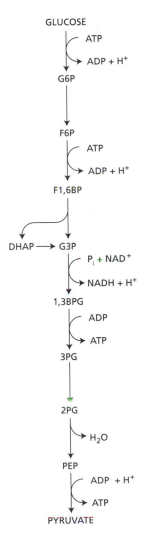

**Table 2–4 The reactions of glycolysis in red blood cells and their associated $\Delta G°$ and $\Delta G$ values** (Problem 2–97).

| STEP | REACTION | $\Delta G°$ | $\Delta G$ |
|---|---|---|---|
| 1 | GLC + ATP → G6P + ADP + $H^+$ | –4.0 | –8.0 |
| 2 | G6P → F6P | +0.4 | –0.6 |
| 3 | F6P + ATP → F1,6BP + ADP + $H^+$ | –3.4 | –5.3 |
| 4 | F1,6BP → DHAP + G3P | +5.7 | –0.3 |
| 5 | DHAP → G3P | +1.8 | +0.6 |
| 6 | G3P + $P_i$ + $NAD^+$ → 1,3BPG + NADH + $H^+$ | +1.5 | –0.4 |
| 7 | 1,3BPG + ADP → 3PG + ATP | –4.5 | +0.3 |
| 8 | 3PG → 2PG | +1.1 | +0.2 |
| 9 | 2PG → PEP + $H_2O$ | +0.4 | –0.8 |
| 10 | PEP + ADP + $H^+$ → PYR + ATP | –7.5 | –4.0 |

GLC = glucose, G6P = glucose 6-phosphate, F6P = fructose 6-phosphate, F1,6BP = fructose 1,6-bisphosphate, DHAP = dihydroxyacetone phosphate, G3P = glyceraldehyde 3-phosphate, 1,3BPG = 1,3-bisphosphoglycerate, 3PG = 3-phosphoglycerate, 2PG = 2-phosphoglycerate, PEP = phosphoenolpyruvate, PYR = pyruvate.

GLUCOSE
ATP
ADP + $H^+$
G6P
F6P
ATP
ADP + $H^+$
F1,6BP
DHAP → G3P
$P_i$ + $NAD^+$
NADH + $H^+$
1,3BPG
ADP
ATP
3PG
2PG
$H_2O$
PEP
ADP + $H^+$
ATP
PYRUVATE

**Figure 2–27** The glycolytic pathway (Problem 2–97). See Table 2–4 for the key to abbreviations.

values are calculated from the $\Delta G°$ values and actual measurements of concentrations of the intermediates in red cells.

Despite the assertion in Problem 2–83 that *all* $\Delta G$ values must be negative, three reactions in red-cell glycolysis have slightly positive $\Delta G$ values. What do you suppose is the explanation for the results in the table?

## HOW CELLS OBTAIN ENERGY FROM FOOD

TERMS TO LEARN

| | | |
|---|---|---|
| citric acid cycle | fermentation | nitrogen fixation |
| electron-transport chain | glycogen | oxidative phosphorylation |
| FAD/$FADH_2$ | glycolysis | starch |
| fat | GTP | |

### DEFINITIONS

Match each definition below with its term from the list above.

**2–98** Central metabolic pathway found in aerobic organisms, which oxidizes acetyl groups derived from food molecules to $CO_2$ and $H_2O$. In eucaryotic cells it occurs in the mitochondria.

**2–99** Energy-storage lipids in cells that are composed of triacylglycerols (triglycerides), which are fatty acids esterified with glycerol.

**2–100** Polysaccharide composed exclusively of glucose units used to store energy in animal cells. Granules of it are especially abundant in liver and muscle cells.

**2–101** Process in bacteria and mitochondria in which ATP formation is driven by the transfer of electrons from food molecules to molecular oxygen.

**2–102** Series of electron carrier molecules along which electrons move from a higher to a lower energy level to a final acceptor molecule, with the associated production of ATP.

**2–103** Ubiquitous metabolic pathway in the cytosol in which sugars are partially metabolized to produce ATP.

### TRUE/FALSE

Decide whether each of these statements is true or false, and then explain why.

**2–104** Because glycolysis is only a prelude to the oxidation of glucose in mitochondria, which yields 15-fold more ATP, glycolysis is not really important for human cells.

**2–105** The reactions of the citric acid cycle do not directly require the presence of oxygen.

### THOUGHT PROBLEMS

**2–106** Match the polymeric molecules in food with the monomeric subunits into which they are digested before they can be oxidized to produce energy.

| | |
|---|---|
| A. fats | 1. amino acids |
| B. polysaccharides | 2. fatty acids |
| C. proteins | 3. glycerol |
| | 4. sugars |

**2–107** From a chemical perspective, the glycolytic pathway (see Figure 2–27) can be thought of as occurring in two stages. The first stage from glucose to glyceraldehyde 3-phosphate (G3P) prepares glucose so its cleavage yields

G3P and an equivalent three-carbon fragment, which is then converted into G3P. The second stage—from G3P to pyruvate—harvests energy in the form of ATP and NADH.

A. Write the balanced equation for the first stage of glycolysis (glucose → G3P).
B. Write the balanced equation for the second stage of glycolysis (G3P → pyruvate).
C. Write the balanced equation for the overall pathway (glucose → pyruvate).

**2–108** The 10 reaction steps that make up the glycolytic pathway in Figure 2–27 are found in most living cells, from bacteria to humans. One could envision countless alternative reaction pathways for sugar oxidation that could, in principle, have evolved to take the place of glycolysis. Discuss the near universality of glycolysis in cells in the context of evolution.

**2–109** At first glance, fermentation of pyruvate to lactate appears to be an optional add-on reaction to glycolysis (Figure 2–28). After all, couldn't cells growing in the absence of oxygen simply discard pyruvate as a waste product? In the absence of fermentation, which products derived from glycolysis would accumulate in cells under anaerobic conditions? Could the metabolism of glucose via the glycolytic pathway continue in the absence of oxygen in cells that cannot carry out fermentation? Why or why not?

**2–110** In the absence of oxygen, cells consume glucose at a high, steady rate. When oxygen is added, glucose consumption drops precipitously and is then maintained at the lower rate. Why is glucose consumed at a high rate in the absence of oxygen and at a low rate in its presence?

**2–111** Arsenate ($AsO_4^{3-}$) is chemically very similar to phosphate ($PO_4^{3-}$) and is used as an alternative substrate by many phosphate-requiring enzymes. In contrast to phosphate, however, the anhydride bond between arsenate and a carboxylic acid group is very quickly hydrolyzed in water. Knowing this, suggest why arsenate is a compound of choice for murderers, but not for cells. Formulate your explanation in terms of the step in glycolysis at which 1,3-bisphosphoglycerate is converted to 3-phosphoglycerate, generating ATP (Figure 2–29).

**2–112** The liver provides glucose to the rest of the body between meals. It does so by breaking down glycogen, forming glucose 6-phosphate in the penultimate step. Glucose 6-phosphate is converted to glucose by splitting off the phosphate ($\Delta G° = -3.3$ kcal/mole). Why do you suppose the liver removes the phosphate by hydrolysis, rather than reversing the reaction by which glucose 6-phosphate is formed from glucose (glucose + ATP → G6P + ADP, $\Delta G° = -4.0$ kcal/mole)? By reversing this reaction, the liver could generate both glucose *and* ATP.

**2–113** What, if anything, is wrong with the following statement: "The oxygen consumed during the oxidation of glucose in animal cells is returned as $CO_2$ to the atmosphere." How might you support your answer experimentally?

GLUCOSE
ATP → ADP
G6P
F6P
ATP → ADP
F1,6BP
DHAP → G3P
$P_i$ + $NAD^+$ → NADH
1,3BPG
ADP → ATP
3PG
2PG
PEP
ADP → ATP
PYRUVATE
NADH → $NAD^+$
NADH, CoA → $NAD^+$, $CO_2$
AcCoA — citric acid cycle (aerobic)
LACTATE — fermentation pathway (anaerobic)

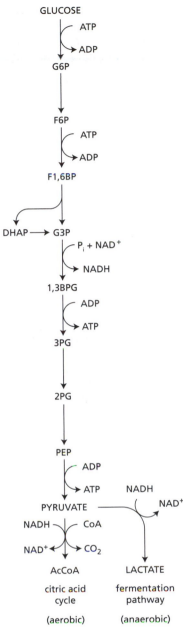

**Figure 2–28** Fermentation of pyruvate to lactate (Problem 2–109). $H^+$ ions and $H_2O$ are not shown.

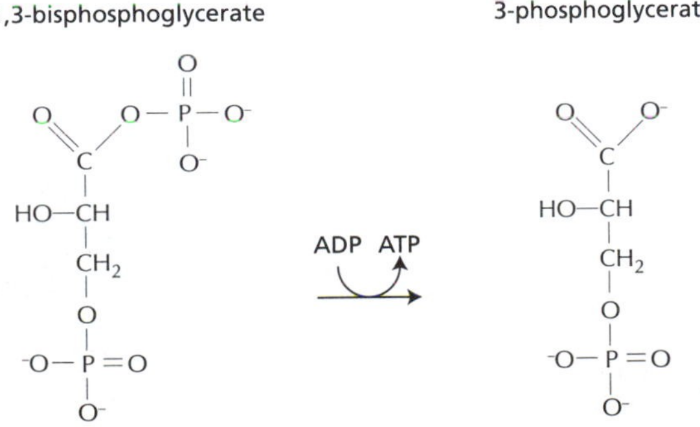


**Figure 2–29** Conversion of 1,3-bisphosphoglycerate and ADP to 3-phosphoglycerate and ATP (Problem 2–111).

**2–114** Oxidation of a sugar yields only about half as much energy as the oxidation of the same weight of a fatty acid. By comparing the structures of the sugar glucose and the fatty acid palmitate in Figure 2–30, can you give a general explanation for this observation?

**2–115** Humans are unable to synthesize all 20 amino acids. In the course of a day, when do you suppose humans make net amounts of new protein? Explain your answer.

**2–116** A cyclic reaction pathway requires that the starting material be regenerated at the end of each cycle. Intermediates in the citric acid cycle (see Figure 2–35), however, are siphoned off for use in a variety of other metabolic reactions. Why does the citric acid cycle not quickly cease to exist?

**2–117** An exceedingly sensitive instrument (yet to be devised) shows that one of the carbon atoms in Charles Darwin's last breath is resident in your bloodstream, where it forms part of a hemoglobin molecule. Suggest how this carbon atom might have traveled from Darwin to you, and list some of the molecules it could have passed through.

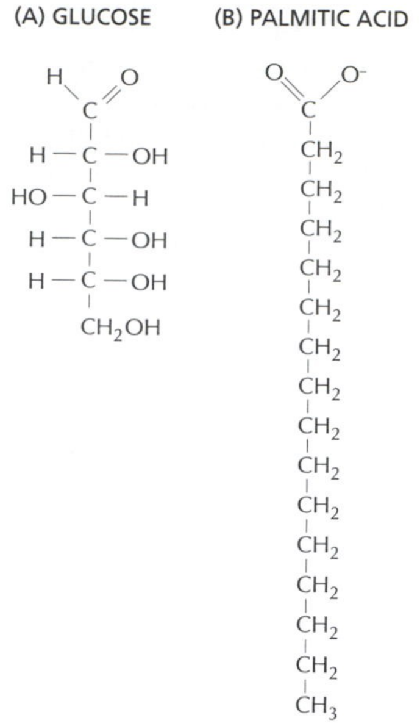

(A) GLUCOSE

H–C(=O)–
H–C–OH
HO–C–H
H–C–OH
H–C–OH
$CH_2OH$

(B) PALMITIC ACID

$O=C(–O^-)$–
$CH_2$–$CH_2$–$CH_2$–$CH_2$–$CH_2$–$CH_2$–$CH_2$–$CH_2$–$CH_2$–$CH_2$–$CH_2$–$CH_2$–$CH_2$–$CH_2$–$CH_3$

**Figure 2–30** Structures of glucose and palmitate (Problem 2–114).

## CALCULATIONS

**2–118** If a cell hydrolyzes and replaces $10^9$ ATP molecules per minute, how long will it take for a cell to consume its own volume (1000 $\mu m^3$) of oxygen? Roughly 90% of the ATP in the cell is regenerated by oxidative phosphorylation (the remainder is regenerated by substrate-level phosphorylation, which is described in Problem 2–122). Assume that 5 molecules of ATP are regenerated by each molecule of oxygen ($O_2$) that is converted to water. Recall that a mole of a gas occupies 22.4 L.

**2–119** Assuming that there are $5 \times 10^{13}$ cells in the human body and that ATP is turning over at a rate of $10^9$ ATP per minute in each cell, how many watts is the human body consuming? (A watt is a joule per second, and there are 4.18 joules/calorie.) Assume that hydrolysis of ATP yields 12 kcal/mole.

**2–120** Does a Snickers™ candy bar (65 g, 325 kcal) provide enough energy to climb from Zermatt (elevation 1660 m) to the top of the Matterhorn (4478 m, Figure 2–31), or might you need to stop at Hörnli Hut (3260 m) to eat another one? Imagine that you and your gear have a mass of 75 kg, and that all of your work is done against gravity (that is, you're just climbing straight up).

$$\text{work (J)} = \text{mass (kg)} \times g\ (\text{m/sec}^2) \times \text{height gained (m)}$$

where $g$ is acceleration due to gravity (9.8 m/sec$^2$). One joule is 1 kg m$^2$/sec$^2$ and there are 4.18 kJ per kcal.

What assumptions made here will greatly underestimate how much candy you need?

**Figure 2–31** The Matterhorn (Problem 2–120).

**2–121** One of the two ATP-generating steps in glycolysis is outlined in Figure 2–32. This sequence of reactions yields ATP and produces pyruvate, which is subsequently converted to acetyl CoA and oxidized to $CO_2$ in the citric acid cycle. Under anaerobic conditions ATP production from phosphoenolpyruvate (PEP) accounts for half of a cell's ATP supply. These 'substrate-level' phosphorylation events (so named to distinguish them from oxidative phosphorylation in mitochondria) were the first to be understood. Consider the conversion of 3-phosphoglycerate (3PG) to PEP, which constitutes the first two reactions in Figure 2–32. Recall that

$$\Delta G° = -2.3\ RT \log K$$
$$= -1.41\ \text{kcal/mole} \log K$$

where $K$ is the equilibrium ratio of the products over the reactants.

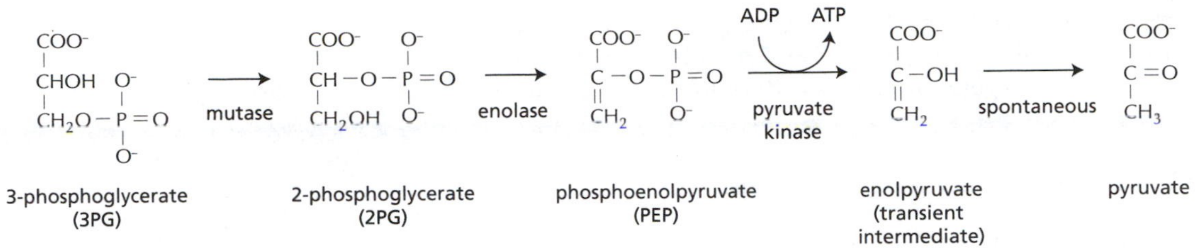

mutase enolase ADP ATP pyruvate kinase spontaneous

3-phosphoglycerate (3PG) → 2-phosphoglycerate (2PG) → phosphoenolpyruvate (PEP) → enolpyruvate (transient intermediate) → pyruvate

**Figure 2–32** Conversion of 3-phosphoglycerate to pyruvate during glycolysis (Problem 2–121).

A. If 10 mM 3PG is mixed with phosphoglycerate mutase, which catalyzes its conversion to 2-phosphoglycerate (2PG) (Figure 2–32), the equilibrium concentrations at 37°C are 8.3 mM 3PG and 1.7 mM 2PG. How would the ratio of equilibrium concentrations change if 1 M 3PG had been added initially? What is the equilibrium constant $K$ for the conversion of 3PG into 2PG, and what is the $\Delta G°$ for the reaction?

B. If 10 mM PEP is mixed with enolase, which catalyzes its conversion to 2PG, the equilibrium concentrations at 37°C are 2.9 mM 2PG and 7.1 mM PEP. What is the $\Delta G°$ for conversion of PEP to 2PG? What is the $\Delta G°$ for the reverse reaction?

C. What is the $\Delta G°$ for the conversion of 3PG to PEP?

**2–122** The study of substrate-level phosphorylation events such as that in Figure 2–32 led to the concept of the 'high-energy' phosphate bond. The term 'high-energy' bond is somewhat misleading since it refers not to the strength of the bond, but rather, to the free-energy change ($\Delta G$) upon hydrolysis. In the conversion of 3PG to PEP (the first two reactions in Figure 2–32), a low-energy phosphate bond is turned into a high-energy phosphate bond. The standard free-energy change ($\Delta G°$) for hydrolysis of the phosphate group in 3PG is about –3.3 kcal/mole, whereas hydrolysis of the phosphate group in PEP (converting it to pyruvate) has a $\Delta G°$ of –14.8 kcal/mole. How is it that moving the phosphate to the 2 position of glycerate and removing water, which has an overall $\Delta G°$ of 0.4 kcal/mole, can have such an enormous effect on the free energy for the subsequent hydrolysis of the phosphate bond?

Consider the set of reactions shown in Figure 2–33.

A. What is the $\Delta G°$ for the conversion of 3PG to pyruvate by way of PEP?

B. What is the $\Delta G°$ for the conversion of 3PG to pyruvate by way of glycerate? What is the $\Delta G°$ for the conversion of glycerate to pyruvate? (Recall that thermodynamic quantities are state functions; that is, they describe differences between the initial and final states—they are independent of the pathway between the states.)

C. Propose an explanation for why the phosphate bond in PEP is a high-energy bond, whereas the one in 3PG is a low-energy bond. (Assume that the removal of water from glycerate has a $\Delta G°$ of –0.5 kcal/mole.)

D. In cells the conversion of PEP to pyruvate is linked to the synthesis of ATP ($\Delta G°$ = 7.3 kcal/mole) as shown in Figure 2–32. What is the $\Delta G°$ for the linked reactions?

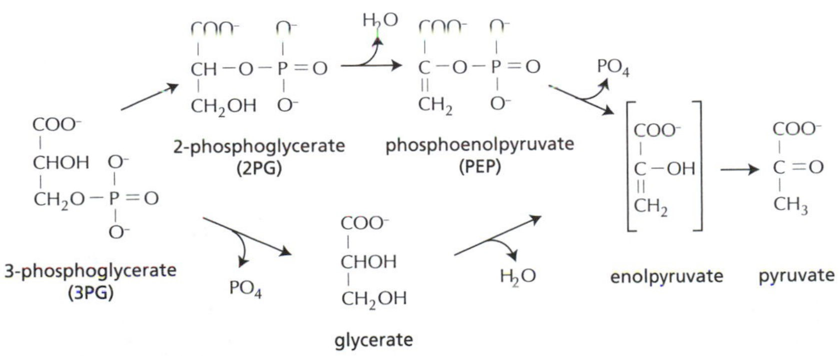


**Figure 2–33** Conversion of 3PG to pyruvate and phosphate by two routes (Problem 2–122). The *bracketed* compound, enolpyruvate, is a transient intermediate.

**2–123** Muscles contain creatine phosphate (CP) as an energy buffer to maintain the levels of ATP in the initial stages of exercise. Creatine phosphate can transfer its phosphate to ADP to generate creatine (C) and ATP, with a $\Delta G°$ of –3.3 kcal/mole.

$$\text{CP} + \text{ADP} \rightarrow \text{C} + \text{ATP} \qquad \Delta G° = -3.3 \text{ kcal/mole}$$

A. In a resting muscle [ATP] = 4 mM, [ADP] = 0.013 mM, [CP] = 25 mM, and [C] = 13 mM. What is the $\Delta G$ for this reaction in resting muscle? Does this value make sense to you? Why or why not?

B. Consider an initial stage in vigorous exercise, when 25% of the ATP has been converted to ADP. Assuming that no other concentrations have changed, what is the $\Delta G$ for the reaction at this stage in exercising muscle? Does this value make sense?

C. If the ATP in muscle could be completely hydrolyzed (in reality it never is), it would power an all-out sprint for about 1 second. If creatine phosphate could be completely hydrolyzed to regenerate ATP, how long could a sprint be powered? Where do you suppose the energy comes from to allow a runner to finish a 200-meter sprint?

**2–124** The last reaction of the citric acid cycle, which regenerates oxaloacetate (OAA) from malate (MAL), has a very positive $\Delta G°$ of 7.1 kcal/mole.

$$\text{MAL} + \text{NAD}^+ \rightarrow \text{OAA} + \text{NADH} \qquad \Delta G° = 7.1 \text{ kcal/mole}$$

Despite its unfavorable equilibrium position, material must flow through this reaction quite readily in mitochondria—otherwise the cycle could not turn.

A. How is flow through the cycle accomplished in the face of such an overwhelmingly positive $\Delta G°$?

B. If the ratio of [$NAD^+$] to [NADH] is maintained at about 10 in mitochondria, what is the minimum ratio of [MAL] to [OAA] when the cycle is turning?

## DATA HANDLING

**2–125** In 1904 Franz Knoop performed what was probably the first successful labeling experiment to study metabolic pathways. He fed many different fatty acids labeled with a terminal benzene ring to dogs and analyzed their urine for excreted benzene derivatives. Whenever the fatty acid had an even number of carbon atoms, phenylacetate was excreted (Figure 2–34A). Whenever the fatty acid had an odd number of carbon atoms, benzoate was excreted (Figure 2–34B).

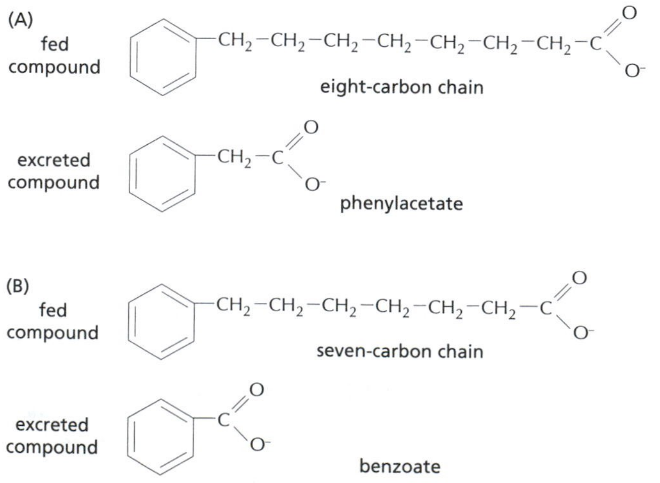

(A)
fed compound
$C_6H_5-CH_2-CH_2-CH_2-CH_2-CH_2-CH_2-CH_2-COO^-$
eight-carbon chain

excreted compound
$C_6H_5-CH_2-COO^-$
phenylacetate

(B)
fed compound
$C_6H_5-CH_2-CH_2-CH_2-CH_2-CH_2-CH_2-COO^-$
seven-carbon chain

excreted compound
$C_6H_5-COO^-$
benzoate

**Figure 2–34** The original labeling experiment to analyze fatty acid oxidation (Problem 2–125). **(A)** Fed and excreted derivatives of an even-number fatty acid chain. **(B)** Fed and excreted derivatives of an odd-number fatty acid chain.

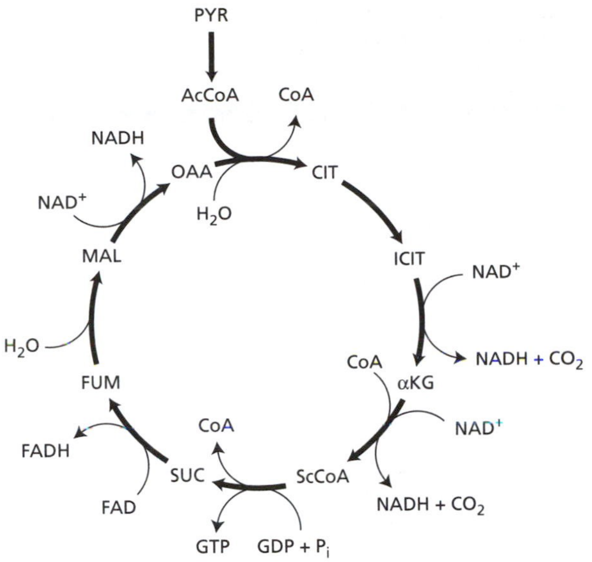

**Figure 2–35** The citric acid cycle (Problem 2–126). PYR = pyruvate, AcCoA = acetyl coenzyme A, CIT = citrate, ICIT = isocitrate, αKG – α-ketoglutarate, ScCoA = succinyl coenzyme A, SUC = succinate, FUM = fumarate, MAL = malate, and OAA = oxaloacetate.

From these experiments Knoop deduced that oxidation of fatty acids to $CO_2$ and $H_2O$ involved the removal of two-carbon fragments from the carboxylic acid end of the chain. Can you explain the reasoning that led him to conclude that two-carbon fragments, as opposed to any other number, were removed, and that degradation was from the carboxylic acid end, as opposed to the other end?

**2–126** In 1937 Hans Krebs deduced the operation of the citric acid cycle (Figure 2–35) from careful observations on the oxidation of carbon compounds in minced preparations of pigeon flight muscle. (Pigeon breast is a rich source of mitochondria, but the function of mitochondria was unknown at the time.) The consumption of $O_2$ and the production of $CO_2$ were monitored with a manometer, which measures changes in volume of a closed system at constant pressure and temperature. Standard chemical methods were used to determine the concentrations of key metabolites. (Remember, radioactive isotopes were not available then.)

In one set of experiments Krebs measured the rate of consumption of $O_2$ during the oxidation of endogenous carbohydrates in the presence or absence of citrate. As shown in Table 2–5, addition of a small amount of citrate resulted in a large increase in the consumption of oxygen. Szent-Györgyi (1924, 1937) and Stare and Baumann (1936) had previously shown that fumarate, oxaloacetate, and succinate also stimulated respiration in extracts of pigeon breast muscle.

When metabolic poisons, such as arsenite or malonate (whose modes of action were undefined), were added to the minced muscles, the results were very different. In the presence of arsenite, 5.5 mmol of citrate were converted into about 5 mmol of α-ketoglutarate. In the presence of malonate an equivalent conversion of citrate into succinate occurred. Furthermore, in the presence of malonate roughly 5 mmol of oxygen were consumed (above background levels in the absence of citrate), which was twice as much as in the presence of arsenite.

**Table 2–5 Respiration in minced pigeon breast in the presence and absence of citrate** (Problem 2–126).

| TIME (minutes) | OXYGEN CONSUMPTION (mmol) | | |
|---|---|---|---|
| | NO CITRATE | 3 mmol CITRATE | DIFFERENCE |
| 30 | 29 | 31 | 2 |
| 60 | 47 | 68 | 21 |
| 90 | 51 | 87 | 36 |
| 150 | 53 | 93 | 40 |

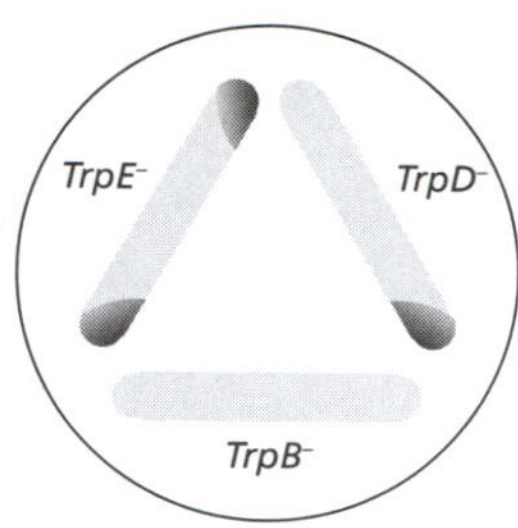


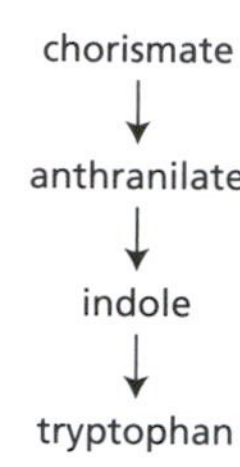


**Figure 2–36** Defining the pathway for tryptophan synthesis using cross-feeding experiments (Problem 2–127). (A) Results of a cross-feeding experiment among mutants defective for steps in the tryptophan biosynthetic pathway. *Dark areas* on the Petri dish show regions of cell growth. (B) The tryptophan biosynthetic pathway. Several steps precede chorismate in the pathway and there are several steps between anthranilate and indole.

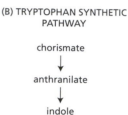

Finally, Krebs showed that the minced muscles were capable of synthesizing citrate if oxaloacetate was added and all traces of oxygen were excluded. None of the other intermediates in the cycle led to a net synthesis of citrate in the absence of oxygen.

A. If citrate ($C_6H_8O_7$) were completely oxidized to $CO_2$ and $H_2O$, how many molecules of $O_2$ would be consumed per molecule of citrate? What is it about the results in Table 2–5 that caught Krebs's attention?

B. Why is the consumption of oxygen so low in the presence of arsenite or malonate? If citrate were oxidized to α-ketoglutarate ($C_5H_6O_5$), how much oxygen would be consumed per molecule of citrate? If citrate were oxidized to succinate ($C_4H_6O_4$), how much oxygen would be consumed per molecule of citrate? Does the observed stoichiometry agree with the expectations based on these calculations?

C. Why, in the absence of oxygen, does oxaloacetate alone cause an accumulation of citrate? Would any of the other intermediates in the cycle cause an accumulation of citrate in the presence of oxygen?

D. Toward the end of the paper Krebs states, "While the citric acid cycle thus seems to occur generally in animal tissues, it does not exist in yeast or in *E. coli*, for yeast and *E. coli* do not oxidize citric acid at an appreciable rate." Why do you suppose Krebs got this point wrong?

**2–127** Pathways for synthesis of amino acids in microorganisms were worked out in part by cross-feeding experiments among mutant organisms that were defective for individual steps in the pathway. Results of cross-feeding experiments for three mutants defective in the tryptophan pathway—*TrpB*⁻, *TrpD*⁻, and *TrpE*⁻—are shown in Figure 2–36A. The mutants were streaked on a Petri dish and allowed to grow briefly in the presence of a very small amount of tryptophan, producing three pale streaks. As shown, heavier growth was observed at points where some streaks were close to other streaks. These spots of heavier growth indicate that one mutant can cross-feed (supply an intermediate) to the other one.

A. From the pattern of cross-feeding shown in Figure 2–36A, deduce the order of the steps controlled by the products of the *TrpB*, *TrpD*, and *TrpE* genes. Explain your reasoning.

B. If accumulated intermediates at the block are responsible for the cross-feeding phenomenon, it should be possible to grow individual mutants on some intermediates. The three mutants were tested for growth on tryptophan and intermediates in the pathway (Figure 2–36B), with the results shown in Table 2–6. Use this information to arrange the defective genes relative to the tryptophan pathway.

**Table 2–6 Growth of mutants on intermediates in the pathway for tryptophan biosynthesis** (Problem 2–127).

| | GROWTH ON MINIMAL MEDIUM SUPPLEMENTED WITH | | | | |
|---|---|---|---|---|---|
| STRAIN | NONE | CHORISMATE | ANTHRANILATE | INDOLE | TRYPTOPHAN |
| Wild type | + | + | + | + | + |
| *TrpB*⁻ | – | – | – | – | + |
| *TrpD*⁻ | – | – | – | + | + |
| *TrpE*⁻ | – | – | + | + | + |

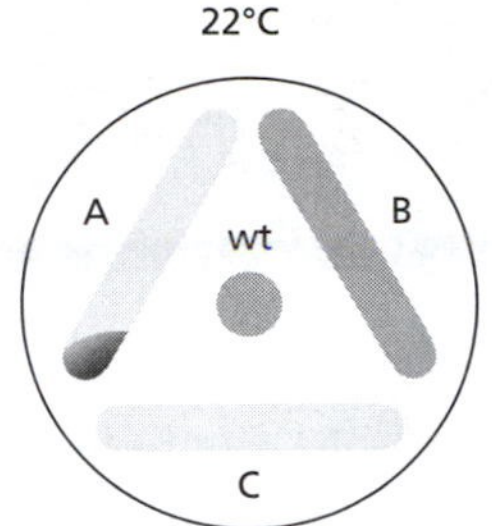

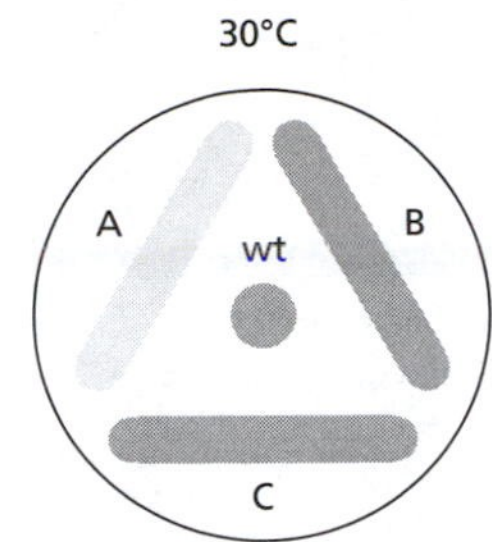

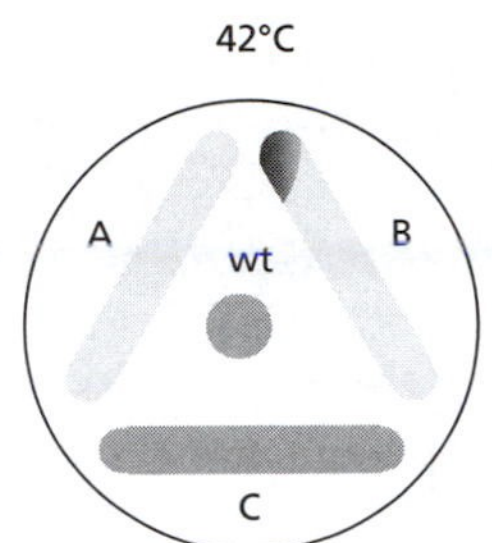


**Figure 2–37** Results of cross-feeding experiments with three strains defective in proline biosynthesis (Problem 2–128). *Dark areas* show regions of cell growth. wt, wild type.

**2–128** You have isolated three different strains of bacteria—*ProA*⁻, *ProB*⁻, and *ProC*⁻—that require added proline for growth. One is cold-sensitive, one is temperature-sensitive (heat-sensitive), and one has a deletion of the gene. You carry out cross-feeding experiments by streaking the strains out on agar plates containing minimal medium supplemented with a very low level of proline. After growth at three temperatures, you observe the results shown in Figure 2–37.

A. Identify the types of mutations—cold-sensitive, temperature-sensitive, or deletion—in each of the strains.

B. Deduce the order in which the gene products act in the pathway for proline biosynthesis.

C. Does the identification of three different genes that affect proline biosynthesis mean that there are three steps in the biosynthetic pathway? Explain your answer.

D. Why do you suppose there was no cross-feeding of *ProA*⁻ by strain *ProC*⁻ at 30°C or 42°C, or between the wild-type bacteria and the mutant strains under any conditions?

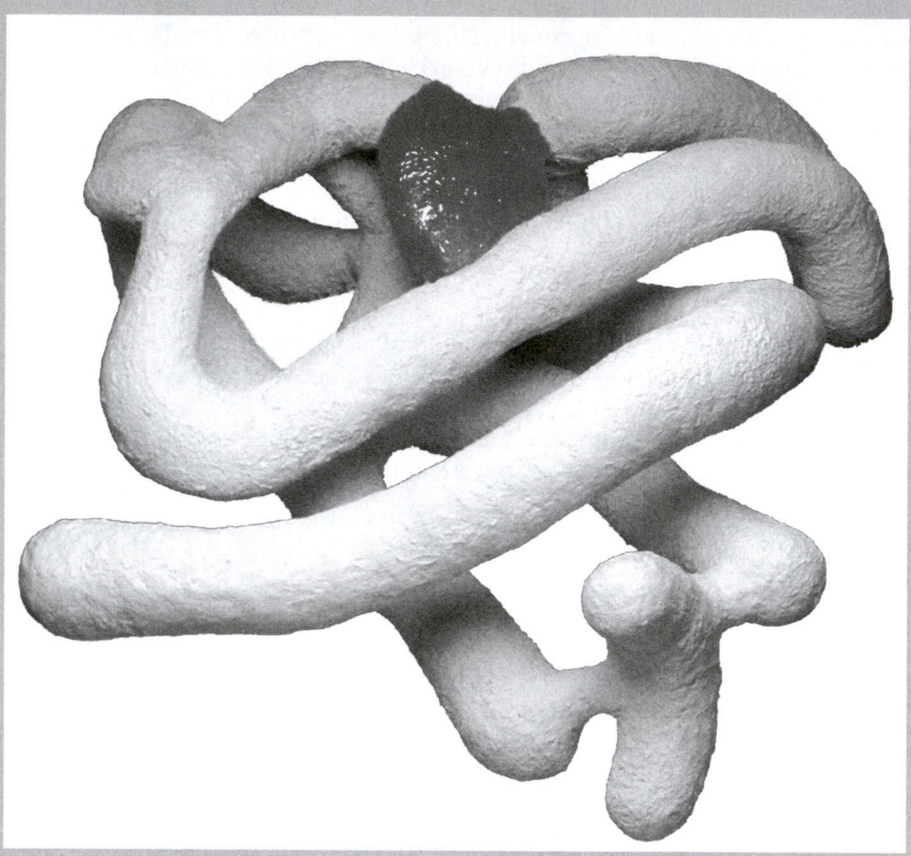

**John Kendrew's original 'sausage' model of myoglobin.** It came as a great surprise to Kendrew and Perutz when the first successful X-ray diffraction map of the electron density of the myoglobin molecule showed no regular features whatsoever. The dark patch at the top of the model represents the heme group that binds oxygen reversibly.

# Proteins

3

## THE SHAPE AND STRUCTURE OF PROTEINS

### In This Chapter


| | |
|---|---|
| THE SHAPE AND STRUCTURE OF PROTEINS | 39 |
| PROTEIN FUNCTION | 47 |


TERMS TO LEARN

| | | |
|---|---|---|
| α helix | polypeptide backbone | quaternary structure |
| β sheet | primary structure | secondary structure |
| binding site | protein | side chain |
| coiled-coil | protein domain | tertiary structure |
| conformation | protein subunit | |

### DEFINITIONS

Match the definition below with its term from the list above.

**3–1** Three-dimensional relationship of the different polypeptide chains in a multisubunit protein or protein complex.

**3–2** Common folding pattern in proteins in which a linear sequence of amino acids folds into a right-handed coil stabilized by internal hydrogen bonding between backbone atoms.

**3–3** The amino acid sequence of a protein.

**3–4** A region on the surface of a protein that can interact with another molecule through noncovalent bonding.

**3–5** Complex three-dimensional form of a folded protein.

**3–6** The chain of repeating carbon and nitrogen atoms, linked by peptide bonds, in a protein.

**3–7** Common structural motif in proteins in which different sections of the polypeptide chain run alongside each other and are joined together by hydrogen bonding between atoms of the polypeptide backbone.

**3–8** Portion of a protein that has a tertiary structure of its own.

**3–9** Regular local folding patterns in a protein, including α helix and β sheet.

### TRUE/FALSE

Decide whether each of these statements is true or false, and then explain why.

**3–10** A protein is at a near entropy minimum (point of lowest disorder, or greatest order) when it is completely stretched out like a string and when it is properly folded up.

**3–11** Each strand in a β sheet is a helix with two amino acids per turn.

**3–12** Loops of polypeptide that protrude from the surface of a protein often form the binding sites for other molecules.

## THOUGHT PROBLEMS

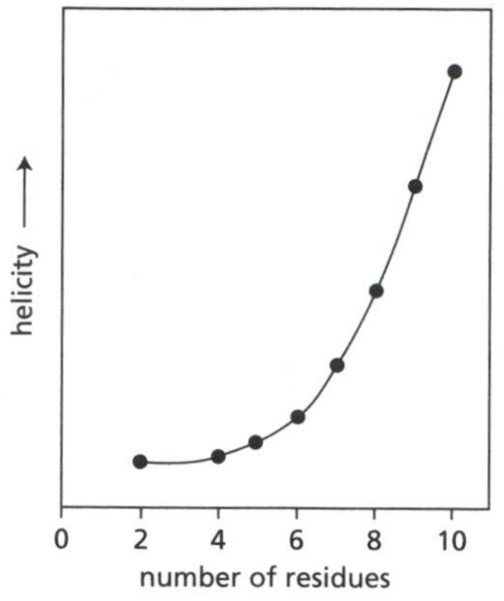


**Figure 3–1** Helicity of various peptides of increasing length (Problem 3–16).

**3–13** Amino acids are commonly grouped as nonpolar (hydrophobic) or as polar (hydrophilic), based on the properties of the amino acid side chains. Quantification of these properties is important for a variety of protein structure predictions, but they cannot readily be determined from the free amino acids themselves. Why are the properties of the side chains difficult to measure using the free amino acids? How do you suppose you might measure the chemical properties of just that portion of an amino acid that corresponds to the side chain? How do you suppose the hydrophilicity or hydrophobicity of a molecule might be determined?

**3–14** What are the four weak (noncovalent) interactions that determine the conformation of a protein?

**3–15** When egg white is heated, it hardens. This cooking process cannot be reversed, but hard-boiled egg white can be dissolved by heating it in a solution containing a strong detergent (such as sodium dodecyl sulfate) together with a reducing agent, like 2-mercaptoethanol. Neither reagent alone has any effect.

A. Why does boiling an egg white cause it to harden?

B. Why does it require both a detergent and a reducing agent to dissolve the hard-boiled egg white?

**3–16** Although α helices are common components of polypeptide chains, they need to be of a certain minimum length. To find out how chain length affects α-helix formation, you measure the circular dichroism (a measure of helicity) for a series of peptides of increasing length (Figure 3–1). Why is there essentially no helix formation until the chain is at least six amino acids long?

**3–17** The uniform arrangement of the backbone carbonyl oxygens and amide nitrogens in an α helix gives the helix a net dipole, so that it carries a partial positive charge at the amino end and a partial negative charge at the carboxyl end. Where would you expect the ends of α helices to be located in a protein? Why?

**3–18** α Helices are often embedded in a protein so that one side faces the surface and the other side faces the interior. Such helices are often termed amphiphilic because the surface side is hydrophilic and the interior side is hydrophobic. A simple way to decide whether a sequence of amino acids might form an amphiphilic helix is to arrange the amino acids around what is known as a 'helix-wheel projection' (Figure 3–2). If the hydrophobic and hydrophilic amino acids are segregated on opposite sides of the wheel, the helix is amphiphilic. Using the helix-wheel projection, decide which of the three peptides in Figure 3–2 might form an amphiphilic helix. (The mnemonic 'FAMILY VW' will help you recognize hydrophobic amino acids.)

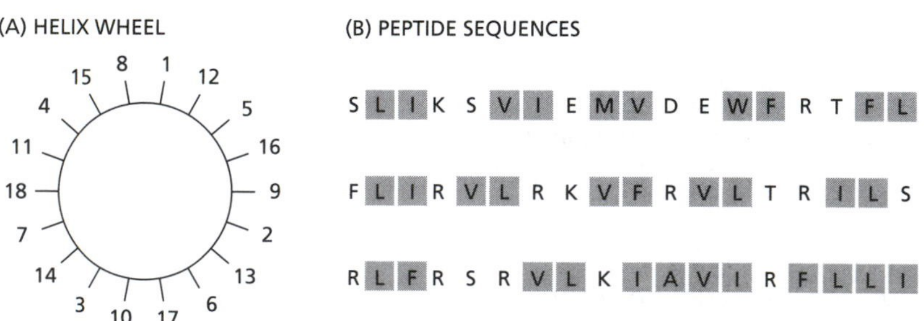


**Figure 3–2** Helix-wheel projection (Problem 3–18). (A) Helix wheel. The *circle* (wheel) represents the helix as viewed from one end. *Numbers* show the positions of the amino acid side chains, as projected on the wheel. The positions of the first 18 amino acids are shown; amino acid 19 would occupy the same position as amino acid 1. Amino acid 1 is closest to the reader; amino acid 18 is farthest away. (B) Peptide sequences. The N-termini are shown at the *left*; hydrophobic amino acids are *shaded*; hydrophilic amino acids are *unmarked*. (See inside back cover for one-letter amino acid code.)

**3–19** Examine the segment of β sheet shown in Figure 3–3. For each strand of the sheet decide whether it is parallel or antiparallel to each of its neighbors.

**3–20** Like α helices, β sheets often have one side facing the surface of the protein and one side facing the interior, giving rise to an amphiphilic sheet with one hydrophobic surface and one hydrophilic surface. From the sequences listed below pick the one that could form a strand in an amphiphilic β sheet. (See inside back cover for one-letter amino acid code; the mnemonic in Problem 3–18 might also be helpful.)

A. A L S C D V E T Y W L I
B. D K L V T S I A R E F M
C. D S E T K N A V F L I L
D. T L N I S F Q M E L D V
E. V L E F M D I A S V L D

**3–21** Several different protein folds are represented in schematic form in Figure 3–4. These diagrams preserve the topology of the protein and allow one to decide, for example, whether a protein is folded in a new way or is an example of a protein fold that is already known. These diagrams also permit a ready demonstration of a fundamental principle of protein folding. For each of these folds, imagine that you could grasp the N and C termini and pull them apart. Would any of the illustrated folds produce a knot when fully stretched out?

**3–22** It is a common observation that antiparallel strands in a β sheet are connected by short loops, but that parallel strands are connected by α helices. Why do you think this is?

**3–23** In 1968 Cyrus Levinthal pointed out a complication in protein folding that is widely known as the Levinthal paradox. He argued that because there are astronomical numbers of conformations open to a protein in the denatured state, it would take a very long time for a protein to search through all the possibilities to find the correct one, even if it tested each possible conformation exceedingly rapidly. Yet denatured proteins typically take less than a second to fold inside the cell or in the test tube. How do you suppose that proteins manage to fold so quickly?

**3–24** Comparison of a homeodomain protein from yeast and *Drosophila* shows that only 17 of 60 amino acids are identical. How is it possible for a protein to change over 70% of its amino acids and still fold in the same way?

**3–25** Neither yeast nor *Drosophila* has been around for more than a few hundred million years, yet they are separated by more than a billion years of evolution. How can that possibly be true?

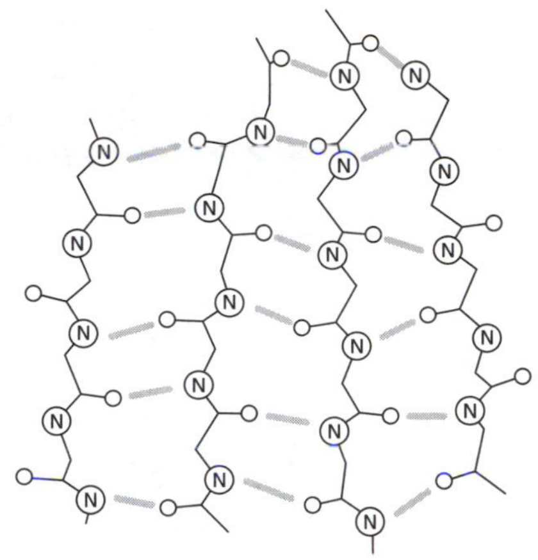

**Figure 3–3** A segment of β sheet from the interior of thioredoxin (Problem 3–19). Amide nitrogens are indicated by *circled* Ns; hydrogen bonds are shown as *gray lines*.

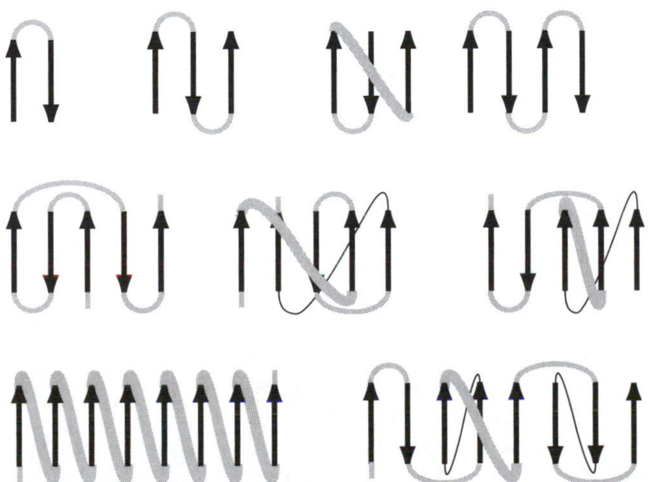

**Figure 3–4** Topological representations of several protein folds (Problem 3–21). *Vertical arrows* represent strands in β sheets; *gray connectors* may be loops or helices. *Thick gray diagonal lines* are above the plane of the page; *thin black lines* lie below the plane of the page.

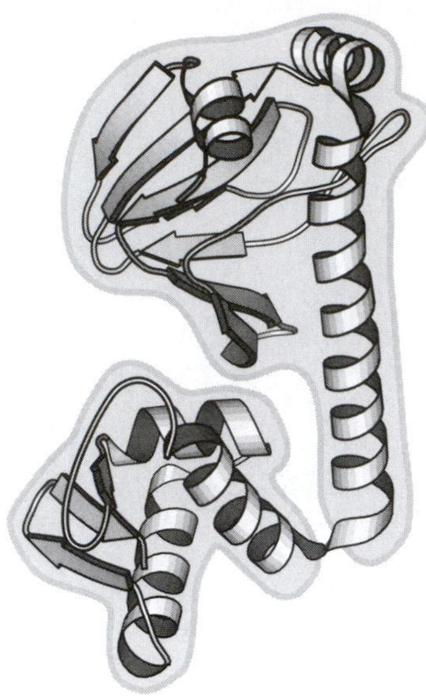

Figure 3–5 Catabolite activator protein from *E. coli* (Problem 3–26). *Shading* indicates its domain structure.

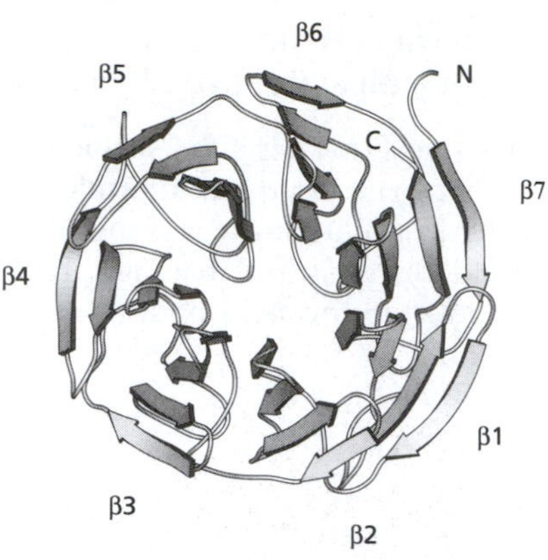


**Figure 3–6** The kelch repeat domain of galactose oxidase from *D. dendroides* (Problem 3–29). The seven individual β propellers are indicated. The N- and C-termini are indicated by N and C.

**3–26** Often, the hard part of protein structure determination by x-ray diffraction is getting good crystals. In difficult cases, there are two common approaches for obtaining crystals: (1) using fragments of the protein and (2) trying homologous proteins from different species.

A. Examine the protein in Figure 3–5. Where would you cleave this protein to obtain fragments that might be expected to fold properly and perhaps form crystals?

B. Why do you suppose that homologous proteins from different species might differ in their ability to form crystals?

**3–27** Some 1000 different protein folds are now known, and it is estimated that there may only be around 2000 in total. Protein folds seem to stay the same as genes evolve, giving rise to large families of similarly folded proteins with related functions. Does this mean that the last common ancestor to all life on Earth had 1000–2000 different genes?

**3–28** A common strategy for identifying distantly related homologous proteins is to search the database using a short signature sequence indicative of the particular protein function. Why is it better to search with a short sequence than with a long sequence? Don't you have more chances for a 'hit' in the database with a long sequence?

**3–29** The so-called kelch motif consists of a four-stranded β sheet, which forms what is known as a β propeller. It is usually found to be repeated four to seven times, forming a kelch repeat domain in a multidomain protein. One such kelch repeat domain is shown in Figure 3–6. Would you classify this domain as an 'in-line' or 'plug-in' type domain?

**3–30** Examine the three protein monomers in Figure 3–7. From the arrangement of complementary binding surfaces, which are indicated by similarly shaped protrusions and invaginations, decide which monomer would assemble into a ring, which would assemble into a chain, and which would assemble into a sheet.

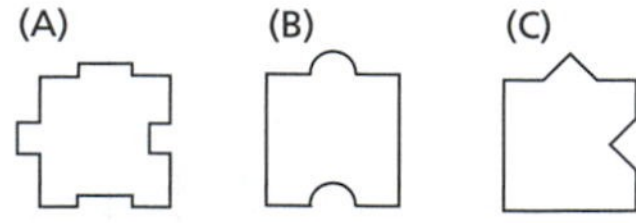


**Figure 3–7** Three protein monomers (Problem 3–30).

**3–31** Cro is a bacterial gene regulatory protein that binds to DNA to turn genes off. It is a symmetrical 'head-to-head' dimer. Each of the two subunits of the dimer recognizes a particular short sequence of nucleotides in DNA. If the sequence of nucleotides recognized by one subunit is represented as an arrow (→), so that the 'head' of the arrow corresponds to DNA recognized by

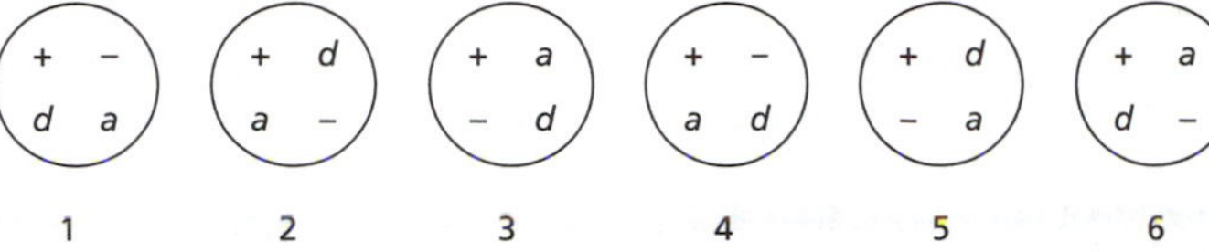


**Figure 3–8** Binding surfaces for six different proteins (Problem 3–33). In each case the bulk of the protein is below the plane of the page.

the 'head' of the subunit, which of the following sequences in DNA represents the binding site for the Cro dimer?

A. →→
B. →←
C. ←←
D. ←→
E. Could be more than one of the above.

**3–32** Why is it that there are numerous examples of 'head-to-head' and 'tail-to-tail' dimers, but few, if any, examples of 'head-to-tail' dimers?

**3–33** Proteins bind to one another via weak interactions across complementary surfaces. Oppositely charged amino acids are apposed, as are hydrogen-bond donors and acceptors, and protrusions match invaginations so that van der Waals contacts can be maximized. When two copies of a protein bind to form a 'head-to-head' dimer, they use the same binding surface. Examine the binding surfaces of the six proteins shown in Figure 3–8, where charged amino acids are indicated by + and –, and hydrogen-bond donors and acceptors are indicated by *d* and *a*. (Protrusions and invaginations—three-dimensional shapes—are not represented in the binding surfaces in Figure 3–8 just because it is difficult to do so, but their absence does not change the general principles derived from this problem.) In which cases could two copies of one protein form a 'head-to-head' dimer in which the charges and hydrogen-bonding groups are appropriately matched? Can you spot any common feature of the surfaces that allows such dimers to form?

**3–34** Nuclear lamin C is a member of the intermediate filament family. Thus, it should show regions of the coiled-coil heptad repeat motif AbcDefg, where A and D are hydrophobic amino acids and b, c, e, f, and g can be almost any amino acid. The sequence of nuclear lamin C is shown in Figure 3–9 with potential coiled-coil regions highlighted. Examine the segment marked 'coil 1A.' Does it conform to the heptad repeat? (Don't forget the mnemonic FAMILY VW.)

coil 1A
METPSQRRATRSGAQASSTPLSPTRITRLQEKED**LQELNDRLAVYIDRVRSLETENA**
coil 1B
**GLRLRITESEEVV**SREVSGIKAA**YEAELGDARKTLDSVAKERARLQLELSKVREEFK**

**ELKARNTKKEGDLIAAQARLKDLEALLNSKEAALSTALSEKRTLEGELHDLRGQVAK**

**LEAALGEAKKQLQDEMLRRVDAENRLQTMKEELDFQKNIYSEELRET**KRRHETRLVE
coil 2
IDNGKQREFESRLAD**ALQQLRAQHEDQVEQYKKELEKTYSAKLDNARQSAERNSNLV**

**GAAHEELQQSRIRIDSLSAQLSQLQKQLAAKEAKLRDLEDSLARERDTSRRLLAEKE**

**REMAEMRARMQQQLDEYQQLLDIKLALDMQIHAYRKLLEGEEER**LRLSPSPTSQRSR

GRASSHSSQTQGGGSVTKKRKLESTESRSSPSQHARTSGRVAVEEVDEEGKFVRLRN

KSNEDQSMGNWQIKRQNGDDPLLTYRFPPKFTLKAGQVVTIWAAGAGATHSPPTDLV

WKAQNTWGCGNSLRTALINSTGEEVAMRKLVRSVTVVEDDEDEDGDDLLHHHHVSGS

RR

**Figure 3–9** The amino acid sequence of nuclear lamin C (Problem 3–34).

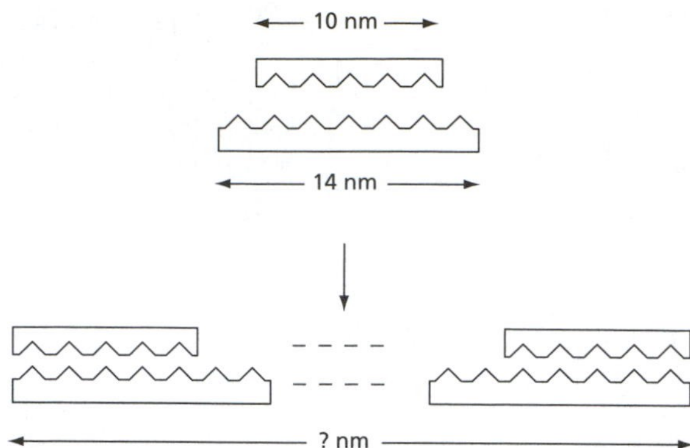


**Figure 3–10** Vernier assembly of two proteins into a fiber of defined length (Problem 3–35).

**3–35** In a vernier type of assembly, rodlike proteins of different lengths form a staggered complex that grows until their ends exactly match.

A. Consider the assembly of a long fibrous complex from two sets of rodlike proteins, one 10 nm in length and the other 14 nm (Figure 3–10). How long will the final fiber be if assembly proceeds until both ends are flush?

B. Initiation of growth of such fibers usually begins with the binding of one kind of subunit to a subunit of the other type. Does the extent of 'stagger' in the very first complex formed make any difference in the length of the final fiber?

## CALCULATIONS

**3–36** Typical proteins have a stability ranging from 7 to 15 kcal/mole at 37°C. Stability is a measure of the equilibrium between the folded and unfolded forms of the protein:

$$\text{folded [F]} \rightleftharpoons \text{unfolded [U]},\ K = [\text{U}]/[\text{F}]$$

For a protein with a stability of 9.9 kcal/mole, calculate the fraction of unfolded protein that would exist at equilibrium at 37°C. At equilibrium,

$$\Delta G^\circ = -RT\ln K = -2.3RT\log K,$$

where $R = 1.98 \times 10^{-3}$ kcal/(K mole) and $T$ is temperature in K (37°C = 310 K).

**3–37** You wish to try your hand at predicting the structure of lysozyme. Because lysozyme has several hundred weak interactions that contribute to its folded conformation, you have decided first to measure their overall contribution, so that you will know how much leeway there is when it comes time to assign values to individual interactions. Since the stability of lysozyme is

$$\Delta G^\circ = G^\circ_{\text{unfolded}} - G^\circ_{\text{folded}}$$

and

$$G^\circ = H^\circ - TS^\circ$$

you determine the standard enthalpy, $H^\circ$, a measure of bond strength, and the standard entropy, $S^\circ$, a measure of disorder, for both the folded and unfolded states. Your values (kcal/mole) are shown below for measurements made at 37°C.

| | $H^\circ$ | $TS^\circ$ |
|---|---|---|
| Unfolded state | 128 | 119 |
| Folded state | 75 | 76 |

A. What is the stability of lysozyme at 37°C?

B. Does this calculation give you hope that you will be able to predict the structure of lysozyme? Why or why not?

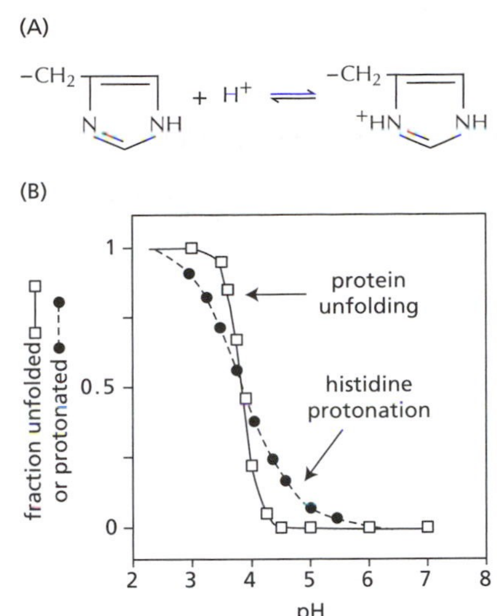


**Figure 3–11** Denaturation of proteins (Problem 3–41). (A) Histidine protonation. (B) Titration curves for protein unfolding and histone protonation.

**3–38** Consider the following statement. "To produce one molecule of each possible kind of polypeptide chain, 300 amino acids in length, would require more atoms than exist in the universe." Given the size of the universe, do you suppose this statement could possibly be correct? Since counting atoms is a tricky business, consider the problem from the standpoint of mass. The mass of the observable universe is estimated to be about $10^{80}$ grams, give or take an order of magnitude or so. Assuming that the average mass of an amino acid is 110 daltons, what would be the mass of one molecule of each possible kind of polypeptide chain 300 amino acids in length? Is this greater than the mass of the universe?

**3–39** The error rate for protein synthesis is estimated to be about 1/10,000; that is, the synthesis machinery incorporates one incorrect amino acid for each 10,000 it inserts. At this error rate, what fraction of proteins will be synthesized correctly for proteins 1000 amino acids, 10,000 amino acids, and 100,000 amino acids in length? [The probability of a correct sequence, $P_C$, equals the fraction correct for each operation, $f_C$, raised to a power equal to the number of operations, $n$: $P_C = (f_C)^n$. For an error rate of 1/10,000, $f_C$ = 0.9999.]

**3–40** It is often said that protein complexes are made from subunits (that is, individually synthesized proteins) rather than as one long protein because it is more likely to give a correct final structure.

A. Assuming the same error rate as in the previous problem, what fraction of bacterial ribosomes would be constituted correctly if the proteins were synthesized as one large protein versus assembled from individual proteins? For the sake of calculation assume that the ribosome is composed of 50 proteins, each 200 amino acids in length, and that the subunits—correct and incorrect—are assembled with equal likelihood into the complete ribosome.

B. Is the assumption that correct and incorrect subunits assemble equally well likely to be true? Why or why not? How would a change in that assumption affect the calculation in part A?

## DATA HANDLING

**3–41** Most proteins denature at both high and low pH. At high pH, the ionization of internal tyrosines is thought to be the main destabilizing influence, whereas at low pH, the protonation of buried histidines (Figure 3–11A) is the likely culprit. A titration curve for the unfolding of the enzyme ribonuclease is shown in Figure 3–11B. Superimposed on it is the expected titration curve for the ionization of a histidine side chain with a p$K$ of about 4, which is typical for a buried histidine (the p$K$ for the side chain of the free amino acid is 6). The titration curve for denaturation is clearly much steeper than that for the side chain. Given the discrepancy between the titration curves for protein unfolding and histidine protonation, how can it be true that protonation of histidine causes protein unfolding?

**3–42** Small proteins may have only one or two amino acid side chains that are totally inaccessible to solvent. Even in large proteins, only about 15% of the amino acids are fully buried. A list of buried side chains from a study of 12 proteins is shown in Table 3–1. The list is ordered by the proportion of each amino acid that is fully buried. What types of amino acids are most commonly buried? Least commonly buried? Are there any surprises on this list?

**Table 3–1 Proportions of amino acids that are inaccessible to solvent in a study of 12 proteins** (Problem 3–42). **See inside back cover for one-letter amino acid code.**

| AMINO ACID SIDE CHAIN | PROPORTION BURIED |
|---|---|
| I | 0.60 |
| V | 0.54 |
| C | 0.50 |
| F | 0.50 |
| L | 0.45 |
| M | 0.40 |
| A | 0.38 |
| G | 0.36 |
| W | 0.27 |
| T | 0.23 |
| S | 0.22 |
| E | 0.18 |
| P | 0.18 |
| H | 0.17 |
| D | 0.15 |
| Y | 0.15 |
| N | 0.12 |
| Q | 0.07 |
| K | 0.03 |
| R | 0.01 |

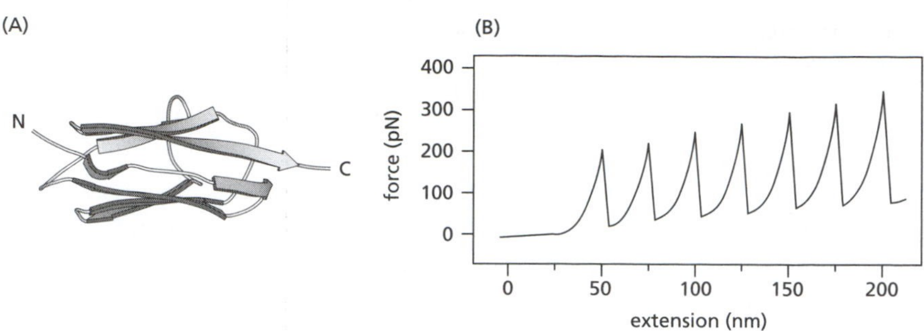


**Figure 3–12** Springlike behavior of titin (Problem 3–43). (A) The structure of an individual Ig domain. (B) Force in piconewtons versus extension in nanometers obtained by atomic force microscopy.

**Problem 16–90** discusses the location of titin in striated muscle.

**3–43** Titin, which has a molecular weight of $3 \times 10^6$, is the largest polypeptide yet described. Titin molecules extend from muscle thick filaments to the Z disc; they are thought to act as springs to keep the thick filaments centered in the sarcomere. Titin is composed of a large number of repeated immunoglobulin (Ig) sequences of 89 amino acids, each of which is folded into a domain about 4 nm in length (Figure 3–12A).

You suspect that the springlike behavior of titin is caused by the sequential unfolding (and refolding) of individual Ig domains. You test this hypothesis using the atomic force microscope, which allows you to pick up one end of a protein molecule and pull with an accurately measured force. For a fragment of titin containing seven repeats of the Ig domain, this experiment gives the sawtooth force-versus-extension curve shown in Figure 3–12B. If the experiment is repeated in a solution of 8 M urea (a protein denaturant) the peaks disappear and the measured extension becomes much longer for a given force. If the experiment is repeated after the protein has been cross-linked by treatment with glutaraldehyde, once again the peaks disappear but the extension becomes much smaller for a given force.

A. Are the data consistent with your hypothesis that titin's springlike behavior is due to the sequential unfolding of individual Ig domains? Explain your reasoning.
B. Is the extension for each putative domain-unfolding event the magnitude you would expect? (In an extended polypeptide chain, amino acids are spaced at intervals of 0.34 nm.)
C. Why is each successive peak in Figure 3–12B a little higher than the one before?
D. Why does the force collapse so abruptly after each peak?

**3–44** You are skeptical of the blanket statement that cysteines in intracellular proteins are not involved in disulfide bonds, while in extracellular proteins they are. To test this statement you carry out the following experiment. As a source of intracellular protein you use reticulocytes, which have no internal membranes and, thus, no proteins from the ER or other membrane-enclosed compartments. As examples of extracellular proteins, you use bovine serum albumin (BSA), which has 37 cysteines, and insulin, which has 6. You denature the soluble proteins from a reticulocyte lysate and the two extracellular proteins so that all cysteines are exposed. To probe the status of cysteines, you treat the proteins with *N*-ethylmaleimide (NEM), which reacts covalently with the –SH groups of free cysteines, but not with sulfur atoms in disulfide bonds. In the first experiment you treat the denatured proteins with radiolabeled NEM, then break any disulfide bonds with dithiothreitol (DTT) and react a second time with unlabeled NEM. In the second experiment you do the reverse: you first treat the denatured proteins with unlabeled NEM, then break disulfide bonds with DTT and treat with radiolabeled NEM. The proteins are separated according to size by electrophoresis on a polyacrylamide gel (Figure 3–13).

A. Do any cytosolic proteins have disulfide bonds?
B. Do the extracellular proteins have any free cysteine –SH groups?
C. How do you suppose the results might differ if you used lysates of cells that have internal membrane-enclosed compartments?

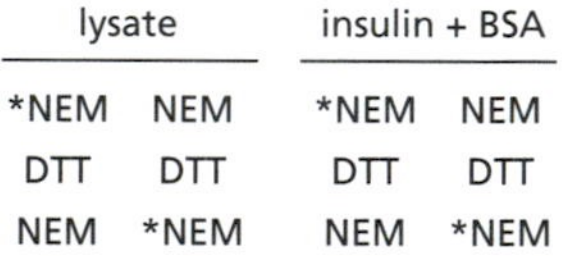

lysate
insulin + BSA
*NEM NEM *NEM NEM
DTT DTT DTT DTT
NEM *NEM NEM *NEM


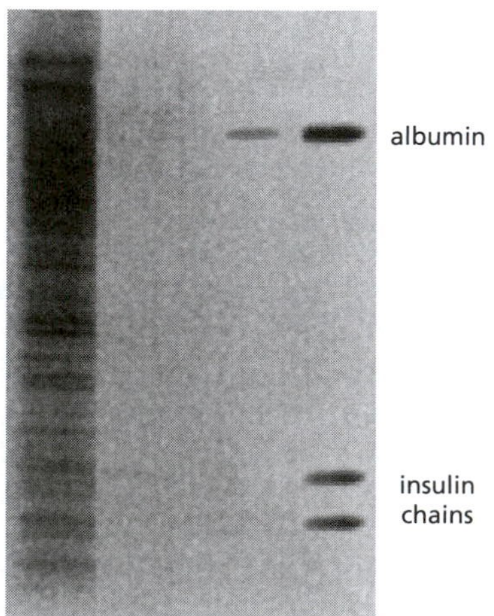


**Figure 3–13** Test for disulfide bonds in cytosolic and extracellular proteins (Problem 3–44). The order of treatment with NEM and DTT is indicated at the top of each lane; *NEM indicates radiolabeled NEM.

# PROTEIN FUNCTION

TERMS TO LEARN

| | | |
|---|---|---|
| active site | feedback inhibition | protein phosphatase |
| allosteric protein | GTP-binding protein (GTPase) | proteomics |
| antibody | ligand | regulatory site |
| antigen | linkage | scaffold protein |
| catalyst | lysozyme | SCF ubiquitin ligase |
| coenzyme | motor protein | substrate |
| equilibrium constant (*K*) | protein kinase | transition state |
| enzyme | | |

## DEFINITIONS

Match the definition below with its term from the list above.

**3–45** A protein that links together a set of activating, inhibiting, adaptor, and substrate proteins at a specific location in a cell.

**3–46** Type of metabolic regulation in which the activity of an enzyme acting near the beginning of a reaction pathway is reduced by a product of the pathway.

**3–47** Protein produced by the immune system in response to a foreign molecule or invading microorganism.

**3–48** Region of an enzyme surface to which a substrate molecule binds in order to undergo a catalyzed reaction.

**3–49** A protein catalyst that speeds up a reaction, often by a factor of a million or more, without itself being changed.

**3–50** Enzyme that removes a phosphate group from a protein by hydrolysis.

**3–51** Mutual effect of the binding of one ligand on the binding of another that is a central feature of the behavior of all allosteric proteins.

**3–52** Enzyme that transfers the terminal phosphate group of ATP to a specific amino acid in a target protein.

**3–53** Rate-limiting structure that forms transiently in the course of a chemical reaction and has the highest free energy of any reaction intermediate.

**3–54** Protein complex that binds to specific target proteins at different times during the cell cycle and adds multiubiquitin chains to mark them for rapid destruction in the proteasome.

**3–55** Protein that changes its conformation (and often its activity) when it binds a regulatory molecule or when it is covalently modified.

**3–56** A term often used to describe research focused on the simultaneous analysis of large numbers of proteins.

**3–57** Small molecule that is tightly associated with a protein catalyst and participates in the chemical reaction, often by forming a covalent bond to the substrate.

## TRUE/FALSE

Decide whether each of these statements is true or false, and then explain why.

**3–58** The tendency for an amino acid side-chain group such as –COOH to release a proton, its p*K*, is the same for the amino acid in solution and for the amino acid in a protein.

**3–59** For a family of related genes that do not match genes of known function in the sequence database, it should be possible to deduce their function by using 'evolutionary tracing' to see where conserved amino acids cluster on their surfaces.

**3–60** Higher concentrations of enzyme give rise to a higher turnover number.

**3–61** Enzymes such as aspartate transcarbamoylase that undergo cooperative allosteric transitions invariably contain multiple identical subunits.

**3–62** Continual addition and removal of phosphates by protein kinases and protein phosphatases is wasteful of energy—since their combined action consumes ATP—but it is a necessary consequence of effective regulation by phosphorylation.

**3–63** Conformational changes in proteins never exceed a few tenths of a nanometer.

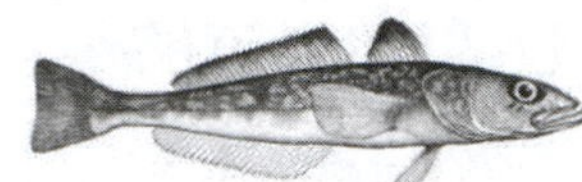

*Dissostichus eleginoides,* the Chilean sea bass

*Pagothenia borchgrevinki*

**Figure 3–14** Two notothenioid fish (Problem 3–64). The notothenioid family now dominates Antarctica's continental shelf, accounting for 50% of the species and 95% of the biomass of fish. The Chilean sea bass is commonly served in restaurants.

## THOUGHT PROBLEMS

**3–64** Antarctic notothenioid fish (Figure 3–14) avoid freezing in their perpetually icy environment because of an antifreeze protein that circulates in their blood. This evolutionary adaptation has allowed the Notothenioidei suborder to rise to dominance in the freezing Southern Ocean. It is said that all proteins function by binding to other molecules. To what ligand do you suppose antifreeze proteins bind to keep the fish from freezing? Or do you think this might be an example of a protein that functions in the absence of any molecular interaction?

**3–65** Aminoacyl-tRNA synthetases attach specific amino acids to their appropriate tRNAs in preparation for protein synthesis. The synthetase that attaches valine to $tRNA^{Val}$ must be able to discriminate valine from threonine, which differ only slightly in structure: valine has a methyl group where threonine has a hydroxyl group (Figure 3–15). Valyl-tRNA synthetase achieves this discrimination in two steps. In the first it uses a binding pocket whose contours allow valine or threonine (but not other amino acids) to bind, but the binding of valine is preferred. This site is responsible for coupling the amino acid to the tRNA. In the second step the enzyme checks the newly made aminoacyl tRNA using a second binding site that is very specific for threonine and hydrolyzes it from the tRNA. How do you suppose it is that the second binding site can be very specific for threonine, whereas the first binding site has only a moderate specificity for valine?

**3–66** Which pair(s) of proteins in Figure 3–8 (see Problem 3–33, p. 43) could bind to one another to form a heterodimer in such a way that all binding groups are satisfied? (Charged amino acids are indicated by + or – and hydrogen-bond donors and acceptors are indicated by *d* and *a*.)

**3–67** You have raised a specific, high-affinity monoclonal antibody against the enzyme you are working on, and have identified its interaction site as a stretch of six amino acids in the enzyme. Your advisor suggests that you could use the antibody to purify the enzyme by affinity chromatography. This technique would involve attaching the antibody to the inert matrix of a column, passing a crude cell lysate over the column, allowing the antibody to bind your enzyme but not other proteins, and finally eluting your enzyme by washing the column with a solution containing the six amino acid peptide corresponding to the binding site. The principal advantage of affinity chromatography is that it allows a rapid, one-step purification under mild conditions that retain enzyme activity.

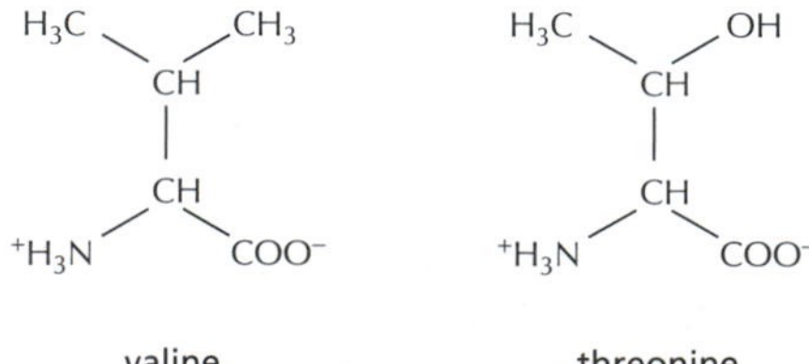


**Figure 3–15** Structures of valine and threonine (Problem 3–65).

In a preliminary experiment you show that if you incubate the antibody with the peptide corresponding to the binding site it will no longer bind to your enzyme, demonstrating that the antibody binds the peptide. Encouraged, you bind the antibody to the column and show that it completely removes your enzyme from the crude cell lysate. When you try to elute your enzyme with a solution containing a high concentration of the peptide, however, you find that none of your enzyme comes off the column. What could have gone wrong? (Think about what must happen for the enzyme to come off the column.)

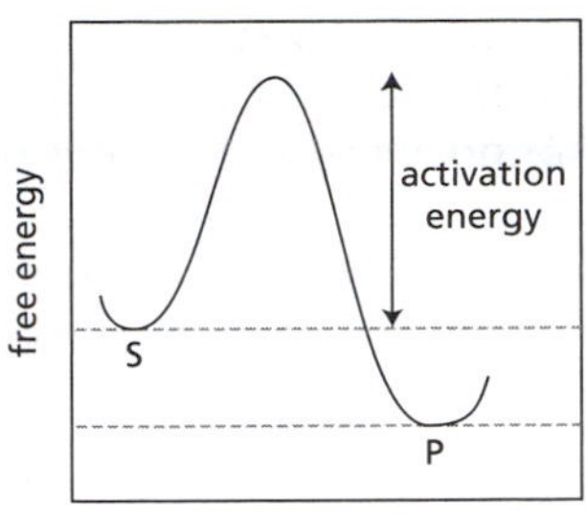


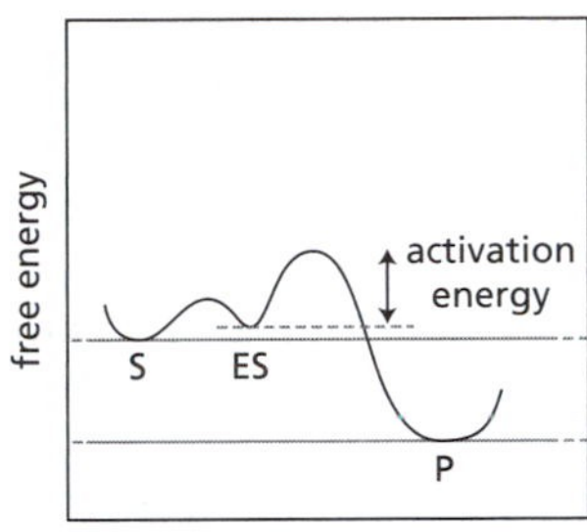


**Figure 3–16** Catalyzed and uncatalyzed reactions showing the free energy at various stages in the progress of the reaction (Problem 3–68).

**3–68** Examine Figure 3–16, which compares the energetics of a catalyzed and uncatalyzed reaction during the progress of the reaction from substrate (S) to product (P). The highest peak in such a diagram corresponds to the transition state, which is an unstable, high-energy arrangement of substrate atoms that is intermediate between substrate and product. The free energy required to surmount this barrier to the reaction is termed the activation energy. Enzymes function by lowering the activation energy, thereby allowing a more rapid approach to equilibrium.

With this diagram in mind, consider the following question. Suppose the enzyme in the diagram were mutated in such a way that its affinity for the substrate was increased by a factor of 100. Assume that there was no other effect beyond increasing the depth of the trough labeled ES (enzyme–substrate complex) in Figure 3–16. Would you expect the rate of the reaction catalyzed by the altered enzyme to be faster, slower, or equal to the reaction rate catalyzed by the normal enzyme?

**3–69** Which one of the following properties of an enzyme is responsible for its saturation behavior; that is, a maximum rate insensitive to increasing substrate concentration?

A. The enzyme does not change the overall equilibrium constant for a reaction.
B. The enzyme lowers the activation energy of a chemical reaction.
C. The enzyme is a catalyst that is not consumed by the reaction.
D. The enzyme has a fixed number of active sites where substrate binds.
E. The product of the enzyme reaction usually inhibits the enzyme.

**3–70** The Michaelis constant, $K_m$, is often spoken of as if it were a measure of the affinity of the enzyme for the substrate: the lower the $K_m$, the higher the binding affinity. This would be true if $K_m$ were the same as $K_d$ (the equilibrium constant for the dissociation reaction), but it is not. For an enzyme-catalyzed reaction

$$E + S \underset{k_{-1}}{\overset{k_1}{\rightleftharpoons}} ES \xrightarrow{k_{cat}} E + P$$

$$K_m = \frac{(k_{-1} + k_{cat})}{k_1}$$

A. In terms of these rate constants, what is $K_d$ for dissociation of the ES complex to E + S?
B. Under what conditions is $K_m$ approximately equal to $K_d$?
C. Does $K_m$ consistently overestimate or underestimate the binding affinity? Or does it sometimes overestimate and sometimes underestimate the binding affinity?

**3–71** You are trying to determine whether it is better to purify an enzyme from its natural source or to express the gene in bacteria and then purify it. You purify the enzyme in the same way from both sources and show that each preparation gives a single band by denaturing gel electrophoresis, a common measure of purity. When you compare the kinetic parameters, you find that both enzymes have the same $K_m$ but the enzyme from bacteria has a 10-fold lower $V_{max}$. Propose possible explanations for this result.

**3–72** The enzyme hexokinase adds a phosphate to D-glucose but ignores its mirror image, L-glucose. Suppose that you were able to synthesize hexokinase entirely from D-amino acids, which are the mirror image of the normal L-amino acids.

A. Assuming that the 'D' enzyme would fold to a stable conformation, what relationship would you expect it to bear to the normal 'L' enzyme?

B. Do you suppose the 'D' enzyme would add a phosphate to L-glucose, and ignore D-glucose?

**3–73** In 1948 Linus Pauling proposed what is now considered to be a key aspect of enzyme function.

"I believe that an enzyme has a structure closely similar to that found for antibodies, but with one important difference, namely, that the surface configuration of the enzyme is not so closely complementary to its specific substrate as is that on an antibody, but is instead complementary to an unstable molecule with only transient existence—namely, the 'activated complex' [transition state, in modern parlance] for the reaction that is catalyzed by the enzyme. The mode of action of an enzyme would then be the following: the enzyme would show a small power of attraction for the substrate molecule or molecules, which would become attached to it in its active surface region. This substrate molecule, or these molecules, would then be strained by the forces of attraction to the enzyme, which would tend to deform it into the configuration for the activated complex, for which the power of attraction by the enzyme is the greatest.... The assumption made above that the enzyme has a configuration complementary to the activated complex, and accordingly has the strongest power of attraction for the activated complex, means that the activation energy for the reaction is less in the presence of the enzyme than in its absence, and accordingly that the reaction would be speeded up by the enzyme."

The enzyme triosephosphate isomerase catalyzes the interconversion of glyceraldehyde 3-phosphate and dihydroxyacetone phosphate through a *cis*-enediolate intermediate (Figure 3–17). Phosphoglycolate (Figure 3–17) is a competitive inhibitor of triosephosphate isomerase with a $K_d$ of 7 μM. The normal substrates for the enzyme have a $K_d$ of about 100 μM. Do you think that phosphoglycolate is a transition-state analog? Why or why not?

**3–74** The mechanism for lysozyme cleavage of its polysaccharide substrate requires Glu 35 in its nonionized form, whereas the nearby Asp 52 must be ionized (Figure 3–18). The p*K* values for the side-chain carboxyl groups on the two amino acids in solution are virtually identical.

A. How can one carboxyl group be charged and the other uncharged in the active site of lysozyme?

B. The pH optimum for lysozyme is about 5. Why do you suppose that the activity decreases above and below this optimum?

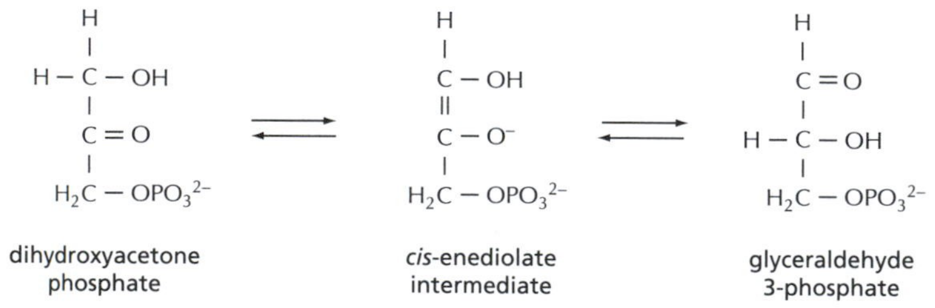

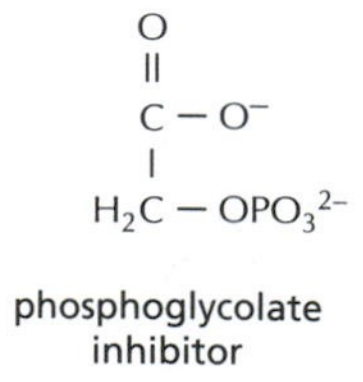

dihydroxyacetone phosphate ($H–C(H)–OH$, $C=O$, $H_2C–OPO_3^{2-}$) ⇌ *cis*-enediolate intermediate ($H–C–OH$ ‖ $C–O^-$, $H_2C–OPO_3^{2-}$) ⇌ glyceraldehyde 3-phosphate ($H–C=O$, $H–C–OH$, $H_2C–OPO_3^{2-}$)

phosphoglycolate inhibitor ($O=C–O^-$, $H_2C–OPO_3^{2-}$)

**Figure 3–17** The reaction catalyzed by triosephosphate isomerase, and the enzyme inhibitor, phosphoglycolate (Problem 3–73).

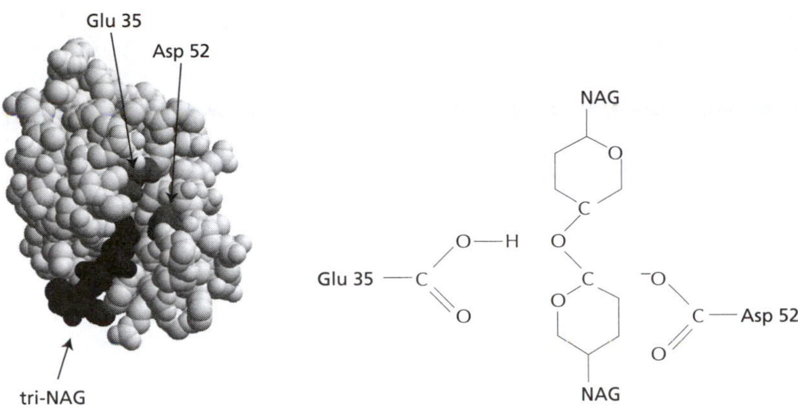


**Figure 3–18** Forms of Glu 35 and Asp 52 required for polysaccharide cleavage by lysozyme (Problem 3–74). In the space-filling model the positions of Glu 35 and Asp 52 are shown relative to the trisaccharide, tri-NAG, which is not quite long enough to be cleaved. In the schematic diagram the positions of Glu 35 and Asp 52 are shown relative to the glycosidic bond to be cleaved in a polysaccharide composed of NAG (*N*-acetylglucosamine) residues.

**3–75** Egg whites, a rich source of nutrients, can be left out at room temperature for days or weeks and nothing grows in them. Egg white provides three main defenses against microorganisms. One is lysozyme, which cleaves bacterial cell walls. A second is a protein called avidin, which binds the essential vitamin biotin with extremely high affinity, making it unavailable to microorganisms. A clue to the third is the observation that washing eggs in water from rusty pipes can cause them to go bad. What necessity for the life of microorganisms might be provided by water from rusty pipes, and how might the egg 'defend' against it?

**3–76** How do you suppose that a molecule of hemoglobin is able to bind oxygen efficiently in the lungs, and yet release it efficiently in the tissues?

**3–77** Assume that two enzymes catalyze successive steps in a metabolic pathway (that is, the product of one enzyme is the substrate for the next) and that the rate of reaction of each enzyme is limited by the rate of diffusion of its substrate. The rate of diffusion is a physical property of the medium and cannot be altered by any changes to the enzymes. Why is it, then, that linking the two enzymes into a complex results in an increase in the metabolic flux through the linked enzymes relative to the unassociated enzymes?

**3–78** If you were in charge of enzyme design for a cell, for what circumstances might you design an enzyme that had a $K_m$ much, much lower than the prevailing substrate concentration ([S] >> $K_m$)? A $K_m$ around the prevailing substrate concentration ([S] ≅ $K_m$)? A $K_m$ much, much higher than the prevailing substrate concentration ([S] <<$K_m$)?

**3–79** Which of the following does NOT describe a mechanism that cells use to regulate enzyme activities?

A. Cells control enzyme activity by phosphorylation and dephosphorylation.
B. Cells control enzyme activity by the binding of small molecules.
C. Cells control the rates of diffusion of substrates to enzymes
D. Cells control the rates of enzyme degradation.
E. Cells control the rates of enzyme synthesis.
F. Cells control the targeting of enzymes to specific organelles.

**3–80** Synthesis of the purine nucleotides AMP and GMP proceeds by a branched pathway starting with ribose 5-phosphate (R5P), as shown schematically in Figure 3–19. Using the principles of feedback inhibition, propose a regulatory strategy for this pathway that ensures an adequate supply of both AMP and GMP and minimizes the buildup of the intermediates (*A–I*) when supplies of AMP and GMP are adequate.

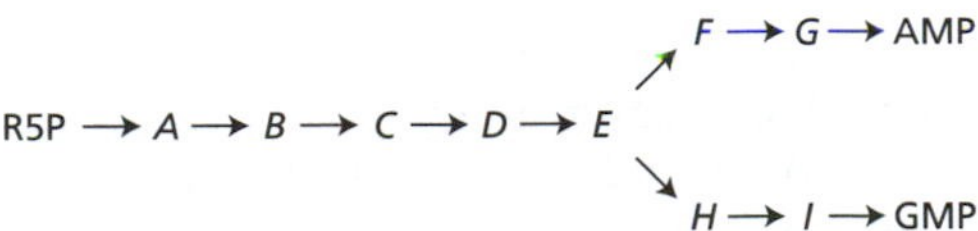


**Figure 3–19** Schematic diagram of the metabolic pathway for synthesis of AMP and GMP from R5P (Problem 3–80).

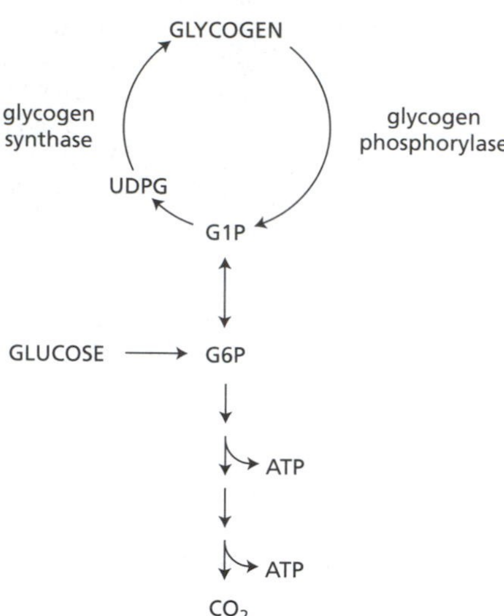


**Figure 3–20** Pathways for glycogen synthesis and its breakdown to glucose 6-phosphate, which is an intermediate along the pathway for glucose metabolism to $CO_2$ (Problem 3–81). G6P stands for glucose 6-phosphate, G1P for glucose 1-phosphate, and UDPG for uridine diphosphoglucose.

**3–81** Pathways devoted to the synthesis of specific bioproducts such as purines are commonly regulated via feedback inhibition by the final product. By contrast, the flow of metabolites through the web of pathways devoted to overall energy metabolism—production and utilization of ATP as well as the buildup and breakdown of internal fuel reserves—is regulated by metabolites whose concentrations reflect the energy status of the cell. ATP-like signal metabolites (ATP, NADH, etc.) tend to accumulate when the cell is slowly consuming ATP to meet its energy needs; AMP-like signal metabolites (ADP, AMP, $P_i$, $NAD^+$, etc.) tend to accumulate when the cell is rapidly using ATP.

Consider the pathways for the synthesis and breakdown of glycogen, the main fuel reserve in muscle cells (Figure 3–20). The synthetic pathway is controlled by glycogen synthase, whereas the breakdown pathway is controlled by glycogen phosphorylase. In resting muscle, which type of signal metabolite would be expected to accumulate? How would those signal metabolites be expected to affect the activity of the two regulated enzymes of glycogen metabolism? What about in exercising muscle?

**3–82** The enzyme glycogen phosphorylase uses phosphate as a substrate to split off glucose 1-phosphate from glycogen, which is a polymer of glucose. Glycogen phosphorylase in the absence of any ligands is a dimer that exists in two conformations: a predominant one with low enzymatic activity and a rarer one with high activity. Both phosphate, a substrate that binds to the active site, and AMP, an activator that binds to an allosteric site, alter the conformational equilibrium by binding preferentially to one conformation. To which conformation of the enzyme would you expect phosphate to bind, and why? To which conformation would you expect AMP to bind, and why? How does the binding of either molecule alter the activity of the enzyme?

**3–83** Monod, Wyman, and Changeux (MWC) originally explained the kinetic behavior of allosteric enzymes by four postulates, as summarized below.

1. All subunits are identical and are arranged symmetrically in the protein.
2. Each subunit carries a binding site for each ligand.
3. The protein can exist in at least two conformations that conserve the symmetry of the protein. The different conformations may have very different affinities for the ligands. (Ligands may be asymmetrically distributed among the subunits.)
4. The binding affinity of a ligand depends only on the conformational state of the enzyme and not on the occupancy of neighboring sites.

A subset of all the possible arrangements of subunits in a cooperative enzyme composed of four identical subunits, each with two conformations, and with different numbers of bound ligands is shown in Figure 3–21. Assuming that the ligand binds much more tightly to one conformation of

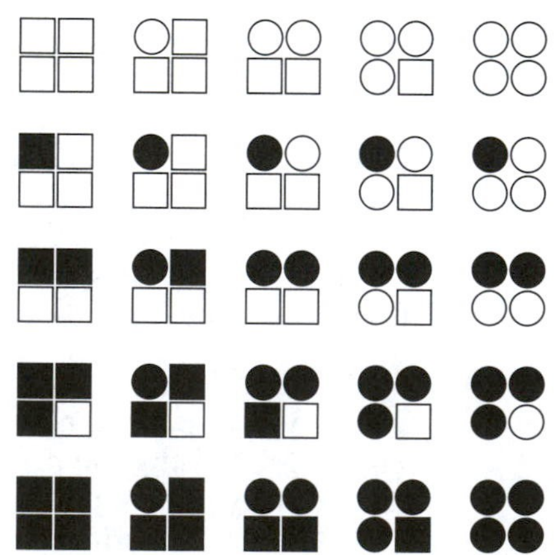

**Figure 3–21** A subset of all the possible arrangements of a tetramer, composed of subunits with either of two conformations, and different numbers of bound ligands (Problem 3–83). *Circles* and *squares* represent the two conformations of the subunits; *black* indicates a subunit with a bound ligand.

subunit (circle), decide which of the tetrameric species are consistent with the MWC postulates. Black subunits in this diagram indicate those with a bound ligand. What would your answer be if the ligand bound equally well to each of the two conformations of subunit?

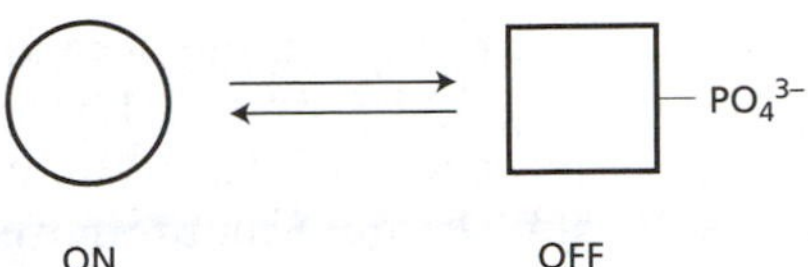


**Figure 3–22** Phosphorylated and nonphosphorylated states of a metabolic enzyme (Problem 3–84).

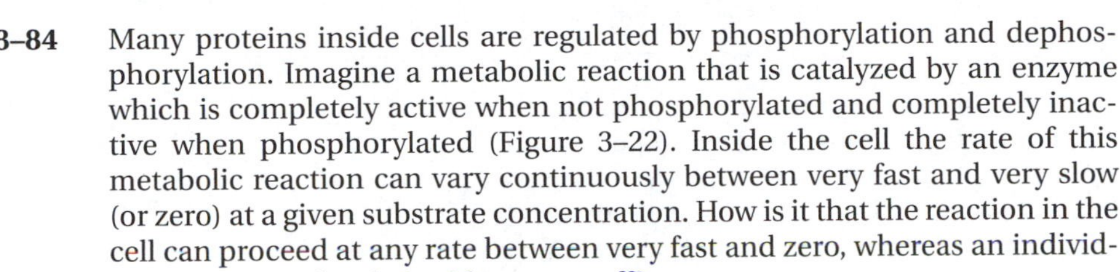

**3–84** Many proteins inside cells are regulated by phosphorylation and dephosphorylation. Imagine a metabolic reaction that is catalyzed by an enzyme which is completely active when not phosphorylated and completely inactive when phosphorylated (Figure 3–22). Inside the cell the rate of this metabolic reaction can vary continuously between very fast and very slow (or zero) at a given substrate concentration. How is it that the reaction in the cell can proceed at any rate between very fast and zero, whereas an individual enzyme molecule is either on or off?

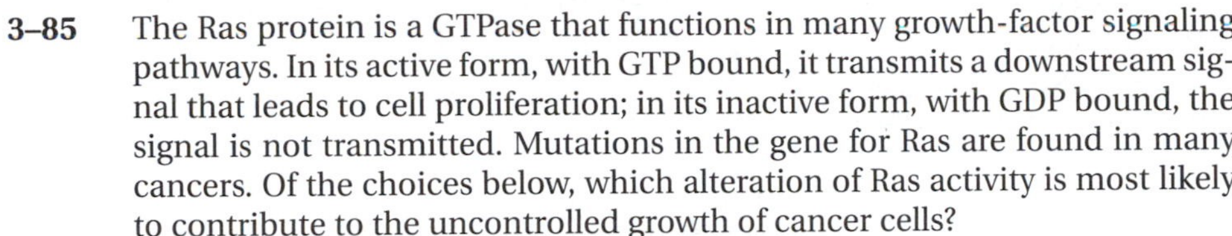

**3–85** The Ras protein is a GTPase that functions in many growth-factor signaling pathways. In its active form, with GTP bound, it transmits a downstream signal that leads to cell proliferation; in its inactive form, with GDP bound, the signal is not transmitted. Mutations in the gene for Ras are found in many cancers. Of the choices below, which alteration of Ras activity is most likely to contribute to the uncontrolled growth of cancer cells?

A. A mutation that prevents Ras from being made.
B. A mutation that increases the affinity of Ras for GDP.
C. A mutation that decreases the affinity of Ras for GTP.
D. A mutation that decreases the affinity of Ras for its downstream targets.
E. A mutation that decreases the rate of hydrolysis of GTP by Ras.

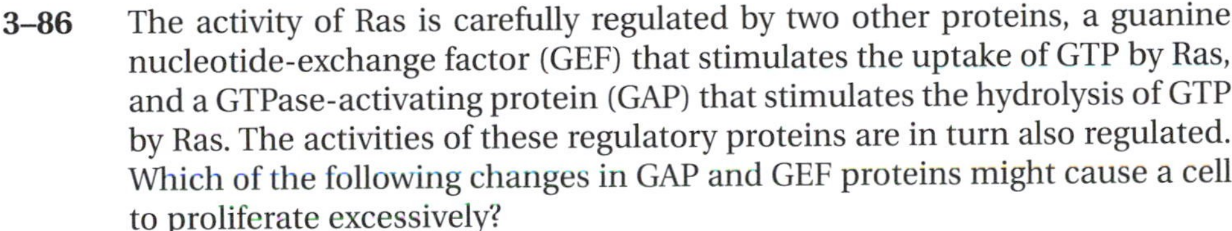

**3–86** The activity of Ras is carefully regulated by two other proteins, a guanine nucleotide-exchange factor (GEF) that stimulates the uptake of GTP by Ras, and a GTPase-activating protein (GAP) that stimulates the hydrolysis of GTP by Ras. The activities of these regulatory proteins are in turn also regulated. Which of the following changes in GAP and GEF proteins might cause a cell to proliferate excessively?

A. A nonfunctional GAP
B. A permanently active GAP
C. A nonfunctional GEF
D. A permanently active GEF

**3–87** Motor proteins generally require ATP (or GTP) hydrolysis to ensure unidirectional movement.

A. In the absence of ATP would you expect a motor protein to stop moving, to wander back and forth, to move in reverse, or to continue moving forward but more slowly?
B. Assume that the concentrations of ATP, ADP, and phosphate were adjusted so that the free-energy change for ATP hydrolysis by the motor protein was equal to zero (instead of very negative, as it is normally). Under these conditions would you expect a motor protein to stop moving, to wander back and forth, to move in reverse, or to continue moving forward but more slowly?

## CALCULATIONS

**3–88** Protein A binds to protein B to form a complex, AB. A cell contains an equilibrium mixture of protein A at a concentration of 1 μM, protein B at a concentration of 1 μM, and the complex AB also at 1 μM.

A. What is the equilibrium constant, $K$, for the reaction $A + B \rightarrow AB$?
B. What would the equilibrium constant be if A, B, and AB were each present in equilibrium at a concentration of 1 nM?
C. At this lower concentration, about how many extra hydrogen bonds would be needed to hold A and B together tightly enough to form the same proportion of the AB complex? (Free-energy change is related to the equilibrium constant by the equation $\Delta G° = -2.3\ RT \log K$, where $R$ is $1.98 \times 10^{-3}$ kcal/(K mole) and $T$ is 310 K. Assume that the formation of one hydrogen bond is accompanied by a favorable free-energy change of about 1 kcal/mole.)

**3–89** An antibody binds to another protein with an equilibrium constant, $K$, of $5 \times 10^9$ M$^{-1}$. When it binds to a second, related protein, it forms three fewer hydrogen bonds, reducing its binding affinity by 2.8 kcal/mole. What is the $K$ for its binding to the second protein? (Free-energy change is related to the equilibrium constant by the equation $\Delta G° = -2.3\ RT \log K$, where $R$ is $1.98 \times 10^{-3}$ kcal/(K mole) and $T$ is 310 K.)

**3–90** The equilibrium constant for a reaction like that of antibody (Ab) binding to a protein (Pr) to form an antibody–protein complex (Ab–Pr) is equal to the ratio of the association rate constant, $k_{on}$, to the dissociation rate constant, $k_{off}$ ($K$ = [Ab–Pr]/([Ab][Pr]) = $k_{on}/k_{off}$). Recall that the association rate ($k_{on}$[Ab][Pr]) equals the dissociation rate ($k_{off}$[Ab–Pr]) at equilibrium. Consider two such reactions. The first has an on rate constant of $10^5$/M sec and an off rate constant of $10^{-3}$/sec at 37°C. The second has an on rate constant of $10^3$/M sec and an off rate of $10^{-5}$/sec at 37°C.

A. What are the equilibrium constants for these two reactions?
B. At equal concentrations of antibody and protein, which of these reactions will reach its equilibrium point more quickly?
C. You wish to use an antibody to purify the protein you are studying. You are concerned that the complex may fall apart in the time it takes you to isolate it, and you are unsure how the off rates relate to the half-time for dissociation; that is, the time at which half the complex will have dissociated. It can be shown the fraction of complex remaining at time $t$ ([Ab–Pr]$_t$) relative to that present initially ([Ab–Pr]$_0$) is

$$\frac{[\text{Ab–Pr}]_t}{[\text{Ab–Pr}]_0} = e^{-k_{off}t}$$

an equation more easily dealt with in its logarithmic form,

$$2.3 \log \frac{[\text{Ab–Pr}]_t}{[\text{Ab–Pr}]_0} = -k_{off}t$$

Using this relationship decide how long it will take for half of the Ab–Pr complexes in each of the above two reactions to dissociate. Neglect any contribution from new complex being formed in the association reaction.

**3–91** Consider an uncatalyzed reaction A ⇌ B. The rate constants for the forward and reverse reactions are $k_f = 10^{-4}$/sec and $k_r = 10^{-7}$/sec. Thus, the rates or velocities ($v$) of the forward and reverse reactions are

$$v_f = k_f\,[A] \qquad \text{and} \qquad v_r = k_r\,[B]$$

The overall reaction rate is

$$v = v_f - v_r = k_f\,[A] - k_r\,[B]$$

A. What is the overall reaction rate at equilibrium?
B. What is the value of the equilibrium constant, $K$?
C. You now add an enzyme that increases $k_f$ by a factor of $10^9$. What will the value of the equilibrium constant be with the enzyme present? What will the value of $k_r$ be?

**3–92** Many enzymes obey simple Michaelis–Menten kinetics, which are summarized by the equation

$$\text{rate} = \frac{V_{max}\,[S]}{[S] + K_m}$$

where $V_{max}$ = maximum velocity, [S] = concentration of substrate, and $K_m$ = the Michaelis constant.

It is instructive to plug a few values of [S] into the equation to see how rate is affected. What are the rates for [S] equal to zero, equal to $K_m$, and equal to infinite concentration?

**3–93** Suppose that the enzyme you are studying is regulated by phosphorylation. When the enzyme is phosphorylated, the $K_m$ for its substrate increases by a factor of 3, but $V_{max}$ is unaltered. At a concentration of substrate equal to the $K_m$ for the unphosphorylated enzyme, decide whether phosphorylation activates the enzyme or inhibits it? Explain your reasoning.

**3–94** For an enzyme that follows Michaelis–Menten kinetics, by what factor does the substrate concentration have to increase to change the rate of the reaction from 20% to 80% $V_{max}$?

A. A factor of 2
B. A factor of 4
C. A factor of 8
D. A factor of 16
E. The factor required cannot be calculated without knowing $K_m$.

**3–95** The 'turnover number,' or $k_{cat}$, for an enzyme is the number of substrate molecules converted into product by an enzyme molecule per unit time when the enzyme is fully saturated with substrate. The maximum rate of a reaction, $V_{max}$, equals $k_{cat}$ times the concentration of enzyme. (Remember that the maximum rate occurs when all of the enzyme is present as the ES complex.) Carbonic anhydrase catalyzes the hydration of $CO_2$ to form $H_2CO_3$. Operating at its maximum rate, 10 μg of pure carbonic anhydrase ($M_r$ 30,000) in 1 mL hydrates 0.90 g of $CO_2$ in 1 minute. What is the turnover number for carbonic anhydrase?

**3–96** It is sometimes difficult to decide whether a molecule is a transition-state analog, especially in reactions with multiple substrates. *N*-(phosphonacetyl)-L-aspartate (PALA) is a very effective inhibitor of the reaction catalyzed by aspartate transcarbamoylase (Figure 3–23). PALA resembles the intermediate in the reaction and it binds with a $K_d$ of 27 nM. The $K_d$ for carbamoyl phosphate is 27 μM and that for aspartate is 11 mM. Although PALA binds more tightly than either substrate, it would be expected to because it combines elements of both substrates into the same molecule and therefore benefits from all the binding contacts made by either substrate separately. One quick way to evaluate the binding of the substrates relative to the inhibitor is by comparing the numerical value for the product of the $K_d$ values for the substrates with that for the $K_d$ for the inhibitor (all $K_d$ values must be in the same units). If the $K_d$ for the inhibitor is less than the product of the $K_d$ values for the two substrates, the inhibitor is likely to be a transition-state analog. By this criterion is PALA a transition-state analog or a 'bisubstrate' analog?

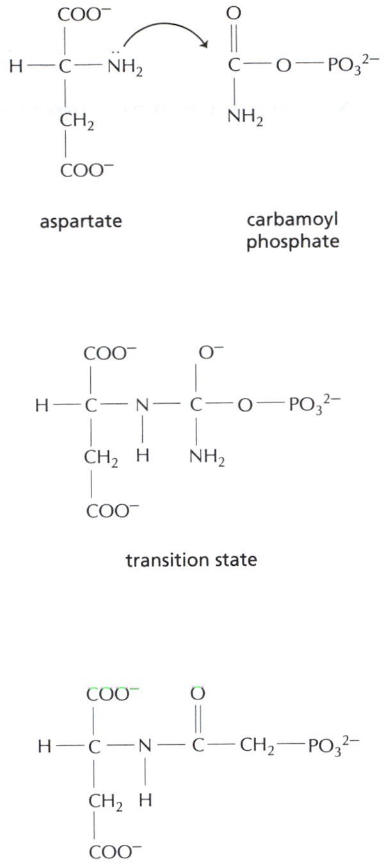

**Figure 3–23** Reaction catalyzed by aspartate transcarbamoylase and the inhibitor, PALA (Problem 3–96).

**3–97** Aspartate transcarbamoylase catalyzes the second step in the synthesis of the pyrimidine nucleotides UMP and CMP. When PALA is added to the growth medium at 2.7 μM (100-fold more than its $K_d$ for binding to the enzyme), cultured cells are very effectively killed, as might be expected from its ability to inhibit this critical enzyme. At a low frequency, cells arise that grow in the presence of PALA. Analysis of these resistant cells shows that they make an aspartate transcarbamoylase identical to the one in normal cells, but that they synthesize 100-times more of it.

A. How does producing more enzyme allow cells to grow in the presence of PALA?
B. Many enzymes become resistant to the effects of an inhibitor by mutational changes that reduce the binding of the inhibitor, and therefore reduce inhibition. Why do you suppose cells with this type of resistance to PALA were never found?

**3–98** Rous sarcoma virus (RSV) carries an oncogene called *Src*, which encodes a continuously active protein tyrosine kinase that leads to unchecked cell proliferation. Normally, Src carries an attached fatty acid (myristoylate) group that allows it to bind to the cytoplasmic side of the plasma membrane. A mutant version of Src that does not allow attachment of myristoylate does not bind to the membrane. Infection of cells with RSV encoding either the normal or the mutant form of Src leads to the same high level of protein tyrosine kinase activity, but the mutant Src does not cause cell proliferation.

A. Assuming that the normal Src is all bound to the plasma membrane and that the mutant Src is distributed throughout the cytoplasm, calculate their relative concentrations in the neighborhood of the plasma membrane. For the purposes of this calculation, assume that the cell is a sphere with a radius of 10 μm and that the mutant Src is distributed throughout, whereas the normal Src is confined to a 4-nm-thick layer immediately beneath the membrane. [For this problem, assume that the membrane has no thickness. The volume of a sphere is $(4/3)\pi r^3$.]

B. The target (X) for phosphorylation by Src resides in the membrane. Explain why the mutant Src does not cause cell proliferation.

pentapeptides with tyrosine

684 708 719 731 739 743 746 755

pentapeptides with phosphotyrosine

684 708 719 731 739 743 746 755

**Figure 3–24** Assay for PI 3-kinase in immunoprecipitates of the PDGF receptor (Problem 3–99). Pentapeptides are indicated by numbers that refer to their position in the PDGF receptor. *Black circles* indicate incorporation of $^{32}$P-phosphate into the substrate for PI 3-kinase.

## DATA HANDLING

**3–99** The binding of platelet-derived growth factor (PDGF) to the PDGF receptor stimulates phosphorylation of 8 tyrosines in the receptor's cytoplasmic domain. The enzyme phosphatidylinositol 3′-kinase (PI 3-kinase) binds to one or more of the phosphotyrosines through its SH2 domains and is thereby activated. To identify the activating phosphotyrosines, you synthesize 8 pentapeptides that contain the critical tyrosines (at the N-terminus) in their phosphorylated or unphosphorylated forms. You then mix an excess of each of the various pentapeptides with phosphorylated PDGF receptor and PI 3-kinase. Immunoprecipitation of the PDGF receptor will bring down any bound PI 3-kinase, which can be assayed by its ability to add $^{32}$P-phosphate to its substrate (Figure 3–24).

A. Do your results support the notion that PI 3-kinase binds to phosphotyrosines in the activated PDGF receptor? Why or why not?

B. The amino acid sequences of the PDGF receptor pentapeptides tested above (numbered according to the position of tyrosine in the PDGF receptor) and of peptide segments that are known to bind to PI 3-kinase in other activated receptors are shown below.

| | | | |
|---|---|---|---|
| 684 | YSNAL | YMMMR | (FGF receptor) |
| 708 | YMDMS | YTHMN | (insulin receptor) |
| 719 | YVPML | YEVML | (hepatocyte growth factor receptor) |
| 731 | YADIE | YMDMK | (steel factor receptor) |
| 739 | YMAPY | YVEMR | (CSF-1 receptor) |
| 743 | YDNYE | | |
| 746 | YEPSA | | |
| 755 | YRATL | | |

What are the common features of peptide segments that form binding sites for PI 3-kinase?

C. Which of the three common types of protein–protein interaction—surface–string, helix–helix, or surface–surface—does the binding of PI 3-kinase with the PDGF receptor most likely illustrate?

**3–100** Antibodies are often used to identify the location of a protein within a cell to gain clues to its function. But one must be careful in interpreting the results. A case in point is the *Brca1* gene, which was identified as the mutated gene in one type of familial breast cancer. Its sequence failed to identify a homolog with a known function. Two groups of scientists raised antibodies to segments of the Brca1 protein (predicted from the sequence to have a molecular mass of about 220 kd) and then reacted them with breast cancer cells using a method that allowed the bound antibodies to be visualized by fluorescence microscopy. One group used an antibody (C20) raised against a 20-amino acid peptide in the C-terminal region, and reported that Brca1 was located in secretory vesicles and on the plasma membrane. A second group, which used antibodies raised against an N-terminal fragment (BPA1) and antibodies raised against a C-terminal fragment (BPA2), reported that

Brca1 was localized to the nucleus. To clear up this contradiction, an immunoblot was performed using all three antibodies against the proteins in lysates from three different breast cancer cell lines (Figure 3–25). (An immunoblot is performed by separating the proteins according to their molecular mass by SDS polyacrylamide-gel electrophoresis, blotting them onto a filter paper, and then reacting them with antibodies in a way that allows the places where antibodies bind to be readily visualized.)

Based on the results of the immunoblot, propose an explanation for the contradictory results obtained by the two groups. Where do you think Brca1 is located in cells?

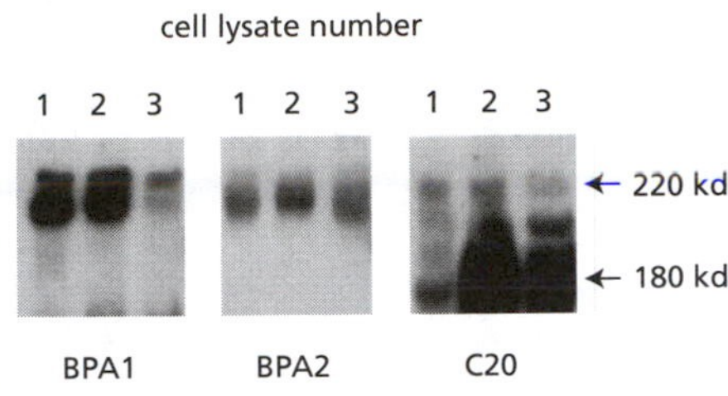


**Figure 3–25** Immunoblot of three antibodies raised against Brca1 (Problem 3–100).

**3–101** Equilibrium dialysis provides a simple method for determining the equilibrium constant for binding of a ligand (L) by a protein (Pr). The protein is confined inside a dialysis sac, formed by an artificial membrane with pores too tiny for the protein to enter, but which the much smaller ligand can freely permeate. The ligand, usually radiolabeled, is added to the solution surrounding the dialysis sac; after equilibrium has been established, the concentration of the ligand is measured in both compartments. The concentration in the external compartment is the concentration of free (unbound) ligand and the concentration in the dialysis sac is the sum of the bound (Pr–L) plus free ligand. By measuring these values after various initial ligand concentrations, the value of the equilibrium constant can be determined. By convention, the equilibrium is usually considered for the dissociation reaction (Pr–L → Pr + L), rather than the association reaction (Pr + L → Pr–L), and thus the equilibrium constant is referred to as the dissociation constant, $K_d$.

It is useful to look at the transformation of the standard equilibrium relationship into the form commonly used to analyze the data. At equilibrium,

$$K_d = \frac{[\text{Pr}][\text{L}]}{[\text{Pr–L}]}$$

Problem 16–60 uses a Scatchard plot to investigate the role of γ-tubulin in microtubule assembly.

Given that the total protein concentration, $[\text{Pr}]_{\text{TOT}}$, is the sum of the concentration of bound protein [Pr–L] and free protein [Pr], we can substitute $[\text{Pr}]_{\text{TOT}}$ – [Pr–L] for [Pr] and rearrange to give

$$[\text{Pr–L}] = \frac{[\text{Pr}]_{\text{TOT}}[\text{L}]}{K_d + [\text{L}]}$$

This is an equation for a rectangular hyperbola. It can be rearranged to give a linear form, as was first done by George Scatchard in 1947 (hence, graphs of such data are commonly known as Scatchard plots):

$$\frac{[\text{Pr–L}]}{[\text{L}]} = \frac{-[\text{Pr–L}]}{K_d} + \frac{[\text{Pr}]_{\text{TOT}}}{K_d}$$

At constant protein concentration and a variety of ligand concentrations, a plot of bound over free ligand ([Pr–L]/[L]) against bound ligand ([Pr–L]) gives a line with slope equal to $-1/K_d$ and an $x$-intercept equal to $[\text{Pr}]_{\text{TOT}}$, which is the total concentration of binding sites.

In the early 1960s, when the nature of the genetically identified repressors of bacterial gene expression had not been defined, Walter Gilbert and Benno Muller-Hill used equilibrium dialysis to measure the binding of an inducer of gene expression (IPTG) to the Lac repressor protein. They used radiolabeled IPTG and two partially purified preparations of Lac repressor protein: one from wild-type cells and the other from mutant cells in which induction of the lactose operon occurred at lower concentrations of IPTG. The mutant cells were assumed to carry a Lac repressor that bound IPTG more tightly. Their data are shown as a Scatchard plot in Figure 3–26.

A. What are the $K_d$ values for the two lines shown in the figure?

B. Which line corresponds to the wild-type Lac repressor and which to the mutant (tighter IPTG-binding) repressor?

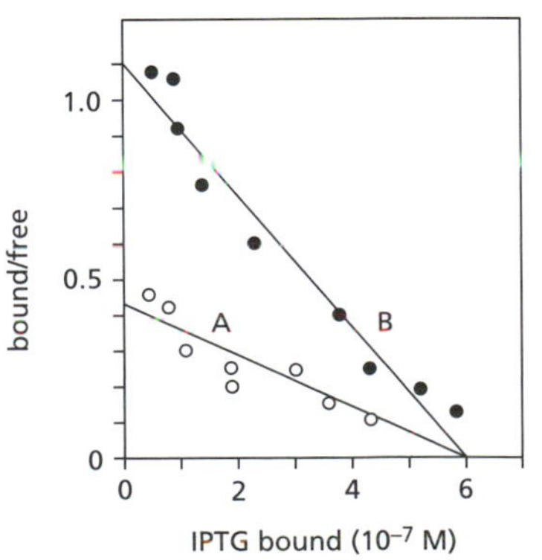


**Figure 3–26** Scatchard plot of equilibrium dialysis data for the binding of IPTG to the Lac repressor (Problem 3–101).

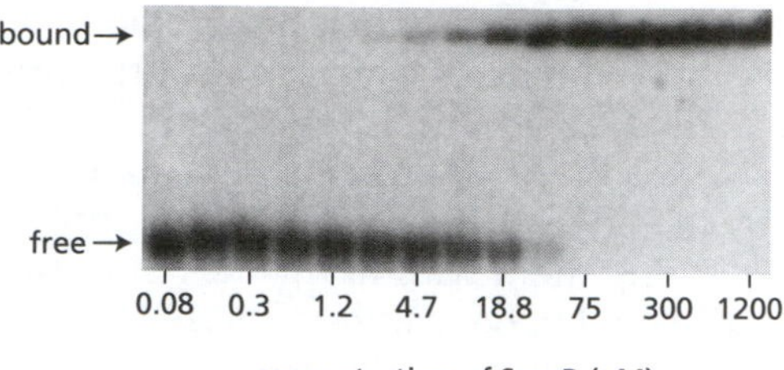


**Figure 3–27** Assay of the binding of purified SmpB protein to $^{32}$P-labeled tmRNA (Problem 3–102). From left to right across the gel, the experiment in each successive lane used a 2-fold increase in concentration of SmpB protein; concentrations in every other lane are indicated.

**3–102** Another common method for determining $K_d$ is to use gel electrophoresis to separate the bound and free forms of the ligand. (Note that this method requires that the off rate be slow enough that the complex will not dissociate during the time it takes to carry out the electrophoresis, usually a few hours.) This method was used to show that the protein, SmpB, binds specifically to a special species of tRNA (tmRNA) that is used in bacteria to eliminate the incomplete proteins made from truncated mRNAs. In this experiment tmRNA was labeled and included in the binding reactions at a concentration of 0.1 nM ($10^{-10}$ M). Purified SmpB protein was included at a range of concentrations and the mixture was incubated until the binding reaction was at equilibrium. Free and bound tmRNA were then separated by electrophoresis and made visible by autoradiography (Figure 3–27).

A. Examine the equation for $K_d$ in the previous problem. When the concentrations of bound and free ligand are equal, what is the relationship between the concentration of free protein and the $K_d$?

B. By visual inspection of Figure 3–27, estimate the $K_d$. Do you have to worry about the concentration of bound protein in this experiment? Why or why not?

C. The concentration of labeled tmRNA in these experiments was 100 pM. Would the results have been the same if tmRNA had been used at 100 nM? At 100 μM?

**Problem 6–92** looks at the mechanism of action of SmpB.

**3–103** If the data in Figure 3–27 are plotted as fraction tmRNA bound versus SmpB concentration, one obtains a symmetrical S-shaped curve as shown in Figure 3–28. This curve is a visual display of a very useful relationship between $K_d$ and concentration, which has broad applicability. The expression for fraction of ligand bound is derived from the equation for $K_d$ by substituting ($[L]_{TOT}$ – [L]) for [Pr–L] and rearranging. Because the total concentration of ligand ($[L]_{TOT}$) is equal to the free ligand ([L]) plus bound ligand ([Pr–L]),

$$\text{fraction bound} = \frac{[L]}{[L]_{TOT}} = \frac{[Pr]}{[Pr] + K_d}$$

(An equivalent relationship in terms of the fraction of protein bound can be derived in an analogous way; it is $[Pr]/[Pr]_{TOT} = [L]/([L] + K_d)$.)

Using this relationship, calculate the fraction of ligand bound for protein concentrations expressed in terms of $K_d$, using Table 3–2.

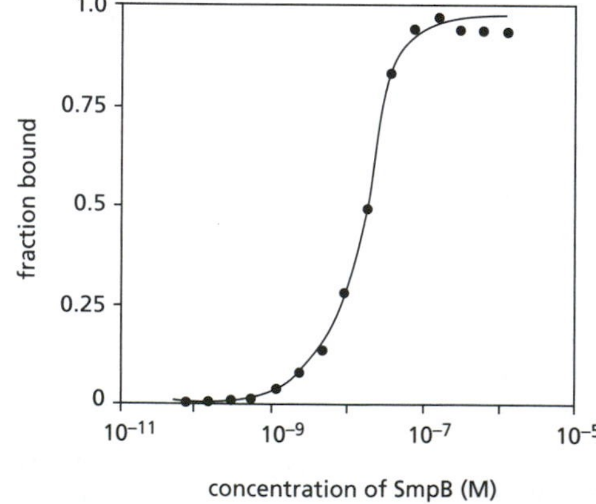


**Figure 3–28** Fraction of tmRNA bound versus SmpB concentration (Problem 3–103).

**Table 3–2 Fraction of ligand bound versus protein concentration** (Problem 3–103).

| PROTEIN CONCENTRATION | FRACTION BOUND (%) |
|---|---|
| $10^4\ K_d$ | |
| $10^3\ K_d$ | |
| $10^2\ K_d$ | |
| $10^1\ K_d$ | |
| $K_d$ | |
| $10^{-1}\ K_d$ | |
| $10^{-2}\ K_d$ | |
| $10^{-3}\ K_d$ | |
| $10^{-4}\ K_d$ | |

**3–104** The Lac repressor regulates the expression of a set of genes for lactose metabolism, which are adjacent to its binding site on the bacterial chromosome. In the absence of lactose in the medium, the binding of the repressor turns the genes off. When lactose is added, an inducer is generated that binds to the repressor and prevents it from binding to its DNA target, thereby turning on gene expression.

Inside *E. coli* there are about 10 molecules of Lac repressor ($10^{-8}$ M) and 1 binding site ($10^{-9}$ M) on the bacterial genome. The dissociation constant, $K_d$, for binding of the repressor to its binding site is $10^{-13}$ M. When an inducer of gene expression binds to the repressor, the $K_d$ for repressor binding to its DNA binding sites increases to $10^{-10}$ M. Use the relationships developed in the previous problem to answer the questions below.

A. In a population of bacteria growing in the absence of lactose, what fraction of the binding sites would you expect to be bound by repressor?

B. In bacteria growing in the presence of lactose, what fraction of binding sites would you expect to be bound by the repressor?

C. Given the information in this problem, would you expect the inducer to turn on gene expression? Why or why not?

D. The Lac repressor binds nonspecifically to any sequence of DNA with a $K_d$ of about $10^{-6}$ M, which is a very low affinity. Can you suggest in a qualitative way how such low-affinity, nonspecific binding might alter the calculations in parts A and B and your conclusion in part C?

**Table 3–3 Initial rates of product formation at various substrate concentrations** (Problem 3–105).

| RATE (μmol/min) | [S] (μM) |
|---|---|
| 0.15 | 0.08 |
| 0.21 | 0.12 |
| 0.7 | 0.54 |
| 1.1 | 1.23 |
| 1.3 | 1.82 |
| 1.5 | 2.72 |
| 1.7 | 4.94 |
| 1.8 | 10.00 |
| **1/RATE (min/μmol)** | **1/[S] (1/μM)** |
| 6.7 | 12.5 |
| 4.8 | 8.3 |
| 1.4 | 1.9 |
| 0.91 | 0.81 |
| 0.77 | 0.55 |
| 0.67 | 0.37 |
| 0.59 | 0.20 |
| 0.56 | 0.10 |

**3–105** The rates of production of product, P, from substrate, S, catalyzed by enzyme, E, were measured under conditions in which very little product was formed. The results are summarized in Table 3–3.

A. Why is it important to measure rates of product formation under conditions in which very little product is formed?

B. Plot these data as rate versus substrate concentration. Is this plot a rectangular hyperbola as expected for an enzyme that obeys Michaelis–Menten kinetics? What would you estimate as the $K_m$ and $V_{max}$ values for this enzyme?

C. To obtain more accurate values for the kinetic constants, the Lineweaver–Burk transformation of the Michaelis–Menten equation is often used so that the data can be plotted as a straight line.

Michaelis–Menten equation:

$$\text{rate} = \frac{V_{max}\,[S]}{[S] + K_m}$$

Lineweaver–Burk equation:

$$\frac{1}{\text{rate}} = \left(\frac{K_m}{V_{max}}\right)\frac{1}{[S]} + \frac{1}{V_{max}}$$

This equation has the form of a straight line, $y = ax + b$. Thus, when 1/rate ($y$) is plotted versus 1/[S] ($x$), the slope of the line equals $K_m/V_{max}$ *(a)* and the $y$ intercept is $1/V_{max}$ *(b)*. Furthermore, it can be shown that the $x$ intercept is equal to $-1/K_m$.

Plot 1/rate versus 1/[S] and determine the kinetic parameters $K_m$ and $V_{max}$. (The values for 1/rate and 1/[S] are shown in Table 3–3.)

**3–106** Lysozyme achieves its antibacterial effect by cleaving the polysaccharide chains that form the bacterial cell wall. In the absence of this rigid mechanical support, the bacterial cell literally explodes due to its high internal osmotic pressure. The cell wall polysaccharide is made up of alternating sugars, *N*-acetylglucosamine (NAG) and *N*-acetylmuramate (NAM), linked together by glycosidic bonds (Figure 3–29). Lysozyme normally cleaves after NAM units in the chain (that is, between NAM and NAG), but will also cleave artificial substrates composed entirely of NAG units. When the crystal structure of lysozyme bound to a chain of three NAG units (tri-NAG) was solved,

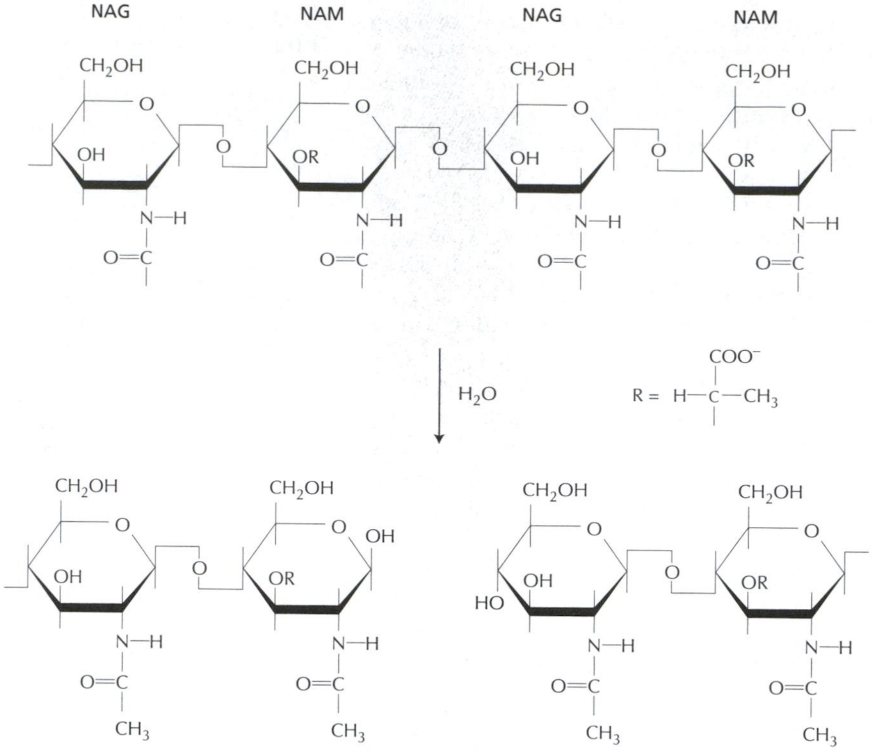

Figure 3–29 Arrangement of NAM and NAG in the bacterial cell-wall polysaccharide (Problem 3–106).

it was discovered that the binding cleft in lysozyme comprised six sugar binding sites, A through F, and that tri-NAG filled the first three of these sites. From the crystal structure it was not apparent, however, which of the five bonds between the six sugars was the one that was normally cleaved. Tri-NAG is not cleaved by lysozyme, although longer NAG polymers are. It was clear from modeling studies that NAM is too large to fit into site C. Where are the catalytic groups responsible for cleavage located relative to the six sugar-binding sites?

A. Between sites A and B
B. Between sites B and C
C. Between sites C and D
D. Between sites D and E
E. Between sites E and F

**3–107** Aspartate transcarbamoylase (ATCase) is an allosteric enzyme with six catalytic and six regulatory subunits. It exists in two conformations: one with low enzymatic activity and the other with high activity. In the absence of any ligands the low-activity conformation predominates. Malate is an inhibitor of ATCase that binds in the active site at the position where the substrate aspartate normally binds. A very peculiar effect of malate is observed when the activity of ATCase is measured at low aspartate concentrations: there is an *increase* in ATCase activity at very low malate concentrations, but then the activity decreases at higher concentrations (Figure 3–30).

A. How is it that malate, a bona fide inhibitor, can increase ATCase activity under these conditions?
B. Would you expect malate to have the same peculiar effect if the measurements were made in the presence of a high concentration of aspartate? Why or why not?

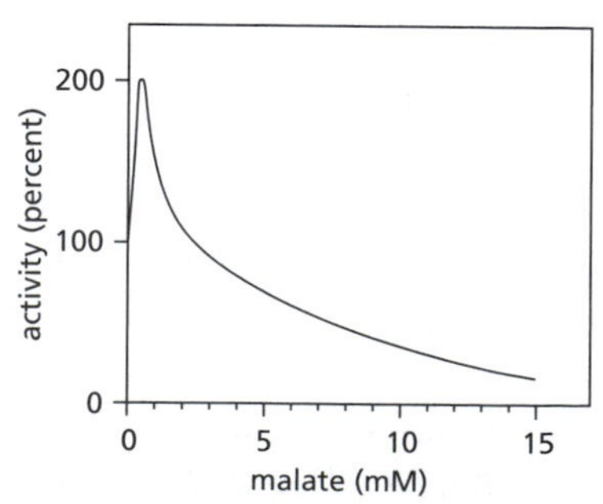


Figure 3–30 The activity of ATCase with increasing concentrations of the inhibitor malate (Problem 3–107). These measurements were made at an aspartate concentration well below its $K_m$.

**3–108** Using a chemically modified version of ATCase, it is possible to monitor binding at active sites due to local changes that occur upon ligand binding. It is also possible to measure global changes in ATCase conformation by effects on the protein's sedimentation rate in a centrifugal field. You want to know how binding at the active sites of ATCase relates to global conformational changes. You incubate the modified ATCase with increasing concentrations

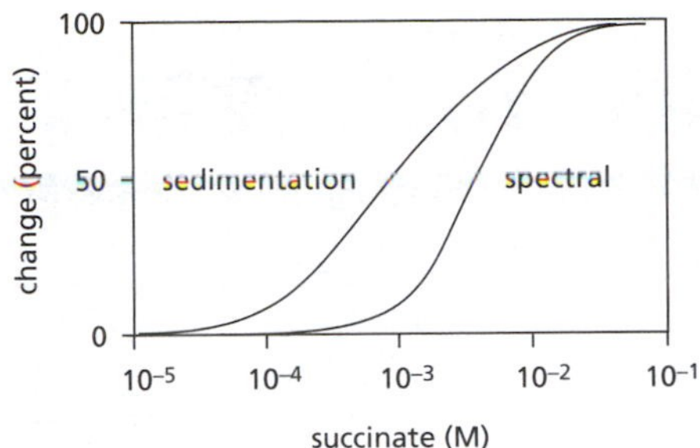


**Figure 3–31** Changes in binding and global conformation of ATCase with increasing succinate (Problem 3–108).

of succinate (a substrate analog that can bind to the active site in each subunit). You measure succinate binding spectrally and global conformational changes by sedimentation. Your results are shown in Figure 3–31. Are these results what you might expect for an allosteric protein like ATCase? Why or why not?

**3–109** Cyclin-dependent protein kinase 2 (Cdk2) regulates critical events in the progression of the cell cycle in mammalian cells. Cdk2 can form a complex with cyclin A and can be phosphorylated by another protein kinase, Civ1, to produce P-Cdk2. To determine the roles of cyclin A and phosphorylation in the function of Cdk2, you purify nonphosphorylated and Civ1-phosphorylated Cdk2. You mix these two forms of Cdk2 and cyclin A in various combinations with $^{32}$P-ATP and assay for phosphorylation of histone H1 (Figure 3–32). You also measure the binding affinity of various forms of Cdk2 for ATP, ADP, cyclin A, and histone H1 (Table 3–4).

A. From Figure 3–32, what is required for Cdk2 to phosphorylate histone H1 efficiently?

B. How do the requirements identified in part A specifically affect the function of Cdk2 relative to its target, histone H1 (Table 3–4 and Figure 3–32)?

C. The usual intracellular concentrations of ATP and ADP are in the range 0.1 to 1 mM. Assume that the binding of cyclin A to Cdk2 or P-Cdk2 does not alter the affinities of either form of Cdk2 for ATP and ADP. Is it likely that the observed changes in affinity for ATP and ADP are important for Cdk2 function? Why or why not?

**3–110** Genome sequencing has revealed a surprisingly large number of protein tyrosine phosphatases (PTPs), very few of which have known roles in the life of a cell. PTP1B was the very first member of the PTP family to be discovered, and its three-dimensional structure and catalytic mechanism are well defined. PTP1B will dephosphorylate almost any phosphotyrosine-containing protein or peptide in the test tube, but in the cell it is likely to have better-defined targets. One strategy for identifying a binding target is to incubate the known protein with a cell lysate, and then isolate the known protein and identify any proteins that are bound to it. An enzyme such as PTP1B, however, binds and releases its target as part of its catalytic cycle. You reason that if you interfere with catalysis you might be able to increase the dwell time of the substrate, making it stable enough for isolation. To this end, you make several mutants of PTP1B by changing specific amino acids in the active site, and measure their kinetic parameters (Table 3–5).

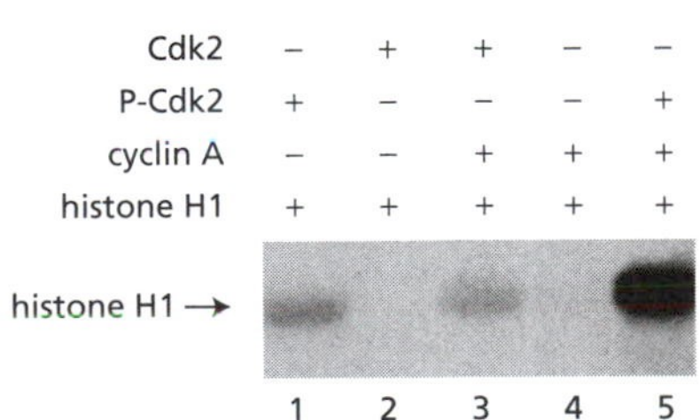


**Figure 3–32** Phosphorylation of histone H1 by various combinations of Cdk2 and cyclin A (Problem 3–109). The amount of radioactive phosphate attached to histone H1 in lanes 1 and 3 is 0.3% and 0.2%, respectively, of that in lane 5.

**Table 3–4 The observed dissociation constants ($K_d$ values) of Cdk2 for ATP, ADP, cyclin A, and histone H1** (Problem 3–109).

| | $K_d$ (μM) | | | |
|---|---|---|---|---|
| | ATP | ADP | CYCLIN A | HISTONE H1 |
| Cdk2 | 0.25 | 1.4 | 0.05 | not detected |
| P-Cdk2 | 0.12 | 6.7 | 0.05 | 100 |
| Cdk2 + cyclin A | | | | 1.0 |
| P-Cdk2 + cyclin A | | | | 0.7 |

**Table 3–5 Kinetic parameters of purified PTP1B mutants** (Problem 3–110).

| ENZYME | $V_{max}$ [nmol/(min/mg)] | $K_M$ (nM) | $k_{cat}$ (min$^{-1}$) |
|---|---|---|---|
| Wild type | 60,200 | 102 | 2244 |
| Tyr-46 → Leu | 4160 | 1700 | 155 |
| Glu-115 → Ala | 5700 | 45 | 212 |
| Lys-120 → Ala | 19,000 | 80 | 708 |
| Asp-181 → Ala | 0.61 | 126 | 0.023 |
| His-214 → Ala | 700 | 20 | 26 |
| Cys-215 → Ser | no activity | | |
| Arg-221 → Lys | 11 | 80 | 0.41 |
| Arg-221 → Met | 3.3 | 1060 | 0.12 |

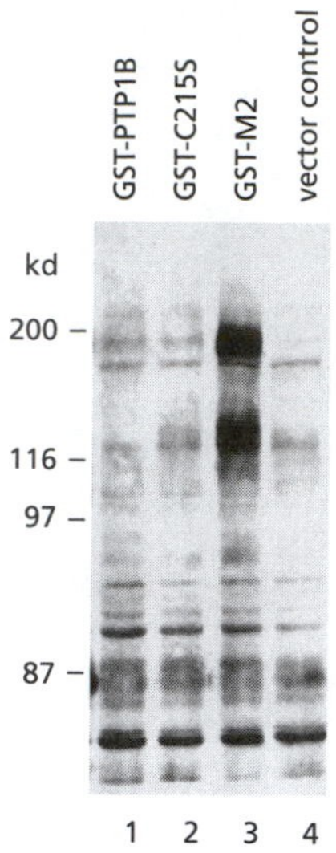


**Figure 3–33** Phosphotyrosine-containing proteins precipitated with GST-tagged PTP1B enzymes (Problem 3–110). GST-PTP1B is the fused wild-type protein; GST-M2 is the fused mutant-2 protein. The positions of 'marker' proteins of known molecular masses are indicated on the *left*.

You pick mutant Cys-215 → Ser (C215S) for study because the –SH group of Cys-215 initiates a nucleophilic attack on the phosphotyrosine, releasing the phosphate. You pick a second mutant—mutant 2—because it has promising kinetic properties. You use genetic engineering tricks to fuse these two mutants and the wild-type enzyme to glutathione *S*-transferase (GST), so that each of the three proteins can be rapidly purified by precipitation with glutathione-Sepharose. You express the GST-tagged proteins, precipitate them from cell lysates, separate the precipitated proteins by gel electrophoresis, and identify phosphotyrosine-containing proteins using anti-phosphotyrosine antibodies. As shown in Figure 3–33, GST-mutant 2 bound two phosphotyrosine-containing proteins (the two dark bands in lane 3), whereas GST-C215S bound none.

A. Which mutant PTP1B in Table 3–5 is likely to correspond to GST-mutant 2? Why do you think this protein gave a successful result?

B. Why might GST-C215S have failed to precipitate any phosphotyrosine-containing proteins?

# DNA, Chromosomes, and Genomes

4

## THE STRUCTURE AND FUNCTION OF DNA

TERMS TO LEARN

| | | |
|---|---|---|
| antiparallel | deoxyribonucleic acid (DNA) | genome |
| base pair | double helix | template |
| complementary | gene | |

### DEFINITIONS

Match each definition below with its term from the list above.

4–1 The totality of the genetic information carried in the DNA of a cell or an organism.

4–2 The three-dimensional structure of DNA, in which two DNA chains held together by hydrogen bonds between the bases are coiled around one another.

4–3 Information-containing element that controls a discrete hereditary characteristic.

4–4 Describes the relative orientation of the two stands in a DNA helix; the polarity of one strand is oriented in the opposite direction to that of the other.

4–5 Two nucleotides in an RNA or DNA molecule that are held together by hydrogen bonds.

### TRUE/FALSE

Decide whether this statement is true or false, and then explain why.

4–6 Human cells do not contain any circular DNA molecules.

### THOUGHT PROBLEMS

4–7 The start of the coding region for the human β-globin gene reads 5′-ATGGTGCAC-3′. What is the sequence of the complementary strand for this segment of DNA?

4–8 Upon returning from a recent trip abroad, you explain to the customs agent that you are bringing in a sample of DNA, deoxyribonucleic acid. He is aghast that you want to bring an acid into his country. What is the acid in DNA? Should the customs agent be wary?

4–9 The chemical structures for an AT and a GC base pair are shown in Figure 4–1, along with their points of attachment to the sugar-phosphate backbones.

A. Indicate the positions of the major and minor grooves of the DNA helix on these representations.

B. Draw the structure of a TA base pair in the same way as in Figure 4–1.

## In This Chapter


| | |
|---|---|
| THE STRUCTURE AND FUNCTION OF DNA | 63 |
| CHROMOSOMAL DNA AND ITS PACKAGING IN THE CHROMATIN FIBER | 65 |
| THE REGULATION OF CHROMATIN STRUCTURE | 72 |
| THE GLOBAL STRUCTURE OF CHROMOSOMES | 76 |
| HOW GENOMES EVOLVE | 81 |


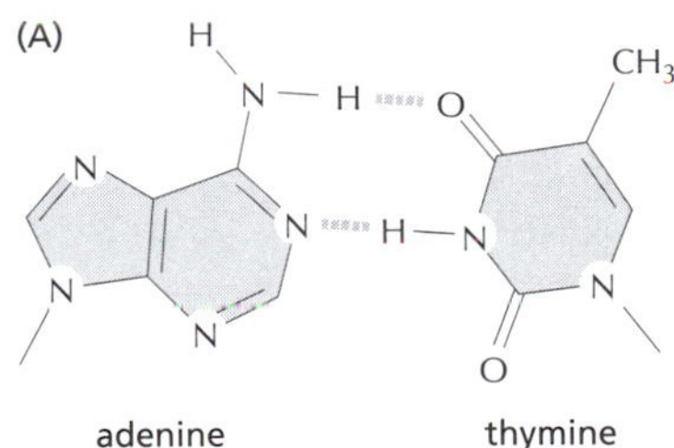


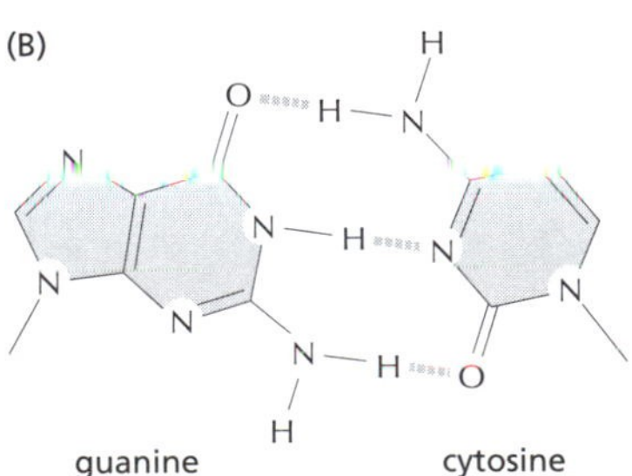


**Figure 4–1** An AT and a GC base pair, as viewed along the helix axis (Problem 4–9). Each base is attached to its deoxyribose sugar via the line extending from the ring nitrogen.

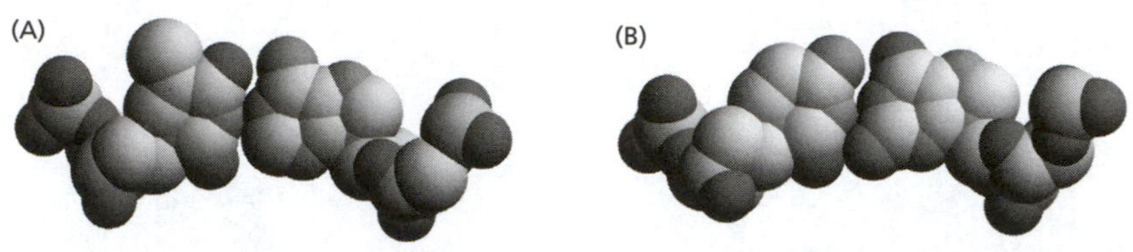


Figure 4–2 Space-filling models of two base pairs (Problem 4–10). Carbon and phosphorus atoms are *light gray*, nitrogen atoms are *intermediate*, and oxygen atoms are *dark gray*. No hydrogen atoms are shown.

C. Do the same chemical moieties (for example, the methyl group of thymine) always project into the same groove?

D. With a sufficiently powerful microscope, do you think it would be possible to read directly the sequence of DNA? Why or why not?

**4–10** Examine the space-filling models of the base pairs shown in Figure 4–2. Each base pair includes two bases, two deoxyribose sugars, and two phosphates. Can you identify the locations of the purine base, the pyrimidine base, the sugars, and the phosphates? Identify each base pair as CG, GC, TA, or AT.

**4–11** DNA isolated from the bacterial virus M13 contains 25% A, 33% T, 22% C, and 20% G. Do these results strike you as peculiar? Why or why not? How might you explain these values?

**4–12** A segment of DNA from the interior of a single strand is shown in Figure 4–3. What is the polarity of this DNA from top to bottom?

**4–13** DNA forms a right-handed helix. Pick out the right-handed helix from those shown in Figure 4–4.

## CALCULATIONS

**4–14** Human DNA contains 20% C on a molar basis. What are the mole percents of A, G, and T?

**4–15** The diploid human genome comprises $6.4 \times 10^9$ bp and fits into a nucleus that is 6 μm in diameter.

A. If base pairs occur at intervals of 0.34 nm along the DNA helix, what is the length of DNA in a human cell?

B. If the diameter of the DNA helix is 2.4 nm, what fraction of the volume of the nucleus is occupied by DNA? [Volume of a sphere is $(4/3)\pi r^3$ and volume of a cylinder is $\pi r^2 h$.]

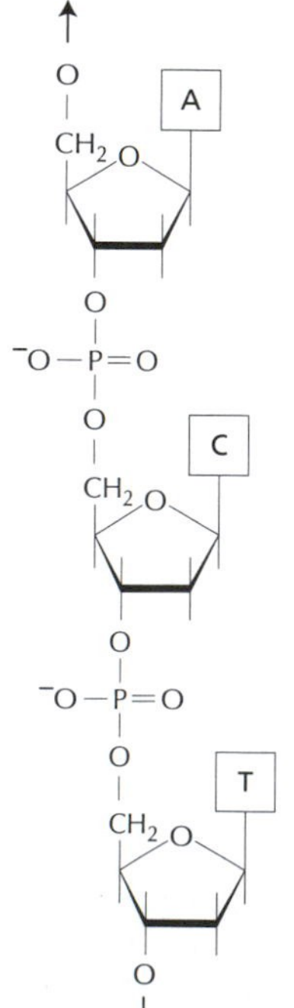


Figure 4–3 Three nucleotides from the interior of a single strand of DNA (Problem 4–12). *Arrows* at the ends of the DNA strand indicate that the structure continues in both directions.

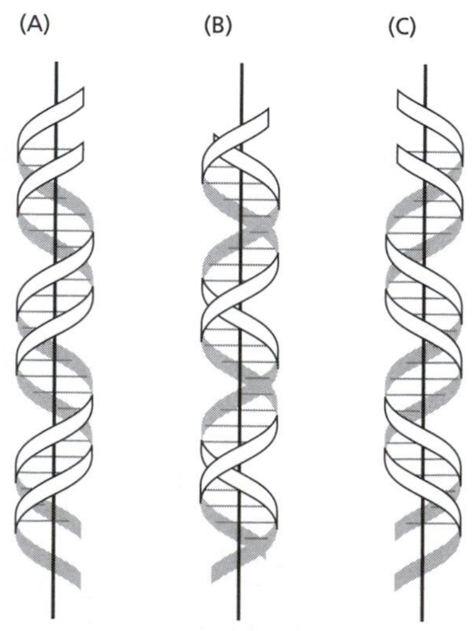


Figure 4–4 Three 'DNA' helices (Problem 4–13).

**4–16** One gram of cultured human cells contains about $10^9$ cells and occupies roughly 1 mL. If the average molecular mass of a base pair is 660 daltons and each cell contains $6.4 \times 10^9$ bp, what mass of DNA is present in this one-gram sample? If all the DNA molecules in the sample were laid end to end to form a single thread, would it be long enough to reach from the Earth to the Moon (385,000 kilometers)?

### DATA HANDLING

**4–17** Bacteriophage T4 attaches to its bacterial host and injects its DNA to initiate an infection that ultimately releases hundreds of progeny virus. In 1952 Alfred Hershey and Martha Chase radiolabeled the DNA of bacteriophage T4 with $^{32}PO_4^{3-}$ and the proteins with $^{35}S$-methionine. They then mixed the labeled bacteriophage with bacteria and after a brief time agitated the mixture vigorously in a blender to detach T4 from the bacteria. They then separated the phage from the bacteria by centrifugation. They demonstrated that the bacteria contained 30% of the $^{32}P$ label but virtually none of the $^{35}S$ label. When new bacteriophage were released from these bacteria, they were also found to contain $^{32}P$ but no $^{35}S$. How does this experiment demonstrate that DNA rather than protein is the genetic material? (Note that bacteriophage T4 contains only protein and DNA.)

## CHROMOSOMAL DNA AND ITS PACKAGING IN THE CHROMATIN FIBER

| TERMS TO LEARN | | |
|---|---|---|
| cell cycle | histone | karyotype |
| centromere | histone H1 | nucleosome |
| chromatin | homologous chromosome (homolog) | replication origin |
| chromosome | intron | telomere |
| exon | | |

### DEFINITIONS

Match each definition below with its term from the list above.

**4–18** Full set of chromosomes of a cell arranged with respect to size, shape, and number.

**4–19** Constricted region of a mitotic chromosome that holds sister chromatids together.

**4–20** Any one of a group of small abundant proteins, rich in arginine and lysine, that form the primary level of chromatin organization.

**4–21** Structure composed of a very long DNA molecule and associated proteins that carries part (or all) of the hereditary information of an organism.

**4–22** The orderly sequence of events by which a cell duplicates its contents and divides into two.

**4–23** Complex of DNA, histones, and nonhistone proteins found in the nucleus of a eucaryotic cells.

**4–24** One of the two copies of a particular chromosome in a diploid cell, each copy being derived from a different parent.

**4–25** Beadlike structure in eucaryotic chromatin, composed of a short length of DNA wrapped around a core of histone proteins.

## TRUE/FALSE

Decide whether each of these statements is true or false, and then explain why.

**4–26** Human females have 23 different chromosomes, whereas human males have 24.

**4–27** The majority of human DNA is thought to be unimportant junk.

**4–28** In a comparison between the DNAs of related organisms such as humans and mice, identifying the conserved DNA sequences facilitates the search for functionally important regions.

**4–29** In the living cell, chromatin usually adopts the extended 'beads-on-a-string' form.

**4–30** The four core histones are relatively small proteins with a very high proportion of positively charged amino acids; the positive charge helps the histones bind tightly to DNA, regardless of its nucleotide sequence.

**4–31** Nucleosomes bind DNA so tightly that they cannot move from the positions where they are first assembled.

## THOUGHT PROBLEMS

**4–32** In the 1950s the techniques for isolating DNA from cells all yielded molecules of about 10,000 to 20,000 base pairs. We now know that the DNA molecules in all cells are very much longer. Why do you suppose such short pieces were originally isolated?

**4–33** Consider the following statement: A human cell contains 46 molecules of DNA in its nucleus. Do you agree with it? Why or why not?

**4–34** An abnormal human karyotype is shown in Figure 4–5. This particular karyotype is found in the cancer cells of more than 90% of patients with chronic myelogenous leukemia. Arrows indicate two abnormal chromosomes. Describe the event that led to this abnormal karyotype. Is this patient male or female?

**4–35** Your advisor, the brilliant bioinformatician, has a high regard for your intellect and industry. She suggests that you write a computer program that will identify the exons of protein-encoding genes directly from the sequence of

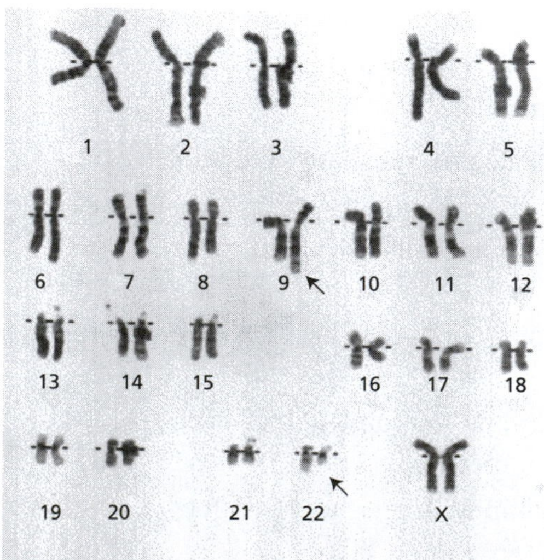


**Figure 4–5** Karyotype illustrating the typical alteration seen in chronic myelogenous leukemia (Problem 4–34).

the human genome. In preparation for that task, you decide to write down a list of the features that might distinguish coding sequences from intronic DNA and sequences outside of genes. What features would you list? (You may wish to try this problem after you have reviewed the basic elements of gene expression in *MBoC* Chapter 6.)

**4–36** Why do you expect to encounter a STOP codon about every 20 codons, or so, on average in a random sequence of DNA? (See the Genetic Code, inside back cover.)

**4–37** Chromosome 3 in orangutans differs from chromosome 3 in humans by two inversion events (Figure 4–6). Draw the intermediate chromosome that resulted from the first inversion and explicitly indicate the segments included in each inversion.

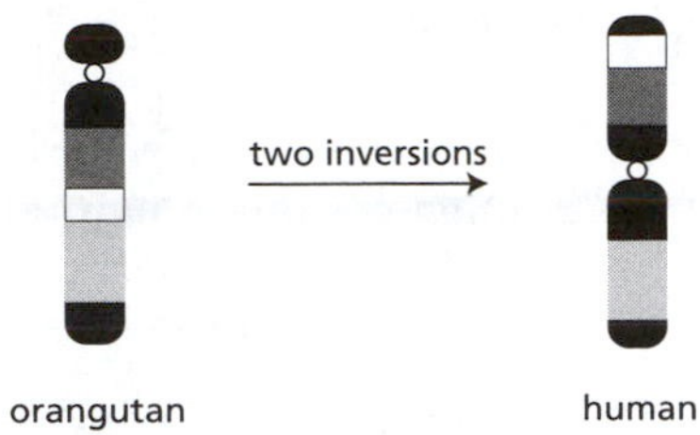


**Figure 4–6** Chromosome 3 in orangutans and humans (Problem 4–37). Differently *shaded blocks* indicate segments of the chromosomes that were derived by previous fusions.

**4–38** Define a 'gene.'

**4–39** List the three specialized DNA sequences and their functions that act to ensure that the number and morphology of chromosomes are constant from one generation of a cell to the next.

**4–40** Describe the consequences that would arise if a eucaryotic chromosome (150 Mb in length) had one of the following features:

A. A single replication origin located in the middle of the chromosome. (DNA replication in animal cells proceeds at about 150 nucleotide pairs per second.)
B. A telomere at only one end of the chromosome.
C. No centromere.

**4–41** Early in development, most human cells turn off expression of an essential component of telomerase, the enzyme responsible for addition of telomere repeat sequences (5′-TTAGGG) to the ends of chromosomes. Thus, as our cells proliferate their telomeres get shorter and shorter, but are normally not lost over the course of a lifetime. If cells are removed from the body and grown in culture, they ultimately enter a state of replicative senescence and stop dividing when their telomeres get too short. By contrast, most tumor cells, in humans and in cell culture, express active telomerase, allowing them to maintain their telomeres and grow beyond the normal limit imposed by senescence—good for them, bad for us.

Anticipating a universal cure for cancer, you set up a company to screen chemical 'libraries' for telomerase inhibitors. The company share price takes a tumble, when a rival group generates a strain of mice from which the telomerase genes have been deleted. These mice breed happily for several generations, but by the sixth generation (when their telomeres are much shorter than those of normal mice) they have a greatly increased tendency to die of tumors. The tumors tend to arise in tissues that show high proliferation rates, such as testis, skin, and blood. Why has this observation shaken the confidence of your investors? Is there a flaw in your hypothesis?

**4–42** Histone proteins are among the most highly conserved proteins in eucaryotes. Histone H4 proteins from a pea and a cow, for example, differ in only 2 of 102 amino acids. However, comparison of the two *gene* sequences shows many more differences. These observations indicate that mutations that change amino acids must be selected against. Why do you suppose that amino acid-altering mutations in histone genes are deleterious?

**4–43** Duplex DNA composed entirely of CTG/CAG trinucleotide repeats (5′-CTG in one strand and 5′-CAG in the other strand) is unusually flexible. If 75 CTG/CAG repeats are incorporated into a much longer DNA molecule and mixed with histone octamers, the first nucleosome that assembles nearly always includes the CTG/CAG repeat region. Can you suggest a reason why CTG/CAG repeats might be such effective elements for positioning nucleosomes? (Consider what the energy of the binding interaction between the histone octamers and the DNA must accomplish.)

## CALCULATIONS

**4–44** Human chromosome 1 contains about $2.8 \times 10^8$ bp. At mitosis this chromosome measures 10 μm in length. Relative to its fully extended length, how compacted is the DNA molecule in chromosome 1 at mitosis? (Recall that a DNA bp is 0.34 nm in length.)

**4–45** The total number of protein-coding genes in the human genome can be calculated in several ways. It is important to remember that all such numbers are estimates at present, because it is still difficult to identify a gene from the DNA sequence. Chromosome 22 has about 700 genes in 48 Mb of sequence, which represents 1.5% of the estimated 3200 Mb in the haploid genome. Using these numbers, how many genes would you estimate for the haploid human genome? If your estimate is significantly larger or smaller than the accepted value of approximately 25,000 genes, suggest possible explanations for the discrepancy.

**4–46** The 700 genes on chromosome 22 (48 Mb) average 19,000 bp in length and contain an average of 5.4 exons, each of which averages 266 bp. On average, what fraction of each gene sequence is converted to mRNA? What fraction of the whole chromosome do genes occupy?

**4–47** A single nucleosome is 11 nm long and contains 146 bp of DNA (0.34 nm/bp). What packing ratio (DNA length to nucleosome length) has been achieved by wrapping DNA around the histone octamer? Assuming that there are an additional 54 bp of extended DNA in the linker between nucleosomes, how condensed is 'beads-on-a-string' DNA relative to fully extended DNA? What fraction of the 10,000-fold condensation that occurs at mitosis does this first level of packing represent?

**4–48** Assuming that the histone octamer forms a cylinder 9 nm in diameter and 5 nm in height and that the human genome forms 32 million nucleosomes, what volume of the nucleus (6 μm in diameter) is occupied by histone octamers? [Volume of a cylinder is $\pi r^2 h$; volume of a sphere is $(4/3)\pi r^3$.] What fraction of the nuclear volume do the DNA and the histone octamers occupy (see Problem 4–15)?

**4–49** Assuming that the 30-nm chromatin fiber contains about 20 nucleosomes (200 bp/nucleosome) per 50 nm of length, calculate the degree of compaction of DNA associated with this level of chromatin structure. What fraction of the 10,000-fold condensation that occurs at mitosis does this level of DNA packing represent?

## DATA HANDLING

**4–50** One way to demonstrate that a chromosome contains a single DNA molecule is to use a technique called pulsed-field gel electrophoresis, which can separate DNA molecules up to $10^7$ bp in length. Ordinary gel electrophoresis cannot separate such long molecules because the steady electric field stretches them out so they travel end-first through the gel matrix at a rate that is independent of their length. If the electric field is changed periodically, however, the DNA molecules are forced to reorient to the new field before continuing their snakelike movement through the gel. The time for reorientation is dependent on length, so that longer molecules move more slowly through the gel.

The results of pulsed-field gel electrophoresis of the DNA from the yeast *Saccharomyces cerevisiae* are shown in Figure 4–7. How many chromosomes does *S. cerevisiae* have?

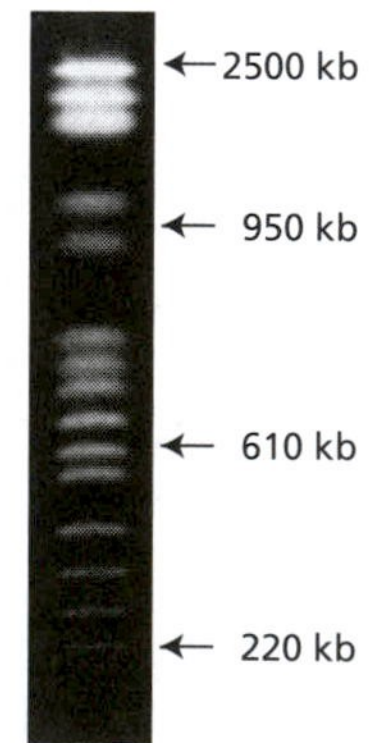


**Figure 4–7** Pulsed-field gel electrophoresis of *S. cerevisiae* chromosomes (Problem 4–50). To minimize the handling of DNA, which would surely break it, the cells themselves are placed at the top of the gel and gently opened by addition of a lysis buffer. The DNA molecules in this gel have been exposed to the dye ethidium bromide, which fluoresces under ultraviolet light when it is bound to DNA. This treatment allows the DNA—otherwise invisible—to be seen.

**4–51** The human U2 small nuclear RNA (U2 snRNA), which is present at thousands of copies per nucleus, plays an important role in mRNA processing. You have made a bacteriophage lambda clone that is 43 kb long and carries

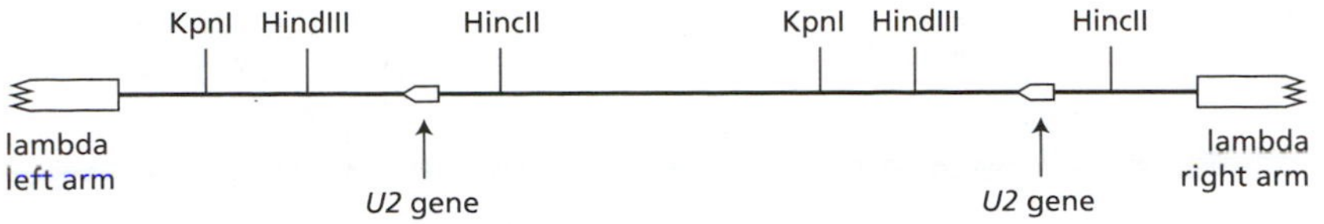


**Figure 4–8** Restriction map of a bacteriophage lambda clone carrying two *U2* genes (Problem 4–51). There are no restriction sites for EcoRI, BglII or XbaI in the stretch of DNA shown. The bacteriophage lambda arms are not shown to scale.

two copies of the *U2* gene, 6 kb apart. The restriction map of this clone is shown in Figure 4–8. (The basics of restriction-enzyme technology are covered in *MBoC* Chapter 8.)

When you cut human genomic DNA to completion with HindIII, HincII (H2), or KpnI (K) and analyze the restriction digest by blot hybridization against the *U2* gene, you detect a single intense band at 6 kb (Figure 4–9, lanes 9 to 11). If you cut with BglII (B), EcoRI (R), or XbaI (X), which do not cut the cloned segment (Figure 4–8), you also detect a single intense band, but of a size greater than 50 kb (Figure 4–9, lanes 1 to 3). If you incubate the genomic DNA with HindIII and remove samples at various times, you see a ladder of bands (lanes 4 to 9). If you cut 2 ng of the bacteriophage lambda cloned DNA with KpnI and run it alongside 10 μg of the genomic KpnI digest, two bands are visible—each of equal intensity to the 6-kb band from the genomic digest (compare lanes 11 and 12).

A. What is the organization of the *U2* genes in the human genome. Explain how the restriction digests define it.

B. Why are two bands visible in the digest of the cloned DNA (lane 12), whereas only one is visible in the digest of genomic DNA (lane 11)?

C. Given that 2 ng of cloned DNA produces a band of equal intensity to that from 10 μg of genomic DNA (lanes 11 and 12), calculate the number of *U2* genes in the human genome. (The bacteriophage lambda clone is 43 kb, and the haploid human genome is $3.2 \times 10^6$ kb.)

D. The sequence of the human genome in this region identifies only 3 genes encoding U2 snRNA. What do you suppose is the basis for the difference in the number of *U2* genes identified by sequencing and by the calculation in part C?

**4–52** About 5% of the human genome consists of duplicated segments of chromosomes, many of which are highly homologous, indicating a relatively recent origin. The high degree of homology occasionally allows inappropriate recombination events to occur between the duplications, which can decrease or increase the number of duplicated segments. Such events are responsible for several human diseases, including the red–green color blindness that affects 8% of the male population. The genes for the red and green visual pigments lie near one another on the X chromosome, one in each copy of the duplicated segment. They are 98% identical throughout most of their length, in both exons and introns; however, the genes can be distinguished by the chance presence of an extra RsaI cleavage site in one of the two genes. As a result, digestion with RsaI gives a longer fragment for the *RsaI-A* gene than for the *RsaI-B* gene, and this so-called restriction fragment length polymorphism (RFLP) can be used to track the two genes (Figure 4–10).

A. To determine which gene encodes which pigment, several normal, red-blind, and green-blind males were screened using a hybridization probe specific for the RsaI RFLP (Figure 4–10). Which gene encodes the red visual pigment, and which encodes the green visual pigment?

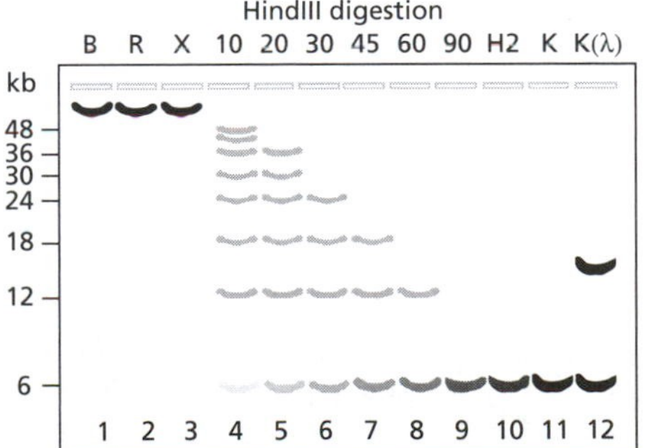


**Figure 4–9** Autoradiograph of various restriction digests of human genomic DNA probed with a radiolabeled *U2* gene (Problem 4–51). Numbers under HindIII indicate time of digestion in minutes. B = BglII; R = EcoRI; X = XbaI; H2 = HincII; K = KpnI. K(λ) indicates a KpnI digest of the bacteriophage DNA carrying the two *U2* genes.

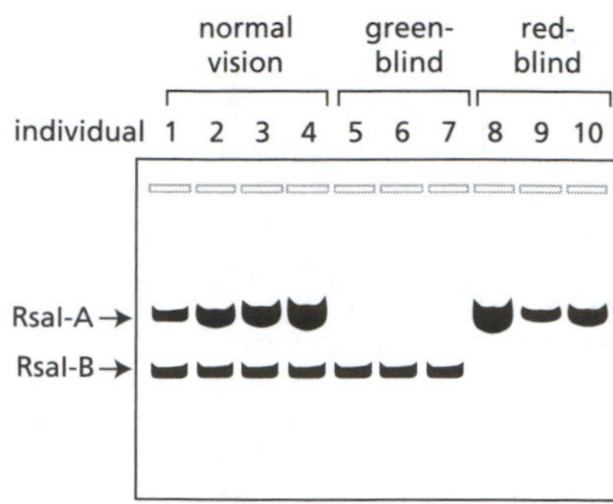


**Figure 4–10** RsaI RFLPs in normal, green-blind, and red-blind males (Problem 4–52). RsaI-A refers to the RFLP characteristic of the *RsaI-A* gene; RsaI-B refers to the RFLP characteristic of the *RsaI-B* gene. Individual males are indicated by number.

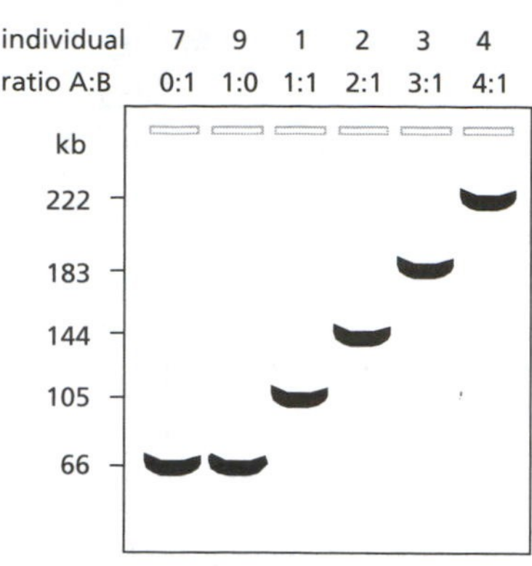


**Figure 4–11** NotI digests of DNA from selected normal and color-blind individuals (Problem 4–52). Numbers for individuals correspond to the numbers in Figure 4–10. The ratios of *RsaI-A* genes to *RsaI-B* genes (A:B) were estimated from the intensity of hybridization in Figure 4–10. The sizes of NotI fragments are indicated in kb.

B. The intensity of hybridization in normal individuals was constant for the *RsaI-B* gene, but surprisingly variable for the *RsaI-A* gene. This anomaly was investigated by digesting the DNA from selected individuals with NotI, which cleaves once within the *RsaI-A* gene but does not cleave the *RsaI-B* gene. The restriction fragments were separated by pulsed-field gel electrophoresis and hybridized with a probe that recognizes both genes (Figure 4–11). What is the basis for the variable intensity of hybridization of *RsaI-A* genes in males with normal color vision? Can your explanation account for the high frequency of color blindness?

C. What is the size of the duplicated chromosomal segment at this site in the human genome?

**4–53** A classic experiment linked telomeres from *Tetrahymena* to a linearized yeast plasmid, allowing the plasmid to grow as a linear molecule—that is, as an artificial chromosome (Figure 4–12). A circular 9-kb plasmid was constructed to contain a yeast origin of replication *(Ars1)* and the yeast *Leu2* gene. Cells that are missing the chromosomal *Leu2* gene, but have taken up the plasmid, can grow in medium lacking leucine. The plasmid was linearized with BglII, which cuts once (Figure 4–12), and then mixed with 1.5-kb *Tetrahymena* telomere fragments generated by cleavage with BamHI (Figure 4–12). The mixture was incubated with DNA ligase and the two restriction nucleases, BglII and BamHI. The ligation products included molecules of 10.5 kb and 12 kb in addition to the original components. The 12-kb band was purified and transformed into yeast, which were then selected for growth in the absence of leucine. Samples of DNA from one transformant were digested with HpaI, PvuII, or PvuI and the fragments were separated by gel electrophoresis and visualized after hybridization to a plasmid-specific probe (Figure 4–13).

A. How do the results of the analysis in Figure 4–13 distinguish between the intended linear form of the plasmid in the transformed yeast and the more standard circular form?

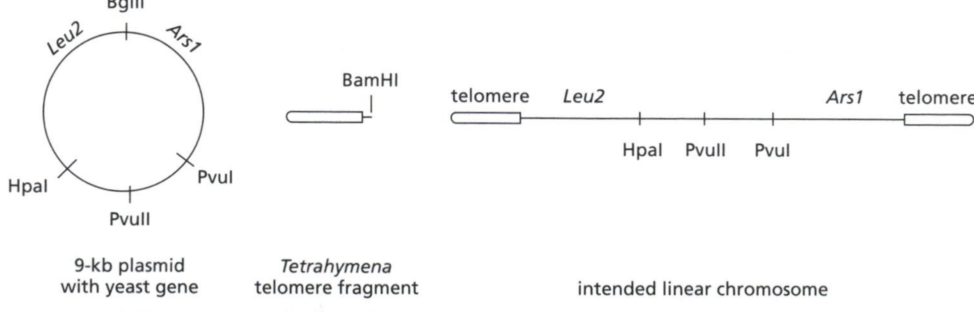


**Figure 4–12** Structure of 9-kb plasmid, telomere fragment, and the intended linear chromosome with *Tetrahymena* telomeres (Problem 4–53). The sites of unique cutting by restriction enzymes are indicated.

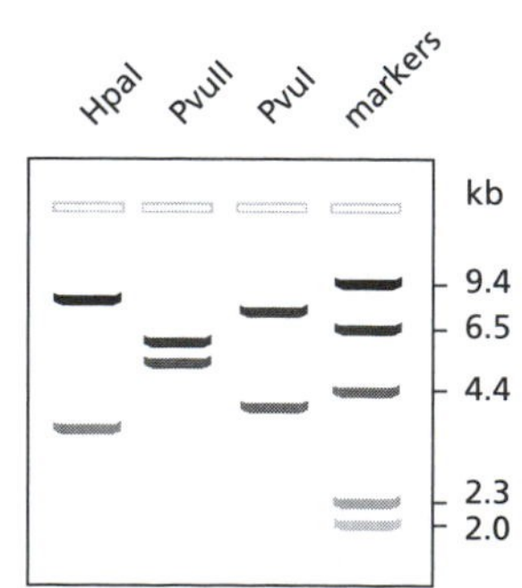


**Figure 4–13** Autoradiograph of restriction analysis of plasmid structure (Problem 4–53). Marker DNAs of known sizes are shown on the *right*.

B. Explain how ligation of the DNA fragments in the presence of the restriction nucleases BamHI and BglII ensures that you get predominantly the linear construct you want. The recognition site for BglII is 5′-A*GATCT-3′ and for BamHI is 5′-G*GATCC-3′, where the asterisk (*) is the site of cutting.

**4–54** You are studying chromatin structure in rat liver DNA. When you digest rat liver nuclei briefly with micrococcal nuclease, extract the DNA, and run it on an agarose gel, it forms a ladder of broad bands spaced at about 200-nucleotide intervals. If you use the enzyme DNase I instead, there is a much more continuous smear of DNA on the gels with only the haziest suggestion of a 200-nucleotide repeat. If you denature the DNase I-treated DNA before fractionating it by gel electrophoresis, you find a new ladder of bands with a regular spacing of about 10 nucleotides.

You are puzzled by the different results with these two enzymes. When you describe the experiments to the rest of your research group, one colleague suggests that the difference derives from the steric properties of the DNA-binding sites on the two enzymes: micrococcal nuclease can only bind and cleave DNA that is free, whereas DNase I can bind and cut free DNA and DNA that is bound to the surface of a nucleosome. Your colleague predicts that if DNA is bound to any surface and digested with DNase I, it will generate a 10-nucleotide ladder. You test this prediction by binding DNA to polylysine-coated plastic dishes and digesting with the two enzymes. Micrococcal nuclease causes minimal digestion, but DNase I generates a 10-nucleotide ladder, verifying your friend's prediction.

A. Why does brief digestion of nuclei with micrococcal nuclease yield a ladder of bands spaced at intervals of about 200 nucleotides?

B. If you digested nuclei extensively with micrococcal nuclease, what pattern would you expect to see after fractionation of the DNA by gel electrophoresis?

C. Explain how your colleague's suggestion accounts for the generation of a 10-nucleotide ladder when nuclei are digested with DNase I.

**4–55** Nucleosomes can be assembled onto defined DNA segments. When a particular 225-bp segment of human DNA was used to assemble nucleosomes and then digested with micrococcal nuclease, uniform fragments 146-bp in length were generated. Subsequent digestion of these fragments with a restriction enzyme that cuts once within the original 225-bp sequence produced two well-defined bands at 37 and 109 bp. Why do you suppose two well-defined fragments were generated by restriction digestion, rather than a range of fragments of different sizes? How would you interpret this result?

**4–56** You have been sent the first samples of a newly discovered Martian microorganism for analysis of its chromatin. The cells resemble Earthly eucaryotes and are composed of similar molecules, including DNA, which is located within a nucleuslike structure in the cell. One member of your team has identified two basic histonelike proteins associated with the DNA in roughly an equal mass ratio with the DNA. You isolate nuclei from the cells and treat them with micrococcal nuclease for various times. You then extract the DNA and run it on an agarose gel alongside DNA from rat-liver nuclei that had been briefly digested with micrococcal nuclease. As shown in Figure 4–14, the digest of rat-liver nuclei gives a standard ladder of nucleosomes, but the digest of the Martian organism gives a smear of products with a nuclease-resistant limit of about 300 nucleotides. As a control, you isolate the Martian DNA free of all protein and digest it with micrococcal nuclease: it is completely susceptible, giving predominantly mono- and dinucleotides.

Do these results suggest that the Martian organism has nucleosomelike structures in its chromatin? If so, what can you deduce about their spacing along the DNA?

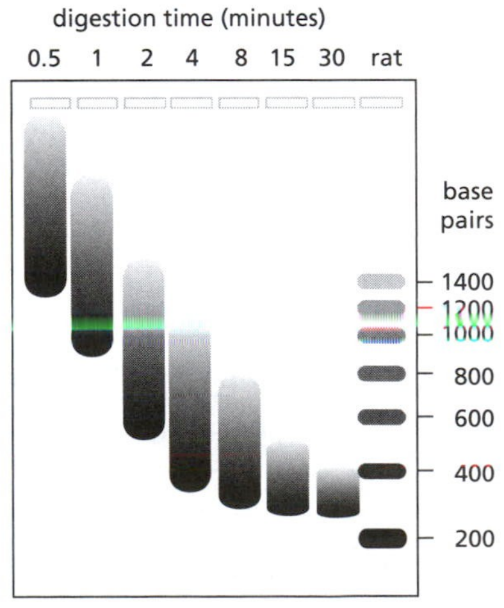


**Figure 4–14** Micrococcal digest of chromatin from a Martian organism (Problem 4–56). The results of digestion of rat-liver chromatin are shown on the *right*.

**4–57** Moving nucleosomes out of the way is important for turning genes on. In yeast the 11-subunit SWI/SNF complex, which is the founding member of the ATP-dependent chromatin remodeling complexes, is required for both

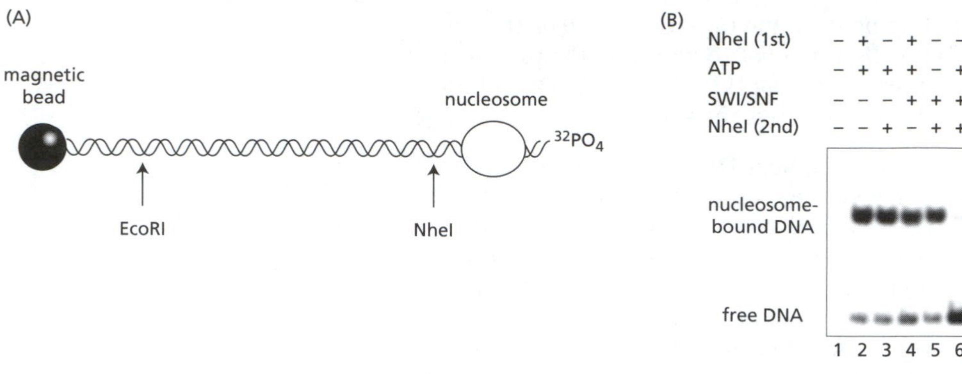


**Figure 4–15** The action of the SWI/SNF complex on a nucleosome (Problem 4–57). (A) Nucleosome-containing substrate tethered to a magnetic bead. Sites of cleavage by the restriction enzymes NheI and EcoRI are indicated. (B) Results of incubation with the SWI/SNF complex before (NheI 2nd) or after (NheI 1st) cleavage with NheI. In the absence of cleavage (lane 1), the labeled substrate does not enter the gel.

activating and repressing gene transcription. How does it work? In principle, it could slide nucleosomes along the chromosome, knock them off the DNA, or transfer them from one duplex to another.

To investigate this problem, you assemble a nucleosome on a 189-bp segment of labeled DNA, which you then ligate to a longer piece of DNA tethered to a magnetic bead, as shown in Figure 4–15A. You incubate this substrate with the SWI/SNF complex either before or after you cut the DNA with NheI, which cleaves near the nucleosome, or with EcoRI, which cleaves near the bead. You use a magnet to separate the DNA that is still attached to the bead from the released DNA. If the nucleosome is present, the released DNA will run with a slower mobility on an agarose gel than if it is absent. Incubation with the SWI/SNF complex in the presence of ATP, followed by NheI digestion (NheI 2nd), releases most of the label as the free DNA fragment (Figure 4–15B, lane 6). By contrast, incubation with SWI/SNF after cleavage with NheI (NheI 1st) releases most of the label as nucleosome-bound DNA (lane 4). Similar experiments with EcoRI digestion showed that incubation with SWI/SNF released most of the label as nucleosome-bound DNA regardless of whether EcoRI cleavage preceded or followed the incubation.

Do your results distinguish among the three possible mechanisms for SWI/SNF action—nucleosome sliding, release, or transfer? Explain how your results argue for or against each of these mechanisms.

# THE REGULATION OF CHROMATIN STRUCTURE

TERMS TO LEARN

| | | |
|---|---|---|
| epigenetic inheritance | heterochromatin | position effect |
| euchromatin | histone code | |

## DEFINITIONS

Match each definition below with its term from the list above.

**4–58** Combination of nucleosomal modifications that determines how and when the DNA packaged in the nucleosome is accessed.

**4–59** Less condensed region of an interphase chromosome that stains diffusely.

**4–60** Form of transmission of information from cell to cell, or from parent to progeny, that is not encoded in DNA.

**4–61** Difference in gene expression that depends on the location of the gene on the chromosome.

## TRUE/FALSE

Decide whether each of these statements is true or false, and then explain why.

**4–62** Deacetylation of histone tails allows nucleosomes to pack together into tighter arrays, which usually reduces gene expression.

**4–63** A modified lysine in a histone can carry one, two, or three methyl groups or an acetyl group in combination with one or two methyl groups.

**4–64** Histone variants are often inserted into already formed chromatin.

## THOUGHT PROBLEMS

**4–65** Phosphorylation of serines and methylation and acetylation of lysines in histone tails affect the stability of chromatin structure above the nucleosome level and have important consequences for gene expression. Draw the structures for serine and phosphoserine, as well as those for lysine, monomethylated lysine, and acetylated lysine. Which modifications alter the net charge on a histone tail? Would you expect the changes in charge to increase or decrease the ability of the tails to interact with DNA?

**4–66** In contrast to histone acetylation, which always correlates with gene activation, histone methylation can lead to either transcriptional activation or repression. How do you suppose that the same modification—methylation—can mediate different biological outcomes?

**4–67** Why is a chromosome with two centromeres (a dicentric chromosome) said to be unstable? Wouldn't a back-up centromere be a good thing for a chromosome, giving it two chances to form a kinetochore and attach to microtubules during mitosis? Wouldn't that help to ensure that the chromosome didn't get left behind at mitosis—sort of like using a belt and braces to keep your pants up?

## DATA HANDLING

**4–68** The first paper to demonstrate different chromatin structures in active and inactive genes used nucleases to probe the globin loci in chicken red blood cells, which express globin mRNAs, and in chicken fibroblasts, which do not. Isolated nuclei from these cells were treated with either micrococcal nuclease or DNase I, and then DNA was prepared and hybridized in vast excess to a $^3$H-thymidine-labeled globin cDNA. (A cDNA is a DNA molecule made as a copy of mRNA.) If the nuclear DNA has not been degraded, it will hybridize to the globin cDNA and protect it from digestion by S1 nuclease, which is specific for single strands of DNA.

Digestion of red-cell nuclei or fibroblast nuclei with micrococcal nuclease (so that about 50% of the DNA was degraded) yielded DNA samples that still protected greater than 90% of the cDNA from subsequent digestion with S1 nuclease. Similarly, digestion of fibroblast nuclei with DNase I (so that less than 20% was degraded) yielded DNA that protected greater than 90% of the cDNA. An identical digestion of red-cell nuclei with DNase I, however, yielded DNA that protected only about 25% of the cDNA. These results are summarized in Table 4–1.

When nucleosome monomers were isolated from red blood cells by digestion with micrococcal nuclease, their DNA protected more than 90% of

**Table 4–1 Protection of globin cDNA by untreated and nuclease-treated chromatin samples** (Problem 4–68).

| | PROTECTED GLOBIN cDNA AFTER TREATMENT | | |
|---|---|---|---|
| SOURCE OF DNA | NONE | MICROCOCCAL | DNase I |
| Fibroblast nuclei | 93% | 91% | 91% |
| Red-cell nuclei | 95% | 92% | 25% |
| Red-cell nucleosomes | | 91% | 25% |
| Red-cell nucleosomes (trypsin) | | 25% | |

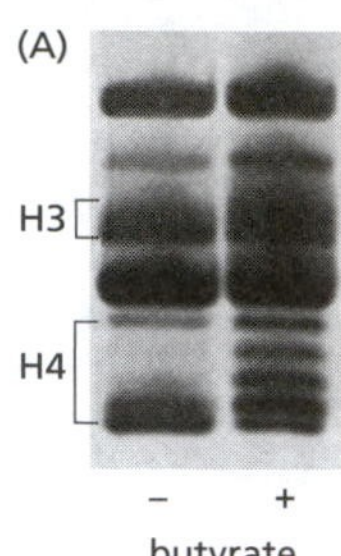


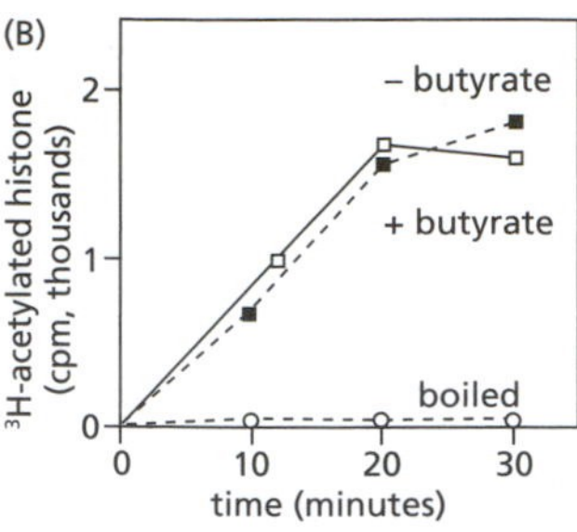


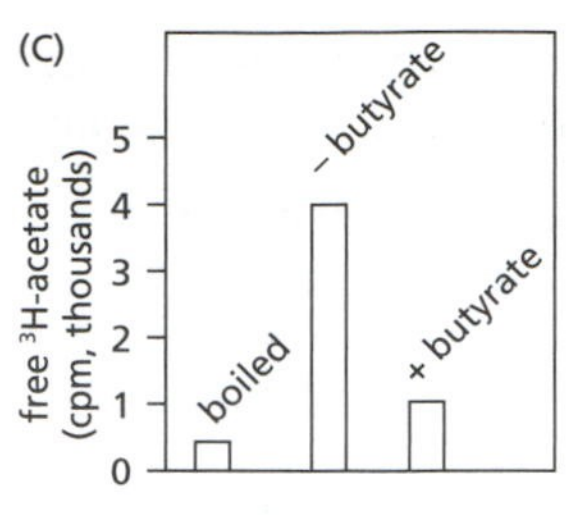


**Figure 4–16** Effects of sodium butyrate on histone acetylation (Problem 4–69). (A) Histone profiles from untreated and butyrate-treated Friend cells. Acetylated histones migrate more slowly than unmodified histones. (B) HAT activity in extracts of Friend cells that were boiled, left untreated (*open squares*), or treated with butyrate (*filled squares*). Extracts were incubated with $^3$H-acetyl CoA, which is the donor used by HATs to transfer acetate to histones, and radioactivity in histones was assayed. (C) HDAC activity in extracts of Friend cells that were boiled, left untreated, or treated with butyrate. Extracts were incubated with $^3$H-acetylated histones and the release of free $^3$H-acetate was measured.

globin cDNA. If the monomers were first treated with DNase I, the isolated DNA protected only 25% of globin cDNA. If the monomers were briefly treated with trypsin to remove 20–30 amino acids from the N-terminus of each histone, digestion of the modified nucleosomes with *micrococcal nuclease* yielded DNA that protected only 25% of globin cDNA (Table 4–1).

A. Which nuclease—micrococcal nuclease or DNase I—digests chromatin that is being expressed (active chromatin)? How can you tell?
B. Does trypsin treatment of nucleosome monomers appear to render a random population or a specific population of nucleosomes sensitive to micrococcal nuclease? How can you tell?
C. Is the alteration that distinguishes active chromatin from bulk chromatin a property of individual nucleosomes, or is it related to the way nucleosome monomers are packaged into higher-order structures within the cell nucleus?

**4–69** When Friend erythroleukemic cells are incubated with sodium butyrate, they differentiate into nondividing, hemoglobin-synthesizing cells. Butyrate-induced differentiation is accompanied by accumulation of acetylated forms of H3 and H4 histones, as can be seen most clearly for H4 in Figure 4–16A. In principle, butyrate treatment could increase the activity of histone acetyl transferases (HATs) or decrease the activity of histone deacetylase complexes (HDACs). To distinguish between these alternatives, HAT and HDAC activities were measured in the presence and absence of butyrate (Figure 4–16B and C). From these results decide how butyrate treatment causes accumulation of acetylated histones.

**4–70** HP1 proteins, a family of proteins found in heterochromatin, are implicated in gene silencing and chromatin structure. The three proteins in humans—HP1α, HP1β, and HP1γ—share a highly conserved chromodomain, which is thought to direct chromatin localization. To determine whether these proteins could bind to the histone H3 N-terminus, you have covalently attached to separate beads various versions of the H3 N-terminal peptide—unmodified, Lys-9-dimethylated (K9-Me), and Ser-10-phosphorylated (S10-P)—along with an unmodified tail from histone H4. This arrangement allows you to incubate the beads with various proteins, wash away unbound proteins, and then elute bound proteins for assay by Western blotting. The results of your 'pull-down' assay for the HP1 proteins are shown in Figure 4–17, along with the results from several control proteins, including Pax5, which is a gene regulatory protein, polycomb protein Pc1, which is known to bind to histones, and Suv39h1, a histone methyltransferase.

Based on these results, which of the proteins tested bind to the unmodified tails of histones? Do any of the HP1 proteins and control proteins selectively bind to any of the various histone N-terminal peptides? What histone modification would you predict would be found in heterochromatin?

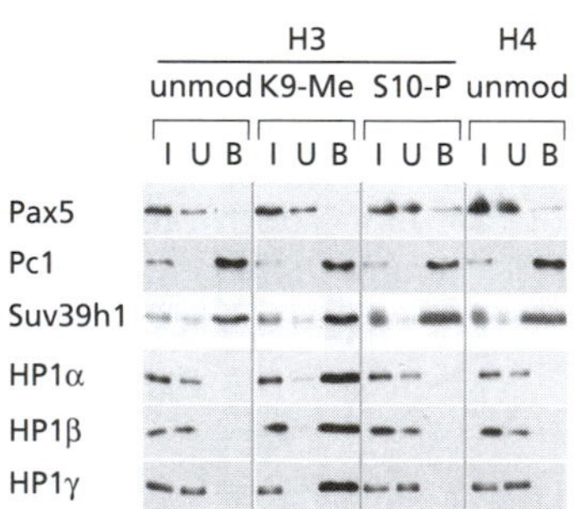


**Figure 4–17** Pull-down assays to determine binding specificity of HP1 proteins (Problem 4–70). Each protein at the *left* was detected by immunoblotting, using a specific antibody after separation by SDS-polyacrylamide gel electrophoresis. For each histone N-terminal peptide, the total input protein (I), the unbound protein (U), and the bound protein (B) are indicated.

**4–71** Look at the two yeast colonies in Figure 4–18. Each of these colonies contains about 100,000 cells descended from a single yeast cell, originally somewhere in the middle of the clump. A white colony arises when the *Ade2* gene is expressed from its normal chromosomal location. When the *Ade2* gene is moved to a location near a telomere, it is packed into heterochromatin and inactivated in most cells, giving rise to colonies that are mostly red. In these

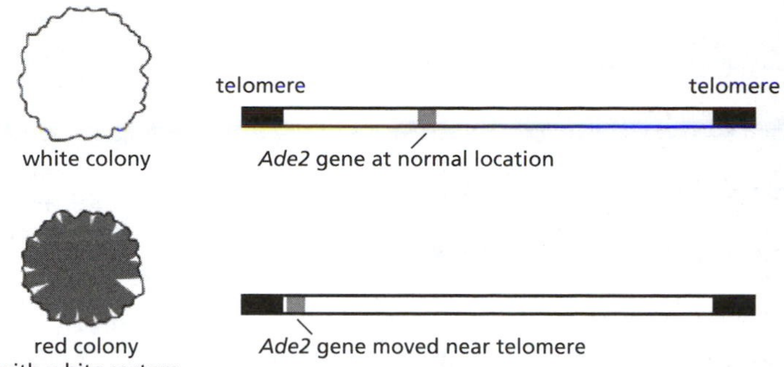


**Figure 4–18** Position effect on expression of the yeast *Ade2* gene (Problem 4–71). The *Ade2* gene codes for one of the enzymes of adenosine biosynthesis, and the absence of the *Ade2* gene product leads to the accumulation of a red pigment. Therefore a colony of cells that express *Ade2* is *white*, and one composed of cells in which the *Ade2* gene is not expressed is *red*.

largely red colonies, white sectors fan out from the middle of the colony. In both the red and white sectors, the *Ade2* gene is still located near telomeres. Explain why white sectors have formed near the rim of the red colony. Based on the existence of these white sectors, what can you conclude about the propagation of the transcriptional state of the *Ade2* gene from mother to daughter cells?

**4–72** High-density DNA microarrays can be used to analyze changes in expression of all the genes in the yeast genome in response to various perturbations. The effects of depletion of histone H4 and deletion of the *Sir3* gene on expression of all yeast genes were analyzed in this way, as summarized on the yeast chromosomes illustrated in Figure 4–19. Depletion of histone H4 was achieved in a strain in which the gene was engineered to respond to galactose. In the absence of galactose the gene is turned off and depletion of histone H4 is evident within 6 hours, leading to a decreased density of nucleosomes throughout the genome. Deletion of *Sir3* removes a critical

**Figure 4–19** Changes in expression of the genes on chromosomes I–XVI of yeast (Problem 4–72). (A) In response to depletion of histone H4. (B) In response to deletion of *Sir3*. *Black bars* indicate genes whose expression was increased relative to wild-type yeast threefold or more. The very light *gray bars* show genes whose expression was decreased by threefold or more; they are not relevant to this problem but have been added for the sake of completeness. Chromosomes were split at their centromeres so that all their telomeres could be aligned at the left; *brackets* indicate the pairs of arms that make up individual chromosomes. Three chromosome arms have been shortened to fit into the figure, as indicated by diagonal lines.

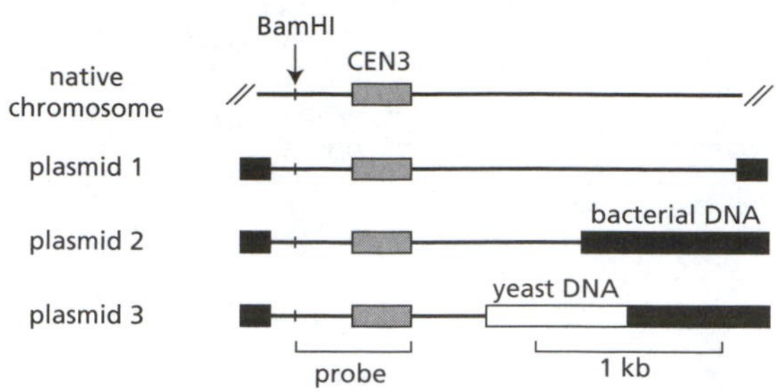


**Figure 4–20** Diagrams of the structures of the native chromosome and three plasmids (Problem 4–73). The native chromosome is linear; its true ends extend well beyond the positions marked by the diagonal lines. The plasmids, which are circular, are shown here as linears for ease of comparison. The native yeast sequences around the centromere are shown as *thin lines*. Bacterial DNA sequences in the plasmids are shown as *black rectangles*. The yeast DNA in plasmid 3, which is shown as a *white rectangle*, is a segment of yeast chromosomal DNA far removed from the centromere.

component of the Sir protein complex that binds to telomeres and is responsible for deacetylation of telomeric nucleosomes.

A. Depletion of histone H4 significantly increased expression of 15% of all yeast genes (Figure 4–19A, black bars). Does loss of histone H4 increase expression of a greater fraction of genes near telomeres than in the rest of the genome? Explain how you arrived at your conclusion.

B. Deletion of the *Sir3* gene significantly increased expression of 1.5% of all yeast genes (Figure 4–19B, black bars). Does the absence of Sir3 protein increase expression of a greater fraction of genes near telomeres than in the rest of the genome? Explain how you arrived at your conclusion.

C. If you concluded that either the depletion of histone H4, the deletion of the *Sir3* gene, or both, preferentially increased expression of genes near telomeres, propose a mechanism for how that might happen.

**4–73** A classic paper examined the arrangement of nucleosomes around the centromere (CEN3) of yeast chromosome III. Because centromeres are the chromosome attachment sites for microtubules, it was unclear whether they would have the usual arrangement of nucleosomes. This study used plasmids into which were cloned various lengths of the native chromosomal DNA around the centromere (Figure 4–20). Chromatin from native yeasts and from yeasts that carried individual plasmids was treated briefly with micrococcal nuclease, and then the DNA was deproteinized and digested with the restriction enzyme BamHI, which cuts the DNA only once (Figure 4–20). The digested DNA was fractionated by gel electrophoresis and analyzed by Southern blotting using a segment of radiolabeled centromeric DNA as a hybridization probe (Figure 4–20). This procedure (called indirect end labeling) allows visualization of all the DNA fragments that include the DNA immediately to the right of the BamHI-cleavage site. As a control, a sample of naked DNA from the same region was treated with micrococcal nuclease and subjected to the same analysis. An autoradiogram of the results is shown in Figure 4–21.

A. When the digestion with BamHI was omitted, a regular, though much less distinct, set of dark bands was apparent. Why does digestion with BamHI make the pattern so much clearer and easier to interpret?

B. Draw a diagram showing the micrococcal-nuclease-sensitive sites on the chromosomal DNA and the arrangement of nucleosomes along the chromosome. What is special about the centromeric region?

C. What is the purpose of including a naked DNA control in the experiment?

D. The autoradiogram in Figure 4–21 shows that the native chromosomal DNA yields a regularly spaced pattern of bands beyond the centromere; that is, the bands at 600 nucleotides and above are spaced at 160-nucleotide intervals. Does this regularity result from the lining up of nucleosomes at the centromere, like cars at a stoplight? Or, is the regularity an intrinsic property of the DNA sequence itself? Explain how the results with plasmids 1, 2, and 3 decide the issue.

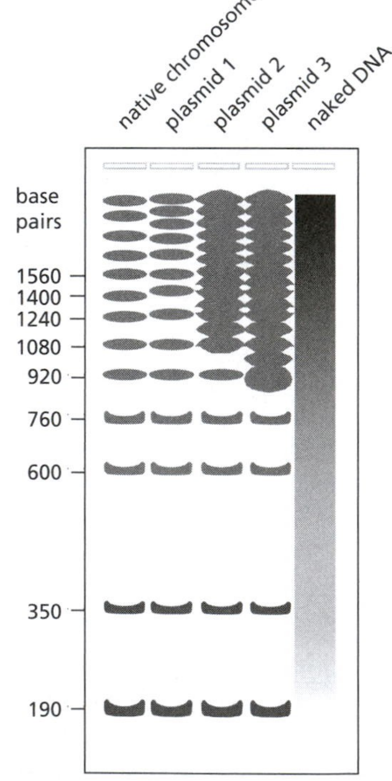


**Figure 4–21** Results of micrococcal-nuclease digestion of DNA around CEN3 (Problem 4–73). Approximate lengths of DNA fragments in nucleotide pairs are indicated on the left of the autoradiogram.

## THE GLOBAL STRUCTURE OF CHROMOSOMES

TERMS TO LEARN

| lampbrush chromosome | mitotic chromosome | polytene chromosome |
|---|---|---|

## DEFINITIONS

Match each definition below with its term from the list above.

**4–74** Giant chromosome in which the DNA has undergone repeated replication without separation into new chromosomes.

**4–75** Paired chromosomes in meiosis in immature amphibian eggs, in which the chromatin forms large stiff loops extending out from the linear axis of the chromosome.

**4–76** Highly condensed duplicated chromosome with the two new chromosomes still held together at the centromere as sister chromatids.

## TRUE/FALSE

Decide whether each of these statements is true or false, and then explain why.

**4–77** In lampbrush chromosomes of amphibian oocytes, most of the DNA is in the loops, which are actively transcribed, while the rest remains highly condensed in the chromomeres, which are generally not transcribed.

**4–78** At the final level of condensation each chromatid of a mitotic chromosome is organized into loops of chromatin that emanate from a central axis.

## THOUGHT PROBLEMS

**4–79** Imagine that a human interphase chromosome could be transferred intact into an amphibian oocyte and that it could form a lampbrush chromosome. What might be learned from knowing the DNA and RNA sequences in the loops, and how might you determine their identity?

**4–80** Although mammalian chromosomes apparently do not form lampbrush chromosomes in amphibian oocytes, chromosomes from different amphibians do. When demembranated sperm heads are injected into oocytes, the sperm chromosomes gradually swell and take on the general appearance of typical lampbrush chromosomes. When *Xenopus laevis* (African clawed toad) sperm heads were injected into *Xenopus laevis* oocytes, they formed lampbrush chromosomes like those in the oocyte, except that they had unpaired loops—as expected for the single chromatids of sperm—instead of the paired loops formed by the sister chromatids in the oocyte chromosomes.

When sperm heads from *Rana pipiens* (Northern leopard frog), which forms large loops in its own oocyte chromosomes, were injected in *X. laevis* oocytes, the resulting lampbrush chromosomes had the small loops typical of those in *X. laevis* oocytes. Similarly, when sperm heads from *X. laevis* were injected into *Notophthalmus viridescens* (red spotted newt) oocytes, the resulting lampbrush chromosomes had the very large loop structure typical of *N. viridescens*.

Do these heterologous injection experiments support the idea that loop structure is an intrinsic property of a chromosome? Why or why not? What do these results imply about experiments designed to map the natural loop domains in mammalian chromosomes by forming them into lampbrush chromosomes in amphibians?

**4–81** Each interphase chromosome tends to occupy a discrete and relatively small area within the nucleus. Does this mean that the particular site a chromosome occupies is critical for cell function? Why or why not?

## DATA HANDLING

**4–82** One of the earliest studies of transcription in lampbrush chromosomes used oocytes from the newt *Triturus*. $^3$H-uridine was injected into the oocytes, and after various times radioactive RNA was detected by autoradiography.

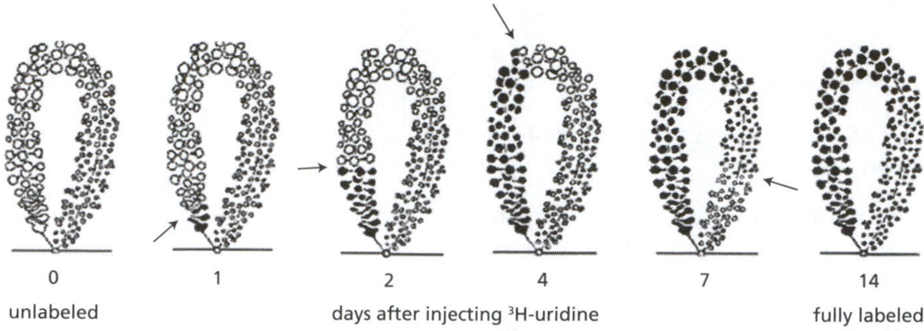


Figure 4–22 Autoradiographs of a giant chromatin loop from a lampbrush chromosome of the newt (Problem 4–82). *Arrows* show forward progress of labeled regions (*black areas*) around the loop at various times after injection of labeled uridine into the oocytes. These giant loops have identical partners on the other side of the chromosome axis, but they are not shown.

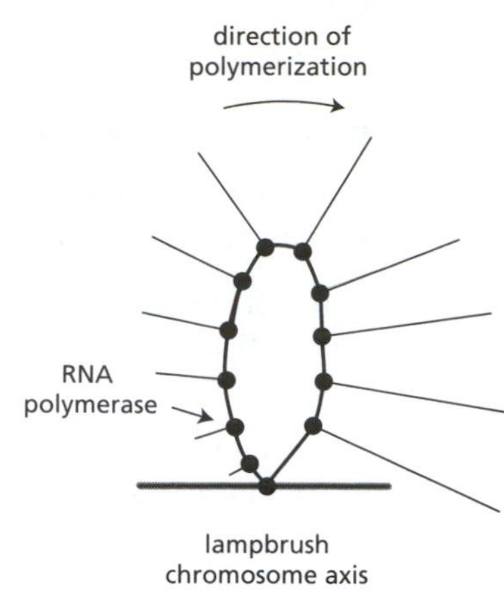


Figure 4–23 Diagrammatic representation of a chromatin loop that is a single transcription unit (Problem 4–82). The progress of RNA polymerase (shown as *black circles*) around the loop is illustrated along with the attendant growth of the nascent RNA chain associated with each polymerase.

Small loops, which are the most common type, were labeled throughout, even at the shortest times of labeling. By contrast, giant loops, which are rare, incorporated label progressively around the loop, beginning after about 1 day and continuing for 14 days before the entire loop was labeled (Figure 4–22).

If loops in lampbrush chromosomes represent single transcription units, in which RNA polymerase initiates and terminates synthesis at the base of the loop, as shown in Figure 4–23, which pattern of loop labeling would you expect: uniform, as in the small loops, or progressive, as in the giant loops?

**4–83** The characteristic banding patterns of the giant polytene chromosomes of *Drosophila melanogaster* provide a visible map of the genome that has proven an invaluable aid in genetic studies for decades. The molecular basis for the more intense staining of bands relative to interbands remains a puzzle. In principle, bands might stain more darkly because they contain more DNA than interbands due to overreplication, or the amount of DNA may be the same in the bands and interbands, but the DNA stains more prominently in the bands because it is more condensed or contains more proteins. These two possibilities—differential replication or differential staining—were distinguished by the experiments described below.

A series of radiolabeled segments spanning 315 kb of a *Drosophila* chromosome were used as hybridization probes to estimate the amount of corresponding DNA present in normal diploid tissues versus DNA from salivary glands, which contain polytene chromosomes. DNA samples from diploid and polytene chromosomes were digested with combinations of restriction enzymes. The fragments were then separated by gel electrophoresis and transferred to nitrocellulose filters for hybridization analysis. In every case the restriction pattern was the same for the DNA from diploid chromosomes and polytene chromosomes, as illustrated for two examples in Figure 4–24. The intensities of many specific restriction fragments were measured and expressed as the ratio of the intensity of the fragment from polytene chromosomes to the intensity of the corresponding fragment from diploid chromosomes (Figure 4–25).

How do these results distinguish between differential replication and differential staining as the basis for the difference between bands and interbands? Explain your reasoning.

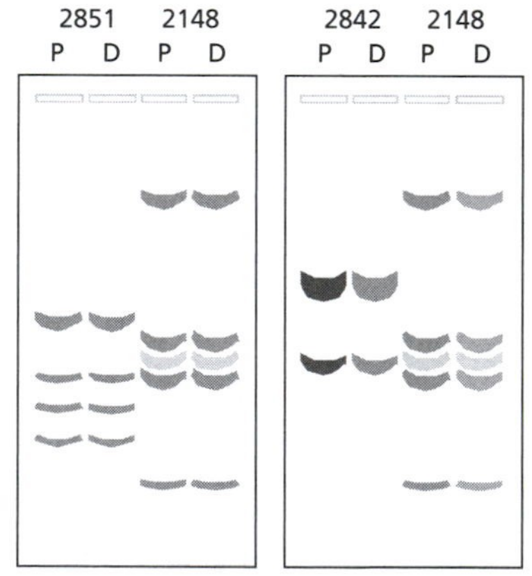


Figure 4–24 Autoradiograph of blot-hybridization analysis of polytene and diploid chromosomes (Problem 4–83). P and D refer to polytene and diploid, respectively. Numbers at the top refer to cloned DNA segments used as probes: 2851 and 2842 are from the 315-kb region under analysis (see Figure 4–25); 2148 is from elsewhere in the genome and was used in all hybridizations to calibrate the amount of DNA added to the gels.

**4–84** The typical coiled phone cord provides an everyday example of the phenomenon of supercoiling. Invariably, the cord becomes coiled about itself forming a tangled mess. These coiled coils are supercoils. Dangling the receiver and letting it spin until it stops can remove them. Similarly, supercoils can be reintroduced by twisting the receiver, which of course is how they get there in the first place.

DNA is coiled into a double helix that exhibits the same phenomenon of supercoiling (Figure 4–26). A relaxed circular DNA, with 10.5 bp per turn of

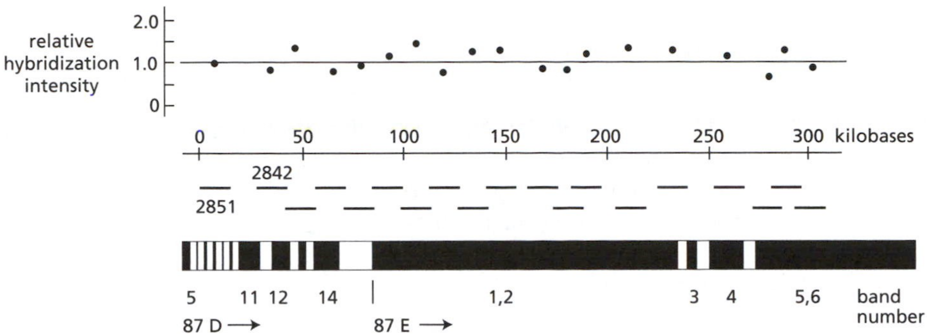


Figure 4–25 Relative amounts of DNA in diploid and polytene chromosomes at different points along the chromosome (Problem 4–83). The chromosomal segment covered by the cloned restriction fragments is shown at the *bottom*, along with the cytological designations for the chromosome regions and bands. The cloned fragments are shown *above* the chromosomes, and the positions of 2851 and 2842 are indicated. The ratio of hybridization of each restriction fragment to DNA from polytene chromosomes versus diploid chromosomes is plotted above each fragment.

the helix, will assume a more or less circular form when laid onto a surface. If one strand of the DNA is broken and wound around its partner two extra times (overwound—an increase in linking number of +2) and then rejoined, the molecule will twist on itself, forming two *positive* supercoils. If one strand in a relaxed circular DNA is broken and rejoined with two fewer turns (underwound—a decrease in linking number of –2), the molecule will twist to form two *negative* supercoils. Positive and negative supercoils each can assume two forms termed plectonemic and solenoidal, although plectonemic supercoils are the only ones that are stable in naked DNA. The effect of supercoiling is to preserve the preferred local winding of DNA at 10.5 bp per turn. In cells the degree of supercoiling of DNA is carefully controlled by special enzymes called topoisomerases that break and rejoin strands of DNA.

Circular plasmid DNA isolated from *E. coli* is highly supercoiled, as is evident when the DNA is separated by electrophoresis on an agarose gel (Figure 4–27, lane 1). When incubated for increasing times with *E. coli* topoisomerase I (which breaks and reseals a single DNA strand in negatively supercoiled DNA but not in positively supercoiled DNA), several new bands appear between the supercoiled DNA and the relaxed DNA (lanes 2–5).

A. In the untreated DNA sample isolated from bacteria, why do you suppose a small fraction of the plasmid molecules are relaxed (Figure 4–27, lane 1)?
B. What are the discrete bands between the highly supercoiled and relaxed bands in Figure 4–27 that appear with increasing times of incubation with topoisomerase I? Why do they move at rates intermediate between relaxed and highly supercoiled DNA?
C. Estimate the number of supercoils that were present in the original plasmid molecules.
D. Did the bacterial plasmid originally contain positive or negative supercoils? Explain your answer.

**4–85** Imagine that you assemble a single nucleosome on a closed circular, relaxed DNA molecule; that is, a circular duplex DNA with no breaks in either strand and zero supercoiling. Wrapping the DNA molecule around the histone octamer forms solenoidal supercoils, which are compensated for by plectonemic supercoils in another part of the molecule.

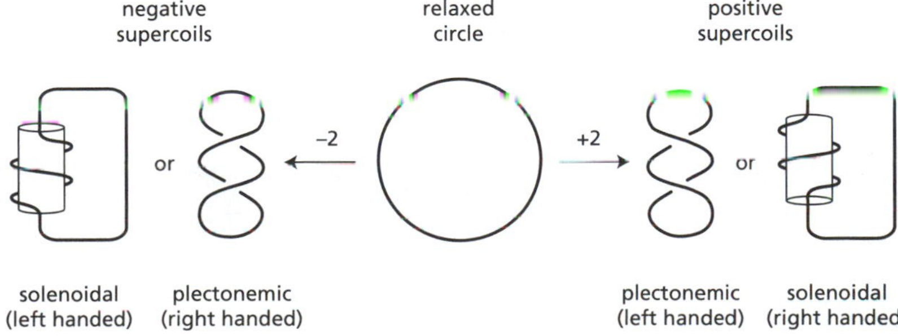


Figure 4–26 Relaxed and supercoiled circular DNA molecules (Problem 4–84). Duplexes of DNA are indicated by single lines. These DNA molecules differ only in the number of times one strand is wound around the other, a quantity known as the linking number. Solenoidal supercoils are shown as wrapped around a cylinder for illustrative purposes.

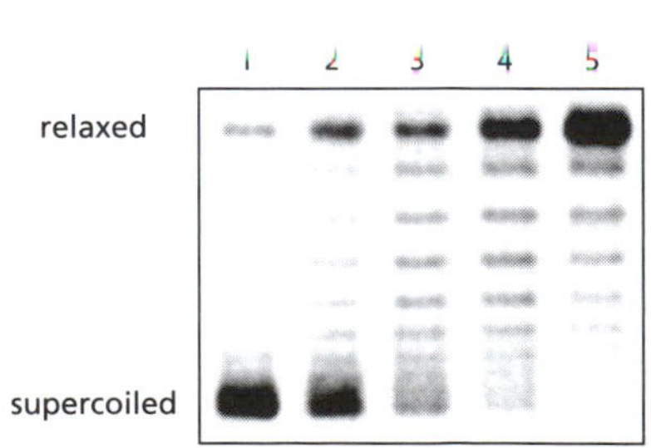


Figure 4–27 Plasmid DNA treated with *E. coli* topoisomerase I for increasing times (Problem 4–84).

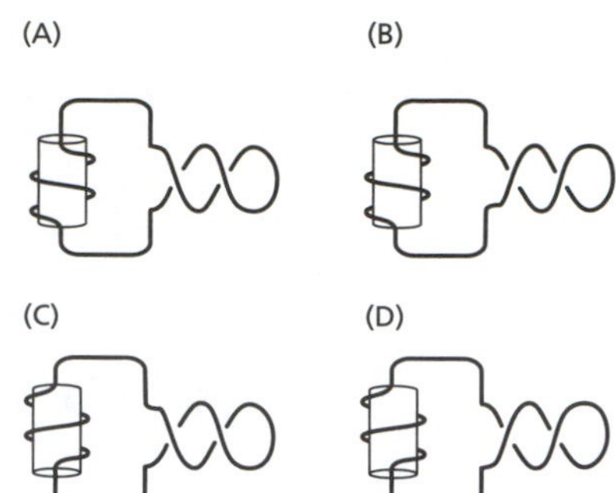


**Figure 4–28** Four possible arrangements of circular DNA molecules with two solenoidal and two plectonemic supercoils (Problem 4–85). The position of the nucleosome is indicated by the cylinder.

Of the four possible arrangements of solenoidal and plectonemic supercoils shown in Figure 4–28, which have a net supercoiling of zero? (Since no breaks were introduced into the DNA in the process of forming the nucleosome, it must retain an overall supercoiling of zero.) Indicate the sign of the supercoiling (positive or negative) on the structures you select.

**4–86** Which of the two alternative arrangements of compensating solenoidal and plectonemic supercoils, generated by the formation of a nucleosome (see Problem 4–85), represents the true biological situation? These alternatives were distinguished by incubating the nucleosome-bound DNA with either *E. coli* topoisomerase I, which can remove only negative plectonemic supercoils, or with calf thymus topoisomerase I, which can remove both negative and positive plectonemic supercoils. Histones were removed after the incubation with a topoisomerase, and the presence of supercoils in the naked DNA was assayed by gel electrophoresis (see Figure 4–27, Problem 4–84). (The sign of the plectonemic supercoils in the naked DNA can be determined by subsequent incubation with *E. coli* topoisomerase I, which relaxes negative supercoils but not positive ones.)

It was found that incubation of nucleosomal DNA with *E. coli* topoisomerase I gave DNA molecules with zero supercoils. By contrast, incubation with calf thymus topoisomerase I gave DNA molecules with two negative supercoils. Are the solenoidal supercoils around biological nucleosomes positive (right handed) or negative (left handed)? What results would you have expected for the other alternative?

**4–87** Condensins use the energy of ATP hydrolysis to drive the coiling of interphase chromosomes into the highly condensed chromosomes visible at mitosis (Figure 4–29). You realize that if condensins operate this way they may change the supercoiling of the DNA, since the chromosomal coils are just large solenoids. To test this hypothesis, you incubate relaxed, closed circular DNA with condensin and ATP. Then you incubate them with either *E. coli* or calf thymus topoisomerase I, remove condensin and assay for supercoils, as shown in Figure 4–30.

A. Do you expect that supercoils present after topoisomerase treatment (Figure 4–30, lanes 5 and 9) will be positive or negative? Why? If the supercoils were generated as a result of condensin-mediated formation of solenoids, were the solenoids right handed (positive) or left handed (negative)?

B. How do you know that condensin isn't simply an ATP-driven topoisomerase; that is, a topoisomerase that uses the energy of ATP to underwind or overwind the DNA, to introduce plectonemic supercoils? What would have been the outcome of the experiment if condensin acted in this way to introduce positive plectonemic supercoils? What would be the outcome if it introduced negative plectonemic supercoils?

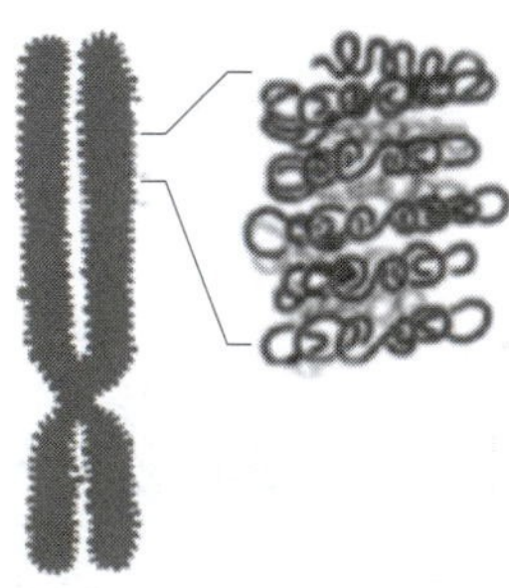

**Figure 4–29** Schematic diagram of the large coils in mitotic chromosomes (Problem 4–87).

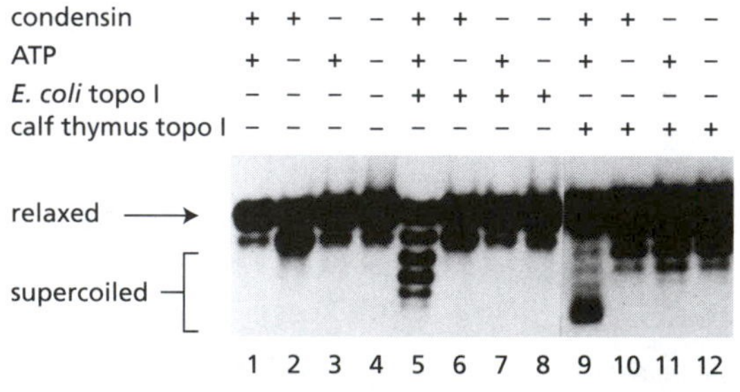


**Figure 4–30** Experiment to test for condensin-mediated coiling of DNA (Problem 4–87). Relaxed, closed circular DNA was incubated with condensin and ATP, as indicated, and then the deproteinized DNA was incubated with topoisomerase I from *E. coli* or calf thymus.

# HOW GENOMES EVOLVE

| TERMS TO LEARN | | |
|---|---|---|
| homologous | pseudogene | single-nucleotide polymorphism (SNP) |
| polymorphic | purifying selection | |

## DEFINITIONS

Match each definition below with its term from the list above.

**4–88** Describes a site in the genome at which two individuals have a reasonable probability—generally greater than 1%—of being different.

**4–89** Gene that has accumulated multiple mutations that have rendered it inactive or nonfunctional.

**4–90** Evolutionary process that eliminates individuals carrying mutations that interfere with important genetic functions.

**4–91** Variation between individuals at a certain nucleotide position in the genome.

## TRUE/FALSE

Decide whether each of these statements is true or false, and then explain why.

**4–92** Many human genes so closely resemble their homologs in yeast that the protein-coding portion of the yeast gene can be substituted with its human homolog.

**4–93** The portion of the human genome subjected to purifying selection corresponds to the protein-coding sequences.

**4–94** Gene duplication and divergence is thought to have played a critical role in the evolution of increased biological complexity.

## THOUGHT PROBLEMS

**4–95** Suppose that you are unable to repair the damage to DNA caused by the loss of purine bases. This defect causes the accumulation of about 5000 mutations per day in the DNA of each of your cells. As the average difference in DNA sequences between humans and chimpanzees is about 1%, it is only a matter of time until you turn into a chimp. What is wrong with this argument?

**4–96** Mobile pieces of DNA—transposable elements—that insert themselves into chromosomes and accumulate during evolution make up more than 40% of the human genome. Transposable elements of four types—long interspersed elements (LINES), short interspersed elements (SINES), LTR retrotransposons, and DNA transposons—are inserted more or less randomly throughout the human genome. These elements are conspicuously rare at the four homeobox gene clusters, *HoxA*, *HoxB*, *HoxC*, and *HoxD*, as illustrated for *HoxD* in Figure 4–31, along with an equivalent region of chromosome 22, which lacks a *Hox* cluster. Each *Hox* cluster is about 100 kb in length and contains 9–11 genes, whose differential expression along the

**Figure 4–31** Transposable elements and genes in 1 Mb regions of chromosomes 2 and 22 (Problem 4–96). Lines that project *upward* indicate exons of known genes. Lines that project *downward* indicate transposable elements; they are so numerous (constituting more than 40% of the human genome) that they merge into nearly a solid block outside the *Hox* clusters.

**Table 4–2 The difference matrix for the first 30 amino acids of the hemoglobin α chains from five species** (Problem 4–97).

| | HUMAN | FROG | CHICKEN | WHALE | FISH |
|---|---|---|---|---|---|
| Human | 0 | ? | 11 | 8 | 17 |
| Frog | | 0 | ? | 17 | 20 |
| Chicken | | | 0 | ? | 20 |
| Whale | | | | 0 | ? |
| Fish | | | | | 0 |

```
Human    VLSPADKTNVKAAWGKVGAHAGEYGAEALE
Frog     LLSADDKKHIKAIMPAIAAHGDKFGGEALY
Chicken  VLSAADKNNVKGIFTKIAGHAEEYGAETLE
Whale    VLSPTDKSNVKATWAKIGNHGAEYGAEALE
Fish     SLSDKDKAAVRALWSKIGKSADAIGNDALS
```

**Figure 4–32** Alignment of the first 30 amino acids of the hemoglobin α chains from five species (Problem 4–97). Amino acids are represented by the one-letter code (see inside back cover).

anteroposterior axis of the developing embryo establishes the basic body plan for humans (and for other animals). Why do you suppose that transposable elements are so rare in the *Hox* clusters?

## CALCULATIONS

**4–97** Nucleotide sequence comparisons are fundamental to our current conception of the tree of life, to our understanding of how mitochondria and chloroplasts were acquired and their subsequent evolution, to the importance and magnitude of horizontal gene transfer, and to the notion that by focusing on a few model organisms we can gain valid insights into all of biology. For these reasons we've designed this problem and the following one to introduce the common methods and assumptions that underlie the art of nucleotide sequence comparison.

A phylogenetic tree represents the history of divergence of species from common ancestors. Construction of such trees from DNA or protein sequences can really only be done with computers: the data sets are enormous and the algorithms are subtle. Nevertheless, some of the fundamental principles of tree construction can be illustrated with a simple example. Consider the first 30 amino acids of the hemoglobin α chains for the five species shown in Figure 4–32.

A. In a common approach, known as the distance-matrix method, the first step is to construct a table of all pairwise differences between the sequences. A partially filled-in example is shown in Table 4–2. Complete the table by filling in the blanks indicated by question marks.

B. According to the information in the completed Table 4–2, which pair of species is most closely related? What is the assumption that underlies your choice?

C. The information in Table 4–2 can be used to arrange species on the phylogenetic tree shown in Figure 4–33. The branching order is determined using a simple kind of cluster analysis. The two most similar species are placed on the adjacent branches at the upper left in Figure 4–33. The species with the fewest average differences relative to this pair is placed on the next branch. In the next step these three species are combined and the average differences from the remaining species are calculated and used to fill in the next branch, and so on. Use this method to arrange the species on the tree in Figure 4–33.

D. Is the branching order you determined in part C the same as you would get by simply using the number of differences relative to human to place the other species on the tree? Why is the method of cluster analysis superior?

**4–98** In the previous question the branching order, or topology, of the tree was established, but actual distances (number of differences) were not assigned to the line segments that make up the tree (Figure 4–34). To calculate distances for line segments is, once again, tedious by hand but easy by computer. The following exercise gives a feeling for how such calculations are done.

A. Using the numbers from the completed distance matrix in Table 4–2 and the branching order determined in the previous problem, write down all the equations for the differences between species in terms of the line segments

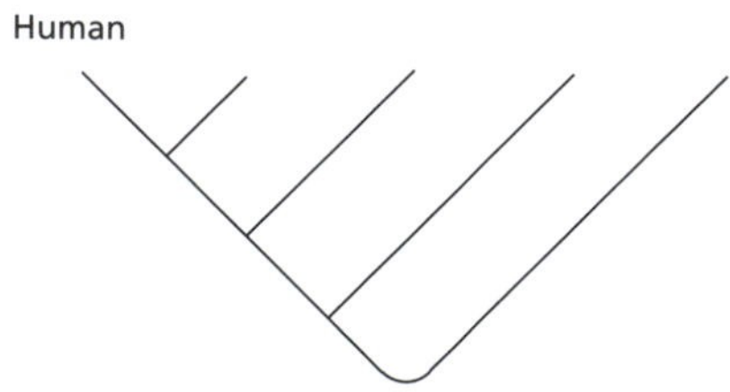


**Figure 4–33** A general phylogenetic tree for five species (Problem 4–97).

that make up the tree (Figure 4–34). Are there enough equations to solve for the lengths of the seven line segments?

B. One straightforward, not too exhausting method for solving these equations is to consider them three at a time; for example,

$$I \rightarrow II = a + b$$
$$I \rightarrow III = a + c + d$$
$$II \rightarrow III = b + c + d$$

There are 10 such three-at-a-time equations for five species. Using the information in the distance-matrix table (see Table 4–2), solve two of these sets of equations—human/whale/chicken and human/whale/frog—for $a$ and $b$. Are the values for $a$ and $b$ the same in the two solutions?

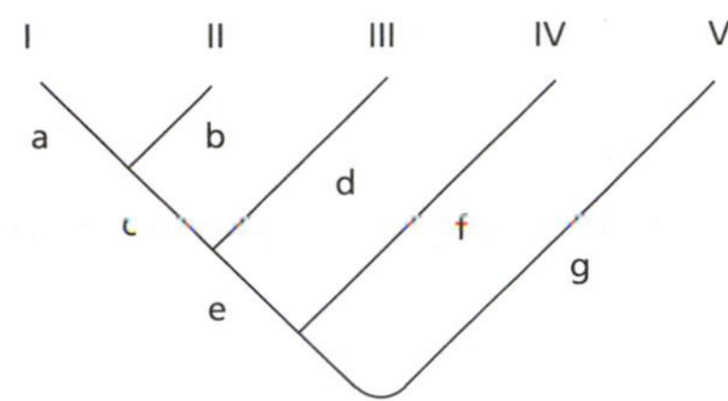

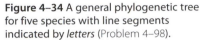

**Figure 4–34** A general phylogenetic tree for five species with line segments indicated by *letters* (Problem 4–98).

## DATA HANDLING

**4–99** The earliest graphical method for comparing nucleotide sequences—the so-called diagon plot—still yields one of the best visual comparisons of sequence relatedness. An example is illustrated in Figure 4–35, where the human β-globin gene is compared with the human cDNA for β globin (Figure 4–35A) and with the mouse β-globin gene (Figure 4–35B). (A cDNA is a DNA molecule made as a copy of mRNA and therefore lacking the introns that are present in genomic DNA.) Diagon plots are generated by comparing blocks of sequence, in this case blocks of 11 nucleotides at a time. If 9 or more of the nucleotides match, a dot is placed on the diagram at the coordinates corresponding to the blocks being compared. A comparison of all possible blocks generates diagrams such as the ones shown in Figure 4–35, in which sequence homologies show up as diagonal lines.

A. From the comparison of the human β-globin gene with the human β-globin cDNA (Figure 4–35A), deduce the positions of exons and introns in the β-globin gene.

B. Are the entire exons of the human β-globin gene (indicated by shading in Figure 4–35B) homologous to the mouse β-globin gene? Identify and explain any discrepancies.

C. Is there any homology between the human and mouse β-globin genes that lies outside the exons? If so, identify its location and offer an explanation for its preservation during evolution.

D. Has either of the genes undergone a change of intron length during their evolutionary divergence? How can you tell?

**4–100** Your first foray into archaeological DNA studies ended in embarrassment. The dinosaur DNA sequences that you so proudly announced to the world later proved to be derived from contaminating modern human cells—probably your own. Setting your sights slightly lower, you decide to try to amplify residual mitochondrial DNA from a well-preserved Neanderthal skeleton. You also redesign your laboratory to minimize the possibility of stray contamination. You carefully prepare three different samples (A, B, and C) of bone from the femur and perform separate polymerase chain reactions

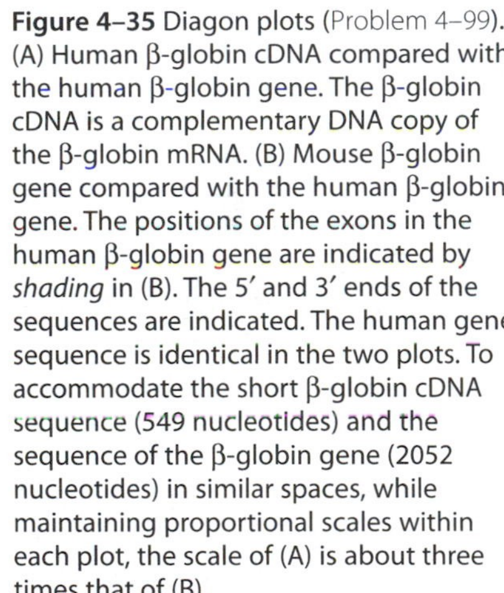

**Figure 4–35** Diagon plots (Problem 4–99). (A) Human β-globin cDNA compared with the human β-globin gene. The β-globin cDNA is a complementary DNA copy of the β-globin mRNA. (B) Mouse β-globin gene compared with the human β-globin gene. The positions of the exons in the human β-globin gene are indicated by *shading* in (B). The 5′ and 3′ ends of the sequences are indicated. The human gene sequence is identical in the two plots. To accommodate the short β-globin cDNA sequence (549 nucleotides) and the sequence of the β-globin gene (2052 nucleotides) in similar spaces, while maintaining proportional scales within each plot, the scale of (A) is about three times that of (B).

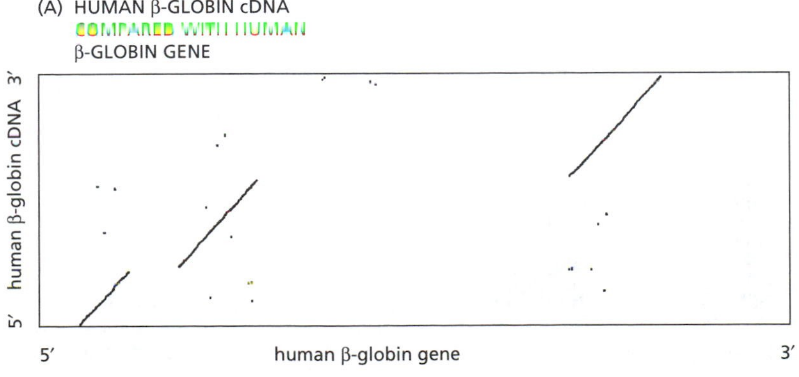


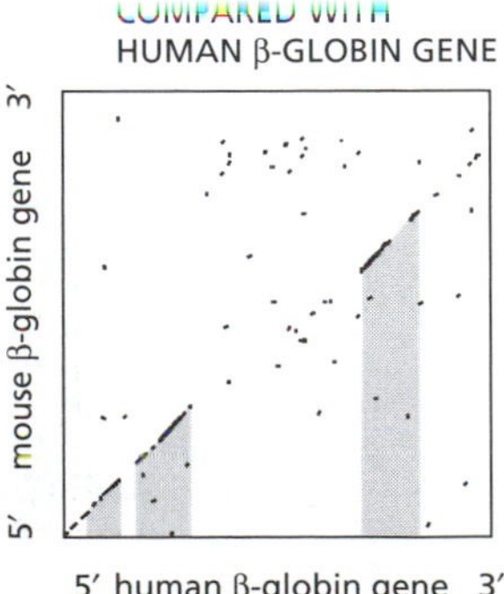

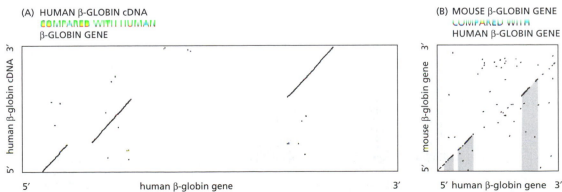

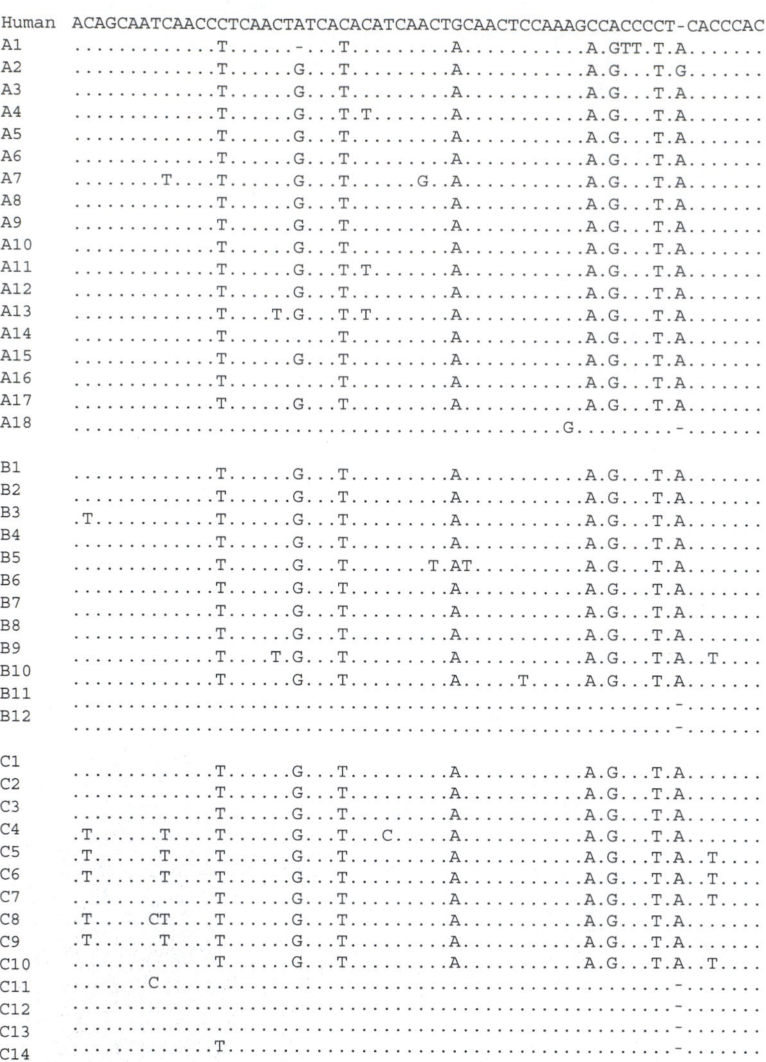

Human ACAGCAATCAACCCTCAACTATCACACATCAACTGCAACTCCAAAGCCACCCCT-CACCCAC
A1 .............T......-...T.........A...........A.GTT.T.A.......
A2 .............T......G...T.........A...........A.G...T.G.......
A3 .............T......G...T.........A...........A.G...T.A.......
A4 .............T......G...T.T.......A...........A.G...T.A.......
A5 .............T......G...T.........A...........A.G...T.A.......
A6 .............T......G...T.........A...........A.G...T.A.......
A7 ........T....T......G...T......G..A...........A.G...T.A.......
A8 .............T......G...T.........A...........A.G...T.A.......
A9 .............T......G...T.........A...........A.G...T.A.......
A10 .............T......G...T.........A...........A.G...T.A.......
A11 .............T......G...T.T.......A...........A.G...T.A.......
A12 .............T......G...T.........A...........A.G...T.A.......
A13 .............T....T.G...T.T.......A...........A.G...T.A.......
A14 .............T..........T.........A...........A.G...T.A.......
A15 .............T......G...T.........A...........A.G...T.A.......
A16 .............T..........T.........A...........A.G...T.A.......
A17 .............T......G...T.........A...........A.G...T.A.......
A18 ............................................G.........-.......
B1 .............T......G...T.........A...........A.G...T.A.......
B2 .............T......G...T.........A...........A.G...T.A.......
B3 .T...........T......G...T.........A...........A.G...T.A.......
B4 .............T......G...T.........A...........A.G...T.A.......
B5 .............T......G...T.......T.AT..........A.G...T.A.......
B6 .............T......G...T.........A...........A.G...T.A.......
B7 .............T......G...T.........A...........A.G...T.A.......
B8 .............T......G...T.........A...........A.G...T.A.......
B9 .............T....T.G...T.........A...........A.G...T.A..T....
B10 .............T......G...T.........A.....T.....A.G...T.A.......
B11 ......................................................-.......
B12 ......................................................-.......
C1 .............T......G...T.........A...........A.G...T.A.......
C2 .............T......G...T.........A...........A.G...T.A.......
C3 .............T......G...T.........A...........A.G...T.A.......
C4 .T......T....T......G...T...C.....A...........A.G...T.A.......
C5 .T......T....T......G...T.........A...........A.G...T.A..T....
C6 .T......T....T......G...T.........A...........A.G...T.A..T....
C7 .............T......G...T.........A...........A.G...T.A..T....
C8 .T.....CT....T......G...T.........A...........A.G...T.A.......
C9 .T......T....T......G...T.........A...........A.G...T.A.......
C10 .............T......G...T.........A...........A.G...T.A..T....
C11 .......C..............................................-.......
C12 ......................................................-.......
C13 ......................................................-.......
C14 .............T........................................-.......
......................................................-.......


**Figure 4–36** Sequences of mitochondrial DNA derived from Neanderthal samples (Problem 4–100). The sequence across the top is a reference human sequence. *Dots* indicate matches to the human sequence; *dashes* indicate missing nucleotides.

(PCR) on them, one in the laboratory of a foreign collaborator. Sure enough, clear products of the expected size are seen in all three reactions. Cloned products from each PCR reaction are individually sequenced with the results shown in Figure 4–36. The sequence of the corresponding region of mitochondrial DNA from a human is shown at the top. Dots indicate matches to the human sequence; dashes indicate missing DNA.

To determine whether the common sequence differences you observe could be due to normal variation within the human population, you make pairwise comparisons of your consensus (most common) Neanderthal sequence with a large number of individual human sequences. You do the same for individual human sequences versus one another and versus chimpanzee sequences. Your pairwise comparisons are shown in Figure 4–37.

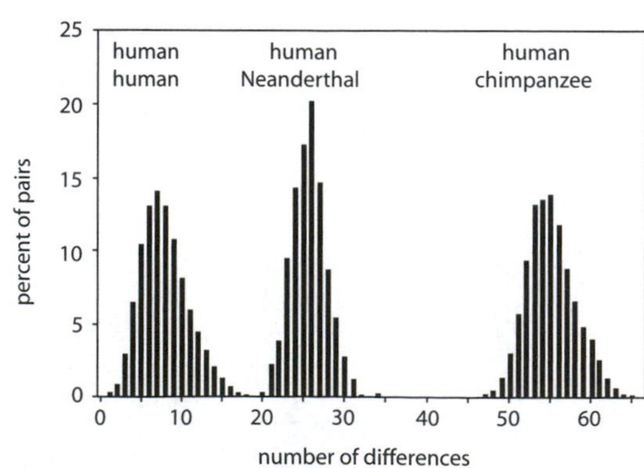


**Figure 4–37** Pairwise comparisons of DNA sequences (Problem 4–100). The human–human distribution compared 994 individual, and distinct, human sequences with one another. These 994 sequences represent contemporary human mitochondrial lineages; that is, distinct sequences occurring in one or more individuals. The fraction of pairs with a given number of differences is plotted. The human–Neanderthal comparison is one Neanderthal sequence against the 994 contemporary human sequences. The human–chimpanzee comparison involved 986 contemporary human lineages with 16 contemporary chimpanzee lineages.

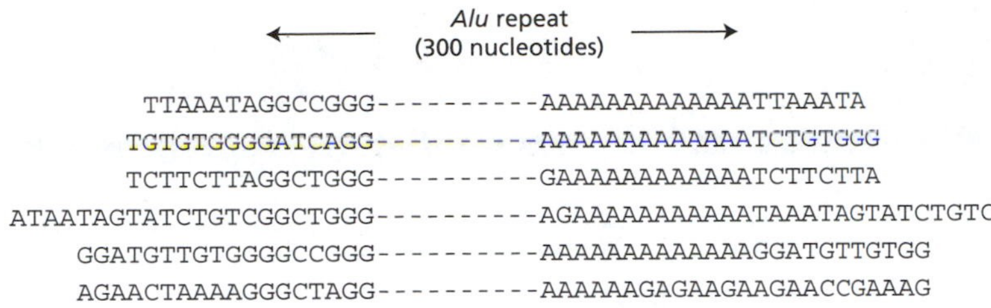


**Figure 4–38** Nucleotide sequences of the six *Alu* inserts in the human albumin-gene family (Problem 4–101). *Dashed lines* indicate nucleotides in the internal part of the *Alu* sequences.

A. Have you successfully identified a Neanderthal mitochondrial DNA sequence? Explain your reasoning.
B. Did your extensive precautions in handling the sample eliminate human contamination?
C. What is the reason for choosing mitochondrial DNA for archaeological DNA studies? Wouldn't nuclear DNA sequences be more informative?
D. What would you consider to be the most important way to confirm or refute your findings?

**4–101** *Alu* sequences are present at six sites in the introns of the human serum albumin and α-fetoprotein genes. (These genes are evolutionary relatives that are located side by side in mammalian genomes.) The same pair of genes in the rat contain no *Alu* sequences. The lineages of rats and humans diverged more than 85 million years ago at the time of the mammalian radiation. Does the presence of *Alu* sequences in the human genes and their absence from the corresponding rat genes mean that *Alu* sequences invaded the human genes only recently, or does it mean that the *Alu* sequences have been removed in some way from the rat genes?

To examine this question, you have sequenced all six of the *Alu* sequences in the human albumin-gene family. The sequences around the ends of the inserted *Alu* elements are shown in Figure 4–38.

A. *Alu* sequences create duplications of several nucleotides on each side of the target site where they insert. Mark the left and right boundaries of the inserted *Alu* sequences and underline the nucleotides in the flanking chromosomal DNA that have been altered by mutation.
B. The rate of nucleotide substitution in introns has been measured at about 3 $\times 10^{-3}$ mutations per million years at each site. Assuming the same rate of substitution into the intron sequences that were duplicated by these *Alu* sequences, calculate how long ago the *Alu* sequences inserted into these genes. (Lump all the *Alu* sequences together to make this calculation; that is, treat them as if they inserted at about the same time.)
C. Why are these particular flanking sequences used in the calculation? Why were larger segments of the intron not included? Why were the mutations in the *Alu* sequences themselves not used?
D. Did these *Alu* sequences invade the human genes recently (after the time of the mammalian radiation), or have they been removed from the rat genes?

**4–102** There has been a colossal snafu in the maternity ward at your local hospital. Four sets of male twins, born within an hour of each other, were inadvertently shuffled in the excitement occasioned by that unlikely event. You have been called in to set things right. As a first step, you want to get the twins matched up. To that end you analyze a small blood sample from each infant using a hybridization probe that detects polymorphic differences in the numbers of simple sequence repeats such as $(CA)_n$ located in widely scattered regions of the genome. The results are shown in Figure 4–39.

A. Which infants are brothers?
B. How will you match brothers to the correct parents?

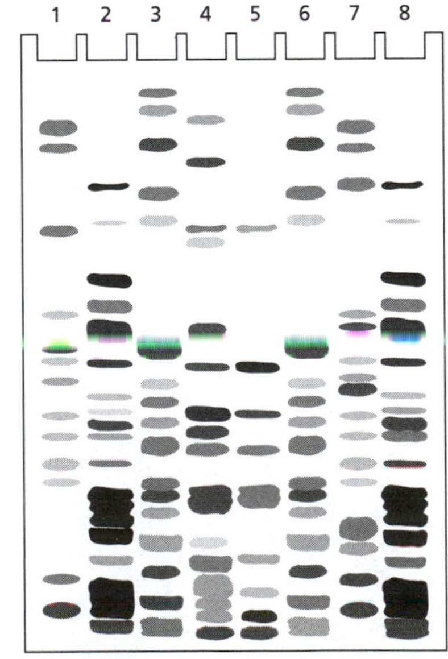


**Figure 4–39** DNA fingerprint analysis of shuffled twins (Problem 4–102).

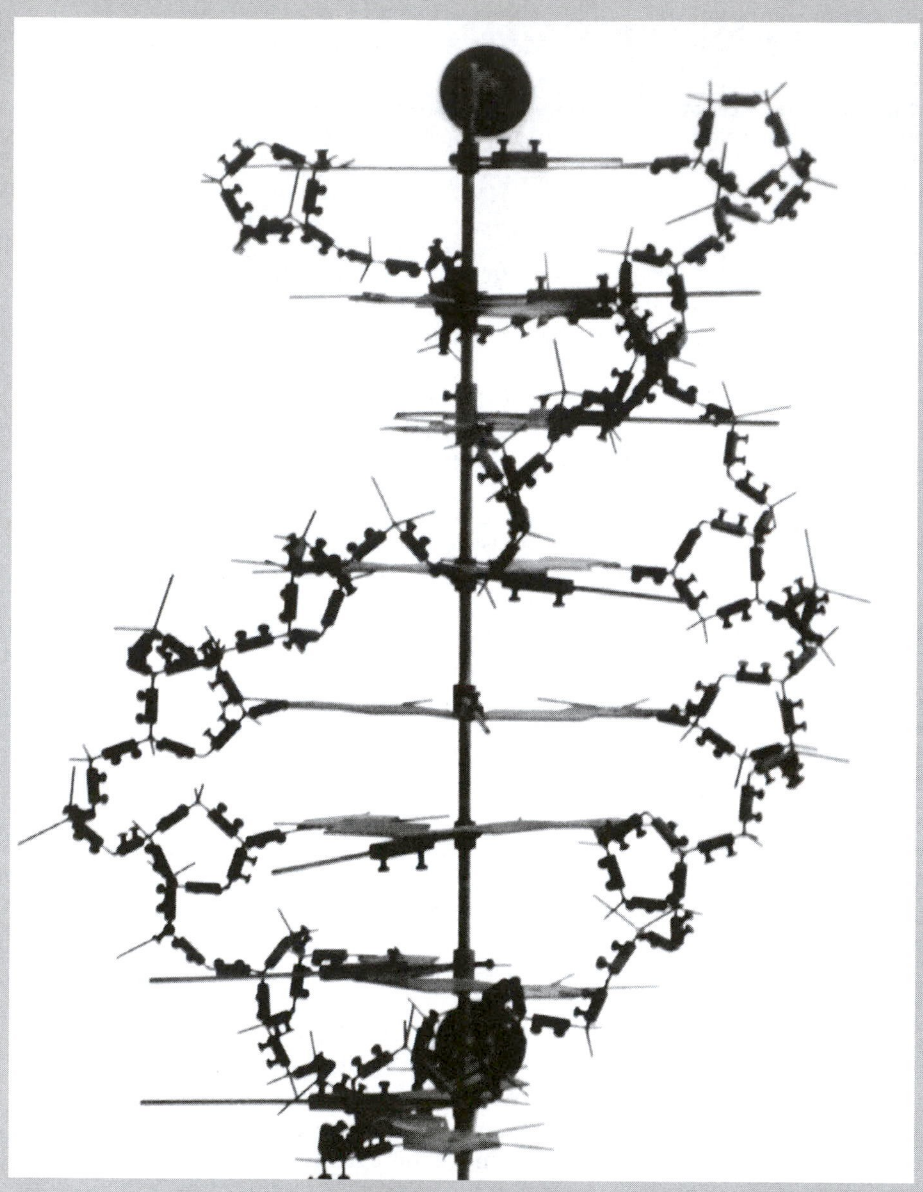

**An early wire model of DNA from the Randall Institute in London where Rosalind Franklin and Maurice Wilkins worked.** It is not so easy to see that the two strands run in opposite directions, but you should be able to mark the 5′ to 3′ polarity if you look closely.

# DNA Replication, Repair, and Recombination

# 5

## In This Chapter


THE MAINTENANCE OF DNA SEQUENCES 87

DNA REPLICATION MECHANISMS 88

THE INITIATION AND COMPLETION OF DNA REPLICATION IN CHROMOSOMES 96

DNA REPAIR 103

HOMOLOGOUS RECOMBINATION 111

TRANSPOSITION AND CONSERVATIVE SITE-SPECIFIC RECOMBINATION 114


## THE MAINTENANCE OF DNA SEQUENCES

TERMS TO LEARN

| | | |
|---|---|---|
| germ cell | mutation rate | somatic cell |
| mutation | | |

### DEFINITIONS

Match each definition below with its term from the list above.

**5–1** A randomly produced, heritable change in the nucleotide sequence of a chromosome.

**5–2** Cell type in a diploid organism that carries only one set of chromosomes and is specialized for sexual reproduction. A sperm or an egg.

**5–3** Any cell of a plant or animal other than a germ cell or germ-line precursor.

### TRUE/FALSE

Decide whether each of these statements is true or false, and then explain why.

**5–4** Both germ-cell DNA stability and somatic-cell DNA stability are essential for the survival of the species.

### THOUGHT PROBLEMS

**5–5** You infected an *E. coli* culture with a virulent bacteriophage. Most of the cells lysed, but a few survived: $1 \times 10^{-4}$ in your sample. You wonder where the resistant bacteria came from. Were they caused by the bacteriophage infection, or did they already exist in the bacterial culture? Earlier, for a different experiment, you had spread a dilute suspension of *E. coli* onto solid medium in a large Petri dish, and, after seeing that about $10^5$ colonies were growing up, you made an imprint of the colonies on that plate and transferred it to three other plates—a process known as replica plating—which creates the same pattern of colonies on the three plates. You realize that you can use these plates to distinguish between the two possibilities. You pipette a suspension of the bacteriophage onto each of the three replica plates so the bacteria can be infected. What result do you expect if the bacteriophage cause resistance? What result do you expect if the resistant bacteria preexist?

**5–6** The following statement sounds patently false. “No two cells in your body have the identical nucleotide sequence.” Provide an argument for why it might be true.

**5–7** Why do calculations based on amino acid differences between the same protein in different species tend to underestimate the actual mutation rate, even after correcting for silent mutations? Is this true for estimates of rates

**Table 5–1 Frequencies of mutant cells in multiple cultures** (Problem 5–10).

| EXPERIMENT | CULTURE (mutant cells/$10^6$ cells) | | | | | | | | | |
|---|---|---|---|---|---|---|---|---|---|---|
| | 1 | 2 | 3 | 4 | 5 | 6 | 7 | 8 | 9 | 10 |
| 1 | 4 | 0 | 257 | 1 | 2 | 32 | 0 | 0 | 2 | 1 |
| 2 | 128 | 0 | 1 | 4 | 0 | 0 | 66 | 5 | 0 | 2 |

based on the fibrinopeptides, which are 20 amino-acid fragments that are discarded from the protein fibrinogen when it is activated to form fibrin during blood clotting? Why or why not?

**5–8** Individual organisms that carry harmful mutations tend to be eliminated from a population by natural selection. It is easy to see how deleterious mutations in bacteria, which have a single copy of each gene, are eliminated by natural selection; the affected bacteria die and the mutation is thereby lost from the population. Eucaryotes, however, have two copies of most genes because they are diploid. It is often the case that an individual with two normal copies of the gene (homozygous, normal) is indistinguishable in phenotype from an individual with one normal copy and one defective copy of the gene (heterozygous). In such cases, natural selection can operate only on an individual with two copies of the defective gene (homozygous, defective). Imagine the situation in which a defective form of the gene is lethal when homozygous, but without effect when heterozygous. Can such a mutation ever be eliminated from the population by natural selection? Why or why not?

## CALCULATIONS

**5–9** Mutations are introduced into the *E. coli* genome at the rate of 1 mutation per $10^9$ base pairs per generation. Imagine that you start with a population of $10^6$ *E. coli*, none of which carry any mutations in your gene of interest, which is 1000 nucleotides in length and not essential for bacterial growth and survival. In the next generation, after the population doubles in number, what fraction of the cells, on average, would you expect to carry a mutation in your gene? After the population doubles again, what would you expect the frequency of mutants in the population to be? What would the frequency be after a third doubling?

## DATA HANDLING

**5–10** To determine the reproducibility of mutation frequency measurements, you do the following experiment. You inoculate each of 10 cultures with a single *E. coli* bacterium, allow the cultures to grow until each contains $10^6$ cells, and then measure the number of cells in each culture that carry a mutation in your gene of interest. You were so surprised by the initial results that you repeated the experiment to confirm them. Both sets of results display the same extreme variability, as shown in Table 5–1. Assuming that the rate of mutation is constant, why do you suppose there is so much variation in the frequencies of mutant cells in different cultures?

# DNA REPLICATION MECHANISMS

TERMS TO LEARN

| | | |
|---|---|---|
| clamp loader | DNA topoisomerase | RNA primer |
| DNA helicase | lagging strand | single-strand DNA-binding (SSB) protein |
| DNA ligase | leading strand | sliding clamp |
| DNA polymerase | replication fork | strand-directed mismatch repair |
| DNA primase | | |

## DEFINITIONS

Match each definition below to its term from the list above.

5–11 Short length of RNA synthesized on the lagging strand during DNA replication and subsequently removed.

5–12 Enzyme that joins two adjacent DNA strands together.

5–13 DNA repair process that replaces incorrect nucleotides inserted during DNA replication.

5–14 Enzyme that opens the DNA helix by separating the single strands.

5–15 One of the two newly made strands of DNA found at a replication fork. It is made by continuous synthesis in the 5′-to-3′ direction.

5–16 A protein complex that encircles the DNA double helix and binds to DNA polymerase, keeping it firmly bound to the DNA while it is moving.

5–17 Enzyme that binds to DNA and reversibly breaks a phosphodiester bond in one or both strands, allowing the DNA to rotate at that point.

5–18 Y-shaped region of a replicating DNA molecule at which the two daughter strands are formed.

5–19 One of the two newly made strands of DNA found at a replication fork. It is made in discontinuous segments that are later joined covalently.

## TRUE/FALSE

Decide whether each of these statements is true or false, and then explain why.

5–20 When read in the same direction (5′-to-3′), the sequence of nucleotides in a newly synthesized DNA strand is the same as in the parental strand used as the template for its synthesis.

5–21 Each time the genome is replicated, half the newly synthesized DNA is stitched together from Okazaki fragments.

5–22 In *E. coli*, where the replication fork travels at 500 nucleotide pairs per second, the DNA ahead of the fork must rotate at nearly 3000 revolutions per minute.

5–23 The mismatch proofreading system in *E. coli* can distinguish the parental strand from the progeny strand as long as one or both are methylated, but not if both strands are unmethylated.

5–24 Topoisomerase I does not require ATP to break and rejoin DNA strands because the energy of the phosphodiester bond is stored transiently in a phosphotyrosine linkage in the enzyme's active site.

## THOUGHT PROBLEMS

5–25 The nucleotide sequence of one DNA strand of a DNA double helix is 5′-GGATTTTTGTCCACAATCA-3′. What is the sequence of the complementary strand?

5–26 The DNA fragment in Figure 5–1 is double stranded at each end but single stranded in the middle. The polarity of the top strand is indicated. Is the phosphate ($PO_4^-$) shown on the bottom strand at the 5′ end or the 3′ end of the fragment to which it is attached?

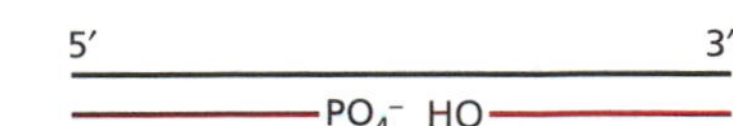


**Figure 5–1** A DNA fragment with a single-stranded gap on the bottom strand (Problem 5–26).

5–27 Look carefully at the structures of the molecules in Figure 5–2.

A. What would you expect to happen if dideoxycytidine triphosphate (ddCTP) were added to a DNA replication reaction in large excess over the concentra-

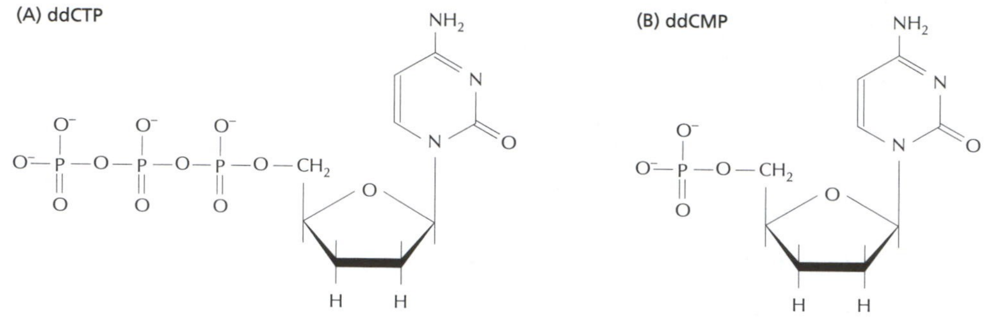

(A) ddCTP

(B) ddCMP

Figure 5–2 Potential replication substrates (Problem 5–27). (A) Dideoxycytidine triphosphate (ddCTP). (B) Dideoxycytidine monophosphate (ddCMP).

tion of deoxycytidine triphosphate (dCTP)? Would it be incorporated into the DNA? If it were, what would happen after that? Give your reasoning.

B. What would happen if ddCTP were added at 10% of the concentration of dCTP?

C. What effects would you expect if dideoxycytidine monophosphate (ddCMP) were added to a DNA replication reaction in large excess, or at 10% of the concentration of dCTP?

**5–28** How would you expect the loss of the 3′-to-5′ proofreading exonuclease activity of DNA polymerase in *E. coli* to affect the fidelity of DNA synthesis? How would its loss affect the rate of DNA synthesis? Explain your reasoning.

**5–29** You have discovered a novel organism that thrives in the ocean depths in the hostile environment of hydrothermal vents. In characterizing its replication, you are astounded to discover that it replicates both strands continuously, using two DNA polymerases: one that synthesizes DNA in the usual 5′-to-3′ direction and a second that synthesizes DNA in the 3′-to-5′ direction. Both polymerases use the standard nucleoside 5′-triphosphates for addition of nucleotides to growing DNA chains. You are surprised to find that *both* newly synthesized DNA strands are made with the same high degree of fidelity that characterizes DNA synthesis in *E. coli.*

A. Briefly describe the four processes that contribute to the high fidelity of DNA replication in *E. coli.*

B. Explain why it is surprising that both strands in this novel organism are replicated with high fidelity.

C. Suggest at least two ways by which high fidelity might be accomplished. If you need to invent additional enzymes to accomplish a specific task, describe their activities.

**5–30** Discuss the following statement: "Primase is a sloppy enzyme that makes many mistakes. Eventually, the RNA primers it makes are replaced with DNA made by a polymerase with higher fidelity. This is wasteful. It would be more energy efficient if a DNA polymerase made an accurate copy in the first place."

**5–31** SSB proteins bind to single-stranded DNA at the replication fork and prevent the formation of short hairpin helices that would otherwise impede DNA synthesis. What sorts of sequences in single-stranded DNA might be able to form a hairpin helix? Write out an example of a sequence that could form a 5-nucleotide hairpin helix, and show the helix.

**5–32** Conditional lethal mutations have proven indispensible in genetic and biochemical analyses of complex processes such as DNA replication. Temperature-sensitive (ts) mutations, which are one form of conditional lethal mutation, allow growth at one temperature (for example, 30°C) but not at a higher temperature (for example, 42°C).

A large number of temperature-sensitive replication mutants have been isolated in *E. coli.* These mutant bacteria are defective in DNA replication at 42°C but not at 30°C. If the temperature of the medium is raised from 30°C to 42°C, these mutants stop making DNA in one of two characteristic ways. The 'quick-stop' mutants halt DNA synthesis immediately, whereas the 'slow-stop' mutants stop DNA synthesis only after many minutes.

A. Predict which of the following proteins, if temperature sensitive, would display a quick-stop phenotype and which would display a slow-stop phenotype. In each case explain your prediction.
   1. DNA topoisomerase I
   2. A replication initiator protein
   3. Single-strand DNA-binding protein
   4. DNA helicase
   5. DNA primase
   6. DNA ligase

B. Cell-free extracts of the mutants show essentially the same patterns of replication as the intact cells. Extracts from quick-stop mutants halt DNA synthesis immediately at 42°C, whereas extracts from slow-stop mutants do not stop DNA synthesis for several minutes after a shift to 42°C. Suppose extracts from a temperature-sensitive DNA helicase mutant and a temperature-sensitive DNA ligase mutant were mixed together at 42°C. Would you expect the mixture to exhibit a quick-stop phenotype, a slow-stop phenotype, or a nonmutant phenotype?

**5–33** DNA repair enzymes preferentially repair mismatched bases on the newly synthesized DNA strand, using the old DNA strand as a template. If mismatches were repaired instead without regard for which strand served as template, would mismatch repair reduce replication errors? Would such an indiscriminate mismatch repair result in fewer mutations, more mutations, or the same number of mutations as there would have been without any repair at all? Explain your answers.

**5–34** If DNA polymerase requires a perfectly paired primer in order to add the next nucleotide, how is it that any mismatched nucleotides 'escape' the polymerase and become substrates for mismatch repair enzymes?

**5–35** DNA damage can interfere with DNA replication. X-rays, for example, generate highly reactive hydroxyl radicals that can break one or both strands of DNA. UV light commonly generates cyclobutane dimers between adjacent T bases in the same DNA strand, which blocks progression of DNA polymerase. If such damage is not repaired, it can have serious consequences when a replication fork encounters it. See if you can predict the appearance of the replication fork after it encounters a nick or a thymine-dimer block in the templates for the leading and lagging strands. Replication forks just before they encounter the damage are shown in Figure 5–3.

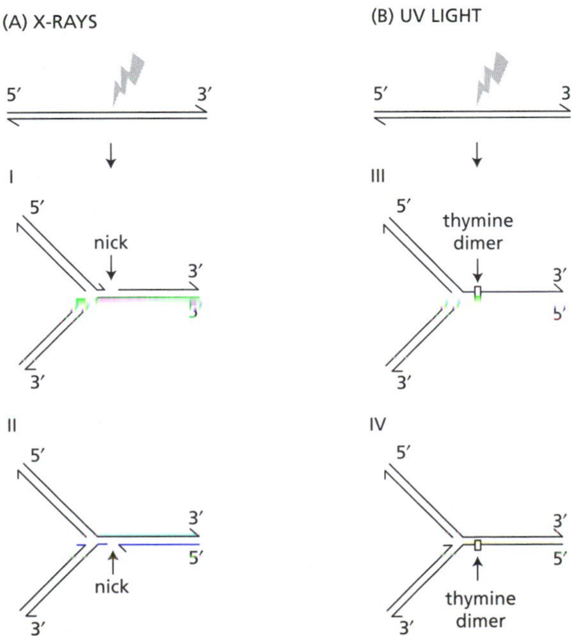


**Figure 5–3** Damage to the templates for the leading and lagging strands (Problem 5–35). (A) X-ray-induced nicks. (B) UV-light-induced thymine dimers. Structures I and III have damage in the template for lagging-strand synthesis. Structures II and IV have damage in the template for leading-strand synthesis.

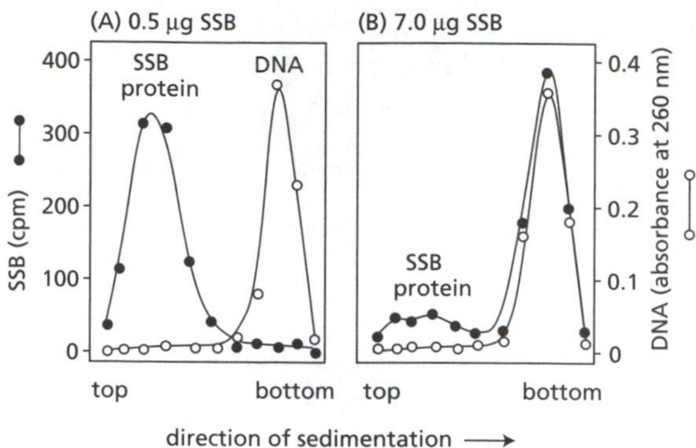


**Figure 5–4** Binding of T4 SSB protein to single-stranded DNA (Problem 5–37). The binding of SSB protein to DNA was analyzed by centrifugation through sucrose gradients, on which the much more massive DNA sediments more rapidly than protein and is consequently found closer to the bottom of the gradient.

**5–36** At the completion of replication of the circular genome of the animal virus SV40, the two daughter circles are interlocked like links in a chain. How do you suppose such interlinked molecules might then separate?

## CALCULATIONS

**5–37** Like all organisms, bacteriophage T4 encodes an SSB protein that is important for removing secondary structure in the single-stranded DNA ahead of the replication fork. The T4 SSB protein is an elongated monomeric protein with a molecular weight of 35,000. It binds tightly to single-stranded, but not double-stranded, DNA. Binding saturates at a 1:12 weight ratio of DNA to protein. The binding of SSB protein to DNA shows a peculiar property that is illustrated in Figure 5–4. In the presence of excess single-stranded DNA (10 μg), virtually no binding is detectable at 0.5 μg SSB protein (Figure 5–4A), whereas nearly all the SSB protein is bound to DNA at 7.0 μg (Figure 5–4B).

A. At saturation, what is the ratio of nucleotides of single-stranded DNA to molecules of SSB protein? (The average mass of a single nucleotide is 330 daltons.)

B. When the binding of SSB protein to DNA reaches saturation, are adjacent monomers of SSB protein likely to be in contact? Assume that a monomer of SSB protein extends for 12 nm along the DNA upon binding and that the spacing of bases in single-stranded DNA after binding is the same as in double-stranded DNA (that is, 10.4 nucleotides per 3.4 nm).

C. Why do you think that the binding of SSB protein to single-stranded DNA depends so strongly on the amount of SSB protein?

**5–38** Approximately how many high-energy bonds are used to replicate the *E. coli* chromosome? How many molecules of glucose would *E. coli* need to consume to provide enough energy to copy its DNA once? How does this mass of glucose compare with the mass of *E. coli*, which is about $10^{-12}$ g? (There are $4.6 \times 10^6$ base pairs in the *E. coli* genome. Oxidation of one glucose molecule yields about 30 high-energy phosphate bonds. Glucose has a molecular mass of 180 daltons, and there are $6 \times 10^{23}$ daltons/g.)

## DATA HANDLING

**5–39** In the electron microscope, it is possible to observe the replication fork directly and, for small DNA molecules, to see the entire replicating structure. In addition, by using appropriate techniques of sample preparation, one can distinguish double-stranded DNA from single-stranded DNA.

A series of hypothetical replicating molecules is illustrated schematically in Figure 5–5, with regions of single-stranded DNA shown as thin lines. In an important early electron microscopic study of bacteriophage lambda replication, some of these structures were observed commonly and others were never seen.

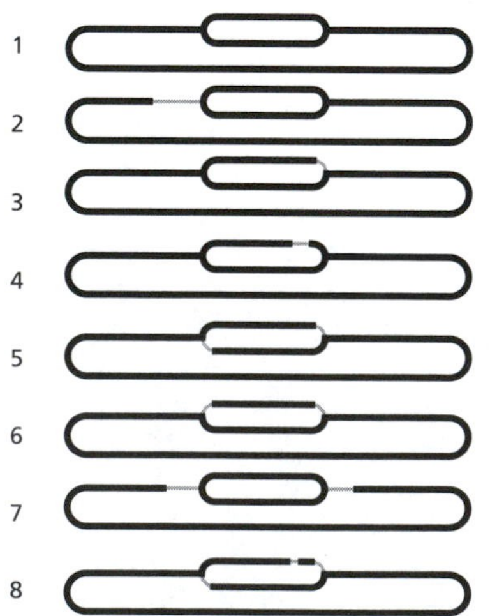


**Figure 5–5** Hypothetical structures of replicating DNA molecules (Problem 5–39). Double-stranded DNA is shown as *thick lines*; single-stranded DNA is shown as *thin lines*.

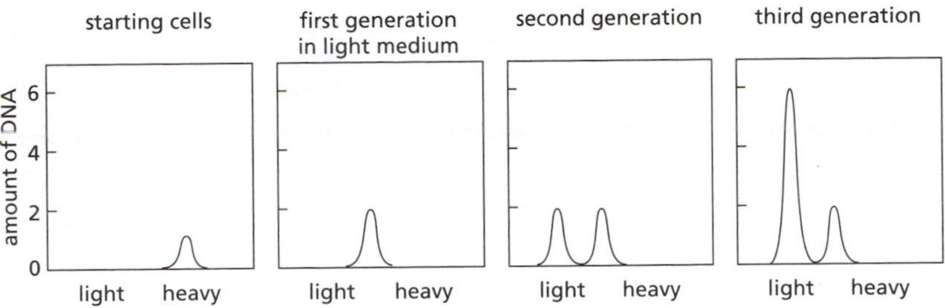


**Figure 5–6** Density of DNAs isolated from cells that were grown for different times in 'light' medium after initial growth in medium enriched for heavy isotopes of nitrogen and carbon (Problem 5–40). Equal culture volumes were analyzed for each time point. Amount of DNA is in arbitrary units, with the peak amount of DNA in the sample containing starting cells set equal to 1.

A. Draw a diagram of a replication structure with two forks moving in opposite directions. Label the ends of all strands (5′ or 3′), and indicate the leading and lagging strands at each replication fork.

B. Based on your knowledge of DNA replication, indicate the structure in Figure 5–5 you would expect to be observed most commonly. In the actual experiment, four of these structures were seen. Which ones do you think they were?

**5–40** A born skeptic, you plan to confirm for yourself the results of a classic experiment originally performed in the 1960s by Meselson and Stahl. They concluded that each daughter cell inherits only one strand of its mother's DNA. To check their results, you 'synchronize' a culture of growing cells, so that virtually all cells begin and then complete DNA synthesis at the same time. You first grow the cells in a medium that contains nutrients highly enriched in heavy isotopes of nitrogen and carbon ($^{15}$N and $^{13}$C in place of the naturally abundant $^{14}$N and $^{12}$C). Cells growing in this 'heavy' medium use the heavy isotopes to build all of their macromolecules, including nucleotides and nucleic acids. You then transfer the cells to a normal, 'light' medium containing $^{14}$N and $^{12}$C nutrients. Finally, you isolate DNA from cells that have grown for different numbers of generations in the light medium and determine the density of their DNA by density-gradient centrifugation. Your data, plotting the amount of DNA isolated versus its density, are shown in Figure 5–6. Are these results in agreement with your expectations? Explain the results.

Problem 8–23 presents data from the original Meselson–Stahl experiment.

**5–41** To study the 3′-to-5′ proofreading exonuclease activity of DNA polymerase I of *E. coli*, you prepare an artificial substrate with a poly(dA) strand as template and a poly(dT) strand as primer. The poly(dT) strand contains a few $^{32}$P-labeled dT nucleotides followed by a few $^{3}$H-labeled dC nucleotides at its 3′ end, as shown in Figure 5–7. You measure the loss of the labeled dTs and dCs either without any dTTP present, so that no DNA synthesis is possible, or with dTTP present, so that DNA synthesis can occur. The results are shown in Figure 5–8.

A. Why were the Ts and Cs labeled with different isotopes?

B. Why did it take longer for the Ts to be removed in the absence of dTTP than the Cs ?

C. Why were none of the Ts removed in the presence of dTTP, whereas the Cs were lost regardless of whether or not dTTP was present?

D. Would you expect different results in Figure 5–8B if you had included dCTP along with the dTTP?

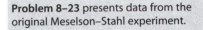

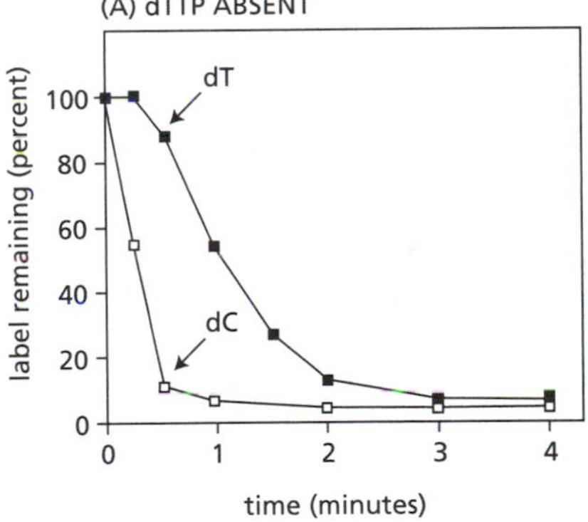


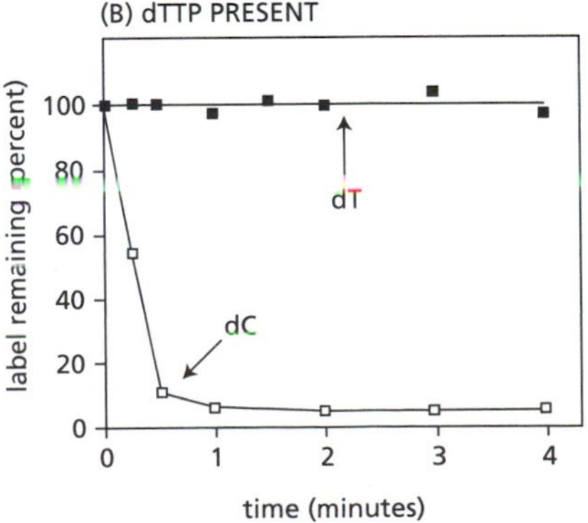


**Figure 5–8** Proofreading by DNA polymerase I (Problem 5–41). (A) In the absence of dTTP. (B) In the presence of dTTP.

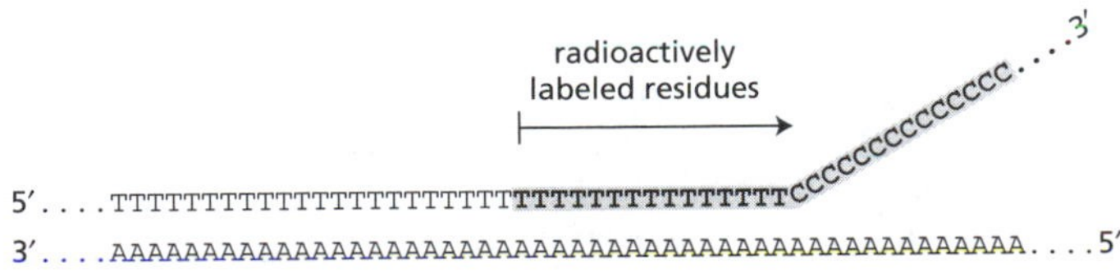


**Figure 5–7** Artificial substrate for studying proofreading by DNA polymerase I of *E. coli* (Problem 5–41). *Shaded* letters indicate nucleotides that are radioactively labeled.

| site | M13 template sequences 5′ → 3′ | DNA sequences linked to RNA primer 5′ → 3′ |
|---|---|---|
| 1 | A T C C T T G C G T T G A A A T | A G G A T |
| 2 | T C T T G T T T G C T C C A G A | C A A G A |
| 3 | A T T C T C T T G T T T G C T C | A G A A T |
| 4 | A C A T G C T A G T T T T A C G | C A T G T |
| 5 | A T T G A C A T G C T A G T T T | T C A A T |
| 6 | A T C T T C C T G T T T T T G G | A A G A T |
| 7 | A A A T A T T T G C T T A T A C | T A T T T |
| 8 | C T A G A A C G G T T A C C C T | T C T A G |

**Figure 5–9** RNA priming during M13 replication (Problem 5–42). M13 template sequences at several sites of RNA priming are shown adjacent to the DNA sequences that were found to be linked to RNA primers at each site. The RNA primers were removed from these DNA chains prior to sequencing.

**5–42** Does RNA priming occur at specific sites or at random sites on the template? The M13 viral DNA, which is a circular single strand, is an ideal template for studying this question. The M13 circle was copied in the presence of DNA polymerase, primosome (a complex of a helicase and an RNA primase), rNTPs, and dNTPs to make a double-stranded circle that still retained RNA primers. Under these conditions the 5′ ends of the RNA primers are not removed, leaving nicks in the newly synthesized strand that correspond to the beginning of each RNA primer. The double-stranded circles were then digested with a restriction nuclease that makes a double-strand cut at a single location in the M13 DNA. When this linear DNA was denatured and separated by gel electrophoresis, several discrete single-stranded segments of newly synthesized DNA were observed. If the linear DNA was first treated with RNase to remove the RNA primers, the single-stranded segments all migrated slightly faster, consistent with their each having originally had a 5-nucleotide RNA primer.

A. Do these results argue that RNA priming occurs at specific sites or random sites on the M13 template? Explain your reasoning.

B. Sequences of the newly synthesized DNA immediately adjacent to the RNA primers were determined for eight single-stranded segments. The first five nucleotides of those DNA segments—the ones that were linked to the RNA primers—are shown in Figure 5–9, along with the corresponding template sequences in the M13 DNA. From these data, deduce the site on each template sequence at which synthesis of the RNA primer began. What is the likely signal for starting the RNA primase reaction on M13 DNA?

**5–43** The *DnaB* gene of *E. coli* encodes a helicase (DnaB) that unwinds DNA at the replication fork. Its properties have been studied using artificial substrates like those shown in Figure 5–10. In such substrates DnaB binds preferentially to the longest single-stranded region (the largest target) available. The experimental approach is to incubate the substrates under a variety of conditions and then subject a sample to electrophoresis on agarose gels. The short single strand (substrates 1 and 2) or strands (substrate 3) will move slowly if still annealed to the longer DNA strand, but will move much faster if unwound and detached. The migration of these short single strands can be followed selectively by making them radioactive and examining their positions in the gel by autoradiography. The migration of the labeled single strands in the three different substrates is shown in Figure 5–11. In the absence of any treatment, the labeled strands move slowly (lanes 4, 8, and 12); when the substrates are heated to 100°C, the labeled strands are detached and migrate more rapidly (lanes 3, 7, and 11).

The results of several experiments are shown in Figure 5–11. Substrate 1, the substrate without tails, was not unwound by DnaB and ATP at 37°C (Figure 5–11, lanes 1 and 2). When either substrate with tails was incubated at 37°C with DnaB and ATP, a significant amount of small fragment was released by unwinding (lanes 6 and 10). For substrate 3, only the 3′ fragment was unwound (lane 10). All unwinding was absolutely dependent on ATP hydrolysis.

Unwinding was considerably enhanced by adding single-stranded DNA-binding protein (SSB) (compare Figure 5–11, lanes 5 and 6 and lanes 9 and

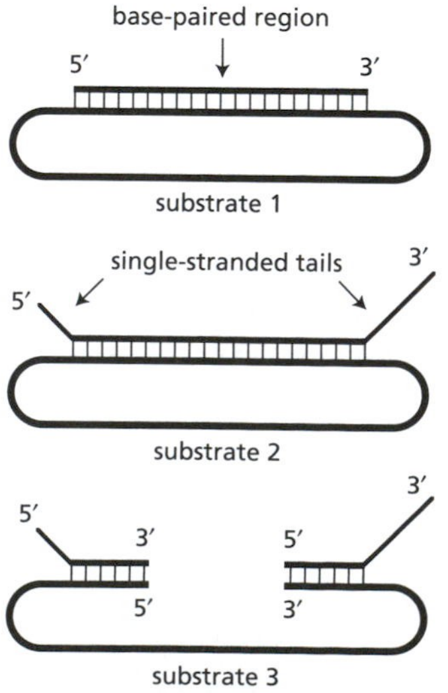

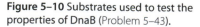

**Figure 5–10** Substrates used to test the properties of DnaB (Problem 5–43).

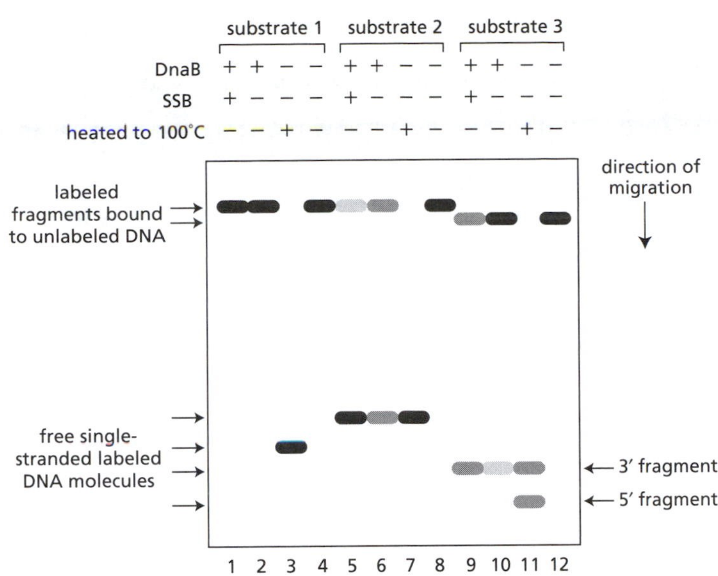


**Figure 5–11** Results of several experiments to measure unwinding by DnaB (Problem 5–43). Only the single-stranded fragments were radioactively labeled. Their positions are shown by the bands in this schematic diagram. ATP was included with DnaB in all incubations, which were carried out at 37°C.

10). Interestingly, SSB had to be added about 3 minutes after DnaB; otherwise it inhibited unwinding.

A. Why do you suppose ATP hydrolysis is required for unwinding?

B. In what direction does DnaB move along the long single-stranded DNA? Is this direction more consistent with its movement on the template for the leading strand or on the template for the lagging strand at a replication fork?

C. Why do you suppose SSB inhibits unwinding when it is added before DnaB, but stimulates unwinding when added after DnaB?

**5–44** The different ways in which DNA synthesis occurs on the leading and lagging strands raises the question as to whether synthesis occurs with equal fidelity on the two strands. One clever approach used reversion of specific mutations in the *E. coli LacZ* gene to address this question. *E. coli* is a good choice for such a study because the same polymerase (DNA pol III) synthesizes both the leading and the lagging strands.

The *LacZ* CC106 allele can regain its function (revert) by converting the mutant AT base pair to the normal GC base pair. This allele was inserted into the *E. coli* chromosome on one side of the normal origin of replication (Figure 5–12A). Two *E. coli* strains were isolated: one with the allele in the 'L' orientation, and the other with it in the opposite, 'R,' orientation. As shown in Figure 5–12B, misincorporation of G opposite T on one strand, or of C opposite A on the other strand, could lead to reversion. Previous studies had shown that C is very rarely misincorporated opposite A. Thus, the most common source of reversion is from misincorporation of G opposite T.

To eliminate the complicating effects of mismatch repair, the experiments were done in two mutant strains of bacteria. One was defective for mismatch repair, which eliminates it from consideration; the other was defective in the proofreading exonuclease, and introduces so many mismatches that it overwhelms the mismatch-repair machinery.

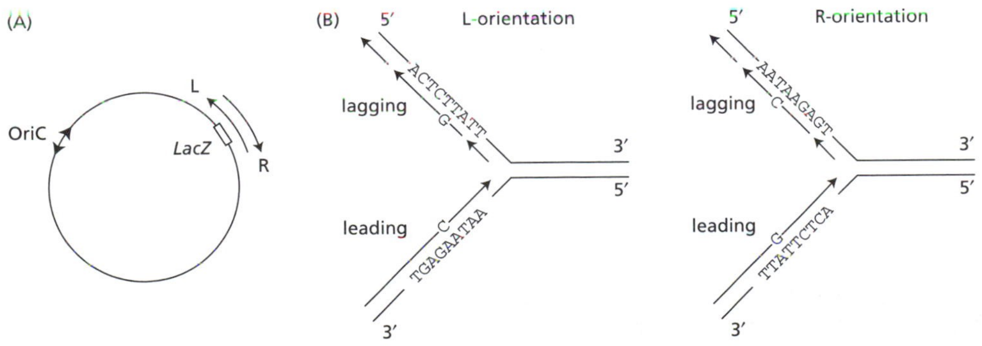


**Figure 5–12** Fidelity of synthesis of the leading and lagging strands (Problem 5–44). (A) Site of insertion of *LacZ* into the *E. coli* chromosome. Orientations are denoted R and L. The *arrows* at OriC represent the two forks initiated at that site. (B) Arrangement of sequences relative to the leading and lagging strands in the L- and R-orientations. The G and C misincorporations that could lead to reversion are indicated.

**Table 5–2 Frequencies (per $10^6$ cells) of revertants of *LacZ* and mutants of *Rif* in mismatch-repair and proofreading deficient strains of *E. coli* (Problem 5–44).**

| *LacZ* ALLELE (mutation measured) | *LacZ* ORIENTATION | MISMATCH-REPAIR DEFICIENT | | PROOFREADING DEFICIENT | |
|---|---|---|---|---|---|
| | | $Lac^- \rightarrow Lac^+$ | $Rif^s \rightarrow Rif^r$ | $Lac^- \rightarrow Lac^+$ | $Rif^s \rightarrow Rif^r$ |
| CC106 (AT → GC) | L | 0.27 | 7.0 | 2.7 | 82 |
| CC106 (AT → GC) | R | 0.51 | 6.4 | 6.7 | 80 |

Accurate frequencies of *LacZ* reversion were measured in the two strains, along with the frequencies of mutation at the *Rif* gene, whose orientation in the chromosome was constant (Table 5–2).

A. On which strand, leading or lagging, does DNA synthesis appear to be more accurate? Explain your reasoning.

B. Can you suggest a reason why DNA synthesis might be more accurate on the strand you have chosen?

## THE INITIATION AND COMPLETION OF DNA REPLICATION IN CHROMOSOMES

TERMS TO LEARN

| | | |
|---|---|---|
| histone chaperone | replication origin | telomerase |
| origin recognition complex (ORC) | S phase | |

### DEFINITIONS

Match each definition below with its term from the list above.

**5–45** Period during a eucaryotic cell cycle in which DNA is synthesized.

**5–46** Large multimeric protein structure that is bound to the DNA at origins of replication in eucaryotic chromosomes throughout the cell cycle.

**5–47** Special DNA sequence on a bacterial or viral chromosome at which DNA replication begins.

**5–48** Enzyme that elongates telomeres, the repetitive nucleotide sequences found at the ends of eucaryotic chromosomes.

### TRUE/FALSE

Decide whether each of these statements is true or false, and then explain why.

**5–49** In a replication bubble, a single parental DNA strand serves as the template strand for leading-strand synthesis in one replication fork and as the template for lagging-strand synthesis in the other fork.

**5–50** When bidirectional replication forks from adjacent origins meet, a leading strand always runs into a lagging strand.

**5–51** In mammalian cells different regions of the genome are replicated in a specified order.

**5–52** If an origin of replication is deleted from a eucaryotic chromosome, the DNA on either side will ultimately be lost, as well, because it cannot be replicated.

### THOUGHT PROBLEMS

**5–53** The laboratory you joined is studying the life cycle of an animal virus that uses a circular, double-stranded DNA as its genome. Your project is to define the location of the origin(s) of replication and to determine whether replication proceeds in one or both directions away from an origin (unidirectional or bidirectional replication). To accomplish your goal, you isolated replicating molecules, cleaved them with a restriction nuclease that cuts the viral genome at one site to produce a linear molecule from the circle, and examined the resulting molecules in the electron microscope. Some of the molecules you observed are illustrated schematically in Figure 5–13. (Note that it is impossible to distinguish the orientation of one DNA molecule from another in the electron microscope.)

You must present your conclusions to the rest of the lab tomorrow. How will you answer the two questions your advisor had posed for you? Is there a single, unique origin of replication or several origins? Is replication unidirectional or bidirectional?

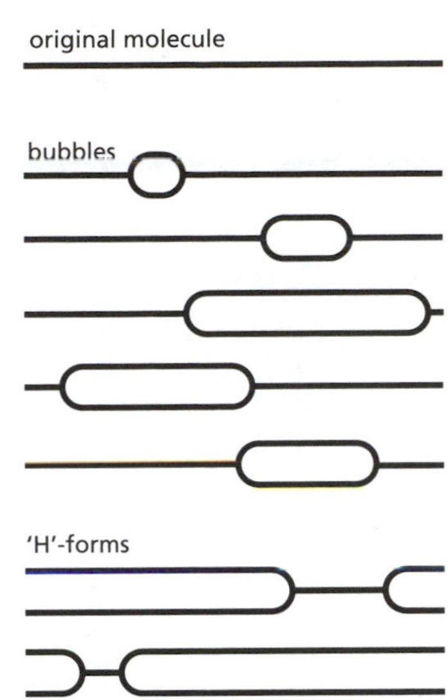


**Figure 5–13** Parental and replicating forms of an animal virus (Problem 5–53).

**5–54** Which one of the following statements about the newly synthesized strand of a human chromosome is correct?

A. It was synthesized from a single origin solely by continuous DNA synthesis.
B. It was synthesized from a single origin solely by discontinuous DNA synthesis.
C. It was synthesized from a single origin by a mixture of continuous and discontinuous DNA synthesis.
D. It was synthesized from multiple origins solely by continuous DNA synthesis.
E. It was synthesized from multiple origins solely by discontinuous DNA synthesis.
F. It was synthesized from multiple origins by a mixture of continuous and discontinuous DNA synthesis.
G. It was synthesized from multiple origins by either continuous or discontinuous DNA synthesis, depending on which specific daughter chromosome is being examined.

**5–55** The mechanism of DNA replication gives rise to the 'end-replication problem' for linear chromosomes. Over time, this problem leads to loss of DNA from the ends of chromosomes. In cells such as yeast, loss of nucleotides during replication is balanced by addition of nucleotides by telomerase. In humans, however, telomerase is turned off in most somatic cells early in development, so that chromosomes become shorter with increasing rounds of replication. Consider one round of replication in a human somatic cell. Which one of the following statements correctly describes the status of the two daughter chromosomes relative to the parent chromosome?

A. One daughter chromosome will be shorter at one end; the other daughter chromosome will be normal at both ends.
B. One daughter chromosome will be shorter at both ends; the other daughter chromosome will be normal at both ends.
C. One daughter chromosome will be shorter at both ends; the other daughter chromosome will be shorter at only one end.
D. Both daughter chromosomes will be shorter at one end, which is the same end in the two chromosomes.
E. Both daughter chromosomes will be shorter at one end, which is the opposite end in the two chromosomes.
F. Both daughter chromosomes will be shorter at both ends.

## CALCULATIONS

**5–56** In the early embryo of *Drosophila* many replication origins are active so that several can be observed in a single electron micrograph, as shown in Figure 5–14.

A. Identify the four replication bubbles in Figure 5–14. Indicate the approximate locations of the origins at which each replication bubble was initiated, and label the replication forks 1 through 8 from left to right across the figure.

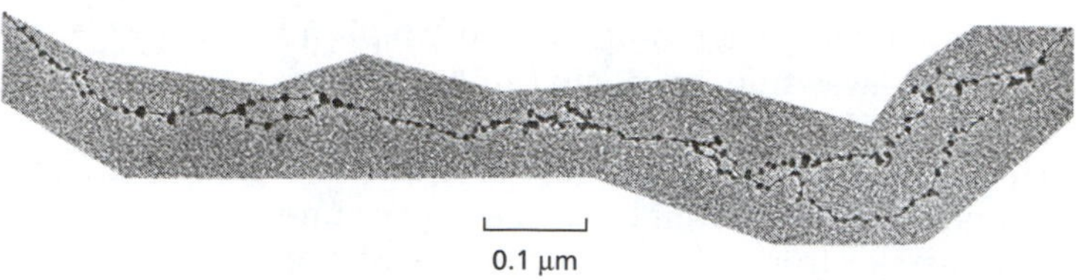


**Figure 5–14** Electron micrograph showing four replication bubbles in a chromosome from the early embryo of *Drosophila* (Problem 5–56).

B. Estimate how long it will take until forks 4 and 5 collide with each other. How long will it take until forks 7 and 8 collide? The distance between nucleotides in DNA is 0.34 nm, and eucaryotic replication forks move at about 50 nucleotides/second. For this problem disregard the nucleosomes evident in Figure 5–14 and assume that the DNA is fully extended.

**5–57** Assuming that there were no time constraints on replication of the genome of a human cell, what would be the minimum number of origins that would be required? If replication had to be accomplished in an 8-hour S phase and replication forks moved at 50 nucleotides/second, what would be the minimum number of origins required to replicate the human genome? (Recall that the human genome comprises a total of $6.4 \times 10^9$ nucleotides on 46 chromosomes.)

## DATA HANDLING

**5–58** You are investigating DNA synthesis in a line of tissue-culture cells using a classic protocol. In this procedure $^3$H-thymidine is added to the cells, which incorporate it at replication forks. Then the cells are gently lysed in a dialysis bag to release the DNA. When the bag is punctured and the solution slowly drained, some of the DNA strands adhere to the walls and are stretched in the general direction of drainage. This method allows very long DNA strands to be isolated intact and examined; however, the stretching collapses replication bubbles so that daughter duplexes lie side by side and cannot be distinguished. The support with its adhered DNA is fixed to a glass slide, overlaid with a photographic emulsion, and exposed for 3 to 6 months. The labeled DNA shows up as tracks of silver grains.

You pretreat the cells to synchronize them at the beginning of S phase. In one experiment you release the synchronizing block and add $^3$H-thymidine immediately. After 30 minutes you wash the cells and change the medium so that the concentration of thymidine is the same as it was, but this time only a third of it is labeled. After an additional 15 minutes you prepare DNA for autoradiography. The results of this experiment are shown in Figure 5–15A. In the second experiment you release the synchronizing block and then wait 30 minutes before adding $^3$H-thymidine. After 30 minutes in the presence of $^3$H-thymidine, you once again change the medium to reduce the concentration of labeled thymidine and incubate the cells for an additional 15 minutes. The results of the second experiment are shown in Figure 5–15B.

A. Explain why in both experiments some regions of the tracks are dense with silver grains (dark), whereas others are less dense (light).

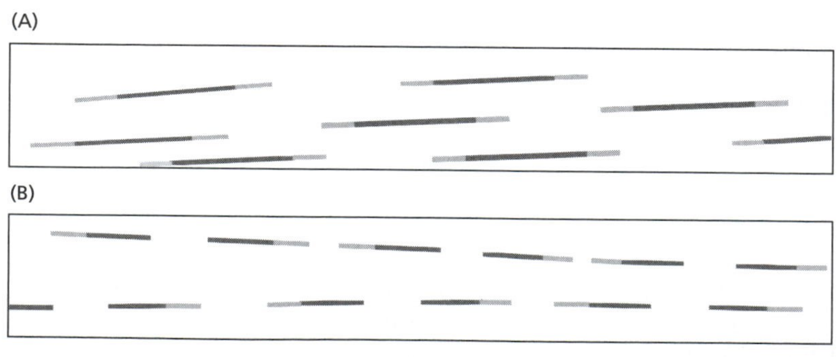


**Figure 5–15** Autoradiographic investigation of DNA replication in cultured cells (Problem 5–58). (A) Addition of labeled thymidine immediately after release from the synchronizing block. (B) Addition of labeled thymidine 30 minutes after release from the synchronizing block.

B. In the first experiment each track has a central dark section with light sections at each end. In the second experiment the dark section of each track has a light section at only one end. Explain the reason for this difference between the two experiments.

C. Estimate the rate of fork movement (μm/min) in these experiments. Do the estimates from the two experiments agree? Can you use this information to gauge how long it would take to replicate the entire genome?

**5–59** One important rule for eucaryotic DNA replication is that no chromosome or part of a chromosome should be replicated more than once per cell cycle. Eucaryotic viruses must evade or break this rule if they are to produce multiple copies of themselves during a single cell cycle. The animal virus SV40, for example, generates 100,000 copies of its genome during a single cycle of infection. In order to accomplish this feat, it synthesizes a special protein, termed T-antigen (because it was first detected immunologically). T-antigen binds to the SV40 origin of replication and in some way triggers initiation of DNA replication.

The mechanism by which T-antigen initiates replication has been investigated *in vitro*. When purified T-antigen, ATP, and SSB protein were incubated with a circular plasmid DNA carrying the SV40 origin of replication, partially unwound structures, like the one shown in Figure 5–16, were observed in the electron microscope. In the absence of any one of these components, no unwound structures were seen. Furthermore, no unwinding occurred in an otherwise identical plasmid that carried a six-nucleotide deletion at the origin of replication, when it was incubated with purified T-antigen, ATP, and SSB protein. If care was taken in the isolation of the plasmid so that it contained no nicks (that is, it was a covalently closed circular DNA), no unwound structures were observed unless topoisomerase I was also present in the mixture.

A. What activity in addition to site-specific DNA binding must T-antigen possess? How might this activity lead to initiation of DNA synthesis?

B. These experiments suggest that the structures observed in the electron microscope are unwound at the SV40 origin of replication. The location of the origin on the SV40 genome is precisely known. How might you use restriction nucleases in addition to the electron microscope to prove this point and to determine whether unwinding occurs in one or both directions away from the origin?

C. Why is there a requirement for topoisomerase I when the plasmid DNA is a covalently closed circle, but not when the plasmid DNA is linear? (Topoisomerase I introduces single-strand breaks into duplex DNA and then rapidly recloses them, so that the breaks have only a transient existence.)

D. Draw an example of the kind of structure that would result if T-antigen repeatedly initiated replication at an SV40 origin that was integrated into a chromosome.

**5–60** Fertilized frog eggs are especially useful for studying the cell-cycle regulation of DNA synthesis. Foreign DNA can be injected into the eggs and followed independently of chromosomal DNA replication. For example, in one study $^{3}H$-labeled viral DNA was injected. The eggs were then incubated in a medium supplemented with $^{32}P$-dCTP and nonradioactive bromodeoxyuridine triphosphate (BrdUTP), which is a thymidine analog that increases the density of DNA into which it is incorporated. Incubation was continued for

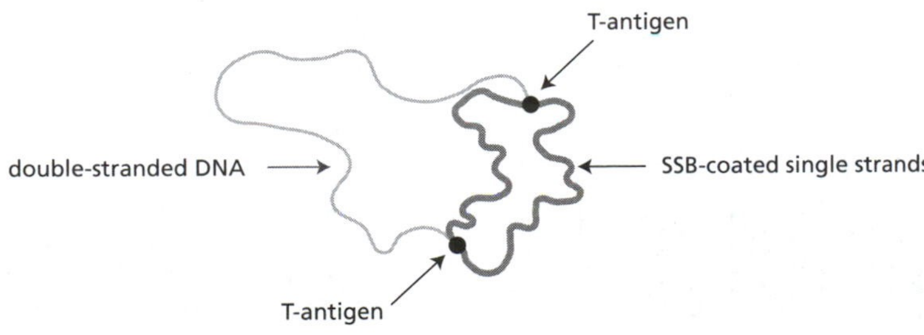


**Figure 5–16** A typical example of a plasmid molecule carrying an SV40 origin of replication after incubation with T-antigen, SSB protein, and ATP (Problem 5–59).

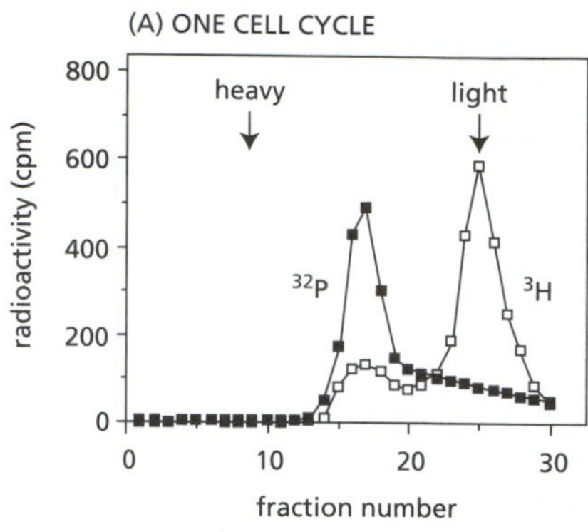


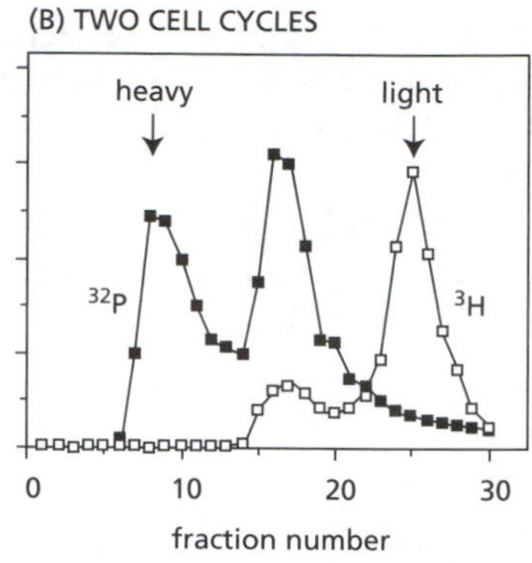


**Figure 5–17** Density distribution of viral DNA after injection into fertilized frog eggs (Problem 5–60). (A) Results after one cell cycle. (B) Results after two cell cycles. The more dense end of the gradient is shown to the *left* (heavy), and the less dense end of the gradient is shown to the *right* (light).

long enough to allow either one or two cell cycles to occur; then the viral DNA was extracted from the eggs and analyzed on CsCl density gradients, which can separate DNA with 0, 1, or 2 BrdU-containing strands. Figure 5–17A and B show the density distribution of viral DNA after incubation for one and two cell cycles, respectively. If the eggs are bathed in cycloheximide (an inhibitor of protein synthesis) during the incubation, the results after incubation for one cycle or for two cycles are like those in Figure 5–17A. (The ability to look specifically at the viral DNA depends on a technical trick: the eggs were heavily irradiated with UV light before the injection in order to block chromosome replication.)

A. Explain how the three density peaks in Figure 5–17 are related to replication of the injected DNA. Why was no $^{32}$P radioactivity associated with the light peak, and why was no $^{3}$H radioactivity associated with the heavy peak?
B. Does the injected DNA mimic the behavior that you would expect for the chromosomal DNA?
C. Why do you think that cycloheximide prevents the appearance of the most dense peak of DNA?

**5–61** Autonomous replication sequences (ARSs), which confer stability on plasmids in yeast, are thought to function as origins of replication. Proving that an ARS is an origin of replication is difficult, mainly because it is very hard to obtain enough well-defined replicating DNA to analyze. This problem can be addressed using a two-dimensional gel-electrophoretic analysis that separates DNA molecules by mass in the first dimension and by shape in the second dimension. Because they have branches, replicating molecules migrate more slowly in the second dimension than do linear molecules of equal mass. By cutting replicating molecules with restriction nucleases, it is possible to generate a continuum of different branched forms that together give characteristic patterns on two-dimensional gels (Figure 5–18).

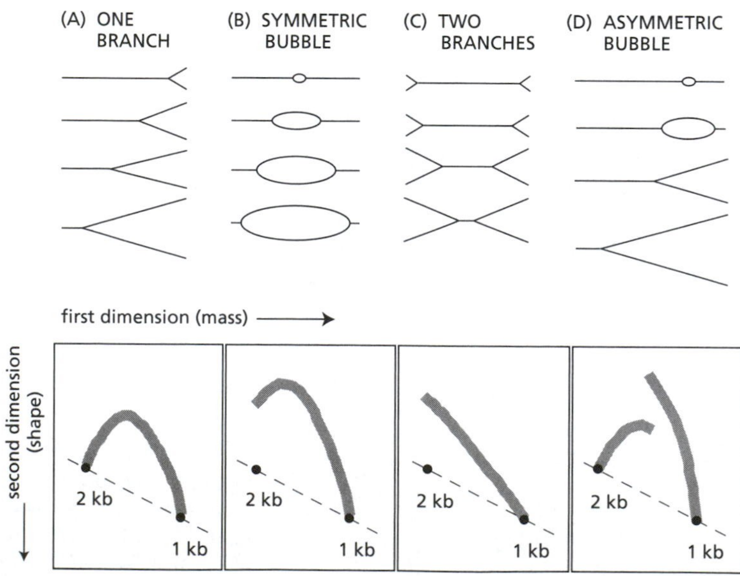


**Figure 5–18** Expected patterns on two-dimensional gels for various DNA molecules (Problem 5–61). (A) Molecules with a single branch. (B) Molecules with a symmetrically located replication bubble. (C) Molecules with two branches. (D) Molecules with an asymmetrically located replication bubble. In the upper half of the figure, 1-kb molecules are shown at progressive stages of replication to 2-kb molecules. In the lower half are shown the corresponding gel patterns that would result from a continuum of such intermediates.

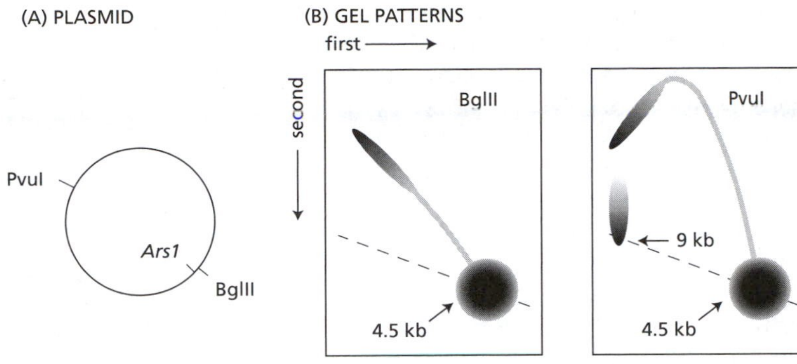


**Figure 5–19** Analysis of replication of an *Ars1*-containing plasmid (Problem 5–61). (A) Structure of the *Ars1* plasmid. (B) The two-dimensional gel patterns resulting from cleavage of replicating plasmids with BglII or PvuI.

You apply this technique to the replication of a plasmid that contains *Ars1*. To maximize the fraction of plasmid molecules that are replicating, you synchronize a yeast culture and isolate DNA from cells in S phase. You then digest the DNA with BglII or PvuI, which cut the plasmid as indicated in Figure 5–19A. You separate the DNA fragments by two-dimensional electrophoresis and visualize the plasmid sequences by autoradiography after blot hybridization to radioactive plasmid DNA (Figure 5–19B).

A. What is the source of the intense spot of hybridization at the 4.5-kb position in both gels in Figure 5–19B?

B. Do the results of this experiment indicate that *Ars1* is an origin of replication? Explain your answer.

C. There is a gap in the arc of hybridization in the PvuI gel pattern in Figure 5–19B. What is the basis for this discontinuity?

**5–62** The shell of a *Drosophila* egg is made from more than 15 different chorion proteins, which are synthesized at a late stage in egg development by follicle cells surrounding the egg. The various chorion genes are grouped in two clusters, one on chromosome 3 and the other on the X chromosome. In each cluster the genes are closely spaced with only a few hundred nucleotides separating adjacent genes. During egg development the number of copies of the chorion genes increases by overreplication of a segment of the surrounding chromosome. The level of amplification around the chorion cluster on chromosome 3 is maximal in the region of the chorion genes, but extends for nearly 50 kb on each side (Figure 5–20).

The DNA sequence responsible for amplification of the cluster on chromosome 3 has been narrowed to a 510-nucleotide segment immediately upstream of one of the chorion genes. When this segment is moved to different places in the genome, those new sites are also amplified in follicle cells. No RNA or protein product seems to be synthesized from this amplification-control element.

A. Sketch what you think the DNA from an amplified cluster would look like in the electron microscope.

B. How many rounds of replication would be required to achieve a 60-fold amplification?

C. How do you think the 510-nucleotide amplification-control element promotes the overreplication of the chorion gene cluster?

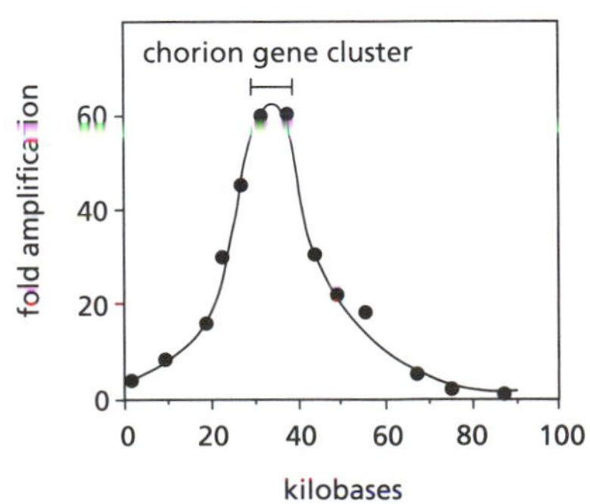

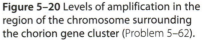

**Figure 5–20** Levels of amplification in the region of the chromosome surrounding the chorion gene cluster (Problem 5–62).

**5–63** In yeast, origin selection is initiated by the origin recognition complex (ORC). ORC is a six-protein DNA-binding complex that recognizes DNA sequences within the yeast origin of replication. One of the components of ORC, Orc1, contains a protein motif (the Walker motif) that is commonly associated with binding and hydrolysis of ATP.

To determine whether binding of ATP, its hydrolysis, or both are required for the recognition of origin DNA by ORC, you carry out the following set of experiments. You first mutate the *Orc1* gene to change an amino acid in the Walker motif of Orc1. You then isolate two versions of ORC: the wild-type form with normal Orc1 and the mutant form with a defective Walker motif. Finally, you measure the binding of these two forms of ORC to origin DNA in the presence of different concentrations of ATP. Binding of the two forms of ORC, as revealed by a DNase I protection assay (DNase footprinting), is shown in Figure 5–21. Exactly the same footprints were obtained when the nonhydrolyzable analog, ATPγS, was used in place of ATP.

A. Indicate the location of ORC binding on the origin DNA in Figure 5–21.
B. Is ATP required for ORC to bind to origin DNA? How can you tell?
C. Is ATP hydrolysis required for ORC binding to origin DNA? How can you tell?
D. Is the Walker motif important to the function of ORC? Explain your answer.

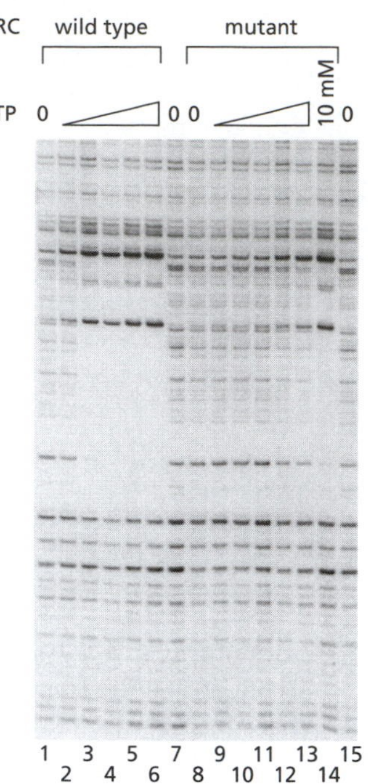


**Figure 5–21** DNase footprinting assay to detect ORC binding to origin DNA (Problem 5–63). *Wedges* indicate increasing concentration of ATP in factors of ten, from 10 nM to 100 μM. In the footprinting assay one strand of the origin-containing DNA was labeled at one end. After ORC was allowed to bind, the complex was treated briefly with DNase I, which breaks the DNA at characteristic places, except where it is protected by ORC. Lanes 7 and 15 omit both ATP and ORC.

**5–64** Origins of replication in mammalian cells have been difficult to define because the sequences required for proper origin function appear to extend over very large stretches of DNA. It may be that replication begins, not at fixed positions, but anywhere within an 'initiation zone.' You wish to characterize origin function within the initiation zone known to exist near the *Dhfr* gene in hamster cells.

You synchronize a population of cells at the $G_1$/S boundary and then release them from the block in the presence of bromodeoxyuridine (BrdU). This base analog is incorporated into the newly synthesized DNA, tagging it for isolation. Your goal is to isolate nascent leading strands. You harvest the DNA 15 minutes after release of the block in order to enrich for nascent strands that are near their origins (Figure 5–22A). To avoid Okazaki fragments, which are about 100 nucleotides in length, you select BrdU-tagged single strands that are about 800 nucleotides long. You then quantify the amount of nascent strands at many sites in a region of 120 kb that includes the initiation zone. You find that nascent strands are concentrated in a 10-kb region, as shown in Figure 5–22B.

A. What is your interpretation of the results shown in Figure 5–22B? Why are there peaks, and what does the double peak indicate?
B. Do you agree with the statement that replication begins anywhere within an initiation zone? Why or why not?

**5–65** You have developed an assay for assembly of nucleosomes onto DNA in order to define the role of chromosome assembly factor 1 (CAF1). You replicate SV40 DNA in a cell-free system in the presence of a labeled nucleotide to tag the replicated molecules, so you can follow them specifically in the assembly assay. After separating the DNA from soluble components in the cell-free replication system, you incubate it with purified CAF1 and a source of histones. You then assay for nucleosome assembly by its effects on the supercoiling status of the circular SV40 genome. Genomes without nucleosomes remain relaxed, whereas genomes with nucleosomes become supercoiled. You separate different supercoiled forms of the DNA by electrophoresis on an agarose gel, which was stained to reveal all forms of DNA and subjected to autoradiography to identify replicated DNA (Figure 5–23).

A. Is most of the SV40 DNA replicated or unreplicated? How can you tell?
B. Does CAF1 assemble nucleosomes on replicated DNA, unreplicated DNA, or both? Explain your answer.

**Problem 8–46** uses monoclonal antibodies to analyze members of the *Xenopus* ORC complex.

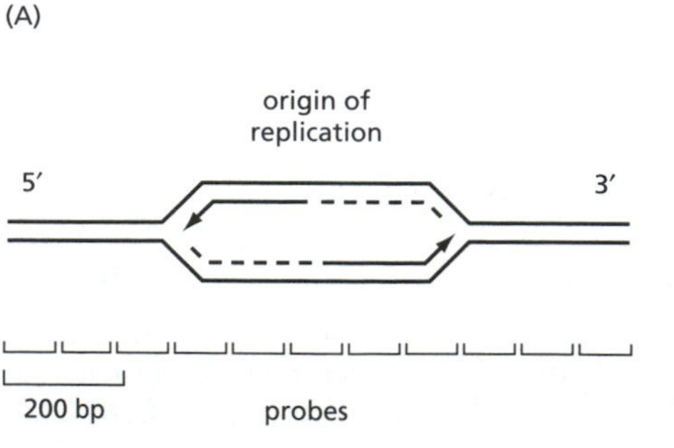


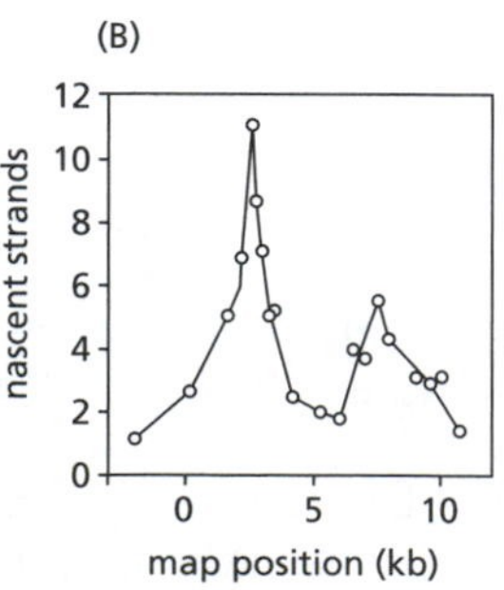


**Figure 5–22** Analysis of an origin in the *Dhfr* gene (Problem 5–64). (A) Arrangement of PCR probes around an origin of replication. The PCR probes—each about 100 bp long—are distributed not only around the origin (as shown), but also throughout the 120-kb initiation zone. (B) PCR-based quantification of nascent strands in the initiation zone.

C. How do you suppose that CAF1 recognizes the DNA it assembles into nucleosomes?

**5–66** You have recently purified and partially sequenced a protein from a ciliated protozoan that seems to be the catalytic subunit of telomerase. You then identify the homologous gene in fission yeast, so you can perform genetic studies that are impossible in the protozoan. You make a targeted deletion of one copy of the gene in a diploid strain of the yeast and then induce sporulation to produce haploid organisms. All four spores germinate perfectly, and you are able to grow colonies on nutrient agar plates. Every 3 days, you re-streak colonies onto fresh plates. After four such serial transfers, the descendants of two of the original four spores grow poorly, if at all. You take cells from the 3-, 6-, and 9-day master plates, prepare DNA from them, and cleave the samples at a chromosomal site about 35 nucleotides away from the start of the telomere repeats. You separate the fragments by gel electrophoresis, and hybridize them to a radioactive telomere-specific probe (Figure 5–24).

A. What is the average length of telomeres in fission yeast?
B. Do the data support the idea that you have identified yeast telomerase? If so, which spores lack telomerase?
C. Assuming that the generation time of this yeast is about 6 hours when growing on plates, by how much do the chromosomes shorten in each generation in the absence of telomerase?
D. If you were to examine the yeast cells that stop dividing, do you suppose they would be smaller, larger, or about the same size as normal yeast cells?

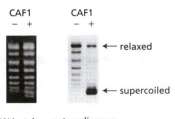

CAF1 – + (DNA stain); CAF1 – + (autoradiogram); ← relaxed; ← supercoiled

**Figure 5–23** CAF1 nucleosome assembly assay (Problem 5–65). The presence or absence of CAF1 in the assembly assay is indicated by + or –, respectively.

# DNA REPAIR

TERMS TO LEARN

| | | |
|---|---|---|
| base excision repair | homologous recombination | nucleotide excision repair |
| DNA repair | nonhomologous end-joining | |

## DEFINITIONS

Match each definition below with its term from the list above.

**5–67** A means for repairing double-strand DNA breaks that links two ends with little regard for sequence homology.

**5–68** Collective term for the enzymatic processes that correct deleterious changes affecting the continuity or sequence of a DNA molecule.

**5–69** Genetic exchange between a pair of identical, or nearly identical, DNA sequences, typically those located on two copies of the same chromosome: either sister chromatids or homologs.

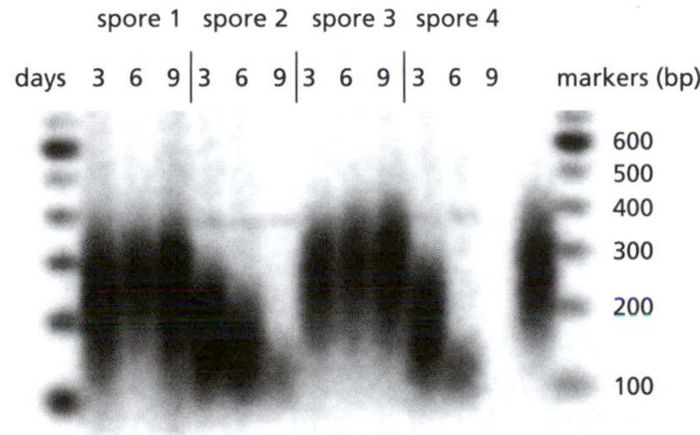


**Figure 5–24** Analysis of telomeres from four fission-yeast spores (Problem 5–66). The results from normal diploid yeast are shown on the *right*, adjacent to the markers.

## TRUE/FALSE

Decide whether each of these statements is true or false, and then explain why.

**5–70** DNA repair mechanisms all depend on the existence of two copies of the genetic information, one in each of the two homologous chromosomes.

**5–71** Spontaneous depurination and the removal of a deaminated C by uracil DNA glycosylase leave identical substrates, which are recognized by AP endonuclease.

**5–72** Only the initial steps in DNA repair are catalyzed by enzymes that are unique to the repair process; the later steps are typically catalyzed by enzymes that play more general roles in DNA metabolism.

## THOUGHT PROBLEMS

**5–73** Discuss the following statement: "The DNA repair enzymes that correct damage introduced by deamination and depurination must preferentially recognize such defects on newly synthesized DNA strands."

**5–74** If you compare the frequency of the sixteen possible dinucleotide sequences in *E. coli* and human cells, there are no striking differences except for one dinucleotide, 5′-CG-3′. The frequency of CG dinucleotides in the human genome is significantly lower than in *E. coli* and significantly lower than expected by chance. Why do you suppose that CG dinucleotides are under-represented in the human genome?

**5–75** Haploid yeast cells that preferentially repair double-strand breaks by homologous recombination are especially sensitive to agents that cause double-strand breaks in DNA. If the breaks occur in the $G_1$ phase of the cell cycle, most yeast cells die; however, if the breaks occur in the $G_2$ phase, a much higher fraction of cells survive. Explain these results.

**5–76** Bacteria have a potent repair system for dealing with pyrimidine dimers. This phenomenon was discovered by careful observation to define an uncontrolled variable in an investigation into the effects of UV light on bacteria—not unlike the scenario described below.

You and your advisor are trying to isolate mutants in *E. coli* using UV light as the mutagenic agent. To get plenty of mutants, you find it necessary to use a dose of irradiation that kills 99.99% of the bacteria. You have been getting much more consistent results than your advisor, who also requires 10-fold to a 100-fold higher doses of irradiation to achieve the same degree of killing. He wonders about the validity of your results since you always do your experiments at night after he has left. When he insists that you come in the morning to do the experiments in parallel, you are surprised when you get exactly the same results as he does. You are a bit chagrined because you had confidently assumed you were better at doing lab work than your advisor, who is rarely seen in a lab coat these days. When you both repeat the experiments in parallel at night, your advisor is surprised at the results, which are exactly as you had originally described them.

Now that you believe one another's observations, you make rapid progress. You find that you need a higher dose of UV light in the afternoon than in the morning to get the same degree of killing. Even higher doses are required on sunny days than on overcast days. Your laboratory faces west. What is the variable that has been plaguing your experiments?

## CALCULATIONS

**5–77** Ku70 and Ku80, two key proteins used in nonhomologous end-joining (NHEJ), form heterodimers that bind to broken DNA ends, helping to align them for joining. How difficult is it for a Ku dimer to find a double-strand break? One way to approach this question is to estimate the average distance between Ku dimers in the nucleus: a break in the DNA will be within half that distance from a Ku dimer. If there are $4 \times 10^5$ Ku dimers in a typical nucleus, what is their average separation? Assume that the Ku dimers are randomly distributed, that a Ku dimer can be approximated as a cube, 8 nm on a side, and that the nucleus is 6 μm in diameter. [Volume of a sphere is $(4/3)\pi r^3$; volume of a cube is $l^3$.]

**5–78** With age, somatic cells are thought to accumulate genomic 'scars' as a result of the inaccurate repair of double-strand breaks by nonhomologous end-joining (NHEJ). Estimates based on the frequency of breaks in primary human fibroblasts suggest that by age 70 each human somatic cell may carry some 2000 NHEJ-induced mutations due to inaccurate repair. If these mutations were distributed randomly around the genome, how many genes would you expect to be affected? Would you expect cell function to be compromised? Why or why not? (Assume that 2% of the genome—1.5% coding and 0.5% regulatory—is crucial information.)

## DATA HANDLING

**5–79** Several genes in *E. coli*, including *UvrA, UvrB, UvrC*, and *RecA*, are involved in repair of UV damage. Strains of *E. coli* that are defective in any one of these genes are much more sensitive to killing by UV light than are nonmutant (wild-type) cells, as shown for *UvrA* and *RecA* strains in Figure 5–25A. Cells mutant for pairs of these genes display a wide range of sensitivity to UV light. The combinations of *Uvr* mutations with one another show little increase in sensitivity relative to the *Uvr* single mutants. By contrast, the combination of a *RecA* mutation with any of the *Uvr* mutations gives a strain that is exquisitely sensitive to UV light, as shown for *UvrARecA* on an expanded scale in Figure 5–25B.

A. Why do combinations of a *RecA* mutation with a *Uvr* mutation give an extremely UV-sensitive strain of bacteria, whereas combinations of mutations in different *Uvr* genes are no more UV sensitive than the individual mutants?

B. According to the Poisson distribution, when a population of bacteria receives an overall average of one lethal 'hit,' 37% (which is $e^{-1}$) will survive because they receive no hits. For the double mutant *UvrARecA*, a dose of 0.04 joules/m$^2$ gives 37% survival (Figure 5–25B). Calculate the number of pyrimidine dimers that constitutes a lethal hit for the *UvrARecA* strain. *E. coli* has

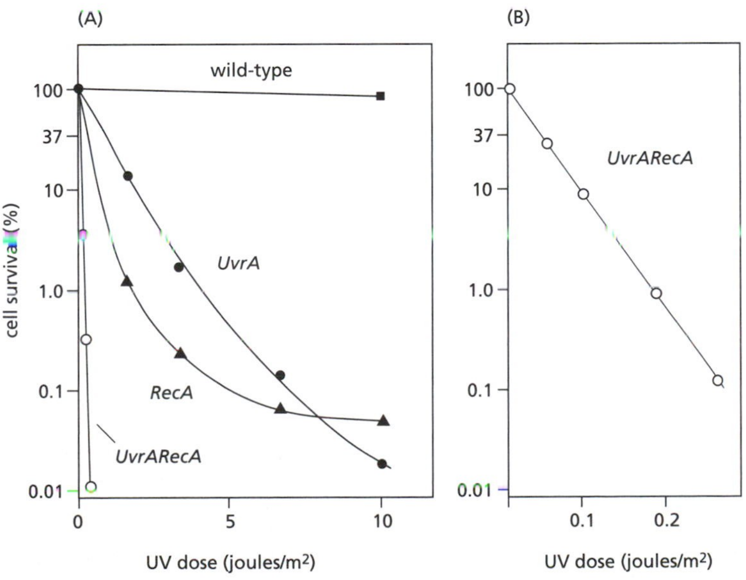


**Figure 5–25** Cell survival as a function of UV dose (Problem 5–79). (A) Survival of wild-type cells, a *UvrA* mutant, a *RecA* mutant, and a *UvrARecA* double mutant. (B) An expanded scale for *UvrARecA* survival.

**Table 5–3 Frequency of UV-induced mutations in various strains of *E. coli*** (Problem 5–80).

| STRAIN | SURVIVAL (%) | MUTATIONS/$10^{10}$ SURVIVORS |
|---|---|---|
| Wild type | 100 | 400 |
| *RecA* | 10 | 1 |
| *UvrA* | 10 | 40,000 |

$4.6 \times 10^6$ base pairs in its genome (assume 50% GC). Exposure of DNA to UV light at 400 joules/$m^2$ converts 1% of the total pyrimidine pairs (TT, TC, CT, plus CC) to pyrimidine dimers.

**5–80** In addition to killing bacteria, UV light causes mutations. You have measured the UV-induced mutation frequency in wild-type *E. coli* and in strains defective in either the *UvrA* gene or the *RecA* gene. The results are shown in Table 5–3. Surprisingly, these strains differ dramatically in their mutability by UV.

A. Assuming that the *RecA* and *UvrA* gene products participate in different pathways for repair of UV damage, decide which pathway is more error prone. Which pathway predominates in wild-type cells?

B. The error-prone pathway uses a specialized DNA polymerase to incorporate nucleotides opposite a site of unrepaired damage. This polymerase tends to incorporate an adenine nucleotide opposite a pyrimidine dimer. Is this a good strategy for dealing with UV damage? Calculate the frequency of base changes (mutations) using A insertion versus random incorporation (each nucleotide with equal probability) for *E. coli*, in which pyrimidine dimers are approximately 60% TT, 30% TC and CT, and 10% CC.

**5–81** The patterns of UV-induced mutations occurring in the *E. coli LacI* gene have been extensively analyzed. Figure 5–26 shows the number of independently isolated missense mutations (amino acid substitutions) and frameshift mutations (changes in the reading frame). There are almost equal numbers of mutations in each category. Missense mutations were identified by scoring for loss of function of the Lac repressor, the protein specified by the *LacI* gene; frameshift mutations were scored by a gene-fusion assay, which is independent of the function of the Lac repressor because it measures the activity of a protein fused to the C-terminus of the Lac repressor.

A. Why do you suppose there are so many more missense mutations near the ends of the gene than in the middle? Why do you suppose the frameshift mutations are more or less evenly distributed across the gene (except for one or two 'hot spots')?

B. Analysis of the hot spot labeled I in Figure 5–26 revealed that its wild-type sequence was TTTTTC and that the mutated sequence was TTTTC. The next most common mutation (labeled II in Figure 5–26) was the change from GTTTTC to GTTTC. Analysis of other frameshifts indicated that they

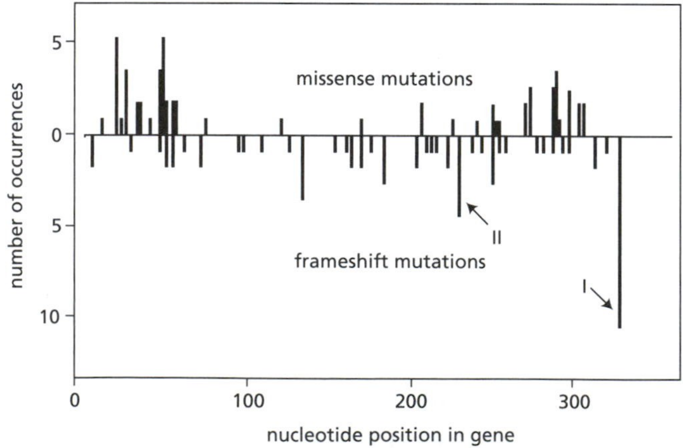


**Figure 5–26** Pattern of UV-induced mutations in the *E. coli LacI* gene (Problem 5–81). Missense mutations are shown *above the line*; frameshift mutations are shown *below the line*.

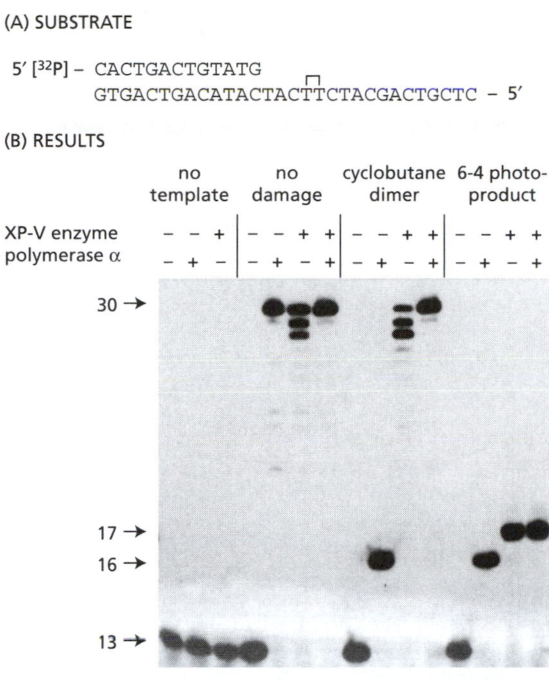


**Figure 5–27** Comparison of DNA polymerase α and the XP-V enzyme (Problem 5–82). (A) Substrate for DNA polymerization. The adjacent Ts indicated by the *bracket* can be selectively modified to produce a cyclobutane dimer or a 6-4 photoproduct. (B) Results obtained with DNA polymerase α and the XP-V enzyme on damaged and undamaged templates. Numbers on the *left* indicate lengths of single-stranded DNA in nucleotides.

resulted most commonly from the loss of one base: no insertions were found. Based on what you know of the nature of UV damage, can you suggest a molecular mechanism for the loss of a single base pair?

**5–82** Humans with the rare genetic disease xeroderma pigmentosum (XP) are extremely sensitive to sunlight and are prone to developing malignant skin cancers. Defects in any one of eight genes can cause XP. Seven XP genes encode proteins involved in nucleotide excision repair (NER). The eighth gene is associated with the XP variant (XP-V) form of the disease. Cells from XP-V patients are proficient for NER but do not accurately replicate UV-damaged DNA. Using a clever assay, you manage to purify from a normal cell extract the enzyme that is missing from XP-V cells. You test its ability to synthesize DNA from the simple template in Figure 5–27A, which contains a TT sequence. This template can be modified to contain either a cyclobutane thymine dimer or the somewhat rarer 6-4 photoproduct (two Ts linked in a different way). You compare the ability of DNA polymerase α (a normal replicative polymerase) and the XP-V enzyme to synthesize DNA from the undamaged template, the template with a cyclobutane dimer, and the template with a 6-4 photoproduct. By labeling the primer (the shorter DNA strand) at its 5′ end and denaturing the reaction products, it is possible to determine whether DNA synthesis has occurred (Figure 5–27B).

A. Is the XP-V enzyme a DNA polymerase? Why or why not?
B. How does the XP-V enzyme differ from DNA polymerase α on an undamaged template? On a template with a cyclobutane dimer? On a template with a 6-4 photoproduct?
C. How accurately do you suppose that the XP-V enzyme copies normal DNA? Would you guess it to be error-prone or faithful?
D. If NER is normal in patients with XP-V, why are they sensitive to sunlight and prone to skin cancers?

**5–83** Mutagens such as *N*-methyl-*N′*-nitro-*N*-nitrosoguanidine (MNNG) and methyl nitrosourea (MNU) are potent DNA-methylating agents and are extremely toxic to cells. Nitrosoguanidines are used in research as mutagens and clinically as drugs in cancer chemotherapy because they preferentially kill cells in the act of replication.

The original experiment that led to the discovery of the alkylation repair system in bacteria was designed to assess the long-term effects of exposure

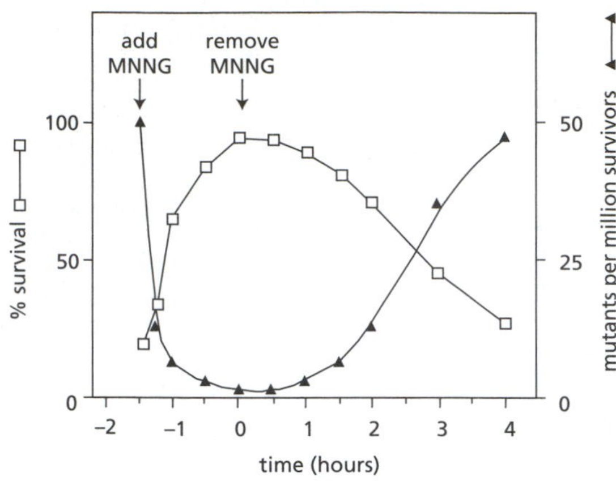


Figure 5–28 Adaptive response of *E. coli* to low doses of MNNG (Problem 5–83). MNNG at 1 μg/mL was present from –1.5 to 0 hours, as indicated by *arrows*. Samples were removed at various times and treated briefly with 100 μg/mL MNNG to assess the number of survivors and the frequency of mutants.

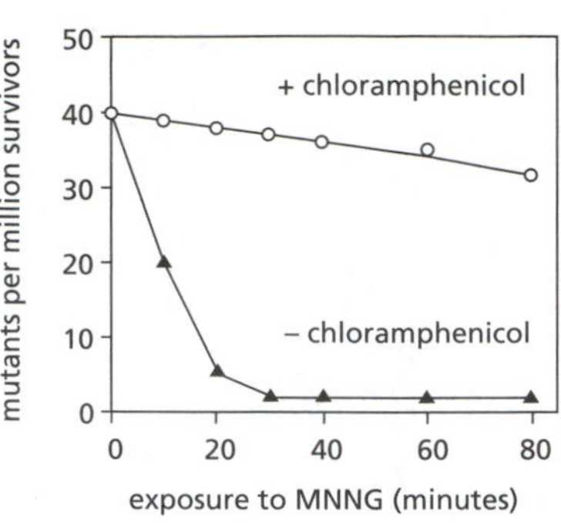


Figure 5–29 Effects of chloramphenicol on the adaptive response to low doses of MNNG (Problem 5–83). After different times of exposure to 1 μg/mL MNNG, samples were removed and treated with 100 μg/mL MNNG to measure susceptibility to mutagenesis.

to low doses of MNNG (as in chemotherapy), in contrast to brief exposure to large doses (as in mutagenesis). Bacteria were placed first in a low concentration of MNNG (1 μg/mL) for 1.5 hours and then in fresh medium lacking MNNG. At various times during and after exposure to the low dose of MNNG, samples of the culture were treated with a high concentration of MNNG (100 μg/mL) for 5 minutes and then tested for viability and the frequency of mutants. As shown in Figure 5–28, exposure to low doses of MNNG temporarily increased the number of survivors and decreased the frequency of mutants among the survivors. As shown in Figure 5–29, the adaptive response to low doses of MNNG was prevented if chloramphenicol (an inhibitor of protein synthesis) was included in the incubation.

A. Does the adaptive response of *E. coli* to low levels of MNNG require activation of preexisting protein or the synthesis of new protein?

B. Why do you think the adaptive response to a low dose of MNNG is so short-lived?

**5–84** The nature of the mutagenic lesion introduced by MNNG and the mechanism of its removal from DNA were identified in the following experiments. To determine the nature of the mutagenic lesion, untreated bacteria and bacteria that had been exposed to low doses of MNNG were incubated with 50 μg/mL $^3$H-MNNG for 10 minutes. Their DNA was isolated and hydrolyzed to nucleotides, and the radioactive purines were then analyzed by paper chromatography as shown in Figure 5–30.

The mechanism of removal of the mutagenic lesion was investigated by first purifying the enzyme responsible. The kinetics of removal were studied by incubating different amounts of the enzyme (molecular weight 19,000) with DNA containing 0.26 pmol of the mutagenic base, which was radioactively labeled with $^3$H. At various times samples were taken, and the DNA was analyzed to determine how much of the mutagenic base

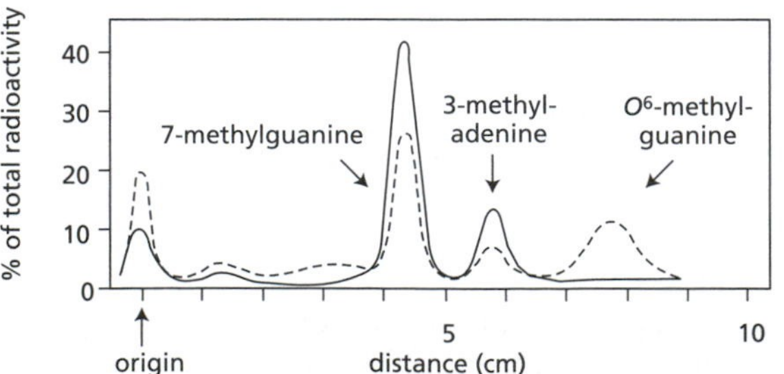


Figure 5–30 Chromatographic separation of labeled methylated purines in the DNA of untreated bacteria and bacteria treated with low doses of MNNG (Problem 5–84). *Solid line* indicates methylated purines from untreated bacteria; *dashed line* indicates those from MNNG-treated bacteria.

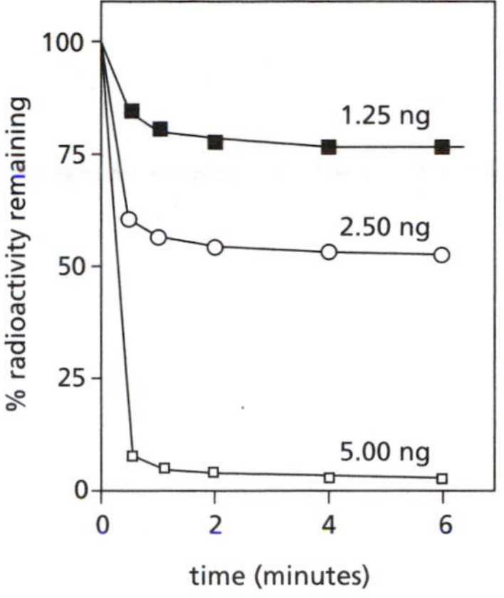


**Figure 5–31** Removal of $^3$H-labeled methyl groups from DNA by the purified enzyme (Problem 5–84). The quantities of purified enzyme are indicated.

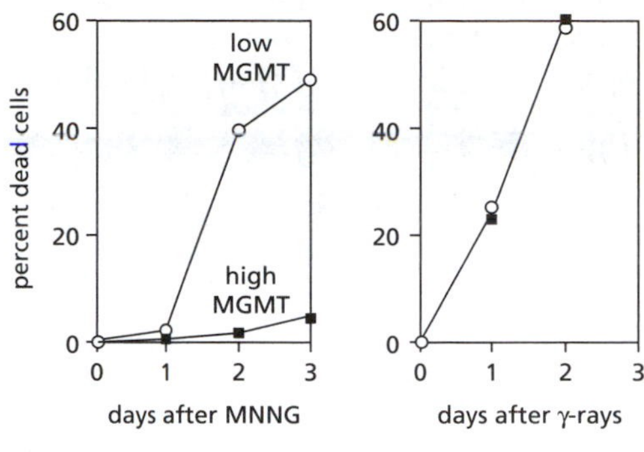


**Figure 5–32** Killing by MNNG and γ-rays of cells that express high or low levels of MGMT (Problem 5–85). Results for cells expressing low levels of MGMT are shown in *open circles*; results for cells expressing high levels of MGMT are shown in *filled squares*.

remained (Figure 5–31). When the experiment was repeated at 5°C instead of 37°C, the initial rates of removal were slower, but exactly the same end points were achieved.

A. Which methylated purine is responsible for the mutagenic action of MNNG?

B. What is peculiar about the kinetics of removal of the methyl group from the mutagenic base? Is this peculiarity due to an unstable enzyme?

C. Calculate the number of methyl groups that are removed by each enzyme molecule. Does this calculation help to explain the peculiar kinetics?

**5–85** Alkylating agents, which are highly cytotoxic, are commonly used as anticancer drugs. The mechanism by which they kill cells is not clear. Do they kill cells by damaging DNA, or by reacting with protein or lipid components of the cell? You know that $O^6$-methylguanine is highly mutagenic and you wonder whether it might also be responsible for cell killing. To test this possibility you construct a cell line that expresses high levels of $O^6$-methylguanine methyltransferase (MGMT), which efficiently removes such groups from the DNA. You then measure the sensitivity of these cells, and of control cells that express very little MGMT, to killing by the alkylating agent MNNG. As a control, you compare sensitivity to γ-irradiation, which kills cells by introducing double-strand breaks and thus should not be dependent on cellular levels of MGMT. The results of your experiments are shown in Figure 5–32. Is $O^6$-methylguanine responsible for the cytotoxic effects of alkylating agents? Explain your reasoning.

**5–86** Nonhomologous end-joining (NHEJ) is responsible for linking DNA sequences that are not homologous to one another. Junctions formed by NHEJ are commonly found at sites of DNA rearrangements, including translocations, inversions, duplications, and deletions. Surveys of large numbers of such junctions have revealed the presence of so-called 'microhomology' at the junctions, typically ranging from 0–5 nucleotides. Examples of junctions with 0 and 2 nucleotides of microhomology are shown in Figure 5–33.

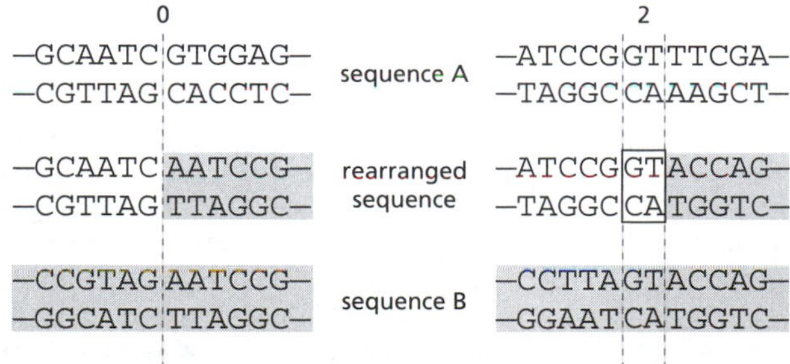


**Figure 5–33** Microhomology at rearrangement junctions (Problem 5–86). In the rearrangement junction with 2 nucleotides of microhomology, the central GT cannot be uniquely assigned to either sequence A or sequence B; it could have come from either.

**Table 5–4 Observed distribution of microhomologies at rearrangement junctions along with the distribution expected by chance** (Problem 5–86).

| | MICROHOMOLOGY | | | | | |
|---|---|---|---|---|---|---|
| Distribution | 0 | 1 | 2 | 3 | 4 | 5 |
| Observed | 47 | 21 | 14 | 13 | 9 | 6 |
| Expected | 62 | 31 | 12 | 4 | 1 | 0 |

The distribution of microhomology expected by chance was calculated on the assumption that the random joining of blunt-ended DNA segments generated the junctions.

Are these short homologies relevant to the mechanism of NHEJ, or are they present by chance? One way to decide the issue is to compare the distribution of observed junctional microhomology with the distribution expected by chance. Microhomologies at 110 NHEJ junctions are shown along with the distribution expected by chance in Table 5–4. One way to determine whether these two distributions are different is to use the statistical method known as chi-square analysis. The readout of a chi-square analysis is a *P* value, which measures the probability that the observed distribution is the same as the expected distribution. Commonly, a *P* value of less than 0.05 is taken to mean that the two distributions are different. (A *P* value of 0.05 means that there is a 5% chance that the observed distribution is the same as the expected distribution.)

Using chi-square analysis, decide whether the observed distribution in Table 5–4 is the same or different from the distribution expected by chance. If you don't have a statistics package available, search for 'chi square calculator' on the Internet.

**5–87** Most eucaryotic cells use two different mechanisms to repair double-strand breaks in DNA: homologous recombination and nonhomologous end-joining. The recombination pathway is favored by a homolog of the Rad52 protein, first defined in yeast. Relatives of the protein known as Ku stimulate the end-joining pathway.

You have purified the human Rad52 protein and are studying its properties. When you mix Rad52 with linear DNA that has short (300- nucleotide) single-strand tails and examine the mixture in the electron microscope, you see structures like the one shown in Figure 5–34A. Such structures are much rarer when the linear DNA is blunt ended. You then expose Rad52-bound, uniformly radiolabeled DNA to nucleases and measure digestion of DNA by release of soluble fragments. As shown in Figure 5–34B, Rad52 protects the DNA from an exonuclease but not from an endonuclease.

A. Where does Rad52 bind on linear DNA? What features of the DNA are important for Rad52 binding?
B. How does Rad52 protect against digestion by the exonuclease, but not against digestion by the endonuclease?
C. How do the properties of Rad52 revealed by these observations fit into its role in recombinational repair of double-strand breaks?

(A) ELECTRON MICROGRAPH

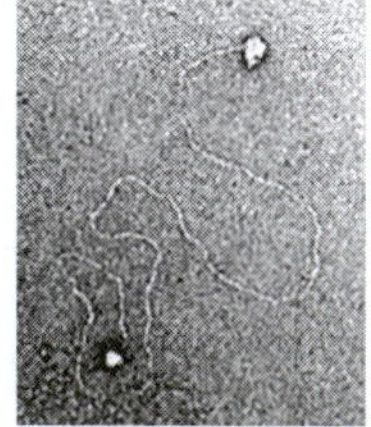

(B) NUCLEASE DIGESTION

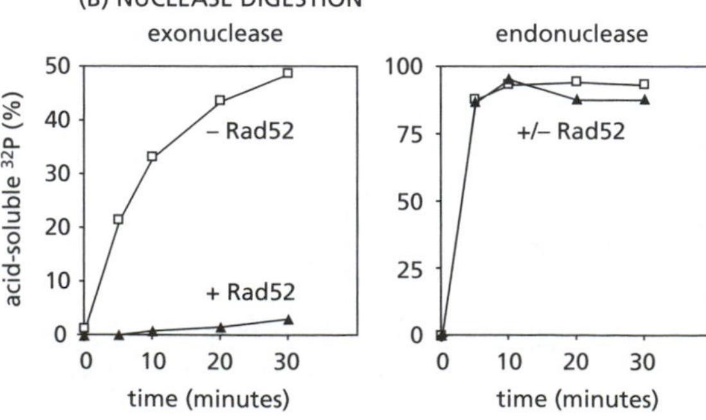


**Figure 5–34** Analysis of the role of Rad52 in repair of double-strand breaks (Problem 5–87). (A) Binding of Rad52 to linear DNA. A schematic representation is shown adjacent to the micrograph. (B) Nuclease digestion of DNA in the presence (+) and absence (–) of bound Rad52.

# HOMOLOGOUS RECOMBINATION

TERMS TO LEARN

| | | |
|---|---|---|
| allele | Holliday junction | Rad51 |
| gene conversion | homologous recombination | RecA protein |
| heteroduplex | hybridization | |

## DEFINITIONS

Match each definition below with its term from the list above.

**5–88** One of a set of alternative forms of a gene. In a diploid cell each gene will have two of these, each occupying the same position (locus) on homologous chromosomes.

**5–89** Process by which DNA sequence information can be transferred from one DNA helix (which remains unchanged) to another DNA helix whose sequence is altered.

**5–90** Experimental process in which two complementary nucleic acid strands form a double helix; a powerful technique for detecting specific nucleotide sequences.

**5–91** X-shaped structure observed in DNA undergoing recombination, in which the two DNA molecules are held together at the site of crossing-over, also called a cross-strand exchange.

## TRUE/FALSE

Decide whether each of these statements is true or false, and then explain why.

**5–92** Homologous recombination requires relatively long regions of homologous DNA on both partners in the exchange.

**5–93** All known mechanisms of gene conversion require a limited amount of DNA synthesis.

## THOUGHT PROBLEMS

**5–94** Using Figure 5–35A as a guide, draw the products of a crossover recombination event between the homologous regions of the molecules represented in Figure 5–35B. In the figure single lines represent the DNA duplex and arrows represent the targets for homologous recombination.

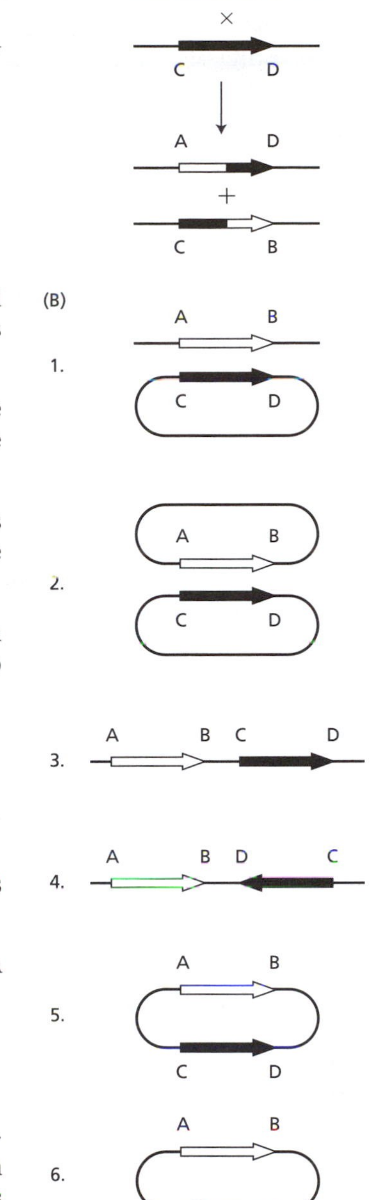

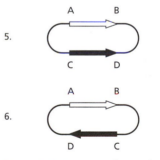

**Figure 5–35** A variety of recombination substrates (Problem 5–94).

**5–95** When plasmid DNA is extracted from *E. coli* and examined by electron microscopy, the majority are monomeric circles, but there are a variety of other forms, including dimeric and trimeric circles. In addition, about 1% of the molecules appear as figure-8 forms, in which the two loops are equal (Figure 5–36A).

You suspect that the figure 8s are recombination intermediates in the formation of a dimer from two monomers (or two monomers from a dimer). To

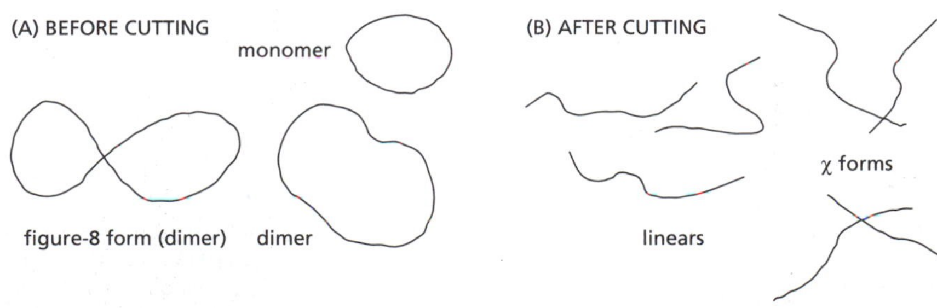


**Figure 5–36** Structures of plasmid molecules (Problem 5–95). (A) Before digestion. (B) After digestion with a single-cut restriction nuclease.

rule out the possibility that they might represent twisted dimers or touching monomers, you digest the DNA sample with a restriction nuclease that cuts at a single site in the monomer, and then examine the molecules. After cutting, only two forms are seen: 99% of the DNA molecules are linear monomers, and 1% are $\chi$ (chi) forms (Figure 5–36B). You note that the $\chi$ forms have an interesting property: the two longer arms are the same length, as are the two shorter arms. In addition, the sum of the lengths of a long arm and a short arm is equal to the length of the monomer plasmid. The position of the crossover point, however, is completely random.

Unsure of yourself and feeling you are probably missing some hidden artifact, you show your pictures to a friend. She points out that your observations prove you are looking at recombination intermediates that arose by random pairing at homologous sites.

A. Is your friend correct? What is her reasoning?
B. How would you expect your observations to differ if you repeated the experiments in a strain of *E. coli* that carried a nonfunctional *RecA* gene?
C. How would the $\chi$ forms have differed from those you observed if the figure 8s had arisen as intermediates in a site-specific recombination between the monomers?
D. What would the $\chi$ forms have looked like if the figure 8s had arisen as intermediates in a totally random, nonhomologous recombination between the monomers?

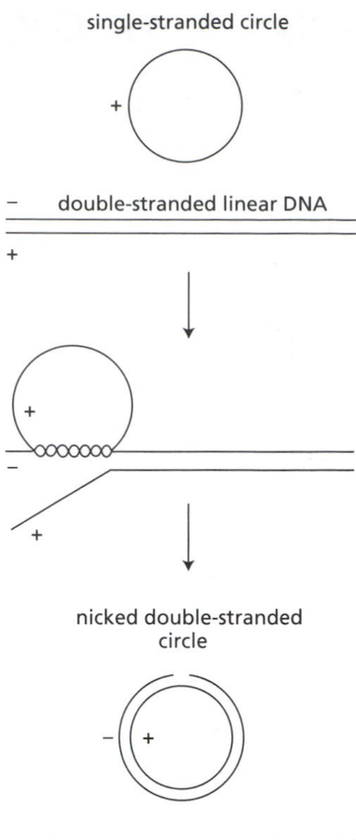


**Figure 5–37** Strand assimilation assay for RecA (Problem 5–96). The single-stranded (+) circle is complementary to the minus (–) strand and identical to the plus strand of the duplex.

**5–96** In *E. coli*, RecA protein catalyzes both the initial pairing step of recombination and subsequent branch migration. It promotes recombination by binding to single-stranded DNA and catalyzing the pairing of such coated single strands to homologous double-stranded DNA. One assay for the action of RecA is the formation of double-stranded DNA circles from a mixture of double-stranded linear molecules and homologous single-stranded circles, as illustrated in Figure 5–37. This reaction proceeds in two steps: circles pair with linears at an end and then they branch migrate until a single-stranded linear DNA is displaced.

One important question about the RecA reaction is whether branch migration is directional. This question has been studied in the following way. Single-stranded circles, which were uniformly labeled with $^{32}P$, were mixed with unlabeled double-stranded linear molecules in the presence of RecA. As the single-stranded DNA pairs with the linear DNA, it becomes sensitive to cutting by restriction nucleases, which do not cut single-stranded DNA. By sampling the reaction at various times, digesting the DNA with a restriction nuclease, and separating the labeled fragments by electrophoresis, you obtain the pattern shown in Figure 5–38A.

A. By comparing the time of appearance of labeled fragments with the restriction map of the circular DNA in Figure 5–38B, deduce which end (5′ or 3′) of

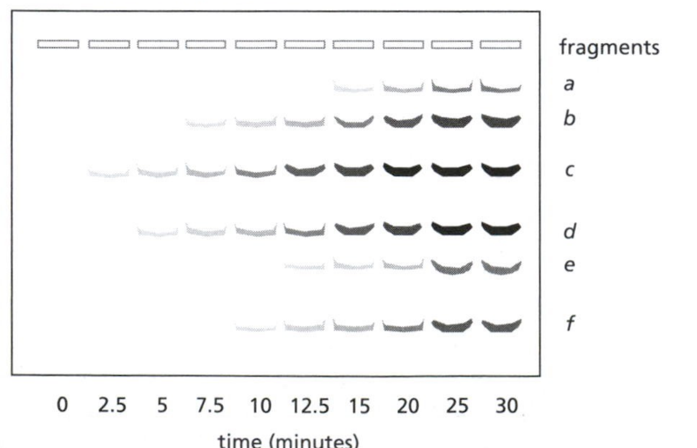


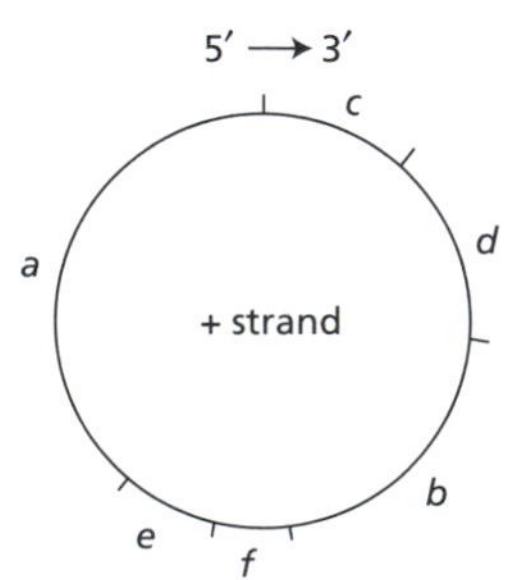


**Figure 5–38** Analysis of branch migration catalyzed by RecA (Problem 5–96). (A) Electrophoretic separation of labeled restriction fragments as a function of time of incubation with RecA. (B) Sites of cleavage represented on the single-stranded DNA circle. Clockwise around the circle is 5′-to-3′.

the minus strand of the linear DNA the circular plus strand pairs with initially. Also deduce the direction of branch migration along the minus strand. (The linear double-stranded DNA was cut at the boundary between fragments *a* and *c* on the restriction map.)

B. Estimate the rate of branch migration, given that the length of this DNA is 7 kb.

C. What would you expect to happen if the linear, double-stranded DNA carried an insertion of 500 nonhomologous nucleotides between restriction fragments *e* and *a*?

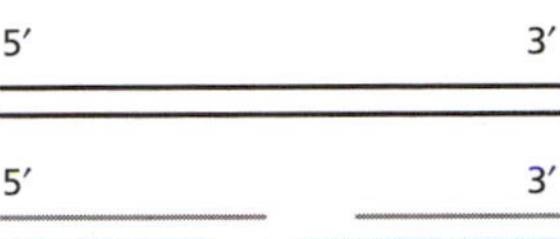

**Figure 5–39** A broken duplex with single-strand tails ready to invade an intact homologous duplex (Problem 5–98).

**5–97** Discuss the following statement: "The Holliday junction contains two distinct pairs of strands (crossing strands and noncrossing strands), which cannot be interconverted without breaking the phosphodiester backbone of at least one strand."

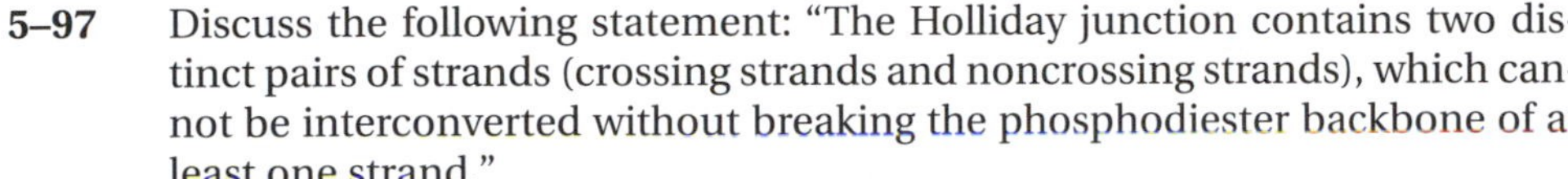

**5–98** Draw the structure of the double Holliday junction that would result from strand invasion by both ends of the broken duplex into the intact homologous duplex shown in Figure 5–39. Label the left end of each strand in the Holliday junction 5′ or 3′ so that the relationship to the parental and recombinant duplexes is clear. Indicate how DNA synthesis would be used to fill in any single-strand gaps in your double Holliday junction.

**5–99** Why is it that recombination between similar, but nonidentical, repeated sequences poses a problem for human cells? How does the mismatch-repair system protect against such recombination events?

## DATA HANDLING

**5–100** Specific DNA sequences known as Chi sites locally stimulate RecBCD-mediated homologous recombination in *E. coli*. Interaction between the RecBCD protein and the Chi site stimulates a rate-limiting step in the recombination pathway. To study this interaction in detail, RecBCD was purified and incubated with a linear, double-stranded DNA fragment containing a Chi site (Figure 5–40).

Different samples of the linear DNA were labeled specifically at the 5′ end on the left (5′L), at the 5′ end on the right (5′R), at the 3′ end on the left (3′L), and at the 3′ end on the right (3′R). Each sample was then incubated in a reaction buffer containing RecBCD. As a control, a separate aliquot of labeled DNA was incubated in the reaction buffer without RecBCD. After 1 hour of reaction the DNA was denatured by boiling, and the resulting single strands were separated by electrophoresis through a polyacrylamide gel. The pattern of radioactively labeled DNA fragments is shown in Figure 5–41. As a control, a sample of 3′R that had been incubated with RecBCD was run on the gel without first denaturing it (Figure 5–41, lane 6).

A. What is the evidence that RecBCD cuts the DNA at the Chi site? Does it cut one or both strands? If you decide it cuts only one strand, specify the strand and indicate your reasoning.

B. What is the evidence that RecBCD can act as a helicase—that is, what is the evidence that it can separate the strands of duplex DNA?

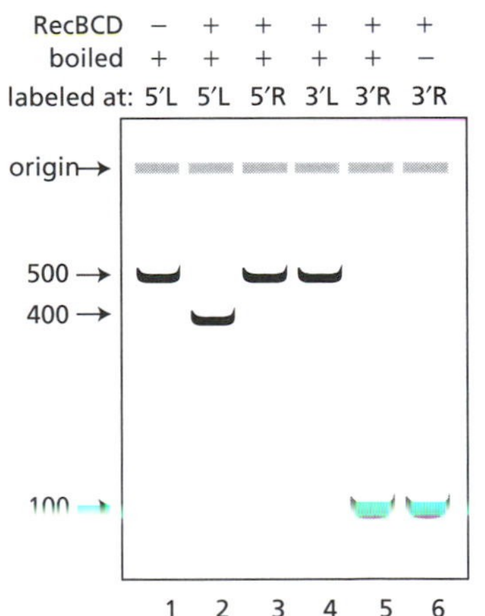

**Figure 5–41** Results of incubating RecBCD with Chi-containing DNA fragments labeled at the ends of defined strands (Problem 5–100). Numbers adjacent to bands indicate the length of the labeled fragment in nucleotides. All samples were denatured before electrophoresis, except the unboiled sample in lane 6.

**5–101** The mechanism of Holliday-junction cleavage by RuvC has been investigated using artificial junctions created by annealing oligonucleotides together (Figure 5–42A). Each of the duplexes involved in these junctions has unique sequences at its ends, which allowed the oligonucleotides to

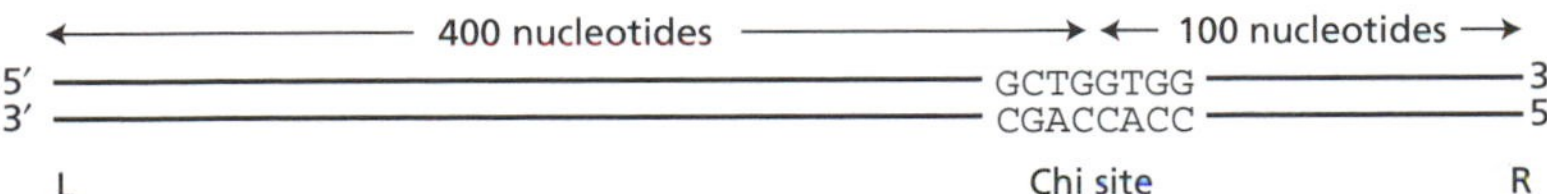

**Figure 5–40** A linear DNA fragment containing a Chi site (Problem 5–100). The sequence of the Chi site is shown. L and R indicate the left and right ends of the fragment.

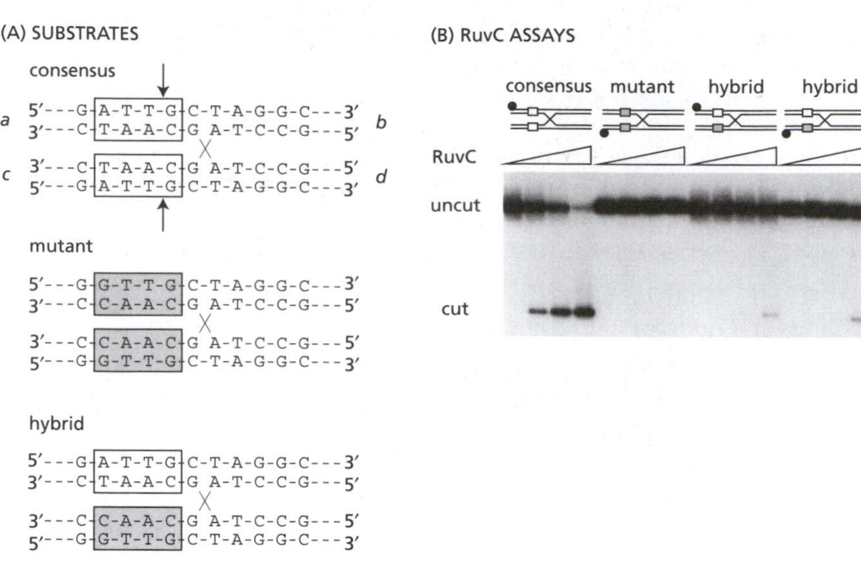


**Figure 5–42** Analysis of Holliday junction cleavage (Problem 5–101). (A) Effects of DNA sequence on resolution of Holliday junctions. Cleavable sequences are shown in *open boxes;* uncleavable sequences are shown in *shaded boxes.* (B) Resolution of Holliday junctions by RuvC. Increasing concentrations of RuvC in each lane are indicated schematically by *triangles.* The locations of the labeled 5′ ends are indicated by *black dots.*

anneal to form the indicated Holliday junctions; they also possess a core of 11 nucleotides that are homologous. The Holliday junction can branch migrate within the core region. A dimer of RuvC binds to Holliday junctions and cleaves a pair of adjacent strands between nucleotides 3 and 4 in the sequences 5′-ATTG, 5′-ATTC, 5′-TTTG, or 5′-TTTC , as shown for 5′-ATTG in Figure 5–42A. These four sequences are represented by the consensus sequence $5'\text{-}^{A}/_{T}TT^{G}/_{C}$.

You want to know whether the two subunits of the RuvC dimer coordinate the cleavages on the two strands or can act independently. To investigate this you make the three Holliday junctions shown in Figure 5–42A: one with two cleavable sequences, one with two uncleavable sequences, and one with one cleavable and one uncleavable sequence. You label the 5′ end of one strand in each, incubate them with RuvC, and analyze cleavage by electrophoresis on denaturing gels, so that the oligonucleotides separate from one another. The results are shown in Figure 5–42B.

A. What fraction of all possible four-nucleotide sequences are cleaved by RuvC?

B. Does the requirement for resolution at a limited set of specific sequences seriously restrict the sites in the *E. coli* genome at which Holliday junctions can be resolved?

C. Do the two subunits of the RuvC dimer coordinate their cleavage of the two strands or can they act independently? Explain your reasoning.

D. Draw out the duplex products generated by RuvC resolution of the 'consensus' Holliday junction in Figure 5–42A, the one with its ends marked by letters.

# TRANSPOSITION AND CONSERVATIVE SITE-SPECIFIC RECOMBINATION

TERMS TO LEARN

| | | |
|---|---|---|
| bacteriophage | nonretroviral retrotransposon | transposable element |
| conservative site-specific recombination | retroviral-like retrotransposon | transposition |
| DNA-only transposon | retrovirus | transposon |
| | reverse transcriptase | |

## DEFINITIONS

Match each definition below with its term from the list above.

**5–102** Length of DNA that moves from a donor site to a target site either by cut-and-paste transposition or by replicative transposition.

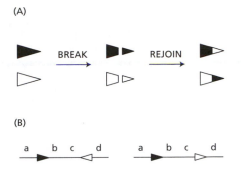

**Figure 5–43** Cre recombinase-mediated site-specific recombination (Problem 5–108). (A) Schematic representation of Cre/LoxP site-specific recombination. The LoxP sequences in the DNA are represented by *triangles* that are colored so that the site-specific recombination event can be followed more readily. (B) DNA substrates containing two arrangements of LoxP sites.

(A) BREAK REJOIN

(B) a b c d a b c d

**5–103** Enzyme that makes a double-stranded DNA copy from a single-stranded RNA template molecule.

**5–104** RNA-containing virus that replicates in a cell by first making a double-stranded DNA intermediate.

**5–105** Any virus that infects bacteria.

**5–106** Rearrangement of DNA that depends on the breakage and rejoining of two DNA helices at specific sequences on each DNA molecule.

## TRUE/FALSE

Consider the following statement and explain your answer.

**5–107** When transposable elements move around the genome, they rarely integrate into the middle of a gene because gene disruption—a potentially lethal event to the cell and the transposon—is selected against by evolution.

## THOUGHT PROBLEMS

**5–108** Cre recombinase is a site-specific enzyme that catalyzes recombination between two LoxP DNA recognition sequences. Cre recombinase pairs two LoxP sites in the same orientation, breaks both duplexes at the same point in the LoxP sites, and joins the ends with new partners so that each LoxP site is regenerated, as shown schematically in Figure 5–43A. Based on this mechanism, predict the arrangement of sequences that will be generated by Cre-mediated site-specific recombination for each of the two DNAs shown in Figure 5–43B.

**5–109** You are investigating the properties of two site-specific recombination systems, Cre/LoxP and FLP/FRT, to determine which might be better for the chromosome engineering project you have in mind. You make the two sets of substrates shown in Figure 5–44, one with LoxP sites and one with FRT sites. You compare the ability of the respective recombinases to promote recombination between adjacent directly repeated recognition sites on a chromosome and recombination between a site on a plasmid and a site on a chromosome (Figure 5–44). You are surprised and puzzled by the results. When you introduce Cre into cells, you get a very high frequency of chromosomal recombination ($10^{-1}$) but a very low frequency of plasmid integration ($10^{-6}$). When you introduce FLP recombinase, you get a lower frequency of chromosomal recombination ($10^{-2}$) but a higher frequency of plasmid integration ($10^{-5}$). You had expected that chromosomal recombination might be more efficient than plasmid integration because the recognition sites are adjacent on the chromosome; however, that consideration alone cannot account for your results. Can you offer an explanation for the outcome of your experiments?

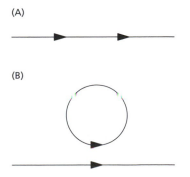

## DATA HANDLING

**5–110** You are studying the procaryotic transposon Tn10 and have just figured out an elegant way to determine whether Tn10 replicates during transposition or moves by a cut-and-paste mechanism. Your idea is based on the key difference between these two mechanisms: both parental strands of the Tn10

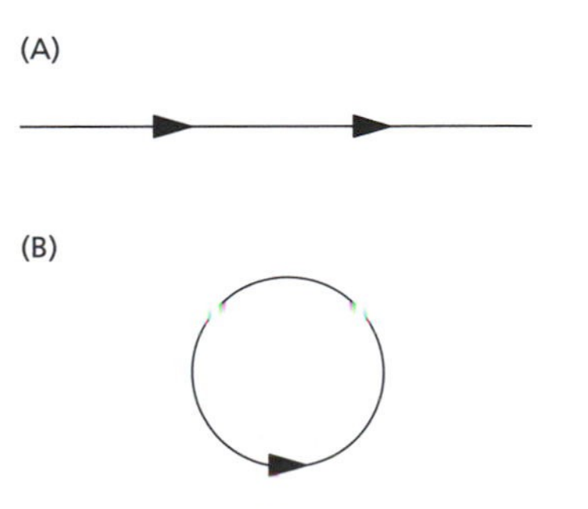


**Figure 5–44** Comparison of Cre/LoxP and FLP/FRT site-specific recombination (Problem 5–109). (A) Chromosomal site-specific recombination between directly repeated sites. (B) Site-specific plasmid integration into a chromosome.

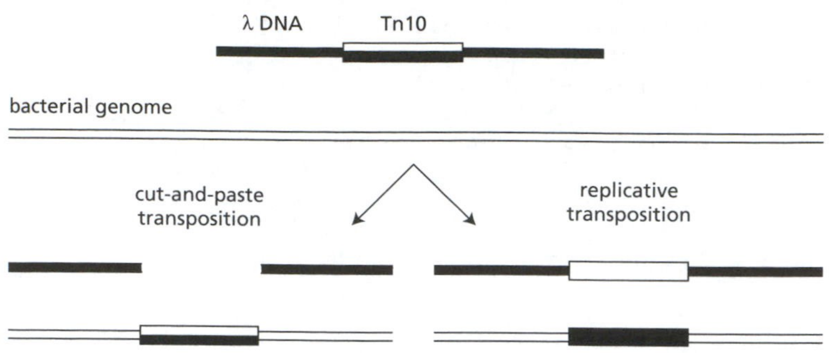


**Figure 5–45** Replicative and cut-and-paste transposition of a transposable element (Problem 5–110). The transposable element is shown as a heteroduplex, which is composed of two genetically different strands—one *black* and one *white*. During replicative transposition, one strand stays with the donor DNA and one strand is transferred to the bacterial genome. In cut-and-paste transposition, the transposable element is cut out of the donor DNA and transferred entirely to the bacterial genome.

move in cut-and-paste transpostion, whereas only one parental strand moves in replicative transposition (Figure 5–45).

You plan to mark the individual strands by annealing strands from two different Tn10s. Both Tn10s contain a gene for tetracycline resistance and a gene for lactose metabolism (*LacZ*), but in one, the *LacZ* gene is inactivated by a mutation. This difference provides a convenient way to follow the two Tn10s since *LacZ*$^+$ bacterial colonies (when incubated with an appropriate substrate) turn blue, but *LacZ*$^-$ colonies remain white. You denature and reanneal a mixture of the two transposon DNAs, which produces an equal mixture of heteroduplexes and homoduplexes (Figure 5–46). You introduce them into *LacZ*$^-$ bacteria, and spread the bacteria onto Petri dishes that contain tetracycline and the color-generating substrate.

Once inside a bacterium, the transposon will move (at very low frequency) into the bacterial genome, where it confers tetracycline resistance on the bacterium. The rare bacterium that gains a Tn10 survives the selective conditions and forms a colony. When you score a large number of such colonies, you find that roughly 25% are white, 25% are blue, and 50% are mixed, with one blue sector and one white sector.

A. Explain the source of each kind of bacterial colony and decide whether the results support a replicative or a cut-and-paste mechanism for Tn10 transposition.
B. You performed these experiments using a recipient strain of bacteria that was incapable of repairing mismatches in DNA. How would you expect the results to differ if you used a bacterial strain that could repair mismatches?

**5–111** The Ty elements of the yeast *Saccharomyces cerevisiae* move to new locations in the genome by transposition through an RNA intermediate. Normally, the Ty-encoded reverse transcriptase is expressed at such a low level that transposition is very rare. To study the transposition process, you engineer a cloned version of the Ty element so that the gene for reverse transcriptase is linked to the galactose control elements. You also 'mark' the element with a segment of bacterial DNA so that you can detect it specifically and thus distinguish it from other Ty elements in the genome. As a target gene to detect transposition, you use a defective histidine (*His*$^-$) gene whose expression is dependent on the insertion of a Ty element near its 5′ end. You show that yeast cells carrying a plasmid with your modified Ty element generate *His*$^+$ colonies at a frequency of $5 \times 10^{-8}$ when grown on glucose. When the same cells are grown on galactose, the frequency of *His*$^+$ colonies is $10^{-6}$, a 20-fold increase.

You notice that cultures of cells with the Ty-bearing plasmid grow normally on glucose but very slowly on galactose. To investigate this phenomenon, you isolate individual colonies that arise under three different conditions: *His*$^-$ colonies grown in the presence of glucose, *His*$^-$ colonies grown in the presence of galactose, and *His*$^+$ colonies grown in the presence of galactose. You eliminate the plasmid from each colony, isolate DNA from each culture, and analyze it by gel electrophoresis and blot hybridization, using the bacterial marker DNA as a probe. Your results are shown in Figure 5–47.

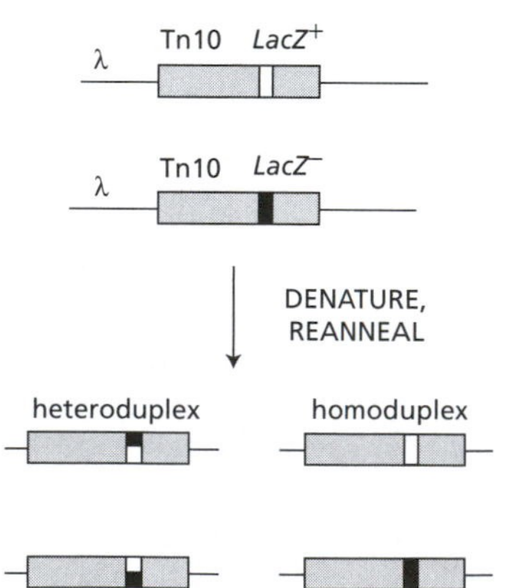


**Figure 5–46** Formation of a mixture of heteroduplexes and homoduplexes by denaturing and reannealing two different Tn10 genomes (Problem 5–110). The Tn10 DNA is shown as a *box*. The sequence differences between the two Tn10s are indicated by the *black* and *white segments*.

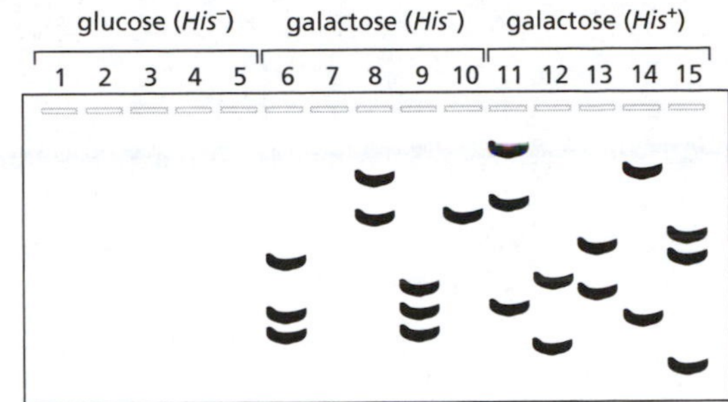


**Figure 5–47** Analysis of cells that harbored a plasmid carrying a modified Ty element (Problem 5–111). Cells were initially grown on glucose or galactose as a carbon source. *His*$^-$ cells were grown in the presence of histidine; *His*$^+$ cells were grown in its absence. Bands indicate restriction fragments that hybridized to the marker DNA originally present in the Ty element carried by the plasmid.

A. Why does transposition occur so much more frequently in cells grown on galactose than it does in cells grown on glucose?

B. As shown in Figure 5–47, *His*$^-$ cells isolated after growth on galactose have about the same number of marked Ty elements in their chromosomes as *His*$^+$ cells that were isolated after growth on galactose. If transposition is independent of histidine selection, why is the frequency of Ty-induced *His*$^+$ colonies so low ($10^{-6}$)?

C. Why do you think it is that cells with the Ty-bearing plasmid grow so slowly on galactose?

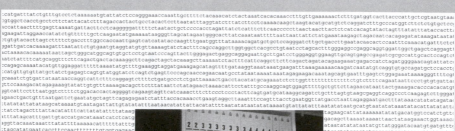

**Reading the Genome.** The complete sequence of the human genome printed out occupies 7 shelves of 16 volumes each containing hundreds of pages like the one shown here, with almost 10 million characters in each volume, printed in 4 point courier type, without any index. Try to see if you can find the gene on this page (*hint*—this comes from human chromosome 11, open on the shelf; to find where you are, try a BLAST search using at least 21 characters, see http://www.ncbi.nlm.nih.gov/BLAST/ and choose nucleotide BLAST). Photograph of the exhibit taken at the opening of the Wellcome Museum in July 2007.

# How Cells Read the Genome: From DNA to Protein

# 6

In This Chapter


| | |
|---|---|
| FROM DNA TO RNA | 119 |
| FROM RNA TO PROTEIN | 134 |
| THE RNA WORLD AND THE ORIGINS OF LIFE | 147 |


## FROM DNA TO RNA

### TERMS TO LEARN

consensus nucleotide sequence
DNA supercoiling
general transcription factor
exon
exosome
intron
mRNA (messenger RNA)
nuclear pore complex
promoter
RNA polymerase
RNA polymerase holoenzyme
RNA splicing
rRNA gene
rRNA (ribosomal RNA)
snoRNA (small nucleolar RNA)
snRNA (small nuclear RNA)
spliceosome
TATA box
terminator
transcription
trans-splicing

### DEFINITIONS

Match the definition below with its term from the list above.

**6–1** Helps to position the RNA polymerase correctly at the promoter, to aid in pulling apart the two strands of DNA to allow transcription to begin, and to release RNA polymerase from the promoter into the elongation mode once transcription has begun.

**6–2** Small RNA molecules that are complexed with proteins to form the ribonucleoprotein particles involved in RNA splicing.

**6–3** Nucleotide sequence in DNA to which RNA polymerase binds to begin transcription.

**6–4** A large protein complex that contains multiple 3′ to 5′ RNA exonucleases which degrade improperly processed mRNAs, introns, and other RNA debris retained in the nucleus.

**6–5** Any one of a number of specific RNA molecules that form part of the structure of the macromolecular complex responsible for synthesis of proteins; often distinguished by their sedimentation coefficient, such as 28S or 5S.

**6–6** RNA molecule that specifies the amino acid sequence of a protein.

**6–7** Process in which intron sequences are excised from RNA transcripts in the nucleus during the formation of messenger and other RNAs.

**6–8** Signal in bacterial DNA that halts transcription.

**6–9** Type of RNA splicing that occurs in a few eucaryotic organisms in which exons from two separate RNA molecules are joined together to form an mRNA.

**6–10** Segment of a eucaryotic gene consisting of a sequence of nucleotides that will be represented in mRNA or other functional RNAs.

**6–11** Large multiprotein structure forming a channel through the nuclear envelope that allows selected molecules to move between nucleus and cytoplasm.

## TRUE/FALSE

Decide whether each of these statements is true or false, and then explain why.

**6–12** The consequences of errors in transcription are less than those of errors in DNA replication.

**6–13** The σ subunit is a permanent component of the RNA polymerase holoenzyme from *E. coli*, allowing it to initiate at appropriate promoters in the bacterial genome.

**6–14** Eucaryotic mRNA molecules carry 3′ ribosyl OH groups at both their 3′ and 5′ ends.

**6–15** Since introns are largely genetic 'junk,' they do not have to be removed precisely from the primary transcript during RNA splicing.

**6–16** RNA polymerase II generates the end of a pre-mRNA transcript when it ceases transcription and releases the transcript; a poly-A tail is then quickly added to the free 3′ end.

## THOUGHT PROBLEMS

**6–17** Consider the expression 'central dogma,' which refers to the proposition that genetic information flows from DNA to RNA to protein. Is the word 'dogma' appropriate in this scientific context?

**6–18** In the electron micrograph in Figure 6–1, are the RNA polymerase molecules moving from right to left or from left to right? How can you tell? Why do the RNA transcripts appear so much shorter than the length of DNA that encodes them?

**6–19** Match the following list of RNAs with their functions.

| | |
|---|---|
| A. mRNA | 1. blocks translation of selected mRNAs |
| B. rRNA | 2. modification and processing of rRNA |
| C. snoRNA | 3. modification of snoRNA and snRNA |
| D. snRNA | 4. components of ribosome |
| E. tRNA | 5. splicing of RNA transcripts |
| F. scaRNA | 6. directs degradation of selected mRNAs |
| G. miRNA | 7. codes for proteins |
| H. siRNA | 8. adaptor for protein synthesis |

**6–20** An RNA polymerase is transcribing a segment of DNA that contains the sequence

5′-GTAACGGATG-3′
3′-CATTGCCTAC-5′

If the polymerase transcribes this sequence from left to right, what will the sequence of the RNA be? What will the RNA sequence be if the polymerase moves right to left?

**6–21** What are the roles of general transcription factors in RNA polymerase II-mediated transcription, and why are they referred to as 'general?'

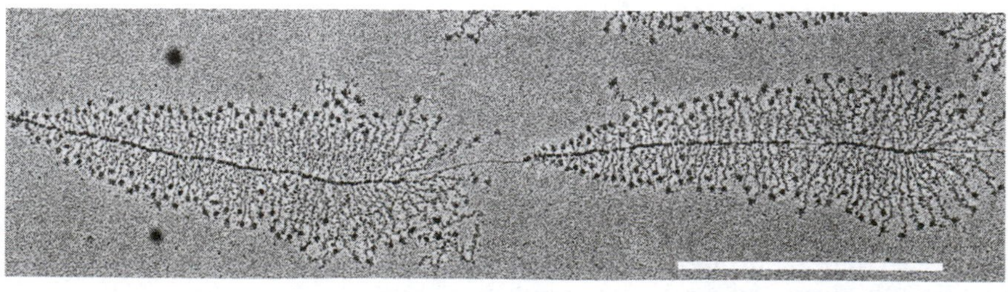

**Figure 6–1** Transcription of two adjacent rRNA genes (Problem 6–18). The scale bar is 1 μm.

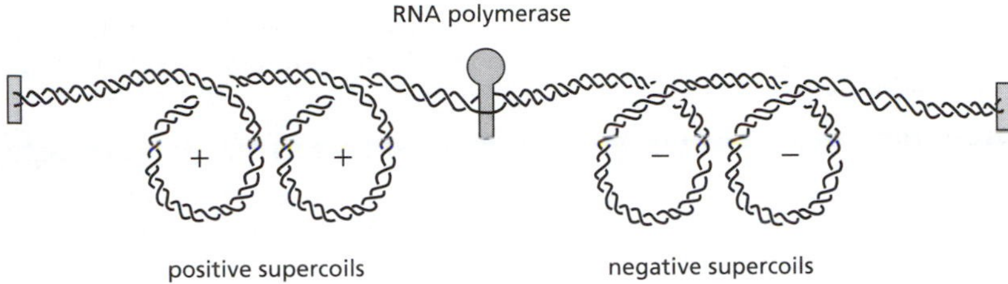


**Figure 6–2** Supercoils around a moving RNA polymerase (Problem 6–22).

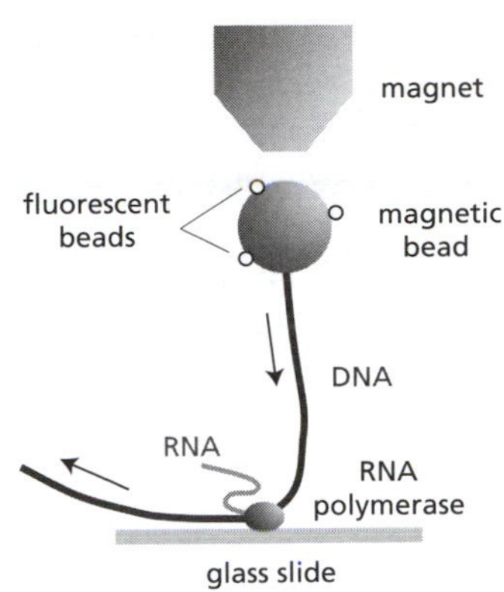


**Figure 6–3** System for measuring the rotation of DNA caused by RNA polymerase (Problem 6–23). The magnet holds the bead upright (but doesn't interfere with its rotation) and the attached fluorescent beads allow the direction of motion to be visualized. RNA polymerase is held in place by attachment to the glass slide.

**6–22** In which direction along the template must the RNA polymerase in Figure 6–2 be moving to have generated the supercoiled structures that are shown? Would you expect supercoils to be generated if the RNA polymerase were free to rotate about the axis of the DNA as it progressed along the template?

**6–23** You have attached an RNA polymerase molecule to a glass slide and have allowed it to initiate transcription on a template DNA that is tethered to a magnetic bead as shown in Figure 6–3. If the DNA with its attached magnetic bead moves relative to the RNA polymerase as indicated in the figure, in which direction will the bead rotate?

**6–24** Why doesn't transcription cause a hopeless tangle? If the RNA polymerase does not revolve around the DNA as it moves, it will induce two DNA supercoils—one in front and one behind—for every 10 nucleotides it transcribes. If, instead, RNA polymerase revolves around the DNA—avoiding DNA supercoiling—then it will coil the RNA around the DNA duplex, once for every 10 nucleotides it transcribes. Thus, for any reasonable size gene the act of transcription should result in hundreds of coils or supercoils…and that's for every single RNA polymerase! So why doesn't transcription lead to a complete snarl?

**Problems 4–84** to **4–87** introduce the principles of supercoiling in DNA.

**6–25** Phosphates are attached to the CTD (C-terminal domain) of RNA polymerase II during transcription. What are the various roles of RNA polymerase II CTD phosphorylation?

**6–26** You are studying a DNA virus that makes a set of abundant proteins late in its infectious cycle. An mRNA for one of these proteins maps to a restriction fragment from the middle of the linear genome. To determine the precise location of this mRNA you anneal it with the purified restriction fragment under conditions where only DNA–RNA hybrid duplexes are stable and DNA strands do not reanneal. When you examine the reannealed duplexes by electron microscopy, you see structures such as that in Figure 6–4. Why are there single-stranded tails at the ends of the DNA–RNA duplex region?

**6–27** Smilin is a (hypothetical) protein that causes people to be happy. It is inactive in many chronically unhappy people. The mRNA isolated from a number of different unhappy individuals in the same family was found to lack an internal stretch of 173 nucleotides that are present in the Smilin mRNA isolated from a control group of generally happy people. The DNA sequences of the Smilin genes from the happy and unhappy persons were determined and compared. They differed by just one nucleotide change—and no nucleotides were deleted. Moreover, the change was found in an intron.

A. Can you hypothesize a molecular mechanism by which a single nucleotide change in a gene could cause the observed *internal* deletion in the mRNA?

B. What consequences for the Smilin protein would result from removing a 173-nucleotide-long internal stretch from its mRNA? Assume that the 173 nucleotides are deleted from the coding region of the Smilin mRNA.

C. What can you say about the molecular basis of unhappiness in this family?

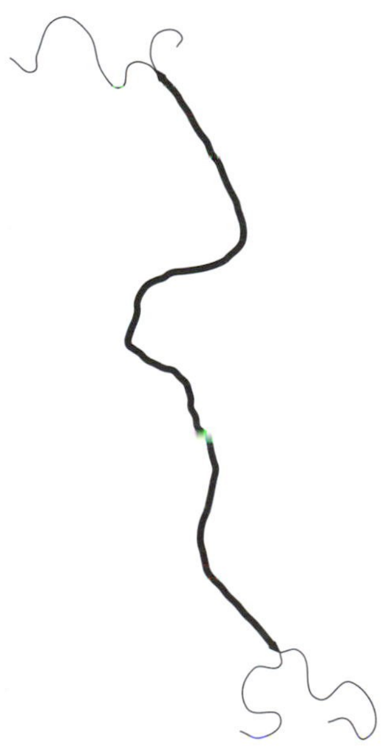

**Figure 6–4** DNA–RNA hybrid between an mRNA and a restriction fragment from adenovirus (Problem 6–26).

**6–28** The human α-tropomyosin gene is alternatively spliced to produce several forms of α-tropomyosin mRNA in various cell types (Figure 6–5). For all forms of the mRNA, the encoded protein sequence is the same for exons 1

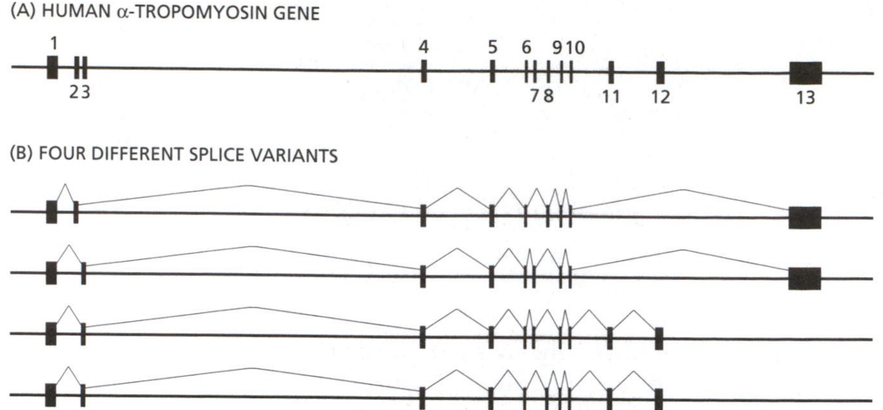


**Figure 6–5** Alternatively spliced mRNAs from the human α-tropomyosin gene (Problem 6–28). (A) Exons in the human α-tropomyosin gene. The locations and relative sizes of exons are shown by the *black rectangles*. (B) Splicing patterns for four α-tropomyosin mRNAs. Splicing is indicated by *lines* connecting the exons that are included in the mRNA.

and 10. Exons 2 and 3 are alternative exons used in different mRNAs, as are exons 7 and 8. Which of the following statements about exons 2 and 3 is the most accurate? Is that statement also the most accurate one for exons 7 and 8? Explain your answers.

A. Exons 2 and 3 must have the same number of nucleotides.
B. Exons 2 and 3 must each contain an integral number of codons (that is, the number of nucleotides divided by 3 must be an integer).
C. Exons 2 and 3 must each contain a number of nucleotides that when divided by 3 leaves the same remainder (that is, 0, 1, or 2).

**6–29** You have printed out a set of DNA sequences around the intron–exon boundaries for genes in the β-globin family, and have taken the thick file to the country to study for the weekend. When you look at the printout, you discover to your annoyance that there's no indication of where in the gene you are. You know that the sequences in Figure 6–6 come from one of the exon/intron or intron/exon boundaries and that the boundaries lie on the dotted line, but you don't know the order of the intron and exon. You know that introns begin with the dinucleotide sequence GT and end with AG, but you realize that these particular sequences would fit *either* as the start *or* the finish of an intron.

If you cannot decide which side is the intron, you will have to cut your weekend short and return to the city (or find a neighbor with Internet access). In desperation, you consider the problem from an evolutionary perspective. You know that introns evolve faster (suffering more nucleotide changes) than exons because they are not constrained by function. Does this perspective allow you to identify the intron, or will you have to pack your bags?

**6–30** Many eucaryotic genes contain a large number of exons. Correct splicing of such genes requires that neighboring exons be ligated to one other; if they are not, exons will be left out. Since 5′ splice sites look alike, as do 3′ splice

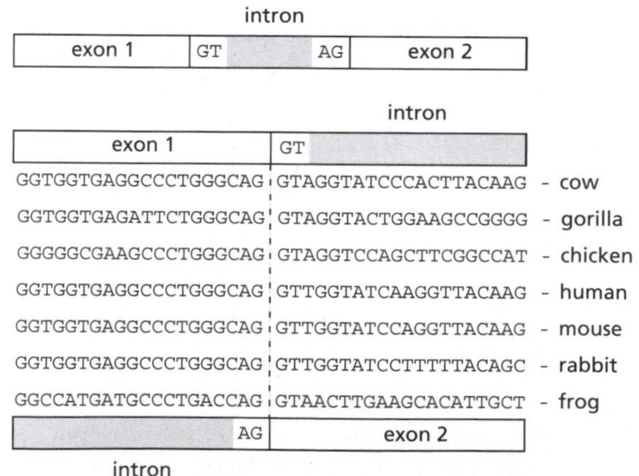


**Figure 6–6** Aligned DNA sequences from the β-globin genes in different species (Problem 6–29). As indicated by the gene structures shown *above* and *below*, the DNA sequences could come from the boundary of exon 1 with the intron or from the boundary of the intron with exon 2.

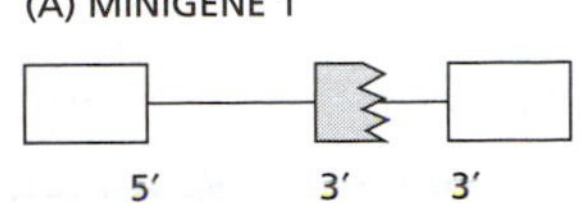


(B) MINIGENE 2
5′ 5′ 3′

**Figure 6–7** A test for intron scanning during RNA splicing (Problem 6–30). (A) Minigene with two 3′ splice sites. (B) Minigene with two 5′ splice sites. *Boxes* represent complete (*open*) or partial (*shaded*) exons; 5′ and 3′ splice junctions are indicated.

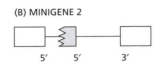

sites, it is remarkable that skipping an exon occurs so rarely. Some mechanism must keep track of neighboring exons and ensure that they are brought together.

One early proposal suggested that the splicing machinery bound to a splice site at one end of an intron and scanned through the intron to find the splice site at the other end. Such a scanning mechanism would guarantee that an exon was never skipped. This hypothesis was tested with two minigenes: one with a duplicated 5′ splice site and the other with a duplicated 3′ splice site (Figure 6–7). These minigenes were transfected into cells and their RNA products were analyzed to see which 5′ and 3′ splice sites were selected during splicing.

A. Draw a diagram of the products you expect from each minigene if the splicing machinery binds to a 5′ splice site and scans toward a 3′ splice site. Diagram the expected products if the splicing machinery scans in the opposite direction.

B. When the RNA products from the transfected minigenes were analyzed, it was found that each minigene generated a mixture of the two possible 5′-to-3′ splice products. Based on these results, would you conclude that neighboring exons are brought together by intron scanning? Why or why not?

**6–31** Self-splicing introns use two distinct strategies to accomplish splicing. Group I introns bind a G nucleotide from solution and activate it for attack on the phosphodiester bond that links the intron to the terminal nucleotide of the upstream exon (Figure 6–8A). Group II introns activate a particularly reactive A nucleotide within the intron sequence and use it to attack the phosphodiester bond that links the intron to the terminal nucleotide of the upstream exon (Figure 6–8B). For both types of intron the next step joins the two exons, releasing the 3′ end of the intron. What are the structures of the excised introns in both cases? Which mechanism more closely resembles pre-mRNA splicing catalyzed by the spliceosome?

**6–32** What does 'export ready' mRNA mean, and what distinguishes an 'export ready' mRNA from a bit of excised intron that needs to be degraded?

**6–33** The nucleolus disappears at each mitosis and then reappears during $G_1$ of the next cell cycle. How is this reversible process thought to be accomplished?

## CALCULATIONS

**6–34** The trypanosome, which is the microorganism that causes sleeping sickness, can vary its surface glycoprotein coat and thus evade the immune defenses of its host. The promoter for the variable surface glycoprotein (VSG) gene proved difficult to locate, but was mapped by measuring the sensitivity of the transcript to UV irradiation. Since RNA polymerases cannot

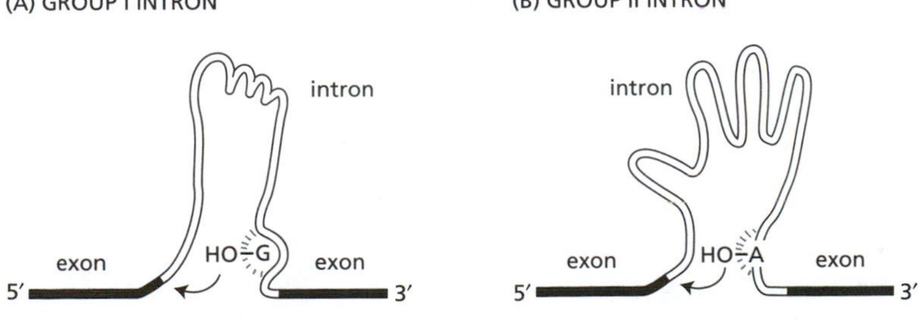


**Figure 6–8** Initial step in self-splicing (Problem 6–31). (A) Reaction catalyzed by a group I self-splicing intron. (B) Reaction catalyzed by a group II self-splicing intron.

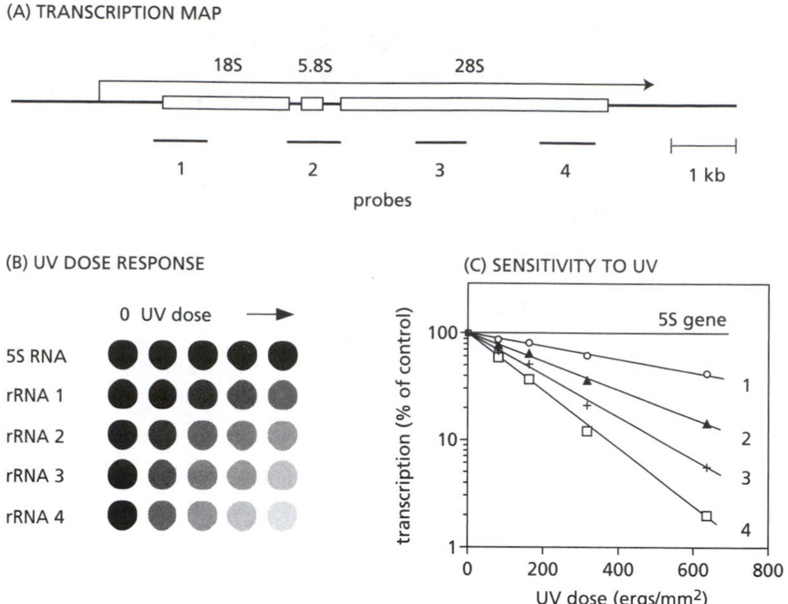


**Figure 6–9** UV mapping of the promoter in trypanosomes (Problem 6–34). (A) Structure of the ribosomal RNA transcription unit. The positions of the hybridization probes along with a scale marker are shown. The transcription unit begins at the left end of the *arrow*. (B) A dot blot of transcripts from the 5S RNA and rRNA genes. The dot blot is done by placing an excess of the DNA probe in spots on a filter paper and then hybridizing radiolabeled RNA to it. (C) Sensitivities of the transcription units to increasing doses of UV irradiation.

transcribe through pyrimidine dimers (the major damage produced by UV irradiation), the sensitivity of transcription to UV irradiation is a measure of the distance between the start of transcription (the promoter) and the point at which transcription is assayed (*Vsg* gene).

Transcription through rRNA genes was used to calibrate the system. The 5S RNA transcription unit is just over 100 nucleotides long, whereas the 18S, 5.8S, and 28S RNAs are part of a single transcription unit that is about 8 kb long (Figure 6–9A). Trypanosomes were exposed to increasing doses of UV irradiation; their nuclei were then isolated and incubated with $^{32}$P-dNTPs. RNA isolated from the nuclei was hybridized to DNA probes corresponding to the 5S RNA gene and various parts of the rRNA gene (Figure 6–9B). Plots of the logarithm of the counts in each spot against the UV dose gave straight lines (Figure 6–9C), with slopes that were proportional to the distance from the hybridization probe to the promoter.

When the experiment was done with a probe from the beginning of the *Vsg* gene, transcription was found to be inactivated about seven times faster for the *Vsg* gene than for probe 4 from the ribosomal transcription unit.

A. Why does RNA transcription increase in sensitivity to UV irradiation with increasing distance from the promoter?
B. Roughly how far is the *Vsg* gene from its promoter? What assumption do you have to make in order to estimate this distance?
C. Transcription through another gene, located about 10 kb upstream of the *Vsg* gene, was about 20% less sensitive to UV irradiation than transcription through the *Vsg* gene. Is this measurement consistent with the possibility that these two genes are transcribed from the same promoter? Explain your reasoning.

**6–35** You have established a transcription assay in which transcripts initiate specifically at an adenovirus promoter in a plasmid (see Problem 6–39). Each transcript is 400 nucleotides long and has an overall composition of $C_2AU$. These transcripts accumulate linearly for about an hour and then reach a plateau. Your assay conditions use a 25-μL reaction volume containing 16 μg/mL of DNA template (the plasmid, which is 3.5 kb in length) with all other components in excess. From the specific activity of the $^{32}$P-CTP and the total radioactivity in transcripts, you calculate that at the plateau 2.4 pmol of CMP was incorporated. (The mass of a nucleotide pair is 660 daltons.)

A. How many transcripts are produced per reaction?
B. How many templates are present in each reaction?
C. How many transcripts are made per template in the reaction?

**Table 6–1 Rate of incorporation of correct and incorrect nucleotides** (Problem 6–36).

| INCORPORATED NUCLEOTIDE (bold) | $V_{max}$ (units/sec) |
|---|---|
| CU**A** | 0.20 |
| CU**G** | 0.0015 |
| CUA**C** | 0.17 |
| CUG**C** | 0.036 |

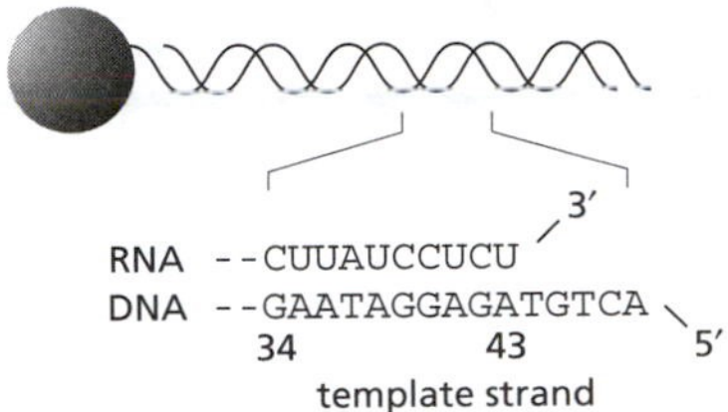


(B) EXPERIMENTAL DATA

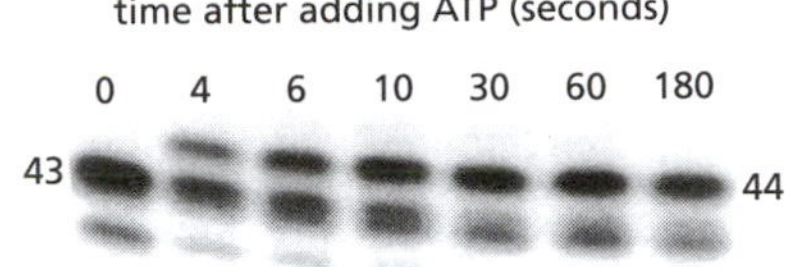


**Figure 6–10** Determination of kinetic parameters for RNA polymerase (Problem 6–36). (A) DNA template attached to an agarose bead. (B) Rate of incorporation of the correct A nucleotide in the nascent RNA chain. (The bands in adjacent lanes form a curve because of an artifact of gel electrophoresis.)

## DATA HANDLING

**6–36** If RNA polymerase proofreads its product in a manner analogous to DNA polymerase, it will slow its rate of nucleotide incorporation after adding an incorrect base to the end of the growing RNA chain. This will allow time for removal of the mismatched nucleotide before the next one is added.

To investigate this possibility, scientists devised a clever technique to measure the rate of nucleotide incorporation by RNA polymerase at a defined point in a DNA template. The template contained a promoter and was covalently attached to agarose beads, so that it (and the attached RNA polymerase) could be removed from solution and washed, and then reincubated in a new mixture of nucleotides (Figure 6–10A). By using a series of solutions containing just one or a couple of nucleotides, the RNA polymerase was 'walked' along the template to the site shown in Figure 6–10A. At that point the RNA polymerase and template were resuspended in a solution containing one nucleotide, and the rate of incorporation of that nucleotide was measured as shown in Figure 6–10B. The rates of incorporation of A and G at position +44 in the RNA strand are shown in Table 6–1. In addition, the incorporation of the following C at position +45 was measured after A or G was incorporated at position +44 (Table 6–1).

A. Imagine that the RNA polymerase was stopped so that the last nucleotide in the RNA chain was the C at position +34. Describe how you might 'walk' the polymerase from there to position +43 to do these experiments.

B. Is the correct nucleotide preferred over the incorrect nucleotide? If so, by what factor is it preferred? Does the presence of an incorrect nucleotide influence the rate of addition of the next nucleotide? Explain your answer.

**6–37** In Figure 6–11 the sequences of 13 promoters recognized by the $\sigma_{70}$ factor of RNA polymerase have been aligned. Deduce the consensus sequences for the –10 and –35 regions of these promoters.

**6–38** Deletion analysis of protein-binding sequences in a promoter can be difficult to interpret because altered spacing between elements can critically affect their function. The 'linker-scanning' method eliminates this potential difficulty by replacing 10-nucleotide segments throughout the promoter with oligonucleotide linkers. A classic paper described this method in an analysis of the promoter for the thymidine kinase (*Tk*) gene. Plasmids

```
                                                           +1
tyrosine tRNA promoter   TCTCAACGTAACACTTTACAGCGGCG..CGTCATTTGATATGATGC.GCCCCGCTTCCCGATAAGGG
ribosomal RNA            GATCAAAAAAATACTTGTGCAAAAAA..TTGGGATCCCTATAATGCGCCTCCGTTGAGACGACAACG
gene promoters           ATGCATTTTTCCGCTTGTCTTCCTGA..GCCGACTCCCTATAATGCGCCTCCATCGACACGGCGGAT
                         CCTGAAATTCAGGGTTGACTCTGAAA..GAGGAAAGCGTAATATAC.GCCACCTCGCGACAGTGAGC
                         CTGCAATTTTTCTATTGCGGCCTGCG..GAGAACTCCCTATAATGCGCCTCCATCGACACGGCGGAT
                         TTTTAAATTTCCTCTTGTCAGGCCGG..AATAACTCCCTATAATGCGCCACCACTGACACGGAACAA
                         GCAAAAATAAATGCTTGACTCTGTAG..CGGGAAGGCGTATTATGC.ACACCCCGCGCCGCTGAGAA
bacteriophage            TAACACCGTGCGTGTTGACTATTTTA.CCTCTGGCGGTGATAATGG..TTGCATGTACTAAGGAGGT
promoters                TATCTCTGGCGGTGTTGACATAAATA.CCACTGGCGGTGATACTGA..GCACATCAGCAGGACGCAC
                         GTGAAACAAAACGGTTGACAACATGA.AGTAAACACGGTACGATGT.ACCACATGAAACGACAGTGA
                         TATCAAAAAGAGTATTGACTTAAAGT.CTAACCTATAGGATACTTA.CAGCCATCGAGAGGGACACG
                         ACGAAAAACAGGTATTGACAACATGAAGTAACATGCAGTAAGATAC.AAATCGCTAGGTAACACTAG
                         GATACAAATCTCCGTTGTACTTTGTT..TCGCGCTTGGTATAATCG.CTGGGGGTCAAAGATGAGTG
```

**Figure 6–11** Sequences recognized by $\sigma_{70}$ factor (Problem 6–37). *Dots* represent spaces that have been added to maximize alignment of sequences in the –10 and –35 regions.

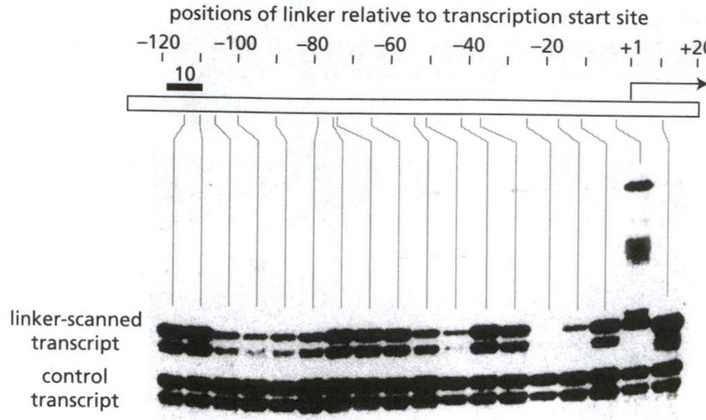


**Figure 6–12** Linker-scanning analysis of the *Tk* promoter (Problem 6–38). Transcripts from linker-scanned plasmids and control plasmids were analyzed by primer extension, using a radiolabeled primer corresponding to sequences about 80 nucleotides from the 5′ end of the transcript. These primers were extended to the 5′ ends of the transcripts and the products were displayed on a denaturing polyacrylamide gel. Two bands are present for both the control and linker-scanned transcripts because of inefficient extension to the very end of the transcript. The position of each linker is indicated at the *center* of the segment it replaced. A 10-base-pair bar—the length of the replacements—is shown at the *top left* of the figure.

in which 10-nucleotide segments had been replaced with linkers were injected into *Xenopus laevis* oocytes along with control plasmids to measure injection efficiency. Results of these injections are shown in Figure 6–12.

A. Estimate from these experiments the locations of sequences that are critical for promoter function, and rank their relative importance.
B. Which, if any, of these elements do you suppose corresponds to the TATA box?

**6–39** Purification of a transcription factor typically requires a rapid assay to prevent its being inactivated before the fraction containing it can be identified. One key technical advance was to use a promoter linked to a 400-nucleotide DNA sequence that contained no C nucleotides. When GTP is omitted from the assay mixture (but CTP, UTP, and ATP are included), the only long RNA transcript is made from the synthetic DNA sequence because all other transcripts terminate when a G is required. This set-up allows a rapid assay of specific transcription simply by measuring the incorporation of a radioactive nucleotide into the long transcript.

To test this idea, two plasmids were constructed that carried the synthetic sequence: one with a promoter from adenovirus (pML1), the other without a promoter (pC1) (Figure 6–13A). Each plasmid was mixed with pure RNA polymerase II, transcription factors, UTP, ATP, and $^{32}$P-CTP. In addition, various combinations of GTP, RNase T1 (which cleaves RNA adjacent to each G nucleotide), and 3′ *O*-methyl GTP (which terminates transcription whenever G is incorporated) were added. The products were measured by gel electrophoresis with the results shown in Figure 6–13B.

A. Why is the 400-nucleotide transcript absent from lane 4 but present in lanes 2, 6, and 8?
B. Can you guess the source of the synthesis in lane 3 when the promoterless pC1 plasmid is used?

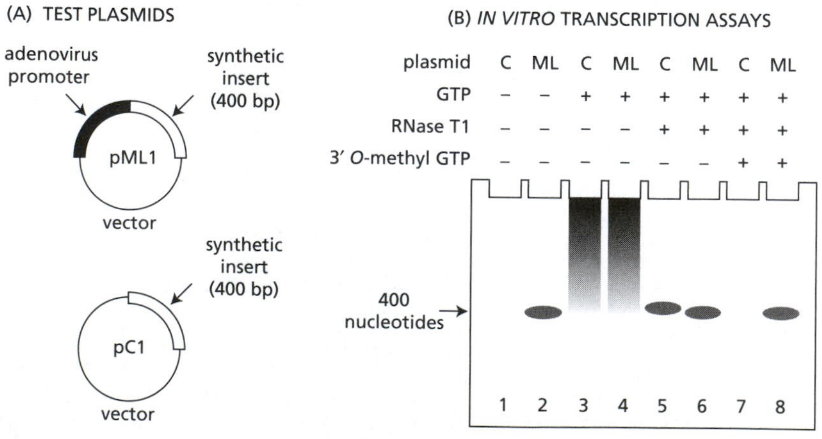


**Figure 6–13** Characterization of transcription using a template without C nucleotides (Problem 6–39). (A) Structures of test plasmids. (B) Results of transcription assays under various conditions. All reactions contain RNA polymerase II, transcription factors, UTP, ATP, and $^{32}$P-CTP. Other components are listed *above* each lane. C is plasmid pC1; ML is plasmid pML1.

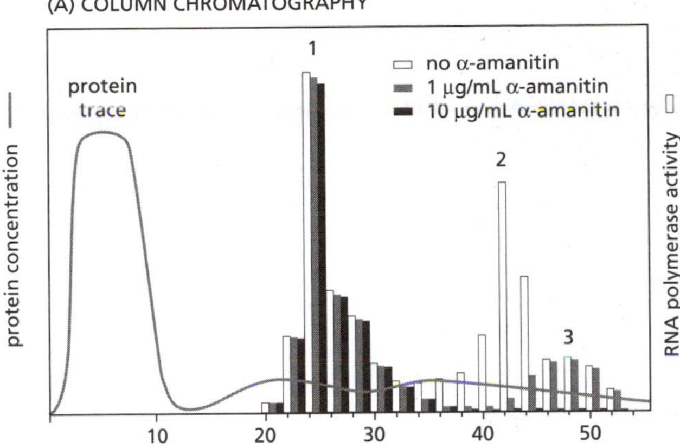

**Figure 6–14** Characterization of RNA polymerase activities (Problem 6–40). (A) Peaks of RNA polymerase activity after column chromatography. Activities were measured in the presence of various concentrations of α-amanitin. (B) The *Amanita phalloides* mushroom.

C. Why is a transcript of about 400 nucleotides present in lane 5 but not in lane 7?

D. The goal in developing this ingenious assay was to aid the purification of transcription factors; however, that process will begin with crude cell extracts, which will contain GTP. How do you suppose you might assay specific transcription in crude extracts?

**6–40** When RNA polymerases were first being characterized in eucaryotes, three peaks of polymerizing activity (1, 2, and 3) were commonly obtained by fractionating cell extracts on chromatography columns (Figure 6–14A). It was unclear whether these peaks corresponded to different RNA polymerases or just to different forms of one polymerase. Incubating the three polymerase fractions in the presence of 1 μg/mL or 10 μg/mL α-amanitin (from *Amanita phalloides,* the world's deadliest mushroom; Figure 6–14B) gave the results shown in Figure 6–14A. Do these results argue for different RNA polymerases or different forms of the same RNA polymerase? Explain your answer.

**6–41** The large subunit of eucaryotic RNA polymerase II in yeast has a CTD (C-terminal domain) that comprises 27 near-perfect repeats of the sequence YSPTSPS. If the normal RNA polymerase II gene is replaced with one that encodes a CTD with only 11 repeats, the cells are viable at 30°C but are unable to grow at 12°C. This cold sensitivity allows revertants to be selected for growth at 12°C. Some of these revertants proved to be dominant mutations in previously unknown genes such as *Srb2.*

Extracts prepared from yeast that are lacking the *Srb2* gene cannot transcribe added DNA templates, but they can be activated for transcription by the addition of Srb2 protein. To test the role of Srb2 in transcription, plasmid DNAs with either a short or a long G-free sequence downstream of a promoter (Figure 6–15A) were incubated separately in the presence or absence of a *limiting* amount of recombinant Srb2 [and in the absence of added nucleotide triphosphates (NTPs) so that transcription could not begin]. The reactions were then mixed and transcription was initiated at various times afterward by adding a mixture of all four NTPs that included $^{32}$P-CTP (Figure 6–15B). After a brief incubation (so that transcription did not have time to reinitiate) the products were displayed by gel electrophoresis. Whichever template was preincubated with Srb2 was the one that was transcribed (Figure 6–15C). By contrast, if an *excess* of Srb2 was mixed with one template during the preincubation, transcription was observed from both templates after mixing.

A. Did Srb2 show a preference for either template when the preincubation was carried out with the individual templates or the mixture (Figure 6–15C, lanes 1 to 3)?

B. Do the results indicate that Srb2 acts stoichiometrically or catalytically? How so?

C. Does Srb2 form part of the preinitiation complex or does it act after transcription has begun? How can you tell?

(A) TEMPLATES

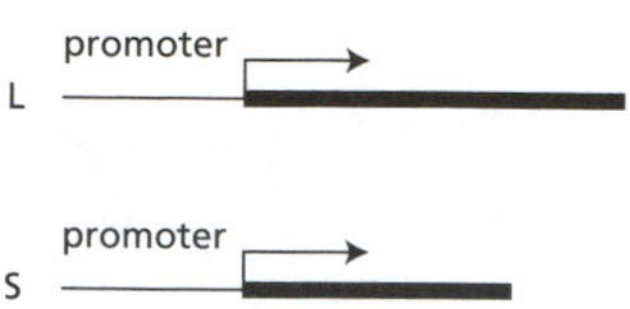


(B) DESIGN OF THE EXPERIMENT

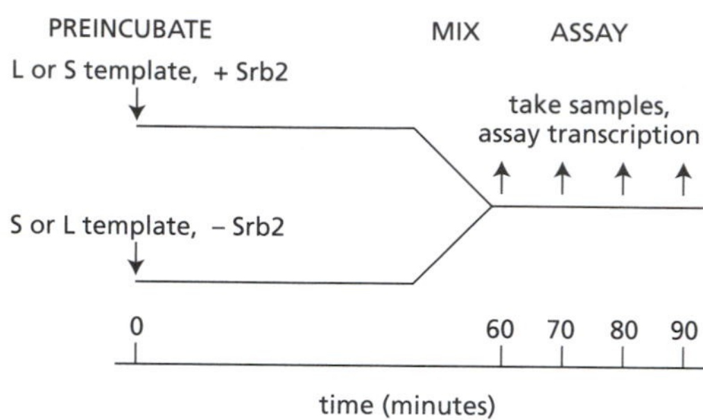


(C) EXPERIMENTAL RESULTS

TEMPLATES

| | | | | | | | | | | | | |
|---|---|---|---|---|---|---|---|---|---|---|---|---|
| preincubation + Srb2 | L | S | L+S | – | L | L | L | L | S | S | S | S |
| preincubation – Srb2 | – | – | – | – | S | S | S | S | L | L | L | L |
| time of assay (min) | 60 | 60 | 60 | 60 | 60 | 70 | 80 | 90 | 60 | 70 | 80 | 90 |

L transcript →

S transcript →

1 2 3 4 5 6 7 8 9 10 11 12

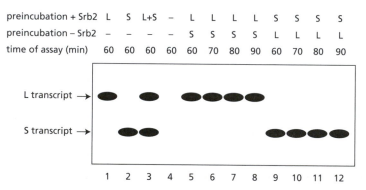

**Figure 6–15** Experiments to test the role of Srb2 in transcription (Problem 6–41). (A) Short (S) and long (L) G-free templates. (B) Experimental design. (C) Experimental results. The transcripts from the S and L templates were visualized by autoradiography.

D. What do you think happens during the preincubation that so strongly favors transcription from the template that was included in the preincubation?

E. Do these results indicate that Srb2 binds to the CTD of RNA polymerase II?

**6–42** Detailed features of the active site of RNA polymerase have been probed using modified RNA nucleotides that can react with nearby targets. In one study an analog of U carrying a reactive chemical group (Figure 6–16A) was incorporated at one of two positions in a radiolabeled RNA and then 'walked'

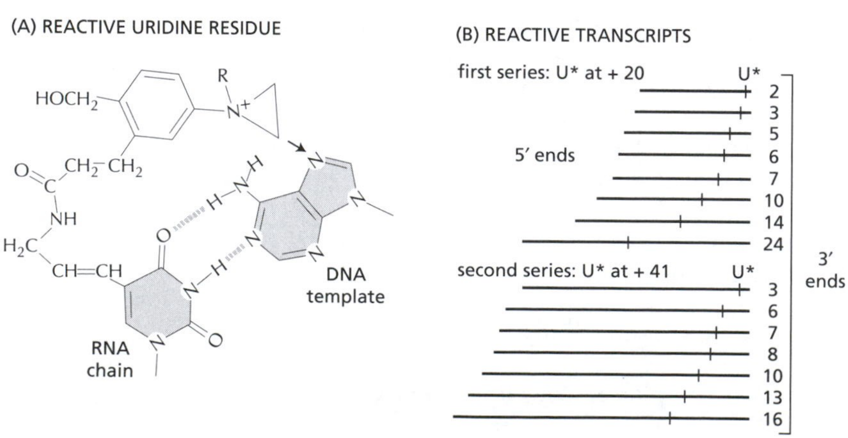


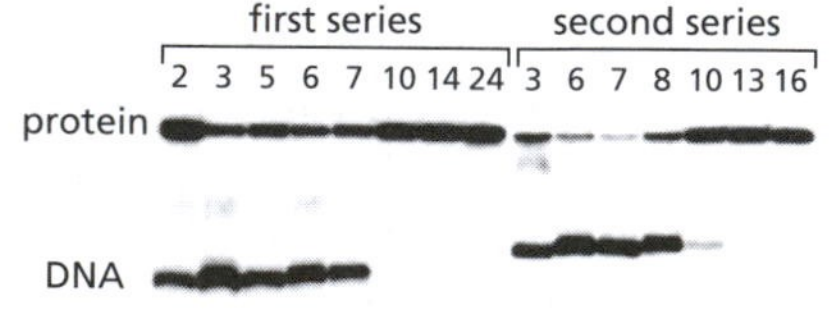


**Figure 6–16** Probing the active site of RNA polymerase (Problem 6–42). (A) The reactivity of a U analog with adenosine in the template DNA. The *arrow* indicates the site of attack. (B) Locations of the U analog in the RNA chain. *Numbers* refer to the number of nucleotides from the reactive U to the 3′ end. (C) The pattern of reaction of the reactive U with DNA and protein.

(A) ELECTRON MICROGRAPH

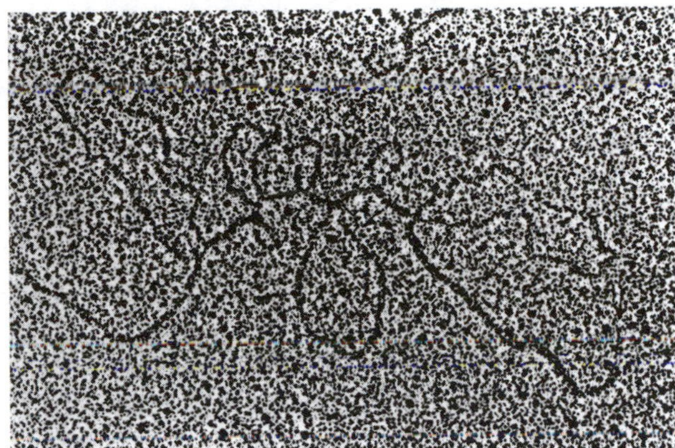

(B) INTERPRETATION

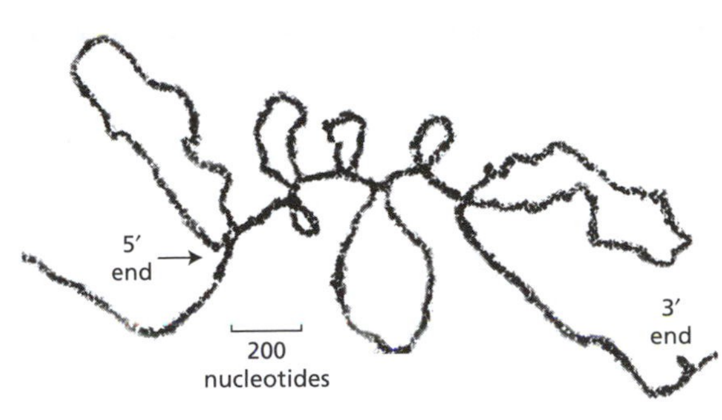


**Figure 6–17** Hybrid between ovalbumin mRNA and a DNA segment containing its gene (Problem 6–43). (A) An electron micrograph. (B) The hybrid with background eliminated. Each of the loops is formed by single-stranded DNA emanating from a central RNA–DNA duplex.

to specific locations relative to the 3′ end of the RNA chain using a technique like that described in Problem 6–36 (Figure 6–16B). When the analog was then activated, it reacted almost exclusively either with the adenine it was paired with in the DNA template strand or with the polymerase (Figure 6–16C). In both series of experiments, the pattern of reactivity with DNA and protein varied depending on the position of the reactive analog relative to the 3′ end of the RNA. Why do you suppose that the activated U reacts so strongly with DNA up to a point and then abruptly begins to react more strongly with protein?

**6–43** The intron–exon structure of eucaryotic genes came as a shock. In the early, skeptical days the most convincing demonstration was visual, as shown, for example, by the electron micrograph in Figure 6–17A. This image was obtained by hybridizing ovalbumin mRNA to a long segment of DNA that contained the gene. To those used to looking at single- and double-stranded nucleic acids in the electron microscope, the structure was clear: a set of single-stranded tails and loops emanating from a central duplex segment whose ends corresponded to the ends of the mRNA (Figure 6–17B). To the extent possible, describe the intron–exon structure of this gene.

**6–44** The interaction of U1 snRNP with the sequences at the 5′ ends of introns is usually shown to involve pairing between the nucleotides in the pre-mRNA and those in the RNA component of the U1 snRNP, as illustrated in Figure 6–18A. But given the relatively small number of interactions, that arrangement is hardly convincing. A series of experiments tested the hypothesis that base-pairing was critical to the function of U1 snRNP in splicing. As shown in Figure 6–18B, a mutant pre-mRNA was generated that could not be spliced. Several mutant U1 RNAs were then tested for their ability to promote splicing of the mutant pre-mRNA, with the results shown in Figure 6–18C. Do these experiments argue that base-pairing is critical to the role of U1 snRNP in splicing? If so, how?

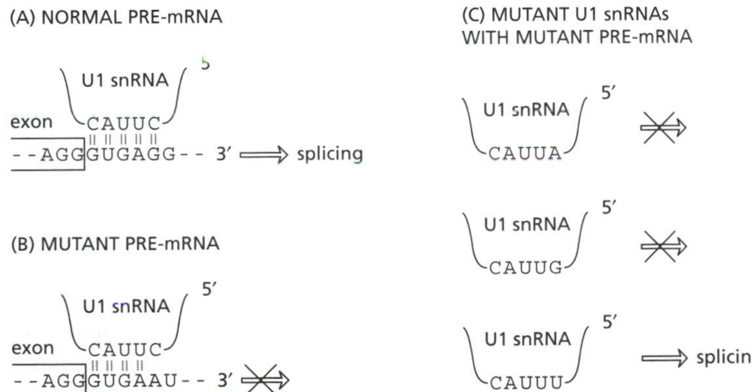


**Figure 6–18** Characterization of the pairing of U1 snRNP with pre-mRNA (Problem 6–44). (A) Interaction of U1 snRNP with a normal pre-mRNA. (B) Interaction of U1 snRNP with a mutant pre-mRNA. (C) Mutant U1 snRNAs.

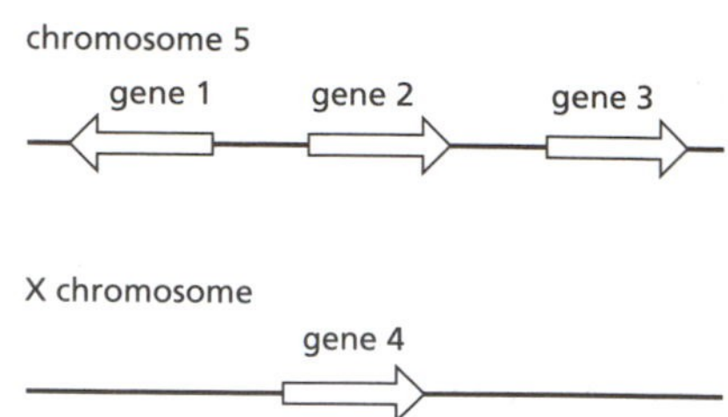

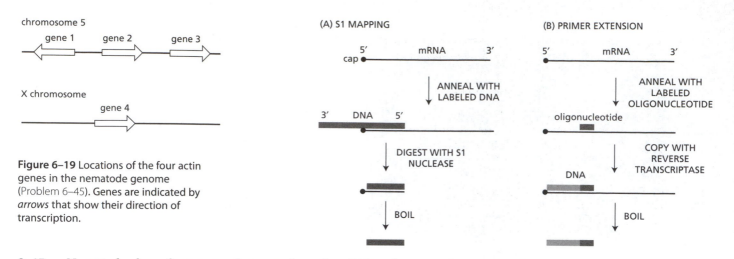

**Figure 6–19** Locations of the four actin genes in the nematode genome (Problem 6–45). Genes are indicated by *arrows* that show their direction of transcription.

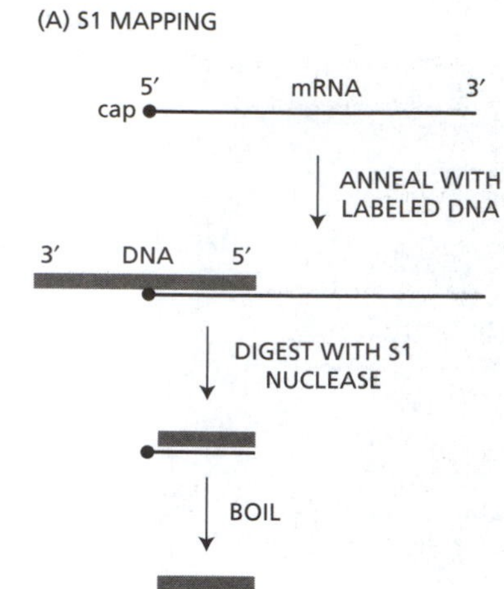


**Figure 6–20** Mapping the 5′ ends of actin mRNAs (Problem 6–45). (A) S1 mapping. (B) Primer extension.

**6–45** Nematodes have four genes that encode actin mRNAs: three are clustered on chromosome 5, and one is located on the X chromosome (Figure 6–19). To identify the start site of transcription you employ S1 mapping and primer extension. For S1 mapping you anneal a radioactive single-stranded segment from the 5′ end of the gene to the mRNAs, digest with S1 nuclease to remove all single strands, and analyze the protected fragments of radioactive DNA on a sequencing gel to determine their length (Figure 6–20A). For primer extension you hybridize specific oligonucleotides to the mRNAs, extend them to the 5′ end, and analyze the resulting DNA segments on sequencing gels (Figure 6–20B).

For gene 4 the two techniques agree, as they usually do. However, for genes 1, 2, and 3, the mRNAs appear to be 20 nucleotides longer when assayed by primer extension. When you compare the mRNA sequences with the sequences for the gene, you find that each has an identical 20-nucleotide segment at its 5′ end, which does not match the sequence of the gene (Figure 6–21).

Using an oligonucleotide complementary to this leader RNA segment, you discover that the corresponding DNA is repeated about 100 times in a cluster on chromosome 5, but it's a long way from the actin genes on chromosome 5. This leader gene encodes an RNA about 100 nucleotides long. The 5′ end of the leader RNA is identical to the segment found at the 5′ ends of the actin mRNAs (Figure 6–21).

A. Assuming that the leader RNA and the actin RNAs are joined by splicing according to the usual rules, indicate on Figure 6–21 the most likely point at which the RNAs are joined.

B. The leader gene and actin genes 1, 2, and 3 are all on the same chromosome. Why is it *not* possible that transcription begins at a leader gene and extends

ACTIN GENE 1

DNA: 5′ TATTATCAATTTAATTTTTCAGGTACATTAAAAACTAATCAAAATG
RNA: 5′ XGUUUAAUUACCCAAGUUUGAGGUACAUUAAAAACUAAUCAAAAUG

ACTIN GENE 2

DNA: 5′ ATAATTCATAATTATTTTGTAGGCTAAGTTCCTCCTAATCTAATAAATCATG
RNA: 5′ XGUUUAAUUACCCAAGUUUGAGGCUAAGUUCCUCCUAAUCUAAUAAAUCAUG

ACTIN GENE 3

DNA: 5′ TATTATCAATTTAATTTTTCAGGTACATTAAAAACTAATCAAAATG
RNA: 5′ XGUUUAAUUACCCAAGUUUGAGGUACAUUAAAAACUAAUCAAAAUG

LEADER RNA GENE

DNA: 5′ GGTTTAATTACCCAAGTTTGAGGTAAACATTCAAACTGA
RNA: 5′ XGUUUAAUUACCCAAGUUUGAGGUAAACAUUCAAACUGA

**Figure 6–21** RNA and DNA sequences of actin genes and leader RNA gene (Problem 6–45). The start site for translation of the actin genes (ATG) is *underlined* in the DNA sequence. The leader RNA segments that are present at the 5′ ends of the actin genes and the leader RNA gene are *underlined* in the RNA sequences. The three nucleotides immediately adjacent to the underlined leader RNA segment are identical in all four genes; thus it is unclear whether they were derived from actin RNA or leader gene RNA. The 5′ nucleotide on the RNAs (X) cannot be determined by primer extension.

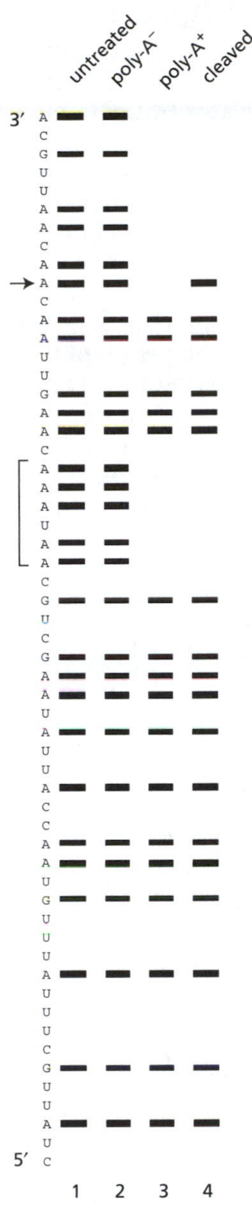


**Figure 6–22** Autoradiographic analysis of experiments to define the purines important in RNA cleavage and polyadenylation (Problem 6–46). The sequence of the precursor RNA is shown at the *left* with the 5′ end at the *bottom* and the 3′ end at the *top*. All RNAs were modified by reaction with diethylpyrocarbonate. RNA that was not treated with extract (untreated) is shown in lane 1. RNA that was treated with extract but was not polyadenylated (poly-A⁻) is shown in lane 2. RNA that was polyadenylated (poly-A⁺) in the extract is shown in lane 3. RNA that was cleaved but not polyadenylated (cleaved) is shown in lane 4. Shorter RNA fragments run faster; that is, they travel farther toward the bottom of the gel during electrophoresis.

through the actin genes to give a precursor RNA that is subsequently spliced to form the individual actin mRNAs?

C. How do you think the actin mRNAs acquire their common leader segment?

**6–46** The AAUAAA sequence just 5′ of the polyadenylation site is a critical signal for polyadenylation, as has been verified in many ways. One elegant confirmation used chemical modification to interfere with specific protein interactions. RNA molecules containing the signal sequence were radiolabeled at one end and then treated with diethylpyrocarbonate, which can modify A and G nucleotides, rendering the RNA sensitive to breakage by subsequent aniline treatment. The experimental conditions employed were such that the RNA molecules contained about one modification each. They were then cleaved with aniline. The resulting series of fragments have lengths that correspond to the positions of As and Gs in the RNA (Figure 6–22, lane 1).

To define critical A and G nucleotides, the modified, but still intact, RNAs were incubated with an extract capable of cleavage and polyadenylation. The RNAs were then separated into those that had acquired a poly-A tail (poly-A⁺) and those that had not (poly-A⁻). These two fractions were treated with aniline and the fragments were analyzed by gel electrophoresis (Figure 6–22, lanes 2 and 3). In a second reaction EDTA was added to the extract along with the modified RNAs; this does not affect RNA cleavage but prevents addition of the poly-A tail. These cleaved RNAs were isolated, treated with aniline, and examined by electrophoresis (lane 4).

A. At which end were the starting RNA molecules labeled?

B. Explain why the bands corresponding to the AAUAAA signal (bracket in Figure 6–22) are missing from the poly-A⁺ RNA and the cleaved RNA.

C. Explain why the band at the arrow (the normal nucleotide to which poly A is added) is missing from the poly-A⁺ RNA but is present in the cleaved RNA.

D. Which A and G nucleotides are important for cleavage, and which A and G nucleotides are important for addition of the poly-A tail?

E. What additional information might be obtained by labeling the RNA molecules at the other end?

**6–47** Unlike most mRNAs, histone mRNAs do not have poly-A tails. They are processed from a longer precursor by cleavage just 3′ of a stem-loop structure in a reaction that depends on a conserved sequence near the cleavage site (Figure 6–23A). A classic paper defined the role of U7 snRNP in cleavage of the histone precursor RNA.

**Figure 6–23** Processing of histone precursor RNAs (Problem 6–47). (A) Nucleotide sequences of histone precursor RNAs. *Horizontal arrows* indicate the inverted repeat sequences capable of forming a stem-loop structure in the precursor, the *vertical arrow* indicates the site of cleavage, and the *underline* shows the position of the conserved region. (B) DNA oligonucleotides used in the experiments. (C) Sequence of human U7 snRNA. The trimethylated cap, $m_3G$, is characteristic of 'U' RNAs. N refers to nucleotides whose identity was unknown.

When nuclear extracts from human cells were treated with a nuclease to digest RNA, they lost the ability to cleave the histone precursor. Adding back a crude preparation of snRNPs restored the activity. The extract could also be inactivated by treatment with RNase H (which cleaves RNA in an RNA–DNA hybrid) in the presence of DNA oligonucleotides containing the conserved sequence—the suspected site of snRNP interaction in the histone precursor (Figure 6–23B). Somewhat surprisingly, the mouse and mammalian consensus oligonucleotides completely blocked histone processing, but the human oligonucleotide had no effect. The two inhibitory oligonucleotides also caused the disappearance of a 63-nucleotide snRNA, which was then purified and partially sequenced (Figure 6–23C).

A. Explain the design of the oligonucleotide experiment. What were these scientists trying to accomplish by incubating the extract with a DNA oligonucleotide in the presence of RNase H?

B. Since a human extract was used, do you find it surprising that the human oligonucleotide did not inhibit processing, whereas the mouse and consensus oligonucleotides did? Can you offer an explanation for this result?

**6–48** In eucaryotes, two distinct classes of snoRNAs, which are characterized by conserved sequence motifs termed boxes, are responsible for 2′-*O*-methylation and pseudouridylation. Box C/D snoRNAs direct 2′-*O*-ribose methylation, and box H/ACA snoRNAs direct pseudouridylation of target RNAs. Recently a novel snoRNA, called U85, was shown to contain both box elements (Figure 6–24). You want to know how these box elements participate in the genesis and function of U85 snoRNA. You begin with the effects of the box elements on the synthesis of U85 snoRNA.

In vertebrates most snoRNAs are processed from introns in pre-mRNA. To test whether the processing of U85 snoRNA occurs in this fashion, you deposit the human U85 gene into the second intron of the human β-globin gene to make the plasmid construct G/U85 (Figure 6–25A). You make the

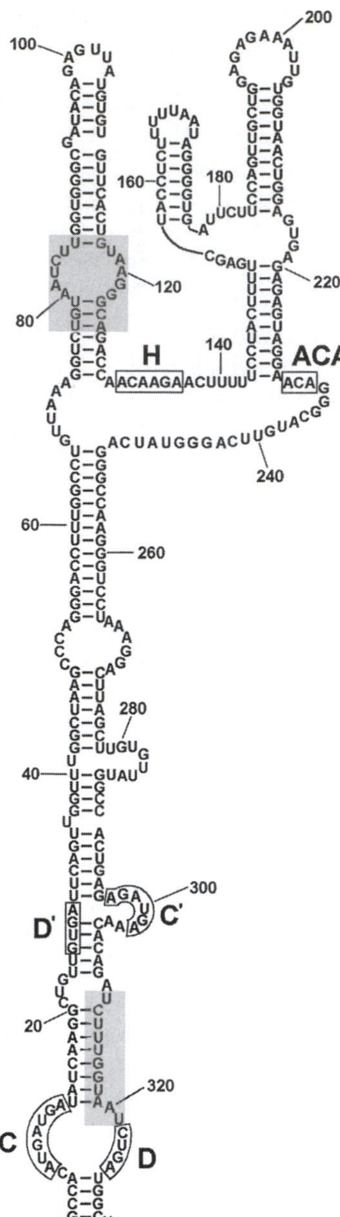


Figure 6–24 Proposed secondary structure of the human U85 snoRNA (Problem 6–48). Box elements are indicated. Regions that are thought to be important for targeting 2′-*O*-methylation and pseudouridylation are *shaded*.

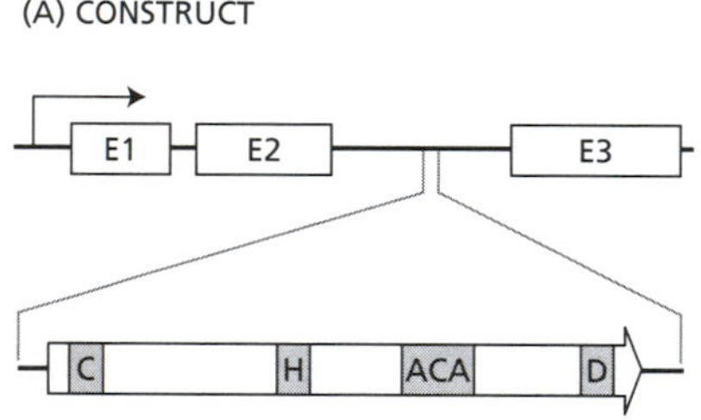


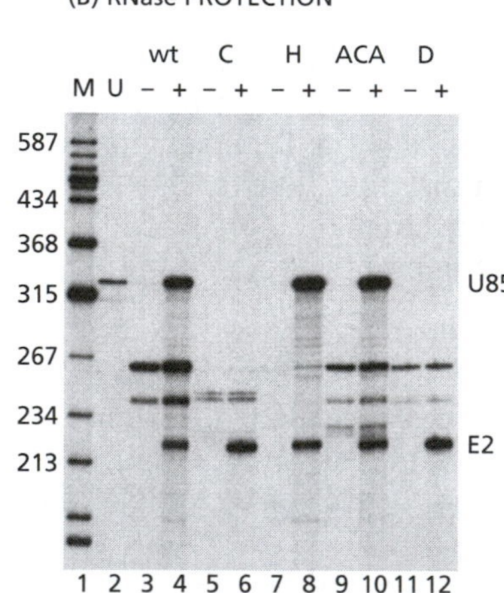


**Figure 6–25** Processing of human U85 snoRNA from the β-globin intron (Problem 6–48). (A) Structure of the G/U85 construct showing the location of U85 in the second intron of the β-globin gene. The positions of the box elements, whose sequences were altered in other, analogous constructs, are indicated in the expanded view of the U85 gene shown *below* the β-globin intron. (B) RNase protection by RNA extracted from cells. Radioactively labeled antisense RNA corresponding to U85 RNA and to β-globin exon RNAs were hybridized to RNA extracted from transfected (+) and nontransfected (–) cells, and treated with RNase. Protected RNAs were separated by electrophoresis. Bands corresponding to U85 RNA and exon 2 (E2) of the β-globin gene are indicated. Bands that show up in both transfected and nontransfected cells are artifacts. The lane marked U contains pure U85 RNA. Wild type (wt) refers to the G/U85 construct with no mutational alterations. Marker RNAs (M) are shown on the *left*; their sizes are indicated in base pairs.

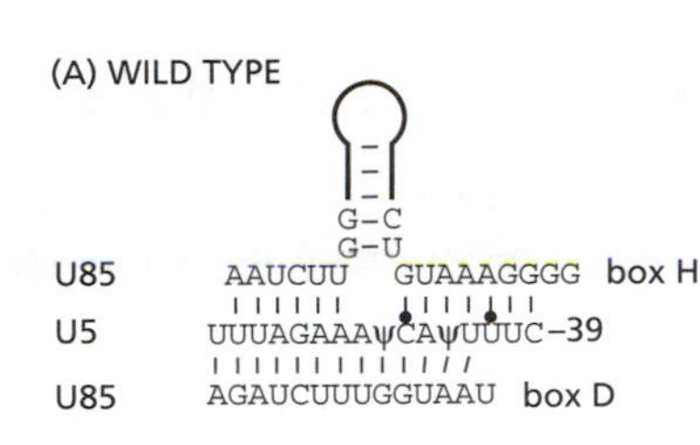


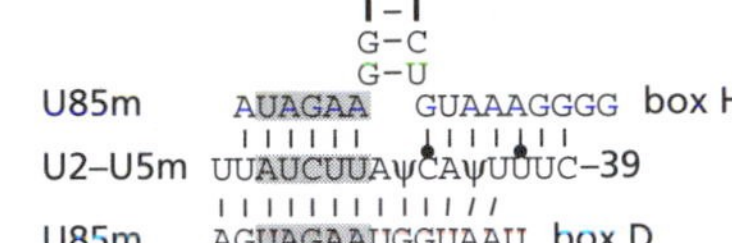


**Figure 6–26** Analysis of potential RNA targets for modification by U85 snoRNA (Problem 6–49). (A) Potential pairing between the guide sequences in U85 snoRNA and a segment of U5 snRNA. Pairing with the guide sequence adjacent to box H, which typically directs pseudouridylation, is shown *above* U5. Pairing with the guide sequence adjacent to box D, which typically directs 2′-*O*-methylation, is shown *below* U5. Only one pairing could occur at a time. Pseudouridines are indicated with ψ, and sites of 2′-*O*-methylation are indicated with *dots*. (B) Potential pairing between the mutant guide sequences in U85m snoRNA and the mutant segment of U5 in U2–U5m.

additional constructs G/U85-C, G/U85-H, G/U85-ACA, and G/U85-D, in which the indicated box element of U85 was replaced with a stretch of C nucleotides. You transfect these constructs into monkey cells. You then analyze the accumulated RNAs by RNase protection, using U85 and β-globin probes (Figure 6–25B). Are any of the conserved box elements important for the accumulation of U85 snoRNA? Explain your reasoning.

**6–49** To identify potential substrate RNAs for the U85 snoRNA, sequences of all known stable cellular RNAs were carefully examined. Sites of 2′-*O*-methylation and pseudouridylation were looked for in sequences that could pair with the putative guide sequences in U85 responsible for such modifications (see Figure 6–24, shaded areas). The search revealed that U85 snoRNA could potentially pair with a region of the U5 spliceosomal snRNA that carries two 2′-*O*-methylated nucleotides, U41 and C45, and two pseudouridines, ψ43 and ψ46 (Figure 6–26A).

To determine whether these U5 snRNA sequences were true substrates for modification by U85 snoRNA, the relevant segment of U5 was inserted into a region of the U2 snRNA gene to create a distinctive molecule U2–U5 that could be readily followed. In addition to the normal sequence, a mutant segment of U5 was inserted into U2 to make U2–U5m. A mutant, U85m, was also constructed with compensating changes in the guide sequences adjacent to the box H and box D regions, so that it could pair with U2–U5m (Figure 6–26B). Expression vectors for U2–U5, U2–U5m, and U2–U5m along with U85m were transfected into cells. All the encoded RNAs were shown to accumulate normally. 2′-*O*-Methylation and pseudouridylation in the critical region in the U2–U5 and U2–U5m molecules were detected by sequencing, as summarized in Table 6–2.

What are the expectations of these experiments if bases in U5 snRNA serve as bona fide targets for U85 snoRNA-dependent modification? Which, if any, of the naturally modified nucleotides in U5—methylated riboses at U41 and C45, and the pseudouridines ψ43 and ψ46—are dependent on U85 snoRNA? Explain your reasoning.

**Table 6–2 Modification of potential target nucleotides in various transfections (Problem 6–49).**

| | | TRANSFECTED MOLECULES | | | MODIFICATIONS[b] | | | |
|---|---|---|---|---|---|---|---|---|
| | | | | | PSEUDOURIDINE | | 2′-*O*-METHYL | |
| | U85[a] | U85m | U2–U5 | U2–U5m | U43 | U46 | U41 | C45 |
| 1 | + | | + | | – | ψ | m | m |
| 2 | + | | | + | – | – | m | – |
| 3 | + | + | | + | – | ψ | m | m |

[a]Endogenous U85 is present in all cells.
[b]Nucleotides that have been converted to pseudouridine are indicated by ψ; nucleotides that have been 2′-*O*-methylated are indicated by an m.

# FROM RNA TO PROTEIN

| TERMS TO LEARN | | |
|---|---|---|
| aminoacyl-tRNA synthetase | molecular chaperone | ribosome |
| anticodon | nonsense-mediated mRNA decay | ribozyme |
| codon | prion disease | rRNA (ribosomal RNA) |
| eucaryotic initiation factor (eIF) | proteasome | translation |
| genetic code | reading frame | tRNA (transfer RNA) |
| initiator tRNA | | |

## DEFINITIONS

Match the definition below with its term from the list above.

**6–50** Large protein complex in the cytosol and nucleus with proteolytic activity that is responsible for degrading the proteins marked for destruction.

**6–51** Set of rules specifying the correspondence between nucleotide triplets in DNA or RNA and amino acids in proteins.

**6–52** Special tRNA that carries methionine and is used to begin translation.

**6–53** Sequence of three nucleotides in a tRNA that is complementary to a three-nucleotide sequence in an mRNA molecule.

**6–54** RNA with catalytic activity.

**6–55** Surveillance system that eliminates defective mRNAs before they can be translated into protein.

**6–56** The three-nucleotide phase in which nucleotides in an mRNA are translated into amino acids in a protein.

**6–57** Enzyme that attaches the correct amino acid to a tRNA molecule to form the activated intermediate used in protein synthesis.

**6–58** Transmissible spongiform encephalopathy such as Creutzfeld–Jacob disease in humans that is apparently caused and transmitted by abnormal forms of proteins.

**6–59** Protein that helps other proteins avoid misfolding pathways that produce inactive or aggregated polypeptides.

## TRUE/FALSE

Decide whether each of these statements is true or false, and then explain why.

**6–60** Wobble pairing occurs between the first position in the codon and the third position in the anticodon.

**6–61** The only significant difference between simple proteases and the proteasome is that the proteasome targets ubiquitylated proteins.

## THOUGHT PROBLEMS

**6–62** For the RNA sequence below indicate the amino acids that are encoded in the three reading frames. If you were told that this segment of RNA was in the middle of an mRNA that encoded a large protein, would you know which reading frame was used? How so? (The genetic code is on the inside back cover.)

AGUCUAGGCACUGA

**6–63** After treating cells with a chemical mutagen, you isolate two mutants. One carries alanine and the other carries methionine at a site in the protein that normally contains valine (Figure 6–27). After treating these two mutants again with the mutagen, you isolate mutants from each that now carry threonine at the site of the original valine (Figure 6–27). Assuming that all mutations involve single-nucleotide changes, deduce the codons that are used for valine, methionine, threonine, and alanine at the affected site. Would you expect to be able to isolate a valine-to-threonine mutant in one step?

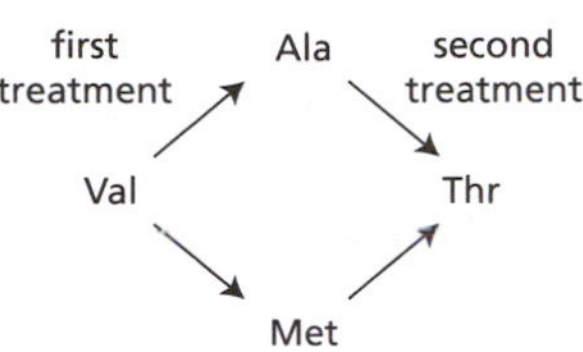


**Figure 6–27** Two rounds of mutagenesis and the altered amino acids at a single position in a protein (Problem 6–63).

**6–64** The genetic code was deciphered in part by experiments in which polynucleotides of repeating sequences were used as mRNAs to direct protein synthesis in cell-free extracts. In the test tube, artificial conditions were used that allowed ribosomes to start protein synthesis anywhere on an RNA molecule, without the need for a translation start codon, as required in a living cell. What types of polypeptides would you expect to be synthesized if the following polynucleotides were used as templates in such a cell-free extract?

A. UUUUUUUUUUUU...
B. AUAUAUAUAUAU...
C. AUCAUCAUCAUC...

**6–65** In *B. lichenformis* a few amino acids are removed from the C-terminus of the β-lactamase enzyme after it is synthesized. The sequence of the original C-terminus can be deduced by comparing it to a mutant in which the reading frame is shifted by the insertion or deletion of a nucleotide and the mutant β-lactamase escapes cleavage. The amino acid sequences of the purified wild-type enzyme and the frameshift mutant from amino acid 263 to the C-terminal end are given below.

```
                                  ↓ site of protein cleavage
wild type:           N M N G K
frameshift mutant:   N M I W Q I C V M K D
```

A. What was the mutational event that gave rise to the frameshift mutant?
B. Deduce the number of amino acids in the synthesized form of the wild-type enzyme and, as far as possible, the sequence of the deleted C-terminus.

**6–66** Which of the following mutational changes would you predict to be the most deleterious to gene function? Explain your answer.

1. Insertion of a single nucleotide near the end of the coding sequence.
2. Removal of a single nucleotide near the beginning of the coding sequence.
3. Deletion of three consecutive nucleotides in the middle of the coding sequence.
4. Deletion of four consecutive nucleotides in the middle of the coding sequence.
5. Substitution of one nucleotide for another in the middle of the coding sequence.

**6–67** Consider the properties of two hypothetical genetic codes constructed with the four common nucleotides: A, G, C, and T.

A. Imagine that the genetic code is constructed so that pairs of nucleotides are used as codons. How many different amino acids could such a code specify?
B. Imagine that the genetic code is a triplet code; that is, it uses three nucleotides to specify each amino acid. In this code, the amino acid specified by each codon depends only on the composition of the codon—not the sequence. Thus, for example, CCA, CAC, and ACC, which all have the composition $C_2A$, would encode the same amino acid. How many different amino acids could such a code specify?
C. Would you expect the genetic codes in A and B to lead to difficulties in the process of translation, using mechanisms analogous to those used in translating the standard genetic code?

**Problem 1–14** considers the amazing mutational resistance of the natural genetic code.

**6–68** One remarkable feature of the genetic code is that amino acids with similar chemical properties often have similar codons. Codons with U or C as the second nucleotide, for example, tend to specify hydrophobic amino acids.

Can you suggest a possible explanation for this phenomenon in terms of the early evolution of the protein synthesis machinery?

**6–69** The rules for wobble pairing in bacteria and eucaryotes are shown in Table 6–3. On the left side of the table the rules are expressed as a wobble codon base and its recognition by possible anticodon bases. [The anticodon base I (inosine) is a common modification in tRNAs; it is generated by deamination of A.] Reformulate these rules as particular anticodon bases and their recognition by possible codon bases, as suggested by the partial information on the right side of the table.

**6–70** Given the wobble rules for codon–anticodon pairing in bacteria, the minimum number of different tRNAs that would be required to recognize all 61 codons is 31. What is the minimum number of different tRNAs that is consistent with the wobble rules used in eucaryotes (see Table 6–3)?

**6–71** A mutation in a bacterial gene generates a UGA stop codon in the middle of the mRNA coding for the protein product. A second mutation in the cell leads to a single nucleotide change in a tRNA that allows the correct translation of the protein; that is, the second mutation 'suppresses' the defect caused by the first. The altered tRNA translates the UGA codon as tryptophan. What nucleotide change has probably occurred in the mutant tRNA molecule? What consequences would the presence of such a mutant tRNA have for the translation of the normal genes in this cell?

**6–72** In a clever experiment performed in 1962, a cysteine that was already attached to $tRNA^{Cys}$ was chemically converted to an alanine. These alanyl-$tRNA^{Cys}$ molecules were then added to a cell-free translation system from which the normal cysteinyl-$tRNA^{Cys}$ molecules had been removed. When the resulting protein was analyzed, it was found that alanine had been inserted at every point in the protein chain where cysteine was supposed to be. Discuss what this experiment tells you about the role of aminoacyl-tRNA synthetases during the normal translation of the genetic code.

**6–73** The charging of a tRNA with an amino acid occurs according to the reaction

$$\text{amino acid} + \text{tRNA} + \text{ATP} \rightarrow \text{aminoacyl-tRNA} + \text{AMP} + \text{PP}_i$$

where $PP_i$ is pyrophosphate, the linked phosphates that were cleaved from ATP to generate AMP. In the aminoacyl-tRNA, the amino acid and tRNA are linked by a high-energy bond. Thus, a large portion of the energy derived from the hydrolysis of ATP is stored in this bond and is available to drive peptide-bond formation at the later stages of protein synthesis. The free-energy change ($\Delta G°$) for the charging reaction shown above is close to zero,

**Table 6–3 Rules for wobble base-pairing between codon and anticodon** (Problem 6–69).

| | WOBBLE CODON BASE | POSSIBLE ANTICODON BASE | | WOBBLE ANTICODON BASE | POSSIBLE CODON BASE |
|---|---|---|---|---|---|
| Bacteria | U | A, G, or I | Bacteria | U | |
| | C | G or I | | C | |
| | A | U or I | | A | |
| | G | C or U | | G | |
| | | | | I | |
| Eucaryotes | U | G or I | Eucaryotes | U | |
| | C | G or I | | C | |
| | A | U | | A | |
| | G | C | | G | |
| | | | | I | |

so that attachment of the amino acid to tRNA would not be expected to be dramatically favored. Can you suggest a further step that could help drive the charging reaction to completion?

**6–74** The protein you are studying contains five leucines and consists of a single polypeptide chain. One leucine is C-terminal and another is N-terminal. In a suspension of cells the average time required to synthesize this polypeptide is 8 minutes. At time zero, radioactive leucine is added to five different suspensions of cells that are *already* in the process of synthesizing the protein. You isolate the complete protein from individual suspensions at 2, 4, 6, 8, and 80 minutes. (Any incomplete polypeptide chains are eliminated at this step.) The proteins are then analyzed for N-terminal and total radioactive leucine. With increasing time of exposure of the cells to the radioactive leucine, the ratio of N-terminal radioactivity to total radioactivity in the isolated protein should:

A. Increase to a final value of 0.2.
B. Remain constant at a value of 0.2.
C. Decrease to a final value of 0.2.
D. An answer cannot be determined from this information.

**6–75** It is commonly reported that 30–50% of a cell's energy budget is spent on protein synthesis. How do you suppose such a measurement might be made?

**6–76** One strand of a section of DNA isolated from *E. coli* reads

5′-GTAGCCTACCCATAGG-3′

A. Suppose that an mRNA were transcribed using the complement of this DNA strand as the template. What would the sequence of the mRNA in this region be?
B. How many different peptides could potentially be made from this sequence of RNA? Would the same peptides be made if the other strand of the DNA served as the template for transcription?
C. What peptide would be made if translation started exactly at the 5′ end of the mRNA in part A? When $tRNA^{Ala}$ leaves the ribosome, what tRNA will be bound next? When the amino group of alanine forms a peptide bond, what bonds, if any, are broken, and what happens to $tRNA^{Ala}$?

**6–77** The elongation factor EF-Tu introduces two short delays between codon–anticodon base-pairing and formation of the peptide bond. These delays increase the accuracy of protein synthesis. Describe these delays and explain how they improve the fidelity of translation.

**6–78** Polycistronic mRNAs are common in procaryotes but extremely rare in eucaryotes. Describe the key differences in protein synthesis that underlie this observation.

**6–79** Procaryotes and eucaryotes both protect against the dangers of translating broken mRNAs. What dangers do partial mRNAs pose for the cell?

**6–80** You have isolated an antibiotic, named edeine, from a bacterial culture. Edeine inhibits protein synthesis but has no effect on either DNA synthesis or RNA synthesis. When added to a reticulocyte lysate, edeine stops protein synthesis after a short lag, as shown in Figure 6–28. By contrast, cycloheximide stops protein synthesis immediately (Figure 6–28). Analysis of the edeine-inhibited lysate by density-gradient centrifugation showed that no polyribosomes remained at the time protein synthesis had stopped. Instead, all the globin mRNA accumulated in an abnormal 40S peak, which contained equimolar amounts of the small ribosomal subunit and initiator tRNA.

A. What step in protein synthesis does edeine inhibit?
B. Why is there a lag between addition of edeine and cessation of protein synthesis? What determines the length of the lag?

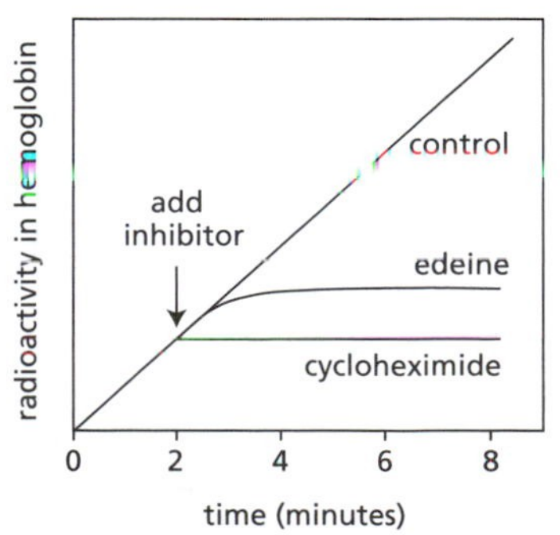


**Figure 6–28** Effects of the inhibitors edeine and cycloheximide on protein synthesis in reticulocyte lysates (Problem 6–80).

C. Would you expect the polyribosomes to disappear if you added cycloheximide at the same time as edeine?

**6–81** In a reticulocyte lysate the polynucleotide 5′-AUGUUUUUUUUU directs the synthesis of Met–Phe–Phe–Phe. In the presence of farsomycin, a new antibiotic perfected by Fluhardy Pharmaceuticals, this polymer directs synthesis of Met–Phe only. From this information, which of the following deductions could you make about farsomycin?

A. It prevents formation of the 80S initiation complex, which contains the initiator tRNA and both ribosomal subunits.
B. It inhibits binding of aminoacyl-tRNAs to the A-site in the ribosome.
C. It inactivates peptidyl transferase activity of the large ribosomal subunit.
D. It blocks translocation of peptidyl-tRNA from the A-site to the P-site of the ribosome.
E. It interferes with chain termination and release of the peptide.

**6–82** Both hsp60-like and hsp70 molecular chaperones share an affinity for exposed hydrophobic patches on proteins, using them as indicators of incomplete folding. Why do you suppose hydrophobic patches serve as critical signals for the folding status of a protein?

**6–83** Most proteins require molecular chaperones to assist in their correct folding. How do you suppose the chaperones themselves manage to fold correctly?

**6–84** Describe the roles of E1, E2, and E3 proteins in conjugating ubiquitin to target proteins. Which components provide the specificity to allow particular proteins or sets of proteins to be targeted for destruction?

## CALCULATIONS

**6–85** Rates of peptide chain growth can be estimated from data such as those shown in Figure 6–29. In this experiment, a tobacco mosaic virus (TMV) mRNA, which encodes a 116,000-dalton protein, was translated in a rabbit-reticulocyte lysate in the presence of $^{35}$S-methionine. Samples were removed at 1-minute intervals and subjected to electrophoresis on SDS-polyacrylamide gels. The separated translation products were visualized by

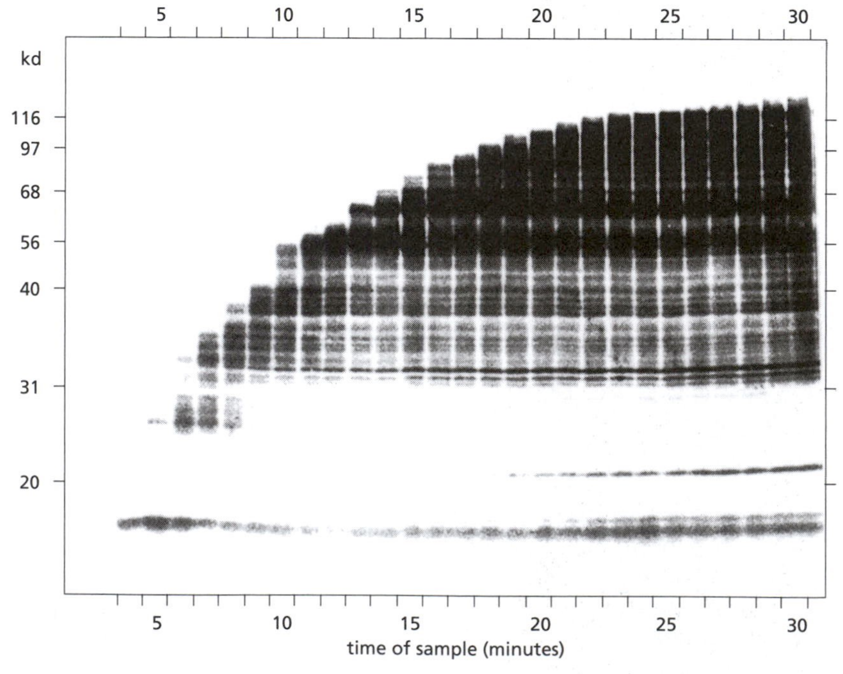


**Figure 6–29** Time course of synthesis of a TMV protein in a rabbit-reticulocyte lysate (Problem 6–85). No radioactivity was detected during the first 3 minutes because the short chains ran off the bottom of the gel. SDS denatures proteins so that they run approximately according to their molecular masses. A scale of molecular masses in kilodaltons is shown on the *left*.

autoradiography. As is apparent in Figure 6–29, the largest detectable polypeptides get larger with time, until the full-length protein appears at about 25 minutes.

A. Is the rate of synthesis linear with time? One simple way to answer this question is to determine the molecular mass of the largest peptide in each sample, as determined by reference to the standards shown on the left in Figure 6–29, and then plot each of these masses against the time at which the relevant sample was taken.

B. What is the rate of protein synthesis (in amino acids/minute) in this experiment? Assume the average molecular mass of an amino acid is 110 daltons.

C. Why does the autoradiograph have so many bands in it rather than just a few bands that get larger as time passes; that is, why does the experiment produce the 'actual' result (Figure 6–30A) rather than the 'theoretical' result (Figure 6–30B)? Can you think of a way to manipulate the experimental conditions to produce the theoretical result?

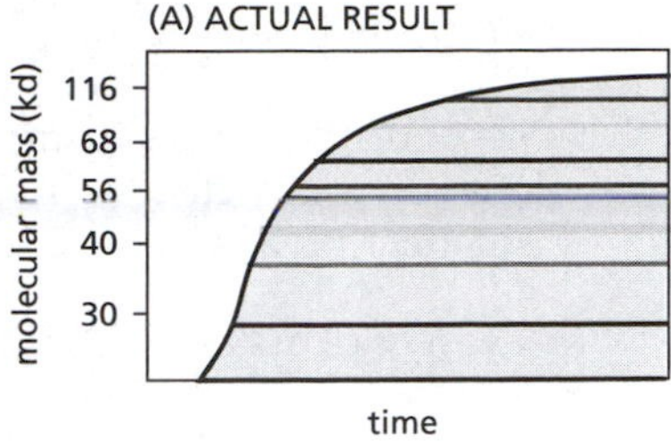


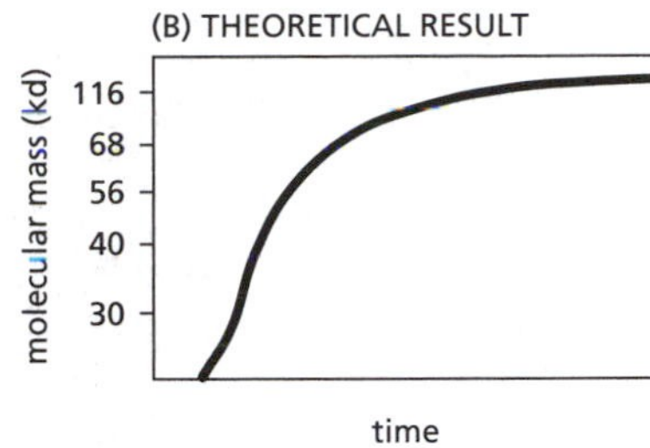


**Figure 6–30** Potential outcomes of experiments on rates of protein synthesis (Problem 6–85).

**6–86** The average molecular weight of proteins encoded in the human genome is about 50,000. A few proteins are very much larger that this average. For example, the protein called titin, which is made by muscle cells, has a molecular weight of 3,000,000.

A. Estimate how long it will take a muscle cell to translate an mRNA coding for an average protein and one coding for titin. The average molecular mass of amino acids is about 110 daltons. Assume that the translation rate is two amino acids per second.

B. If the nucleotides in the coding portion of the mRNA constitute 5% of the total that are transcribed, how long will it take a muscle cell to transcribe a gene for an average protein versus the titin gene. Assume that the transcription rate is 20 nucleotides per second.

**6–87** Protein synthesis consumes four high-energy phosphate bonds per added amino acid. Transcription consumes two high-energy phosphate bonds per added nucleotide. Calculate how many protein molecules will have been made from an individual mRNA at the point when the energy cost of translation is equal to the energy cost of transcription. Assume that the nucleotides in the coding portion of the mRNA constitute 5% of the total that are transcribed.

**6–88** The overall accuracy of protein synthesis is difficult to measure because mistakes are very rare. One ingenious approach used flagellin (molecular weight 40,000). Flagellin is the sole protein in bacterial flagella and thus easy to purify. Because flagellin contains no cysteine, it allows for sensitive detection of cysteine that has been misincorporated into the protein.

To generate radioactive cysteine, bacteria were grown in the presence of $^{35}SO_4^{2-}$ (specific activity $5.0 \times 10^3$ cpm/pmol) for exactly one generation with excess unlabeled methionine in the growth medium (to prevent the incorporation of $^{35}S$ label into methionine). Flagellin was purified and assayed: 8 μg of flagellin was found to contain 300 cpm of $^{35}S$ radioactivity.

**Problem 3–39** examines the relationship between error rate, length, and fraction of proteins correctly synthesized.

A. Of the flagellin molecules that were synthesized during the labeling period, what fraction contained cysteine? Assume that the mass of flagellin doubles during the labeling period and that the specific activity of cysteine in flagellin is equal to the specific activity of the $^{35}SO_4^{2-}$ used to label the cells.

B. In flagellin, cysteine is misincorporated at the arginine codons CGU and CGC. In terms of anticodon–codon interaction, what mistake is made during the misincorporation of cysteine for arginine?

C. Given that there are 18 arginines in flagellin and assuming that all arginine codons are equally represented, what is the frequency of misreading of each sensitive (CGU and CGC) arginine codon?

D. Assuming that the error frequency per codon calculated above applies to all amino acid codons equally, calculate the percentage of molecules that are correctly synthesized for proteins 100, 1000, and 10,000 amino acids in length. The probability of synthesizing a correct protein is $P = (1 - E)^n$, where $E$ is the error frequency and $n$ is the number of amino acids added.

## DATA HANDLING

**6–89** Many of the errors in protein synthesis occur because tRNA synthetases have difficulty discriminating between related amino acids. For example, isoleucyl-tRNA synthetase (IleRS) normally activates isoleucine.

$$\text{IleRS} + \text{Ile} + \text{ATP} \rightarrow \text{IleRS(Ile-AMP)} + \text{PP}_i$$

At a frequency about 1/180 of the correct activation, IleRS misactivates valine.

$$\text{IleRS} + \text{Val} + \text{ATP} \rightarrow \text{IleRS(Val-AMP)} + \text{PP}_i$$

Protein synthesis is more accurate than this frequency might suggest because the synthetase subsequently edits out most of its mistakes, in a reaction that depends on the presence of $\text{tRNA}^{\text{Ile}}$.

$$\text{IleRS(Val-AMP)} + \text{tRNA}^{\text{Ile}} \rightarrow \text{IleRS} + \text{Val} + \text{AMP} + \text{tRNA}^{\text{Ile}} \qquad \text{EDITING}$$

The $\text{tRNA}^{\text{Ile}}$ is, of course, also required for proper aminoacylation (charging) by isoleucine to make Ile-$\text{tRNA}^{\text{Ile}}$.

$$\text{IleRS(Ile-AMP)} + \text{tRNA}^{\text{Ile}} \rightarrow \text{Ile-tRNA}^{\text{Ile}} + \text{IleRS} + \text{AMP} \qquad \text{CHARGING}$$

Are the parts of $\text{tRNA}^{\text{Ile}}$ that are required for isoleucine charging the same as those that are required for valine editing?

One approach to this question is to make changes in the $\text{tRNA}^{\text{Ile}}$ to see whether the two activities track with one another. Rather than change the sequence nucleotide by nucleotide, blocks of sequence changes were made, using $\text{tRNA}^{\text{Val}}$ as a donor. $\text{tRNA}^{\text{Val}}$ by itself does not stimulate isoleucine charging or valine editing by IleRS. Changing its anticodon from 5′-CAU to 5′-GAU, however, allows it to be charged fairly efficiently by IleRS. A variety of chimeric tRNAs were made by combining bits of $\text{tRNA}^{\text{Ile}}$ and $\text{tRNA}^{\text{Val}}$. The ability of each chimera to stimulate isoleucine charging and valine editing was then tested, as shown in Figure 6–31.

A. What (at a minimum) must be inserted into $\text{tRNA}^{\text{Val}}$ to permit isoleucine charging by IleRS?

B. What (at a minimum) must be inserted into $\text{tRNA}^{\text{Val}}$ to permit valine editing by IleRS?

**Problem 3–65** looks at editing and charging from the standpoint of the isoleucyl-tRNA synthetase.

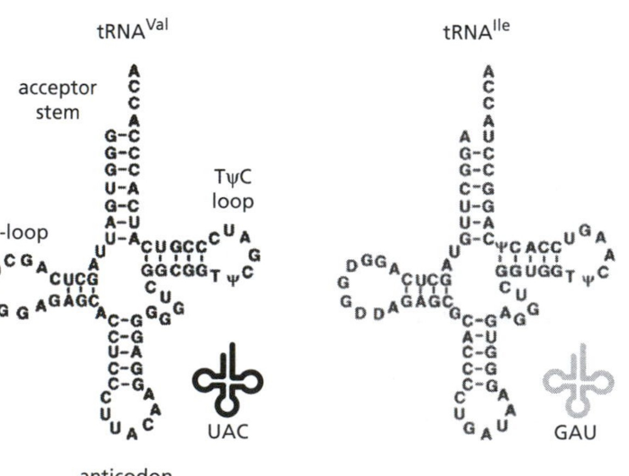


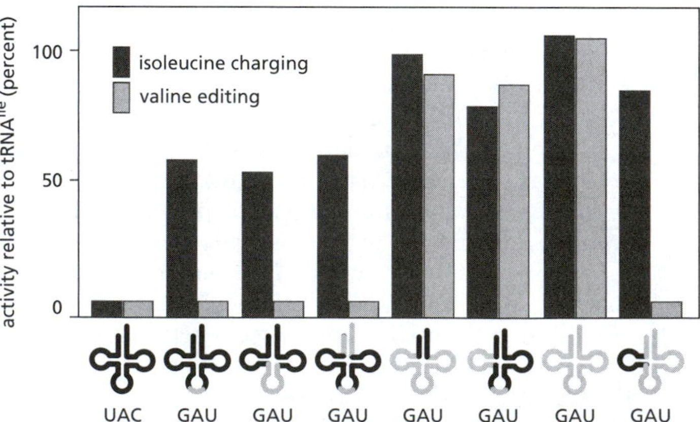


**Figure 6–31** Portions of $\text{tRNA}^{\text{Ile}}$ responsible for charging and editing (Problem 6–89). (A) $\text{tRNA}^{\text{Val}}$ and $\text{tRNA}^{\text{Ile}}$. (B) Isoleucine charging and valine editing. Chimeric tRNAs composed of bits from $\text{tRNA}^{\text{Val}}$ (*black*) and $\text{tRNA}^{\text{Ile}}$ (*gray*) are shown *below* the results for isoleucine charging (*black bars*) and valine editing (*gray bars*).

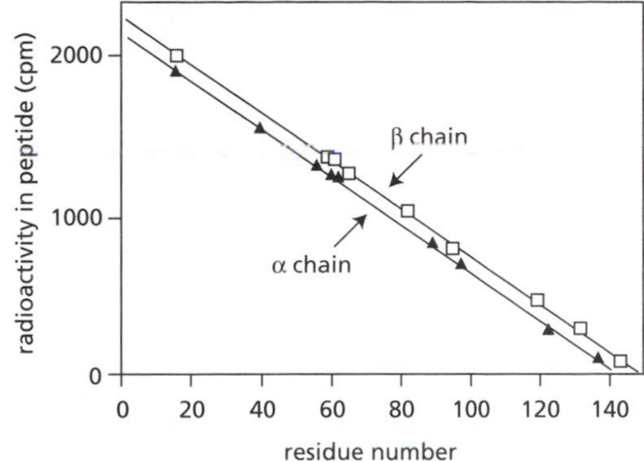


**Figure 6–32** Synthesis of α- and β-globin chains (Problem 6–90).

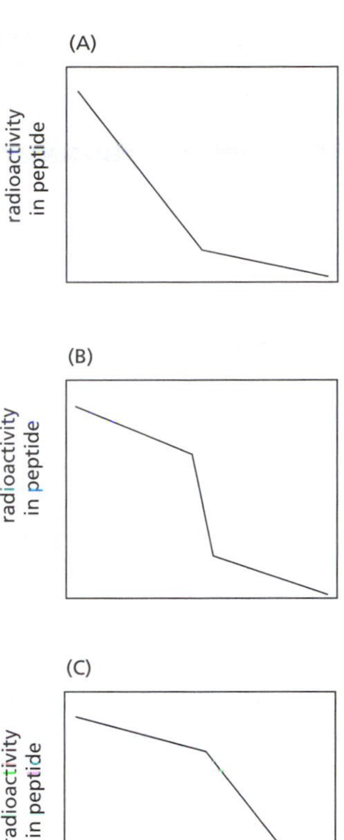


**Figure 6–33** Hypothetical curves for globin synthesis with a roadblock to ribosome movement at the midpoint of the mRNA (Problem 6–90). These schematic diagrams are analogous to the graph in Figure 6–32.

C. Does IleRS recognize the same features of $tRNA^{Ile}$ when it catalyzes isoleucine charging that it does when it carries out valine editing? Explain your answer.

**6–90** Consider the following experiment on the coordinated synthesis of the α and β chains of hemoglobin. Rabbit reticulocytes were labeled with $^3$H-lysine for 10 minutes, which is very long relative to the time required for the synthesis of a single globin chain. The ribosomes, with attached nascent chains, were then isolated by centrifugation to give a preparation free of soluble (finished) globin chains. The nascent globin chains were digested with trypsin, which gives peptides ending in C-terminal lysine or arginine. The peptides were then separated by high-performance liquid chromatography (HPLC), and their radioactivity was measured. A plot of the radioactivity in each peptide versus the position of the lysines in the chains (numbered from the N-terminus) is shown in Figure 6–32.

A. Do these data allow you to decide which end of the globin chain (N- or C-terminus) is synthesized first? How so?
B. In what ratio are the two globin chains produced? Can you estimate the relative numbers of α- and β-globin mRNA molecules from these data?
C. How long does a protein chain stay attached to the ribosome once the termination codon has been reached?
D. It was once suggested that heme is added to nascent globin chains during their synthesis and, furthermore, that ribosomes must wait for the insertion of heme before they can proceed. The straight lines in Figure 6–32 indicate that ribosomes do not pause significantly, and heme is now thought to be added after synthesis. From among the graphs shown in Figure 6–33, choose the one that would have resulted if there were a significant roadblock to ribosome movement halfway down the globin mRNA.

**6–91** Termination codons in bacteria are decoded by one of two proteins. Release factor 1 (RF1) recognizes UAG and UAA, whereas RF2 recognizes UGA and UAA. For RF2, a comparison of the nucleotide sequence of the gene with the amino acid sequence of the protein revealed a startling surprise, which is contained within the sequences shown below the gene in Figure 6–34. Sequences of the gene and protein were checked carefully to rule out any artifacts.

A. What is the surprise?
B. What hypothesis concerning the regulation of RF2 expression is suggested by this observation?

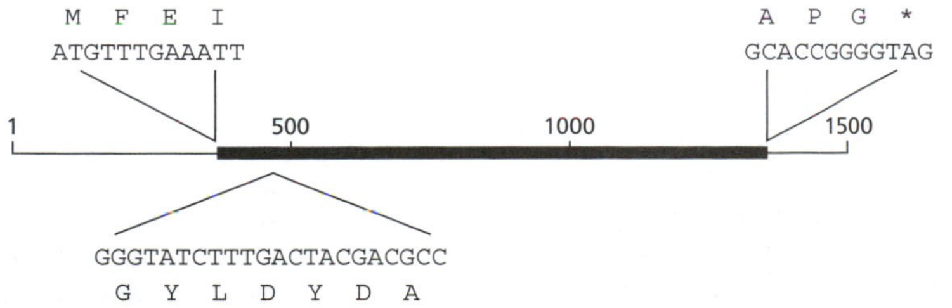


**Figure 6–34** Schematic representation of the gene for RF2 (Problem 6–91). The coding sequence is shown as a *thick line*, with sequences at the start and finish shown for reference.

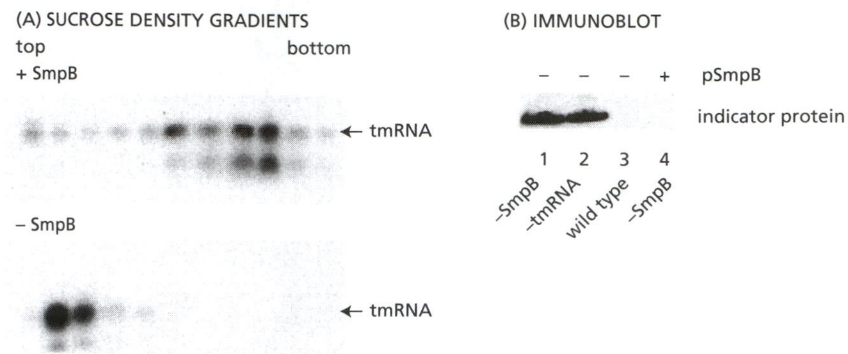


**Figure 6–35** Analysis of the role of SmpB in tmRNA-mediated protein tagging and destruction (Problem 6–92).
(A) Association of tmRNA with ribosomes in normal and SmpB-deficient cells. Cell components were separated by centrifugation after layering the cell extract on top of a sucrose gradient. Under the conditions used in these experiments, free tmRNA remains near the top of the gradient and polysomes move toward the bottom. The position of tmRNA in the gradient was determined by hybridization to tmRNA-specific oligonucleotides. (B) Destruction of a protein fragment in wild-type cells and in SmpB-deficient (–SmpB) or tmRNA-deficient (–tmRNA) cells. An mRNA that encodes the indicator protein but does not carry a stop codon was included in all lanes. Plasmid pSmpB expresses SmpB; its presence (+) or absence (–) is indicated *above* the lanes.

**6–92** In procaryotic cells the SmpB protein binds to tmRNA, thereby allowing it to attach a C-terminal tag to proteins whose translation is stalled. You wish to find out the point at which SmpB operates. tmRNA is initially charged with alanine and then it binds to the A-site of a ribosome in which mRNA translation has stalled. The alanine is attached to the end of the protein along with 10 additional amino acids that tmRNA encodes. Finally, the tagged protein is released from the ribosome for destruction by a special protease.

Several experiments were carried out to determine where SmpB acts in this sequence of events. Alanyl-tRNA synthetase was shown to charge tmRNA with alanine equally well in the presence and absence of SmpB. The association of tmRNA with ribosomes, however, was different in the presence (+) and absence (–) of SmpB (Figure 6–35A). Destruction of proteins generated by stalled ribosomes, which was detected using a convenient indicator protein, was also different in the presence (+) and absence (–) of SmpB (Figure 6–35B).

To the extent that these experiments allow, describe the point at which SmpB is needed to permit tmRNA to carry out its function.

**Problem 3–102** characterizes the tight binding between SmpB and tmRNA.

**6–93** You are studying protein synthesis in *Tetrahymena*, which is a unicellular ciliate. You have good news and bad news. The good news is that you have the first bit of protein and nucleic acid sequence data for the C-terminus of a *Tetrahymena* protein, as shown below:

```
 I   M   Y   K   Q   V   A   Q   T   Q   L   *
AUU AUG UAU AAG UAG GUC GCA UAA ACA CAA UUA UGA GAC UUA
```

The bad news is that you have been unable to translate a preparation of *Tetrahymena* mRNA in a reticulocyte lysate, which is a standard system for analyzing protein synthesis *in vitro*. The mRNA preparation looks good by all criteria, but the translation products are mostly small polypeptides (Figure 6–36, lane 1).

To figure out what is wrong, you do a number of control experiments using a pure mRNA from tobacco mosaic virus (TMV) that encodes a 116-kd protein. TMV mRNA alone is translated just fine in the *in vitro* system, giving a major band at 116 kd—the expected product—and a very minor band about 50 kd larger (Figure 6–36, lane 2). When *Tetrahymena* RNA is added, there is a decrease in the smaller of the two bands and a significant increase in the larger one (lane 3). When some *Tetrahymena* cytoplasm (minus the ribosomes) is added, the TMV mRNA now gives mostly the higher molecular mass product (lane 4); furthermore, much to your delight, the previously inactive *Tetrahymena* mRNA now appears to be translated (lane 4). You confirm this by leaving out the TMV mRNA (lane 5).

A. What is unusual about the sequence data for the *Tetrahymena* protein?
B. How do you think the more minor of the two higher molecular mass bands is produced from pure TMV mRNA in the reticulocyte lysate (lane 2)?
C. Explain the basis for the shift in proportions of the major and minor TMV proteins upon addition of *Tetrahymena* RNA alone and in combination with *Tetrahymena* cytoplasm. What *Tetrahymena* components are likely to be required for the efficient translation of *Tetrahymena* mRNA?
D. Comment on the evolutionary implications of your results.

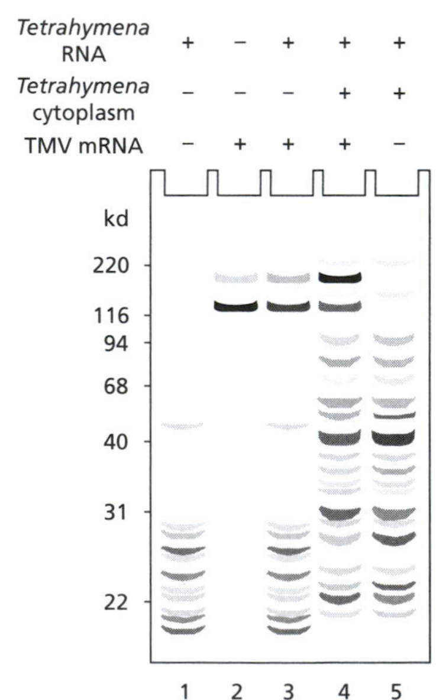


**Figure 6–36** Translation of TMV and *Tetrahymena* mRNA in the presence and absence of various components from *Tetrahymena* (Problem 6–93). The molecular masses of marker proteins are indicated in kilodaltons on the *left*.

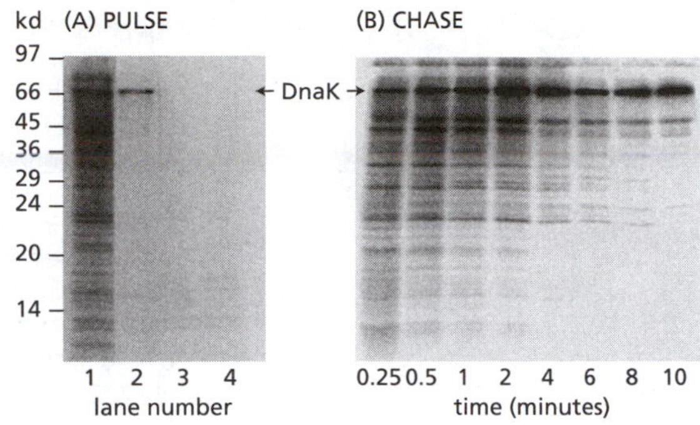


**Figure 6–37** Association of DnaK with nascent proteins (Problem 6–94). (A) Pulse-labeled proteins immunoprecipitated by antibodies to DnaK. Wild-type bacteria (lane 1); SDS-treated wild-type bacteria (lane 2); *DnaK*-deletion strain (lane 3); and mixture of labeled *DnaK*-deletion strain and unlabeled wild-type strain (lane 4). Size markers are indicated on the *left* and the position of DnaK is indicated on the *right*. (B) Pulse-chase experiment. Wild-type bacteria labeled for 15 seconds with $^{35}$S-methionine were then incubated for varying times in the presence of an excess of unlabeled methionine before immunoprecipitation by antibodies against DnaK.

**6–94** Hsp70 molecular chaperones are thought to bind to hydrophobic regions of nascent polypeptides on ribosomes. This binding was difficult to demonstrate for DnaK, which is one of the two major Hsp70 chaperones in *E. coli.* In one approach nascent proteins were labeled with a 15-second pulse of $^{35}$S-methionine, isolated in the absence of ATP, and then incubated with antibodies against DnaK. A collection of labeled proteins was precipitated as shown in Figure 6–37A, lane 1. The proteins were not precipitated if they were treated beforehand with the strong detergent SDS (lane 2), or if they were isolated from a mutant strain missing DnaK (*DnaK*-deletion, lane 3). If labeled *DnaK*-deletion cells were mixed with unlabeled wild-type cells before the proteins were isolated, DnaK antibodies did not precipitate labeled proteins (lane 4). Finally, if unlabeled methionine was added in excess after the pulse of $^{35}$S-methionine, the labeled proteins disappeared with time (Figure 6–37B).

A. Do the series of control experiments in Figure 6–37A, lanes 2 to 4, argue that DnaK is bound to the labeled proteins in a meaningful way (as opposed to a random aggregation, for example)?
B. When ATP was present during the isolation of the proteins, antibodies against DnaK did not precipitate any proteins. How do you suppose ATP might interfere with precipitation of labeled proteins?
C. Why do you suppose that the labeled proteins disappeared with time in the presence of excess unlabeled methionine?
D. Do these experiments show that DnaK binds to proteins as they are being synthesized on ribosomes? Why or why not?

**6–95** DnaK and trigger factor (TF) are the two major Hsp70 chaperones in *E. coli*, yet neither gene is essential for growth at 37°C. To investigate this seeming paradox, you construct three mutant strains for comparison with the wild-type (wt) parent. One carries a deletion of the TF gene (Δ*Tig*); the second carries an altered version of *DnaK* that is expressed from an isopropyl β-D-thiogalactopyranoside (IPTG)-inducible promoter (I-*DnaK*); and the third contains both mutations. You spot various dilutions of the four types of bacteria on plates supplemented with IPTG (which turns on the *DnaK* gene) or lacking IPTG (which turns off the *DnaK* gene) and grow them at a variety of temperatures (Figure 6–38).

A. Compare the growth properties of the strains (Δ*Tig* and I-*DnaK*) that express a single Hsp70.

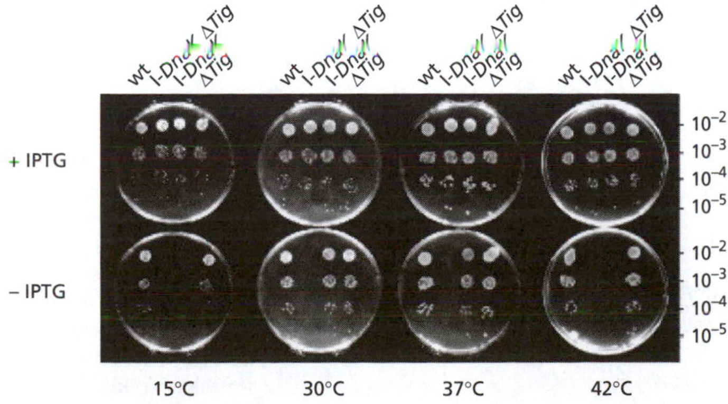


**Figure 6–38** Analysis of growth of bacterial strains defective for one or both hsp70 proteins (Problem 6–95). Dilutions (shown on the *right*) of bacterial cultures (labeled at the *top*) were spotted on agar plates in the presence and absence of IPTG (indicated on the *left*) and grown at various temperatures (shown at the *bottom*). *White spots* indicate bacterial growth.

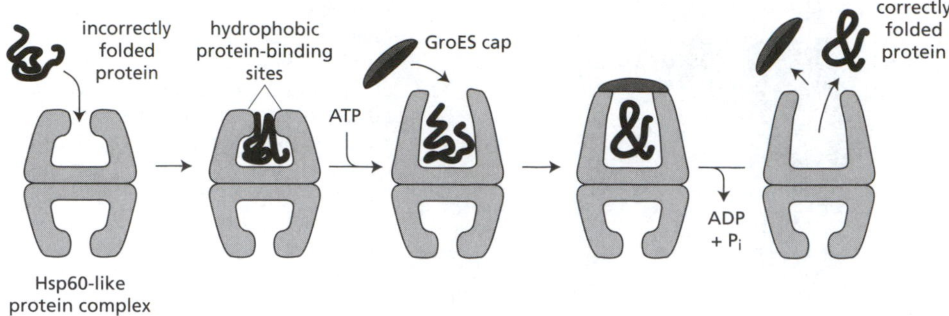


**Figure 6–39** Protein refolding by the bacterial GroEL chaperone (Problem 6–96). A misfolded protein is initially captured by hydrophobic interactions along one rim of the barrel. The subsequent binding of ATP plus the GroES cap increases the size of the cavity and confines the protein in the enclosed space, where it has a new opportunity to fold. After about 15 seconds, ATP hydrolysis ejects the protein, whether folded or not, and the cycle repeats.

B. Contrast the growth properties of the single mutants with those of the double mutant (I-*DnaK* Δ*Tig*), which expresses neither heat shock protein. Suggest an explanation for any significant differences in the growth properties of the single and double mutants.

**6–96** Hsp60-like molecular chaperones provide a large central cavity in which misfolded proteins can attempt to refold. Two models, which are not mutually exclusive, can be considered for the role of the Hsp60-like chaperones in the refolding process. They might act passively to provide an isolation chamber that aids protein folding by preventing aggregation with other proteins. Alternatively, Hsp60-like chaperones might actively unfold misfolded proteins to remove stable, but incorrect, intermediate structures that block proper folding. The involvement of ATP binding and hydrolysis and the associated conformational changes of the Hsp60-like chaperones could be used in favor of either model.

In bacteria the Hsp60-like chaperone GroEL binds to a misfolded protein, then binds ATP and the GroES cap, and after about 15 seconds hydrolyzes the ATP and ejects the protein (Figure 6–39). To distinguish between a passive and an active role for GroEL in refolding, you label a protein by denaturing it in tritiated water, $^3H_2O$. When the denaturant is removed and the protein is transferred to normal water, $^1H_2O$, most of the radioactivity is lost within 10 minutes, but a stable core of 12 tritium atoms exchanges on a much longer time scale—a behavior typical of amide hydrogen atoms involved in stable hydrogen bonds. Disruption of these bonds would allow their exchange within a few milliseconds.

You prepare the radioactive substrate and mix it immediately with a slight molar excess of GroEL, and then wait 10 minutes for the rapidly exchanging tritium atoms to be lost. The addition of GroES or ATP alone has no effect on the exchange, as shown by the upper curve in Figure 6–40. The addition of GroES and ATP together causes a rapid loss of tritium (Figure 6–40, lower curve). Addition of GroES plus ADP has no effect, but GroES plus AMPPNP, a nonhydrolyzable analog of ATP, promotes a rapid exchange that is indistinguishable from GroES plus ATP.

A. After the addition of components to the complex of tritiated protein and GroEL, it took a minimum of 45 seconds to separate the protein from the freed tritium label. Did the exchange of tritium occur within one cycle of binding and ejection by GroEL, which takes about 15 seconds, or might it have required more than one cycle? Explain your answer.

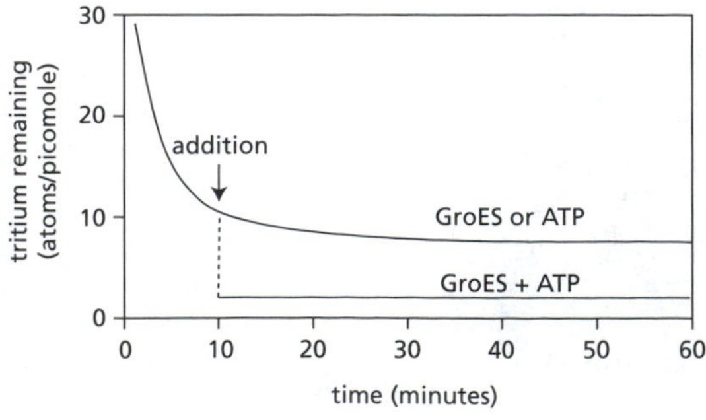


**Figure 6–40** Effects of chaperone components on rates of tritium exchange (Problem 6–96).

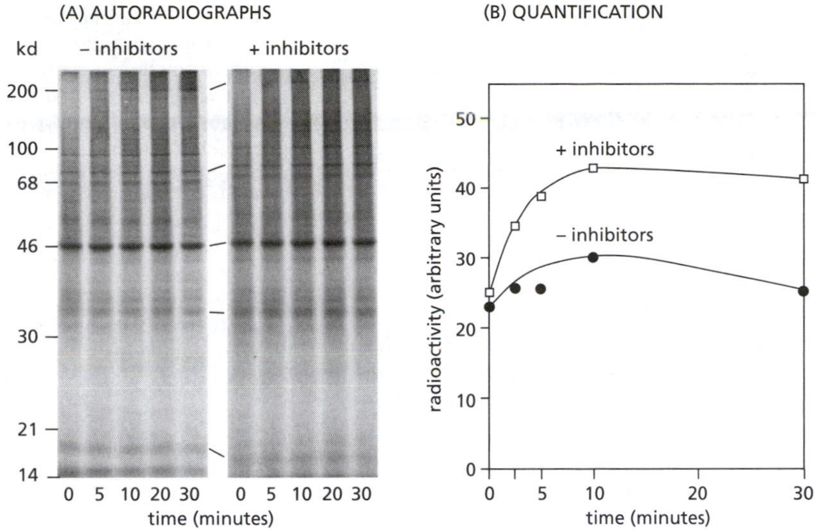


**Figure 6–41** Analysis of newly synthesized proteins in the presence and absence of proteasome inhibitors (Problem 6–97). (A) Autoradiographs of SDS-polyacrylamide gel electrophoretic separation of cellular proteins. A combination of inhibitors—carbobenzoxyl-Leu-Leu-leucinal (zLLL), lactacystin, and clasto-lactacystin β-lactone—was used to block proteasome activity. Samples were taken at various times up to 30 minutes after the addition of $^{35}$S-methionine. Equal amounts of total cellular protein were loaded into each lane. (B) Quantification of newly synthesized proteins in the presence and absence of proteasome inhibitors. The amount of radioactivity in each entire lane in (A) was used as a basis for comparison. Radioactivity was quantified by Phosphorimager analysis.

B. Do the results support a passive isolation-chamber model, or an active-unfolding model, for GroEL action? Explain your reasoning.

**6–97** You wish to measure the fraction of newly synthesized proteins that are degraded by proteasomes. Your strategy is to assay newly synthesized proteins in the presence and in the absence of inhibitors of proteasome function. You pulse-label mouse lymph node cells with $^{35}$S-methionine and then chase with excess unlabeled methionine for 30 minutes, all in the presence or absence of proteasome inhibitors. At various times during the chase you take aliquots of cells and boil them in SDS to denature all proteins. For each sample, you load identical amounts of cellular protein onto SDS-polyacrylamide gels, separate them by electrophoresis, and quantify the radioactivity in each lane. Autoradiographs of the gels and the results of quantification are shown in Figure 6–41.

A. What fraction of newly synthesized proteins is degraded by proteasomes? Explain your reasoning.
B. All proteins appear to be equally affected by proteasome inhibitors; that is, there appears to be a general increase in intensity of radioactivity throughout the gel. How do you suppose that blocking proteasomes might lead to such a generalized increase in intensity of radioactivity?

**Problem 15–148** illustrates the role of the proteasome in regulating the activity of β-catenin.

**6–98** Fission yeast cyclin B, known as Cdc13, must be destroyed at the metaphase–anaphase transition to allow normal progression through the cell cycle. Destruction of Cdc13 requires a nine amino acid sequence—the so-called destruction box or D-box—near its N-terminus. The D-box mediates an interaction with the anaphase-promoting complex (APC), which is the E3 component of an E2–E3 ubiquitin ligase that adds multiubiquitin chains and targets Cdc13 for destruction by the proteasome.

You reason that overexpression of the N-terminal 70 amino acids (N70) of Cdc13, which contains the D-box, might block the ability of APC to trigger the destruction of Cdc13 and stop cell proliferation. Sure enough, overexpression of N70 proves to be lethal. As a check on your hypothesis, you assay for accumulation of Cdc13 and another D-box protein (Cut2), which you expect to be protected from destruction, as well as two non-D-box proteins, Rum1 and Cdc18, which you expect to be unaffected because their ubiquitylation is independent of APC. Much to your surprise, all four proteins accumulate when N70 is overexpressed (Figure 6–42). To clarify the situation, you prepare two other versions of N70: one in which all lysines have been replaced by arginines (K0-N70) and one that carries a mutant D-box (dm-N70). Overexpression of these two proteins gives the results shown in Figure 6–42.

Propose an explanation for the differences in the patterns of accumulation of Cdc13, Cut2, Rum1, and Cdc18 after overexpression of dm-N70, K0-N70, and N70.

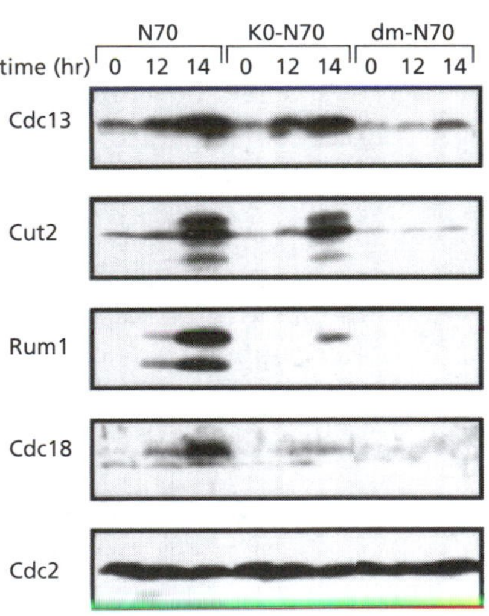


**Figure 6–42** Levels of specific proteins after expression of various N-terminal segments of Cdc13 (Problem 6–98). Each version of the Cdc13 N-terminus was expressed in fission yeast from the *Nmt1* promoter, which can be induced by adding thiamine to the medium. Induction takes about 12 hours. Proteins were detected by immunoblotting with specific antibodies. Cdc2 is not subject to ubiquitylation and thus serves as a loading control.

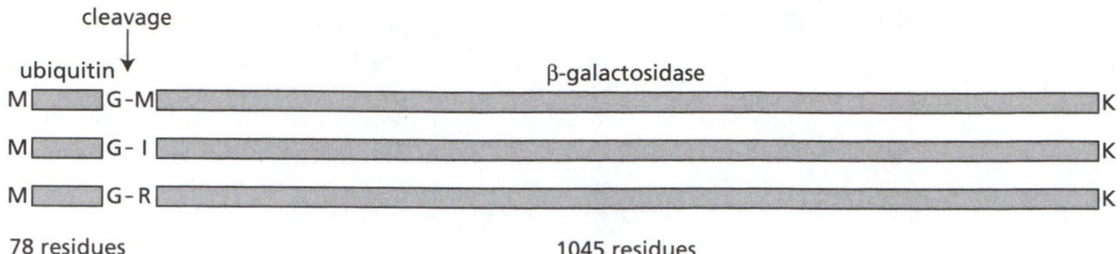


**Figure 6–43** Fusion proteins encoded by three yeast plasmids (Problem 6–99). Upon expression in yeast, the fusion proteins are cleaved at the peptide bonds indicated by the *arrow*. The β-galactosidase molecules liberated by cleavage differ only at their N-termini.

**Problem 17–100** considers the consequences of an indestructible cyclin on progress through the cell cycle.

**6–99** The life-spans of proteins are appropriate to their *in vivo* tasks; for example, structural proteins tend to be long-lived, whereas regulatory proteins are usually short-lived. In eucaryotic cells life-spans are strongly influenced by the N-terminal amino acid. The first experiments that revealed these effects used hybrid proteins consisting of ubiquitin fused to β-galactosidase, as shown in Figure 6–43. When plasmids encoding these proteins were introduced into yeast, the hybrid proteins were synthesized but the ubiquitin was cleaved off exactly at the junction with β-galactosidase, generating proteins with different N-termini (Figure 6–43).

To measure the half-lives of these β-galactosidase molecules, yeast cells were grown for several generations in the presence of a radioactive amino acid. Protein synthesis was then blocked with the inhibitor cycloheximide (CHX). The rate of degradation of β-galactosidase was determined by removing samples from the cultures at various times, purifying β-galactosidase and measuring the amount of associated radioactivity after SDS-gel electrophoresis. The results at the 5-minute time point are shown in Figure 6–44A, and a graph depicting results for all time points is shown in Figure 6–44B.

A. Using recombinant DNA techniques, it would have been straightforward to generate a series of plasmids in which the first codon in the β-galactosidase gene was changed. Why do you suppose this more direct approach was not tried?

B. Estimate the half-life (time at which half the material has been degraded) of each of the three species of β-galactosidase.

**6–100** Two aspects of the work in the previous problem might have made you curious. First, how did the investigators know that ubiquitin was actually removed from the N-terminus? Second, what are the bands above the position of β-galactosidase in Figure 6–44A? Any fragments of β-galactosidase due to degradation should run below β-galactosidase.

To address these questions, the investigators used β-galactosidase antibodies to isolate nonradioactive β-galactosidase from cells transfected with the same three plasmids (see Figure 6–43). These samples were separated by electrophoresis as before, then transferred to a filter paper and reacted with radioactive antibodies specific for ubiquitin. As shown in Figure 6–45, antibodies against ubiquitin did not react with material at the position of

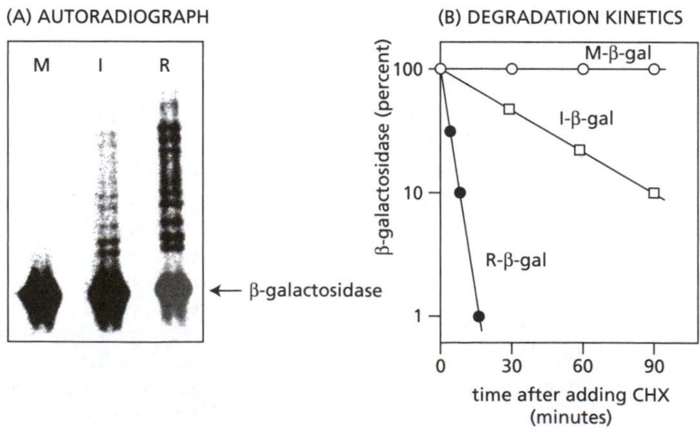


**Figure 6–44** Analysis of half-lives of proteins with different N-termini (Problem 6–99). (A) Electrophoretic separation of radioactive β-galactosidase. Antibodies directed against β-galactosidase were used to precipitate the protein. The N-terminal amino acids are indicated *above* the lanes.
(B) Disappearance of β-galactosidases with time after termination of protein synthesis by the addition of cycloheximide (CHX). The level of β-galactosidase is expressed as a percentage of that present immediately after protein synthesis was blocked.

β-galactosidase but did react with a ladder of bands above β-galactosidase, which are at the same positions as those in Figure 6–44A.

A. Do these experiments demonstrate that the ubiquitin at the N-terminus of the hybrid protein is removed?

B. Offer an explanation for the presence of ubiquitin above the position of β-galactosidase in the experiments with isoleucine (I) and arginine (R) at the N-terminus, but not in the experiments with methionine (M) at the N-terminus.

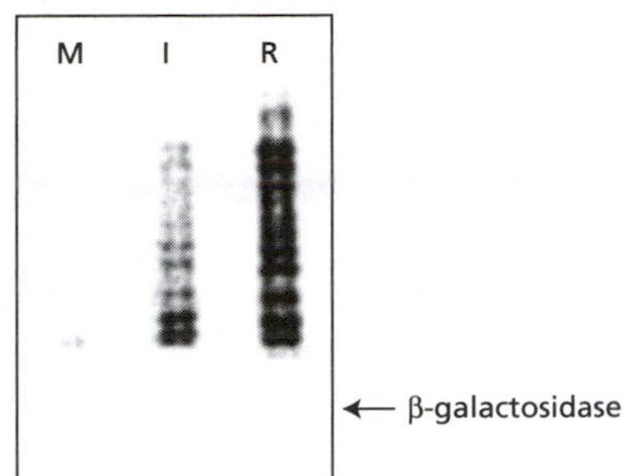


**Figure 6–45** Electrophoresis of unlabeled protein expressed from the ubiquitin β-galactosidase fusion gene, followed by reaction with labeled antibodies against ubiquitin (Problem 6–100). The N-terminal amino acids are indicated *above* the lanes. The position at which intact β-galactosidase runs is marked by an *arrow*.

# THE RNA WORLD AND THE ORIGINS OF LIFE

TERMS TO LEARN

RNA world

## DEFINITIONS

Match the definition below with its term from the list above.

**6–101** A hypothetical state of evolution that existed on Earth before modern cells arose, in which RNA both stored genetic information and catalyzed chemical reactions in primitive cells.

## TRUE/FALSE

Decide whether each of these statements is true or false, and then explain why.

**6–102** Protein enzymes are thought to greatly outnumber ribozymes in modern cells because they catalyze a much greater variety of reactions at much faster rates than ribozymes.

## THOUGHT PROBLEMS

**6–103** What is so special about RNA that makes it such an attractive evolutionary precursor to DNA and protein?

**6–104** Discuss the following statement: "During the evolution of life on Earth, RNA has been demoted from its glorious position as the first replicating catalyst. Its role now is as a mere messenger in the information flow from DNA to protein."

**6–105** Imagine a warm pond on the primordial Earth. Chance processes have just assembled a single copy of an RNA molecule with a catalytic site that can carry out RNA replication. This RNA molecule folds into a structure that is capable of linking nucleotides according to instructions in an RNA template. Given an adequate supply of nucleotides, will this RNA molecule be able to catalyze its own replication? Why or why not?

**6–106** If an RNA molecule could form a hairpin with a symmetric internal loop, as shown in Figure 6–46, could the complement of this RNA form a similar structure? If so, would there be any regions of the two structures that are identical? Which ones?

**6–107** Why are compartments thought to have been necessary for evolution in the RNA world?

**6–108** What is it about DNA that makes it a better material than RNA for the storage of genetic information?

**6–109** An RNA molecule with the ability to catalyze RNA replication—the linkage of RNA nucleotides according to the information in an RNA template—would have been a key ribozyme in the RNA world. Your advisor wants to

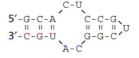

```
           C-U
          /   \
5'-G-C-A       C-C-G
   | | |       | | | \
3'-C-G-U       G-G-C- U
          \   /
           A-C
```

**Figure 6–46** An RNA hairpin with a symmetric internal loop (Problem 6–106).

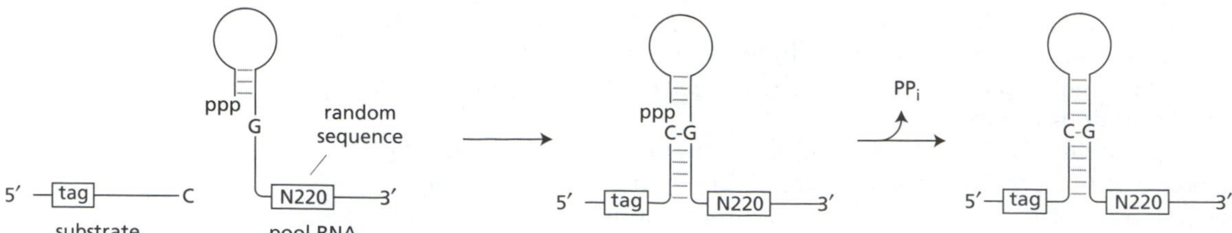


Figure 6–47 Ribozyme-catalyzed reaction selected for in this scheme for *in vitro* evolution (Problem 6–109). The random sequence in the pool RNA is 220 nucleotides in length (N220). The 3′ end of the substrate oligonucleotide is complementary to the constant 5′ end of the pool RNA molecules, so that it can pair with the pool RNAs as shown.

evolve such an RNA replicase *in vitro* by selection and amplification of rare functional molecules from a pool of random RNA sequences. The difficulty has been to devise a strategy that models a reasonable first step and that can be fit into an *in vitro* selection scheme. When you arrived in the lab today, you found the scheme shown in Figures 6–47 and 6–48 on your desk with a note from your advisor that he wants to talk with you about it tomorrow when he returns from an out of town trip. He's even left you some questions to 'focus' the discussion.

A. As shown in Figure 6–47, the activity being selected for is the attachment of an oligonucleotide 'tag' to the end of the catalytic RNA molecule. How is this reaction an analog of nucleotide addition to the end of a primer on a template RNA?

B. Why is it important in the selection scheme in Figure 6–48 that the 'tag' RNA be linked to the same RNA that catalyzed its attachment?

C. Why does the starting pool of RNA molecules have constant regions at each end and a random segment in the middle? What specific roles do these segments play in the overall scheme?

D. How is a catalytic RNA molecule selected from the pool and specifically amplified?

E. Why do you suppose it is necessary to repeat the cycles of selection and amplification? Why not simply purify the ribozymes at the end of the first cycle?

## CALCULATIONS

**6–110** Curses! Your advisor has sent you an email with still more questions about his selection scheme for evolving an RNA replicase (see Figures 6–47 and 6–48).

A. He thinks you will be able to generate about a milligram of RNA to begin the selection. How many molecules will be present in this amount of RNA? (Assume that an RNA nucleotide has a mass of 330 daltons and that the RNA molecules are 300 nucleotides in length.)

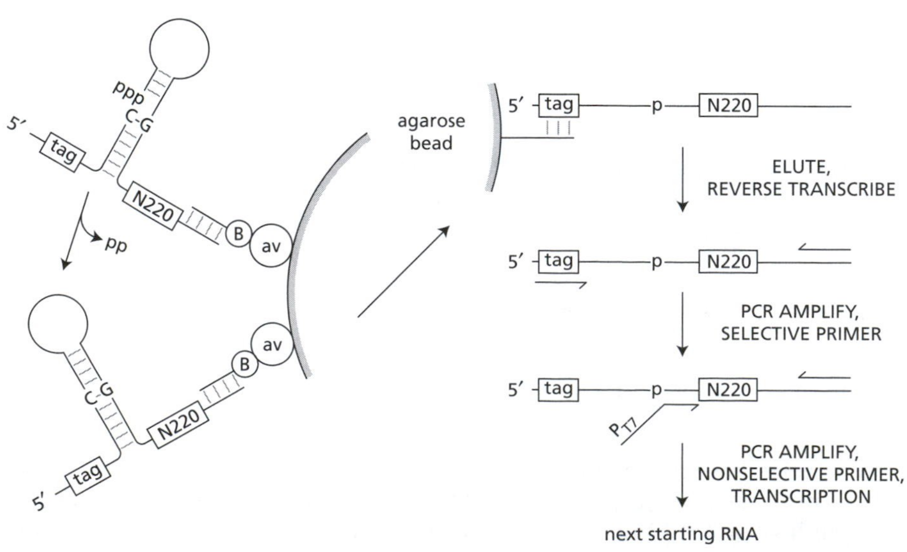


Figure 6–48 One round in the cyclic selection scheme to amplify individual ribozymes from a random pool (Problem 6–109). Each pool RNA is linked at one end to the substrate RNA molecule and at the other end to a complementary DNA oligonucleotide linked to an agarose bead for ease of manipulation. In the final PCR amplification with a nonselective primer, the noncomplementary portion of the oligonucleotide carries the promoter for T7 RNA polymerase ($P_{T7}$), which allows the final DNA product to be transcribed back into RNA.

**Table 6–4 Summary of rounds of selection for a ribozyme with RNA ligation activity** (Problem 6–111).

| ROUND | ERROR-PRONE PCR | LIGATION CONDITIONS | | LIGATION RATE (per hour) |
|---|---|---|---|---|
| | | $MgCl_2$ (mM) | TIME (hours) | |
| 0 | | | | 0.000003 |
| 1 | No | 60 | 16 | <0.000004 |
| 2 | No | 60 | 16 | 0.0008 |
| 3 | No | 60 | 16 | 0.0094 |
| 4 | No | 60 | 16 | 0.027 |
| 5 | Yes | 60 | 0.50 | 0.16 |
| 6 | Yes | 60 | 0.17 | 0.40 |
| 7 | Yes | 60 | 0.02 | 0.86 |
| 8 | No | 4 | 0.12 | 3.2 |
| 9 | No | 2.5 | 0.17 | 4.5 |
| 10 | No | 2.5 | 0.17 | 8.0 |

B. How many different molecules are possible if the central 220-nucleotide segment is completely random? What fraction of all possible molecules will be present in your 1-mg sample?

C. If the ligation reaction could only be catalyzed by a single, *unique* 50-nucleotide RNA sequence, what do you suppose your chances of success would be? What does the general success of such selection schemes imply about the range of RNA molecules that are capable of catalysis?

## DATA HANDLING

**6–111** Once you and your advisor had ironed out the details of the selection scheme, the work went fairly quickly. You have now carried out 10 rounds of selection with the results shown in Table 6–4. You have cloned and sequenced 15 individual RNA molecules from pool 10: no two are the same, although 11 of the 15 have very similar sequences. Your advisor is very excited by your results and has asked you to give the next departmental seminar. You know from your conversations with other students that you will need to prepare careful explanations for the questions listed below.

A. Why did you use error-prone PCR, which can introduce mutations, in some of the rounds of amplification?

B. Why did you reduce the time and $Mg^{2+}$ concentration—both of which increase the difficulty of ligation—in successive rounds of selection?

C. How much of an improvement in ligation rate have you found in your 10 rounds of selection and amplification?

D. Why is there still such diversity among the RNA molecules after 10 rounds of selection and amplification?

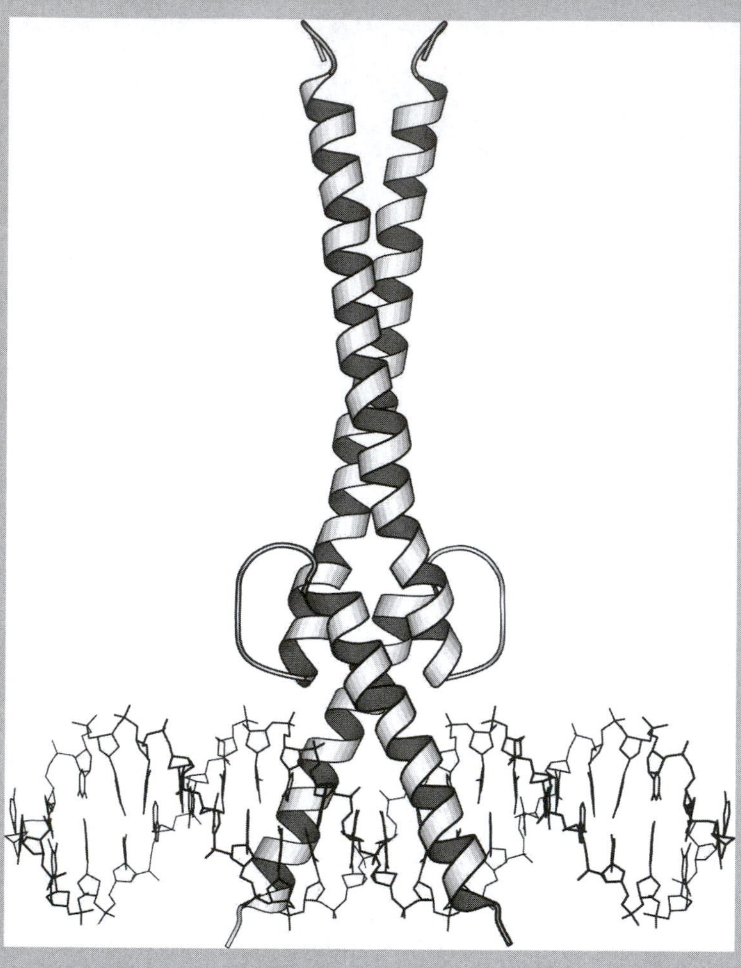

**The MAX homodimer bound to DNA.** A perfect illustration of how helix-loop-helix proteins bind to DNA by forming dimers. Coordinates from 1AN2 determined by AR Ferre-D'Amare, GC Prendergast, EB Ziff, and SK Burley, *Nature* 363:38–45, 1993.

# Control of Gene Expression

7

In This Chapter

AN OVERVIEW OF GENE CONTROL 151

DNA-BINDING MOTIFS IN GENE REGULATORY PROTEINS 153

HOW GENETIC SWITCHES WORK 160

THE MOLECULAR GENETIC MECHANISMS THAT CREATE SPECIALIZED CELL TYPES 172

POST-TRANSCRIPTIONAL CONTROLS 182

## AN OVERVIEW OF GENE CONTROL

| TERMS TO LEARN | | |
|---|---|---|
| mRNA degradation control | RNA processing control | transcriptional control |
| protein activity control | RNA transport control | translational control |
| RNA localization control | | |

### DEFINITIONS

Match each definition below with its term from the list above.

**7–1** Regulates which RNAs are exported from the nucleus.

**7–2** Regulates when and how often a given gene sequence is made into RNA.

**7–3** Regulates which mRNA molecules are selectively destabilized in the cytoplasm.

**7–4** Regulates which mRNAs are selected to be used for protein synthesis by ribosomes.

**7–5** Regulates the splicing and modification of RNA transcripts.

### TRUE/FALSE

Decide whether each of these statements is true or false, and then explain why.

**7–6** When the nucleus of a fully differentiated carrot cell is injected into a frog egg whose nucleus has been removed, the injected donor nucleus is capable of programming the recipient egg to produce a normal carrot.

**7–7** The differences in the patterns of proteins produced in different specialized cell types are accurately reflected in the patterns of expressed mRNAs.

### THOUGHT PROBLEMS

**7–8** A small portion of a two-dimensional display of proteins from human brain is shown in Figure 7–1. These proteins were separated on the basis of size in one dimension and electrical charge (isoelectric point) in the other. Not all

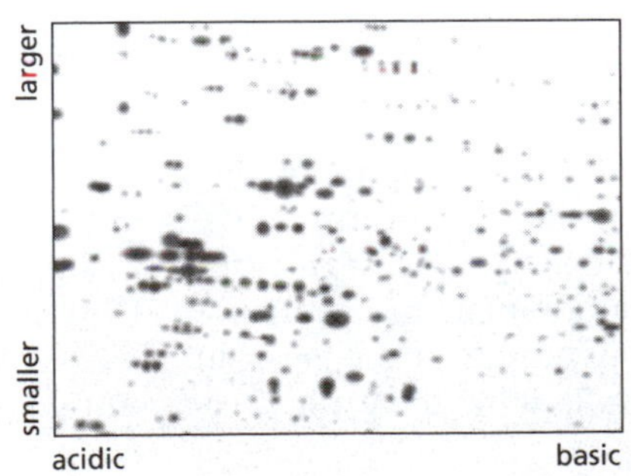


**Figure 7–1** Two-dimensional separation of proteins from the human brain (Problem 7–8). The proteins were displayed using two-dimensional gel electrophoresis. Only a small portion of the protein spectrum is shown.

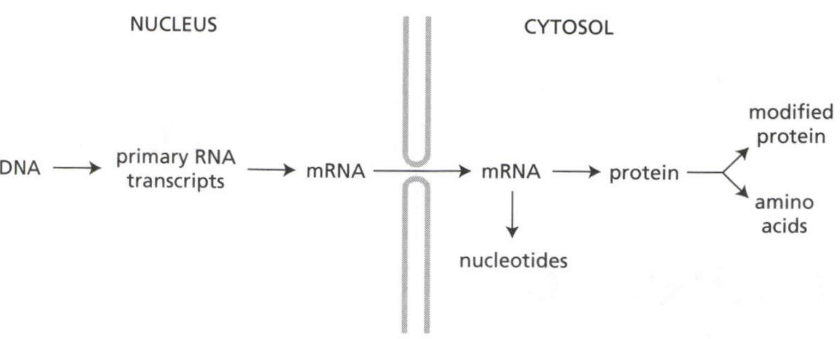


**Figure 7–2** Six steps at which the pathway for eucaryotic gene expression can be controlled (Problem 7–10).

protein spots on such displays are products of different genes; some represent modified forms of a protein that migrate to different positions. Pick out a couple of sets of spots that could represent proteins that differ by the number of phosphates they carry. Explain the basis for your selection.

**7–9** DNA microarray analysis of the patterns of mRNA abundance in different human cell types shows that the level of expression of almost every active gene is different. The patterns of mRNA abundance are so characteristic of cell type that they can be used to determine the tissue of origin of cancer cells, even though the cells may have metastasized to different parts of the body. By definition, however, cancer cells are different from their noncancerous precursor cells. How do you suppose then that patterns of mRNA expression might be used to determine the tissue source of a human cancer?

**7–10** In principle, a eucaryotic cell can regulate gene expression at any step in the pathway from DNA to the active protein (Figure 7–2).

A. Place the types of control listed below at appropriate places on the diagram in Figure 7–2.
   1. mRNA degradation control
   2. protein activity control
   3. RNA processing control
   4. RNA transport and localization control
   5. transcriptional control
   6. translational control

B. Which of the types of control listed above are unlikely to be used in bacteria?

## DATA HANDLING

**7–11** In the original cloning of sheep from somatic cells, the success rate was very low. For example, only 1 lamb (Dolly) was born from 277 zygotes that were reconstructed using nuclei derived from breast cells and enucleated, unfertilized eggs. Other experiments using nuclei from embryonic or fetal lamb cells had a higher success rate, albeit still a low one. Given the rarity of successful events, it was critical to eliminate inadvertent mating of either the oocyte donor or the surrogate mother as the source of newborn lambs. To determine whether the cloned animals were derived from the donor nuclei, the researchers analyzed DNA microsatellites (short, repetitive DNA sequences) at four loci in surrogate mothers and donor cells (Figure 7–3). These loci were chosen because many different lengths are present in sheep populations.

A. Do the results in Figure 7–3 argue that the lambs were derived from the transplanted nuclei, or from an inadvertent mating? Explain your answer.

B. What would the results have looked like for the alternative you did not choose in question A?

**7–12** Developmentally programmed genome rearrangements occur in mammals during the generation of diversity in the immune system. In B cells, for example, the variable (V) and constant (C) segments of the immunoglobulin gene are juxtaposed by deletion of a long segment of DNA that separates them in other cells. Digestion of unrearranged germ-line DNA with a restriction nuclease that cuts in DNA flanking the V and C segments generates two

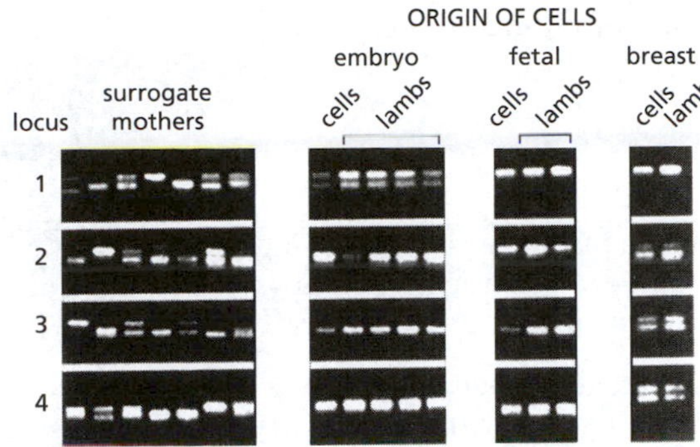


Figure 7–3 Microsatellite analysis of seven surrogate mothers, the three different nuclear donor cell types, and the seven lambs that were born (Problem 7–11). Four polymorphic loci were used in the analysis. The surrogate mothers are arranged *left* to *right* in the same order as the lambs they gave birth to. Nuclear donor cells were derived from embryo, fetus, or breast. At each of the four polymorphic loci, flanking PCR primers were used to amplify DNA that included a particular microsatellite. Microsatellites with different numbers of repeats give rise to different length PCR products.

bands on a Southern blot when hybridized to radioactive probes specific for the V and the C segments (Figure 7–4). Would you expect the V- and C-segment probes to hybridize to the same or different DNA fragments after digestion of B-cell DNA with the same restriction nuclease? Sketch a possible pattern of hybridization to B-cell DNA that is consistent with your expectations. Explain the basis for your pattern of hybridization. (Without a lot more information you cannot predict the exact pattern, so focus on the general features of the pattern.)

# DNA-BINDING MOTIFS IN GENE REGULATORY PROTEINS

TERMS TO LEARN

| | | |
|---|---|---|
| chromatin immunoprecipitation | gene regulatory protein | homeodomain |
| combinatorial control | helix–loop–helix (HLH) motif | leucine zipper motif |
| DNA affinity chromatography | helix–turn–helix | zinc finger |
| gel-mobility shift assay | | |

## DEFINITIONS

Match each definition below with its term from the list above.

**7–13** Detection of the binding of a protein to DNA by the altered migration of a labeled DNA fragment in an electric field.

**7–14** Any protein that binds to a specific DNA sequence to alter expression of a gene.

**7–15** DNA-binding motif in which an α helix and a β sheet are held together by a metal ion.

**7–16** Regulation of a step in a cellular process by defined assortments of different proteins, rather than by individual proteins.

**7–17** Structural motif in many DNA-binding proteins in which two α helices from separate proteins are joined together in a coiled-coil.

**7–18** Method based on cross-linking proteins to DNA in living cells that is used to determine the sites on DNA that a specific protein occupies.

## TRUE/FALSE

Decide whether each of these statements is true or false, and then explain why.

**7–19** Because the individual contacts are weak, the interactions between regulatory proteins and DNA are among the weakest in biology.

**7–20** In terms of its biochemical function, the helix–loop–helix motif is more closely related to the leucine zipper motif than it is to the helix–turn–helix motif.

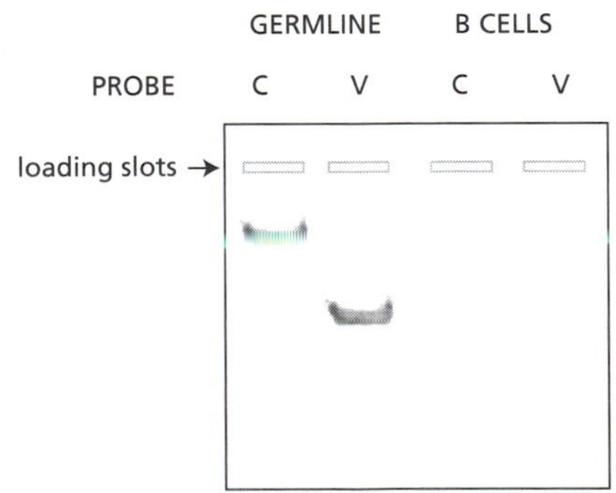


Figure 7–4 Southern blot of DNAs from germ line and B cells (Problem 7–12). Germ-line DNA and B-cell DNA were digested with the same restriction nuclease. Only the hybridization to germ-line DNA is shown.

7–21 DNA affinity chromatography allows purification of unlimited amounts of DNA-binding proteins from a cell extract.

## THOUGHT PROBLEMS

7–22 Figure 7–5 shows a short stretch of a DNA helix displayed as a space-filling model. Indicate the major and minor grooves and provide a scale. Is it possible to tell the polarity of each of the strands in this figure?

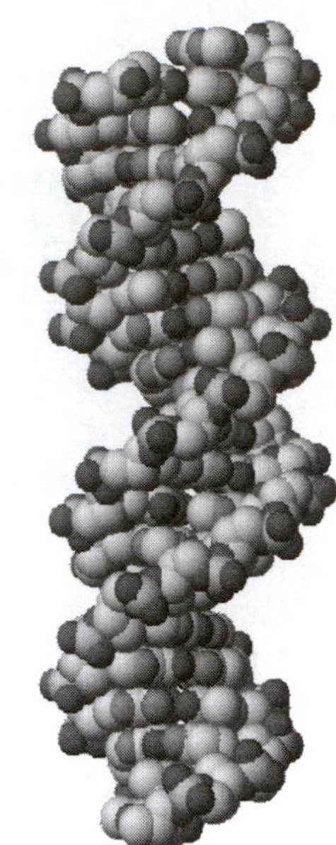

Figure 7–5 A space-filling model of a DNA duplex (Problem 7–22).

7–23 Explain how DNA-binding proteins can make sequence-specific contacts to a double-stranded DNA molecule without breaking the hydrogen bonds that hold the bases together. Indicate how, by making such contacts, a protein can distinguish a C–G from a T–A base pair. Use Figure 7–6 to indicate what sorts of noncovalent bonds (hydrogen bonds, electrostatic attractions, or hydrophobic interactions) could be used to discriminate between C–G and T–A. (You do not need to specify any particular amino acid on the protein.)

7–24 What are the two fundamental components of a genetic switch?

7–25 The nucleus of a eucaryotic cell is much larger than a bacterium, and it contains much more DNA. As a consequence, a DNA-binding protein in a eucaryotic cell must be able to select its specific binding site from among many more unrelated sequences than does a DNA-binding protein in a bacterium. Does this present any special problems for eucaryotic gene regulation?

Consider the following situation. Assume that the eucaryotic nucleus and the bacterial cell each have a single copy of the same DNA-binding site. In addition, assume that the nucleus is 500 times the volume of the bacterium, and has 500 times as much DNA. If the concentration of the gene regulatory protein that binds the site were the same in the nucleus and in the bacterium, would the regulatory protein find its binding site equally as well in the eucaryotic nucleus as it does in the bacterium? Explain your answer.

7–26 One type of zinc finger motif consists of an α helix and a β sheet held together by a zinc ion (Figure 7–7). When this motif binds to DNA, the α helix is positioned in the major groove, where it makes specific contacts with the bases. This type of zinc finger is often found in a cluster with additional zinc fingers, an arrangement that allows strong and specific DNA-protein interactions to be built up through a repeating basic structural unit. Why do you suppose this motif is thought to enjoy a particular advantage over other DNA-binding motifs when the strength and specificity of the DNA–protein interaction need to be adjusted during evolution?

**Problem 3–24** considers how homeodomain proteins with very different sequences can still fold in the same way.

**Problem 3–31** examines the symmetry match between the subunits of the Cro repressor and its binding half-sites in DNA.

7–27 Many gene regulatory proteins form dimers of identical or slightly different subunits on the DNA. Suggest two advantages of dimerization?

7–28 The lambda repressor binds as a dimer to critical sites on the bacteriophage lambda genome to keep the lytic genes turned off, which stabilizes the prophage (integrated) state. Each molecule of the repressor consists of an

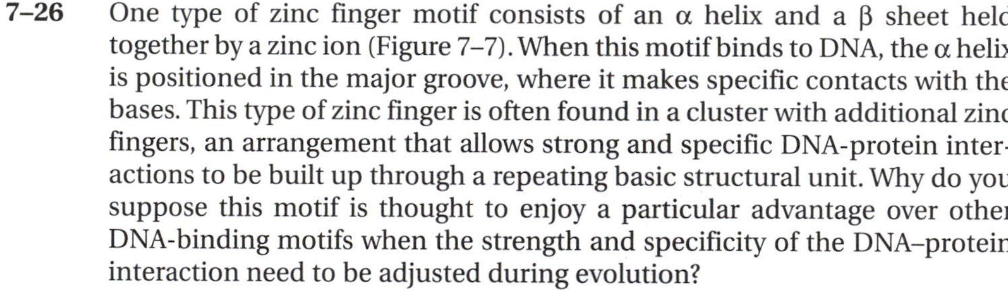

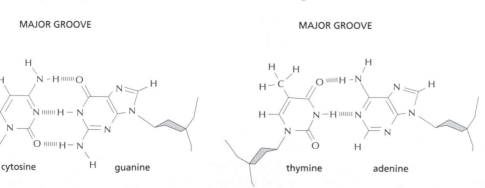

Figure 7–6 C–G and T–A base pairs (Problem 7–23).

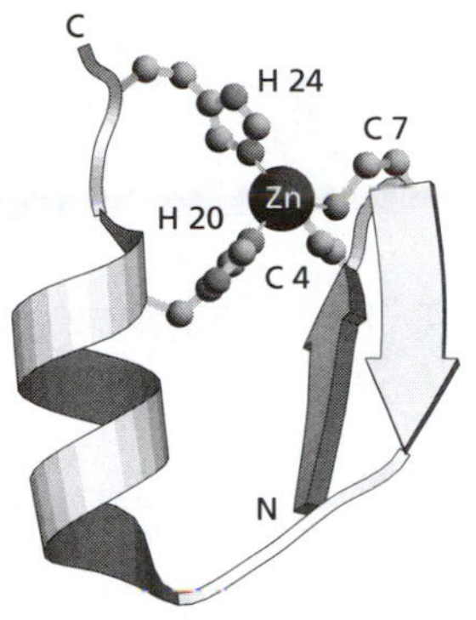


**Figure 7–7** One type of zinc finger motif (Problem 7–26). The zinc ion interacts with Cys (C) and His (H) side chains so that the α helix is held tightly to one end of the β sheet.

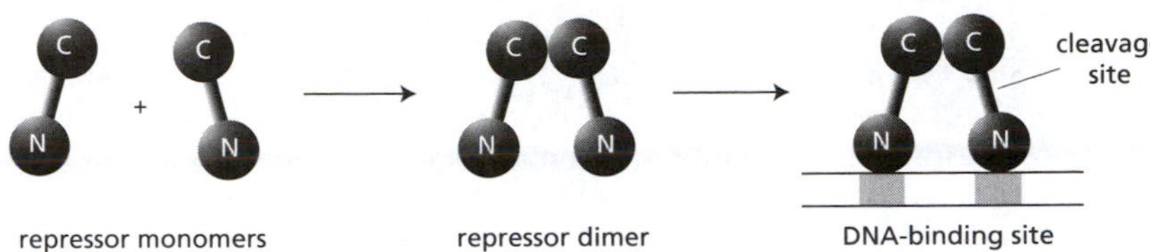


**Figure 7–8** Domains of the lambda repressor and the binding of repressor dimers to DNA (Problem 7–28).

N-terminal DNA-binding domain and a C-terminal dimerization domain (Figure 7–8). Upon induction (for example, by irradiation with UV light), the genes for lytic growth are expressed, lambda progeny are produced, and the bacterial cell lyses to release the viral progeny. Induction is initiated by cleavage of the lambda repressor at a site between the DNA-binding domain and the dimerization domain. In the absence of bound repressor, RNA polymerase initiates transcription of the lytic genes, triggering lytic growth. Given that the number (concentration) of DNA-binding domains is unchanged by cleavage of the repressor, why do you suppose its cleavage results in its removal from the DNA?

**7–29** The differentiation of muscle cells from the somites of the developing embryo is controlled by myogenin, an HLH gene regulatory protein that functions as a heterodimer with another member of the Myod family of HLH proteins (Figure 7–9A). The activity of myogenin must be carefully controlled lest it trigger premature expression of the muscle program of cell differentiation. The *myogenin* gene is turned on in advance of the time when it is needed, but myogenin is prevented from functioning by its tight binding to Id, an HLH protein that lacks a DNA-binding domain, and by phosphorylation of its DNA-binding domain (Figure 7–9B). Explain how dimerization with Id and phosphorylation of the DNA-binding domain might act to keep myogenin nonfunctional.

## CALCULATIONS

**7–30** One common method for determining the DNA sites bound by a gene regulatory protein is to mix random sequences of DNA with the binding protein, separate the bound sequences, amplify the bound sequences by PCR, and then repeat this binding-release-amplification cycle until a consensus

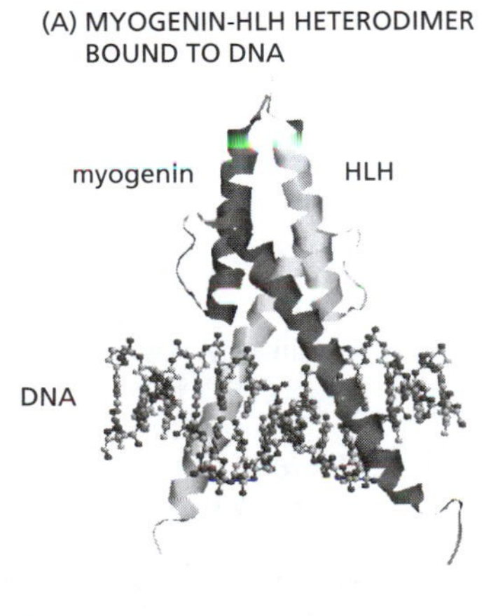


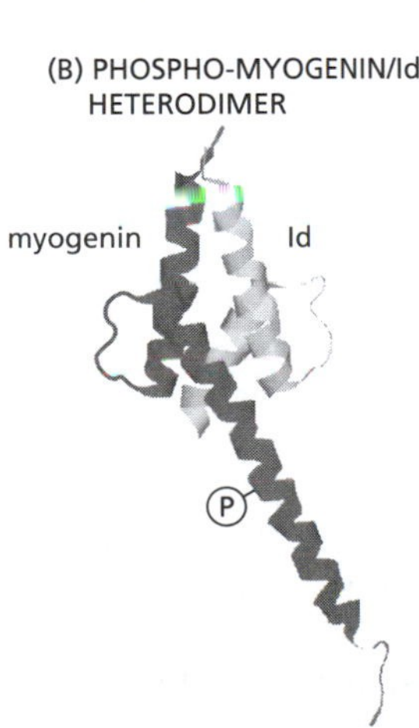


**Figure 7–9** The gene regulatory protein myogenin (Problem 7–29). (A) Myogenin as part of a heterodimer bound to DNA. (B) An inactive form of phosphorylated myogenin bound to Id.

sequence emerges. But is it reasonable to expect that all possible consensus-site binding sequences will actually be present in the starting sample of oligonucleotide?

Consider the following specific example. You wish to test all possible consensus sequences that are 14 nucleotides long. You synthesize a population of oligonucleotides that carry a central 26-nucleotide long random sequence, flanked on either side by 25-nucleotide-long defined sequences to serve as primer sites for PCR amplification. You convert the single-stranded oligonucleotides to double-stranded ones, using one of the PCR primers. You begin the first cycle of binding with an 0.4-ng sample of the synthesized population of double-stranded oligonucleotides.

A. How many double-stranded oligonucleotides are present in the 0.4-ng sample with which you start the experiment? (The average mass of a nucleotide pair is 660 daltons.)

B. Assuming that the starting population of oligonucleotides was truly random in the 26 central nucleotides, would you expect to find all possible 14-base-pair-long sequences represented in the starting sample?

## DATA HANDLING

**7–31** When Jacob and Wollman tried to check the genetic linkage between the *Gal* gene and an integrated bacteriophage lambda genome (termed a lambda prophage), they discovered a surprising phenomenon they referred to as 'erotic induction' (which was later called zygotic induction for publication). In a bacterial mating used to determine genetic linkage, a portion of the chromosome is transferred via a narrow tube from the donor bacterium to the recipient. Jacob and Wollman found that if the donated chromosome carried a lambda prophage, but the recipient cell did not, lambda growth was induced in the recipient cell, which then lysed, producing lambda phage. If the recipient cell carried the lambda prophage, however, no lysis was observed. A summary of results from all their matings is shown in Figure 7–10.

A. Are these results consistent with the notion that a repressor encoded by the prophage normally keeps the bacteriophage's lytic genes turned off. Why or why not?

B. Suppose that the prophage prevented lytic growth by expressing a gene regulatory protein that turned on a gene for an anti-lysis protein. Would the results of the matings have been the same or different? Explain your answer.

**7–32** DNA-binding proteins often find their specific sites much faster than would be anticipated by simple three-dimensional diffusion. The Lac repressor, for example, associates with the *Lac* operator—its DNA-binding site—more than 100 times faster than expected from this model. Clearly, the repressor must find the operator by mechanisms that reduce the dimensionality or volume of the search in order to hasten target acquisition.

Several techniques have been used to investigate this problem. One of the most elegant used strongly fluorescent RNA polymerase molecules that could be followed individually. An array of DNA molecules was aligned in parallel by an electrophoretic technique and anchored to a glass slide. Fluorescent RNA polymerase molecules were then allowed to flow across them at an oblique angle (Figure 7–11A). Traces of individual RNA polymerases showed that about half flowed in the same direction as the bulk and about half deviated from the bulk flow in a characteristic manner (Figure 7–11B). If the RNA polymerase molecules were first incubated with short DNA fragments containing a strong promoter, all the traces followed the bulk flow.

A. Offer an explanation for why some RNA polymerase molecules deviated from the bulk flow as shown in Figure 7–11B. Why did incubation with short DNA fragments containing a strong promoter eliminate traces that deviated from the bulk flow?

B. Do these results suggest an explanation for how site-specific DNA-binding molecules manage to find their sites faster than expected by diffusion?

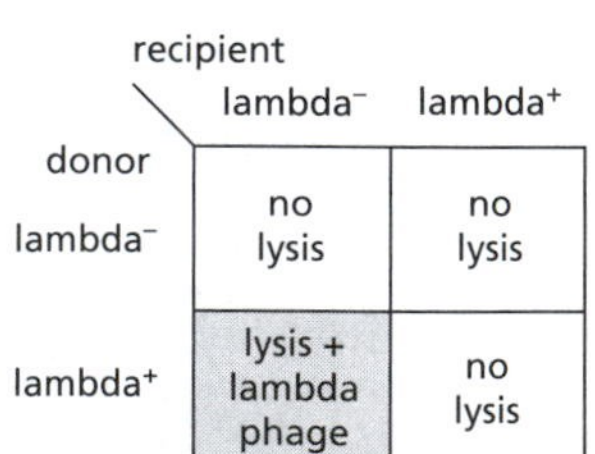


**Figure 7–10** Results of matings between bacteria with and without lambda prophages (Problem 7–31). Lambda$^-$ indicates the absence of a prophage; lambda$^+$ indicates its presence.

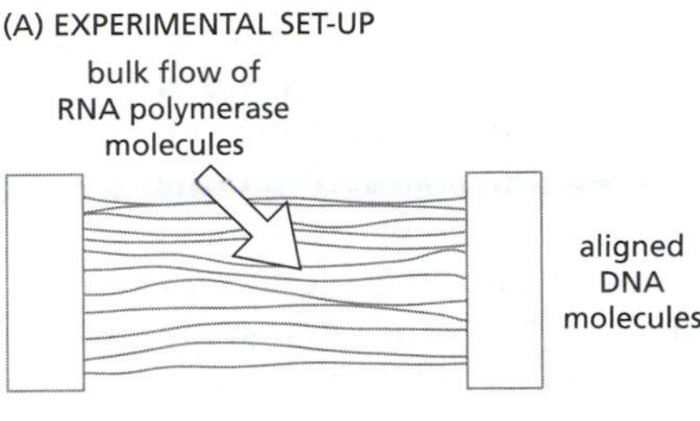


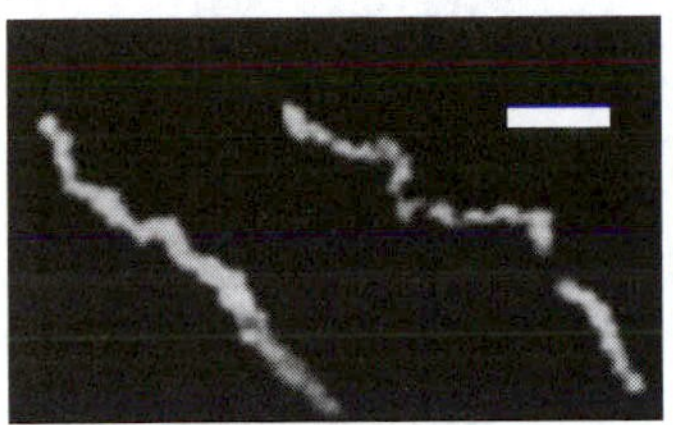


**Figure 7–11** Interactions of individual RNA polymerase molecules with DNA (Problem 7–32). (A) Experimental set-up. DNA molecules are aligned and anchored to glass slide at their ends, and highly fluorescent RNA polymerase molecules are allowed to flow across them. (B) Traces of two individual RNA polymerase molecules. The one on the *left* has traveled with the bulk flow, and the one on the *right* has deviated from it. The scale bar is 10 μm.

C. Based on your explanation, would you expect a site-specific DNA-binding molecule to find its target site faster in a population of short DNA molecules or in a population of long DNA molecules? Assume that the concentration of target sites is identical and that there is one target site per DNA molecule.

**7–33** The binding of a protein to a DNA sequence can cause the DNA to bend to make appropriate contacts with groups on the surface of the protein. Such protein-induced DNA bending can be readily detected by the way the protein–DNA complexes migrate through polyacrylamide gels. The rate of migration of bent DNA depends on the average distance between its ends as it gyrates in solution: the more bent the DNA, the closer together the ends are on average and the more slowly it migrates. If there are two sites of bending, the end-to-end distance depends on whether the bends are in the same (*cis*) or opposite (*trans*) directions (Figure 7–12A).

You have shown that the catabolite activator protein (CAP) causes DNA to bend by more than 90° when it binds to its regulatory site. You wish to know the details of the bent structure. Specifically, is the DNA at the center of the CAP-binding site bent so that the minor groove of the DNA helix is on the inside, or is it bent so that the major groove is on the inside? To answer this,

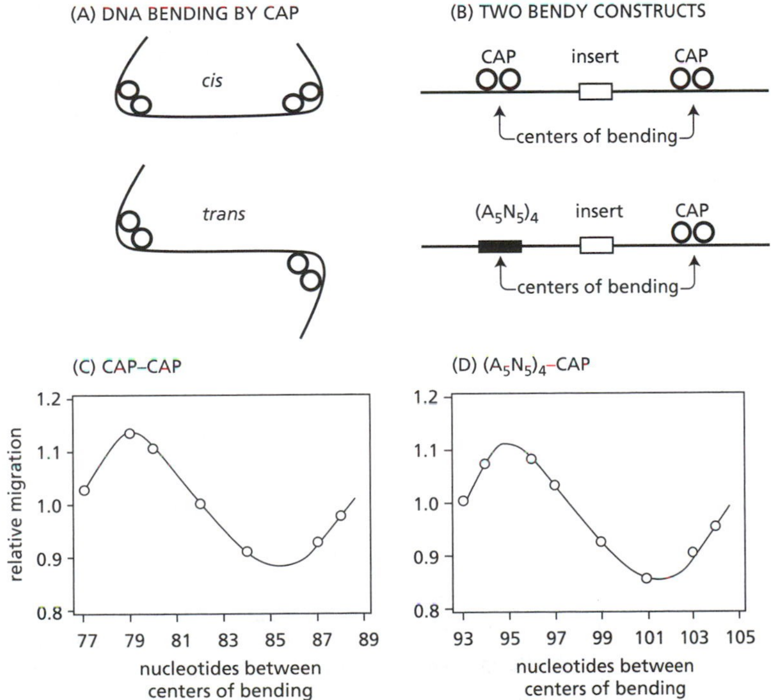


**Figure 7–12** Bending of DNA by CAP binding (Problem 7–33). (A) *Cis* and *trans* configurations of a pair of bends. (B) Two constructs used to investigate DNA bending by CAP binding. (C) Relative migration as a function of the number of nucleotides between the centers of bending in the CAP–CAP construct. (D) Relative migration as a function of the number of nucleotides between the centers of bending in the $(A_5N_5)_4$–CAP construct.

you prepare two kinds of constructs, as shown in Figure 7–12B. In one, you place two CAP-binding sequences on either side of a central site into which you insert a series of DNA segments that vary from 10 to 20 nucleotides in length. In the other, you flank the insertion site with one CAP-binding sequence and one $(A_5N_5)_4$ sequence, which is known to bend with the major groove on the inside. You measure the migration of the CAP-bound constructs relative to the corresponding CAP-bound DNA with no insert. You then plot the relative migration versus the number of nucleotides between the centers of bending (Figure 7–12C and D).

A. Assuming that there are 10.6 nucleotides per turn of the DNA helix, estimate the number of turns that separate the centers of bending of the two CAP-binding sites at the point of minimum relative migration. How many helical turns separate the centers of bending at the point of maximum relative migration?

B. Is the relationship between the relative migration and the separation of the centers of bending of the CAP sites what you would expect, assuming that the *cis* configuration migrates slower than the *trans* configuration (Figure 7–12C)? Explain your answer.

C. How many helical turns separate the centers of bending at the point of minimum migration of the construct with one CAP site and one $(A_5N_5)_4$ site (Figure 7–12D)?

D. Which groove of the helix faces the inside of the bend at the center of bending of the CAP site?

**7–34** The *Fos* and *Jun* oncogenes encode proteins that form a heterodimeric regulator of transcription. Leucine zipper domains in each protein mediate their dimerization through coiled-coil interactions. Dimerization juxtaposes the DNA-binding domains of each protein, positioning them for interaction with regulatory sites in DNA. The dynamics of Fos–Jun interaction in the presence and absence of the AP-1 DNA-binding site were investigated by fluorescence resonance energy transfer (FRET), which is well suited for the rapid measurements that are necessary in such studies.

Fos was tagged with fluorescein (Fos–F) and Jun was tagged with rhodamine (Jun–R), as shown in Figure 7–13A. Fluorescein absorbs light at 490 nm and emits light at 530 nm, whereas rhodamine absorbs light at 530 nm and emits light at 603 nm. When Fos–F and Jun–R are brought into close proximity through heterodimerization, light energy absorbed by fluorescein at 490 nm is efficiently transferred to rhodamine through nonradiative energy transfer and emitted by rhodamine at 603 nm. Dimerization thus decreases fluorescence by fluorescein at 530 nm and increases fluorescence by rhodamine at 603 nm, as shown in Figure 7–13B. In the presence of AP-1 DNA, FRET is even more efficient (Figure 7–13B), indicating that binding to the DNA brings the two fluorophores into even closer proximity.

FRET was also used to measure the ability of Fos–Jun dimers to exchange subunits with monomers in solution in the presence and absence of AP-1 DNA (Figure 7–13C). Fos–F and Jun–R were preincubated in the absence of

**Figure 7–13** Dynamics of Fos–Jun heterodimerization in the presence and absence of AP-1 DNA (Problem 7–34). (A) Arrangement of fluorophores in Fos–F and Jun–R. In the dimer, the excited fluorescein on Fos transfers energy to the rhodamine on Jun. (B) Emission spectra of Fos–F alone, Fos–F with Jun–R, and Fos–F with Jun–R and AP-1 DNA. The emission spectra were recorded between 500 and 700 nm after excitation at 490 nm. (C) Analysis of the exchange of Fos–F and Jun–R in the presence and absence of DNA. Rhodamine fluorescence at 603 nm was followed over time after excitation of fluorescein at 490 nm. A 10-fold excess of unlabeled Fos was added at the time indicated by the *arrow*.

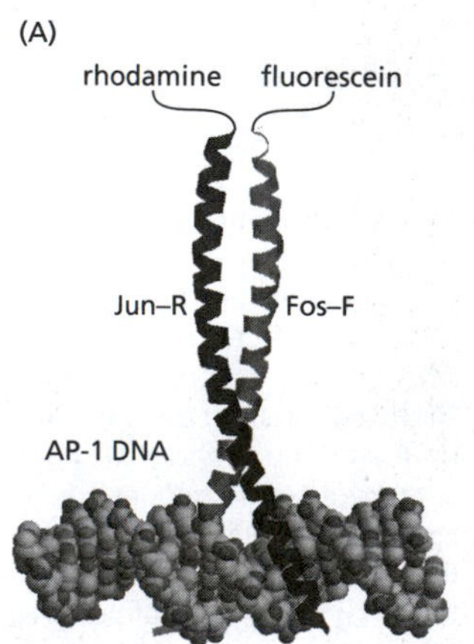


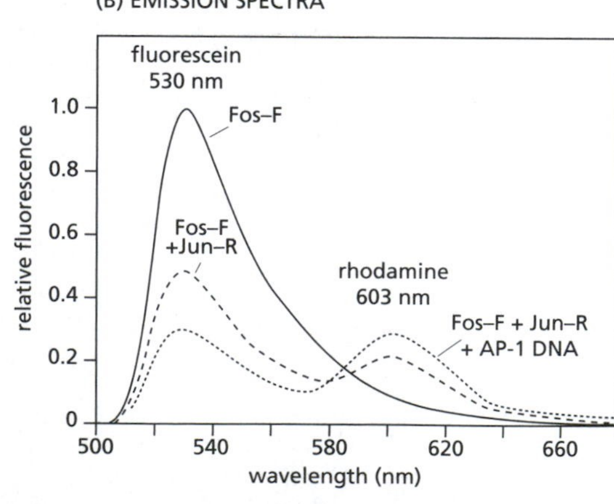


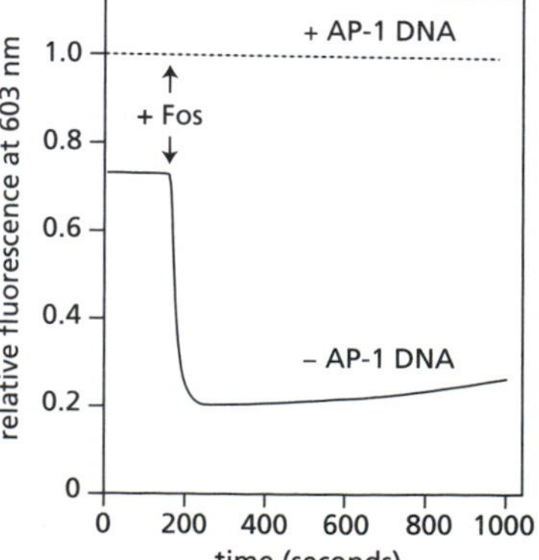

DNA to allow free heterodimers to form, or in its presence to allow DNA-bound heterodimers to form. A 10-fold excess of Fos (without fluorescein) was then added to both solutions, and rhodamine fluorescence at 603 nm was followed after excitation at 490 nm (Figure 7–13C).

A. Do free heterodimers exchange subunits with added unlabeled Fos? Do heterodimers bound to DNA exchange subunits? How can you tell? Explain any significant differences in the behavior of free and DNA-bound heterodimers.

B. In most cells, there are many distinct leucine zipper proteins, several of which can interact to form a variety of heterodimers. If the results in Figure 7–13C were typical of the leucine zipper proteins, what do they imply about the leucine zipper heterodimers in cells?

**7–35** You have cloned four partial cDNAs for a gene regulatory protein to identify the portion of the protein that binds to its DNA recognition sequence. The partial cDNA clones extend for different distances toward the 5′ end of the gene (Figure 7–14). These cDNA clones were transcribed and translated *in vitro,* and then the translation products were mixed with highly radioactive DNA containing the recognition sequence for the gene regulatory protein. When the mixtures were analyzed by polyacrylamide gel electrophoresis, some of the proteins were found to bind to the DNA fragment, slowing its mobility (Figure 7–14, lanes 3, 4, and 5). When cDNA clones 3 and 4 were mixed together before transcription and translation, three bands appeared in the gel-mobility shift assay (lane 6).

A. Why do the radioactive DNA molecules migrate at different rates?

B. Where in the gene regulatory protein is the portion that recognizes DNA?

C. Why were there three shifted bands when cDNA clones 3 and 4 were mixed together? What does that tell you about the structure of the gene regulatory protein?

**7–36** You are working on a gene that encodes a gene regulatory protein, and you wish to know the DNA sequence to which it binds. Your advisor, who is known for her clever imagination and green thumb when it comes to molecular biology, suggests that you might be able to use PCR to amplify rare DNA molecules that are bound by your protein. Her idea consists of four steps, as outlined in Figure 7–15. (1) Synthesize a population of random-sequence oligonucleotides 26 nucleotides long, flanked by defined 25-nucleotide sequences that can serve as primer sites for PCR amplification (Figure 7–15A). (2) Add these oligonucleotides to a crude cell extract that contains

(A) MAP OF THE CLONE

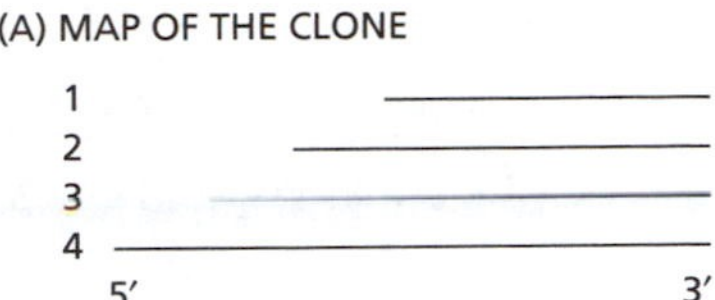

(B) GEL-MOBILITY SHIFT ASSAY

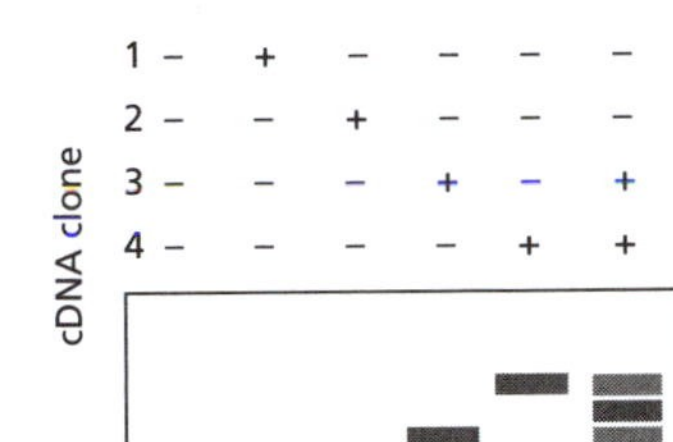

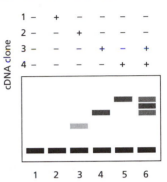


**Figure 7–14** Analysis of gene regulatory protein binding to DNA (Problem 7–35). (A) Structure of partial cDNAs encoding portions of the protein. The 5′ end of the gene corresponds to the N-terminal end of the protein. (B) Gel-mobility shift assays of gene regulatory protein binding to DNA. The grid shows which clones supplied the translation products employed in each of the lanes of the gel. The DNA, which contains a recognition site for the regulatory protein, was radioactively labeled and the bands were visualized by autoradiography.

(A) ORIGINAL RANDOM-SEQUENCE OLIGONUCLEOTIDE

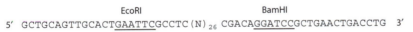

EcoRI BamHI

5′ GCTGCAGTTGCACTGAATTCGCCTC(N)$_{26}$ CGACAGGATCCGCTGAACTGACCTG 3′

(B) SELECTION AND AMPLIFICATION PROTOCOL

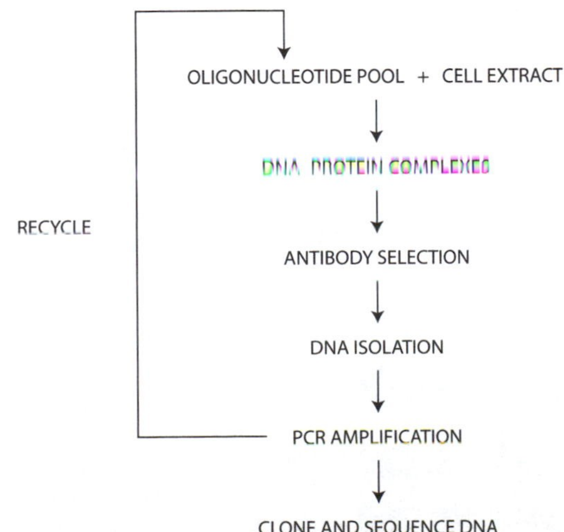


**Figure 7–15** Selection and amplification of specific DNA sequences from a random pool (Problem 7–36). (A) Original random-sequence oligonucleotide used for selection. N stands for any nucleotide. The EcoRI and BamHI sites facilitate cloning of the selected DNA for sequence analysis. (B) Scheme for selecting and amplifying oligonucleotides that bind to a sequence-specific DNA-binding protein.

**Table 7–1 Nucleotide sequences of selected and amplified DNAs** (Problem 7–36).

| |
|---|
| GAATTCGCCTCGAGCACATCATTGCCCATATATGGCACGACAGGATCC |
| GAATTCGCCTCTTCTAATGCCCATATATGGACTTGCTCGACAGGATCC |
| GGATCCTGTCGGTCCTTTATGCCCATATATGGTCATTGAGGCGAATTC |
| GAATTCGCCTCATGCCCATATATGGCAATAGGTGTTTCGACAGGATCC |
| GAATTCGCCTCTATGCCCATATAAGGCGCCACTACCCCGACAGGATCC |
| GAATTCGCCTCGTTCCCAGTATGCCCATATATGGACACGACAGGATCC |
| GGATCCTGTCGACACCATGCCCATATTTGGTATGCTCGAGGCGAATTC |
| GAATTCGCCTCATTTATGAACATGCCCTTATAAGGACCGACAGGATCC |
| GAATTCGCCTCTAATACTGCAATGCCCAAATAAGGAGCGACAGGATCC |
| GAATTCGCCTCATGCCCAAATATGGTCATCACCTACACGACAGGATCC |

Underlined sequences correspond to PCR primer sites. Two sequences are shown starting at the BamHI end so that the binding sites are oriented the same way in all sequences.

the gene regulatory protein, which will bind to those oligonucleotides that contain its recognition site. (3) Isolate the oligonucleotides that are bound to the protein using antibodies against it. (4) PCR amplify the selected oligonucleotides for sequence analysis.

You begin this procedure with the single-stranded random-sequence oligonucleotides, which you convert to double-stranded DNA using one of the PCR primers. After four rounds of selection and amplification, as outlined in Figure 7–15B, you digest the isolated DNA with BamHI and EcoRI, clone the fragments into a plasmid, and sequence 10 individual clones (Table 7–1).

A. What is the consensus sequence to which your gene regulatory protein binds?

B. Does the consensus binding sequence show any signs of symmetry; that is, is any portion of it palindromic? Does this help you decide whether the gene regulatory protein binds to DNA as a monomer or as a dimer?

**7–37** You have determined the DNA-binding site of a protein by DNA footprinting after labeling one strand. To check your results, you repeat the experiment after labeling the other strand of the duplex. You find that the footprints are slightly offset from one another relative to the sequence of the DNA (Figure 7–16). If the protein binds to the same duplex in both cases, how can the footprints on the two strands be different?

**Problem 5–63** demonstrates the use of DNA footprinting to identify protein-binding sites.

# HOW GENETIC SWITCHES WORK

TERMS TO LEARN

| | | |
|---|---|---|
| gene | operator | promoter |
| gene control region | operon | regulatory sequence |
| gene repressor protein | positive control | tryptophan repressor |
| negative control | | |

## DEFINITIONS

Match each definition below with its term from the list above.

**7–38** Type of regulation of gene expression in which the active DNA-binding form of the gene regulatory protein turns the gene off.

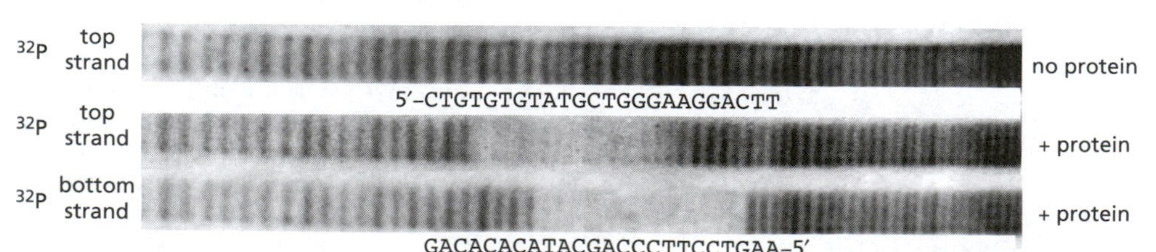


**Figure 7–16** DNA footprints (Problem 7–37). The sequences of the DNA from the top and bottom strands around the footprint are shown. The 5′ ends of the top or bottom strands were labeled with $^{32}$P. Chemical cleavage was used to introduce DNA breaks, which are fairly randomly distributed, as shown by the roughly equal intensities of the individual bands. Each band corresponds to a fragment of DNA that differs from those on either side by one nucleotide.

**7–39** Nucleotide sequence in DNA to which RNA polymerase binds to begin transcription.

**7–40** In a bacterial chromosome, a group of contiguous genes that are transcribed from a single promoter into a single mRNA molecule.

**7–41** General term for a gene regulatory protein that prevents the initiation of transcription.

**7–42** Type of regulation of gene expression in which the active DNA-binding form of the gene regulatory protein turns the gene on.

**7–43** A segment of DNA that is transcribed into RNA, along with its control regions.

**7–44** Short stretch of DNA within a bacterial promoter that contains a binding site for a gene regulatory protein and controls transcription of the adjacent gene.

## TRUE/FALSE

Decide whether each of these statements is true or false, and then explain why.

**7–45** Many gene regulatory proteins in eucaryotes can act even when they are bound to DNA thousands of nucleotide pairs away from the promoter that they influence.

**7–46** It seems likely that the close-packed arrangement of bacterial genes and genetic switches developed from more extended forms of switches in response to the evolutionary pressure to maintain a small genome.

## THOUGHT PROBLEMS

**7–47** Define negative control and positive control in terms of how the active forms of the gene regulatory proteins function. Explain how an 'inducing' ligand turns on a gene that is negatively controlled. How does an inducing ligand turn on a gene that is positively controlled? Explain how an 'inhibitory' ligand turns off a negatively controlled gene, and how it turns off a positively controlled gene.

**7–48** The genes encoding the enzymes for arginine biosynthesis are located at several positions around the genome of *E. coli*. Their expression is coordinated by the ArgR gene regulatory protein. The activity of ArgR is modulated by arginine. Upon binding arginine, ArgR, dramatically changes its affinity for the regulatory sequences in the promoters of the genes for the arginine biosynthetic enzymes. Given that ArgR is a gene repressor, would you expect that ArgR would bind more tightly, or less tightly, to the regulatory sequences when arginine is abundant? If ArgR functioned instead as a gene activator, would you expect the binding of arginine to increase, or to decrease, its affinity for its regulatory sequences? Explain your answers.

**7–49** Bacterial cells can take up the amino acid tryptophan from their surroundings, or, if the external supply is insufficient, they can synthesize tryptophan from small molecules in the cell. The tryptophan repressor inhibits transcription of the genes in the tryptophan operon, which encodes the tryptophan biosynthetic enzymes. Upon binding tryptophan, the tryptophan repressor binds to a site in the promoter of the operon.

A. Why is tryptophan-dependent binding to the operon a useful property for the tryptophan repressor?

B. What would you expect to happen to the regulation of the tryptophan biosynthetic enzymes in cells that express a mutant form of the tryptophan repressor that (i) cannot bind to DNA or (ii) binds to DNA even when no tryptophan is bound to it?

C. What would happen in scenarios (i) and (ii) if the cell produced normal tryptophan repressor from a second unmutated copy of the gene?

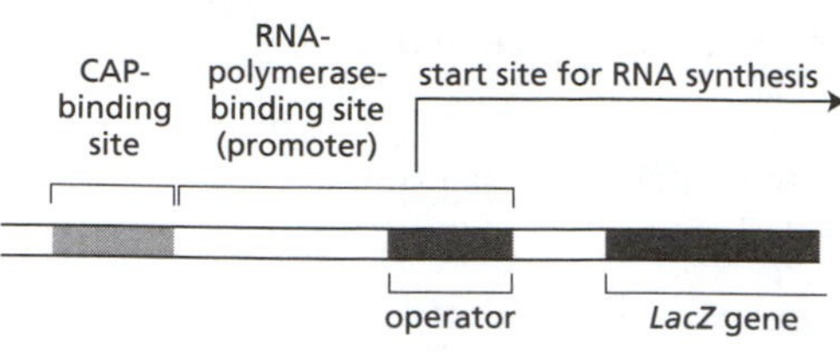


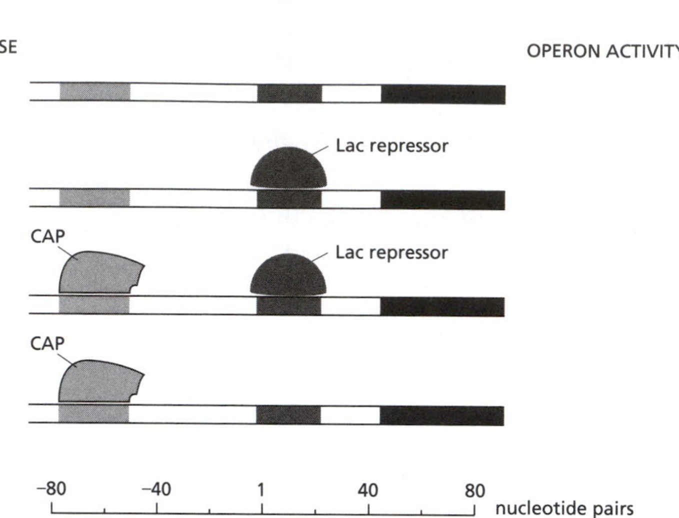

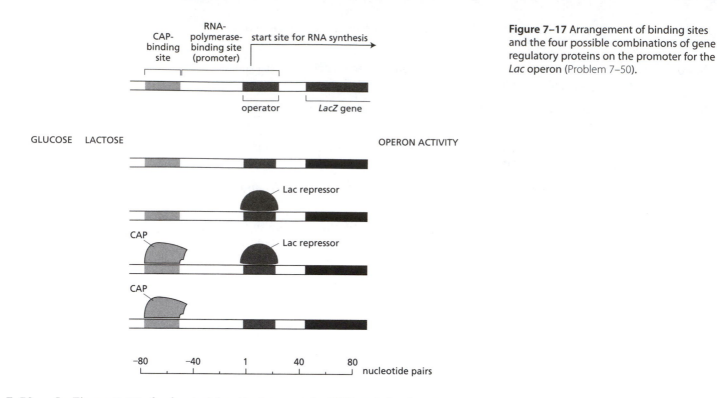

**Figure 7–17** Arrangement of binding sites and the four possible combinations of gene regulatory proteins on the promoter for the *Lac* operon (Problem 7–50).

**7–50** In Figure 7–17, the bacterial activator protein CAP and the Lac repressor have been placed in the four possible combinations on their binding sites in the promoter for the *Lac* operon. Each combination of gene regulatory proteins corresponds to a particular mixture of glucose and lactose. For each of the four combinations, indicate on the left-hand side of the figure which sugars must be present and, on the right-hand side, whether the operon is expected to be turned on or off.

**7–51** Imagine that you have created a fusion between the *Trp* operon, which encodes the enzymes for tryptophan biosynthesis, and the *Lac* operon, which encodes the enzymes necessary for lactose utilization (Figure 7–18). Under which set of conditions (A–F below) will β-galactosidase be expressed in the strain that carries the fused operon?

A. Only when lactose and glucose are both absent.
B. Only when lactose and glucose are both present.
C. Only when lactose is absent and glucose is present.
D. Only when lactose is present and glucose is absent.
E. Only when tryptophan is absent.
F. Only when tryptophan is present.

**7–52** When enhancers were initially found to influence activity at remote promoters, two principal models were invoked to explain this action at a distance. In the 'DNA looping' model, direct interactions between proteins

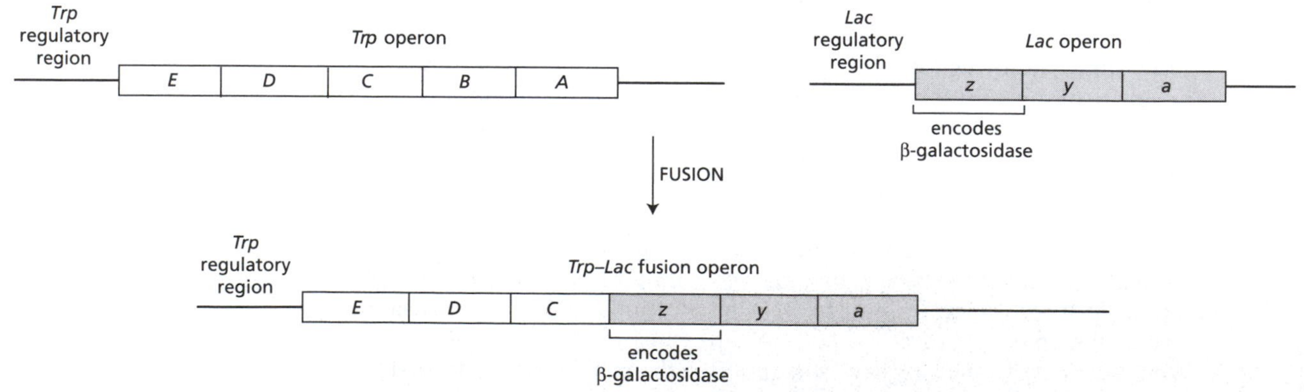


**Figure 7–18** Separated (normal) and fused *Trp* and *Lac* operons (Problem 7–51).

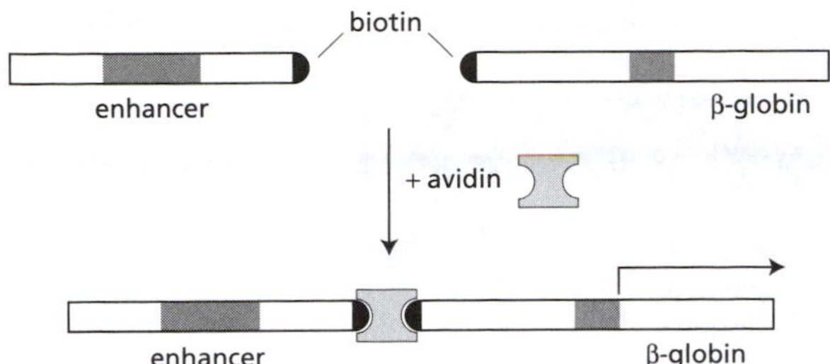


**Figure 7–19** Stimulation of β-globin gene expression by an enhancer linked via a protein bridge (Problem 7–52). Each DNA molecule carries biotin attached to one end, as shown. In the presence of the protein avidin, the two molecules are linked together and transcription occurs as shown by the *arrow above* the β-globin gene.

bound at enhancers and promoters were proposed to stimulate RNA polymerase. In the 'scanning' or 'entry-site' model, RNA polymerase (or a gene regulatory protein) was proposed to bind at the enhancer and then slide along the DNA until it reached the promoter. These two models were distinguished using an enhancer on one piece of DNA and the β-globin gene with its promoter on a separate piece of DNA (Figure 7–19). The β-globin gene was not expressed from the mixture of pieces. When the two segments of DNA were joined via a protein linker, the β-globin gene was expressed.

How does this experiment distinguish between the DNA looping model and the scanning model? Explain your answer.

**7–53** Some gene regulatory proteins bind to DNA and cause the double helix to bend at a sharp angle. Such 'bending proteins' can affect the initiation of transcription without contacting the RNA polymerase, the general transcription factors, or any gene regulatory protein. Can you devise a plausible explanation for how such proteins might work to modulate transcription? Draw a diagram that illustrates your explanation.

**7–54** The yeast Gal4 gene activator protein comprises two domains: a DNA-binding domain and an activation domain. The DNA-binding domain allows it to bind to appropriate DNA sequences located near genes that are required for metabolism of the sugar galactose. The activation domain binds to components of the transcriptional machinery (including RNA polymerase), attracting them to the promoter, so the regulated genes can be turned on. In the absence of Gal4, the galactose genes cannot be turned on. When Gal4 is expressed normally, the genes can be maximally activated. When Gal4 is massively overexpressed, however, the galactose genes are turned off. Why do you suppose that too much Gal4 squelches expression of the galactose genes?

**7–55** How are histone acetylases and chromatin remodeling complexes recruited to unmodified chromatin, and how are they thought to aid in the activation of transcription from previously silent genes?

**Problem 4–57** measures the capacity of chromatin remodeling complexes to move nucleosomes out of the way.

**7–56** How is it that protein–protein interactions that are too weak to cause proteins to assemble in solution can nevertheless allow the same proteins to assemble into complexes on DNA?

**7–57** Consider the following argument: "If the expression of every gene depends on a set of gene regulatory proteins, then the expression of these gene regulatory proteins must also depend on the expression of other gene regulatory proteins, and their expression must depend on the expression of still other gene regulatory proteins, and so on. Cells would therefore need an infinite number of genes, most of which would code for gene regulatory proteins." How does the cell get by without having to achieve the impossible?

**7–58** A locus control region (LCR) that lies far upstream of the cluster of β-like globin genes regulates the entire gene cluster. In skin cells, for example, which do not express the globin genes, the whole gene cluster is tightly packaged into chromatin. By contrast, in erythroid cells the gene cluster is decondensed and individual genes are transcribed in a characteristic developmental sequence. How is the LCR thought to help regulate expression of the globin genes?

**Table 7–2 Responses of normal and mutant bacteria to the presence and absence of arabinose** (Problem 7–59).

| GENOTYPE | *AraA* GENE PRODUCT – ARABINOSE | *AraA* GENE PRODUCT +ARABINOSE |
|---|---|---|
| *AraC*⁺ | 1 | 1000 |
| *AraC*⁻ | 1 | 1 |

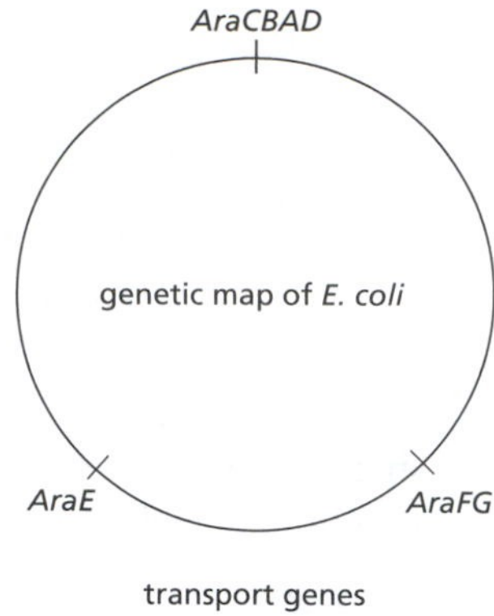


**Figure 7–20** Chromosomal locations of the genes involved in arabinose metabolism (Problem 7–59).

## DATA HANDLING

**7–59** In the absence of glucose, *E. coli* can proliferate using the pentose sugar arabinose for energy, by inducing a set of genes that are arrayed in three groups on the chromosome (Figure 7–20). The *AraA, AraB,* and *AraD* genes encode enzymes for the metabolism of arabinose. The *AraC* gene encodes a gene regulatory protein that binds adjacent to the arabinose promoters and coordinates expression of the genes involved in arabinose metabolism. (The other two groups of genes encode proteins involved in arabinose transport.)

To understand the regulatory properties of the AraC protein, you isolate a mutant bacterium with a deletion of the *AraC* gene. As shown in Table 7–2, the mutant strain does not induce expression of the *AraA* gene when arabinose is added to the medium.

A. Do the results in Table 7–2 indicate that the AraC protein regulates arabinose metabolism by negative control or by positive control? Explain your answer.

B. What would the data in Table 7–2 have looked like if the AraC protein regulated expression of the enzymes of arabinose metabolism by the type of control that you did not choose in part A?

**7–60** *E. coli* proliferates faster on the monosaccharide glucose than it does on the disaccharide lactose for two reasons: (1) lactose is taken up more slowly than glucose and (2) lactose must be hydrolyzed to glucose and galactose (by β-galactosidase) before it can be further metabolized.

When *E. coli* is grown on a medium containing a mixture of glucose and lactose, it proliferates with complex kinetics (Figure 7–21, squares). The bacteria proliferate faster at the beginning than at the end, and there is a lag between these two phases when they virtually stop dividing. Assays of the concentrations of the two sugars in the medium show that glucose falls to very low levels after a few cell doublings (Figure 7–21, circles), but lactose remains high until near the end of the experimental time course (not shown). Although the concentration of lactose is high throughout most of the experiment, β-galactosidase, which is regulated as part of the *Lac* operon, is not induced until more than 100 minutes have passed (Figure 7–21, triangles).

A. Explain the kinetics of bacterial proliferation during the experiment. Account for the rapid initial rate, the slower final rate, and the delay in the middle of the experiment.

**Problem 3–104** analyzes regulation of the *Lac* operon in terms of the quantitative changes in affinity of the Lac repressor upon binding the inducing ligand.

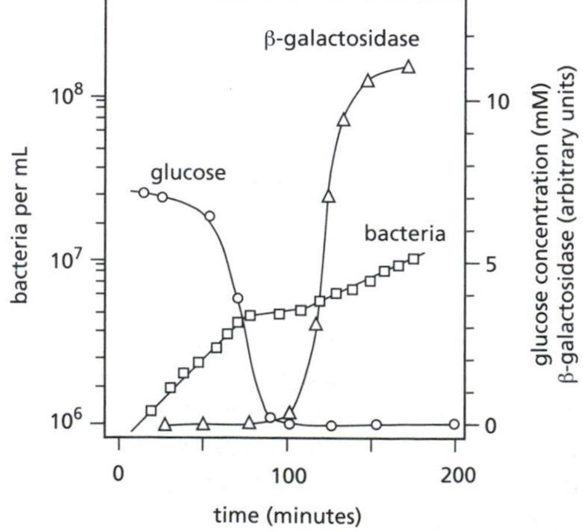


**Figure 7–21** Proliferation of *E. coli* on a mixture of glucose and lactose (Problem 7–60).

B. Explain why the *Lac* operon is not induced by lactose during the rapid initial phase of bacterial proliferation.

**7–61** Transcription of the bacterial gene encoding the enzyme glutamine synthetase is regulated by the availability of nitrogen in the cell. The key transcriptional regulator is the NtrC protein, which stimulates transcription only when it is phosphorylated. Phosphorylation of NtrC is controlled by the NtrB protein, which is both a protein kinase and a protein phosphatase. The balance between its kinase and phosphatase activities—hence the level of phosphorylation of NtrC and transcription of glutamine synthetase—is determined by other proteins that respond to the ratio of α-ketoglutarate and glutamine. (This ratio is a sensitive indicator of nitrogen availability because two nitrogens—as ammonia—must be added to α-ketoglutarate to make glutamine.)

Transcription of the gene for glutamine synthetase can be achieved *in vitro* by adding RNA polymerase, a special sigma factor, and phosphorylated NtrC to a linear DNA template containing the gene and its upstream regulatory region. NtrC binds to five sites upstream of the promoter. Although binding of NtrC is only slightly increased by phosphorylation, transcription is absolutely dependent on phosphorylation, even though RNA polymerase binds strongly to the promoter in the absence of NtrC.

Activation of transcription by NtrC was characterized using three different templates: the normal gene with intact regulatory sequences, a gene with all of the NtrC-binding sites deleted, and a gene with three NtrC-binding sites at its 3′ end (Figure 7–22). In the absence of phosphorylated NtrC, no transcription occurred from any of the templates. In the presence of 100 nM phosphorylated NtrC, all three templates supported maximal transcription. The three templates differed significantly, however, in the concentration of NtrC required for half-maximal rates of transcription: the normal gene (Figure 7–22A) required 5 nM NtrC, the gene with 3′ binding sites (Figure 7–22C) required 10 nM NtrC, and the gene without NtrC-binding sites (Figure 7–22B) required 50 nM NtrC.

A. If RNA polymerase can bind to the promoter of the glutamine synthetase gene in the absence of NtrC, why do you suppose NtrC is needed to activate transcription?
B. If NtrC can bind to its binding sites regardless of its state of phosphorylation, why do you suppose phosphorylation is necessary for transcription?

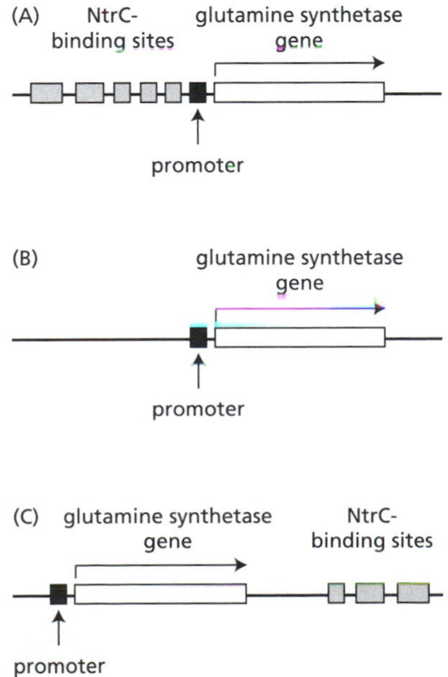


**Figure 7–22** Three templates for studying the role of NtrC in transcription of the glutamine synthetase gene (Problem 7–61). (A) The normal gene with intact upstream regulatory sequences. (B) A gene with all of the NtrC-binding sites deleted. (C) A gene with three NtrC-binding sites at the 3′ end of the gene.

C. If NtrC can activate transcription even when its binding sites are absent, what role do the binding sites play?

**7–62** Regulation of arabinose metabolism in *E. coli* is fairly complex. Not only are the genes scattered around the chromosome (see Figure 7–20), but the regulatory protein, AraC, acts as both a positive and a negative regulator (Figure 7–23A). When AraC binds at site 1 in the presence of arabinose (and the absence of glucose) transcription increases roughly 100-fold over the level measured in the absence of AraC. Binding of AraC at site 2 in the absence of arabinose represses transcription of the *AraBAD* genes about 10-fold below the level measured in the absence of AraC. The combined effects of negative regulation at site 2 (in the absence of arabinose) and positive regulation at site 1 (in the presence of arabinose) means that addition of arabinose causes a 1000-fold increase in transcription of the *AraBAD* genes.

Positive regulation by binding at site 1 seems straightforward since that site lies adjacent to the promoter and presumably facilitates RNA polymerase binding or stimulates initiation of transcription. You are puzzled by the results of binding at site 2. Site 2 lies 280 nucleotides upstream from the start site of transcription. Regulatory effects over such distances seem more reminiscent of enhancers in eucaryotic cells. To study the mechanism of repression at site 2 more easily, you move the whole *AraBAD* regulatory region so that it is immediately in front of the *GalK* gene, whose encoded enzyme, galactokinase, is simple to assay.

To determine the importance of the spacing between the two AraC-binding sites, you insert or delete nucleotides at the insertion point indicated in Figure 7–23A. You then assay the activity of the promoter in the absence of arabinose by growing the bacteria on special indicator plates. If galactokinase is not made, the bacterial colonies are white; if it is, the colonies are red. You streak out bacteria containing the altered spacings against a scale that shows how many nucleotides were deleted or inserted (Figure 7–23B). Much to your surprise, red and white streaks are interspersed.

When you show your results to your advisor, she is very pleased and tells you that these experiments distinguish among three potential mechanisms of repression from a distance. (1) An alteration in the structure of the DNA could propagate from the repression site to the transcription site, making the promoter an unfavorable site for RNA polymerase binding. (2) The protein could bind cooperatively (oligomerize) at the repression site in such a way that additional subunits continue to be added until the growing chain covers the promoter, blocking transcription. (3) The DNA could form a loop

(A) *AraBAD* REGULATORY REGION

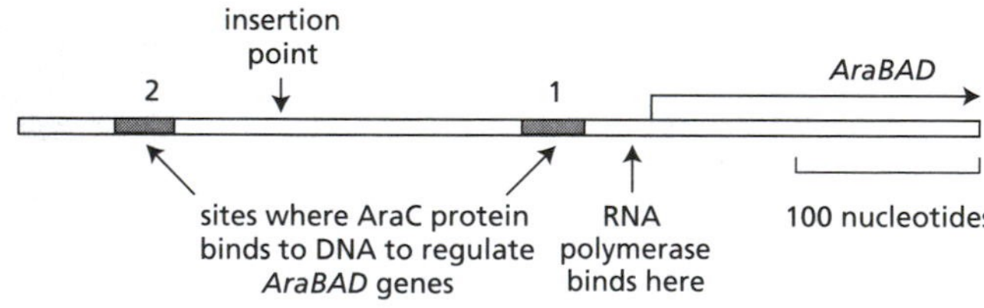


(B) GALACTOKINASE STREAK TEST

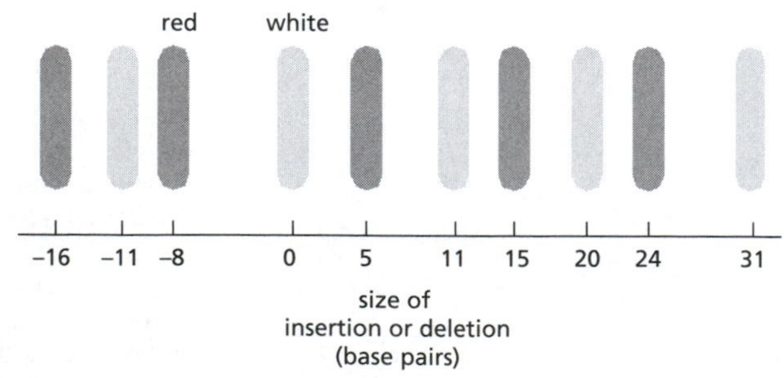


**Figure 7–23** Analysis of regulation of the *AraBAD* genes (Problem 7–62). (A) Arrangement of AraC-binding sites in the *AraBAD* operon. (B) Results of altering the spacing between AraC binding sites 1 and 2. For this streak assay, the *AraBAD* regulatory region was inserted in front of the *GalK* gene. Various numbers of nucleotides were inserted or deleted between the AraC-binding sites at the point indicated by the *arrow* in A. Colonies were streaked out against a scale that showed how many nucleotides had been inserted or deleted. Colonies that do not express galactokinase are *white*; those that do are *red*.

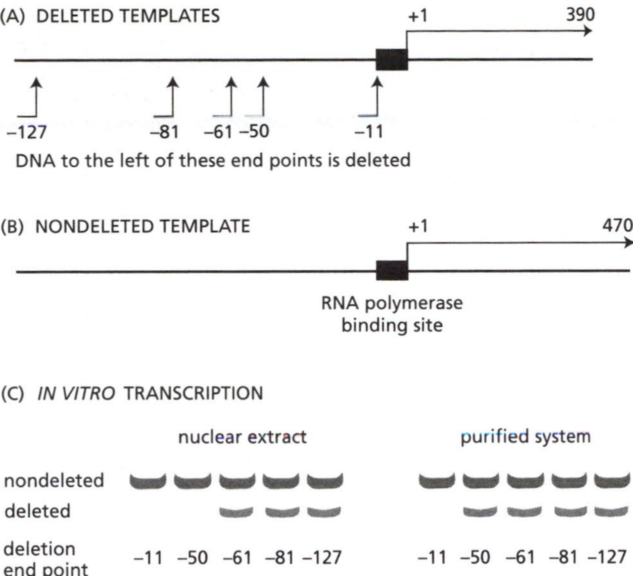


**Figure 7–24** Analysis of transcription from a viral promoter (Problem 7–63). (A) Deletion templates. These templates are all truncated at their 3′ ends so that they give rise to slightly shorter transcripts than the longer, control template in B. The deletions remove DNA to the *left* of the indicated end points. Nucleotides are numbered from the start site of transcription (+1); *negative numbers* indicate nucleotides in front of the start site. (B) Nondeleted control template. This template serves as an internal control; it gives rise to a transcript that is 80 nucleotides longer than the experimental templates in A. (C) Results of transcription of a mixture of control and deleted templates in a crude nuclear extract (*left*) and using purified components (*right*). *Negative numbers* identify the particular deletion template that was included in each mixture.

so that the protein bound at the distant repression site could interact with proteins (or DNA) at the transcription start site.

Which of these general mechanisms do your data support? How do you account for the patterns of red and white streaks?

**7–63** You have developed an *in vitro* transcription system using a defined segment of DNA that is transcribed under the control of a viral promoter. Transcription of this DNA occurs when you add purified RNA polymerase II and the general transcription factors, TFIID, TFIIB, and TFIIE. Transcription in your *in vitro* system using purified components occurs at low efficiency relative to the transcription that occurs if you use a crude nuclear lysate instead. This suggests that an important gene regulatory protein is missing from the purified components. To search for the DNA sequence to which this putative regulatory protein binds, you make a series of deletions upstream of the start site for transcription (Figure 7–24A), and compare their transcriptional activity in the purified system and in the crude extract. As an internal control, you mix each of the deletions with a nondeleted template that encodes a slightly longer transcript (Figure 7–24B). The results of these assays are shown in Figure 7–24C. Roughly ten times less material from the reactions in the nuclear extracts was run on the gel so that the intensities of the bands would be approximately equal. Deletions up to –61 have no effect on transcription, and the –11 deletion inactivates transcription in both the purified system and the crude extract. Surprisingly, the –50 deletion is transcribed as efficiently as the control template in the purified system, but not in the nuclear extract.

You purify the gene regulatory protein that is responsible for this effect and show that it stimulates transcription approximately 10-fold from the control template and from the –61 deletion template; however, it does not stimulate transcription from the –50 deletion template. The purified protein binds very transiently in the absence of TFIID (a 20-second half-life), but binds stably (with a half-life greater than 5 hours) in its presence.

A. Where is the binding site for the gene regulatory protein located?

B. In the nuclear extract, which contains the stimulatory gene regulatory protein, no transcript from the –50 deletion template is evident on the gel (Figure 7–24C). By contrast, in the purified system, which is missing the stimulatory protein, the transcript from the –50 deletion template is clearly visible. How would you explain this difference?

C. How would you account for the marked difference in stability of binding of the regulatory protein in the presence and absence of TFIID?

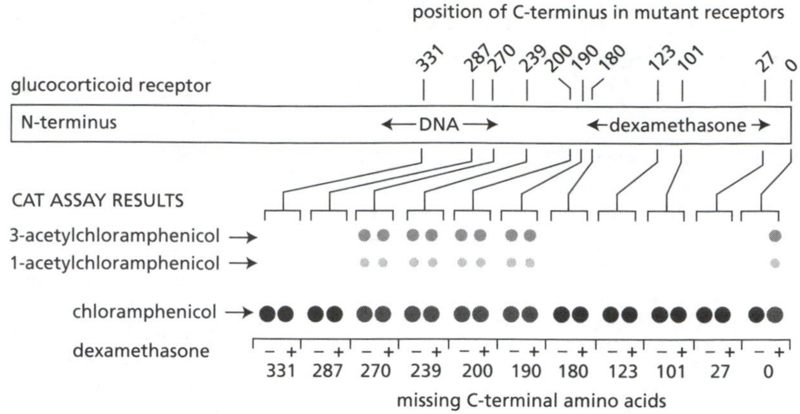


**Figure 7–25** Effect of C-terminal deletions on the activity of the glucocorticoid receptor (Problem 7–64). The schematic diagram at the *top* illustrates the position of the DNA-binding site (DNA) and the glucocorticoid-binding site (dexamethasone) in the receptor, as well as the positions of the C-terminal deletions. The *lower* diagram shows the results of a standard CAT assay obtained by mixing cell extracts with $^{14}$C-chloramphenicol. The *lowest* spot is unreacted chloramphenicol; the *upper* spots show the attachment of acetyl groups to one or the other of two positions on chloramphenicol. The presence (+) or absence (–) of dexamethasone is indicated *below* appropriate lanes.

**7–64** Hormone receptors for glucocorticoids, upon hormone binding, become DNA-binding proteins that activate a specific set of responsive genes. Genetic and molecular studies indicate that the DNA- and hormone-binding sites occupy distinct regions of the C-terminal half of the glucocorticoid receptor. Hormone binding could generate a functional DNA-binding protein in either of two ways: by altering receptor conformation to create a DNA-binding domain or by altering the conformation to uncover a preexisting DNA-binding domain.

These possibilities were investigated by comparing the activities of a series of C-terminal deletions (Figure 7–25). Segments of the cDNA that encode N-terminal portions of the receptor were expressed by transfection into appropriate cells. The capacity of the receptor fragments to activate responsive genes was tested by co-transfections with a reporter plasmid carrying a glucocorticoid response element linked to the chloramphenicol acetyltransferase *(Cat)* gene. As shown in Figure 7–25, cells co-transfected with a cDNA for the full-length receptor (that is, with its C-terminus at position 0) responded as expected. In the presence of the glucocorticoid dexamethasone, CAT activity was readily detected, which showed that the *Cat* gene was functioning; in the absence of dexamethasome, no CAT activity was detected. Six mutant receptors, lacking 27, 101, 123, 180, 287, and 331 C-terminal amino acids, failed to activate CAT expression in the presence or absence of dexamethasone. In contrast, four mutant receptors, lacking 190, 200, 239, and 270 C-terminal amino acids, activated CAT expression in the presence and absence of dexamethasone. Separate experiments indicated that the mutant receptors were synthesized equally.

How do these experiments distinguish between the proposed models for hormone-dependent conversion of the normal receptor to a DNA-binding form? Does hormone binding create a DNA-binding site or does it uncover a preexisting one?

**7–65** An early attempt to understand how the packing of DNA into chromatin affects transcription in eucaryotes used a defined template, purified RNA polymerase II, and the general transcription factors—TFIIA, TFIIB, TFIID, and TFIIE. If the template was first assembled into nucleosomes, no transcription was detected when the transcription components were added (Figure 7–26, lane 2). If the template was preincubated with the transcription components (in the absence of NTPs), and then assembled into chromatin and the chromatin template purified, transcription proceeded just as well as it does on the naked DNA template when the transcription components were added (compare lanes 1 and 3). This suggested that the transcription components, when bound to the template, keep the promoter accessible.

This phenomenon was investigated in more detail in two additional kinds of experiment. In one, individual components were omitted during the preincubation (Figure 7–26, lanes 4 to 8). In the second, individual components were left out during the transcription assay (lanes 9 to 13).

A. Which of the transcription components must be present during the preincubation in order for the template to be active after chromatin assembly?

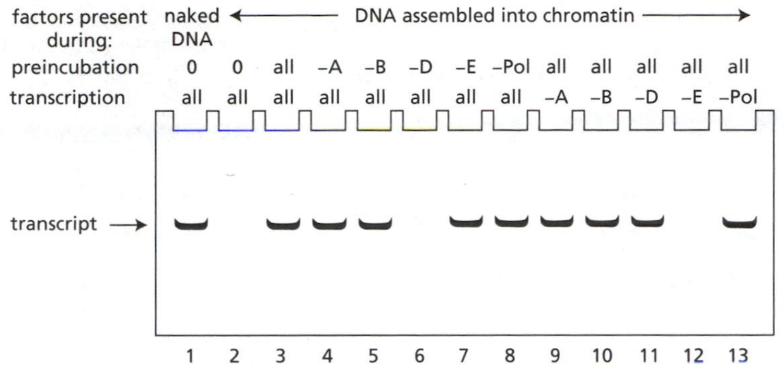


**Figure 7–26** Effects of chromatin assembly on transcription (Problem 7–65). Transcription templates were preincubated with none, all, or all minus one of the transcription components (for example, –A means that TFIIA was left out, and –Pol means that RNA polymerase II was left out). After assembly into chromatin, the template with bound proteins was purified away from the unbound components in the preincubation mixture. The transcription assay was also carried out in the presence of all, or all minus one of the transcription components.

B. Which of the transcription components can form a complex with the template that is stable to chromatin formation and subsequent purification?

C. Which of the transcription components must be added during the assay in order to produce a transcript?

**7–66** Coactivators do not bind to DNA, but instead serve as bridging molecules that link gene-specific activators to general transcription factors at the promoter. The SAGA complex in yeast is a large multiprotein complex that is required for transcription of many genes. SAGA contains a variety of gene regulatory proteins, including a histone acetyltransferase and a subset of TATA-binding protein associated factors (TAFs). If SAGA functions as a coactivator, it should be physically present at the promoters it regulates, and its recruitment should be governed by an activation signal.

To test this idea, you focus on the regulation of the *Gal1* and *Gal10* genes by the Gal4 activator, which binds to DNA sequences (UAS) adjacent to their promoters (Figure 7–27A, shaded boxes). *Gal4* requires components of the SAGA complex for activation. When cells are grown on galactose, the activation domain of Gal4 is exposed, allowing the transcription machinery to assemble at the adjacent galactose promoters. If the cells are grown instead on the sugar raffinose, Gal4 remains bound to the UAS, but its activation domain is occluded and transcription does not occur.

To determine whether the SAGA complex associates with Gal4 on the chromosome, you use chromatin immunoprecipitation. You prepare two strains of yeast that each expresses a derivative of a protein in the SAGA complex—either Spt3 or Spt20—that are fused to a stretch of amino acids from hemagglutinin (HA). These so-called HA tags do not interfere with the function of these proteins, and they allow very efficient immunoprecipitation using commercial antibodies against HA. You prepare small chromatin

(A) MAP OF THE *Gal1-10* LOCUS

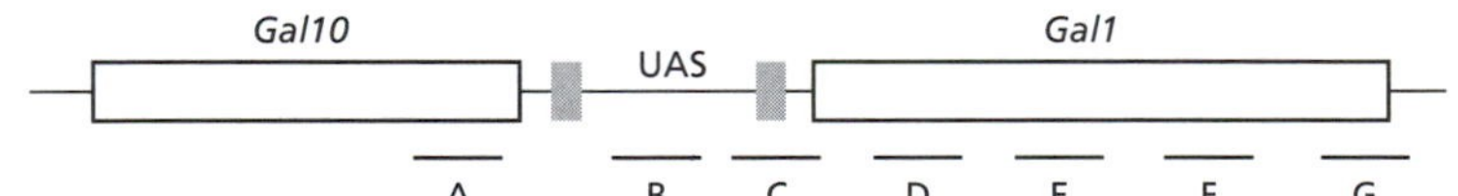


(B) PCR PRODUCTS FROM CHROMATIN IMMUNOPRECIPITATES

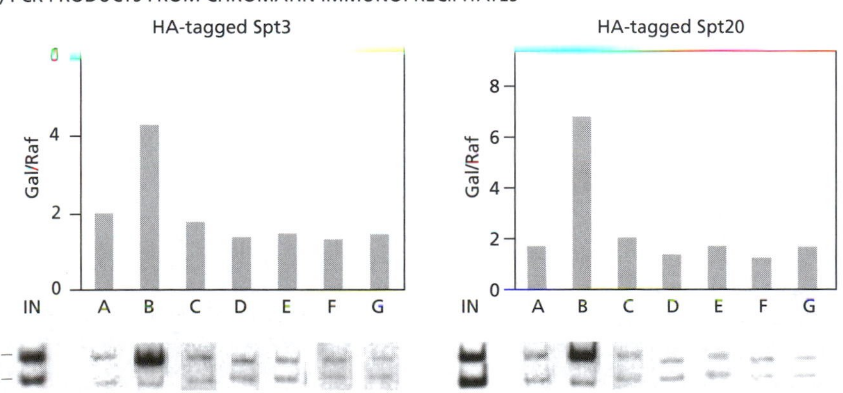


**Figure 7–27** Analysis of SAGA association with the *Gal* promoter (Problem 7–66). (A) Organization of the *Gal10* and *Gal1* genes. A common UAS serves both promoters, which are indicated by *shaded regions*. The positions of the PCR fragments that were used for quantification are shown. (B) Results of chromatin immunoprecipitation in strains expressing HA–Spt3 or HA–Spt20. The two strains were grown in medium containing raffinose (Raf) or galactose (Gal), their chromatin was fragmented and immunoprecipitated with HA antibodies, and quantitative PCR was carried out on the isolated DNA. The PCR products from strains grown on raffinose were loaded onto the gels and run for a few minutes before the PCR products from strains grown on galactose, so that the products could be compared within the same lanes. Finally, the amounts of PCR products were measured, normalized to the loading control (IN), and the ratio (Gal/Raf) was calculated. IN refers to the input chromatin before immunoprecipitation; PCR products in those lanes were from an unaffected part of the genome and were used as controls for loading differences. About 50- to 100-fold less input chromatin (IN) was loaded per lane than immunoprecipitated chromatin, reflecting the inefficiency of chromatin immunoprecipitation.

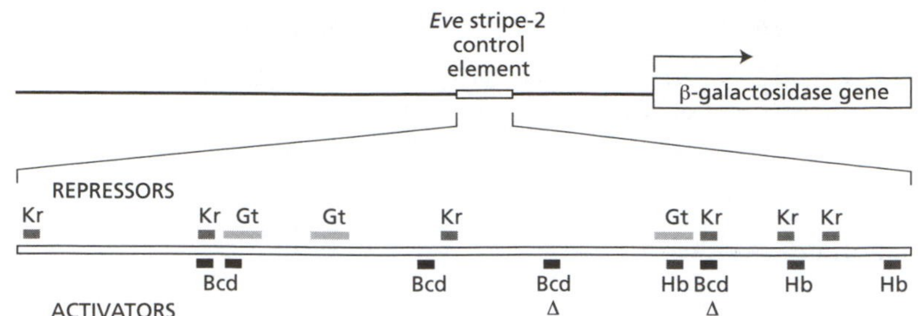


**Figure 7–28** Binding sites for repressors and activators of *Eve* stripe 2 (Problem 7–67).

fragments from the two strains after growth on raffinose or galactose, and then precipitate them with HA antibodies. You strip the precipitated chromatin of protein and analyze the DNA by quantitative PCR to measure the amounts of specific DNA segments in the precipitates. The ratio of immunoprecipitated chromatin from galactose-grown cells to that from raffinose-grown cells (Gal/Raf) is shown for several segments around the galactose promoter in Figure 7–27B.

A. Which of the tested fragments would you have expected to be enriched if SAGA behaved as a coactivator? Explain your answer.

B. Does SAGA meet the criteria for a coactivator? Is it physically present at the promoter, and does its recruitment depend on an activation signal?

**7–67** The protein encoded by the *Even-skipped (Eve)* gene of *Drosophila* is a transcriptional regulator required for proper segmentation in the middle of the body. It first appears about 2 hours after fertilization at a uniform level in all the embryonic nuclei. Not long after that, it forms a pattern of seven stripes across the embryo. Each stripe is under the control of a separate module in the promoter, which provides binding sites for both repressors and activators of *Eve* transcription. In the case of stripe 2, there are multiple, overlapping binding sites for activators and repressors (Figure 7–28). Two activators, Hunchback (Hb) and Bicoid (Bcd), and two repressors, Giant (Gt) and Krüppel (Kr), are required to give the normal pattern. The binding sites for these proteins have been mapped onto the 670-nucleotide segment shown in Figure 7–28; deletion of this upstream segment abolishes *Eve* expression in stripe 2. The patterns of expression of Hunchback, Bicoid, Giant, and Krüppel in the embryo are shown in Figure 7–29. It seems that *Eve* expression in stripe 2 occurs only in the region of the embryo that expresses both activators, but neither repressor: a simple enough rule.

To check if this rule is correct, you construct a β-galactosidase reporter gene driven by a 5-kb upstream segment from the *Eve* promoter. (This segment also includes the controlling elements for stripes 3 and 7.) In addition to the normal upstream element, you make three mutant versions in which several of the binding sites in the *Eve* stripe-2 control segment have been deleted. (Note, however, that because many of the binding sites overlap it is not possible to delete all of one kind of site without affecting some of the other sites.)

Construct 1. Deletion of all the Krüppel-binding sites
Construct 2. Deletion of all the Giant-binding sites

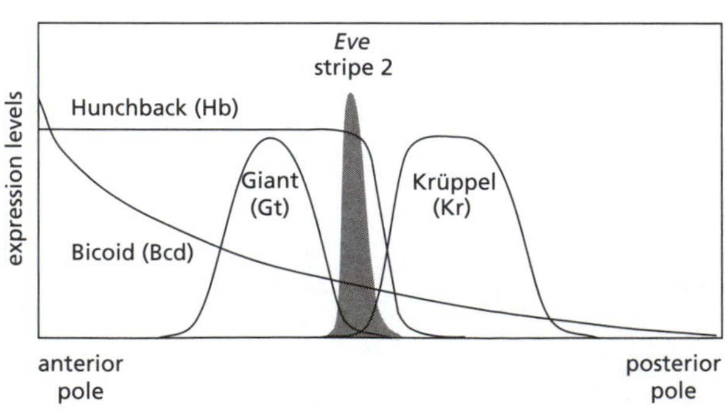


**Figure 7–29** Expression of repressors and activators of *Eve* stripe 2 in the *Drosophila* embryo (Problem 7–67).

Construct 3. Deletion of two Bicoid-binding sites (indicated by Δs in Figure 7–28)

You make flies containing these novel genetic constructs integrated into their chromosomes and determine the patterns of β-galactosidase expression in their embryos, which are shown in Figure 7–30.

A. Match the mutant embryos to the mutant constructs.
B. You began these experiments to test the simple rule that *Eve* expression in stripe 2 occurs in the embryo where the two activators are present and the two repressors are absent. Do the results with the mutant embryos confirm this rule? Explain your answer.
C. Offer a plausible explanation for why there is no expression of β-galactosidase at the anterior pole of mutant embryo D in Figure 7–30.
D. In the *Eve* stripe-2 control segment, the binding sites for the two activators do not overlap, nor do the binding sites for the two repressors; however, it is often the case that binding sites for activators overlap binding sites for repressors. What does this overlap suggest about the mode of genetic control of *Eve* in stripe 2, and what might be the consequences for stripe morphology?

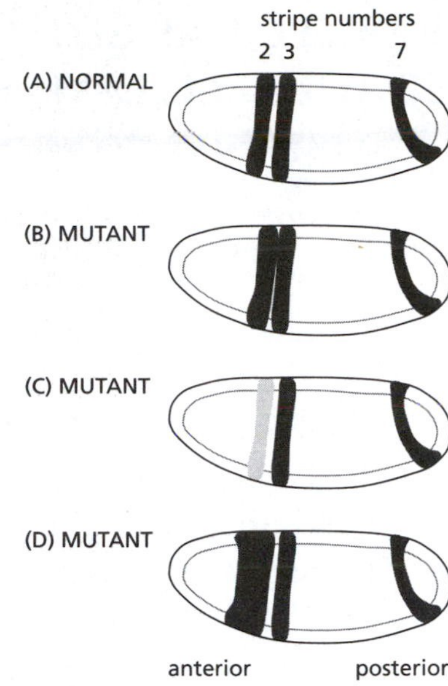


**Figure 7–30** Embryonic expression of β-galactosidase from constructs with normal or mutated *Eve* stripe-2 control elements (Problem 7–67).

**7–68** Genes present at the expressed *Mat* locus on chromosome III determine the mating type of a haploid yeast cell. Identical mating-type genes also exist at the *Hml* and *Hmr* loci on the same chromosome (Figure 7–31A). At these two silent loci, the mating-type genes are not expressed, even though all the signals for expression are present. Each silent mating type locus is bracketed by two short DNA sequences, designated E and I, which bind a variety of proteins and serve to establish and maintain repression of genes within each locus.

To investigate the function of these elements, you insert the *Ura3* gene at different locations between E and I and outside the *Hml* locus, as shown in Figure 7–31B. You then test the growth of these strains in complete medium, in the absence of uracil (–uracil), and in the presence of 5-fluoroorotic acid (+FOA) (Figure 7–31B). These growth conditions test for the activity of the *Ura3* gene as follows: in complete medium, expression of the Ura3 protein is irrelevant; in the absence of uracil, expression of Ura3 is required for growth; and in the presence of FOA, expression of Ura3 is lethal to the cell.

Do these control elements specifically repress the mating-type genes, or do they act like insulators that control the expression of any gene placed between them?

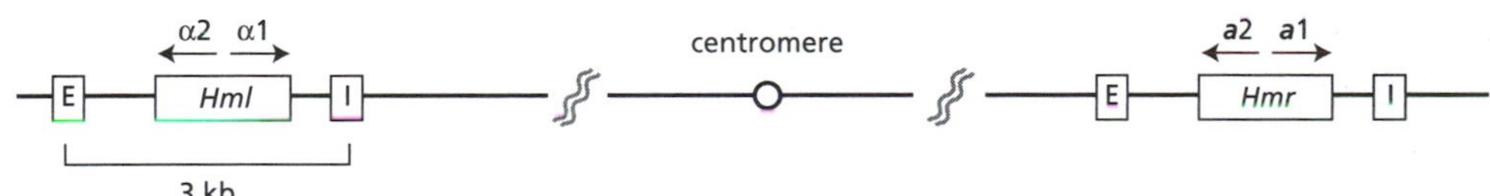


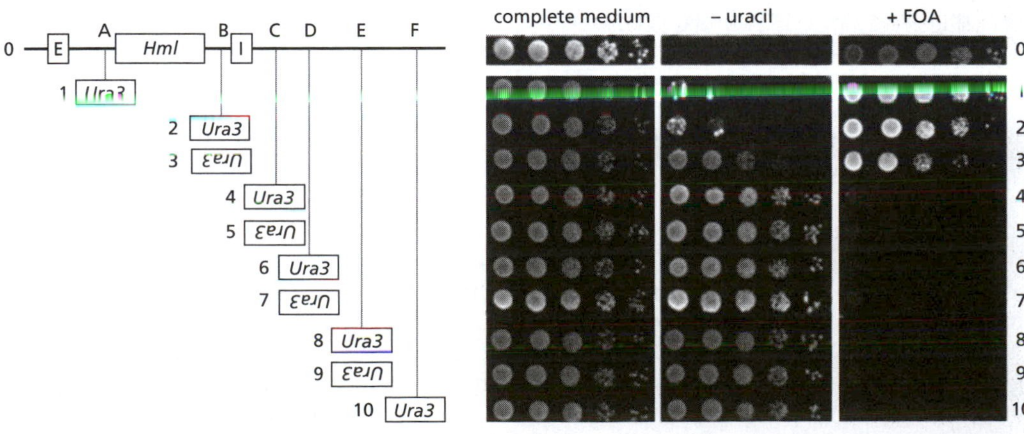


**Figure 7–31** Effects of E and I elements at *Hml* on the expression of *Ura3* genes inserted nearby (Problem 7–68). (A) Arrangement of E and I elements around *Hml* and *Hmr*. The mating-type genes are shown as α1, α2, a1, and a2. (B) Locations of *Ura3* genes inserted around the *Hml* locus. Orientations of the *Ura3* genes are indicated by the direction of *lettering*. Growth phenotypes of each strain are indicated on the *right*. Strain 0 is the parental strain with no *Ura3* gene; strains 1 to 10 have a *Ura3* gene inserted as indicated. Serial dilutions of each strain were spotted on agar plates (with the highest concentration on the left). The agar contained complete medium, complete medium minus uracil (–uracil), or complete medium plus FOA (+FOA). *White areas* indicate growth.

# THE MOLECULAR GENETIC MECHANISMS THAT CREATE SPECIALIZED CELL TYPES

TERMS TO LEARN

| | | |
|---|---|---|
| CG island | genomic imprinting | phase variation |
| Cro protein | lambda repressor protein | X-inactivation |
| DNA methylation | mating | X-inactivation center (XIC) |
| epigenetic inheritance | mating-type (*Mat*) locus | |

## DEFINITIONS

Match each definition below with its term from the list above.

**7–69** Transcriptional silencing of gene expression on one of the two X chromosomes in the somatic cells of female mammals.

**7–70** Addition of a $-CH_3$ group to DNA; restricted to cytosines in CG sequences in vertebrate DNA.

**7–71** Heritable difference in the phenotype of a cell or an organism that does not result from changes in the nucleotide sequence of the DNA.

**7–72** Control of expression of genes for some cell-surface proteins in bacteria by reversible gene rearrangements.

**7–73** Situation where a gene is either expressed or not expressed in the embryo, depending on which parent it is inherited from.

**7–74** Long region of DNA with a much greater than average density of CG sequences, which usually remain unmethylated.

## TRUE/FALSE

Decide whether each of these statements is true or false, and then explain why.

**7–75** Reversible genetic rearrangements are a common way of regulating gene expression in procaryotes and mammalian cells.

**7–76** The fibroblasts and other cell types that are converted to muscle cells by expression of myogenic proteins have probably already accumulated a number of gene regulatory proteins that can cooperate with the myogenic proteins to switch on muscle-specific genes.

**7–77** CG islands are thought to have arisen during evolution because they were associated with portions of the genome that remained active, hence unmethylated, in the germ line.

## THOUGHT PROBLEMS

**7–78** Bacteriophage lambda can replicate as a prophage or lytically. In the prophage state, the viral DNA is integrated into the bacterial chromosome and is copied once per cell division. In the lytic state, the viral DNA is released from the chromosome and replicates many times. This viral DNA then produces viral coat proteins that enclose the replicated viral genomes to form many new virus particles, which are released when the bacterial cell bursts.

These two states are controlled by the gene regulatory proteins cI and Cro, which are encoded by the virus. In the prophage state, cI is expressed; in the lytic state, Cro is expressed. In addition to regulating the expression of other genes, cI is a repressor of transcription of the gene that encodes Cro, and Cro is a repressor of the gene that encodes cI (Figure 7–32). When bacteria containing a lambda prophage are briefly irradiated with UV light, cI protein is degraded.

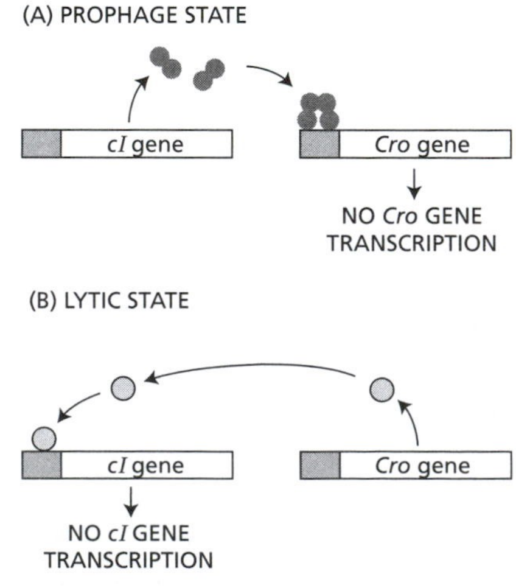


**Figure 7–32** Regulation of bacteriophage lambda replication by cI and Cro (Problem 7–78). (A) The prophage state. (B) The lytic state.

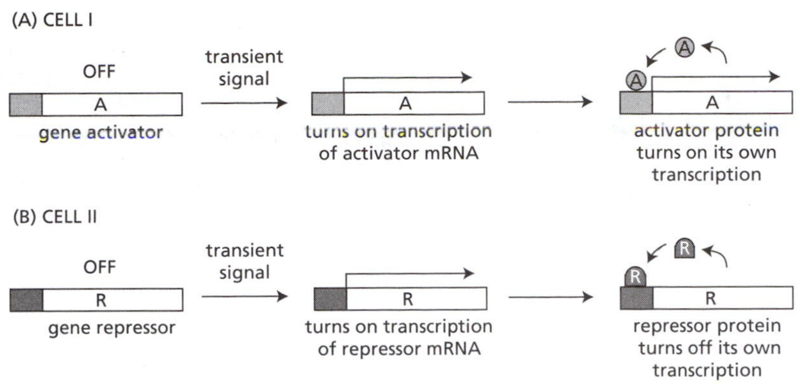


**Figure 7–33** Gene regulatory circuits and cell memory (Problem 7–79). (A) Induction of synthesis of gene activator A by a transient signal. (B) Induction of synthesis of gene repressor R by a transient signal.

A. What will happen next?
B. Will the change in (A) be reversed when the UV light is switched off?
C. How is this mechanism beneficial to the virus?

**7–79** Imagine the two situations shown in Figure 7–33. In cell I, a transient signal induces the synthesis of protein A, which is a gene activator that turns on many genes including its own. In cell II, a transient signal induces the synthesis of protein R, which is a gene repressor that turns off many genes including its own. In which, if either, of these situations will the descendants of the original cell 'remember' that the progenitor cell had experienced the transient signal? Explain your reasoning.

**7–80** Most people who are completely blind have circadian rhythms that are 'free-running;' that is, their circadian rhythms are not synchronized to environmental time cues and they oscillate on a cycle of about 24.5 hours. Why do you suppose the circadian clocks of blind people are not entrained to the same 24-hour clock as the majority of the population? Can you guess what symptoms might be associated with a free-running circadian clock? Do you suppose that blind people have trouble sleeping?

**7–81** Figure 7–34 shows a simple scheme by which three gene regulatory proteins might be used during development to create eight different cell types. How many cell types could you create, using the same rules, with four different gene regulatory proteins? Myod is a gene regulatory protein that can trigger the entire program of muscle differentiation when expressed in fibroblasts.

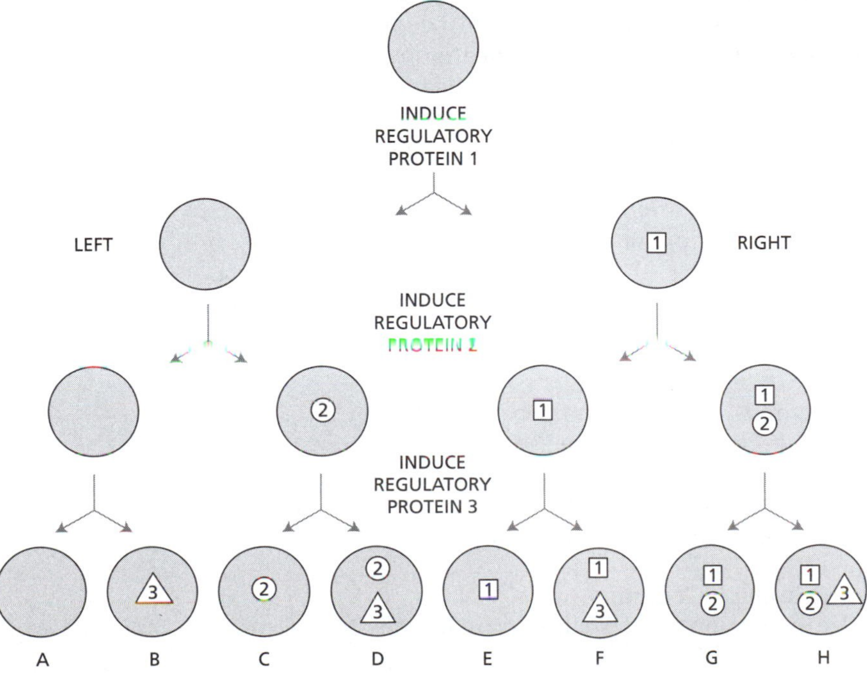


**Figure 7–34** An illustration of combinatorial gene control for development (Problem 7–81). In this simple, idealized scheme a 'decision' to make a new gene regulatory protein *(numbered symbols)* is made after each cell division. In this scheme, the daughter cell on the *right* is induced to make the new regulatory protein. Each gene regulatory protein is assumed to continue to be expressed after it is induced, thereby allowing different combinations of regulatory proteins to be built up. In this example, eight different cell types have been created.

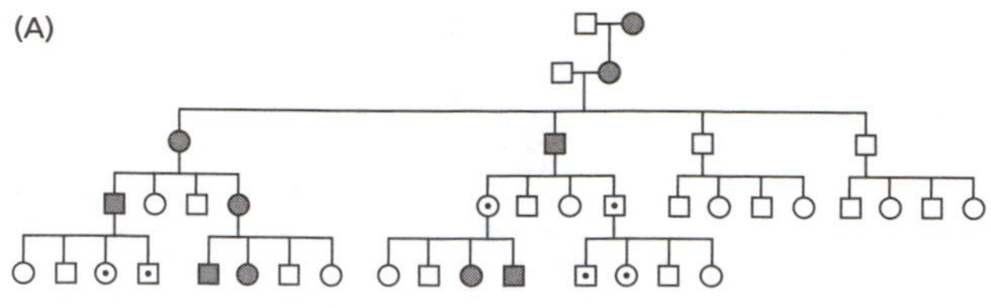


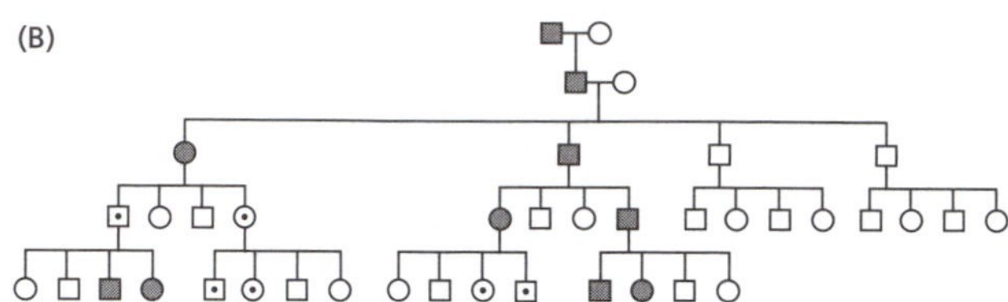


**Figure 7–35** Pedigrees reflecting maternal and paternal imprinting (Problem 7–83). In one pedigree the gene is paternally imprinted; in the other it is maternally imprinted. In generations 3 and 4 only one of the two parents in the indicated matings is shown; the other parent is a normal individual from outside this pedigree. Affected individuals are represented by *shaded circles* for females and *shaded squares* for males. *Dotted symbols* indicate individuals that carry the deletion but do not display the phenotype.

How might you accomodate this observation into the scheme shown in Figure 7–34?

**7–82** Maintenance methyltransferase, *de novo* DNA methyltransferases, and demethylating enzymes play crucial roles in the changes in methylation patterns during development. Starting with the unfertilized egg, describe in a general way how these enzymes bring about the observed changes in genomic DNA methylation.

**7–83** Examine the two pedigrees shown in Figure 7–35. One results from deletion of a maternally imprinted, autosomal gene. The other pedigree results from deletion of a paternally imprinted autosomal gene. In both pedigrees, affected individuals (shaded symbols) are heterozygous for the deletion. These individuals are affected because one copy of the chromosome carries an imprinted, inactive gene, while the other carries a deletion of the gene. Dotted symbols indicate individuals that carry the deleted locus, but do not display the mutant phenotype. Decide which of the pedigrees involves paternal imprinting and which involves maternal imprinting. Explain your answer.

**7–84** Imprinting occurs only in mammals, and why it should exist at all is a mystery. One idea is that it represents an evolutionary endpoint in a tug of war between the sexes. In most mammalian species, a female can mate with multiple males, generating multiple embryos with different fathers. If one father could cause more rapid growth of his embryo, it would prosper at the expense of the other embryos. While this would be good for the father's genes (in an evolutionary sense), it would drain the resources of the mother, potentially putting her life at risk (not good for her genes). Thus, it is in the mother's interest to counter these paternal effects with maternal changes that limit the growth of the embryo.

Based on this scenario, decide whether the *Igf2* gene, whose product—insulin-like growth factor-2—is required for prenatal growth, is more likely to be imprinted in the male or in the female.

**7–85** Sexually reproducing organisms use a variety of strategies to compensate for the gene-dosage differences arising from the different numbers of X chromosomes in the two sexes.

A. From the list below, select the methods of dosage compensation that are used in mice, *Drosophila*, and nematodes, and indicate the sex in which it occurs.
   1. Two-fold up-regulation of gene expression from one X chromosome.
   2. Two-fold down-regulation of gene expression from both X chromosomes.
   3. Elimination of gene expression from one X chromosome.

B. Which of these strategies equalize expression of the X chromosome genes in the two sexes?

## DATA HANDLING

**7–86** It is relatively common for pathogenic organisms to change their coats periodically in order to evade the immune surveillance of their host. *Salmonella* (a bacterium that can cause food poisoning) can exist in two antigenically distinguishable forms, or phases as their discoverer called them in 1922. Bacteria in the two different phases synthesize different kinds of flagellin, which is the protein that makes up the flagellum. Phase 1 bacteria switch to phase 2 and vice versa about once per thousand cell divisions. Two kinds of explanation were originally considered for the switch mechanism: a DNA rearrangement, such as insertion or inversion, and a DNA modification, such as methylation.

The unlinked genes *H1* and *H2* encode the two flagellins responsible for phase variation. The genetic element that enables the bacteria to switch phases is very closely linked to the *H2* gene. To distinguish between the proposed mechanisms of switching, a segment of DNA containing the *H2* gene was cloned. When this segment was introduced into *E. coli*, which has no flagella, most bacteria that picked up the plasmid acquired the ability to swim, indicating that they were synthesizing the flagellin encoded by the *H2* gene. A few colonies, however, were nonmotile even though they carried the plasmid. When DNA was prepared from cultures grown from these nonmotile colonies and introduced into a fresh culture of *E. coli*, some of the transformed bacteria were able to swim, indicating that H2 flagellin synthesis had been switched on.

Plasmid DNA was prepared from these switching cultures, digested with a restriction nuclease, heated to separate the DNA strands, and then slowly cooled to allow DNA strands to reanneal. The DNA molecules were then examined by electron microscopy. About 5% of the molecules contained a bubble, formed by two equal-length single-stranded DNA segments, at a unique position near one end, as shown in Figure 7–36.

A. All *Salmonella* can swim no matter which type of flagellin they are synthesizing. In contrast, the transformed *E. coli* switched between a form that was able to swim and one that was immotile. Why?

B. Explain how these results distinguish between mechanisms of switching that involve DNA rearrangement and ones that involve DNA modification.

C. How do these results distinguish among site-specific DNA rearrangements that involve deletion of DNA, addition of DNA, or inversion of DNA?

**7–87** One of the key regulatory proteins produced by the yeast mating-type locus is a repressor protein known as α2. In haploid cells of the α mating type, α2 is essential for turning off a set of genes that are specific for the **a** mating type. In **a**/α diploid cells, the α2 repressor collaborates with the product of the **a**1 gene to turn off a set of haploid-specific genes in addition to the **a**-specific genes. Two distinct but related types of conserved DNA sequences are found upstream of these two sets of regulated genes: one in front of the **a**-specific genes and the other in front of the haploid-specific genes. Given the relatedness of these upstream sequences, it is most likely that α2 binds to both; however, its binding properties must be modified in some way by the **a**1 protein before it can recognize the haploid-specific sequence. You wish to understand the nature of this modification. Does **a**1 catalyze covalent modification of α2, or does it modify α2 by binding to it stoichiometrically?

To study these questions, you perform three types of experiment. In the first, you measure the binding of **a**1 and α2, alone and together, to the two kinds of upstream regulatory DNA sites. As shown in Figure 7–37, **a**1 alone does not bind DNA fragments that contain either regulatory site (Figure 7–37, lane 2), whereas α2 binds to **a**-specific fragments, but not to haploid-specific fragments (lane 3). The mixture of **a**1 and α2 binds to **a**-specific *and* haploid-specific fragments (lane 4).

In the second series of experiments, you add a vast excess of unlabeled DNA containing the **a**-specific sequence, along with the mixture of **a**1 and

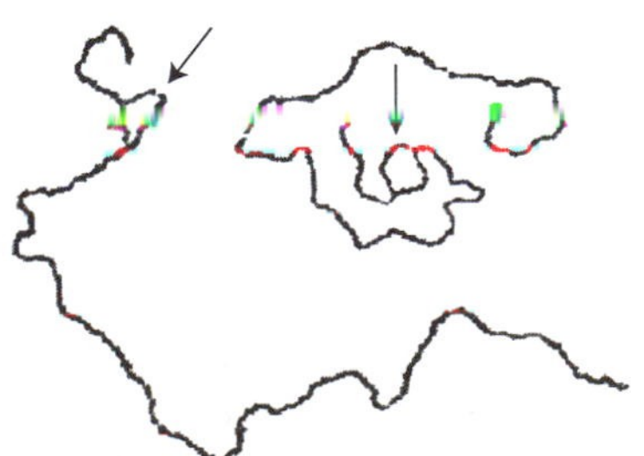

**Figure 7–36** Reannealed DNA fragments from cultures of switching *E. coli* (Problem 7–86). *Arrows* indicate single-stranded bubbles.

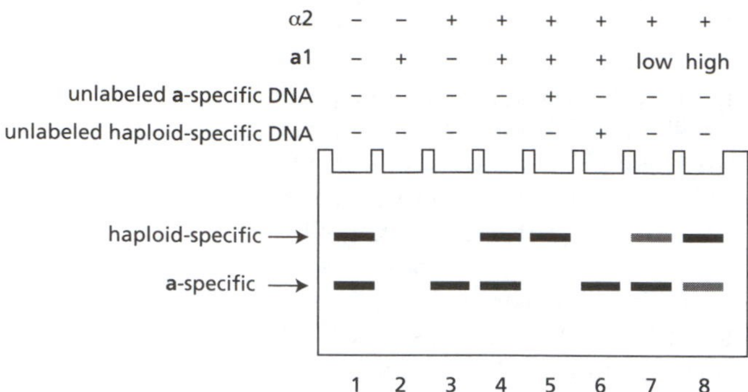


**Figure 7–37** Binding of regulatory proteins to fragments of DNA containing the **a**-specific or haploid-specific regulatory sequences (Problem 7–87). Various combinations of regulatory proteins were incubated with a mixture of **a**-specific and haploid-specific radioactive DNA fragments (shown in lane 1). At the end of the incubation, the samples were precipitated with antibody against the proteins, and the DNA fragments in the precipitate were run on the gel. The gel was then placed against x-ray film to expose the positions of the radioactive DNA fragments.

α2 proteins. Under these conditions, the haploid-specific fragment is still bound (Figure 7–37, lane 5). Similarly, if you add an excess of unlabeled haploid-specific DNA, the **a**-specific fragment is still bound (lane 6).

In the third set of experiments, you vary the ratio of **a**1 relative to α2. When α2 is in excess, binding to the haploid-specific fragment is decreased (Figure 7–37, lane 7); when **a**1 is in excess, binding to the **a**-specific fragment is decreased (lane 8).

A. In the presence of **a**1, is α2 present in two forms with different binding specificities, or in one form that can bind to both regulatory sequences? How do your experiments distinguish between these alternatives?

B. An α2 repressor with a small deletion in its DNA-binding domain does not bind to DNA fragments containing the haploid-specific sequence. When this mutant protein is expressed in a diploid cell along with normal α2 and **a**1 proteins, the haploid-specific genes are turned on. (These genes are normally off in a diploid.) In the light of this result and your other experiments, do you consider it more likely that **a**1 catalyzes a covalent modification of α2, or that **a**1 modifies α2 by binding to it stoichiometrically to form an **a**1–α2 complex?

**7–88** You have discovered a new strain of yeast with a novel mating system. There are two haploid mating types, called M and F. Cells of opposite mating type can mate to form M/F diploid cells. These diploids can undergo meiosis and sporulate, but they cannot mate with each other or with either haploid mating type.

Your genetic analysis shows there are four genes that control mating type. When one pair of genes, *M1* and *M2*, are at the mating-type locus, the cells are mating-type M; when a second pair of genes, *F1* and *F2*, are at the mating-type locus, the cells are mating-type F. You have also identified three sets of regulated genes: one that is expressed specifically in M-type haploids (Msg), one in F-type haploids (Fsg), and one in sporulating cells (Ssg). You obtain viable mutants ($M1^-$, $M2^-$, $F1^-$, $F2^-$) that are defective in each of the mating-type genes, and study their effects—individually and in combination—on the mating phenotype and on the expression of their regulated genes. Your results with haploid and diploid cells containing different combinations of mutants are shown in Table 7–3.

Suggest a regulatory scheme to explain how the *M1*, *M2*, *F1*, and *F2* gene products control the expression of the M-specific, F-specific, and sporulation-specific sets of genes. Indicate which gene products are activators and which are repressors of transcription, and decide whether the gene products act alone or in combination.

**7–89** Expression of the *Clock* gene in zebrafish shows a strong circadian oscillation in many tissues *in vivo* and in culture, indicating that endogenous circadian oscillators exist in peripheral tissues. A defining feature of circadian clocks is that they can be set or entrained to local time, usually by the environmental light–dark cycle. An important question is whether peripheral oscillators are entrained by signals from central pacemakers in the brain or are themselves directly light-responsive.

**Table 7–3 Phenotypes of mutants that affect mating in a new strain of yeast** (Problem 7–88).

| | MUTANT | MATING PHENOTYPE | GENES EXPRESSED |
|---|---|---|---|
| Haploid cells | wild-type M | M | Msg |
| | $M1^-$ | nonmating | Msg, Fsg |
| | $M2^-$ | M | Msg |
| | $M1^-M2^-$ | nonmating | Msg, Fsg |
| | wild-type F | F | Fsg |
| | $F1^-$ | F | Fsg |
| | $F2^-$ | nonmating | Msg, Fsg |
| | $F1^-F2^-$ | nonmating | Msg, Fsg |
| Diploid cells | wild-type M/F | nonmating | Ssg |
| | $M1^-/F1^-$ | F | Fsg |
| | $M1^-/F2^-$ | nonmating | Msg, Fsg, Ssg |
| | $M2^-/F1^-$ | nonmating | none |
| | $M2^-/F2^-$ | M | Msg |

Msg = M-specific genes; Fsg = F-specific genes; Ssg = sporulation-specific genes

To address this question, you isolate hearts from zebrafish and culture them over a period of 5 days. You expose one group to a light–dark cycle (14 hours light–10 hours dark) and the other to a dark–light cycle (10 hours dark–14 hours light). Periodically, you measure the mRNA levels of two circadian rhythm proteins: Clock, whose mRNA is known to oscillate, and Timeless, whose mRNA does not. Your results are shown in Figure 7–38.

A. Do cycles of light and dark entrain the circadian clock in isolated hearts? Explain your answer.
B. How do you suppose light influences isolated cells and organs?

**7–90** An elegant early study examined whether a transcriptional complex could remain bound to the DNA during DNA replication, to serve, perhaps, as a basis for inheritance of the parental pattern of gene expression. An active transcription complex was assembled on the *Xenopus* 5S RNA gene on a plasmid. The plasmid was then replicated *in vitro* and subsequently tested for its ability to support transcription.

Replicated and unreplicated templates were distinguished by the restriction nucleases DpnI, MboI, and Sau3A, which have different sensitivities to the methylation state of their recognition sequence GATC (Figure 7–39A). This sequence is present once at the beginning of the 5S RNA gene, and when it is cleaved, no transcription occurs. When the plasmid is grown in wild-type *E. coli*, GATC is methylated at the A on both strands by the bacterial *Dam* methylase. Replication of fully methylated DNA *in vitro* generates

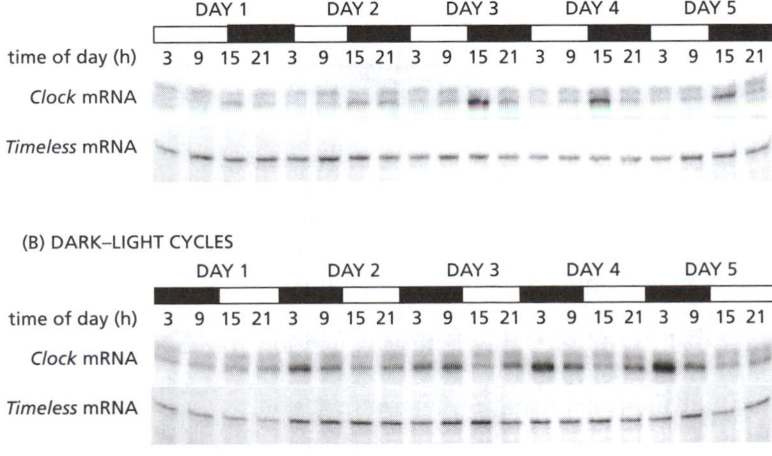


**Figure 7–38** Influence of cycles of light and dark on mRNAs for two circadian rhythm proteins in cultured zebrafish hearts (Problem 7–89). (A) Light–dark cycles. (B) Dark–light cycles. Hearts were cultured in light–dark or dark–light cycles for 5 days. The positions of mRNAs for Clock and Timeless are indicated.

(A) CUTTING PATTERNS

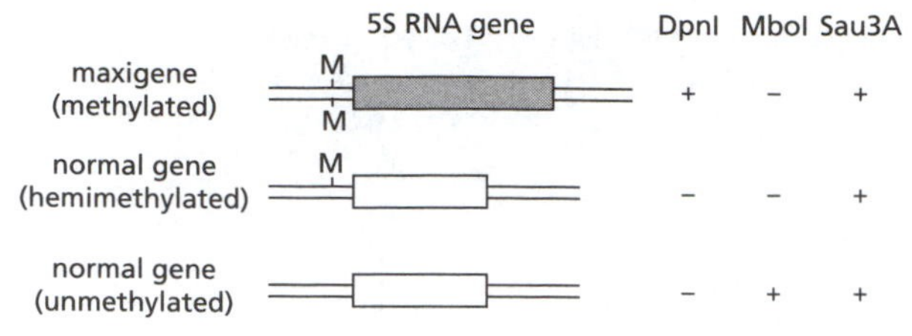

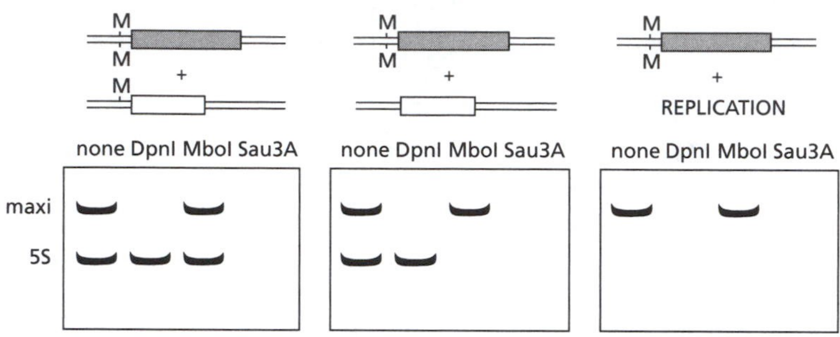


**Figure 7–39** Analysis of stability of transcription complexes during DNA replication (Problem 7–90). (A) Sensitivity of 5S RNA genes in different methylation states to digestion with restriction enzymes. The 5S RNA maxigene is *shaded* and the normal gene is *white*. M indicates that a strand is methylated. Sensitivity to cleavage by a restriction nuclease is indicated by (+); resistance is indicated by (–). (B) Transcriptional activity in the presence and absence of replication.

daughter duplexes that are methylated on only one strand (hemimethylated) in the first round and are unmethylated after subsequent rounds. By starting with fully methylated DNA and inducing its replication *in vitro*, transcription could be assayed specifically from the replicated DNA by treating the DNA with DpnI, which selectively cuts unreplicated (methylated) DNA, thus preventing its transcription.

Two versions of the 5S RNA gene—a slightly longer 'maxigene' and the normal gene—were constructed so that their transcripts could be distinguished (Figure 7–39B). Mixtures of the fully methylated maxigene with either the hemimethylated or unmethylated normal gene served as controls for the specificity of the restriction digestions (Figure 7–39B). To test the effect of replication on transcription, transcription complexes were assembled on the fully methylated maxigene, replication was induced, and transcriptional activity was assayed before and after cleavage with restriction nucleases (Figure 7–39B).

A. Does the methylation status of the 5S RNA gene affect its level of transcriptional activity? How can you tell?
B. Does the pattern of transcription after cleavage of the mixtures of maxigenes and normal genes with the various restriction nucleases match your expectations? Explain your answer.
C. In your experiment about half of the DNA molecules were replicated. Does the pattern of transcription after replication and cleavage indicate that the transcription complex remains bound to the 5S RNA gene during replication?
D. In these experiments the authors were careful to show that greater than 90% of the molecules were assembled into active transcription complexes and that 50% of the molecules were replicated. How would the pattern have changed if only 10% of the molecules were assembled into active transcription complexes, and only 10% were replicated? Would the conclusions have changed?

**7–91** You are studying the role of DNA methylation in the control of gene expression, using the human γ-globin gene as a test system (Figure 7–40A). Globin mRNA can be detected when this gene is integrated into the genome of mouse fibroblasts, even though it is expressed at much lower levels than in red cells. If the gene is methylated at all 27 of its CG sites before integration, its expression is blocked completely. You are using this system to decide whether a single critical methylation site is sufficient to determine globin expression.

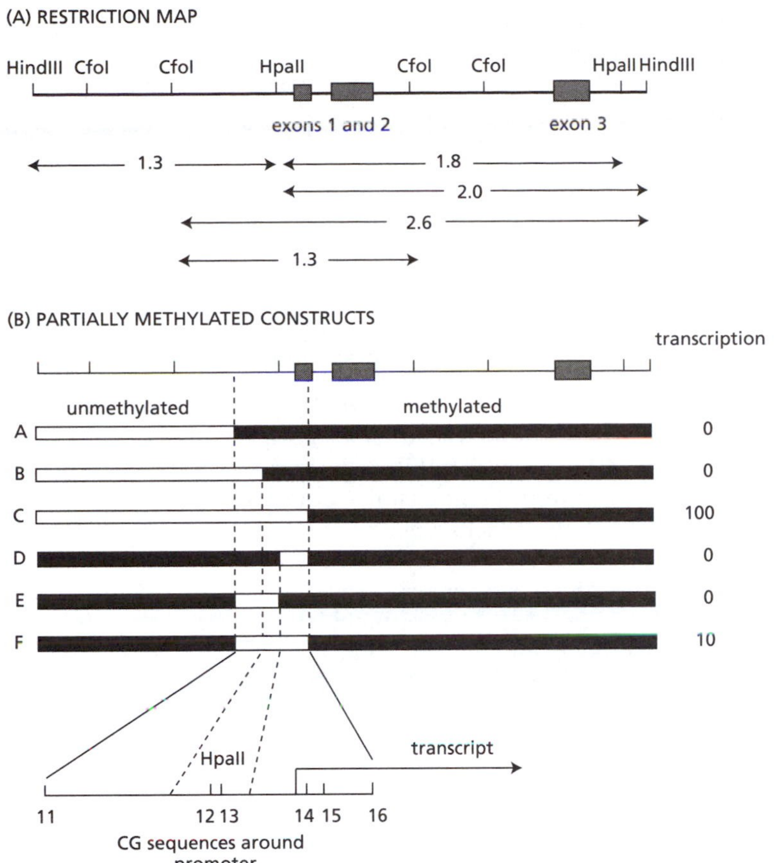


**Figure 7–40** Effects of methylation on transcription of the γ-globin gene (Problem 7–91). (A) HindIII fragment containing the γ-globin gene. Sites of cleavage by the methylation-sensitive restriction nucleases, CfoI and HpaII, are indicated along with the sizes of the larger fragments that are observed on the gel in Figure 7–41. (B) Methylated constructs of the γ-globin gene. The methylated segments of the gene are shown in *black*. The six CG sites around the promoter are shown in more detail *below* the constructs. The level of expression of γ-globin RNA from each construct is shown on the *right* as a percentage of the expression from a fully unmethylated construct.

You use a combination of site-directed mutagenesis and primed synthesis in the presence of 5-methyl dCTP to create several different γ-globin constructs that are unmethylated in specific regions of the gene. These constructs are illustrated in Figure 7–40B, with the methylated regions shown in black. The arrangement of six methylation sites around the promoter is shown below the constructs. Sites 11, 12, and 13 are unmethylated in construct E, sites 14, 15, and 16 are unmethylated in construct D, and all six sites are unmethylated in construct F. You integrate these constructs in mouse fibroblasts, grow cell lines containing individual constructs, and measure γ-globin RNA synthesis relative to cell lines containing the fully unmethylated construct (Figure 7–40B).

To check whether the methylation patterns were correctly inherited, you isolate DNA samples from cell lines containing constructs B, C, or F and digest them with HindIII plus CfoI or HpaII. CfoI and HpaII do not cleave if their recognition sites are methylated. The cleavage sites for these nucleases are shown in Figure 7–40A, along with the sizes of relevant restriction fragments larger than 1 kb. You separate the cleaved DNA samples on a gel and visualize them by hybridization to the radiolabeled HindIII fragment (Figure 7–41).

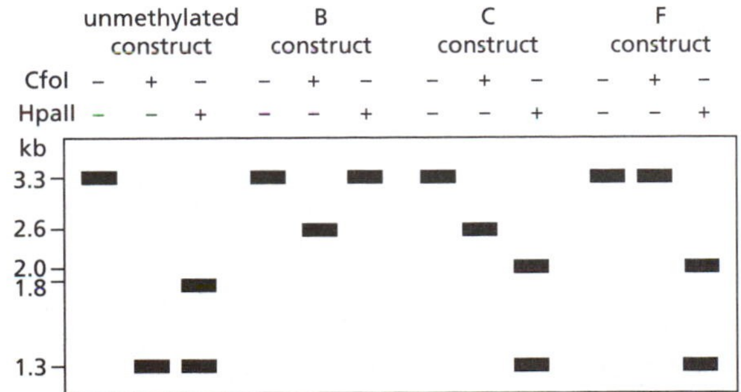


**Figure 7–41** Restriction patterns from cell lines containing constructs B, C, or F (Problem 7–91). HindIII was included in all digests. Fragments less than 1 kb are not shown.

A. To create some of the methylated DNA substrates you used a single-stranded version of the gene as a template and primed synthesis of the second strand using 5-methyl dCTP instead of dCTP in the reaction. This method creates a DNA molecule containing 5-methyl C in place of C in one strand. After isolation of colonies containing such constructs, you found that they have 5-methyl C on both strands, but only at CG sequences. Explain the retention of 5-methyl C in CG sequences and their loss elsewhere.

B. Do the restriction patterns of the constructs in the isolated cell lines (Figure 7–41) indicate that the CG sequences that were methylated during creation of the constructs (Figure 7–40) were maintained in the cell?

C. Does the γ-globin transcription associated with the cell lines containing the constructs (Figure 7–40) indicate that a single critical site of methylation determines whether the gene is expressed?

**7–92** The embryonic mouse fibroblast cell line, 10T½ (pronounced 10 T and a half), is a stable line with cells that look and behave like fibroblasts. If these cells are exposed to medium containing 5-azacytidine (5-aza C) for 24 hours, they will then differentiate into cartilage, fat, or muscle cells when they grow to a high cell density. (Treatment with 5-aza C reduces the general level of DNA methylation, allowing some previously inactive genes to become active.) If the cells are grown at low cell density after the treatment, they retain their original fibroblastlike shape and behavior, but even after many generations, they still differentiate when they reach a high cell density. 10T½ cells that have not been exposed to 5-aza C do not differentiate, no matter what the cell density is.

When the treated cells differentiate, about 25% turn into muscle cells (myoblasts). The high frequency of myoblast formation led your advisor to hypothesize that a single master regulatory gene, which is normally repressed by methylation, may trigger the entire transformation. Accordingly, he persuades you to devise a strategy to clone the gene. You assume the gene is off before treatment with 5-aza C and on in the induced myoblasts. If this assumption is valid, you should be able to find sequences corresponding to the gene among the cDNA copies of mRNAs that are synthesized after 5-aza C treatment. You decide to screen an existing cDNA library from normal myoblasts (which according to your assumption will also express the gene) using a set of radioactive probes to identify likely cDNA clones. You prepare three radioactive probes.

**Probe 1.** You isolate RNA from myoblasts induced by 5-aza C treatment of fibroblasts, and prepare radioactive cDNA copies.

**Probe 2.** You hybridize the radioactive cDNA from the 5-aza-C-induced myoblasts with RNA from untreated 10T½ cells and discard all the RNA:DNA hybrids.

**Probe 3.** You isolate RNA from normal myoblasts and prepare radioactive cDNA copies, which you then hybridize to RNA from untreated 10T½ cells; you discard the RNA:DNA hybrids.

The first probe hybridizes to a large number of clones from the myoblast cDNA library, but only about 1% of those clones hybridize to probes 2 and 3. Overall, you find four distinct patterns of hybridization (Table 7–4).

A. What is the purpose of hybridizing the radioactive cDNA from 5-aza-C-induced and normal myoblasts to the RNA from untreated 10T½ cells? In other words, why are probes 2 and 3 useful?

B. What general kinds of genes might you expect to find in each of the four classes of cDNA clones (A, B, C, and D in Table 7–4)? Which class of cDNA clone is most likely to include sequences corresponding to the putative muscle regulatory gene you are seeking?

**Table 7–4 Patterns of myoblast cDNA hybridization with radioactive probes** (Problem 7–92).

| CLASS | PROBE 1 | PROBE 2 | PROBE 3 |
|---|---|---|---|
| A | + | – | – |
| B | + | – | + |
| C | + | + | – |
| D | + | + | + |

**7–93** In female mammals, one X chromosome in each cell is chosen at random for inactivation early in development. X-inactivation, which involves more than 1000 genes in humans, is crucial for equalizing expression of X-chromosome genes in males and females. A critical clue to the mechanism of X-inactivation came from the isolation of a large number of cDNAs for genes on the human X chromosome. Their expression patterns were characterized in cells from normal males and females, in cells from individuals with abnormal numbers of X chromosomes, and in rodent:human hybrid cell lines that retained either one inactive human X chromosome ($X_i$) or one active human X chromosome ($X_a$). Among all these cDNAs, there were three patterns of expression, as illustrated in Figure 7–42 for cDNAs A, B, and C.

A. For each pattern of expression, decide whether the gene is expressed from the active X, the inactive X, or both. Which pattern do you expect to be the most common? Which pattern is the most surprising?
B. From the results with cells from abnormal individuals, formulate a rule as to how many chromosomes are inactivated, and how many remain active during X inactivation.

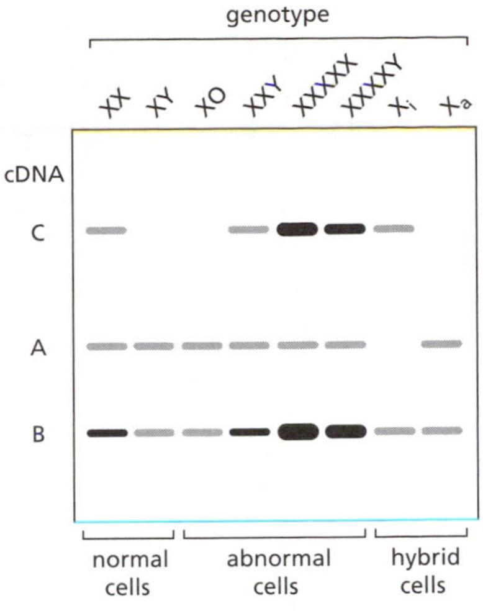


**Figure 7–42** Northern analysis of gene expression from cells with different numbers and types of X chromosomes (Problem 7–93). RNA from cells was run out on gels, blotted onto nitrocellulose, probed with a mixture of radioactive cDNAs A, B, and C, and visualized by autoradiography. The positions of the RNA bands that correspond to genes A, B, and C were determined in a separate experiment.

**7–94** One research group directly tested the hypothesis that the *Xist* gene in mice is required for X-inactivation by introducing a targeted deletion of *Xist* into one X chromosome. Using female embryonic stem (ES) cells in which genes on the two X chromosomes could be distinguished due to polymorphisms, they followed X-inactivation during differentiation of ES cells in culture. ES cells normally maintain both X chromosomes in the active state; however, when they are induced to differentiate, they randomly inactivate one.

The scientists considered three hypotheses. (1) ES cells mutant for one *Xist* gene would fail to register the presence of two X chromosome and thus fail to undergo X-inactivation. (2) The *Xist* knockout would prevent X-inactivation of the targeted X chromosome, thus predisposing the normal X chromosome to preferential X-inactivation. (3) The mutation would have no effect on X-inactivation at all.

Using allele-specific (X chromosome-specific) oligonucleotide probes, they were able to determine which allele of an X-chromosome gene was expressed in individual cells. As shown in Figure 7–43, they examined cells from mutant ES cells that were undifferentiated, and from mutant and non-mutant ES cells that had undergone differentiation. Only a few cells are shown in Figure 7–43, but analysis of many more confirmed the patterns shown. The A allele marks the X chromosome whose *Xist* gene is intact in the mutant ES cells; the B allele marks the X chromosome from which the *Xist* gene was deleted.

Which of the three hypotheses do these results support? Explain your reasoning.

**7–95** Using strand-specific probes of transcription in the X-inactivation center, a second gene was found to be transcribed in the opposite direction to *Xist* and was named *Tsix* to indicate its antisense orientation. The *Tsix* transcript, like the *Xist* transcript, has no significant open reading frame and is thought to function as an RNA molecule. As shown in Figure 7–44, the *Tsix* transcript extends all the way across the *Xist* gene. To determine whether *Tsix* plays a role in counting X chromosomes, choosing which one to inactivate, or in silencing the inactive X, female ES cells were generated in which the promoter for *Tsix* had been deleted from one X chromosome. When these ES cells were allowed to differentiate, it was found that the X chromosome with the *Tsix* deletion was always inactivated.

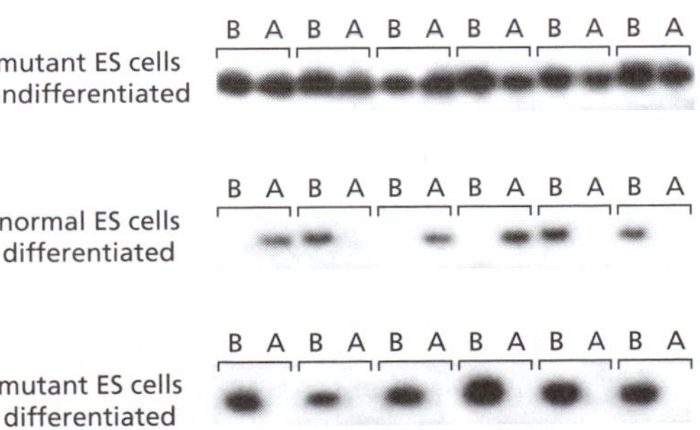


**Figure 7–43** Analysis of expression of X chromosome-specific alleles in undifferentiated and differentiated ES cells (Problem 7–94). Analysis of the A and B alleles from individual cells are paired as indicated by the *brackets*.

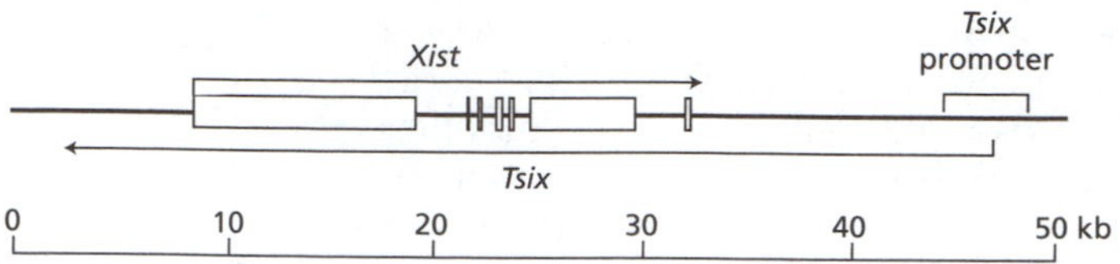


**Figure 7–44** Arrangement of *Xist* and *Tsix* transcripts in the X-inactivation center (Problem 7–95). *Boxes* indicate the exons of *Xist*; exons are undefined for *Tsix*. The promoter deletion in the *Tsix* mutant is indicated.

A. Is *Tsix* important for the counting, choice, or silencing of X chromosomes? Explain your answer.
B. Before X-inactivation, *Tsix* is expressed from both alleles, as is *Xist*. At the onset of X-inactivation *Tsix* expression becomes confined to the future active X, whereas *Xist* expression is restricted to the future inactive X. Can you suggest some possible ways that *Tsix* might regulate *Xist*?

# POST-TRANSCRIPTIONAL CONTROLS

| TERMS TO LEARN | | |
|---|---|---|
| alternative RNA splicing | regulated nuclear transport | small interfering RNA (siRNA) |
| internal ribosome entry site (IRES) | RNA editing | translational control |
| microRNA (miRNA) | RNA interference (RNAi) | transcription attenuation |
| post-transcriptional control | | |

## DEFINITIONS

Match each definition below with its term from the list above.

**7–96** Natural defense mechanism in many organisms that is directed against foreign RNA molecules, especially those that occur in double-stranded form.

**7–97** Production of a functional RNA by insertion or alteration of individual nucleotides in an RNA transcript after it has been synthesized.

**7–98** Inhibition of gene expression in bacteria by the premature termination of transcription.

**7–99** A way to generate different proteins from the same gene by combining different segments of the initial RNA transcript to make distinct mRNAs.

**7–100** General term for a regulatory event that occurs after RNA polymerase has bound to the gene's promoter and begun RNA synthesis.

**7–101** A class of short noncoding RNAs that regulate gene expression; roughly one-third of human genes are thought to be regulated in this way.

## TRUE/FALSE

Decide whether each of these statements is true or false, and then explain why.

**7–102** If the site of transcript cleavage and polyadenylation of a particular RNA in one cell is downstream of the site used for cleavage and polyadenylation of the same RNA in a second cell, the protein produced from the longer polyadenylated RNA will necessarily contain additional amino acids at its C-terminus.

**7–103** The single *Drosophila Dscam* gene, which has the potential to produce more than 38,000 different proteins by alternative splicing, is as complex as the whole human genome.

## THOUGHT PROBLEMS

**7–104** RNA polymerase II commonly terminates transcription of the HIV (the human AIDS virus) genome a few hundred nucleotides after it begins, unless helped along by a virus-encoded protein called Tat. In the absence of Tat, the CTD (C-terminal domain) of RNA polymerase is hypophosphorylated, but in the presence of Tat it becomes hyperphosphorylated and is converted to a form that can efficiently transcribe the entire HIV genome. Tat apparently aids this transition by binding to a specific hairpin structure in the nascent viral RNA. Tat then recruits a collection of proteins that enhance the ability of RNA polymerase to continue transcribing. Among the Tat-recruited proteins is the positive transcription elongation factor b, P-TEFb, which is composed of cyclin T1 and the cyclin-dependent protein kinase Cdk9.

Flavopiridol is the most potent inhibitor of Cdk9 yet discovered. Flavopiridol binds at the site on Cdk9 where ATP would normally bind and blocks Cdk9-mediated phosphorylation of target proteins. Would you expect flavopiridol to interfere with HIV transcription? Why or why not?

**7–105** Transcripts from the mitochondrial DNA of trypanosomes are heavily edited by insertion and deletion of U nucleotides at numerous sites. The information for the editing process is provided by a large number of small guide RNAs that pair at specific places in the transcript. Typically, the guide RNAs pair perfectly at their 5′ ends (as shown by the boxed nucleotides in Figure 7–45), but imperfectly at their 3′ ends. Examine the duplex structure formed by the pairing of the small guide RNAs with the pre-edited transcript (shown in the upper half of Figure 7–45), and then compare the sequences of the pre-edited and edited transcripts. Formulate a set of rules for RNA editing of mitochondrial transcripts in trypanosomes.

**Problems 6–48 and 6–49** examine a related type of RNA editing in which modifications are introduced into U5 snRNA by U85 snoRNA.

**7–106** Several distinct mechanisms for mRNA localization have been discovered. They all require specific signals in the mRNA itself, usually in the 3′ untranslated region (UTR). Briefly outline three mechanisms by which cellular mRNAs might become localized in the cell.

**7–107** Regulation of ferritin translation is controlled by the interaction between a hairpin structure in the mRNA, termed an iron-response element (IRE), and the iron-response protein (IRP) that binds to it. When the IRE is bound by the IRP, translation is inhibited. The location of the IRE in the mRNA is critical for its function. To work properly, it must be positioned near the 5′ end of the mRNA. If it is moved more than 60 nucleotides downstream of the cap structure, the IRE no longer inhibits translation. Why do you suppose there is such a critical position dependence for regulation of translation by IRE and IRP?

**7–108** Although essential, iron is also potentially toxic. It is maintained at optimal levels in mammalian cells by the actions of three proteins. Transferrin binds to extracellular iron ions and delivers them to cells; the transferrin receptor binds iron-loaded transferrin and brings it into the cell; and ferritin binds iron (up to 4500 atoms in the internal cavity in each complex) to provide an intracellular storage site.

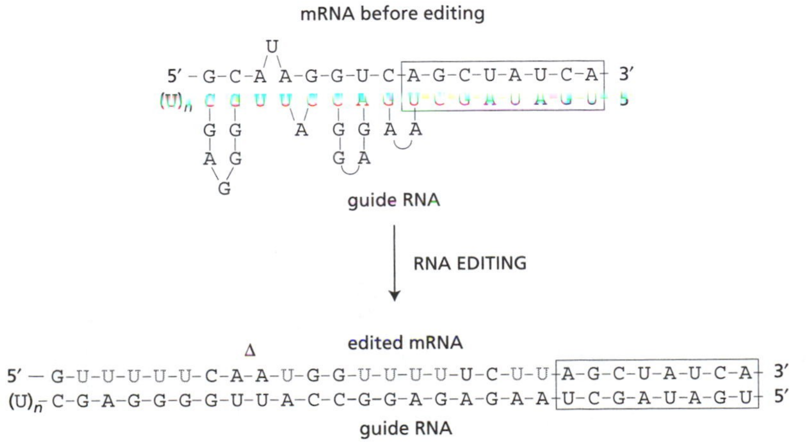

mRNA before editing

5′ – G-C-A-U-A-G-G-U-C-A-G-C-U-A-U-C-A- 3′

guide RNA

RNA EDITING

edited mRNA

5′ — G-U-U-U-U-U-C-A-A-U-G-G-U-U-U-U-U-C-U-U-A-G-C-U-A-U-C-A- 3′

$(U)_n$- C-G-A-G-G-G-G-U-U-A-C-C-G-G-A-G-A-G-A-A-U-C-G-A-U-A-G-U- 5′

guide RNA

**Figure 7–45** RNA editing of mitochondrial transcripts in trypanosomes (Problem 7–105). Only a small portion of the pre-edited RNA is shown, whereas most of the guide RNA is shown, except for its 3′ end, which consists of a string of U nucleotides ($U_n$). The *boxed* nucleotides indicate the portion of the initial structure that is perfectly paired. In the edited mRNA, inserted U nucleotides are shown in *light gray*, and the deleted U nucleotide is indicated by the Δ.

The regulation of the transferrin receptor, like that of ferritin (see Problem 7–107), is accomplished by IREs and IRPs, and in both cases, iron binding to IRPs prevents their binding to IREs. Nevertheless, the mechanism of transferrin receptor regulation is quite different from that of ferritin. The transferrin receptor mRNA contains five IREs that are all located in the 3′ untranslated region of the mRNA (instead of at the 5′ end, as in ferritin). In addition, binding of IRPs to these IREs increases the translation of transferrin receptor (as opposed to a decrease for ferritin).

A. Does this opposite regulation of ferritin and transferrin receptor in response to iron levels make biological sense? Consider the consequences of high and low iron levels.

B. In the presence of iron, the transferrin receptor mRNA is rapidly degraded; in the absence of iron, it is stable. Can you suggest a mechanism for how iron levels might be linked to the stability of transferrin receptor mRNA?

**7–109** Vg1 mRNA encodes a member of the TGFβ family of growth factors, which is important for mesoderm induction. In *Xenopus*, Vg1 mRNA is localized to the vegetal pole of the oocyte. Translation of Vg1 mRNA does not occur until a late stage, after localization is complete. Analysis of the 3′ untranslated region identified two elements: a localization element and a UA-rich translation control element. Translational regulation was shown to be independent of the status of the poly-A tail: it remains unchanged in length throughout oogenesis, and translational regulation could be demonstrated on injected mRNAs that had no poly-A tails. By contrast, translational repression was abolished when an internal ribosome entry site (IRES) sequence was inserted into the mRNA upstream of the coding sequence. How do you suppose translation of Vg1 mRNA is controlled?

## CALCULATIONS

**7–110** Ferritin stores iron in many tissues, and its synthesis increases up to 2-fold in the presence of iron. You wish to define the molecular mechanism for the induction of ferritin synthesis. You carry out a series of experiments in which you inject rats with a ferric salt solution or saline, each with or without actinomycin D, which inhibits RNA synthesis. Three hours later you kill the rats, homogenize their livers, and prepare polysome and supernatant fractions. You extract RNA from each fraction and translate it in a cell-free system in the presence of radioactive amino acids. You measure total protein synthesis by incorporation of label, and the synthesis of ferritin by precipitation with ferritin-specific antibodies (Table 7–5).

A. For each sample, calculate the percentage of total protein synthesis that is due to synthesis of ferritin. Given that 85% of the bulk mRNA is bound to ribosomes in the polysome fraction and 15% is free (in the supernatant fraction), determine the percentage of total ferritin mRNA in the polysome and supernatant fractions under each condition of treatment. (Assume that the percentages of polysome and supernatant mRNA are not changed by treatment with actinomycin D.)

**Table 7–5 Synthesis of ferritin in the rat after various treatments** (Problem 7–110).

| INJECTION | ACTINOMYCIN D | FRACTION | TOTAL PROTEIN SYNTHESIS | FERRITIN SYNTHESIS |
|---|---|---|---|---|
| Saline | absent | polysomes | 750,000 | 700 |
| | | supernatant | 255,000 | 1400 |
| Iron | absent | polysomes | 500,000 | 900 |
| | | supernatant | 400,000 | 500 |
| Saline | present | polysomes | 800,000 | 800 |
| | | supernatant | 600,000 | 3000 |
| Iron | present | polysomes | 780,000 | 1380 |
| | | supernatant | 550,000 | 700 |

Numbers show radioactivity (cpm) incorporated into total proteins or into ferritin.

B. Do your results distinguish between transcriptional and post-transcriptional control of ferritin synthesis by iron? If so, how?

C. How would you account for the 2-fold increase in ferritin synthesis in the presence of iron?

**7–111** A very active cell-free protein synthesis system can be prepared from reticulocytes. These immature red blood cells have already lost their nuclei but still contain ribosomes. A reticulocyte lysate can translate each globin mRNA many times provided heme is added (Figure 7–46). Heme serves two functions: it is required for assembly of globin chains into hemoglobin, and, surprisingly, it is required to maintain a high rate of protein synthesis. If heme is omitted, protein synthesis stops after a brief lag (Figure 7–46).

The first clue to the molecular basis for the effect of heme on globin synthesis came from a simple experiment. A reticulocyte lysate was incubated for several hours in the absence of heme. When 5 μL of this preincubated lysate was added to 100 μL of a fresh lysate in the presence of heme, protein synthesis was rapidly inhibited (Figure 7–46). When further characterized, the inhibitor (termed heme-controlled repressor, or HCR) was shown to be a protein with a molecular weight of 180,000. Pure preparations of HCR at a concentration of 1 μg/mL completely inhibit protein synthesis in a fresh, heme-supplemented lysate.

A. Calculate the ratio of HCR molecules to ribosomes and globin mRNA at the concentration of HCR that inhibits protein synthesis. Reticulocyte lysates contain 1 mg/mL ribosomes (molecular weight, 4 million), and the average polysome contains four ribosomes per globin mRNA.

B. Do the results of this calculation favor a catalytic or a stoichiometric mechanism for HCR inhibition of protein synthesis?

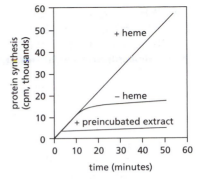

**Figure 7–46** Protein synthesis in a reticulocyte lysate (Problem 7–111). The lysate to which preincubated extract was added also contained added heme.

## DATA HANDLING

**7–112** A repeated hexanucleotide element, TGCATG, has been shown to regulate the tissue-specific splicing of the fibronectin alternative exon EIIIB. You wonder if it might also regulate alternative splicing in the calcitonin/CGRP gene (Figure 7–47). In thyroid cells, an mRNA containing exons 1, 2, 3, and 4 encodes calcitonin. In neuronal cells, an mRNA containing exons 1, 2, 3, 5, and 6 encodes calcitonin gene related peptide (CGRP).

When you examine the calcitonin/CGRP gene you find five copies of a related repeat, GCATG, within 500 nucleotides of the exon-4 splice site (Figure 7–47). To analyze their potential role in alternative splicing, you make several calcitonin/CGRP constructs in which the repeats have been altered singly or in combination. You transfect the constructs into HeLa cells, which normally give the calcitonin splicing pattern, and into F9 cells, which normally give the CGRP splicing pattern. You find that no single mutation alters the pattern seen with wild type (not shown); however, various combinations of the mutations have dramatic effects (Figure 7–47).

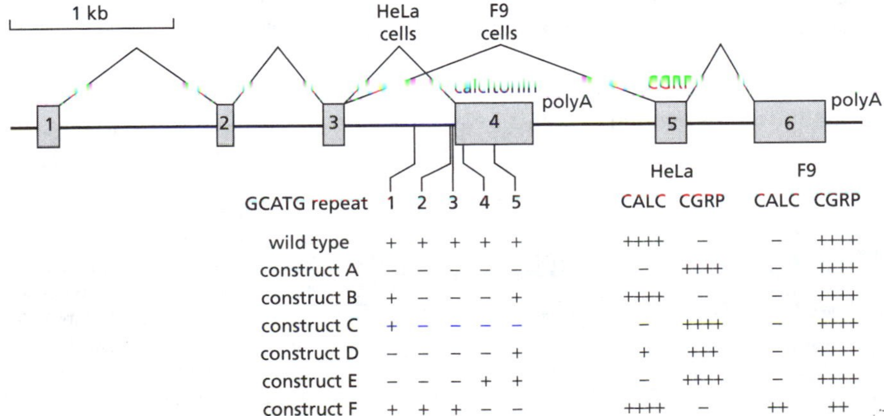


| GCATG repeat | 1 | 2 | 3 | 4 | 5 | HeLa CALC | HeLa CGRP | F9 CALC | F9 CGRP |
|---|---|---|---|---|---|---|---|---|---|
| wild type | + | + | + | + | + | ++++ | – | – | ++++ |
| construct A | – | – | – | – | – | – | ++++ | – | ++++ |
| construct B | + | – | – | – | + | ++++ | – | – | ++++ |
| construct C | + | – | – | – | – | – | ++++ | – | ++++ |
| construct D | – | – | – | – | + | + | +++ | – | ++++ |
| construct E | – | – | – | + | + | – | ++++ | – | ++++ |
| construct F | + | + | + | – | – | ++++ | – | ++ | ++ |

**Figure 7–47** Effects of mutations in GCATG repeats on alternative splicing of calcitonin/CGRP pre-mRNA (Problem 7–112). In HeLa cells, exon 3 is spliced to exon 4 to make calcitonin mRNA; in F9 cells, exon 3 is spliced to exon 5 to make CGRP mRNA. The positions of the GCATG repeats around exon 4 are indicated. In the various constructs the presence of a GCATG repeat is indicated by a *plus*, and its absence by a *minus*. Production of calcitonin- (CALC) and CGRP-specific spliced RNA is indicated relative to wild type as follows: ++++ = 80–100%, +++ = 60–80%, ++ = 40–60%, + = 20–40%, and – = 0–20%.

A. Why do you suppose the most dramatic changes were seen in HeLa cells rather than in F9 cells?

B. Is there a particular combination of GCATG repeats that is critical for proper calcitonin mRNA production in HeLa cells? If so, what is it?

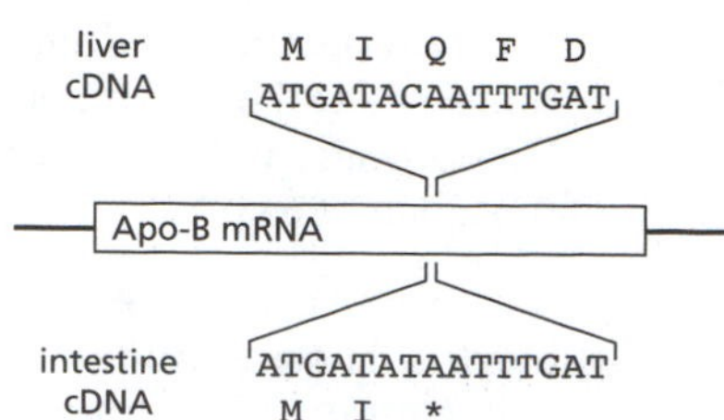


**Figure 7–48** Location of the sequence differences in cDNA clones from ApoB RNA isolated from liver and intestine (Problem 7–113). The encoded amino acid sequences, in the one-letter code, are shown aligned with the cDNA sequences. The *asterisk* indicates a stop codon.

**7–113** In humans, two closely related forms of apolipoprotein B (ApoB) are found in blood as constituents of the plasma lipoproteins. ApoB48, which has a molecular mass of 48 kilodaltons, is synthesized by the intestine and is a key component of chylomicrons, the large lipoprotein particles responsible for delivery of dietary triglycerides to adipose tissue for storage. ApoB100, which has a molecular mass of 100 kilodaltons, is synthesized in the liver for formation of the much smaller, very low-density lipoprotein particles used in the distribution of triglycerides to meet energy needs. A classic set of studies defined the surprising relationship between these two proteins.

Sequences of cloned cDNA copies of the mRNAs from these two tissues revealed a single difference: cDNAs from intestinal cells had a T, as part of a stop codon, at a point where the cDNAs from liver cells had a C, as part of a glutamine codon (Figure 7–48). To verify the differences in the mRNAs and to search for corresponding differences in the genome, RNA and DNA were isolated from intestinal and liver cells and then subjected to PCR amplification, using oligonucleotides that flanked the region of interest. The amplified DNA segments from the four samples were tested for the presence of the T or C by hybridization to oligonucleotides containing either the liver cDNA sequence (oligo-Q) or the intestinal cDNA sequence (oligo-STOP). The results are shown in Table 7–6.

Are the two forms of ApoB produced by transcriptional control from two different genes, by a processing control of the RNA transcript from a single gene, or by differential cleavage of the protein product from a single gene? Explain your reasoning.

**7–114** The level of β-tubulin gene expression is established in cells by an unusual regulatory pathway, in which the intracellular concentration of free tubulin dimers (composed of one α-tubulin and one β-tubulin subunit) regulates the rate of new β-tubulin synthesis. The initial evidence for such an autoregulatory pathway came from studies with drugs that cause assembly or disassembly of all cellular tubulin. For example, treatment of cells with colchicine, which causes microtubule depolymerization into tubulin dimers, represses β-tubulin synthesis 10-fold. Autoregulation of β-tubulin synthesis by tubulin dimers is accomplished at the level of β-tubulin mRNA stability. The first 12 nucleotides of the coding portion of the mRNA were found to contain the site responsible for this autoregulatory control.

Since the critical segment of the mRNA involves a coding region, it is not clear whether the regulation of mRNA stability results from the interaction of tubulin dimers with the RNA or with the nascent protein. Either interaction might plausibly trigger a nuclease that would destroy the mRNA.

These two possibilities were tested by mutagenizing the regulatory region on a cloned version of the gene. The mutant genes were then transfected into cells, and the stability of their mRNAs was assayed in the presence of excess free tubulin dimers. The results from a dozen mutants that affect a short region of the mRNA are shown in Figure 7–49.

Does the regulation of β-tubulin mRNA stability result from an interaction with the RNA or from an interaction with the encoded protein? Explain your reasoning.

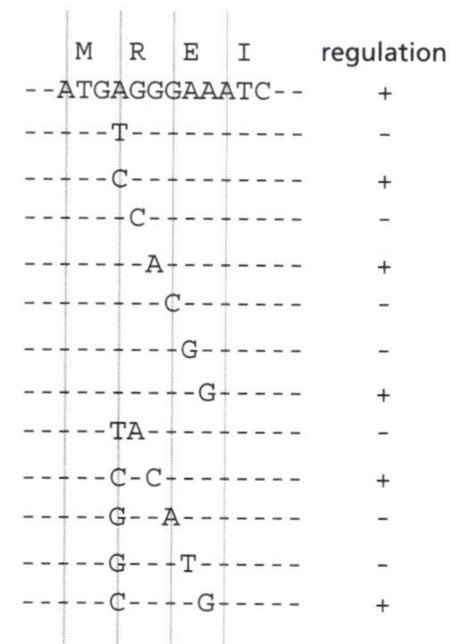


**Figure 7–49** Effects of mutations on the regulation of β-tubulin mRNA stability (Problem 7–114). The wild-type sequence for the first 12 nucleotides of the coding portion of the gene is shown at the *top*, and the first four amino acids beginning with methionine (M) are indicated *above* the codons. The nucleotide changes in the 12 mutants are shown *below*; only the altered nucleotides are indicated. Regulation of mRNA stability is shown on the *right*: + indicates wild-type response to changes in intracellular tubulin concentration and – indicates no response to changes. *Vertical dotted lines* mark the position of the first nucleotide in each codon.

**Table 7–6 Hybridization of specific oligonucleotides to the amplified segments from liver and intestine RNA and DNA** (Problem 7–113).

| | RNA | | DNA | |
|---|---|---|---|---|
| | LIVER | INTESTINE | LIVER | INTESTINE |
| Oligo-Q | + | – | + | + |
| Oligo-STOP | – | + | – | – |

Hybridization is indicated by +; absence of hybridization is indicated by –.

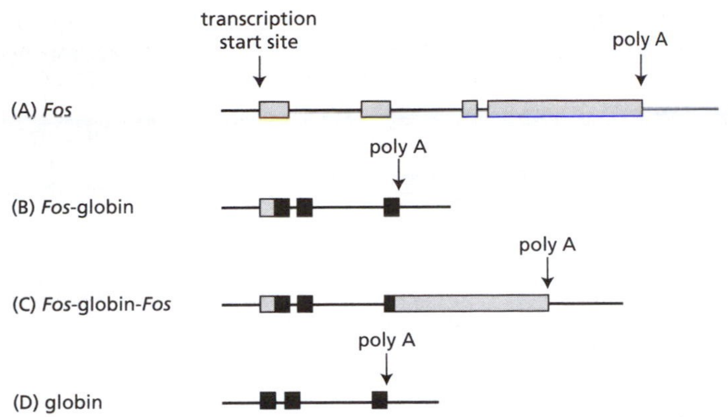


**Figure 7–50** Structures of the human *Fos* gene and the hybrid genes of *Fos* with the human β-globin gene (Problem 7–115). (A) The cloned *Fos* gene. (B) A hybrid gene, *Fos*–globin, with *Fos* sequences at the 5′ end. (C) A hybrid gene, *Fos*–globin–*Fos*, with *Fos* sequences at the 5′ and 3′ ends. (D) The normal β-globin gene. *Gray boxes* denote *Fos* exons; *black boxes* are β-globin exons. The junctions between *Fos* and globin sequences in the hybrid genes are located in exons. The site for polyadenylation is indicated by 'poly A.'

**7–115** The *Fos* gene encodes a gene regulatory protein, whose activation is one of the earliest transcriptional responses to serum growth factors. The *Fos* transcription rate in mouse cells increases markedly within 5 minutes, reaches a maximum by 10 to 15 minutes, and abruptly decreases thereafter.

To study the mechanism of transient induction of the human *Fos* gene, you clone a DNA fragment that contains the complete transcription unit, starting 750 nucleotides upstream of the transcription start site and ending 1.5 kb downstream of the poly-A addition site (Figure 7–50A). When you transfer this cloned segment into mouse cells (so you can distinguish the human *Fos* mRNA from the endogenous mRNA) and stimulate cell growth one day later by adding serum (a rich source of growth factors), you observe the same sort of transient induction (Figure 7–51A), although the timescale is slightly longer than for the mouse gene.

By analyzing a series of deletion mutants, you show that an enhancerlike element (SRE, serum response element), 300 nucleotides upstream of the transcription start site, is necessary for increased transcription in response to serum. You also investigate the stability of the *Fos* mRNA by creating two hybrid genes containing portions of the human β-globin gene, which encodes a stable mRNA. The structures of the globin gene and the hybrid genes are shown in Figure 7–50. When the hybrid genes are transfected into mouse cells in low serum, they respond to serum added 24 hours later as indicated in Figure 7–51B and C.

A. Which portion of the *Fos* gene confers instability on *Fos* mRNA?

B. The *Fos*–globin hybrid gene carries all the normal *Fos* regulatory elements, including the SRE, and yet the mRNA is present at time zero (24 hours after transfection, but before serum addition), and does not increase appreciably after serum addition. Can you account for this behavior in terms of mRNA stability?

**7–116** In nematodes the choice between spermatogenesis and oogenesis in the hermaphrodite germ line depends on translational regulation of the *Tra2* and *Fem3* genes, as shown in Figure 7–52. When they are expressed, *Tra2* promotes oogenesis and *Fem3* promotes spermatogenesis. Translation of each gene's mRNA is regulated by the binding of proteins to elements within

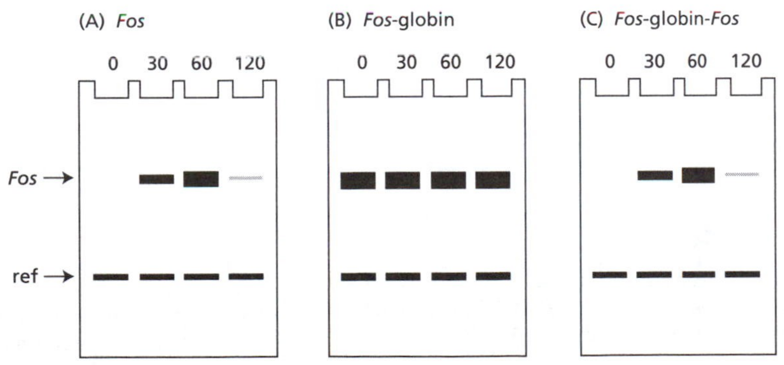


**Figure 7–51** Responses of *Fos* and hybrid genes to addition of serum (Problem 7–115). (A) The cloned human *Fos* gene. (B) The *Fos*–globin hybrid gene with *Fos* sequences at the 5′ end. (C) The *Fos*–globin–*Fos* hybrid gene with *Fos* sequences at the 5′ and 3′ ends. The *upper* band in each gel is the transcript containing *Fos* sequences. The *lower* band is a reference transcript (to control for recoveries of RNA) from a gene that does not respond to serum addition. Sampling times (in minutes) are indicated *above* each lane.

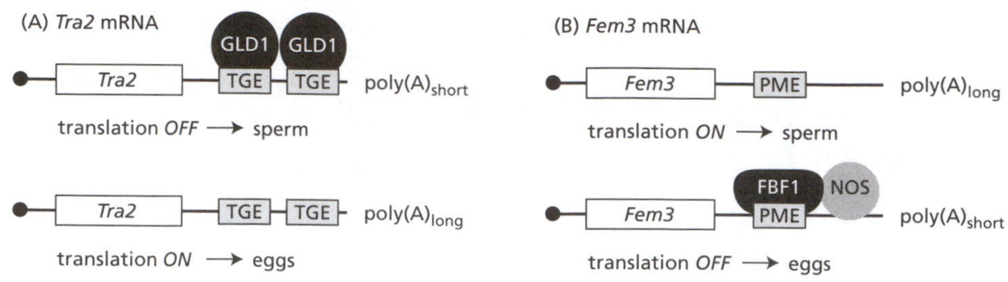


**Figure 7–52** Translational control of the choice between spermatogenesis and oogenesis (Problem 7–116). (A) *Tra2* mRNA. (B) *Fem3* mRNA. Control elements within the 3′ untranslated regions of the two genes are shown, along with the specific proteins that bind to those elements.

their 3′ untranslated regions. In each case, the protein-bound mRNA, although stable, is not efficiently translated and has a short poly-A tail. In each case, the mRNA in its unbound form is translationally active and has a long poly-A tail. How do you suppose that the lengths of the poly-A tails might affect the efficiency of translation of these mRNAs?

**7–117** To investigate the molecular mechanism by which the *Tra2* regulatory elements (TGEs) control translation, you are injecting *Tra2* mRNAs into *Xenopus* oocytes and following their fate. You have shown that an oocyte protein binds to the TGEs and inhibits translation of injected *Tra2* mRNA. Now you wish to determine how mRNAs with bound proteins come to have short poly-A tails. To investigate this question, you prepare two kinds of radiolabeled *Tra2* mRNA: one with the normal pair of TGE elements, and one with those elements deleted (see Figure 7–52A). Each RNA has a tail of 65 A nucleotides. You inject these mRNAs into one-cell *Xenopus* embryos and then reisolate the mRNAs at various cell stages thereafter. The reisolated radiolabeled RNA was analyzed by gel electrophoresis and autoradiography (Figure 7–53).

A. Do the TGEs alter the overall stability of the mRNAs; that is, are the mRNAs destroyed more rapidly in the presence or absence of the TGEs? How can you tell?

B. Do the TGEs influence the length of the poly-A tails? How can you tell?

**7–118** Nematode mutants with defects in the germ-line-specific protein GLD1, which binds to the TGE sequences in *Tra2* mRNA (see Figure 7–52A), have phenotypes that suggest that GLD1 may bind more target mRNAs than just the one from *Tra2*. To try to locate additional mRNA targets of GLD1, the researchers constructed a gene encoding a GLD1 protein fused to a short stretch of amino acids, termed a FLAG tag (which can be precipitated readily by a commercially available antibody). The modified *Gld1* gene was introduced stably into a nematode carrying a nonfunctional *Gld1* gene. The strategy for isolating new GLD1 targets is shown in Figure 7–54. RNAs were isolated using FLAG antibodies, or nonspecific mouse IgG antibodies, and then converted to cDNAs. Potential GLD1 targets among the FLAG cDNAs were enriched by removing the cDNAs that were also present in the IgG cDNAs prepared using nonspecific antibodies—a technique called subtractive

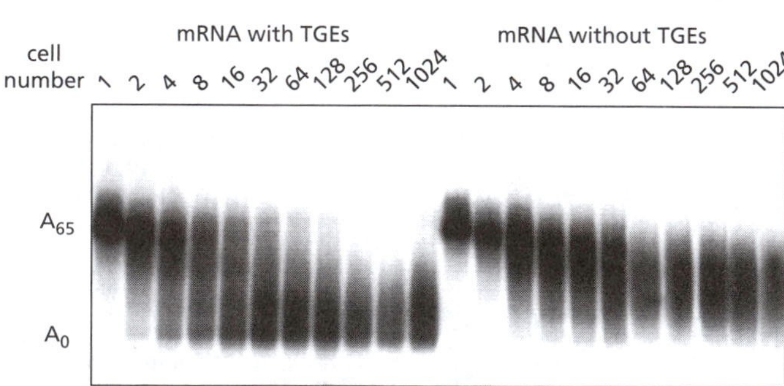


**Figure 7–53** Injection of radiolabeled mRNAs with and without TGEs into *Xenopus* one-cell embryos (Problem 7–117). The lengths of the poly-A tails are indicated on the *left*. Cell number refers to the number of cells in the embryos when they were harvested.

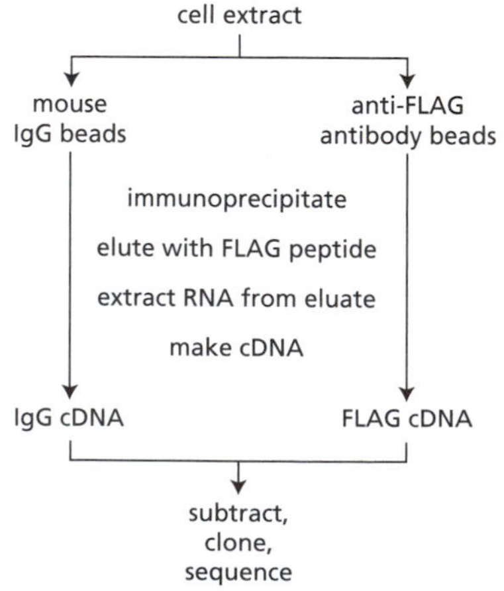


**Figure 7–54** Scheme for isolating mRNA targets of GLD1 (Problem 7–118). The common cDNAs in the two preparations were removed from the FLAG cDNA population by subtractive hybridization.

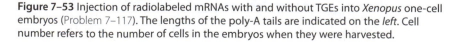

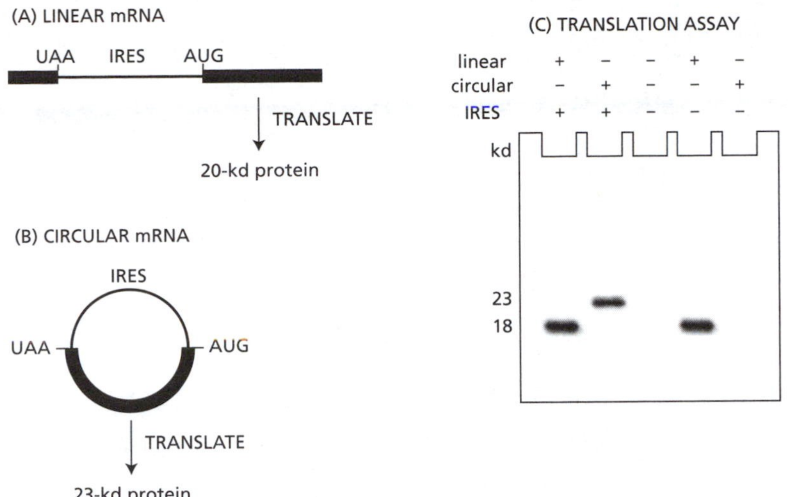


**Figure 7–55** Analysis of effects of IRESs on translation (Problem 7–119). (A) Linear mRNA that contains an IRES. The structure of the mRNA without the IRES was the same. (B) Circular mRNA that contains an IRES. The structure of the mRNA without the IRES was the same. Circular RNAs were prepared by ligating the ends together; they were purified from the linear starting molecules by gel electrophoresis. (C) Display of translation products from various species of mRNA.

hybridization. From one subtractive hybridization, 211 clones were sequenced and found to correspond to 94 different genes. From an independent subtractive hybridization, 198 clones were found to correspond to 89 different genes. Comparison of the two data sets indicated that 17 genes were found in common.

A. Why did the investigators focus on the cDNAs that were common to the two independent subtractive hybridizations? Shouldn't all the cDNAs present after subtraction be targets for GLD1?

B. For one specific gene, *Rme2*, the investigators (1) showed that the Rme2 protein was expressed specifically in the germ line; (2) demonstrated that the locations of GLD1 and Rme2 in the germ line were nonoverlapping; and (3) identified GLD1-binding sites in *Rme2* mRNA. How do these observations support their conclusion that *Rme2* mRNA is a bona fide target of GLD1?

**7–119** You are skeptical that IRESs really allow direct binding of the eucaryotic translation machinery to the interior of an mRNA. As a critical test of this notion, you prepare a set of linear and circular RNA molecules, with and without IRESs (Figure 7–55A and B). You translate these various RNAs in rabbit reticulocyte lysates and display the translation products by SDS gel electrophoresis (Figure 7–55C). Do these results support or refute the idea that IRESs allow ribosomes to initiate translation of mRNAs in a cap-independent fashion? Explain your answer.

**Problem 6–91** considers the translational regulation of *E. coli* release factor 2, a protein required for termination of translation.

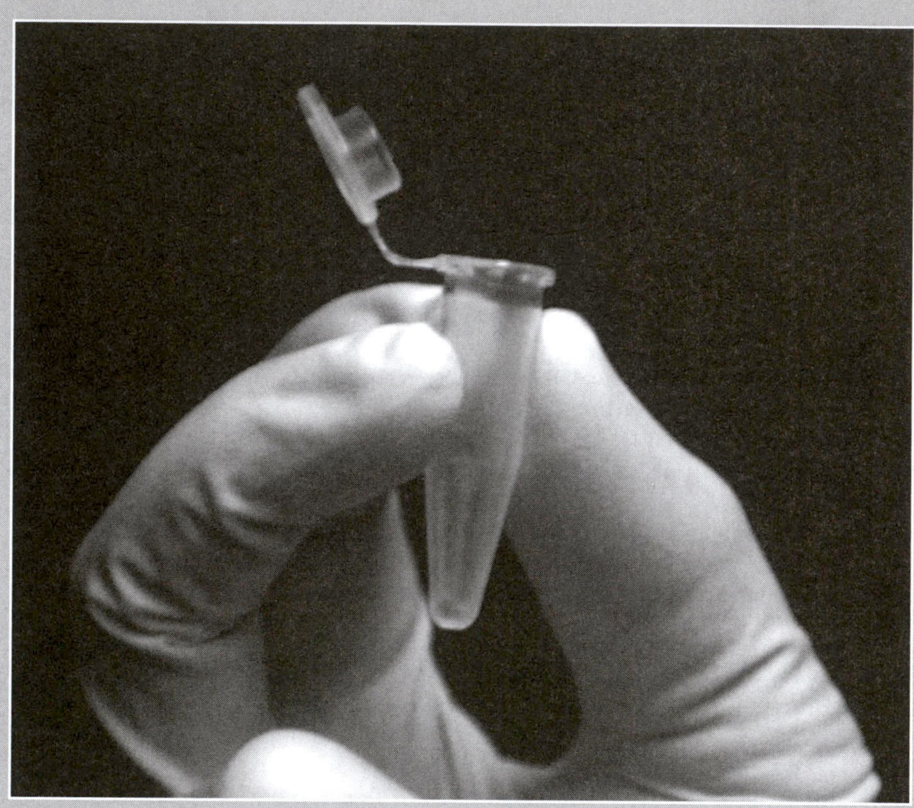

**An Eppendorf tube containing 10 microliters of solution.** Most molecular biology reactions are carried out in these tiny plastic tubes, which were invented in 1961 at the Eppendorf hospital laboratories in the suburbs of Hamburg, Germany, as part of a system for handling very small volumes of clinical samples. These tubes are inert, robust, and inexpensive.

# Manipulating Proteins, DNA, and RNA

8

## In This Chapter


| | |
|---|---|
| ISOLATING CELLS AND GROWING THEM IN CULTURE | 191 |
| PURIFYING PROTEINS | 192 |
| ANALYZING PROTEINS | 195 |
| ANALYZING AND MANIPULATING DNA | 203 |
| STUDYING GENE EXPRESSION AND FUNCTION | 214 |


## ISOLATING CELLS AND GROWING THEM IN CULTURE

TERMS TO LEARN

hybridoma    monoclonal antibody

### DEFINITIONS

Match each definition below with its term from the list above.

**8–1** Antibody secreted by a hybridoma cell line.

**8–2** Cell line used in the production of monoclonal antibodies; obtained by fusing antibody-secreting B cells with cells of a lymphocyte tumor.

### TRUE/FALSE

Decide whether each of these statements is true or false, and then explain why.

**8–3** Because a monoclonal antibody recognizes a specific antigenic site (epitope), it binds only to the specific protein against which it was made.

**8–4** For both therapeutic cloning and reproductive cloning a somatic cell nucleus is first transplanted into an egg, whose nucleus has been removed, and then allowed to develop into a very early embryo stage *in vitro*.

### THOUGHT PROBLEMS

**8–5** A common step in the isolation of cells from a sample of animal tissue is to treat it with trypsin, collagenase, and EDTA. Why is such a treatment necessary, and what does each component accomplish? And why doesn't this treatment kill the cells?

**8–6** Isolation of cells from tissues, fluorescence-activated cell sorting, and laser capture microdissection are just a few of the ways for generating homogeneous cell populations. Why do you suppose it is important to have a homogeneous cell population for many experiments?

**8–7** Distinguish among the terms 'primary culture,' 'secondary culture,' and 'cell line.'

**8–8** Why are human embryonic stem cell lines thought to be especially promising from a medical perspective?

**8–9** Consider the following two statements. "The most important advantage of the hybridoma technique is that monoclonal antibodies can be made against molecules that constitute only a minor component of a complex mixture." "The most important advantage of the hybridoma technique is that antibodies that may be present as only minor components in conventional antiserum can be produced in quantity in pure form as monoclonal antibodies." Are these two statements equivalent? Why or why not?

**Problem 3–67** examines the use of monoclonal antibodies for affinity purification of proteins.
**Problem 12–98** uses monoclonal antibodies to elucidate the mechanism of protein import into peroxisomes.
**Problems 19–66** and **19–67** investigate integrin binding and signaling using monoclonal antibodies with special properties.

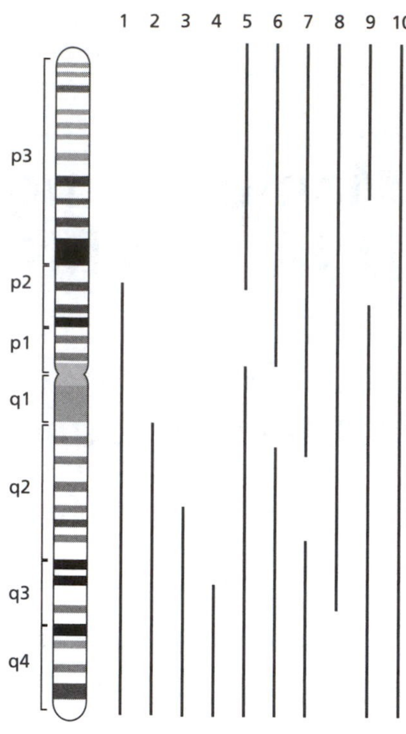


**Figure 8–1** Mapping the gene for the FeLV-C receptor using human–rodent hybrid cell lines that carry portions of human chromosome 1 (Problem 8–12). The hybrid cell lines that could be infected by FeLV-C are shown, with the portions of human chromosome 1 retained in the individual hybrid cell lines indicated as *vertical lines*. The p and q designations refer to a standard convention (the Paris nomenclature) for describing chromosome positions. Short arm locations are labeled p (*petit*) and long arms q (*queue*). Each chromosome arm is divided into regions labeled p1, p2, p3, q1, q2, q3, etc., counting outwards from the centromere. Regions are delimited by specific landmarks, which are distinct morphological features, including the centromere and certain prominent bands. Regions are divided into bands labeled p11 (one-one, not eleven), p12, etc. Sub-bands are designated p11.1, p11.2, etc., and sub-sub bands are designated p11.11, p11.12, etc. In all cases the numbers increase from the centromere toward the telomere.

**8–10** Do you suppose it would be possible to raise an antibody against another antibody? Explain your answer.

## CALCULATIONS

**8–11** You want to isolate rare cells that are present in a population at a frequency of 1 in $10^5$ cells, and you need 10 of those cells to do an experiment. If your fluorescence-activated cell sorter can sort cells at the rate of 1000 per second, how long would it take to collect enough rare cells for your experiment?

## DATA HANDLING

**8–12** Panels of human–rodent cell hybrids that retain one or a few human chromosomes, parts of human chromosomes, or radiation-induced fragments of human chromosomes have proven enormously useful in mapping genes to defined locations. Now that the human genome has been sequenced, it is a trivial matter to know a gene's location if you have a bit of sequence from the gene, for example, from a cDNA clone. Nevertheless, there are many instances in which such panels of cells are still invaluable.

You wish to map the location of the receptor for feline leukemia virus type C (FeLV-C), which infects human cells but not rodent cells. Using a panel of hybrid cells carrying whole chromosomes, you have shown that FeLV-C infects only those hybrids carrying human chromosome 1. Using a second panel carrying portions of chromosome 1, you show that FeLV-C infects several of the hybrid cell lines. The segments of chromosome 1 that are present in the infectable hybrid cell lines are shown in Figure 8–1. Where on chromosome 1 is the gene for the FeLV-C receptor located?

**Problem 7–93** uses hybrid cell lines carrying either an active or an inactive X chromosome to investigate the mechanism of X-inactivation.

# PURIFYING PROTEINS

TERMS TO LEARN

affinity chromatography
column chromatography
fusion protein
high-performance liquid chromatography (HPLC)
purified cell-free system

## DEFINITIONS

Match each definition below with its term from the list above.

**8–13** General term for purification technique in which a mixture of proteins is passed through a column containing a porous solid matrix.

**8–14** Type of chromatography that uses columns packed with special chromatography resins composed of tiny spheres that attain a high degree of resolution, even at very fast flow rates.

**8–15** Purification technique in which a protein mixture is passed over a matrix to

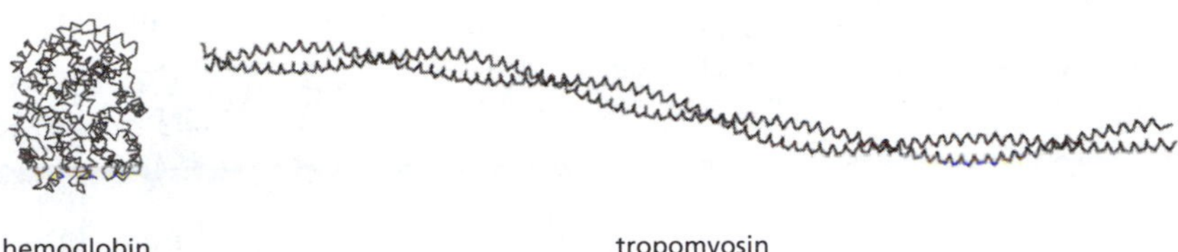


Figure 8–2 Backbone models of tropomyosin and hemoglobin (Problem 8–20).

which ligands specific for the desired protein have been attached, so that the protein is retained on the matrix.

## TRUE/FALSE

Decide whether each of these statements is true or false, and then explain why.

**8–16** It is possible to pellet hemoglobin by centrifugation at sufficiently high speed.

**8–17** If the beads used in gel-filtration chromatography had pores of a uniform size, proteins would either be excluded from the pores or included in them, but would not be further fractionated.

## THOUGHT PROBLEMS

**8–18** Describe how you would use preparative centrifugation to purify mitochondria from a cell homogenate.

**8–19** Distinguish between velocity sedimentation and equilibrium sedimentation. For what general purpose is each technique used? Which do you suppose might be best suited for separating two proteins of different size?

**8–20** Tropomyosin, at 93 kd, sediments at 2.6S, whereas the 65-kd protein, hemoglobin, sediments at 4.3S. (The sedimentation coefficient S is a linear measure of the rate of sedimentation: both increase or decrease in parallel.) These two proteins are shown as α-carbon backbone models in Figure 8–2. How is it that the bigger protein sediments more slowly than the smaller one? Can you think of an analogy from everyday experience that might help you with this problem?

**Problems 5–37, 6–80, 6–92,** and **15–51** give examples of the use of velocity sedimentation.

**8–21** Distinguish among ion-exchange chromatography, hydrophobic chromatography, gel-filtration chromatography, and affinity chromatography in terms of the column material and the basis for separation of a mixture of proteins.

## CALCULATIONS

**8–22** The purification of a protein usually requires multiple steps and often involves several types of column chromatography. A key component of any purification is an assay for the desired protein. The assay can be a band on a gel, a structure in the electron microscope, the ability to bind to another molecule, or an enzymatic activity. The purification of an enzyme is particularly instructive because the assay allows one to quantify the extent of purification at each step. Consider the purification of the enzyme shown in Table 8–1. The total volume, total protein, and total enzyme activity are shown at each step.

**Table 8–1 Purification of an enzyme** (Problem 8–22).

| PROCEDURE | TOTAL VOLUME (mL) | TOTAL PROTEIN (mg) | TOTAL ACTIVITY (units) | SPECIFIC ACTIVITY (units/mg) |
|---|---|---|---|---|
| 1. Crude extract | 2000 | 15,000 | 150,000 | |
| 2. Ammonium sulfate precipitation | 320 | 4000 | 140,000 | |
| 3. Ion-exchange chromatography | 100 | 550 | 125,000 | |
| 4. Gel-filtration chromatography | 85 | 120 | 105,000 | |
| 5. Affinity chromatography | 8 | 5 | 75,000 | |

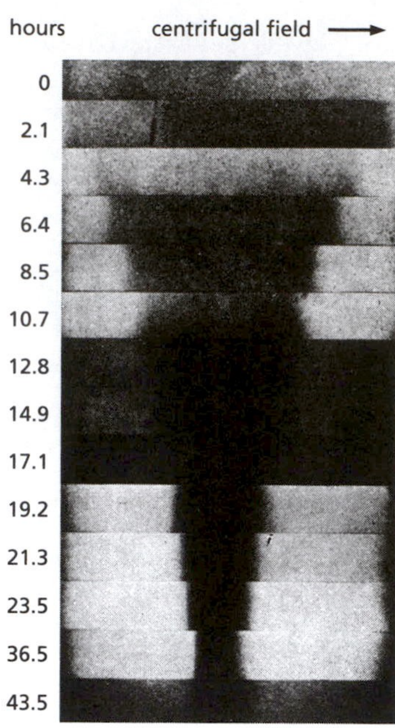


**Figure 8–3** Ultraviolet absorption photographs showing successive stages in the banding of *E. coli* DNA (Problem 8–23). DNA, which absorbs UV light, shows up as *dark regions* in the photographs. The bottom of the centrifuge tube is on the *right*.

A. For each step in the purification procedure, calculate the specific activity of the enzyme (units of activity per mg of protein). How can you tell that purification has occurred at each step?
B. Which of the purification steps was most effective? Which was least effective?
C. If you were to carry the purification through additional steps, how would the specific activity change? How could you tell from specific activity measurements that the enzyme was pure? How might you check on that conclusion?
D. If the enzyme is pure at the end of the purification scheme in Table 8–1, what proportion of the protein in the starting cell does it represent?

## DATA HANDLING

**8–23** In the classic paper that demonstrated the semi-conservative replication of DNA, Meselson and Stahl began by showing that DNA itself will form a band when subjected to equilibrium sedimentation. They mixed randomly fragmented *E. coli* DNA with a solution of CsCl so that the final solution had a density of 1.71 g/mL. As shown in Figure 8–3, with increasing length of centrifugation at 70,000 times gravity, the DNA, which was initially dispersed throughout the centrifuge tube, became concentrated over time into a discrete band in the middle.

A. Describe what is happening with time and explain why the DNA forms a discrete band.
B. What is the buoyant density of the DNA? (The density of the solution at which DNA 'floats' at equilibrium defines the 'buoyant density' of the DNA.)
C. Even if the DNA were centrifuged for twice as long—or even longer—the width of the band remains about what is shown at the bottom of Figure 8–3. Why doesn't the band become even more compressed? Suggest some possible reasons to explain the thickness of the DNA band at equilibrium.

**Problems 5–40, 5–60, 10–30,** and **14–112** give examples of the use of equilibrium sedimentation.

**8–24** The result of gel-filtration chromatography of six, roughly spherical proteins is shown in Figure 8–4. The identities of the proteins, their molecular masses, and their elution volumes are indicated in Table 8–2. (The elution volume identifies when each protein came off the column.)

A. Why do the smaller proteins come off the column later than the larger proteins?
B. Plot molecular mass versus elution volume. Now plot the log of the molecular mass versus the elution volume. Which plot gives a straight line? What do you suppose is the basis for that result?

**8–25** In preliminary studies you've determined that your partially purified protein is stable (retains activity) between pH 5.0 and pH 7.5. On either side of that pH range the protein is no longer active. Your advisor now wants you to do a quick experiment to determine conditions for ion-exchange chromatography. He

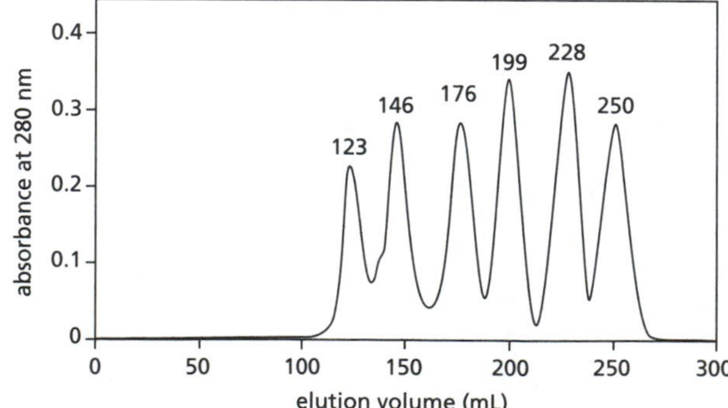


**Figure 8–4** Elution profile for proteins fractionated by gel-filtration chromatography (Problem 8–24). The absorbance at 280 nm is a measure of protein concentration. Each of the peaks is identified by its elution volume.

**Table 8–2 Proteins separated by gel-filtration chromatography** (Problem 8–24).

| PROTEIN | MOLECULAR MASS (kd) | MOLECULAR MASS (log) | ELUTION VOLUME (mL) |
|---|---|---|---|
| Ribonuclease A | 13 | 4.14 | 250 |
| Chymotrypsinogen | 25 | 4.40 | 228 |
| Ovalbumin | 43 | 4.63 | 199 |
| Bovine serum albumin | 67 | 4.83 | 176 |
| Aldolase | 158 | 5.20 | 146 |
| Catalase | 232 | 5.37 | 123 |

has left instructions for you. First, you're supposed to mix a bit of the crude preparation with a small amount of the ion-exchange resin DEAE-Sepharose in a series of buffer solutions that have a pH between 5.0 and 7.5. Next, you are to pellet the resin and assay the supernatant for the presence of your protein. Finally, he tells you to use this information to pick the proper pH to do the ion-exchange chromatography. You have completed the first two steps and have obtained the results shown in Figure 8–5. But you are a little uncertain as to how to use the information to pick the pH for the chromatography.

A. At which end of the pH range is the charge on your protein more positive and at which end is it more negative? [Over this pH range the positively charged amine groups on the DEAE-Sepharose beads (Figure 8–5A) are unaffected.]

B. For the chromatography, should you pick a pH at which the protein binds to the beads (pH 6.5 to 7.5) or a pH where it does not bind (pH 5.0 to 6.0)? Explain your choice.

C. Should you pick a pH close to the boundary (that is, pH 6.0 or 6.5) or far away from the boundary (that is, pH 5.0 or pH 7.5)? Explain your reasoning.

D. How will you carry out ion-exchange chromatography of your protein? What are the various steps you will use to accomplish the separation of your protein from others via ion-exchange chromatography?

**Problem 6–40** gives an example of column chromatography.

**Problems 3–67 and 15–148** show examples of affinity chromatography for protein purification.

# ANALYZING PROTEINS

TERMS TO LEARN

| | |
|---|---|
| chemical biology | surface plasmon resonance (SPR) |
| fluorescence resonance energy transfer (FRET) | two-dimensional gel electrophoresis |
| nuclear magnetic resonance (NMR) spectroscopy | two-hybrid system |
| SDS polyacrylamide-gel electrophoresis (SDS-PAGE) | Western blotting (immunoblotting) |
| | x-ray crystallography |

## DEFINITIONS

Match each definition below with its term from the list above.

**8–26** Analysis of the release of electromagnetic radiation by atomic nuclei in a magnetic field, due to flipping of the orientation of their magnetic dipole moments.

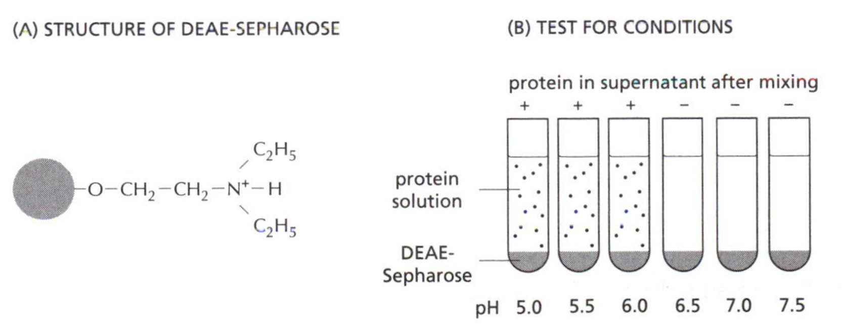


**Figure 8–5** Preliminary test to determine conditions for ion-exchange chromatography (Problem 8–25). (A) Structure of the charged amine groups attached to Sepharose beads. (B) Results of mixing your protein with DEAE-Sepharose beads. Samples of the protein were mixed with DEAE-Sepharose beads in buffers at a range of pH values, and then the mixtures were centrifuged to pellet the beads. The presence of the protein in the supernatant is indicated by a +; its absence is indicated by a –.

**8–27** Technique for identifying interacting proteins using genetically engineered yeast cells.

**8–28** Technique for protein separation in which the protein mixture is run first in one direction and then in a direction at right angles to the first.

**8–29** Technique in which a protein mixture is separated by running it through a gel containing a detergent that binds to and unfolds the proteins.

**8–30** Technique for monitoring the closeness of two fluorescently labeled molecules (and thus their interactions) in cells.

**8–31** Technique by which proteins are separated by electrophoresis, immobilized on a paper sheet, and then analyzed, usually by means of a labeled antibody.

## TRUE/FALSE

Decide whether each of these statements is true or false, and then explain why.

**8–32** Given the inexorable progress of technology, it seems inevitable that the sensitivity of detection of molecules will ultimately be pushed beyond the yoctomole level ($10^{-24}$ mole).

**Problems 6–23, 14–31,** and **16–78** describe the analysis of single molecules of RNA polymerase, ATP synthase, and a kinesin motor, respectively.

**8–33** Surface plasmon resonance (SPR) measures association ($k_{on}$) and dissociation ($k_{off}$) rates between molecules in real time, using small amounts of unlabeled molecules, but it does not give the information needed to determine the binding constant ($K$).

## THOUGHT PROBLEMS

**8–34** How is it that smaller molecules move through a gel-filtration column more slowly than larger molecules, whereas in SDS polyacrylamide-gel electrophoresis (SDS-PAGE) the opposite is true: larger molecules move more slowly than small molecules?

**8–35** You are all set to run your first SDS polyacrylamide-gel electrophoresis. You have boiled your samples of protein in SDS in the presence of mercaptoethanol and loaded them into the wells of a polyacrylamide gel. You are now ready to attach the electrodes. Uh oh, does the positive electrode (the anode) go at the top of the gel, where you loaded your proteins, or at the bottom of the gel?

**8–36** You hate the smell of mercaptoethanol. Since there are no disulfide bonds in intracellular proteins (see Problem 3–44), you have convinced yourself that it is not necessary to treat a cytoplasmic homogenate with mercaptoethanol prior to SDS-PAGE. You heat a sample of your homogenate in SDS and subject it to electrophoresis. Much to your surprise, your gel looks horrible; it is an ugly smear! You show your result to a fellow student with a background in chemistry, and she suggests that you treat your sample with *N*-ethyl maleimide (NEM), which reacts with free sulfhydryls. You run another sample of your homogenate after treating it with NEM and SDS. Now the gel looks perfect!

If intracellular proteins don't have disulfide bonds—and they don't—why didn't your original scheme work? And how does treatment with NEM correct the problem?

**Problems 7–8, 16–39, and 16–108** use two-dimensional gel electrophoresis to separate proteins.

**Problem 5–61** illustrates the use of two-dimensional gel electrophoresis of DNA in the analysis of origins of DNA replication.

**8–37** For separation of proteins by two-dimensional polyacrylamide-gel electrophoresis, what are the two types of electrophoresis that are used in each dimension? Do you suppose it makes any difference which electrophoretic method is applied first? Why or why not?

**8–38** Discuss the following statement: "With the ever-expanding databases of protein sequences and structures, it will soon be possible to input an amino acid sequence of an unknown protein and, by analogy to known proteins, determine its structure and function. Thus, it will not be long before biochemists are put out of work."

**8–39** Hybridoma technology allows one to generate monoclonal antibodies to virtually any protein. Why is it, then, that tagging proteins with epitopes is such a commonly used technique, especially since an epitope tag has the potential to interfere with the function of the protein?

**8–40** Phage display is a powerful method for screening large collections of peptides for binding to selected targets. Random DNA sequences encoding the peptides are introduced into a virus that infects *E. coli* (a bacteriophage or phage) so that they are fused with a gene encoding one of proteins that forms the viral coat. This collection of phage is then screened for binding to a purified protein of interest (see Figure 8–12). Bound phage can be amplified by growth in *E. coli* and their genomes can be sequenced to determine the encoded peptide. Because phage display seems to be just what you need to find the binding site recognized by your protein, you look into the method in more detail.

In the supplier's information package on phage display, you find that they use a 'reduced' genetic code with only 32 codons (Figure 8–6). They synthesize mixtures of oligonucleotides with all four nucleotides in the first two positions of a codon, but they include only T and G in the third position. In addition, they grow their phage on a strain of bacteria that inserts glutamine (Q) at a TAG codon, which is normally a stop codon.

A. Why do you suppose they use a reduced genetic code? Can you see any advantages to including just T and G at the third position of codons and growing the phage in the special strain of bacteria that inserts glutamine at TAG codons?

B. If all four nucleotides were used at each position, all possible codons would be produced equally (assuming no biases). Amino acids would then be encoded by such a mixture in the same proportions as their codons are represented in the standard genetic code (see inside back cover). That is to say, cysteine, which has two codons, would be encoded half as often as alanine, which has four codons, and one-third as often as leucine, which has six codons. Are the relative frequencies of the various amino acids the same for the reduced genetic code as for the standard genetic code? If not, how do they differ?

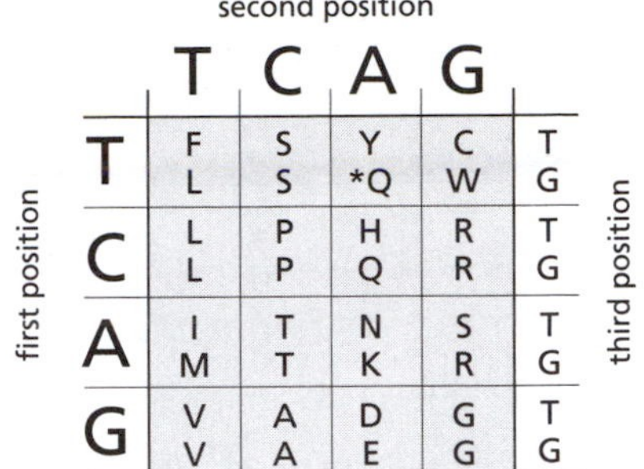


**Figure 8–6** Reduced genetic code (Problem 8–40). Compare this code with the standard genetic code (see inside back cover). *Q indicates that the special strain of bacteria in which the phage are grown inserts glutamine at TAG codons.

## CALCULATIONS

**8–41** How many copies of a protein need to be present in a cell in order for it to be visible as a band on a gel? Assume that you can load 100 µg of cell extract onto a gel and that you can detect 10 ng in a single band by silver staining. The concentration of protein in cells is about 200 mg/mL (see Problem 2–58), and a typical mammalian cell has a volume of about 1000 $\mu m^3$ and a typical bacterium a volume of about 1 $\mu m^3$. Given these parameters, calculate the number of copies of a 120-kd protein that would need to be present in a mammalian cell and in a bacterium in order to give a detectable band on a gel. You might try an order-of-magnitude guess before you make the calculations.

**8–42** You want to know the sensitivity for detection of immunoblotting (Western blotting), using an enzyme-linked second antibody to detect the antibody directed against your protein (Figure 8–7A). You are using the mouse monoclonal antibody 4G10, which is specific for phosphotyrosine residues, to detect phosphorylated proteins. You first phosphorylate the myelin basic protein *in vitro* using a tyrosine protein kinase that adds one phosphate per molecule. You then prepare a dilution series of the phosphorylated protein and subject the samples to SDS-PAGE. The protein is then transferred (blotted) onto a nitrocellulose filter, incubated with the 4G10 antibody, and

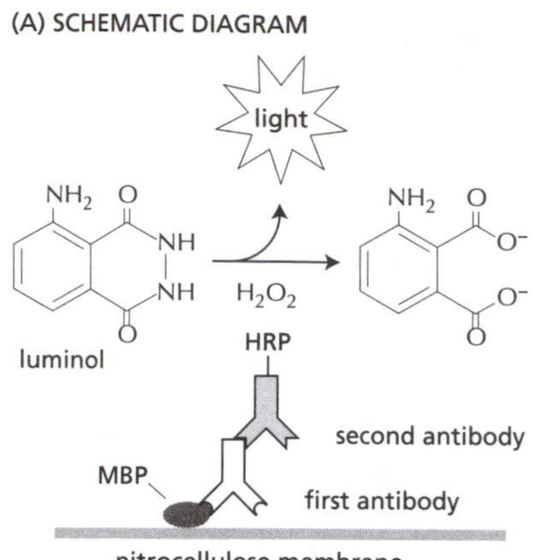


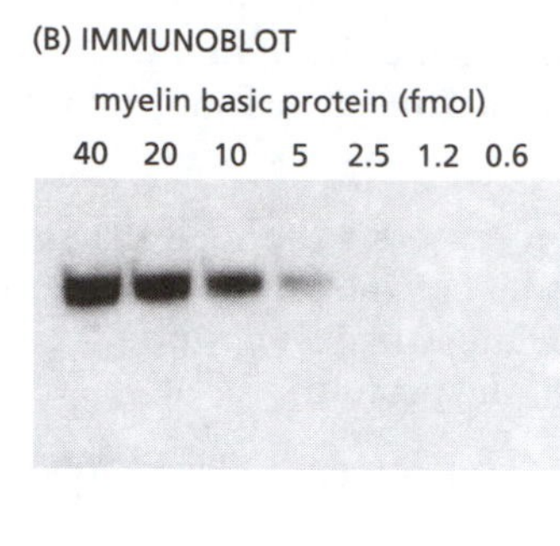


**Figure 8–7** Sensitivity of detection of immunoblotting (Problem 8–42). (A) Schematic diagram of experiment. MBP stands for myelin basic protein. In the presence of hydrogen peroxide, HRP converts luminol to a chemiluminescent molecule that emits light, which is detected by exposure of an x-ray film. (B) Exposed film of an immunoblot. The number of femtomoles of myelin basic protein in each band is indicated.

washed to remove unbound antibody. The blot is then incubated with a second goat anti-mouse antibody that carries horseradish peroxidase (HRP) conjugated to it, and any excess unbound antibody is again washed away. You place the blot in a thin plastic bag, add reagents that chemiluminesce when they react with HRP (Figure 8–7A), and place the bag against a sheet of x-ray film. When the film is developed you see the picture shown in Figure 8–7B.

A. Given the amounts of phosphorylated myelin basic protein indicated in each lane in Figure 8–7B, calculate the detection limit of this method in terms of molecules of protein per band.
B. Assuming that you were using monoclonal antibodies to detect proteins, would you expect that the detection limit would depend on the molecular mass of the protein? Why or why not?

**Problems 3–100, 6–98, 12–86, 12–98, and 15–50** give examples of various uses of immunoblotting in cell biological experiments.

**8–43** In preparing for a phage-display experiment (see Problems 8–40 and 8–50) you notice that the supplier created the phage-display libraries by inserting random oligonucleotides of appropriate length into phage genomes. The initial libraries contained $3 \times 10^9$ different phages. Is this a sufficient number of phages so that the library could contain all possible peptides 7 amino acids in length? Could it contain all possible peptides 12 amino acids in length? Could it contain all the possible peptides 20 amino acids in length?

## DATA HANDLING

**8–44** Figure 8–8 shows an autoradiograph of an SDS-PAGE separation of radiolabeled proteins in a cell-free extract of sea urchin eggs. Alongside are shown a set of radiolabeled marker proteins of defined molecular mass. Two bands that contain known proteins—the small subunit of ribonucleotide reductase and cyclin B—are indicated.

A. Do the standard set of proteins migrate at a rate that is inversely proportional to their molecular masses? That is to say, would you expect a protein of 35 kd, for example, to migrate twice as far down the gel as a protein of 70 kd? Do you suppose a plot of log molecular mass versus migration would give a more linear relationship?
B. How would you use the standard set of proteins to estimate the molecular masses of ribonucleotide reductase and cyclin B? What would you estimate the molecular masses of these two proteins to be?
C. The sequences of the genes for these two proteins give molecular masses of 44 kd for ribonucleotide reductase and 46 kd for cyclin B. Can you offer some possible reasons why the SDS-PAGE estimate for the molecular mass of cyclin B is so far off?

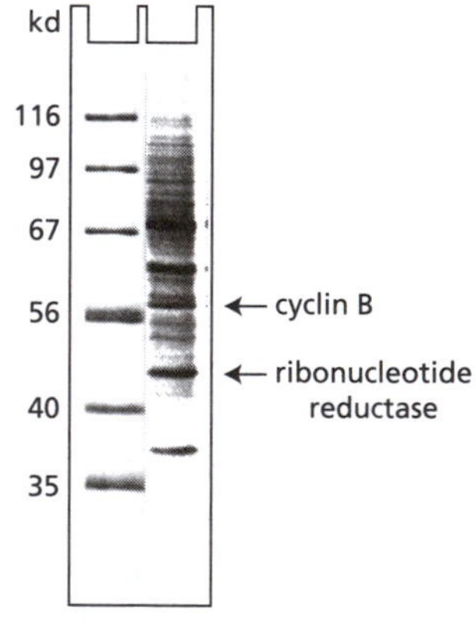


**Figure 8–8** Autoradiograph of radiolabeled proteins separated by SDS-PAGE (Problem 8–44). A set of radiolabeled marker proteins with known molecular masses is shown in the *left hand* lane, along with their molecular masses in kilodaltons. Radiolabeled proteins from a sea urchin egg extract are shown in the *right hand* lane. *Arrows* mark the bands that correspond to cyclin B and the small subunit of ribonucleotide reductase.

**8–45** You have isolated the proteins from two adjacent spots after two-dimensional polyacrylamide-gel electrophoresis and digested them with trypsin. When the masses of the peptides were measured by MALDI-TOF mass spectrometry, the peptides from the two proteins were found to be identical

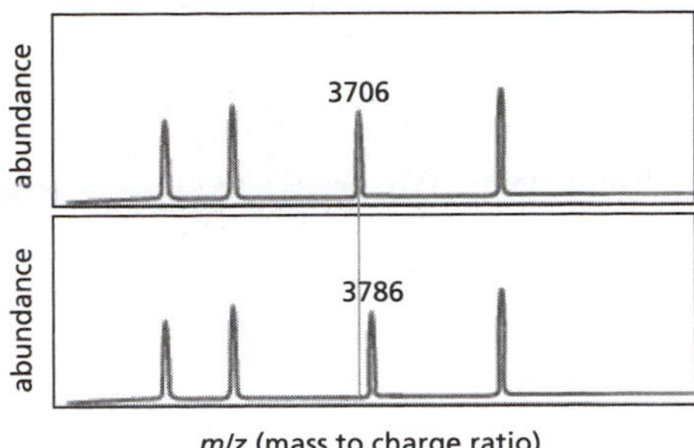


**Figure 8–9** Masses of peptides measured by MALDI-TOF mass spectrometry (Problem 8–45).

except for one (Figure 8–9). For this peptide, the mass to charge ($m/z$) values differed by 79.97, a value that does not correspond to a difference in amino acid sequence. (For example, glutamic acid instead of valine at one position would give an $m/z$ difference of around 30.) Can you suggest a possible difference between the two peptides that might account for the observed $m/z$ difference?

**Problems 3–99, 6–94, 7–66** and **10–67** demonstrate the use of antibodies to analyze associated proteins by co-immunoprecipitation.

**Problems 3–110, 13–32,** and **15–119** accomplish the same objective using GST 'pull downs.'

**8–46** You have raised four different monoclonal antibodies to *Xenopus* Orc1, which is a component of the DNA replication origin recognition complex (ORC) found in eucaryotes. You want to use the antibodies to immunopurify other members of ORC. To decide which of your monoclonal antibodies—TK1, TK15, TK37, or TK47—is best suited for this purpose, you covalently attach them to beads, incubate them with a *Xenopus* egg extract, spin the beads down and wash them carefully, and then solubilize the bound proteins with SDS. You use SDS-PAGE to separate the solubilized proteins and stain them, as shown in Figure 8–10.

A. From these results which bands do you think arise from proteins that are present in ORC?

B. Why do you suppose the various monoclonal antibodies give such different results?

C. Which antibody do you think is the best one to use in future studies of this kind? Why?

D. How might you determine which band on this gel is Orc1?

**8–47** The yeast two-hybrid system depends on the modular nature of many transcription factors, which have one domain that binds to DNA and another domain that activates transcription. Domains can be interchanged by recombinant DNA methods, allowing hybrid transcription factors to be constructed. Thus, the DNA-binding domain of the *E. coli* LexA repressor can be combined with the powerful VP16 activation domain from herpesvirus to activate transcription of genes downstream of a LexA DNA-binding site (Figure 8–11).

If the two domains of the transcription factor can be brought into proximity by protein–protein interactions, they will activate transcription. This is the key feature of the two-hybrid system. Thus, if one member of an interacting pair of proteins is fused to the DNA-binding domain of LexA (to form the 'bait') and the other is fused to the VP16 activation domain (to form the 'prey'), transcription will be activated when the two hybrid proteins interact inside a yeast cell. It is possible to design powerful screens for protein–protein interactions, if the gene whose transcription is turned on is essential for growth or can give rise to a colored product.

To check out the ability of the system to find proteins with which Ras interacts, hybrid genes were constructed that contained the LexA DNA-binding domain, one fused to Ras (LexA–Ras) and the other fused to nuclear lamin (LexA–lamin). A second pair of constructs contained the VP16 activation

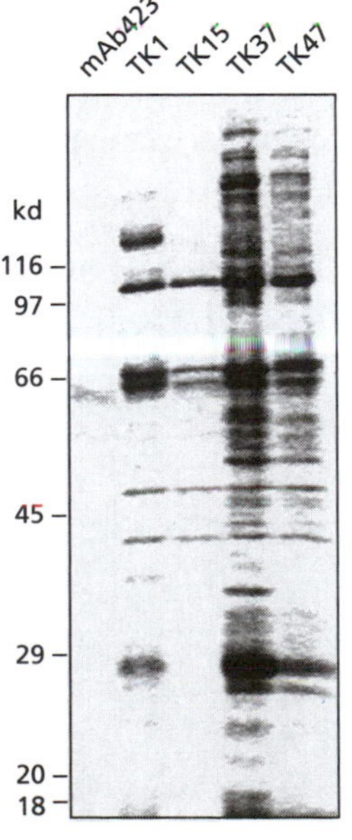

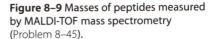

**Figure 8–10** Immunoaffinity purification of *Xenopus* ORC (Problem 8–46). The monoclonal antibody mAb423 is specific for an antigen not found in *Xenopus* extracts and thus serves as a control. The positions of marker proteins are shown at the *left* with their masses indicated in kilodaltons.

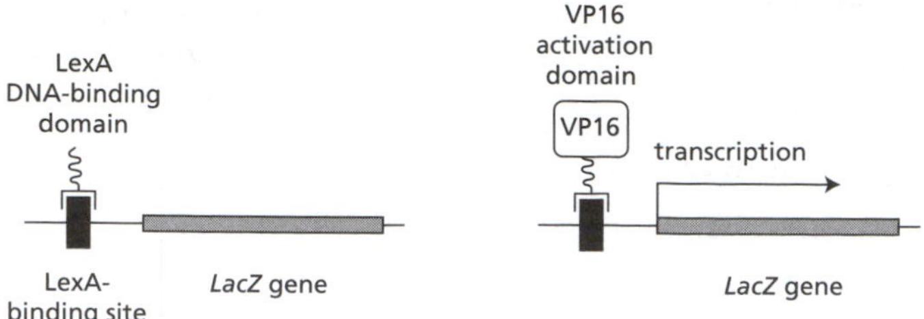


**Figure 8–11** Activation of transcription by a hybrid transcription factor (Problem 8–47).

domain alone (VP16) or fused to the adenylyl cyclase gene (VP16–CYR). Adenylyl cyclase is known to interact with Ras and serves as a positive control; nuclear lamins do not interact with Ras and serve as a negative control. These plasmid constructs were introduced into a strain of yeast containing copies of the *His3* gene and the *LacZ* gene, both with LexA-binding sites positioned immediately upstream. Individual transformed colonies were tested for the ability to grow on a plate lacking histidine, which requires expression of the *His3* gene. In addition, they were tested for ability to form blue colonies (as compared to the normal white colonies) when grown in the presence of an appropriate substrate (XGAL) for β-galactosidase. The setup for the experiment is outlined in Table 8–3.

A. Fill in Table 8–3 with your expectations. Use a plus sign to indicate growth on plates lacking histidine and a minus sign to indicate no growth. Write 'blue' or 'white' to indicate the color of colonies grown in the presence of XGAL.
B. For any entries in the table that you expect to grow in the absence of histidine and form blue colonies with XGAL, sketch the structure of the active transcription factor on the *LacZ* gene.
C. If you want two proteins to be expressed in a single polypeptide chain, what must you be careful to do when you fuse the two genes together?

**Problems 7–62, 7–64, 7–67, 12–43, 12–58, 15–47, 16–38, 17–49, 17–95, 17–146, 18–23, and 18–24** give examples of the use of reporter genes to provide information on the timing and location of gene expression.

**8–48** To use the two-hybrid system to screen for proteins that interact with Ras, a cDNA library was made with the cDNA inserts positioned at the C-terminus of the *Vp16* gene segment. The library was transfected into yeast that already contained the LexA–Ras plasmid described in Problem 8–47. The transformed cells were then grown on plates in the presence of XGAL and in the absence of histidine, and blue colonies were isolated for further testing.

The LexA–Ras plasmid was eliminated from such cells by genetic selection, and the 'cured' cells, which contained only a VP16–cDNA plasmid, were checked for growth in the absence of histidine and for color when grown in the presence of XGAL. Cured cells that did not grow in the absence of histidine and formed white colonies in the presence of XGAL were then transformed with the LexA–lamin plasmid (see Problem 8–47) and tested again for growth in the absence of histidine and colony color in the presence of XGAL. Only cells that did not grow in the absence of histidine and formed white colonies in the presence of XGAL were analyzed further.

**Table 8–3 Experiments to test the two-hybrid system** (Problem 8–47).

| PLASMID CONSTRUCTS | | GROWTH ON PLATES LACKING HISTIDINE | COLOR ON PLATES WITH XGAL |
|---|---|---|---|
| BAIT | PREY | | |
| LexA–Ras | | | |
| LexA–lamin | | | |
| | VP16 | | |
| | VP16-CYR | | |
| LexA–Ras | VP16 | | |
| LexA–Ras | VP16-CYR | | |
| LexA–lamin | VP16 | | |
| LexA–lamin | VP16-CYR | | |

Since its introduction in 1989, the two-hybrid system has been modified considerably to deal with the problem of false positives; that is, to eliminate cDNA clones that do not really encode a protein that interacts with the protein of interest. Several of these modifications were built into the selection scheme used here.

A. When the cDNA library was initially transfected into yeast containing the LexA–Ras plasmid, many white colonies (in addition to blue colonies) were observed when the cells were grown in the presence of XGAL and in the absence of histidine. What type of VP16–cDNA might give rise to such white colonies?

B. Among the colonies that were cured of the LexA–Ras plasmid—and thus contained only a VP16–cDNA plasmid—some grew in the absence of histidine and turned blue when grown in the presence of XGAL. What type of VP16–cDNA might give this result in the absence of the LexA–Ras plasmid?

C. When the LexA-lamin plasmid was introduced into cells that contained a VP16–cDNA plasmid (ones that alone did not grow in the absence of histidine and did not turn blue in the presence of XGAL), some formed blue colonies in the absence of histidine and in the presence of XGAL. What type of VP16–cDNA might give these results in the presence of the LexA–lamin plasmid?

**8–49** Out of $1.4 \times 10^6$ original transformants only 19 clones met the stringent criteria set up in Problem 8–48. The cDNA inserts downstream of the *Vp16* gene segment were sequenced in each. Nine clones had cDNA inserts that corresponded to the N-terminal domain of known Raf serine/threonine protein kinases. This was an exciting finding because other work had shown that immunoprecipitates of Raf could phosphorylate and activate MAP kinase kinase, suggesting that Raf was the missing link between Ras and the MAP-kinase cascade (see *MBoC* Chapter 15).

The scientists that identified Raf by the two-hybrid system did not stop there. They made fusion proteins between the maltose-binding protein and Raf (MBP–Raf) and between glutathione-*S*-transferase and Ras (GST–Ras). They then produced the two fusion proteins in bacteria and tested their ability to bind one another by passing the mixture over an amylose affinity column, which will bind MBP–Raf. When the bound proteins were eluted with maltose, a significant fraction of the input GST–Ras protein was found to elute with the MBP–Raf protein. This result provided a biochemical confirmation of the genetic results with the two-hybrid system.

Why do you think these scientists felt it necessary to demonstrate a direct interaction of Ras and Raf by biochemical studies?

**8–50** The surfaces of different varieties of cyclin A proteins all contain a hydrophobic cleft that is known to bind the sequence PSACRNLFG. This sequence is found close to the N-terminus of the cyclin-dependent kinase inhibitor p27. Looking for potential anti-cancer drugs, you decide to use phage display to hunt for peptides with very high affinity for cyclin A. You attach cyclin A to the bottoms of plastic dishes and 'pan' for phage that will bind (Figure 8–12). You use three different phage M13 libraries that bear randomized 7-, 12-, or 20-amino acid sequences on one of their coat proteins. You isolate phages that bind to the immobilized cyclin A and sequence the segment of their coat protein gene that encodes the peptide. Sequences for

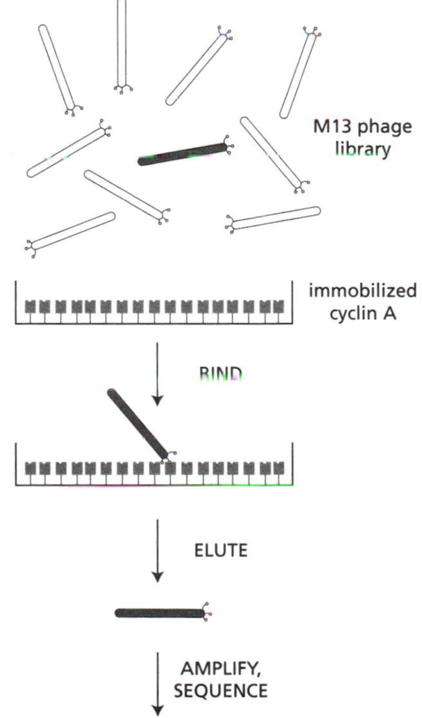


**Figure 8–12** Panning a phage-display library for phages that bind to immobilized cyclin A (Problem 8–50). The M13 phage library contains a randomized segment of the coat protein gene so that each phage displays one of a large number of possible amino acid sequences on its surface. The phage library is incubated with the immobilized protein in the plate, unbound phage are washed away, and bound phage are eluted and amplified by growth in *E. coli*. This cycle is repeated 2–3 more times and then individual phages are isolated, amplified, and analyzed by sequencing.

**Table 8–4 Peptides encoded by phages that were selected by panning of a phage-display library against immobilized cyclin A** (Problem 8–50).

| CLONE | LIBRARY | PEPTIDE SEQUENCE |
|---|---|---|
| 1 | 7-mer | LEPRMLF |
| 2 | 7-mer | TLPRQLF |
| 3 | 7-mer | LKPTKLF |
| 4 | 7-mer | LIPKNLF |
| 5 | 7-mer | FLPRALF |
| 1 | 12-mer | NVRVELFPPTKV |
| 2 | 12-mer | KSSVVRSLFVPT |
| 3 | 12-mer | ERPSAQRSLVFW |
| 4 | 12-mer | NLFYPRNLFPEF |
| 5 | 12-mer | YPSPARNLLPMF |
| 6 | 12-mer | ATIRELFPPTLP |
| 1 | 20-mer | HQPESVKRSLFKPAHSALEP |
| 2 | 20-mer | EVARRELFADHSLVHVGHVR |
| 3 | 20-mer | EHKALPGKAVTGPKRELVFQ |
| p27 cyclin A-binding sequence | | PSACRNLFGP |

14 clones are shown in Table 8–4. By comparing all 15 sequences, decide which amino acid residues are likely to be the most critical for binding to cyclin A.

**8–51** You want to use surface plasmon resonance (SPR) to measure the rate of dissociation of your protein from its ligand. You immobilize the ligand on the biosensor surface and allow a solution of your pure protein to flow across it until the binding reaches equilibrium, as indicated by the plateau value for the resonance angle (Figure 8–13). At this point you replace the protein solution with buffer. Now the resonance angle decreases as the protein dissociates from the ligand. When you put these data through the computer algorithm supplied with the machine, you are surprised to find that the dissociation curve has two components. The initial part of the curve is characterized by a high off rate, whereas the final part has a low off rate (Figure 8–13).

A. Why are you surprised that a protein–ligand complex should have two distinct dissociation rates?

B. Can you suggest some possible explanations for your results? What kinds of explanations might lead to two off rates, where only one is expected?

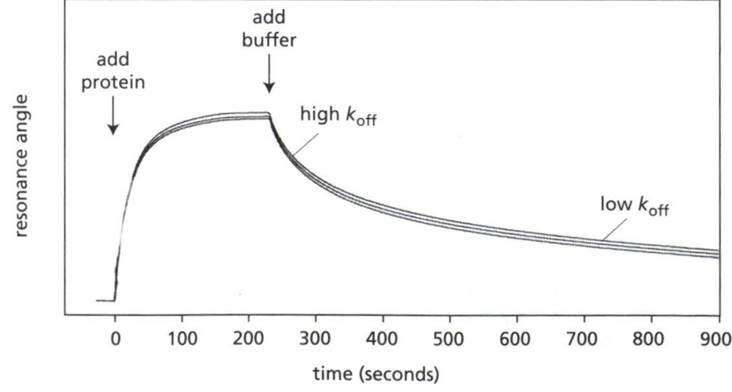


**Figure 8–13** SPR measurement of protein association with and dissociation from a ligand immobilized on a biosensor (Problem 8–51).

# ANALYZING AND MANIPULATING DNA

TERMS TO LEARN

| | | |
|---|---|---|
| bacterial artificial chromosome (BAC) | genetic engineering | polymerase chain reaction (PCR) |
| cDNA clone | genomic DNA clone | probe |
| cDNA library | genomic DNA library | recombinant DNA |
| dideoxy method | hybridization | recombinant DNA technology |
| DNA cloning | Northern blotting | restriction nuclease |
| expression vector | plasmid vector | Southern blotting |

## DEFINITIONS

Match each definition below with its term from the list above.

**8–52** A virus or plasmid that carries a DNA sequence into a suitable host cell and there directs the abundant synthesis of the protein encoded by the sequence.

**8–53** Any DNA molecule formed by joining DNA segments from different sources.

**8–54** Small circular DNA molecule that replicates independently of the genome and can be used for DNA cloning.

**8–55** A collection of clones that contain a variety of DNA segments from the genome of an organism.

**8–56** One of a large number of enzymes that can cleave a DNA molecule at any site where a specific short sequence of nucleotides occurs.

**8–57** Manipulation of DNA with precision in a test tube or an organism.

**8–58** A DNA clone of a DNA copy of an mRNA molecule.

**8–59** Technique in which RNA fragments, separated by electrophoresis, are immobilized on a paper sheet and then detected by hybridization with a labeled nucleic acid probe.

**8–60** Technique for generating multiple copies of specific regions of DNA by the use of sequence-specific primers and multiple cycles of DNA synthesis.

**8–61** Procaryotic cloning vector that can accommodate large pieces of DNA up to 1 million base pairs.

**8–62** The process whereby two complementary nucleic acid strands form a double helix.

## TRUE/FALSE

Decide whether each of these statements is true or false, and then explain why.

**8–63** Bacteria that make a specific restriction nuclease for defense against viruses have evolved in such a way that their own genome does not contain the recognition sequence for that nuclease.

**8–64** Pulsed-field gel electrophoresis uses a strong electric field to separate very long DNA molecules, stretching them out so that they travel end-first through the gel at a rate that depends on their length.

**8–65** Imagine that RNA or DNA molecules in a crude mixture are separated by electrophoresis and then hybridized to a probe. If molecules of only one size become labeled, one can be fairly certain that the hybridization was specific.

**8–66** By far the most important advantage of cDNA clones over genomic clones is that they can contain the complete coding sequence of a gene.

**8–67** If each cycle of PCR doubles the amount of DNA synthesized in the previous cycle, then 10 cycles will give a $10^3$-fold amplification, 20 cycles will give a $10^6$-fold amplification, and 30 cycles will give a $10^9$-fold amplification.

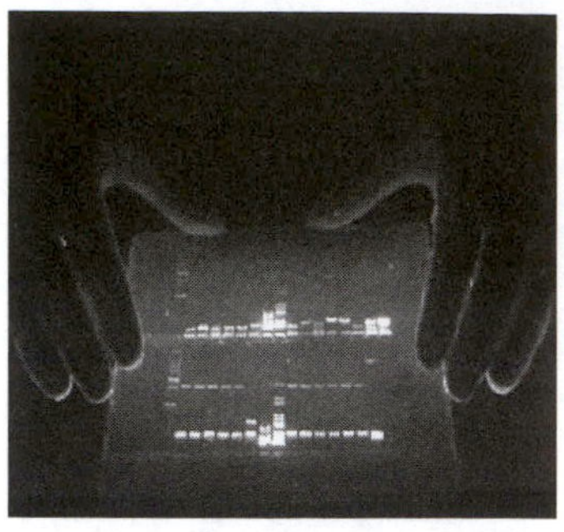

**Figure 8–14** DNA fragments separated by gel electrophoresis and stained with ethidium bromide (Problem 8–68). Each of the *bright bands* on the gel represents a site where fragments of DNA have migrated during electrophoresis. Ethidium intercalates between base pairs in double-stranded DNA. Removal of ethidium from the aqueous environment and fixing its orientation in the nonpolar environment of DNA enhance its fluorescence dramatically. When irradiated with long-wavelength UV light, it fluoresces a bright orange.

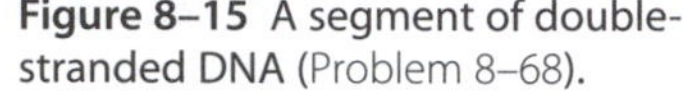


**Figure 8–15** A segment of double-stranded DNA (Problem 8–68).

## THOUGHT PROBLEMS

**8–68** Figure 8–14 shows a picture of DNA fragments that have been separated by gel electrophoresis and then stained by ethidium bromide, a molecule that fluoresces intensely under long wavelength UV light when it is bound to DNA. Such gels are a standard way of detecting the products of cleavage by restriction nucleases. For the DNA fragment shown in Figure 8–15, decide whether it will be cut by the restriction nucleases EcoRI (5′-GAATTC), AluI (5′-AGCT), and PstI (5′-CTGCAG). For those that cut the DNA, how many products will be produced?

**8–69** The restriction nucleases BamHI and PstI cut their recognition sequences as shown in Figure 8–16.

A. Indicate the 5′ and 3′ ends of the cut DNA molecules.

B. How would the ends be modified if you incubated the cut molecules with DNA polymerase in the presence of all four dNTPs?

C. After the reaction in part B, could you still join the BamHI ends together by incubation with T4 DNA ligase? Could you still join the PstI ends together? (T4 DNA ligase will join blunt ends together as well as cohesive ends.)

D. Will joining of the ends in part C regenerate the BamHI site? Will it regenerate the PstI site?

**8–70** The restriction nuclease EcoRI recognizes the sequence 5′-GAATTC and cleaves between the G and A to leave 5′ protruding single strands (like BamHI, see Figure 8–16). PstI, on the other hand, recognizes the sequence 5′-CTGCAG and cleaves between the A and G to leave 3′ protruding single strands (see Figure 8–16). These two recognition sites are displayed on the helical representations of DNA in Figure 8–17.

A. For each restriction site indicate the position of cleavage on each strand of the DNA.

B. From the positions of the cleavage sites decide for each restriction nuclease whether you expect it to approach the recognition site from the major-groove side or from the minor-groove side.

**8–71** Which, if any, of the restriction nucleases listed in Table 8–5 will *definitely* cleave a segment of cDNA that encodes the peptide KIGPACF? (See inside back cover for the genetic code.)

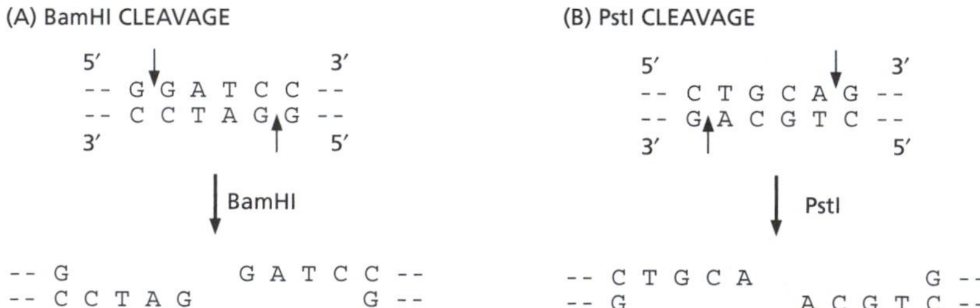


**Figure 8–16** Restriction nuclease cleavage of DNA (Problem 8–69). (A) BamHI cleavage. (B) PstI cleavage. Only the nucleotides that form the recognition sites are shown.

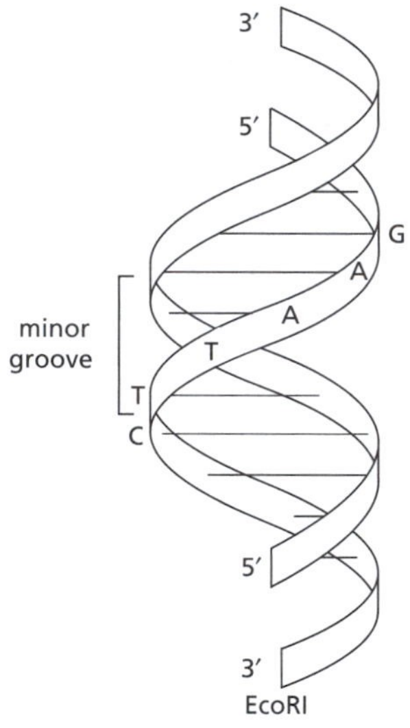


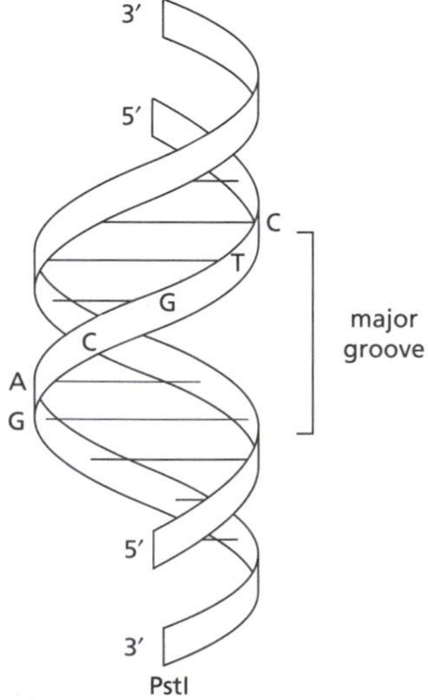


**Figure 8–17** Restriction sites on helical DNA (Problem 8–70).

**Table 8–5 A set of restriction nucleases and their recognition sequences** (Problem 8–71).

| RESTRICTION NUCLEASE | RECOGNITION SEQUENCE |
|---|---|
| AluI | AGCT |
| Sau96I | GGNCC |
| HindIII | AAGCTT |

N stands for any nucleotide.

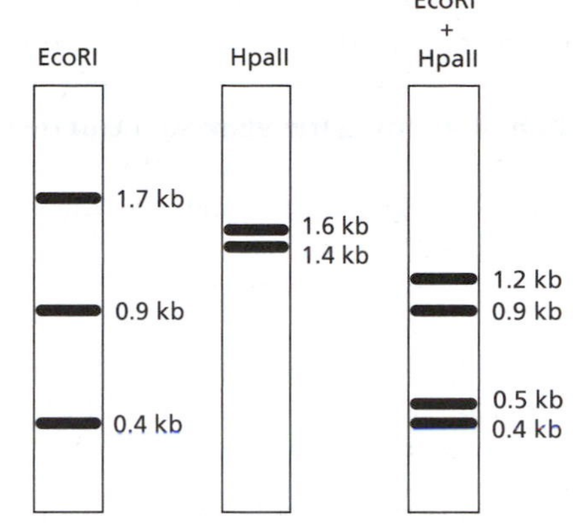

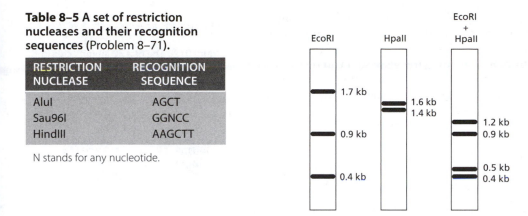

**Figure 8–18** Sizes of DNA bands produced by digestion of a 3.0-kb fragment by EcoRI, HpaII, and a mixture of the two (Problem 8–72). Sizes of the fragments are shown in kilobases.

**8–72** You wish to make a restriction map of a 3.0-kb BamHI restriction fragment. You digest three samples of the fragment with EcoRI, HpaII, and a mixture of EcoRI and HpaII. You then separate the fragments by gel electrophoresis and visualize the DNA bands by staining with ethidium bromide (Figure 8–18). From these results, prepare a restriction map that shows the relative positions of the EcoRI and HpaII recognition sites and the distances in kilobases (kb) between them.

**8–73** If you add DNA to wells at the top of a gel, should you place the positive electrode (anode) at the top or at the bottom of the gel? Explain your choice.

**8–74** You want to clone a DNA fragment that has KpnI ends into a vector that has BamHI ends. The problem is that BamHI and KpnI ends are not compatible: BamHI leaves a 5′ overhang and KpnI leaves a 3′ overhang (Figure 8–19). A friend suggests that you try to link them with an oligonucleotide 'splint' as shown in Figure 8–19. It is not immediately clear to you that such a scheme will work because ligation requires an adjacent 5′ phosphate and 3′ hydroxyl. Although molecules that are cleaved with restriction nucleases have appropriate ends, oligonucleotides are synthesized with hydroxyl groups at both ends. Also, although the junction shown in Figure 8–23 is BamHI–KpnI, the other junction is KpnI–BamHI, and you are skeptical that the same oligonucleotide could splint both junctions.

A. Draw a picture of the KpnI–BamHI junction and the oligo splint that would be needed. Is this oligonucleotide the same or different from the one shown in Figure 8–19?

B. Draw a picture of the molecule after treatment with DNA ligase. Indicate which if any of the nicks will be ligated.

C. Will your friend's scheme work?

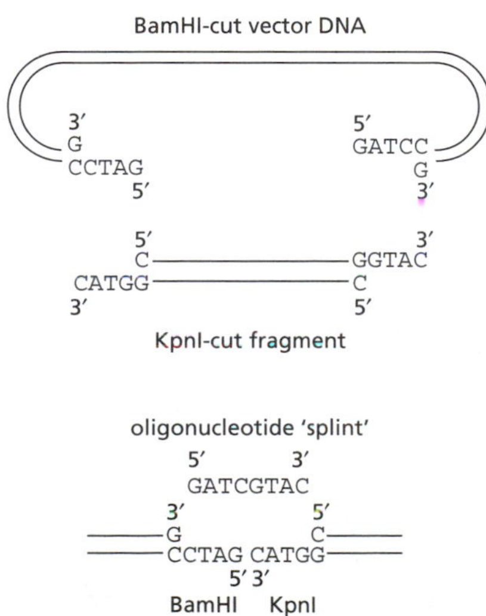


**Figure 8–19** Scheme to use an oligonucleotide 'splint' to link incompatible restriction ends (Problem 8–74).

**8–75** It's midnight. Your friend has awakened you with yet another grandiose scheme. He has spent the last two years purifying a potent modulator of the immune response. Tonight he got the first 30 amino acids of the sequence (Figure 8–20). He wants your help in cloning the gene so it can be expressed at high levels in bacteria. He argues that this protein, by stimulating the immune system, could be the ultimate cure for the common cold. He's already picked out a trade name—Immustim.

Even though he gets carried away at times, he is your friend, and you are intrigued by this idea. You promise to call him back in 15 minutes as soon as you have checked out the protein sequence. What two sets of degenerate 20-nucleotide-long oligonucleotide probes will you recommend to your friend as the best hybridization probes for screening a genomic DNA library?

```
        10         20         30
MFYWMIGRST EDWMPLYMKD FWAKHSLICE
```

**Figure 8–20** The first 30 amino acids in your friend's protein (Problem 8–75).

**Problems 7-92, 11-93, 15-45, 15-119, 17-12, and 17-36** provide examples of cloning from cDNA libraries

**8–76** You wish to know whether the cDNA you have isolated and sequenced is the product of a unique gene or is made by a gene that is a member of a family of related genes. To address this question, you digest cell DNA with a restriction nuclease that cleaves the genomic DNA but not the cDNA, separate the fragments by gel electrophoresis, and visualize bands using radioactive cDNA as a probe. The Southern blot shows two bands, one of which hybridizes more strongly to the probe than the other.

You interpret the stronger hybridizing band as the gene that encodes your cDNA and the weaker band as a related gene. When you explain your result to your advisor, she cautions that you have not proven that there are two genes. She suggests that you repeat the Southern blot in duplicate, probing one with a radioactive segment from the 5′ end of the cDNA and the other with a radioactive segment from the 3′ end of the cDNA.

A. How might you get two hybridizing bands if the cDNA was the product of a unique gene?

B. What results would you expect from the experiment your advisor proposed if there were a single unique gene? If there were two related genes?

**8–77** How would a DNA sequencing reaction be affected if the ratio of dideoxynucleoside triphosphates (ddNTPs) to deoxynucleoside triphosphates (dNTPs) were increased? What would the consequences be if the ratio were decreased?

**8–78** Discuss the following statement: "From the nucleotide sequence of a cDNA clone, the complete amino acid sequence of a protein can be deduced by applying the genetic code. Thus, protein biochemistry has become superfluous because there is nothing more that can be learned by studying the protein."

**8–79** DNA sequencing of your own two β-globin genes (one from each of your two copies of chromosome 11) reveals a mutation in one of the genes. Given this information alone, how much should you worry about being a carrier of an inherited disease that could be passed on to your children? What other information would you like to have to assess your risk?

**8–80** You want to amplify the DNA between the two stretches of sequence shown in Figure 8–21. Of the listed primers choose the pair that will allow you to amplify the DNA by PCR.

DNA to be amplified

```
5′-GACCTGTGGAAGC ———————————— CATACGGGATTGA-3′
3′-CTGGACACCTTCG ———————————— GTATGCCCTAACT-5′
```

primers

| | |
|---|---|
| (1) 5′-GACCTGTCCAAGC-3′ | (5) 5′-CATACGGGATTGA-3′ |
| (2) 5′-CTGGACACCTTCG-3′ | (6) 5′-GTATGCCCTAACT-3′ |
| (3) 5′-CGAAGGTGTCCAG-3′ | (7) 5′-TGTTAGGGCATAC-3′ |
| (4) 5′-GCTTCCACAGGTC-3′ | (8) 5′-TCAATCCCGTATG-3′ |

**Figure 8–21** DNA to be amplified and potential PCR primers (Problem 8–80).

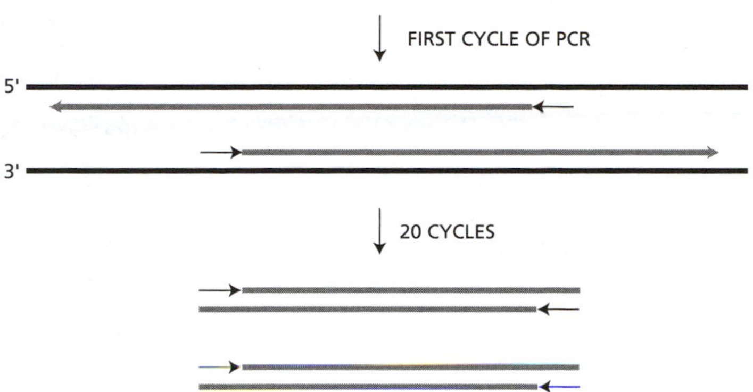


**Figure 8–22** Products of PCR after 1 and 20 cycles (Problem 8–81).

**8–81** In the very first round of PCR using genomic DNA, the DNA primers prime synthesis that terminates only when the cycle ends (or when a random end of DNA is encountered). Yet, by the end of 20 to 30 cycles—a typical amplification—the only visible product is defined precisely by the ends of the DNA primers (Figure 8–22). In what cycle is a double-stranded fragment of the correct size first generated?

**8–82** You want to express a rare human protein in bacteria so that you can make large quantities of it. To aid in its purification, you decide to add a stretch of six histidines to the N-terminus or the C-terminus of the protein. Such histidine-tagged proteins bind tightly to $Ni^{2+}$ columns but can be readily eluted with a solution of EDTA or imidazole. This procedure allows an enormous purification in one step.

The nucleotide sequence that encodes your protein is shown in Figure 8–23. Design a pair of PCR primers, each with 18 nucleotides of homology to the gene, that will amplify the coding sequence and add an initiation codon followed by six histidine codons to the N-terminus. Design a pair of primers that will add six histidine codons followed by a stop codon to the C-terminus.

**8–83** You have now cloned in an expression vector both versions of the histidine-tagged protein you created in Problem 8–82. Neither construct expresses particularly strongly in bacteria, but the product is soluble. You pass the crude extract over a $Ni^{2+}$ affinity column, which binds histidine-tagged proteins specifically. After washing the column extensively, you elute your protein from the column using a solution containing imidazole (Figure 8–24), which releases your protein.

When you subject the eluted protein to electrophoresis and stain the gel for protein, you are pleased to find bands in the eluate that are not present when control bacteria are treated similarly. But you are puzzled to see that the construct tagged at the N-terminus gives a ladder of shorter proteins below the full-length protein, whereas the C-terminally tagged construct yields exclusively the full-length protein. The amount of full-length protein is about the same for each construct.

**Problems 3–110, 7–66, 12–98, 13–32, 15–119,** and **15–148** illustrate experimental uses for a variety of protein tags.

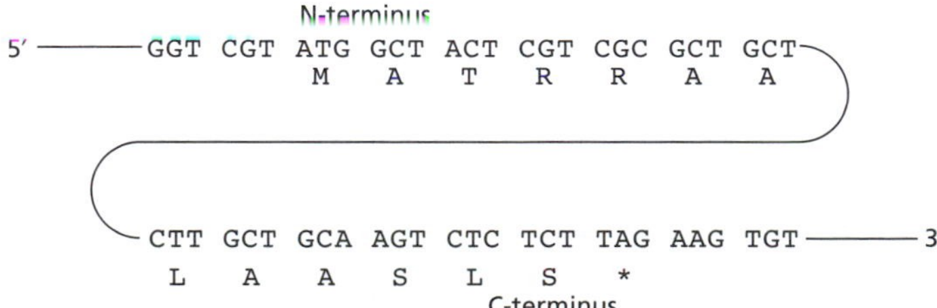

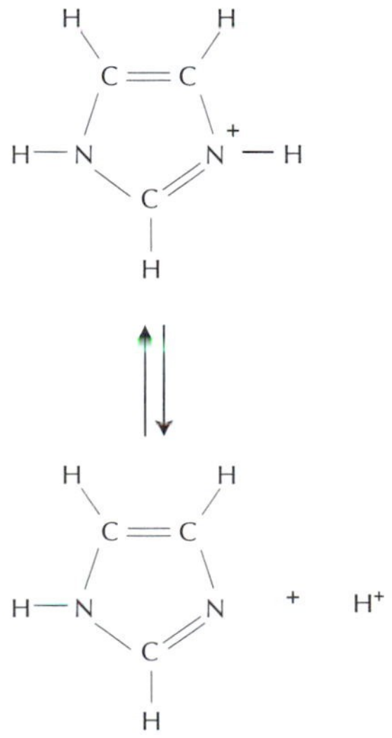

**Figure 8–23** Nucleotide sequence around the N- and C-termini of the protein you want to modify (Problem 8–82). The encoded amino acid sequence is indicated *below* each codon using the one-letter code. The *asterisk* (*) indicates the stop codon. Only the top strand of the double-stranded DNA is shown.

**Figure 8–24** Structure of imidazole (Problem 8–83).

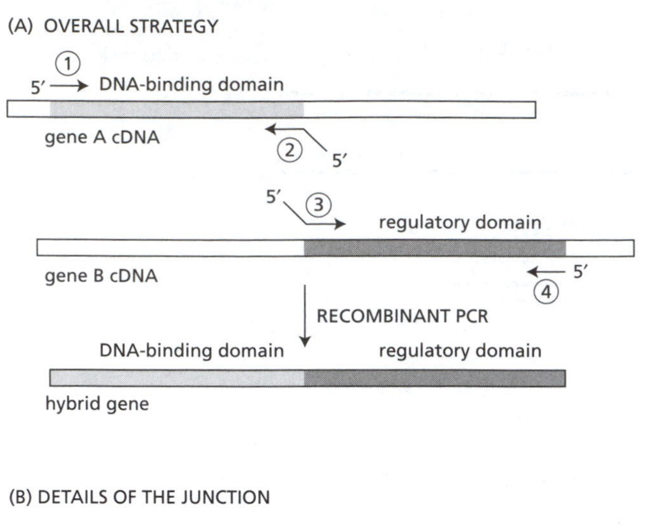


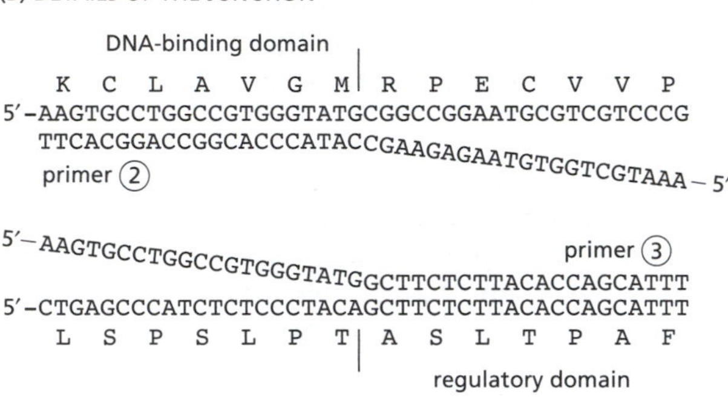


**Figure 8–25** Recombinant PCR (Problem 8–84). (A) Overall strategy. (B) Details of the junction. Note that primers 2 and 3 are complementary to one another over their entire length. Only one strand of the cDNA is shown; it is the strand with the same sequence as the mRNA. Primer 2 will pair with the cDNA strand that is shown. Primer 3 will pair with the complement of that strand.

**Problems 12–52 and 12–132** present two of the earliest uses of synthetic gene fusions to investigate biological questions.

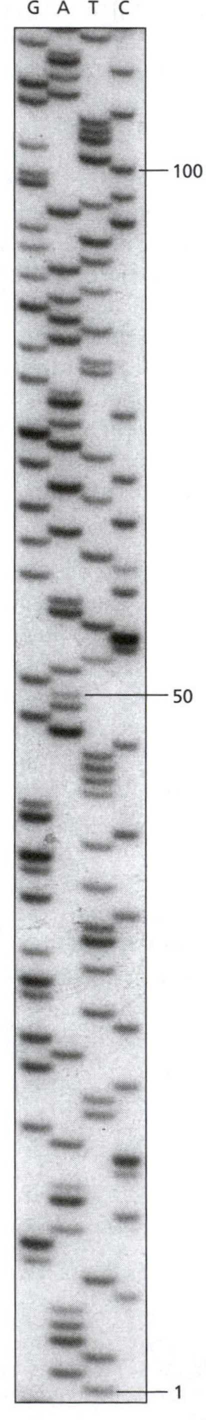


A. Why does a solution of imidazole release a histidine-tagged protein from the $Ni^{2+}$ column?
B. Offer an explanation for the difference in the products generated by the two constructs.

**8–84** An adaptation of standard PCR, called recombinant PCR, allows virtually any two nucleotide sequences to be joined any way you want. Imagine, for example, that you want to combine the DNA-binding domain of protein A to the regulatory domain of protein B in order to test your conjectures about how these proteins work. The target domains in the cDNAs and the arrangement of PCR primers needed to join the domains are shown in Figure 8–25.

Recombinant PCR is usually carried out in two steps. In the first step, PCR primers 1 and 2 are used to amplify the target segment of gene A, and in a separate reaction, primers 3 and 4 are used to amplify the target sequence in gene B. In the second step, the individually amplified products, separated from their primers, are mixed together and amplified using primers 1 and 4. *Voila!* The desired hybrid gene is the major product.

A. Explain how recombinant PCR manages to link the two gene segments together.
B. Illustrate schematically the structure and arrangement of primers you would use to put the regulatory domain of gene B at the N-terminus of the hybrid protein.

**8–85** A clear example of a dideoxy sequencing gel is shown in Figure 8–26. Try reading it. As read from the bottom of the gel to the top, the sequence corresponds to the mRNA for a protein. Can you find the open reading frame in this sequence? What protein does it code for?

**Figure 8–26** A dideoxy sequencing gel of a cloned segment of DNA (Problem 8-85). The lanes are labeled G, A, T, and C to indicate which ddNTP was included in the reaction.

**8–86** After decades of work, an old classmate of yours has isolated a small amount of attractase, an enzyme producing a powerful human pheromone, from hair samples of Hollywood celebrities. To produce attractase for his personal use, he obtained a complete genomic clone of the attractase gene, connected it to a strong bacterial promoter on an expression plasmid, and introduced the plasmid into *E. coli* cells. He was devastated to find that no attractase was produced in the cells. What is the likely explanation for his failure?

## CALCULATIONS

**8–87** The restriction nuclease Sau3A recognizes the sequence 5′-GATC and cleaves on the 5′ side (to the left) of the G. (Since the top and bottom strands of most restriction sites read the same in the 5′-to-3′ direction, only one strand of the site need be shown.) The single-stranded ends produced by Sau3A cleavage are identical to those produced by BamHI cleavage (see Figure 8–16), allowing the two types of ends to be joined together by incubation with DNA ligase. (You may find it helpful to draw out the product of this ligation to convince yourself that it is true.)

A. What fraction of BamHI sites can be cut with Sau3A? What fraction of Sau3A sites can be cut with BamHI?

B. If two BamHI ends are ligated together, the resulting site can be cleaved again by BamHI. The same is true for two Sau3A ends. Suppose you ligate a Sau3A end to a BamHI end. Can the hybrid site be cut with Sau3A? Can it be cut with BamHI?

C. What do you suppose is the average size of DNA fragments produced by digestion of chromosomal DNA with Sau3A? What's the average size with BamHI?

**8–88** You are constructing a cDNA library in a high-efficiency cloning vector called λYES (Yeast–*E. coli* Shuttle). This vector links a bacteriophage lambda genome to a plasmid that can replicate in *E. coli* and yeast. It combines the advantages of viral and plasmid cloning vectors. cDNAs can be inserted into the plasmid portion of the vector, which can then be packaged into a virus coat *in vitro*. The packaged vector DNA infects *E. coli* much more efficiently than plasmid DNA on its own. Once inside *E. coli*, the plasmid sequence in λYES can be induced to recombine out of the lambda genome and replicate on its own. This allows cDNAs to be isolated on plasmids, which are ideal for subsequent manipulations.

To maximize the efficiency of cloning, both the vector and the cDNAs are prepared in special ways. A preparation of double-stranded cDNAs with blunt ends is ligated to the blunt end of a double-stranded oligonucleotide adaptor composed of paired 5′-CGAGATTTACC and 5′-GGTAAATC oligonucleotides, each of which carries a phosphate at its 5′ end. The vector DNA is cut at its unique XhoI site (5′-C*TCGAG) and then incubated with DNA polymerase in the presence of dTTP only. The vector DNA and cDNAs are then mixed and ligated together. This procedure turns out to be very efficient. Starting with 2 μg of vector and 0.1 μg of cDNA, you make a library consisting of $4 \times 10^7$ recombinant molecules.

A. Given that the vector is 43 kb long and the average size of the cDNAs is about 1 kb, estimate the ratio of vector molecules to cDNA molecules in the ligation mixture.

B. What is the efficiency with which vector molecules are converted to recombinant molecules in this procedure? (The average mass of a nucleotide pair is 660 daltons.)

C. Explain how the treatments of the vector molecules and the cDNAs allow them to be ligated together. Can the treated vector molecules be ligated to one another? Can the cDNAs be ligated to one another?

D. How does the treatment of the vector and cDNAs improve the efficiency of generating recombinant DNA molecules?

E. Can a cDNA be cut out of a recombinant plasmid with XhoI?

protein sequence

M Q K F N

ATGCA(A/G)AA(A/G)TT(T/C)AA(T/C)

degenerate oligonucleotide

**Figure 8–27** A degenerate oligonucleotide probe for the Factor VIII gene based on a stretch of amino acids from the protein (Problem 8–90). Because more than one DNA triplet can encode each amino acid, a number of different nucleotide sequences are possible for each amino acid sequence. Although only one of these sequences will actually code for the protein in the genomic DNA, it is impossible to tell in advance which one it is. Therefore, a mixture of the possible sequences—called a degenerate oligonucleotide probe—is used to search a genomic library for the gene.

**8–89** To prepare a genomic DNA library, it is necessary to fragment the genome so that it can be cloned in a vector. A common method is to use a restriction nuclease.

A. How many different DNA fragments would you expect to obtain if you cleaved human genomic DNA with Sau3A (5′-GATC)? (Recall that there are $3.2 \times 10^9$ base pairs in the haploid human genome.) How many would you expect to get with EcoRI (5′-GAATTC)?

B. Human genomic libraries are often made from fragments obtained by cleaving human DNA with Sau3A in such a way that the DNA is only partially digested; that is, so that not all the Sau3A sites have been cleaved. What is a possible reason for doing this?

**8–90** A degenerate set of oligonucleotide probes for the Factor VIII gene for blood clotting is shown in Figure 8–27. Each of these probes is only 15 nucleotides long. On average, how many exact matches to any single 15-nucleotide sequence would you expect to find in the human genome ($3.2 \times 10^9$ base pairs)? How many matches to the collection of sequences in the degenerate oligonucleotide probe would you expect to find? How might you determine that a match corresponds to the Factor VIII gene?

## DATA HANDLING

**8–91** You have purified two DNA fragments that were generated by BamHI digestion of recombinant DNA plasmids. One fragment is 400 nucleotide pairs, and the other is 900 nucleotide pairs. You want to join them together as shown in Figure 8–28 to create a hybrid gene, which, if your speculations are right, will have amazing new properties.

You mix the two fragments together and incubate them in the presence of DNA ligase. After 30 minutes and again after 8 hours, you remove samples and analyze them by gel electrophoresis. You are surprised to find a complex pattern of fragments instead of the 1.3-kb recombinant molecule of interest (Figure 8–29A). You notice that with longer incubation the smaller fragments diminish in intensity and the larger ones become more intense. If you cut the ligated mixture with BamHI, you regenerate the starting fragments (Figure 8–13A).

Puzzled, but undaunted, you purify the 1.3-kb fragment from the gel (arrow in Figure 8–29A) and check its structure by digesting a sample of it with BamHI. As expected, the original two bands are regenerated (Figure 8–29B). Just to be sure it is the structure you want, you digest another sample with EcoRI. You expected this digestion to generate two fragments of 300 nucleotides and one fragment of 700 nucleotides. Once again you are surprised by the complexity of the gel pattern (Figure 8–29B).

A. Why are there so many bands in the original ligation mixture?

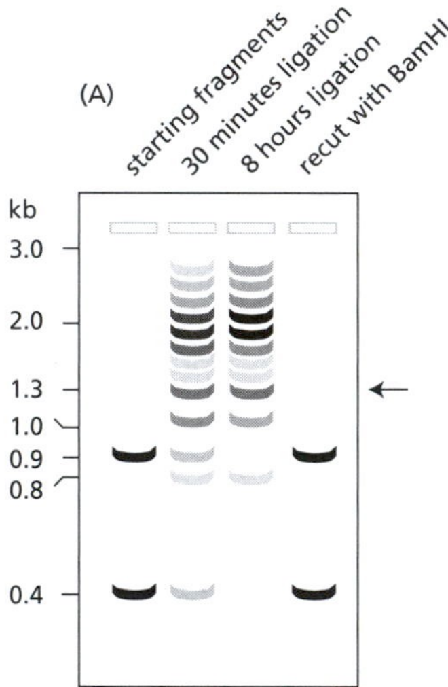


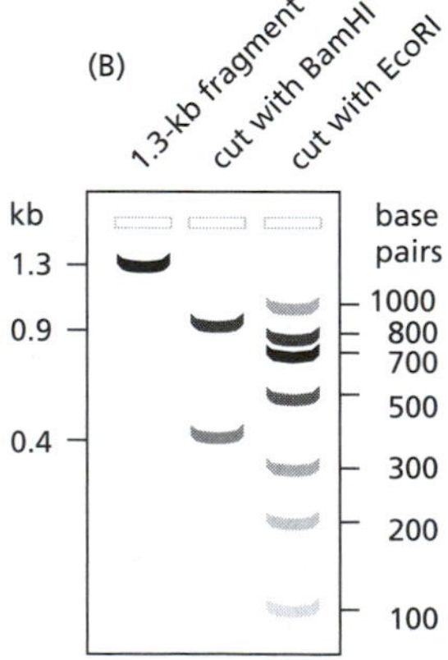


**Figure 8–29** Construction of the hybrid gene (Problem 8–91). (A) Ligation of pure DNA 0.9-kb and 0.4-kb fragments. The position of the 1.3-kb fragment is marked by an *arrow*. (B) Diagnostic digestion of the purified 1.3-kb fragment.

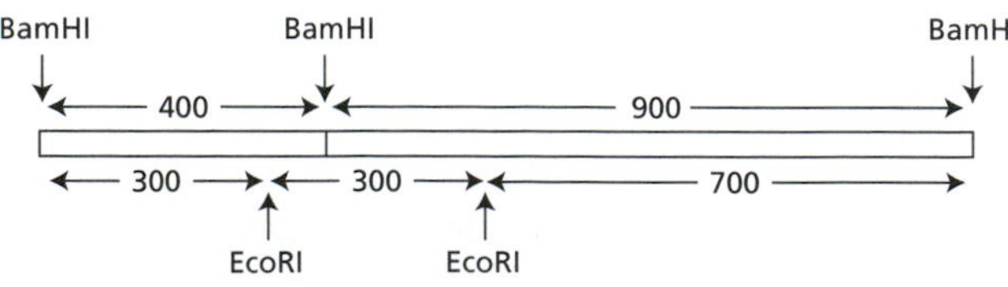


**Figure 8–28** Final structure of the desired hybrid gene (Problem 8–91).

B. Why are so many fragments produced by EcoRI digestion of the pure 1.3-kb fragment?

**8–92** *Tetrahymena* is a ciliated protozoan with two nuclei. The smaller nucleus (the micronucleus) maintains a master copy of the cell's chromosomes. The micronucleus participates in sexual conjugation, but not in day-to-day gene expression. The larger nucleus (the macronucleus) maintains a 'working' copy of the cell's genome in the form of a large number of gene-sized double-stranded DNA fragments (minichromosomes), which are actively transcribed. The minichromosome that contains the ribosomal RNA genes is present in many copies. It can be separated from the other minichromosomes by centrifugation and studied in detail.

When examined by electron microscopy, each ribosomal minichromosome is a linear structure 21 kb in length. Ribosomal minichromosomes also migrate at 21 kb when subjected to gel electrophoresis (Figure 8–30, lane 1). If the minichromosome is cut with the restriction nuclease BglII, the two fragments that are generated (13.4 kb and 3.8 kb) do not sum to 21 kb (lane 2). When the DNA is cut with other restriction enzymes, the sizes of the fragments always sum to less than 21 kb; moreover, the fragments in each digest add up to different overall lengths.

If uncut minichromosomes are first denatured and reannealed before they are run on a gel, the 21-kb fragments are replaced by double-stranded fragments exactly half their length, 10.5 kb (Figure 8–30, lane 3). Similarly, if the BglII-cut minichromosomes are denatured and reannealed, the 13.4-kb fragments are replaced by double-stranded fragments half their length, 6.7 kb (lane 4).

Explain why the restriction fragments do not appear to add up to 21 kb. Why does the electrophoretic pattern change when the DNA is denatured and reannealed? What do you think might be the overall organization of sequences in the ribosomal minichromosome?

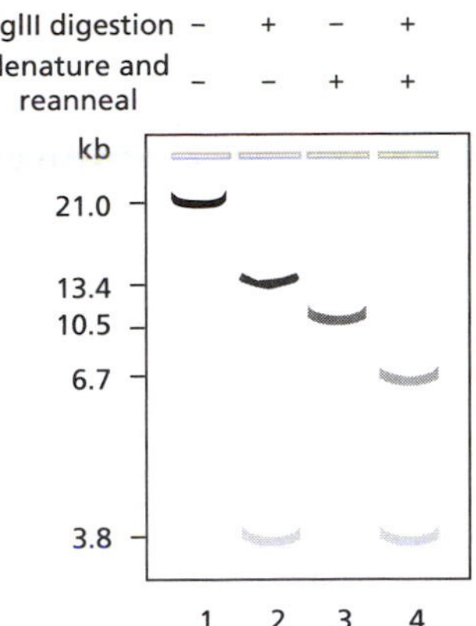


**Figure 8–30** Restriction analysis of the *Tetrahymena* ribosomal minichromosome (Problem 8–92). Numbers at the *left* indicate the sizes of the bands in kilobases.

**8–93** You have cloned a 4-kb segment of a gene into a plasmid vector (Figure 8–31) and now wish to prepare a restriction map of the gene in preparation for other DNA manipulations. Your advisor left instructions on how to do it, but she is now on vacation, so you are on your own. You follow her instructions, as outlined below.

1. Cut the plasmid with EcoRI.
2. Add a radioactive label to the EcoRI ends.
3. Cut the labeled DNA with BamHI.
4. Purify the insert away from the vector.
5. Digest the labeled insert briefly with a restriction nuclease so that on average each labeled molecule is cut about one time.
6. Repeat step 5 for several different restriction nucleases.
7. Run the partially digested samples side by side on an agarose gel.
8. Place the gel against x-ray film so that fragments with a radioactive end can expose the film to produce an autoradiograph.
9. Draw the restriction map.

Your biggest problem thus far has been step 5; however, by decreasing the amounts of nuclease and lowering the temperature, you were able to find conditions for partial digestion. You have now completed step 8, and your autoradiograph is shown in Figure 8–32.

Unfortunately, your advisor was not explicit about how to construct a map from the data in the autoradiograph. She is due back tomorrow. Will you figure it out in time?

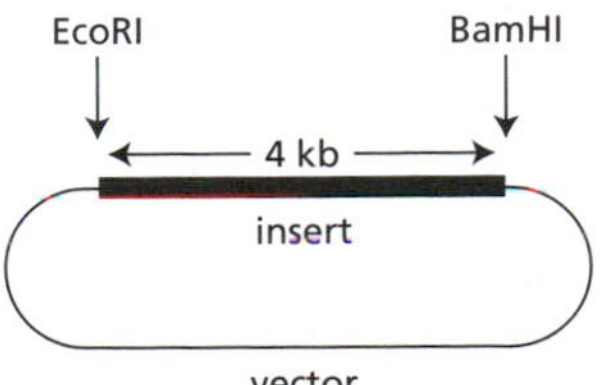


**Figure 8–31** Recombinant plasmid containing a cloned DNA segment (Problem 8–93).

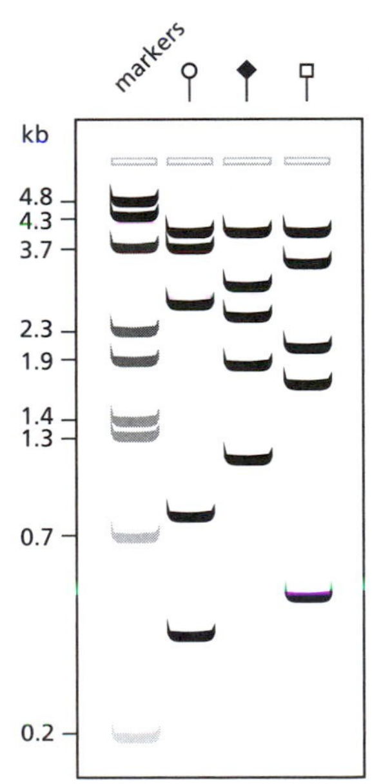


**Figure 8–32** Autoradiograph showing the electrophoretic separation of the labeled fragments after partial digestion with the three restriction nucleases represented by the *symbols* (Problem 8–93). Numbers at the *left* indicate the sizes of a set of marker fragments in kilobases.

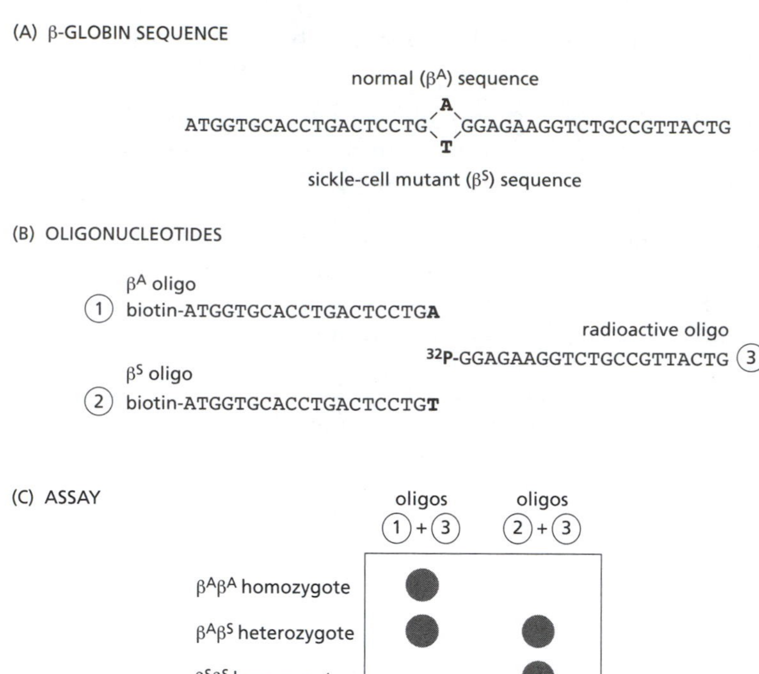


**Figure 8–33** Oligonucleotide-ligation assay (Problem 8–94). (A) Sequence of the β-globin gene around the site of the sickle-cell ($\beta^S$) mutation. The normal $\beta^A$ sequence carries an A at the central position; the sickle cell mutant $\beta^S$ sequence has a T instead. (B) Specific oligonucleotides for ligation assay. (C) Assays to detect the single-nucleotide difference between the $\beta^A$ and $\beta^S$ sequences. After hybridization to patient DNA, biotinylated oligonucleotides were collected in a spot on a sheet of filter paper and exposed to x-ray film to detect radioactivity, which turns the film black.

**8–94** Many mutations that cause human genetic diseases involve the substitution of one nucleotide for another, as is the case for sickle-cell anemia (Figure 8–33A). An assay based on ligation of oligonucleotides provides a rapid way to detect such specific single-nucleotide differences. This assay uses pairs of oligonucleotides: for each pair, one oligonucleotide is labeled with biotin and the other with a radioactive (or fluorescent) tag. In the assay shown in Figure 8–33B for the detection of the mutation responsible for sickle-cell anemia, two pairs of oligonucleotides are hybridized to DNA from an individual and incubated in the presence of DNA ligase. Biotinylated oligonucleotides are then bound to streptavidin on a solid support and any associated radioactivity is visualized by autoradiography, as shown in Figure 8–33C.

A. Do you expect the $\beta^A$ and $\beta^S$ oligonucleotides to hybridize to both $\beta^A$ and $\beta^S$ DNA?

B. How does this assay distinguish between $\beta^A$ and $\beta^S$ DNA?

**Problems 4–73, 7–12, and 20–43** provide additional examples of Southern blotting.

**8–95** The DNA of certain animal viruses can integrate into a cell's DNA as shown schematically in Figure 8–34. You want to know the structure of the viral genome as it exists in the integrated state. You digest samples of viral DNA and DNA from cells that contain the integrated virus with restriction nucleases that cut the viral DNA at known sites (Figure 8–35A). Subsequently you separate the fragments by electrophoresis on agarose gels and visualize the bands that contain viral DNA by Southern blotting, using radioactive viral DNA as a hybridization probe. You obtain the patterns shown in Figure 8–35B.

From this information, decide in which of the five segments of the viral genome (labeled *a* to *e* in Figure 8–35A) the integration event occurred.

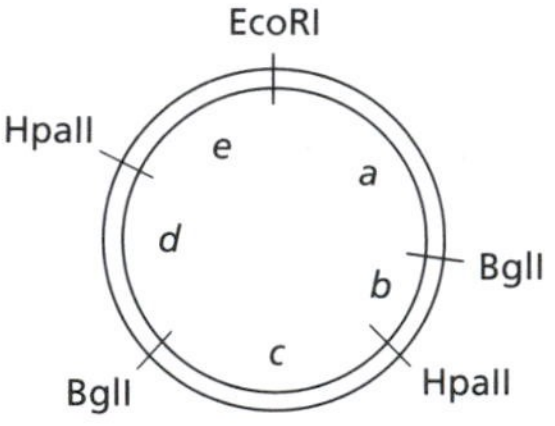


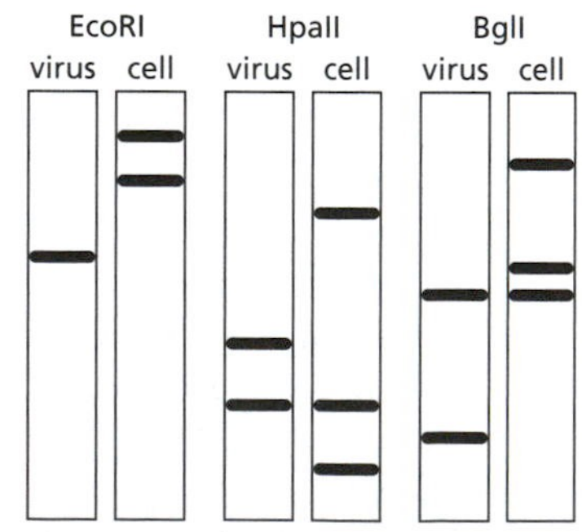


**Figure 8–35** Southern blots of viral and cell DNA digested with various restriction nucleases and incubated with a viral DNA probe (Problem 8–95). (A) Restriction sites on the viral genome. The DNA segments defined by these sites are indicated by the letters *a* to *e*. (B) Restriction digests of viral DNA and cellular DNA. Agarose gels separate DNA fragments on the basis of size—the smaller the fragment, the farther it moves toward the *bottom* of the gel.

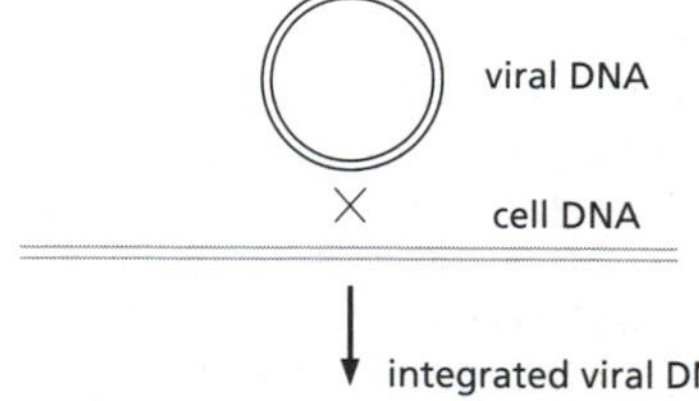


**Figure 8–34** Integration of viral DNA into cell DNA (Problem 8–95).

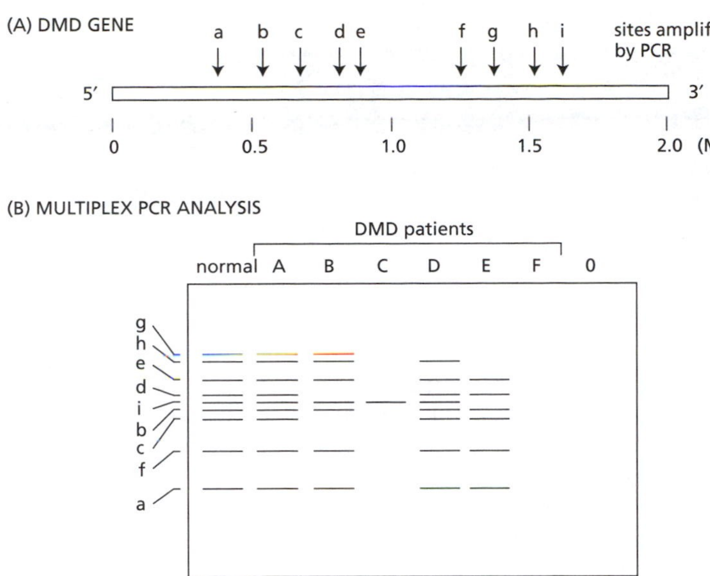


**Figure 8–36** Multiplex PCR analysis of six DMD patients (Problem 8–96). (A) The DMD gene with the nine sites amplified by PCR indicated by *arrows*. The sizes of the PCR products are so small on this scale that their location is simply indicated. (B) Agarose gel display of amplified PCR products. 'Normal' indicates a normal male. The lane marked '0' shows a negative control with no added DNA.

**8–96** Duchenne's muscular dystrophy (DMD) is among the most common human genetic diseases, affecting approximately 1 in 3500 male births. One-third of all new cases arise via new mutations. The DMD gene, which is located on the X chromosome, is greater than 2 million base pairs in length and contains at least 70 exons. Large deletions account for about 60% of all cases of the disease, and they tend to be concentrated around two regions of the gene.

The very large size of the DMD gene complicates the analysis of mutations. One rapid approach, which can detect about 80% of all deletions, is termed multiplex PCR. It uses multiple pairs of PCR primers to amplify nine different segments of the gene in the two most common regions for deletions (Figure 8–36A). By arranging the PCR primers so that each pair gives a different size product, it is possible to amplify and analyze all nine segments in one PCR reaction. An example of multiplex PCR analysis of six unrelated DMD males is shown in Figure 8–36B.

A. Describe the extent of the deletions, if any, in each of the six DMD patients.
B. What additional control might you suggest to confirm your analysis of patient F?

**8–97** You want to clone a cDNA into an expression vector so you can make large amounts of the encoded protein in *E. coli.* The cDNA is flanked by BamHI sites and you plan to insert it at the BamHI site in the vector. This is your first experience with cloning, so you carefully follow the procedures in the cloning manual.

The manual recommends that you cleave the vector DNA and then treat it with alkaline phosphatase to remove the 5′ phosphates. The next step is to mix the treated vector with the BamHI-cut cDNA fragment and incubate with DNA ligase. After ligation the DNA is mixed with bacterial cells that have been treated to make them competent to take up DNA. Finally, the mixture is spread onto culture dishes filled with a solid growth medium that contains an antibiotic that kills all cells that have not taken up the vector. The vector allows cells to survive because it carries a gene for resistance to the antibiotic.

The cloning manual also suggests four controls.

**Control 1.** Plate bacterial cells that have not been exposed to any vector onto the culture dishes.

**Control 2.** Plate cells that have been transfected with vector that has not been cut.

**Control 3.** Plate cells that have been transfected with vector that has been cut (but not treated with alkaline phosphatase) and then incubated with DNA ligase (in the absence of the cDNA fragment).

**Table 8–6 Results of your cDNA cloning endeavor** (Problem 8–97).

| PREPARATION OF SAMPLE | RESULTS OF EXPERIMENTS | | |
|---|---|---|---|
| | 1 | 2 | 3 |
| Control 1: cells alone | TMTC | 0 | 0 |
| Control 2: uncut vector | TMTC | 0 | >1000 |
| Control 3: omit phosphatase, omit cDNA | TMTC | 0 | 435 |
| Control 4: omit cDNA | TMTC | 0 | 25 |
| Experimental sample | TMTC | 0 | 34 |

TMTC means too many to count.

**Control 4.** Plate cells that have been transfected with vector that has been cut and treated with alkaline phosphatase and then incubated with DNA ligase (in the absence of the cDNA fragment).

For your first attempt at the experiment and all its controls, you borrow a fellow student's competent cells, but all the plates have too many colonies to count (Table 8–6). For your second attempt, you use cells you have prepared yourself, but this time you get no colonies on any plate (Table 8–6). Ever the optimist, you try again, and this time you are rewarded with more encouraging results (Table 8–6). You pick 12 colonies from the experimental sample, prepare plasmid DNA from them, and digest the DNA with BamHI. Nine colonies yield a single band the same size as the linearized vector, but three colonies have, in addition, a fragment the size of the cDNA you wanted to clone. Success is sweet!

A. What do you think happened in the first experiment? What is the point of control 1?
B. What do you think happened in the second experiment? What is the point of control 2?
C. What is the point of doing controls 3 and 4?
D. Why does the cloning manual recommend treating the vector with alkaline phosphatase?

# STUDYING GENE EXPRESSION AND FUNCTION

TERMS TO LEARN

| | | |
|---|---|---|
| allele | genetics | quantitative RT-PCR |
| complex trait | genotype | reverse genetics |
| complementation test | haplotype block | RNA interference (RNAi) |
| conditional mutation | haplotype map (hapmap) | site-directed mutagenesis |
| Cre/lox | *in situ* hybridization | totipotent |
| DNA microarray | knockout mouse | transgene |
| epistasis analysis | phenotype | transgenic organism |
| genetic screen | polygenic | |

## DEFINITIONS

Match each definition below with its term from the list above.

**8–98** Describes an inherited characteristic that is influenced by multiple genes, each of which makes a small contribution to the phenotype.

**8–99** A search through a large collection of mutants for a mutant with a particular phenotype.

**8–100** Description of a cell that has the ability to give rise to all parts of the organism.

**8–101** One of a set of alternative forms of a gene.

**8–102** The observable character of a cell or an organism.

**8–103** A mouse in which both copies of a gene have been inactivated by gene targeting.

**8–104** Method for inactivating genes that introduces into cells a double-stranded RNA that is processed and hybridized to a targeted mRNA, directing its degradation.

**8–105** Ancestral chromosome segment that has been inherited with little genetic rearrangement across the generations.

**8–106** Animal or plant that has been permanently engineered by gene deletion, gene insertion, or gene replacement.

**8–107** The genetic constitution of an individual cell or organism.

## TRUE/FALSE

Decide whether each of these statements is true or false, and then explain why.

**8–108** In an organism whose genome has been sequenced and is publically available, identifying the mutant gene responsible for an interesting phenotype is as easy for mutations induced by chemical mutagenesis as it is for those generated by insertional mutagenesis.

**8–109** Loss-of-function mutations are usually recessive.

**8–110** If two mutations have a synthetic phenotype, it usually means that the mutations are in genes whose products operate in the same pathway.

## THOUGHT PROBLEMS

**8–111** Distinguish between the following genetic terms:

A. Locus and allele
B. Homozygous and heterozygous
C. Genotype and phenotype
D. Dominant and recessive

**8–112** Explain the difference between a gain-of-function mutation and a dominant-negative mutation. Why are both these types of mutation usually dominant?

**Problems 12–53, 12–54, 12–77, 12–96,** and **12–131** describe various genetic screens and genetic selections that have been used to find mutants defective in particular cellular processes.

**8–113** Discuss the following statement: "We would have no idea today of the importance of insulin as a regulatory hormone if its absence were not associated with the devastating human disease diabetes. It is the dramatic consequences of its absence that focused early efforts on the identification of insulin and the study of its normal role in physiology."

**8–114** What are single-nucleotide polymorphisms (SNPs), and how can they be used to locate a mutant gene by linkage analysis?

**8–115** Fatty acid synthase in mammalian cells is encoded by a single gene. This remarkable protein carries out seven distinct biochemical reactions. The mammalian fatty acid synthase gene is homologous to seven different *E. coli* genes, each of which encodes one of the functions of the mammalian protein. Do you think it is likely that the proteins in *E. coli* function together as a complex? Why or why not?

**8–116** How does reverse genetics differ from standard genetics?

**8–117** One of the first organisms that was genetically engineered using modern DNA technology was a bacterium that lives on the surface of strawberry plants. This bacterium normally makes a protein, called ice-protein, which causes the efficient formation of ice crystals around it when the temperature drops to just below freezing. Thus, strawberries harboring this bacterium are particularly susceptible to frost damage because the ice crystals destroy their cells. Strawberry farmers have a strong financial interest in preventing such damage.

A genetically engineered version of this bacterium was constructed with the ice-protein gene knocked out. The mutant bacteria were then introduced in large numbers into strawberry fields, where they displaced the normal bacteria. This approach has been successful: strawberries bearing the mutant bacteria show a much reduced susceptibility to frost damage.

At the time they were carried out, the open-field trials triggered an intense debate because they represented the first release into the environment of an organism that had been genetically engineered using recombinant DNA techniques. All preliminary experiments were carried out with extreme caution and in strict containment.

Discuss some of the issues that arise from such applications of DNA technology. Do you think that bacteria lacking the ice-protein could be isolated without the use of modern DNA technology? Is it likely that such mutations have already occurred in nature? Would the use of a mutant bacterial strain isolated from nature be of lesser concern? Should we be concerned about the risks posed by the application of genetic engineering techniques in agriculture, medicine, and technology? Explain your answers.

```
   L  R  D  P  Q  G  G  V  I
5'-CTTAGAGACCCGCAGGGCGGCGTCATC-3'
```

**Figure 8–37** Sequence of DNA and the encoded protein (Problem 8–119).

**8–118** The cells in an individual animal contain nearly identical genomes. In an experiment, a tissue composed of multiple cell types is fixed and subjected to *in situ* hybridization with a DNA probe to a particular gene. To your surprise, the hybridization signal is much stronger in some cells that in others. Explain this result.

**Problem 15–152** employs *in situ* hybridization in *Drosophila* embryos to investigate cell signaling in development.

**8–119** From previous work, you suspect that the glutamine (Q) in the protein segment in Figure 8–37 plays an important role at the active site. Your advisor wants you to alter the protein in three ways: change the glutamine to lysine (K), change the glutamine to glycine (G), and delete the glutamine from the protein. You plan to accomplish these mutational alterations on a version of your gene that is cloned into M13 viral DNA. You want to hybridize an appropriate oligonucleotide to the M13 viral DNA, so that when DNA polymerase extends the oligonucleotide around the single-stranded M13 circle, it will complete a strand that encodes the complement of the desired mutant protein. Design three 20-nucleotide-long oligonucleotides that could be hybridized to the cloned gene on single-stranded M13 viral DNA as the first step in effecting the mutational changes.

**8–120** A mutation engineered *in vitro* as in Problem 8–119 introduces a mismatch into the DNA. It the mismatched DNA were introduced into cells and immediately replicated, it would be expected to generate an equal mixture of wild-type and mutant genes (Figure 8–38). If the mismatch DNA were not

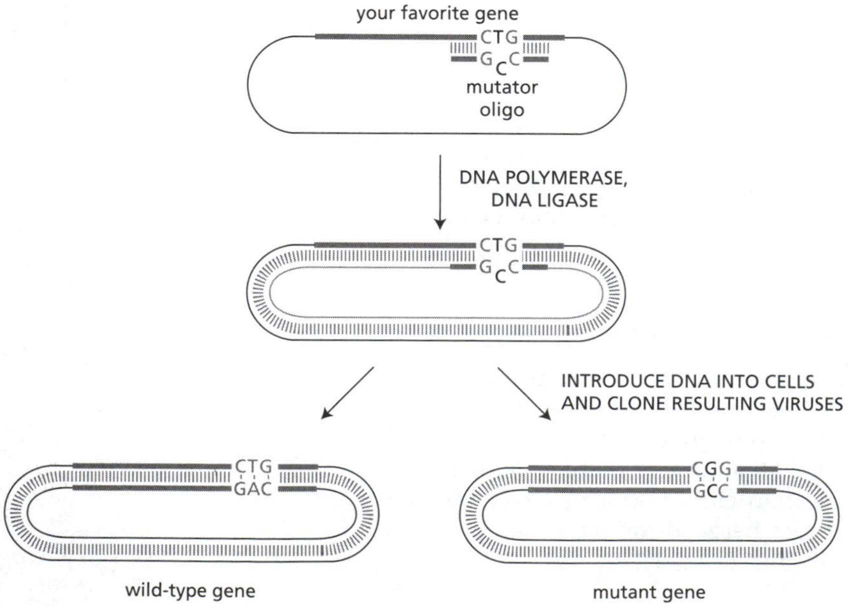

**Figure 8–38** Site-directed mutagenesis (Problem 8–120).

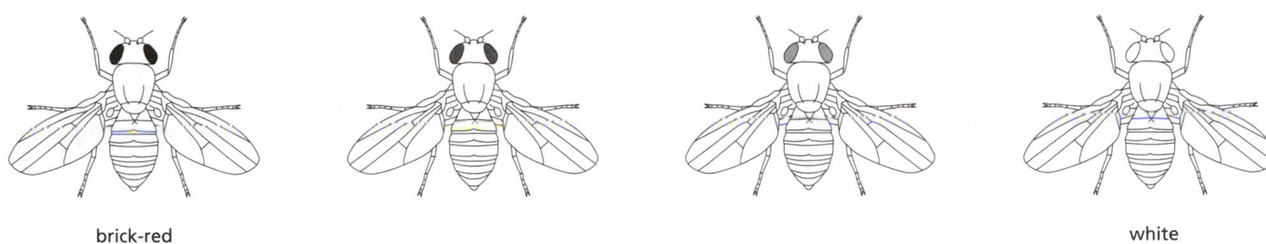


**Figure 8–39** *Drosophila* with different color eyes (Problem 8–121). Wild-type flies with brick-red eyes are shown on the *left* and white-eyed flies are shown on the *right*. Flies with eye colors between red and white are shown in between.

replicated immediately, would you expect the mismatched DNA to be recognized and repaired by DNA mismatch repair enzymes? If so, would you expect mismatch repair to increase, reduce, or not affect the frequency of mutants? Explain your answer.

## DATA HANDLING

**8–121** Early genetic studies in *Drosophila* laid the foundation for our current understanding of genes. *Drosophila* geneticists were able to generate mutant flies with a variety of easily observable phenotypic changes. Alterations from the fly's normal brick-red eye color have a venerable history because the very first mutant found by Thomas Hunt Morgan was a white-eyed fly (Figure 8–39). Since that time a large number of mutant flies with intermediate eye colors have been isolated and given names that challenge your color sense: garnet, ruby, vermilion, cherry, coral, apricot, buff, and carnation. The mutations responsible for these eye-color phenotypes are recessive. To determine whether the mutations affected the same or different genes, flies homozygous for each mutation were bred to one another in pairs and the eye colors of their progeny were noted. In Table 8–7, brick-red wild-type eyes are shown as (+) and other colors are indicated as (–).

A. How is it that flies with two different eye colors—ruby and white, for example—give rise to progeny that all have brick-red eyes?

B. Which mutations affect different genes and which mutations are alleles of the same gene?

C. How can alleles of the same gene give different eye colors? That is to say, why don't all the mutations in the same gene give the same phenotype?

**8–122** You have designed and constructed a DNA microarray that carries 20,000 allele-specific oligonucleotides (ASOs). These ASOs correspond to the wild-type and mutant alleles associated with 1000 human diseases. You have designed the microarray so the ASO that hybridizes to the mutant allele is located right below the ASO that hybridizes to the same site in the wild-type sequence. This arrangement is illustrated in Figure 8–40 for ASOs that are specific for the sickle-cell allele ($\beta^S$) and the corresponding site in the wild-type allele ($\beta^A$). ASO$^{\beta S}$ hybridizes to the sickle-cell mutation, and ASO$^{\beta A}$ hybridizes to the corresponding position in the wild-type allele. As a test of

**Table 8–7 Complementation analysis of *Drosophila* eye-color mutations** (Problem 8–121).

| MUTATION | WHITE | GARNET | RUBY | VERMILION | CHERRY | CORAL | APRICOT | BUFF | CARNATION |
|---|---|---|---|---|---|---|---|---|---|
| White | – | + | + | + | – | – | – | – | + |
| Garnet | | – | + | + | + | + | + | + | + |
| Ruby | | | – | + | + | + | + | + | + |
| Vermilion | | | | – | + | + | + | + | + |
| Cherry | | | | | – | – | – | – | + |
| Coral | | | | | | – | – | – | + |
| Apricot | | | | | | | – | – | + |
| Buff | | | | | | | | – | + |
| Carnation | | | | | | | | | – |

Brick-red eyes are indicated as (+). Other colors are indicated as (–).

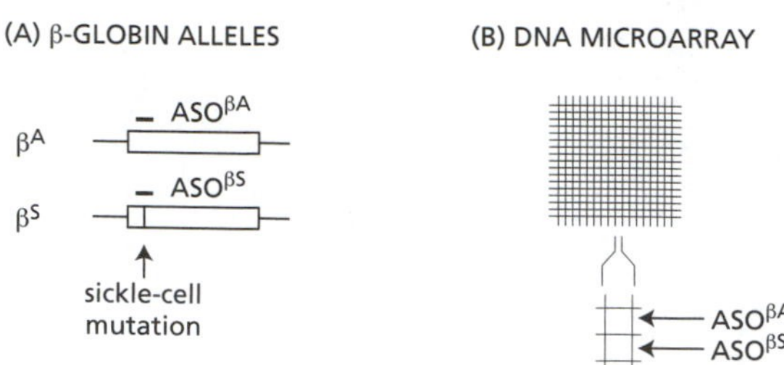


**Figure 8–40** DNA microarray for detection of disease alleles (Problem 8–122). (A) The β-globin alleles. The wild-type β-globin gene ($\beta^A$) and the sickle-cell allele ($\beta^S$) are shown. The position of the sickle-cell mutation is shown by a *vertical line*. The ASOs are shown as *short horizontal lines* arranged above the sites in the gene to which they hybridize. Because the ASOs are located at corresponding positions, each is specific for its allele: the wild-type ASO will not hybridize to the sickle-cell allele, nor will the sickle-cell ASO hybridize to the wild-type allele. (B) DNA microarray. A tiny section of the microarray is enlarged to illustrate the locations of the wild-type and sickle-cell ASOs.

your microarray, you carry out hybridizations of DNA isolated from individuals who are homozygous for the wild-type allele, homozygous for the sickle-cell allele, or heterozygous for the wild-type and sickle-cell alleles. For each DNA sample draw the expected patterns of hybridization to the globin ASOs on your microarray.

**8–123** Now that news of your disease-specific DNA microarray has gotten around, you are being inundated with requests to analyze various samples. Just today you received requests from four physicians for help in the prenatal diagnosis of the same disease. Each of the pregnant mothers has a family history of this disease. You included on your array the five alleles known to cause this disease (Figure 8–41). Each of these alleles is recessive. You agree to help. You prepare samples of fetal DNA gotten by amniocentesis and hybridize them to your microarrays. Your data are shown in Figure 8–41C. Assuming that the five disease alleles shown in Figure 8–41A are the only ones in the human population, decide for each sample of DNA whether the individual will have the disease or not. Explain your reasoning.

**8–124** The *Raf1* gene encodes a serine/threonine protein kinase that may be important in vertebrate development. In *Drosophila,* when the homolog of this gene is defective, embryos develop through the blastula stage normally, but from then on development is abnormal, leading to a truncated embryo that is lacking the posterior four segments. In vertebrates, fibroblast growth factor (FGF) stimulates mesoderm induction and posterior development, and it is thought that Raf1 protein participates in the signaling pathway triggered by FGF. Raf1 consists of a serine/threonine kinase domain that is the C-terminal half of the molecule and an N-terminal regulatory domain that inhibits the kinase activity until a proper signal is received.

To test the role of Raf1, you construct what you hope will be a dominant-negative *Raf1* cDNA by changing a key residue in the ATP-binding domain. This mutated cDNA encodes a kinase-defective protein, which you term Naf (not a functional Raf). You prepare RNA from the *Raf1* cDNA and from the *Naf* cDNA and inject them individually or as a mixture (in equal amounts) into both cells of the two-cell frog embryo to determine their effects on frog

**Problems 15–92** and **18–22** illustrate the uses of dominant-negative mutations in analyzing cell signaling pathways.

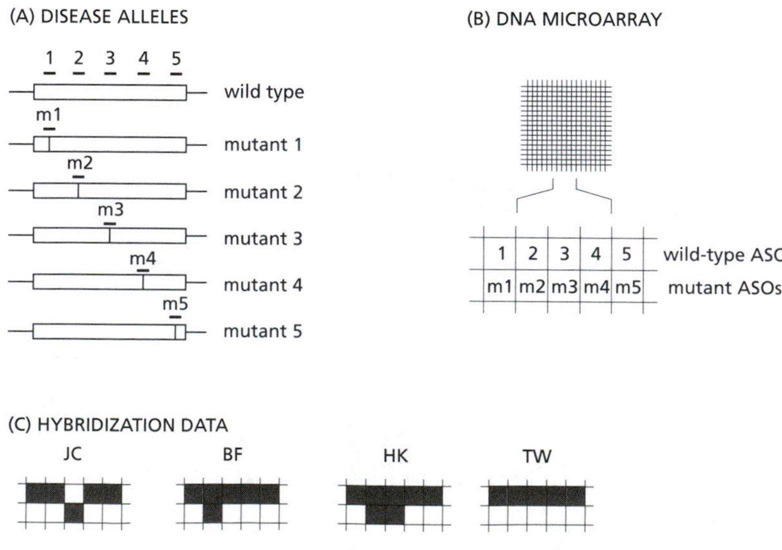


**Figure 8–41** DNA microarray analysis of alleles present in prenatal samples (Problem 8–123). (A) Wild-type and disease alleles. *Vertical lines* indicate the sites of the mutations in the disease-causing alleles. The ASOs specific for the disease mutations are shown as m1, m2, etc. The corresponding ASOs that hybridize to the wild-type gene at sites that correspond to the position of the mutations are labeled 1, 2, etc. (B) DNA microarray. The arrangement of wild-type and mutant ASOs is indicated. (C) Hybridization data. Samples of fetal DNA were hybridized to DNA arrays. *Dark spots* indicate sites where hybridization occurred. *Letters* identify the patients.

**Table 8–8 Injection of *Raf1* and *Naf* RNA into two-cell embryos** (Problem 8–124).

| INJECTION | TOTAL SURVIVORS | PHENOTYPE OF TADPOLE NORMAL | TRUNCATED TAIL | OTHER ABNORMALITIES |
|---|---|---|---|---|
| *Raf1* RNA | 94 | 75% | 0% | 25% |
| *Naf* RNA | 80 | 40% | 36% | 24% |
| *Raf1* + *Naf* RNA | 93 | 73% | 5% | 22% |
| water | 101 | 92% | 0% | 8% |
| uninjected | 80 | 99% | 0% | 1% |

development. These experiments are summarized in Table 8–8. The truncated-tail phenotype produced in some injections is very striking (Figure 8–42).

A. Do your experiments provide evidence in support of the hypothesis that Raf1 participates in posterior development in vertebrates? If so, how?

B. Given the description of the Raf1 protein, suggest a way that Naf might act as a dominant-negative protein.

C. Why does the co-injection of *Raf1* RNA with *Naf* RNA result in a low frequency of the truncated-tail phenotype?

**8–125** You've heard about this cool technique for targeted recombination into embryonic stem cells that allows you to create a null allele or a conditional allele more or less at the same time. As your friend explained it to you, you first carry out standard gene targeting into embryonic stem (ES) cells by homologous recombination, as shown in Figure 8–43. The *Neo* gene, which codes for resistance to the antibiotic neomycin, allows selection for ES cells that have incorporated the vector. These cells can then be screened by Southern blotting for those that have undergone a targeted event. The really cool part is to flank the *Neo* gene and an adjacent exon or two with lox sites. This technique is commonly referred to as 'floxing.' Once the modified ES cells have been identified, they can be exposed to the Cre recombinase, which promotes site-specific recombination between pairs of lox sites. One advantage is that it allows you to get rid of the *Neo* gene and any bacterial DNA segments, which can sometimes influence the phenotype.

A. What possible products might you get from expression of Cre in modified ES cells that carry three lox sites, as indicated in Figure 8–43?

B. Which product(s) would be a null allele?

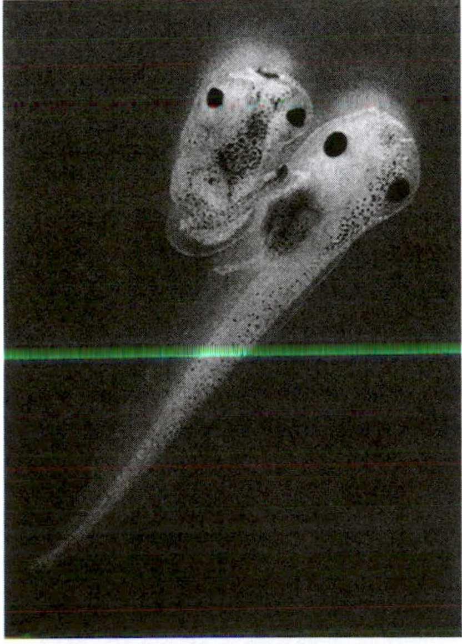

**Figure 8–42** A normal tadpole and one with a truncated tail (Problem 8–124).

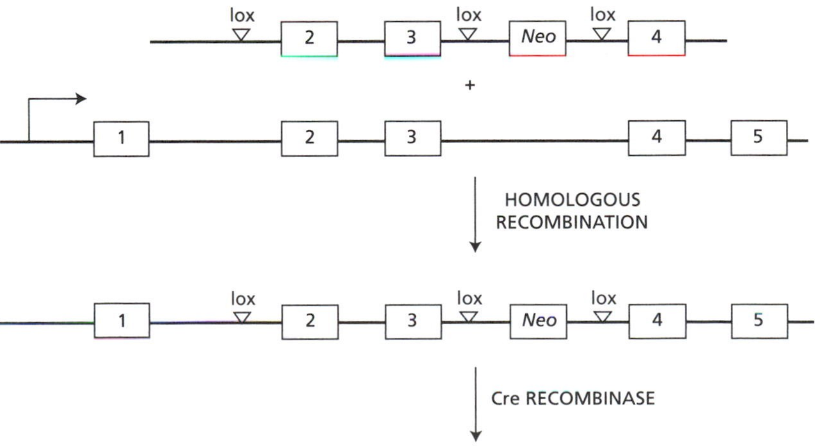


**Figure 8–43** Targeted modification of a gene in mouse ES cells (Problem 8–125). Exons are shown as *boxes*. The targeting vector is fully homologous to the gene, except for the presence of the three lox sites and the *Neo* gene, all of which are in introns. Cre recombinase, which can be introduced by transfection of a Cre-expression vector, catalyzes a site-specific recombination event between a pair of lox sites in about 20% of transfected cells.

C. Which product(s) would have a pair of lox sites but be an otherwise normal allele?

D. If you had one mouse that expressed Cre under the control of a tissue-specific promoter, can you use the allele in part C (after you've put it into the germ line of a mouse) as a conditional allele; that is, one whose defect is expressed only in a particular tissue?

# Visualizing Cells

9

## In This Chapter

LOOKING AT CELLS IN THE LIGHT MICROSCOPE 221

LOOKING AT CELLS AND MOLECULES IN THE ELECTRON MICROSCOPE 228

## LOOKING AT CELLS IN THE LIGHT MICROSCOPE

TERMS TO LEARN

| | |
|---|---|
| bright-field microscope | image processing |
| cell doctrine | ion-sensitive indicator |
| confocal microscope | light microscope |
| dark-field microscope | limit of resolution |
| differential-interference-contrast microscope | microelectrode |
| fluorescence microscope | optical tweezers |
| fluorescence recovery after photobleaching (FRAP) | phase-contrast microscope |
| fluorescence resonance energy transfer (FRET) | photoactivation |
| green fluorescent protein (GFP) | two-photon effect |

### DEFINITIONS

Match each definition below with its term from the list above.

**9–1** Fluorescent protein (from a jellyfish) that is widely used as a marker for monitoring the movements of proteins in living cells.

**9–2** The minimal separation between two objects at which they appear distinct.

**9–3** The normal light microscope in which the image is obtained by simple transmission of light through the object being viewed.

**9–4** Computer treatment of images gained from microscopy that reveal information not immediately visible to the eye.

**9–5** Similar to a light microscope but the illuminating light is passed through one set of filters before the specimen, to select those wavelengths that excite the dye, and through another set of filters before it reaches the eye, to select only those wavelengths emitted when the dye fluoresces.

**9–6** Type of light microscope that produces a clear image of a given plane within a solid object. It uses a laser beam as a pinpoint source of illumination and scans across the plane to produce a two-dimensional optical section.

**9–7** Technique for monitoring the closeness of two fluorescently labeled molecules (and thus their interaction) in cells.

**9–8** Piece of fine glass tubing, pulled to an even finer tip, that is used to inject electric current into cells.

### TRUE/FALSE

Decide whether each of these statements is true or false, and then explain why.

**9–9** Computer-assisted image processing makes it possible to see clear images of objects such as single microtubules (0.025 μm) that are well below the limit of resolution (0.2 μm).

**9–10** Because the DNA double helix is only 10 nm wide—well below the resolution of the light microscope—it is impossible to see chromosomes in living cells without special stains.

**9–11** Caged molecules can be introduced into a cell and then activated by a strong pulse of laser light at the precise time and cellular location chosen by the experimenter.

**Problems 13–119, 16–106, and 16–113** give examples of the use of caged $Ca^{2+}$, caged ATP, and caged rhodamine, respectively.

## THOUGHT PROBLEMS

**9–12** Examine the diagram of the light microscope in Figure 9–1. Identify and label the eyepiece, the condenser, the light source, the objective, and the specimen. At what two points in the light path is the image of the specimen magnified?

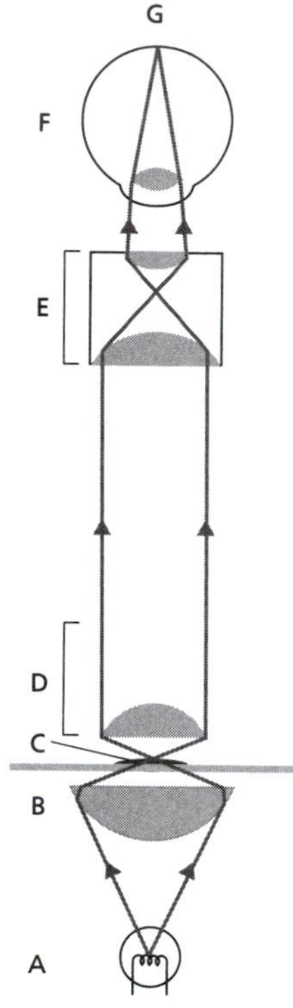


**Figure 9–1** Schematic diagram of a light microscope (Problem 9–12).

**9–13** Why is it important to keep dust, fingerprints, and other smudges off the lenses of a light microscope?

**9–14** When light enters a medium with a different optical density, it bends in a direction that depends on the refractive indices of the two media. Air and glass, for example, have refractive indices of 1.00 and 1.51, respectively. When light enters glass—the medium with the higher refractive index—it bends *toward* a line drawn normal to the surface (Figure 9–2A). Conversely, when light exits glass into air, it bends *away* from the normal line (Figure 9–2B). Using these principles, draw the paths of two parallel light rays that pass through the hemispherical glass lens shown in Figure 9–2C. Will the two rays converge or diverge? Would the result be any different if the glass lens were flipped so that light entered the flat surface?

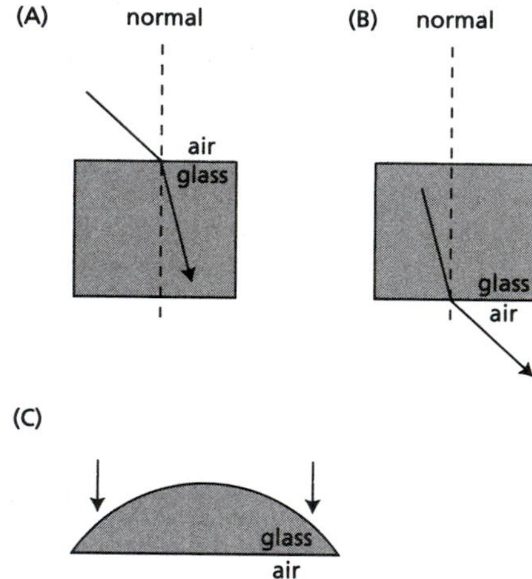


**Figure 9–2** Refraction of light at air–glass interfaces (Problem 9–14). (A) A ray of light passing from air to glass. (B) A ray of light passing from glass to air. (C) Two parallel rays of light entering a glass lens.

**9–15** The diagrams in Figure 9–3 show the paths of light rays passing through a specimen with a dry lens and with an oil-immersion lens. Offer an explanation for why oil-immersion lenses should give better resolution. Air, glass, and oil have refractive indices of 1.00, 1.51, and 1.51, respectively.

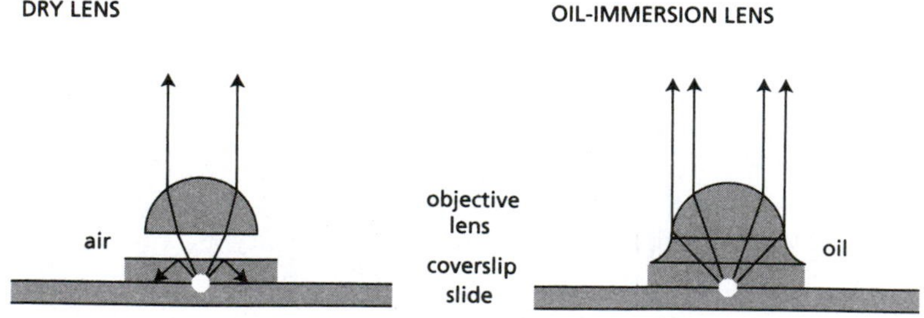


**Figure 9–3** Paths of light rays through dry and oil-immersion lenses (Problem 9–15). The *white circle* at the origin of the light rays is the specimen.

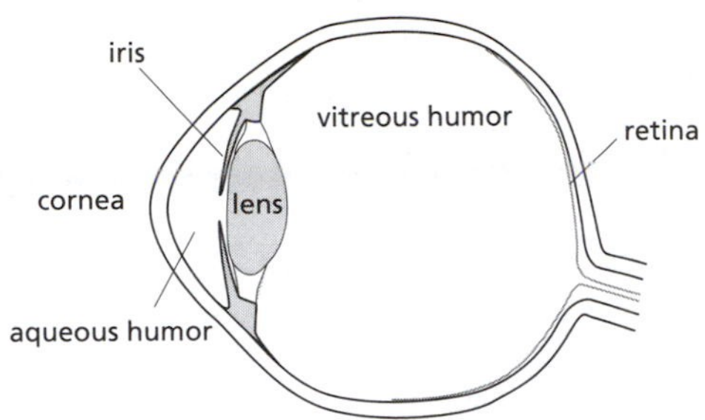

**Figure 9–4** Diagram of the human eye (Problem 9–16).

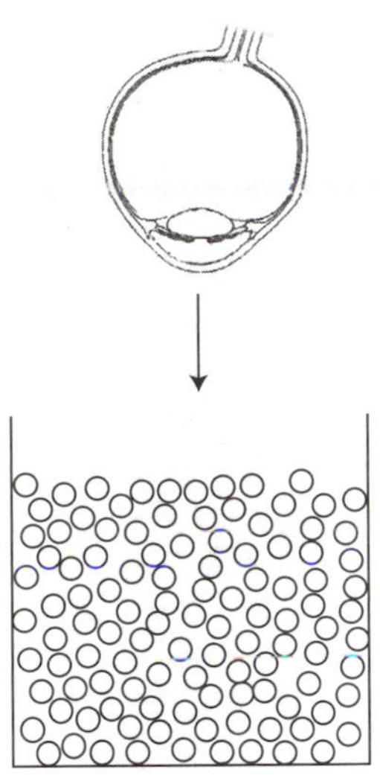

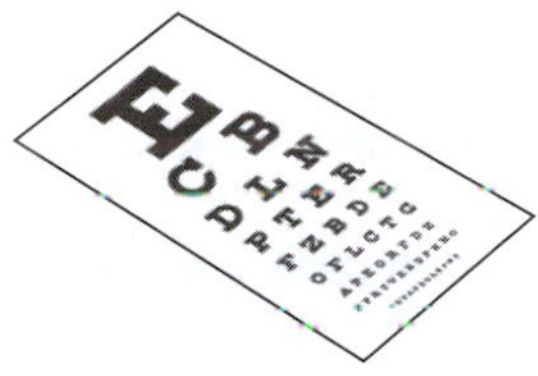

**Figure 9–5** Viewing an eye chart through a beaker filled with clear glass balls (Problem 9–18).

**9–16** Figure 9–4 shows a diagram of the human eye. The refractive indices of the components in the light path are: cornea 1.38, aqueous humor 1.33, crystalline lens 1.41, and vitreous humor 1.38. Where does the main refraction—the main focusing—occur? What role do you suppose the lens plays?

**9–17** Why do humans see so poorly underwater? And why do goggles help?

**9–18** Reading through a beaker filled with clear glass balls is impossible (Figure 9–5). Do you suppose it would help to fill the beaker with water? With immersion oil? Explain your reasoning.

**9–19** Examine the four photomicrographs of the same cell in Figure 9–6. Match each image to the technique listed below.

1. Bright-field microscopy
2. Dark-field microscopy
3. Nomarski differential-interference-contrast microscopy
4. Phase-contrast microscopy

**9–20** Explain the difference between resolution and magnification.

**9–21** Many fluorescent dyes that stain DNA require excitation by ultraviolet light. Hoechst 33342, for example, binds to DNA and absorbs light at 352 nm and emits at 461 nm. If you want to use this membrane-permeant dye in living cells, do you have to worry about ultraviolet light damage to the DNA, which absorbs light maximally at around 260 nm? Why or why not?

**9–22** Why is it, do you suppose, that a fluorescent molecule, having absorbed a single photon of light at one wavelength, *always* emits a photon at a longer wavelength?

**Problems 16–76, 17–66,** and **17–116** show examples of pictures taken by Nomarski differential-interference-contrast microscopy.

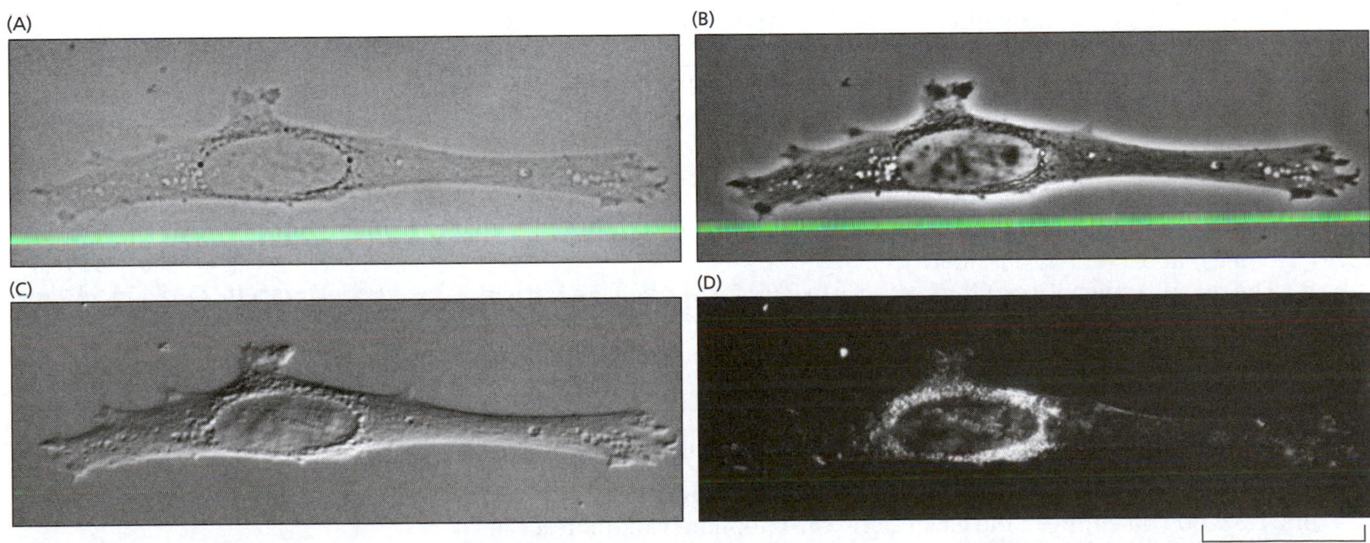

**Figure 9–6** Four photomicrographs of the same cell (Problem 9–19).

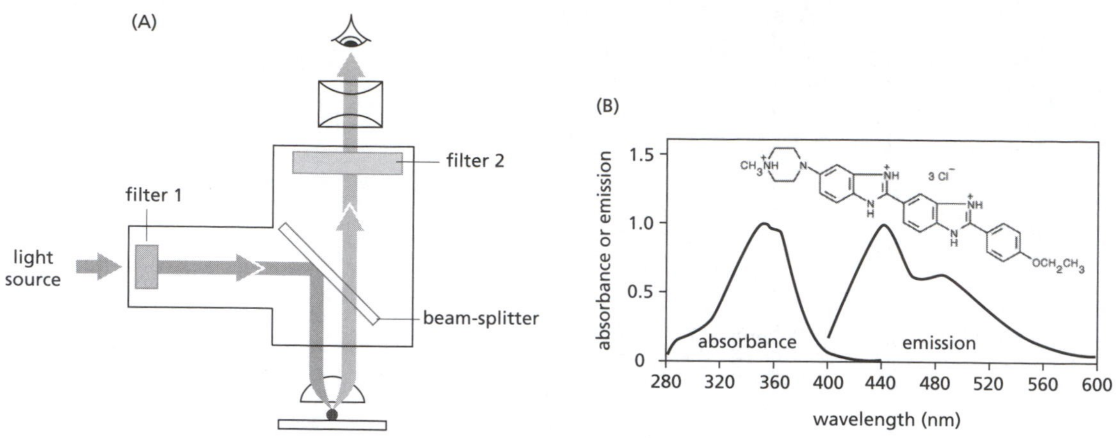


**Figure 9–7** Fluorescence microscopy (Problem 9–23). (A) The light path through a fluorescence microscope. (B) Absorption and fluorescence emission spectra of Hoechst 33342 bound to DNA. The structure of Hoechst 33342 is shown in the inset.

**9–23** A fluorescence microscope, which is shown schematically in Figure 9–7A, uses two filters and a beam-splitting (dichroic) mirror to excite the sample and capture the emitted fluorescent light. Imagine that you wish to view the fluorescence of a sample that has been stained with Hoechst 33342, a common stain for DNA. The absorption and fluorescence emission spectra for Hoechst 33342 are shown in Figure 9–7B.

A. Which of the following commercially available barrier filters would you select to place between the light source and the sample? Which one would you place between the sample and the eyepiece?
   1. A filter that passes wavelengths between 300 nm and 380 nm.
   2. A filter that passes all wavelengths above 420 nm.
   3. A filter that passes wavelengths between 450 nm and 490 nm.
   4. A filter that passes all wavelengths above 515 nm.
   5. A filter that passes wavelengths between 510 nm and 560 nm.
   6. A filter that passes all wavelengths above 590 nm.

B. How would you design your beam-splitting mirror? Which wavelengths would be reflected and which would be transmitted?

**Problems 7-32, 12-48, 12-49, 12-50, 14-31, 16-41, 16-63, 16-65, 16-111, 17-97,** and **17-98** provide examples of the use of fluorescence microscopy in various cell biological experiments.

**9–24** Antibodies that bind to specific proteins are important tools for defining the locations of molecules in cells. The sensitivity of the primary antibody—the antibody that reacts with the target molecule—is often enhanced by using labeled secondary antibodies that bind to it. What are the advantages and disadvantages of using secondary antibodies that carry fluorescent tags versus those that carry bound enzymes?

**Problems 12-92** and **13-118** illustrate the use of fluorescently tagged primary and secondary antibodies.

**9–25** The green fluorescent protein (GFP) was isolated as a cDNA from a species of jellyfish that glows green. When the cDNA for GFP was introduced into bacteria, the colonies they formed glowed pale green under UV light. In these early studies the following pertinent observations provided important insights into how GFP becomes fluorescent.
   1. When bacteria are grown anaerobically, they express normal amounts of GFP but it is not fluorescent.
   2. The denatured GFP found in insoluble protein aggregates (inclusion bodies) in bacteria is not fluorescent.
   3. The rate of appearance of fluorescence follows first-order kinetics, with a time constant that is independent of the concentration of GFP.
   4. Random mutations introduced into the cDNA coding for GFP produced some proteins with appreciably brighter fluorescence and some with different colors.

Comment on what each of these observations says about GFP fluorescence.

**Problems 12-43, 12-44, 12-58, 16-38, 17-49, 17-146, 18-23,** and **18-24** illustrate the usefulness of GFP for tagging proteins and following biological processes in cells.

**9–26** Figure 9–8 shows a series of modified GFPs that emit light in a range of colors. How do you suppose the exact same chromophore can fluoresce at so many different wavelengths?

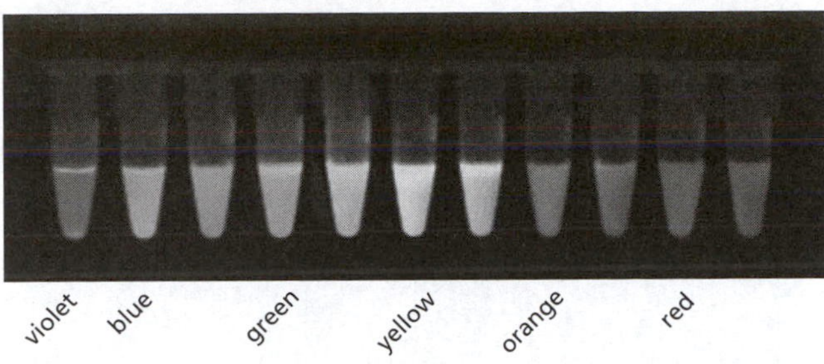


Figure 9–8 A rainbow of colors produced by modified GFPs (Problem 9–26). The colors are arranged like those in a rainbow, from blue on the *left* to red at the *right*. (See Figure Q9–3, p. 614, in *MBoC* for a color version of this image.)

**9–27** You just developed an autoradiograph after a two-week exposure. You had incubated your protein with a cell-cycle kinase in the presence of $^{32}$P-ATP in hopes of demonstrating that it was indeed a substrate for the kinase. You see the hint of a band on the gel at the right position, but it is just too faint to be convincing. You show your result to your advisor and tell him that you've put the blot against a fresh sheet of film, which you plan to expose for a longer period of time. He gives you a sideways look and tells you to do the experiment over again and use more radioactivity. What's wrong with your plan to reexpose the blot for a longer time?

## CALCULATIONS

**9–28** The resolving power of a light microscope depends on the width of the cone of light that illuminates the specimen, the wavelength of the light used, and the refractive index of the medium separating the specimen from the objective and condenser lenses, according to the formula

$$\text{resolution} = \frac{0.61\ \lambda}{n \sin \theta}$$

where $\lambda$ equals the wavelength of light used, $n$ is the refractive index, and $\theta$ is half the angular width of the cone of rays collected by the objective lens. Assuming an angular width of 120° ($\theta$ = 60°), calculate the various resolutions you would expect if the sample were illuminated with violet light ($\lambda$ = 0.4 μm) or red light ($\lambda$ = 0.7 μm) in a refractive medium of air ($n$ = 1.00) or oil ($n$ = 1.51). Which of these conditions would give you the best resolution?

**9–29** At a certain critical angle, which depends on the refractive indices of the two media, light from the optically denser medium will be bent sufficiently that it cannot escape and will be reflected back into the denser medium. The formula that describes refraction is

$$n_i \sin \theta_i = n_t \sin \theta_t$$

where $n_i$ and $n_t$ are the refractive indices of the incident and transmitting media, respectively, and $\theta_i$ and $\theta_t$ are the incident and transmitted angles, respectively, as shown in Figure 9–9A. For a glass–air interface calculate the incident angle for a light ray that is bent so that it is parallel to the surface (Figure 9–9B). The refractive indices for air and glass are 1.00 and 1.51, respectively.

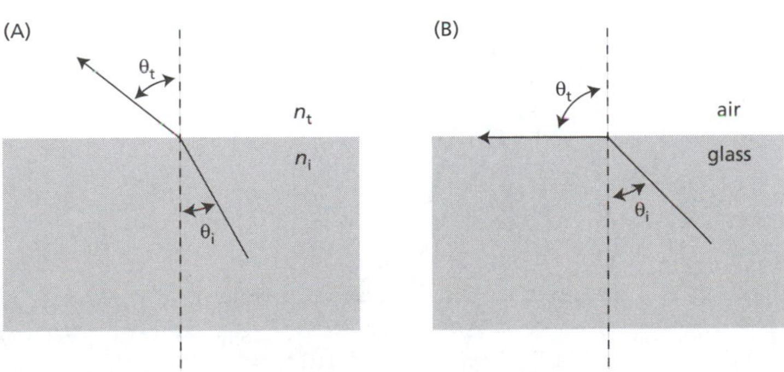


Figure 9–9 Refraction at the interface between media with different refractive indices (Problem 9–29). (A) The relationship between incident and transmitted angles for materials with different refractive indices such that $n_i > n_t$. (B) The incident angle for a glass–air interface such that the transmitted (bent) ray is parallel to the interface.

**Table 9–1 Excitation properties of a few common biological fluorophores** (Problem 9–30).

| FLUOROPHORE | OPE[a] ABSORPTION MAXIMUM (nm) | TPE[b] ABSORPTION MAXIMUM (nm) | EMISSION MAXIMUM (nm) |
|---|---|---|---|
| DAPI | 358 | 685 | 461 |
| ER-Tracker | 374 | 728 | 575 |
| Mito-Tracker | 579 | 1133 | 599 |
| Alexa Fluor 488 | 491 | 985 | 515 |
| FITC-IgG | 490 | 947 | 525 |

[a]One-photon excitation; [b]Two-photon excitation.

## DATA HANDLING

**9–30** Fluorescent molecules can be excited by a single high-energy photon or by multiple lower-energy photons. A list of commonly used biological fluorophores is given in Table 9–1. Why do you suppose that the absorption maximum for two-photon excitation is about twice that for one-photon excitation?

**9–31** You wish to attach fluorescent tags to two different proteins so that you can follow them independently. The excitation and emission spectra for cyan, green, and yellow fluorescent proteins (CFP, GFP, and YFP) are shown in Figure 9–10. Can you pick any pair of these proteins? Or are some pairs better than others? Explain your answer.

**9–32** Consider a fluorescent detector designed to report the cellular location of active protein tyrosine kinases. A blue (cyan) fluorescent protein (CFP) and a yellow fluorescent protein (YFP) were fused to either end of a hybrid protein domain. The hybrid protein segment consisted of a substrate peptide recognized by the Abl protein tyrosine kinase and a phosphotyrosine binding domain (Figure 9–11A). Stimulation of the CFP domain does not cause emission by the YFP domain when the domains are separated. When the CFP and YFP domains are brought close together, fluorescence resonance energy transfer (FRET) allows excitation of CFP to stimulate emission by YFP. FRET

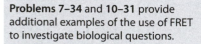

**Problems 7–34** and **10–31** provide additional examples of the use of FRET to investigate biological questions.

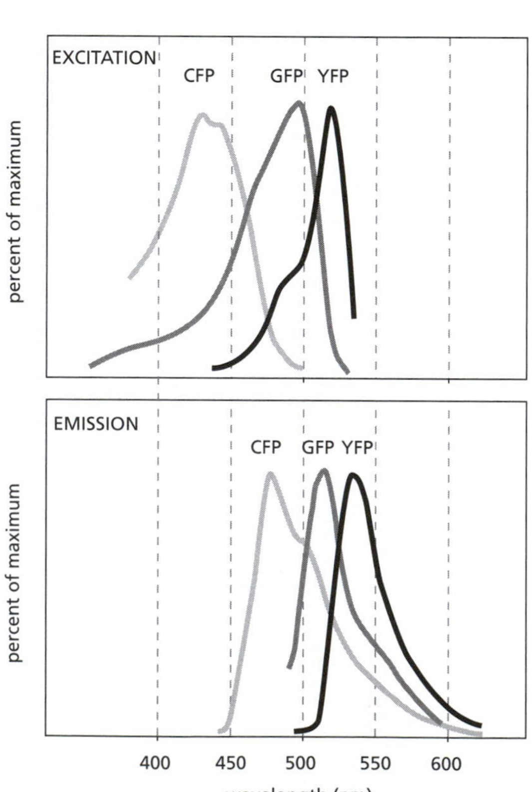


**Figure 9–10** Excitation and emission spectra for CFP, GFP, and YFP (Problem 9–31).

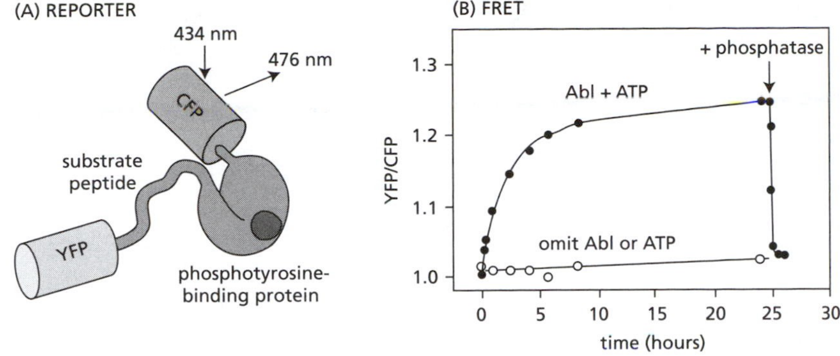


**Figure 9–11** Fluorescent reporter protein designed to detect tyrosine phosphorylation (Problem 9–32). (A) Domain structure of reporter protein. Four domains are indicated: CFP, YFP, tyrosine kinase substrate peptide, and a phosphotyrosine-binding domain. (B) FRET assay. YFP/CFP is normalized to 1.0 at time zero. The reporter was incubated in the presence (or absence) of Abl and ATP for the indicated times. *Arrow* indicates time of addition of a tyrosine phosphatase.

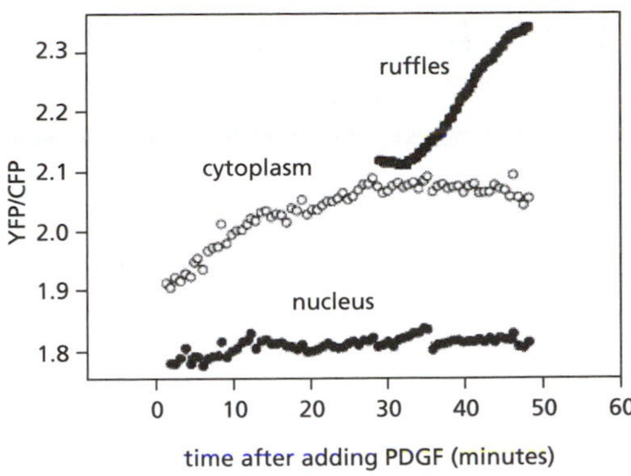


**Figure 9–12** Time course of FRET in various parts of the cell after addition of PDGF (Problem 9–33). FRET was measured as the increase in the ratio of emission at 526 nm to that at 476 nm (YFP/CFP) when CFP was excited by 434-nm light.

shows up experimentally as an increase in the ratio of emission at 526 nm versus 476 nm (YFP/CFP) when CFP is excited by 434-nm light.

Incubation of the reporter protein with Abl protein tyrosine kinase in the presence of ATP gave an increase in YFP/CFP emission (Figure 9–11B). In the absence of ATP or the Abl protein, no FRET occurred. FRET was also eliminated by addition of a tyrosine phosphatase (Figure 9–11B). Describe as best you can how the reporter protein detects active Abl protein tyrosine kinase.

**9–33** Cells activate Abl protein tyrosine kinase in response to platelet-derived growth factor (PDGF). PDGF binds to the PDGF receptor, which activates Src, which then activates Abl. It is unclear where in the cell Abl is active, but one of the consequences of PDGF stimulation is the appearance of membrane ruffles. To investigate this question, the reporter construct described in Problem 9–32 was transfected into cells, which were then stimulated by addition of PDGF. Using fluorescence microscopy, YFP emission in response to CFP excitation (FRET) was followed in different parts of the cell, with the results shown in Figure 9–12. What can you infer about the cellular distribution of active Abl protein tyrosine kinase in response to PDGF?

**9–34** You are using a cameleon indicator to measure intracellular concentrations of $Ca^{2+}$. The indicator is composed of a central calmodulin domain, which converts from an extended form to a much more compact form upon calcium binding, and two flanking fluorescent proteins, with CFP attached at one end and YFP attached at the other. You have expressed this indicator in cells and now wish to measure intracellular changes in $Ca^{2+}$ concentration in response to the biological process you are studying. The instructions say to excite the cameleon at 440 nm and to measure emission at 535 nm. How do you suppose the cameleon indicator works?

**Problem 11–42** describes the use of a fluorescent indicator of intracellular pH.

**9–35** How many molecules of your labeled protein are required for detection by autoradiography? You added 1 μL containing 10 μCi of $\gamma$-$^{32}$P-ATP (a negligible amount of ATP) to 9 μL of cell extract that had an ATP concentration of 1 mM. You incubated the mixture to allow transfer of phosphate to proteins in the extract. You then subjected 1 μL of the mixture to SDS-PAGE, dried the gel, and placed it against a sheet of x-ray film. After an overnight exposure you saw a barely detectable band in the location of your protein. You know from previous experience that a protein labeled at 1 count per minute per band (1 cpm equals 1 disintegration per minute, dpm, for $^{32}$P) will form such a band after an overnight exposure. How many molecules of labeled protein are in the band, if you assume 1 phosphate per molecule? (Some useful conversion factors for radioactivity are shown on the inside of the front cover.)

# LOOKING AT CELLS AND MOLECULES IN THE ELECTRON MICROSCOPE

TERMS TO LEARN

cryoelectron microscopy
electron microscope (EM)
electron-microscope tomography
immunogold electron microscopy
negative staining
scanning electron microscope (SEM)
single-particle reconstruction

## DEFINITIONS

Match each definition below with its term from the list above.

**9–36** A contrast-enhancing technique for the electron microscope in which a heavy-metal salt is used to create a reverse, or negative, image of the object.

**9–37** Type of microscope that uses a beam of electrons to create an image.

**9–38** Electron microscopy technique in which the objects to be viewed, such as macromolecules and viruses, are rapidly frozen.

**9–39** Type of electron microscope that produces an image of the surface of an object.

**9–40** Electron microscopy technique in which cellular structures or molecules of interest are labeled with antibodies tagged with electron-dense gold particles, which show up as black spots on the image.

## TRUE/FALSE

Decide whether each of these statements is true or false, and then explain why.

**9–41** Transmission electron microscopy (TEM) and scanning electron microscopy (SEM) can both be used to examine a structure in the interior of a thin section; TEM provides a projection view, while SEM captures electrons scattered from the structure and gives a more three-dimensional view.

## THOUGHT PROBLEMS

**9–42** A major challenge to electron microscopists from the beginning was to convince others that what they observed in micrographs truly reflected structures that were originally present in the living cell. Outline a current approach to this problem.

**9–43** The techniques of metal shadowing and negative staining both use heavy metals such as platinum and uranium to provide contrast. If these metals don't actually bind to defined biological structures (which they don't), how is it that they can help to make such structures visible?

**9–44** It is sometimes difficult to tell bumps from pits just by looking at the pattern of shadows. Consider Figure 9–13, which shows a set of shaded circles. In Figure 9–13A the circles appear to be bumps; however, when the picture is simply turned upside down (Figure 9–13B), the circles seem to be pits. This is a classic illusion. The same illusion is present in metal shadowing, as shown in the two electron micrographs in Figure 9–13. In one the membrane appears to be covered in bumps, while in the other the membrane looks heavily pitted. Is it possible for an electron microscopist to be sure that one view is correct, or is it all arbitrary? Explain your reasoning.

**9–45** Nuclear pore complexes, which are large assemblages of more than 30 different proteins, mediate the exchange of macromolecules between the nucleus and cytoplasm. You have gently isolated nuclei from *Dictyostelium discoideum* and frozen them in vitreous ice for examination by cryoelectron tomography. You obtain a tomogram of a nucleus and combine the images

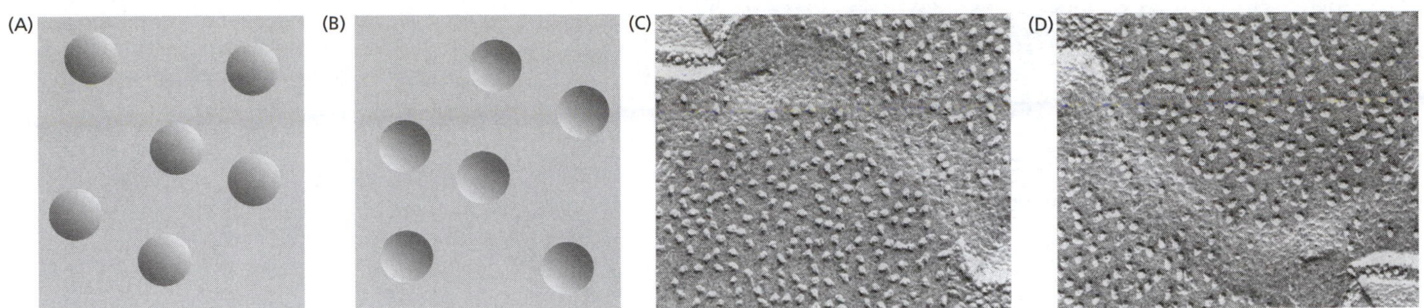


Figure 9–13 Bumps and pits (Problem 9–44). (A) Shaded circles that look like bumps. (B) Shaded circles that look like pits. (C) An electron micrograph oriented so that it appears to be covered with bumps. (D) An electron micrograph oriented so that it appears to be covered with pits.

from 267 nuclear pore complexes. You expect that individual nuclei will have been arrested in different states of transport because the isolated nuclei were shown to be competent for transport. Assuming that parts of the structure flex and move during the transport process, how do you suppose that averaging structures in different states will affect your final picture? Can you think of a way to improve the quality of the image you would get from this dataset?

## CALCULATIONS

**9–46** The practical resolving power of modern electron microscopes is around 0.1 nm. The major reason for this constraint is the small numerical aperture ($n \sin \theta$), which is limited by θ (half the angular width of rays collected at the objective lens). Assuming that the wavelength (λ) of the electron is 0.004 nm and that the refractive index ($n$) is 1.0, calculate the value for θ. How does that value compare with a θ of 60°, which is typical for light microscopes?

$$\text{resolution} = \frac{0.61\ \lambda}{n \sin \theta}$$

## DATA HANDLING

**9–47** You are studying two proteins that you think may be components of gap junctions, which are structures that allow adjacent cells to exchange small molecules. When cells are frozen and fractured and viewed in the electron microscope, gap junctions show up clearly as specialized areas densely populated with membrane particles. To decide whether your proteins are components of gap junctions, you have prepared antibodies against each of them. To one antibody you've attached 15-nm gold particles; to the other, 10-nm gold particles. You prepare freeze-fractured cells for electron microscopy and then incubate them with your gold-tagged antibodies, with the results shown in Figure 9–14. Do these results indicate that both proteins are part of gap junctions? Why or why not?

**Problem 16–75** shows how gold-tagged antibodies can be used to identify one end of a microtubule.

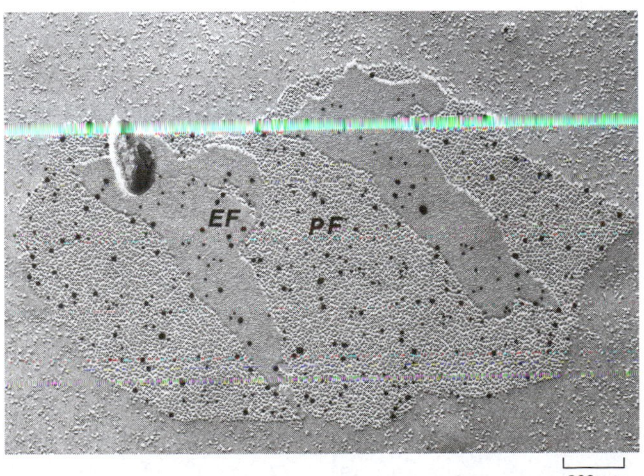


Figure 9–14 A freeze-fracture electron micrograph of a gap junction (Problem 9–47). The central densely packed area is the gap junction. The two leaflets of the plasma membrane are indicated as EF (for external face) and PF (for protoplasmic face). Gold particles show up as *black dots*.

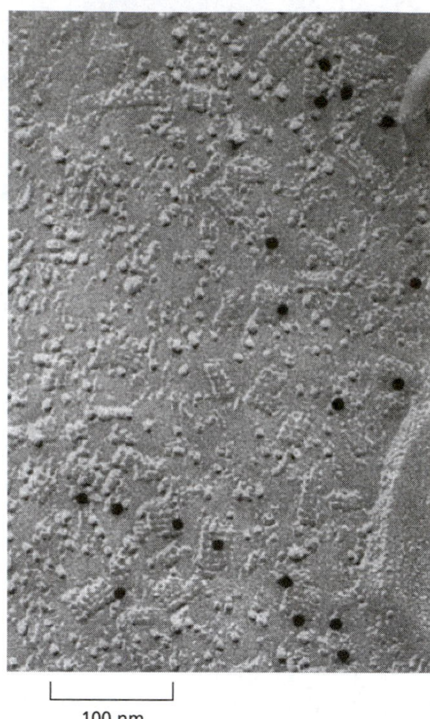


**Figure 9–15** Freeze-fracture micrograph of an astrocyte membrane labeled with gold-tagged antibodies against AQP4 (Problem 9–48).

**9–48** Aquaporin water channels play a major role in water metabolism and osmoregulation in many cells, including nerve cells of the brain and spinal cord. Although the aquaporin channels are known to be located in the plasma membrane, their structural organization is unknown. You have prepared a highly specific antibody against AQP4, which forms the water channels in the astrocytes of the brain. You prepare a freeze-fractured sample from the brain, incubate it with gold-tagged antibodies against AQP4, and examine it by electron microscopy (Figure 9–15).

A. Are the gold particles (black dots) consistently associated with any particular structure?

B. Are there any examples of black dots that are not associated with these structures? Are there any examples of structures that do not have black dots? How do your answers to these questions affect your confidence that the structure you've identified is the aquaporin water channel?

# Membrane Structure

# 10

## In This Chapter

THE LIPID BILAYER 231

MEMBRANE PROTEINS 238

## THE LIPID BILAYER

TERMS TO LEARN

| | | |
|---|---|---|
| amphiphilic | hydrophilic | liposome |
| black membrane | hydrophobic | phosphoglyceride |
| cholesterol | lipid bilayer | phospholipid |
| ganglioside | lipid droplet | plasma membrane |
| glycolipid | lipid raft | |

### DEFINITIONS

Match the definition below with its term from the list above.

**10–1** Artificial planar lipid bilayer formed across a hole in a partition between two compartments.

**10–2** Artificial phospholipid bilayer vesicle formed from an aqueous suspension of phospholipid molecules.

**10–3** Describes a nonpolar molecule or part of a molecule that cannot form energetically favorable interactions with water molecules and therefore does not dissolve in water.

**10–4** Small region of the plasma membrane enriched in sphingolipids and cholesterol.

**10–5** Any glycolipid having one or more sialic acid residues in its structure; especially abundant in the plasma membranes of nerve cells.

**10–6** Having both hydrophobic and hydrophilic regions, as in a phospholipid or a detergent molecule.

**10–7** The main type of phospholipid in animal cell membranes, with two fatty acids and a polar head group attached to a three-carbon glycerol backbone.

**10–8** Lipid molecule with a characteristic four-ring steroid structure that is an important component of the plasma membranes of animal cells.

### TRUE/FALSE

Decide whether each of these statements is true or false, and then explain why.

**10–9** Although lipid molecules are free to diffuse in the plane of the bilayer, they cannot flip-flop across the bilayer unless enzyme catalysts called phospholipid translocators are present in the membrane.

**10–10** All of the common phospholipids—phosphatidylcholine, phosphatidylethanolamine, phosphatidylserine, and sphingomyelin—carry a positively charged moiety on their head group, but none carry a net positive charge.

**10–11** Glycolipids are never found on the cytoplasmic face of membranes in living cells.

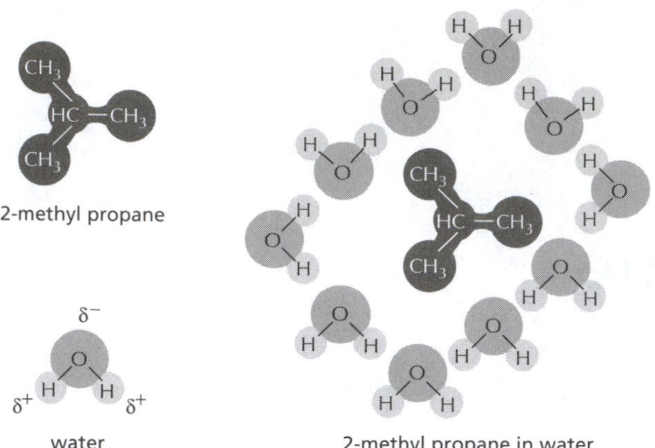


**Figure 10–1** Icelike cage of water molecules around a hydrophobic solute (Problem 10–12).

## THOUGHT PROBLEMS

**10–12** Hydrophobic solutes are said to "force adjacent water molecules to reorganize into icelike cages" (Figure 10–1). This statement seems paradoxical because water molecules do not interact with hydrophobic solutes. How could water molecules 'know' about the presence of a hydrophobic solute and change their behavior to interact differently with one another? Discuss this seeming paradox and develop a clear concept of what is meant by an 'icelike' cage. How does it compare to ice? Why would such a cagelike structure be energetically unfavorable relative to pure water?

**10–13** When a lipid bilayer is torn, why doesn't it seal itself by forming a 'hemi-micelle' cap at the edges, as shown in Figure 10–2?

**10–14** Five students in your class always sit together in the front row. This could be because (1) they really like each other or (2) nobody else in your class wants to sit next to them. Which explanation holds for the assembly of a lipid bilayer? Explain your answer. If the lipid bilayer assembled for the opposite reason, how would its properties differ?

**10–15** The properties of a lipid bilayer are determined by the structures of its lipid molecules. Predict the properties of the lipid bilayers that would result if the following were true:

A. Phospholipids had only one hydrocarbon chain instead of two.
B. The hydrocarbon chains were shorter than normal, say about 10 carbon atoms long.
C. All of the hydrocarbon chains were saturated.
D. All of the hydrocarbon chains were unsaturated.
E. The bilayer contained a mixture of two kinds of lipid molecule, one with two saturated hydrocarbon tails and the other with two unsaturated hydrocarbon tails.
F. Each lipid molecule were covalently linked through the end carbon atom of one of its hydrocarbon chains to a lipid molecule in the opposite monolayer.

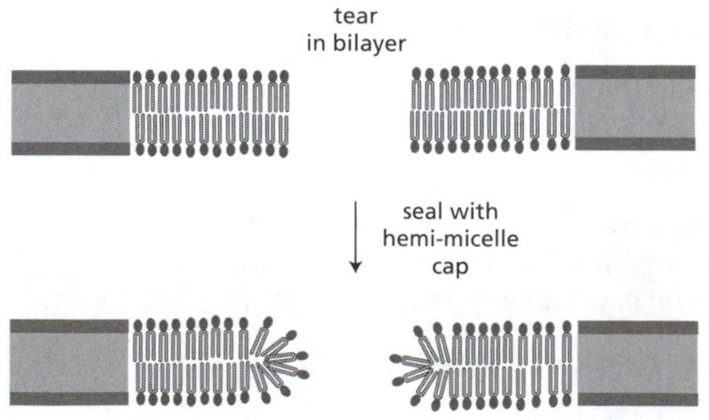


**Figure 10–2** A torn lipid bilayer sealed with a hypothetical 'hemi-micelle' cap (Problem 10–13).

**10–16** What is meant by the term 'two-dimensional fluid'?

**10–17** Margarine is made from vegetable oil by a chemical process. Do you suppose this process converts saturated fatty acids to unsaturated ones, or vice versa? Explain your answer.

**10–18** Which one of the phospholipids listed below is present in very small quantities in the plasma membranes of mammalian cells, despite its crucial role in cell signaling?

A. Phosphatidylcholine
B. Phosphatidylethanolamine
C. Phosphatidylinositol
D. Phosphatidylserine
E. Sphingomyelin

**10–19** Predict which one of the following organisms will have the highest percentage of unsaturated fatty acid chains in their membranes. Explain your answer.

A. Antarctic fish
B. Desert iguana
C. Human being
D. Polar bear
E. Thermophilic bacterium

**10–20** Lipid rafts are rich in both sphingolipids and cholesterol, and it is thought that cholesterol plays a central role in raft formation since lipid rafts apparently do not form in its absence. Why do you suppose that cholesterol is essential for formation of lipid rafts? (Hint: sphingolipids other than sphingomyelin have large head groups composed of several linked sugar molecules.)

**10–21** If lipid rafts form because sphingolipid and cholesterol molecules preferentially associate, why do you think it is that they aggregate into multiple tiny rafts instead of into a single large one?

**10–22** Why are lipid rafts thicker than other parts of the bilayer?

**10–23** The lipid bilayers found in cells are fluid, yet asymmetrical in the composition of the monolayers. Is this a paradox? Explain your answer.

**10–24** Phosphatidylserine, which is normally confined to the cytoplasmic monolayer of the plasma membrane lipid bilayer, is redistributed to the outer monolayer during apoptosis. How is this redistribution accomplished?

## CALCULATIONS

**10–25** Within a monolayer, lipid molecules exchange places with their neighbors every $10^{-7}$ seconds. It takes about 1 second for a lipid molecule to diffuse from one end of a bacterium to the other, a distance of 2 μm.

A. Are these numbers in agreement? Assume that the diameter of a lipid head group is 0.5 nm. Explain why or why not.
B. To gain an appreciation for the great speed of molecular motions, assume that a lipid molecule is the size of a ping-pong ball (4-cm diameter) and that the floor of your living room (6 m × 6 m) is covered wall to wall in a monolayer of balls. If two neighboring balls exchanged positions once every $10^{-7}$ seconds, how fast would they be moving in kilometers per hour? How long would it take for a ball to move from one end of the room to the other?

**10–26** If a lipid raft is typically 70 nm in diameter and each lipid molecule has a diameter of 0.5 nm, about how many lipid molecules would there be in a lipid raft composed entirely of lipid? At a ratio of 50 lipid molecules per protein molecule (50% protein by mass), how many proteins would be in a typical raft? (Neglect the loss of lipid from the raft that would be required to accommodate the protein.)

## DATA HANDLING

**10–27** While crossing the Limpopo River on safari in Africa, a friend of yours was bitten by a poisonous water snake and nearly died from extensive hemolysis. A true biologist at heart, he captured the snake before he passed out and has asked you to analyze the venom to discover the basis of its hemolytic activity. You find that the venom contains a protease, a neuraminidase (which removes sialic acid residues from gangliosides), and a phospholipase (which cleaves bonds in phospholipids). Treatment of isolated red blood cells with these purified enzymes gave the results shown in Table 10–1. Analysis of the products of hemolysis produced by phospholipase treatment showed an enormous increase in free phosphorylcholine (choline with a phosphate group attached) and diacylglycerol (glycerol with two fatty acid chains attached).

A. What is the substrate for the phospholipase, and where is it cleaved?

B. In light of what you know about the structure of the plasma membrane, can you suggest why the phospholipase causes lysis of the red blood cells, but the protease and neuraminidase do not.

**Table 10–1 Results of treatment of red blood cells with enzymes isolated from snake venom** (Problem 10–27).

| PURIFIED ENZYME | HEMOLYSIS |
|---|---|
| Protease | no |
| Neuraminidase | no |
| Phospholipase | yes |

**10–28** A classic paper studied the behavior of lipids in the two monolayers of a membrane by labeling individual molecules with nitroxide groups, which are stable free radicals (Figure 10–3). These spin-labeled lipids can be detected by electron spin resonance (ESR) spectroscopy, a technique that does not harm living cells. Spin-labeled lipids are introduced into small lipid vesicles, which are then fused with cells, thereby transferring the labeled lipids into the plasma membrane.

The two spin-labeled phospholipids shown in Figure 10–3 were incorporated into intact human red blood cell membranes in this way. To determine whether they were introduced equally into the two monolayers of the bilayer, ascorbic acid (vitamin C), which is a water-soluble reducing agent that does not cross membranes, was added to the medium to destroy any nitroxide radicals exposed on the outside of the cell. The ESR signal was followed as a function of time in the presence and absence of ascorbic acid as indicated in Figure 10–4A and B.

A. Ignoring for the moment the difference in extent of loss of ESR signal, offer an explanation for why phospholipid 1 (Figure 10–4A) reacts faster with ascorbate than does phospholipid 2 (Figure 10–4B). Note that phospholipid 1 reaches a plateau in about 15 minutes, whereas phospholipid 2 takes almost an hour.

B. To investigate the difference in extent of loss of ESR signal with the two phospholipids, the experiments were repeated using red cell ghosts that had been resealed to make them impermeable to ascorbate (Figure 10–4C and D). Resealed red cell ghosts are missing all of their cytoplasm, but have an intact plasma membrane. In these experiments the loss of ESR signal for both phospholipids was negligible in the absence of ascorbate and reached a plateau at 50% in the presence of ascorbate. What do you suppose might account for the difference in extent of loss of ESR signal in experiments with red cell ghosts (Figure 10–4C and D) versus those with normal red cells (Figure 10–4A and B).

C. Were the spin-labeled phospholipids introduced equally into the two monolayers of the red cell membrane?

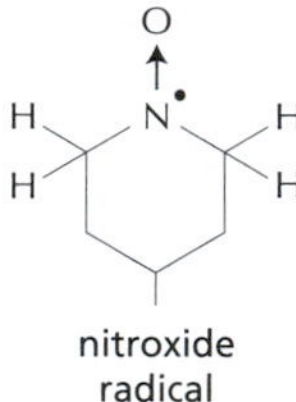


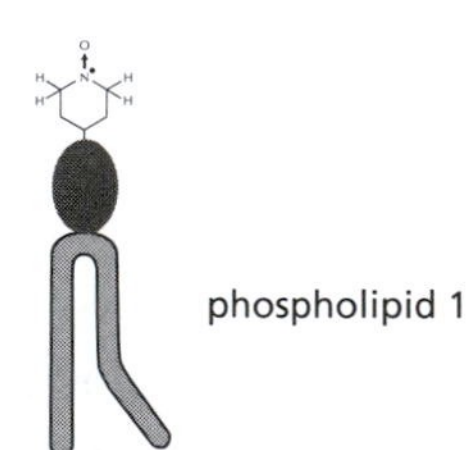


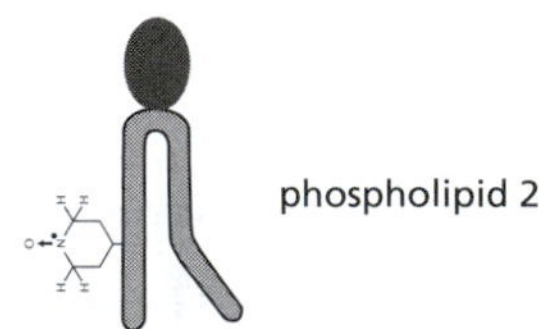


**Figure 10–3** Structures of two nitroxide-labeled lipids (Problem 10–28). The nitroxide radical is shown at the *top*, and its position of attachment to the phospholipids is shown *below*.

**10–29** You wish to determine the distribution of the phospholipids in the plasma membrane of the human red blood cell. Phospholipids make up 60% of the lipids in the red cell bilayer, with cholesterol (23%) and glycolipids (3%) accounting for most of the rest. To measure the distribution of individual phospholipids, you treat intact red cells and permeable red cell ghosts (1) with two different phospholipases and (2) with a fluorescent reagent, abbreviated SITS, which specifically labels primary amine groups, but will not penetrate an intact membrane.

As summarized in Table 10–2, treatment with sphingomyelinase degrades most of the sphingomyelin in intact red cells (without causing lysis) and in

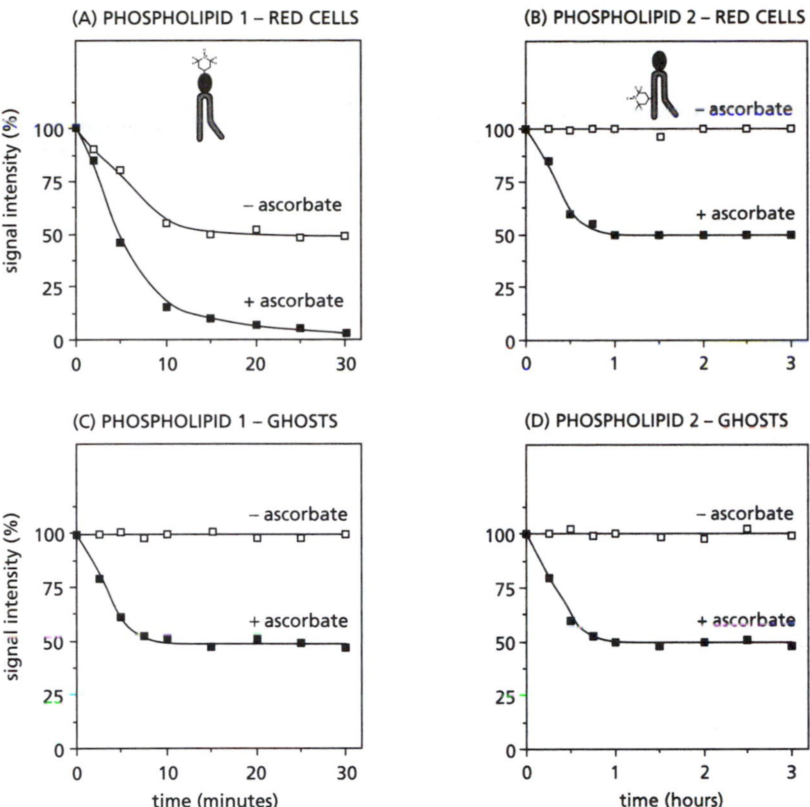


**Figure 10–4** Decrease in ESR signal intensity as a function of time in red cells and red cell ghosts in the presence and absence of ascorbate (Problem 10–28). (A and B) Phospholipid 1 and phospholipid 2 in red cells. (C and D) Phospholipid 1 and phospholipid 2 in red cell ghosts.

permeable red cell ghosts. The phospholipases in sea snake venom degrade only phosphatidylcholine in intact red cells (without causing lysis), but they degrade phosphatidylserine and phosphatidylethanolamine, as well, in permeable red cell ghosts. SITS labels all the phosphatidylethanolamine and phosphatidylserine in permeable red cell ghosts but labels virtually none in intact red blood cells.

A. From these results deduce the distribution of the four principal phospholipids in red cell membranes. Which of the phospholipids, if any, are located in both monolayers of the membrane?

B. Why did you use red blood cells and not other animal cells for these experiments?

**10–30** GPI-anchored proteins are insoluble in ice-cold solutions of the detergent Triton X-100. This is surprising since the hydrophobic portions of the GPI anchors might be expected to partition efficiently in detergent micelles, as indeed they do at higher temperatures. Because GPI anchors are added after protein synthesis during transport to the plasma membrane, you carry out a pulse-labeling experiment to determine when the proteins become insoluble. Cells that express a typical GPI-anchored protein called PLAP were labeled for 5 minutes in $^{35}$S-methionine and quickly transferred to nonradioactive medium. At intervals, the cells were treated with ice-cold

**Table 10–2 Sensitivity of phospholipids in intact human red cells and permeable red cell ghosts to phospholipases and to a fluorescent label that cannot penetrate an intact membrane** (Problem 10–29).

| PHOSPHOLIPID | SPHINGOMYELINASE | | SEA SNAKE VENOM | | SITS FLUORESCENCE | |
|---|---|---|---|---|---|---|
| | RED CELLS | GHOSTS | RED CELLS | GHOSTS | RED CELLS | GHOSTS |
| Phosphatidylcholine | – | – | + | + | – | – |
| Phosphatidylethanolamine | – | – | – | + | – | + |
| Phosphatidylserine | – | – | – | + | – | + |
| Sphingomyelin | + | + | – | – | – | – |

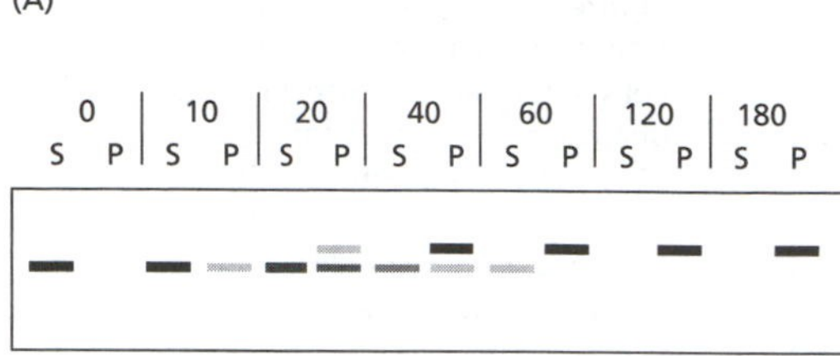


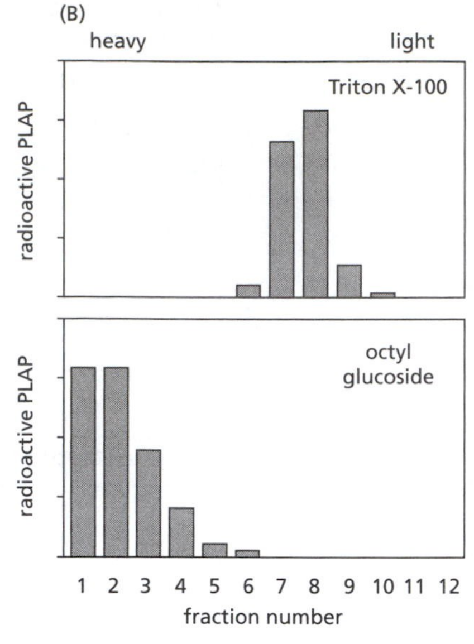


**Figure 10–5** Analysis of a GPI-anchored protein (Problem 10–30). (A) Time course of appearance of PLAP in the supernatant and pellet fractions. Numbers indicate minutes after transfer to a nonradioactive medium. (B) Density of PLAP in Triton X-100 and octyl glucoside. Fractions are removed from the bottom of the tube; thus, the *bottom* of the gradient is at the *left* (heavy) and the *top* of the gradient is at the *right* (light).

Triton X-100 and centrifuged. The supernatant (soluble) and pellet (insoluble) fractions were analyzed by immunoprecipitation with an antibody specific for PLAP and electrophoresis in an SDS polyacrylamide gel. The time course for recovery of PLAP in the supernatant (S) and pellet (P) fractions is shown in Figure 10–5A.

A. When does PLAP become insoluble in Triton X-100? Why is there a shift in its mobility by SDS polyacrylamide-gel electrophoresis? Why do you suppose that there is a time delay before PLAP becomes insoluble?

B. When the composition of the insoluble fraction from Triton X-100 was analyzed by centrifugation to equilibrium in a sucrose density gradient, PLAP was found to be located near the top of the gradient (low density) (Figure 10–5B). When solubilized by a different detergent, octyl glucoside, PLAP was found near the bottom of the gradient (high density) (Figure 10–5B). Why do you suppose PLAP is located at the top of the sucrose gradient in the Triton X-100-insoluble fraction but at the bottom of the gradient when extracted with octyl glucoside?

C. Analysis of the lipids in the visibly milky band corresponding to fractions 7 and 8 from the Triton X-100-insoluble PLAP (Figure 10–5B) showed that they were enriched for sphingomyelin, other sphingolipids, and cholesterol, but depleted for phosphatidylethanolamine, phosphatidylcholine, and phosphatidylserine. Do these results support the idea of lipid rafts in the membrane? Why or why not?

D. When fractions 7 and 8 from the Triton X-100-insoluble PLAP were analyzed by electron microscopy, they were found to contain vesicles ranging from 100 to 1000 nm in diameter, with PLAP embedded in their membranes. What is the surface area of a 100-nm sphere (area = $4\pi r^2$)? How does this compare with the surface area of a lipid raft 70 nm in diameter (area = $\pi r^2$)? Does this microscopic analysis support the idea of lipid rafts in membranes? Why or why not?

**10–31** Sphingolipids, cholesterol, and some proteins form detergent-insoluble complexes in ice-cold Triton X-100, which was the first evidence in favor of lipid rafts in membranes. By itself, such evidence is unsatisfying because insoluble complexes could be artifacts of the extraction method. A sensitive form of fluorescence resonance energy transfer (FRET) has been used to investigate the existence of lipid rafts in living cells in a way that avoids this potential artifact.

When a fluorophore on a membrane protein is illuminated with plane-polarized light, its fluorescent emission will have the same polarization as the incident light because the protein in the membrane rotates slowly relative to the time for emission. However, if the absorbed energy is first transferred to a nearby fluorophore (by FRET) the subsequently emitted fluorescent light will be oriented differently than the incident light because the nearby protein will nearly always have a different orientation in the membrane than the original molecule. Thus, the loss of polarization is a sensitive measure of FRET between fluorescently tagged membrane proteins.

To use this technique to investigate lipid rafts, two cell lines were obtained that expressed different forms of the monomeric folate receptor in their plasma membranes: a GPI-anchored form and a transmembrane-anchored form. The folate receptors in each cell line were made fluorescent by addition of a fluorescent folate analog. Cells tagged in this way showed variation in fluorescence intensity over their surface because of chance variations in the distributions of the labeled receptors. These random variations allowed different densities of receptors to be analyzed within the same cell.

Polarization of the emitted light from individual pixels (1 $\mu m^2$, tiny areas of the cell surface, but much larger than individual rafts) was measured with two filters: one parallel to the incident light to capture direct emission, and one perpendicular to the incident light to capture emission after FRET. The density (concentration) of receptors in the pixel was measured by the total fluorescence intensity. Polarization of the fluorescent light was measured as intensity detected via the parallel filter ($I_{par}$) minus intensity detected via the perpendicular filter ($I_{perp}$), divided by the total intensity $[(I_{par} - I_{perp})/(I_{par} + I_{perp})]$.

A. The expectations for dispersed receptors versus receptors in lipid rafts are shown in Figure 10–6. Explain these expectations.

B. The actual experiments showed that transmembrane-anchored folate receptors followed the expectations shown in Figure 10–6A, whereas the GPI-anchored folate receptors followed those in Figure 10–6B. Do these experiments provide evidence for the existence of lipid rafts in the plasma membrane? Why or why not?

C. When these same cells were grown in the presence of compactin, which inhibits cholesterol synthesis, both types of folate receptor gave results like those in Figure 10–6A. What bearing does this observation have on the existence of lipid rafts? Explain your reasoning.

**10–32** The asymmetric distribution of phospholipids in the two monolayers of the plasma membrane implies that very little spontaneous flip-flop occurs or, alternatively, that any spontaneous flip-flop is rapidly corrected by phospholipid translocators that return phospholipids to their appropriate monolayer. The rate of phospholipid flip-flop in the plasma membrane of intact red blood cells has been measured to decide between these alternatives.

One experimental measurement used the same two spin-labeled phospholipids described in Problem 10–28 (see Figure 10–3). To measure the rate of flip-flop from the cytoplasmic monolayer to the outer monolayer, red cells with spin-labeled phospholipids exclusively in the cytoplasmic monolayer were incubated for various times in the presence of ascorbate and the loss of ESR signal was followed. To measure the rate of flip-flop from the

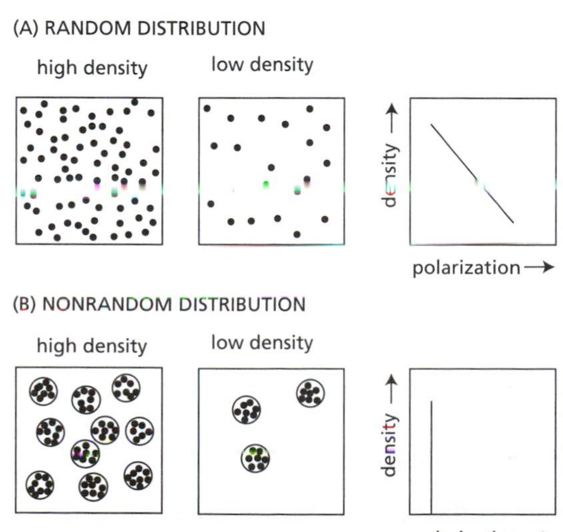


**Figure 10–6** Expectations for polarization of fluorescence at different densities of receptors (Problem 10–31). (A) Randomly distributed receptors. The *boxed area* represents a pixel and the *solid dots* represent fluorescent receptors. (B) Receptors clustered in microdomains. *Circles* represent microdomains such as lipid rafts. *Solid dots* represent fluorescent receptors.

outer to the cytoplasmic monolayer, red cells with spin-labeled phospholipids exclusively in the outer monolayer were incubated for various times in the absence of ascorbate and the loss of ESR signal was followed. The results of these experiments are illustrated in Figure 10–7.

A. From the results in Figure 10–7, estimate the rate of flip-flop from the cytoplasmic to the outer monolayer and from the outer to the cytoplasmic monolayer. A convenient way to express such rates is as the half-time of flip-flop—that is, the time it takes for half the phospholipids to flip-flop from one monolayer to the other.

B. From what you learned about the behavior of the two spin-labeled phospholipids in Problem 10–28, deduce which one was used to label the cytoplasmic monolayer of the intact red blood cells, and which one was used to label the outer monolayer.

C. Using the information in this problem, propose a method to generate intact red cells that contain spin-labeled phospholipids exclusively in the cytoplasmic monolayer, and a method to generate cells spin labeled exclusively in the outer monolayer.

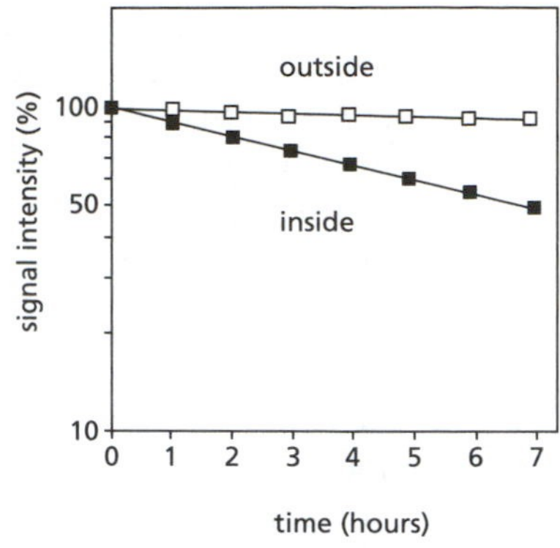


**Figure 10–7** Decrease in ESR signal intensity of red blood cells containing spin-labeled phospholipids in the outer monolayer (outside) and cytoplasmic monolayer (inside) of the plasma membrane (Problem 10–32).

## MEMBRANE PROTEINS

TERMS TO LEARN

| | |
|---|---|
| bacteriorhodopsin | lectin |
| carbohydrate layer | multipass transmembrane protein |
| cortex | peripheral membrane protein |
| detergent | single-pass transmembrane protein |
| electron crystallography | spectrin |
| glycosylphosphatidylinositol (GPI) anchor | transmembrane protein |
| integral membrane protein | |

### DEFINITIONS

Match the definition below with its term from the list above.

**10–33** Protein that binds tightly to a specific sugar.

**10–34** The outer coat of a eucaryotic cell, composed of oligosaccharides linked to intrinsic plasma membrane glycoproteins and glycolipids, as well as proteins that have been secreted and reabsorbed onto the cell surface.

**10–35** Abundant protein associated with the cytosolic side of the plasma membrane in red blood cells, forming a rigid network that supports the membrane.

**10–36** Protein whose polypeptide chain crosses the lipid bilayer more than once.

**10–37** Pigmented protein found in the plasma membrane of *Halobacterium halobium*, where it pumps protons out of the cell in response to light.

**10–38** The complicated cytoskeletal network in the cytosol just beneath the plasma membrane.

**10–39** Protein that is attached to one face of a membrane by noncovalent interactions with other membrane proteins and can be removed by relatively gentle treatments that leave the bilayer intact.

**10–40** Type of lipid linkage, formed as proteins pass through the endoplasmic reticulum, by which some proteins are attached to the noncytosolic surface of the membrane.

### TRUE/FALSE

Decide whether each of these statements is true or false, and then explain why.

**10–41** The basic structure of biological membranes is determined by the lipid bilayer, but their specific functions are carried out largely by proteins.

**10–42** Hydropathy plots are useful for identifying hydrophobic polypeptide segments that are long enough to span a membrane as an α helix or as a β sheet.

**10–43** Whereas all the carbohydrate in the plasma membrane faces outward on the external surface of the cell, all the carbohydrate on internal membranes faces toward the cytosol.

**10–44** Human red blood cells contain no internal membranes other than the nuclear membrane.

**10–45** Each molecule of bacteriorhodopsin contains a single chromophore called retinal, which, when activated by a photon of light, causes a series of small conformational changes in the protein that results in the transfer of protons from the inside to the outside of the cell.

**10–46** Although membrane domains with different protein compositions are well known, there are at present no examples of membrane domains that differ in lipid composition.

## THOUGHT PROBLEMS

**10–47** Which of the arrangements of membrane-associated proteins indicated in Figure 10–8 have been found in biological membranes?

**10–48** Compare the hydrophobic forces that hold a membrane protein in the lipid bilayer with those that help proteins fold into a unique three-dimensional structure.

**10–49** Which one of the following statements correctly describes the mass ratio of lipids to proteins in membranes?

A. The mass of lipids greatly exceeds the mass of proteins.
B. The mass of proteins greatly exceeds the mass of lipids.
C. The masses of lipids and proteins are about equal.
D. The mass ratio of lipids to proteins varies widely in different membranes.

**10–50** Name the three general types of lipid anchors that are used to attach proteins to membranes.

**10–51** Monomeric single-pass transmembrane proteins span a membrane with a single α helix that has characteristic chemical properties in the region of the bilayer. Which of the three 20-amino acid sequences listed below is the most likely candidate for such a transmembrane segment? Explain the reasons for your choice. (See inside back cover for one-letter amino acid code; FAMILY VW is a convenient mnemonic for hydrophobic amino acids.)

A. I T L I Y F G V M A G V I G T I L L I S
B. I T P I Y F G P M A G V I G T P L L I S
C. I T E I Y F G R M A G V I G T D L L I S

**10–52** Consider a transmembrane protein complex that forms a hydrophilic pore across the plasma membrane of a eucaryotic cell. The pore is made of five

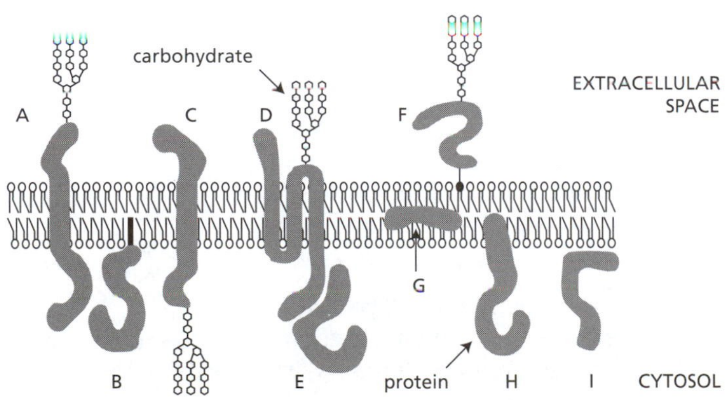


**Figure 10–8** A variety of possible associations of proteins with a membrane (Problem 10–47).

similar protein subunits, each of which contributes a membrane-spanning α helix to form the pore. Each α helix has hydrophilic amino acid side chains on one side of the helix and hydrophobic amino acid side chains on the opposite side. Propose a possible arrangement of these five α helices in the membrane.

**10–53** Proteins that form a β-barrel pore in the membrane have several β strands that span the membrane. The pore-facing side of each strand carries hydrophilic amino acid side chains, whereas the bilayer-facing sides carry hydrophobic amino acid side chains. Which of the three 10-amino acid sequences listed below is the most likely candidate for a transmembrane β strand in a β-barrel pore? Explain the reasons for your choice. (See inside back cover for one-letter amino acid code.)

A. A D F K L S V E L T
B. A F L V L D K S E T
C. A F D K L V S E L T

**10–54** Why is it that membrane-spanning protein segments are almost always α helices or β barrels, but never disordered chains?

**10–55** Why do you suppose that α helices are more common than β barrels in transmembrane proteins?

**10–56** You are studying the binding of proteins to the cytoplasmic face of cultured neuroblastoma cells and have found a method that gives a good yield of inside-out vesicles from the plasma membrane. Unfortunately, your preparations are contaminated with variable amounts of right-side-out vesicles. Nothing you have tried avoids this problem. A friend suggests that you pass your vesicles over an affinity column made of lectin coupled to solid beads. What is the point of your friend's suggestion?

**10–57** Why is it that intrachain (and interchain) disulfide (S–S) bonds form readily between cysteine side chains (–SH) outside the cell but not in the cytosol?

**10–58** Detergents are small amphiphilic molecules—one end hydrophobic and the other hydrophilic—that tend to form micelles in water. Examine the structures of SDS and Triton X-100 in Figure 10–9 and explain why the black portions are hydrophilic and the gray sections are hydrophobic.

**10–59** Why does a red blood cell membrane need proteins?

**10–60** Pythagoras forbade his followers to eat fava beans. Beyond the political implications (Greeks voted with beans), there turns out to be a rational basis for this proscription. In the Middle East defective forms of the gene encoding glucose-6-phosphate dehydrogenase (G6PD) are common. These mutant forms of the gene typically reduce G6PD activity to about 10% of normal. They have been selected for in the Middle East, and in other areas of the world where malaria is common, because they afford protection against the malarial parasite. G6PD controls the first step in the pathway for NADPH production. A lower-than-normal level of NADPH in red blood cells creates an environment unfavorable for growth of the protist *Plasmodium falciparum*, which causes malaria.

Although somewhat protected against malaria, G6PD-deficient individuals occasionally have other problems. NADPH is the principal agent required to keep the red cell cytosol in a properly reduced state, constantly converting transient disulfide bonds (–S–S–) back to sulfhydryls (–SH HS–). When a G6PD-deficient individual eats raw or undercooked fava beans, an oxidizing substance in the beans overwhelms the reducing capacity of the red cells, leading to a severe—sometimes life-threatening—hemolytic anemia. How do you suppose eating fava beans leads to anemia?

**10–61** Glycophorin, a protein in the plasma membrane of the red blood cell, normally exists as a homodimer that is held together entirely by interactions between its transmembrane domains. Since transmembrane domains are

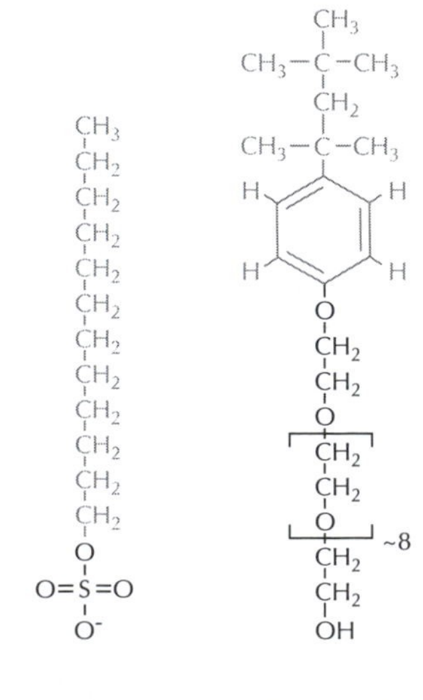

sodium dodecyl sulfate (SDS)

Triton X-100

**Figure 10–9** The structures of SDS and Triton X-100 (Problem 10–58).

**Table 10–3 Proportion of stain associated with three membrane-associated proteins** (Problem 10–65).

| PROTEIN | MOLECULAR WEIGHT | PERCENT OF STAIN |
|---|---|---|
| Spectrin | 250,000 | 25 |
| Band 3 | 100,000 | 30 |
| Glycophorin | 30,000 | 2.3 |

hydrophobic, how is it that they can associate with one another so specifically?

**10–62** Describe the different methods that cells use to restrict proteins to specific regions of the plasma membrane. Is a membrane with many anchored proteins still fluid?

**10–63** Many cells interact with the extracellular matrix. Why is there any ambiguity in 'where the plasma membrane ends and the extracellular matrix begins'?

## CALCULATIONS

**10–64** In the membrane of a human red blood cell the ratio of the mass of protein (average molecular weight 50,000) to phospholipid (average molecular weight 800) to cholesterol (molecular weight 386) is about 2:1:1. How many lipid molecules (phospholipid + cholesterol) are there for every protein molecule?

**10–65** Estimates of the number of membrane-associated proteins per cell and the fraction of the plasma membrane occupied by such proteins provide a useful quantitative basis for understanding the structure of the plasma membrane. These calculations are straightforward for proteins in the plasma membrane of a red blood cell because red cells are readily isolated from blood and they contain no internal membranes to confuse the issue. Plasma membranes are prepared, the membrane-associated proteins are separated by SDS polyacrylamide-gel electrophoresis, and then they are stained with a dye (Coomassie Blue). Because the intensity of color is roughly proportional to the mass of protein present in a band, quantitative estimates can be made as shown in Table 10–3.

A. From the information in Table 10–3, calculate the number of molecules of spectrin, band 3, and glycophorin in an individual red blood cell. Assume that 1 mL of red cell ghosts contains $10^{10}$ cells and 5 mg of total membrane protein.

B. Calculate the fraction of the plasma membrane that is occupied by band 3. Assume that band 3 is a cylinder 3 nm in radius and 10 nm in height and is oriented in the membrane as shown in Figure 10–10. The total surface area of a red cell is $10^8$ nm$^2$.

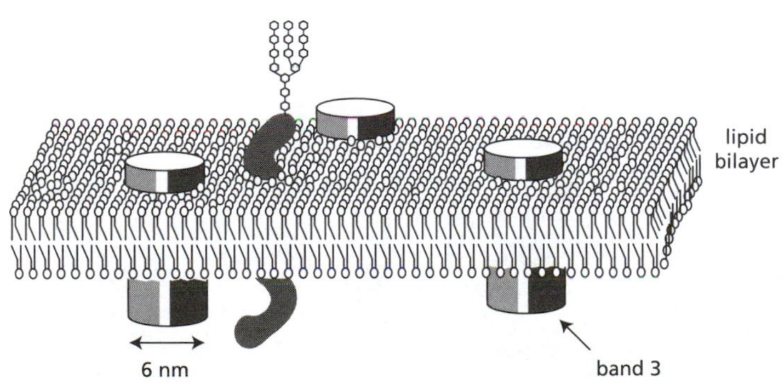


**Figure 10–10** Schematic diagram of band 3, represented as a cylinder, in the plasma membrane (Problem 10–65).

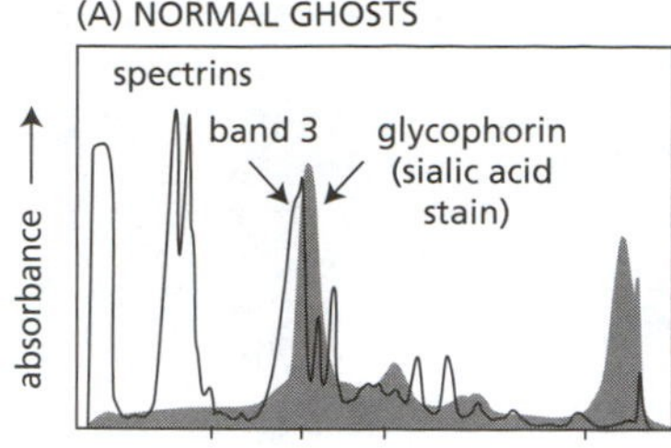


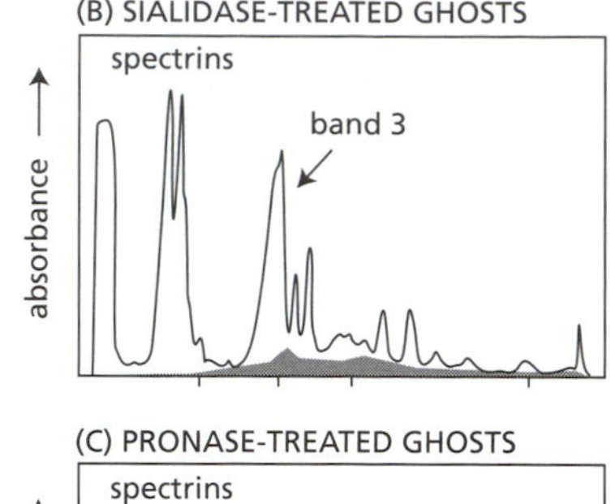


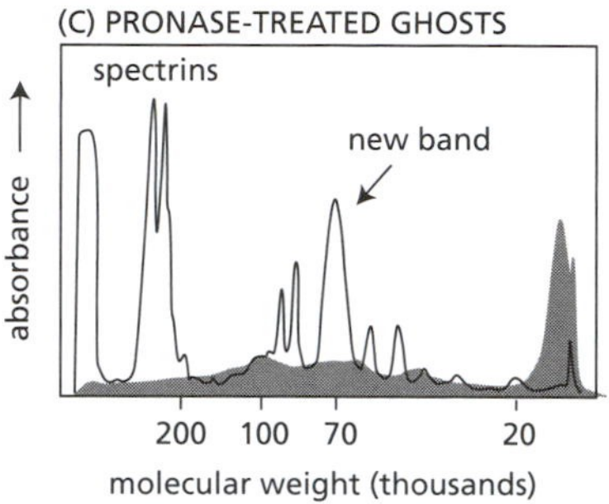


**Figure 10–11** Analysis of proteins in the plasma membrane of red cell ghosts (Problem 10–66). (A) Untreated red cell ghosts. (B) Sialidase-treated red cell ghosts. (C) Pronase-treated red cell ghosts. Membrane-associated proteins were separated by SDS polyacrylamide-gel electrophoresis and then stained for protein and for carbohydrate. *Lines* indicate the distribution of proteins, and *shaded regions* indicate the distribution of carbohydrate.

## DATA HANDLING

**10–66** Enzymatic digestion of sealed right-side-out red cell ghosts was originally used to determine the sidedness of the major membrane-associated proteins: spectrin, band 3, and glycophorin. These experiments made use of sialidase, which removes sialic acid from protein, and pronase, which cleaves peptide bonds. The proteins from normal ghosts and enzyme-treated ghosts were separated by SDS polyacrylamide-gel electrophoresis and then stained for protein and carbohydrate (Figure 10–11).

A. How does the information in Figure 10–11 allow you to decide whether the carbohydrate of glycophorin is on the cytoplasmic or the external surface, and how does it allow you to decide which of the red cell proteins are exposed on the external side of the cell?

B. When you show your deductions to a colleague, she challenges your conclusion that some proteins are not exposed on the external surface and suggests instead that these proteins may be resistant to pronase digestion. What control experiment can you propose to test this possibility?

C. How would you modify this enzymatic approach in order to determine which red cell proteins span the plasma membrane?

**10–67** One difficult problem in molecular biology is to define the associations between different proteins in complex assemblies. The associations involving spectrin, ankyrin, band 3, and actin, which generate the filamentous meshwork on the cytoplasmic surface of the red blood cell plasma membrane, have been investigated in several ways. One general method is to use antibodies that are specific for individual proteins. A mixture of two proteins is incubated together, and then an antibody specific for one of them is added. The resulting antibody–protein complexes are then precipitated and analyzed. This technique, when applied to pairwise mixtures of spectrin, ankyrin, band 3, and actin, yields the results summarized in Table 10–4. From the information in the table, deduce the associations between these proteins.

**10–68** Look carefully at the transmembrane band 3 proteins in Figure 10–12A. Imagine that you could mark all the band 3 proteins specifically with a fluorescent group and measure their mobility by fluorescence recovery after photobleaching (FRAP). Sketch the recovery curve you would expect to see with time after photobleaching a small spot in the membrane (Figure 10–12B).

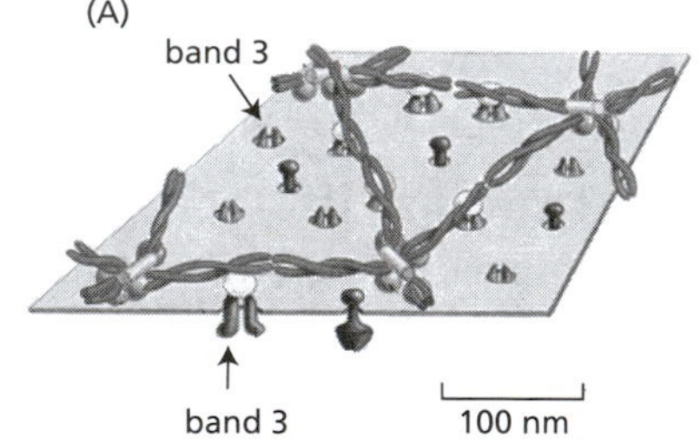


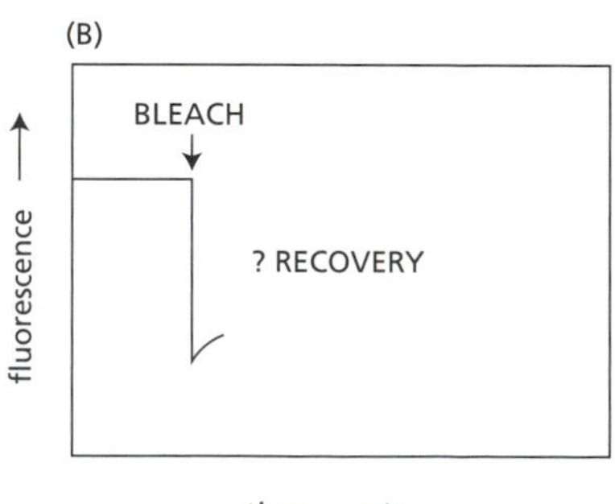


**Figure 10–12** Mobility of membrane proteins (Problem 10–68). (A) Model of red blood cell membrane. (B) Recovery of fluorescence after photobleaching a red cell plasma membrane containing band 3 protein tagged with a fluorescent group.

**Table 10–4 Precipitation of red blood cell plasma membrane proteins by antibodies specific for individual proteins** (Problem 10–67).

| PROTEIN MIXTURE | ANTIBODY SPECIFICITY | PROTEINS IN PELLET |
|---|---|---|
| 1. Band 3 + actin | actin | actin |
| 2. Band 3 + spectrin | spectrin | spectrin |
| 3. Band 3 + ankyrin | ankyrin | band 3 + ankyrin |
| 4. Actin + spectrin | spectrin | actin + spectrin |
| 5. Actin + ankyrin | ankyrin | ankyrin |
| 6. Spectrin + ankyrin | spectrin | spectrin + ankyrin |

# Membrane Transport of Small Molecules and the Electrical Properties of Membranes

# 11

## In This Chapter

PRINCIPLES OF MEMBRANE TRANSPORT 243

TRANSPORTERS AND ACTIVE MEMBRANE TRANSPORT 246

ION CHANNELS AND THE ELECTRICAL PROPERTIES OF MEMBRANES 252

## PRINCIPLES OF MEMBRANE TRANSPORT

| TERMS TO LEARN | |
|---|---|
| active transport | membrane transport protein |
| channel | passive transport (facilitated diffusion) |
| electrochemical gradient | transporter |

### DEFINITIONS

Match each definition below with its term from the list above.

**11–1** An aqueous pore in a lipid membrane, with walls made of protein, through which selected ions or molecules can pass.

**11–2** The movement of a small molecule or ion across a membrane due to a difference in concentration or electrical charge.

**11–3** General term for a membrane-embedded protein that serves as a carrier of ions or small molecules from one side of the membrane to the other.

**11–4** Movement of a molecule across a membrane that is driven by ATP hydrolysis or other form of metabolic energy.

**11–5** Driving force for ion movement that is due to differences in ion concentration and electrical charge on either side of the membrane.

### TRUE/FALSE

Decide whether each of these statements is true or false, and then explain why.

**11–6** The plasma membrane is highly impermeable to all charged molecules.

**11–7** Transport by transporters can be either active or passive, whereas transport by channels is always passive.

### THOUGHT PROBLEMS

**11–8** Order the molecules on the following list according to their ability to diffuse through a lipid bilayer, beginning with the one that crosses the bilayer most readily. Explain your order.

1. $Ca^{2+}$
2. $CO_2$
3. Ethanol
4. Glucose
5. RNA
6. $H_2O$

**11–9** Why are the maximum rates of transport by transporters and channels thought to be so different?

**11–10** Transporters were once imagined to operate like revolving doors, transporting their ligands while maintaining a sealed lipid bilayer. Why do you suppose this mechanism is no longer considered likely?

**11–11** A simple enzyme reaction can be represented by the equation

$$E + S \rightleftharpoons ES \rightarrow E + P$$

where E is the enzyme, S is the substrate, P is the product, and ES is the enzyme–substrate complex.

A. Write a corresponding equation describing a transporter (T) that mediates transport of a solute (S) down its concentration gradient.

B. The Michaelis–Menten equation for the simple enzyme reaction above is

$$\text{rate} = V_{max} \frac{[S]}{[S] + K_m}$$

Where 'rate' is the initial rate of the reaction, $V_{max}$ is the maximum rate of the enzyme-catalyzed reaction, and $K_m$ is the Michaelis constant. Write the corresponding Michaelis–Menten equation for the process of solute transport by a transporter. What do rate, $V_{max}$, and $K_m$ mean in the equation for transport?

C. Would these equations provide an appropriate description for channels? Why or why not?

**Problems 3–70, 3–92, 3–93,** and **3–94** introduce and elaborate on the equations for enzyme kinetics.

**11–12** How is it possible for some molecules to be at equilibrium across a biological membrane and yet not be at the same concentration on both sides?

**11–13** Ionophores are small hydrophobic molecules that dissolve in lipid bilayers and increase the permeability of the bilayer to specific inorganic ions. There are two classes of ionophore—mobile ion carriers, which move within the bilayer, and channel formers, which span the bilayer. Both types operate by shielding the charge on the transported ion so that the ion can penetrate the hydrophobic interior of the lipid bilayer. How would you expect the activities of a channel-forming ionophore and mobile ion carrier to change as you lowered the temperature of a lipid bilayer, increasing its viscosity?

## CALCULATIONS

**11–14** Brain cells, which depend on glucose for energy, use the glucose transporter GLUT3, which has a $K_m$ of 1.5 mM. Liver cells, which store glucose (as glycogen) after a meal and release glucose between meals, use the glucose transporter GLUT2, which has a $K_m$ of 15 mM.

A. Calculate the rate (as a percentage of $V_{max}$) of glucose uptake in brain cells and in liver cells at circulating glucose concentrations of 3 mM (starvation conditions), 5 mM (normal levels), and 7 mM (after a carbohydrate-rich meal). Rearranging the Michaelis–Menten equation gives

$$\frac{\text{rate}}{V_{max}} = \frac{[S]}{[S] + K_m}$$

B. Although the concentration of glucose in the general circulation normally doesn't rise much above 7 mM, the liver is exposed to much higher concentrations after a meal. The intestine delivers glucose into the portal circulation, which goes directly to the liver. In the portal circulation the concentration of glucose can be as high as 15 mM. At what fraction of the maximum rate ($V_{max}$) do liver cells import glucose at this concentration?

C. Do these calculations fit with the physiological functions of brain and liver cells? Why or why not?

**11–15** Cells use transporters to move nearly all metabolites across membranes. But how much faster is a transporter than simple diffusion? There is sufficient information available for glucose transporters to make a comparison. The normal circulating concentration of glucose in humans is 5 mM, whereas the intracellular concentration is usually very low. (For this problem assume the internal concentration of glucose is 0 mM.)

A. At what rate (molecules/sec) would glucose diffuse into a cell if there were no transporter? The permeability coefficient for glucose is $3 \times 10^{-8}$ cm/sec. Assume a cell is a sphere with a diameter of 20 µm. The rate of diffusion equals the concentration difference multiplied by the permeability coefficient and the total surface area of the cell (surface area = $4\pi r^2$). (Remember to convert everything to compatible units so that the rate is molecules/sec.)

B. If in the same cell there are $10^5$ GLUT3 molecules ($K_m$ = 1.5 mM) in the plasma membrane, each of which can transport glucose at a maximum rate of $10^4$ molecules per second, at what rate (molecules/sec) will glucose enter the cell? How much faster is transporter-mediated uptake of glucose than entry by simple diffusion?

## DATA HANDLING

**11–16** Some bacterial cells can grow on either ethanol ($CH_3CH_2OH$) or acetate ($CH_3COO^-$) as their only carbon source. Initial rates of entry of these two molecules as a function of external concentration are shown in Table 11–1, but the identity of the molecules is not shown.

A. Plot the data from the table as initial rate versus concentration.

B. What can you say about the transport of the two molecules from your plot of the data? From the graphs deduce which molecule is ethanol and which is acetate. Rationalize your choice.

C. If it is possible to do so from the data you have plotted, estimate the $V_{max}$ and $K_m$ values for ethanol and acetate.

**11–17** Cytochalasin B, which is often used as an inhibitor of actin-based motility systems, is also a very potent competitive inhibitor of D-glucose uptake into mammalian cells. When red blood cell ghosts (red cells, emptied of their cytoplasm) are incubated with $^3$H-cytochalasin B and then irradiated with ultraviolet light, the cytochalasin becomes cross-linked to the glucose transporter, GLUT1. Cytochalasin is not cross-linked to the transporter if an excess of D-glucose is present during the labeling reaction; however, an excess of L-glucose (which is not transported) does not interfere with labeling.

If membrane proteins from labeled ghosts are separated by SDS polyacrylamide-gel electrophoresis, GLUT1 appears as a fuzzy radioactive band extending from 45,000 to 70,000 daltons. If labeled ghosts are treated with an enzyme that removes attached sugars before electrophoresis, the fuzzy band disappears and a much sharper band at 46,000 daltons takes its place.

A. Why does an excess of D-glucose, but not L-glucose, prevent cross-linking of cytochalasin to GLUT1?

B. Why does GLUT1 appear as a fuzzy band on SDS polyacrylamide gels?

**11–18** Insulin is a small protein hormone that binds to receptors in the plasma membranes of many cells. In fat cells this binding dramatically increases the rate of uptake of glucose into the cells. The increase occurs within minutes and is not blocked by inhibitors of protein synthesis or of glycosylation. These results suggest that insulin increases the activity of the glucose

**Table 11–1 Initial rates of entry of ethanol and acetate** (Problem 11–16).

| CARBON SOURCE (mM) | INITIAL RATE OF ENTRY (µmol/min) | |
|---|---|---|
| | MOLECULE A | MOLECULE B |
| 0.1 | 2.0 | 18 |
| 0.3 | 6.0 | 46 |
| 1.0 | 20 | 100 |
| 3.0 | 60 | 150 |
| 10.0 | 200 | 182 |

**Table 11–2 Amount of GLUT4 associated with the plasma membrane and internal membranes in the presence and absence of insulin** (Problem 11–18).

| MEMBRANE FRACTION | BOUND $^3$H-CYTOCHALASIN B (cpm/mg protein) | |
|---|---|---|
| | UNTREATED CELLS (– INSULIN) | TREATED CELLS (+ INSULIN) |
| Plasma membrane | 890 | 4480 |
| Internal membranes | 4070 | 80 |

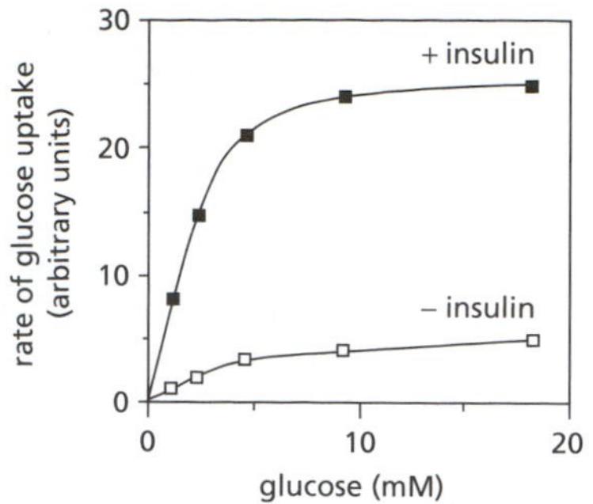


**Figure 11–1** Rate of glucose uptake into cells in the presence and absence of insulin (Problem 11–18).

transporter, GLUT4, in the plasma membrane, without increasing the total number of GLUT4 molecules in the cell.

The two experiments described below suggest a possible mechanism for this insulin effect. In the first experiment, the initial rate of glucose uptake in control and insulin-treated cells was measured, with the results shown in Figure 11–1. In the second experiment, the concentration of GLUT4 in fractionated membranes from control and insulin-treated cells was measured, using the binding of radioactive cytochalasin B as the assay (see Problem 11–17), as shown in Table 11–2.

A. Deduce the mechanism by which glucose transport through GLUT4 increases in insulin-treated cells.

B. Does insulin stimulation alter either the $K_m$ or the $V_{max}$ of GLUT4? How can you tell from these data?

**11–19** Gramicidin A is a channel-forming ionophore, one of the two classes of small hydrophobic molecules that enhance the passage of inorganic ions across the lipid bilayer (see Problem 11–13). Two gramicidin molecules come together end to end across the bilayer to form a pore that is selective for cations (Figure 11–2A). If a minute amount of gramicidin A is added to an artificial bilayer, movement of cations across the bilayer can be detected as changes in current, as shown in Figure 11–2B. What do the step-wise changes in current tell you about the gramicidin channel? Why are some peaks twice as high as others?

# TRANSPORTERS AND ACTIVE MEMBRANE TRANSPORT

TERMS TO LEARN

| | | |
|---|---|---|
| ABC transporter | multidrug resistance (MDR) protein | P-type pump |
| antiporter | $Na^+$-$K^+$ ATPase | symporter |
| $Ca^{2+}$ pump ($Ca^{2+}$ ATPase) | $Na^+$-$K^+$ pump ($Na^+$ pump) | transcellular transport |
| F-type pump | osmolarity | uniporter |
| lactose permease | | |

## DEFINITIONS

Match each definition below with its term from the list above.

**11–20** The concentration of a solute expressed in terms of the osmotic pressure it can exert.

**11–21** Large superfamily of membrane transport proteins that use the energy of ATP hydrolysis to transfer peptides and a variety of small molecules across membranes.

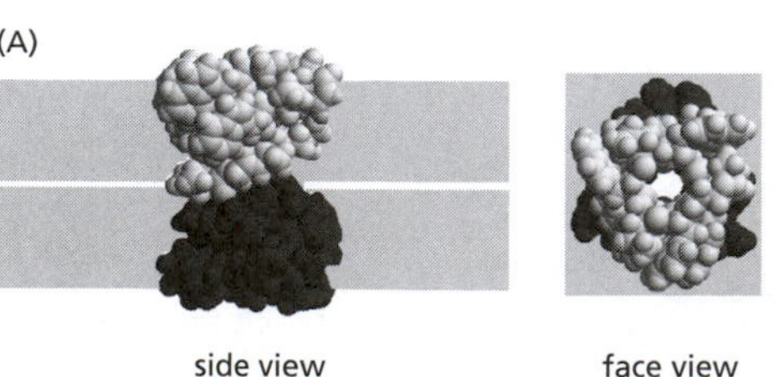


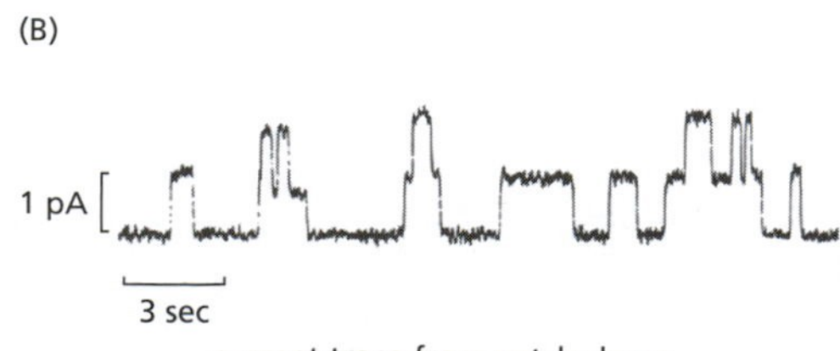


**Figure 11–2** Channels induced in a membrane by gramicidin A (Problem 11–19). (A) Space-filling model of gramicidin A in side and face views. (B) Currents across a lipid bilayer exposed to a minute amount of gramicidin A. The timescale of the trace is shown in seconds (sec) and the current is shown in picoamperes (pA). A 1-pA current corresponds to about $6 \times 10^6$ ions per second.

**11–22** Type of ABC transporter protein that can pump hydrophobic drugs (such as some anticancer drugs) out of the cytoplasm of eucaryotic cells.

**11–23** Membrane carrier protein that transports two different ions or small molecules across a membrane in opposite directions, either simultaneously or in sequence.

**11–24** Transport of solutes across an epithelium, by means of membrane transport proteins in the apical and basal surfaces of the epithelial cells.

**11–25** Carrier protein that transports two types of solute across the membrane in the same direction.

**11–26** Carrier protein that transports a single solute from one side of the membrane to the other.

## TRUE/FALSE

Decide whether each of these statements is true or false, and then explain why.

**11–27** A symporter would function as an antiporter if its orientation in the membrane were reversed (that is, if the portion of the protein normally exposed to the cytosol faced the outside of the cell instead).

**11–28** The co-transport of $Na^+$ and a solute into a cell, which harnesses the energy in the $Na^+$ gradient, is an example of primary active transport.

**11–29** In response to depolarization of the muscle cell plasma membrane, the $Ca^{2+}$ pumps in the sarcoplasmic reticulum (SR) use the energy of ATP hydrolysis to move $Ca^{2+}$ from the lumen of the SR to the cytosol to initiate muscle contraction.

## THOUGHT PROBLEMS

**11–30** What are the three main ways in which cells carry out active transport? Briefly describe each.

**11–31** Which of the ions listed in Table 11–3 could be used to drive an electrically neutral coupled transport of a solute across the plasma membrane? Indicate the direction of movement of the listed ions (inward or outward) and indicate what sort of ion would be co-transported to preserve electrical neutrality.

Incidentally, there is a glaring intracellular deficiency of anions relative to cations in Table 11–3, yet cells are electrically neutral. What anions do you suppose are missing from this table?

**Table 11–3 A comparison of ion concentrations inside and outside a typical mammalian cell** (Problem 11–31).

| COMPONENT | INTRACELLULAR CONCENTRATION (mM) | EXTRACELLULAR CONCENTRATION (mM) |
|---|---|---|
| Cations | | |
| $Na^+$ | 5–15 | 145 |
| $K^+$ | 140 | 5 |
| $Mg^{2+}$ | 0.5 | 1–2 |
| $Ca^{2+}$ | $10^{-4}$ | 1–2 |
| $H^+$ | $7 \times 10^{-5}$ ($10^{-7.2}$ M or pH 7.2) | $4 \times 10^{-5}$ ($10^{-7.4}$ M or pH 7.4) |
| Anions | | |
| $Cl^-$ | 5–15 | 110 |

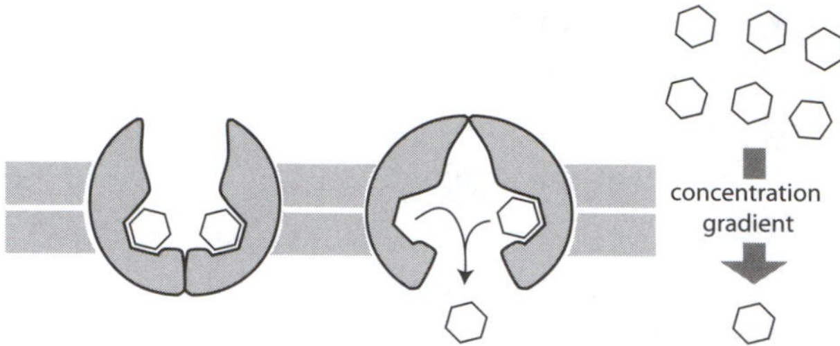


**Figure 11–3** Hypothetical model showing how a conformational change in a transporter could mediate passive transport of glucose (Problem 11–33). The transition between the two conformational states is proposed to occur randomly and to be completely reversible, regardless of binding site occupancy.

**11–32** A transmembrane protein has the following properties: it has two binding sites, one for solute A and one for solute B. The protein can undergo a conformational change to switch between two states: either both binding sites are exposed exclusively on one side of the membrane, or both are exposed exclusively on the other side of the membrane. The protein can switch between the two conformational states only if both binding sites are occupied or if both binding sites are empty, but cannot switch if only one binding site is occupied.

A. What kind of a transporter do these properties define?

B. Do you need to specify any additional properties to turn this protein into a transporter that couples the movement of solute A up its concentration gradient to the movement of solute B down its electrochemical gradient?

C. Write a set of rules like those in the body of this problem that defines the properties of an antiporter.

**11–33** A model for a uniporter that could mediate passive transport of glucose down its concentration gradient is shown in Figure 11–3. How would you need to change the diagram to convert the transporter into a pump that transports glucose up its concentration gradient by hydrolyzing ATP? Explain the need for each of the steps in your new illustration.

**11–34** Ion transporters are 'linked' together—not physically, but as a consequence of their actions. For example, cells can raise their intracellular pH, when it becomes too acidic, by exchanging external $Na^+$ for internal $H^+$, using a $Na^+$-$H^+$ antiporter. The change in internal $Na^+$ is then redressed using the $Na^+$-$K^+$ pump.

A. Can these two transporters, operating together, normalize both the $H^+$ and the $Na^+$ concentrations inside the cell?

B. Does the linked action of these two pumps cause imbalances in either the $K^+$ concentration or the membrane potential?

**11–35** Why does export of $HCO_3^-$ out of a cell via the $Cl^-$-$HCO_3^-$ exchanger lower the intracellular pH?

**11–36** Cells continually generate $CO_2$ as a consequence of the oxidation of glucose and fatty acids that is required to meet their energy needs.

A. If unopposed by any transport process, what effect would the continual production of $CO_2$ have on intracellular pH?

B. How does the $Na^+$-driven $Cl^-$-$HCO_3^-$ exchanger counteract the changes in intracellular pH?

C. Does the combination of metabolic oxidation and $Na^+$-driven $Cl^-$-$HCO_3^-$ exchange increase the total concentration of $CO_2$ + $HCO_3^-$ in the cell, decrease it, or keep it constant?

D. How do you suppose $CO_2$ exits from the cell?

**11–37** $CO_2$ is removed from the body through the lungs in a process that is mediated by red blood cells, as summarized in Figure 11–4. Transport of $CO_2$ is coupled to the transport of $O_2$ through hemoglobin. Upon release of $O_2$ in the tissues, hemoglobin undergoes a conformational change that raises the p$K$ of a histidine side chain, allowing it to bind an $H^+$, which is generated by hydration of $CO_2$. This process occurs in reverse in the lungs when $O_2$ is bound to hemoglobin.

A. To what extent does the intracellular pH of the red blood cell vary during its movement from the tissues to the lungs and back, and why is this so?

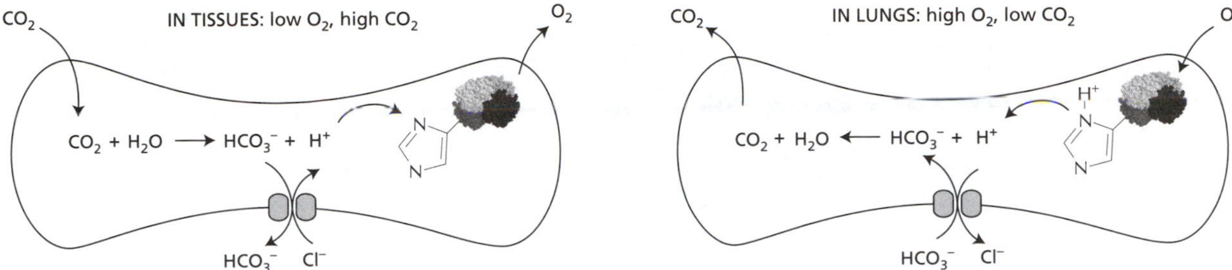


**Figure 11–4** Red blood cell-mediated transport of $CO_2$ from the tissues to the lungs (Problem 11–37). Low and high $O_2$ and $CO_2$ refer to their concentrations, or partial pressures.

B. In what form, and where, is the $CO_2$ during its movement from the tissues to the lungs?

C. How is it that the $Cl^-$-$HCO_3^-$ exchanger operates in one direction in the tissues and in the opposite direction in the lungs?

**11–38** A rise in the intracellular $Ca^{2+}$ concentration causes muscle cells to contract. In addition to an ATP-driven $Ca^{2+}$ pump, heart muscle cells, which contract quickly and regularly, have an antiporter that exchanges $Ca^{2+}$ for extracellular $Na^+$ across the plasma membrane. This antiporter rapidly pumps most of the entering $Ca^{2+}$ ions back out of the cell, allowing the cell to relax. Ouabain and digitalis, drugs that are used in the treatment of patients with heart disease, make the heart contract more strongly. Both drugs function by partially inhibiting the $Na^+$-$K^+$ pump in the membrane of the heart muscle cell. Can you propose an explanation for the effects of these drugs in patients? What will happen if too much of either drug is taken?

**11–39** You have prepared lipid vesicles (spherical lipid bilayers) that contain $Na^+$-$K^+$ pumps as the sole membrane protein. Assume for the sake of simplicity that each pump transports one $Na^+$ one way and one $K^+$ the other way in each pumping cycle, as illustrated in Figure 11–5. All of the $Na^+$-$K^+$ pumps are oriented so that the portion of the molecule that normally faces the cytosol faces the outside of the vesicle. Predict what would happen under each of the following conditions.

A. The solution inside and outside the vesicles contains both $Na^+$ and $K^+$ ions, but no ATP.

B. The solution inside the vesicles contains both $Na^+$ and $K^+$ ions; the solution outside contains both ions, as well as ATP.

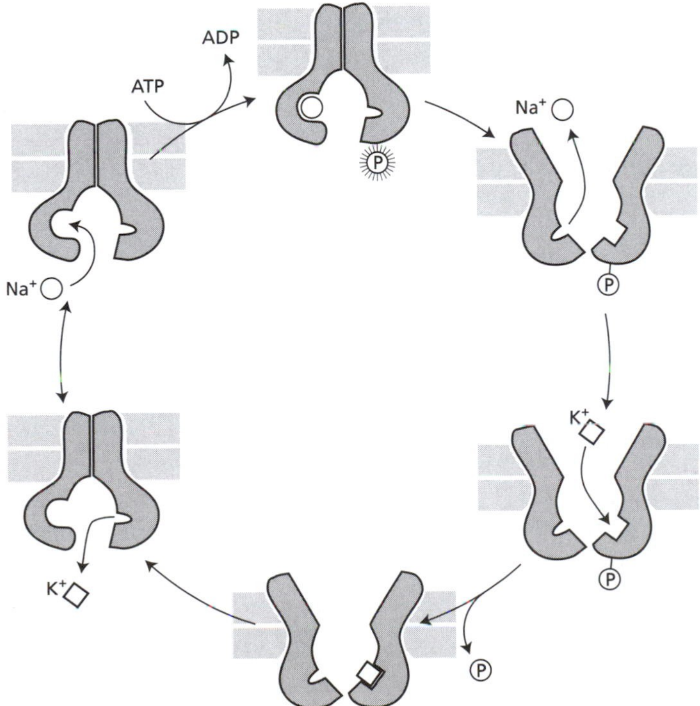


**Figure 11–5** The $Na^+$-$K^+$ pump (Problem 11–39).

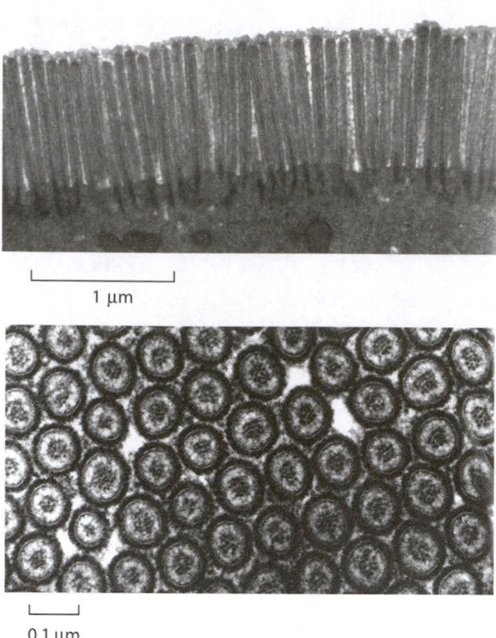


**Figure 11–6** Microvilli of intestinal epithelial cells in profile and cross section (Problem 11–40).

C. The solution inside contains $Na^+$; the solution outside contains $Na^+$ and ATP.
D. The solution is as in B, but the $Na^+$-$K^+$ pump molecules are randomly oriented, some facing one direction, some the other.

## CALCULATIONS

**11–40** A principal function of the plasma membrane is to control the entry of nutrients into the cell. This function is especially important for the epithelial cells that line the gut because they are responsible for absorbing virtually all the nutrients that enter the body. As befits this critical role, their plasma membranes are specialized so that the surface facing the gut is folded into numerous fingerlike projections, termed microvilli. Microvilli increase the surface area of the intestinal cells, providing more efficient absorption of nutrients. Microvilli are shown in profile and cross section in Figure 11–6. From the dimensions given in the figure, estimate the increase in surface area that microvilli provide (for the portion of the plasma membrane in contact with the lumen of the gut) relative to the corresponding surface of a cell with a 'flat' plasma membrane.

**11–41** How much energy does it take to pump substances across membranes? Or, to put it another way, since active transport is usually driven directly or indirectly by ATP, how steep a gradient can ATP hydrolysis maintain for a particular solute? For transport into the cell the free-energy change ($\Delta G_{in}$) per mole of solute moved across the plasma membrane is

$$\Delta G_{in} = -2.3RT \log \frac{C_o}{C_i} + zFV$$

Where $R$ = the gas constant, $1.98 \times 10^{-3}$ kcal/K mole
$T$ = the absolute temperature in K (37°C = 310 K)
$C_o$ = solute concentration outside the cell
$C_i$ = solute concentration inside the cell
$z$ = the valence (charge) on the solute
$F$ = Faraday's constant, 23 kcal/V mole
$V$ = the membrane potential in volts (V)

Since $\Delta G_{in} = -\Delta G_{out}$, the free-energy change for transport out of the cell is

$$\Delta G_{out} = 2.3RT \log \frac{C_o}{C_i} - zFV$$

At equilibrium, where $\Delta G = 0$, the equations can be rearranged to the more familiar form known as the Nernst equation.

$$V = 2.3\ \frac{RT}{zF}\ \log\ \frac{C_o}{C_i}$$

For the questions below, assume that hydrolysis of ATP to ADP and $P_i$ proceeds with a $\Delta G$ of –12 kcal/mole; that is, ATP hydrolysis can drive active transport with a $\Delta G$ of +12 kcal/mole. Assume that $V$ is –60 mV.

A. What is the maximum concentration gradient that can be achieved by the ATP-driven active transport into the cell of an uncharged molecule such as glucose, assuming that 1 ATP is hydrolyzed for each solute molecule that is transported?

B. What is the maximum concentration gradient that can be achieved by active transport of $Ca^{2+}$ from the inside to the outside of the cell? How does this maximum compare with the actual concentration gradient observed in mammalian cells (see Table 11–3)?

C. Calculate how much energy it takes to drive the $Na^+$-$K^+$ pump. This remarkable molecular device transports five ions for every molecule of ATP that is hydrolyzed: 3 $Na^+$ out of the cell and 2 $K^+$ into the cell. The pump typically maintains internal $Na^+$ at 10 mM, external $Na^+$ at 145 mM, internal $K^+$ at 140 mM, and external $K^+$ at 5 mM. As shown in Figure 11–7, $Na^+$ is transported against the membrane potential, whereas $K^+$ is transported with it. (The $\Delta G$ for the overall reaction is equal to the sum of the $\Delta G$ values for transport of the individual ions.)

D. How efficient is the $Na^+$-$K^+$ pump? That is, what fraction of the energy available from ATP hydrolysis is used to drive transport?

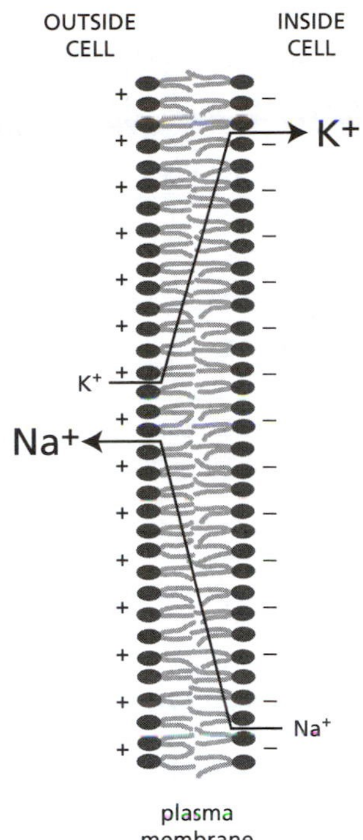


**Figure 11–7** $Na^+$ and $K^+$ gradients and direction of pumping across the plasma membrane (Problem 11–41). *Large letters* symbolize high concentrations and *small letters* symbolize low concentrations. Both $Na^+$ and $K^+$ are pumped against chemical concentration gradients; however, $Na^+$ is pumped up the electrical gradient, whereas $K^+$ is pumped down the electrical gradient.

## DATA HANDLING

**11–42** If you have ever used the standard probe on a pH meter, you may well wonder how pH could possibly be measured in the tiny volumes inside cellular compartments. The recent development of pH-sensitive fluorophores has simplified this difficult task immensely. One such fluorescent indicator is a hydrophobic ester of SNARF-1, which can enter cells by passive diffusion and then is trapped inside after intracellular enzymes hydrolyze the ester bonds to liberate SNARF-1 (Figure 11–8). SNARF-1 absorbs light at 488 nm and emits fluorescent light with peaks at 580 nm and 640 nm. Emission spectra for SNARF-1 at pH 6.0 and pH 9.0 are shown in Figure 11–9. The p$K$ of SNARF-1 is 7.5.

A. Explain why the ester of SNARF-1 diffuses through membranes, whereas the cleaved form stays inside cells.

B. Why do you think there are two peaks of fluorescence (at 580 nm and at 640 nm) that change so dramatically in intensity with a change in pH (see Figure 11–9)? What features of SNARF-1 might be important in this?

C. What forms of SNARF-1 are present at pH 6.0 and what are their relative proportions? At pH 9.0? The Henderson–Hasselbalch equation describing the dissociation of a weak acid is pH = p$K$ + log ([salt]/[acid]).

**Figure 11–8** Structure of the ester and free forms of SNARF-1 (Problem 11–42). The blocking ester groups are shown as R. The acid (HA) and salt ($A^-$) forms of SNARF-1 are indicated. The two different resonance structures for the salt form of SNARF-1 are shown in *brackets*.

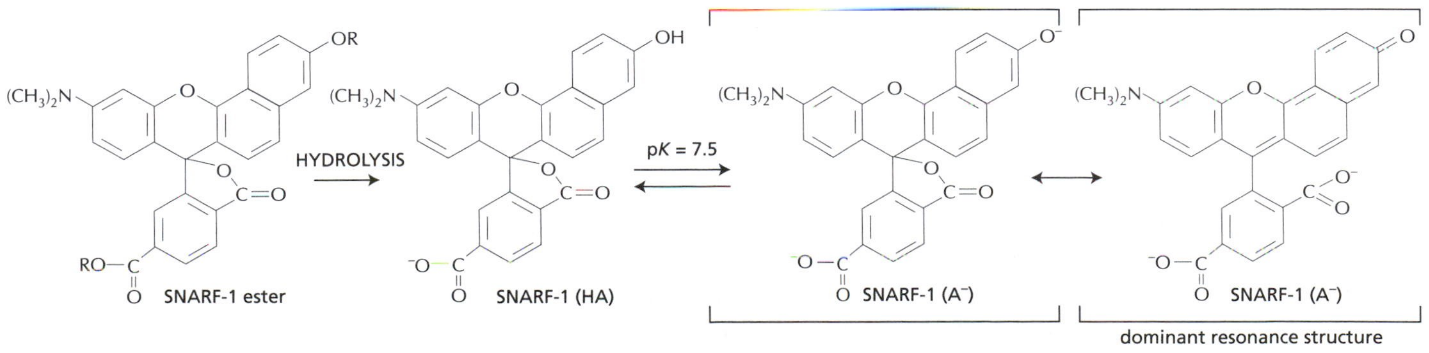

SNARF-1 ester — HYDROLYSIS → SNARF-1 (HA) ⇄ (p$K$ = 7.5) [ SNARF-1 ($A^-$) ⟷ SNARF-1 ($A^-$) ]

dominant resonance structure

D. Sketch an approximate curve for the SNARF-1 emission spectrum inside a cell at pH 7.2. (All such curves pass through the point where the two curves in Figure 11–9 cross.)

E. Why do you suppose indicators such as SNARF-1 that have emission spectra with two peaks are preferred to those that have a single peak?

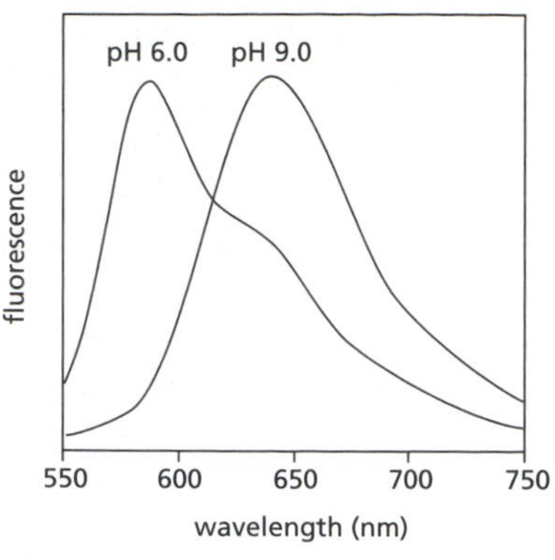


**Figure 11–9** Emission spectra of SNARF-1 at pH 6.0 and pH 9.0 (Problem 11–42). SNARF-1 in solution at pH 6.0 or pH 9.0 was excited by light at 488 nm and the intensity of fluorescence was determined at wavelengths from 550 nm to 750 nm.

# ION CHANNELS AND THE ELECTRICAL PROPERTIES OF MEMBRANES

| TERMS TO LEARN | | |
|---|---|---|
| acetylcholine receptor | initial segment | oligodendrocyte |
| action potential | ion channel | patch-clamp recording |
| adaptation | $K^+$ leak channel | resting membrane potential |
| AMPA receptor | long-term depression (LDP) | rapidly inactivating $K^+$ channel |
| aquaporin (water channel) | long-term potentiation (LTP) | Schwann cell |
| axon | membrane potential | selectivity filter |
| $Ca^{2+}$-activated $K^+$ channel | myelin sheath | synapse |
| delayed $K^+$ channel | Nernst equation | transmitter-gated ion channel |
| dendrite | neuron (nerve cell) | voltage-gated cation channel |
| excitatory neurotransmitter | neuromuscular junction | voltage-gated $K^+$ channel |
| glial cell | neurotransmitter | voltage-gated $Na^+$ channel |
| inhibitory neurotransmitter | NMDA receptor | |

## DEFINITIONS

Match each definition below with its term from the list above.

**11–43** The adjustment in sensitivity of a cell or organism following repeated stimulation that allows a response even when there is a high background level of stimulation.

**11–44** Extension of a nerve cell, typically branched and relatively short, that receives stimuli from other nerve cells.

**11–45** The long-lasting increase (days to weeks) in the sensitivity of certain synapses in the hippocampus that is induced by a short burst of repetitive firing in the presynaptic neurons.

**11–46** Rapid, transient, self-propagating electrical signal in the plasma membrane of a cell such as a neuron or muscle cell: a nerve impulse.

**11–47** Quantitative expression that relates the equilibrium ratio of concentrations of an ion on either side of a permeable membrane to the voltage difference across the membrane.

**11–48** Voltage difference across a membrane due to the slight excess of positive ions on one side and of negative ions on the other (typically –60 mV, inside negative, for an animal cell).

**11–49** General term for a membrane protein that selectively allows cations such as $Na^+$ to cross a membrane in response to changes in membrane potential.

**11–50** That part of an ion channel structure that determines which ions it can transport.

**11–51** Insulating layer of specialized cell membrane wrapped around vertebrate axons.

**11–52** A $K^+$-transporting ion channel in the plasma membrane of animal cells that remains open even in a 'resting' cell.

**11–53** Long thin nerve cell process capable of rapidly conducting nerve impulses over long distances so as to deliver signals to other cells.

**11–54** Transmembrane protein complex that forms a water-filled channel across the lipid bilayer through which specific inorganic ions can diffuse down their electrochemical gradients.

**11–55** Cell with long processes specialized to receive, conduct, and transmit signals in the nervous system.

**11–56** Specialized junction between a nerve cell and another cell, across which the nerve impulse is transferred, usually by a neurotransmitter, which is secreted by the nerve cell and diffuses to the target cell.

**11–57** Small signaling molecule such as acetylcholine, glutamate, GABA, or glycine, secreted by a nerve cell at a chemical synapse to signal to the postsynaptic cell.

**11–58** Technique in which the tip of a small glass electrode is sealed onto an area of cell membrane, thereby making it possible to record the flow of current through individual ion channels.

**11–59** Supporting cell of the nervous system, including oligodendrocytes and astrocytes in the vertebrate central nervous system and Schwann cells in the peripheral nervous system.

## TRUE/FALSE

Decide whether each of these statements is true or false, and then explain why.

**11–60** Transporters saturate at high concentrations of the transported molecule when all their binding sites are occupied; channels, on the other hand, do not bind the ions they transport and thus the flux of ions through a channel does not saturate.

**11–61** The resting membrane potential of a typical animal cell arises predominantly through the action of the $Na^+$-$K^+$ pump, which in each cycle transfers 3 $Na^+$ ions out of the cell and 2 $K^+$ into the cell, leaving an excess of negative charges inside the cell.

**11–62** The membrane potential arises from movements of charge that leave ion concentrations practically unaffected, causing only a very slight discrepancy in the number of positive and negative ions on the two sides of the membrane.

**11–63** Upon stimulation of a nerve cell, two processes limit the entry of $Na^+$ ions: (1) the membrane potential reaches the $Na^+$ equilibrium potential, which stops further net entry of $Na^+$, and (2) the $Na^+$ channels are inactivated and cannot reopen until the original resting potential has been restored.

**11–64** The aggregate current crossing the membrane of an entire cell indicates the degree to which individual channels are open.

**11–65** Transmitter-gated ion channels open in response to specific neurotransmitters in their environment but are insensitive to the membrane potential; therefore, they cannot by themselves (in the absence of ligand) generate an action potential.

**11–66** When an action potential depolarizes the muscle cell membrane, the $Ca^{2+}$ pump is responsible for pumping $Ca^{2+}$ from the sarcoplasmic reticulum into the cytosol to initiate muscle contraction.

## THOUGHT PROBLEMS

**11–67** According to Newton's laws of motion, an ion exposed to an electric field in a vacuum would experience a constant acceleration from the electric driving force, just as a falling body in a vacuum constantly accelerates due to gravity. In water, however, an ion moves at constant velocity in an electric field. Why do you suppose that is?

**11–68** What two properties distinguish an ion channel from a simple aqueous pore?

**11–69** Name the three ways in which an ion channel can be gated.

**11–70** $K^+$ channels typically transport $K^+$ ions 10,000-times better than $Na^+$ ions, yet these two ions carry the same charge and are about the same size (0.133 nm for $K^+$ and 0.095 nm for $Na^+$). Briefly describe how a $K^+$ channel manages to discriminate against $Na^+$ ions.

**11–71** You have prepared lipid vesicles that contain molecules of the $K^+$ leak channel, all oriented so that their cytosolic surface faces the outside of the vesicle. Predict how $K^+$ ions will move under the following conditions and what sort of membrane potential will develop.

A. Equal concentrations of $K^+$ ion are present inside and outside the vesicle.
B. $K^+$ ions are present only inside the vesicle.
C. $K^+$ ions are present only outside the vesicle.

**11–72** If a frog egg and a red blood cell are placed in pure water, the red blood cell will swell and burst, but the frog egg will remain intact. Although a frog egg is about one million times larger than a red cell, they both have nearly identical internal concentrations of ions so that the same osmotic forces are at work in each. Why do you suppose red blood cells burst in water, while frog eggs do not?

**11–73** Aquaporins allow water to move across a membrane, but prevent the passage of ions. How does the structure of the pore through which the water molecules move prevent passage of ions such as $K^+$, $Na^+$, $Ca^+$, and $Cl^-$? $H^+$ ions present a different problem because they move by relay along a chain of hydrogen-bonded water molecules (Figure 11–10). How does the pore prevent the relay of $H^+$ ions across the membrane?

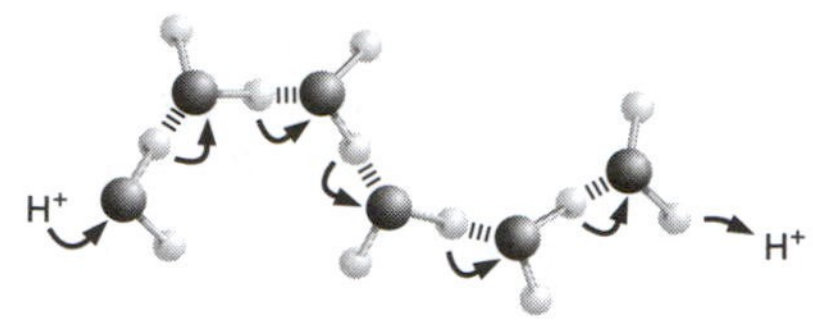


**Figure 11–10** Rapid diffusion of $H^+$ ions by a molecular relay system involving the making and breaking of hydrogen bonds between adjacent water molecules (Problem 11–73).

**Problem 9–48** looks at aquaporins in the membranes of brain astrocytes, using freeze-fracture electron microscopy.

**11–74** Explain in 100 words or less how an action potential is passed along an axon.

**11–75** The myelin sheath, which insulates many vertebrate axons, changes the way the action potential is conducted, allowing it to jump from node to node in a process called saltatory conduction. What are the two main advantages of saltatory conduction?

**11–76** The neurotransmitter acetylcholine is made in the cytosol and then transported into synaptic vesicles, where its concentration is more than 100-fold higher than in the cytosol. Synaptic vesicles isolated from neurons can take up additional acetylcholine if it is added to the solution in which they are suspended, but only in the presence of ATP. $Na^+$ ions are not required for acetylcholine uptake, but, curiously, raising the pH of the solution in which the synaptic vesicles are suspended increases acetylcholine uptake. Furthermore, transport is inhibited in the presence of drugs that make the membrane permeable to $H^+$ ions. Suggest a mechanism that is consistent with all these observations.

**11–77** Excitatory neurotransmitters open $Na^+$ channels, while inhibitory neurotransmitters open either $Cl^-$ or $K^+$ channels. Rationalize this observation in terms of the effects of these ions on the firing of an action potential.

**11–78** Acetylcholine-gated cation channels do not discriminate among $Na^+$, $K^+$, and $Ca^{2+}$ ions, allowing all to pass through freely. How is it, then, that when acetylcholine receptors in muscle cells open there is a large net influx principally of $Na^+$?

**11–79** In the disease myasthenia gravis, the human body makes—by mistake—antibodies to its own acetylcholine receptors. These antibodies bind to and inactivate acetylcholine receptors on the plasma membranes of muscle cells. The disease leads to a progressive weakening of the patient's motor responses. Early on, they may have difficulty opening their eyelids, for example. As the disease progresses, most muscles weaken and patients have difficulty speaking and swallowing. Eventually, impaired breathing can cause death. Explain which step of muscle function is affected.

**11–80** To make antibodies against the acetylcholine receptor from the electric organ of electric eels, you inject the purified receptor into mice. You note an interesting correlation: mice with high levels of antibodies against the receptor appear weak and sluggish; those with low levels are lively. You suspect that the antibodies against the eel acetylcholine receptors are reacting with the mouse acetylcholine receptors, causing many of the receptors to be destroyed. Since a reduction in the number of acetylcholine receptors is also the basis for the human autoimmune disease myasthenia gravis, you wonder whether an injection of the drug neostigmine into the mice might give them a temporary restoration of strength, as it does for myasthenic patients. Neostigmine inhibits acetylcholinesterase, the enzyme responsible for hydrolysis of acetylcholine in the synaptic cleft. Sure enough, when you inject your mice with neostigmine, they immediately become very active. Propose an explanation for how neostigmine restores temporary function to a neuromuscular synapse with a reduced number of acetylcholine receptors.

**11–81** The ion channels for neurotransmitters such as acetylcholine, serotonin, GABA, and glycine have similar overall structures. Each class comprises an extremely diverse set of channel subtypes, with different ligand affinities, different channel conductances, and different rates of opening and closing. Why is such extreme diversity a good thing from the standpoint of the pharmaceutical industry?

## CALCULATIONS

**11–82** The 'ball-and-chain' model for the rapid inactivation of voltage-gated $K^+$ channels has been elegantly confirmed for the *shaker* $K^+$ channel from *Drosophila melanogaster.* (The *shaker* $K^+$ channel in *Drosophila* is named after a mutant form that causes excitable behavior—even anesthetized flies keep twitching.) Deletion of the N-terminal amino acids from the normal *shaker* channel gives rise to a channel that opens in response to membrane depolarization but stays open (Figure 11–11A, 0 μM) instead of rapidly closing as the normal channel does. A peptide (MAAVAGLYGLGEDRQHRKKQ) that corresponds to the deleted N-terminus can partially inactivate the open channel at 50 μM and completely inactivate it at 100 μM (Figure 11–11A).

Is the concentration of free peptide (100 μM) required to inactivate the defective $K^+$ channel anywhere near the normal local concentration of the tethered ball on a normal channel? Assume that the tethered ball can explore a hemisphere [volume = $(2/3)\ \pi r^3$] with a radius of 21.4 nm, the length of the polypeptide 'chain' (Figure 11–11B). Calculate the concentration for one ball in this hemisphere. How does that value compare with the concentration of free peptide needed to inactivate the channel?

**11–83** If the resting membrane potential of a cell is –70 mV and the thickness of the lipid bilayer is 4.5 nm, what is the strength of the electric field across the membrane in V/cm? What do you suppose would happen if you applied this voltage to two metal electrodes separated by a 1-cm air gap?

**11–84** The squid giant axon occupies a unique position in the history of our understanding of cell membrane potentials and nerve action. Its large size (0.2–1.0 mm in diameter and 5–10 cm in length) allowed electrodes, large by modern

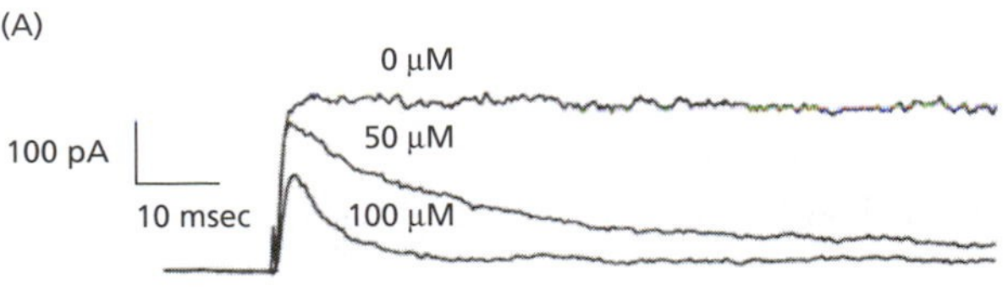


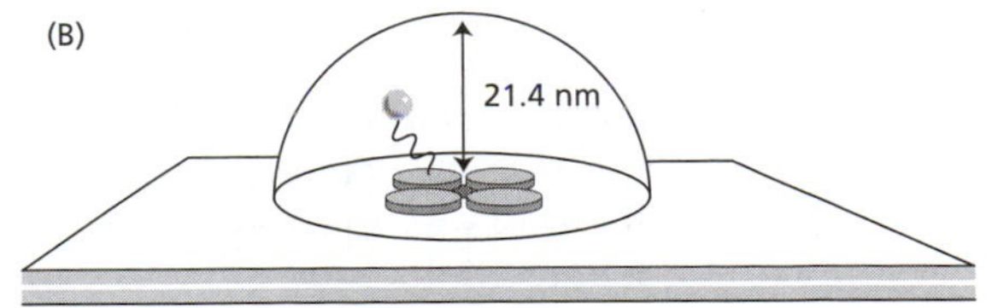


**Figure 11–11** Inactivation of voltage-gated $K^+$ channels (Problem 11–82). (A) Patch clamp recording of a defective *shaker* $K^+$ channel in the absence and presence of inactivating peptide. Current through the channel is indicated in picoamps (pA). (B) A 'ball' tethered by a 'chain' to a normal channel.

standards, to be inserted so that intracellular voltages could be measured. When an electrode is stuck into an intact giant axon, the membrane potential registers –70 mV. When the axon, suspended in a bath of seawater, is stimulated to conduct a nerve impulse, the membrane potential changes transiently from –70 mV to +40 mV.

The Nernst equation relates equilibrium ionic concentrations to the membrane potential.

$$V = 2.3\ \frac{RT}{zF}\ \log\ \frac{C_o}{C_i}$$

For univalent ions and 20°C (293°K),

$$V = 58\ \text{mV} \times \log \frac{C_o}{C_i}$$

A. Using this equation, calculate the potential across the resting membrane (1) assuming that it is due solely to $K^+$ and (2) assuming that it is due solely to $Na^+$. (The $Na^+$ and $K^+$ concentrations in axon cytoplasm and in seawater are given in Table 11–4.) Which calculation is closer to the measured resting potential? Which calculation is closer to the measured action potential? Explain why these assumptions approximate the measured resting and action potentials.

B. If the solution bathing the squid giant axon is changed from seawater to artificial seawater in which NaCl is replaced with choline chloride, there is no effect on the resting potential, but the nerve no longer generates an action potential upon stimulation. What would you predict would happen to the magnitude of the action potential if the concentration of $Na^+$ in the external medium were reduced to a half or a quarter of its normal value, using choline chloride to maintain osmotic balance?

**Table 11–4 Ionic compositions of seawater and of cytoplasm from the squid giant axon** (Problem 11–84).

| ION | CYTOPLASM | SEAWATER |
|---|---|---|
| $Na^+$ | 65 mM | 430 mM |
| $K^+$ | 344 mM | 9 mM |

**11–85** Intracellular changes in ion concentration often trigger dramatic cellular events. For example, when a clam sperm contacts a clam egg, it triggers ionic changes that result in the breakdown of the egg nuclear envelope, condensation of chromosomes, and initiation of meiosis. Two observations confirm that ionic changes initiate these cellular events: (1) suspending clam eggs in seawater containing 60 mM KCl triggers the same intracellular changes as do sperm; (2) suspending eggs in artificial seawater lacking calcium prevents activation by 60 mM KCl.

A. How does 60 mM KCl affect the resting potential of eggs? The intracellular $K^+$ concentration is 344 mM and that of normal seawater is 9 mM. Remember from Problem 11–84 that

$$V = 58\ \text{mV} \times \log \frac{C_o}{C_i}$$

B. What does the lack of activation by 60 mM KCl in calcium-free seawater suggest about the mechanism of KCl activation?

C. What would you expect to happen if the calcium ionophore, A23187, (which allows selective entry of $Ca^{2+}$) was added to a suspension of eggs (in the absence of sperm) in (1) regular seawater and (2) calcium-free seawater?

**11–86** The number of $Na^+$ ions entering the squid giant axon during an action potential can be calculated from theory. Because the cell membrane separates positive and negative charges, it behaves like a capacitor. From the known capacitance of biological membranes, the number of ions that enter during an action potential can be calculated. Starting from a resting potential of –70 mV, it can be shown that $1.1 \times 10^{-12}$ moles of $Na^+$ must enter the cell per $cm^2$ of membrane during an action potential.

To determine experimentally the number of entering $Na^+$ during an action potential, a squid giant axon (1 mm in diameter and 5 cm in length) was suspended in a solution containing radioactive $Na^+$ (specific activity = $2 \times 10^{14}$ cpm/mole) and a single action potential was propagated down its length. When the cytoplasm was analyzed for radioactivity, a total of 340 cpm were found to have entered the axon.

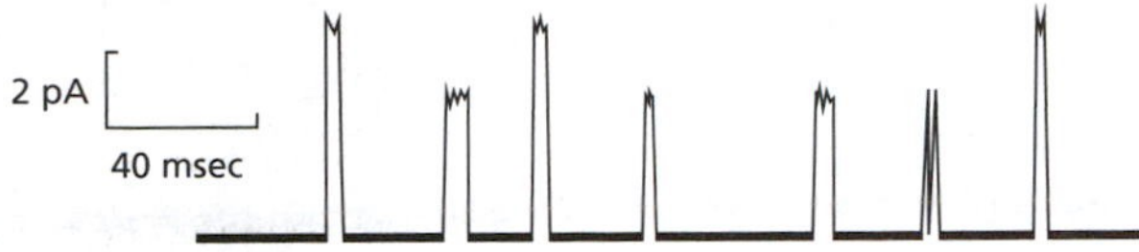


**Figure 11–12** Patch-clamp measurements of acetylcholine-gated cation channels in young rat muscle (Problem 11–88).

A. How well does the experimental measurement match the theoretical calculation?
B. How many moles of $K^+$ must cross the membrane of the axon, and in which direction, to reestablish the resting potential after the action potential is over?
C. Given that the concentration of $Na^+$ inside the axon is 65 mM, calculate the fractional increase in internal $Na^+$ concentration that results from the passage of a single action potential down the axon.
D. At the other end of the spectrum of nerve sizes are small dendrites about 0.1 μm in diameter. Assuming the same length (5 cm), the same internal $Na^+$ concentration (65 mM), and the same resting and action potentials as for the squid giant axon, calculate the fractional increase in internal $Na^+$ concentration that would result from the passage of a single action potential down a dendrite.
E. Is the $Na^+$-$K^+$ pump more important for the continuing performance of a giant axon, or of a dendrite?

**11–87** Cytosolic $Ca^{2+}$ concentrations typically rise about 50-fold when $Ca^{2+}$ channels open in the plasma membrane. Assume that 1000 $Ca^{2+}$ channels open in a cell with a volume of 1000 μm$^3$ and an internal $Ca^{2+}$ concentration of 100 nM. Each $Ca^{2+}$ channel passes $10^6$ $Ca^{2+}$ ions per second. For how long would the channels need to stay open in order to raise the cytosolic $Ca^{2+}$ concentration 50-fold to 5 μM?

**11–88** Acetylcholine-gated cation channels at the neuromuscular junction open in response to acetylcholine released by the nerve terminal and allow $Na^+$ ions to enter the muscle cell, which causes membrane depolarization and ultimately leads to muscle contraction.
A. Patch-clamp measurements show that young rat muscles have cation channels that respond to acetylcholine (Figure 11–12). How many kinds of channel are there? How can you tell?
B. For each kind of channel, calculate the number of ions that enter in one millisecond. (One ampere is a current of one coulomb per second; one pA equals $10^{-12}$ ampere. An ion with a single charge such as $Na^+$ carries a charge of $1.6 \times 10^{-19}$ coulomb.)

## DATA HANDLING

**11–89** The *shaker* $K^+$ channel in *Drosophila* opens in response to membrane depolarization and then rapidly inactivates via a ball-and-chain mechanism. The *shaker* $K^+$ channel assembles as a tetramer composed of four subunits, each with its own ball and chain. Do multiple balls in the tetrameric channel act together to inactivate the channel, or is one ball sufficient?

This question has been answered by mixing subunits from two different forms of the $K^+$ channel: the normal subunits with balls, and scorpion toxin-resistant subunits without balls (Figure 11–13A). Scorpion toxin prevents the opening of normal (toxin-sensitive) $K^+$ channels, but not channels composed entirely of toxin-resistant subunits. Moreover, hybrid channels containing even a single toxin-sensitive subunit fail to open in the presence of toxin.

Mixed $K^+$ channels were created by injecting a mixture of mRNAs for the two types of subunit into frog oocytes. After expression, patches of membrane containing several hundred channels were studied by patch-clamp recording after subjecting the membranes to depolarization in the absence or presence of scorpion toxin (Figure 11–13B).

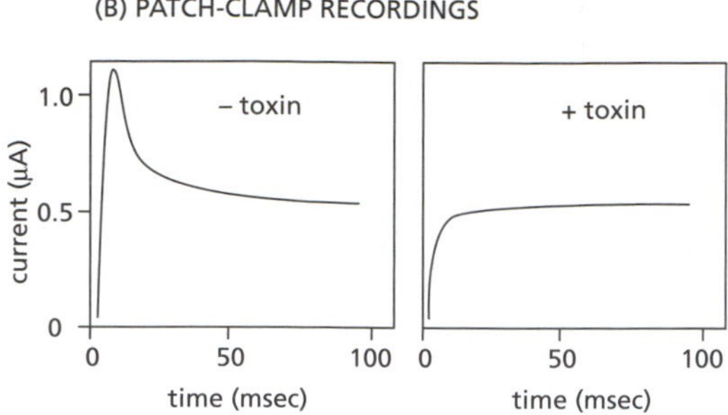


**Figure 11–13** Analysis of the inactivation of the *shaker* $K^+$ channel (Problem 11–89). (A) Mixture of normal subunits with balls (*white*) and toxin-resistant subunits without balls (*black*). If the two subunits were expressed in equal amounts, the different forms of the $K^+$ channel would be present in the ratio 1:4:6:4:1; however, higher levels of toxin-resistant subunit mRNA were used, significantly skewing the ratios in favor of channels with toxin-resistant subunits. (B) Patch-clamp recordings in the presence and absence of scorpion toxin.

A. Sketch the expected patch-clamp recordings, in the presence and absence of toxin, for a pure population of $K^+$ channels composed entirely of toxin-resistant subunits without balls.
B. Sketch the expected patch-clamp recordings, in the presence and absence of toxin, for a pure population of $K^+$ channels composed entirely of normal (toxin-sensitive) subunits with balls.
C. Sketch the expected patch-clamp recordings, in the presence and absence of toxin, for a mixed population of $K^+$ channels, 50% composed entirely of normal (toxin-sensitive) subunits with balls and 50% composed entirely of toxin-resistant subunits without balls.
D. The key observation in Figure 11–13B is that the final plateau values of the currents in the absence and presence of toxin are the same. Does this observation argue that a single ball can close a channel or that multiple balls must act in concert? (Think about what the curves would look like if one, two, three, or four balls were required to close a channel.)

**11–90** A recording from a patch of membrane (Figure 11–14A) measured current through acetylcholine-gated ion channels, which open when acetylcholine is bound (Figure 11–14B). To obtain the recording, acetylcholine was added to the solution in the micropipette (Figure 11–14A). Describe what you can learn about the channels from this recording. How would the recording differ if acetylcholine were (a) omitted or (b) added only to the solution outside the micropipette?

**11–91** One important parameter for understanding any particular membrane transport process is to know the number of copies of the specific transport protein present in the cell membrane. To measure the number of voltage-gated $Na^+$ channels in the rabbit vagus nerve, you use a potent toxin, saxitoxin, a shellfish poison that specifically inactivates the voltage-gated $Na^+$ channels in these nerve cells. Assuming that each channel binds one toxin molecule, the number of $Na^+$ channels in a segment of vagus nerve will be equal to the maximum number of bound toxin molecules.

You incubate identical segments of nerve for 8 hours with increasing amounts of $^{125}$I-labeled toxin. You then wash the segments to remove unbound toxin and measure the radioactivity associated with the nerve segments to determine the toxin-binding curve (Figure 11–15, upper curve). You are puzzled because you expected to see binding reach a maximum

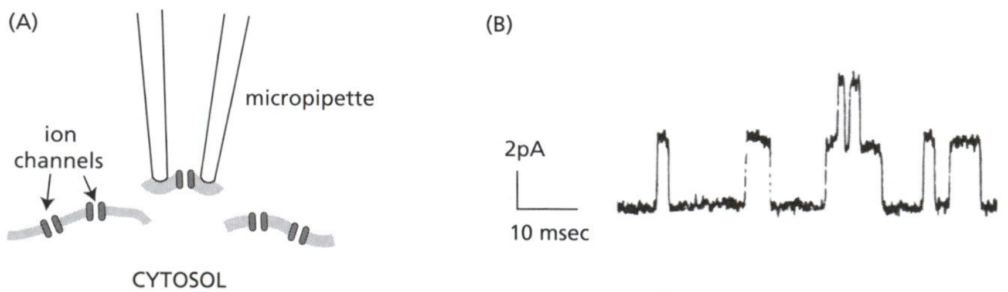


**Figure 11–14** Analysis of acetylcholine-gated ion channels (Problem 11–90). (A) A micropipette with a patch of membrane. (B) Patch-clamp recording of current through acetylcholine-gated ion channels.

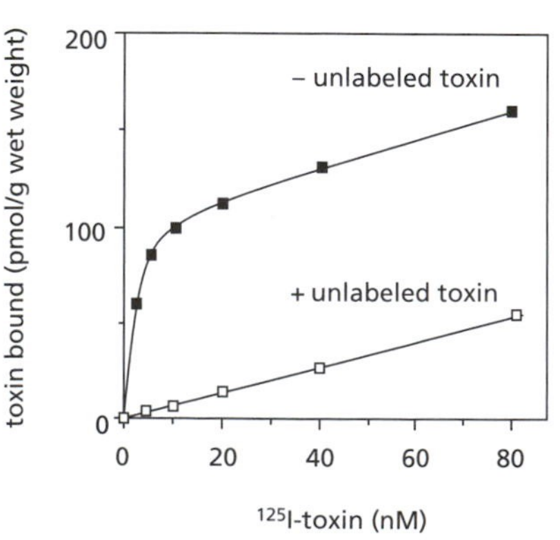


**Figure 11–15** Toxin-binding curves in the presence and absence of unlabeled saxitoxin (Problem 11–91).

(saturate) at high concentrations of toxin; however, no distinct end point was reached, even at higher concentrations of toxin than those shown in the figure. After careful thought, you design a control experiment in which the binding of labeled toxin is measured in the presence of a large molar excess of unlabeled toxin. The results of this experiment, which are shown in the lower curve in Figure 11–15, make everything clear and allow you to calculate the number of $Na^+$ channels in the membrane of the vagus nerve axon.

A. Why does binding of the labeled toxin not saturate? What is the point of the control experiment, and how does it work?

B. Given that 1 gram of vagus nerve has an axonal membrane area of 6000 $cm^2$ and assuming that the $Na^+$ channel is a cylinder with a diameter of 6 nm, calculate the number of $Na^+$ channels per square micrometer of axonal membrane and the fraction of the cell surface occupied by the channel. (Use 100 pmol as the amount of toxin specifically bound to the receptor.)

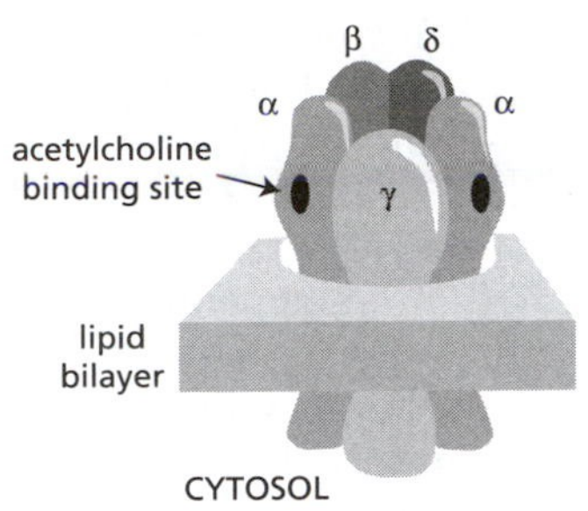


**Figure 11–16** A model for the structure of the acetylcholine receptor (Problem 11–92).

**11–92** The acetylcholine-gated cation channel is a pentamer of homologous proteins containing two copies of the α subunit and one copy each of the β, γ, and δ subunits (Figure 11–16). The two distinct acetylcholine-gated channels in young rat muscle (see Problem 11–88) differ in their γ subunits. The 6-pA channel contains the $\gamma_1$ subunit, whereas the 4-pA channel contains the $\gamma_2$ subunit.

To pinpoint the portions of the $\gamma_1$ and $\gamma_2$ subunits responsible for the conductance differences of the two channels, you prepare a series of chimeric cDNAs in which different portions of the $\gamma_1$ and $\gamma_2$ cDNAs have been swapped (Figure 11–17). Since it seems likely that the transmembrane segments determine channel conductance, you have designed the chimeric

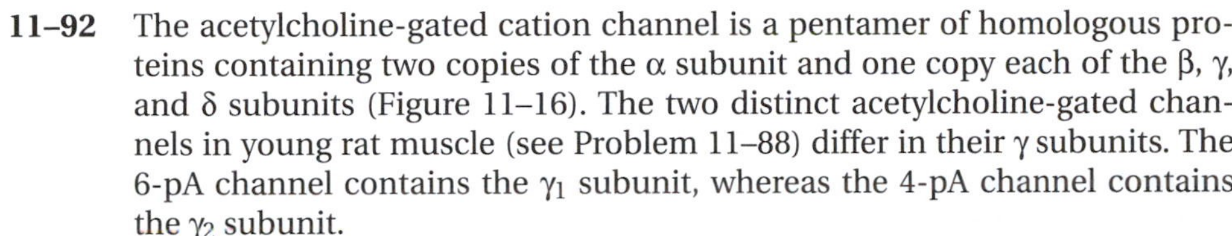


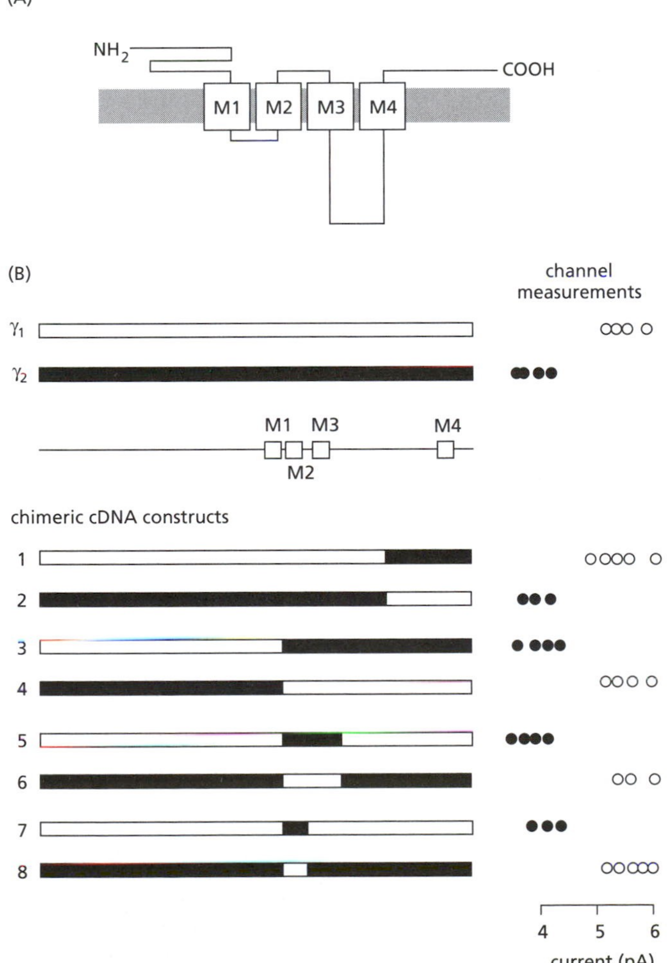


**Figure 11–17** Analysis of the structural basis for conductance differences in two acetylcholine-gated channels (Problem 11–92). (A) Arrangement of a γ subunit in the membrane. (B) Current through channels made with the $\gamma_1$ subunit, the $\gamma_2$ subunit, or chimeric subunits. The current through the different channels is indicated on the *right*. *Open circles* indicate channels with conductance like those made from the $\gamma_1$ subunit; *filled circles* indicate channels with conductance like those made from the $\gamma_2$ subunit. M1, M2, M3, and M4 indicate the four transmembrane segments in the γ subunits.

cDNAs to test different combinations of the four transmembrane segments (M1 to M4 in Figure 11–17A). You mix mRNA made from each of the chimeric cDNAs with mRNA from the α, β, and δ subunits, inject the mRNAs into *Xenopus* oocytes, and measure the current through individual channels by patch clamping (Figure 11–17B).

A. Which transmembrane segment is responsible for the differences in conductance through the two types of acetylcholine-gated cation channels?

B. Why do you suppose that the difference in channel conductance is all due to one transmembrane segment?

C. Assume that a threonine in the critical transmembrane segment contributes to the constriction that forms the ion-selectivity filter for the channel. How would the current through the channel change if glycine were substituted at that position? If leucine were substituted at that position? How might you account for the different currents through channels constructed with $\gamma_1$ and $\gamma_2$ subunits?

**11–93** Glutamate is a major excitatory neurotransmitter in the central nervous system. Glutamate binds to a variety of receptors, some of which control intracellular signaling via G proteins and others that control cation channels. One of the latter, the NMDA receptor, opens a $Ca^{2+}$ channel in response to *N*-methyl-D-aspartate (NMDA), which is an analog of glutamate.

The NMDA receptor was cloned from mRNA isolated from rat forebrain. The mRNA was first size-fractionated on a sucrose gradient, and individual fractions were then injected into *Xenopus* oocytes. A 3–5 kb fraction conferred NMDA responsiveness, as assessed by electrophysiological measurements on the injected oocytes, which normally do not respond to NMDA. A cDNA library was prepared from the active mRNA fraction and then subdivided into 10 aliquots containing 1000 clones each. RNA was made from each aliquot by *in vitro* transcription and then injected into oocytes. Aliquots whose RNA conferred NMDA responsiveness were further divided into aliquots containing fewer cDNA clones and retested for the ability of the transcribed RNA to confer NMDA responsiveness to injected oocytes. Repeated subdivision of NMDA-responsive cDNA aliquots allowed the cDNA for the NMDA receptor to be cloned as outlined in Figure 11–18.

The identity of the NMDA receptor was confirmed using a variety of pharmacological agents that are known agonists (activators) or antagonists (inhibitors) of the NMDA receptor or other glutamate receptors (Table 11–5).

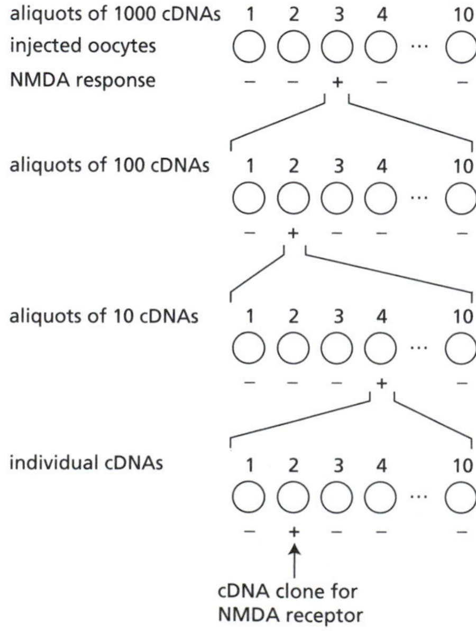


**Figure 11–18** Pooling strategy for isolating the cDNA encoding the NMDA receptor (Problem 11–93).

**Table 11–5 Pharmacology of the cloned NMDA receptor** (Problem 11–93).

| AGONIST | | ANTAGONIST | | RESPONSE |
|---|---|---|---|---|
| FOR NMDA RECEPTOR | FOR OTHER GLUTAMATE RECEPTORS | FOR NMDA RECEPTOR | FOR OTHER GLUTAMATE RECEPTORS | |
| NMDA (100 μM) | | | | 100% |
| Ibotenate (100 μM) | | | | 71% |
| L-HCA (100 μM) | | | | 85% |
| | Kainate (500 μM) | | | <5% |
| | AMPA (50 μM) | | | <5% |
| NMDA (100 μM) | | D-APV (10 μM) | | 27% |
| NMDA (100 μM) | | 7-Cl-KYNA (50 μM) | | 2% |
| NMDA (100 μM) | | | GAMS (100 μM) | 97% |
| NMDA (100 μM) | | | JSTX (100 μM) | 83% |

Oocytes injected with the putative cDNA for the cloned NMDA receptor were tested with a variety of pharmacological agents whose receptor specificities and effective concentrations were already defined from previous work.

L-HCA = homocysteate; AMPA = α-amino-3-hydroxy-5-methyl-4-isoxazolepropionate; D-APV = 2-amino-5-phosphonovalerate; 7-Cl-KYNA = 7-chlorokynurenate; GAMS = γ-D-glutamylaminomethyl sulphonate; JSTX = Joro spider toxin.

A. The pooling strategy outlined in Figure 11–18 is an efficient way to find one particular cDNA in a complex mixture. How many oocyte injections and analyses for NMDA responsiveness would have been required if the original cDNA library were tested one cDNA at a time? How many were required using the pooling strategy?

B. How do the data in Table 11–5 confirm that the cDNA for the NMDA receptor was cloned?

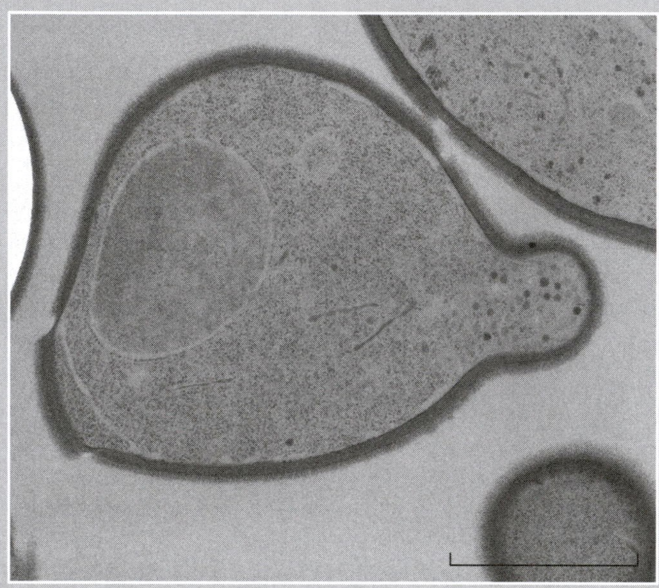

**A snap-frozen electron microscope section of a yeast cell captured in the act of forming a growing bud.** The Golgi apparatus and secretory vesicles are clearly visible. Scale bar is 1.6 microns. (Micrograph courtesy of Heinz Schwartz at the MPI Tübingen.)

# Intracellular Compartments and Protein Sorting

# 12

## In This Chapter

THE COMPARTMENTALIZATION OF CELLS 263

THE TRANSPORT OF MOLECULES BETWEEN THE NUCLEUS AND THE CYTOSOL 266

THE TRANSPORT OF PROTEINS INTO MITOCHONDRIA AND CHLOROPLASTS 274

PEROXISOMES 279

THE ENDOPLASMIC RETICULUM 282

## THE COMPARTMENTALIZATION OF CELLS

| TERMS TO LEARN | | |
|---|---|---|
| cytoplasm | organelle | sorting signal |
| cytosol | signal patch | transmembrane transport |
| gated transport | signal peptidase | vesicular transport |
| lumen | signal sequence | |

### DEFINITIONS

Match the definition below with its term from the list above.

**12–1** Movement of proteins across a bilayer from the cytosol to a topologically distinct compartment.

**12–2** Contents of a cell that are contained within its plasma membrane but, in the case of eucaryotic cells, outside the nucleus.

**12–3** Protein sorting signal that consists of a specific three-dimensional arrangement of atoms on the folded protein's surface.

**12–4** Contents of the main compartment of the cell, excluding the nucleus and membrane-bounded compartments such as endoplasmic reticulum and mitochondria.

**12–5** Movement of proteins through nuclear pore complexes between the cytosol and the nucleus.

**12–6** Membrane-enclosed compartment in a eucaryotic cell that has a distinct structure, macromolecular composition, and function.

**12–7** Protein sorting signal that consists of a short continuous sequence of amino acids.

### TRUE/FALSE

Decide whether each of these statements is true or false, and then explain why.

**12–8** The biological membranes that partition the cell into functionally distinct compartments are impermeable.

**12–9** Like the lumen of the ER, the interior of the nucleus is topologically equivalent to the outside of the cell.

**12–10** Membrane-bound and free ribosomes, which are structurally and functionally identical, differ only in the proteins they happen to be making at a particular time.

**12–11** Each signal sequence specifies a particular destination in the cell.

## THOUGHT PROBLEMS

**12–12** Discuss the following statement: "The plasma membrane is only a minor component of most eucaryotic cells."

**12–13** In a typical animal cell, which of the following compartments is the largest? Which compartment is present in the greatest number?

A. Cytosol
B. Endoplasmic reticulum
C. Endosome
D. Golgi apparatus
E. Lysosome
F. Mitochondria
G. Nucleus
H. Peroxisome

**12–14** Is it really true that *all* human cells contain the same basic set of membrane-enclosed organelles? Do you know of any examples of human cells that do not have a complete set of organelles?

**12–15** When cells are treated with drugs that depolymerize microtubules, the Golgi apparatus is fragmented into small vesicles and dispersed throughout each cell. When such drugs are removed, cells typically recover and grow normally. If cells that have recovered from such treatment are examined by electron microscopy, they are found to contain a perfectly normal looking Golgi apparatus. Does this mean that the Golgi apparatus has been synthesized anew from scratch? If not, how do you suppose it might have happened?

**12–16** Why do eucaryotic cells require a nucleus as a separate compartment when procaryotic cells manage perfectly well without?

**12–17** What is the fate of a protein with no sorting signal?

**12–18** Protein synthesis in a liver cell occurs nearly exclusively on free ribosomes in the cytosol and on ribosomes that are bound to the ER membrane. (A small fraction of total protein synthesis is directed by the mitochondrial genome and occurs on ribosomes in the mitochondrial matrix.) Which type of protein synthesis—in the cytosol or on the ER—do you think is responsible for the majority of protein synthesis in a liver cell? Assume that the average density and lifetimes of proteins are about the same in all compartments. Explain the basis for your answer. Would your answer change if you took into account that some proteins are secreted from liver cells?

**12–19** List the organelles in an animal cell that obtain their proteins via gated transport, via transmembrane transport, or via vesicular transport.

**12–20** Imagine that you have engineered a set of genes, each encoding a protein with a pair of conflicting signal sequences that specify different compartments. If the genes were expressed in a cell, predict which signal would win out for the following combinations. Explain your reasoning.

A. Signals for import into nucleus and import into ER.
B. Signals for import into peroxisomes and import into ER.
C. Signals for import into mitochondria and retention in ER.
D. Signals for import into nucleus and export from nucleus.

**12–21** If you think of the protein as a traveler, what kind of vehicle would best describe the sorting receptor: a private car, a taxi, or a bus? Explain your choice.

**12–22** If membrane proteins are integrated into the ER membrane by means of protein translocators, which are themselves membrane proteins, how do the first protein translocators become incorporated into the ER membrane?

## CALCULATIONS

**12–23** A typical animal cell is said to contain some 10 billion protein molecules that need to be sorted into their proper compartments. That's a lot of proteins. Can 10 billion protein molecules even fit into a cell? An average protein encoded by the human genome is 450 amino acids in length, and the average mass of an amino acid in a protein is 110 daltons. Given that the average density of a protein is 1.4 g/cm$^3$, what fraction of the volume of a cell would 10 billion protein molecules occupy? Consider a liver cell, which has a volume of about 5000 μm$^3$, and a pancreatic exocrine cell, which has a volume of about 1000 μm$^3$.

**12–24** The lipid bilayer, which is 5 nm thick, occupies about 60% of the volume of a typical cell membrane. (Lipids and proteins contribute equally on a mass basis, but lipids are less dense and therefore account for more of the volume.) For liver cells and pancreatic exocrine cells, the total area of all cell membranes is estimated at about 110,000 μm$^2$ and 13,000 μm$^2$, respectively. What fraction of the total volumes of these cells is accounted for by lipid bilayers? The volumes of liver cells and pancreatic exocrine cells are about 5000 μm$^3$ and 1000 μm$^3$, respectively.

**12–25** A decrease in surface-to-volume ratio relative to bacteria is one rationale for why large eucaryotic cells evolved to possess such a profusion of internal membranes. How much different are the surface-to-volume ratios for bacteria and eucaryotic cells? A liver cell has about 110,000 μm$^2$ of membrane, 2% of which is plasma membrane, and a volume of about 5000 μm$^3$. A pancreatic exocrine cell has about 13,000 μm$^2$ of membrane, 5% of which is plasma membrane, and a volume of about 1000 μm$^3$. The rod-shaped bacterium, *E. coli*—essentially a cylinder 2 μm long and 1 μm in diameter—has a surface area of 7.85 μm$^2$ and a volume of 1.57 μm$^3$.

What are the surface-to-volume ratios for these cells? How much smaller are they in the eucaryotic cells versus *E. coli?* What would the surface-to-volume ratio for *E. coli* be if its diameter and length were increased by a factor of 10, so that its volume was increased 1000-fold to 1570 μm$^3$—about the same size as the eucaryotic cells? (For a cylinder, volume is $\pi r^2 h$ and surface area is $2\pi r^2 + 2\pi rh$.)

**12–26** The rough ER is the site of synthesis of many different classes of membrane protein. Some of these proteins remain in the ER, whereas others are sorted to compartments such as the Golgi apparatus, lysosomes, and the plasma membrane. One measure of the difficulty of the sorting problem is the degree of 'purification' that must be achieved during transport from the ER to the other compartments. For example, if membrane proteins bound for the plasma membrane represented 90% of all proteins in the ER, then only a small degree of purification would be needed (and the sorting problem would appear relatively easy). On the other hand, if plasma membrane proteins represented only 0.01% of the proteins in the ER, a very large degree of purification would be required (and the sorting problem would appear correspondingly more difficult).

What is the magnitude of the sorting problem? What fraction of the membrane proteins in the ER are destined for other compartments? A few simple considerations allow one to answer these questions. Assume that all proteins on their way to other compartments remain in the ER for 30 minutes on average before exiting, and that the ratio of proteins to lipids in the membranes of all compartments is the same.

A. In a typical growing cell that is dividing once every 24 hours, the equivalent of one new plasma membrane must transit the ER every day. If the ER membrane is 20 times the area of a plasma membrane, what is the ratio of plasma membrane proteins to other membrane proteins in the ER?

B. If in the same cell the Golgi membrane is three times the area of the plasma membrane, what is the ratio of Golgi membrane proteins to other membrane proteins in the ER?

C. If the membranes of all other compartments (lysosomes, endosomes, inner nuclear membrane, and secretory vesicles) that receive membrane proteins from the ER are equal in total area to the area of the plasma membrane, what fraction of the membrane proteins in the ER of this cell are permanent residents of the ER membrane?

## DATA HANDLING

**12–27** Although the vast majority of transmembrane proteins insert into membranes with the help of dedicated protein-translocation machines, a few proteins can insert into membranes on their own. Such proteins may provide a window into how membrane insertion occurred in the days before complex translocators had evolved.

You are studying a protein that inserts itself into the bacterial membrane independent of the normal translocation machinery. This protein has an N-terminal, 18-amino acid hydrophilic segment that is located on the outside of the membrane, a 19-amino acid hydrophobic transmembrane segment flanked by negatively and positively charged amino acids, and a C-terminal domain that resides inside the cell (Figure 12–1A). If the protein is properly inserted in the membrane, the N-terminal segment is exposed externally where it can be clipped off by a protease, allowing you to quantify insertion.

To examine the roles of the hydrophobic segment and its flanking charges, you construct a set of modified genes that express mutant proteins with altered charges in the N-terminal segment, altered lengths of the hydrophobic segment, and combinations of the two (Figure 12–1B). For each gene you measure the fraction of the total protein that is cleaved by the protease, which is the fraction that was inserted correctly (Figure 12–1B). To assess the contribution of the normal membrane potential (positive outside, negative inside), you repeat the measurements in the presence of CCCP, an ionophore that eliminates the charge on the membrane (Figure 12–1B).

A. Which of the two N-terminal negative charges is the more important for insertion of the protein in the presence of the normal membrane potential (minus CCCP)? Explain your reasoning.

B. In the presence of the membrane potential (minus CCCP), is the hydrophobic segment or the N-terminal charge more important for insertion of the protein into the membrane? Explain your reasoning.

C. In the absence of the membrane potential (plus CCCP), is the hydrophobic segment or the N-terminal charge more important for insertion of the protein into the membrane? Explain your reasoning.

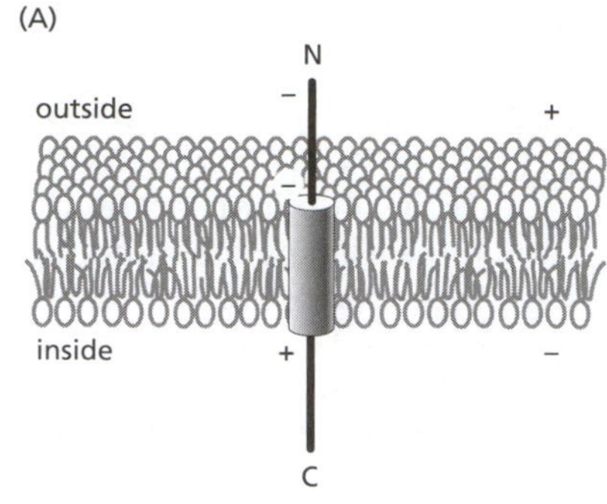


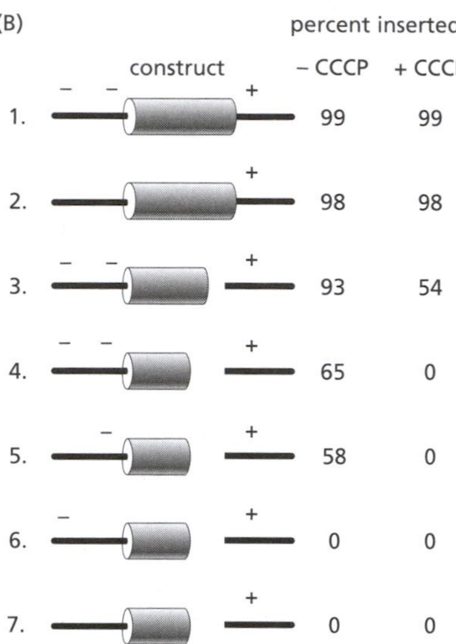


**Figure 12–1** Insertion of a small protein into the bacterial membrane (Problem 12–27). (A) Normal orientation of the protein in the membrane. (B) Mutant proteins used to investigate the contributions of the N-terminal negative charges and length of the hydrophobic segment to membrane insertion. The presence of negative charges is indicated by –; *cylinders* indicate the length of α-helical, transmembrane hydrophobic segments; deletions of the hydrophobic segments are shown as *gaps*. Percent inserted refers to the proportion of the total protein whose N-termini are sensitive to protease digestion.

# THE TRANSPORT OF MOLECULES BETWEEN THE NUCLEUS AND THE CYTOSOL

| TERMS TO LEARN | | |
|---|---|---|
| inner nuclear membrane | nuclear import receptor | nuclear pore complex (NPC) |
| NPC protein | nuclear lamin | nuclear transport receptor |
| nuclear envelope | nuclear lamina | outer nuclear membrane |
| nuclear export receptor | nuclear localization signal | Ran |
| nuclear export signal | | |

## DEFINITIONS

Match the definition below with its term from the list above.

**12–28** Sorting signal contained in the structure of macromolecules and complexes that are transported from the nucleus to the cytosol through nuclear pore complexes.

**12–29** Large multiprotein structure forming a channel through the nuclear envelope that allows selected molecules to move between nucleus and cytoplasm.

**12–30** Double membrane surrounding the nucleus.

**12–31** Monomeric GTPase present in both cytosol and nucleus that is required for the active transport of macromolecules into and out of the nucleus through nuclear pore complexes.

**12–32** Fibrous meshwork of proteins on the inner surface of the inner nuclear membrane.

**12–33** Protein that binds nuclear localization signals and facilitates the transport of proteins with these signals from the cytosol into the nucleus through nuclear pore complexes.

**12–34** Sorting signal found in proteins destined for the nucleus and which enable their selective transport into the nucleus from the cytosol through the nuclear pore complexes.

**12–35** The portion of the nuclear envelope that is continuous with the endoplasmic reticulum and is studded with ribosomes on its cytosolic surface.

## TRUE/FALSE

Decide whether each of these statements is true or false, and then explain why.

**12–36** The nuclear membrane is freely permeable to ions and other small molecules under 5000 daltons.

**12–37** To avoid the inevitable congestion that would occur if two-way traffic through a single pore were allowed, nuclear pore complexes are specialized so that some mediate import while others mediate export.

**12–38** Some proteins are kept out of the nucleus, until needed, by inactivating their nuclear localization signals by phosphorylation.

**12–39** All cytosolic proteins have nuclear export signals that allow them to be removed from the nucleus when it reassembles after mitosis.

## THOUGHT PROBLEMS

**12–40** As shown in Figure 12–2, the inner and outer nuclear membranes form a continuous sheet, connecting through the nuclear pores. Continuity implies that membrane proteins can move freely between the two nuclear membranes by diffusing through the bilayer at the nuclear pores. Yet the inner and outer nuclear membranes have different protein compositions, as befits their different functions. How do you suppose this apparent paradox is reconciled?

**12–41** How is it that a single nuclear pore complex can efficiently transport proteins that possess different kinds of nuclear localization signal?

**12–42** How do you suppose that proteins with a nuclear export signal get into the nucleus?

**12–43** You have modified the green fluorescent protein (GFP) to carry a nuclear localization signal at its N-terminus and a nuclear export signal at its C-terminus. When this protein is expressed in cells, you find that about 60% of it is located in the nucleus and the rest in the cytoplasm. You assume that the protein is shuttling between the nucleus and the cytoplasm and that its rates of import and export determine the observed distribution. Nevertheless, you are concerned that the modified GFP might not fold properly even though it retains its fluorescent properties. Specifically, you worry that the 40% in the cytoplasm is not being transported at all because neither signal is exposed, and the 60% in the nucleus has only the nuclear import signal available. How might addition of an inhibitor of nuclear export to such cells resolve this issue?

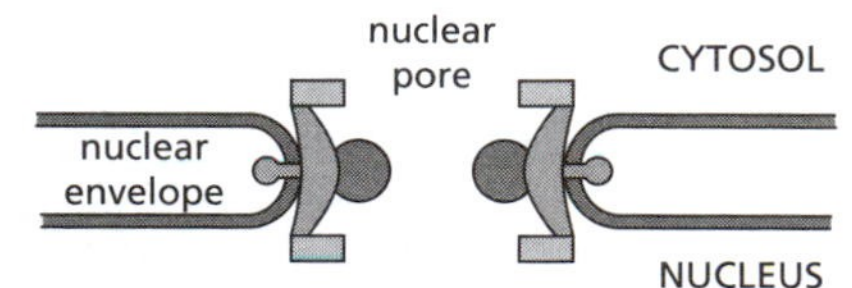


**Figure 12–2** Cross-section through a nuclear pore complex, showing continuity of inner and outer nuclear membranes (Problem 12–40).

**12–44** Your advisor is explaining his latest results in your weekly lab meeting. By fusing his protein of interest to GFP, he has shown that it is located entirely in the nucleus. But he wonders if it is a true nuclear protein or a shuttling protein that just spends most of its time in the nucleus. He is unsure how to resolve this issue. Having just read an article about how a similar problem was answered, you suggest that he make a heterocaryon by fusing cells that are expressing his tagged protein with an excess of cells that are not expressing it. You tell him that in the presence of a protein synthesis inhibitor to block new synthesis of the tagged protein he can resolve the issue by examining fused cells with two nuclei. He gives you a puzzled look and asks how does that help. You tell him what he has so often told you: "Think about it."

A. How would examining the two nuclei in a heterocaryon answer the question? What results would you expect if the protein were a true nuclear protein? What would you expect if it were a shuttling protein?

B. Why did you suggest that a protein synthesis inhibitor would be needed in this experiment?

**12–45** Nuclear localization signals are not cleaved off after transport into the nucleus, whereas the signal sequences for import into other organelles are often removed after import. Why do you suppose it is critical that nuclear localization signals remain attached to their proteins?

## CALCULATIONS

**12–46** A nuclear pore can dilate to accommodate a gold particle 26 nm in diameter. If it could accommodate a spherical protein of the same dimensions, what would the protein's molecular mass (g/mole) be? [Assume that the density of the protein is 1.4 g/cm$^3$. The volume of a sphere is $(4/3)\pi r^3$.]

**12–47** Assuming that 32 million histone octamers are required to package the human genome, how many histone molecules must be transported per second per pore complex in cells whose nuclei contain 3000 nuclear pores and is dividing once per day?

**12–48** To test the hypothesis that the directionality of transport across the nuclear membrane is determined primarily by the gradient of the Ran-GDP outside the nucleus and Ran-GTP inside the nucleus, you decide to reverse the gradient to see if you can force the import of a protein that is normally exported from the nucleus. You add a well-defined nuclear export substrate, fluorescent BSA coupled with a nuclear export signal (NES-BSA), to the standard permeabilized cell assay. Sure enough, it is excluded from the nuclei (Figure 12–3A). Now you add Crm1, the nuclear export receptor that recognizes the export signal, and RanQ69L-GTP, a mutant form of Ran that cannot hydrolyze GTP. With these additions, the tagged BSA now enters the nuclei (Figure 12–3B). Unlike conventional nuclear import, which concentrates imported proteins in the nucleus, the concentration of NES-BSA in the nucleus in the import assay is no higher than in the surrounding cytoplasm.

A. Why doesn't NES-BSA accumulate to a higher concentration in the nucleus than in the cytoplasm in these experiments?

B. In a standard nuclear import assay with added cytoplasm and GTP, proteins with a nuclear localization signal accumulate to essentially 100% in the nucleus (see Figure 12–6, Problem 12–54). How is it that the standard assay allows 100% accumulation in the nucleus?

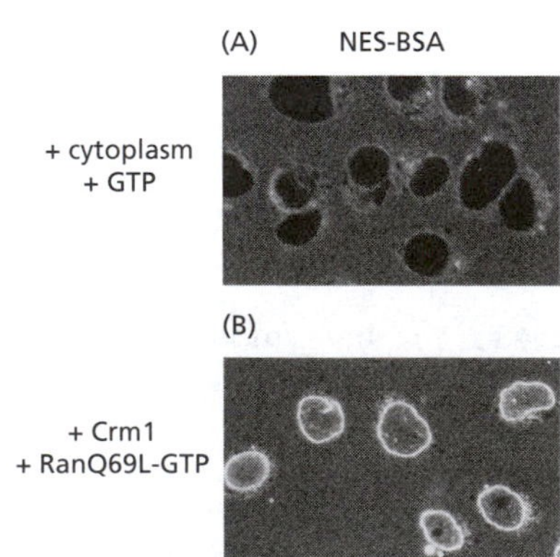


**Figure 12–3** Directionality of nuclear transport (Problem 12–48). (A) Exclusion of fluorescent NES-BSA from the nucleus in a standard import assay. (B) Import of fluorescent NES-BSA in the presence of Crm1 and RanQ69L-GTP.

**12–49** The bright ring of fluorescent NES-BSA around the nuclei in Figure 12–3B suggests that the transport substrate accumulates at the nuclear periphery. How many molecules of NES-BSA would need to be bound per nuclear pore complex to give such a bright ring? As a basis for estimating the answer, calculate the concentration of NES-BSA in the nuclear membrane, assuming that one NES-BSA is bound per pore complex, that there are 3000 pores per nucleus, and that the volume of the nuclear membrane is 16 fL ($16 \times 10^{-15}$ L). How does this calculated concentration compare with the concentration of

NES-BSA used in these experiments (0.3 μM)? If the ring of fluorescence in Figure 12–3B is ten times brighter than the background, how many NES-BSA molecules would you calculate are bound per nuclear pore complex? How does this estimate fit with the structure of the pore complex?

**12–50** By following the increase in nuclear fluorescence over time in the forced nuclear import experiments in Figure 12–3B, the authors were able to show that nuclear fluorescence reached half its maximal value in 60 seconds. Since the added concentration of NES-BSA was 0.3 μM, its concentration in the nucleus after 60 seconds was 0.15 μM. Given that the volume of the nucleus is 500 fL and that each nucleus contains 3000 nuclear pores, calculate the rate of import of NES-BSA per pore in this experiment. (For this calculation, neglect export of NES-BSA.) Is your answer physiologically reasonable? Why or why not?

## DATA HANDLING

**12–51** Before nuclear pore complexes were well understood, it was unclear whether nuclear proteins diffused passively into the nucleus and accumulated there by binding to residents of the nucleus such as chromosomes, or were actively transported and accumulated regardless of their affinity for nuclear components.

A classic experiment that addressed this problem used several forms of radioactive nucleoplasmin, which is a large pentameric protein involved in chromatin assembly. In this experiment, either the intact protein or the nucleoplasmin heads, tails, or heads with a single tail were injected into the cytoplasm of a frog oocyte, or into the nucleus (Figure 12–4). All forms of nucleoplasmin, except heads, accumulated in the nucleus when injected into the cytoplasm, and all forms were retained in the nucleus when injected there.

A. What portion of the nucleoplasmin molecule is responsible for localization in the nucleus?

B. How do these experiments distinguish between active transport, in which a nuclear localization signal triggers transport by the nuclear pore complex, and passive diffusion, in which a binding site for a nuclear component allows accumulation in the nucleus?

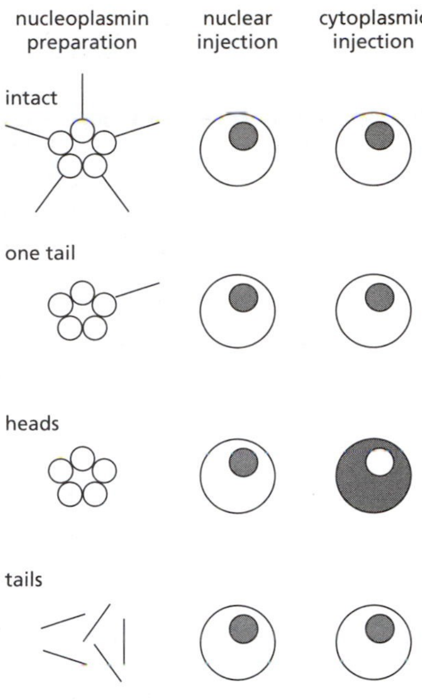


**Figure 12–4** Cellular location of injected nucleoplasmin and nucleoplasmin components (Problem 12–51). Schematic diagrams of autoradiographs show the cytoplasm and nucleus with the location of nucleoplasmin indicated by the *shaded* areas.

**12–52** You have just joined a laboratory that is analyzing the nuclear transport machinery in yeast. Your advisor, who is known for her extraordinarily clever ideas, has given you a project with enormous potential. In principle, it would allow a genetic selection for conditional-lethal mutants in the nuclear transport apparatus.

She gave you two plasmids. Each plasmid consists of a hybrid gene under the control of a regulatable promoter (Figure 12–5). The hybrid gene is a fusion between a gene whose product is normally imported into the nucleus and the gene for the restriction nuclease EcoRI. The plasmid $pNL^+$ contains a functional nuclear localization signal (NLS); the plasmid $pNL^-$ does not have an NLS. The promoter, which is from the yeast *Gal1* gene, allows transcription of the hybrid gene only when the sugar galactose is present in the growth medium.

Following her instructions, you introduce the plasmids into yeast (in the absence of galactose) and then assay the transformed yeast in medium containing glucose and in medium containing galactose. Your results are shown in Table 12–1. You don't remember what your advisor told you to expect, but you know you will be expected to explain these results at the weekly lab

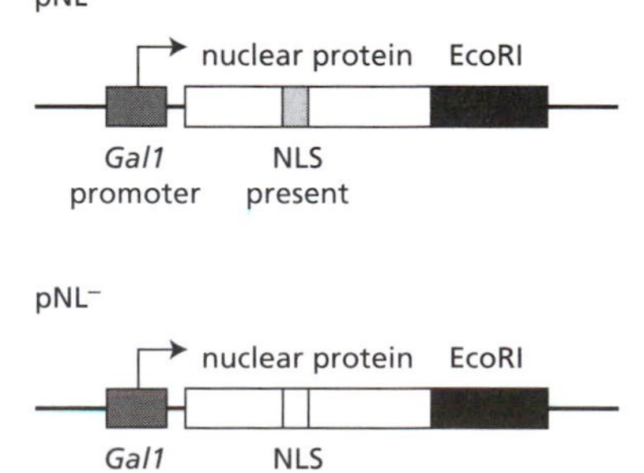


**Figure 12–5** Two plasmids for investigating nuclear localization in yeast (Problem 12–52). Plasmids are shown as linear molecules for clarity. *Arrows* indicate direction of transcription from the *Gal1* promoter.

**Table 12–1 Results of proliferation experiments with yeast carrying plasmids $pNL^+$ or $pNL^-$ (Problem 12–52).**

| PLASMID | GLUCOSE MEDIUM | GALACTOSE MEDIUM |
|---|---|---|
| $pNL^+$ | proliferation | death |
| $pNL^-$ | proliferation | proliferation |

meeting. Why do yeasts with the pNL$^+$ plasmid proliferate in the presence of glucose but die in the presence of galactose, whereas yeasts with the pNL$^-$ plasmid proliferate in both media?

**12–53** Now that you understand why plasmid pNL$^+$ (see Figure 12–5) kills cells in galactose-containing medium, you begin to understand how your advisor intends to exploit its properties to select mutants in the nuclear transport machinery. You also understand why she emphasized that the desired mutants would have to be *conditionally* lethal. Since nuclear import is essential to the cell, a fully defective mutant could never be grown and thus would not be available for study. By contrast, conditional-lethal mutants can be grown perfectly well under one set of conditions (permissive conditions). Under a different set of conditions (restrictive conditions), the cells exhibit the defect, which can then be studied.

With this overall strategy in mind, you design a scheme to select for temperature-sensitive (*ts*) mutants in the nuclear translocation machinery. You want to find mutants that proliferate well at low temperature (the permissive condition) but are defective at high temperature (the restrictive condition). You plan to mutagenize cells containing the pNL$^+$ plasmid at low temperature and then shift them to high temperature in the presence of galactose. You reason that at the restrictive temperature the nuclear transport mutants will not take up the killer protein encoded by pNL$^+$ and therefore will not be killed. Normal cells will transport the killer protein into the nucleus and die. After one or two hours at the high temperature to allow selection against normal cells, you intend to lower the temperature and put the surviving cells on nutrient agar plates containing glucose. You expect that the nuclear translocation mutants will recover at the low temperature and form colonies.

When you show your advisor your scheme, she is very pleased at your initiative. However, she sees a critical flaw in your experimental design that would prevent isolation of nuclear translocation mutants, but she won't tell you what it is—she hints that it has to do with the killer protein made during the high-temperature phase of the experiment.

A. What is the critical flaw in your original experimental protocol?
B. How might you modify your protocol to correct this flaw?
C. Assuming for the moment that your original protocol would work, can you think of any other types of mutants (not involved in nuclear transport) that would survive your selection scheme?

**12–54** Nuclear import and export are typically studied in cells treated with mild detergents under carefully controlled conditions, so that the plasma membrane is freely permeable to proteins, but the nuclear envelope remains intact. Because the cytoplasm leaks out, nuclei take up proteins only when the cell remnants are incubated with fresh cytoplasm as a source of critical soluble proteins. One common substrate for uptake studies is serum albumin that is cross-linked to a nuclear localization signal peptide and tagged with a fluorescent label such as fluorescein, so that its uptake into nuclei can be observed by microscopy.

The small GTPase, Ran, was shown to be required for nuclear uptake by using this assay, and not surprisingly, it requires GTP to function. Since neither Ran-GDP nor Ran-GTP binds to nuclear localization signals, other factors must be responsible for identifying proteins to be imported into the nucleus.

Using this assay system, you purify a protein, which you name importin, that in the presence of Ran and GTP promotes uptake of the labeled substrate into nuclei to about the same extent as crude cytosol in the presence of GTP (Figure 12–6A). No uptake occurs in the absence of GTP, and Ran alone is unable to promote nuclear uptake. Importin by itself causes a GTP-independent accumulation of substrate at the nuclear periphery, but does not promote nuclear uptake (Figure 12–6A).

To define the steps in the uptake pathway, you first incubate nuclei with substrate in the presence of importin. You then wash away free importin and substrate and incubate a second time with Ran and GTP (Figure 12–6B).

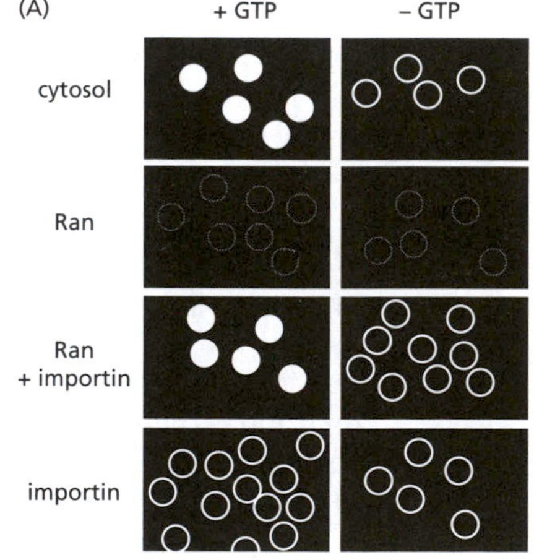


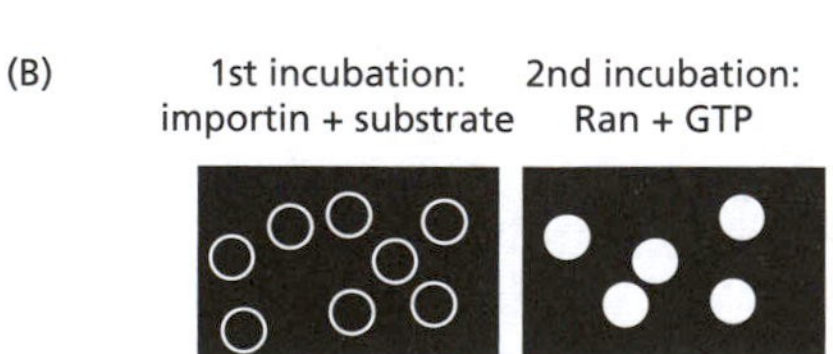


**Figure 12–6** Effects of Ran, importin, and GTP on nuclear uptake of fluorescein-labeled substrate (Problem 12–54). (A) Comparison of various combinations of Ran and importin relative to cytosol in the presence and absence of GTP. (B) Two-stage incubation of cell remnants with importin and substrate and then with Ran and GTP. *Circles* are the nuclei; *very light circles* are nuclei without bound substrate.

A. Why do you think the substrate accumulates at the nuclear periphery, as is seen in the absence of GTP or with importin alone in the presence of GTP?
B. From these data decide whether importin is a nuclear import receptor, a Ran-GAP, or a Ran-GEF (the two Ran-specific regulatory proteins that trigger conversion between Ran-GDP and Ran-GTP). Explain your reasoning.
C. To the extent these data allow, define the order of events that leads to uptake of substrate into the nucleus.

**12–55** The structures of Ran-GDP and Ran-GTP (actually Ran-GppNp, a stable GTP analog) are strikingly different, as shown in Figure 12–7. Not surprisingly, Ran-GDP binds to a different set of proteins than does Ran-GTP.

To look at the uptake of Ran itself into the nuclei of permeabilized cells, you attach a red fluorescent tag to a cysteine side chain in Ran to make it visible. This modified Ran supports normal nuclear uptake. Fluorescent Ran-GDP is taken up by nuclei only if cytoplasm is added, whereas a mutant form, RanQ69L-GTP, which is unable to hydrolyze GTP, is not taken up in the presence or absence of cytoplasm. To identify the cytoplasmic protein that is crucial for Ran-GDP uptake, you construct affinity columns with bound Ran-GDP or bound RanQ69L-GTP and pass cytoplasm through them. Cytoplasm passed over a Ran-GDP column no longer supports nuclear uptake of Ran-GDP, whereas cytoplasm passed over a column of RanQ69L-GTP retains this activity. You elute the bound proteins from each column and analyze them on an SDS polyacrylamide gel, looking for differences that might identify the factor that is required for nuclear uptake of Ran (Figure 12–8).

A. Why did you use RanQ69L-GTP instead of Ran-GTP in these experiments? Could you have used Ran-GppNp instead of RanQ69L-GTP to achieve the same purpose?
B. Which of the many proteins eluted from the two different affinity columns is a likely candidate for the factor that promotes nuclear import of Ran-GDP?
C. What other protein or proteins would you predict the Ran-GDP import factor would be likely to bind in order to carry out its function?
D. How might you confirm that the factor you have identified is necessary for promoting the nuclear uptake of Ran?

**12–56** The original experiments on nuclear uptake uniformly—but incorrectly—concluded that GTP hydrolysis was essential for import. This conclusion was largely based on results with nonhydrolyzable analogs of GTP such as GppNp, which efficiently blocked nuclear import. It is now known that a Ran-GEF, which converts Ran-GDP to Ran-GTP, exists in the nucleus and a Ran-GAP, which converts Ran-GTP to Ran-GDP, exists in the cytosol, often associated with the pore complexes. These two proteins ensure that Ran-GTP is present primarily in the nucleus and that Ran-GDP is present primarily in the cytoplasm.

The experiments that cleared up this situation—and helped define our present view of nuclear import—showed that some Ran-GEF (a protein known as RCC1) leaks out of nuclei during preparation of cell remnants. How might the presence of Ran-GEF outside the nucleus lead to a spurious requirement for GTP hydrolysis in nuclear import? Why does a nonhydrolyzable analog of GTP such as GppNp block import under these conditions while GTP does not?

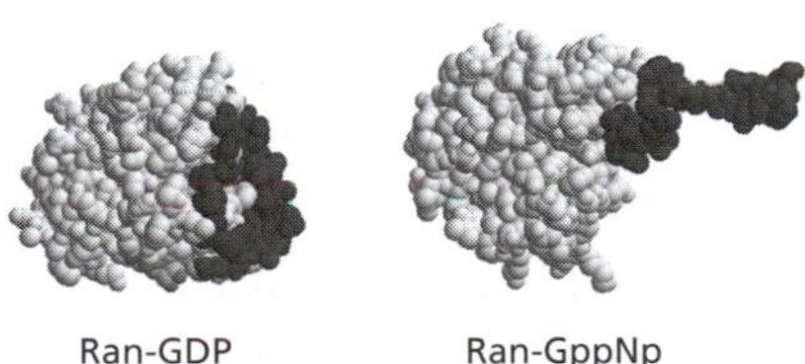


**Figure 12–7** Structures of Ran-GDP and Ran-GppNp (Problem 12–55). The *shaded portions* of the structures show the segment of Ran that differs most dramatically when GDP or the GTP analog is bound.

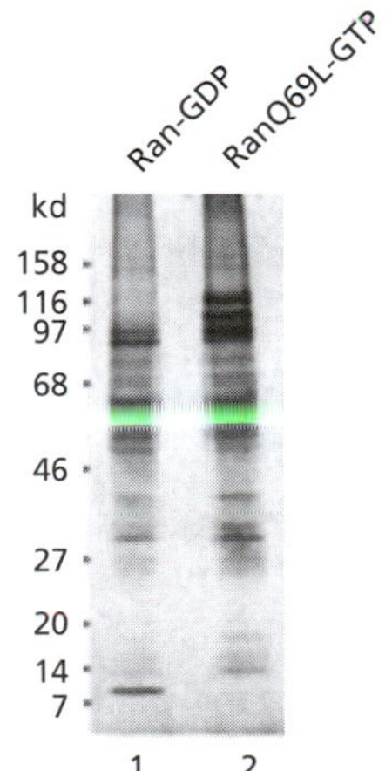


**Figure 12–8** Proteins eluted from Ran-GDP (lane 1) and RanQ69L-GTP (lane 2) affinity columns (Problem 12–55). The molecular masses of marker proteins are shown on the *left*.

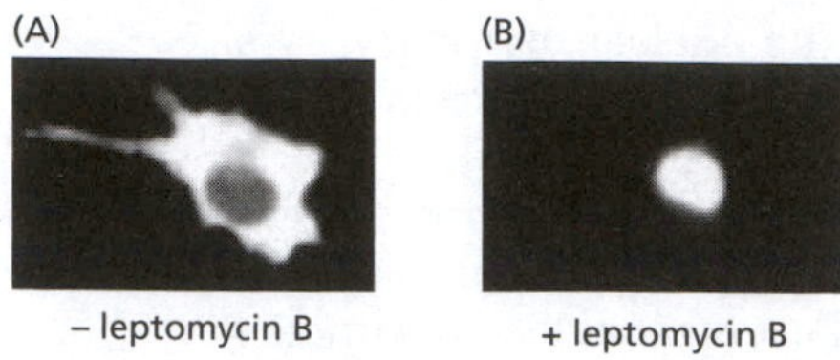


**Figure 12–9** The cellular distribution of Rev (Problem 12–57). (A) Absence of leptomycin B. (B) Presence of leptomycin B.

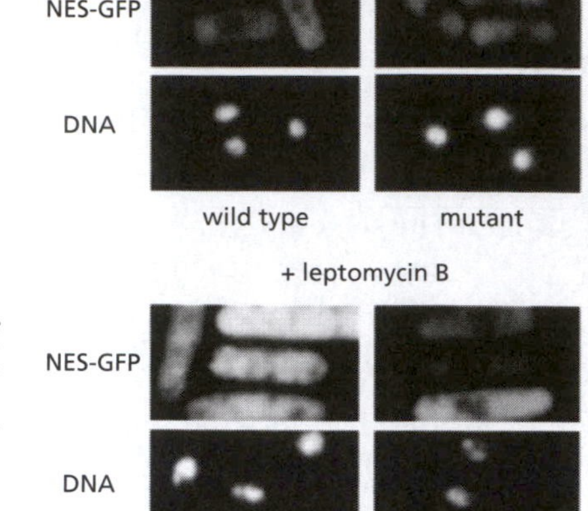


**Figure 12–10** Distribution of NES-GFP in *S. pombe* in the presence and absence of leptomycin B (Problem 12–58). *Light areas* in the NES-GFP panels show the position of GFP. *Light areas* in the DNA panels result from a stain that binds to DNA and marks the position of the nuclei in the cells in the NES-GFP panels.

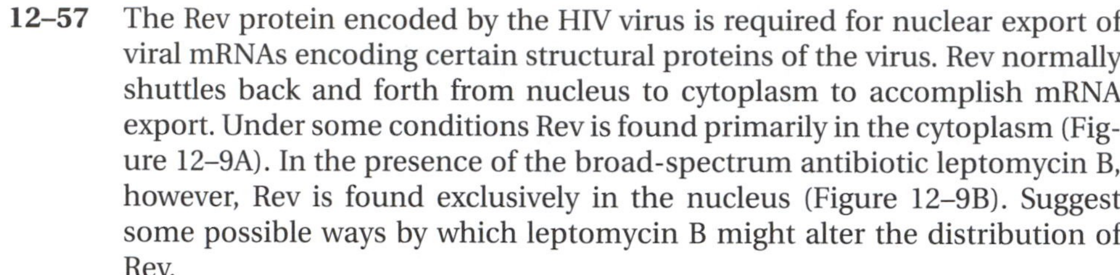

**12–57** The Rev protein encoded by the HIV virus is required for nuclear export of viral mRNAs encoding certain structural proteins of the virus. Rev normally shuttles back and forth from nucleus to cytoplasm to accomplish mRNA export. Under some conditions Rev is found primarily in the cytoplasm (Figure 12–9A). In the presence of the broad-spectrum antibiotic leptomycin B, however, Rev is found exclusively in the nucleus (Figure 12–9B). Suggest some possible ways by which leptomycin B might alter the distribution of Rev.

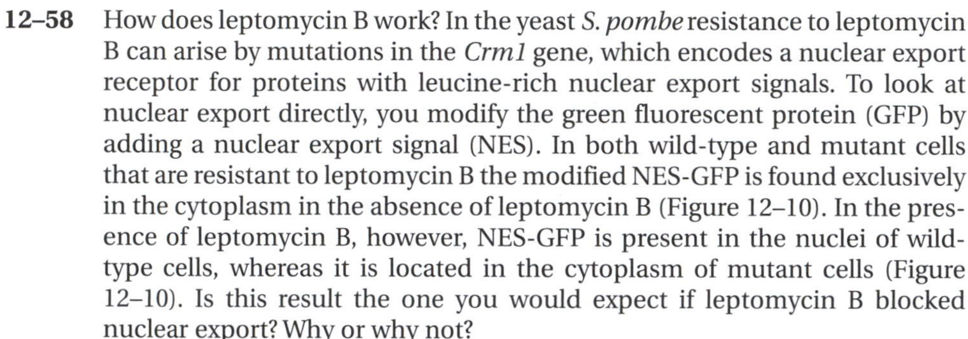

**12–58** How does leptomycin B work? In the yeast *S. pombe* resistance to leptomycin B can arise by mutations in the *Crm1* gene, which encodes a nuclear export receptor for proteins with leucine-rich nuclear export signals. To look at nuclear export directly, you modify the green fluorescent protein (GFP) by adding a nuclear export signal (NES). In both wild-type and mutant cells that are resistant to leptomycin B the modified NES-GFP is found exclusively in the cytoplasm in the absence of leptomycin B (Figure 12–10). In the presence of leptomycin B, however, NES-GFP is present in the nuclei of wild-type cells, whereas it is located in the cytoplasm of mutant cells (Figure 12–10). Is this result the one you would expect if leptomycin B blocked nuclear export? Why or why not?

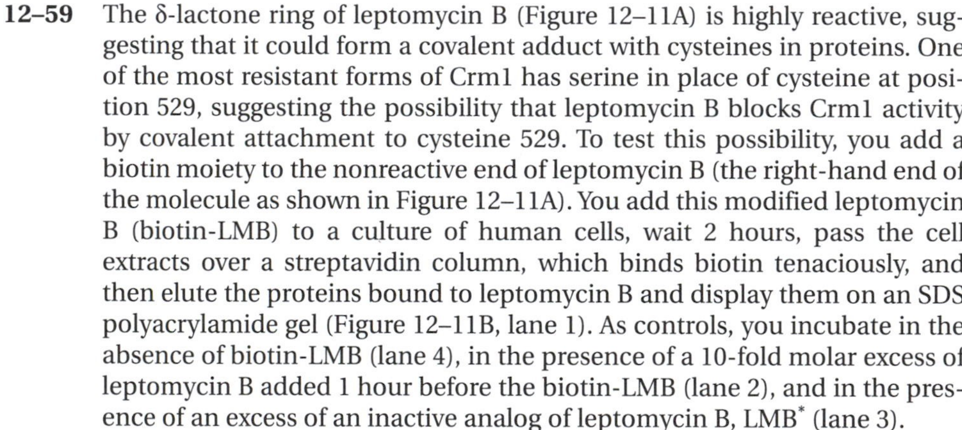

**12–59** The δ-lactone ring of leptomycin B (Figure 12–11A) is highly reactive, suggesting that it could form a covalent adduct with cysteines in proteins. One of the most resistant forms of Crm1 has serine in place of cysteine at position 529, suggesting the possibility that leptomycin B blocks Crm1 activity by covalent attachment to cysteine 529. To test this possibility, you add a biotin moiety to the nonreactive end of leptomycin B (the right-hand end of the molecule as shown in Figure 12–11A). You add this modified leptomycin B (biotin-LMB) to a culture of human cells, wait 2 hours, pass the cell extracts over a streptavidin column, which binds biotin tenaciously, and then elute the proteins bound to leptomycin B and display them on an SDS polyacrylamide gel (Figure 12–11B, lane 1). As controls, you incubate in the absence of biotin-LMB (lane 4), in the presence of a 10-fold molar excess of leptomycin B added 1 hour before the biotin-LMB (lane 2), and in the presence of an excess of an inactive analog of leptomycin B, LMB* (lane 3).

A. Figure 12–11, lanes 1 and 3, show a strong new band of 102 kd that is retained on the affinity column when biotin-LMB is added to the cell culture medium. How might you show that this protein is Crm1?

B. Why was the 102-kd protein not seen in Figure 12–11, lanes 2 and 4? What was the point of performing these particular control experiments?

C. Several proteins that bind to the streptavidin column are present in all four lanes in Figure 12–11B. Can you suggest what kinds of proteins they might be? (Think about why they might have been retained on the streptavidin affinity column.)

**12–60** Frog oocytes are a useful experimental system for studying nuclear export because they are large cells with large nuclei. It is easy (with practice) to

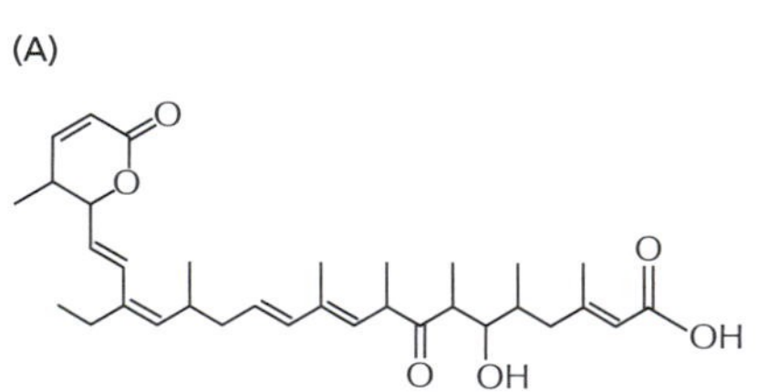


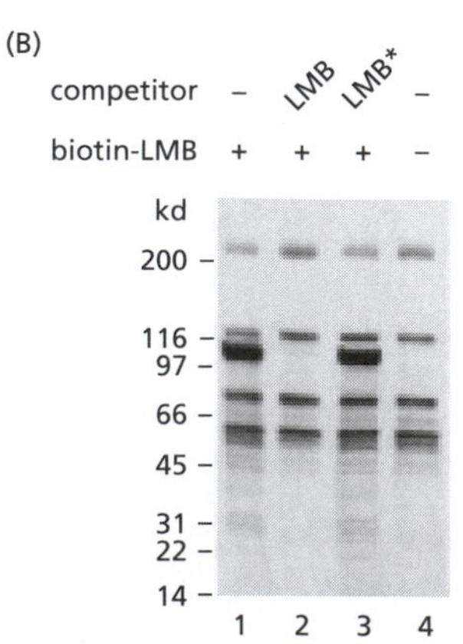


**Figure 12–11** Identifying a protein that reacts with leptomycin B (Problem 12–59). (A) Structure of leptomycin B. (B) Proteins eluted from a streptavidin affinity column after various cell treatments. Molecular masses of marker proteins are shown at the *left* in kilodaltons.

inject oocytes with labeled RNA and to separate the nucleus and cytoplasm to follow the fate of the injected label. You inject a mixture of various $^{32}$P-labeled RNA molecules into the nucleus in the presence and absence of leptomycin B to study its effect on nuclear export of RNA. Immediately after injection and three hours later, you analyze total (T), cytoplasmic (C), and nuclear (N) contents by polyacrylamide-gel electrophoresis and autoradiography (Figure 12–12).

A. How good was your injection technique? Did you actually inject into the nucleus? Did you rip apart the nuclear envelope when you injected the RNAs? How do you know?
B. Which, if any, of the RNAs are normally exported from the nucleus?
C. Is the export of any of the RNAs inhibited by leptomycin B? What does your answer imply about export of this collection of RNAs?

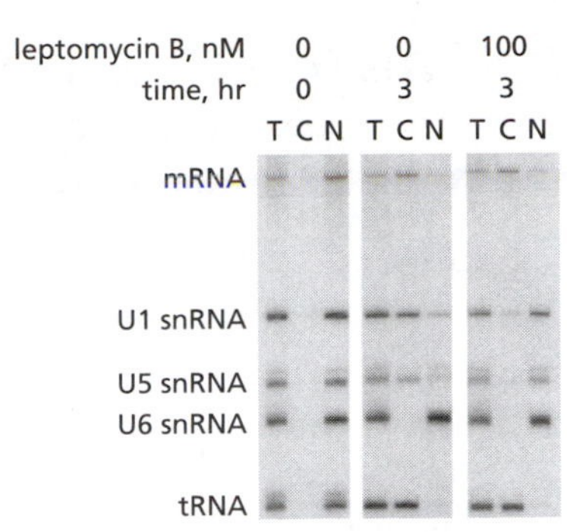


**Figure 12–12** Effects of leptomycin B on export of various RNAs from frog oocytes (Problem 12–60). Total (T), cytoplasmic (C), and nuclear (N) fractions are indicated.

**12–61** The gene for the κ-light chain of immunoglobulins contains a DNA sequence element (an enhancer) that regulates its expression. A protein called NFκB binds to this enhancer and can be found in nuclear extracts of B cells, which actively synthesize immunoglobulins. By contrast, NFκB activity is absent from precursor B cells (pre-B cells), which do not express immunoglobulins. You want to know how NFκB regulates expression of κ-light-chain genes.

If pre-B cells are treated with phorbol ester (which activates protein kinase C), NFκB DNA-binding activity appears within a few minutes in parallel with the transcriptional activation of the κ-light-chain gene. This activation of NFκB is not blocked by cycloheximide, which is an inhibitor of protein synthesis.

You find that cytoplasmic extracts of unstimulated pre-B cells contain an inactive form of NFκB that can be activated for DNA binding by treatment with mild denaturants. (Denaturants disrupt noncovalent bonds but do not break covalent bonds.) To understand the relationship between NFκB activation and transcriptional regulation of κ-light-chain genes, you isolate nuclear (N) and cytoplasmic (C) fractions from pre-B cells before and after stimulation by phorbol ester. You then measure NFκB activity, before and after treatment with mild denaturants, by its ability to form a complex with labeled DNA fragments that carry the enhancer. The results are shown in Figure 12–13.

A. How does the subcellular localization of NFκB change in pre-B cells in response to phorbol ester treatment?
B. Do you think that the activation of NFκB in response to phorbol ester occurs because NFκB is phosphorylated?
C. Outline a molecular mechanism to account for activation of NFκB by treatment with phorbol ester.

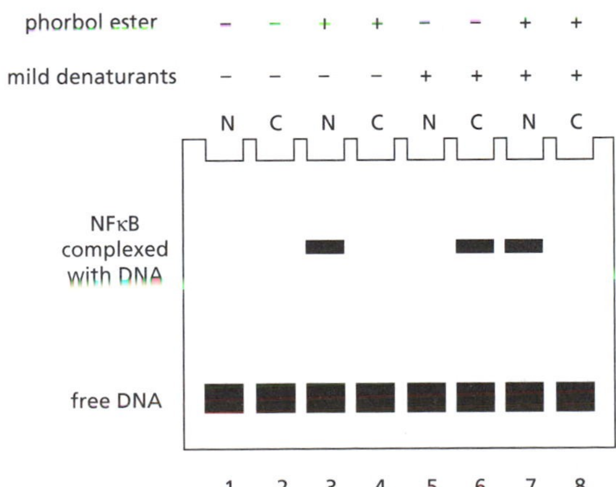


**Figure 12–13** Gel retardation assay of NFκB activity under various conditions (Problem 12–61). N and C refer to nuclear and cytoplasmic fractions, respectively. The mild denaturants are a combination of formamide (27%) and sodium deoxycholate (0.2%). When NFκB binds to the DNA, it forms a complex that migrates more slowly on the gel.

# THE TRANSPORT OF PROTEINS INTO MITOCHONDRIA AND CHLOROPLASTS

TERMS TO LEARN

| | | |
|---|---|---|
| chloroplast | mitochondrial Hsp70 | SAM complex |
| inner membrane | mitochondrial precursor protein | stroma |
| intermembrane space | outer membrane | thylakoid |
| matrix space | OXA complex | TIM complex |
| mitochondria | protein translocator | TOM complex |

## DEFINITIONS

Match the definition below with its term from the list above.

**12–62** Membrane-enclosed organelles, about the size of bacteria, that carry out oxidative phosphorylation and produce most of the ATP in eucaryotic cells.

**12–63** Part of a multisubunit protein assembly that is bound to the matrix side of the TIM23 complex and acts as a motor to pull the precursor protein into the matrix space.

**12–64** Multisubunit protein assembly that transports proteins across the mitochondrial outer membrane.

**12–65** The matrix space of a chloroplast.

**12–66** The membrane of a mitochondrion that encloses the matrix and is folded into cristae.

**12–67** Multisubunit protein assembly that transports proteins across the mitochondrial inner membrane.

**12–68** Central subcompartment of a mitochondrion, enclosed by the inner mitochondrial membrane.

**12–69** Flattened sac of membrane in a chloroplast that contains the protein subunits of the photosynthetic system and of the ATP synthase.

**12–70** Protein encoded by a nuclear gene, synthesized in the cytosol, and subsequently transported into mitochondria.

## TRUE/FALSE

Decide whether each of these statements is true or false, and then explain why.

**12–71** The TOM complex is required for the import of all nucleus-encoded mitochondrial proteins.

**12–72** The two signal sequences required for transport of nucleus-encoded proteins into the mitochondrial inner membrane via the TIM23 complex are cleaved off the protein in different mitochondrial compartments.

**12–73** Import of proteins into mitochondria and chloroplasts is very similar; even the individual components of their transport machinery are homologous, as befits their common evolutionary origin.

## THOUGHT PROBLEMS

**12–74** To aid your studies of protein import into mitochondria, you treat yeast cells with cycloheximide, which blocks ribosome movement along mRNA. When you examine these cells in the electron microscope, you are surprised to find cytosolic ribosomes attached to the outside of the mitochondria. You have never seen attached ribosomes in the absence of cycloheximide. To investigate this phenomenon further, you prepare mitochondria from cells that

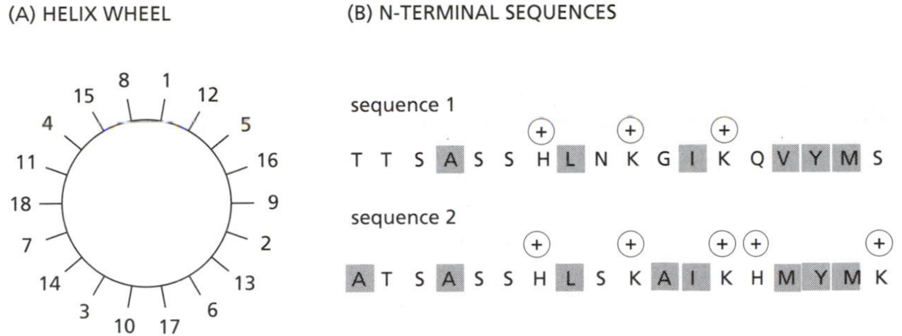


**Figure 12–14** Analysis of the N-terminal sequences for human and chicken glutamine synthetase (Problem 12–75). (A) Helix-wheel projection of an α helix. *Numbers* show the positions of the first 18 amino acids of an α helix; amino acid 19 would occupy the same position as amino acid 1. (B) N-terminal amino acid sequences for glutamine synthetases from humans and chickens. The N-termini are shown at the *left*; hydrophobic amino acids are *shaded*; positively charged amino acids are indicated by +.

have been treated with cycloheximide and then extract the mRNA that is bound to the ribosomes associated with the mitochondria. You translate this mRNA *in vitro* and compare the protein products with similarly translated mRNA from the cytosol. The results are clear-cut: the mitochondria-associated ribosomes are translating mRNAs that encode mitochondrial proteins.

You are astounded! Here, clearly visible in the electron micrographs, seems to be proof that protein import into mitochondria occurs during translation. How might you reconcile this result with the prevailing view that mitochondrial proteins are imported only after they have been synthesized and released from ribosomes?

**12–75** Glutamine synthetase is an enzyme involved in ammonia detoxification. In chickens the enzyme is targeted to mitochondria, whereas in humans it is cytosolic. Amphiphilic α helices with positively charged amino acids on one side and hydrophobic amino acids on the other are the key features of signal peptides used for protein import into mitochondria. Using the helix-wheel projection in Figure 12–14A, decide which of the N-terminal sequences in Figure 12–14B is from the chicken enzyme and which is from the human enzyme.

**Problem 3–18** describes the use of a helix-wheel projection.

**12–76** You have made a peptide that contains a functional mitochondrial import signal. Would you expect the addition of an excess of this peptide to affect the import of mitochondrial proteins? Why or why not?

**12–77** Components of the TIM complex, the multisubunit protein translocator in the mitochondrial inner membrane, are much less abundant than those of the TOM complex. They were initially identified using a genetic trick. The yeast *Ura3* gene, whose product is normally located in the cytosol where it is essential for synthesis of uracil, was modified to carry an import signal for the mitochondrial matrix. A population of cells carrying the modified *Ura3* gene was then grown in the absence of uracil. Most cells died, but the rare cells that grew were shown to be defective for mitochondrial import. Explain how this selection identifies cells with defects in components required for mitochondrial import. Why don't normal cells with the modified *Ura3* gene grow in the absence of uracil? Why do cells that are defective for mitochondrial import grow in the absence of uracil?

**12–78** Mitochondria normally provide cells with most of the ATP they require to meet their energy needs. Mitochondria that cannot import proteins are defective for ATP synthesis. How is it that cells with import-defective mitochondria can survive at all? How do they get the ATP they need to function?

**12–79** Five protein translocator complexes direct proteins to various compartments in mitochondria. Indicate on Table 12–2 the mitochondrial compartment(s) into which each translocator inserts its transport substrates.

**Table 12–2 Mitochondrial compartments and translocator complexes** (Problem 12–79).

| COMPARTMENT | SAM | TOM | TIM22 | TIM23 | OXA |
|---|---|---|---|---|---|
| Outer membrane | | | | | |
| Intermembrane space | | | | | |
| Inner membrane | | | | | |
| Matrix | | | | | |

**12–80** Describe in a general way how you might use radiolabeled proteins and proteases to study import processes in isolated, intact mitochondria. What sorts of experimental controls might you include to ensure that the results you obtain mean what you think they do?

**12–81** If the enzyme dihydrofolate reductase (DHFR), which is normally located in the cytosol, is engineered to carry a mitochondrial targeting sequence at its N-terminus, it is efficiently imported into mitochondria. If the modified DHFR is first incubated with methotrexate, which binds tightly to the active site, the enzyme remains in the cytosol. How do you suppose that the binding of methotrexate interferes with mitochondrial import?

**12–82** Why do mitochondria need a fancy translocator to import proteins across the outer membrane? Their outer membranes already have large pores formed by porins.

## CALCULATIONS

**12–83** The vast majority of mitochondrial proteins are imported through the outer membrane by the multisubunit protein translocators known as TOM complexes. Since the number of TOM complexes in yeast mitochondria is known (10 pmole/mg of mitochondrial protein), it is possible to calculate whether a co-translational mechanism could account for the bulk of mitochondrial protein import. If mitochondrial proteins were all imported co-translationally, then 10 pmol of TOM complexes would need to import 1 mg of protein each generation. Given that mitochondria double every 3 hours and that the rate of protein synthesis is 3 amino acids per second (which is therefore the maximum rate of import through a single TOM complex), how many mg of mitochondrial protein could 10 pmol of TOM complexes import in one generation? (On average an amino acid has a mass of 110 daltons.)

## DATA HANDLING

**12–84** Barnase is a 110-amino acid bacterial ribonuclease that is much used as a model for studies of protein folding and unfolding. It forms a compact folded structure that has a high energy of activation for unfolding (about 20 kcal/mole). Can such a protein be imported into mitochondria? To the N-terminus of barnase, you add 35, 65, or 95 amino acids from the N-terminus of pre-cytochrome $b_2$, all of which include the cytochrome's mitochondrial import signal. N35-barnase is not imported, N65-barnase is imported at a low rate, and N95-barnase is imported very efficiently into isolated mitochondria. None of these N-terminal extensions have any measurable effect on the stability of the barnase domain. If these proteins are denatured before testing for import, they are all imported at the same high rate. How do you suppose that longer N-terminal extensions facilitate the import of barnase?

**12–85** Are proteins imported into mitochondria as completely unfolded polypeptide chains, or can the translocation apparatus accommodate fully or partially folded structures? That is, is the protein sucked up like a noodle, or is it swallowed whole, as a python devours its prey? It is possible to engineer cysteine amino acids into barnase and then cross-link them to make disulfide bonds either between C5 and C78 or between C43 and C80 (Figure 12–15). Import of N95-barnase (see Problem 12–84) was tested in the presence and absence of disulfide cross-links at these two positions. Its import was unaffected by either cross-link. By contrast, import of N65-barnase was blocked by the C5–C78 cross-link but unaffected by the C43–C80 cross-link. Do these results allow you to distinguish between import of extended polypeptide chains or of folded structures? Why or why not?

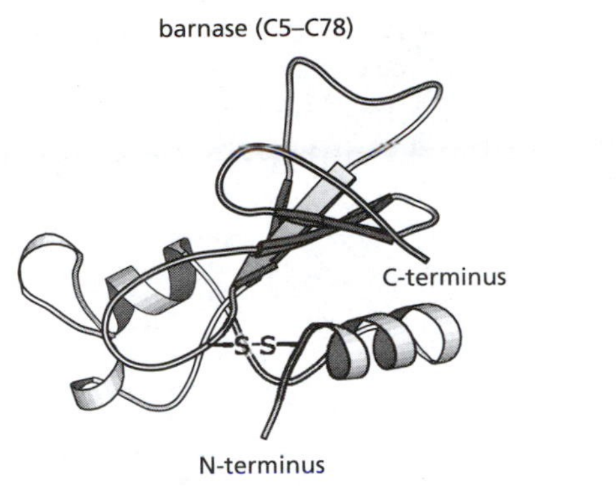


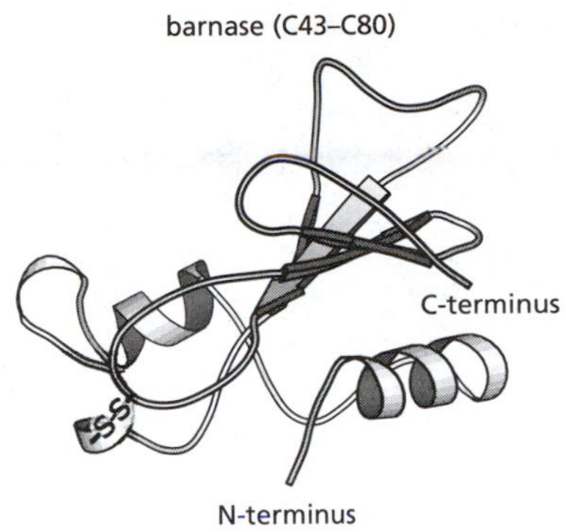


**Figure 12–15** Structures of barnase molecules with disulfide bonds between cysteines at positions 5 and 78 (C5–C78) or between positions 43 and 80 (C43–C80) (Problem 12–85).

**12–86** Tim23 is a key component of the mitochondrial TIM23 protein translocator complex. The N-terminal half of Tim23 is hydrophilic, while the C-terminal half is hydrophobic and probably spans the membrane four times, as suggested by the hydropathy plot in Figure 12–16A. To determine the arrangement of Tim23 in mitochondrial membranes, a protease was added to intact mitochondria or to mitoplasts, which are mitochondria from which the outer membranes have been removed. The mobility of Tim23 was detected on SDS polyacrylamide gels by immunoblotting with antibodies specific for the N-terminal half or for the extreme C-terminus of Tim23 (Figure 12–16B). Normal-sized Tim23 was present in both mitochondria and mitoplasts (as shown for mitochondria in Figure 12–16B), but Tim23 was partially digested when mitochondria and mitoplasts were treated with a protease (Figure 12–16B).

A. Is Tim23 an integral component of the inner or outer mitochondrial membrane? Explain your reasoning.

B. To the extent the information in this problem allows, diagram the arrangement of Tim23 in mitochondrial membranes.

**12–87** Chloroplasts contain six compartments—outer membrane, intermembrane space, inner membrane, stroma, thylakoid membrane, and thylakoid lumen (Figure 12–17)—each of which is populated by specific sets of proteins. To investigate the import of nucleus-encoded proteins into chloroplasts, you have chosen to study ferredoxin (FD), which is located in the stroma, and plastocyanin (PC), which is located in the thylakoid lumen, as well as two hybrid genes: ferredoxin with the plastocyanin signal peptide (pcFD) and plastocyanin with the ferredoxin signal peptide (fdPC). You translate mRNAs

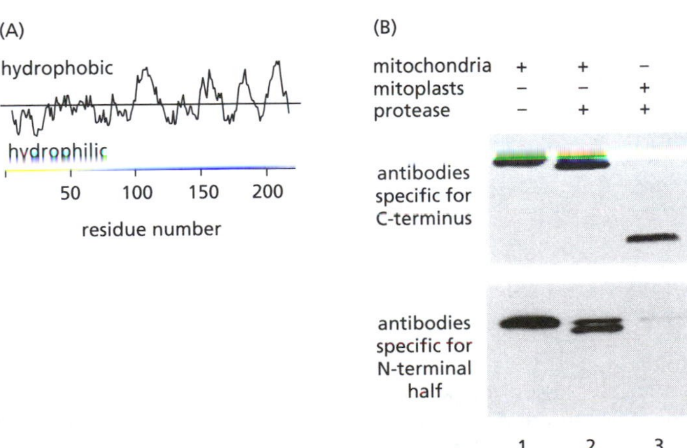


**Figure 12–16** Arrangement of Tim23 in mitochondrial membranes (Problem 12–86). (A) Hydropathy plot for Tim23. (B) Sensitivity of Tim23 in mitochondria and mitoplasts to digestion with a protease.

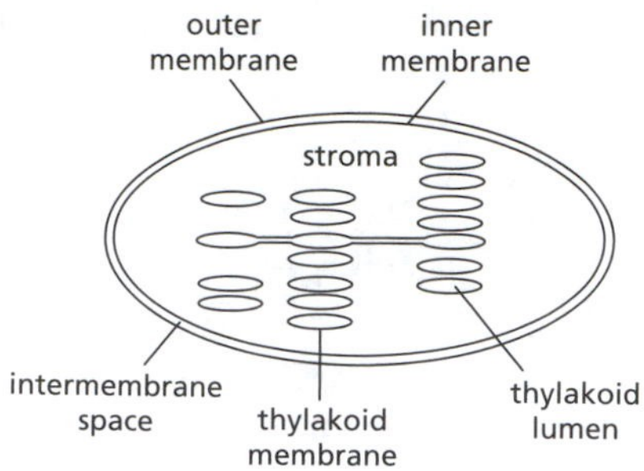


**Figure 12–17** The six compartments of a chloroplast (Problem 12–87).

from these four genes *in vitro*, mix the translation products with isolated chloroplasts for a few minutes, reisolate the chloroplasts after protease treatment, and fractionate them to find which compartments the proteins have entered (Figure 12–18A). The status of the normal and hybrid proteins at each stage of the experiment is shown in Figure 12–18B: each lane in the gels corresponds to a stage of the experiment as indicated alongside the experimental protocol in Figure 12–18A.

A. How efficient is chloroplast uptake of ferredoxin and plastocyanin in your *in vitro* system? How can you tell?

B. Are ferredoxin and plastocyanin localized to their appropriate chloroplast compartments in these experiments? How can you tell?

C. Are the hybrid proteins imported as you would expect if the N-terminal signal peptides determined their final location? Comment on any significant differences.

D. Why are there three bands in experiments with plastocyanin and pcFD but only two bands in experiments with ferredoxin and fdPC? To the extent you can, identify the molecular species in the bands and their relationships to one another.

E. Based on your experiments, propose a model for the import of proteins into the stroma and thylakoid lumen.

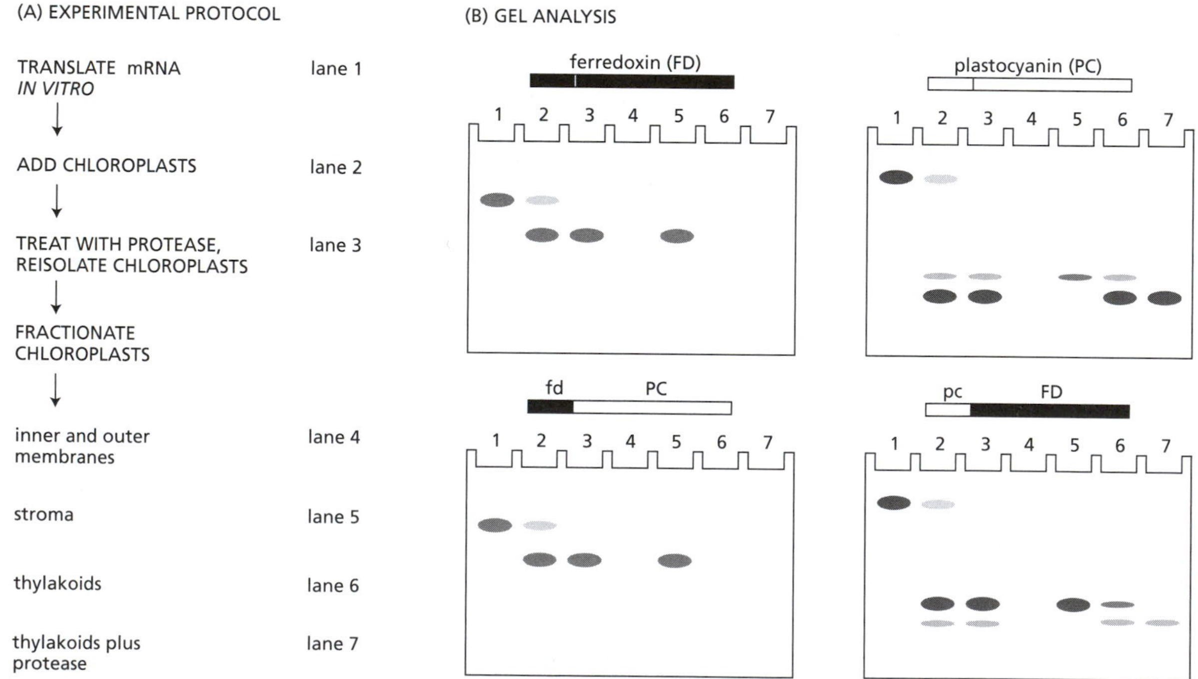


**Figure 12–18** Import of ferredoxin and plastocyanin into chloroplast compartments (Problem 12–87). (A) Experimental protocol. (B) Gel analysis. Samples from each stage in the experimental protocol were analysed by gel electrophoresis. Each lane in (B) corresponds to a particular experimental treatment in (A). Ferredoxin gene segments are *black*; plastocyanin gene segments are *white*. The signal peptides in all genes are located at the *left* (N-terminal) end. The ferredoxin signal peptide in the hybrid gene is shown as fd; the plastocyanin signal peptide in the hybrid gene is shown as pc.

# PEROXISOMES

TERMS TO LEARN

peroxin    peroxisome

## DEFINITIONS

Match the definition below with its term from the list above.

**12–88** Small membrane-bounded organelle that uses molecular oxygen to oxidize organic molecules.

**12–89** One of several proteins that are involved in protein import into peroxisomes.

## TRUE/FALSE

Decide whether this statement is true or false, and then explain why.

**12–90** Peroxisomes are found in only a few specialized types of eucaryotic cell.

## THOUGHT PROBLEMS

**12–91** Primary hyperoxaluria type 1 (PH1) is a lethal autosomal recessive disease caused by a deficiency of the liver-specific peroxisomal enzyme alanine:glyoxylate aminotransferase (AGT). About one-third of PH1 patients possess significant levels of AGT protein and enzyme activity. Analysis of these patients shows that their AGT contains two critical single amino acid changes: one that interferes with peroxisomal targeting and a second that allows the N-terminus to form an amphiphilic α helix with a positively charged side. Where do you suppose this mutant AGT is found in cells from these patients? Explain your reasoning.

**12–92** Catalase, an enzyme normally found in peroxisomes, is present in normal amounts in cells that do not have visible peroxisomes. It is possible to determine the location of catalase in such cells using immunofluorescence microscopy with antibodies specific for catalase. Fluorescence micrographs of normal cells and peroxisome-deficient cells are shown in Figure 12–19. Where is catalase located in cells without peroxisomes (Figure 12–19B)? Why does catalase show up as small dots of fluorescence in normal cells (Figure 12–19A)?

**12–93** Cells with functional peroxisomes incorporate 9-(1′-pyrene)nonanol (P9OH) into membrane lipids. Exposure of such cells to UV light causes cell death because excitation of the pyrene moiety generates reactive oxygen species, which are toxic to cells. Cells that do not make peroxisomes lack a critical enzyme responsible for incorporating P9OH into membrane lipids. How might you make use of P9OH to select for cells that are missing peroxisomes?

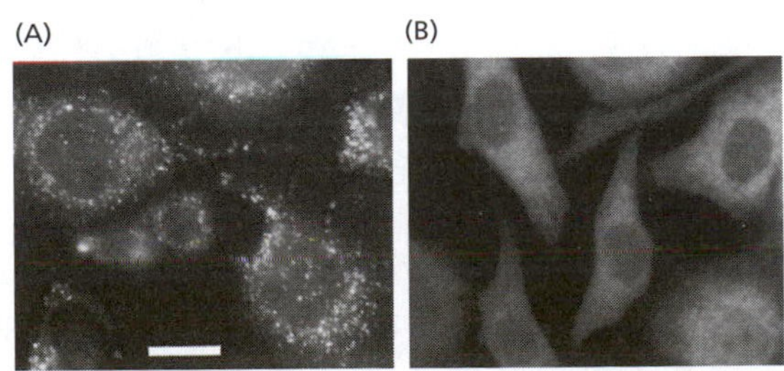


**Figure 12–19** Location of catalase in cells as determined by immunofluorescence microscopy (Problem 12–92). (A) Normal cells. (B) Peroxisome-deficient cells. Cells were reacted with antibodies specific for catalase, washed, and then stained with a fluorescein-labeled second antibody that is specific for the catalase-specific antibody. The two panels are the same size; a 20-μm scale bar is shown in (A).

**12–94** If peroxisomes come from preexisting peroxisomes by growth and fission, how is it that mutant yeast cells with no visible peroxisomes, peroxisome ghosts, or peroxisome remnants, can suddenly form fully functional peroxisomes when the missing gene is introduced?

## DATA HANDLING

**12–95** Trypanosomes are single-celled parasites that cause sleeping sickness when they infect humans. Trypanosomes from humans carry the enzymes for a portion of the glycolytic pathway in a peroxisomelike organelle, termed the glycosome. By contrast, trypanosomes from the tsetse fly—the intermediate host—carry out glycolysis entirely in the cytosol. This intriguing difference has alerted the interest of the pharmaceutical company that employs you. Your company wishes to exploit this difference to control the disease.

You decide to study the enzyme phosphoglycerate kinase (PGK) because it is in the affected portion of the glycolytic pathway. Trypanosomes from the tsetse fly express PGK entirely in the cytosol, whereas trypanosomes from humans express 90% of the total PGK activity in glycosomes and only 10% in the cytosol. When you clone PGK genes from trypanosomes, you find three forms that differ slightly from one another. Exploiting these small differences, you design three oligonucleotides that hybridize specifically to the mRNAs from each gene. Using these oligonucleotides as probes, you determine which genes are expressed by trypanosomes from humans and from tsetse flies. The results are shown in Figure 12–20.

A. Which PGK genes are expressed in trypanosomes from humans? Which are expressed in trypanosomes from tsetse flies?
B. Which PGK gene probably encodes the glycosomal form of PGK?
C. Do you think that the minor cytosolic PGK activity in trypanosomes from humans is due to inaccurate sorting into glycosomes? Explain your answer.

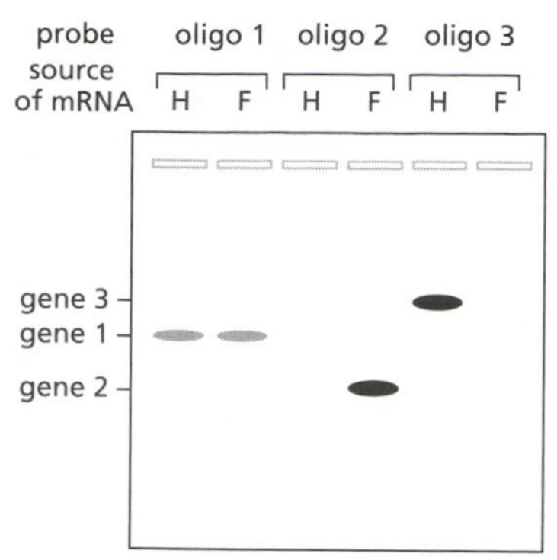


**Figure 12–20** Hybridization of specific oligonucleotide probes (oligos) to mRNA isolated from trypanosomes from humans (H) and tsetse flies (F) (Problem 12–95). The intensity of the bands on the autoradiograph reflects the concentrations of the mRNAs.

**12–96** You have developed an assay for peroxisomal function. Mutagenized cell colonies are incubated with a soluble radioactive precursor that is converted into a readily detectable insoluble product by a peroxisomal enzyme. Your laborious screen of 25,000 colonies is finally rewarded by the discovery of two colonies that do not have the insoluble radioactive product. Sure enough, these mutant cells lack typical peroxisomes as judged by electron microscopy.

When you test these cells for peroxisomal enzymes, you find that catalase activity is virtually the same as in normal cells. By contrast, acyl CoA oxidase activity is absent in both mutant cell lines. To investigate the acyl CoA oxidase deficiency, you perform a pulse-chase experiment: you grow cells for 1 hour in medium containing $^{35}$S-methionine, then transfer them to unlabeled medium and immunoprecipitate acyl CoA oxidase at various times after transfer (Figure 12–21). You observe two forms of oxidase in normal cells, but only one in the mutant cell lines. To clarify the relationship between the 75-kd and 53-kd forms of the oxidase, you isolate mRNA from wild-type and the mutant cell lines, translate it *in vitro*, and immunoprecipitate acyl CoA oxidase. All three sources of mRNA give similar levels of the 75-kd form, but none of the 53-kd form.

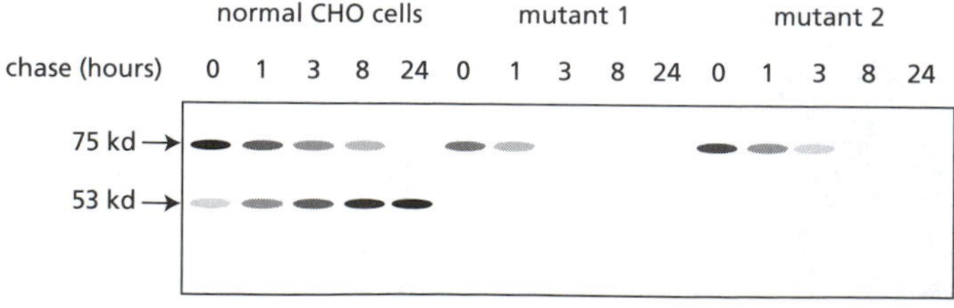


**Figure 12–21** Pulse-chase experiments with normal and mutant cells (Problem 12–96).

A. How do you think the two forms of acyl CoA oxidase in normal cells are related? Which one, if either, do you suppose is the active enzyme?

B. Why do the mutant cells have only the 75-kd form of acyl CoA oxidase, and why do you think it disappears during the pulse-chase experiment? If you had done a similar experiment with catalase, do you suppose it would have behaved the same way?

**12–97** Through cleverness and hard work, you have isolated 14 cell lines that are missing peroxisomes, as judged by electron microscopy. You wish to know how many different genes are represented in this collection of mutant cell lines. You carry out two types of complementation analysis—fusion of cells to form heterocaryons and transfection with cDNAs for defined peroxisomal genes—and test for the appearance of peroxisomes. Your results are shown in Table 12–3, where the presence of peroxisomes is indicated by + and their absence by –.

As is apparent from the blank spaces in Table 12–3, you haven't done all the possible heterocaryon analyses (nor all the possible transfection analyses, for that matter). Is your present data set adequate to allow you to group all 14 mutant cell lines unambiguously into complementation groups? To the extent the data allow, sort mutant cell lines into complementation groups and indicate which groups are mutant in the *Pex2*, *Pex5*, and *Pex6* genes.

**12–98** Proteins that are imported into the peroxisome matrix using a C-terminal tripeptide signal are recognized by the cytosolic receptor Pex5, which docks at a complex of proteins in the peroxisomal membrane. Delivery by a cytosolic receptor distinguishes peroxisomal import from import into mitochondria, chloroplasts, and the endoplasmic reticulum. Moreover, unlike import into those organelles, peroxisomes can import fully folded proteins and protein oligomers. You wish to distinguish three potential modes of action for Pex5: (1) it delivers its cargo to the peroxisomal membrane but remains cytosolic, (2) it enters the peroxisome along with its cargo; or (3) it cycles between the cytosol and the peroxisomal matrix.

To define the mechanism for Pex5-mediated import, you modify the *Pex5* gene to encode an N-terminal peptide segment that includes a cleavage site for a protease localized exclusively to the matrix of the peroxisome (Figure 12–22A). Immediately adjacent to the cleavage site is a sequence of amino acids (the so-called FLAG tag) that can be recognized by commercial antibodies. One antibody, mAb2, binds the FLAG tag in any context, whereas another, mAb1, binds the FLAG tag only when it is at the N-terminus and thus will detect only cleaved Pex5. By preparing a whole-cell extract (WCE) and fractionating it into a pellet (P), which contains the peroxisomes, and supernatant (S), which contains the cytosol, you can distinguish the three

**Table 12–3 Complementation analysis of peroxisome-deficient cell lines** (Problem 12–97).

| | HETEROCARYON ANALYSIS | | | | | | | | cDNA TRANSFECTION | | |
|---|---|---|---|---|---|---|---|---|---|---|---|
| MUTANTS | Z65 | ZP105 | ZP92 | Z24 | ZP109 | ZP110 | ZP114 | ZP165 | *Pex2* | *Pex5* | *Pex6* |
| ZP109 | | | | + | | | | | – | – | – |
| ZP110 | | | | + | + | | | | – | – | – |
| ZP114 | | | | + | + | + | | | – | – | – |
| ZP116 | – | + | + | | + | | | | + | – | – |
| ZP119 | + | + | + | + | + | + | + | – | – | – | – |
| ZP160 | – | | | | | | | | + | – | – |
| ZP161 | | | | | | – | | | – | – | – |
| ZP162 | | – | | | | | | | – | + | – |
| ZP164 | | | – | | | | | | – | – | + |
| ZP165 | | | | + | | | | – | – | – | – |

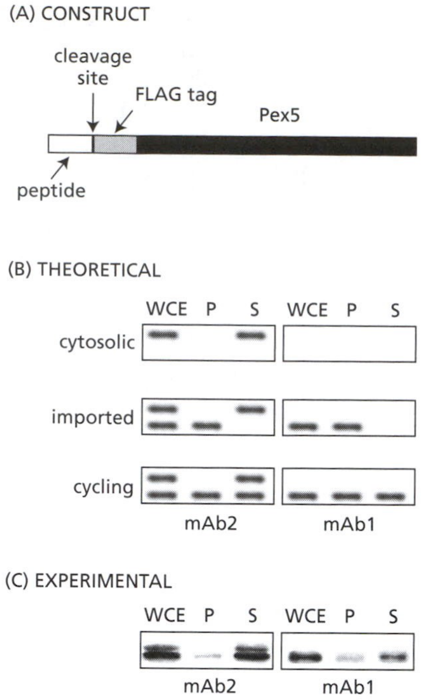

**Figure 12–22** Mechanism of Pex5-mediated peroxisomal import (Problem 12–98). (A) Modified Pex5. (B) Expectations for different mechanisms of Pex5-mediated peroxisomal import. (C) Analysis of transfection of the modified *Pex5* gene into cells. 'WCE' stands for whole-cell extract, 'P' for pellet, and 'S' for supernatant. Proteins were detected by immunoblotting using mAb1 or mAb2, followed by reaction with a second antibody that binds mAb1 and mAb2 and also carries bound horseradish peroxidase, which then converts an added precursor into a light-emitting molecule that can be detected on photographic film. *Black bands* correspond to sites where the film has been exposed by the light-emitting molecule. In (B) uncleaved Pex5 (*upper band*) and cleaved Pex5 (*lower band*) are separated by a larger distance than they are in the experiment (C) for clarity.



possible mechanisms of Pex5-mediated import—cytosolic, imported, cycling—by using the mAb1 and mAb2 antibodies, as shown in Figure 12–22B.

When you express the modified *Pex5* gene in cells, prepare cell fractions, separate the proteins by electrophoresis, and react them with mAb1 and mAb2 antibodies, you obtain the results shown in Figure 12–22C.

A. Explain the theoretical results (Figure 12–22B) expected for each of the three possible mechanisms of Pex5-mediated import.
B. Based on the results in Figure 12–22C, how does Pex5 mediate the import of proteins into the matrix of the peroxisome? Explain your reasoning.
C. Pex5-mediated import into peroxisomes resembles most closely import into what other cellular organelle?

# THE ENDOPLASMIC RETICULUM

TERMS TO LEARN

| | | |
|---|---|---|
| BiP | free ribosome | Sec61 complex |
| calnexin | glycoprotein | signal-recognition particle (SRP) |
| calreticulin | GPI anchor | single-pass transmembrane protein |
| co-translational | membrane-bound ribosome | smooth ER |
| dolichol | microsome | SRP receptor |
| endoplasmic reticulum (ER) | multipass transmembrane protein | start-transfer signal |
| ER lumen | post-translational | stop-transfer signal |
| ER resident protein | protein glycosylation | unfolded protein response |
| ER retention signal | rough ER | |
| ER signal sequence | | |

## DEFINITIONS

Match the definition below with its term from the list above.

**12–99** Region of the ER not associated with ribosomes.

**12–100** Type of lipid linkage by which some proteins are bound to the membrane.

**12–101** Labyrinthine membrane-enclosed compartment in the cytoplasm of eucaryotic cells, where lipids are synthesized and membrane-bound proteins and secretory proteins are made.

**12–102** Ribonucleoprotein complex that binds an ER signal sequence on a partially synthesized polypeptide chain and directs the polypeptide and its attached ribosome to the ER.

**12–103** Hydrophobic amino acid sequence that halts translocation of a polypeptide chain through the ER membrane, thus, anchoring the protein chain in the membrane.

**12–104** Describes import of a protein into the ER before the polypeptide chain is completely synthesized.

**12–105** Short amino acid sequence on a protein that keeps it in the ER.

**12–106** Cellular action triggered by an accumulation of misfolded proteins in the ER.

**12–107** The protein translocator that forms a water-filled pore in the ER membrane, allowing passage of a polypeptide chain as it is being synthesized by membrane-bound ribosomes.

**12–108** Describes any process involving a protein that occurs after protein synthesis is completed.

**12–109** Small vesicle that is derived from fragmented ER produced when cells are homogenized.

**12–110** Any protein with one or more oligosaccharide chains covalently linked to amino acid side chains.

**12–111** Ribosome in the cytosol that is unattached to any membrane.

## TRUE/FALSE

Decide whether each of these statements is true or false, and then explain why.

**12–112** The signal peptide binds to a hydrophobic site on the ribosome causing a pause in protein synthesis, which resumes when SRP binds to the signal peptide.

**12–113** Nascent polypeptide chains are transferred across the ER membrane through a pore in the Sec61 protein translocator complex.

**12–114** In multipass transmembrane proteins the odd-numbered transmembrane segments (counting from the N-terminus) act as start-transfer signals and the even-numbered segments act as stop-transfer signals.

**12–115** The ER lumen contains a mixture of thiol-containing reducing agents that prevent the formation of S—S linkages (disulfide bonds) by maintaining the cysteine side chains of lumenal proteins in reduced (–SH) form.

**12–116** Some membrane proteins are attached to the cytoplasmic surface of the plasma membrane through a C-terminal linkage to a glycosylphosphatidylinositol (GPI) anchor.

## THOUGHT PROBLEMS

**12–117** Explain how an mRNA molecule can remain attached to the ER membrane while the individual ribosomes translating it are released and rejoin the cytosolic pool of ribosomes after each round of translation.

**12–118** Why are cytosolic Hsp70 chaperone proteins required for import of proteins into mitochondria and chloroplasts, but not for co-translational import into the ER?

**12–119** Where would you expect to see microsomes in an electron micrograph of a liver cell?

**12–120** If smooth ER and rough ER are continuous with each other, how is it that rough microsomes (derived from rough ER) can have more than 20 proteins that are not present in smooth microsomes (derived from smooth ER)?

**12–121** You are studying post-translational translocation of a protein into purified microsomes, but you find that import is very inefficient. Which one of the following might be expected to increase the efficiency of import if added to the mixture of protein and microsomes? Explain your answer.

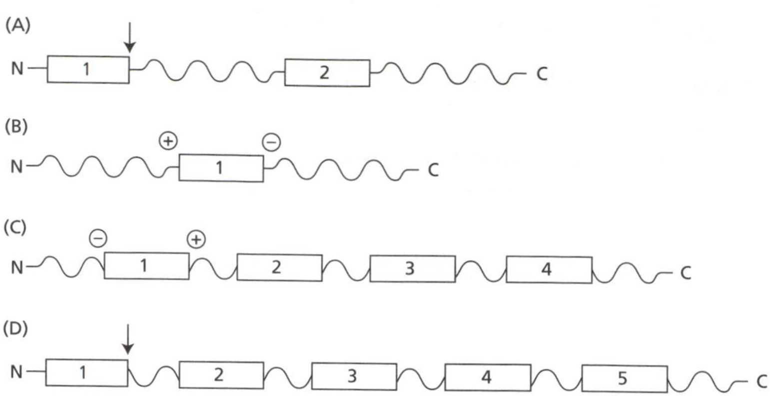


**Figure 12–23** Distribution of the membrane-spanning segments in proteins to be inserted into the ER membrane (Problem 12–123). *Boxes* represent membrane-spanning segments and *arrows* indicate sites at which signal sequences are cleaved. The *pluses* and *minuses* indicate the charges at the ends of some transmembrane segments.

A. BiP
B. Cytosolic Hsp70
C. Free ribosomes
D. Sec61 complex
E. SRP

**12–122** Compare and contrast protein import into the ER and into the nucleus. List at least two major differences in the mechanisms and speculate on why the nuclear mechanism might not work for ER import and vice versa.

**12–123** Four membrane proteins are represented schematically in Figure 12–23. The boxes represent membrane-spanning segments and the arrows represent sites for cleavage of the signal sequence. Predict how each of the mature proteins will be arranged across the membrane of the ER. Indicate clearly the N- and C-termini relative to the cytosol and the lumen of the ER, and label each box as a start-transfer or stop-transfer signal.

**12–124** Examine the multipass transmembrane protein shown in Figure 12–24. What would you predict would be the effect of converting the first hydrophobic transmembrane segment to a hydrophilic segment? Sketch the arrangement of the modified protein in the ER membrane.

**12–125** Why might it be advantageous to add a preassembled block of 14 sugars to a protein in the ER, rather than building the sugar chains step-by-step on the surface of the protein by the sequential addition of sugars by individual enzymes?

**12–126** Discuss the following statement: "The reason why the sugar tree on dolichol phosphate is synthesized in such a baroque fashion, with extensive flipping of dolichol-phosphate-sugar intermediates across the ER membrane, is because there are no nucleoside triphosphates in the ER lumen. The activated forms of the sugars are all generated by reaction with UTP or GTP in the cytosol."

**12–127** Outline the steps by which misfolded proteins in the ER trigger synthesis of additional ER chaperone proteins. How does this response benefit the cell?

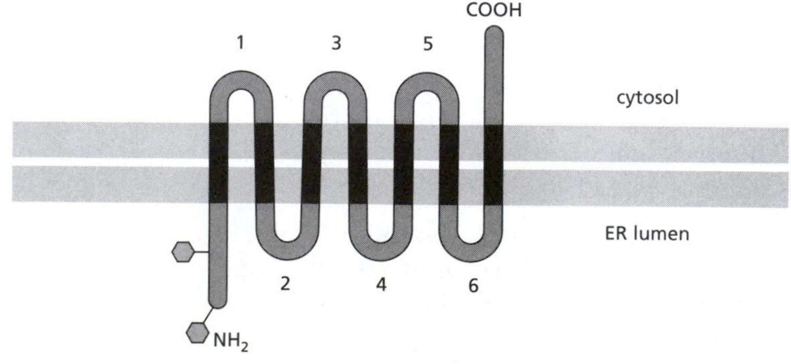


Figure 12–24 Arrangement of a multipass transmembrane protein in the ER membrane (Problem 12–124). *Hexagons* represent covalently attached oligosaccharides.

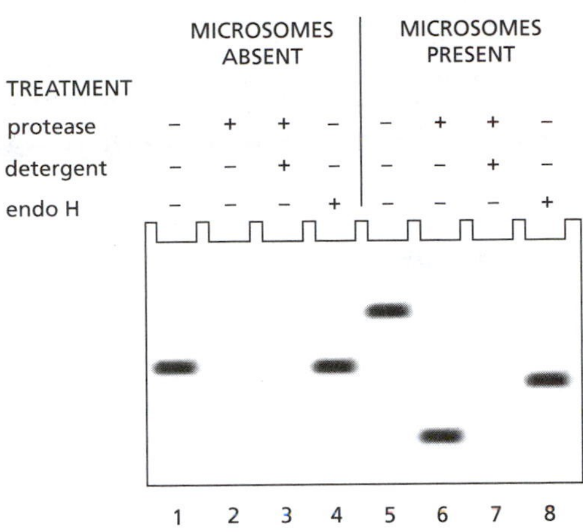


**Figure 12–25** Results of translation of a pure mRNA in the presence and absence of microsomal membranes (Problem 12–129). Treatments of the products of translation before electrophoresis are indicated at the *top* of each lane. Electrophoresis was on an SDS polyacrylamide gel, which separates proteins on the basis of size, with lower molecular weight proteins migrating farther down the gel.

**12–128** All new phospholipids are added to the cytoplasmic leaflet of the ER membrane, yet the ER membrane has a symmetrical distribution of different phospholipids in its two leaflets. By contrast, the plasma membrane, which receives all its membrane components ultimately from the ER, has a very asymmetrical distribution of phospholipids in the two leaflets of its lipid bilayer. How is symmetry generated in the ER membrane, and how is asymmetry generated in the plasma membrane?

## DATA HANDLING

**12–129** Translocation of proteins across rough microsomal membranes can be judged by several experimental criteria: (1) the newly synthesized proteins are protected from added proteases, unless detergents are present to solubilize the lipid bilayer; (2) the newly synthesized proteins are glycosylated by oligosaccharide transferases, which are localized exclusively to the lumen of the ER; (3) the signal peptides are cleaved by signal peptidase, which is active only on the lumenal side of the ER membrane.

Use these criteria to decide whether a protein is translocated across rough microsomal membranes. The mRNA is translated into protein in a cell-free system in the absence or presence of microsomes. Samples of newly synthesized proteins are treated in four different ways: (1) no treatment, (2) addition of a protease, (3) addition of a protease and detergent, and (4) disruption of microsomes and addition of endoglycosidase H (endo H), which removes *N*-linked sugars that are added in the ER. An electrophoretic analysis of these samples is shown in Figure 12–25.

A. Explain the experimental results that are seen in the absence of microsomes (Figure 12–25, lanes 1 to 4).

B. Using the three criteria outlined in the problem, decide whether the experimental results in the presence of microsomes (Figure 12–25, lanes 5 to 8) indicate that the protein is translocated across microsomal membranes. How would you account for the migration of the proteins in Figure 12–25, lanes 5, 6, and 8?

C. Is the protein anchored in the membrane, or is it translocated all the way through the membrane?

**12–130** An elegant early study of co-translational import of protein into the ER focused on the source of energy for translocation: does the ribosome 'push' the protein across or does the translocation machinery 'pull' it? The authors reasoned that if the ribosome pushed the protein across, they should not be able to uncouple translation from translocation, whereas if the translocation machinery pulled the protein across, they might be able to.

They cloned a gene for an ER protein onto a plasmid so they could transcribe it *in vitro* by adding RNA polymerase (Figure 12–26). By cutting the

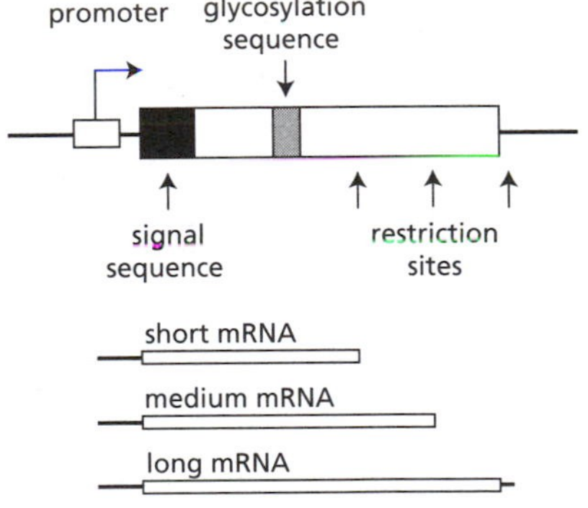


**Figure 12–26** A cloned gene for testing the coupling of translation and translocation (Problem 12–130). The plasmid is shown as a linear sequence for simplicity. The protein-coding segment is shown by the *large rectangle*; promoter sequences are indicated by the *small rectangle*. Cleavage sites for restriction nucleases used to truncate the transcription template are indicated. The three different mRNA products of transcription from the truncated templates are indicated *below* the map of the gene.

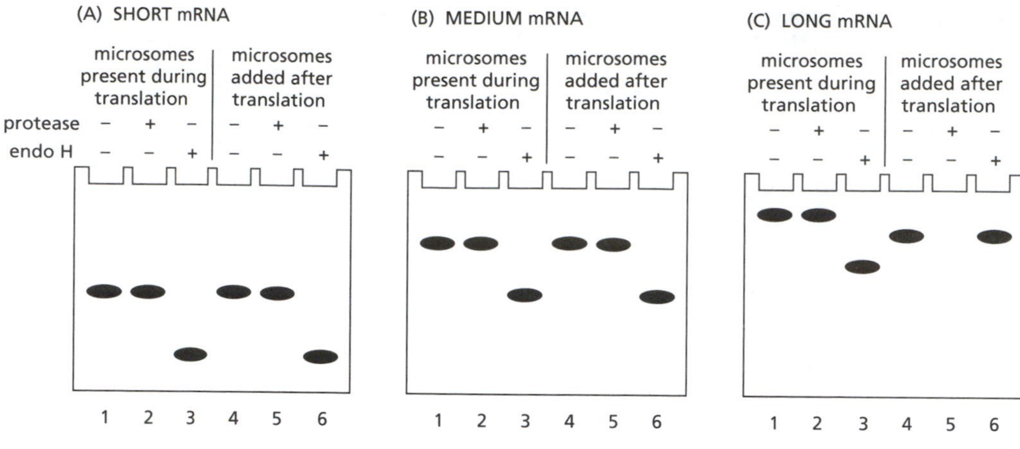


**Figure 12–27** Results of experiments to test the coupling of translation and translocation (Problem 12–130). (A) Short mRNA. (B) Medium mRNA. (C) Long mRNA. Treatments of samples before electrophoresis are indicated *above* the gels. Endo H treatment was applied after the microsomes were disrupted. Endo H removes sugars of the type added in the ER.

plasmid at different positions within the gene, they could also generate two mRNAs that were shorter than normal (Figure 12–26). These three mRNAs were translated *in vitro*. In one experiment rough microsomes were added before translation began. In a second experiment microsomes were added after translation was complete (along with cycloheximide to inhibit any additional protein synthesis). Translocation was assessed by comparing samples that were untreated, treated directly with protease, or treated with endoglycosidase H (endo H) after disruption of the microsomes. The products were then displayed by SDS-polyacrylamide gel electrophoresis (Figure 12–27).

A. Were the proteins from each of the mRNAs translocated when the microsomes were present during translation? How can you tell?
B. Was translocation uncoupled from translation in any of the experiments? Explain your answer.
C. From these results the authors concluded that the translocation machinery pulls proteins across the membrane. What was their reasoning?
D. It is now well accepted that ribosomes push their proteins across the ER membrane during co-translational insertion. How do you suppose the authors might reinterpret their results in hindsight, with the benefit of nearly two decades of additional experimentation?

**12–131** Things are not going well. You've just had a brief but tense meeting with your research advisor that you will never forget, and it is clear that your future in his lab is in doubt. He is not fond of your habit of working late and sleeping late; he thinks it is at the root of your lack of productivity (as he perceives it). A few days after the meeting, you roll in at the bright and early hour of noon, to be informed by your student colleagues that your advisor is "looking all over for you." You are certain the axe is going to fall. But it turns out he is excited, not upset. He's just heard a seminar that reminded him of the note you'd left on his desk the month before, describing a selection scheme you had crafted (late one night, of course) for isolating mutants in the ER translocation machinery. You are completely flabbergasted.

You quickly settle on the details for the selection, which involves fusing an ER import signal to the N-terminus of the *His4* gene product. His4 is a cytosolic enzyme that converts histidinol to histidine. Yeast strains that are defective for an early step in the histidine biosynthetic pathway can grow on added histidinol, if His4 is present. You decide to look for temperature-sensitive (*ts*) mutants, which are normal at 24°C but defective at 37°C. Using a strain that expresses His4 with an ER import signal, you select for cells that grow on histidinol at 30°C and screen them for ones that die at 37°C. The first mutant you isolate is in the *Sec61* gene (later shown to encode a principal component of the translocator through which ribosomes insert nascent proteins across the ER membrane.) You are back in your advisor's good graces!

A. Why is it that normal cells with the modified His4 cannot grow on histidinol, whereas cells with a defective ER-import apparatus can?

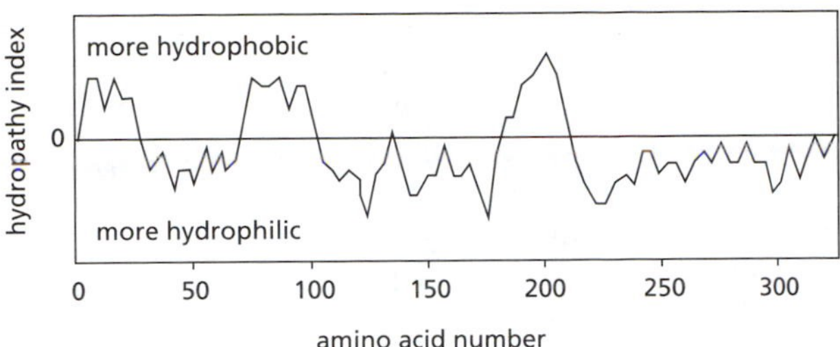


**Figure 12–28** Hydropathy plot of a membrane protein (Problem 12–132). The three hydrophobic peaks indicate the positions of three potential membrane-spanning segments.

B. Why did the selection scheme set out to find ts mutants? Why was selection applied at the intermediate temperature of 30°C, rather than at 24°C or 37°C?

**12–132** A classic paper describes a genetic method for determining the organization of a bacterial protein in the membrane of *E. coli.* The hydropathy plot of the protein in Figure 12–28 indicated three potential membrane-spanning segments. Hybrid fusion proteins of different lengths, some with internal deletions, were made with the membrane protein at the N-terminus and alkaline phosphatase at the C-terminus (Figure 12–29). Alkaline phosphatase is easy to assay in whole cells and has no significant hydrophobic stretches. Moreover, when it is on the cytoplasmic side of the membrane its activity is low, and when it is on the external side of the membrane (in the periplasmic space) its activity is high. The assayed levels of alkaline phosphatase activity are indicated (HIGH or LOW) in Figure 12–29.

A. How is the protein organized in the membrane? Explain how the results with the fusion proteins indicate this arrangement.

B. How is the organization of the membrane protein altered by the deletion? Are your measurements of alkaline phosphatase activity in the internally deleted plasmids consistent with the altered arrangement?

**12–133** Mitochondria and peroxisomes, as opposed to most other cellular membranes, acquire new phospholipids via phospholipid exchange proteins. One such protein, PC exchange protein, specifically transfers phosphatidylcholine (PC) between membranes. Its activity can be measured by mixing red blood cell ghosts (intact plasma membranes with cytoplasm removed) with synthetic phospholipid vesicles containing radioactively labeled PC in

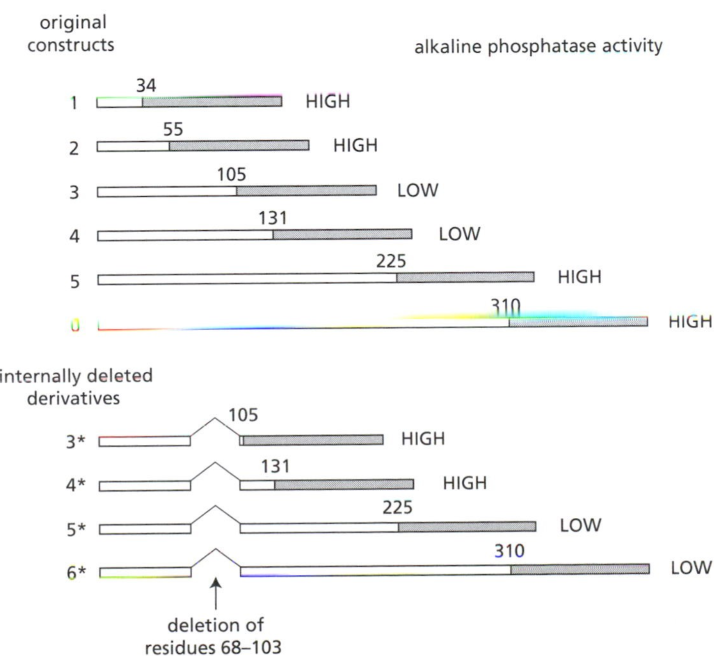


**Figure 12–29** Structures of hybrid proteins used to determine the organization of a membrane protein (Problem 12–132). The membrane protein (*unshaded segment*) is at the N-terminus and alkaline phosphatase (*shaded segment*) is at the C-terminus of the protein. The *inverted V* indicates the site from which amino acids were deleted from modified hybrid proteins. The most C-terminal amino acid of the membrane protein is *numbered* in each hybrid protein. The activity of alkaline phosphatase in each hybrid protein is shown on the *right.*

both monolayers of the vesicle bilayer. After incubation at 37°C, the mixture is centrifuged briefly so that ghosts form a pellet, whereas the vesicles stay in the supernatant. The amount of exchange is determined by measuring the radioactivity in the pellet.

Figure 12–30 shows the result of an experiment along these lines, using labeled (donor) vesicles with an outer radius of 10.5 nm and a bilayer 4.0 nm in thickness. No transfer occurred in the absence of the exchange protein, but in its presence up to 70% of the labeled PC in the vesicles could be transferred to the red cell membranes.

Several control experiments were performed to explore the reason why only 70% of the label in donor vesicles was transferred.

1. Five times as many membranes from red cell ghosts were included in the incubation: the transfer still stopped at the same point.
2. Fresh exchange protein was added after 1 hour: it caused no further transfer.
3. The labeled lipids remaining in donor vesicles at the end of the reaction were extracted and made into fresh vesicles: 70% of the label in these vesicles was exchangeable.

When the red cell ghosts that were labeled in this experiment were used as donor membranes in the reverse experiment (that is, transfer of PC from red cell membranes to synthetic vesicles), 96% of the label could be transferred to the acceptor vesicles.

A. What possible explanations for the 70% limit do each of the three control experiments eliminate?
B. What do you think is the explanation for the 70% limit? (Hint: the area of the outer surface of these small donor vesicles is about 2.6 times larger than the area of the inner surface.)
C. Why do you think that almost 100% of the label in the red cell membrane can be transferred back to the vesicle?

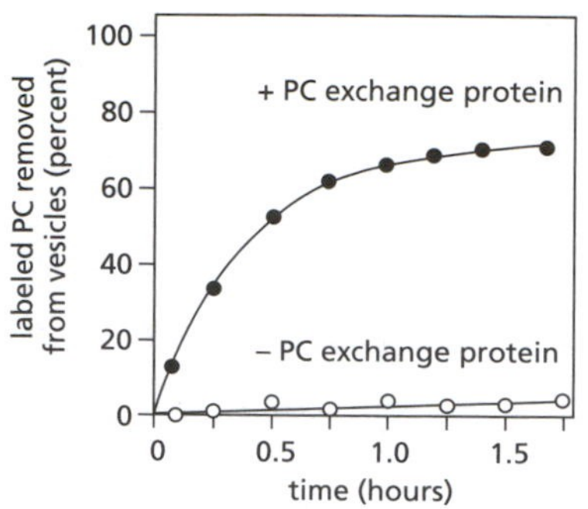


**Figure 12–30** Transfer of labeled PC from donor vesicles to red cell membranes by PC exchange protein (Problem 12–133).

# Intracellular Vesicular Traffic

# 13

## In This Chapter

THE MOLECULAR MECHANISMS OF MEMBRANE TRANSPORT AND THE MAINTENANCE OF COMPARTMENT DIVERSITY 289

TRANSPORT FROM THE ER THROUGH THE GOLGI APPARATUS 296

TRANSPORT FROM THE *TRANS* GOLGI NETWORK TO LYSOSOMES 302

TRANSPORT INTO THE CELL FROM THE PLASMA MEMBRANE: ENDOCYTOSIS 305

TRANSPORT FROM THE *TRANS* GOLGI NETWORK TO THE CELL EXTERIOR: EXOCYTOSIS 310

## THE MOLECULAR MECHANISMS OF MEMBRANE TRANSPORT AND THE MAINTENANCE OF COMPARTMENT DIVERSITY

TERMS TO LEARN

| | | |
|---|---|---|
| adaptor protein | COPII-coated vesicle | retromer |
| Arf protein | dynamin | Sar1 protein |
| clathrin | NSF | SNARE protein (SNARE) |
| clathrin-coated vesicle | phosphatidylinositide (PIP) | transport vesicle |
| coated vesicle | Rab effector | t-SNARE |
| coat-recruitment GTPase | Rab protein | v-SNARE |
| COPI-coated vesicle | | |

### DEFINITIONS

Match the definition below with its term from the list above.

**13–1** General term for a membrane-enclosed container that moves material between membrane-enclosed compartments within the cell.

**13–2** Any of a large family of monomeric GTPases present in the plasma membrane and organelle membranes that confer specificity on vesicle docking.

**13–3** A protein that mediates binding between the clathrin coat and transmembrane proteins, including transmembrane cargo receptors.

**13–4** Cytosolic GTPase that binds to the neck of a clathrin-coated vesicle and helps it to pinch off from the membrane.

**13–5** The protein that catalyzes the disassembly of the helical domains of paired SNARE proteins.

**13–6** Coated vesicle that transports material from the plasma membrane, and between endosomal and Golgi compartments.

**13–7** The coat-recruitment GTPase responsible for both COPI coat assembly and clathrin coat assembly at Golgi membranes.

**13–8** Protein that facilitates vesicle transport, docking, and membrane fusion once it is bound by an activated Rab protein.

**13–9** General term for a member of the large family of proteins that catalyze the membrane fusion reactions in membrane transport.

**13–10** A multiprotein complex that assembles on endosomal membranes only when it can bind the cytoplasmic tail of a receptor in a curved membrane that has phosphorylated phosphotidylinositol lipid head groups.

**13–11** General term for a transport vesicle that carries a distinctive cage of proteins covering its cytosolic surface.

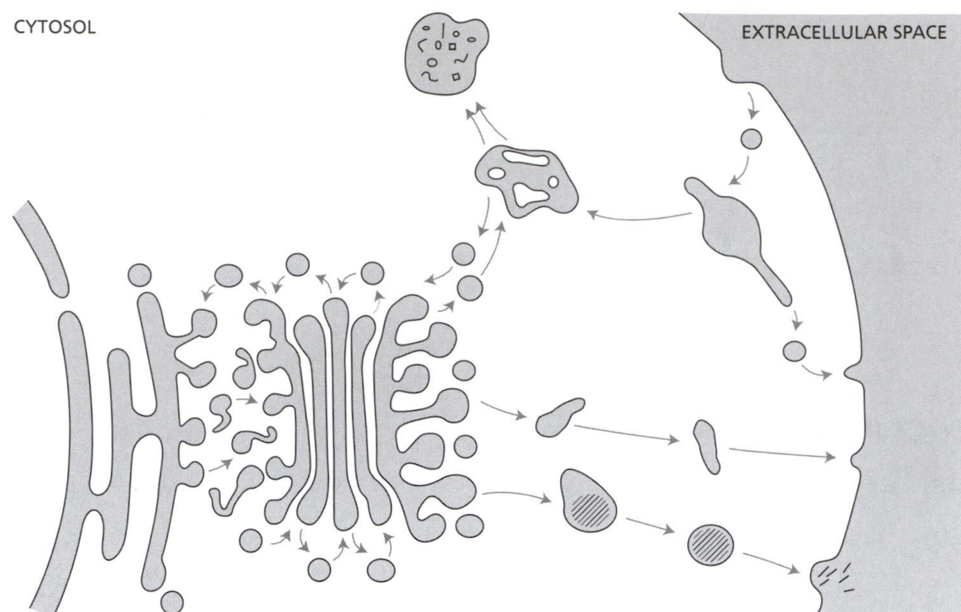


**Figure 13–1** The intracellular compartments in the biosynthetic–secretory, endocytic, and retrieval pathways, with flow between compartments indicated (Problem 13–17).

**13–12** The coat-recruitment GTPase responsible for COPII coat assembly at the ER membrane.

## TRUE/FALSE

Decide whether each of these statements is true or false, and then explain why.

**13–13** In all events involving fusion of a vesicle to a target membrane, the cytosolic leaflets of the vesicle and target bilayers always fuse together, as do the leaflets that are not in contact with the cytosol.

**13–14** Complementary Rab proteins on transport vesicles and target membranes bind to one another to allow transport vesicles to dock selectively at their appropriate target membranes.

## THOUGHT PROBLEMS

**13–15** In a nondividing cell such as a liver cell, why must the flow of membrane between compartments be balanced, so that the retrieval pathways match the outward flow? Would you expect the same balanced flow in a gut epithelial cell, which is actively dividing?

**13–16** At least three different coats form around transport vesicles. What two principal functions do these different coats have in common?

**13–17** The diagram in Figure 13–1 shows the various intracellular compartments involved in the biosynthetic–secretory, endocytic, and retrieval pathways.

A. Label the various compartments in the diagram.

B. Indicate on the arrows whether the indicated flow is part of the biosynthetic–secretory pathway, the endocytic pathway, or a retrieval pathway.

**13–18** Discuss the following analogy: "Cargo receptors competing to be transported by the coated pit system can be compared to skiers joining a cable-car network. Entry is permitted to ticket holders only, but there is no guarantee of who is found with whom in a particular cable car, although all travelers, hopefully, will reach the next station."

**13–19** The clathrin coat on a vesicle is made up of numerous triskelions that form a cage 60–200 nm in diameter, composed of both pentagonal and hexagonal

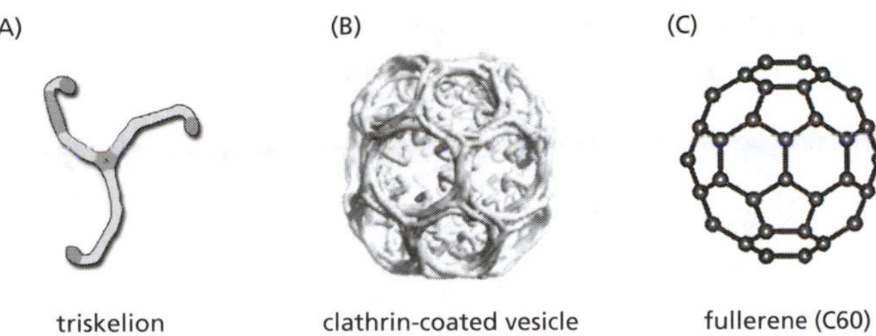


**Figure 13–2** Structure of a clathrin coat (Problem 13–19). (A) A triskelion subunit. (B) A clathrin-coated vesicle. (C) A C60 fullerene.

faces, just like C60 fullerene (Figure 13–2). Sketch the location of an individual triskelion in the clathrin-coated vesicle in Figure 13–2B. At what point in its structure does a triskelion have to be most flexible to accommodate changes in size of the vesicle? At what point does it have to be most flexible to fit into both the pentagonal and hexagonal faces?

**13–20** Yeast, and many other organisms, make a single type of clathrin heavy chain and a single type of clathrin light chain; thus, they make a single kind of clathrin coat. How is it then that a single clathrin coat can be used for three different transport pathways—Golgi to late endosomes, plasma membrane to early endosomes, and immature secretory vesicles to Golgi—that each involves different specialized cargo proteins?

**13–21** Clathrin-coated vesicles bud from eucaryotic plasma membrane fragments when adaptor proteins, clathrin, and dynamin-GTP are added. What would you expect to observe if the following modifications were made to the experiment? Explain your answers.

A. Adaptor proteins were omitted.
B. Clathrin was omitted.
C. Dynamin was omitted.
D. Procaryotic membrane fragments were used.

**13–22** The molecular details of how dynamin mediates the final membrane-fusion step in clathrin-coated vesicle formation (Figure 13–3) are controversial. Although dynamin has a GTPase domain, it is not clear how GTP is used to accomplish membrane fusion. One view is that dynamin uses the energy of GTP hydrolysis to pinch off the neck of the vesicle; that is, that dynamin itself is a mechanochemical 'pinchase' powered by GTP hydrolysis. An alternative view is that the GTPase domain of dynamin behaves more like a conventional small GTPase, regulating the activities of other proteins, which are the true pinchases. In this view, the binding of GTP to dynamin serves as the 'ON' signal to recruit the proteins that pinch off the vesicle.

One attempt to distinguish between these alternatives used a mutant form of dynamin in which a threonine at position 65, which is in the GTPase domain, was changed to alanine. This T65A mutant dynamin cannot hydrolyze GTP, although it is normal by all other criteria. It was then tested for its ability to support formation of clathrin-coated vesicles in an *in vitro*

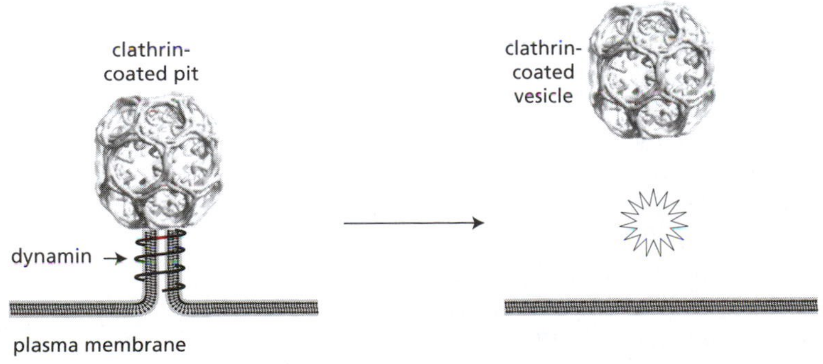


**Figure 13–3** Dynamin-mediated step in clathrin-coated vesicle formation (Problem 13–22).

system. The mutant dynamin blocked formation of clathrin-coated vesicles. Which model for dynamin function does this result support? What result would have been expected according to the alternative hypothesis? Explain your answers.

**13–23** Correct the following description, as necessary. "Sar1 protein is a COPII-recruitment GTPase that facilitates the unidirectional transfer of COPII vesicles from the ER membrane to the Golgi membrane. A unique directionality is imposed on the transfer by the locations of a guanine-nucleotide exchange factor (GEF) and a GTPase-activating protein (GAP). The GEF, which is located in the ER membrane, stimulates vesicle formation by mediating attachment of Sar1-GTP to the ER membrane, where it recruits COPII subunits. The GAP, which is located in the Golgi membrane, stimulates vesicle docking by stimulating hydrolysis of Sar1-bound GTP. Sar1-GDP causes disassembly of the COPII coat, which prepares the vesicle for fusion with the Golgi membrane."

**13–24** Imagine that Arf1 protein was mutated so that it could not hydrolyze GTP, regardless of its binding partners. Would you expect COPI-coated vesicles to form normally? How would you expect transport mediated by COPI-coated vesicles to be affected? If this were the only form of Arf1 in a cell, would you expect it to be lethal? Explain your answers.

**13–25** How can it possibly be true that complementary pairs of specific SNAREs uniquely mark vesicles and their target membranes? After vesicle fusion, the target membrane will contain a mixture of t-SNAREs and v-SNAREs. Initially, these SNAREs will be tightly bound to one another, but NSF can pry them apart, reactivating them. What do you suppose prevents target membranes from accumulating a population of v-SNAREs equal to or greater than their population of t-SNAREs?

**13–26** Consider the v-SNAREs that direct transport vesicles from the *trans* Golgi network to the plasma membrane. They, like all other v-SNAREs, are membrane proteins that are integrated into the membrane of the ER during their biosynthesis, and are then transported by vesicles to their destination. Thus, transport vesicles budding from the ER contain at least two kinds of v-SNAREs—those that target the vesicles to the *cis* Golgi network and those that are in transit to the *trans* Golgi network to be packaged into different transport vesicles destined for the plasma membrane.

A. Why do you suppose this might be a problem?

B. How do you suppose the cell might solve this problem?

**13–27** Viruses are the ultimate scavengers—a necessary consequence of their small genomes. Wherever possible they make use of the cell's machinery to accomplish the steps involved in their own reproduction. Many different viruses have membrane coverings. These so-called enveloped viruses gain access to the cytosol by fusing with a cell membrane. Why do you suppose that each of these viruses encodes its own special fusion protein, rather than making use of a cell's SNAREs?

## CALCULATIONS

**13–28** For fusion of a vesicle with its target membrane to occur, the membranes have to be brought to within 1.5 nm so that the two bilayers can join (Figure 13–4). Assuming that the relevant portions of the two membranes at the fusion site are circular regions 1.5 nm in diameter, calculate the number of water molecules that would remain between the membranes. (Water is 55.5 M and the volume of a cylinder is $\pi r^2 h$.) Given that an average phospholipid occupies a membrane surface area of 0.2 $nm^2$, how many phospholipids would be present in each of the opposing monolayers at the fusion site? Are there sufficient water molecules to bind to the hydrophilic head groups of this number of phospholipids? (It is estimated that 10–12 water molecules

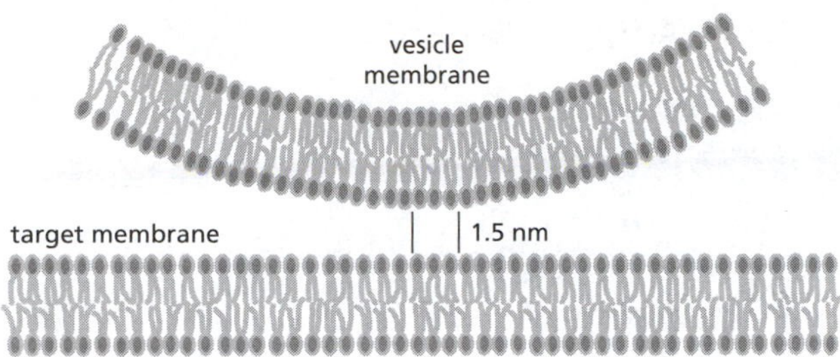


**Figure 13–4** Close approach of a vesicle and its target membrane in preparation for fusion (Problem 13–28).

are normally associated with each phospholipid head group at the exposed surface of a membrane.)

## DATA HANDLING

**13–29** When the fungal metabolite brefeldin A is added to cells, the Golgi apparatus largely disappears and the Golgi proteins intermix with those in the ER. Brefeldin-A treatment also causes the rapid dissociation of some Golgi-associated peripheral membrane proteins, including subunits of the COPI coat. These observations imply that brefeldin A prevents transport involving COPI-coated vesicles by blocking the assembly of coats, and thus the budding of transport vesicles. In principle, brefeldin A could block formation of COPI-coated vesicles at any point in the normal scheme for assembly, which is shown in Figure 13–5. The following observations identify the point of action of brefeldin A.

1. Arf with bound GTPγS (a nonhydrolyzable analog of GTP) causes COPI-coated vesicles to form when added to Golgi membranes. Formation of vesicles in this way is not affected by brefeldin A.
2. Arf with bound GDP exchanges GDP for GTP when added to Golgi membranes. This exchange reaction does not occur in the presence of brefeldin A. The essential component in the Golgi membrane is sensitive to trypsin digestion, suggesting that it is a protein.

Given these experimental observations, how do you think brefeldin A blocks formation of COPI-coated vesicles?

**13–30** Small GTPases are generally active in the GTP-bound state and inactive when the GTP is hydrolyzed to GDP. In the absence of a GTPase activating protein (GAP), small GTPases typically hydrolyze GTP very slowly. The mechanism by which GAP stimulates GTP hydrolysis is known for the small GTPase Ras. When Ras–GAP binds to Ras, it alters the conformation of Ras and provides a critical, catalytic arginine 'finger' that stabilizes the transition state for GTP hydrolysis, thereby stimulating hydrolysis by several orders of magnitude.

During assembly of COPI-coated vesicles, Arf1—a small GTPase—binds to Arf1–GAP, which locks Arf1 into its active catalytic conformation but does not supply the catalytic arginine. Since COPI subunits also bind to Arf1, you

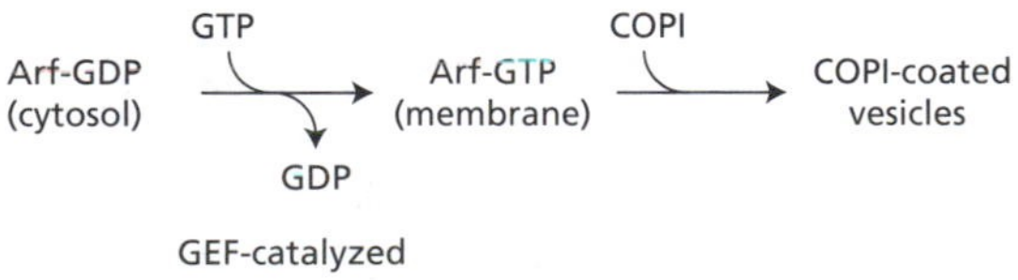

Arf-GDP (cytosol) —GTP in, GDP out (GEF-catalyzed)→ Arf-GTP (membrane) —COPI→ COPI-coated vesicles

**Figure 13–5** Normal pathway for formation of COPI-coated vesicles (Problem 13–29). The small GTPase Arf carries a bound GDP in its cytosolic form. In response to a GEF, Arf releases GDP and picks up GTP. Binding of GTP causes a conformational change that exposes a fatty acid tail on Arf, which promotes binding of Arf-GTP to the membrane. COPI subunits bind to Arf-GTP to form COPI-coated vesicles.

**Table 13–1 Rates of GTP hydrolysis by various combinations of Arf1, Arf1–GAP, and COPI subunits** (Problem 13–30).

| COMPONENTS ADDED | RATE OF GTP HYDROLYSIS |
|---|---|
| Arf1 | 0 |
| Arf1 + Arf1–GAP | 1 |
| Arf1 + COPI subunits | 0 |
| Arf1 + Arf1–GAP + COPI subunits | 1000 |

wonder if they might affect GTP hydrolysis. To test this possibility, you mix Arf1, Arf1–GAP, and COPI subunits in various combinations and measure GTP hydrolysis (Table 13–1).

How would you interpret these results? How might you further test your conclusions?

**13–31** SNAREs exist as complementary partners that carry out membrane fusions between appropriate vesicles and their target membranes. In this way, a vesicle with a particular variety of v-SNARE will fuse only with a membrane that carries the complementary t-SNARE. In some instances, however, fusions of identical membranes (homotypic fusions) are known to occur. For example, when a yeast cell forms a bud, vesicles derived from the mother cell's vacuole move into the bud where they fuse with one another to form a new vacuole. These vesicles carry both v-SNAREs and t-SNAREs. Are both types of SNARE essential for this homotypic fusion event?

To test this point, you have developed an ingenious assay for fusion of vacuolar vesicles. You prepare vesicles from two different mutant strains of yeast: strain B has a defective gene for vacuolar alkaline phosphatase (Pase); strain A is defective for the protease that converts the precursor of alkaline phosphatase (pro-Pase) into its active form (Pase) (Figure 13–6A). Neither strain has active alkaline phosphatase, but when extracts of the strains are mixed, vesicle fusion generates active alkaline phosphatase, which can be easily measured (Figure 13–6).

Now you delete the genes for the vacuolar v-SNARE, t-SNARE, or both in each of the two yeast strains. You prepare vacuolar vesicles from each and test them for their ability to fuse, as measured by the alkaline phosphatase assay (Figure 13–6B).

What do these data say about the requirements for v-SNAREs and t-SNAREs in fusion of vacuolar vesicles? Does it matter which kind of SNARE is on which vesicle?

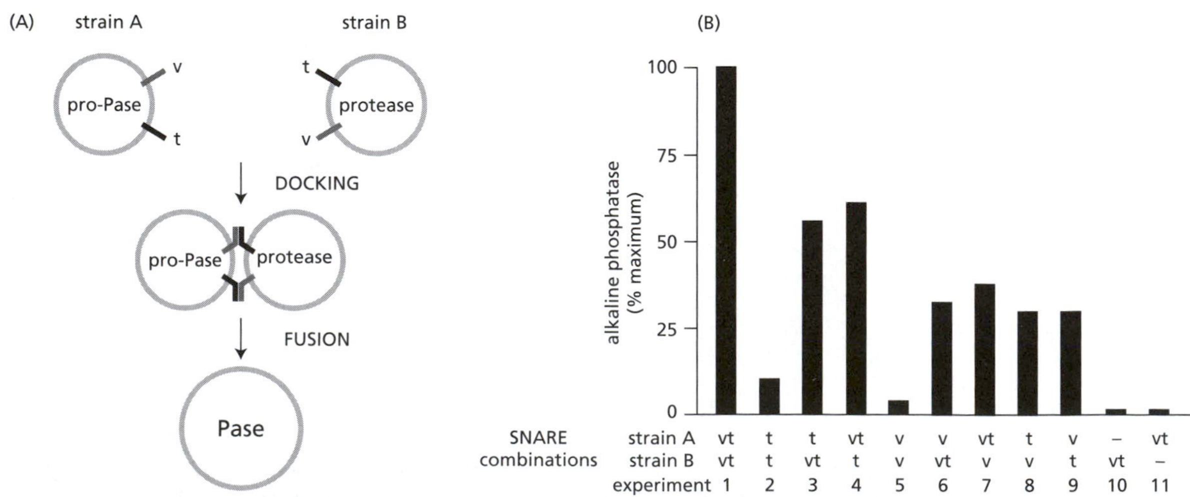


**Figure 13–6** SNARE requirements for vesicle fusion (Problem 13–31). (A) Scheme for measuring fusion of vacuolar vesicles. (B) Results of fusions of vesicles with different combinations of v-SNAREs and t-SNAREs. The SNAREs present on the vesicles of the two strains are indicated as v (v-SNARE) and t (t-SNARE).

**13–32** A complex of syntaxin (t-SNARE), Snap25 (t-SNARE), and synaptobrevin (v-SNARE) are responsible for docking synaptic vesicles at the plasma membrane of a nerve terminal. But what is the next step in fusion?

You have discovered a protein (synaphin) that interacts with syntaxin and wonder if it also binds to the complex. By making a fusion of synaphin to glutathione *S*-transferase (GST), you can rapidly test this possibility. The GST provides a convenient tag that allows you to rapidly isolate GST–synaphin, along with any other proteins bound to it, out of a complex mixture of proteins (see *MBoC* Chapter 8). You incubate GST–synaphin with purified Snap25, synaptobrevin, or both, in the presence of increasing concentrations of syntaxin. You then add beads with attached glutathione—the substrate for GST—to the mixture, and allow GST–synaphin and its associated proteins, to bind via GST to the beads. You wash the beads to eliminate unbound proteins, and then incubate the beads with free glutathione to release GST–synaphin and its associated proteins. (This technique of using a GST fusion protein to detect interactions of a fused protein with other proteins is commonly known as a GST 'pull down.') You measure the amount of syntaxin that was bound under the various conditions by Western blotting using syntaxin-specific antibodies, as shown in Figure 13–7.

A. In this experiment, can you tell whether synaphin binds to Snap25 or to synaptobrevin in the absence of syntaxin? Why or why not?
B. How does the binding of synaphin to syntaxin compare with binding of synaphin to syntaxin complexed with other molecules?
C. What do you suppose is the target for synaphin binding at a synaptic terminal?

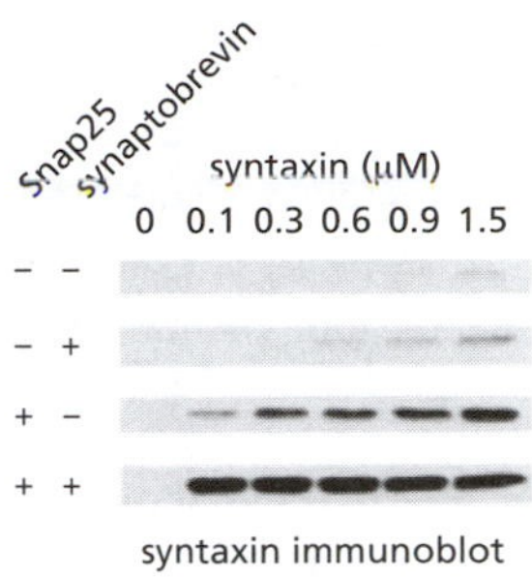


**Figure 13–7** Detection of syntaxin in a GST–synaphin pull down (Problem 13–32). After the GST pull down, samples were boiled in SDS to disrupt protein–protein interactions and then separated by electrophoresis on an SDS-containing gel. Syntaxin was detected by reaction with an antibody specific for it.

**13–33** To learn more about the role of synaphin in synaptic vesicle fusion, you incubate the complete SNARE complex (consisting of the two t-SNAREs, syntaxin and Snap25, and the v-SNARE, synaptobrevin) with increasing concentrations of synaphin. You then separate the products by electrophoresis under conditions where the SNARE complex remains intact, and probe a blot of the gel with antibodies specific for syntaxin (Figure 13–8). In the absence of synaphin, the SNARE complex runs predominantly at about 60 kd, as expected for the intact complex. In the presence of high concentrations of synaphin, however, the 60-kd band disappears and syntaxin is detected in slower-migrating (larger) complexes. When a 15-amino acid peptide, corresponding to the portion of synaphin that binds to syntaxin, is added to the incubation mixture at high concentration, only the 60-kd complex is detected (this result is not shown in Figure 13–8).

A. What do you suppose the larger complexes are, and how do you think synaphin promotes their formation?
B. How do you suppose the peptide interferes with formation of the larger complexes?
C. Injection of the peptide into a squid giant axon causes a complete block of synaptic transmission. What do you suppose this means in terms of the normal role of synaphin in synaptic vesicle fusion?
D. Can you formulate a model of how synaphin might promote synaptic vesicle fusion?

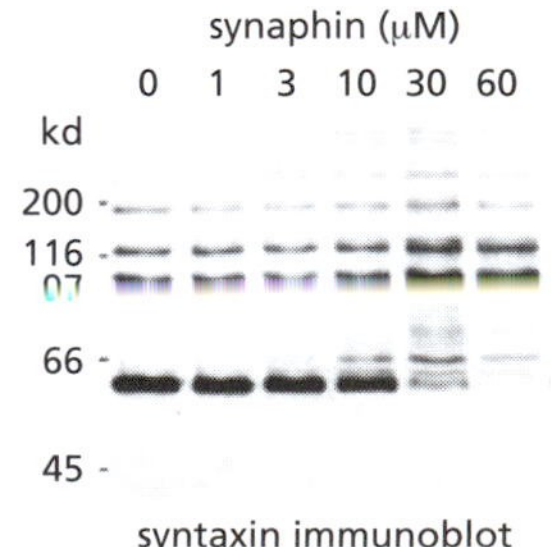


**Figure 13–8** Results of incubating increasing concentrations of synaphin with the SNARE complex from nerve terminals (Problem 13–33). Gels were run under conditions that keep the SNARE complex intact. Syntaxin was detected using antibodies specific for it.

**13–34** You wish to identify the target proteins that are bound by NSF and its two accessory proteins. You incubate purified NSF and its accessory proteins with a crude detergent extract of synaptic membranes, and then add NSF-specific antibodies that are attached to beads. By centrifuging the mixture, you can readily separate the beads, and any attached proteins, from the rest of the crude extract. The proteins attached to the beads can be analyzed by SDS polyacrylamide-gel electrophoresis. When the incubation is carried out in the presence or absence of ATP, you find that NSF alone is present on the beads. If you incubate in the presence of ATPγS, a nonhydrolyzable analog of ATP, the beads bring down NSF, its accessory proteins, syntaxin, Snap25, and synaptobrevin. What is the substrate for NSF and its accessory proteins? Why does the experiment work when ATPγS is present, but not in the presence or absence of ATP?

**13–35** The *Sec4* gene of budding yeast encodes a small GTPase that plays an essential role in the secretion pathway that forms the daughter bud. Normally, about 80% of the Sec4 protein is found on the cytosolic surface of transport vesicles and 20% is free in the cytosol. When temperature-sensitive *Sec4* mutants of yeast (*Sec4*$^{ts}$) are incubated at high temperature, growth ceases and small vesicles accumulate in the daughter bud.

To define the role of Sec4 in secretion, you engineer two specific *Sec4* mutants based on the way other small GTPases work. One mutant, *Sec4*-ccΔ, lacks two cysteines at its C-terminus, which you expect will prevent attachment of the fatty acid required for membrane binding. The second mutant, *Sec4N133I*, encodes an isoleucine in place of the normal asparagine at position 133; you expect that this protein will be locked into its active state, even though it should not be able to bind GTP or GDP.

You find that Sec4-ccΔ binds GTP but remains entirely cytosolic with none bound to vesicles. When expressed at high levels in yeast that also have a normal *Sec4* gene, it does not inhibit their growth. In contrast, Sec4N133I is located almost entirely on vesicles, and when it is expressed at high levels in normal yeast, it completely inhibits growth, and the yeast are found to be packed with small vesicles.

A. Do you think Sec4 is required for formation of vesicles, for vesicle fusion with target membranes, or for both? Based on its function, would you guess it was analogous to mammalian Arf, Sar1, or Rab proteins?

B. Using your knowledge of the way the analogous mammalian protein works, outline how you think normal Sec4 functions in vesicle formation and fusion. Why is some Sec4 free in the cytosol of wild-type cells? How does removal of the C-terminal cysteines prevent Sec4-ccΔ from carrying out its function?

C. Why do you think expression of Sec4N133I inhibits growth of the yeast that also express normal Sec4?

## TRANSPORT FROM THE ER THROUGH THE GOLGI APPARATUS

TERMS TO LEARN

| | | |
|---|---|---|
| *cis*-face | Golgi apparatus (Golgi complex) | *trans*-face |
| *cis* Golgi network (CGN) | high-mannose oligosaccharide | *trans* Golgi network (TGN) |
| cisternal maturation model | *O*-linked glycosylation | vesicular transport model |
| complex oligosaccharide | proteoglycan | |

### DEFINITIONS

Match the definition below with its term from the list above.

**13–36** The hypothesis that new cisternae form continuously at the *cis* face of the Golgi and then migrate through the stack as they mature.

**13–37** Molecule consisting of one or more glycosaminoglycan chains attached to a core protein.

**13–38** The side of the Golgi stack at which material enters the organelle.

**13–39** Chain of sugars attached to a glycoprotein that is generated by initially trimming the original oligosaccharide attached in the ER and by then adding other sugars.

**13–40** Membrane-bounded organelle in eucaryotic cells in which proteins and lipids transferred from the ER are modified and sorted.

**13–41** Chain of sugars attached to a glycoprotein that contains many mannose residues.

**13–42** Meshwork of interconnected cisternae and tubules on the side of the Golgi stack at which material is transferred out of the Golgi.

## TRUE/FALSE

Decide whether each of these statements is true or false, and then explain why.

**13–43** There is one strict requirement for the exit of a protein from the ER: it must be correctly folded.

**13–44** All of the glycoproteins and glycolipids in intracellular membranes have their oligosaccharide chains facing the lumenal side, and all those in the plasma membrane have their oligosaccharide chains facing the outside of the cell.

**13–45** The Golgi apparatus confers the heaviest glycosylation of all on proteoglycan core proteins, which are converted into proteoglycans by the addition of one or more *O*-linked glycosaminoglycan chains.

## THOUGHT PROBLEMS

**13–46** How is it that soluble proteins in the ER can be selectively recruited into vesicles destined for the Golgi?

**13–47** Isn't quality control always a good thing? How can quality control in the ER be detrimental to cystic fibrosis patients?

**13–48** In the assay for the homotypic fusion of yeast vacuolar vesicles described in Problem 13–31, it was necessary to incubate the vesicles from the two strains with NSF and ATP before they were mixed together. If they were not pretreated in this way, the vesicles would not fuse. Why do you suppose that step was necessary? Would you expect that such a treatment would be required for vesicles that each carried just one kind of SNARE (a t-SNARE or a v-SNARE)?

**13–49** The C-terminal 40 amino acids of three ER-resident proteins—calnexin, calreticulin, and HMG CoA reductase—are shown in Figure 13–9. Decide for each protein whether it is likely to be transmembrane or soluble. Explain your answer.

**13–50** If you were to remove the ER retrieval signal from protein disulfide isomerase (PDI), which is normally a soluble resident of the ER lumen, where would you expect the modified PDI to be located?

**13–51** The KDEL receptor must shuttle back and forth between the ER and the Golgi apparatus in order to accomplish its task of ensuring that soluble ER proteins are retained in the ER lumen. In which compartment does the KDEL receptor bind its ligands more tightly? In which compartment does it bind its ligands more weakly? What is thought to be the basis for its different binding affinities in the two compartments? If you were designing the system, in which compartment would you have the highest concentration of KDEL receptor? Would you predict that the KDEL receptor, which is a transmembrane protein, would itself possess an ER retrieval signal?

**13–52** When the KDEL retrieval signal is added to rat growth hormone or human chorionic gonadotropin, two proteins that are normally secreted, the proteins are still secreted, but about six times more slowly. If the C-terminal L in

Calnexin C-terminus

...KDKGDEEEEGEEKLEEKQKSDAEEDGGTVSQEEEDRKPKAEEDEILNRSPRNRKPRRE

Calreticulin

...KQDEEQRLKEEEEDKKRKEEEEAEDKEDDEDKDEDEEDEEDKEEDEEEDVPGQAKDEL

HMG CoA reductase

...PGENARQLARIVCGTVMAGELSLMAALAAGHLVKSHMIHNRSKINLQDLQGACTKKTA

**Figure 13–9** C-terminal amino acids of proteins that are residents of the ER (Problem 13–49).

the signal is changed to V, the proteins are once again secreted at their normal rate. By contrast, bona fide ER resident proteins rarely, if ever, are secreted from the cell; they are usually captured and returned very efficiently. How is it, do you suppose, that normal resident proteins with a KDEL signal are efficiently retained in the ER, whereas secreted proteins to which a KDEL signal has been added are not efficiently retained? Is this what you would expect if the KDEL signal and the KDEL receptor accounted entirely for retention of soluble proteins in the ER?

**13–53** Processing of *N*-linked oligosaccharides is not uniform among species. Most mammals, with the exception of humans and Old World primates, occasionally add galactose, in place of an *N*-acetylneuraminic acid, to a galactose, forming a terminal Gal(α1–3)Gal disaccharide on some branches of an *N*-linked oligosaccharide. How does this explain the preferred use of Old World primates for production of recombinant proteins for therapeutic use in humans?

**13–54** Cells have evolved a set of complicated pathways for addition of carbohydrates to proteins, implying that carbohydrates serve important functions. List three functions that carbohydrates on proteins are known to carry out.

**13–55** Most exported proteins move across the Golgi apparatus in 5–15 minutes, but very large proteins, like procollagen type I (PC), can take more than an hour. How might you account for this observation—different rates of protein movement—in the vesicle transport and cisternal maturation models for protein transport through the Golgi apparatus?

## DATA HANDLING

**13–56** The vesicular stomatitis virus (VSV) G protein is a typical membrane glycoprotein. In addition to its signal peptide, which is removed after import into the ER, the G protein contains a single membrane-spanning segment that anchors the protein in the plasma membrane. The membrane-spanning segment consists of 20 uncharged and mostly hydrophobic amino acids that are flanked by basic amino acids (Figure 13–10). Twenty amino acids arranged in an α helix is just sufficient to span the 3-nm thickness of the lipid bilayer of the membrane.

To test the length requirements for membrane-spanning segments, you modify a cloned version of the VSV G protein to generate a series of mutants in which the membrane-spanning segment is shorter, as indicated in Figure 13–10. When you introduce the modified plasmids into cultured cells, roughly the same amount of G protein is synthesized from each mutant as from wild-type cells. You analyze the cellular distribution of the altered G proteins in several ways.

1. You examine the cellular location of the modified VSV G proteins by immunofluorescence microscopy, using G-specific antibodies tagged with fluorescein.

| | | ← membrane-spanning segment → | | | | | | | | | | | | | | | | | | | |
|---|---|---|---|---|---|---|---|---|---|---|---|---|---|---|---|---|---|---|---|---|---|
| pMS20 | **K** | S | S | I | A | S | F | F | F | I | I | G | L | I | I | G | L | F | L | V | L | **R** |
| pMS18 | **K** | S | S | I | A | S | F | F | F | I | I | G | L | - | - | G | L | F | L | V | L | **R** |
| pMS16 | **K** | S | S | I | A | S | F | F | F | I | I | G | - | - | - | - | L | F | L | V | L | **R** |
| pMS14 | **K** | S | S | I | A | S | F | F | F | I | I | - | - | - | - | - | - | F | L | V | L | **R** |
| pMS12 | **K** | S | S | I | A | S | F | F | F | I | - | - | - | - | - | - | - | - | L | V | L | **R** |
| pMS8 | **K** | S | S | I | A | - | - | - | - | - | - | - | - | - | - | - | - | F | L | V | L | **R** |
| pMS0 | **K** | - | - | - | - | - | - | - | - | - | - | - | - | - | - | - | - | - | - | - | - | **R** |

**Figure 13–10** The membrane-spanning domains of normal and mutant VSV G proteins (Problem 13–56). Plasmid numbers indicate the number of amino acids in the membrane-spanning segment; for example, pMS20 contains the wild type, 20-amino acid segment. *Dashed lines* indicate amino acids that are missing in the other plasmids. *Boxed letters* indicate the basic amino acids that flank the membrane-spanning segment.

**Table 13–2 Results of experiments characterizing the cellular distribution of G proteins from normal and mutant cells** (Problem 13–56).

| PLASMID | CELLULAR LOCATION | ENDO H TREATMENT | PROTEASE TREATMENT |
|---|---|---|---|
| pMS20 | plasma membrane | resistant | sensitive |
| pMS18 | plasma membrane | resistant | sensitive |
| pMS16 | plasma membrane | resistant | sensitive |
| pMS14 | plasma membrane | resistant | sensitive |
| pMS12 | intracellular | +/– resistant | sensitive |
| pMS8 | intracellular | sensitive | sensitive |
| pMS0 | intracellular | sensitive | resistant |

2. You characterize the attached oligosaccharide chains by digestion with endoglycosidase H (endo H), which cleaves off *N*-linked oligosaccharides until the first mannose is removed in the medial portion of the Golgi apparatus (see Figure 13–11, Problem 13–57).
3. You determine whether the altered VSV G proteins retain the small C-terminal cytoplasmic domain (which characterizes the normal G protein) by treating isolated microsomes with a protease. In the normal VSV G protein this domain is sensitive to protease treatment and removed.

The results of these experiments are summarized in Table 13–2.

A. To the extent these data allow, deduce the intracellular location of each altered VSV G protein that fails to reach the plasma membrane.

B. For the VSV G protein, what is the minimum length of the membrane-spanning segment that is sufficient to anchor the protein in the membrane?

C. What is the minimum length of the membrane-spanning segment that is consistent with proper sorting of the G protein? How is it that transmembrane segments shorter than this can make it to the Golgi apparatus, but then not be able to exit?

**13–57** You have isolated several mutant cell lines that are defective in their ability to add carbohydrate to exported proteins. Using an easily purified protein that carries only *N*-linked complex oligosaccharides, you have analyzed the sugar monomers that are added in the different mutant cells. Each mutant is unique in the kinds and numbers of different sugars contained in its *N*-linked oligosaccharides (Table 13–3).

A. Arrange the mutants in the order that corresponds to the steps in the pathway for processing *N*-linked oligosaccharides (Figure 13–11). (Assume that each mutant cell line is defective for a single enzyme required to construct the *N*-linked oligosaccharide.)

**Table 13–3 Analysis of the sugars present in the *N*-linked oligosaccharides from wild-type and mutant cell lines defective in oligosaccharide processing** (Problem 13–57).

| CELL LINE | Man | GlcNAc | Gal | NANA | Glc |
|---|---|---|---|---|---|
| Wild type | 3 | 5 | 3 | 3 | 0 |
| Mutant A | 3 | 5 | 0 | 0 | 0 |
| Mutant B | 5 | 3 | 0 | 0 | 0 |
| Mutant C | 9 | 2 | 0 | 0 | 3 |
| Mutant D | 9 | 2 | 0 | 0 | 0 |
| Mutant E | 5 | 2 | 0 | 0 | 0 |
| Mutant F | 3 | 3 | 0 | 0 | 0 |
| Mutant G | 8 | 2 | 0 | 0 | 0 |
| Mutant H | 9 | 2 | 0 | 0 | 2 |
| Mutant I | 3 | 5 | 3 | 0 | 0 |

Abbreviations: Man = mannose; GlcNAc = *N*-acetylglucosamine; Gal = galactose; NANA = *N*-acetylneuraminic acid, or sialic acid; Glc = glucose. Numbers indicate the number of sugar monomers in the oligosaccharide.

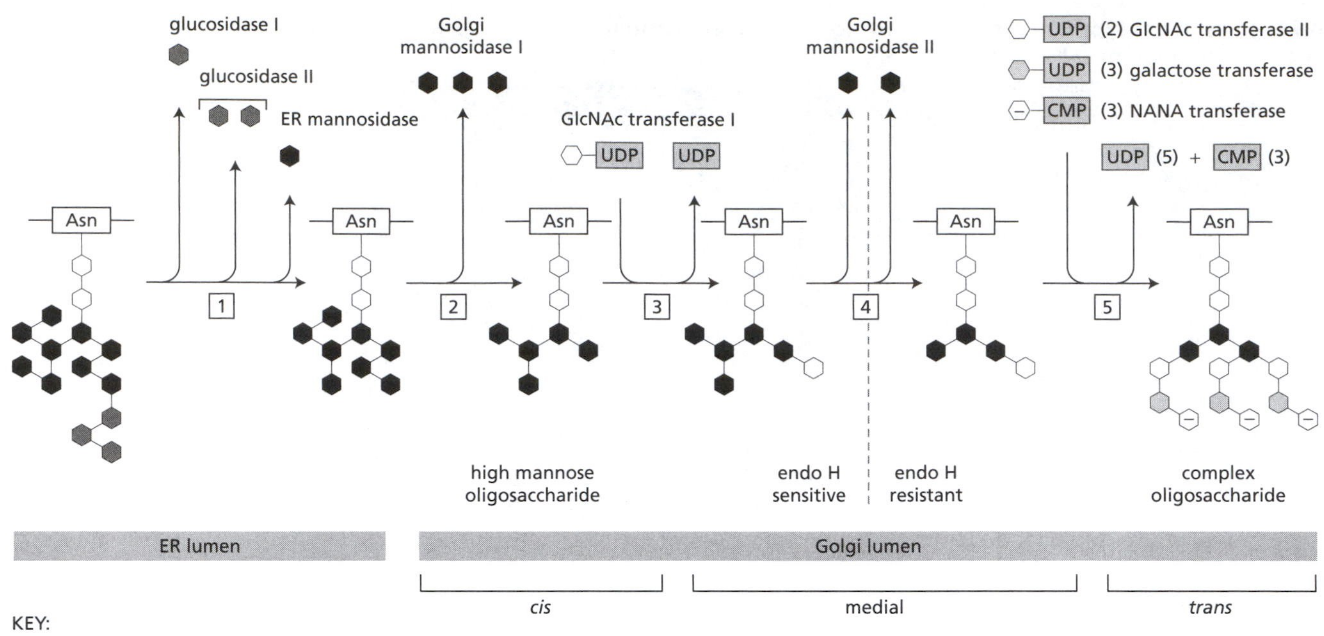


**Figure 13–11** Oligosaccharide processing in the ER and the Golgi apparatus (Problem 13–57). The reactions listed in step 5 occur in two different compartments; addition of GlcNAc occurs in the medial compartment, whereas addition of Gal and NANA occurs in the *trans* compartment.

B. Which of these mutants are defective in processing events that occur in the ER? Which mutants are defective in processing steps that occur in the Golgi?

C. Which of the mutants are likely to be defective in a processing enzyme that is directly responsible for modifying *N*-linked oligosaccharides? Which mutants might not be defective in a processing enzyme, but rather in another enzyme that affects oligosaccharide processing indirectly?

**13–58** Two extreme models—vesicular transport and cisternal maturation—have been proposed to account for the movement of molecules across the polarized structure of the Golgi apparatus. In the vesicular transport model, the individual Golgi cisternae remain in place as proteins move through them (Figure 13–12A). By contrast, in the cisternal maturation model, the individual

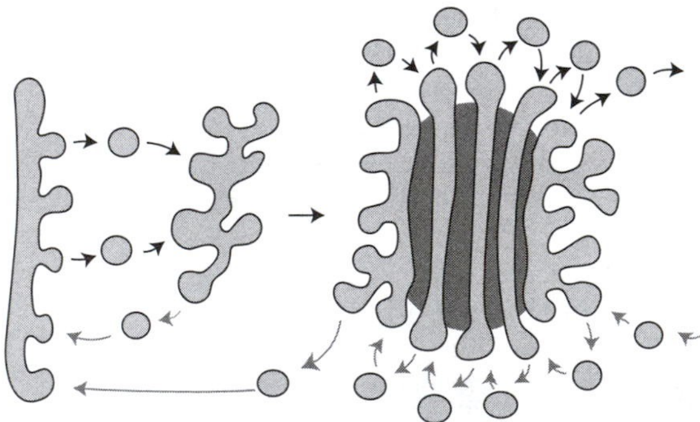


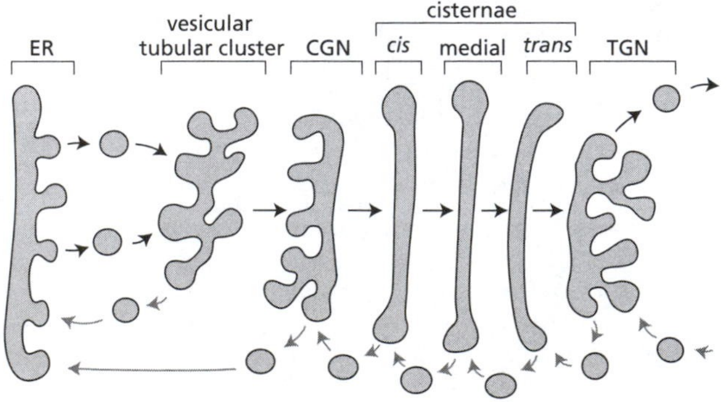


**Figure 13–12** Two models for the movement of molecules through the Golgi apparatus (Problem 13–58). (A) The vesicular transport model. (B) The cisternal maturation model. In (B) the individual cisternae have been separated for illustration purposes.

**Table 13–4 Addition of galactose to VSV G protein after fusion of VSV-infected cells with uninfected cells** (Problem 13–59).

| INFECTED CELLS | UNINFECTED CELLS | PRECIPITATE | SUPERNATANT |
|---|---|---|---|
| Mutant cells | wild-type cells | 45% | 55% |
| Mutant cells | mutant cells | 5% | 95% |
| Wild-type cells | wild-type cells | 85% | 15% |

Golgi cisternae move across the stack, carrying the proteins with them (Figure 13–12B). Transport vesicles serve critical functions in both models, but their roles are distinctly different. Describe the roles of the transport vesicles in each of the two models. Comment specifically on the roles of vesicles in the forward movement of proteins across the Golgi stack, in the retention of Golgi resident proteins in individual cisternae, and on the return of escaped ER proteins to the ER.

**13–59** One early test of the vesicular transport and cisternal maturation models (see Figure 13–12) looked for the movement of a protein between Golgi cisternae. This study made use of mutant cells that cannot add galactose to proteins, which normally occurs in the *trans* compartment of the Golgi (see Figure 13–11). The mutant cells were infected with vesicular stomatitis virus (VSV) to provide a convenient marker protein, the viral G protein. At an appropriate point in the infection, an inhibitor of protein synthesis was added to stop further synthesis of G protein. The infected cells were then incubated briefly with a radioactive precursor of GlcNAc, which is added only in the medial cisterna of the Golgi (see Figure 13–11). Next, the infected mutant cells were fused with uninfected wild-type cells to form a common cytoplasm containing both wild-type and mutant Golgi stacks. After a few minutes, the cells were dissolved with detergent and all the VSV G protein was captured using G-specific antibodies. After separation from the antibodies, the G proteins carrying galactose were precipitated, using a lectin that binds galactose. The radioactivity in the precipitate and in the supernatant was measured. The results of this experiment along with control experiments (which used mutant cells only or wild-type cells only) are shown in Table 13–4.

A. Between which two compartments of the Golgi apparatus is the movement of proteins being tested in this experiment? Explain your answer.
B. If proteins moved through the Golgi apparatus by cisternal maturation, what would you predict for the results of this experiment? If proteins moved through the Golgi via vesicular transport, what would you predict?
C. Which model is supported by the results in Table 13–4?

**13–60** A second test of the vesicular transport and cisternal maturation models used electron microscopy to follow movement of procollagen type I (PC) across the Golgi apparatus. PC is a long (300 nm) rodlike protein that is composed of three chains coiled into an uninterrupted triple helix. During its assembly, proline amino acids within the individual chains are hydroxylated in the ER. This allows the chains to form the helix, which is the signal that the PC is ready to exit the ER. The assembled PC is too large to fit into transport vesicles, which are typically 70–90 nm in diameter.

PC can be detected by electron microscopy using antibodies specific for the triple helix, which detect only the assembled form, or using antibodies specific for its globular head groups, which detect both the individual chains and the fully assembled form. Its movement through the Golgi apparatus was followed after two synchronization procedures, both of which yielded the same result. When assembly in the ER was blocked by a reversible inhibitor of proline hydroxylase, PC that was already in the Golgi apparatus was seen to disappear first from the *cis* cisterna, then the medial cisterna, and finally from the *trans* cisterna. If the block was maintained until the Golgi apparatus was empty of PC, and then the block was removed, PC appeared first in the *cis* cisterna, then in the medial cisterna, and finally in the *trans* cisterna. Both antibodies yielded the same result, and no PC was observed except within Golgi cisternae. Three-dimensional reconstructions from serial sections show that PC was always located in Golgi cisternae, never in vesicles or in budding vesicles.

A. Why was it important in these experiments to use two antibodies, one specific for triple-helical PC and the other that could also detect individual chains? What possibility were the authors concerned about?

B. Which model for movement of proteins through the Golgi apparatus—vesicular transport or cisternal maturation—do these results support? Explain your reasoning.

# TRANSPORT FROM THE *TRANS* GOLGI NETWORK TO LYSOSOMES

TERMS TO LEARN

| | | |
|---|---|---|
| acid hydrolase | lysosome | M6P receptor protein |
| autophagy | lysosomal storage disease | vacuole |

## DEFINITIONS

Match the definition below with its term from the list above.

**13–61** Digestion of obsolete parts of the cell by the cell's own lysosomes.

**13–62** Very large fluid-filled vesicle found in most plant and fungal cells, typically occupying more than 30% of the cell volume.

**13–63** Membrane-bounded organelle in eucaryotic cells that contains digestive enzymes, which are typically most active at the acid pH found in the lumen.

## TRUE/FALSE

Decide whether each of these statements is true or false, and then explain why.

**13–64** Lysosomal membranes contain a proton pump that utilizes the energy of ATP hydrolysis to pump protons out of the lysosome, thereby maintaining the lumen at a low pH.

**13–65** Late endosomes are converted to mature lysosomes by the loss of distinct endosomal membrane proteins and a further decrease in their internal pH.

**13–66** If cells were treated with a weak base such as ammonia or chloroquine, which raises the pH of organelles toward neutrality, M6P receptors would be expected to accumulate in the Golgi because they could not bind to the lysosomal enzymes.

## THOUGHT PROBLEMS

**13–67** How does the low pH of lysosomes protect the rest of the cell from lysosomal enzymes in case the lysosome breaks?

**13–68** Imagine that an autophagosome is formed by engulfment of a mitochondrion by the ER membrane. How many layers of membrane separate the matrix of the mitochondrion from the cytosol outside the autophagosome? Identify the source of each membrane and the spaces between the membranes.

**13–69** The principal pathway for transport of lysosomal hydrolases from the *trans* Golgi network (pH 6.6) to the late endosomes (pH 6) and for the recycling of M6P receptors back to the Golgi depends on the pH difference between those two compartments. From what you know about M6P receptor binding and recycling and the pathways for delivery of material to lysosomes, describe the consequences of changing the pH in those two compartments.

A. What do you suppose would happen if the pH in late endosomes were raised to pH 6.6?

B. What do you suppose would happen if the pH in the *trans* Golgi network were lowered to pH 6?

**Figure 13–13** A normal mouse and the *Mocha* mouse (Problem 13–71). In addition to its light coat color, the *Mocha* mouse has a poor sense of balance.

**13–70** Patients with I-cell disease are missing the enzyme GlcNAc phosphotransferase, which catalyzes the first of the two steps required for addition of phosphate to mannose to create the M6P marker (see Figure 13–14). In the absence of the M6P marker, the M6P receptor cannot bind to the protein and deliver it to a lysosome. How do you suppose that the lysosomes in some cells from these patients—liver cells, for example—acquire a normal complement of lysosomal enzymes?

**13–71** Melanosomes are specialized lysosomes that store pigments for eventual release by exocytosis. Various cells such as skin and hair cells then take up the pigment, which accounts for their characteristic pigmentation. Mouse mutants that have defective melanosomes often have pale or unusual coat colors. One such light-colored mouse, the *Mocha* mouse (Figure 13–13), has a defect in the gene for one of the subunits of the adaptor protein complex AP3, which is associated with coated vesicles budding from the *trans* Golgi network. How might the loss of AP3 cause a defect in melanosomes?

## CALCULATIONS

**13–72** Most lysosomal hydrolases are tagged with several oligosaccharide chains that can each acquire multiple M6P groups. Multiple M6P groups substantially increase the affinity of these hydrolases for M6P receptors, and markedly improve the efficiency of sorting to lysosomes. The reasons for the increase in affinity are twofold: one relatively straightforward and the other more subtle. Both can be appreciated at a conceptual level by comparing a hydrolase with a single M6P group to one with four equivalent M6P groups.

The binding of a hydrolase (H) to the M6P receptor (R) to form a hydrolase–receptor complex (HR) can be represented as

$$\mathrm{H + R} \underset{k_{\mathrm{off}}}{\overset{k_{\mathrm{on}}}{\rightleftharpoons}} \mathrm{HR}$$

At equilibrium the rate of association of hydrolase with the receptor ($k_{\mathrm{on}}$[H][R]) equals the rate of dissociation ($k_{\mathrm{off}}$[HR])

$$k_{\mathrm{on}}\,[\mathrm{H}][\mathrm{R}] = k_{\mathrm{off}}\,[\mathrm{HR}]$$

$$\frac{k_{\mathrm{on}}}{k_{\mathrm{off}}} = \frac{[\mathrm{HR}]}{[\mathrm{H}][\mathrm{R}]} = K$$

where $K$ is the equilibrium constant for the association. Because the equilibrium constant for association is a measure of the strength of binding between two molecules, it is sometimes called the affinity constant: the larger the value of the affinity constant, the stronger the binding.

A. Consider first the hypothetical situation in which the hydrolase and the M6P receptor are both soluble—that is, the receptor is not in a membrane. Assume that the M6P receptor has a single binding site for M6P groups, and think about the interaction between a single receptor and a hydrolase molecule. How do you think the rate of association will change if the hydrolase has one M6P group or four equivalent M6P groups? How will the rate at which the hydrolase dissociates from a single receptor change if the hydrolase has one M6P group or four? Given the effect on association and dissociation, how will the affinity constants differ for a hydrolase with one M6P group versus a hydrolase with four M6P groups?

B. Consider the situation in which a hydrolase with four equivalent M6P groups has bound to one receptor already. Assuming that the first receptor is locked in place, how will the binding to the first receptor influence the affinity constant for binding to a second receptor? (For simplicity, assume that the binding to the first receptor does not interfere with the ability of other M6P groups on the hydrolase to bind to a second receptor.)

C. In the real situation, as it occurs during sorting, the M6P receptors are in the Golgi membrane, whereas the hydrolases are initially free in the lumen of the Golgi. Consider the situation in which a hydrolase with four M6P groups has bound to one receptor already. In this case how do you think the binding of the hydrolase to the first receptor will influence the affinity constant for binding to a second receptor? (Think about this question from the point of view of how the binding changes the distribution of the hydrolase with respect to the lumen and the membrane.)

**Problem 19–19** looks at multivalent interactions in the context of the attachment of phage T4 to *E. coli*.

**Problem 3–98** examines the effect of attachment of Src to the membrane on its affinity for a membrane-bound target.

## DATA HANDLING

**13–73** Patients with Hunter's syndrome or with Hurler's syndrome rarely live beyond their teens. These patients accumulate glycosaminoglycans in lysosomes due to the lack of specific lysosomal enzymes necessary for their degradation. When cells from patients with the two syndromes are fused, glycosaminoglycans are degraded properly, indicating that the cells are missing different degradative enzymes. Even if the cells are just cultured together, they still correct each other's defects. Most surprising of all, the medium from a culture of Hurler's cells corrects the defect in Hunter's cells (and vice versa). The corrective factors in the media are inactivated by treatment with proteases, by treatment with periodate, which destroys carbohydrate, and by treatment with alkaline phosphatase, which removes phosphates.

A. What do you suppose the corrective factors are? Beginning with the donor patient's cells, describe the route by which the factors reach the medium and subsequently enter the recipient cells to correct the lysosomal defects.

B. Why do you suppose the treatments with protease, periodate, and alkaline phosphatase inactivate the corrective factors?

C. Would you expect a similar sort of correction scheme to work for mutant cytosolic enzymes?

**13–74** Children with I-cell disease synthesize perfectly good lysosomal enzymes, but they are secreted outside the cell instead of being sorted to lysosomes. The mistake occurs because the cells lack GlcNAc phosphotransferase, which is required to create the M6P marker that is essential for proper sorting (Figure 13–14). In principle, I-cell disease could also be caused by deficiencies in GlcNAc phosphoglycosidase, which removes GlcNAc to expose M6P (Figure 13–14), or in the M6P receptor itself. Thus, there are three potential kinds of I-cell disease, which could be distinguished by the ability of various culture supernatants to correct defects in mutant cells. Imagine that you have three cell lines (A, B, and C), each of which derives from a patient with one of the three hypothetical I-cell diseases. Experiments with supernatants from these cell lines give the results below.

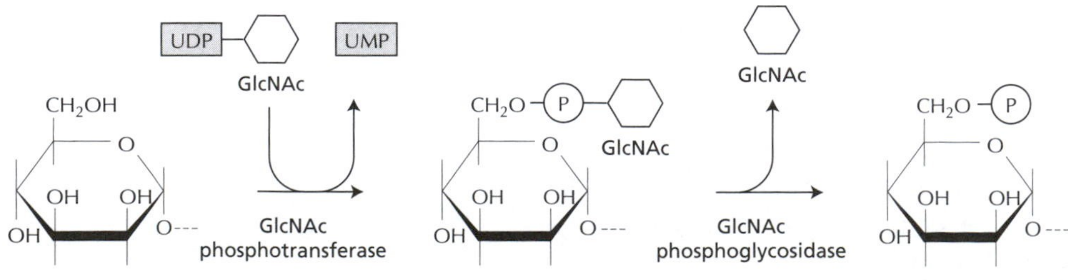


**Figure 13–14** Synthesis of M6P marker on a lysosomal hydrolase (Problem 13–74).

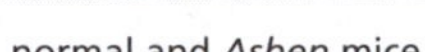

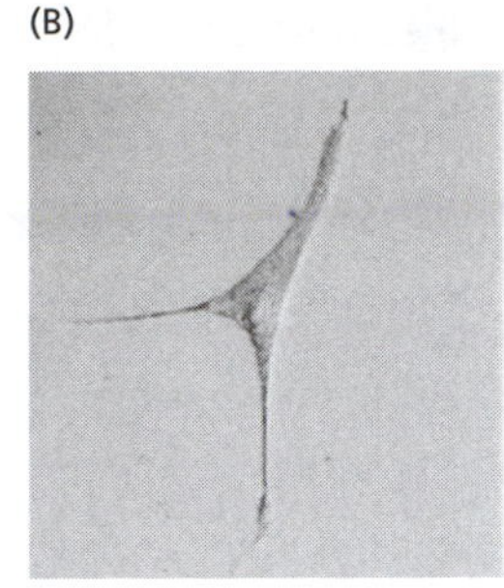

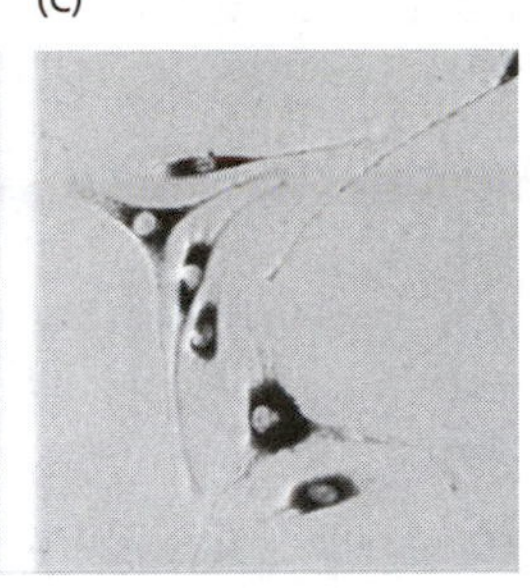


**Figure 13–15** Pigmentation defects in *Ashen* mice (Problem 13–75). (A) Normally pigmented mice and pale *Ashen* mice. (B) A melanocyte from a normal mouse. (C) Melanocytes from an *Ashen* mouse.

1. The supernatant from normal cells corrects the defects in B and C but not the defect in A.
2. The supernatant from A corrects the defect in Hurler's cells, which are missing a specific lysosomal enzyme, but the supernatants from B and C do not.
3. If the supernatants from the mutant cells are first treated with phosphoglycosidase to remove GlcNAc, then the supernatants from A and C correct the defect in Hurler's cells, but the supernatant from B does not.

From these results deduce the nature of the defect in each of the mutant cell lines.

**13–75** More than 50 different genes are known to affect coat color in mice. Three of them—*Dilute, Leaden,* and *Ashen*—are grouped together because of their highly similar phenotypes. Although these mice have normal melanosomes in their melanocytes, the pigment in the melanosomes is not delivered correctly to hair cells, giving rise to pale coats, as shown for *Ashen* mice in Figure 13–15A. *Dilute* mice lack an unconventional myosin heavy chain, MyoVa, which interacts with a microtubule-based transport motor. *Ashen* mice carry a mutation in the gene for the Rab protein, Rab27a, which associates with melanosomes. *Leaden* mice are missing melanophilin (Mlph), which is a modular protein with individual domains that bind to MyoVa, to Rab27a, and to actin filaments in the cell cortex.

Melanocytes from normal mice have a characteristic branch morphology (Figure 13–15B) and normally discharge their melanosomes near the tips of the branches. As shown in Figure 13–15C, melanocytes from *Ashen* mice have a normal morphology but their melanosomes surround the nucleus. Melanocytes from *Dilute* mice and *Leaden* mice have the same appearance as those from *Ashen* mice. Try to put these observations together to formulate a hypothesis to account for the normal delivery of melanosomes to the tips of the melanocyte branches, and for the defects in melanosome function in these mice.

# TRANSPORT INTO THE CELL FROM THE PLASMA MEMBRANE: ENDOCYTOSIS

TERMS TO LEARN

caveola
caveolin
clathrin-coated pit
early endosome
endocytosis
late endosome
low-density lipoprotein (LDL)
macrophage
multivesicular body
neutrophil
phagocytosis
phagosome
pinocytosis
receptor-mediated endocytosis
recycling endosome
transcytosis
transferrin receptor

## DEFINITIONS

Match the definition below with its term from the list above.

**13–76** General term for the process by which cells take up macromolecules, particulate substances, and even other cells into membrane-bounded vesicles.

**13–77** Complex vesicle with invaginating buds and internal vesicles involved in the maturation of early endosomes into late endosomes.

**13–78** Phagocytic cell, derived from a hematopoietic stem cell, that ingests invading microorganisms and plays an important role in scavenging senescent cells and apoptotic cells.

**13–79** Type of endocytosis in which soluble materials are taken up from the environment and incorporated into vesicles for digestion.

**13–80** The uptake of material at one face of a cell by endocytosis, its transfer across a cell in vesicles, and its discharge from another face by exocytosis.

**13–81** Invagination that forms from lipid rafts at the cell surface and buds off internally to form a pinocytotic vesicle.

**13–82** Region of plasma membrane of animal cells that is covered with the protein clathrin on its cytosolic face; it will bud off from the membrane to form an intracellular vesicle.

**13–83** Process by which macromolecules bind to complementary transmembrane receptor proteins, accumulate in coated pits, and then enter the cells as receptor–macromolecule complexes in clathrin-coated vesicles.

**13–84** One of a family of structural proteins in caveolae that are unusual because they extend multiple hydrophobic loops into the membrane from the cytosolic side, but do not cross the membrane.

**13–85** Membrane-bounded compartment just beneath the plasma membrane, to which external molecules are first delivered by endocytosis.

**13–86** Specialized form of endocytosis in which a cell uses large endocytic vesicles to ingest large particles such as microorganisms and dead cells.

## TRUE/FALSE

Decide whether each of these statements is true or false, and then explain why.

**13–87** Any particle that is bound to the surface of a phagocyte will be ingested by phagocytosis.

**13–88** Like the LDL receptor, most of the more than 25 different receptors known to participate in receptor-mediated endocytosis enter coated pits only after they have bound their specific ligands.

**13–89** All the molecules that enter early endosomes ultimately reach late endosomes, where they become mixed with newly synthesized acid hydrolases and end up in lysosomes.

**13–90** During transcytosis, vesicles that form from coated pits on the apical surface fuse with the plasma membrane on the basolateral surface, and in that way transport molecules across the epithelium.

## THOUGHT PROBLEMS

**13–91** A macrophage ingests the equivalent of 100% of its plasma membrane each half hour by endocytosis. What is the rate at which membrane is returned by exocytosis?

**13–92** The electron micrograph in Figure 13–16A is illustrated schematically by the drawing in Figure 13–16B. Name the structures that are labeled in the drawing.

**13–93** Caveolae are thought to form from lipid rafts, which are patches of the plasma membrane that are especially rich in cholesterol and glycosphingolipids. Caveolae may collect cargo proteins by virtue of the lipid composition of their

**Problems 10–20, 10–21, 10–22, 10–26, 10–30,** and **10–31** examine various properties of lipid rafts in membranes.

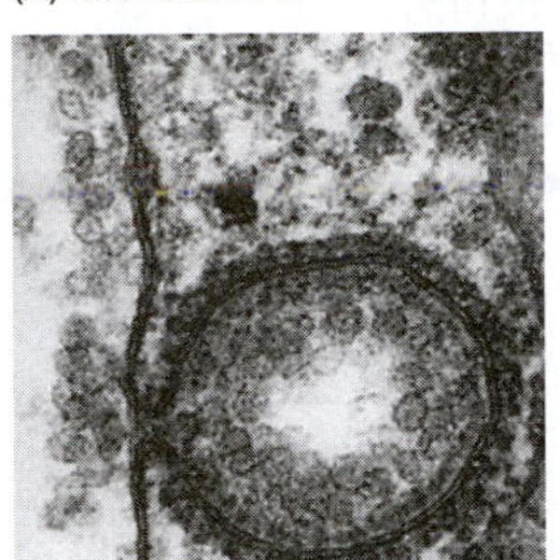

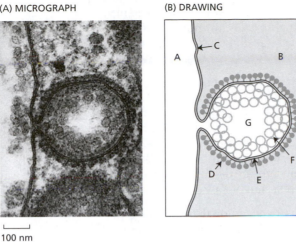

Figure 13–16 A coated pit about to bud from the membrane (Problem 13–92). (A) An electron micrograph. (B) A schematic drawing.

membrane, rather than by assembly of a cytosolic protein coat. What might you predict would be a characteristic of the structure of transmembrane proteins that collect in caveolae?

**13–94** Influenza viruses are surrounded by a membrane that contains a fusion protein, which is activated by acidic pH. Upon activation, the protein causes the viral membrane to fuse with cell membranes. An old folk remedy for the flu recommends that one should spend a night in a horse stable. Odd as it may sound, there is a rational basis for this advice. Air in stables contains ammonia ($NH_3$) generated by bacteria from the horses' urine. Sketch a diagram showing the pathway (in detail) by which flu virus enters cells, and speculate how $NH_3$ might protect cells from virus infection. (Hint: $NH_3$ can neutralize acidic solutions by the reaction $NH_3 + H^+ \rightarrow NH_4^+$.)

**13–95** Iron (Fe) is an essential trace metal that is needed by all cells. It is required, for example, for the synthesis of the heme groups that are part of cytochromes and hemoglobin. Iron is taken into cells via a two-component system. The soluble protein transferrin circulates in the bloodstream, and the transferrin receptor is a membrane protein that is continually endocytosed and recycled to the plasma membrane. Fe ions bind to transferrin at neutral pH but not at acidic pH. Transferrin binds to the transferrin receptor at neutral pH only when it has bound an Fe ion, but it binds to the receptor at acidic pH even in the absence of bound iron. From these properties, describe how iron is taken up, and discuss the advantages of this elaborate scheme.

## CALCULATIONS

**13–96** Cells take up extracellular molecules by receptor-mediated endocytosis and by fluid-phase endocytosis. A classic paper compared the efficiencies of these two pathways by incubating human cells for various periods of time in a range of concentrations of either $^{125}$I-labeled epidermal growth factor (EGF), to measure receptor-mediated endocytosis, or horseradish peroxidase (HRP), to measure fluid-phase endocytosis. Both EGF and HRP were found to be present in small vesicles with an internal radius of 20 nm. The uptake of HRP was linear (Figure 13–17A), while that of EGF was initially linear but reached a plateau at higher concentrations (Figure 13–17B).

A. Explain why the shapes of the curves in Figure 13–17 are different for HRP and EGF.

B. From the curves in Figure 13–17, estimate the difference in the uptake rates for HRP and EGF when both are present at 40 nM. What would the difference be if both were present at 40 μM?

C. Calculate the average number of HRP molecules that get taken up by each endocytic vesicle (radius 20 nm) when the medium contains 40 μM HRP. [The volume of a sphere is $(4/3)\pi r^3$.]

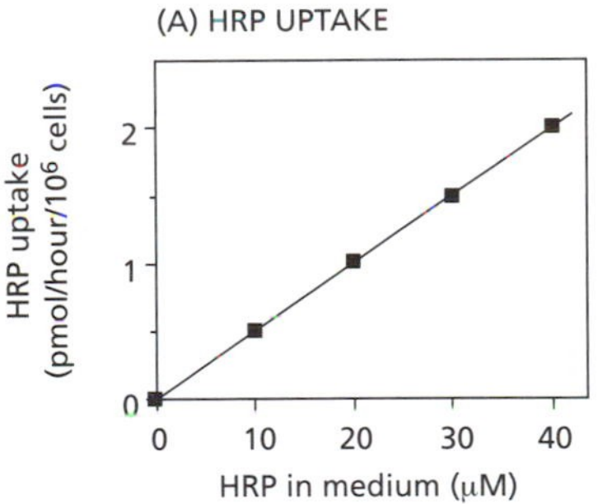


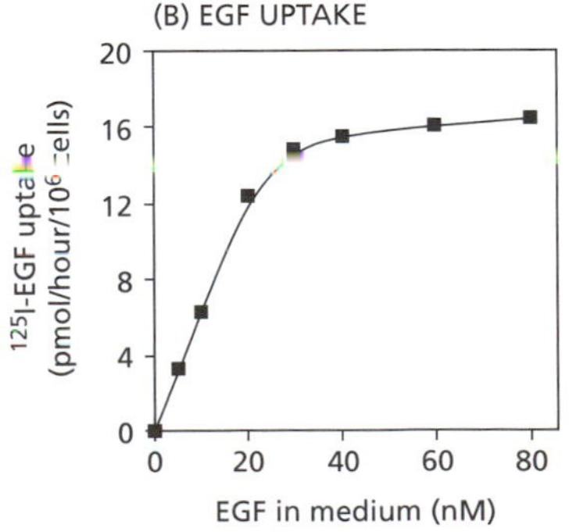


Figure 13–17 Uptake of HRP and EGF as a function of their concentration in the medium (Problem 13–96).

D. The scientists who did these experiments said at the time, "These calculations clearly illustrate how cells can internalize EGF by endocytosis while excluding all but insignificant quantities of extracellular fluid." What do you think they meant?

**13–97** A ligand for receptor-mediated endocytosis circulates at a concentration of 1 nM ($10^{-9}$ M). It is taken up in coated vesicles with a volume of $1.66 \times 10^{-18}$ L (about 150 nm in diameter). On average, there are 10 of its receptors in each coated vesicle. If all the receptors were bound to the ligand, how much more concentrated would the ligand be in the vesicle than it was in the extracellular fluid? What would the dissociation constant ($K_d$) for the receptor–ligand binding need to be in order to concentrate the ligand 1000-fold in the vesicle? (You may wish to review the discussion of $K_d$ in Problem 3–103.)

**13–98** The recycling of transferrin receptors has been studied by labeling the receptors on the cell surface and following their fate at 0°C and 37°C. A sample of intact cells at 0°C was reacted with radioactive iodine under conditions that label cell-surface proteins. If these cells were kept on ice and incubated in the presence of trypsin, which destroys the receptors without damaging the integrity of the cell, the radioactive transferrin receptors were completely degraded. If the cells were first warmed to 37°C for 1 hour and then treated with trypsin on ice, about 70% of the initial radioactivity was resistant to trypsin. At both temperatures, most of the receptors were not labeled and most remained intact, as apparent from a protein stain.

A second sample of cells that had been surface-labeled at 0°C and incubated at 37°C for 1 hour was analyzed with transferrin-specific antibodies, which identify transferrin receptors via their linkage to Fe–transferrin complexes. If intact cells were reacted with antibody, 0.54% of the labeled proteins were bound by antibody. If the cells were first dissolved in detergent, 1.76% of the labeled proteins were bound by antibody.

A. When the cells were kept on ice, why did trypsin treatment destroy the labeled transferrin receptors, but not the majority of receptors? Why did most of the labeled receptors become resistant to trypsin when the cells were incubated at 37°C?

B. What fraction of all the transferrin receptors is on the cell surface after a 1-hour incubation at 37°C? Do the two experimental approaches agree?

**13–99** The average time for transferrin receptors to cycle between the cell surface and endosomes has been determined by labeling cell-surface receptors with radioactive iodine at 0°C, and then following their fate at 37°C. At various times after shifting labeled cells to 37°C, samples were diluted into ice-cold medium that contained trypsin. The amount of radioactivity in trypsin-resistant transferrin receptors was measured after separation from other membrane components by two-dimensional polyacrylamide-gel electrophoresis. The results are shown in Figure 13–18.

A. The initial rate of internalization of labeled transferrin receptors is indicated by the dashed line. Why does the rate of internalization decline with time?

B. Using the initial rate of internalization, estimate the fraction of surface receptors that were internalized each minute.

C. What fraction of the *total* receptor population was internalized each minute?

D. At the rate determined in part C, how many minutes would it take for the equivalent of the entire population of receptors to be internalized? Explain why this time equals the average time for a receptor to cycle from the cell surface through the endosomal compartment and back to the cell surface.

E. On average, how long does each transferrin receptor spend on the cell surface?

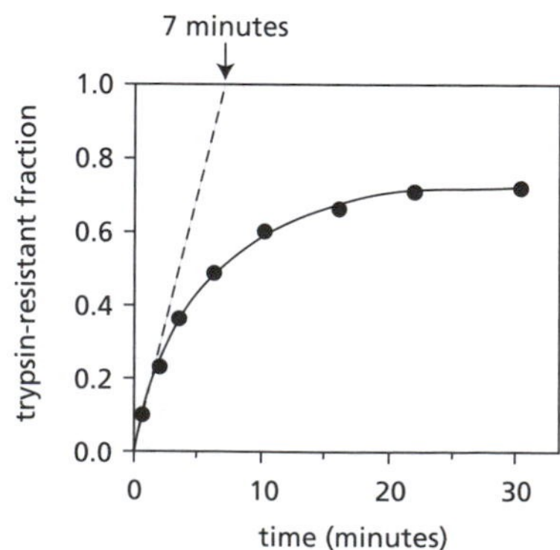


**Figure 13–18** Fraction of labeled transferrin receptor that was trypsin resistant as a function of time after labeling (Problem 13–99). The *dashed line* indicates the initial rate of internalization.

## DATA HANDLING

**13–100** Cholesterol is an essential component of the plasma membrane, but people who have very high levels of cholesterol in their blood (hypercholes-

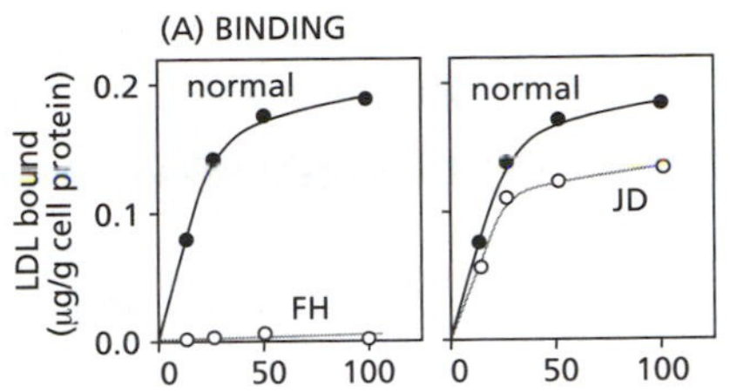


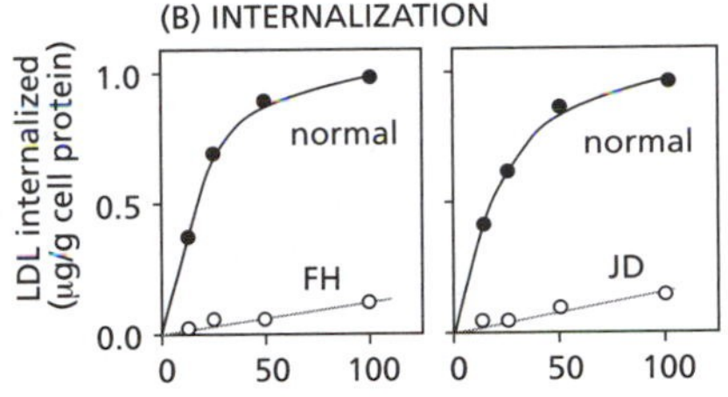


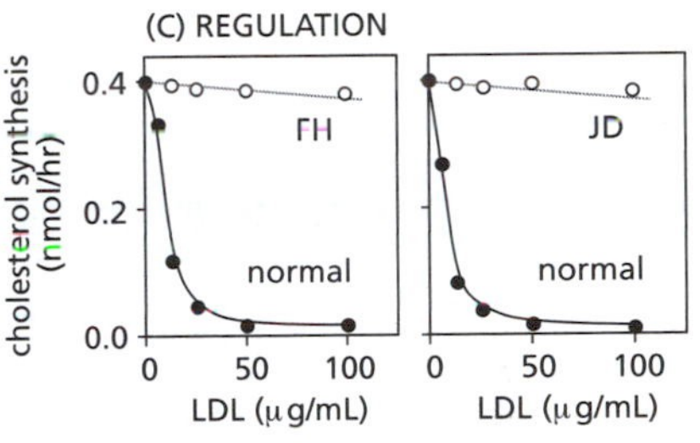


**Figure 13–19** LDL metabolism in normal cells and in cells from patients with severe familial hypercholesterolemia (Problem 13–100). (A) Surface binding of LDL. Assays at 4°C allow binding but not internalization. (B) Internalization of LDL. After binding at 4°C, the cells are warmed to 37°C. Binding and uptake of LDL can be followed by labeling LDL either with ferritin particles, which can be seen by electron microscopy, or with radioactive iodine, which can be measured in a gamma counter. (C) Regulation of cholesterol synthesis by LDL.

terolemia) tend to have heart attacks. Blood cholesterol is carried in the form of cholesterol esters in low-density lipoprotein (LDL) particles. LDL binds to a high-affinity receptor on the cell surface, enters the cell via a coated pit, and ends up in lysosomes. There its protein coat is degraded, and cholesterol esters are released and hydrolyzed to cholesterol. The released cholesterol enters the cytosol and inhibits the enzyme HMG CoA reductase, which controls the first unique step in cholesterol biosynthesis. Patients with severe hypercholesterolemia cannot remove LDL from the blood. As a result, their cells do not turn off normal cholesterol synthesis, which makes the problem worse.

LDL metabolism can be conveniently divided into three stages experimentally: binding of LDL to the cell surface, internalization of LDL, and regulation of cholesterol synthesis by LDL. Skin cells from a normal person and two patients suffering from severe familial hypercholesterolemia were grown in culture and tested for LDL binding, LDL internalization, and LDL regulation of cholesterol synthesis. The results are shown in Figure 13–19.

A. In Figure 13–19A, the surface binding of LDL by normal cells is compared with LDL binding by cells from patients FH and JD. Why does binding by normal cells and by JD's cells reach a plateau? What explanation can you suggest for the lack of LDL binding by FH's cells?
B. In Figure 13–19B, internalization of LDL by normal cells increases as the external LDL concentration is increased, reaching a plateau 5-fold higher than the amount of externally bound LDL. Why does LDL enter cells from patients FH or JD at such a slow rate?
C. In Figure 13–19C, the regulation of cholesterol synthesis by LDL in normal cells is compared with that in cells from FH and JD. Why does increasing the external LDL concentration inhibit cholesterol synthesis in normal cells, but affect it only slightly in cells from FH or JD?
D. How would you expect the rate of cholesterol synthesis to be affected if normal cells and cells from FH or JD were incubated with cholesterol itself? (Free cholesterol crosses the plasma membrane by diffusion.)

**13–101** What is wrong with JD's metabolism of LDL? As discusssed in Problem 13–100, JD's cells bind LDL with the same affinity as normal cells and in almost the same amounts, but the binding does not lead to internalization of LDL. Two classes of explanation could account for JD's problem:

1. JD's LDL receptors are defective in a way that prevents internalization, even though the LDL-binding domains on the cell surface are unaffected.
2. JD's LDL receptors are entirely normal, but there is a mutation in the cellular internalization machinery such that loaded LDL receptors cannot be brought in.

To distinguish between these explanations, JD's parents were studied. It is known that an autosomal gene encodes the receptor that binds LDL. Thus, each parent must have donated one defective gene to JD. JD's mother suffered from mildly elevated blood cholesterol. Her cells bound only half as much LDL as normal cells, but the bound LDL was internalized at the same rate as in normal cells. JD's father also had mild hypercholesterolemia, but his cells bound even more LDL than normal cells. Of the bound LDL, less than half the label could be internalized; the rest remained on the cell surface.

The association of this family's LDL receptors with coated pits was studied by electron microscopy, using LDL that was labeled with ferritin. The results are shown in Table 13–5.

A. Why does JD's mother have mild hypercholesterolemia? Based on the LDL-binding and internalization studies, and on the EM observations, decide what kind of defective LDL-receptor gene she passed to JD?

B. Why does JD's father have mild hypercholesterolemia? Based on the LDL-binding and internalization studies, and on the EM observations, decide what kind of defective LDL-receptor gene he passed to JD?

C. Can you account for JD's hypercholesterolemia from the behavior of the LDL receptors in his parents? In particular, how is it that JD binds nearly a normal amount of LDL, but has severe hypercholesterolemia.

D. At the beginning of this problem, two possible explanations—defective receptor or defective internalization machinery—were proposed to account for the lack of internalization by JD's LDL receptors in the face of nearly normal LDL binding. Do these studies allow you to decide between these alternative explanations?

**Table 13–5 Distribution of LDL receptors on the surface of cells from JD and his parents as compared with normal individuals** (Problem 13–101).

| INDIVIDUAL | NUMBER OF LDL RECEPTORS | |
|---|---|---|
| | IN PITS | OUTSIDE PITS |
| Normal male | 186 | 195 |
| Normal female | 186 | 165 |
| JD | 10 | 342 |
| JD's father | 112 | 444 |
| JD's mother | 91 | 87 |

## TRANSPORT FROM THE *TRANS* GOLGI NETWORK TO THE CELL EXTERIOR: EXOCYTOSIS

TERMS TO LEARN

| | | |
|---|---|---|
| constitutive secretory pathway | immature secretory vesicle | secretory vesicle |
| default pathway | regulated secretory pathway | synaptic vesicle |
| exocytosis | | |

### DEFINITIONS

Match the definition below with its term from the list above.

**13–102** Specialized class of tiny secretory vesicles that store neurotransmitter molecules.

**13–103** Pathway for exocytosis that operates continuously in all cells.

**13–104** Membrane-bounded organelle in which molecules destined to be exported are stored prior to release.

**13–105** Process involving fusion of vesicles with the plasma membrane.

**13–106** Pathway for exocytosis that operates mainly in cells specialized for secreting products rapidly on demand.

### TRUE/FALSE

Decide whether each of these statements is true or false, and then explain why.

**13–107** When a foreign gene encoding a secretory protein is introduced into a secretory cell that normally does not make the protein, the alien secretory protein is not packaged into secretory vesicles.

**13–108** Once a secretory vesicle is properly positioned beneath the plasma membrane, it will immediately fuse with the membrane and release its contents to the cell exterior.

### THOUGHT PROBLEMS

**13–109** In a cell capable of regulated secretion, what are the three main classes of protein that must be separated before they leave the *trans* Golgi network?

**13–110** You are interested in exocytosis and endocytosis in a line of cultured liver cells that secrete albumin and take up transferrin. To distinguish between these events, you tag transferrin with colloidal gold and prepare ferritin-labeled antibodies that are specific for albumin. You add the tagged transferrin to the medium, and then after a few minutes you fix the cells, prepare thin sections, and react them with ferritin-labeled antibodies against albumin. Colloidal gold and ferritin are both electron dense and therefore readily visible when viewed by electron microscopy; moreover, they can be easily distinguished from one another on the basis of size.

A. Will this experiment allow you to identify vesicles in the exocytic and endocytic pathways? How?
B. Not all the gold-labeled vesicles are clathrin coated. Why?

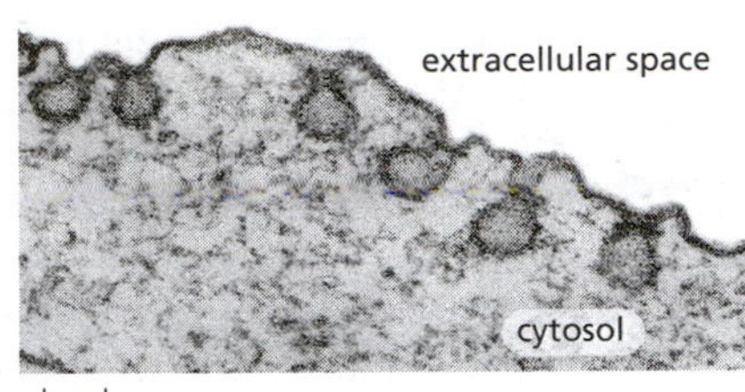


**Figure 13–20** Electron micrograph of a nerve terminal from a *shibire* mutant fly at elevated temperature (Problem 13–113).

**13–111** What would you expect to happen in cells that secrete large amounts of protein through the regulated secretory pathway, if the ionic conditions in the ER lumen could be changed to resemble those of the *trans* Golgi network?

**13–112** Antitrypsin, which inhibits certain proteases, is normally secreted into the bloodstream by liver cells. Antitrypsin is absent from the bloodstream of patients who carry a mutation that results in a single amino acid change in the protein. Antitrypsin deficiency causes a variety of severe problems, particularly in lung tissue, because of uncontrolled protease activity. Surprisingly, when the mutant antitrypsin is synthesized in the laboratory, it is as active as the normal antitrypsin at inhibiting proteases. Why then does the mutation cause the disease? Think of more than one possibility and suggest ways in which you could distinguish among them.

**13–113** Dynamin was first identified as a microtubule-binding protein, and its sequence indicated that it was a GTPase. The key to its function came from neurobiological studies in *Drosophila. Shibire* mutant flies, which carry a mutation in the dynamin gene, are rapidly paralyzed when the temperature is elevated. They recover quickly once the temperature is lowered. The complete paralysis at the elevated temperature suggested that synaptic transmission between nerve and muscle cells was blocked. Electron micrographs of synapses of the paralyzed flies showed a loss of synaptic vesicles and a tremendously increased number of coated pits relative to normal synapses (Figure 13–20).

Suggest an explanation for the paralysis shown by the *shibire* mutant flies, and indicate why signal transmission at a synapse might require dynamin.

## DATA HANDLING

**13–114** Liver cells secrete a broad spectrum of proteins into the blood via the constitutive pathway. You are interested in how long it takes for different proteins to be secreted. Accordingly, you add $^{35}$S-methionine to cultured liver cells to label proteins as they are synthesized. You then sample the medium at various times to measure the appearance of individual labeled proteins. As shown in Figure 13–21, albumin appears after 20 minutes, transferrin appears after 50 minutes, and retinol-binding protein appears after 90 minutes. You are surprised at the variability in secretion rates, which bear no obvious relationship to the size, function, or quantity of the individual proteins.

Why do transferrin and the retinol-binding protein take so much longer than albumin to be secreted? You suspect that the slow step in secretion occurs either in the ER or in the Golgi apparatus. To determine which, you label cells for 4 hours, which is long enough for the labeled proteins to reach the same steady-state distribution as unlabeled proteins. (At steady state the influx into a pathway exactly equals efflux from the pathway.) You then

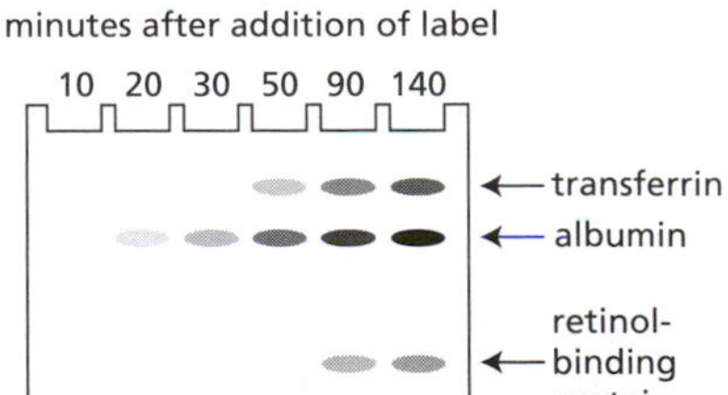


**Figure 13–21** Time of appearance of secreted proteins in the medium (Problem 13–114). At various times after labeling, proteins were immunoprecipitated with specific antibodies, separated by gel electrophoresis, and subjected to autoradiography.

homogenize the cells to break the ER and Golgi into vesicles and separate the vesicles by density on a sucrose gradient. You measure the amount of labeled albumin and transferrin that are associated with the two types of vesicle (Figure 13–22).

Does the slow step in the constitutive secretion of transferrin occur in the ER or in the Golgi? Where does the slow step in the constitutive secretion of albumin occur? How do these experiments allow you to decide?

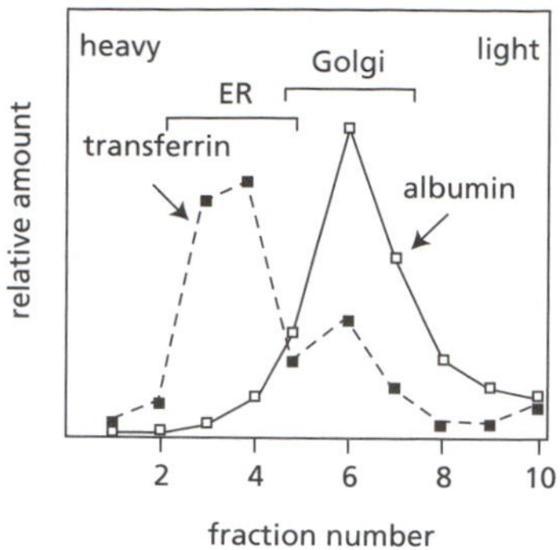


**Figure 13–22** Distribution of albumin and transferrin in vesicles derived from the ER and Golgi (Problem 13–114). Labeled albumin and transferrin were assayed by immunoprecipitation, electrophoresis, and autoradiography.

**13–115** Proteins without special signals are transported between cisternae in Golgi stacks and onward to the plasma membrane via the nonselective constitutive secretory pathway, or the default pathway, as it is commonly known. This transport is also sometimes referred to as bulk transport because the Golgi contents do not become concentrated in the vesicles. Given that transport in clathrin-coated vesicles is so highly concentrating, you are skeptical that no concentration occurs in the default pathway for secretion.

To determine whether vesicles in the default pathway concentrate their contents, you infect cells with vesicular stomatitis virus (VSV) and follow the viral G protein, which is transported by the default pathway. Your idea, an ambitious one, is to compare the concentration of G protein in the lumen of the Golgi stacks with that in the associated transport vesicles. You intend to measure G-protein concentration by preparing thin sections of VSV-infected cells and incubating them with G-specific antibodies tagged with gold particles. Since the gold particles are visible in electron micrographs as small black dots, it is relatively straightforward to count dots in the lumens of transport vesicles (fully formed and just budding) and of the Golgi apparatus. You make two estimates of G-protein concentration: (1) the number of gold particles per cross-sectional area (surface density) and (2) the number of gold particles per linear length of membrane (linear density). Your results are shown in Table 13–6.

Do the vesicles involved in the default pathway concentrate their contents or not? Explain your reasoning.

**13–116** Insulin is synthesized as a pre-pro-protein in the β cells of the pancreas. Its pre-peptide is cleaved off after it enters the ER lumen. To define the cellular location at which its pro-peptide is removed, you have prepared two antibodies: one that is specific for pro-insulin, and one that is specific for insulin. You have tagged the anti-pro-insulin antibody with a red fluorophore and the anti-insulin antibody with green fluorophore, so that you can follow them independently in the same cell. When you incubate a pancreatic β cell with a mixture of your two antibodies, you obtain the results shown in Table 13–7. In what cellular compartment is the pro-peptide removed from pro-insulin?

**13–117** Polarized epithelial cells must make an extra sorting decision since their plasma membranes are divided into apical and basolateral domains, which are populated by distinctive sets of proteins. Proteins destined for the apical or basolateral domain seem to travel there directly from the *trans* Golgi network. One way to sort proteins to these domains would be to use a specific sorting signal for one group of proteins, which would then be actively recognized and directed to one domain, and to allow the rest to travel via a default pathway to the other domain.

**Table 13–6 Relative densities of G protein in Golgi and vesicle lumens and membranes** (Problem 13–115).

| SOURCE OF GOLGI | SITE MEASURED | PARAMETER MEASURED | MEAN DENSITY |
|---|---|---|---|
| Uninfected cells | whole Golgi | surface density | 5/μm$^2$ |
| Infected cells | whole Golgi | surface density | 271/μm$^2$ |
| Infected cells | Golgi buds and vesicles | surface density | 233/μm$^2$ |
| Infected cells | Golgi cisternal membranes | linear density | 6/μm |
| Infected cells | Golgi buds and vesicles | linear density | 4/μm |

**Table 13–7 Fluorescence associated with various compartments of β cells after reaction with fluorescent antibodies directed against pro-insulin and insulin** (Problem 13–116).

| COMPARTMENT | FLUORESCENCE |
|---|---|
| *cis* Golgi network | red |
| Endoplasmic reticulum | red |
| Golgi cisternae | red |
| Immature secretory vesicles | yellow |
| Lysosomes | none |
| Mature secretory vesicles | green |
| Mitochondria | none |
| Nucleus | none |
| *trans* Golgi network | red |

Consider the following experiment to identify the default pathway. The cloned genes for several foreign proteins were engineered by recombinant DNA techniques so that they could be expressed in the polarized epithelial cell line MDCK. These proteins are secreted in other types of cells, but are not normally expressed in MDCK cells. The cloned genes were introduced into the polarized MDCK cells, and their sites of secretion were assayed. Although the cells remained polarized, the foreign proteins were delivered in roughly equal amounts to the apical and basolateral domains.

A. What is the expected result of this experiment, based on the hypothesis that targeting to one domain of the plasma membrane is actively signaled and targeting to the other domain is via a default pathway?

B. Do these results support the concept of a default pathway as outlined above?

**13–118** Neurons are difficult to study because of their excessively branched structure and long thin dendrites, as shown in Figure 13–23. Fluorescently tagged antibodies are powerful tools for investigating certain aspects of neuron structure. Synaptic vesicles, for example, were shown to be concentrated in the presynaptic cells at nerve synapses in this way. A culture of neurons was first exposed for 1 hour to a fluorescently tagged antibody specific for the lumenal domain of synaptotagmin, a transmembrane protein that resides exclusively in the membranes of synaptic vesicles. The culture was then washed thoroughly to remove all synaptotagmin antibodies. When the culture was examined by fluorescence microscopy, dots of color from the synaptotagmin-specific antibody were found to mark the positions of the synaptic vesicles in the nerve terminals.

If antibodies do not cross intact membranes, how do you suppose the synaptic vesicles get labeled? When the procedure was repeated using an antibody specific for the cytoplasmic domain of synaptotagmin, the nerve terminals did not become labeled. Explain the results with the two different antibodies for synaptotagmin.

**13–119** The original version of the SNARE hypothesis suggested that the ATP-dependent disassembly of SNAREs by NSF provided the energy necessary for membrane fusion, and thus that NSF should act at the last step in secretion. More recent evidence suggests that NSF acts at an earlier step to prime the vesicle for secretion, and that SNAREs alone are sufficient to catalyze membrane fusion at the last step in secretion. It is important to know what really happens, and you have the means at hand to answer the question.

Using a whole-cell patch-clamp protocol (Figure 13–24A), you can diffuse cytosolic components into the cell through the pipet and assay exocytosis by changes in capacitance, which is a measure of the increase in the area of the plasma membrane. To control precisely the timing of vesicle fusion, you enclose $Ca^{2+}$ in a photosensitive chemical 'cage,' from which it can be released with a flash of light. In response to $Ca^{2+}$ release, there is a rapid burst of vesicle fusion (indicated by a rapid rise in capacitance), followed by a longer, slower fusion process (Figure 13–24B, –NEM). The initial burst

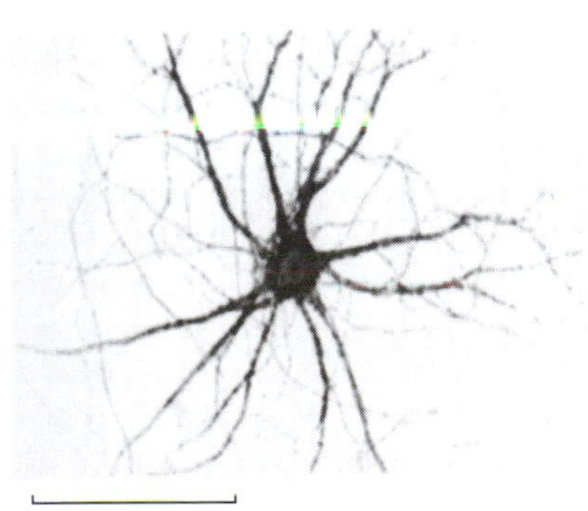


**Figure 13–23** A hippocampal neuron (Problem 13–118).

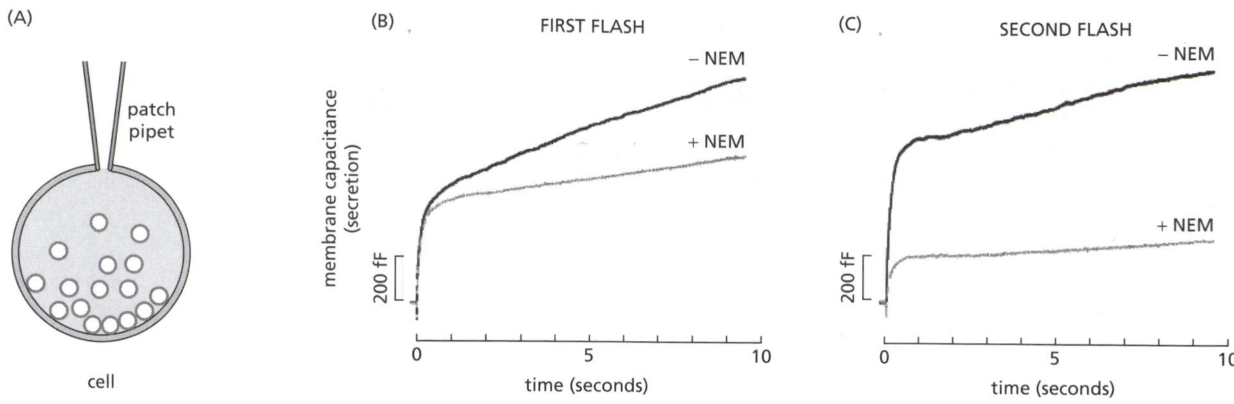


**Figure 13–24** Analysis of the role of NSF in vesicle fusion (Problem 13–119). (A) Whole-cell patch-clamp. (B) Responses to the first flash-mediated release of $Ca^{2+}$. (C) Reponses to the second flash-mediated release of $Ca^{2+}$. Secretion was measured as an increase in membrane capacitance (in femtoFarads, fF) in the absence (–NEM) or presence (+NEM) of an inhibitor of NSF.

represents the fusion of vesicles that were just waiting for the $Ca^{2+}$ trigger. Because $Ca^{2+}$ is rapidly removed from the cell, the procedure can be repeated with a second flash of light 2 minutes later; it yields the same rapid and slow components (Figure 13–24C, –NEM).

To test the role of NSF in vesicle fusion, you first diffuse *N*-ethylmaleimide (NEM) into cells to inhibit NSF. (The name 'NSF' stands for NEM-sensitive factor.) You then repeat the flash protocols as before. In response to the first flash, the rapid component is unaffected, but the slow component is decreased (Figure 13–24B, +NEM). In response to the second flash, both components are inhibited (Figure 13–24C, +NEM).

A. What do you suppose the slow component of the fusion process represents?
B. Why does inhibition of NSF affect the slow component after both flashes, but inhibit the rapid component only after the second flash?
C. Which of the alternatives for the role of NSF in vesicle fusion—acting at the last step or an early step—do these experiments support? Explain your reasoning.
D. Propose a model for the molecular role of NSF in fusion of secretory vesicles.

# Energy Conversion: Mitochondria and Chloroplasts

# 14

## In This Chapter


THE MITOCHONDRION 315

ELECTRON-TRANSPORT CHAINS AND THEIR PROTON PUMPS 322

CHLOROPLASTS AND PHOTOSYNTHESIS 329

THE GENETIC SYSTEMS OF MITOCHONDRIA AND PLASTIDS 336


## THE MITOCHONDRION

TERMS TO LEARN

| | | |
|---|---|---|
| ATP synthase | electron-transport chain | outer membrane |
| chemiosmotic coupling | inner membrane | oxidative phosphorylation |
| citric acid cycle | intermembrane space | proton-motive force |
| cristae | matrix | respiratory chain |
| electrochemical proton gradient | mitochondria | |

### DEFINITIONS

Match the definition below with its term from the list above.

**14–1** General term for a series of electron carrier molecules along which electrons move from a higher to a lower energy level, and ultimately to a final acceptor molecule.

**14–2** The subcompartment formed between the inner and outer mitochondrial membranes.

**14–3** Metabolic pathway that oxidizes acetyl groups to $CO_2$.

**14–4** Electron-transport chain that receives high-energy electrons from the citric acid cycle and generates the proton gradient across the inner mitochondrial membrane that is used to power ATP synthesis.

**14–5** Enzyme in the inner membrane of a mitochondrion that catalyzes the formation of ATP from ADP and inorganic phosphate.

**14–6** Mechanism by which a pH gradient across a membrane is used to drive an energy-requiring process, such as ATP production or the rotation of bacterial flagella.

**14–7** Process in bacteria and mitochondria in which ATP formation is driven by the transfer of electrons from food molecules to molecular oxygen, with the intermediate generation of a proton gradient across a membrane.

**14–8** The result of a combined pH gradient and membrane potential.

**14–9** A sievelike membrane surrounding mitochondria that is permeable to all molecules of 5000 daltons or less.

### TRUE/FALSE

Decide whether each of these statements is true or false, and then explain why.

**14–10** Due to the many specialized transport proteins in the mitochondrial outer membrane, the intermembrane space is chemically equivalent to the cytosol with respect to small molecules.

**14–11** The most important contribution of the citric acid cycle to energy metabolism is the extraction of high-energy electrons during the oxidation of acetyl CoA to $CO_2$.

**14–12** Each respiratory enzyme complex in the electron-transport chain has a greater affinity for electrons than its predecessors, so that electrons pass sequentially from one complex to another until they are finally transferred to oxygen, which has the greatest electron affinity of all.

**14–13** If the flow of protons through ATP synthase were blocked by an inhibitor, addition of a small amount of oxygen into an anaerobic preparation of submitochondrial particles (which have their matrix surface exposed to the surrounding medium) would result in a burst of respiration that would cause the medium to become more basic.

## THOUGHT PROBLEMS

**14–14** Mitochondria in liver cells appear to move freely in the cytosol, whereas those in cardiac muscle are immobilized at positions between adjacent myofibrils. Do you suppose these differences are a trivial consequence of cell architecture or do they reflect some underlying functional advantage? Explain your answer.

**14–15** Electron micrographs show that mitochondria in heart muscle have a much higher density of cristae than mitochondria in skin cells. Why do you suppose this should be?

**14–16** In the 1860s Louis Pasteur noticed that when he added $O_2$ to a culture of yeast growing anaerobically on glucose, the rate of glucose consumption declined dramatically. Explain the basis for this result, which is known as the Pasteur effect.

**14–17** During a marathon, a runner's leg muscles are well supplied with $O_2$ and thus abstract all the energy available from the oxidation of glucose to $CO_2$. By contrast, in a sprint, which requires even more ATP hydrolysis per second, the extreme contraction of the leg muscles restricts blood flow and severely reduces the oxygen available for oxidative phosphorylation. As a result, nearly all of the sprinter's ATP comes from glycolysis alone. Glycolysis releases only about 1/15 of the energy generated by oxidation of glucose. With such a reduced energy yield from glucose, how is it possible for a sprinter to sprint?

**14–18** The citric acid cycle generates NADH and $FADH_2$, which are then used in the process of oxidative phosphorylation to make ATP. If the citric acid cycle, which does not use oxygen, and oxidative phosphorylation are separate processes, as they are, then why is it that the citric acid cycle stops almost immediately upon removal of $O_2$?

**14–19** The respiratory chain is relatively inaccessible to experimental manipulation in intact mitochondria. After disrupting mitochondria with ultrasound, it is possible to isolate functional submitochondrial particles, which consist of broken cristae that have resealed inside out into small closed vesicles. In these vesicles the components that originally faced the matrix are now exposed to the surrounding medium. How do you suppose such an arrangement might aid in the study of electron transport and ATP synthesis?

**14–20** As electrons move down the respiratory chain, protons are pumped across the inner membrane. Are those protons confined to the intermembrane space? Why or why not?

**14–21** You have reconstituted into the same membrane vesicles purified bacteriorhodopsin, which is a light-driven $H^+$ pump from a photosynthetic bacterium, and purified ATP synthase from ox heart mitochondria. Assume that

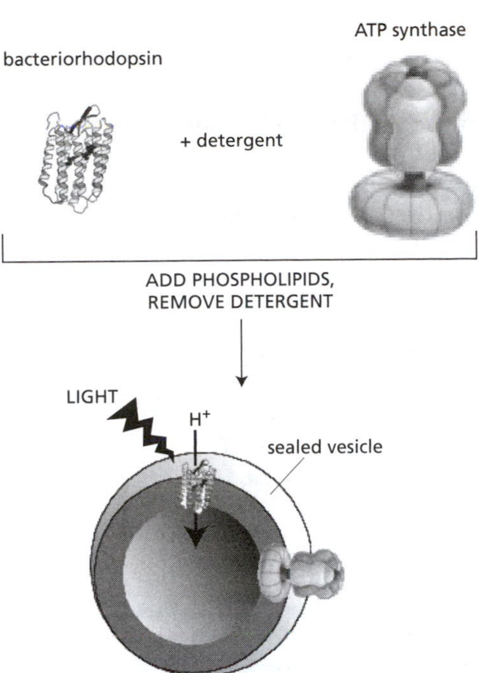


**Figure 14–1** Reconstitution of bacteriorhodopsin and ATP synthase into lipid vesicles (Problem 14–21).

all molecules of bacteriorhodopsin and ATP synthase are oriented as shown in Figure 14–1, so that protons are pumped into the vesicle and ATP synthesis occurs on the outer surface.

A. If you add ADP and phosphate to the external medium and shine light into the suspension of vesicles, would you expect ATP to be synthesized? Why or why not?
B. If you prepared the vesicles without being careful to remove all the detergent, which makes the bilayer leaky to protons, would you expect ATP to be synthesized?
C. If the ATP synthase molecules were randomly oriented so that about half faced the outside of the vesicle and half faced the inside, would you expect ATP to be synthesized? If the bacteriorhodopsin molecules were randomly oriented, would you expect ATP to be synthesized? Explain your answers.
D. You tell a friend over dinner about your new experiments. He questions the validity of an approach that utilizes components from so widely divergent, unrelated organisms. As he so succinctly puts it, "Why would anybody want to mix vanilla pudding with brake fluid?" Defend your approach against his criticism.

**14–22** When dinitrophenol (DNP) is added to mitochondria, the inner membrane becomes permeable to protons. When the drug valinomycin is added to mitochondria, the inner membrane becomes permeable to $K^+$.

A. How will the electrochemical proton gradient change in response to DNP?
B. How will it change in response to valinomycin?

**14–23** An elongated arm—the stator—links the catalytic head group (the $\alpha_3\beta_3$ complex) of the ATP synthase to the membrane-embedded rotor component. Attached to the rotor is a stalk (the axle-like $\gamma$ subunit) that turns inside the head group to force the conformational changes that lead to ATP synthesis. If the stator were missing, would ATP be synthesized in response to the proton flow? Why or why not?

**14–24** A variety of coupled transport processes that occur across the inner mitochondrial membrane are illustrated in Figure 14–2. For each, decide whether transport is with the electrochemical proton gradient, against it, or unaffected by it. For those transport processes that are affected by the gradient, identify which component of the gradient (membrane potential or ΔpH) affects transport.

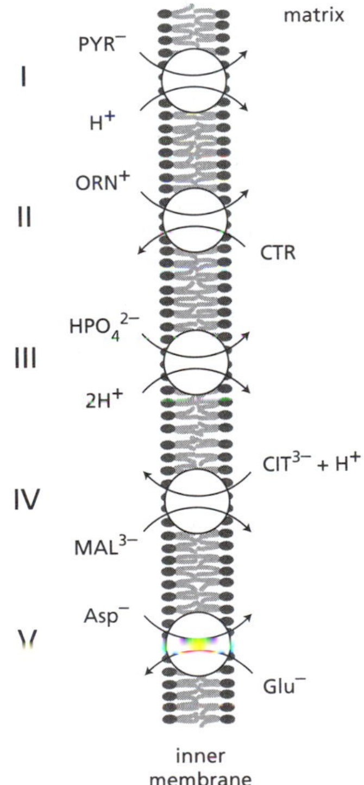


**Figure 14–2** Five coupled transport processes that occur across the inner mitochondrial membrane (Problem 14–24). PYR is pyruvate; ORN is ornithine; CTR is citrulline; CIT is citrate; MAL is malate; Asp is aspartic acid; and Glu is glutamic acid.

## CALCULATIONS

**14–25** In actively respiring liver mitochondria, the pH inside the matrix is about half a pH unit higher than that in the cytosol. Assuming that the cytosol is at pH 7 and the matrix is a sphere with a diameter of 1 μm [$V = (4/3)\ \pi r^3$], calculate the total number of protons in the matrix of a respiring liver mitochondrion. If the matrix began at pH 7 (equal to that in the cytosol), how many protons would have to be pumped out to establish a matrix pH of 7.5 (a difference of 0.5 pH unit)?

**14–26** *Thiobacillus ferrooxidans* is a bacterium that lives on slag heaps at pH 2. It is an important organism in the mining industry, as we will cover in Problems 14–57 and 14–88. Is it possible for *T. ferrooxidans* to make ATP for free using the natural pH gradient supplied by the environment? The $\Delta G$ available from the transport of protons from the outside (pH 2) to the inside (pH 6.5) of *T. ferrooxidans* is given by the Nernst equation

$$\Delta G = 2.3\ RT \log \frac{[H^+]_{in}}{[H^+]_{out}} + nFV$$

where $R = 1.98 \times 10^{-3}$ kcal/K mole, $T$ = temperature in kelvins, $n$ = the number of electrons transferred, $F$ = 23 kcal/V mole, and $V$ = the membrane potential.

In *T. ferrooxidans* grown at pH 2, the membrane potential is zero. Assuming that $T$ = 310 K and that $\Delta G$ = 11 kcal/mole for ATP synthesis under the prevailing intracellular conditions, how many protons (to the nearest integer) would have to enter the cell through the ATP synthase to drive ATP synthesis? In order for proton transport to be coupled to ATP synthesis, could the protons pass through the ATP synthase one at a time, or would they all have to pass through together?

**14–27** Heart muscle gets most of the ATP needed to power its continual contractions through oxidative phosphorylation. When oxidizing glucose to $CO_2$, heart muscle consumes $O_2$ at a rate of 10 μmol/min per g of tissue, in order to replace the ATP used in contraction and give a steady-state ATP concentration of 5 μmol/g of tissue. At this rate, how many seconds would it take the heart to consume an amount of ATP equal to its steady-state levels? (Complete oxidation of one molecule of glucose to $CO_2$ yields 30 ATP, 26 of which are derived by oxidative phosphorylation using the 12 pairs of electrons captured in the electron carriers NADH and $FADH_2$.)

**14–28** The relationship of free-energy change ($\Delta G$) to the concentrations of reactants and products is important because it predicts the direction of spontaneous chemical reactions. Familiarity with this relationship is essential for understanding energy conversions in cells. Consider, for example, the hydrolysis of ATP to ADP and inorganic phosphate ($P_i$):

$$ATP + H_2O \rightarrow ADP + P_i$$

The free-energy change due to ATP hydrolysis is

$$\Delta G = \Delta G^\circ + RT\ \ln \frac{[ADP][P_i]}{[ATP]}$$
$$= \Delta G^\circ + 2.3\ RT \log \frac{[ADP][P_i]}{[ATP]}$$

where the concentrations are expressed as molarities (by convention, the concentration of water is not included in the expression). $R$ is the gas constant ($1.98 \times 10^{-3}$ kcal/K mole), $T$ is temperature (assume 37°C, which is 310 K), and $\Delta G^\circ$ is the standard free-energy change (–7.3 kcal/mole for ATP hydrolysis to ADP and $P_i$).

A. Calculate $\Delta G$ for ATP hydrolysis when the concentrations of ATP, ADP, and $P_i$ are all equal to 1 M. What is $\Delta G$ when the concentrations of ATP, ADP, and $P_i$ are all equal to 1 mM?

B. In a resting muscle, the concentrations of ATP, ADP, and $P_i$ are approximately 5 mM, 1 mM, and 10 mM, respectively. What is $\Delta G$ for ATP hydrolysis in resting muscle?

C. What will $\Delta G$ equal when the hydrolysis reaction reaches equilibrium? At $[P_i]$ = 10 mM, what will be the ratio of [ATP] to [ADP] at equilibrium?

D. Show that, at constant $[P_i]$, $\Delta G$ decreases by 1.4 kcal/mole for every 10-fold increase in the ratio of [ATP] to [ADP], regardless of the value of $\Delta G°$. (For example, $\Delta G$ decreases by 2.8 kcal/mole for a 100-fold increase, by 4.2 kcal/mole for a 1000-fold increase, and so on.)

**14–29** Suspensions of isolated mitochondria will synthesize ATP until the $[ATP]/[ADP][P_i]$ ratio is about $10^4$. If each proton that moves into the matrix down an electrochemical proton gradient of 200 mV liberates 4.6 kcal/mole of free energy ($\Delta G = -4.6$ kcal/mole), what is the minimum number of protons that will be required to synthesize ATP ($\Delta G° = 7.3$ kcal/mole) at 37°C?

## DATA HANDLING

**14–30** It was difficult to define the molecular mechanism by which the electrochemical proton gradient is coupled to ATP synthesis. A fundamental question at the outset was whether ATP was synthesized directly from ADP and inorganic phosphate or by transfer of a phosphate from an intermediate source such as a phosphoenzyme or some other phosphorylated molecule.

One elegant approach to this question analyzed the stereochemistry of the reaction mechanism. As is often done, the investigators studied the reverse reaction (ATP hydrolysis into ADP and phosphate) to gain an understanding of the forward reaction. (A basic principle of enzyme catalysis is that the forward and reverse reactions are precisely the opposite of one another.) All enzyme-catalyzed phosphate transfers occur with inversion of configuration about the phosphate atom; thus one-step mechanisms, in which the phosphate is transferred directly between substrates, result in inversion of the final product (Figure 14–3A).

To analyze the stereochemistry of ATP hydrolysis, the investigators first generated a version of ATP with three distinct atoms (S, $^{16}O$, and $^{18}O$) attached stereospecifically to the terminal phosphorus atom (Figure 14–3B). They then hydrolyzed this compound to ADP and inorganic phosphate using purified ATP synthase in the presence of $H_2O$ that was enriched for $^{17}O$. Using NMR to analyze the resulting inorganic phosphate, they could determine whether the configuration about the phosphorus atom had been inverted or retained (Figure 14–3B).

A. How does this experiment distinguish between synthesis of ATP directly from ADP and inorganic phosphate and synthesis of ATP through a phosphorylated intermediate?

B. Their analysis showed that the configuration had been inverted. Does this result support direct synthesis of ATP or synthesis of ATP through a phosphorylated intermediate?

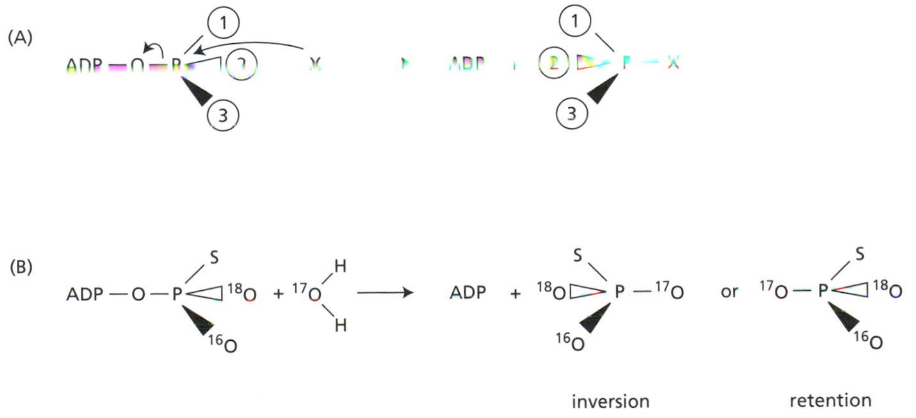

**Figure 14–3** Stereochemistry of phosphate transfer reactions (Problem 14–30). (A) Inversion of configuration by a one-step phosphate transfer reaction. (B) Experimental setup for assaying stereochemistry of ATP synthesis by ATP synthase. *Thin* bonds are in the same plane as the page; *thick white* bonds point behind the page; and *thick black* bonds project out of the page. Oxygen atoms are indicated by their atomic number.

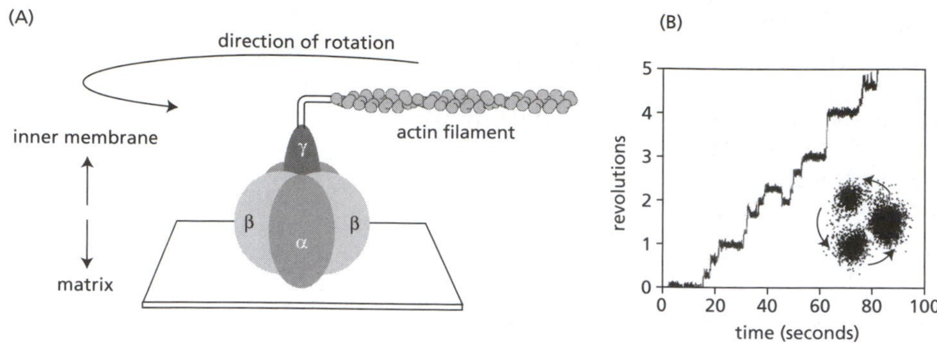


**Figure 14–4** Experimental set-up for observing rotation of the γ subunit of ATP synthase (Problem 14–31). (A) The immobilized $\alpha_3\beta_3\gamma$ complex. The β subunits are anchored and a fluorescent actin filament is attached to the γ subunit. (B) Stepwise revolution of the actin filament. The indicated trace is a typical example from one experiment. The inset shows the positions in the revolution at which the actin filament pauses.

**14–31** ATP synthase is the world's smallest rotary motor. Passage of $H^+$ ions through the membrane-embedded portion of ATP synthase (the $F_0$ component) causes rotation of the single, central axle-like γ subunit inside the head group. The tripartite head is composed of the three αβ dimers, the β subunit of which is responsible for synthesis of ATP. The rotation of the γ subunit induces conformational changes in the αβ dimers that allow ADP and $P_i$ to be converted into ATP. A variety of indirect evidence had suggested rotary catalysis by ATP synthase, but seeing is believing.

To demonstrate rotary motion, a modified form of the $\alpha_3\beta_3\gamma$ complex was used. The β subunits were modified (on their inner membrane-distal ends) so they could be firmly anchored to a solid support. The γ subunit was modified (on the end that normally inserts into the $F_0$ component in the inner membrane) so that a fluorescently tagged, readily visible filament of actin could be attached (Figure 14–4A). This arrangement allows rotations of the γ subunit to be visualized as revolutions of the long actin filament. In these experiments ATP synthase was studied in the reverse of its normal mechanism by allowing it to hydrolyze ATP. At low ATP concentrations the actin filament was observed to revolve in steps of 120° and then pause for variable lengths of time, as shown in Figure 14–4B.

A. Why does the actin filament revolve in steps with pauses in between? What does this rotation correspond to in terms of the structure of the $\alpha_3\beta_3\gamma$ complex?
B. In its normal mode of operation inside the cell, how many ATP molecules do you suppose would be synthesized for each complete 360° rotation of the γ subunit? Explain your answer.

**14–32** The three αβ dimers in each ATP synthase normally exist in three different conformations: one empty, one with ADP and $P_i$ bound, and one with ATP bound. The conformational changes are driven sequentially by rotation of the γ subunit, which in turn is driven by the flow of protons through the ATP synthase. As viewed from the inner membrane, looking at the underside of the $\alpha_3\beta_3$ complex, these sites could be arranged in either of two ways around the central γ subunit (Figure 14–5). The sequential, linked conformational

**Figure 14–5** The two possible arrangements of conformations of the three αβ dimers in ATP synthase, along with the linked conformational changes driven by proton flow (Problem 14–32). One αβ dimer is *shaded* to emphasize that its position remains fixed as it changes conformation in response to proton flow. The perspective illustrated is from the *inner* membrane, looking at the *underside* of the tripartite $\alpha_3\beta_3$ complex.

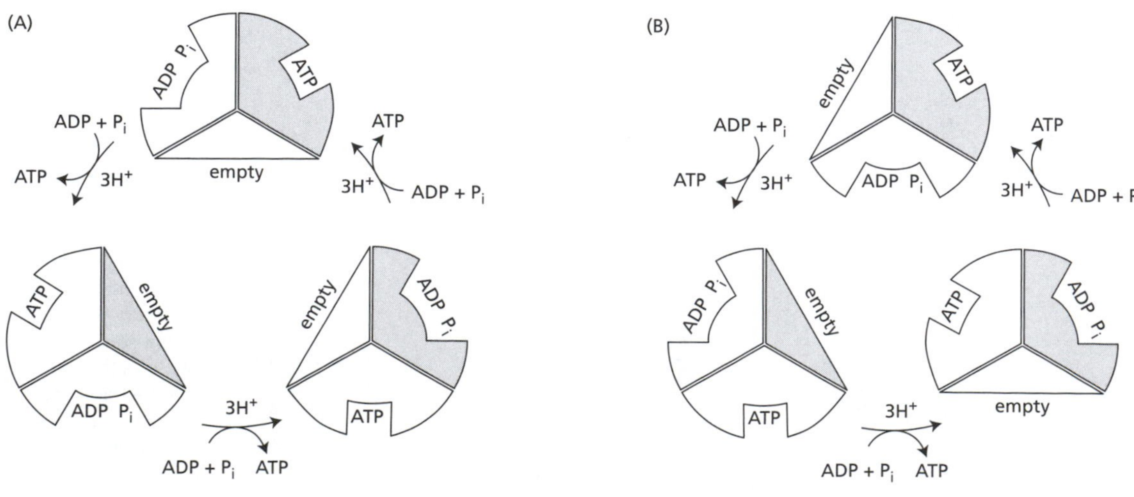

changes driven by proton flow are also shown for the two arrangements in the figure. In Problem 14–31 the revolutions of the attached actin filaments during ATP hydrolysis were shown to be counterclockwise when viewed from the same perspective (see Figure 14–4). Which of the two arrangements of conformations of αβ dimers shown in Figure 14–5 is correct? Explain your answer.

**14–33** A manuscript has been submitted to a prestigious scientific journal. In it the authors describe an experiment using an immobilized $\alpha_3\beta_3\gamma$ complex with an attached actin filament like that shown in Figure 14–4A. The authors show that they can mechanically rotate the γ subunit by applying force to the actin filament. Moreover, in the presence of ADP and phosphate each 120° clockwise rotation of the γ subunit is accompanied by the synthesis of one molecule of ATP. Is this result at all reasonable? What would such an observation imply about the mechanism of ATP synthase? Should this manuscript be considered for publication in one of the best journals?

**14–34** The ADP–ATP antiporter in the mitochondrial inner membrane can exchange ATP for ATP, ADP for ADP, and ATP for ADP. Even though mitochondria can transport both ADP and ATP, there is a strong bias in favor of exchange of external ADP for internal ATP in actively respiring mitochondria. You suspect that this bias is due to the conversion of ADP into ATP inside the mitochondrion. ATP synthesis would continually reduce the internal concentration of ADP and thereby create a favorable concentration gradient for import of ADP. The same process would increase the internal concentration of ATP, thereby creating favorable conditions for export of ATP.

To test your hypothesis, you conduct experiments on isolated mitochondria. In the absence of substrate (when the mitochondria are not respiring and the membrane is uncharged), you find that ADP and ATP are taken up at the same rate. When you add substrate, the mitochondria begin to respire, and ADP enters mitochondria at a much faster rate than ATP. As you expected, when you add an uncoupler (dinitrophenol, which collapses the pH gradient) along with the substrate, ADP and ATP again enter at the same rate. When you add an inhibitor of ATP synthase (oligomycin) along with the substrate, ADP is taken up much faster than ATP. Your results are summarized in Table 14–1. You are puzzled by the results with oligomycin, since your hypothesis predicted that the rates of uptake would be equal.

When you show the results to your advisor, she compliments you on your fine experiments and agrees that they disprove the hypothesis. She suggests that you examine the structures of ATP and ADP (Figure 14–6) if you wish to understand the behavior of the antiporter. What is the correct explanation for the biased exchange by the ADP–ATP antiporter under some of the experimental conditions and an unbiased exchange under others?

**14–35** How many molecules of ATP are formed from ADP + $P_i$ when a pair of electrons from NADH is passed down the electron-transport chain to oxygen? Is the number an integer or not? These deceptively simple questions are difficult to answer from purely theoretical considerations, but they can be measured directly with an oxygen electrode, as illustrated in Figure 14–7. At the indicated time, a suspension of mitochondria was added to a phosphate-

**Table 14–1 Entry of ADP and ATP into isolated mitochondria** (Problem 14–34).

| EXPERIMENT | SUBSTRATE | INHIBITOR | RELATIVE RATES OF ENTRY |
|---|---|---|---|
| 1 | absent | none | ADP = ATP |
| 2 | present | none | ADP > ATP |
| 3 | present | dinitrophenol | ADP = ATP |
| 4 | present | oligomycin | ADP > ATP |

In all cases the initial rates of entry of ATP and ADP were measured.

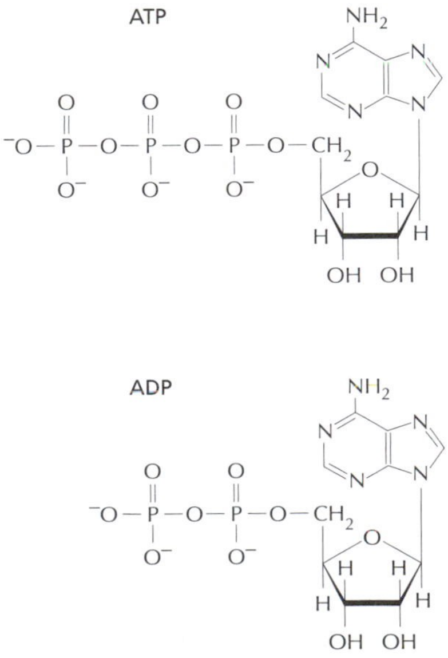

**Figure 14–6** Structures of ATP and ADP (Problem 14–34).

buffered solution containing β-hydroxybutyrate, which can be oxidized by mitochondria to generate NADH + $H^+$. After an initial burst in oxygen consumption, the rate of decline in oxygen concentration slowed to the background rate. When the rate of oxygen consumption stabilized, 500 nmol of ADP were added, causing a rapid increase in the rate of consumption until all the ADP had been converted into ATP, at which point the rate of oxygen consumption again slowed to the background rate.

A. Why did the rate of oxygen consumption increase dramatically over the background rate when ADP was added? Why did oxygen consumption return to the background rate when all the ADP had been converted to ATP?

B. Why do you think it is that mitochondria consume oxygen at a slow background rate in the absence of added ADP?

C. How many ATP molecules were synthesized per pair of electrons transferred down the electron-transport chain to oxygen ($P/2e^-$ ratio)? How many ATP molecules were synthesized per oxygen atom consumed (P/O ratio)? (Remember that $\frac{1}{2}O_2 + 2e^- \rightarrow H_2O$.)

D. In experiments like this, what processes in addition to ATP production are driven by the electrochemical proton gradient?

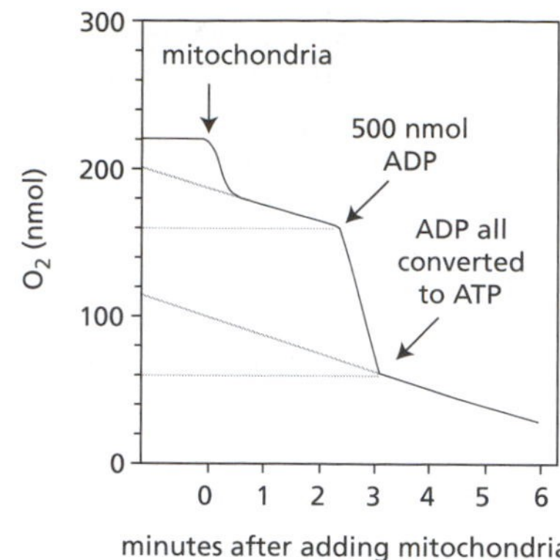


**Figure 14–7** Consumption of oxygen by mitochondria under various experimental conditions (Problem 14–35). The rates of oxygen consumption at various times during the experiment are shown by downward sloping lines, with faster rates shown by steeper lines.

# ELECTRON-TRANSPORT CHAINS AND THEIR PROTON PUMPS

TERMS TO LEARN

| | | |
|---|---|---|
| cytochrome | NADH dehydrogenase complex | redox reaction |
| cytochrome *b*-$c_1$ complex | quinone (Q) | respiratory control |
| cytochrome oxidase complex | redox pairs | respiratory enzyme complex |
| iron–sulfur center | redox potential | |

## DEFINITIONS

Match the definition below with its term from the list above.

**14–36** An electron-driven proton pump in the respiratory chain that accepts electrons from cytochrome *c* and generates water using molecular oxygen as an electron acceptor.

**14–37** Regulatory mechanism that is thought to control the rate of electron transport in the respiratory chain according to need via a direct influence of the electrochemical proton gradient.

**14–38** Colored, heme-containing protein that transfers electrons during cellular respiration.

**14–39** Electron-transporting group consisting of either two or four iron atoms bound to an equal number of sulfur atoms.

**14–40** A reaction in which one component becomes oxidized and the other reduced.

**14–41** Any of the major protein complexes of the mitochondrial respiratory chain that act as electron-driven proton pumps to generate the proton gradient across the inner membrane.

**14–42** The affinity of a redox pair for electrons, generally measured as the voltage difference between an equimolar mixture of the pair and a standard reference.

## TRUE/FALSE

Decide whether each of these statements is true or false, and then explain why.

**14–43** Most cytochromes have a higher redox potential (higher affinity for electrons) than iron–sulfur centers, which is why the cytochromes tend to serve as electron carriers near the $O_2$ end of the respiratory chain.

**14–44** The three respiratory enzyme complexes in the mitochondrial inner membrane exist in structurally ordered arrays that facilitate the correct transfer of electrons between appropriate complexes.

**14–45** Lipophilic weak acids short-circuit the normal flow of protons across the inner membrane, thereby eliminating the proton-motive force, stopping ATP synthesis, and blocking the flow of electrons.

## THOUGHT PROBLEMS

**14–46** Both $H^+$ and $Ca^{2+}$ are ions that move through the cytosol. Why is the movement of $H^+$ ions so much faster than that of $Ca^{2+}$ ions? How do you suppose the speed of these two ions would be affected by freezing the solution? Would you expect them to move faster or slower? Explain your answer.

**14–47** Distinguish between a hydrogen atom, a proton, a hydride ion, and a hydrogen molecule.

**14–48** The half reactions for some of the carriers in the respiratory chain are given in Table 14–2. From their $E_0'$ values what would you guess is their order in the chain? What would you need to know before you were more certain of their order?

**14–49** You mix components of the respiratory chain in a solution as indicated below. Assuming that the electrons must follow the standard path through the electron-transport chain, in which experiments would you expect a net transfer of electrons to cytochrome *c*? Discuss why no electron transfer occurs in the other experiments.

A. Reduced ubiquinone and oxidized cytochrome *c*
B. Oxidized ubiquinone and oxidized cytochrome *c*
C. Reduced ubiquinone and reduced cytochrome *c*
D. Oxidized ubiquinone and reduced cytochrome *c*
E. Reduced ubiquinone, oxidized cytochrome *c*, and cytochrome $b$-$c_1$ complex
F. Oxidized ubiquinone, oxidized cytochrome *c*, and cytochrome $b$-$c_1$ complex
G. Reduced ubiquinone, reduced cytochrome *c*, and cytochrome $b$-$c_1$ complex
H. Oxidized ubiquinone, reduced cytochrome *c*, and cytochrome $b$-$c_1$ complex

**14–50** The cytochrome oxidase complex is strongly inhibited by cyanide, which binds to the $Fe^{3+}$ form of cytochrome $a_3$. Carbon monoxide (CO) also inhibits the cytochrome oxidase complex by binding to cytochrome $a_3$, but it binds to the $Fe^{2+}$ form. Cyanide kills at very low concentration because of its effects on cytochrome oxidase. By contrast, much larger amounts of CO must be inhaled to cause death, and its toxicity is due to its binding to the heme group of hemoglobin, which also carries an $Fe^{2+}$. In a sense, by sopping up CO hemoglobin protects cytochrome oxidase from being inhibited by CO. One treatment for cyanide poisoning—if administered quickly enough—is to give sodium nitrite, which oxidizes $Fe^{2+}$ to $Fe^{3+}$. How do you suppose sodium nitrite protects against the effects of cyanide?

**Table 14–2 Standard redox potentials for electron carriers in the respiratory chain** (Problem 14–48).

| HALF REACTION | $E_0'$ (V) |
|---|---|
| ubiquinone + 2 $H^+$ + 2 $e^-$ → ubiquinol | 0.045 |
| cytochrome *b* ($Fe^{3+}$) + $e^-$ → cytochrome *b* ($Fe^{2+}$) | 0.077 |
| cytochrome $c_1$ ($Fe^{3+}$) + $e^-$ → cytochrome $c_1$ ($Fe^{2+}$) | 0.22 |
| cytochrome *c* ($Fe^{3+}$) + $e^-$ → cytochrome *c* ($Fe^{2+}$) | 0.25 |
| cytochrome *a* ($Fe^{3+}$) + $e^-$ → cytochrome *a* ($Fe^{2+}$) | 0.29 |
| cytochrome $a_3$ ($Fe^{3+}$) + $e^-$ → cytochrome $a_3$ ($Fe^{2+}$) | 0.55 |

**14–51** The two different diffusible electron carriers, ubiquinone and cytochrome *c*, shuttle electrons between the three protein complexes of the electron-transport chain. In principle, could the same diffusible carrier be used for both steps? If not, why not? If it could, what characteristics would it need to possess and what would be the disadvantages of such a situation?

**14–52** The uncoupler dinitrophenol was once prescribed as a diet drug to aid in weight loss. How would an uncoupler of oxidative phosphorylation promote weight loss? Why do you suppose that it is no longer prescribed?

**14–53** If you were to impose an artificially large electrochemical proton gradient across the mitochondrial inner membrane, would you expect electrons to flow up the respiratory chain, in the reverse of their normal direction? Why or why not?

**14–54** Some bacteria have become specialized to live in an alkaline environment at pH 10. They maintain their internal environment at pH 7. Why is it that they cannot exploit the pH difference across their membrane to get ATP for free using a standard ATP synthase? Can you suggest an engineering modification to ATP synthase that would allow it to generate ATP from proton flow in such an environment?

## CALCULATIONS

**14–55** One of the problems in understanding redox reactions is coming to grips with the language. Consider the reduction of pyruvate by NADH:

$$\text{pyruvate} + \text{NADH} + \text{H}^+ \rightleftharpoons \text{lactate} + \text{NAD}^+$$

In redox reactions, oxidation and reduction necessarily occur together; however, it is convenient to list the two halves of a redox reaction separately. By convention, each half reaction is written as a reduction: oxidant + $e^-$ → reductant. For the reduction of pyruvate by NADH the half reactions are

$$\text{pyruvate} + 2\,\text{H}^+ + 2e^- \rightarrow \text{lactate} \qquad E_0' = -0.19\text{ V}$$
$$\text{NAD}^+ + \text{H}^+ + 2e^- \rightarrow \text{NADH} \qquad E_0' = -0.32\text{ V}$$

where $E_0'$ is the standard redox potential and refers to a reaction occurring under standard conditions (25°C or 298 K, all concentrations at 1 M, and pH 7). To obtain the overall equation for reduction of pyruvate by NADH, it is necessary to reverse the NAD$^+$/NADH half reaction and change the sign of $E_0'$:

$$\text{pyruvate} + 2\,\text{H}^+ + 2e^- \rightarrow \text{lactate} \qquad E_0' = -0.19\text{ V}$$
$$\text{NADH} \rightarrow \text{NAD}^+ + \text{H}^+ + 2e^- \qquad E_0' = +0.32\text{ V}$$

Summing these two half reactions and their $E_0'$ values gives the overall equation and its $\Delta E_0'$ value:

$$\text{pyruvate} + \text{NADH} + \text{H}^+ \rightleftharpoons \text{lactate} + \text{NAD}^+ \qquad \Delta E_0' = +0.13\text{ V}$$

When a redox reaction takes place under nonstandard conditions, the tendency to donate electrons ($\Delta E$) is equal to $\Delta E_0'$ modified by a concentration term:

$$\Delta E = \Delta E_0' - \frac{2.3\,RT}{nF}\log\frac{[\text{lactate}][\text{NAD}^+]}{[\text{pyruvate}][\text{NADH}]}$$

where $R = 1.98 \times 10^{-3}$ kcal/K mole, $T$ = temperature in kelvins, $n$ = the number of electrons transferred, and $F$ = 23 kcal/V mole.

$\Delta G$ is related to $\Delta E$ by the equation

$$\Delta G = -nF\Delta E$$

Since the signs of $\Delta G$ and $\Delta E$ are opposite, a favorable redox reaction has a positive $\Delta E$ and a negative $\Delta G$.

A. Calculate $\Delta G$ for reduction of pyruvate to lactate at 37°C with all reactants and products at a concentration of 1 M.
B. Calculate $\Delta G$ for the reaction at 37°C under conditions where the concentrations of pyruvate and lactate are equal and the concentrations of $NAD^+$ and NADH are equal.
C. What would the concentration term need to be for this reaction to have a $\Delta G$ of zero at 37°C?
D. Under normal conditions in vascular smooth muscle (at 37°C), the concentration ratio of $NAD^+$ to NADH is 1000, the concentration of lactate is 0.77 μmol/g, and the concentration of pyruvate is 0.15 μmol/g. What is $\Delta G$ for reduction of pyruvate to lactate under these conditions?

**14–56** All the redox reactions in the citric acid cycle use $NAD^+$ to accept electrons except for the oxidation of succinate to fumarate, where FAD is used.

$$\text{succinate} + \text{FAD} \rightleftharpoons \text{fumarate} + \text{FADH}_2$$

Is FAD a more appropriate electron acceptor for this reaction than $NAD^+$? The relevant half reactions are given below. The half reaction for FAD/$FADH_2$ is for the cofactor as it exists covalently attached to the enzyme that catalyzes the reaction.

$$\text{fumarate} + 2\,H^+ + 2e^- \rightarrow \text{succinate} \qquad E_0' = 0.03\ \text{V}$$
$$\text{FAD} + 2\,H^+ + 2e^- \rightarrow \text{FADH}_2 \qquad E_0' = 0.05\ \text{V}$$
$$\text{NAD}^+ + H^+ + 2e^- \rightarrow \text{NADH} \qquad E_0' = -0.32\ \text{V}$$

A. From these $E_0'$ values decide which, if either, cofactor (FAD or $NAD^+$) would yield a negative $\Delta G°$ value for the reaction?
B. For the FAD reaction, what is the concentration ratio of products to reactants that would give a $\Delta G$ value of zero at 37°C? For the $NAD^+$ reaction, what concentration ratio would give a $\Delta G$ value of zero?
C. Is FAD a more appropriate electron acceptor for this reaction than $NAD^+$? Explain your answer.

**14–57** *Thiobacillus ferrooxidans*, the bacterium that lives on slag heaps at pH 2, is used by the mining industry to recover copper and uranium from low-grade ore by an acid leaching process. The bacteria oxidize $Fe^{2+}$ to produce $Fe^{3+}$, which in turn oxidizes (and solubilizes) these minor components of the ore. It is remarkable that the bacterium can live in such an environment. It does so by exploiting the pH difference between the environment and its cytoplasm (pH 6.5) to drive synthesis of ATP (see Problem 14–26) and NADPH, which it can then use to fix $CO_2$ and nitrogen. In order to keep its cytoplasmic pH constant, *T. ferrooxidans* uses electrons from $Fe^{2+}$ to reduce $O_2$ to water, thereby removing the protons:

$$4\,Fe^{2+} + O_2 + 4\,H^+ \rightarrow 4\,Fe^{3+} + 2\,H_2O$$

What are the energetics of these processes? Is the flow of electrons from $Fe^{2+}$ to $O_2$ energetically favorable? How difficult is it to reduce $NADP^+$ using electrons from $Fe^{2+}$? These are key questions for understanding how *T. ferrooxidans* manages to thrive in such an unlikely niche.

A. What are $\Delta E$ and $\Delta G$ for the reduction of $O_2$ by $Fe^{2+}$, assuming that the reaction occurs under standard conditions? The half reactions are

$$Fe^{3+} + e^- \rightarrow Fe^{2+} \qquad E_0' = 0.77\ \text{V}$$
$$O_2 + 4\,H^+ + 4e^- \rightarrow 2\,H_2O \qquad E_0' = 0.82\ \text{V}$$

B. Write a balanced equation for the reduction of $NADP^+ + H^+$ by $Fe^{2+}$. What is $\Delta G$ for this reaction under standard conditions? The half reaction for $NADP^+$ is

$NADP^+ + H^+ + 2e^- \rightarrow NADPH \qquad E_0' = -0.32\ V$

What is $\Delta G$ for the reduction of $NADP^+ + H^+$ by $Fe^{2+}$, if the concentrations of $Fe^{3+}$ and $Fe^{2+}$ are equal, the concentration of NADPH is10-fold greater than that of $NADP^+$, and the temperature is 310 K? (Note: adjusting the number of atoms and electrons in order to balance the chemical equation does not affect $E_0'$ or $\Delta E_0'$ values. It does, however, affect $\Delta E$ by its influence on the concentration term: each concentration term must be raised to an exponent equal to the number of atoms or molecules used in the balanced equation.)

**14–58** What is the standard free energy change ($\Delta G°$) associated with transfer of electrons from NADH to $O_2$, according to the balanced equation below?

$$O_2 + 2\ NADH + 2\ H^+ \rightleftharpoons 2\ H_2O + 2\ NAD^+$$

The half reactions are

$O_2 + 4\ H^+ + 4e^- \rightarrow 2\ H_2O \qquad E_0' = 0.82\ V$
$NAD^+ + H^+ + 2e^- \rightarrow NADH \qquad E_0' = -0.32\ V$

A common way of writing this equation is

$$\tfrac{1}{2}\ O_2 + NADH + H^+ \rightleftharpoons H_2O + NAD^+$$

What is $\Delta G°$ for this equation? Do the two calculations give the same answer? Explain why they do or don't.

## DATA HANDLING

**14–59** In 1925 David Keilin used a simple spectroscope to observe the characteristic absorption bands of the cytochromes that make up the electron-transport chain in mitochondria. A spectroscope passes a very bright light through the sample of interest and then through a prism to display the spectrum from red to blue. If molecules in the sample absorb light of particular wavelengths, dark bands will interrupt the colors of the rainbow. Keilin found that tissues from a wide variety of animals all showed the pattern in Figure 14–8. (This pattern had actually been observed several decades before by an Irish physician named MacMunn, but he thought all the bands were due to a single pigment. His work was all but forgotten by the 1920s.)

The different heat stabilities of the individual absorption bands and their different intensities in different tissues led Keilin to conclude that the absorption pattern was due to three components, which he labeled cytochromes *a*, *b*, and *c* (Figure 14–8). His key discovery was that the absorption bands disappeared when oxygen was introduced (Figure 14–9A) and then reappeared when the samples became anaerobic (Figure 14–9B). He later confessed, "This visual perception of an intracellular respiratory process was one of the most impressive spectacles I have witnessed in the course of my work."

Keilin subsequently discovered that cyanide prevented the bands from disappearing when oxygen was introduced (Figure 14–9C). When urethane (an inhibitor of electron transport that is no longer used) was added, bands

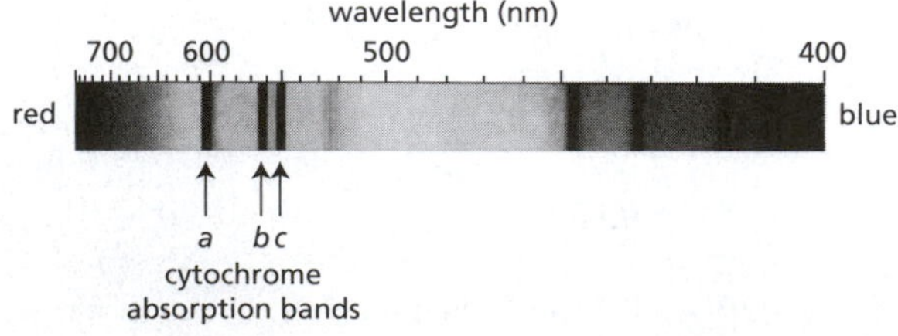


**Figure 14–8** Cytochrome absorption bands (Problem 14–59).

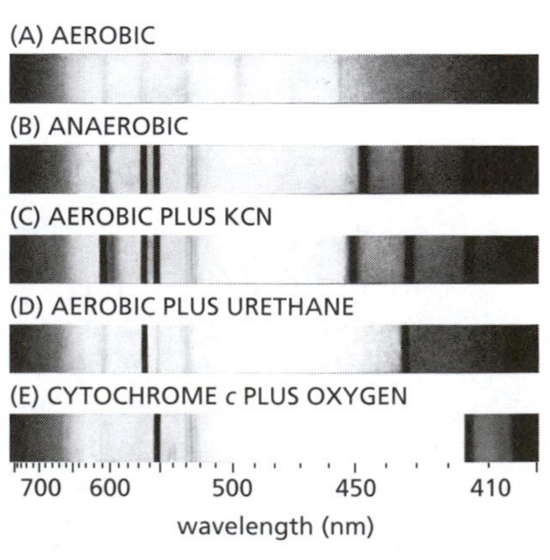


**Figure 14–9** Cytochrome absorption bands under a variety of experimental conditions (Problem 14–59).

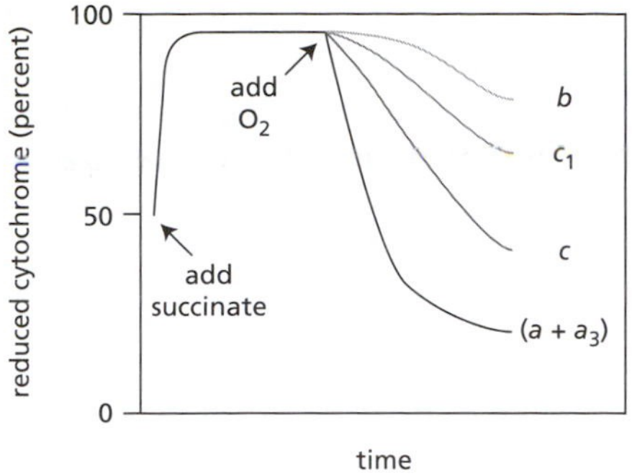


Figure 14–10 Rapid spectrophotometric analysis of the rates of oxidation of electron carriers in the respiratory chain (Problem 14–60). Cytochromes *a* and $a_3$ cannot be distinguished and thus are listed as cytochrome $(a + a_3)$.

*a* and *c* disappeared in the presence of oxygen, but band *b* remained (Figure 14–9D). Finally, using cytochrome *c* extracted from dried yeast, he showed that the band due to cytochrome *c* remained when oxygen was present (Figure 14–9E).

A. Is it the reduced (electron-rich) or the oxidized (electron-poor) forms of the cytochromes that give rise to the bands that Keilin observed?

B. From Keilin's observations, deduce the order in which the three cytochromes carry electrons from intracellular substrates to oxygen.

C. One of Keilin's early observations was that the presence of excess glucose prevented the disappearance of the absorption bands when oxygen was added. How do you think that rapid glucose oxidation to $CO_2$ might explain this observation?

**14–60** If isolated mitochondria are incubated with a source of electrons such as succinate, but without oxygen, electrons enter the respiratory chain, reducing each of the electron carriers almost completely. When oxygen is then introduced, the carriers become oxidized at different rates (Figure 14–10). How does this result allow you to order the electron carriers in the respiratory chain? What is their order?

**14–61** Inhibitors have provided extremely useful tools for analyzing mitochondrial function. Figure 14–11 shows three distinct patterns of oxygen electrode traces obtained using a variety of inhibitors. In all experiments mitochondria were added to a phosphate-buffered solution containing succinate as the sole source of electrons for the respiratory chain. After a short interval ADP was added followed by an inhibitor, as indicated in Figure 14–11. The rates of oxygen consumption at various times during the experiment are shown by downward sloping lines, with faster rates shown by steeper lines (see Figure 14–7).

A. Based on the descriptions of the inhibitors in Table 14–3, assign each inhibitor to one of the oxygen traces in Figure 14–11. All these inhibitors stop ATP synthesis.

B. Using the same experimental protocol indicated in Figure 14–11, sketch the oxygen traces that you would expect for the sequential addition of the pairs of inhibitors in the list below.

1. FCCP followed by cyanide
2. FCCP followed by oligomycin
3. Oligomycin followed by FCCP

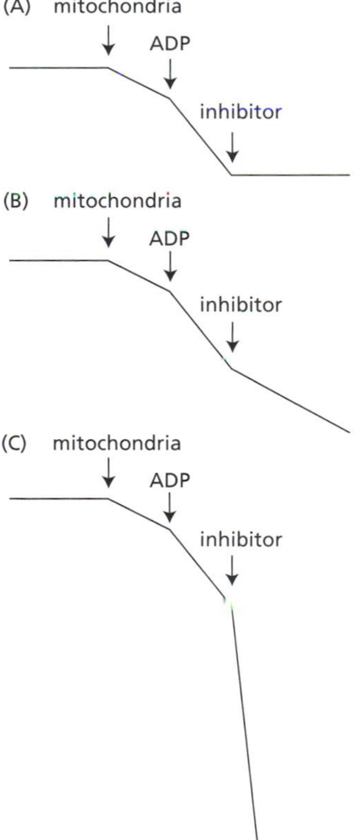


Figure 14–11 Oxygen traces showing three patterns of inhibitor effects on oxygen consumption by mitochondria (Problem 14–61).

**Table 14–3 Effects of a variety of inhibitors of mitochondrial function** (Problem 14–61).

| INHIBITOR | FUNCTION |
|---|---|
| 1. FCCP | makes membranes permeable to protons |
| 2. Malonate | prevents oxidation of succinate |
| 3. Cyanide | inhibits cytochrome oxidase |
| 4. Atractyloside | inhibits the ADP–ATP antiporter |
| 5. Oligomycin | inhibits ATP synthase |
| 6. Butylmalonate | blocks mitochondrial uptake of succinate |

**14–62** During operation of the respiratory chain, cytochrome *c* accepts an electron from the cytochrome *b*-$c_1$ complex and transfers it to the cytochrome oxidase complex. What is the relationship of cytochrome *c* to the two complexes it connects? Does cytochrome *c* simultaneously connect to both complexes, serving like a wire to allow electrons to flow through it from one complex to the other, or does cytochrome *c* move between the complexes like a ferry, picking up an electron from one and handing it over to the next? An answer to this question is provided by the experiments described below.

Individual lysines on cytochrome *c* were modified to replace the positively charged amino group with a neutral group or with a negatively charged group. The effects of these modifications on electron transfer from the cytochrome *b*-$c_1$ complex to cytochrome *c* and from cytochrome *c* to the cytochrome oxidase complex were then measured. Some modifications had no effect (Figure 14–12, open circles), whereas others inhibited electron transfer from the cytochrome *b*-$c_1$ complex to cytochrome *c* (Figure 14–12A, shaded circles) or from cytochrome *c* to the cytochrome oxidase complex (Figure 14–12B, shaded circles). In an independent series of experiments, several lysines that inhibited transfer of electrons to and from cytochrome *c* were shown to be protected from acetylation when cytochrome *c* was bound either to the cytochrome *b*-$c_1$ complex or to the cytochrome oxidase complex.

How do these results distinguish between the possibilities that cytochrome *c* connects the two complexes by binding to them simultaneously or by shuttling between them?

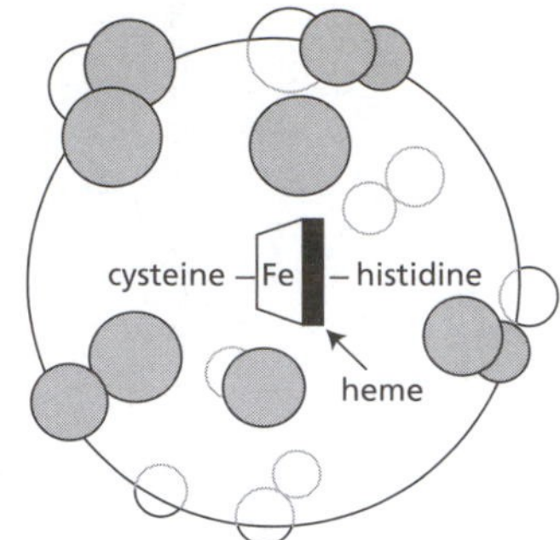


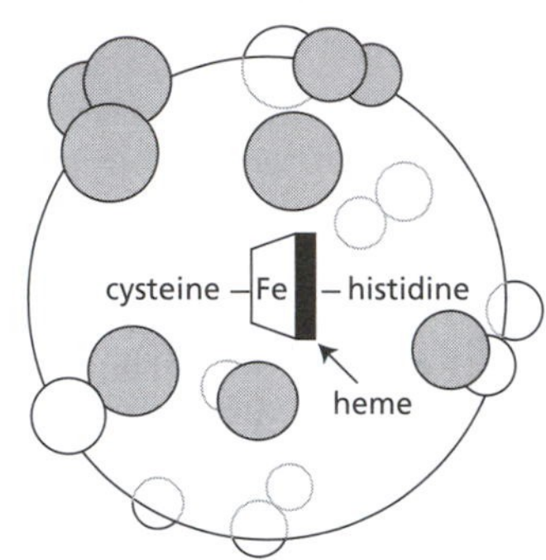


**Figure 14–12** Analysis of electron transfer by cytochrome *c* (Problem 14–62). (A) Modified lysines that inhibit the transfer of electrons from the cytochrome *b*-$c_1$ complex to cytochrome *c*. (B) Modified lysines that inhibit the transfer of electrons from cytochrome *c* to the cytochrome oxidase complex. One edge of the heme group protrudes from the front surface of cytochrome *c* toward the reader. *Circles* show the positions of lysines. *Solid circles* are on the front surface of the molecule; *dashed circles* are on the back surface. The *larger* the circle, the *closer* it is to the reader. *Shaded circles* indicate lysines that inhibit electron transfer when modified; *open circles* indicate lysines that do not inhibit electron transfer when modified.

**14–63** Methanogenic bacteria produce methane as the end-product of electron transport. *Methanosarcina barkeri*, for example, when grown under hydrogen in the presence of methanol, transfers electrons from hydrogen to methanol, producing methane and water:

$$CH_3OH + H_2 \rightarrow CH_4 + H_2O$$

This reaction is analogous to those used by aerobic bacteria and mitochondria, which transfer electrons from carbon compounds to oxygen, producing water and carbon dioxide. Given the peculiar biochemistry involved in methanogenesis, it was initially unclear whether methanogenic bacteria synthesized ATP by electron-transport-driven phosphorylation.

In the experiments depicted in Figure 14–13, methanol was added to cultures of methanogenic bacteria grown under hydrogen. The production of $CH_4$, the magnitude of the electrochemical proton gradient, and the intracellular concentration of ATP were then assayed in the presence of the inhibitors TCS and DCCD. Based on the effects of these inhibitors, and by analogy to mitochondria, decide for each inhibitor if it blocks electron transport, inhibits ATP synthase, or uncouples electron transport from ATP synthesis. Explain your reasoning.

**14–64** The electrochemical proton gradient is responsible not only for ATP production in bacteria, mitochondria, and chloroplasts, but also for powering bacterial flagella. The flagellar motor is thought to be driven directly by the flux of protons through it. To test this idea, you analyze a motile strain of

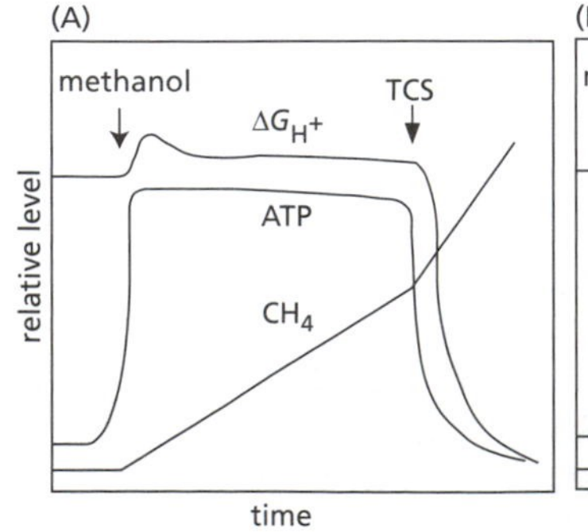


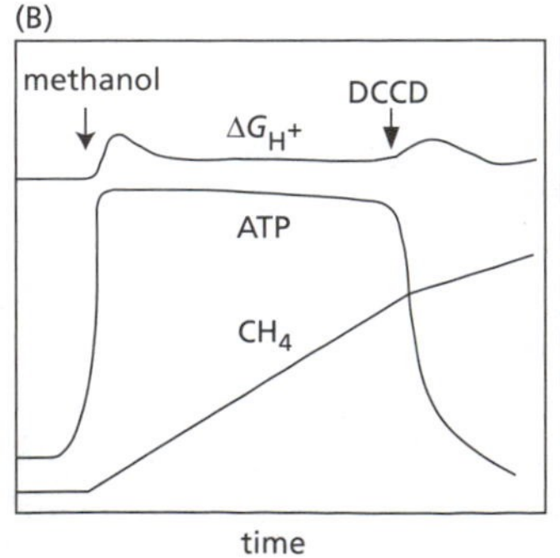


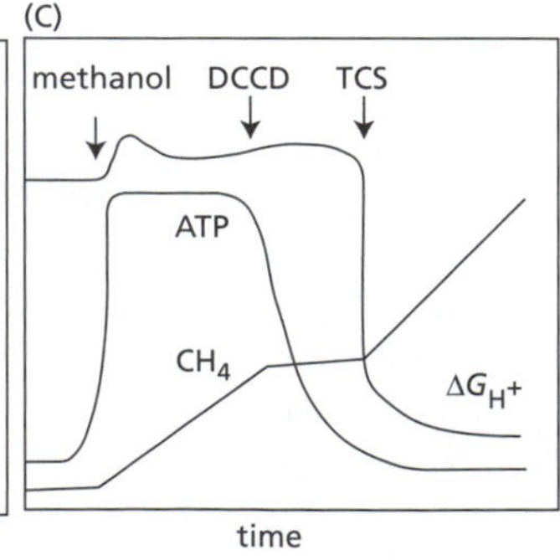


**Figure 14–13** Time course of $CH_4$ production after addition of methanol to a culture of methanogenic bacteria growing under $H_2$ (Problem 14–63). The magnitude of the electrochemical proton gradient ($\Delta G_{H^+}$), the intracellular ATP concentration, and the production of $CH_4$ are expressed in arbitrary units.

**Table 14–4 Effects of ionophores on the swimming of normal bacteria** (Problem 14–64).

| OBSERVATION | GLUCOSE | IONOPHORE | ION IN MEDIUM | EFFECT ON BACTERIA |
|---|---|---|---|---|
| 1 | present | FCCP ($H^+$) | — | stop swimming |
| 2 | present | valinomycin ($K^+$) | $K^+$ | keep swimming |
| 3 | absent | valinomycin ($K^+$) | $K^+$ | remain motionless |
| 4 | absent | valinomycin ($K^+$) | $Na^+$ | swim briefly |

The specificity of each ionophore is shown in parentheses.

*Streptococcus.* These bacteria swim when glucose is available for oxidation, but they do not swim when glucose is absent (and no other substrate is available for oxidation). Using a series of ionophores that alter the pH gradient or the membrane potential (the two components of the electrochemical proton gradient), you make several observations. (Ionophores are small hydrophobic molecules that dissolve in membranes and increase their permeability to specific inorganic ions.)

1. Bacteria that are swimming in the presence of glucose stop swimming upon addition of the proton ionophore FCCP.
2. Bacteria that are swimming in the presence of glucose in a medium containing $K^+$ are unaffected by addition of the $K^+$ ionophore valinomycin.
3. Bacteria that are motionless in the absence of glucose in a medium containing $K^+$ remain motionless upon addition of valinomycin.
4. Bacteria that are motionless in the absence of glucose in a medium containing $Na^+$ swim briefly upon addition of valinomycin and then stop.

These observations are summarized in Table 14–4.

A. Explain how each of these observations is consistent with the idea that the flagellar motor is driven by a flux of protons. (The concentration of $K^+$ inside these bacteria is lower than the concentration of $K^+$ in the medium.)

B. Wild-type bacteria can swim in the presence or absence of oxygen. Mutant bacteria that are missing ATP synthase, which couples proton flow to ATP production, can swim only in the presence of oxygen. How are normal bacteria able to swim in the absence of oxygen when there is no electron flow? How do you think the loss of the ATP synthase prevents swimming in mutant bacteria?

# CHLOROPLASTS AND PHOTOSYNTHESIS

| TERMS TO LEARN | | |
|---|---|---|
| antenna complex | chloroplast | photosystem |
| carbon fixation | noncyclic photophosphorylation | plastid |
| carbon-fixation reactions | photochemical reaction center | starch |
| carbon-fixation cycle (Calvin cycle) | photorespiration | stroma |
| chlorophyll | photosynthetic electron-transfer | sucrose |

## DEFINITIONS

Match the definition below with its term from the list above.

**14–65** Light-driven reactions in photosynthesis in which electrons move along the electron-transport chain in the thylakoid membrane.

**14–66** Part of a photosystem that captures light energy and channels it into the photochemical reaction center.

**14–67** Process by which green plants incorporate carbon atoms from atmospheric carbon dioxide into sugars.

**14–68** The part of a photosystem that converts light energy into chemical energy.

**14–69** Organelle in green algae and plants that contains chlorophyll and carries out photosynthesis.

**14–70** A disaccharide composed of one glucose unit and one fructose unit that is the major form in which sugar is transported between plant cells.

**14–71** Light-absorbing green pigment that plays a central role in photosynthesis.

**14–72** Photosynthetic process that produces both ATP and NADPH in plants and cyanobacteria.

**14–73** Major metabolic pathway by which $CO_2$ is incorporated into carbohydrate during the second stage of photosynthesis in plants.

## TRUE/FALSE

Decide whether each of these statements is true or false, and then explain why.

**14–74** In a general way, one might view the chloroplast as a greatly enlarged mitochondrion in which the cristae have been pinched off to form a series of interconnected submitochondrial particles in the matrix space.

**14–75** When a molecule of chlorophyll in an antenna complex absorbs a photon, the excited electron is rapidly transferred from one chlorophyll molecule to another until it reaches the photochemical reaction center.

## THOUGHT PROBLEMS

**14–76** Both mitochondria and chloroplasts use electron transport to pump protons, creating an electrochemical proton gradient, which drives ATP synthesis. Are protons pumped across the same (analogous) membranes in the two organelles? Is ATP synthesized in the analogous compartments? Explain your answers.

**14–77** Deriving a balanced equation from the numerous individual reactions of the carbon-fixation cycle (Calvin cycle) should be straightforward, but evidently is not. A selection of prominent biochemistry texts list a variety of equations for the balanced reaction (Table 14–5). All agree on the amounts of $CO_2$, NADPH, and ATP required to produce glucose ($C_6H_{12}O_6$), but disagree on water and protons. Part of the difficulty is that explicit chemical structures are not shown for ATP, ADP, and $P_i$. $P_i$ ($HPO_4^{2-}$) incorporates an additional O and H that are not present in ATP, but must be accounted for in a balanced equation. Which equation in Table 14–5 is the balanced equation for the Calvin cycle?

**Problem 1–32** presents a classic experiment from the 1930s that reasoned correctly that photosynthetic fixation of $CO_2$ released $O_2$ by photolysis of $H_2O$, not $CO_2$ as was widely believed.

**14–78** A suspension of the cyanobacterium *Chlamydomonas* is actively carrying out photosynthesis in the presence of light and $CO_2$. If you turned off the light, how would you expect the amounts of ribulose 1,5-bisphosphate and 3-phosphoglycerate to change over the next minute? How about if you left the light on but removed the $CO_2$?

**14–79** Recalling Joseph Priestley's famous experiment in which a sprig of mint saved the life of a mouse in a sealed chamber, you decide to do an analogous experiment to see how a $C_3$ and a $C_4$ plant do when confined together in a sealed environment. You place a corn plant ($C_4$) and a geranium ($C_3$) in a sealed plastic chamber with normal air (300 parts per million $CO_2$) on a

**Table 14–5 A selection of 'balanced' equations for the reactions of the Calvin cycle** (Problem 14–77).

| EQUATIONS FOR THE CALVIN CYCLE |
|---|
| A. $6\,CO_2 + 12\,NADPH + 18\,ATP + 12\,H_2O \rightarrow C_6H_{12}O_6 + 12\,NADP^+ + 18\,ADP + 18\,P_i + 6\,H^+$ |
| B. $6\,CO_2 + 12\,NADPH + 18\,ATP + 12\,H^+ + 12\,H_2O \rightarrow C_6H_{12}O_6 + 12\,NADP^+ + 18\,ADP + 18\,P_i$ |
| C. $6\,CO_2 + 12\,NADPH + 18\,ATP + 12\,H^+ + 12\,H_2O \rightarrow C_6H_{12}O_6 + 12\,NADP^+ + 18\,ADP + 18\,P_i + 6\,O_2$ |
| D. $6\,CO_2 + 12\,NADPH + 18\,ATP + 12\,H_2O \rightarrow C_6H_{12}O_6 + 12\,NADP^+ + 18\,ADP + 18\,P_i$ |
| E. $6\,CO_2 + 12\,NADPH + 18\,ATP + 12\,H^+ \rightarrow C_6H_{12}O_6 + 12\,NADP^+ + 18\,ADP + 18\,P_i + 6\,H_2O$ |

windowsill in your laboratory. What will happen to the two plants? Will they compete or collaborate? If they compete, which one will win and why?

**14–80** Why are plants green?

**14–81** Treatment of chloroplasts with the herbicide DCMU stops $O_2$ evolution and photophosphorylation. If an artificial electron acceptor is added that accepts electrons from plastoquinone (Q), oxygen evolution is restored but not photophosphorylation. Propose a site at which DCMU acts in the flow of electrons through photosystems I and II (Figure 14–14). Explain your reasoning. Why is DCMU a herbicide?

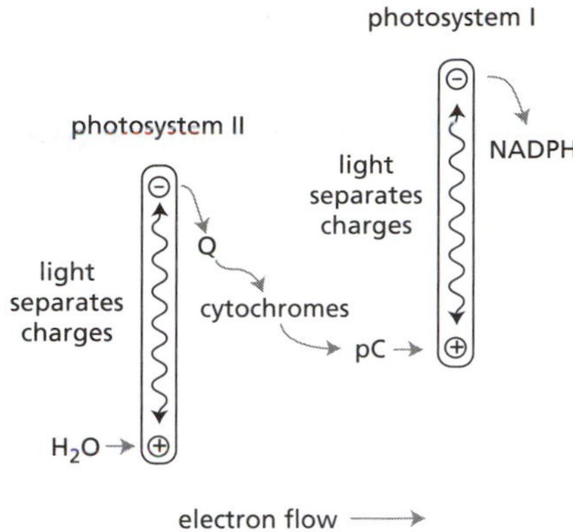


**Figure 14–14** Flow of electrons through photosystems I and II during photosynthesis in chloroplasts (Problem 14–81). Electrons from photosystem II flow to plastoquinone (Q), then to the cytochrome $b_6$-$f$ complex (cytochromes), and then to plastocyanin (pC), after which they enter photosystem I. The protons pumped by the cytochrome $b_6$-$f$ complex generate an electrochemical gradient, which is used to drive ATP synthesis.

**14–82** In chloroplasts, protons are pumped out of the stroma across the thylakoid membrane, whereas in mitochondria, they are pumped out of the matrix across the inner membrane. Explain how this arrangement allows chloroplasts to generate a larger proton gradient across the thylakoid membrane than mitochondria can generate across the inner membrane.

**14–83** Unlike mitochondria, chloroplasts do not have a transporter that allows them to export ATP to the cytosol. How, then, does the rest of the cell get the ATP it needs to survive?

## CALCULATIONS

**14–84** How much energy is available in visible light? How much energy does sunlight deliver to Earth? How efficient are plants at converting light energy into chemical energy? The answers to these questions provide an important backdrop to the subject of photosynthesis.

Each quantum or photon of light has energy $h\upsilon$, where $h$ is Planck's constant ($1.58 \times 10^{-37}$ kcal sec/photon) and $\upsilon$ is the frequency in $sec^{-1}$. The frequency of light is equal to $c/\lambda$, where $c$ is the speed of light ($3.0 \times 10^{17}$ nm/sec) and $\lambda$ is the wavelength in nm. Thus, the energy ($E$) of a photon is

$$E = h\upsilon = hc/\lambda$$

A. Calculate the energy of a mole of photons ($6 \times 10^{23}$ photons/mole) at 400 nm (violet light), at 680 nm (red light), and at 800 nm (near infrared light).

B. Bright sunlight strikes Earth at the rate of about 0.3 kcal/sec per square meter. Assuming for the sake of calculation that sunlight consists of monochromatic light of wavelength 680 nm, how many seconds would it take for a mole of photons to strike a square meter?

C. Assuming that it takes eight photons to fix one molecule of $CO_2$ as carbohydrate under optimal conditions (8–10 photons is the currently accepted value), calculate how long it would take a tomato plant with a leaf area of 1 square meter to make a mole of glucose from $CO_2$. Assume that photons strike the leaf at the rate calculated above and, furthermore, that all the photons are absorbed and used to fix $CO_2$.

D. If it takes 112 kcal/mole to fix a mole of $CO_2$ into carbohydrate, what is the efficiency of conversion of light energy into chemical energy after photon capture? Assume again that eight photons of red light (680 nm) are required to fix one molecule of $CO_2$.

**14–85** What fraction of the free energy of light at 700 nm is captured when a chlorophyll molecule (P700) at the photochemical reaction center in photosystem I absorbs a photon? The equation for calculating the free energy available in one photon of light is given in Problem 14–84. If one assumes standard conditions, the captured free energy ($\Delta G = -nF\Delta E_0'$) can be calculated from the standard redox potential for P700* (excited) → P700 (ground state), which can be gotten from the half reactions:

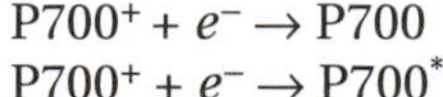
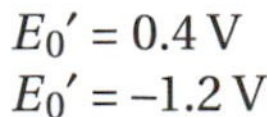

$P700^+ + e^- \rightarrow P700$ $\quad E_0' = 0.4$ V

$P700^+ + e^- \rightarrow P700^*$ $\quad E_0' = -1.2$ V

**14–86** The balanced equation for production of NADPH by noncyclic photophosphorylation is

$$2\ H_2O + 2\ NADP^+ \rightarrow 2\ NADPH + 2\ H^+ + O_2$$

How many photons must be absorbed to generate two NADPH and a molecule of $O_2$? (Assume one photon excites one electron.)

**14–87** What is the standard free-energy change ($\Delta G°$) for the reduction of $NADP^+$ by $H_2O$? The half reactions are

$$NADP^+ + H^+ + 2\ e^- \rightarrow NADPH \qquad E_0' = -0.32\ V$$
$$O_2 + 4\ H^+ + 4\ e^- \rightarrow 2\ H_2O \qquad E_0' = 0.82\ V$$

**14–88** *T. ferrooxidans*, the slag-heap bacterium that lives at pH 2, fixes $CO_2$ like photosynthetic organisms but uses the abundant $Fe^{2+}$ in its environment as a source of electrons instead of $H_2O$. *T. ferrooxidans* oxidizes $Fe^{2+}$ to $Fe^{3+}$ to reduce $NADP^+$ to NADPH, a very unfavorable reaction with a $\Delta E$ of about –1.1 V. It does so by coupling production of NADPH to the energy of the natural proton gradient across its membrane, which has a free-energy change ($\Delta G$) of –6.4 kcal/mole $H^+$. What is the smallest number of protons to the nearest integer that would be required to drive the reduction of $NADP^+$ by $Fe^{2+}$? How do you suppose proton flow is mechanistically coupled to the reduction of $NADP^+$?

**14–89** Conversion of 6 $CO_2$ to one molecule of glucose via the Calvin cycle requires 18 ATP and 12 NADPH. Assuming that both noncyclic and cyclic photophosphorylation generate 1 ATP per pair of electrons, what is the ratio of electrons that travel the noncyclic path to those that travel the cyclic path to meet the needs of the Calvin cycle? What is the ratio of electrons processed in photosystem I to those processed in photosystem II to meet the needs of the Calvin cycle?

**14–90** What is the free energy change ($\Delta G$) for ATP synthesis under typical conditions in which ATP, ADP, and $P_i$ are at 3 mM, 0.1 mM, and 10 mM, respectively, at 37°C? The standard free energy change for ATP synthesis is 7.3 kcal/mole.

## DATA HANDLING

**14–91** Recently, your boss expanded the lab's interest in photosynthesis from algae to higher plants. You have decided to study photosynthetic carbon fixation in the cactus, but so far you have had no success: cactus plants do not seem to fix $^{14}CO_2$, even in direct sunlight. Under the same conditions your colleagues studying dandelions get excellent incorporation of $^{14}CO_2$ within seconds of adding it and are busily charting new biochemical pathways.

Depressed, you leave the lab one day without dismantling the labeling chamber. The following morning you remove the cactus and, much to your surprise, find it has incorporated a great deal of $^{14}CO_2$. Evidently, the cactus fixed carbon during the night. When you repeat your experiments at night in complete darkness, you find that cactus plants incorporate label splendidly.

Although you are forced to shift your work habits, at least now you can make some progress. A brief exposure to $^{14}CO_2$ labels one compound—malate—almost exclusively. During the night, labeled malate builds up to very high levels in specialized vacuoles inside chloroplast-containing cells. In addition, the starch in these same cells disappears. During the day the malate disappears, and labeled starch is formed in a process that requires light. Furthermore, you find that $^{14}CO_2$ reappears in these cells during the day.

These results remind you in some ways of $CO_2$ pumping in $C_4$ plants (Figure 14–15), but they are quite distinct in other ways.

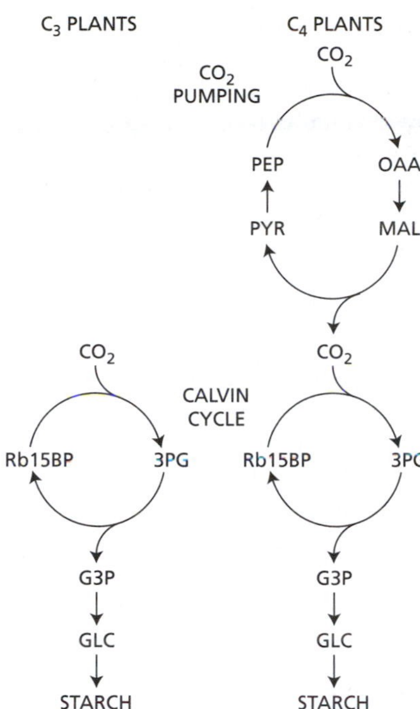


**Figure 14–15** Fixation of $CO_2$ in $C_3$ and $C_4$ plants (Problem 14–91). In these simplified diagrams the usage of ATP and NADPH in various reactions is not shown. PEP is phosphoenolpyruvate; OAA is oxaloacetate; MAL is malate; PYR is pyruvate; Rb15BP is ribulose 1,5-bisphosphate; 3PG is 3-phosphoglycerate; G3P is glyceraldehyde 3-phosphate; and GLC is glucose.

A. Why is light required for starch formation in the cactus? Is it required for starch formation in $C_4$ plants? In $C_3$ plants?

B. Using the reactions of the $CO_2$ pump in $C_4$ plants as a starting point (Figure 14–15), sketch a brief outline of $CO_2$ fixation in the cactus. Which reactions occur during the day and which at night?

C. A cactus depleted of starch cannot fix $CO_2$, but a $C_4$ plant could. Why is starch required for $CO_2$ fixation in the cactus but not in $C_4$ plants?

D. Can you offer an explanation for why this method of $CO_2$ fixation is advantageous for cactus plants?

**14–92** Careful experiments comparing absorption and action spectra of plants ultimately led to the notion that two photosystems cooperate in chloroplasts. The absorption spectrum is the amount of light captured by photosynthetic pigments at different wavelengths. The action spectrum is the rate of photosynthesis (for example, $O_2$ evolution or $CO_2$ fixation) resulting from the capture of photons.

T.W. Engelmann, who used simple equipment and an ingenious experimental design, probably made the first measurement of an action spectrum in 1882. He placed a filamentous green alga into a test tube along with a suspension of oxygen-seeking bacteria. He allowed the bacteria to use up the available oxygen and then illuminated the alga with light that had been passed through a prism to form a spectrum. After a short time he observed the results shown in Figure 14–16. Sketch the action spectrum for this alga and explain how this experiment works.

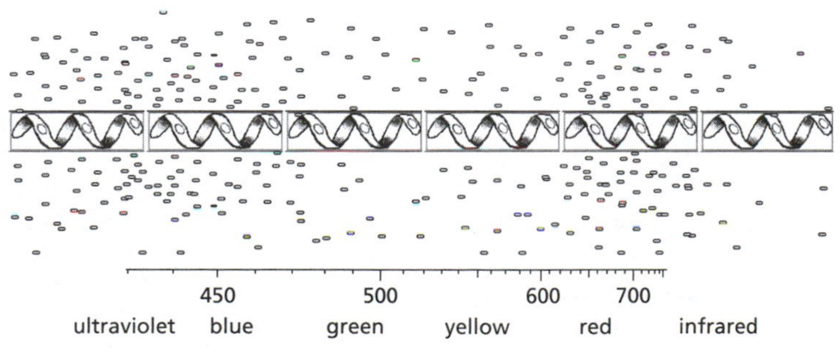


**Figure 14–16** Experiment to measure the action spectrum of a filamentous green alga (Problem 14–92). Bacteria, which are indicated by the *tiny rectangles*, were distributed evenly throughout the test tube at the beginning of the experiment.

**14–93** If all pigments captured energy and delivered it to the photosystem with equal efficiency, then the absorption spectrum and the action spectrum would have the same shape; however, the two spectra differ slightly (Figure 14–17A). When a ratio of the two spectra is displayed (Figure 14–17B), the most dramatic difference is the so-called 'red drop' at long wavelengths. In 1957 Emerson found that if shorter-wavelength light (650 nm) was mixed with the less effective longer-wavelength light (700 nm), the rate of $O_2$ evolution was much enhanced over either wavelength given alone. This result, along with others, suggested that two photosystems (now called photosystem I and photosystem II) were cooperating with one another, and it led to the familiar Z scheme for photosynthesis.

One clue to the order in which the two photosystems are linked came from experiments in which illumination was switched between 650 nm and 700 nm. As shown in Figure 14–18, a shift from 700 nm to 650 nm was accompanied by a transient burst of $O_2$ evolution, whereas a shift from 650 nm to 700 nm was accompanied by a transient depression in $O_2$ evolution.

Using your knowledge of the Z scheme of photosynthesis, explain why these so-called chromatic transients occur, and deduce whether photosystem II, which accepts electrons from $H_2O$, is more responsive to 650-nm light or to 700-nm light.

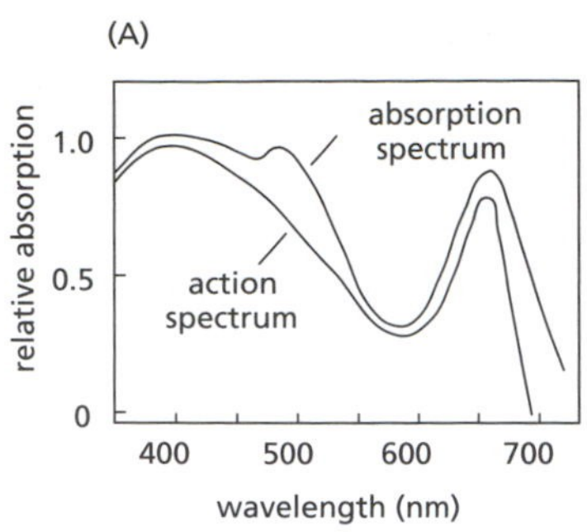


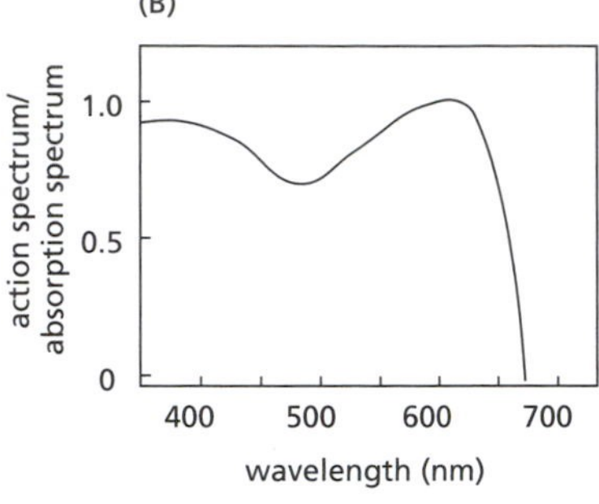


**Figure 14–17** Analysis of photosynthesis in the alga *Chlorella* (Problem 14–93). (A) Absorption and action spectra. The action spectrum shows the evolution of $O_2$ at different wavelengths. (B) The ratio of the action spectrum to the absorption spectrum. The ratio is shown on an arbitrary scale.

**14–94** The most compelling early evidence for the Z scheme of photosynthesis came from measuring the oxidation states of the cytochromes in algae under different regimes of illumination (Figure 14–19). Illumination with light at 680 nm caused oxidation of cytochromes (indicated by the upward trace in Figure 14–19A). Additional illumination with light at 562 nm caused reduction of the cytochromes (indicated by the downward trace in Figure 14–19A). When the lights were then turned off, both effects were reversed (Figure 14–19A). In the presence of the herbicide DCMU (see Problem 14–81), no reduction with 562-nm light occurred (Figure 14–19B).

A. In these algae, which wavelength stimulates photosystem I and which stimulates photosystem II?
B. How do these results support the Z scheme for photosynthesis; that is, how do they support the idea that there are two photosystems that are linked by cytochromes?
C. On which side of the cytochromes does DCMU block electron transport—on the side nearer photosystem I or the side nearer photosystem II?

**14–95** Photosystem II accepts electrons from water, generating $O_2$, and donates them via the electron-transport chain to photosystem I. Each photon absorbed by photosystem II transfers only a single electron, and yet four electrons must be removed from water to generate a molecule of $O_2$. Thus, four photons are required to evolve a molecule of $O_2$:

$$2\,H_2O + 4\,h\nu \rightarrow 4e^- + 4\,H^+ + O_2$$

How do four photons cooperate in the production of $O_2$? Is it necessary that four photons arrive at a single reaction center simultaneously? Can four activated reaction centers cooperate to evolve a molecule of $O_2$? Or is there some sort of 'gear wheel' that collects the four electrons from $H_2O$ and transfers them one at a time to a reaction center?

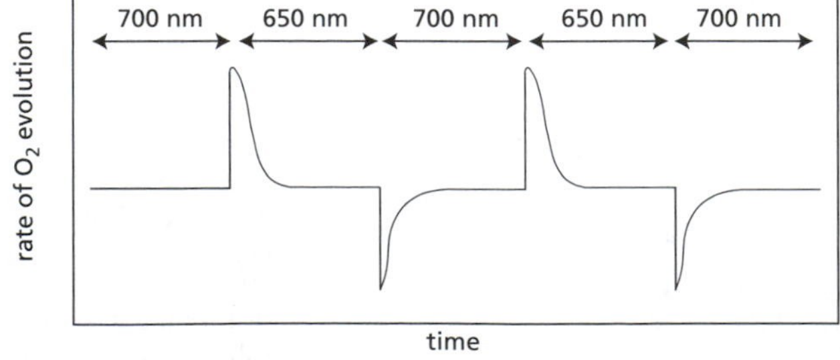


**Figure 14–18** Chromatic transients observed upon switching between 650-nm light and 700-nm light (Problem 14–93). The intensities of light at the two wavelengths were adjusted beforehand so that alone each gave the same rate of $O_2$ evolution.

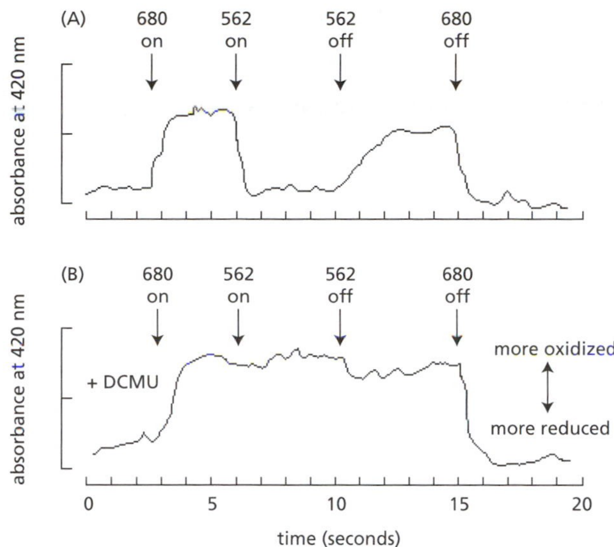


**Figure 14–19** Oxidation state of cytochromes after illumination of algae with light of different wavelengths (Problem 14–94). (A) In the absence of DCMU. (B) In the presence of DCMU. An *upward* trace indicates oxidation of the cytochromes; a *downward* trace indicates reduction of the cytochromes.

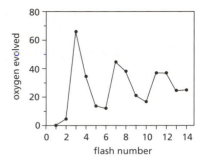

**Figure 14–20** Oxygen evolution by spinach chloroplasts in response to saturating flashes of light (Problem 14–95). The chloroplasts were placed in the dark for 40 minutes prior to the experiment to allow them to come to the same 'ground' state. Oxygen production is expressed in arbitrary units.

To investigate this problem, you expose dark-adapted spinach chloroplasts to a series of brief saturating flashes of light (2 μsec) separated by short periods of darkness (0.3 sec) and measure the evolution of $O_2$ that results from each flash. Under this lighting regime most photosystems capture a photon during each flash. As shown in Figure 14–20, $O_2$ is evolved with a distinct periodicity: the first burst of $O_2$ occurs on the third flash, and subsequent peaks occur every fourth flash thereafter. If you first inhibit 97% of the photosystem II reaction centers with DCMU and then repeat the experiment, you observe the same periodicity of $O_2$ production, but the peaks are only 3% of the uninhibited values.

A. How do these results distinguish among the three possibilities posed at the outset (simultaneous action, cooperation among reaction centers, and a gear wheel)?
B. Why do you think it is that the first burst of $O_2$ occurs after the third flash, whereas additional peaks occur at four-flash intervals? (Consider what this observation implies about the dark-adapted state of the chloroplasts.)
C. Can you suggest a reason why the periodicity in $O_2$ evolution becomes less pronounced with increasing flash number?

**14–96** In an insightful experiment performed in the 1960s, chloroplasts were first soaked in an acidic solution at pH 4, so that the stroma and thylakoid space became acidified (Figure 14–21). They were then transferred to a basic solution (pH 8). This rapidly increased the pH of the stroma to 8, while the thylakoid space temporarily remained at pH 4. A burst of ATP synthesis was observed, and the pH difference between the thylakoid space and the stroma quickly disappeared.

A. Explain why these conditions lead to ATP synthesis.
B. Is light needed for the experiment to work? Why or why not?

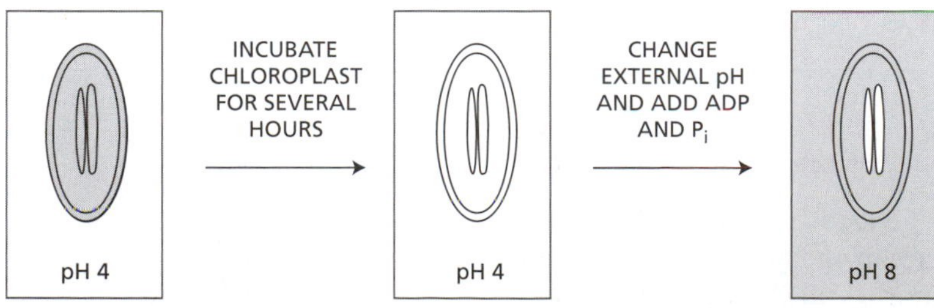


**Figure 14–21** Soaking chloroplasts in acidic and basic solutions (Problem 14–96). *White* areas are at pH 4.

C. What would happen if the solutions were switched so that the first incubation was in the pH 8 solution and the second one was in the pH 4 solution? Explain your answer.
D. Does this experiment support the chemiosmotic model, or raise questions about it?

**14–97** In one of the earliest and most convincing tests of the chemiosmotic model, formation of ATP was assayed in a suspension of thylakoid vesicles (stromal surface facing outward, that is, right-side-out). These vesicles were first acidified to pH 4 and then made alkaline in the presence of ADP and $^{32}PO_4$. As shown in Figure 14–22A, the yield of ATP was greater when the vesicles were acidified with succinic acid than when they were acidified with HCl. Furthermore, the yield of ATP increased with increasing concentration of succinic acid (even though pH 4 solutions were used in all experiments). As shown in Figure 14–22B, the yield of ATP also depended on the pH of the alkaline stage of the experiment, with the yield of ATP increasing up to pH 8.5.
A. Why does acidification with succinic acid yield more ATP than acidification with HCl? And why does the yield of ATP increase with increasing concentrations of succinic acid? (Succinic acid has two carboxylic acid groups with p*K* values of 4.2 and 5.5.)
B. Why is the yield of ATP so critically dependent on the pH of the alkaline stage of the experiment?
C. Predict the effects of treatments with FCCP, an uncoupler of electron transport and ATP synthesis, and DCMU, which blocks electron transport, during the alkaline incubation. Explain the basis for your predictions.

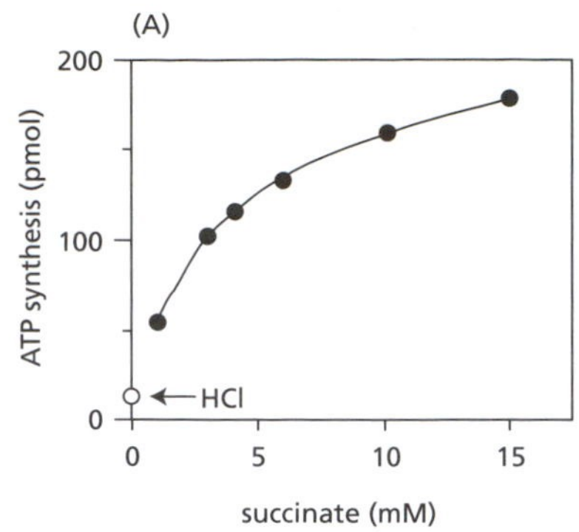


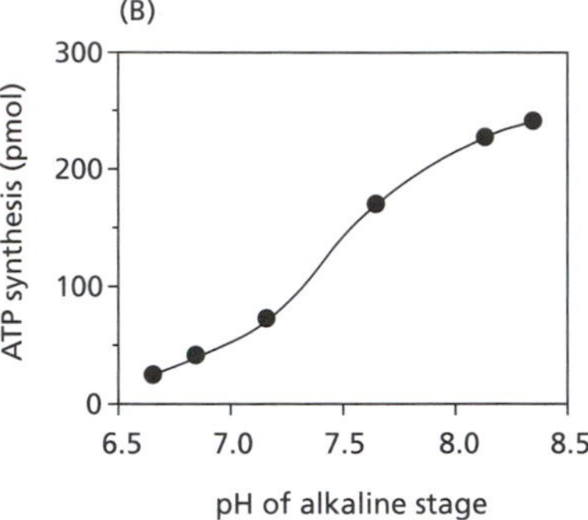


**Figure 14–22** Test of the chemiosmotic model in thylakoid vesicles (Problem 14–97). (A) Yield of ATP with HCl and with increasing concentrations of succinic acid. The yield of ATP with HCl is shown on the *y* axis, where the concentration of succinic acid is zero. In all cases the thylakoid suspension was treated with acid for 60 seconds, then treated with alkali for 15 seconds in the presence of ADP and $^{32}PO_4$, at which point the reaction was stopped and the yield of radioactive ATP was measured. (B) Yield of ATP at different pHs during the alkaline stage of the experiment.

# THE GENETIC SYSTEMS OF MITOCHONDRIA AND PLASTIDS

TERMS TO LEARN

| | |
|---|---|
| cytoplasmic (non-Mendelian) inheritance | maternal inheritance |
| endosymbiont hypothesis | mitotic segregation |

## DEFINITIONS

Match the definition below with its term from the list above.

**14–98** The idea that mitochondria and chloroplasts evolved from bacteria that were endocytosed more than 1 billion years ago.

**14–99** Pattern of mitochondrial inheritance in higher animals that arises because the egg cells always contribute much more cytoplasm to the zygote than does the sperm.

**14–100** Random process whereby cells carrying mitochondria with a mixture of genomes can give rise to cells that carry mitochondrial DNA of only one genotype.

## TRUE/FALSE

Decide whether each of these statements is true or false, and then explain why.

**14–101** Mitochondria and chloroplasts replicate their DNA in synchrony with the nuclear DNA and divide when the cell divides, thereby ensuring a constant number of organellar genomes.

**14–102** The mitochondrial genetic code differs slightly from the nuclear code, but it is identical in mitochondria from all species that have been examined.

**14–103** The presence of introns in organellar genes is not surprising since similar introns have been found in related genes from bacteria whose ancestors are thought to have given rise to mitochondria and chloroplasts.

**14–104** Mutations that are inherited according to Mendelian rules affect nuclear genes; mutations whose inheritance violates Mendelian rules are likely to affect organelle genes.

**14–105** Mitochondria, which divide by fission, can apparently grow and divide indefinitely in the cytoplasm of proliferating eucaryotic cells even in the complete absence of a mitochondrial genome.

## THOUGHT PROBLEMS

**14–106** You have discovered a remarkable, one-celled protozoan that lives in an anaerobic environment. It has caught your attention because it has absolutely no mitochondria, an exceedingly rare condition among eucaryotes. If you could show that this organism derives from an ancient lineage that split off from the rest of eucaryotes before mitochondria were acquired, it would truly be a momentous discovery. You sequence the organism's genome so you can make detailed comparisons. It is clear from sequence comparisons that your organism does indeed derive from an ancient lineage. But here and there, scattered around the genome are bits of DNA that in aggregate resemble the bacterial genome from which mitochondria evolved. Propose a plausible evolutionary history for your organism.

**Problem 1–45** examines the evolutionary history of *Giardia*, a parasitic eucaryote that is missing mitochondria, ER, and the Golgi apparatus.

**14–107** At the cellular level, evolutionary theories are particularly difficult to test since fossil evidence is lacking. The possible evolutionary origins of mitochondria and chloroplasts must be sought in living organisms. Fortunately, living forms resembling the ancestral types required by the endosymbiont hypothesis for the origin of mitochondria and chloroplasts can be found today. For example, the plasma membrane of the free-living aerobic bacterium *Paracoccus denitrificans* contains a respiratory chain that is nearly identical to the respiratory chain of mammalian mitochondria—both in the types of respiratory components present and in its sensitivity to respiratory inhibitors such as antimycin and rotenone. Indeed, no significant feature of the mammalian respiratory chain is absent from *Paracoccus. Paracoccus* effectively assembles in a single organism all those features of the mitochondrial inner membrane that are otherwise distributed at random among other aerobic bacteria.

Imagine that you are a protoeucaryotic cell looking out for your evolutionary future. You have been observing proto-*Paracoccus* and are amazed at its incredibly efficient use of oxygen in generating ATP. With such a source of energy your horizons would be unlimited. You plot to hijack a proto-*Paracoccus* and make it work for you and your descendants. You plan to take it into your cytoplasm, feed it any nutrients it needs, and harvest the ATP. Accordingly, one dark night you trap a philandering proto-*Paracoccus*, surround it with your plasma membrane, and imprison it in a new cytoplasmic compartment. To your relief, the proto-*Paracoccus* seems to enjoy its new environment. After a day of waiting, however, you feel as sluggish as ever. What has gone wrong with your scheme?

**14–108** How many codons specify STOP in human mitochondria?

**14–109** Examine the variegated leaf shown in Figure 14–23. The dark patches are green and the light patches are yellow. Yellow patches surrounded by green

**Figure 14–23** A variegated leaf of *Aucuba japonica* with green (*dark*) and yellow (*light*) patches (Problem 14–109).

are common, but there are no green patches surrounded by yellow. Propose an explanation for this phenomenon.

**14–110** The pedigrees in Figure 14–24 show one example each of the following types of mutation: mitochondrial mutation, autosomal recessive mutation, autosomal dominant mutation, and X-linked recessive mutation. In each family the parents have had nine children. Assign each pedigree to one type of mutation. Explain the basis for your assignments.

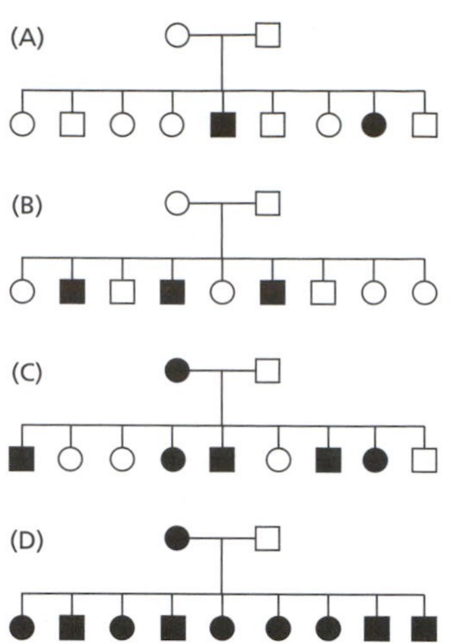


**Figure 14–24** Hypothetical pedigrees representing four patterns of inheritance (Problem 14–110). Males are shown as *squares*; females as *circles*. Affected individuals are shown as *filled symbols*; unaffected individuals are shown as *empty symbols*.

## CALCULATIONS

**14–111** How many genomes are there, on average, in a single mitochondrion in a human liver cell? Assume that the weight of mitochondrial DNA is 1% of the weight of the nuclear DNA and that there are 1000 mitochondria per liver cell. The human mitochondrial genome is 16,569 bp and the diploid human genome is $6.4 \times 10^9$ bp.

## DATA HANDLING

**14–112** Replication of an individual mitochondrial genome takes about an hour in mouse cells, which is about 5% of the 20-hour cell-generation time. Since the average number of mitochondrial genomes per cell is constant, they must replicate, on average, once per cell cycle. Do mitochondrial genomes replicate at random times throughout the cell cycle or is their replication confined to one particular stage, for example, S phase?

Your elegant approach to this question is to label mouse cell mitochondrial DNA briefly (2 hours) with $^3$H-thymidine, to chase with nonradioactive thymidine for various times, and finally to label with 5-bromodeoxyuridine (BrdU) for 2 hours (Figure 14–25). Any DNA that was replicated during both exposures will be radioactive (due to the $^3$H-thymidine) and denser than normal mitochondrial DNA (due to the BrdU). Using density-gradient centrifugation, you separate heavy (BrdU-labeled) mitochondrial DNA from light (non-BrdU-labeled) mitochondrial DNA and measure the radioactivity associated with each (Figure 14–26). Your results show that a relatively constant fraction of $^3$H-labeled mitochondrial DNA was shifted to a higher density, regardless of the time between the labeling periods (the chase times).

A. Do these results fit better with your expectations for mitochondrial replication during a specific part of the cell cycle or at random times? Why?

B. One of your colleagues criticizes the design of these experiments. He suggests that you cannot distinguish between random or timed replication

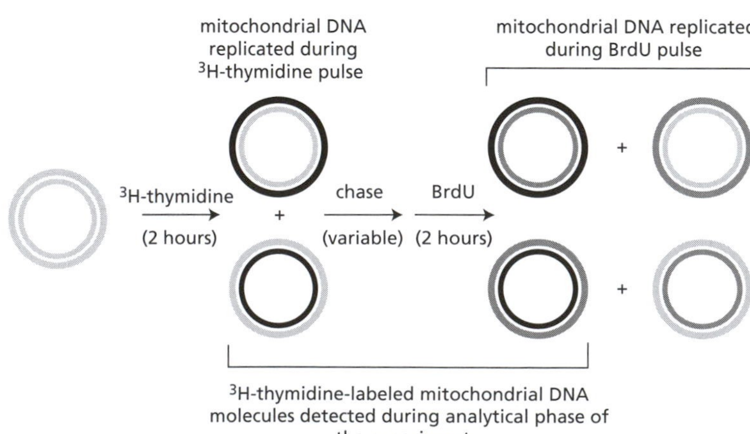


**Figure 14–25** Experimental design to assess the timing of mitochondrial DNA replication (Problem 14–112). Unlabeled mitochondrial DNA is indicated with *light gray lines;* DNA labeled with $^3$H-thymidine is indicated with *black lines;* DNA labeled with BrdU is indicated with *dark gray lines.*

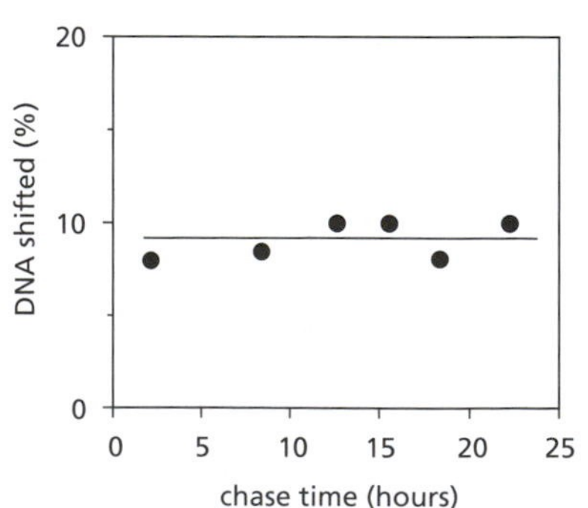


**Figure 14–26** The fraction of mitochondrial DNA labeled with $^3$H-thymidine that is shifted to heavy density after various lengths of chase (Problem 14–112). The fraction of DNA that is density shifted is the radioactivity at the heavy density divided by the total radioactivity. The length of chase is the time from the end of the $^3$H-thymidine labeling to the beginning of the BrdU labeling (see Figure 14–25).

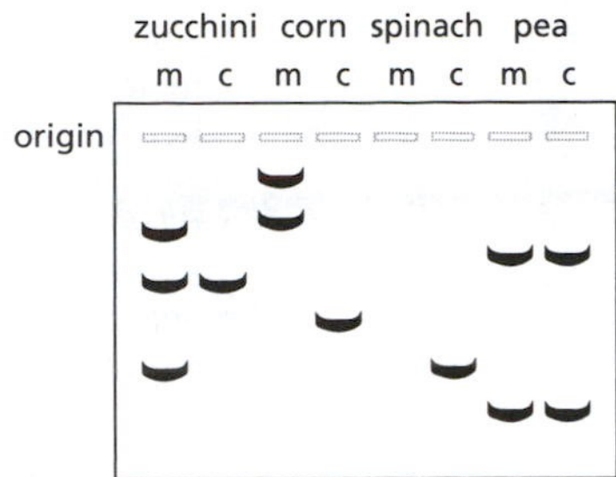


**Figure 14–27** Patterns of hybridization of a probe from spinach chloroplast DNA to mitochondrial and chloroplast DNAs from zucchini, corn, spinach, and pea (Problem 14–113). Lanes labeled *m* contain mitochondrial DNA; lanes labeled *c* contain chloroplast DNA. Restriction fragments to which the probe hybridized are shown as *dark bands*.

because the cells were growing asynchronously (that is, within the cell population all different stages of the cell cycle were represented). How does this concern affect your interpretation?

C. What results would you expect for a similar analysis of nuclear DNA? Assume that the DNA synthesis phase—S phase—of the 20-hour cell cycle is 5 hours long. Explain your answer.

D. What results would you expect if mitochondrial DNA were replicated at all times during the cell cycle, but once an individual molecule was replicated, it had to wait exactly one cell cycle before it was replicated again? Would they be the same as the results in Figure 14–26, the same as you expect for nuclear DNA, or would they give you a new pattern entirely? Why?

**14–113** It is well accepted that transfer of DNA from organellar genomes to nuclear genomes is common during evolution. Do transfers between organellar genomes also occur? One experiment to search for genetic transfers between organellar genomes used a defined restriction fragment from spinach chloroplasts, which carried information for the gene for the large subunit of ribulose bisphosphate carboxylase. This gene has no known mitochondrial counterpart. Thus, if a portion of the chloroplast DNA in the restriction fragment was transferred to the mitochondrial genome, it would show up as a hybridizing band at a novel position. Mitochondrial and chloroplast DNAs were prepared from zucchini, corn, spinach, and pea. All these DNAs were digested with the same restriction nuclease, and the resulting fragments were separated by electrophoresis. The fragments were then transferred to a filter and hybridized to a radioactive preparation of the spinach fragment. A schematic representation of the autoradiograph is shown in Figure 14–27.

A. It is very difficult to prepare mitochondrial DNA that is not contaminated to some extent with chloroplast DNA. How do these experiments control for contamination of the mitochondrial DNA preparation by chloroplast DNA?

B. Which of these plant mitochondrial DNAs appear to have acquired chloroplast DNA?

**Problems 1–47** and **1–48** present evidence for transfer of DNA from mitochondrial to nuclear genomes and reveal some of the mechanisms involved.

**14–114** The majority of mRNAs, tRNAs, and rRNAs in human mitochondria are transcribed from one strand of the genome. These RNAs are all present initially on one very long transcript, which is 93% of the length of the DNA strand. During mitochondrial protein synthesis, these RNAs function as separate, independent species of RNA. The relationship of the individual RNAs to the primary transcript and many of the special features of the mitochondrial genetic system have been revealed by comparing the sequences of the RNAs with the nucleotide sequence of the genome. An overview of the transcription map is shown in Figure 14–28.

Three segments of the nucleotide sequence of the human mitochondrial genome are shown in Figure 14–29 along with the three mRNAs that are

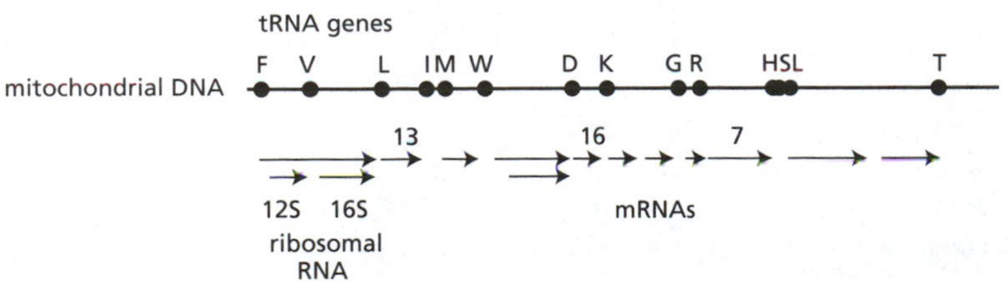


**Figure 14–28** Transcription map of human mitochondrial DNA (Problem 14–114). Individual tRNA genes are indicated by *black circles*; the amino acids they carry are shown in the one-letter code. The three mRNAs whose detailed sequences are shown in Figure 14–29 are indicated by *number*.

```
tRNA^L                                         tRNA^I
TTCTTAACAACATACCCAT.........CTCAAACCTAAGAAATATG        DNA
         ACAUACCCAU.........CUCAAACCUAAAAAAAAAA        mRNA 13
            M  P                E  T  *                protein

tRNA^D                                         tRNA^K
TATATCTTAATGGCACATG.........CTCTAGAGCCCACTGTAAA        DNA
         AUGGCACAUG.........CUCUAGAGCCAAAAAAAAA        mRNA 16
          M  A  H               S  *                   protein

tRNA^R                                         tRNA^H
ATTTACCAAATGCCCCTCA.........TTTTCCTCTTGTAAATATA        DNA
         AUGCCCCUCA.........UUUUCCUCUUAAAAAAAAA        mRNA 7
          M  P  L            F  S  S  *                protein
```

**Figure 14–29** Arrangements of tRNA and mRNA sequences at three places on the human mitochondrial genome (Problem 14–114). *Underlined* sequences indicate tRNA genes. The sequences of the mRNAs are shown *below* the corresponding genes. The middle portions of the mRNAs and their genes are indicated by *dots*. The 5′ ends of the sequences are shown at the *left*. The 5′ ends of the mRNAs are unmodified and the 3′ ends have poly-A tails. The encoded protein sequences are indicated below the mRNAs, with the letter for the amino acid immediately under the center nucleotide of the codon. An asterisk (*) indicates a termination codon.

generated from those regions. The nucleotides that encode tRNA species are underlined; the amino acids encoded by the mRNAs are indicated below the center base of the codon.

A. In terms of codon usage and mRNA structure, in what two ways does initiation of protein synthesis in mitochondria differ from initiation in the cytoplasm?

B. In what two ways are the termination codons for protein synthesis in mitochondria unusual? (The termination codons are shown in Figure 14–29 as asterisks.)

C. Does the arrangement of tRNA and mRNA sequences in the genome suggest a possible mechanism for processing the primary transcript into individual RNA species?

**14–115** A friend of yours has been studying a pair of mutants in the fungus *Neurospora*, which she has whimsically named *poky* and *puny*. Both mutants grow at about the same rate, but much more slowly than wild type. Your friend has been unable to find any supplement that improves their growth rates. Her biochemical analysis shows that each mutant displays a different abnormal pattern of cytochrome absorption. To characterize the mutants genetically, she crossed them to wild type and to each other and tested the growth rates of the progeny. She has come to you because she is puzzled by the results.

She explains that haploid nuclei from the two parents fuse during a *Neurospora* mating and then divide meiotically to produce four haploid spores, which can be readily tested for their growth rates. The parents contribute unequally to the diploid: one parent (the protoperithecial parent) donates a nucleus and the cytoplasm; the other (the fertilizing parent) contributes little more than a nucleus—much like egg and sperm in higher organisms. As shown in Table 14–6, the 'order' of the crosses sometimes makes a difference: this is a result she has not seen before.

Can you help your friend understand these results?

**Table 14–6 Genetic analysis of *Neurospora* mutants** (Problem 14–115).

| | | | SPORE COUNTS | |
|---|---|---|---|---|
| CROSS | PROTOPERITHECIAL PARENT | FERTILIZING PARENT | FAST GROWTH | SLOW GROWTH |
| 1 | *poky* | wild | 0 | 1749 |
| 2 | wild | *poky* | 1334 | 0 |
| 3 | *puny* | wild | 850 | 799 |
| 4 | wild | *puny* | 793 | 801 |
| 5 | *poky* | *puny* | 0 | 1831 |
| 6 | *puny* | *poky* | 754 | 710 |
| 7 | wild | wild | 1515 | 0 |
| 8 | *poky* | *poky* | 0 | 1389 |
| 9 | *puny* | *puny* | 0 | 1588 |

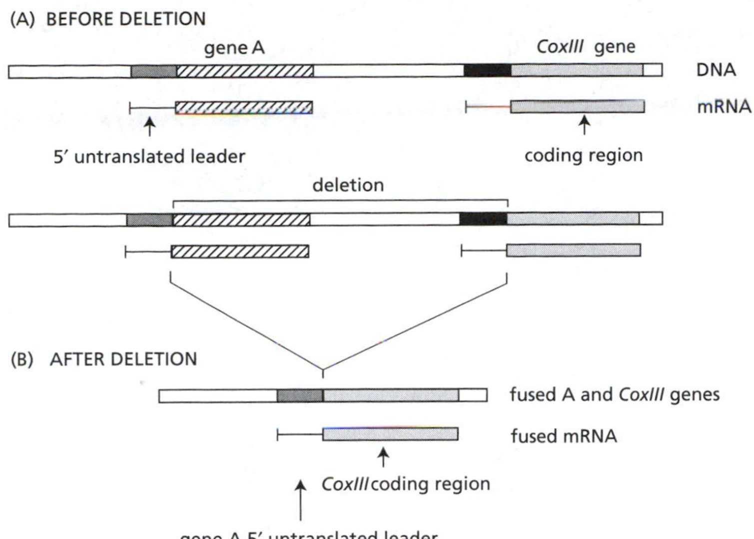


**Figure 14–30** Schematic representation of the mitochondrial deletion mutations that suppress the nuclear mutation *Pet494* (Problem 14–116).

**14–116** Mutants of yeast that are defective in mitochondrial function grow on fermentable substrates such as glucose, but they fail to grow on nonfermentable substrates such as glycerol. Genetic crosses can distinguish nuclear and mitochondrial mutations. Nearly 200 different nuclear genes have been defined by complementation analysis of nuclear petite (*Pet*) mutants. Surprisingly, mutations in about a quarter of these genes affect the expression of single, or restricted sets of, mitochondrial gene products without blocking overall gene expression.

For example, *Pet494* mutants contain normal levels of all the known mitochondrial gene products except subunit III of cytochrome oxidase (CoxIII). However, *Pet494* mutants contain normal levels of CoxIII mRNA, which appears normal in size. Thus, some posttranscriptional step in expression of the gene for CoxIII appears to be regulated by the normal *Pet494* gene product. It could be required to promote translation of CoxIII mRNA or to stabilize CoxIII during assembly of the cytochrome oxidase complex. To distinguish between these possibilities, mitochondrial mutations that suppress the respiratory defect of *Pet494* mutations were selected and analyzed. All these mutations were deletions that fused the CoxIII coding region to the 5′ untranslated region of another gene, as illustrated in Figure 14–30.

A. How do these results distinguish between a requirement for translation of CoxIII mRNA and a requirement for stabilization of the CoxIII protein?

B. Each of the mitochondrial deletions eliminates one or more genes that are essential for mitochondrial function, yet these yeast strains contain all the normal mitochondrial mRNAs and make normal colonies when grown on nonfermentable substrates. How can this be?

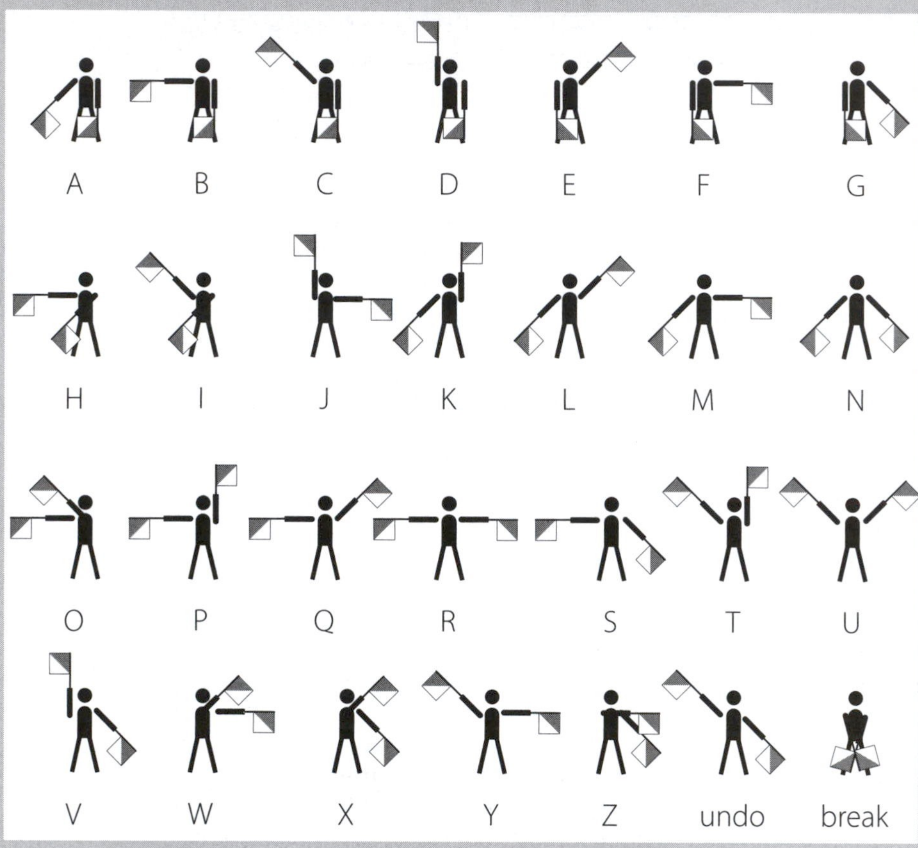


**The semaphore alphabet as seen by the recipient.** You may like to consider the similarities and differences between this human mode of communication and the signaling networks used by cells.

# Mechanisms of Cell Communication

# 15

## In This Chapter

GENERAL PRINCIPLES OF CELL COMMUNICATION 343

SIGNALING THROUGH G-PROTEIN-COUPLED CELL-SURFACE RECEPTORS (GPCRs) AND SMALL INTRACELLULAR MEDIATORS 352

SIGNALING THROUGH ENZYME-COUPLED CELL-SURFACE RECEPTORS 361

SIGNALING PATHWAYS DEPENDENT ON REGULATED PROTEOLYSIS OF LATENT GENE REGULATORY PROTEINS 367

SIGNALING IN PLANTS 370

## GENERAL PRINCIPLES OF CELL COMMUNICATION

TERMS TO LEARN

adaptation (desensitization)
adaptor
contact-dependent signaling
endocrine cell
enzyme-coupled receptor
extracellular signal molecule
gap junction
G-protein-coupled receptor
GTPase-activating protein (GAP)
GTP-binding protein
guanine nucleotide exchange factor (GEF)
hormone
interaction domain
intracellular signaling protein
ion-channel-coupled receptor
local mediator
monomeric GTPase
morphogen
negative feedback loop
neurotransmitter
nitric oxide (NO)
NO synthase (NOS)
nuclear receptor superfamily
paracrine signaling
phosphorylation
phosphorylation cascade
postive feedback loop
primary cilium
protein kinase
protein phosphatase
receptor
scaffold protein
second messenger
serine/threonine kinase
signaling cascade
small intracellular mediator
steroid hormone
synaptic signaling
tyrosine kinase

### DEFINITIONS

Match the definition below with its term from the list above.

**15–1** Protein that binds to a GTP-binding protein and activates it by stimulating release of tightly bound GDP, thereby allowing it to bind GTP.

**15–2** General term for a protein that binds a specific extracellular molecule (ligand) and initiates a response in the cell.

**15–3** Alteration of sensitivity following repeated stimulation, reducing a cell's response to that level of stimulus.

**15–4** Compact protein module that binds to a particular structural motif in another protein (or lipid) molecule with which the signaling protein interacts.

**15–5** Communicating cell–cell junction that allows ions and small molecules to pass from the cytoplasm of one cell to the cytoplasm of the next.

**15–6** An arrangement of signaling interactions within a pathway that counteracts the effect of the stimulus and thereby abbreviates the response and limits its final level.

**15–7** Enzyme that catalyzes the release of nitric oxide by deamination of the amino acid arginine.

**15–8** Hydrophobic signaling molecule with a characteristic four-ringed structure derived from cholesterol.

**15–9** Short-range cell–cell communication via secreted local mediators that act on adjacent cells.

**15–10** A signal relay chain involving multiple protein kinases, each of which is activated by phosphorylation and then phosphorylates the next protein kinase in the sequence.

**15–11** Small molecule that is formed in the cytosol, or released into it, in response to an extracellular signal and that helps to relay the signal to the interior of the cell.

**15–12** Specialized animal cell that secretes a hormone into the blood.

**15–13** General term for a signal relay chain containing multiple amplification steps.

**15–14** Molecule from outside the cell that communicates the behavior or actions of other cells in the environment and elicits an appropriate response.

**15–15** Enzyme that transfers the terminal phosphate group of ATP to a specific amino acid of a target protein.

**15–16** Small signal molecule secreted by the presynaptic nerve cell at a chemical synapse to relay the signal to the postsynaptic cell.

**15–17** Protein that binds to a GTP-binding protein and inactivates it by stimulating its GTPase activity so that its bound GTP is hydrolyzed to GDP.

**15–18** Protein that organizes groups of interacting intracellular signaling proteins into signaling complexes.

**15–19** Cell–cell communication in which the signal molecule remains bound to the signaling cell and only influences cells that physically touch it.

## TRUE/FALSE

Decide whether each of these statements is true or false, and then explain why.

**15–20** There is no fundamental distinction between signaling molecules that bind to cell-surface receptors and those that bind to intracellular receptors.

**15–21** The receptors involved in paracrine, synaptic, and endocrine signaling all have very high affinity for their respective signaling molecules.

**15–22** All small intracellular mediators (second messengers) are water soluble and diffuse freely through the cytosol.

## THOUGHT PROBLEMS

**15–23** Surgeons use succinylcholine, which is an acetylcholine analog, as a muscle relaxant. Care must be taken because some individuals recover abnormally slowly from this paralysis, with life-threatening consequences. Such individuals are deficient in an enzyme called pseudocholinesterase, which is normally present in the blood, where it slowly inactivates succinylcholine by hydrolysis to succinate and choline.

If succinylcholine is an analog of acetylcholine, why do you think it causes muscles to relax and not contract as acetylcholine does?

**15–24** To remain a local response, paracrine signal molecules must be prevented from straying too far from their point of origin. Suggest different ways by which this geographical restriction could be accomplished.

**15–25** Compare and contrast signaling by neurons to signaling by endocrine cells. What are the relative advantages of these two mechanisms for cellular communication?

**15–26** Cells communicate in ways that resemble human communication. Decide which of the following forms of human communication are analogous to autocrine, paracrine, endocrine, and synaptic signaling by cells.

A. A telephone conversation
B. Talking to people at a cocktail party
C. A radio announcement
D. Talking to yourself

**15–27** Why do signaling responses that involve changes in proteins already present in the cell occur in milliseconds to seconds, whereas responses that require changes in gene expression require minutes to hours?

**15–28** It seems counterintuitive that a cell with an abundant supply of nutrients would commit suicide if not constantly stimulated by signals from other cells. What do you suppose might be the purpose of such regulation?

**15–29** How is it that different cells can respond in different ways to exactly the same signaling molecule even when they have identical receptors?

**15–30** A rise in the cyclic GMP levels in smooth muscle cells causes relaxation of the blood vessels in the penis, resulting in an erection. Explain how the natural signal molecule NO and the drug Viagra produce an increase in cyclic GMP.

**15–31** Nuclear receptors carry a binding site for a signal molecule and for a DNA sequence. How is it that identical nuclear receptors in different cells can activate different genes when they bind the same signal molecule?

**15–32** The steroid hormones cortisol, estradiol, and testosterone are all derived from cholesterol by modifications that introduce polar groups such as –OH and =O (Figure 15–1). If cholesterol itself were not normally found in cell membranes, do you suppose it could be used effectively as a hormone, provided that an appropriate intracellular receptor were available?

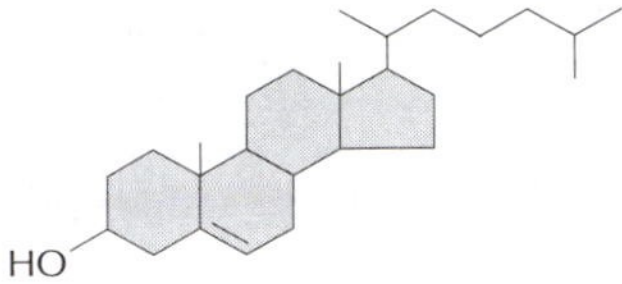


$CH_2OH$
C=O
HO
O
cortisol

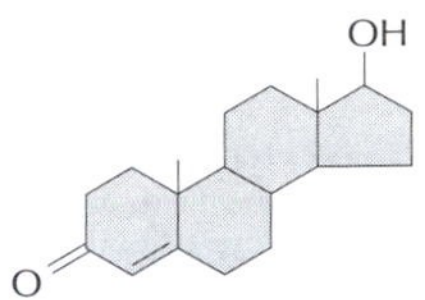


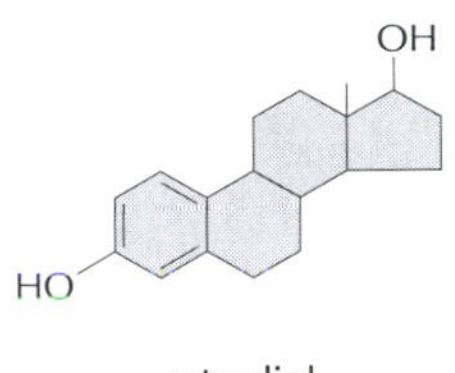


**Figure 15–1** Steroid hormones and their parent molecule, cholesterol (Problem 15–32).

**15–33** The signaling mechanisms used by a steroid hormone receptor and by an ion-channel-coupled receptor have very few components. Can either mechanism lead to an amplification of the initial signal? If so, how?

**15–34** Working out the order in which the individual components in a signaling pathway act is an essential step in defining the pathway. Imagine that two protein kinases, PK1 and PK2, act sequentially in an intracellular signaling pathway. When either kinase is completely inactivated, cells do not respond to the normal extracellular signal. By contrast, cells containing a mutant form of PK1 that is permanently active respond even in the absence of an extracellular signal. Doubly mutant cells that contain inactivated PK2 and permanently active PK1 respond in the absence of a signal.

In the normal signaling pathway, does PK1 activate PK2 or does PK2 activate PK1? What outcome would you have predicted for a doubly mutant cell line with an activating mutation in PK2 and an inactivating mutation in PK1? Explain your reasoning.

**15–35** Proteins play a variety of roles in signaling pathways. Match the following list of signaling proteins with the best description of their functions. Each description should be matched with only one protein.

A. Amplifier proteins
B. Anchoring proteins
C. Integrator proteins
D. Modulator proteins
E. Relay proteins
F. Scaffold proteins
G. Transducer proteins

1. bind multiple signaling proteins together in a functional complex
2. combine signals from two or more pathways before passing it onward
3. convert the signal to a different form
4. greatly increase the signal they receive
5. maintain specific signaling proteins at a precise location in the cell
6. modify the activity of signaling proteins to regulate signal strength
7. pass the message to the next signaling component in the pathway

**15–36** Why do you suppose that phosphorylation/dephosphorylation, as opposed to allosteric binding of small molecules, for example, has evolved to play such a prominent role in switching proteins on and off in signaling pathways?

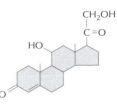

**15–37** The two main classes of molecular switches involve changes in phosphorylation state or changes in guanine nucleotide binding. Comment on the following statement. "In the regulation of molecular switches, protein kinases and guanine nucleotide exchange factors (GEFs) always turn proteins on, whereas protein phosphatases and GTPase-activating proteins (GAPs) always turn proteins off."

Problem 3–84 asks how molecular on/off switches can give rise to intermediate responses

**15–38** Consider a signaling pathway that proceeds through three protein kinases that are sequentially activated by phosphorylation. In one case the kinases are held in a signaling complex by a scaffolding protein; in the other, the kinases are freely diffusing (Figure 15–2). Discuss the properties of these two types of organization in terms of signal amplification, speed, and potential for cross-talk between signaling pathways.

**15–39** Proteins in signaling pathways use a variety of binding domains to assemble into signaling complexes. Match the following domains with their binding targets. (A binding target can be used more than once.)

| | |
|---|---|
| A. PH domain | 1. phosphorylated tyrosines |
| B. PTB domain | 2. proline-rich sequences |
| C. SH2 domain | 3. phosphorylated inositol phospholipids |
| D. SH3 domain | |

Problem 3–99 analyzes the binding of an SH2 domain to its target protein

**15–40** Describe three ways in which a gradual increase in an extracellular signal can be sharpened to produce an abrupt or nearly all-or-none cellular response.

## CALCULATIONS

**15–41** Suppose that the circulating concentration of hormone is $10^{-10}$ M and the $K_d$ for binding to its receptor is $10^{-8}$ M. What fraction of the receptors will have hormone bound? If a meaningful physiological response occurs when 50% of the receptors have bound a hormone molecule, how much will the concentration of hormone have to rise to elicit a response? Recall that the fraction of receptors (R) bound to hormone (H) to form a receptor–hormone complex (R–H) is $[R\text{–}H]/([R] + [R\text{–}H]) = [R\text{–}H]/[R]_{TOT} = [H]/([H] + K_d)$.

**Problems 3–101, 3–102,** and **3–103** develop the relationship for fraction of bound ligand from the expression for the dissociation constant, $K_d$, and illustrate the usefulness in several experimental settings.

**15–42** Radioimmunoassay (RIA) is a powerful tool for quantifying virtually any substance of biological interest because it is sensitive, accurate, and fast. RIA technology arose from studies on adult onset diabetes. Some patients had antibodies with high affinity for insulin, and RIA was developed as a method to distinguish free insulin from antibody-bound insulin.

How can high-affinity antibodies be exploited to measure low concentrations of insulin? When a small amount of insulin-specific antiserum is mixed with an equally small amount of very highly radioactive insulin, some binds and some remains free according to the equilibrium.

$$\text{Insulin (I)} + \text{Antibody (A)} \rightleftharpoons \text{Insulin–Antibody (I–A) Complex}$$

$$K = \frac{[\text{I–A}]}{[\text{I}][\text{A}]}$$

When increasing amounts of unlabeled insulin are added to a fixed amount of labeled insulin and anti-insulin antibody, the ratio of bound to free radioactive insulin decreases as expected from the equilibrium expression. If the concentration of the unlabeled insulin is known, then the resulting curve serves as a calibration against which other unknown samples can be compared (Figure 15–3).

You have three samples of insulin whose concentrations are unknown. When mixed with the same amount of radioactive insulin and anti-insulin antibody used in Figure 15–3, the three samples gave the following ratios of bound to free insulin:

| | |
|---|---|
| Sample 1 | 0.67 |
| Sample 2 | 0.31 |
| Sample 3 | 0.46 |

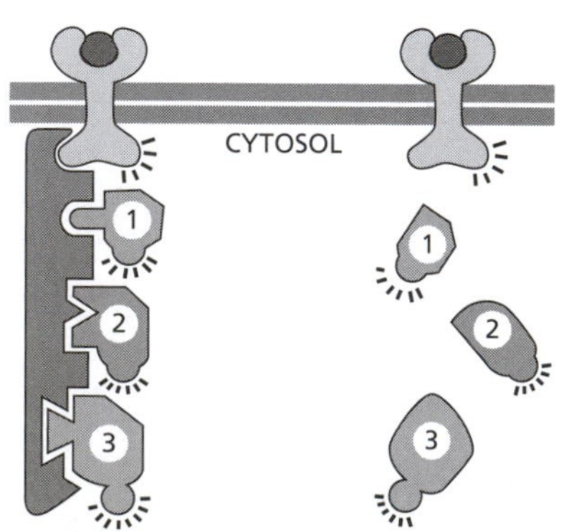


**Figure 15–2** A protein kinase cascade organized by a scaffolding protein or composed of freely diffusing components (Problem 15–38).

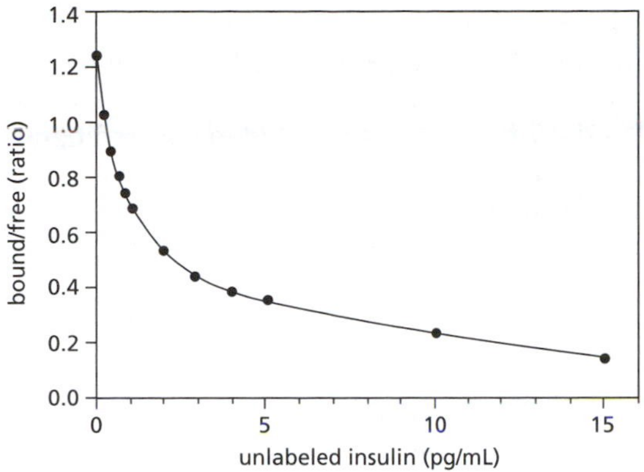


**Figure 15–3** Calibration curve for radioimmunoassay of insulin (Problem 15–42).

A. What is the concentration of insulin in each of these unknown samples?
B. What portion of the standard curve is the most accurate, and why?
C. If the antibodies were raised against pig insulin, which is similar but not identical to human insulin, would the assay still be valid for measuring human insulin concentrations?

**15–43** The optimal sensitivity of radioimmunoassay (RIA) occurs when the concentrations of the unknown and the radioactive tracer are equal. Thus the sensitivity of RIA is limited by the specific activity of the radioactive ligand. The more highly radioactive the ligand, the smaller the amount needed per assay and, therefore, the smaller amount of unknown that can be detected.

You wish to measure an unknown sample of insulin (molecular weight, 11,466) using an insulin tracer labeled with radioactive iodine, which has a half-life of 7 days. Assume that you can attach one atom of radioactive iodine per molecule of insulin, that iodine does not interfere with antibody binding, that each radioactive disintegration has a 50% probability of being detected as a 'count,' and that the limit of detection in the RIA requires a total of at least 1000 counts per minute (cpm) distributed between the bound and free fractions.

A. Given these parameters, calculate how many picograms of radioactive insulin will be required for an assay.
B. At optimal sensitivity, how many picograms of unlabeled insulin will you be able to detect?

**15–44** Two intracellular molecules, A and B, are normally synthesized at a constant rate of 1000 molecules per second per cell. Each molecule of A survives an average of 100 seconds, while each molecule of B survives an average of 10 seconds.

A. How many molecules of A and B will a cell contain?
B. If the rates of synthesis of both A and B were suddenly increased 10-fold to 10,000 molecules per second—without any change in their average lifetimes—how many molecules of A and B would be present after 1 second?
C. Which molecule would be preferred for rapid signaling? Explain your answer.

## DATA HANDLING

**15–45** There are roughly 10 billion protein molecules per cell, but cell-surface receptors are typically present at only 1000 to 100,000 copies per cell. Their low abundance makes them extremely difficult to purify by normal biochemical methods. One common approach is to clone the receptor gene and then express it to make the receptor. Cloning is usually done using a pooling strategy. A cDNA library is made from a cell that expresses the receptor. The library is divided into pools of 1000 plasmid clones and each pool is then transfected into mammalian cells that normally do not express the receptor. Cells that have received plasmids survive and form colonies on a Petri dish. A solution containing a radioactive ligand specific for the receptor is added to the dishes, incubated for a time, and then washed off. Cells are fixed in ice-cold methanol and the dishes are subjected to autoradiography to identify colonies that have bound the ligand (Figure 15–4).

A. Which pool contains a cDNA for the receptor?
B. Why don't all the colonies on the petri dish transfected with the positive pool bind the radioactive ligand?
C. How would you complete the cloning to obtain a pure plasmid containing the receptor cDNA?

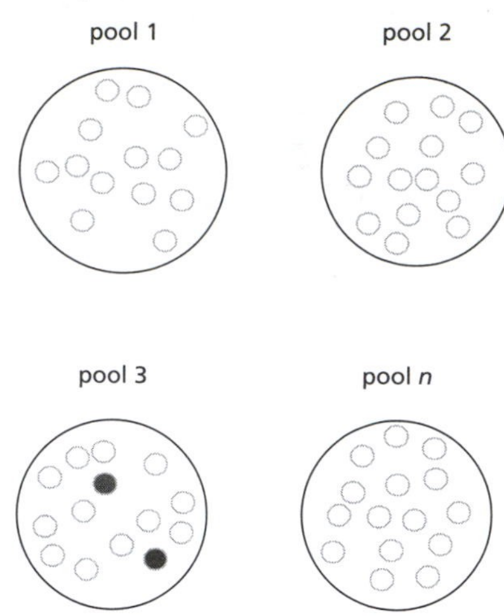


**Figure 15–4** Autoradiograph of colonies transfected with pools from a cDNA library (Problem 15–45). *Dark circles* indicate colonies that have bound the radioactive ligand. In such an autoradiograph, unlabeled colonies would not be visible; they have been included here for completeness.

**15–46** The cellular slime mold *Dictyostelium discoideum* is a eucaryote that lives on the forest floor as independent motile cells called amoebae, which feed on bacteria and yeast. When their food supply is exhausted, the amoebae stop dividing and gather together to form tiny, multicellular, wormlike structures, which crawl about as slugs. How do individual amoebae know when to stop dividing and how to find their way into a common aggregate? A set of classic experiments investigated this phenomenon more than half a century ago.

Amoebae aggregate when placed on a glass coverslip under water, provided that simple salts are present. The center of the aggregation pattern can be removed with a pipette and placed in a field of fresh amoebae, which immediately start streaming toward it. Thus, the center is emitting some sort of attractive signal. Four experiments were designed to determine the nature of the signal. In each, an existing center of aggregation was used as the source of the signal and previously unexposed amoebae served as the target cells. The arrangements of aggregation centers and test amoebae at the beginning and end of the experiments are shown in Figure 15–5.

Do these results show that *Dictyostelium discoideum* aggregates through the action of a secreted chemical signal? Explain your reasoning.

**15–47** A segment of the Moloney murine sarcoma virus carries an enhancer of gene expression, which confers glucocorticoid responsiveness on genes that are linked to it. Figure 15–6 shows the activity of a reporter gene (chloramphenicol acetyl transferase, *Cat*) after transfection into two cell lines in the presence and absence of dexamethasone, a glucocorticoid. You are puzzled by the results with cell line 1 because the viral enhancer increased CAT expression 20-fold in the absence of dexamethasone.

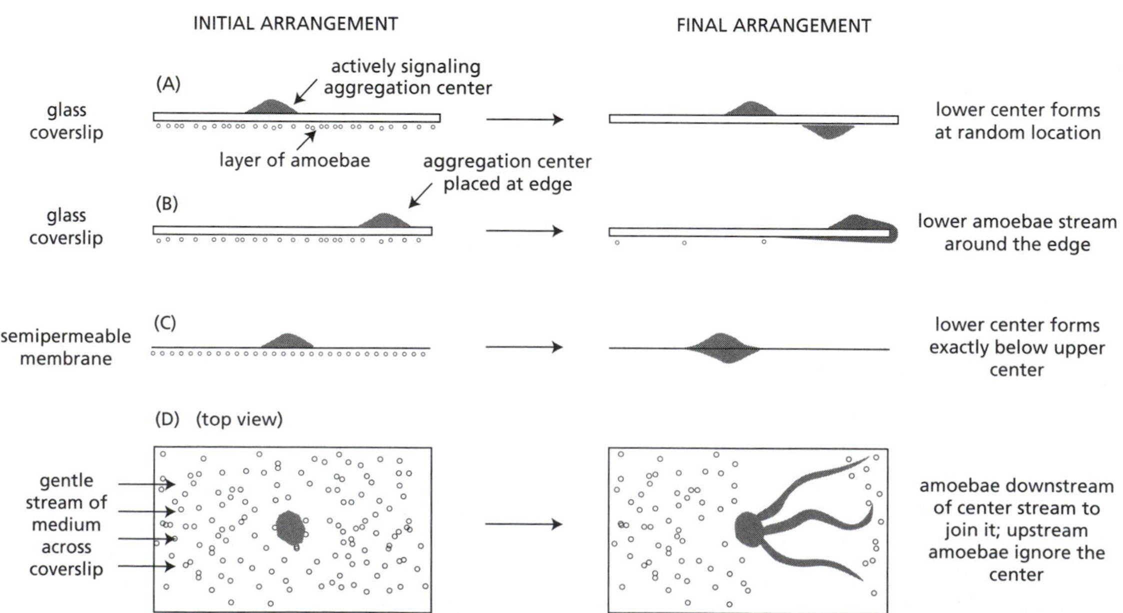


**Figure 15–5** Four experiments to study the nature of the attractive signal generated by aggregation centers (Problem 15–46).

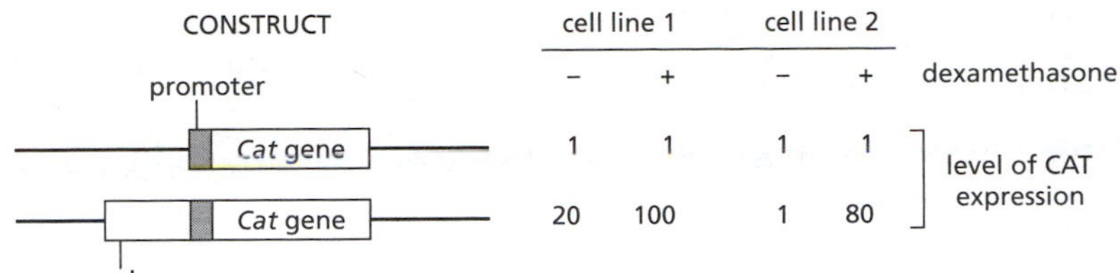


**Figure 15–6** Transfection of constructs into two different cell lines (Problem 15–47). Presence (+) and absence (–) of the hormone (dexamethasone) are indicated. *Numbers* indicate expression levels of the *Cat* gene product; in all cases, the values are expressed relative to a construct without the viral enhancer.

A. Do both cell lines contain glucocorticoid receptors? How can you tell?
B. Propose an explanation for the difference in the CAT activity in cell lines 1 and 2 after transfection with the construct containing the viral enhancer in the absence of dexamethasone.
C. Based on your explanation, predict the outcome of an experiment in which a variety of shorter pieces of the viral enhancer are placed in front of the *Cat* gene and tested for CAT activity in these two cell lines.

**15–48** Glucocorticoids induce transcription of mouse mammary tumor virus (MMTV) genes. Using hybrid constructs containing the MMTV long terminal repeat (LTR) linked to an easily assayable gene, you have shown that the LTR contains regulatory elements that respond to glucocorticoids. To map the glucocorticoid response elements within the LTR, you delete different portions of the LTR (Figure 15–7) and measure the binding of glucocorticoid receptors to the remaining DNA segments. To measure binding, you cut each of the mutant DNAs into fragments, purify the fragments containing LTR sequences, label the ends with $^{32}P$, and then incubate a mixture of the labeled fragments with the purified glucocorticoid receptor. You assess binding by passing the incubation mixture through a nitrocellulose filter, which binds protein (and any attached DNA) but not free DNA. You display the DNA fragments that were bound to the receptor by agarose gel electrophoresis. The electrophoretic patterns of the starting mixture of fragments and the bound fragments are shown in Figure 15–7.

A. Where in the LTR are the glucocorticoid response elements located?
B. A list of 11 MMTV sequences (several from the LTR) that are bound by the glucocorticoid receptor are shown in Figure 15–8. For ease of analysis the sequences have already been aligned via the consensus sequence to which the receptor binds. Identify the consensus sequence for glucocorticoid receptor binding. (A consensus sequence is the average or most typical form

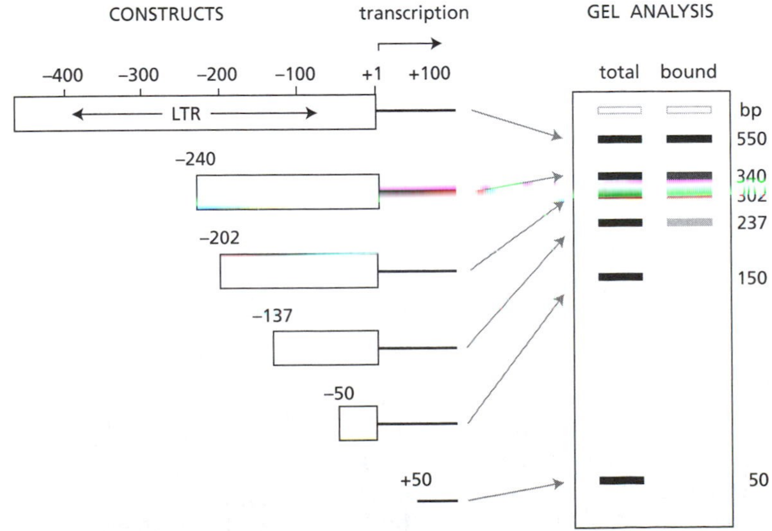


**Figure 15–7** Diagram of the LTR deletions and the electrophoretic patterns of the starting and bound mixtures of DNA fragments (Problem 15–48). Nucleotides are numbered relative to the start site for transcription, which is at +1. Negative numbers are in front of the start site, and positive numbers are after it. The electrophoretic pattern represents an autoradiograph of the agarose gel.

1. CCAAGGAGGGGACAGTGGCTGGACTAATAG
2. GGACTAATAGAACATTATTCTCCAAAAACT
3. TCGTTTTAAGAACAGTTTGTAACCAAAAAC
4. AGGATGTGAGACAAGTGGTTTCCTGACTTG
5. AGGAAAATAGAACACTCAGAGCTCAGATCA
6. CAGAGCTCAGATCAGAACCTTTGATACCAA
7. CATGATTCAGCACAAAAAGAGCGTGTGCCA
8. CTGTTATTAGGACATCGTCCTTTCCAGGAC
9. CCTAGTGTAGATCAGTCAGATCAGATTAAA
10. GATCAGTCAGATCAGATTAAAAGCAAAAAG
11. TTCCAAATAGATCCTTTTTGCTTTTAATCT

**Figure 15–8** DNA sequences that are bound by the glucocorticoid receptor (Problem 15–48). The DNA sequences have been aligned by their consensus binding sequences.

of a sequence that is reproduced with minor variations. It shows the nucleotide most often found at each position.)

**15–49** Studies with the fruit fly *Drosophila* provided initial clues to the complex changes in patterns of gene expression that a simple hormone can trigger. *Drosophila* larvae molt in response to an increase in the concentration of the steroid hormone ecdysone. The polytene chromosomes of the *Drosophila* salivary glands are an excellent experimental system in which to study the pattern of gene activity initiated by the hormone because active genes enlarge into puffs that are visible in the light microscope. Furthermore, the size of a puff is proportional to the rate at which the gene is transcribed. Prior to addition of ecdysone, a few puffs—termed intermolt puffs—are already active. Upon exposure of dissected salivary glands to ecdysone, these intermolt puffs regress, and two additional sets of puffs appear. The early puffs arise within a few minutes after addition of ecdysone; the late puffs arise within 4–10 hours. The concentration of ecdysone does not change during this time period. The pattern of puff appearance and disappearance is illustrated for a typical puff in each category in Figure 15–9A.

Two critical experiments helped to define the relationships between the different classes of puff. In the first, cycloheximide, which blocks protein synthesis, was added at the same time as ecdysone. As illustrated in Figure 15–9B, under these conditions the early puffs did not regress and the late puffs were not induced. In the second experiment, ecdysone was washed out after a 2-hour exposure. As illustrated in Figure 15–9C, this treatment caused an immediate regression of the early puffs and a *premature induction* of the late puffs.

A. Why do you think the early puffs didn't regress and the late puffs weren't induced in the presence of cycloheximide? Why do you think the intermolt puffs were unaffected?
B. Why do you think the early puffs regressed immediately when ecdysone was removed? Why do you think the late puffs arose prematurely under these conditions?
C. Outline a model for ecdysone-mediated regulation of the puffing pattern.

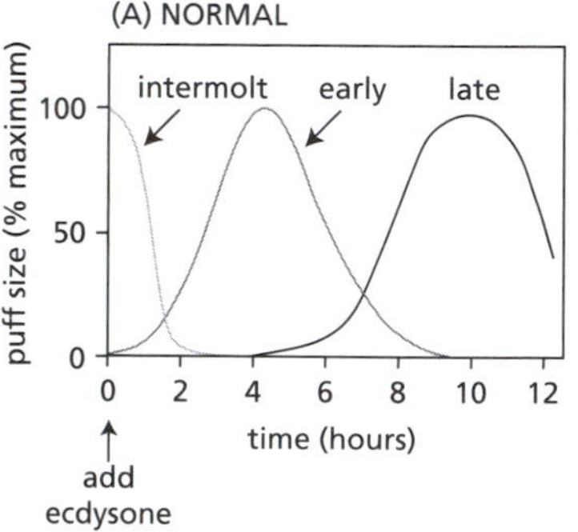


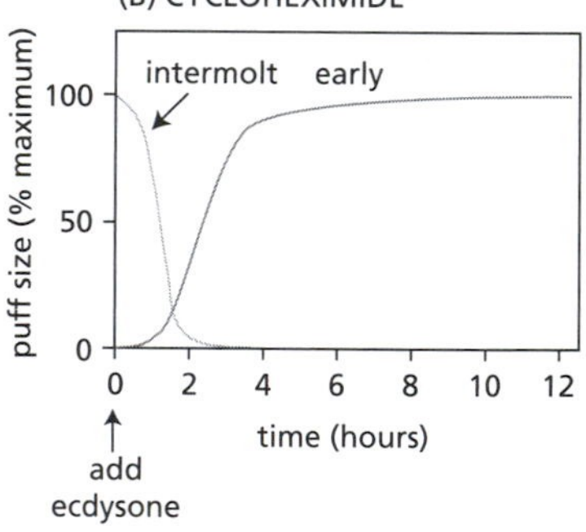


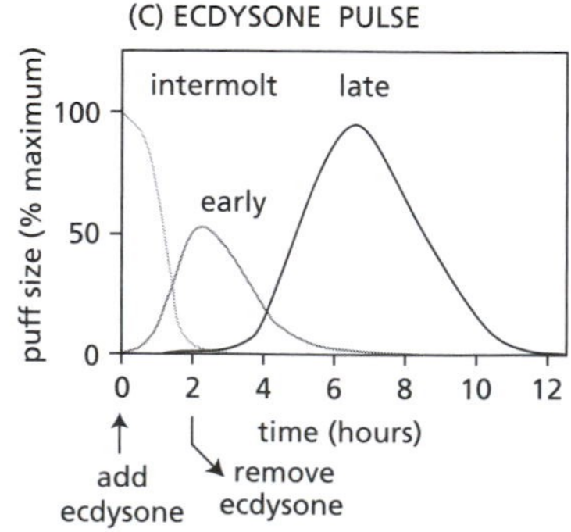


**Figure 15–9** Puffing patterns in salivary gland giant chromosomes (Problem 15–49). (A) Normal puffing pattern. (B) Puffing pattern in the presence of cycloheximide. (C) Puffing pattern after removal of ecdysone.

**15–50** Activation ('maturation') of frog oocytes is signaled through a MAP kinase signaling module. An increase in the hormone progesterone triggers the module by stimulating the translation of the mRNA for Mos, which is the frog's MAP kinase kinase kinase (Figure 15–10). Maturation is easy to score visually by the presence of a white spot in the middle of the brown surface of the oocyte (Figure 15–10). To determine the dose–response curve for progesterone-induced activation of MAP kinase, you place 16 oocytes in each of six plastic dishes and add various concentrations of progesterone. After an overnight incubation, you crush the oocytes, prepare an extract, and determine the state of MAP kinase phosphorylation (hence, activation) by SDS polyacrylamide gel electrophoresis (Figure 15–11A). This analysis shows a graded response of MAP kinase to increasing concentrations of progesterone.

Before you crushed the oocytes, you noticed that not all oocytes in individual dishes had white spots. Had some oocytes undergone partial activation and not yet reached the white-spot stage? To answer this question, you repeat the experiment but this time you analyze MAP kinase activation in

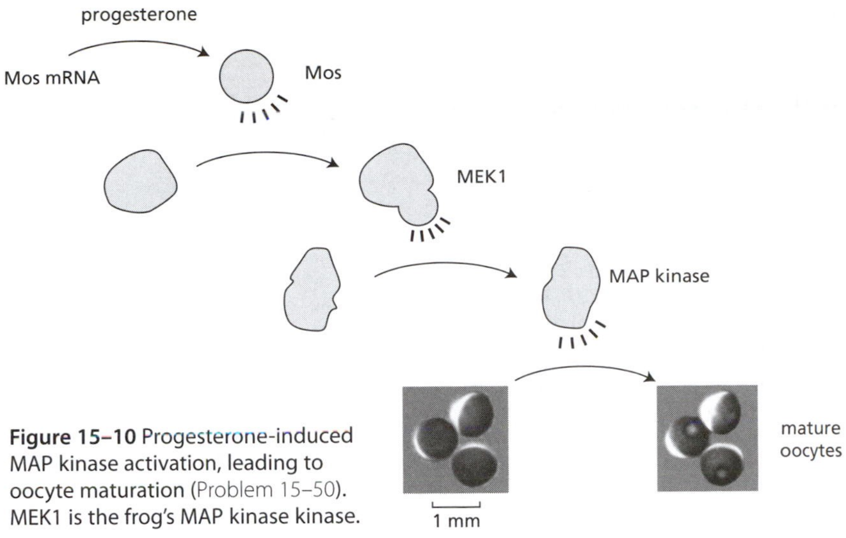


**Figure 15–10** Progesterone-induced MAP kinase activation, leading to oocyte maturation (Problem 15–50). MEK1 is the frog's MAP kinase kinase.

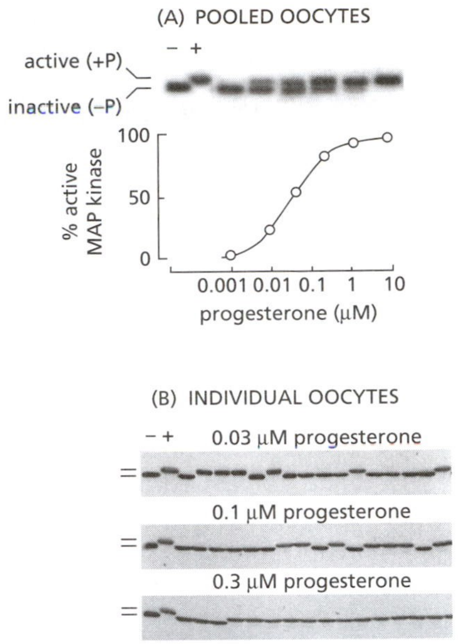

**Figure 15–11** Activation of frog oocytes (Problem 15–50). (A) Phosphorylation of MAP kinase in pooled oocytes. (B) Phosphorylation of MAP kinase in individual oocytes. MAP kinase was detected by immunoblotting using a MAP-kinase-specific antibody. The first two lanes in each gel contain nonphosphorylated, inactive MAP kinase (–) and phosphorylated, active MAP kinase (+).

individual oocytes. You are surprised to find that each oocyte has either a fully activated or a completely inactive MAP kinase (Figure 15–11B). How can an all-or-none response in individual oocytes give rise to a graded response in the population?

**15–51** After prolonged exposure to isoproterenol, which binds to β-adrenergic receptors and activates adenylyl cyclase, cells become refractory and cease responding. To investigate the basis for this desensitization phenomenon, you treat cells for various times with isoproterenol, wash it out, and then assay for β-adrenergic receptors by measuring the binding of dihydroalprenolol, which is hydrophobic, and CGP-12177, which is hydrophilic. You find that CGP-12177 binding decreases in parallel to adenylyl cyclase activity, whereas dihydroalprenolol binding remains high (Figure 15–12). When you lyse isoproterenol-treated (desensitized) cells and fractionate membrane-enclosed vesicles by centrifugation through sucrose-density gradients, you find that CGP-12177 binds to vesicles derived from the plasma membrane (as indicated by the presence of the marker enzyme 5′-nucleotidase), whereas dihydroalprenolol binds not only to these vesicles, but also to an additional population of vesicles (Figure 15–13).

A. Give an explanation for the differences in binding by dihydroalprenolol and CGP-12177.

B. What do you think might be the basis for isoproterenol-induced desensitization in these cells?

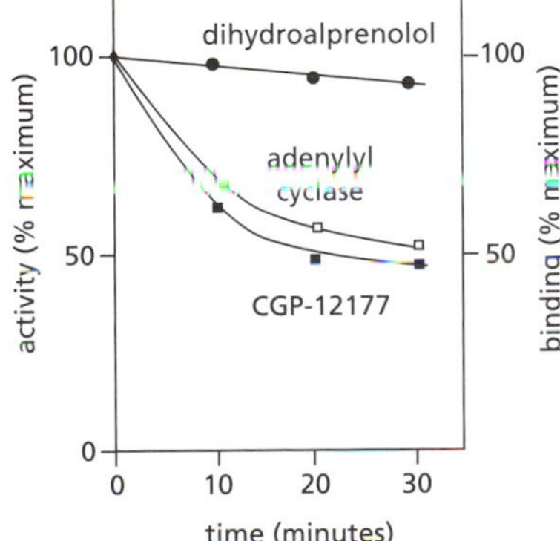


**Figure 15–12** Adenylyl cyclase activity, $^3$H-CGP-12177 binding and $^3$H-dihydroalprenolol binding at various times after treatment with isoproterenol (Problem 15–51). All activities are expressed as a percentage of the values at time zero.

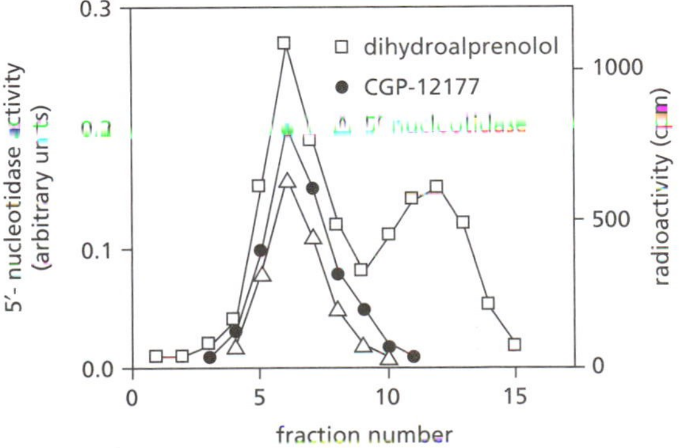


**Figure 15–13** Ligand binding in sucrose-gradient fractions from isoproterenol-treated cells (Problem 15–51).

**15–52** The nicotinic acetylcholine receptor is a neurotransmitter-dependent ion channel, which is composed of four types of subunit. Phosphorylation of the receptor by protein kinase A attaches one phosphate to the $\gamma$ subunit and one phosphate to the $\delta$ subunit. Fully phosphorylated receptors desensitize much more rapidly than unmodified receptors. To study this process in detail, you phosphorylate two preparations of receptor to different extents (0.8 mole phosphate/mole receptor and 1.2 mole phosphate/mole receptor) and measure desensitization over several seconds (Figure 15–14). Both preparations behave as if they contain a mixture of receptors; one form that is rapidly desensitized (the initial steep portion of the curves) and another form that is desensitized at the same rate as the untreated receptor.

A. Assuming that the $\gamma$ and $\delta$ subunits are independently phosphorylated at equal rates, calculate the percentage of receptors that carry zero, one, and two phosphates per receptor at the two extents of phosphorylation.

B. Do these data suggest that desensitization requires one phosphate or two phosphates per receptor? If you decide that desensitization requires only one phosphate, indicate whether the phosphate has to be on one specific subunit or can be on either of the subunits.

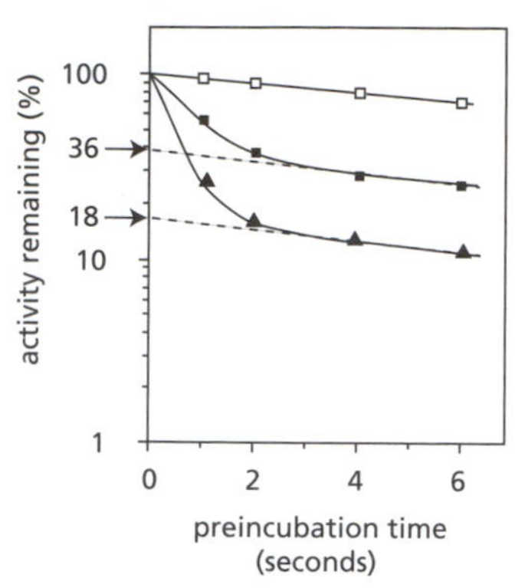


**Figure 15–14** Desensitization rates for untreated acetylcholine receptor and two preparations of phosphorylated receptor (Problem 15–52). *Open squares* represent untreated receptors; *filled squares* represent receptors with 0.8 mole phosphate/mole receptor; and *filled triangles* represent receptors with 1.2 mole phosphate/mole receptor. *Arrows* indicate the fractions of the phosphorylated preparations that behaved like the untreated receptor.

# SIGNALING THROUGH G-PROTEIN-COUPLED CELL-SURFACE RECEPTORS (GPCRs) AND SMALL INTRACELLULAR MEDIATORS

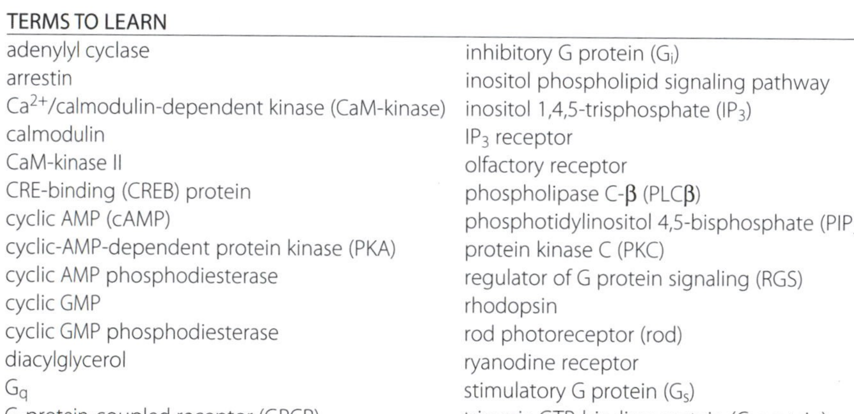

| TERMS TO LEARN | |
|---|---|
| adenylyl cyclase | inhibitory G protein ($G_i$) |
| arrestin | inositol phospholipid signaling pathway |
| $Ca^{2+}$/calmodulin-dependent kinase (CaM-kinase) | inositol 1,4,5-trisphosphate ($IP_3$) |
| calmodulin | $IP_3$ receptor |
| CaM-kinase II | olfactory receptor |
| CRE-binding (CREB) protein | phospholipase C-$\beta$ (PLC$\beta$) |
| cyclic AMP (cAMP) | phosphotidylinositol 4,5-bisphosphate ($PIP_2$) |
| cyclic-AMP-dependent protein kinase (PKA) | protein kinase C (PKC) |
| cyclic AMP phosphodiesterase | regulator of G protein signaling (RGS) |
| cyclic GMP | rhodopsin |
| cyclic GMP phosphodiesterase | rod photoreceptor (rod) |
| diacylglycerol | ryanodine receptor |
| $G_q$ | stimulatory G protein ($G_s$) |
| G-protein-coupled receptor (GPCR) | trimeric GTP-binding protein (G protein) |
| GPCR kinase (GRK) | |

## DEFINITIONS

Match the definition below with its term from the list above.

**15–53** G protein that activates adenylyl cyclase and thereby increases cyclic AMP concentration.

**15–54** Protein composed of three subunits, one of which is activated by the binding of GTP.

**15–55** Ubiquitous calcium-binding protein whose interactions with other proteins is governed by changes in intracellular $Ca^{2+}$ concentration.

**15–56** Enzyme that hydrolyzes cyclic AMP to adenosine 5′-monophosphate (5′-AMP).

**15–57** Cell-surface receptor that associates with an intracellular G protein upon activation by an extracellular ligand.

**15–58** Enzyme that participates in desensitization of GPCRs by phosphorylating them after they have been activated by ligand binding.

**15–59** $Ca^{2+}$-release channel in the ER membrane that is activated by $Ca^{2+}$ binding in the absence of $IP_3$.

**15–60** Enzyme bound to the cytoplasmic surface of the plasma membrane that converts membrane $PI(4,5)P_2$ to diacylglycerol and $IP_3$.

**15–61** Protein that is an α-subunit-specific GTPase-activating protein (GAP).

**15–62** Small intracellular mediator that is released from a phospholipid in the plasma membrane and diffuses to the ER, where it opens $Ca^{2+}$-release channels.

**15–63** Enzyme that phosphorylates target proteins in response to a rise in intracellular cyclic AMP.

**15–64** A $Ca^{2+}$-dependent protein kinase that is activated by diacylglycerol.

**15–65** Light-sensitive GPCR in rod photoreceptor cells of the retina.

**15–66** Protein kinase whose activity is regulated by the binding of $Ca^{2+}$-activated calmodulin, and which indirectly mediates the effects of $Ca^{2+}$ by phosphorylation of other proteins.

**15–67** Protein that binds to the cyclic AMP response elements found in the regulatory region of many genes activated by cyclic AMP.

## TRUE/FALSE

Decide whether each of these statements is true or false, and then explain why.

**15–68** Different isoforms of protein kinase A in different cell types explain why the effects of cyclic AMP vary depending on the target cell.

**15–69** The activity of any protein regulated by phosphorylation depends on the balance at any instant between the activities of the kinases that phosphorylate it and the phosphatases that dephosphorylate it.

**15–70** In contrast to the more direct signaling pathways used by nuclear receptors, catalytic cascades of intracellular mediators provide numerous opportunities for amplifying the responses to extracellular signals.

## THOUGHT PROBLEMS

**15–71** GPCRs activate G proteins by reducing the strength of GDP binding, allowing GDP to dissociate and GTP, which is present at much higher concentrations, to bind. How do you suppose the activity of a G protein would be affected by a mutation that caused its affinity for GDP to be reduced without significantly changing its affinity for GTP?

**15–72** When adrenaline (epinephrine) binds to adrenergic receptors on the surface of a muscle cell, it activates a G protein, initiating a signaling pathway that results in breakdown of muscle glycogen. How would you expect glycogen breakdown to be affected if muscle cells were injected with a nonhydrolyzable analog of GTP, which can't be converted to GDP? Consider what would happen in the absence of adrenaline and after a brief exposure to it.

**15–73** Should RGS (regulator of G protein signaling) proteins be classified as GEFs (guanine nucleotide exchange factors) or GAPs (GTPase-activating proteins)? Explain what role this activity plays in modulating G-protein-mediated responses in animals and yeasts.

**15–74** What is 'cyclic' about cyclic AMP?

**15–75** Explain why cyclic AMP must be broken down rapidly in a cell to allow rapid signaling.

**15–76** You are trying to purify adenylyl cyclase from brain. The assay is based on the conversion of α-$^{32}$P-ATP to cAMP. You can easily detect activity in crude brain homogenates stimulated by isoproterenol, which binds to β-adrenergic receptors, but the enzyme loses activity when low-molecular-weight cofactors are removed by dialysis. What single molecule do you think you could add back to the dialyzed homogenate to restore activity?

**15–77** Propose specific types of mutation in the gene for the regulatory subunit of cyclic AMP-dependent protein kinase (PKA) that could lead to either a permanently active PKA, or to a permanently inactive PKA.

**15–78** During a marathon, runners draw heavily on their internal reserves of glycogen (carbohydrate) and triglycerides (fat) to fuel muscle contraction. Initially, energy is derived mostly from carbohydrates, with increasing amounts of fat being used as the race progresses. If runners use up their muscle glycogen reserves before they finish the race, they hit what is known as 'the wall,' a point of diminished performance that arises because fatty acids from triglyceride breakdown cannot be delivered to the muscles quickly enough to sustain maximum effort. One trick that marathon runners use to avoid the wall is to drink a cup of strong black coffee an hour or so before the race begins. Coffee contains caffeine, which is an inhibitor of cyclic AMP phosphodiesterase. How do you suppose inhibition of this enzyme helps them avoid the wall?

**15–79** How is an $IP_3$-triggered $Ca^{2+}$ response terminated?

**15–80** Why do you suppose cells use $Ca^{2+}$ (intracellular concentration $10^{-7}$ M) for signaling rather than the more abundant $Na^+$ (intracellular concentration $10^{-3}$ M)?

**15–81** EGTA chelates $Ca^{2+}$ with high affinity and specificity. How would microinjection of EGTA affect glucagon-triggered breakdown of glycogen in liver? How would it affect vasopressin-triggered breakdown of glycogen in liver?

**15–82** Phosphorylase kinase integrates signals from the cyclic-AMP-dependent and $Ca^{2+}$-dependent signaling pathways that control glycogen breakdown in liver and muscle cells (Figure 15–15). Phosphorylase kinase is composed of four subunits. One is the protein kinase that catalyzes the addition of phosphate to glycogen phosphorylase to activate it for glycogen breakdown. The other three subunits are regulatory proteins that control the activity of the catalytic subunit. Two contain sites for phosphorylation by PKA, which is activated by cyclic AMP. The remaining subunit is calmodulin, which binds $Ca^{2+}$ when the cytosolic concentration rises. The regulatory subunits control the equilibrium between the active and inactive conformations of the catalytic subunit.

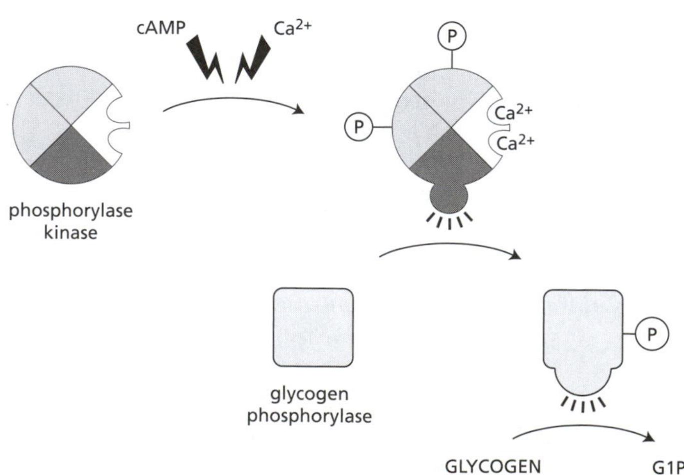


**Figure 15–15** Integration of cyclic-AMP-dependent and $Ca^{2+}$-dependent signaling pathways by phosphorylase kinase in liver and muscle cells (Problem 15–82). G1P is glucose 1-phosphate, the product of the cleavage of a glucose from glycogen using a phosphate.

How does this arrangement allow phosphorylase kinase to serve its role as an integrator protein for the multiple pathways that stimulate glycogen breakdown?

**15–83** CaM-kinase II is a remarkable molecular memory device. How does CaM-kinase II 'remember' its exposure to $Ca^{2+}$/calmodulin and why does it eventually 'forget?'

**15–84** The outer segments of rod photoreceptor cells can be broken off, isolated, and used to study the effects of small molecules on visual transduction because the broken end of each segment remains unsealed. How would you expect the visual response to be affected by the following additions?

A. An inhibitor of cyclic GMP phosphodiesterase.
B. A nonhydrolyzable analog of GTP.
C. An inhibitor of rhodopsin-specific kinase.

**15–85** Patients with Oguchi's disease have an inherited form of nightblindness. After a flash of bright light these individuals recover their night vision (become dark adapted) very slowly. Night vision depends almost entirely on the visual responses of rod photoreceptor cells. What aspect of the visual response in these patients' rod cells do you suppose is defective? What genes, when defective, might give rise to Oguchi's disease?

**15–86** In muscle cells, adrenaline binds to the β-adrenergic receptor to initiate a signaling cascade that leads to the breakdown of glycogen (Figure 15–16). At what points in this pathway is the signal amplified?

**15–87** A critical feature of all signaling cascades is that they must be turned off rapidly when the extracellular signal is removed. Examine the signaling cascade in Figure 15–16. Describe how each component of this signaling pathway is returned to its inactive state when adrenaline is removed.

## CALCULATIONS

**15–88** In a classic paper the number of β-adrenergic receptors on the membranes of frog erythrocytes was determined by using a competitive inhibitor of adrenaline, alprenolol, which binds to the receptors 500 times more tightly than adrenaline. These receptors normally bind adrenaline and stimulate adenylyl cyclase activity. Labeled alprenolol was mixed with erythrocyte membranes, left for 10 minutes at 37°C, and then the membranes were

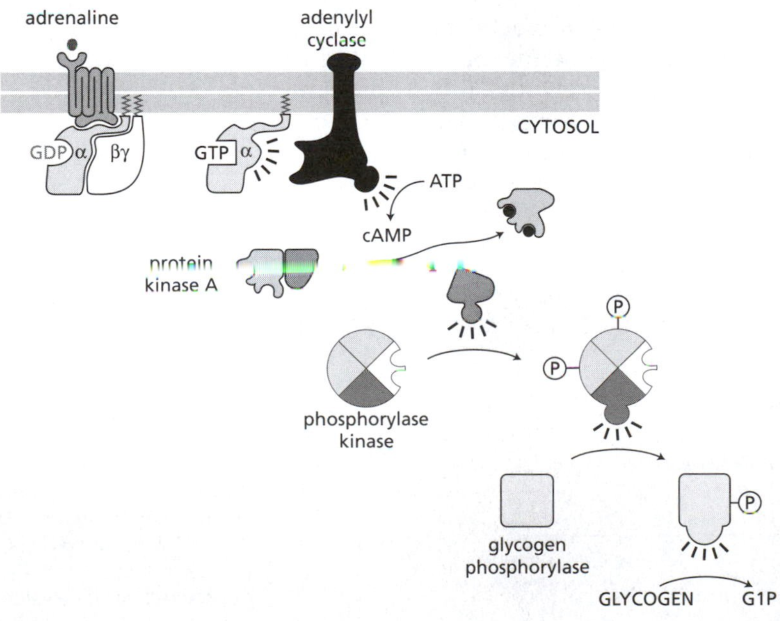


**Figure 15–16** Signaling cascade for activation of glycogen breakdown by adrenaline in muscle cells (Problem 15–86). G1P is glucose 1-phosphate; cAMP bound to the regulatory subunits of PKA is shown as *black balls*.

pelletted by centrifugation and the radioactivity in the pellet was measured. The experiment was done in two ways. The binding of increasing amounts of $^3$H-alprenolol to a fixed amount of erythrocyte membranes was measured to determine total binding. The experiment was repeated in the presence of a vast excess of unlabeled alprenolol to measure nonspecific binding. The results are shown in Figure 15–17.

A. On Figure 15–17 sketch the curve for specific binding of alprenolol to β-adrenergic receptors. Has alprenolol binding to the receptors reached saturation?

B. Assuming that one molecule of alprenolol binds per receptor, calculate the number of β-adrenergic receptors on the membrane of a frog erythrocyte. The specific activity of the labeled alprenolol is $1 \times 10^{13}$ cpm/mmol, and there are $8 \times 10^8$ frog erythrocytes per milligram of membrane protein.

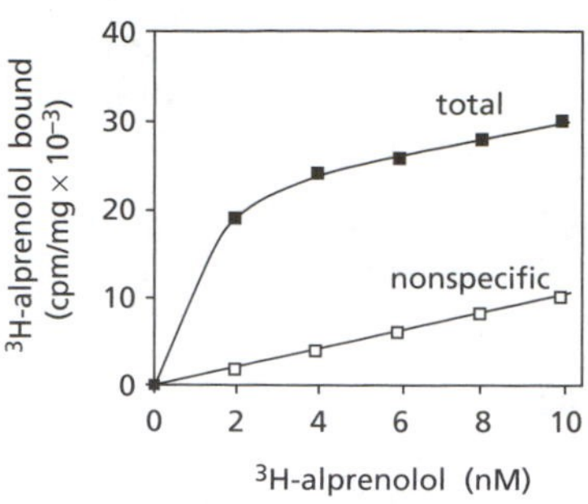


**Figure 15–17** Binding of $^3$H-alprenolol to frog erythrocyte membranes (Problem 15–88).

**15–89** In visual transduction one activated rhodopsin molecule leads to the hydrolysis of $5 \times 10^5$ cyclic GMP molecules per second. One stage in this enormous signal amplification is achieved by cyclic GMP phosphodiesterase, which hydrolyzes 1000 molecules of cyclic GMP per second. The additional factor of 500 could arise because one activated rhodopsin activates 500 transducin ($G_t$) molecules, or because one activated transducin activates 500 cyclic GMP phosphodiesterases, or through a combination of both effects. One experiment to address this question measured the amount of GppNp (a nonhydrolyzable analog of GTP) that is bound by transducin in the presence of different amounts of activated rhodopsin. As indicated in Figure 15–18, 5.5 mmol of GppNp were bound per mole of total rhodopsin when 0.0011% of the rhodopsin was activated.

A. Assuming that each transducin molecule binds one molecule of GppNp, calculate the number of transducin molecules that are activated by each activated rhodopsin molecule. Which mechanism of amplification does this measurement support?

B. Binding studies have shown that transducin-GDP has a high affinity for activated rhodopsin and that transducin-GTP has a low affinity; conversely, transducin-GTP has a high affinity and transducin-GDP has a low affinity for cyclic GMP phosphodiesterase. Are these affinities consistent with the mechanism of amplification you deduced from the above experiment? Explain your reasoning.

## DATA HANDLING

**15–90** The mating behavior of yeast depends on signaling peptides termed pheromones that bind to pheromone GPCRs (Figure 15–19). When the α-factor pheromone binds to a wild-type yeast cell, it blocks cell-cycle progression, arresting proliferation until a mating partner is found. Yeast mutants with defects in one or more of the components of the G protein have characteristic phenotypes in the absence and in the presence of the α-factor pheromone (Table 15–1). Strains with defects in any of these genes cannot undergo the mating response and are therefore termed sterile.

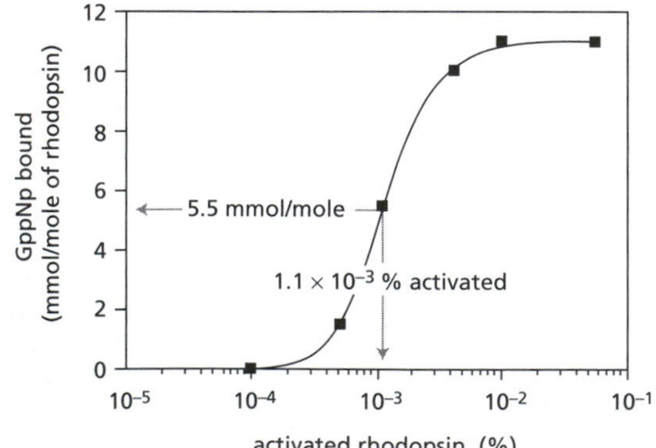


**Figure 15–18** Binding of GppNp to rod cell membranes as a function of the fraction of activated rhodopsin (Problem 15–89). Background binding of GppNp to rod cell membranes in the dark has been subtracted from the values shown.

**Table 15–1 Mating phenotypes of various mutant and nonmutant strains of yeast** (Problem 15–90).

| MUTATION | PHENOTYPE MINUS α FACTOR | PHENOTYPE PLUS α FACTOR |
|---|---|---|
| None (wild type) | normal proliferation | arrested proliferation, mating response |
| α subunit deleted | arrested proliferation | arrested proliferation, sterile |
| β subunit deleted | normal proliferation | normal proliferation, sterile |
| γ subunit deleted | normal proliferation | normal proliferation, sterile |
| α and β deleted | normal proliferation | normal proliferation, sterile |
| α and γ deleted | normal proliferation | normal proliferation, sterile |
| β and γ deleted | normal proliferation | normal proliferation, sterile |

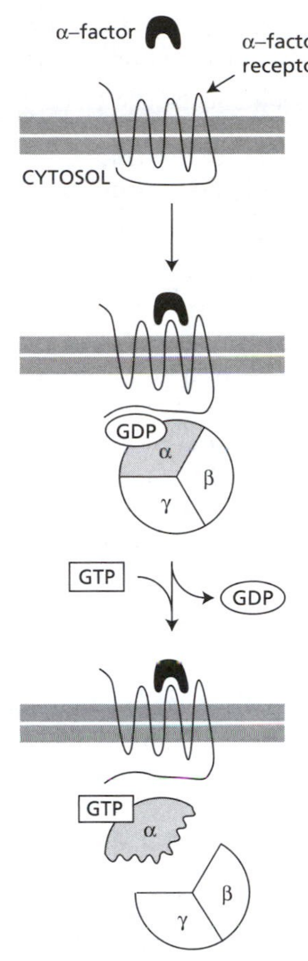


**Figure 15–19** α-Factor pheromone signaling via α-factor GPCR and G protein (Problem 15–90).

A. Based on genetic analysis of the yeast mutants, decide which component of the G protein normally transmits the mating signal to the downstream effector molecules.

B. Predict the proliferation and mating phenotypes in the absence and presence of the α-factor pheromone of strains with the following mutant α subunits:

1. An α subunit that can bind GTP but cannot hydrolyze it.
2. An α subunit with an altered N-terminus to which the fatty acid myristoylate cannot be added, thereby preventing its localization to the plasma membrane.
3. An α subunit that cannot bind to the activated pheromone receptor.

**15–91** You are baffled. You are studying the control of cyclic AMP levels in brain slices. You have confirmed that signal molecules such as isoproterenol that act through β-adrenergic receptors cause a modest increase in cyclic AMP, as expected from G-protein-mediated coupling between the receptor and adenylyl cyclase. You find a puzzling synergy, however, between isoproterenol and a number of pharmacological agents that by themselves have no effect on cyclic AMP levels. What is the basis for this paradoxical augmentation of cyclic AMP levels?

A biochemist friend of yours has suggested a possible explanation: she has found in *in vitro* experiments that βγ subunits from inhibitory trimeric G proteins stimulate type II adenylyl cyclase, which is expressed in brain. To test this idea in cells, you plan to express the cDNAs encoding the component proteins in human kidney cells, which lack the receptors found in brain. In this way you hope to reconstruct the effects you observed in brain slices, but in a much simpler background.

You transfect the kidney cells with various combinations of cDNAs encoding type II adenylyl cyclase, the dopamine receptor (which interacts with an inhibitory G protein), and a mutated (constitutively active) $\alpha_s$* subunit. You measure the levels of cyclic AMP in the resulting cell lines in the absence or presence of quinpirole, which activates the dopamine receptor (Figure 15–20). You also measure the effects of pertussis toxin, which blocks the signal from $G_i$-coupled receptors by modifying the $\alpha_i$ subunit in such a way that it can no longer bind GTP or dissociate from its βγ subunit (Figure 15–20).

A. Did you succeed in reproducing the paradoxical result you observed in brain slices in the transfected kidney cells? How so?

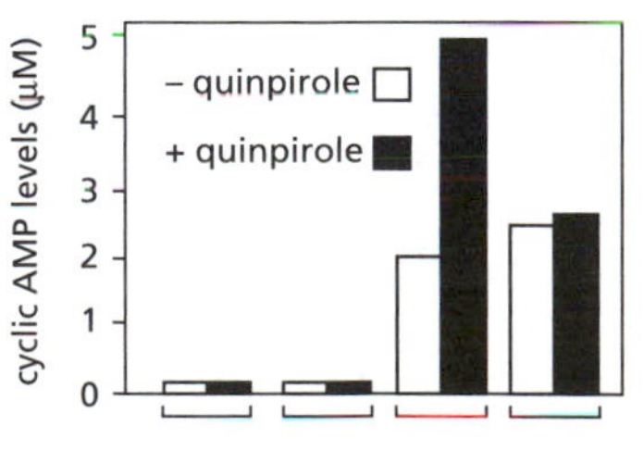


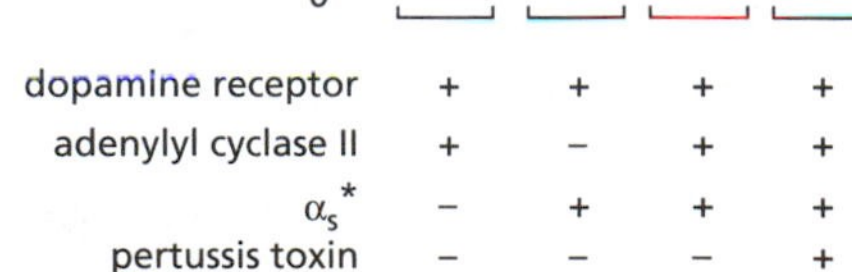

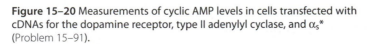

**Figure 15–20** Measurements of cyclic AMP levels in cells transfected with cDNAs for the dopamine receptor, type II adenylyl cyclase, and $\alpha_s$* (Problem 15–91).

B. Explain the effects of pertussis toxin in your experiments.
C. What do your experiments indicate is required for maximal activation of type II adenylyl cyclase? Propose a molecular explanation for the augmented activation of type II adenylyl cyclase.
D. Predict the results of expressing the cDNA for the α subunit of transducin, which does not bind to adenylyl cyclase but binds tightly to free βγ subunits.

**15–92** You are working with a hamster cell line that stops proliferating when intracellular cyclic AMP reaches high levels. By stimulating adenylyl cyclase with cholera toxin and by inhibiting cyclic AMP phosphodiesterase with theophylline, cyclic AMP can be artificially elevated. Under these conditions only cells that are resistant to the effects of cyclic AMP can grow.

In this way you isolate several resistant colonies that grow under the selective conditions and assay them for PKA activity: they are all defective. About 10% of the resistant lines are completely missing PKA activity. The remainder possess PKA activity, but a very high level of cyclic AMP is required for activation. To characterize the resistant lines further, you fuse them with the parental cells and test the hybrids for resistance to cholera toxin. Hybrids between parental cells and PKA-negative cells are sensitive to cholera toxin, which indicates that the mutations in these cells are recessive. By contrast, hybrids between parental cells and resistant cells with altered PKA responsiveness are resistant to cholera toxin, which indicates that these mutations are dominant.

A. Is PKA an essential enzyme in these hamster cells?
B. PKA is a tetramer consisting of two catalytic (protein kinase) and two regulatory (cyclic-AMP-binding) subunits. Propose an explanation for why the mutations in some cyclic-AMP-resistant cell lines are recessive, while others are dominant.
C. Do these experiments support the notion that all cyclic AMP effects in hamster cells are mediated by PKA? Do they rule out the possibility that cyclic AMP-binding proteins play other critical roles in this hamster cell line?

**15–93** A particularly graphic illustration of the subtle, yet important, role of cyclic AMP in the whole organism comes from studies of the fruit fly *Drosophila melanogaster*. In search of the gene for cyclic AMP phosphodiesterase, one laboratory measured enzyme levels in flies with chromosomal duplications or deletions and found consistent alterations in flies with mutations involving bands 3D3 and 3D4 on the X chromosome. Duplications in this region have about 1.5 times the normal activity of the enzyme; deletions have about half the normal activity.

An independent laboratory at the same institution was led to the same chromosomal region through work on behavioral mutants of fruit flies. The researchers had developed a learning test in which flies were presented with two metallic grids, one of which was electrified. If the electrified grid was painted with a strong-smelling chemical, normal flies quickly learned to avoid it, even when it was no longer electrified. The mutant flies, on the other hand, never learned to avoid the smelly grid; they were aptly called *Dunce* mutants. The *Dunce* mutation was mapped genetically to bands 3D3 and 3D4.

Is the learning defect really due to lack of cyclic AMP phosphodiesterase or are the responsible genes simply closely linked? Further experiments showed that the level of cyclic AMP in *Dunce* flies was 1.6 times higher than in normal flies. Furthermore, sucrose gradient analysis of homogenates of *Dunce* and normal flies revealed two cyclic AMP phosphodiesterase activities, one of which was missing in *Dunce* flies (Figure 15–21).

A. Why do *Dunce* flies have higher levels of cyclic AMP than normal flies?
B. Explain why homozygous (both chromosomes affected) duplications of the nonmutant *Dunce* gene cause cyclic AMP phosphodiesterase levels to be elevated 1.5-fold and why homozygous deletions of the gene reduce enzyme activity to half the normal value.
C. What would you predict would be the effect of caffeine, a phosphodiesterase inhibitor, on the learning performance of normal flies?

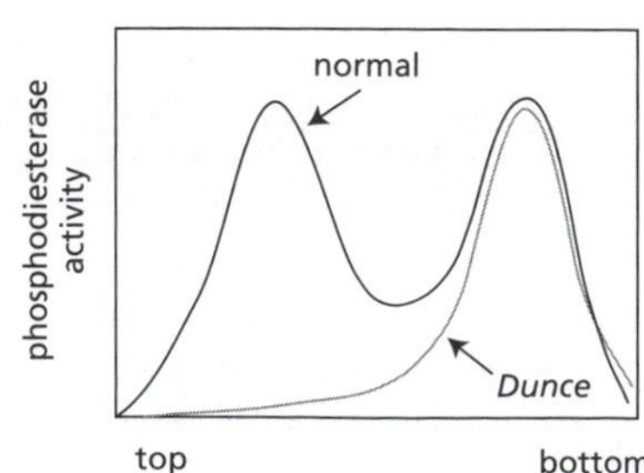


**Figure 15–21** Sucrose gradient analysis of cyclic AMP phosphodiesterase activity in homogenates of normal and *Dunce* flies (Problem 15–93).

**15–94** In a cell line derived from normal rat thyroid, stimulation of the $\alpha_1$-adrenergic receptor increases both inositol trisphosphate ($IP_3$) formation and release of arachidonic acid. $IP_3$ elevates intracellular $Ca^{2+}$, which mediates thyroxine efflux, whereas arachidonic acid serves as a source of prostaglandin $E_2$, which stimulates DNA synthesis. How is arachidonic acid release connected to the adrenergic receptor? Arachidonic acid could arise by cleavage from the diacylglycerol that accompanies $IP_3$ production. Alternatively, it could arise through an independent effect of the receptor on phospholipase $A_2$, which can directly release arachidonic acid from intact phosphoglycerides. Consider the following experimental observations:

1. Addition of noradrenaline (norepinephrine) to cell cultures stimulates the production of both $IP_3$ and arachidonic acid.
2. If the $\alpha_1$-adrenergic receptors are made unresponsive to noradrenaline by treatment with phorbol esters (which act through protein kinase C to cause phosphorylation, hence inactivation, of the receptor), addition of noradrenaline causes no increase in either $IP_3$ or arachidonic acid.
3. When cells are made permeable and treated with GTP$\gamma$S (a nonhydrolyzable analog of GTP), the production of both $IP_3$ and arachidonic acid is increased.
4. If cells are treated with neomycin (which blocks the action of phospholipase C), subsequent treatment of permeabilized cells with GTP$\gamma$S stimulates arachidonic acid production but causes no increase in $IP_3$.
5. If cells are treated with pertussis toxin (which inactivates inhibitory G protein, $G_i$), subsequent treatment of permeabilized cells with GTP$\gamma$S stimulates the production of $IP_3$ but causes no increase in arachidonic acid.

A. Which of the two proposed mechanisms for arachidonic acid production in these cells do the observations support?

B. Describe a molecular pathway for the activation of arachidonic acid production that is consistent with the experimental results.

**15–95** The primary role of platelets is to control blood clotting. When they encounter the exposed basement membrane (collagen fibers) of a damaged blood vessel or a newly forming fibrin clot, they change their shape from round to spiky and stick to the damaged area. At the same time, they begin to secrete serotonin and ATP, which accelerate similar changes in newly arriving platelets, leading to the rapid formation of a clot. The platelet response is regulated by protein phosphorylation. Significantly, platelets contain high levels of two protein kinases: PKC, which initiates serotonin release, and myosin light-chain kinase, which mediates the change in shape.

When platelets are stimulated with thrombin, the light chain of myosin and an unknown protein of 40,000 daltons are phosphorylated. When platelets are treated with a calcium ionophore, which increases membrane permeability to $Ca^{2+}$, only the myosin light chain is phosphorylated; when they are treated with diacylglycerol, only the 40-kd protein is phosphorylated. Experiments using a range of concentrations of diacylglycerol in the presence or absence of calcium ionophore show that the extent of phosphorylation of the 40-kd protein depends only on the concentration of diacylglycerol (Figure 15–22A). Serotonin release, however, depends on diacylglycerol and the calcium ionophore (Figure 15–22B).

A. Based on these experimental observations, describe the normal sequence of molecular events that leads to phosphorylation of the myosin light chain and the 40-kd protein. Indicate how the calcium ionophore and diacylglycerol treatments interact with the normal sequence of events.

B. Why do you think serotonin release requires both calcium ionophore and diacylglycerol?

**15–96** Unlike myosin from skeletal muscle, smooth muscle myosin interacts with actin only when its light chains are phosphorylated. Phosphorylation is controlled by variations in the intracellular concentration of $Ca^{2+}$, which is mediated through calmodulin. If myosin light-chain kinase is purified from

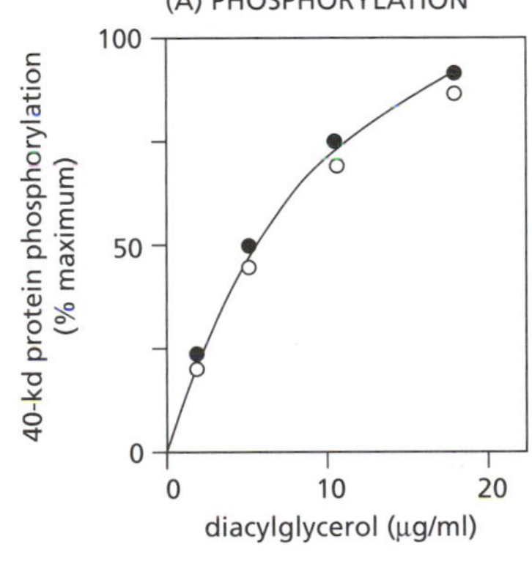


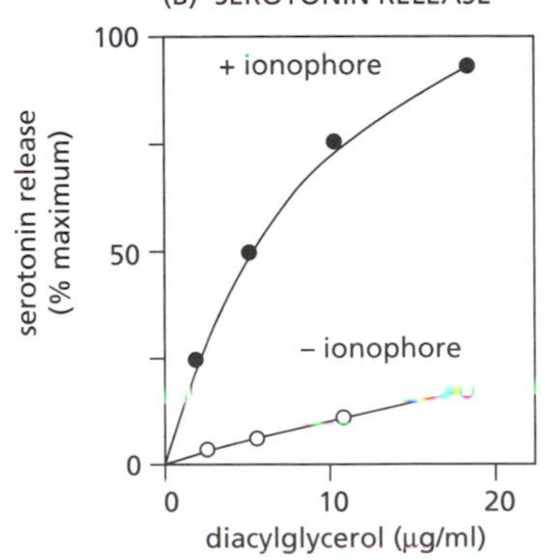


**Figure 15–22** Treatments of platelets with calcium ionophore and diacylglycerol (Problem 15–95). (A) Effects on phosphorylation of the 40-kd protein. (B) Effects on serotonin release. *Filled circles* indicate the presence of calcium ionophore and *open circles* indicate its absence.

**Table 15–2 Activities of myosin light-chain kinase purified in the presence and absence of protease inhibitors** (Problem 15–96).

| PURIFICATION SCHEME | ADDITIONS TO ASSAY MIX | RELATIVE ACTIVITY |
|---|---|---|
| Minus inhibitors | none | 50 |
| Minus inhibitors | $Ca^{2+}$ | 50 |
| Minus inhibitors | calmodulin | 50 |
| Minus inhibitors | $Ca^{2+}$/calmodulin | 50 |
| Plus inhibitors | none | 1 |
| Plus inhibitors | $Ca^{2+}$ | 1 |
| Plus inhibitors | calmodulin | 1 |
| Plus inhibitors | $Ca^{2+}$/calmodulin | 100 |

smooth muscle in the absence of protease inhibitors, its kinase activity is the same in the presence or absence of $Ca^{2+}$/calmodulin. In the presence of protease inhibitors, however, the purified kinase is inactive unless $Ca^{2+}$/calmodulin is present. As shown in Table 15–2, the purified kinase now behaves as expected; it shows an absolute dependence on $Ca^{2+}$/calmodulin.

A. Why do you suppose the original purified enzyme was active independently of $Ca^{2+}$/calmodulin?

B. Propose a sequence of molecular events that leads to the contraction of smooth muscles. Begin with the entry of $Ca^{2+}$ into the cytosol.

**15–97** Acetylcholine acts on muscarinic GPCRs in the heart to open $K^+$ channels, thereby slowing the heart rate. This process can be directly studied using the inside-out membrane patch-clamp technique. The external surface of the membrane is in contact with the solution in the bore of the pipet, and the cytoplasmic surface faces outward and can be exposed readily to a variety of solutions (Figure 15–23). Receptors, G proteins, and $K^+$ channels remain associated with the membrane patch.

When acetylcholine is added to a pipet with a whole cell attached, $K^+$ channels open as indicated by the flow of current (Figure 15–23A). Under similar circumstances with a patch of membrane inserted into a buffered salts solution, no current flows (Figure 15–23B). When GTP is added to the buffer, however, current resumes (Figure 15–23C). Subsequent removal of GTP stops the current (Figure 15–23D). The results of several similar experiments to test the effects of different combinations of components are summarized in Table 15–3.

A. Why do you think it is that $G_{\beta\gamma}$ activated the channel when the complete G protein did not? Is the active component of the G protein in this system the same as the one that activates adenylyl cyclase in other cells?

B. Addition of GppNp (a nonhydrolyzable analog of GTP) causes the $K^+$ channel to open in the absence of acetylcholine (Table 15–3, line 4). The flow of current, however, rose very slowly and reached its maximum only after a minute (compare with the immediate rise in Figure 15–23A and C). How do you suppose GppNp causes the channels to open slowly in the absence of acetylcholine?

C. To the extent that these experiments allow, draw a scheme for the activation of $K^+$ channels in heart cells in response to acetylcholine.

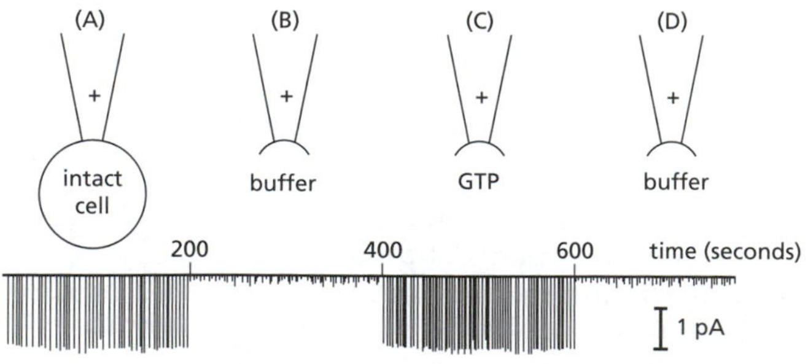


**Figure 15–23** Experimental setup and typical results of patch-clamp analysis of $K^+$ channel activation by acetylcholine (Problem 15–97). The buffer is a salts solution that does not contain nucleotides or $Ca^{2+}$. In all these experiments, acetylcholine is present inside the pipet, as indicated by the *plus* sign. The current through the membrane is measured in picoamps (pA). In (C) the GTP is added to the buffer.

**Table 15–3 Responses of K⁺ channel to various experimental manipulations** (Problem 15–97).

| | ADDITIONS | | | |
|---|---|---|---|---|
| | ACETYLCHOLINE | SMALL MOLECULES | G-PROTEIN COMPONENTS | $K^+$ CHANNEL |
| 1 | + | none | none | closed |
| 2 | + | GTP | none | open |
| 3 | – | GTP | none | closed |
| 4 | – | GppNp | none | open |
| 5 | – | none | G protein | closed |
| 6 | – | none | $G_\alpha$ | closed |
| 7 | – | none | $G_{\beta\gamma}$ | open |
| 8 | – | none | boiled G protein | closed |

# SIGNALING THROUGH ENZYME-COUPLED CELL-SURFACE RECEPTORS

TERMS TO LEARN

Akt
bacterial chemotaxis
Cdc42
chemotaxis receptor
cytokine receptor
cytoplasmic tyrosine kinase
enzyme-coupled receptor
Eph receptor
ephrins
focal adhesion kinase (FAK)
Grb2
histidine-kinase-associated receptor
JAK–STAT signaling pathway
Janus kinase (JAK)
MAP kinase cascade (MAP kinase module)
mTOR
phosphoinositide
phosphoinositide 3-kinase (PI 3-kinase)
phospholipase C-γ (PLCγ)
PI 3-kinase–Akt pathway
pleckstrin homology (PH) domain
protein tyrosine phosphatase
Rac
Ras-GAP
Ras-GEF
Ras-MAP-kinase signaling pathway
Ras
Ras superfamily
receptor serine/threonine kinase
receptor tyrosine kinase (RTK)
Rheb
Rho
Rho family
SH2 domain
Smad family
Src family
STATs
TOR
transforming growth factor-β (TGFβ) superfamily
tyrosine-kinase-associated receptor

## DEFINITIONS

Match the definition below with its term from the list above.

**15–98** The largest class of cell-surface-bound extracellular signal proteins.

**15–99** Large family of structurally related, secreted, dimeric proteins that act as hormones and local mediators to control a wide range of biological functions in all animals.

**15–100** Cell-surface receptor that when activated by ligand binding adds phosphates from ATP to tyrosine side chains in its own cytoplasmic domain.

**15–101** The founding member of a superfamily of monomeric GTPases that help to relay signals from cell-surface receptors to the nucleus.

**15–102** A group of monomeric GTPases that regulate both the actin and microtubule cytoskeletons.

**15–103** Responses of bacteria to attractants and repellents in their environment.

**15–104** Cytoplasmic tyrosine kinase present at cell–matrix junctions in association with the cytoplasmic tails of integrins.

**15–105** A kinase that is involved in intracellular signaling pathways activated by cell-surface receptors and that phosphorylates inositol phospholipids at the 3 position of the inositol ring.

**15–106** Cell-surface receptor in which the cytoplasmic domain either has enzymatic activity itself or is associated with an intracellular enzyme.

**15–107** Cell-surface receptor that activates a tyrosine kinase that is noncovalently bound to the receptor.

**15–108** A three-component signaling module used in various signaling pathways in eucaryotic cells.

**15–109** One of several intracellular signaling pathways that leads from cell-surface receptors to the nucleus it is distinguished by providing one of the more direct routes.

**15–110** Protein domain found in intracellular signaling proteins by which they bind to inositol phospholipids phosphorylated by PI 3-kinase.

**15–111** A protein domain that is homologous to a region in Src, is present in many proteins, and binds to a short amino acid sequence containing a phosphotyrosine.

**15–112** A crucial signaling protein in the PI 3-kinase–Akt signaling pathway, so named because it is the target of rapamycin.

## TRUE/FALSE

Decide whether each of these statements is true or false, and then explain why.

**15–113** Binding of extracellular ligands to receptor tyrosine kinases activates the intracellular catalytic domain by propagating a conformational change across the lipid bilayer through a single transmembrane $\alpha$ helix.

**15–114** PI 3-kinase phosphorylates the inositol head groups of phospholipids at the 3 position of the ring so that they can be cleaved by phospholipase C to produce $IP_3$.

**15–115** Protein tyrosine phosphatases display exquisite specificity for their substrates, unlike serine/threonine protein phosphatases, which have rather broad specificity.

**Problem 3–110** illustrates a strategy used to define target substrates for a protein phosphatase.

## THOUGHT PROBLEMS

**15–116** Antibodies are Y-shaped molecules that carry two identical binding sites. Imagine that you have obtained an antibody that is specific for the extracellular domain of a receptor tyrosine kinase. If cells were exposed to the antibody, would you expect the receptor tyrosine kinase to be activated, inactivated, or unaffected? Explain your reasoning.

**15–117** What is bidirectional signaling and how do ephrins and ephrin receptors mediate it?

**15–118** Genes encoding mutant forms of a receptor tyrosine kinase can be introduced into cells that express the normal receptor from their own genes. If the mutant genes are expressed at considerably higher levels than the normal genes, what will be the consequences for receptor-mediated signaling of introducing genes for the following mutant receptors?

A. A mutant receptor tyrosine kinase that lacks its extracellular domain.

B. A mutant receptor tyrosine kinase that lacks its intracellular domain.

**15–119** The SH3 domain, which comprises about 60 amino acids, recognizes and binds to structural motifs in other proteins. The motif recognized by SH3 domains was found by constructing a fusion protein between an SH3 domain and glutathione-*S*-transferase (GST). GST fusions allow for easy purification using a glutathione-affinity column, which binds GST specifically. After tagging the purified GST-SH3 protein with biotin to make it easy to detect, it was used to screen filters containing *E. coli* colonies expressing a cDNA library. Two different clones were identified that bound to the SH3 domain: in both cases binding was shown to occur at short proline-rich sequences.

A. Could you use biotin-tagged GST-SH2 proteins in the same way to find cDNAs for proteins that bind to SH2 domains? Why or why not?

B. Many proteins bind to short strings of amino acids in other proteins. How do you think these kinds of interactions differ from the kinds of interactions found between the protein subunits of multisubunit enzymes?

**15–120** The Ras protein functions as a molecular switch that is turned on by a guanine-nucleotide exchange factor (GEF) that causes it to bind GTP. A GTPase-activating protein (GAP) turns the switch off by inducing Ras to hydrolyze its bound GTP to GDP much more rapidly than in the absence of the GAP. Thus Ras works like a light switch that one person turns on and another turns off. In a cell line that lacks the Ras-specific GAP, what abnormalities in Ras activity, if any, would you expect to find in the absence of extracellular signals, and in their presence?

**15–121** What are the similarities and differences between the reactions that lead to the activation of G proteins and those that lead to the activation of Ras?

**15–122** In principle, the activated, GTP-bound form of Ras could be increased by activating a guanine-nucleotide exchange factor (GEF) or by inactivating a GTPase-activating protein (GAP). Why do you suppose that Ras-mediated signaling pathways always increase Ras–GTP by activating a GEF rather than inactivating a GAP?

**15–123** A single amino acid change in Ras eliminates its ability to hydrolyze GTP, even in the presence of a GTPase-activating protein (GAP). Roughly 30% of human cancers have this change in Ras. You have just identified a small molecule that prevents the dimerization of a receptor tyrosine kinase that signals via Ras. Would you expect this molecule to be effective in the treatment of cancers that express this common, mutant form of Ras? Why or why not?

**15–124** Phosphorylation of MAP kinase at a threonine (T) and a tyrosine (Y) residue causes a visible conformational change in the enzyme (Figure 15–24). Can you guess where the active enzyme binds ATP and its target proteins?

**15–125** Receptorlike tyrosine phosphatases are single-pass transmembrane proteins whose ligands and functions are largely unknown. Explain in a general

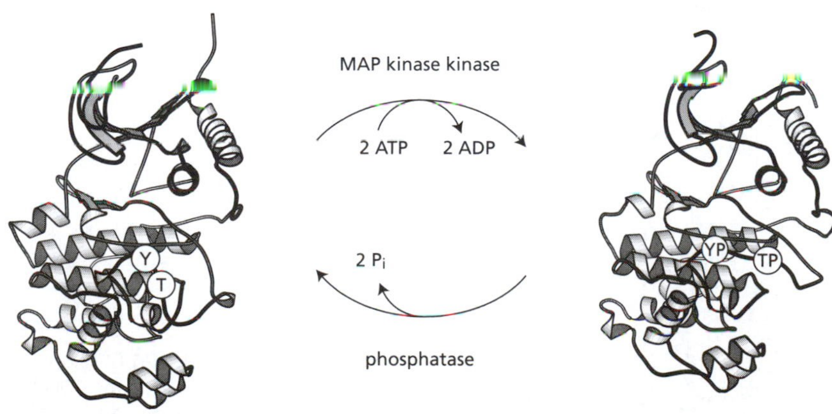


**Figure 15–24** Conformational change of MAP kinase upon phosphorylation (Problem 15–124).

way how the signaling pathway triggered by these putative receptors can be studied by replacing their extracellular domains with the extracellular domain from a growth factor receptor such as the EGF receptor.

## DATA HANDLING

**15–126** What does autophosphorylation mean? When a receptor tyrosine kinase binds its ligand and forms a dimer, do the individual receptor molecules phosphorylate themselves or does one receptor cross-phosphorylate the other, and vice versa? To investigate this question, you've constructed genes for three forms of a receptor tyrosine kinase: the normal form with an active kinase domain and three sites of phosphorylation; a large form that carries an inactivating point mutation in the kinase domain but retains the three phosphorylation sites; and a short version that has an active kinase domain but is lacking the sites of phosphorylation (Figure 15–25A). You express the genes singly and in combination in a cell line that lacks this receptor tyrosine kinase, and then break open the cells and add the ligand for the receptor in the presence of radioactive ATP. You immunoprecipitate the receptors and analyze them for expression levels by staining for protein (Figure 15–25B) and for phosphorylation by autoradiography (Figure 15–25C).

A. What results would you expect on the autoradiograph if individual receptors phosphorylated themselves?

B. What would you expect if receptors cross-phosphorylated each other?

C. Which model for autophosphorylation do your data support?

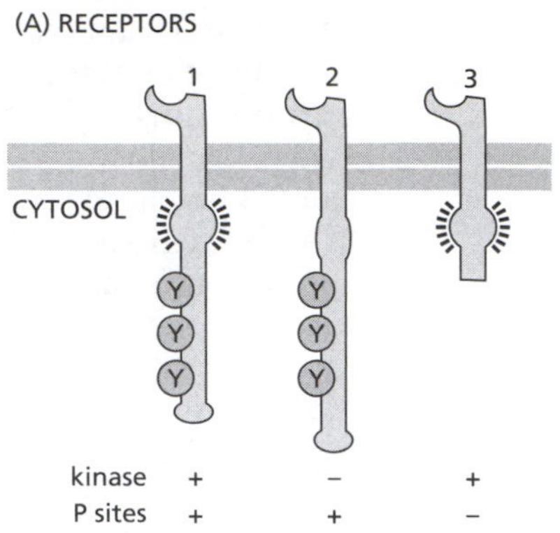


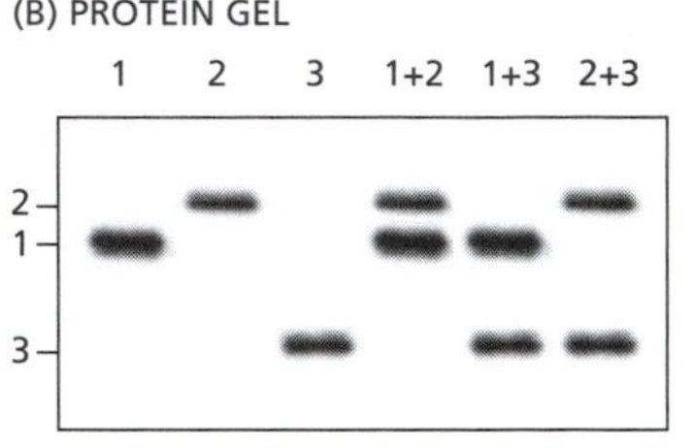


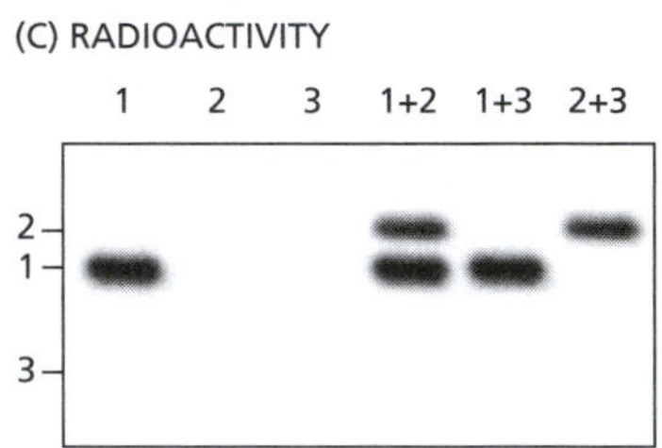


**Figure 15–25** Analysis of autophosphorylation (Problem 15–126). (A) Normal and mutant receptor tyrosine kinases. P sites refers to the sites of phosphorylation. (B) Expression of receptor tyrosine kinases. (C) Phosphorylation of receptor tyrosine kinases.

**15–127** When activated, the platelet-derived growth factor (PDGF) receptor phosphorylates itself on multiple tyrosines. These phosphorylated tyrosines serve as assembly sites for several SH2-domain-containing proteins that include phospholipase C-γ (PLCγ), a Ras-specific GTPase-activating protein (GAP), a subunit of phosphatidylinositol 3-kinase (PI3K), and a phosphotyrosine phosphatase (PTP) (Figure 15–26). PDGF binding stimulates several changes in the target cell, one of which is an increase in DNA synthesis, as measured by incorporation of radioactive thymidine or bromodeoxyuridine into DNA.

To determine which of the bound proteins is responsible for activation of DNA synthesis, you construct several mutant genes for the PDGF receptor that retain individual or combinations of tyrosine phosphorylation sites. When expressed in cells that do not make a PDGF receptor of their own, each of the receptors is phosphorylated at its tyrosines upon binding of PDGF. As shown in Figure 15–27, DNA synthesis is stimulated to different extents in cells expressing the mutant receptors.

What roles do PI3K, GAP, PTP, and PLCγ play in the stimulation of DNA synthesis by PDGF?

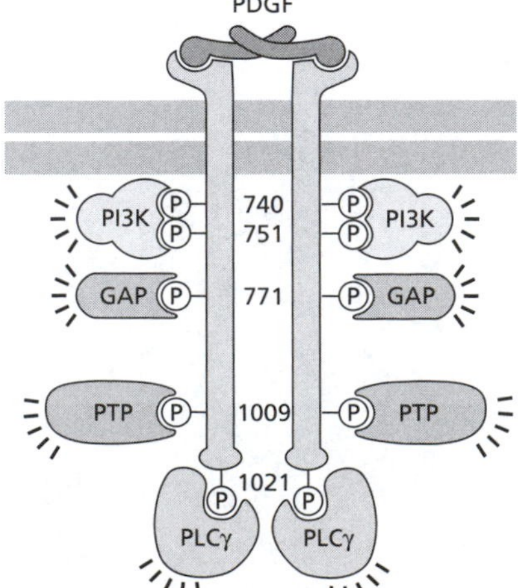


**Figure 15–26** The signaling complex assembled on the PDGF receptor (Problem 15–127).

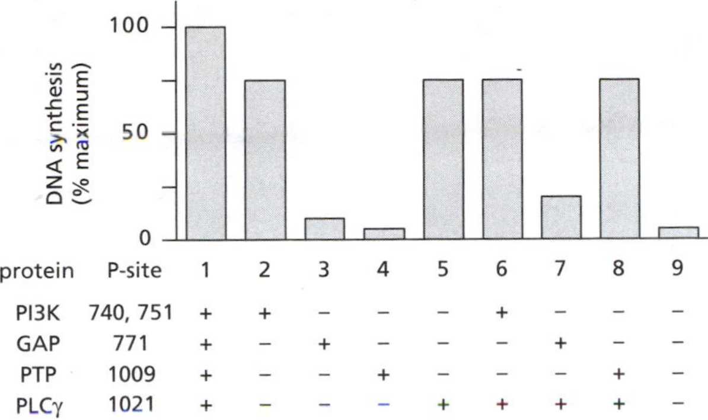


**Figure 15–27** Stimulation of DNA synthesis by the normal PDGF receptor and by receptors missing some phosphorylation sites (Problem 15–127). Stimulation by the normal receptor is set arbitrarily at 100%. The presence of a phosphorylation site is indicated by +; absence of a site, by –.

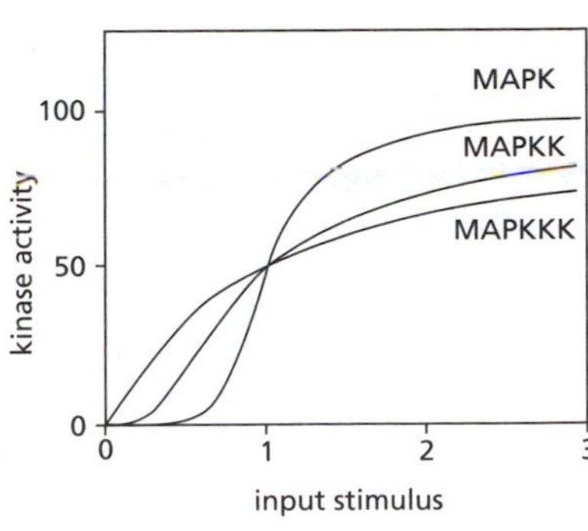


**Figure 15–28** Stimulus–response curves for the components of the MAPK cascade (Problem 15–128). For ease of comparison the curves have been normalized so that an input stimulus of 1 gives 50% activation of the kinases.

**15–128** MAP kinase kinase kinase (MAPKKK) activates MAP kinase kinase (MAPKK) by phosphorylation of two serine side chains. Doubly phosphorylated (active) MAPKK, in turn, activates MAP kinase (MAPK) by the phosphorylation of a threonine and a tyrosine. The doubly phosphorylated MAPK then phosphorylates a variety of target proteins to bring about complex changes in cell behavior. It is possible to write down all of the rate equations for the individual steps in this activation cascade, as well as for the removal of the phosphates (inactivation) by protein phosphatase, and to solve them by making reasonable assumptions about the concentrations of the proteins. The calculated plot of activation of the kinases versus input stimulus is shown in Figure 15–28. Why is the very steep response curve for MAPK a good thing for this signaling pathway?

**15–129** An explicit assumption in the analysis in Problem 15–128 is that the components of the MAP kinase cascade operate independently of one another, so that the dual phosphorylation events that activate MAPKK and MAPK occur one at a time as molecules collide in solution. How do you suppose the curves in Figure 15–28 would change if a scaffold protein held the kinases of the MAP kinase cascade together? Most MAP kinase cascades are scaffolded. What is the advantage of linking these kinases onto scaffold proteins?

**15–130** Akt is a key protein kinase in the signaling pathway that leads to cell growth. Akt is activated by a phosphatidylinositol-dependent protein kinase (PDK1), which phosphorylates threonine 308. At the same time, serine 473 is phosphorylated. Your advisor has been unsuccessful in purifying the protein kinase responsible for the phosphorylation of serine 473, but you think you know what is going on. You construct genes encoding two mutant forms of Akt: one carries a point mutation in the kinase domain, Akt-K179M, which renders it kinase dead, and the other carries a point mutation in the domain required to bind to PDK1 (Akt-T308A), which cannot be activated by PDK1. You transfect each of these constructs, and a construct for wild-type Akt, into cells that do not express their own Akt. You treat a portion of the cells with an insulin-like growth factor (IGF1), which activates PDK1, and analyze the phosphorylation state of the various forms of Akt using antibodies specific for Akt or for particular phosphorylated amino acids (Figure 15–29).

What is the identity of the enzyme that phosphorylates serine 473 on Akt?

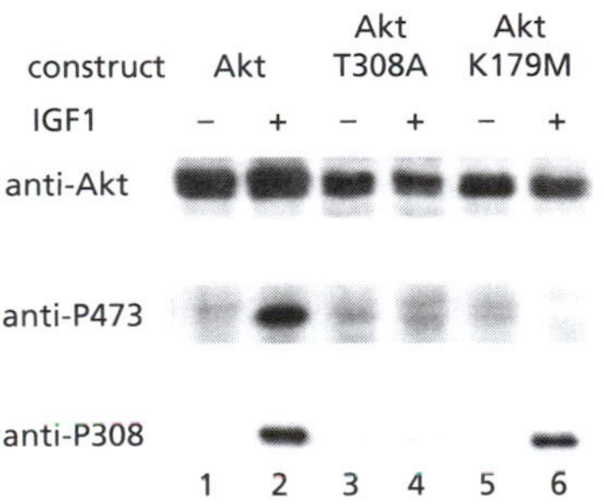


**Figure 15–29** Expression levels of various forms of Akt and their degree of phosphorylation in the presence and absence of IGF1 (Problem 15–130). Anti-Akt recognizes all three forms of Akt regardless of their phosphorylation state; anti-P473 specifically recognizes the phosphorylated serine at position 473; anti-P308 specifically recognizes the phosphorylated threonine at position 308.

**15–131** Interferon-γ (IFNγ) is a cytokine produced by activated T lymphocytes. It binds to surface receptors on macrophages and stimulates their efficient scavenging of invading viruses and bacteria via a JAK–STAT signaling pathway. A number

**Figure 15–30** Sequence elements in the transcription factor that responds to IFNγ (Problem 15–131).

of genes are activated in response to IFNγ binding, all of which contain a DNA sequence element with partial dyad symmetry (TTCCXGTAA) that is required for the IFNγ response.

You have cloned the gene for the STAT transcription factor that is activated in response to INFγ binding. The sequence of the gene indicates that the protein contains several heptad repeat sequences near its N-terminus—a common dimerization domain in many transcription factors—and SH2 and SH3 domains adjacent to a site for tyrosine phosphorylation near the C-terminus (Figure 15–30). By making antibodies to the protein, you show that it is normally located in the cytosol. After 15 minutes exposure to IFNγ, the protein becomes phosphorylated on a tyrosine and moves to the nucleus.

Suspecting that tyrosine phosphorylation is the key to the regulation of this transcription factor, you assay its ability to bind the DNA sequence element in the presence of high concentrations of free phosphotyrosine or when mixed with anti-phosphotyrosine antibodies. Both treatments inhibit binding of the protein to DNA, as does treatment with a protein phosphatase. Finally, you measure the molecular weight of the cytosolic and nuclear forms of the protein, which suggest that the cytosolic form is a monomer and the nuclear form is a dimer.

A. Do you think that phosphorylation of the transcription factor is necessary for the factor to bind to DNA, or do you think phosphorylation is required to create an acidic activation domain to promote transcription?

B. Bearing in mind that SH2 domains bind phosphotyrosine, how do you suppose free phosphotyrosine might interfere with the activity of the transcription factor?

C. How might tyrosine phosphorylation of the protein promote its dimerization? How do you think dimerization enhances its binding to DNA?

**15–132** Four types of chemotaxis receptors have been identified in *E. coli.* These receptors mediate chemotactic responses to two different amino acids, to sugars, and to dipeptides. As part of a practical demonstration in bacterial chemotaxis, your instructor has given you a wild-type strain with all four receptors intact and four mutant strains with one or more of the receptors missing (Table 15–4). Your assignment is to identify the receptor that mediates the response to each attractant. The experimental assay is very simple. You fill a capillary pipet with a solution of the attractant, dip it into a buffered solution containing bacteria, remove it after 5 minutes, and count the number of bacteria in the capillary. Your results are shown in Table 15–4. Identify each attractant and its appropriate receptor.

**15–133** To clarify the relationship between the structure of a chemotaxis receptor and the functions of stimulus recognition, signal transduction, and adaptation in *Salmonella typhimurium,* you have cloned the normal gene for the

**Table 15–4** Chemotaxis in wild-type and mutant strains of *E. coli* (Problem 15–132).

| | | NUMBER OF CELLS (1000s) IN CAPILLARY | | | | |
|---|---|---|---|---|---|---|
| STRAIN | INTACT RECEPTORS | SERINE | ASPARTATE | RIBOSE | PRO-GLY | NONE |
| 1 | Tap, Tar, Trg, Tsr | 59 | 105 | 95 | 6.6 | 0.5 |
| 2 | Tap, Tar, Trg | 0.7 | 84 | 77 | 13 | 0.8 |
| 3 | Trg, Tsr | 34 | 0.7 | 59 | 0.6 | 0.6 |
| 4 | Tap, Trg, Tsr | 55 | 0.6 | 65 | 4.1 | 0.5 |
| 5 | Tar, Trg, Tsr | 70 | 59 | 85 | 0.9 | 0.8 |

aspartate receptor and a mutant that is missing 35 C-terminal amino acids. By introducing the wild-type and truncated forms of the gene back into *Salmonella* that are missing the normal gene, you can test for functional differences between the two cloned genes. Both the wild-type and the truncated receptor are overexpressed about 15-fold above normal—not uncommon for genes expressed from plasmids—but both bind aspartate normally. The only major difference is that the truncated receptor cannot be methylated.

Wild-type bacteria in the absence of an attractant change their direction of rotation (tumble) every few seconds. Upon exposure to an attractant, however, the changes in direction of rotation are suppressed, which leads to a period of smooth swimming. If the concentration of attractant remains constant (even if high), wild-type bacteria quickly adapt and begin again to tumble every few seconds. To observe these behavioral changes experimentally, you tether bacteria by their flagella to coverslips so that you can observe their direction of rotation. You then expose them to aspartate and count the number of bacteria that do not reverse their direction of rotation in 1-minute intervals. The results for wild-type cells and for cells containing the cloned aspartate receptors are shown in Figure 15–31.

A. Is signal transduction by the two cloned receptors normal?
B. Are the adaptive properties of the cloned receptors normal?
C. Suggest molecular explanations for why the cloned normal receptor and the cloned truncated receptor, when introduced into bacteria, respond differently from the normal receptor in wild-type cells (Figure 15–31).

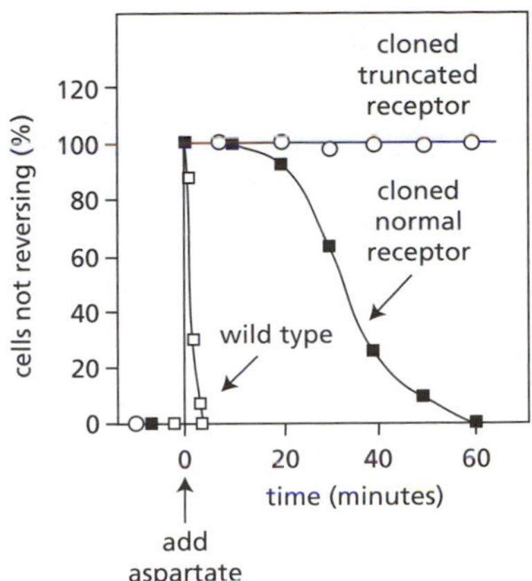


**Figure 15–31** Behavior of wild-type cells and cells containing the cloned receptors (Problem 15–133).

# SIGNALING PATHWAYS DEPENDENT ON REGULATED PROTEOLYSIS OF LATENT GENE REGULATORY PROTEINS

| TERMS TO LEARN | | |
|---|---|---|
| β-catenin | Hedgehog protein | Notch |
| Cubitus interruptus (Ci) | iHog | Patched |
| Delta | IκB | Smoothened |
| Dishevelled | LDL-receptor-related protein (LRP) | Wnt/β-catenin pathway |
| Frizzled | NFκB proteins | Wnt proteins |

## DEFINITIONS

Match the definition below with its term from the list above.

**15–134** Receptor protein involved in what may be the most widely used signaling pathway in animal development; its ligands are cell-surface proteins such as Delta.

**15–135** A family of secreted signal molecules that act as local mediators and morphogens during development; they were initially discovered as the products of the *Wingless* gene in flies and the *Int1* gene in mice.

**15–136** A signaling pathway activated by Wnt binding to both the Frizzled receptor and the LRP co-receptor.

**15–137** A group of secreted signal molecules that act as local mediators and morphogens during development and whose effects are mediated through the cell-surface receptor Patched and its binding partner Smoothened.

**15–138** A target of Hedgehog signaling, this gene regulatory molecule is a full-length gene activator in the presence of Hedgehog and a partially proteolyzed gene repressor in its absence.

**15–139** Latent gene regulatory proteins that are present in most cells in both animals and plants and are central to many stressful, inflammatory, and innate immune responses.

## TRUE/FALSE

Decide whether each of these statements is true or false, and then explain why.

**15–140** Signaling pathways that activate latent gene regulatory proteins depend on regulated proteolysis to control activity and location.

**15–141** Notch is both a cell-surface receptor and a latent gene regulatory protein.

**15–142** Because one of the targets of NFκB activation is the IκBα gene, the cytoplasmic inhibitor of NFκB, a negative feedback loop is established that limits the duration of the NFκB response.

## THOUGHT PROBLEMS

**15–143** One of the most difficult aspects of signaling through pathways that involve proteolysis is keeping track of the names of the components and their functions. Sort the following list of proteins into the Notch, Wnt/β-catenin and Hedgehog signaling pathways, list them in the order that they function, and select one or more appropriate descriptors that identify their role.

| PROTEIN | FUNCTION |
|---|---|
| A. β-Catenin | 1. Amplifier protein |
| B. Cubitus interruptus (Ci) | 2. Anchoring protein |
| C. Delta | 3. Extracellular signal protein |
| D. Dishevelled | 4. Gene regulatory protein |
| E. Frizzled | 5. Integrator protein |
| F. Glycogen synthase kinase 3 (GSK3) | 6. Latent gene regulatory protein |
| G. Hedgehog | 7. Modulator protein |
| H. iHog | 8. Protease |
| I. LRP | 9. Receptor protein |
| J. Notch | 10. Relay protein |
| K. Patched | 11. Scaffold protein |
| L. Presenilin | 12. Transducer protein |
| M. Smoothened | |
| N. Rbpsuh | |
| O. Wnt | |

**15–144** Like Notch, the β-amyloid precursor protein (APP) is cleaved near its transmembrane segment to release an extracellular and an intracellular component. Explain how the fragments of APP relate to the amyloid plaques that are characteristic of Alzheimer's disease.

**15–145** The Wnt planar polarity signaling pathway normally ensures that each wing cell in *Drosophila* has a single hair. Overexpression of the *Frizzled* gene from a heat-shock promoter (hs-*Fz*) causes multiple hairs to grow from many cells (Figure 15–32A). This phenotype is suppressed if hs-*Fz* is combined with a heterozygous deletion ($Dsh^{\Delta}$) of the *Dishevelled* gene (Figure 15–32B). Do these results allow you to order the action of Frizzled and Dishevelled in the signaling pathway? If so, what is the order? Explain your reasoning.

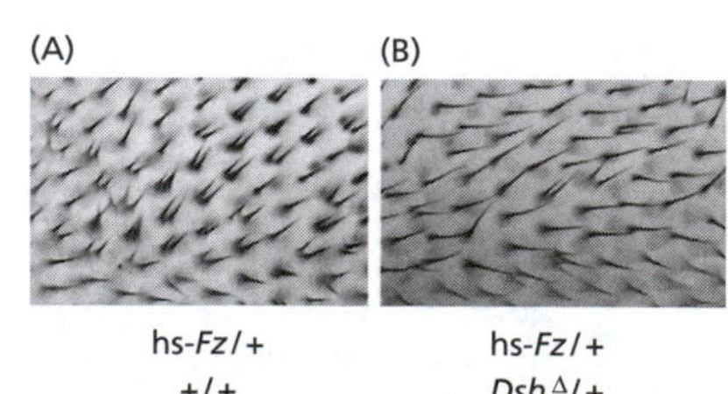


**Figure 15–32** Pattern of hair growth on wing cells in genetically different *Drosophila* (Problem 15–145).

**15–146** There are two mutational routes to the uncontrolled cell proliferation and invasiveness that characterize cancer cells. The first is to make a stimulatory gene (a proto-oncogene) hyperactive: this type of mutation has a dominant effect so that only one of the cell's two gene copies needs to undergo change. The second is to make an inhibitory gene (a tumor suppressor gene) inactive: this type of mutation usually is recessive so that both the cell's gene copies must be inactivated.

Mutations of the *Apc* (adenomatous polyposis coli) gene occur in 80% of human colon cancers. Normal APC increases the affinity of the degradation complex for β-catenin, which in excess can enter the nucleus and promote transcription of key target genes for cell proliferation. Given this information, which category—oncogene or tumor suppressor—would you expect the *Apc* gene to belong to? Why?

**15–147** Latent gene regulatory proteins are prevented from entering the nucleus until the cell receives an appropriate signal. List four ways by which cells keep gene regulatory proteins out of the nucleus, and give an example of a latent gene regulatory protein that is controlled by each mechanism.

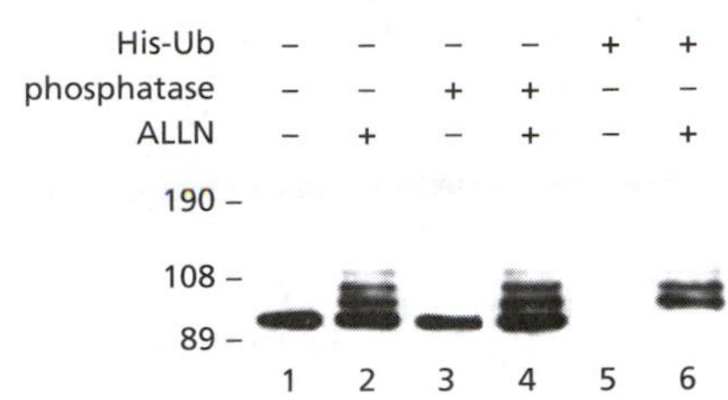


**Figure 15–33** Electrophoretic analysis of β-catenin (Problem 15–148). β-Catenin was detected using β-catenin-specific antibodies. Size markers in kilodaltons are shown on the left.

## DATA HANDLING

**15–148** β-Catenin is a target for phosphorylation by glycogen synthase kinase 3 and also a substrate for degradation in proteasomes. Treatment of mouse fibroblasts with a proteasome inhibitor, ALLN, increases the stability of β-catenin and causes the appearance of new, slower migrating forms of the protein on SDS-polyacrylamide gels (Figure 15–33, lanes 1 and 2). Do these new bands represent phosphorylated proteins, which often run more slowly on such gels (see Figure 15–11), or do they arise by addition of ubiquitin, which would increase their size? To test for phosphorylation, you treat samples with a protein phosphatase that efficiently removes phosphates from other proteins in the same sample (not shown) and run them on gels (lanes 3 and 4). To test for ubiquitylation, you express His-tagged ubiquitin in cells, treat with ALLN (or not), and then purify His-ubiquitylated proteins by $Ni^{2+}$-column chromatography before running them on gels (lanes 5 and 6). In all cases you detect β-catenin specifically, using antibodies directed against it.

Are the slower migrating forms of β-catenin due to phosphorylation or ubiquitylation? Explain your answer.

Problem 12–61 examines the spatial control of the latent gene regulatory protein NFκB

**15–149** β-Catenin can be phosphorylated by glycogen synthase kinase 3 (GSK3) and it can be degraded in proteasomes. β-Catenin could be sensitized for degradation by phosphorylation, it could be protected from degradation by phosphorylation, or its phosphorylation status could be irrelevant for degradation. To distinguish among these possibilities, you generate cell lines that express either a mutant GSK3 that cannot carry out phosphorylation, or a mutant β-catenin that is missing its site of phosphorylation. In the presence and absence of the proteasome inhibitor, ALLN, both cell lines yield β-catenin that migrates as a single band, with no slower migrating bands visible, in contrast to the situation with nonmutant β-catenin and GSK3 (which is shown in Figure 15–33, lanes 1 and 2). What is the relationship between β-catenin phosphorylation and its degradation in proteasomes? Explain your answer.

**15–150** The *Hedgehog* gene encodes the Hedgehog precursor protein, which is 471 amino acids long. The precursor protein (Figure 15–34A) is normally cleaved between glycine 257 (G257) and cysteine 258 (C258) to generate a fragment that is active in local and long-range signaling. Cleavage is essential for signaling. You clone a segment of the *Hedgehog* gene encoding a portion of the protein that includes the cleavage site and the entire C-terminus. When you purify this protein and incubate it in buffer, you observe cleavage over the course of several hours, as shown in Figure 15–34B. If you vary its concentration over a 256-fold range and assay cleavage after 4 hours of incubation, you observe the results shown in Figure 15–34C.

A. Explain how these data support the idea that the Hedgehog precursor protein cleaves itself. How do they rule out the possibility that the purified protein is contaminated with a bacterial protease, for example?

B. Does a molecule of precursor protein cleave itself, or does it cleave another molecule of the precursor; that is, is the reaction intramolecular or intermolecular?

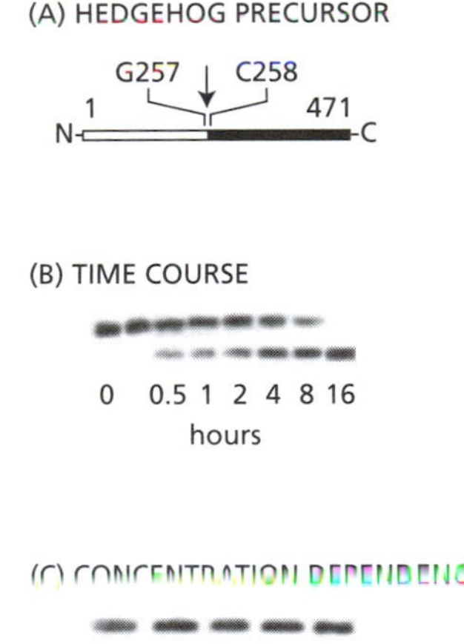

**Figure 15–34** Mechanism of cleavage of the Hedgehog precursor protein (Problem 15–150). (A) Site of cleavage in the Hedgehog precursor protein. (B) Time course of cleavage of the fragment of the precursor protein. (C) Dependence of cleavage on concentration of the precursor protein fragment.

**15–151** To find out what happens to the fragments of Hedgehog after cleavage, you express three versions: wild-type Hedgehog precursor, an uncleavable form, and the N-terminal cleavage product (Figure 15–35A). In fly embryos the

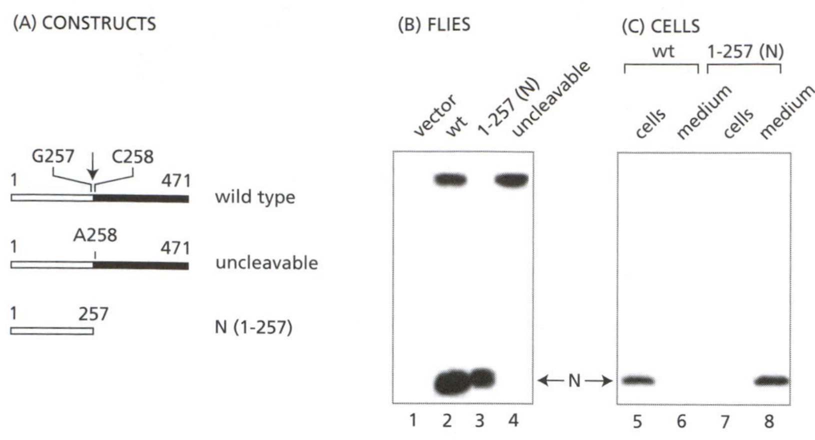


Figure 15–35 Fate of the fragments of Hedgehog after cleavage (Problem 15–151). (A) Constructs encoding different forms of the Hedgehog precursor protein. (B) Results of expression in *Drosophila* embryos. (C) Results of expression in insect cells. Hedgehog fragments were detected using antibodies specific for the N-terminal segment.

constructs behave as expected: wild-type Hedgehog is cleaved, the uncleavable version is not, and the N-terminal segment is expressed (Figure 15–35B). When wild-type Hedgehog and the N-terminal segment are expressed in insect cells, however, the N-terminal segment from wild-type Hedgehog remains associated with the cells, while the synthesized N-terminal segment is secreted into the medium (Figure 15–35C). Can you suggest possible explanations for the difference in localization of the N-terminal segment?

**15–152** If you overexpress various Hedgehog contructs (see Figure 15–35A) in flies and examine the pattern of Wnt expression (a well-characterized target of Hedgehog signaling), you observe a striped pattern of expression in all cases, but some constructs lead to thicker stripes than normal (Figure 15–36).

A. Which part of the Hedgehog molecule is responsible for signaling?

B. All the cells in the embryo are overexpressing the various Hedgehog constructs. Why is it, do you suppose, that you observe the same basic striped pattern of Wnt expression in all of them?

C. Why do you see stripes of Wnt expression even in the absence of Hedgehog overexpression?

## SIGNALING IN PLANTS

| TERMS TO LEARN | |
|---|---|
| auxin | growth regulator (plant hormone) |
| brassinosteroids | leucine-rich repeat (LRR) receptor kinase |
| cryptochrome | phototropin |
| ethylene | phytochrome |

### DEFINITIONS

Match the definition below with its term from the list above.

**15–153** A cytoplasmic serine/threonine kinase in plants that is activated by red light and inactivated by far-red light.

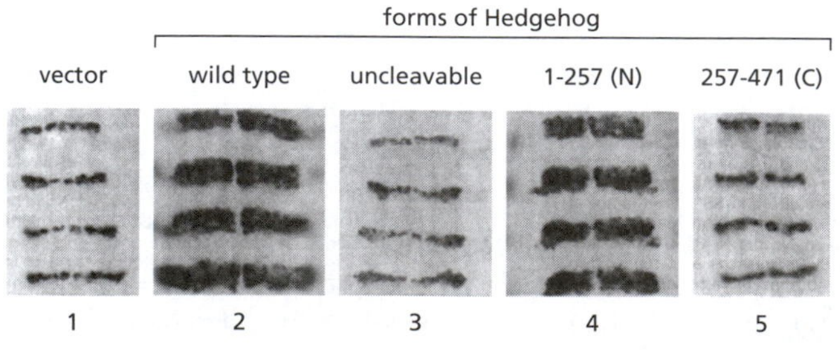


Figure 15–36 Patterns of Wnt expression in *Drosophila* embryos that are overexpressing various Hedgehog constructs (Problem 15–152). Wnt expression was detected by *in situ* hybridization.

**15–154** Small gas molecule influential in various aspects of plant development, including fruit ripening and leaf abscission.

**15–155** General term for a signal molecule that helps coordinate growth and development in plants.

**15–156** Flavoprotein responsive to blue light, found in both plants and animals; in animals it is involved in circadian rhythms.

**15–157** A growth regulator that helps plants grow toward light, grow upward rather than branch out, and extend their roots downward.

**15–158** Common type of receptor serine/threonine kinase in plants, characterized by an extracellular portion rich in repeated segments containing a high proportion of leucine.

## TRUE/FALSE

Decide whether each of these statements is true or false, and then explain why.

**15–159** Even though plants and animals independently evolved multicellularity, they use virtually all the same signaling proteins and second messengers for cell–cell communication.

**15–160** Remarkably, the auxin efflux transporter PIN3 in the cap cells of the root quickly redistribute themselves in response to a change in the direction of the gravity vector, so that they pump auxin toward the side of the root pointing downward.

## THOUGHT PROBLEMS

**15–161** The last common ancestor to plants and animals was a unicellular eucaryote. Thus, it is thought that multicellularity and the attendant demands for cell communication arose independently in these two lineages. This evolutionary viewpoint accounts nicely for the vastly different mechanisms that plants and animals use for cell communication. Fungi use signaling mechanisms and components that are very similar to those used in animals. Which of the phylogenetic trees shown in Figure 15–37 does this observation support?

**15–162** If signaling arose as a solution to the demands of multicellularity, how then do you account for the very similar mechanisms of signaling that are used in animals and the unicellular fungus *Saccharomyces cerevisiae*?

**15–163** How is it that plant growth regulators can be present throughout a plant and yet have specific effects on particular cells and tissues?

## DATA HANDLING

**15–164** The ripening of fruit is a complicated process of development, differentiation, and death (except for the seeds, of course). The process is triggered by minute amounts of ethylene gas. (This was discovered by accident many years ago; the paraffin stoves used to heat greenhouses in the olden days gave off enough ethylene to initiate the process.) The ethylene is normally

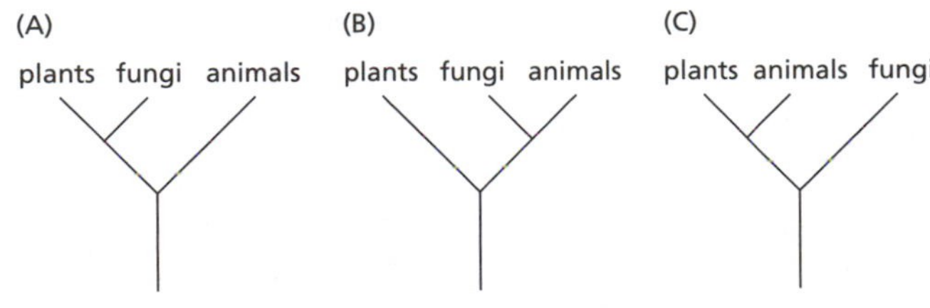


**Figure 15–37** Three possible phylogenetic relationships among plants, animals, and fungi (Problem 15–161).

produced by the fruits themselves in a biochemical pathway, the rate-limiting step of which is controlled by ACC synthase, which converts *S*-adenosylmethionine to a cyclopropane compound that is the immediate precursor of ethylene. Ethylene initiates a program of sequential gene expression that includes the production of several new enzymes, including polygalacturonase, which probably contributes to softening the cell wall.

Your company, Agribucks, is trying to make mutant tomatoes that cannot synthesize their own ethylene. Such fruit could be allowed to stay longer on the vine, developing their flavor while remaining green and firm. They could be shipped in this robust unripe state and exposed to ethylene just before arrival at market. This should allow them to be sold at the peak of perfection, and the procedure involves no artificial additives of any kind.

You decide to use an antisense approach, which works especially well in plants. You place an ACC synthase cDNA into a plant expression vector so that the gene will be transcribed in reverse, introduce it into tomato cells, and regenerate whole tomato plants. Sure enough, ethylene production is inhibited by 99.5% in these transgenic tomato plants, and their fruit fails to ripen. But when placed in air containing a small amount of ethylene, they turn into beautiful, tasty ripe red fruit in about 2 weeks.

A. How do you imagine that transcribing the ACC synthase gene in reverse blocks the production of ethylene?
B. Will you be a millionaire before you are 30?

# The Cytoskeleton

16

In This Chapter

THE SELF-ASSEMBLY AND DYNAMIC STRUCTURE OF CYTOSKELETAL FILAMENTS 373

HOW CELLS REGULATE THEIR CYTOSKELETAL FILAMENTS 382

MOLECULAR MOTORS 388

THE CYTOSKELETON AND CELL BEHAVIOR 391

## THE SELF-ASSEMBLY AND DYNAMIC STRUCTURE OF CYTOSKELETAL FILAMENTS

TERMS TO LEARN

| | | |
|---|---|---|
| cytoskeleton | minus end | protofilament |
| dynamic instability | neurofilament | tubulin |
| intermediate filament | plus end | treadmilling |
| keratin | | |

### DEFINITIONS

Match the definition below with its term from the list above.

**16–1** A linear chain of protein subunits joined end to end, which associates laterally with other such chains to form cytoskeletal components.

**16–2** The property of sudden conversion from growth to shrinkage, and vice versa, in a protein filament such as a microtubule or an actin filament.

**16–3** The end of a microtubule or an actin filament at which addition of monomers occurs most readily; the fast-growing end.

**16–4** General term for the fibrous protein filaments (about 10 nm in diameter) that form ropelike networks in animal cells.

**16–5** The process by which a polymeric protein filament is maintained at constant length by addition of protein subunits at one end and loss of subunits at the other.

**16–6** System of protein filaments in the cytoplasm of a eucaryotic cell that gives the cell its shape and the capacity for directed movement.

### TRUE/FALSE

Decide whether each of these statements is true or false, and then explain why.

**16–7** The structural polarity of all microtubules is such that α-tubulin is exposed at one end and β-tubulin is exposed at the opposite end.

**16–8** The role of ATP hydrolysis in actin polymerization is similar to the role of GTP hydrolysis in tubulin polymerization: both serve to weaken the bonds in the polymer and thereby promote depolymerization.

**16–9** Like actin filaments and microtubules, cytoplasmic intermediate filaments are found in all eucaryotes.

### THOUGHT PROBLEMS

**16–10** In general terms what are the cellular functions of intermediate filaments, microtubules, and actin filaments?

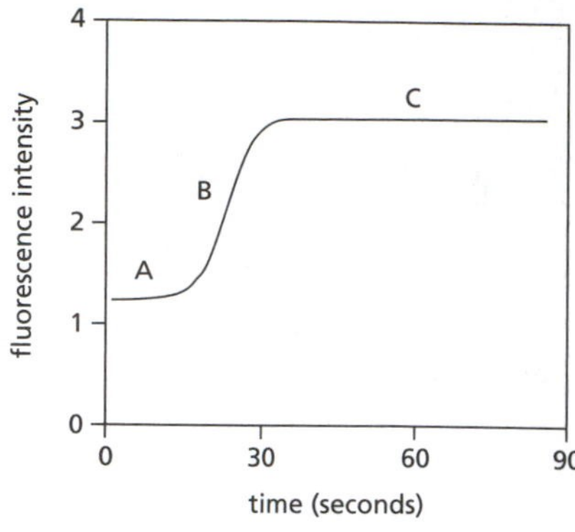


Figure 16–1 Formation of actin filaments over time, starting with purified actin monomers that are labeled with a fluorescent probe (Problem 16–12). Upon polymerization the fluorescence of the probe increases, which allows polymerization to be measured. The intensity of fluorescence at zero seconds is due to the background fluorescence of the actin monomers. The three phases of polymerization are indicated as A, B, and C.

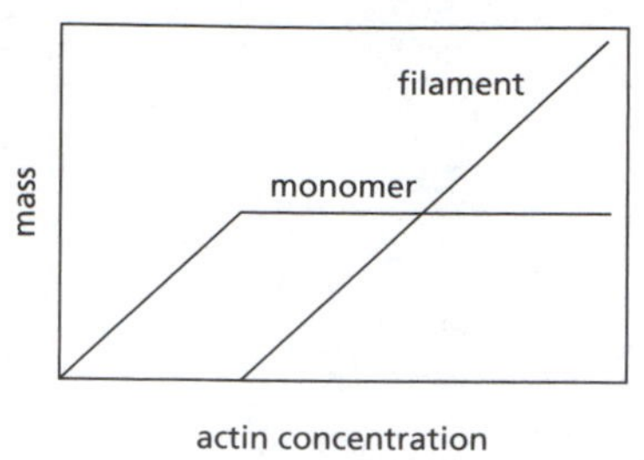


Figure 16–2 Mass of actin monomers and filaments as a function of actin concentration (Problem 16–13).

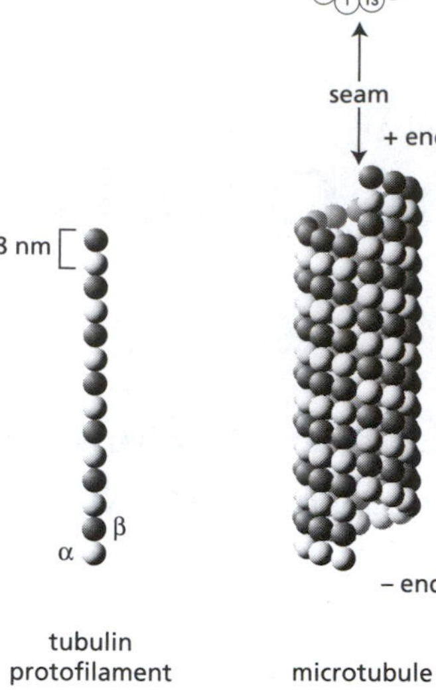


Figure 16–3 Structure of a 13-protofilament microtubule, showing the seam between the first and thirteenth protofilaments (Problem 16–15).

**16–11** If each type of cytoskeletal filament is made up of subunits that are held together by weak noncovalent bonds, how is it possible for a human being to lift heavy objects?

**16–12** A typical time course of polymerization of actin filaments from actin subunits is shown in Figure 16–1.

A. Explain the properties of actin polymerization that account for each of the three phases of the polymerization curve.

B. How would the curve change if you doubled the concentration of actin? Would the concentration of free actin at equilibrium be higher or lower than in the original experiment, or would it be the same in both?

**16–13** Figure 16–2 shows the equilibrium distribution of actin in free subunits (monomers) and in filaments, as a function of actin concentration. Indicate the critical concentration of actin on this diagram.

**16–14** Why do you suppose it is much easier to add tubulin to existing microtubules than to start a new microtubule from scratch?

**16–15** In a 13-filament microtubule the majority of lateral interactions are between like subunits, with α-tubulin binding to α-tubulin and β-tubulin binding to β-tubulin. Between the first and thirteenth protofilaments, however, there is a seam at which α-tubulin interacts with β-tubulin (Figure 16–3). Are these heterotypic interactions (α with β) likely to be stronger than, weaker than, or the same strength as homotypic interactions (α with α or β with β)? Explain your reasoning.

**16–16** Imagine that the polymer in Figure 16–4A can add subunits at either end, just like actin filaments and microtubules. Imagine also three hypothetical types of free subunit, as shown in Figure 16–4B. Each type of subunit can add to the polymer and, once added, it adopts the conformation of the other subunits in the polymer (Figure 16–4A). For each of these subunits, decide which end of the polymer, if either, will grow at the faster rate when the concentration of that subunit is higher than the critical concentration required for polymerization. Explain your reasoning. For any of the subunits, will there be a concentration at which one end will preferentially grow while the other shrinks? Why or why not?

**16–17** The microtubules in Figure 16–5A were obtained from a population that was growing rapidly, whereas the one in Figure 16–5B came from microtubules undergoing catastrophic shrinkage. Comment on any differences between the two images and suggest likely explanations for those you observe.

(A) POLYMERIZATION

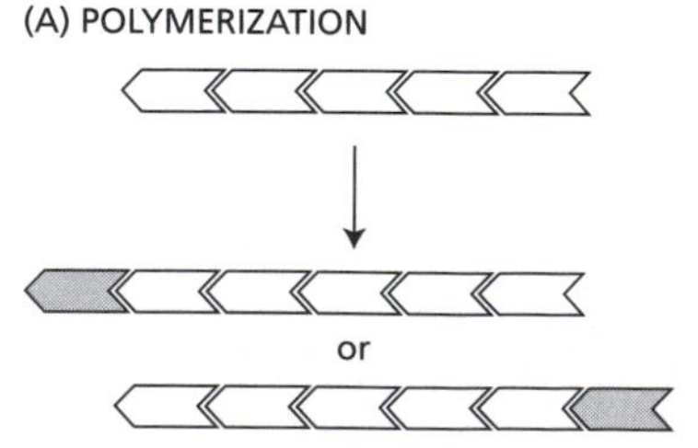


(B) SUBUNIT CONFORMATIONS

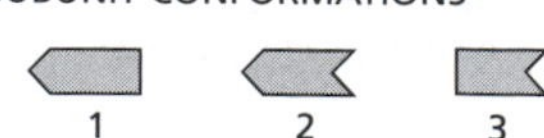


Figure 16–4 Polymerization of a polymer (Problem 16–16). (A) Addition of a subunit to a polymer. (B) Conformations of three hypothetical subunits.

**16–18** Dynamic instability causes microtubules either to grow or to shrink rapidly. Consider an individual microtubule that is in its shrinking phase.

A. What must happen at the end of the microtubule in order for it to stop shrinking and start growing?

B. How would an increase in the tubulin concentration affect this switch from shrinking to growing?

C. What would happen if GDP, but no GTP, were present in the solution?

D. What would happen if the solution contained an analog of GTP that could not be hydrolyzed?

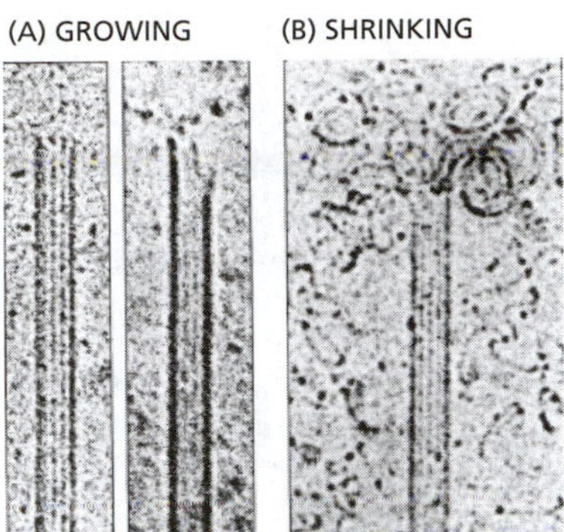


**Figure 16–5** Electron microscopic analysis of microtubule dynamics (Problem 16–17). (A) Rapidly growing microtubules. (B) Catastrophically shrinking microtubule.

**16–19** The β-tubulin subunit of an αβ-tubulin dimer retains its bound GTP for a short time after it has been added to a microtubule, yielding a GTP cap whose size depends on the relative rates of polymerization and GTP hydrolysis. A simple notion about microtubule growth dynamics is that the ends with GTP caps grow, whereas ends without GTP caps shrink. To test this idea, you allow microtubules to form under conditions where you can watch individual microtubules. You then sever one microtubule in the middle using a laser beam. Would you expect the newly exposed plus and minus ends to grow or to shrink? Explain your answer.

**16–20** Tubulin is distantly related to the GTPase family that includes Ras. The β-tubulin in an αβ-tubulin dimer hydrolyzes GTP at a slow intrinsic rate. When it is incorporated into a growing microtubule, its rate of GTP hydrolysis increases dramatically, like Ras when it interacts with a GTPase activating protein (GAP). What is the GAP for β-tubulin?

**16–21** List differences between bacteria and animal cells that could have depended on the appearance during evolution of some or all of the components of the present eucaryotic cytoskeleton. Why do you suppose a cytoskeleton might have been crucial for each of these differences to evolve?

**16–22** The amino acid sequences of actins and tubulins from all eucaryotes are remarkably well conserved, yet the large numbers of proteins that interact with these filaments are no more conserved than most other proteins in different species. How can it be that the filament proteins themselves are highly conserved, while the proteins that interact with them are not?

**Problem 3–34** examines the sequence of an intermediate filament protein for the characteristic coiled-coil heptad repeat motif.

**16–23** Why is it that intermediate filaments have identical ends and lack polarity, whereas actin filaments and microtubules have two distinct ends with a defined polarity?

**16–24** Which of the following types of cell would you expect to contain a high density of cytoplasmic intermediate filaments? Explain your answers.

A. *Amoeba proteus* (a free-living amoeba).

B. Human skin epithelial cell.

C. Smooth muscle cell in the digestive tract of a vertebrate.

D. Nerve cell in the spinal cord of a mouse.

E. Human sperm cell.

F. Plant cell.

**16–25** Disulfide bonds do not form in the cytosol of eucaryotic cells (see Problem 3–44). Yet keratin intermediate filaments in the skin are cross-linked by disulfide bonds. How can that be?

**16–26** Although knockouts of genes for some intermediate filaments have detectable phenotypes in mice, gene knockouts for vimentin or glial fibrillary acid protein (GFAP) appear normal. Mice with knockouts for both vimentin and GFAP, however, exhibit impaired function of their astrocytes, which are accessory cells in the central nervous system. Why do you suppose that individual knockouts for vimentin and GFAP are normal, while the combined knockout has a demonstrable deficiency?

**16–27** The drugs taxol, extracted from the bark of yew trees, and colchicine, an alkaloid from autumn crocus, have opposite effects. Taxol binds tightly to microtubules and stabilizes them. When added to cells, it causes much of the free tubulin to assemble into microtubules. In contrast, colchicine prevents microtubule formation. Taxol and colchicine are equally toxic to dividing cells, and both are used as anticancer drugs. Based on your knowledge of microtubule dynamics, explain why these drugs are toxic to dividing cells despite their opposite modes of action.

**16–28** The common laboratory reagent acrylamide, used as a precursor in making polyacrylamide gels, is a potent neurotoxin. One hypothesis for its toxic effects is that it destroys neurofilament bundles by binding to the subunits and causing their depolymerization. To test this possibility, you compare acrylamide toxicity in normal mice and knockout mice that are lacking neurofilaments. (Surprisingly, these knockout mice have no obvious mutant phenotype.) You find that acrylamide is an equally potent neurotoxin in normal mice and these knockout mice. Is acrylamide toxicity mediated through its effects on neurofilaments?

## CALCULATIONS

**16–29** At 1.4 mg/mL pure tubulin, microtubules grow at a rate of about 2 μm/min. At this growth rate how many αβ-tubulin dimers (8 nm in length) are added to the ends of a microtubule each second?

**16–30** The average time it takes particles to diffuse a distance of $x$ cm is

$$t = x^2/2D$$

where $t$ is the time in seconds and $D$ is the diffusion coefficient, which is a constant that depends on the size and shape of the particle.

A. How long would it take for a small molecule, a protein molecule, and a membrane-enclosed vesicle to diffuse across a cell 10 μm in diameter. A typical diffusion coefficient for a small molecule is $5 \times 10^{-6}$ cm$^2$/sec, for a protein molecule $5 \times 10^{-7}$ cm$^2$/sec, and for a membrane vesicle $5 \times 10^{-8}$ cm$^2$/sec.

B. Why do you suppose a cell relies on the strategy of polymerizing and depolymerizing cytoskeletal filaments, rather than on diffusion of the filaments themselves, to accomplish its cytoskeletal rearrangements?

## DATA HANDLING

**16–31** If you add short actin filaments marked by bound myosin heads (myosin-decorated filaments) to a solution with an excess of actin monomers, wait for a few minutes and then examine the filaments by electron microscopy, you see the picture shown in Figure 16–6.

A. Which is the plus end of the myosin-decorated filament and which is the minus end? Which is the 'barbed' end and which is the 'pointed' end? How can you tell?

B. If you diluted the mixture so that the actin concentration was below the critical concentration, which end would depolymerize more rapidly?

C. When the actin filament depolymerizes, why are subunits removed exclusively from the ends and not from the middle of the filament?

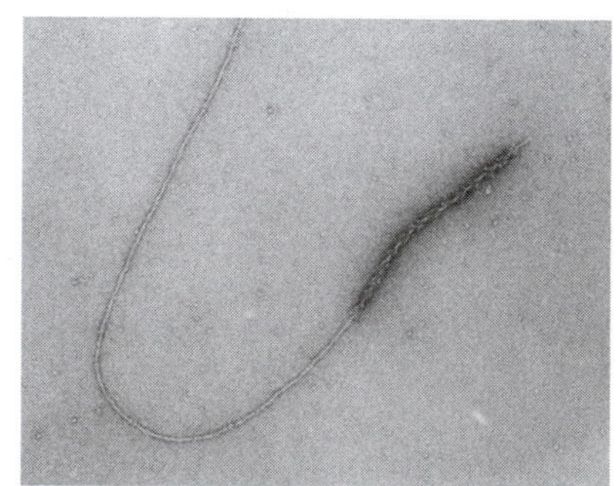

**Figure 16–6** Myosin-decorated actin filament after a few minutes in a solution with excess actin monomers (Problem 16–31). The shorter, thicker segment is the myosin-decorated actin filament.

**16–32** The orientation of the αβ-tubulin dimer in a microtubule was determined in several ways. GTP-coated fluorescent beads, for example, were found to bind exclusively at the plus ends of microtubules. By contrast, gold beads coated with antibodies specific for a peptide of α-tubulin bound exclusively at the minus end. How do these observations define the orientation of the αβ-tubulin dimer in the microtubule? Which tubulin subunit, α or β, is at which end? Explain your reasoning.

**16–33** The complex kinetics of microtubule assembly make it hard to predict the behavior of individual microtubules. Some microtubules in a population

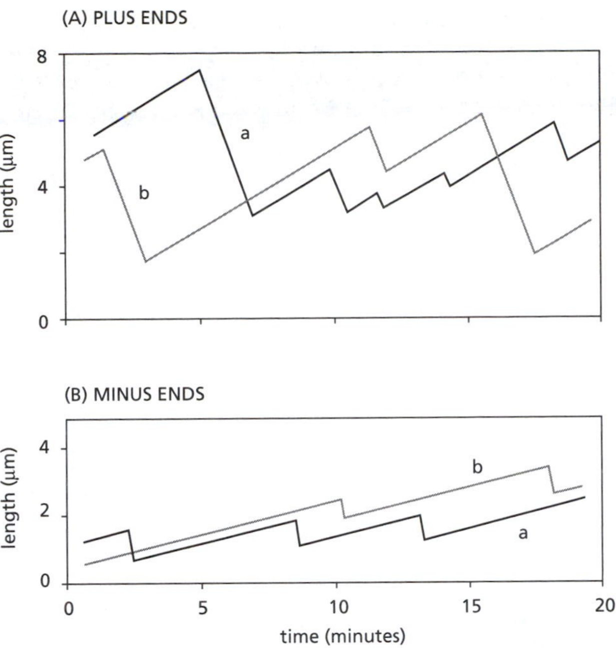


**Figure 16–7** Analysis of growth kinetics of individual microtubules (Problem 16–33). (A) Changes in length at the plus ends. (B) Changes in length at the minus ends. Results from two individual microtubules are identified with a and b.

can grow, even as the majority shrink to nothing. One simple hypothesis proposed to explain this behavior is that a growing end is protected from disassembly by a GTP cap and that a faster growing end has a longer GTP cap. Real-time video observations of changes in length with time are shown for two individual microtubules in Figure 16–7. Measurements of their rates of growth and shrinkage show that the plus end of each microtubule grows three times faster than the minus end, and shrinks at half the rate.

A. Are changes in length at the two ends of a microtubule dependent or independent of one another? How can you tell?

B. What does the GTP-cap hypothesis predict about the rate of switching between growing and shrinking states at the fast-growing end relative to the slow-growing end? Does the outcome of this experiment support the GTP-cap hypothesis?

**16–34** The growth rates at the plus and minus ends of actin filaments as a function of actin concentration are shown in Figure 16–8A and, on an expanded scale, in Figure 16–8B.

A. The data in Figure 16–8A were gathered by measuring initial growth rates at each actin concentration. Similar data gathered for any Michaelis–Menten enzyme would generate a hyperbolic plot, instead of the linear plots shown here. Why does the growth rate of actin filaments continue to increase linearly with increasing actin concentration, whereas an enzyme-catalyzed reaction reaches a plateau with increasing substrate concentration?

B. Figure 16–8B shows the filament growth rates at low actin concentration on an expanded scale. Imagine that you could add actin filaments to a solution of actin subunits at the concentrations indicated as A, B, C, D, and E. For each of these concentrations, decide whether the added actin filament would grow or shrink at its plus and minus ends. What is the critical concentration for the plus end? What is the critical concentration for the minus end? Would treadmilling occur at any of these concentrations?

**16–35** Comparisons of microtubule behavior between species point to differences that raise questions about the biological importance of dynamic instability. Notothenioid fishes, for example, which live in the Southern Ocean at a

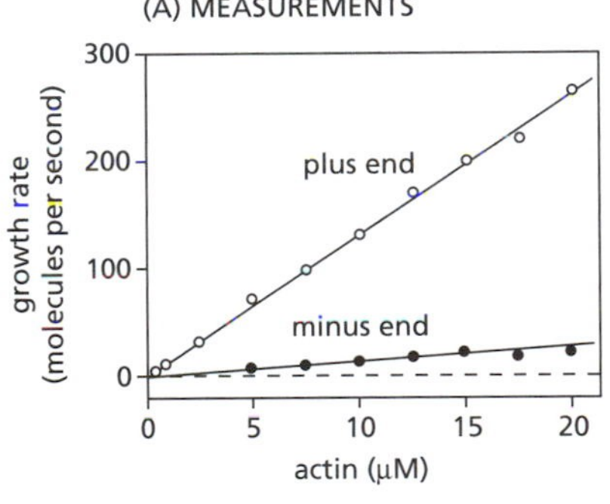


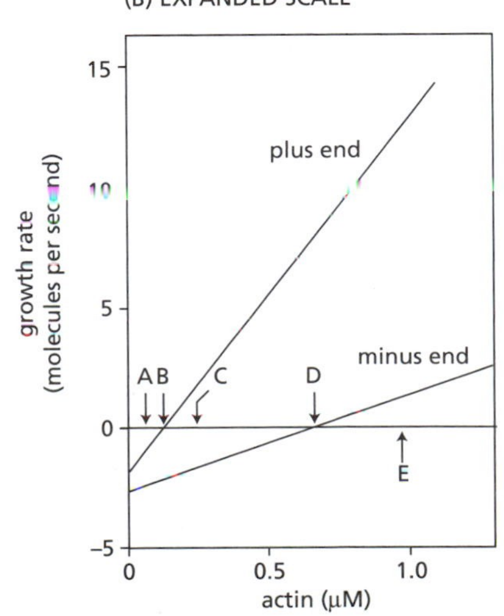


**Figure 16–8** Growth rates at the plus and minus ends of actin filaments as a function of actin concentration (Problem 16–34). (A) Measurements of growth rates over a broad range of actin concentrations. (B) Growth rates at low actin concentrations, shown on an expanded scale.

**Table 16–1 Properties of individual microtubules in notothenioid fish and the domestic cow (Problem 16–35).**

| MICROTUBULES | GROWTH RATE (μm/min) | SHRINKAGE RATE (μm/min) | CATASTROPHE FREQUENCY ($min^{-1}$) | RESCUE FREQUENCY ($min^{-1}$) |
|---|---|---|---|---|
| Notothenioid fish | 0.27 | 0.8 | 0.008 | <0.0004 |
| Domestic cow | 2.18 | 61.2 | 0.52 | 3.1 |

Multiple individual microtubules were observed by videomicroscopy near the body temperature for each species: 5°C for fish and 37°C for cow. Average growth rates were calculated for growing microtubules, and average shrinkage rates were calculated for shrinking microtubules. Changes from growth to shrinkage (catastrophe) and from shrinkage to growth (rescue) were averaged over the observation period and expressed as frequency of events per minute.

constant temperature of –1.8°C, have remarkably stable microtubules compared with warm-blooded vertebrates such as the cow. This is an essential modification for notothenioid fish because normal microtubules disassemble completely into αβ-tubulin dimers at 0°C. Measurements on individual microtubules in solutions of pure tubulin show that fish microtubules grow at a much slower rate, shrink at a much slower rate, and only rarely switch from growth to shrinkage (catastrophe) or from shrinkage to growth (rescue) (Table 16–1).

**Problem 3–64** looks at antifreeze proteins in notothenioid fish, another adaptation to life lived below 0°C.

A. The amino acid sequences of the α- and β-tubulin subunits from notothenioid fish differ from those of the cow at positions and in ways that might reasonably be expected to stabilize the microtubule, in accord with the data in Table 16–1. Would you expect these changes to strengthen the interactions between the α- and β-tubulin subunits in the αβ-dimer, between adjacent dimers in the protofilament, or between tubulin subunits in adjacent protofilaments? Explain your reasoning.

B. Dynamic instability is thought to play a fundamental role in the rapid microtubule rearrangements that occur in cells. How do you suppose cells in these notothenioid fishes manage to alter their microtubule architecture quickly enough to accomplish essential cell functions? Or do you suppose that these cells exist with a stable microtubule cytoskeleton that only slowly rearranges itself?

**16–36** A standard purification scheme for tubulin is to prepare a cell extract, chill it to 0°C, spin it at high speed and save the supernatant. The supernatant is then warmed to 37°C and incubated in the presence of GTP. The mixture is then spun at high speed and the pellet is saved and redissolved. Then the cycle is repeated: chill the dissolved pellet, spin at high speed, save the supernatant, incubate with GTP at 37°C, spin at high speed, save the pellet. After a few cycles one obtains a pure preparation of tubulin. Explain how this procedure yields pure tubulin.

**16–37** Your ultimate goal is to understand human consciousness, but your advisor wants you to understand some basic facts about actin assembly. He tells you that ATP binds to actin monomers and is required for assembly. But, ATP hydrolysis is not necessary for polymerization since ADP can, under certain circumstances, substitute for the ATP requirement. ADP filaments, however, are much less stable than ATP filaments, supporting your secret suspicion that the free energy of ATP hydrolysis really is used to drive actin assembly.

Your advisor suggests that you make careful measurements of the quantitative relationship between the number of ATP molecules hydrolyzed and the number of actin monomers linked into polymer. The experiments are straightforward. To measure ATP hydrolysis, you add $\gamma^{32}$P-ATP to a solution of polymerizing actin, take samples at intervals, and determine how much radioactive phosphate has been produced. To assay polymerization, you measure the increase in light scattering that is caused by formation of the actin filaments. Your results are shown in Figure 16–9. Your light-scattering measurements indicate that 20 μmoles of actin monomers were polymerized. Since the number of polymerized actin monomers matches exactly the number of ATP molecules hydrolyzed, you conclude that one ATP is hydrolyzed as each new monomer is added to an actin filament.

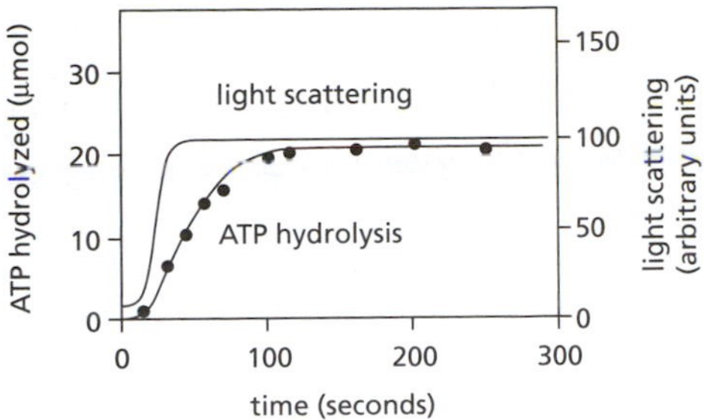


**Figure 16–9** The kinetics of actin polymerization and ATP hydrolysis (Problem 16–37).

When you show your advisor the data and tell him your conclusions, he smiles and very gently tells you to look more closely at the graph. He says your data prove that actin can polymerize without ATP hydrolysis.

A. What does your advisor see in the data that you have overlooked?

B. What do your data imply about the distribution of ATP and ADP in polymerizing actin filaments?

**16–38** The intermediate filament networks in cells must be dealt with in some way when a cell divides. Figure 16–10 shows the vimentin networks (tagged with vimentin–GFP) in kidney cells from baby hamster (BHK-21) and rat kangaroo (PtK2) that are undergoing division. By examining these photographs, decide how each of these cell types handles its vimentin network.

**16–39** Although the mechanism of disassembly of most intermediate filaments is unclear, it is well defined for their ancestors, the nuclear lamins. The nuclear envelope is strengthened by a fibrous meshwork of lamins (the nuclear lamina), which supports the membrane on the nuclear side. When cells enter mitosis, the nuclear envelope breaks down and the nuclear lamina disassembles. Assembly and disassembly of the nuclear lamina may be controlled by reversible phosphorylation of lamins A, B, and C, since the lamins from cells that are in mitosis carry significantly more phosphate than do the lamins from cells that are in interphase.

To investigate the role of phosphorylation, you label cells with $^{35}$S-methionine and purify lamins A, B, and C from mitotic cells and from interphase cells. You then analyze the purified lamins from each source, along with a mixture of them, by two-dimensional gel electrophoresis (Figure 16–11A). You also treat the samples with alkaline phosphatase, which removes phosphates from proteins, and analyze them in the same way (Figure 16–11B).

A. Why does treatment with alkaline phosphatase reduce the number of lamin spots to three, regardless of the number seen in the absence of phosphatase treatment?

B. How many phosphate groups are attached to lamins A, B, and C during interphase? How many are attached during mitosis? How can you tell?

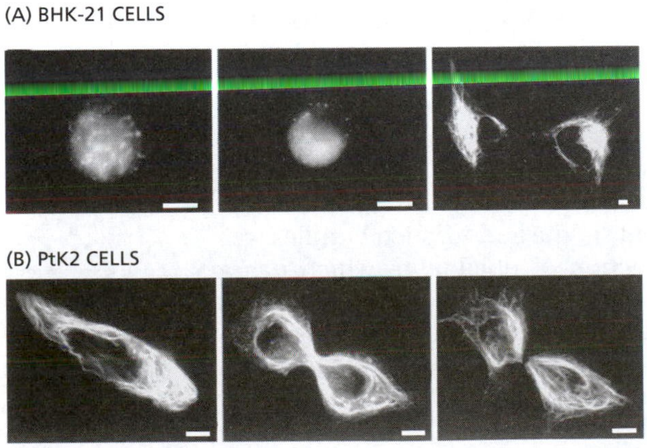

(A) BHK-21 CELLS

(B) PtK2 CELLS

**Figure 16–10** Vimentin networks during cell division (Problem 16–38). (A) BHK-21 cells. The three images from *left* to *right* correspond to prometaphase and anaphase of mitosis and to the daughter cells. Note that the first two images are magnified relative to the third. (B) PtK2 cells. The images from *left* to *right* correspond to prometaphase, telophase, and late cytokinesis. Note that in late cytokinesis the cells are still connected by a bridge of cytoplasm. The scale bar in each picture is 5 μm.

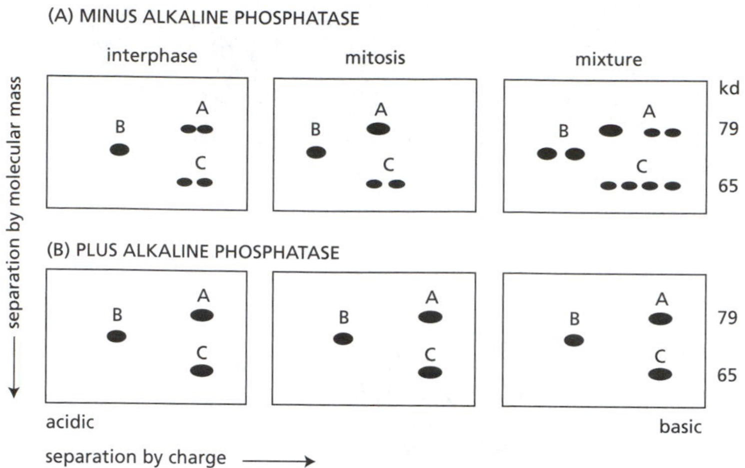


**Figure 16–11** Two-dimensional separation of nuclear lamins from cells in interphase and mitosis (Problem 16–39). (A) No treatment with alkaline phosphatase. (B) Treatment with alkaline phosphatase. *Letters* identify the positions of lamins A, B, and C. The purified lamins from interphase and mitotic cells were added together to create the mixture. Acidic proteins are more negatively charged; basic proteins are more positively charged. Each *rectangle* represents a distinct two-dimensional separation of nuclear lamins.

C. Why was $^{35}$S-methionine rather than $^{32}$P-phosphate used to label lamins in experiments designed to measure phosphorylation differences? How would the autoradiograms have differed if $^{32}$P-phosphate had been used instead?

D. Do you think these results prove that lamin disassembly during mitosis is caused by their reversible phosphorylation? Why or why not?

**16–40** Cytochalasin B strongly inhibits certain forms of cell motility, such as cytokinesis and the ruffling of growth cones, and it dramatically decreases the viscosity of gels formed with mixtures of actin and a wide variety of actin-binding proteins. These observations suggest that cytochalasin B interferes with the assembly of actin filaments. In the classic experiment that defined its mechanism, short lengths of actin filaments were decorated with myosin heads and then mixed with actin subunits in the presence or absence of cytochalasin B. Assembly of actin filaments was measured by assaying the viscosity of the solution (Figure 16–12) and by examining samples by electron microscopy (Figure 16–13).

A. Suggest a plausible mechanism to explain how cytochalasin B inhibits actin filament assembly. Account for the appearance of the filaments in the electron micrographs and the viscosity measurements (both the altered rate and extent).

B. The normal growth characteristics of an actin filament and the actin-binding properties of cytochalasin B argue that actin monomers undergo a conformational change upon addition to an actin filament. How so?

**16–41** Phalloidin, which is a toxic peptide from the mushroom *Amanita phalloides*, binds to actin filaments. Phalloidin tagged with a fluorescent probe is commonly used to stain actin filament assemblies in cells (Figure 16–14A). If phalloidin is attached to a gold particle instead, its binding to actin filaments can be examined at high resolution by scanning transmission electron microscopy. Figure 16–14B shows a micrograph of an actin filament with bound phalloidin and Figure 16–14C shows the same picture with the contrast adjusted so that only the points of highest intensity (the gold particles) are visible. Does phalloidin bind to every actin subunit? How can you tell?

**16–42** Isolated bundles of actin filaments from the acrosomal processes of *Limulus polyphemus* (horseshoe crab) sperm have readily distinguishable plus ends (tapered) and minus ends (blunt). Assembly at the ends of such bundles was used to determine the mechanism of action of phalloidin, which has a marked effect on actin assembly. When phalloidin is mixed with actin in a molar ratio of at least 1:1, the growth rate increases at both ends, as shown for minus ends in Figure 16–15A. Because growth rate = $k_{on}$[actin]$_{initial}$ – $k_{off}$, these plots have the form $y = mx + b$, so that the slope of the line equals $k_{on}$ and the $y$ intercept equals $-k_{off}$.

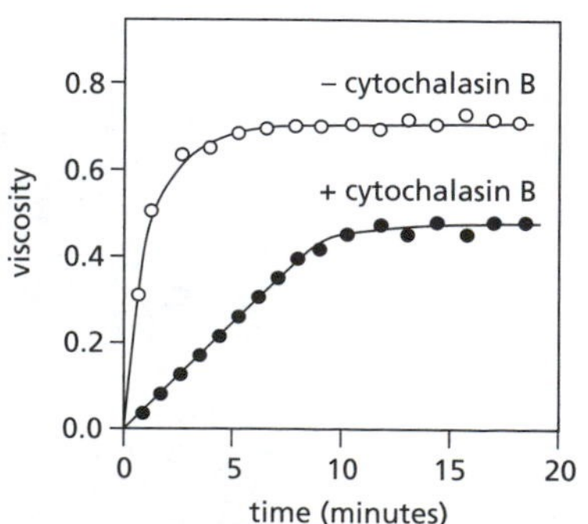


**Figure 16–12** Increase in the viscosity of actin solutions in the presence and absence of cytochalasin B (Problem 16–40).

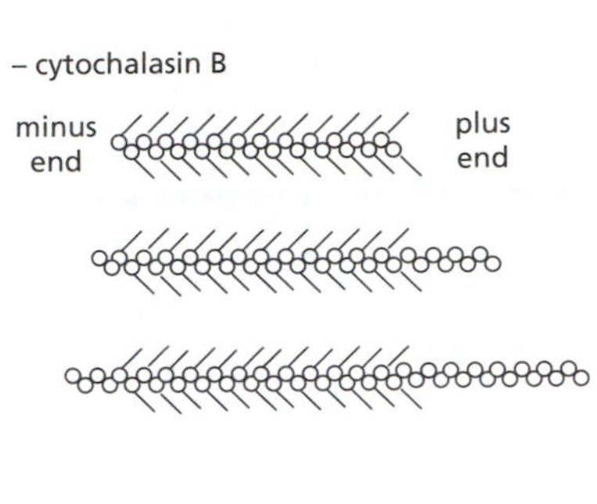


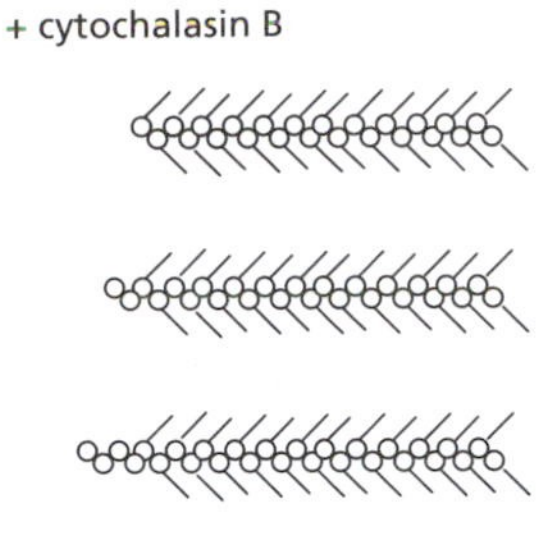


**Figure 16–13** Appearance of typical actin filaments formed in the presence and absence of cytochalasin B (Problem 16–40). The decorated actin filaments present before the addition of actin monomers are shown at the *top* of each set of three. Filaments present after increasing times of incubation with actin monomers are shown *below*.

A. By analyzing the on and off rates, decide how phalloidin increases the growth rate of actin filaments? Explain your reasoning.
B. In Figure 16–15B actin filaments grown in the presence or absence of phalloidin were diluted in the absence of actin monomers and their disassembly was assayed. Do these results confirm or contradict your conclusions from part A? Explain your answer.
C. What is the critical concentration for actin assembly at the minus end in the absence of phalloidin? What is the critical concentration for actin assembly at the minus end in the presence of phalloidin?
D. Propose a molecular mechanism for the effects of phalloidin on actin assembly.

**16–43** Swinholide A is a member of a class of lipophilic compounds termed macrolides, which include a number of useful antibiotics such as erythromycin, that are synthesized by Actinomycetes. Swinholide A is a 'twin' molecule, composed of two identical halves (Figure 16–16A). When added to cells growing in culture, swinholide A disrupts the actin cytoskeleton. Your advisor has shown conclusively that swinholide A binds a pair of actin monomers. She suspects that swinholide A causes actin filaments to depolymerize by sequestering actin subunits in a nonfunctional dimeric form and thus accelerating depolymerization through mass action effects. She wants you to test this hypothesis.

You prepare actin filaments tagged with a probe that fluoresces intensely in the filament but much less so in the free subunits (or swinholide-bound subunits). This allows you to follow depolymerization readily and rapidly as a loss of fluorescence. Just as your advisor predicted, depolymerization increases in the presence of increasing concentrations of swinholide A (Figure 16–16B). But you notice two features of these curves that suggest to you that swinholide A may actually sever actin filaments. One of these features is illustrated in Figure 16–16C, which shows a nonlinear dependence of the initial rate of depolymerization on the concentration of swinholide A. A simple mass-action effect—the sequestering of actin monomers by binding to swinholide A—predicts a linear dependence; however, increasing increments in swinholide A concentration have a progressively greater effect on depolymerization.

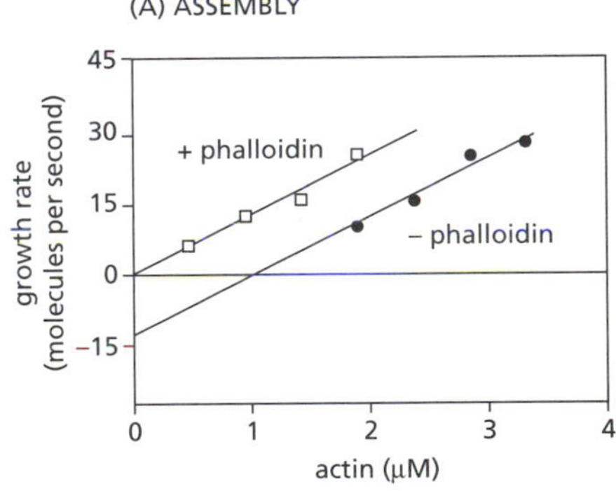


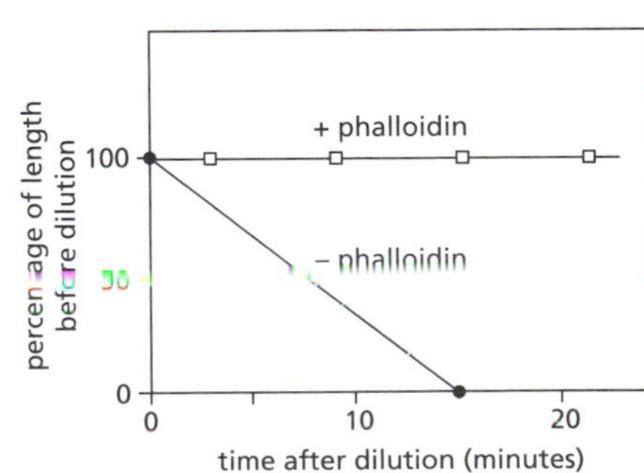


**Figure 16–15** Effects of phalloidin on actin filaments (Problem 16–42). (A) Growth rates at the minus ends of acrosomal bundles in the presence and absence of phalloidin. (B) Disassembly of actin filaments upon dilution in the presence and absence of phalloidin.

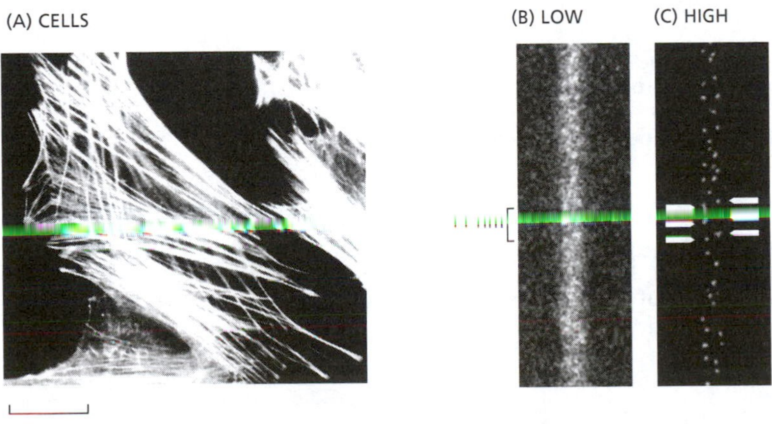


**Figure 16–14** Binding of phalloidin to actin filaments (Problem 16–41). (A) The actin cytoskeleton stained with fluorescent phalloidin. (B) An actin filament bound by gold-tagged phalloidin at low contrast. (C) The same actin filament as in (B), but at high contrast. *Bright bands* mark the positions of six gold particles.

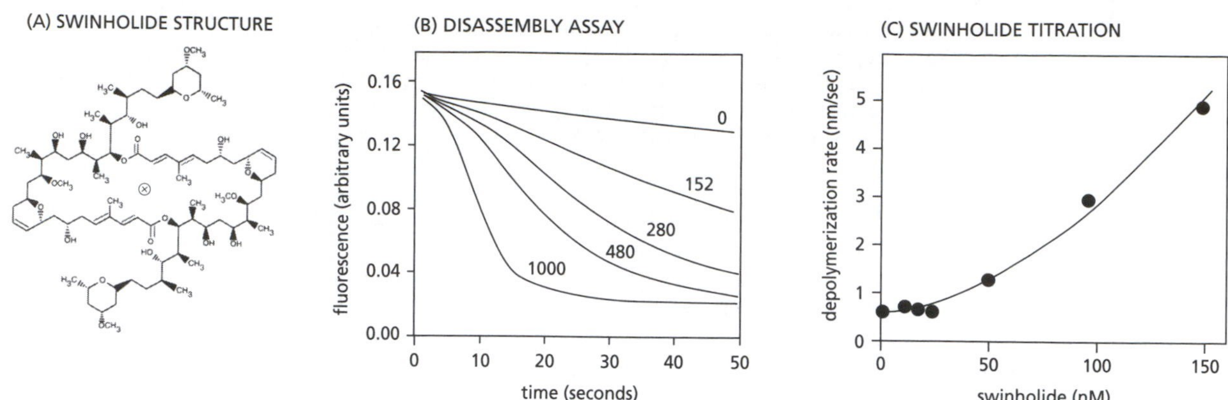


**Figure 16–16** Effects of swinholide A on actin filaments (Problem 16–43). (A) Structure of swinholide A. The identical halves of swinholide are arranged head to tail, so that if the molecule were rotated 180° about the indicated axis (*circle* with an X in it), it would superimpose on itself. For this reason it is said to have a twofold axis of symmetry. (B) Time course of actin filament depolymerization in the presence and absence of swinholide A. *Numbers* indicate the concentration of swinholide A (nM) used in each depolymerization assay. (C) Initial rates of depolymerization as a function of swinholide A concentration.

A. In Figure 16–16B, why does fluorescence reach a plateau value (at about 0.03) instead of decreasing to zero?
B. The other odd feature you noticed about depolymerization in the presence of swinholide A (Figure 16–16B) is that the lines have a 'hump' in them in the first few seconds (before they reach the plateau at later times). Why does this hump suggest that swinholide A severs actin filaments?
C. Assuming that swinholide A does sever actin filaments, is one molecule enough, or are multiple molecules needed? How do you know?

# HOW CELLS REGULATE THEIR CYTOSKELETAL FILAMENTS

TERMS TO LEARN

| | |
|---|---|
| ARP complex | formin |
| cell cortex | γ-tubulin ring complex (γ-TuRC) |
| centriole | microtubule-associated protein (MAP) |
| centrosome | microtubule-organizing center (MTOC) |

## DEFINITIONS

Match the definition below with its term from the list above.

**16–44** Centrally located organelle of animal cells that is the primary microtubule-organizing center and acts as the spindle pole during mitosis.

**16–45** Specialized layer of cytoplasm on the inner face of the plasma membrane, rich in actin filaments.

**16–46** Protein assembly containing a special form of tubulin, along with other proteins, that is an efficient nucleator of microtubule growth.

**16–47** Short cylindrical array of microtubules, a pair of which are embedded in the major microtubule organizing center of an animal cells.

## TRUE/FALSE

Decide whether each of these statements is true or false, and then explain why.

**16–48** All microtubule-organizing centers contain centrioles that help nucleate microtubule polymerization.

**16–49** All proteins that bind to the ends of microtubules or actin filaments cap the ends to prevent further polymerization.

## THOUGHT PROBLEMS

**16–50** A solution of pure αβ-tubulin dimers is thought to nucleate microtubules by forming a linear protofilament about seven dimers in length. At that point, the probabilities that the next αβ-dimer will bind laterally or to the end of the protofilament are about equal. The critical event for microtubule formation is thought to be the first lateral association (Figure 16–17). How does lateral association promote the subsequent rapid formation of a microtubule?

**16–51** How does a centrosome 'know' when it has found the center of the cell?

**16–52** Some actin-binding proteins significantly increase the rate at which the formation of actin filaments is initiated in the cytosol. How might such proteins do this? What must they *not* do when binding the actin monomers?

**16–53** How are γ-TuRC and the ARP complex similar, and how are they different?

**16–54** The concentration of actin in cells is 50–100 times greater than the critical concentration observed for pure actin in a test tube. How is this possible? What prevents the actin subunits in cells from polymerizing into filaments? Why is it advantageous to the cell to maintain such a large pool of actin subunits?

**16–55** The observations on individual microtubules in Figure 16–7 (see Problem 16–33) were made with microtubules polymerized in a solution of pure αβ-tubulin. How do you think the observations would change if centrosomes were used to nucleate microtubule growth? What would happen if you added microtubule-associated proteins (MAPs) to the solution of pure tubulin?

**16–56** Cofilin preferentially binds to older actin filaments and promotes their disassembly. How does cofilin distinguish old filaments from new ones?

**16–57** Filamin cross-links actin filaments at roughly right angles to produce a viscous gel that is required for cells to extend thin sheetlike lamellipodia. Why is the loss of filamin in melanoma cells bad news for the melanoma cells, but good news for the patient?

**16–58** When cells enter mitosis, their existing array of cytoplasmic microtubules has to be rapidly broken down and replaced with the mitotic spindle, which pulls the chromosomes into the daughter cells. The enzyme katanin, named after Japanese samurai swords, is activated during the onset of mitosis and cleaves microtubules into short pieces. What do you suppose is the fate of the microtubule fragments created by katanin?

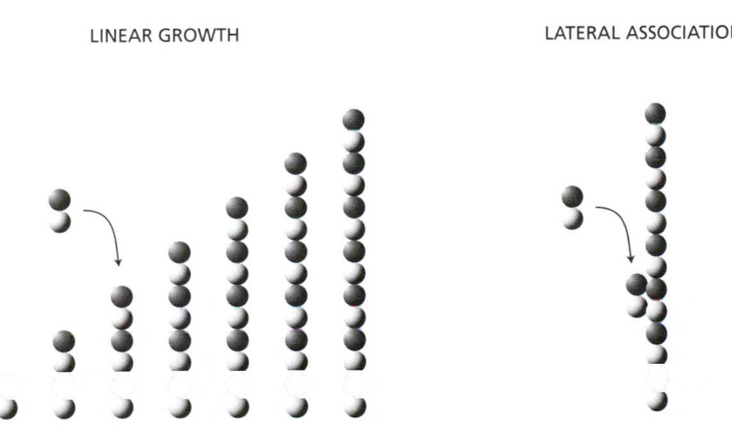


**Figure 16–17** Model for microtubule nucleation by pure αβ-tubulin dimers (Problem 16–50).

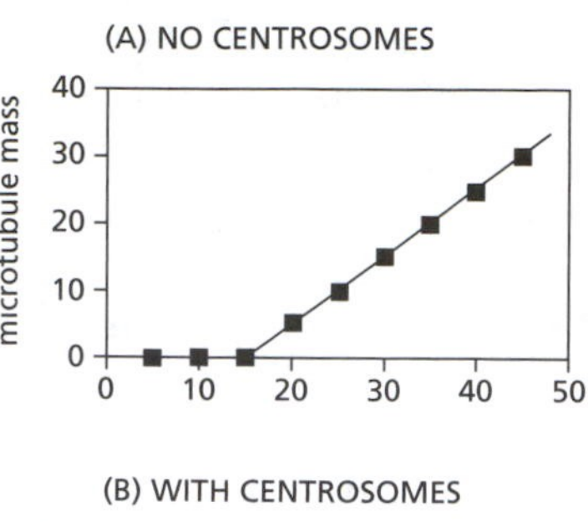


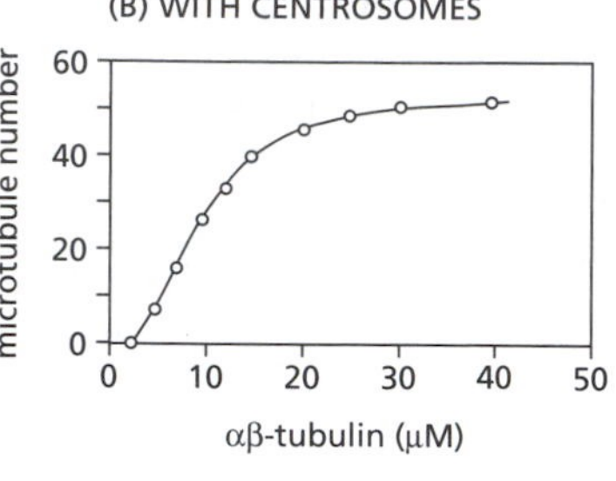


**Figure 16–18** Analysis of microtubule assembly (Problem 16–59). (A) Mass of microtubules assembled in the absence of centrosomes as a function of tubulin concentration. (B) Average number of microtubules per centrosome as a function of tubulin concentration. Concentrations refer to αβ-tubulin dimers, which are the subunits of assembly.

## CALCULATIONS

**16–59** The function of microtubules depends on their specific spatial organization within the cell. How are specific arrangements created, and what determines the formation and disappearance of individual microtubules?

To address these questions, investigators have studied the *in vitro* assembly of αβ-tubulin dimers into microtubules. Below 15 μM αβ-tubulin, no microtubules are formed; above 15 μM, microtubules form readily (Figure 16–18A). If centrosomes are added to the solution of tubulin, microtubules begin to form at less than 5 μM (Figure 16–18B). (Different assays were used in the two experiments—total weight of microtubules in Figure 16–18A and the average number of microtubules per centrosome in Figure 16–18B—but the lowering of the critical concentration for microtubule assembly in the presence of centrosomes is independent of the method of assay.)

A. Why do you think that the concentration at which microtubules begin to form (the critical concentration) is different in the two experiments?

B. Why do you think that the plot in Figure 16–18A increases linearly with increasing tubulin concentration above 15 μM, whereas the plot in Figure 16–18B reaches a plateau at about 25 μM?

C. The concentration of αβ-tubulin dimers (the subunits for assembly) in a typical cell is 1 mg/mL and the molecular weight of a tubulin dimer is 110,000. What is the molar concentration of tubulin dimers in cells? How does the cellular concentration compare with the critical concentrations in the two experiments in Figure 16–18? What are the implications for the assembly of microtubules in cells?

**16–60** The γ-tubulin ring complex (γ-TuRC), which nucleates microtubule assembly in cells, includes γ-tubulin and several accessory proteins. To get at its mechanism of nucleation, you have prepared monomeric γ-tubulin by *in vitro* translation and purification. You measure the effect of adding monomeric γ-tubulin to a solution of αβ-tubulin dimers, as shown in Figure 16–19.

A. In the presence of monomeric γ-tubulin, the lag time for assembly of microtubules decreases, and assembly occurs more rapidly (Figure 16–19A). How would you account for these two effects of γ-tubulin?

B. The critical concentration of αβ-tubulin needed for the assembly of microtubules is reduced from about 3.2 μM in the absence of γ-tubulin to about 1.7 μM in its presence (Figure 16–19B). How do you suppose γ-tubulin lowers the critical concentration? How does this account for the greater extent of polymerization in Figure 16–19A? (Think about the end—plus or minus—at which polymerization occurs in the presence of γ-tubulin.)

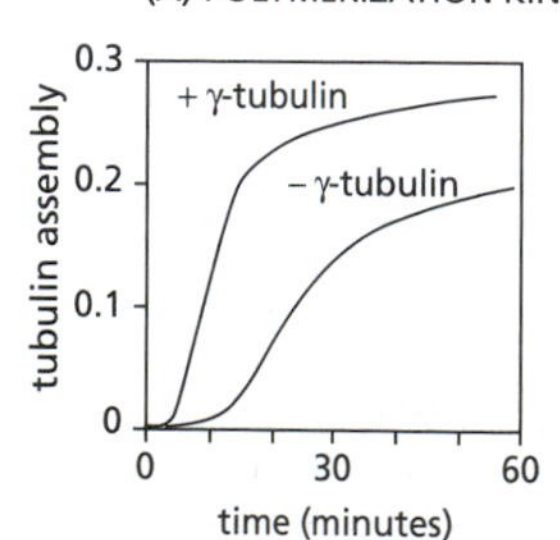


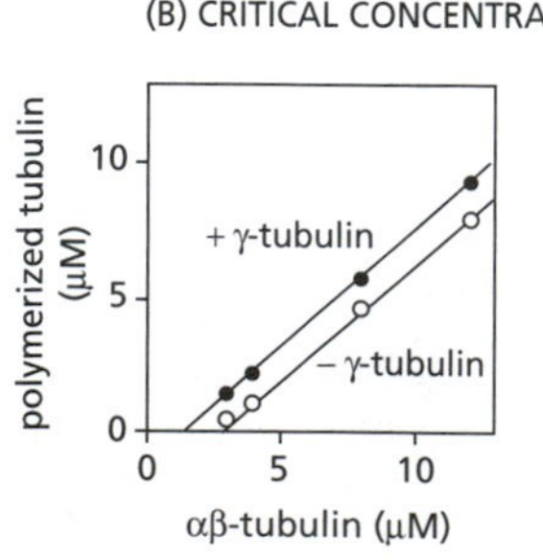


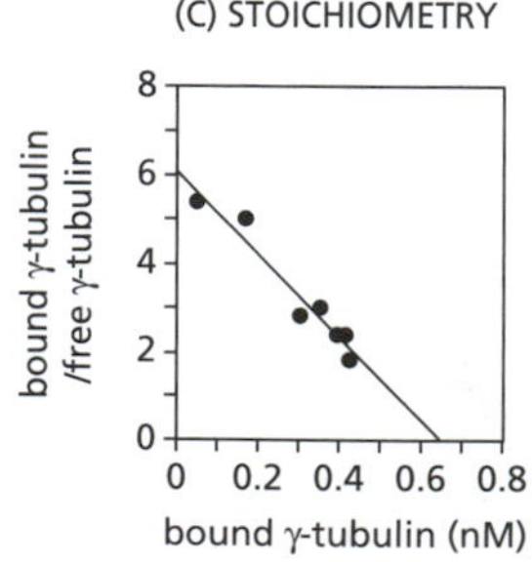


**Figure 16–19** Effects of γ-tubulin on microtubule polymerization (Problem 16–60). (A) Kinetics of polymerization in the presence and absence of γ-tubulin. (B) The critical concentration for microtubule assembly in the presence of 0.6 nM γ-tubulin and in its absence. (C) Stoichiometry of γ-tubulin binding as determined by a Scatchard plot.

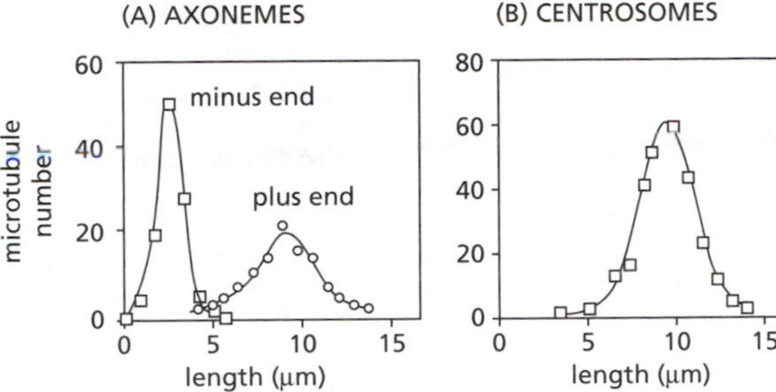


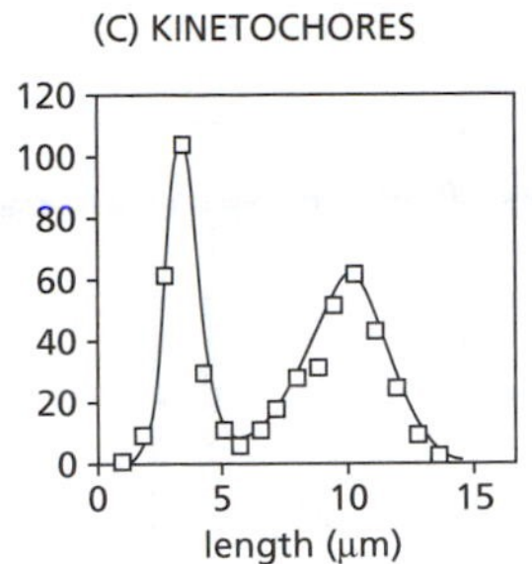


**Figure 16–20** Length distributions of microtubules (Problem 16–61). (A) Nucleation by flagellar axonemes, where the plus and minus ends can be distinguished. (B) Nucleation by centrosomes. (C) Nucleation by kinetochores.

C. A Scatchard plot (see Problem 3–101) of the bound over free γ-tubulin versus bound γ-tubulin in the presence of microtubules at 0.62 nM is shown in Figure 16–19C. What is the $K_d$ for γ-tubulin binding to microtubules? (The slope of the line in a Scatchard plot is $-1/K_d$.) How many γ-tubulin monomers are bound per microtubule in these experiments? (The *x*-intercept in a Scatchard plot is equal to the total number of binding sites.) How many γ-tubulin monomers do you think it takes to nucleate one microtubule?

## DATA HANDLING

**16–61** In addition to centrosomes, flagellar axonemes and kinetochores also can serve as nucleation sites for microtubule assembly. The following experiment was designed to determine whether these two structures nucleate microtubule growth by binding to the plus end or to the minus end of the nascent microtubule. Flagellar axonemes were included as a control since their plus and minus ends can be distinguished. Centrosomes and kinetochores (and flagellar axonemes) were incubated briefly in unlabeled tubulin to nucleate microtubule growth. A high concentration of biotin-labeled tubulin was then added and the incubation was continued for 10 minutes. At that point the preparations were fixed and the biotin-labeled segments were visualized by adding fluorescein-labeled antibodies specific for biotin. The lengths of the biotin-labeled segments were measured and plotted as shown in Figure 16–20.

A. Which end of a newly assembled microtubule is attached to the plus end of the flagellar axoneme?
B. Which end of a microtubule assembled on a flagellar axoneme grows faster?
C. Which end of an assembled microtubule is attached to a centrosome? To a kinetochore? Explain your reasoning.

**16–62** In the presence of monomeric γ-tubulin, αβ-tubulin dimers nucleate microtubules by forming a protofilament about three dimers in length, as opposed to seven dimers in the absence of γ-tubulin (see Problem 16–50). By separating α-tubulin from β-tubulin on polyacrylamide gels, transferring the proteins to nitrocellulose paper, and blotting with $^{35}$S-γ-tubulin, you show that γ-tubulin binds to β-tubulin (Figure 16–21). Based on this information, propose a model for nucleation of microtubules by monomeric γ-tubulin. Incorporate into your model whatever is needed to ensure that γ-tubulin caps the nucleated end (that is, permits growth in only one direction).

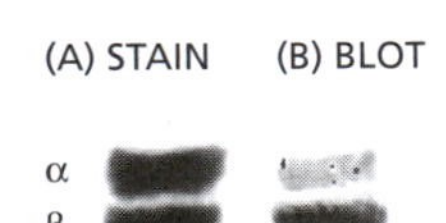


**Figure 16–21** Blot of separated α- and β-tubulin by $^{35}$S-γ-tubulin (Problem 16–62). (A) Protein stain showing positions of α- and β-tubulin. (B) Autoradiograph of blot with $^{35}$S-γ-tubulin.

**16–63** In the paper that defined the γ-tubulin ring complex (γ-TuRC), the authors purified the complex from *Xenopus* oocytes and showed that it dramatically stimulated the nucleation of microtubules. To determine whether nucleation occurred at plus ends or at minus ends, they polymerized microtubules in two steps. In the first step, microtubules nucleated with or without γ-TuRC were allowed to form in the presence of a small amount of αβ-tubulin containing a low proportion of rhodamine-labeled αβ-tubulin, which makes the microtubules fluoresce dimly. In the second step these

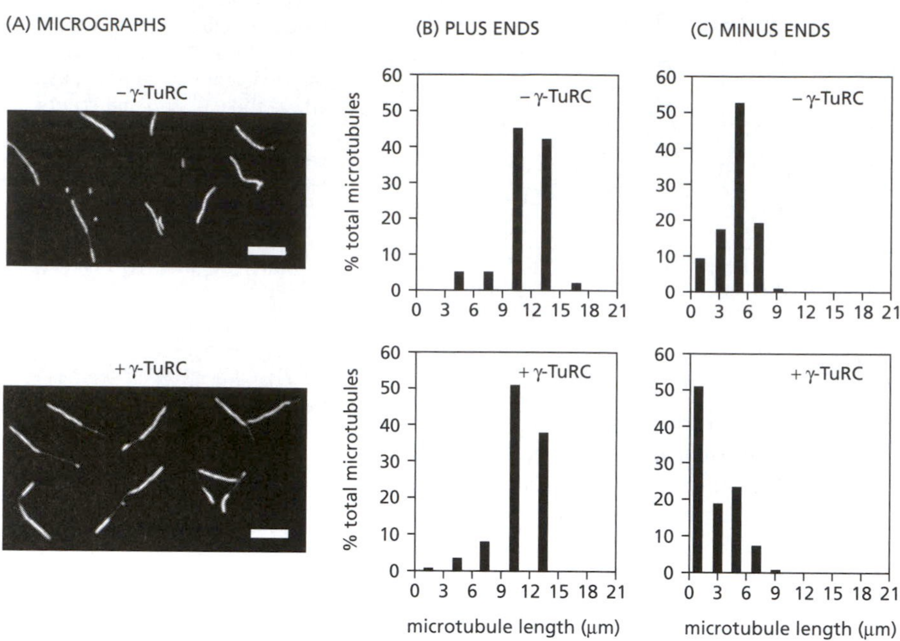


**Figure 16–22** Effects of γ-TuRC on microtubule assembly (Problem 16–63). (A) Example of microtubules grown in the presence and absence of γ-TuRC. Scale bar is 10 μm. (B) Distribution of the lengths of bright segments at microtubule plus ends in the presence and absence of γ-TuRC. (C) Distributions of the lengths of bright segments at microtubule minus ends in the presence and absence of γ-TuRC. In (B) and (C) only microtubules with a defined dim segment and one or two bright terminal segments were counted.

microtubules were allowed to extend at both ends in the presence of a higher proportion of rhodamine-tagged tubulin to label the ends brightly. The longer bright segment identifies the plus end, and the shorter segment the minus end (Figure 16–22A). Measurements of the lengths of a large number of bright segments in individual microtubules yielded the data in Figure 16–22B and C. At which end of the microtubule is γ-TuRC when it nucleates? Explain your reasoning.

**16–64** Accessory proteins that regulate the nucleation of actin filaments promote binding of the ARP complex to actin filaments so that most new filaments form as branches from existing ones. These proteins could stimulate ARP binding to the sides of existing filaments or to the plus end of a growing filament in a way that does not interfere with growth. Both possibilities would yield the final characteristic branched network of filaments. To distinguish between these alternatives, you mix the regulatory proteins with the ARP complex and actin subunits in the presence of actin filaments that are capped at their plus ends. After a short incubation you examine the resulting structures by electron microscopy. How will this experiment distinguish between these alternatives? What structures would you expect to see according to each model for nucleation by the ARP complex?

**16–65** The intracellular pathogenic bacterium *Listeria monocytogenes* propels itself through the cytosol on a comet tail of actin filaments (Figure 16–23). Remarkably, only a single bacterial protein, the transmembrane protein ActA, is required for this motility. ActA is distributed unequally on the surface of the

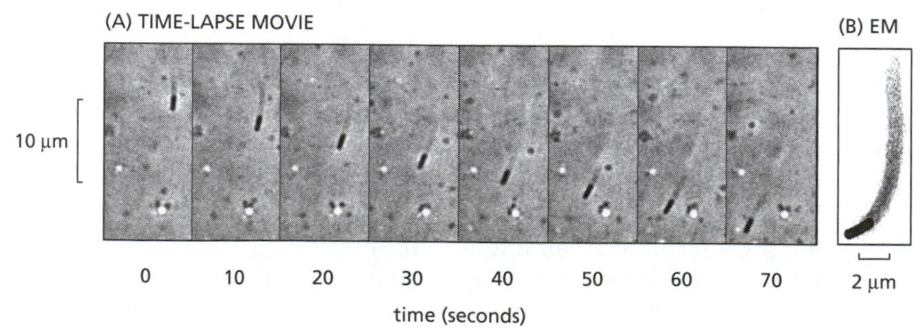


**Figure 16–23** Movement of a bacterium through the cytosol on a comet tail of actin filaments (Problem 16–65). (A) Time-lapse movie. (B) Electron micrograph. The bacterium is 2 μm in length.

bacterium, with maximum concentrations at the pole in contact with the actin tail. The effects of ActA on actin polymerization in the presence and absence of the ARP complex are shown in Figure 16–24A. The first few seconds of the reactions are shown on an expanded scale in Figure 16–24B. Polymerization of actin was followed using pyrene-actin, which exhibits much higher fluorescence intensity when actin is polymerized.

A. What are the effects of ActA and the ARP complex, separately and together, on the rate of nucleation of actin filaments? Explain your answer.
B. How do you suppose that the polymerization of actin by ActA and the ARP complex propels the bacterium across the cell. In the comet tail of actin filaments, which ends—plus or minus—do you suppose are pointed at the bacterium?

**16–66** You have two proteins that you suspect cap the ends of actin filaments. To determine whether they do and, if so, which protein caps which end, you measure filament formation as a function of actin concentration in the absence of either protein, in the presence of protein 1, and in the presence of protein 2 (Figure 16–25). Which protein caps the plus end and which caps the minus end? How can you tell? Give examples of proteins in the cell that you would expect to behave like protein 1 and protein 2.

**16–67** The haploid slime mold, *Dictyostelium discoideum,* lives on the forest floor as independently motile cells called amoebae, which feed on bacteria and yeast. When the food supply is exhausted, the amoebae come together to form multicellular wormlike slugs that crawl around, eventually differentiating into a plantlike structure with a fruiting body full of spores. When growth conditions become favorable once again, the spores germinate to form amoebae, completing the developmental life cycle.

*Dictyostelium* has two major actin filament cross-linking proteins: α-actinin and gelation factor. You have isolated two mutant strains of *Dictyostelium,* one that fails to express α-actinin and one that lacks gelation factor. Surprisingly, neither of these single mutants shows any defect in cell growth, motility, or development. By contrast, the double mutants are clearly defective. Although they move well as amoebae, when they form slugs they stay put, instead of crawling around as they normally would.

A. Your studies with the single mutants, defective in either α-actinin or gelation factor, suggest that neither gene is essential. Why then do you think that the loss of both genes causes such dramatic defects in the cell movements associated with development?
B. Why do you suppose that cells from the double mutant move all right as amoebae, but not as slugs?

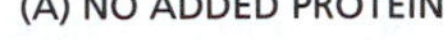


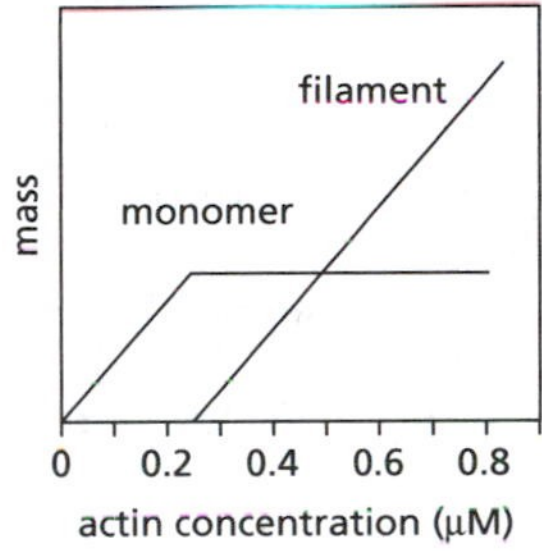


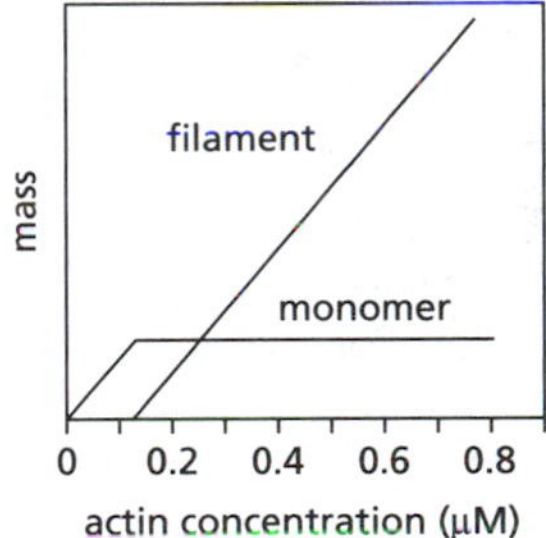


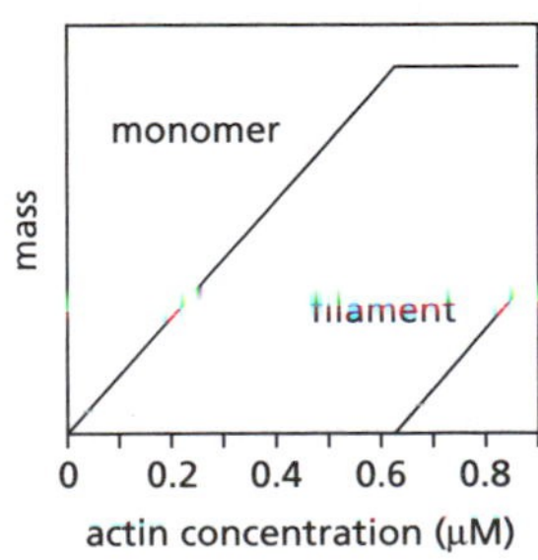


**Figure 16–25** Effects of two proteins on actin polymerization (Problem 16–66). (A) Polymerization of pure actin. (B) Actin polymerization in the presence of protein 1. (C) Actin polymerization in the presence of protein 2. The mass of actin, as monomers or filaments, was determined at equilibrium.

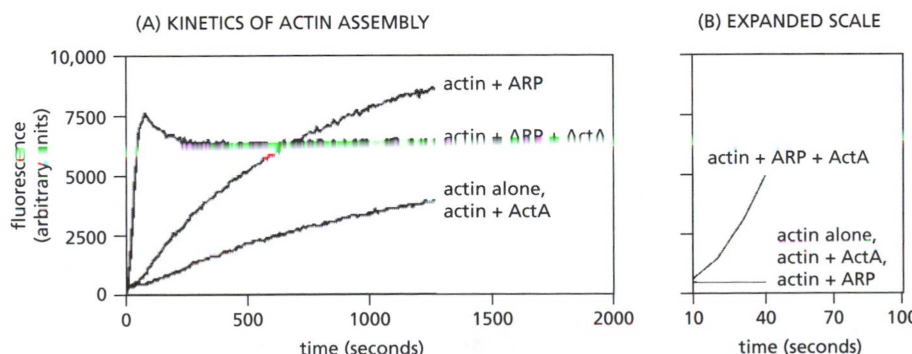


**Figure 16–24** Effects of ActA and the ARP complex on actin polymerization (Problem 16–65). (A) Kinetics of polymerization of actin in the presence of ActA and the ARP complex. (B) Kinetics of polymerization on an expanded scale. In all cases actin was present at 2 μM, and ActA and the ARP complex were present at 30 nM.

# MOLECULAR MOTORS

| TERMS TO LEARN | | |
|---|---|---|
| dynein | motor protein | myosin |
| kinesin | | |

## DEFINITIONS

Match the definition below with its term from the list above.

**16–68** A member of the family of motor proteins that move along microtubules by walking toward the minus end.

**16–69** The motor protein in muscle that generates the force for muscle contraction.

**16–70** A motor protein that moves along microtubules by walking toward the plus end.

## TRUE/FALSE

Decide whether each of these statements is true or false, and then explain why.

**16–71** Myosin II molecules have two motor domains and a rodlike tail that allows them to assemble into bipolar filaments, which are crucial for the efficient sliding of oppositely oriented actin filaments past each other.

**16–72** In most animal cells, minus end-directed microtubule motors deliver their cargo to the periphery of the cell, whereas plus end-directed microtubule motors deliver their cargo to the interior of the cell.

## THOUGHT PROBLEMS

**16–73** There are no known motor proteins that move on intermediate filaments. Suggest an explanation for this observation.

**16–74** A useful technique for studying a microtubule motor is to attach the motor proteins by their tails to a glass coverslip (the tails stick avidly to a clean glass surface) and then to allow microtubules to settle onto them. In the light microscope, the microtubules can be seen to move over the surface of the coverslip as the heads of the motors propel them (Figure 16–26).

A. Since the motor proteins attach in random orientations to the coverslip, how can they generate coordinated movement of individual microtubules, rather than engaging in a tug-of-war?

B. In which direction will microtubules crawl on a bed of dynein motor molecules (that is, will they move plus end first or minus end first)?

C. In the experiment shown in Figure 16–26, some of the microtubules were marked by gold beads that were bound by minus end-specific antibodies. Is the motor protein on the coverslip a plus end or minus end-directed motor? How can you tell?

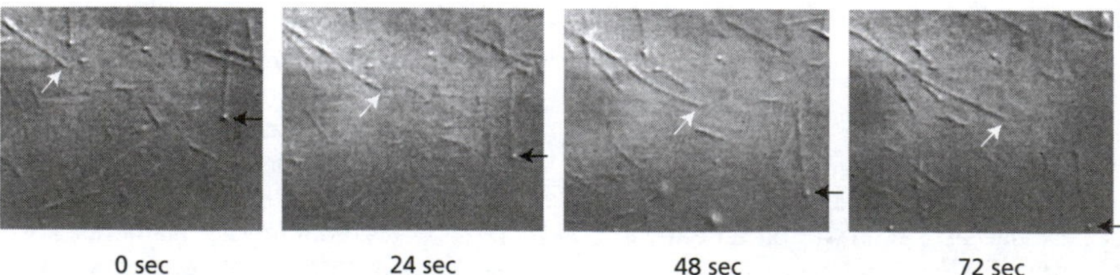


**Figure 16–26** Movement of microtubules on a bed of microtubule motor molecules (Problem 16–74). *Black arrows* mark the movement of a microtubule with a gold bead attached via antibodies to the minus end of a microtubule; *white arrows* mark the movement of a microtubule without an attached bead. Pictures were taken using video-enhanced differential interference contrast (VE-DIC) microscopy.

**16–75** In Problem 16–32 the orientation of the αβ-tubulin dimer in the microtubule was determined by showing that α-tubulin antibody-coated gold beads bound to the minus end. The electron micrographs, however, just showed microtubules with beads at one end (Figure 16–27). How do you suppose the investigators knew which end was which? Design an experiment to determine the orientation of microtubules labeled at one end with a gold bead.

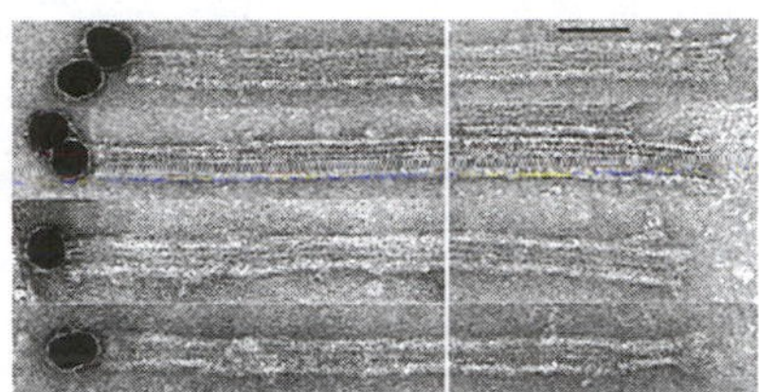

**Figure 16–27** Microtubules with α-tubulin antibody-coated gold beads attached to one end (Problem 16–75). At the *vertical line* a section of each microtubule has been removed so that the two ends can be displayed side by side.

**16–76** Living systems continually transform chemical free energy into motion. Muscle contraction, ciliary movement, cytoplasmic streaming, cell division, and active transport are examples of the ability of cells to transduce chemical free energy into mechanical work. In all these instances, a protein motor harnesses the free energy released in a chemical reaction to drive an attached molecule (the ligand) in a particular direction. Analysis of free-energy transduction in favorable biological systems suggests that a set of general principles governs the process in cells.

1. A cycle of reactions is used to convert chemical free energy into mechanical work.
2. At some point in the cycle a ligand binds very tightly to the protein motor.
3. At some point in the cycle the motor undergoes a major conformational change that alters the physical position of the ligand.
4. At some point in the cycle the binding constant of the ligand markedly decreases, allowing the ligand to detach from the motor.

These principles are illustrated by the two cycles for free-energy transduction shown in Figure 16–28: (1) the sliding of actin and myosin filaments against each other and (2) the active transport of $Ca^{2+}$ from inside the cell, where its concentration is low, to the cell exterior, where its concentration is high. An examination of these cycles underscores the principles of free-energy transduction.

A. What is the source of chemical free energy that powers these cycles, and what is the mechanical work that each cycle accomplishes?

B. What is the ligand that is bound tightly and then released in each of the cycles? Indicate the points in each cycle where the ligand is bound tightly.

C. Identify the conformational changes in the protein motor that constitute the 'power stroke' and 'return stroke' of each cycle.

## DATA HANDLING

**16–77** Kinesin carries vesicles for long distances along microtubule tracks in the cell. Are the two motor domains of a kinesin molecule essential to accomplish this task, or could a one-headed motor protein function just as well? Using recombinant DNA techniques, a version of kinesin was prepared that was identical to normal kinesin except that one motor domain was absent. Wild-type kinesin with two motor domains and recombinant kinesin with

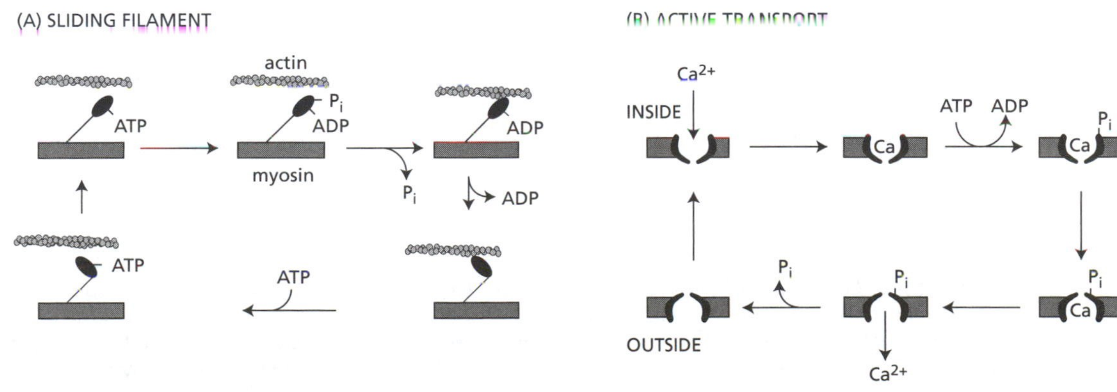


**Figure 16–28** Transduction of chemical free energy into mechanical work (Problem 16–76). (A) Sliding of actin filaments relative to myosin filaments. (B) Active transport of $Ca^{2+}$ from the inside to the outside of the cell. In both cycles *arrows* are drawn in only one direction to emphasize their normal operation. The phosphorylation and dephosphorylation steps in the active transport cycle are catalyzed by enzymes that are not shown in the diagram.

one were attached to coverslips at a variety of densities and the rate at which microtubules were bound and moved (collectively, termed the landing rate) was measured (Figure 16–29).

A. Why do you suppose that the curves at low motor densities are so different?

B. What do these experiments say about the design of the kinesin motor: are two heads required for vesicle transport, or is only one needed? Explain your reasoning.

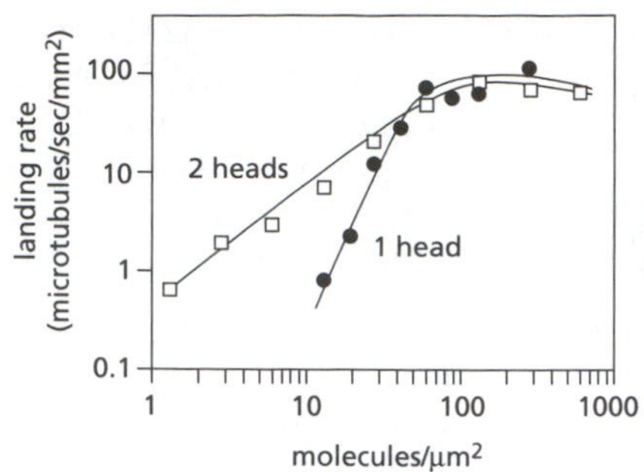


**Figure 16–29** Landing rates—binding and moving—of microtubules as a function of motor protein density (Problem 16–77). Results with wild-type kinesin are shown as *open squares*, while those with recombinant kinesin are shown as *solid circles*.

**16–78** The movements of single motor-protein molecules can be analyzed directly. Using polarized laser light, it is possible to create interference patterns that exert a centrally directed force, ranging from zero at the center to a few piconewtons at the periphery (about 200 nm from the center). Individual molecules that enter the interference pattern are rapidly pushed to the center, allowing them to be captured and moved at the experimenter's discretion.

Using such 'optical tweezers,' single kinesin molecules can be positioned on a microtubule that is fixed to a coverslip. Although a single kinesin molecule cannot be seen optically, it can be tagged with a silica bead and tracked indirectly by following the bead (Figure 16–30A). In the absence of ATP, the kinesin molecule remains at the center of the interference pattern, but with ATP it moves toward the plus end of the microtubule. As kinesin moves along the microtubule, it encounters the force of the interference pattern, which simulates the load kinesin carries during its actual function in the cell. Moreover, the pressure against the silica bead counters the effects of Brownian (thermal) motion, so that the position of the bead more accurately reflects the position of the kinesin molecule on the microtubule.

Traces of the movements of two kinesin molecules along a microtubule are shown in Figure 16–30B.

A. As shown in Figure 16–30B, all movement of kinesin is in one direction (toward the plus end of the microtubule). What supplies the free energy needed to ensure a unidirectional movement along the microtubule?

B. What is the average rate of movement of each kinesin along the microtubule?

C. What is the length of each step that a kinesin takes as it moves along a microtubule?

D. From other studies it is known that kinesin has two globular domains that each can bind to β-tubulin, and that kinesin moves along a single protofilament in a microtubule. In each protofilament the β-tubulin subunit repeats at 8-nm intervals. Given the step length and the interval between β-tubulin subunits, how do you suppose a kinesin molecule moves along a microtubule?

E. Is there anything in the data in Figure 16–30B that tells you how many ATP molecules are hydrolyzed per step?

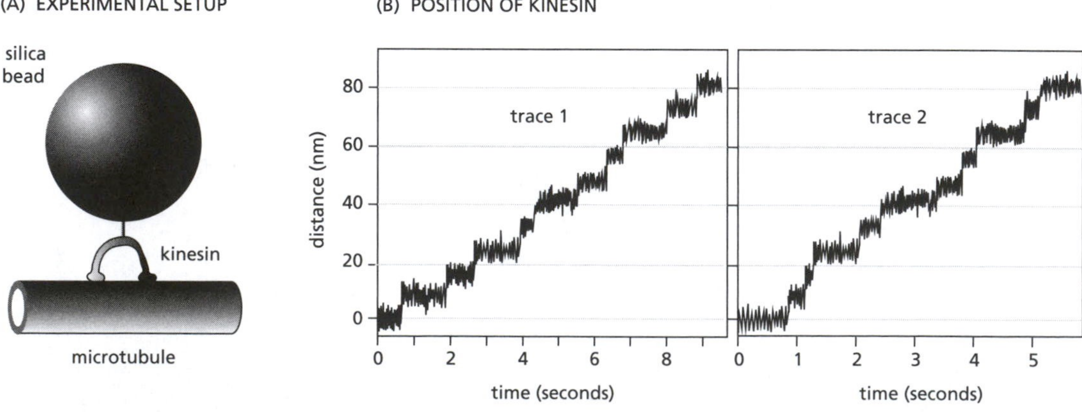


**Figure 16–30** Movement of kinesin along a microtubule (Problem 16–78). (A) Experimental setup with kinesin linked to a silica bead, moving along a microtubule. (B) Position of kinesin (as visualized by the position of the silica bead) relative to the center of the interference pattern, as a function of time of movement along the microtubule. The jagged nature of the trace results from Brownian motion of the bead. The movements of two different kinesin molecules are shown.

# THE CYTOSKELETON AND CELL BEHAVIOR

TERMS TO LEARN

| | | |
|---|---|---|
| axoneme | flagellum | pseudopodium |
| cilium | lamellipodium | Rho protein family |
| filopodium | myofibril | WASp protein |

## DEFINITIONS

Match the definition below with its term from the list above.

**16–79** A group of closely related monomeric GTPases that includes Cdc42, Rac, and Rho.

**16–80** Bundle of microtubules and associated proteins that forms the core of a cilium or flagellum in a eucaryotic cell and is responsible for their movements.

**16–81** Long, highly organized bundle of actin, myosin, and other proteins in the cytoplasm of muscle cells that contracts by a sliding-filament mechanism.

**16–82** Flattened, two-dimensional protrusion of membrane, supported by a meshwork of actin filaments, that is extended from the leading edge of crawling epithelial cells, fibroblasts, and some neurons.

**16–83** Long, hairlike protrusion from the surface of a eucaryotic cell whose undulations drive the cell through a fluid medium.

## TRUE/FALSE

Decide whether each of these statements is true or false, and then explain why.

**16–84** Motor neurons trigger action potentials in muscle cell membranes that open voltage-sensitive $Ca^{2+}$ channels in T-tubules, allowing extracellular $Ca^{2+}$ to enter the cytosol, bind to troponin C, and initiate rapid muscle contraction.

**16–85** When activated by $Ca^{2+}$ binding, troponin C causes troponin I to release its hold on actin, thereby allowing the tropomyosin molecules to shift their positions slightly so that the myosin heads can bind to the actin filaments.

**16–86** Neutrophils move toward a source of bacterial infection by chemotaxis, using receptors on their surface to respond to a gradient of *N*-formylated peptides derived from bacterial proteins.

## THOUGHT PROBLEMS

**16–87** Kinesin-1 motors are highly processive, moving long distances on microtubule tracks without dissociating. By contrast, myosin II motors in skeletal muscle do not move processively; they take only one or a few steps before letting go. How are these different degrees of processivity adapted to the biological functions of kinesin-1 and myosin II?

**16–88** Compare the structure of intermediate filaments with that of the myosin II filaments in skeletal muscle cells. What are the major similarities? What are the major differences? How do the differences in structure relate to their function?

**16–89** Which one of the following changes takes place when a skeletal muscle contracts?

A. Z discs move farther apart.
B. Actin filaments contract.
C. Myosin filaments contract.
D. Sarcomeres become shorter.

(A)

(B)

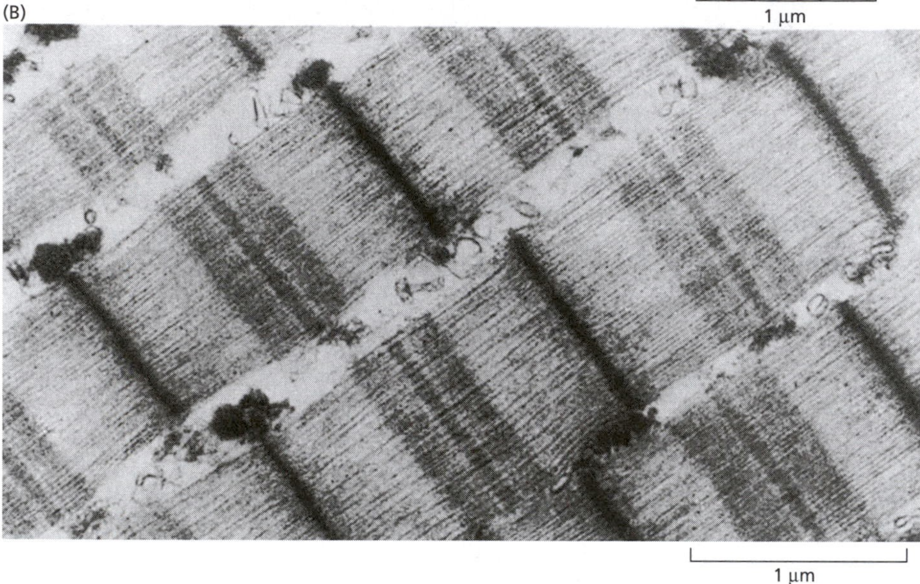


**Figure 16–31** Two electron micrographs of striated muscle in longitudinal section (Problem 16–90). The micrograph in (B) is a much lighter exposure than the one in (A). At the same exposure, the entire space between the thin dark lines in (B) would be as dark as the fat dark band in (A).

**16–90** Two electron micrographs of striated muscle in longitudinal section are shown in Figure 16–31. The sarcomeres in these micrographs are in two different states of contraction.

A. Using the micrograph in Figure 16–31A, identify the locations of the following:
   1. Dark band
   2. Light band
   3. Z disc
   4. Myosin II filaments
   5. Actin filaments (show plus and minus ends)
   6. α-Actinin
   7. Nebulin
   8. Titin

B. Locate the same features on the micrograph in Figure 16–31B. Be careful!

**Problem 3–43** describes the basis for the springlike behavior of titin, which keeps the myosin thick filaments centered in the sarcomere.

**16–91** Troponin molecules are evenly spaced along an actin filament with one troponin bound at every seventh actin molecule. How do you suppose troponin molecules can be positioned this regularly?

**16–92** What two major roles does ATP hydrolysis play in muscle contraction?

**16–93** An electron micrograph of a cross section through a flagellum is shown in Figure 16–32.

A. Assign the following components to the indicated positions on the figure.
   A microtubule

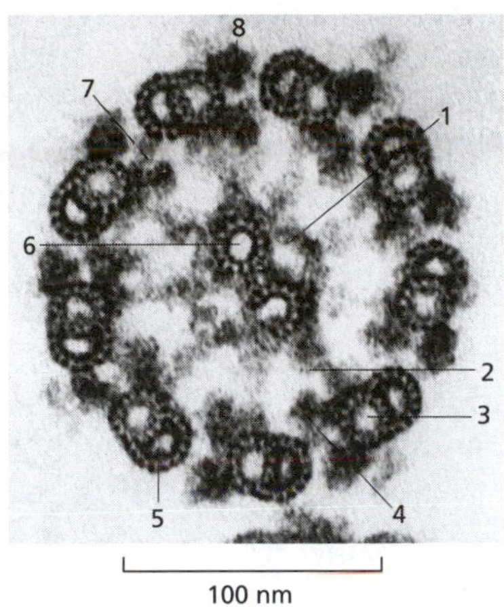


**Figure 16–32** Electron micrograph of a cross section through a flagellum of *Chlamydomonas reinhardtii* (Problem 16–93).

B microtubule
Outer dynein arm
Inner dynein arm
Inner sheath
Nexin
Radial spoke
Singlet microtubule

B. Which of the above structures are composed of tubulin?

**16–94** The sliding-microtubule mechanism for ciliary bending is undoubtedly correct. The consequences of sliding are straightforward when a pair of outer doublets is considered in isolation. The dynein arms are arranged so that, when activated, they push their neighboring outer doublet outward toward the tip of the cilium. If the pair of outer doublets is linked together by nexin molecules, they will bend so that the one that has been pushed toward the tip will define the inside of the curve (see Figure 16–40A). It is confusing, however, to think about sliding in the circular array of outer doublets in the axoneme. If all the dynein arms in a circular array were equally active, there could be no significant relative motion. (The situation is equivalent to a circle of strongmen, each trying to lift his neighbor off the ground; if they all succeeded, the group would levitate.)

Devise a pattern of dynein activity (consistent with axoneme structure and the directional pushing of dynein) that could account for bending of the axoneme in one direction. How would this pattern change for bending in the opposite direction?

**16–95** Distinguish among the three processes—protrusion, attachment, and traction—that make up the crawling movements of cells.

**16–96** The locomotion of fibroblasts in culture is immediately halted by the drug cytochalasin B, which caps actin filaments. Addition of colchicine, which depolymerizes microtubules, causes fibroblasts to cease to move directionally and to begin extending lamellipodia in seemingly random directions. Injection of fibroblasts with antibodies to vimentin intermediate filaments has no discernible effect on their migration. What do these observations suggest to you about the involvement of the three different cytoskeletal filaments in fibroblast locomotion?

**16–97** Actin filaments are said to 'push' on the cell membrane to cause it to form a protrusion. But there are problems with a pushing mechanism at both ends of the filaments. When a plus end reaches the membrane and abuts it, how are new subunits added to extend the filament (allowing it to push)? And how is the minus end of the filament anchored so that the filament isn't simply pushed back into the cell's interior? What do you suppose might be the answers to these questions?

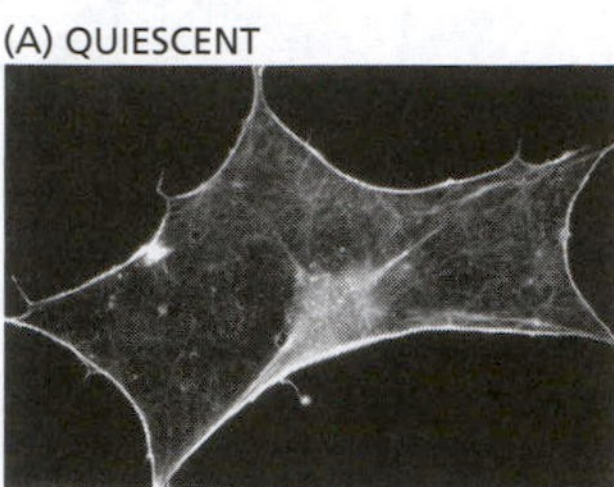

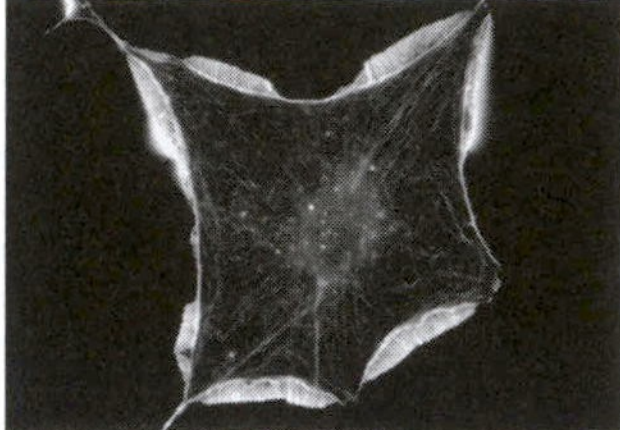


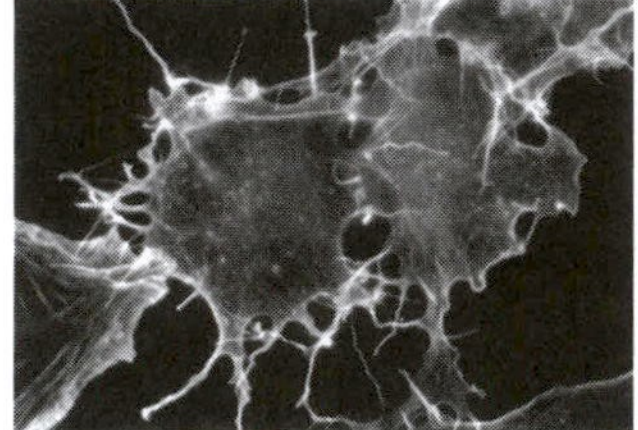


**Figure 16–33** Actin cytoskeleton in different cells (Problem 16–98). (A) Quiescent cells. (B) Cells with prominent stress fibers. (C) Cells with multiple lamellipodia. (D) Cells with many long filopodia. Cells in B, C, and D were injected with an activated form of a monomeric GTPase.

**16–98** The characteristic actin staining in a quiescent cell is shown in Figure 16–33A. When such cells are injected with a constitutively activated form of Rac, Rho, or Cdc42 monomeric GTPases, they dramatically alter their actin cytoskeletons. Which GTPase is associated with formation of stress fibers (Figure 16–33B), lamellipodia (Figure 16–33C), and filopodia (Figure 16–33D)?

**16–99** How is the unidirectional motion of a lamellipodium maintained?

**16–100** In addition to conducting impulses in both directions, nerve axons carry vesicles to and from the cell body along microtubule tracks. Do outbound vesicles move along microtubules that are oriented in one direction and incoming vesicles move along oppositely oriented microtubules? Or are microtubules all oriented in the same direction, with different motor proteins providing the directionality?

To distinguish between these possibilities, you prepare a cross section through a nerve axon and decorate the microtubules with tubulin, which binds to the tubulin subunits of the microtubule to form hooks. The decorated microtubules are illustrated in Figure 16–34. Do all the microtubules run in the same direction or not? How can you tell?

**16–101** Mice that are homozygous for a knockout of the gene for the kinesin motor protein KIF1B die at birth. Heterozygous knockouts survive, but suffer from a progressive muscle weakness similar to human neuropathies. Humans with Charcot–Marie–Tooth disease type 2A have a mutation in one copy of the gene for KIF1B that prevents the protein from binding to ATP. The heterozygous mice and the human patients have very similar progressive neuropathies. How do you suppose that the loss of one copy of a gene for a kinesin motor can have such profound effects on nerve function?

## CALCULATIONS

**16–102** Using the equation for diffusion given in Problem 16–30, calculate the average time it would take for a vesicle to diffuse to the end of an axon 10 cm in length. The diffusion coefficient of a typical vesicle is $5 \times 10^{-8}$ cm$^2$/sec.

**16–103** A mitochondrion 1 μm long can travel the 1 meter length of the axon from the spinal cord to the big toe in a day. The Olympic men's freestyle swimming record for 200 meters is 1.75 minutes. In terms of body lengths per day, who is moving faster: the mitochondrion or the Olympic record holder? (Assume that the swimmer is 2 meters tall.)

## DATA HANDLING

**16–104** As a laboratory exercise, you and your classmates are carrying out experiments on isolated muscle fibers using 'caged' ATP (Figure 16–35). Since caged ATP does not bind to muscle components, it can be added to a muscle fiber without stimulating activity. Then, at some later time it can be split by a flash of laser light to release ATP instantly throughout the muscle fiber.

To begin the experiment, you treat an isolated, striated muscle fiber with glycerol to make it permeable to nucleotides. You then suspend it in a buffer containing ATP in an apparatus that allows you to measure any tension generated by fiber contraction. As illustrated in Figure 16–36, you measure the tension generated after several experimental manipulations: removal of ATP by dilution, addition of caged ATP, and activation of caged ATP by laser light.

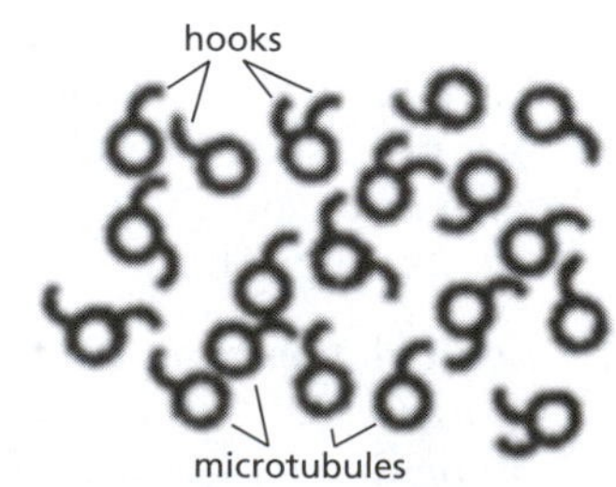


**Figure 16–34** Tubulin-decorated microtubules in a cross section through a nerve axon (Problem 16–100). The hooks represent the tubulin decoration.

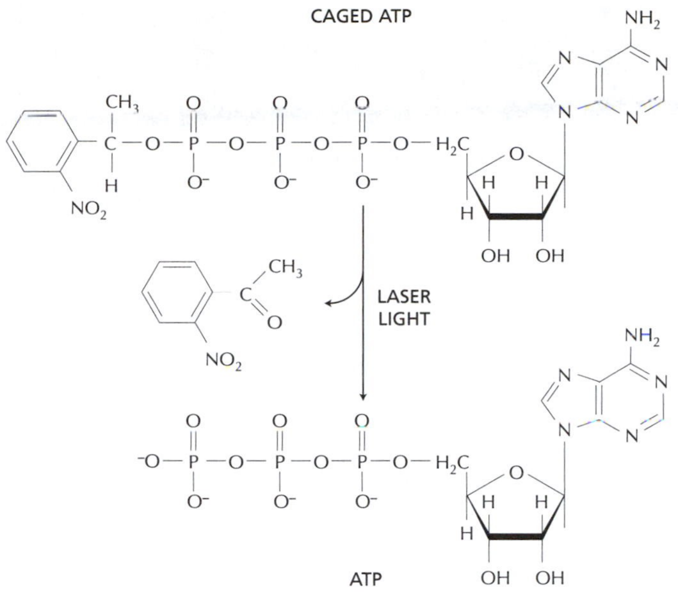

Figure 16–35 Caged ATP (Problem 16–104).

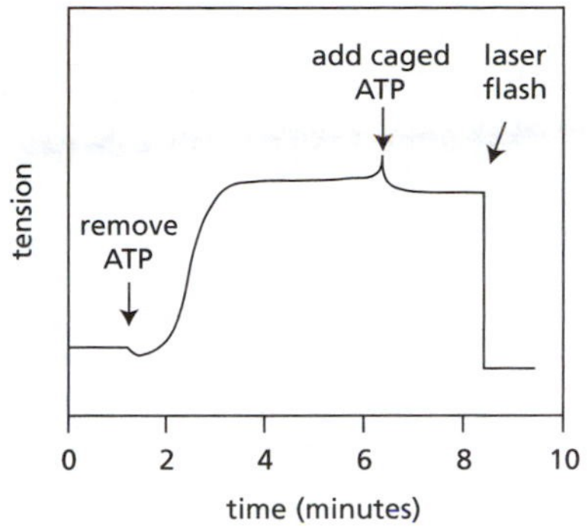

Figure 16–36 Tension in a striated muscle fiber as a result of various experimental manipulations (Problem 16–104).

You are somewhat embarrassed because your results are very different from everyone else's. In checking over your experimental protocol, you realize that you forgot to add $Ca^{2+}$ to your buffers. The teaching assistant in charge of your section tells you that your experiment is actually a good control for the class, but you will have to answer the following questions to get full credit.

A. Why did the ATP in the suspension buffer not cause the muscle fiber to contract?
B. Why did the subsequent removal of ATP generate tension? Why did tension develop so gradually? (If our muscles normally took a full minute to contract, we would move very slowly.)
C. Why did laser illumination of a fiber containing caged ATP lead to relaxation?

**16–105** Detailed measurements of sarcomere length and tension during isometric contraction in striated muscle provided crucial early support for the sliding filament model of muscle contraction. Based on your understanding of the sliding filament model and the structure of a sarcomere, propose a molecular explanation for the relationship of tension to sarcomere length in the portions of Figure 16–37 marked I, II, III, and IV. (In this muscle the length of the myosin filament is 1.6 μm, and the lengths of the actin thin filaments that project from the Z discs are 1.0 μm.)

**16–106** On ciliated cells, the beating of individual cilia is usually coordinated so that the cilia move in the same direction, thereby imparting unidirectional motion to the cell (or to the surrounding fluid). In principle, the coordinated,

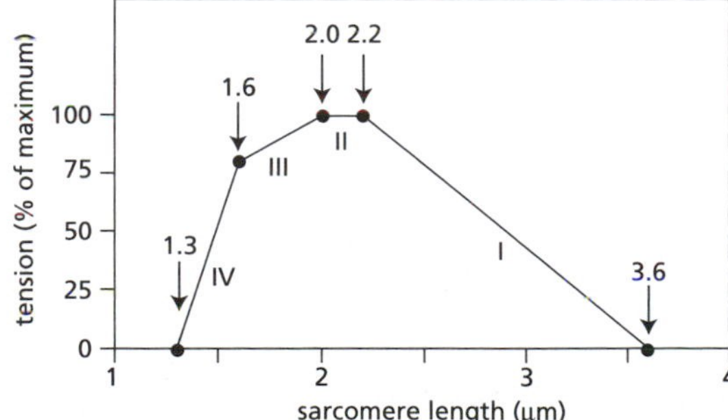


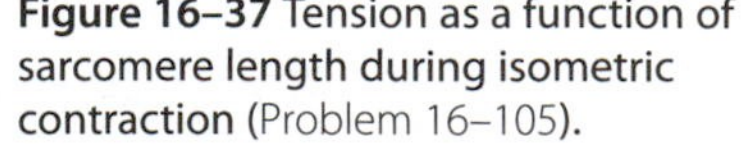
Figure 16–37 Tension as a function of sarcomere length during isometric contraction (Problem 16–105).

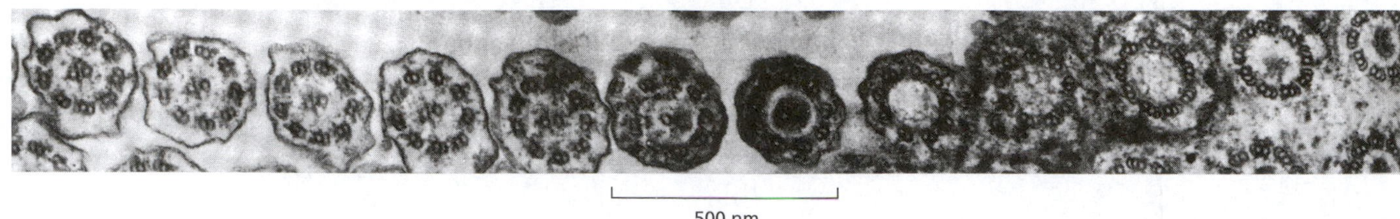


**Figure 16–38** Electron micrograph of a cross section through the cortex of a ciliated protozoan *Tetrahymena* (Problem 16–106).

unidirectional beating of adjacent cilia could be determined by some feature of their structure or, alternatively, by some cellular control mechanism that is independent of ciliary structure.

An electron micrograph of a cross section through the cortex of the ciliated protozoan *Tetrahymena* is shown in Figure 16–38. The plane of sections grazes the surface of the cell, showing in successive sections how the '9 + 2' arrangement of microtubules in the axoneme leads into the nine triplet microtubules of the basal body. Are there any clues in the micrograph that allow you to decide whether axoneme structure or cellular control is the basis for the unidirectional beating of adjacent cilia? Explain your reasoning.

**16–107** When analyzed in detail, the rhythmic beating of a cilium is revealed as a series of precisely repeated movements. In *Chlamydomonas* the flagellar beat cycle is straightforward (Figure 16–39A). The beat cycle begins with a power stroke, which is initiated by a bend near the base of the flagellum (arrow at base of flagellum 1 in Figure 16–39A). The power stroke ends when the bent segment of flagellum extends roughly through half the circumference of a circle (flagellum 5 in Figure 16–39A). The return stroke is formed by the movement of the semicircular segment of the flagellum outward toward the tip, which is accomplished by further bending at the leading edge of the semicircle and relaxation at the trailing edge (flagella 6 to 8 in Figure 16–39A).

A. How much sliding of microtubule doublets against one another is required to account for the observed bending of the flagellum into a semicircle? Calculate how much farther the doublet on the inside of the semicircle protrudes beyond the doublet on the outside of the semicircle at the tip of the flagellum (Figure 16–40A). The width of a flagellum is 180 nm.

B. The elastic nexin molecules that link adjacent outer doublets must stretch to accommodate the bending of a flagellum into a semicircle. If the length of an unstretched nexin molecule at the base of a flagellum is 30 nm, what is the length of a stretched nexin molecule at the tip of a flagellum (Figure 16–40B)? Adjacent doublets are 30 nm apart.

(A) WILD TYPE

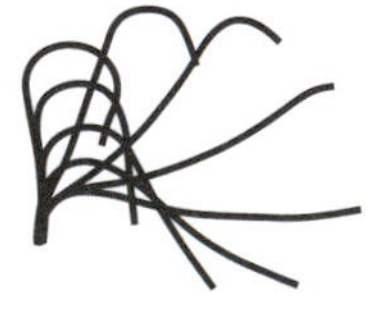

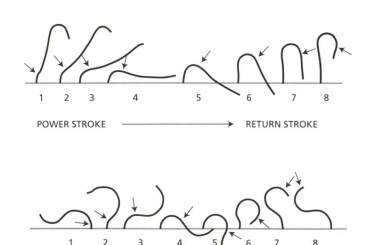

1 2 3 4 5 6 7 8

POWER STROKE ⟶ RETURN STROKE

(B) DOUBLE MUTANT

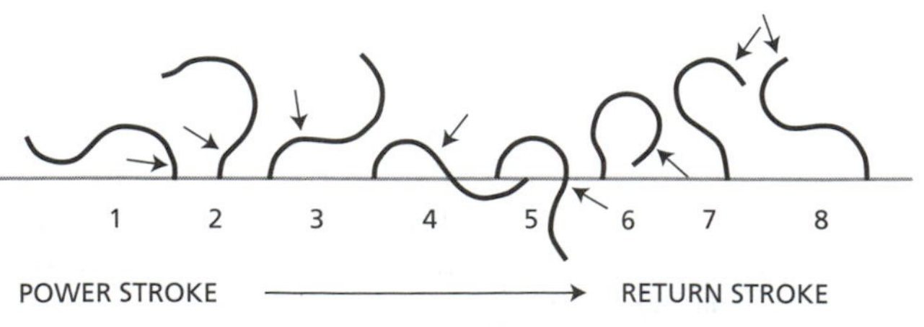


**Figure 16–39** Analysis of flagellar beat cycles in *Chlamydomonas* (Problem 16–107). (A) Wild-type flagella. (B) Double-mutant flagella. The beating flagella shown on the *left* were photographed with stroboscopic illumination under a microscope. Individual frames from these pictures represent flagella at successive stages in the beat cycle. *Arrows* point to the tip-proximal end of a bent segment of flagellum, as it moves from the bottom of the flagellum to the tip during the beat cycle.

C. *Chlamydomonas* mutants that are missing radial spokes have paralyzed flagella. The paralysis can be overcome by mutations in a second gene (called $Sup_{pf}$, for suppressor of paralyzed flagella), which encodes a component of the outer dynein arm. Although the flagella now move, their beat pattern is aberrant (Figure 16–39B). At the gross level, how does the aberrant beat stroke of the double mutant differ from that of the wild type? What does this gross difference suggest for the function mediated by the radial spokes in *Chlamydomonas*?

**16–108** The structure of the ciliary axoneme, which is composed of more than 200 different proteins, is exceedingly complex. *Chlamydomonas reinhardtii*, which bears two flagella, is an extremely useful organism for analyzing axoneme structure. Physiological and microscopic observation are straightforward, but even more important is the ease of genetic and biochemical analysis. It is a simple matter to isolate mutants with paralyzed flagella because they cannot move. These mutants can then be assigned to specific genes by genetic crosses, which are routine with *Chlamydomonas*. Finally, wild-type and mutant flagella can be detached readily (for example, by pH shock) and recovered for easy biochemical analysis.

Many of the mutants with paralyzed flagella are missing one or another of the major substructures of the axoneme, such as the radial spokes, the outer dynein arms, the inner dynein arms, or one or both central microtubules. In most cases, loss of the axonemal substructure is caused by mutation of a single gene, and yet biochemical analysis shows that the defective axoneme is missing multiple proteins. Consider, for example, the mutant *Pf*14 (paralyzed flagella); electron micrographs of its flagella show a complete absence of radial spokes (Figure 16–41A), and two-dimensional electrophoretic analysis shows that it lacks 17 different proteins (Figure 16–41B).

A single-gene defect that results in multiple protein deficiencies could have two underlying explanations: (1) the defect is in a regulatory gene that controls the synthesis of the missing proteins, or (2) the defect is in a gene whose product must be present in the structure before the other proteins can be added. Two approaches have been used to evaluate these possibilities.

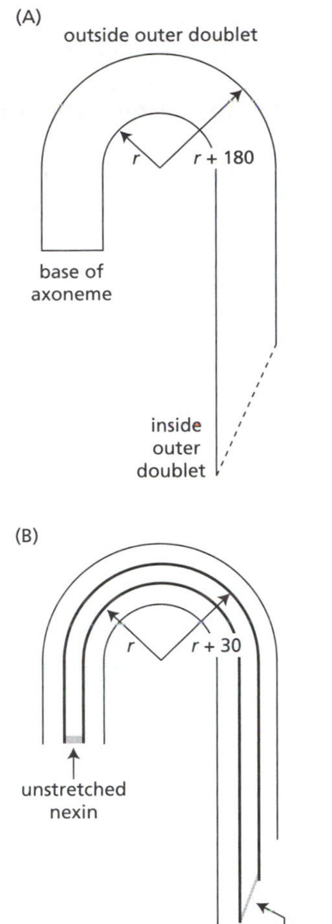


**Figure 16–40** Flagella bent into half circles (Problem 16–107). (A) Representation showing the 'inside' and 'outside' doublets, which are 180-nm apart. (B) Representation showing adjacent doublets, which are 30-nm apart, and the nexin molecules that link them.

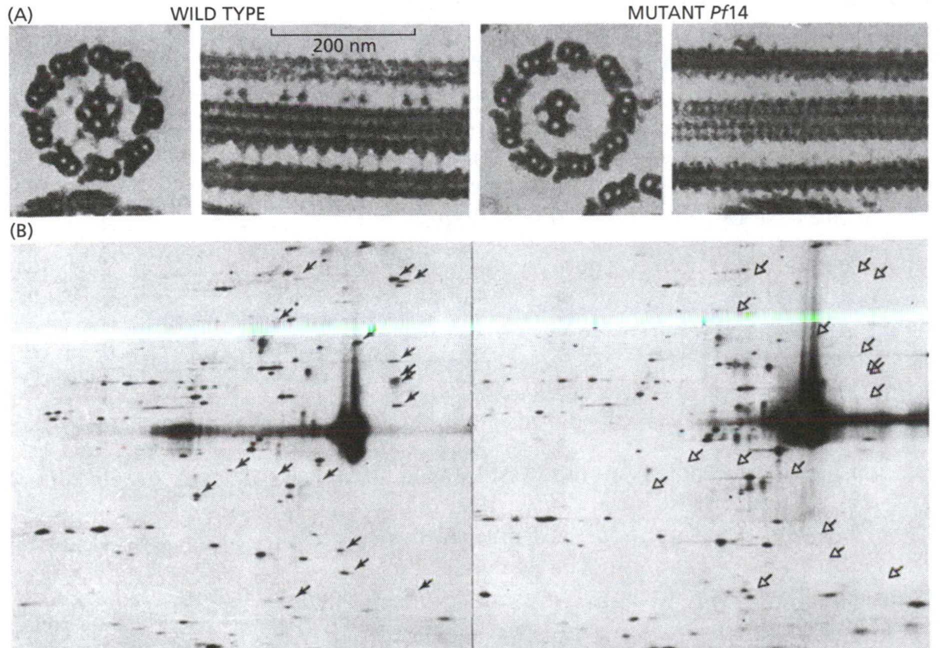


**Figure 16–41** Comparison of flagella from wild-type *Chlamydomonas* with those from the mutant *Pf*14 (Problem 16–108). (A) Electron micrographs showing flagella in transverse and longitudinal sections. Notice the absence of radial spokes in *Pf*14. (B) Analysis of flagella by two-dimensional gel electrophoresis. The first dimension (*horizontal*) is separation by isoelectric focusing with the more acidic proteins to the *right*; the second dimension is separation by molecular mass using SDS-gel electrophoresis. *Arrows* indicate the positions of proteins that are present in wild type but missing in *Pf*14. The highly exposed areas correspond to α- and β-tubulin, which far outnumber the other axonemal proteins.

A. The first method takes advantage of a feature of the regular mating cycle of *Chlamydomonas.* In the mating reaction, biflagellate gametes fuse efficiently to give a population of temporary dikaryons with four flagella. In a mating of *Pf*14 with wild-type gametes, the paralyzed flagella recover function after fusion, indicating that the defective structures can be repaired without completely rebuilding them. (In normal cells there is a pool of flagellar components sufficient to rebuild an entire flagellum in the absence of protein synthesis.) To distinguish between the possible explanations for flagellar defects, investigators labeled mutant cells by growth in $^{35}SO_4$ and then fused the labeled gametes to nonradioactive wild-type gametes in the presence of an inhibitor of protein synthesis. After recovery of function, the flagella were isolated and the radioactive proteins were analyzed by two-dimensional gel electrophoresis followed by autoradiography.

Predict the expected electrophoretic pattern of *radioactive* proteins from the dikaryon if the affected gene controlled the synthesis of the missing proteins. How would it differ from the electrophoretic pattern that would be expected if the affected gene product participated in assembly?

B. The second approach was to expose *Pf*14 to a mutagen to generate revertants that regained flagellar function, not because the original defect was corrected, but because a second alteration within the gene compensated for the first one. (Depending on the gene, such intragenic revertants can be very common.) The proteins from several such revertants were compared with those in the wild type by two-dimensional gel electrophoresis.

How might this method distinguish between the two possible explanations for the defect in *Pf*14?

**16–109** One of the most striking examples of a purely actin-based cellular movement is the extension of the acrosomal process of a sea cucumber sperm. The sperm contains a store of unpolymerized actin in its head. When a sperm makes contact with a sea cucumber egg, the actin polymerizes rapidly to form a long spearlike extension. The tip of the acrosomal process penetrates the egg, and it is probably used to pull the sperm inside.

Are actin monomers added to the base or to the tip of the acrosomal bundle of actin filaments during extension of the process? If the supply of monomers to the site of assembly depends on diffusion, it should be possible to distinguish between these alternatives by measuring the length of the acrosomal process with increasing time. If actin monomers are added to the base of the process, which is inside the head, the rate of growth should be linear because the distance between the site of assembly and the pool of monomers does not change with time. On the other hand, if the subunits are added to the tip, the rate of growth should decline progressively as the acrosomal process gets longer because the monomers must diffuse all the way down the shaft of the process. In this case, the rate of extension should be proportional to the square root of time. Plots of the length of the acrosomal process versus time and the square root of time are shown in Figure 16–42.

A. Are the ascending portions of the plots in Figure 16–42 more consistent with the addition of actin monomers to the base or to the tip of the acrosomal process?

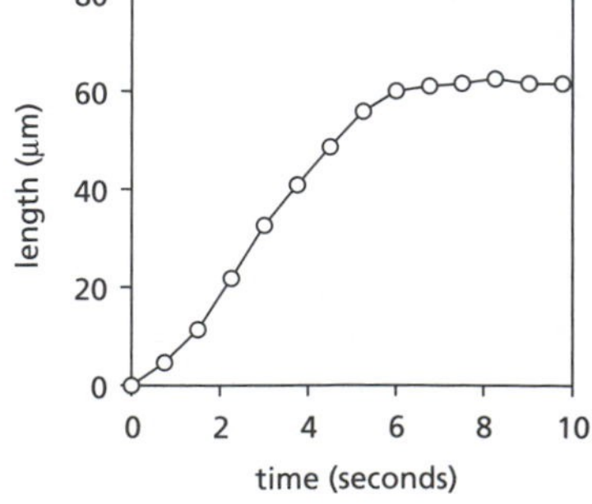


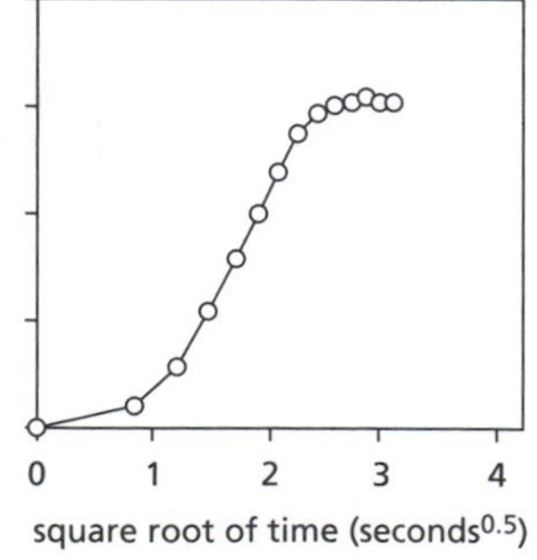


**Figure 16–42** Plots of the length of the acrosome versus time and the square root of time (Problem 16–109).

B. Why do you suppose the process grows so slowly at the beginning and at the end of the acrosomal reaction?

**16–110** Kinesin motors transport oligomers of neurofilament proteins down axonal microtubules to sites where they are used in the construction or repair of neurofilaments. Classic studies that followed pulse-labeled neurofilament proteins during axonal transport agree that the peak of radioactivity broadens markedly during transport. More recent studies demonstrated that unphosphorylated neurofilament proteins bind strongly to kinesin motors and weakly to existing neurofilaments. By contrast, the phosphorylated forms bind weakly to kinesin motors and strongly to neurofilaments.

A. How might the phosphorylation dependence of oligomer binding to kinesin motors and neurofilaments account for the broadening of the transport wave?

B. If you could track the movement of single oligomers down an axon, how would you expect them to move? Consider an oligomer at the leading edge of the transport wave and one at the trailing edge.

**16–111** Nerve growth cones navigate along stereotyped pathways during development by continually extending and retracting slender filopodia to sense directional cues in the environment. The bundled actin cytoskeleton in such a filopodium grows at the tip by actin polymerization, and is pulled back into the cell over time, a phenomenon known as retrograde flow. In principle, extension and retraction of filopodia could be controlled by regulating the rate of actin polymerization or the rate of retrograde flow. The experiments below were carried out to determine which rate—polymerization or retrograde flow—is regulated.

Actin monomers tagged with caged rhodamine were injected into cells and allowed to incorporate into actin filaments. The rhodamine in a narrow segment of the actin bundles near the tip of a single filopodium was uncaged by brief irradiation, yielding a fluorescent mark that allowed retrograde flow to be observed directly over time (Figure 16–43A). Extension and retraction of the tip of the filopodium was followed microscopically (Figure 16–43A). Actin polymerization was taken as the distance between the fluorescent mark and the tip of the filopodium (Figure 16–43A). A summary of the data for this single filopodium is shown in Figure 16–43B. Are extension and contraction of this filopodium regulated by the rate of actin polymerization or by the rate of retrograde flow? Explain your reasoning.

**16–112** Yeast cells choose bud sites on their surface in two distinct spatial patterns: axial for **a** and α haploid cells, and bipolar for **a**/α diploid cells (Figure 16–44A). The selection of a new bud site establishes a cell polarity that involves the cytoskeleton and determines the site of new cell growth (Figure 16–44B). To find the genes responsible for bud-site selection, mutagenized α cells were examined visually to identify mutant cells with altered budding patterns. Five genes, *Bud1* to *5*, were identified in this way: all are nonessential genes that have no effect on cell growth or morphology. Further genetic tests indicated that these genes are involved in a single pathway for bud-site selection. In the absence of *Bud1*, *Bud2*, or *Bud5* the budding pattern is

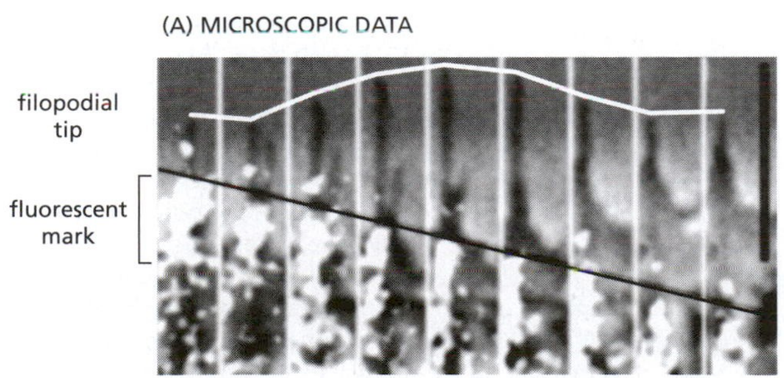


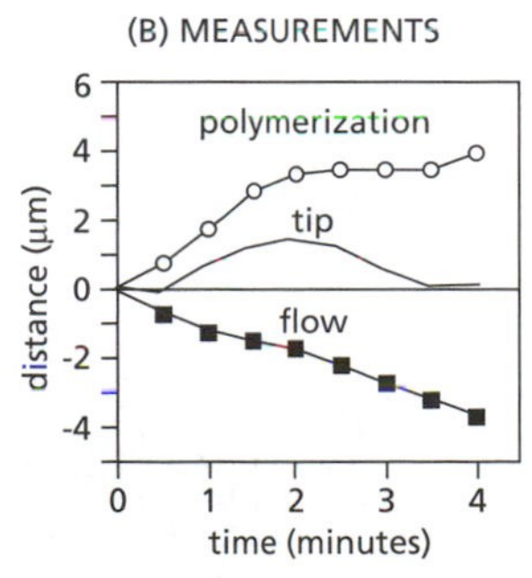


**Figure 16–43** Regulation of filopodial extension and retraction (Problem 16–111). (A) Time-lapse observations of a single filopodium. The *white line* identifies the position of the tip of the filopodium; the *black line* marks the position of the fluorescent segment. The scale bar at the right is 5 μm. (B) Summary of the data. For the data summary the positions of the tip and the fluorescent segment were arbitrarily set at zero at zero minutes, as was the difference between the tip and the fluorescent mark. The position of the filopodium is labeled 'tip;' the position of the fluorescent mark is labeled 'flow' for retrograde flow; and the difference is labeled 'polymerization' for actin polymerization.

randomly either bipolar or axial: in the absence of *Bud3* or *Bud4* the budding pattern is bipolar. All five genes are required for axial budding.

Sequence analysis of the genes and biochemical characterization of the proteins indicate that *Bud1* encodes a monomeric GTPase, *Bud2* encodes a GTPase-activating protein (GAP) that stimulates GTP hydrolysis by the Bud1 protein, and *Bud5* encodes a guanine nucleotide exchange factor (GEF) that catalyzes the exchange of GTP for the GDP bound to the Bud1 monomeric GTPase.

A. By analogy to the use of monomeric GTPases, GAPs, and GEFs in directing vesicular transport between membrane-bounded compartments, design a plausible scheme by which the Bud1, Bud2, and Bud5 proteins might be used to deliver critical cytosolic proteins to a bud site.
B. Does your scheme account for the initial selection of the site for bud formation?

**16–113** Activation of Cdc42, a monomeric GTPase, triggers actin polymerization and bundling to form either filopodia or shorter cell protrusions called microspikes. These effects of Cdc42 could be mediated by N-WASp, which is a multifunctional protein. As shown in Figure 16–45A, N-WASp contains a pleckstrin homology (PH) domain, which binds to $PIP_2$; a Cdc42-binding domain (G); a verprolin homology domain (V), which binds to actin; a cofilin-homology domain (C), which can bind to actin filaments; and a C-terminal acidic domain (A), which binds the ARP complex.

In *Xenopus* egg extracts, a convenient source of components, the addition of Cdc42 charged with GTPγS, a nonhydrolyzable analog of GTP, stimulates actin polymerization (Figure 16–45B). If the extract is depleted of N-WASp using N-WASp-specific antibodies, no actin polymerization is observed when Cdc42-GTPγS is added (Figure 16–45B). Actin polymerization can be restored by the addition of purified N-WASp, but not by the addition of either of two mutant forms of N-WASp: one (H208D) that cannot bind to Cdc42, and a second (Δcof) that eliminates the function of the cofilin domain (Figure 16–45A).

Do these experiments support a role for N-WASp in the rearrangement of actin filaments in response to Cdc42 activation? Explain your reasoning. Include a discussion of why the two mutant forms of N-WASp do not restore actin polymerization.

**16–114** To determine the mechanism by which N-WASp mediates activation by Cdc42, polymerization was measured in the presence of purified components. In the presence of the ARP complex, N-WASp stimulates actin polymerization substantially over the ARP complex or N-WASp alone, but not

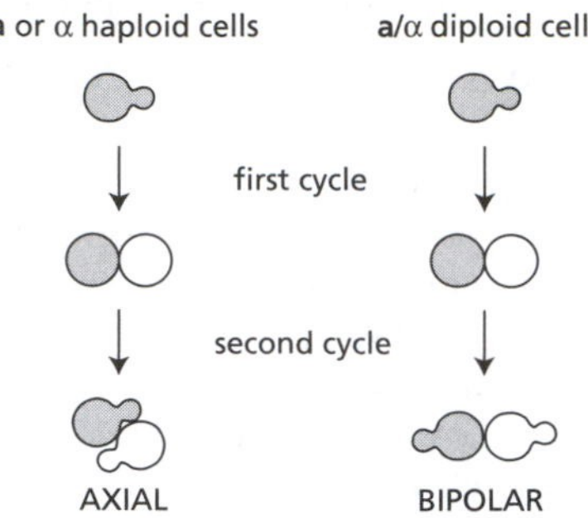


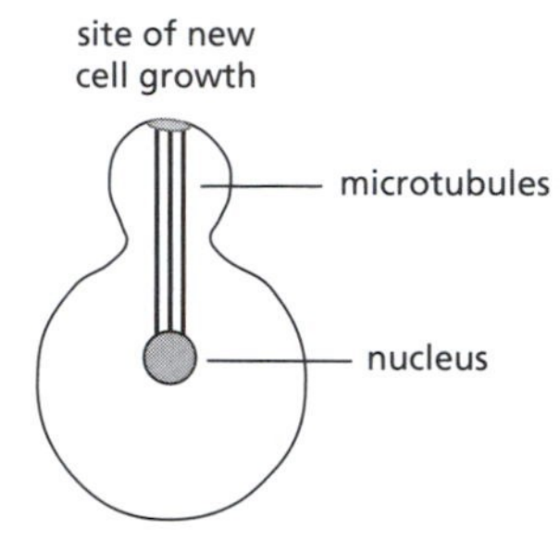


**Figure 16–44** Budding in yeast (Problem 16–112). (A) Natural patterns of budding. (B) Cell polarization during bud formation.

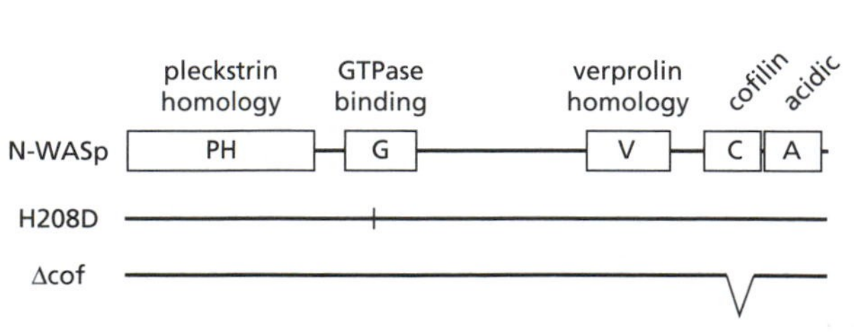


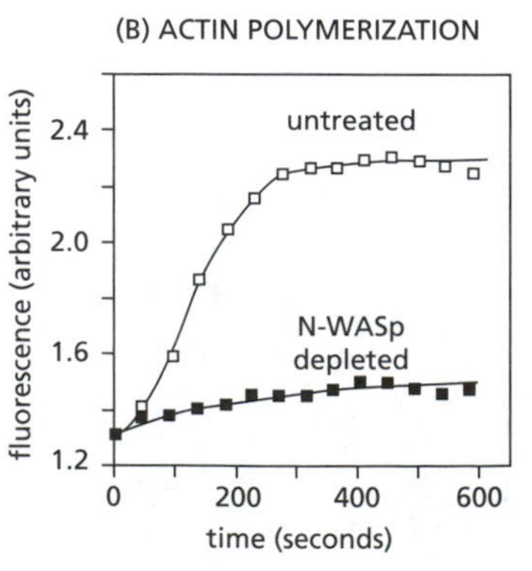


**Figure 16–45** The role of N-WASp in Cdc42-triggered actin polymerization and bundling (Problem 16–113). (A) Domain structure of N-WASp and two mutants defective in individual domains. (B) Actin polymerization in untreated and N-WASp-depleted egg extracts. In the polymerization assays, the extracts were supplemented with pyrene-actin, which fluoresces much more highly in the filament than it does on the subunit.

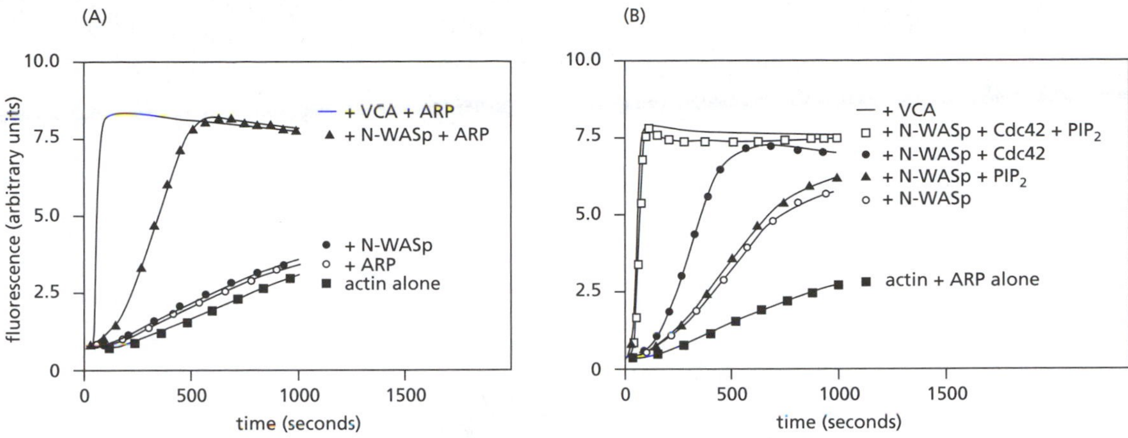


**Figure 16–46** Polymerization of actin in the presence of various purified components (Problem 16–114). (A) Mixtures of actin, N-WASp, the ARP complex, and the VCA C-terminal segment of N-WASp. (B) Mixtures of actin, N-WASp, the ARP complex, Cdc42-GTPγS, and $PIP_2$-containing vesicles. Vesicles without $PIP_2$ do not stimulate in any combination with the other components.

nearly so dramatically as the C-terminal segment of N-WASp that contains just the verprolin (V), cofilin (C), and acidic (A) domains (Figure 16–46A). To account for the difference between N-WASp and its C-terminal VCA segment, N-WASp and the ARP complex were mixed with combinations of Cdc42-GTPγS and vesicles containing $PIP_2$, as shown in Figure 16–46B.

A. What is required for N-WASp to stimulate polymerization as efficiently as its C-terminal VCA segment? Explain your reasoning.

B. Based on these results, propose a model for the activation of N-WASp and its stimulation of actin polymerization.

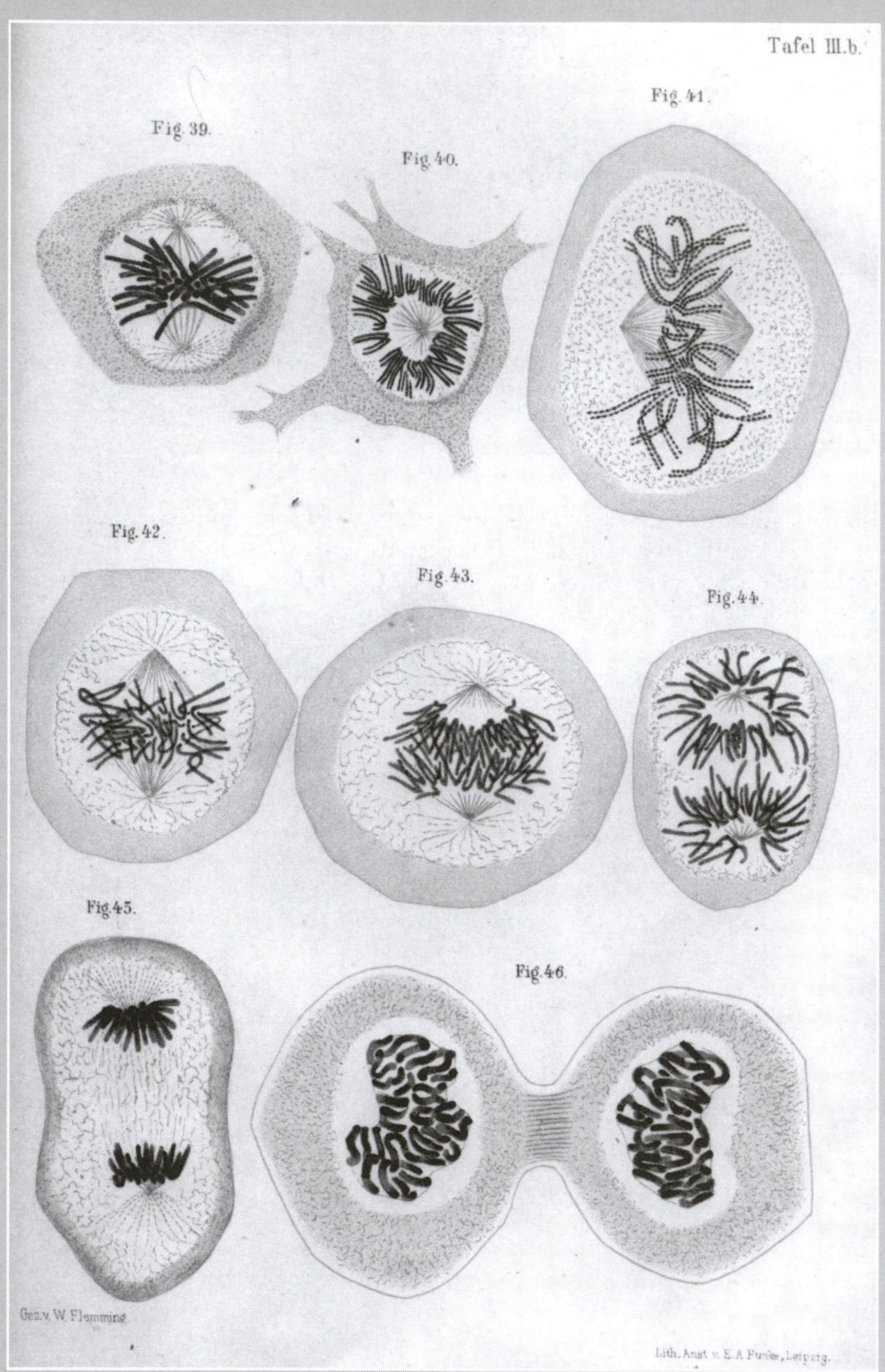


**Table IIIb of Walther Flemming's 1882 magnum opus, Zellsubstanz, Kern und Zellheilung contains these first published pictures of the chromosomes during mitosis.**

# The Cell Cycle

17

## In This Chapter


| | |
|---|---|
| OVERVIEW OF THE CELL CYCLE | 403 |
| THE CELL-CYCLE CONTROL SYSTEM | 407 |
| S PHASE | 410 |
| MITOSIS | 413 |
| CYTOKINESIS | 427 |
| CONTROL OF CELL DIVISION AND CELL GROWTH | 430 |


## OVERVIEW OF THE CELL CYCLE

TERMS TO LEARN

| | | |
|---|---|---|
| budding yeast | $G_1$ phase | restriction point |
| cell cycle | $G_2$ phase | sister chromatid |
| cell-division-cycle gene (*Cdc* gene) | interphase | Start |
| fission yeast | mitotic spindle | |

### DEFINITIONS

Match the definition below with its term from the list above.

**17–1** The long period of the cell cycle between one mitosis and the next.

**17–2** The orderly sequence of events by which a cell duplicates its contents and divides into two.

**17–3** The checkpoint in the cell cycle that governs the cell's commitment to enter S phase.

**17–4** The phase of the eucaryotic cell cycle between the end of cytokinesis and the start of DNA synthesis.

**17–5** Common name often given to the baker's yeast *Saccharomyces cerevisiae,* which divides by producing a small outgrowth that separates to become a new individual.

### TRUE/FALSE

Decide whether each of these statements is true or false, and then explain why.

**17–6** Since there are about $10^{13}$ cells in an adult human, and about $10^{10}$ cells die and are replaced each day, we become new people every three years.

**17–7** Although the lengths of all phases of the cycle are variable to some extent, by far the greatest variation occurs in the duration of $G_1$.

### THOUGHT PROBLEMS

**17–8** If the most basic function of the cell cycle is to duplicate accurately the DNA in the chromosomes and then distribute the copies precisely to the daughter cells, why are there gaps between S phase and M phase?

**17–9** Many cell-cycle genes from human cells function perfectly well when expressed in yeast cells. Why do you suppose that is considered remarkable? After all, many human genes encoding enzymes for metabolic reactions also function in yeast, and no one thinks that is remarkable.

**17–10** What determines the length of S phase? One possibility is that it depends on how much DNA the nucleus contains. As a test, you measure the length of S

**Table 17–1 Correlation between length of S phase and DNA content** (Problem 17–10).

| ORGANISM | DNA CONTENT OF NUCLEUS (pg) | LENGTH OF S PHASE (hr) |
|---|---|---|
| Lizard | 3.2 | 15 |
| Frog | 15 | 26 |
| Newt | 45 | 41 |

phase in dividing cells of a lizard, a frog, and a newt, each one of which has a different amount of DNA. As shown in Table 17–1, the length of S phase does increase with increasing DNA content.

Even though these organisms are similar, they are different species. You recall that it is possible to obtain haploid embryos of frogs and repeat your measurements with haploid and diploid frog cells. Haploid frog cells have the same length S phase as diploid frog cells. Further research in the literature shows that in plants, tetraploid strains of beans and oats have the same length S phase as their diploid cousins.

Propose an explanation to reconcile these apparently contradictory results. Why do you suppose the length of S phase increases with increasing DNA content in different species, but remains constant with increasing DNA content in the same species?

**17–11** The budding yeast *Saccharomyces cerevisiae* and the fission yeast *Schizosaccharomyces pombe* provide facile genetic systems for studying a wide range of eucaryotic cell biological processes. If cell-cycle progression is essential for cell viability, as it is in these yeasts, how is it possible to isolate cells that are defective in cell-cycle genes?

**17–12** You have isolated a new *Cdc* mutant of budding yeast that forms colonies at 25°C but not at 37°C. You would now like to isolate the wild-type gene that corresponds to the defective gene in your *Cdc* mutant. How might you isolate the wild-type gene using a plasmid-based DNA library prepared from wild-type yeast cells?

**17–13** Fertilized eggs from the frog *Xenopus*, which contain 100,000 times more cytoplasm than a typical mammalian cell, are a favorite choice for studying the biochemistry of the cell cycle. Why isn't it just as easy to study these biochemical questions by growing large numbers of mammalian cells, which is a straightforward process?

**17–14** For many experiments, it is desirable to have a population of cells that are traversing the cell cycle synchronously. One of the first, and still often used, methods for synchronizing cells is the so-called double thymidine block. When high concentrations of thymidine are added to the culture fluid, cells in S phase stop DNA synthesis, though other cells are not affected. The excess thymidine blocks the enzyme ribonucleotide reductase, which is responsible for converting ribonucleotides into deoxyribonucleotides. When this enzyme is inhibited, the supply of deoxyribonucleotides falls and DNA synthesis stops. When the excess thymidine is removed by changing the medium, the supply of deoxyribonucleotides rises and DNA synthesis resumes normally.

For a cell line with a 22-hour cell cycle divided so that M phase = 0.5 hour, $G_1$ phase = 10.5 hours, S phase = 7 hours, and $G_2$ phase = 4 hours, a typical protocol for synchronization by a double thymidine block would be as follows:

1. At 0 hours ($t = 0$ hours) add excess thymidine.
2. After 18 hours ($t = 18$ hours) remove excess thymidine.
3. After an additional 10 hours ($t = 28$ hours) add excess thymidine.
4. After an additional 16 hours ($t = 44$ hours) remove excess thymidine.

A. At what point in the cell cycle is the cell population when the second thymidine block is removed?

B. Explain how the times of addition and removal of excess thymidine synchronize the cell population.

## CALCULATIONS

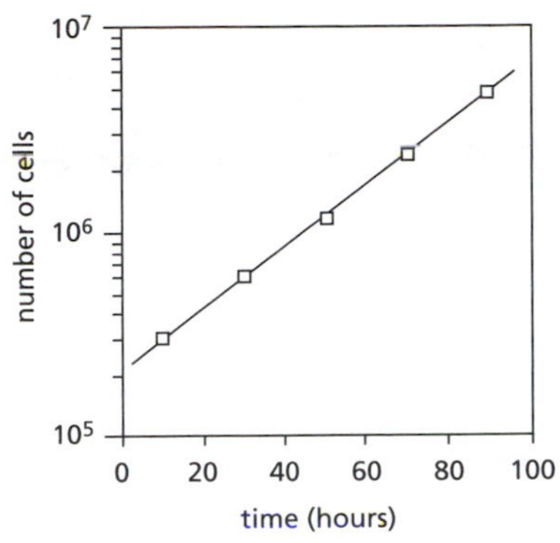


**Figure 17–1** Increase in the number of mouse L cells with time (Problem 17–16).

**17–15** The fraction of cells in a population that are undergoing mitosis (the mitotic index) is a convenient way to estimate the length of the cell cycle. You have decided to measure the cell cycle in the liver of the adult mouse by measuring the mitotic index. Accordingly, you have prepared liver slices and stained them to make cells in mitosis easy to recognize. After 3 days of counting, you have found only 3 mitoses in 25,000 cells. Assuming that M phase lasts 30 minutes, calculate the average length of the cell cycle in the liver of an adult mouse.

**17–16** The overall length of the cell cycle can be measured from the doubling time for a population of exponentially proliferating cells. The doubling time of a population of mouse L cells was determined by counting the number of cells in samples of culture fluid at various times (Figure 17–1). What is the overall length of the cell cycle in mouse L cells?

## DATA HANDLING

**17–17** A common first step in characterizing cell-division-cycle (*Cdc*) mutants is to define the phase of the cell cycle at which the mutation blocks the cell's progress. Temperature-sensitive *Cdc* mutants are particularly useful because they grow and divide normally at one temperature (the permissive temperature) but express a mutant phenotype at a higher temperature (the restrictive temperature). One method for characterizing temperature-sensitive *Cdc* mutants uses the drug hydroxyurea, which blocks DNA synthesis by inhibiting ribonucleotide reductase (which provides deoxyribonucleotide precursors). A useful feature of this method is that the block can be rapidly reversed by changing the incubation medium to one that lacks hydroxyurea. Consider the following results with the hypothetical mutants *Cdc*101 and *Cdc*102.

You incubate a culture of a yeast *Cdc*101 mutant at its restrictive temperature (37°C) for 2 hours (the approximate length of the cell cycle) so that its mutant phenotype is expressed. Then you change the medium to one containing hydroxyurea, and incubate the culture at the permissive temperature (20°C). None of the cells divide.

You now reverse the order of treatment. You incubate *Cdc*101 at 20°C for 2 hours in medium containing hydroxyurea, and then change to a medium without hydroxyurea, and incubate at 37°C. The cells undergo one round of division.

You repeat these two experiments with the *Cdc*102 mutant. The cells do not divide in either case.

A. In what phase of the cell cycle is *Cdc*101 blocked at the restrictive temperature? Explain the results of the two different temperature-shift experiments.
B. In what phase of the cell cycle is *Cdc*102 blocked at the restrictive temperature? Explain the results of the two different temperature-shift experiments.

**17–18** You have isolated a temperature-sensitive mutant of budding yeast. It proliferates well at 25°C, but at 35°C all the cells develop a large bud and then halt their progression through the cell cycle. The characteristic morphology of the cells at the time they stop cycling is known as the landmark morphology.

It is very difficult to obtain synchronous cultures of this yeast, but you would like to know exactly where in the cell cycle the temperature-sensitive gene product must function—its execution point, in the terminology of the field—in order for the cell to complete the cycle. A clever friend, who has a good microscope with a heated stage and a video camera, suggests that you take movies of a field of cells as they experience the temperature increase, and follow the morphology of the cells as they stop cycling. Since the cells do not move much, it is relatively simple to study individual cells. To make sense of what you see, you arrange a circle of pictures of cells at the start of the experiment in order of the size of their daughter buds. You then find the

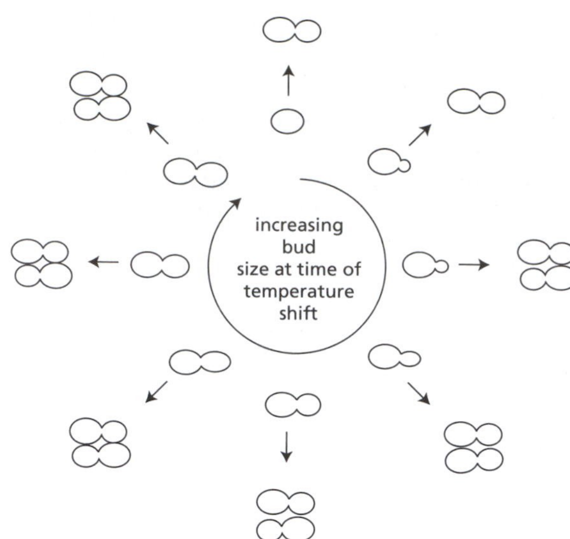


**Figure 17–2** Time-lapse photography of a temperature-sensitive mutant of yeast (Problem 17–18). Cells on the *inner ring* are arranged in order of their bud size, which corresponds to their position in the cell cycle. After 6 hours at 37°C, they have given rise to the cells shown on the *outer ring*. No further growth or division occurs.

corresponding pictures of those same cells 6 hours later, when growth and division has completely stopped. The results with your mutant are shown in Figure 17–2.

A. Indicate on the diagram in Figure 17–2 where the execution point for your mutant lies.

B. Does the execution point correspond to the time at which the cell cycle is arrested in your mutant? How can you tell?

**17–19** Cells that grow and divide in medium containing radioactive thymidine covalently incorporate the thymidine into their DNA during S phase. Consider a simple experiment in which cells were labeled by a brief (30 minute) exposure to radioactive thymidine. The medium was then replaced with one containing unlabeled thymidine, and the cells were allowed to grow and divide for some additional time. At different time points after replacement of the medium, cells were examined in a microscope. Cells in mitosis were identified by their condensed chromosomes. The fraction of mitotic cells that had radioactive DNA was determined by autoradiography and plotted as a function of time after the thymidine labeling (Figure 17–3).

A. Would all the cells in the population be expected to contain radioactive DNA after the labeling procedure?

B. Initially, there were no mitotic cells that contained radioactive DNA (Figure 17–3). Why is this?

C. Explain the rise and fall of the curve in Figure 17–3.

D. Given that mitosis lasts 30 minutes, estimate the lengths of the $G_1$, S, and $G_2$ phases from these data. (Hint: Use the points where the curves correspond to 50% labeled mitoses to estimate the lengths of phases in the cell cycle.)

**17–20** The phases of the cell cycle can also be determined using a continuous labeling protocol, if the overall length of the cell cycle is known (for example, as determined in Problem 17–16). In this method, $^3$H-thymidine is added to an asynchronous culture of cells (randomly distributed throughout the cell cycle). At various times thereafter, cells are stained and prepared for autoradiography. Cells that incorporated $^3$H-thymidine expose the photographic emulsion and are covered by silver grains. In Figure 17–4A, the fraction of mitotic cells that are labeled is plotted as a function of time after addition of $^3$H-thymidine. In Figure 17–4B, the average number of silver grains above mitotic cells is plotted as a function of time after addition of $^3$H-thymidine. Assuming the M phase lasts 30 minutes, and that the overall length of the cell cycle was measured at 20 hours, deduce the duration of the $G_1$, S, and $G_2$ phases in these cells. Explain your reasoning.

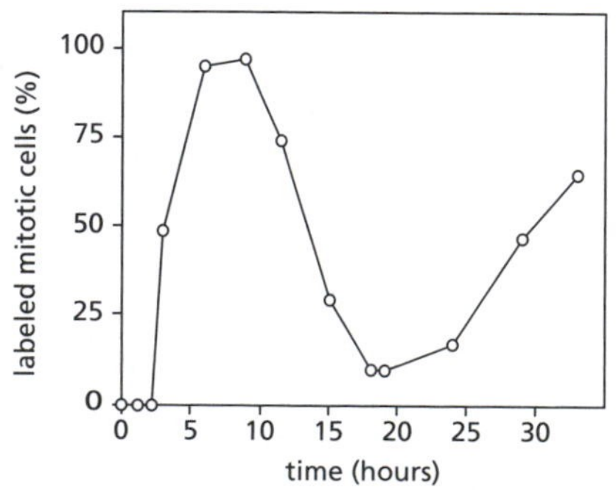


**Figure 17–3** Percentage of mitotic cells that were labeled as a function of time after a brief incubation with radioactive thymidine (Problem 17–19).

**17–21** Hoechst 33342 is a membrane-permeant dye that fluoresces when it binds to DNA. When a population of cells is incubated briefly with Hoechst dye and then sorted in a flow cytometer, which measures the fluorescence of each cell, the cells display various levels of fluorescence as shown in Figure 17–5.

A. Which cells in Figure 17–5 are in the $G_1$, S, $G_2$, and M phases of the cell cycle. Explain the basis for your answer.

B. Sketch the sorting distributions you would expect for cells that were treated with inhibitors that block the cell cycle in the $G_1$, S, or M phase. Explain your reasoning.

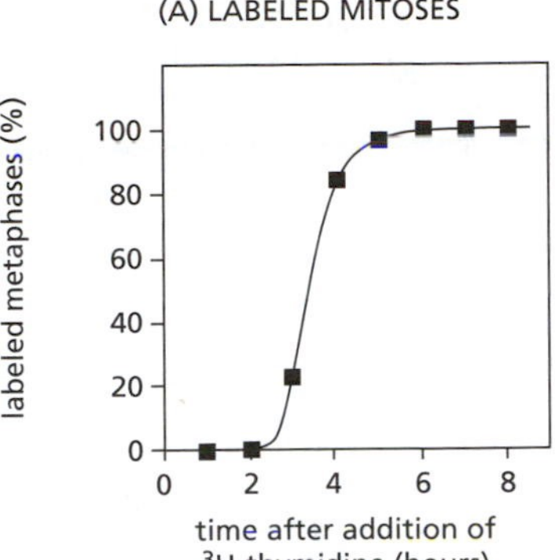

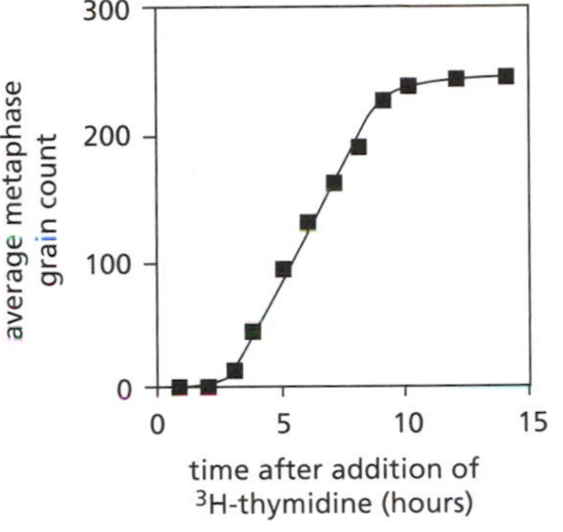


**Figure 17–4** Labeled mitotic cells as a function of time after addition of $^3$H-thymidine (Problem 17–20).
(A) Fraction of labeled mitotic cells.
(B) Average number of silver grains above labeled mitotic cells.

## THE CELL-CYCLE CONTROL SYSTEM

TERMS TO LEARN

| | | |
|---|---|---|
| anaphase-promoting complex or cyclosome (APC/C) | cyclin | M-Cdk |
| Cdc20 | cyclin–Cdk complex | M-cyclin |
| Cdc25 | cyclin-dependent kinase (Cdk) | metaphase-to-anaphase transition |
| Cdh1 | $G_1$-Cdk | S-Cdk |
| Cdk-activating kinase (CAK) | $G_1$-cyclin | SCF |
| Cdk inhibitor protein (CKI) | $G_1$/S-Cdk | S-cyclin |
| cell-cycle control system | $G_1$/S-cyclin | Wee1 |
| checkpoint | $G_2$/M checkpoint | |

### DEFINITIONS

Match the definition below with its term from the list above.

**17–22** A member of the family of protein kinases that have to be complexed with a cyclin protein in order to act.

**17–23** One of the several places in the eucaryotic cell cycle where progress through the cycle can be halted until conditions are suitable for the cell to proceed to the next stage.

**17–24** The ubiquitin ligase that promotes the destruction of a specific set of proteins, thereby promoting the separation of sister chromatids and the completion of M phase.

**17–25** The cyclin-Cdk complex responsible for stimulating entry into mitosis at the $G_2$/M checkpoint.

**17–26** One of a family of proteins that rise and fall in concentration in step with the eucaryotic cell cycle, thereby regulating the activity of the crucial protein kinases that control progression through the cell cycle.

**17–27** The final major checkpoint in the cell cycle, where the control system stimulates sister-chromatid separation, leading to the completion of mitosis and cytokinesis.

**17–28** A timing mechanism that triggers events of the cell cycle in a set sequence, using feedback from the processes it controls to ensure that one stage is complete before the next one begins.

**17–29** General term for one of the several protein assemblies that form periodically during the cell cycle as the level of cyclin increases, and partially activate the cyclin-dependent kinase component.

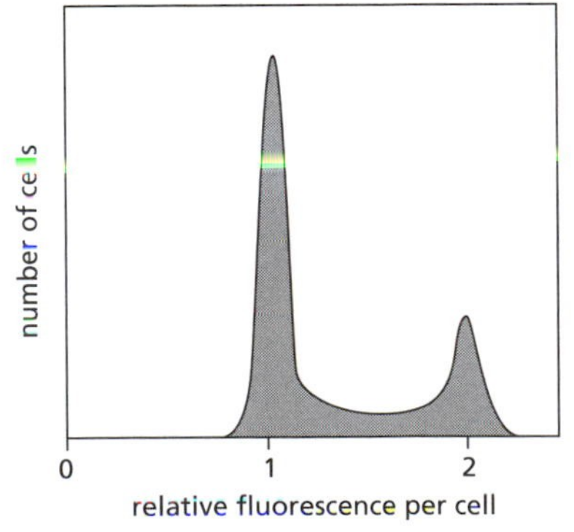

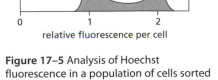

**Figure 17–5** Analysis of Hoechst fluorescence in a population of cells sorted in a flow cytometer (Problem 17–21).

### TRUE/FALSE

Decide whether each of these statements is true or false, and then explain why.

**17–30** The regulation of cyclin–Cdk complexes depends entirely on phosphorylation and dephosphorylation.

**17–31** In order for proliferating cells to maintain a relatively constant size, the length of the cell cycle must match the time it takes for the cell to double in size.

## THOUGHT PROBLEMS

**17–32** Vertebrate cells use several different Cdks to manage the various transitions in the cell cycle, yet budding yeast is able to get by with a single Cdk. How do budding yeast cells manage that neat trick?

**17–33** As director of the Royal Ballet, you are intrigued by the similarity between your nightly show and the cell cycle. Each night the theater fills, the lights go out, the curtain rises, the dancers perform, the curtain falls, the lights go on, and the theater empties—a defined sequence of events, just like the cell cycle. You wonder whether checkpoint mechanisms might enhance theater operations, as they do in the cell cycle. As a first step, you set out to design a system so that when all the seats are filled, the curtains will automatically rise. You have the engineers place sensors in each seat to detect a seated person. But you haven't yet decided how to connect the sensors to the curtains. Do you want the sensors to send out a positive signal that in aggregate will be enough to raise the curtain, or do you want each sensor to send out a negative signal that will stop the curtain going up until every seat is filled? Which do you think would be the more reliable system? Which system do you think a cell would use?

## DATA HANDLING

**17–34** Frog oocytes mature into eggs when incubated with progesterone. Egg maturation is characterized by disappearance of the nucleus (termed germinal vesicle breakdown) and formation of a meiotic spindle. The requirement for progesterone can be bypassed by microinjecting 50 nL of egg cytoplasm directly into a fresh oocyte (1000 nL), which then matures normally (Figure 17–6). Progesterone-independent maturation is triggered by maturation-promoting factor (MPF) activity in the egg cytoplasm—later called mitosis-promoting factor and shown to be M-Cdk.

At early times after progesterone treatment, inhibition of protein synthesis by cyclohexamide blocks egg maturation. However, a few hours before oocytes become eggs—a time that corresponds to the appearance of MPF activity—progesterone-induced maturation can no longer be blocked by cyclohexamide.

Is synthesis of MPF itself the cycloheximide-sensitive event? To test this possibility, you transfer MPF serially from egg to oocyte to test whether its activity diminishes with dilution. You first microinject 50 nL of cytoplasm from an activated egg into an immature oocyte as shown in Figure 17–6; when the oocyte matures into an egg, you transfer 50 nL of its cytoplasm into another immature oocyte; and so on. Surprisingly, you find that you can

**Problem 15–50** analyzes progesterone-induced egg maturation.

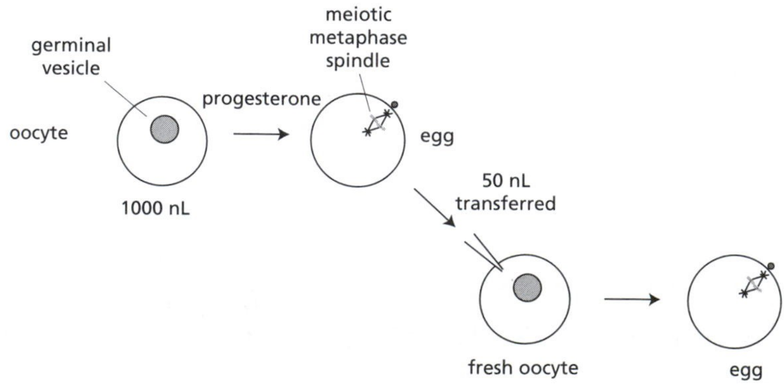


**Figure 17–6** Progesterone- and MPF-induced maturation of oocytes (Problem 17–34).

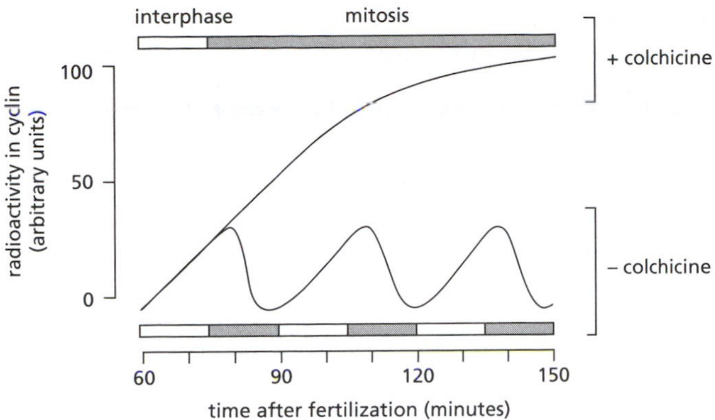


**Figure 17–7** Effects of colchicine treatment on cyclin B destruction and cell-cycle progression (Problem 17–35). *Boxes* indicate position in the cell cycle; *white segments* represent interphase and *gray segments* represent mitosis.

continue this process for at least 10 transfers, even when the recipient oocytes are bathed in cycloheximide! Moreover, the apparent MPF activity in the last egg is equal to that in the first egg.

A. What dilution factor is achieved by 10 serial transfers of 50 nL into 1000 nL? Do you consider it likely that a molecule might have an undiminished biological effect over this concentration range?

B. How do you suppose MPF activity can be absent from immature oocytes yet appear in activated eggs, even when protein synthesis has been blocked by cycloheximide?

C. Propose a means by which MPF activity might be maintained through repeated serial transfers.

**Problem 3–109** describes the activation of Cdk2 by phosphorylation.

**17–35** You are studying the synthesis and destruction of cyclins in dividing clam embryos. You suspect that colchicine, a drug that binds to tubulin and arrests cells in mitosis, might work by inhibiting the normal destruction of cyclin B at the metaphase–anaphase transition. To test this idea, you add $^{35}$S-methionine to a suspension of fertilized clam eggs, divide the suspension in two and add colchicine to one. You take duplicate samples at 5-minute intervals, one for analysis of cyclin B by gel electrophoresis and the other for analysis of mitotic chromosomes by fixing and staining the cells. Untreated cells alternated between interphase and mitosis every 30 minutes, whereas colchicine-treated cells entered mitosis normally but remained there for hours (Figure 17–7). Colchicine treatment also abolished the disappearance of cyclin B that normally precedes the metaphase–anaphase transition (Figure 17–7).

To get a clearer picture of how colchicine inhibits cyclin B destruction, you repeat the experiment, but add the protein synthesis inhibitor emetine just before cells enter mitosis. In the absence of colchicine, emetine-treated cells entered and exited mitosis normally, and divided into two daughter cells, which then remained indefinitely in interphase (Figure 17–8). In the presence of colchicine, the emetine-treated cells stayed in mitosis for about 2 hours and then decondensed their chromosomes and re-formed nuclei without dividing. In the presence and absence of colchicine, the exit from mitosis coincided with the disappearance of cyclin B.

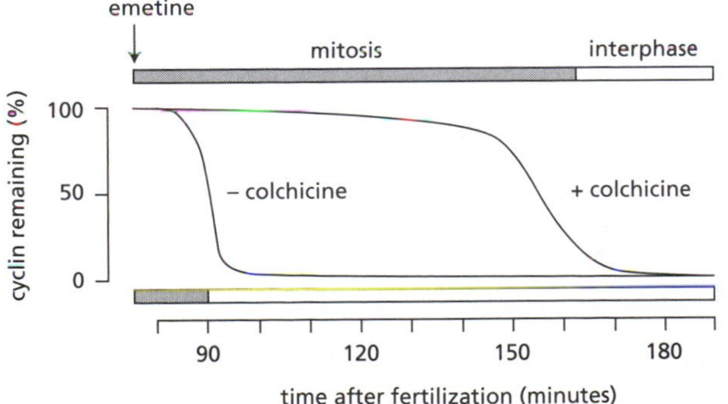


**Figure 17–8** Effects of emetine treatment on cyclin B destruction and cell-cycle progression in normal cells and colchicine-treated cells (Problem 17–35). *Boxes* indicate position in the cell cycle; *white segments* represent interphase and *gray segments* represent mitosis.

A. What effect, if any, does colchicine have on cyclin B synthesis and destruction? Can these effects explain how colchicine causes metaphase arrest?
B. How do you suppose that inhibition of protein synthesis eventually reverses the metaphase arrest produced by colchicine?

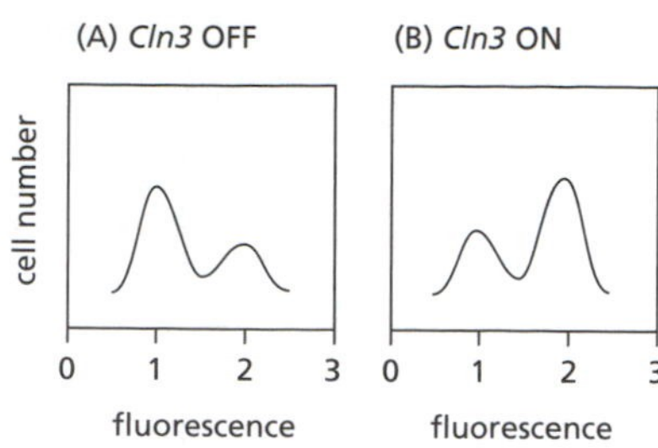


**Figure 17–9** Flow cytometry of wild type yeast cells (Problem 17–37). (A) Cln3 not expressed from plasmid. (B) Cln3 overexpressed from plasmid. Numbers indicate relative fluorescence per cell.

**17–36** You have isolated a temperature-sensitive mutant of the *Cdc28* gene, which encodes Cdk1 in budding yeast. At the restrictive temperature, the mutant Cdk1 binds the $G_1$ cyclins Cln1 and Cln2 so weakly that colonies do not form. When a cDNA library on a high copy-number plasmid was expressed in the mutant strain, three different cDNAs were found that allowed colony formation at the restrictive temperature. One was the cDNA from the wild-type *Cdc28* gene; the other two were cDNAs that encoded cyclins Cln1 and Cln2. If the cDNAs for the cyclin genes were transferred to a plasmid that was maintained at one copy per cell, neither one allowed the *Cdc28* mutant to grow at high temperature. Suggest a mechanism whereby high-level expression of Cln1 or Cln2 would allow survival of mutant *Cdc28* cells at high temperature, whereas low-level expression would not.

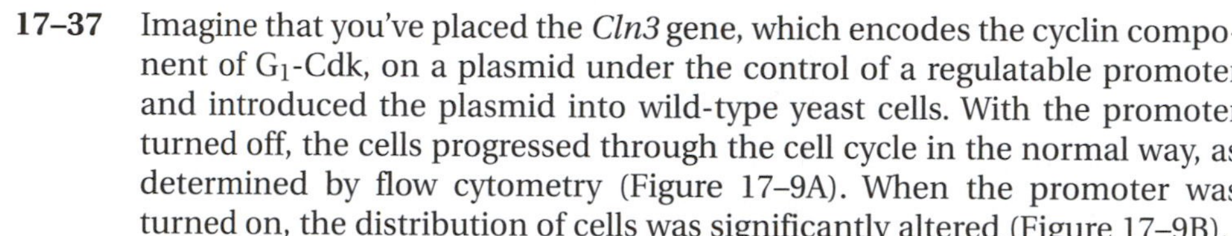

**17–37** Imagine that you've placed the *Cln3* gene, which encodes the cyclin component of $G_1$-Cdk, on a plasmid under the control of a regulatable promoter and introduced the plasmid into wild-type yeast cells. With the promoter turned off, the cells progressed through the cell cycle in the normal way, as determined by flow cytometry (Figure 17–9A). When the promoter was turned on, the distribution of cells was significantly altered (Figure 17–9B).
A. How did the distribution of cells in the phases of the cell cycle change when the cyclin Cln3 was overexpressed? How do you suppose that increased expression of the cyclin Cln3 caused these changes?
B. When Cln3 was overexpressed, would the cells have been smaller than normal, the same size, or larger than normal? Explain your reasoning.

**17–38** You have isolated two temperature-sensitive strains of yeast (which you've named *giant* and *tiny*) that show very different responses to elevated temperature. At high temperature, *giant* cells grow until they become enormous, but no longer divide. By contrast, *tiny* cells have a very short cell cycle and divide when they are very much smaller than usual. You are amazed to discover that these strains arose by different mutations in the very same gene. Based on your understanding of cell-cycle regulation by Cdk1, Wee1, and Cdc25, propose an explanation for how two different mutations in one of these genes might have given rise to the *giant* and *tiny* strains.

# S PHASE

| TERMS TO LEARN | | |
|---|---|---|
| Cdc6 | geminin | preinitiation complex |
| Cdt1 | Mcm protein | prereplicative complex (pre-RC) |
| cohesin | origin recognition complex (ORC) | |

## DEFINITIONS

Match the definition below with its term from the list above.

**17–39** Complex of proteins that holds sister chromatids together along their length until they separate at mitosis.

**17–40** Large protein complex that is bound throughout the cell cycle at origins of replication in eucaryotic chromosomes.

**17–41** One of a group of related proteins in the eucaryotic cell that are loaded onto the origin recognition complexes in DNA in late mitosis and early $G_1$ to form the prereplicative complex and license replication.

**17–42** Protein that binds to and inhibits a key component of the prereplicative complex.

## TRUE/FALSE

Decide whether each of these statements is true or false, and then explain why.

**17–43** Initiation of DNA synthesis is permitted only at origins of replication that contain a prereplicative complex.

**17–44** While other proteins come and go during the cell cycle, the proteins of the origin recognition complex remain bound to the DNA throughout.

## THOUGHT PROBLEMS

**17–45** In budding yeast, a prereplicative complex, consisting of ORC, Cdc6, and Mcm proteins, is established at origins of replication during the $G_1$ phase. S-Cdk then triggers origin firing and helps to prevent re-replication. But not all yeast origins begin replication at the same time: some fire early in S phase, while others fire late. How is it possible for S-Cdk to trigger origin firing at a variety of different times, yet also prevent re-replication at all origins? The details of this process are largely unknown. Propose a scheme that could account for this behavior of S-Cdk.

**17–46** The yeast cohesin subunit Scc1, which is essential for sister-chromatid pairing, can be artificially regulated for expression at any point in the cell cycle. If expression is turned on at the beginning of S phase, all the cells divide satisfactorily and survive. By contrast, if Scc1 expression is turned on only after S phase is completed, the cells fail to divide and they die, even though Scc1 accumulates in the nucleus and interacts efficiently with chromosomes. Why do you suppose that cohesin must be present during S phase for cells to divide normally?

**17–47** If cohesins join sister chromatids all along their length, how is it possible for the cell to generate mitotic chromosomes such as that shown in Figure 17–10, which clearly shows the two sister chromatids as separate domains?

1 μm

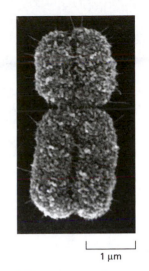

**Figure 17–10** A scanning electron micrograph of a fully condensed mitotic chromosome from vertebrate cells (Problem 17–47).

## DATA HANDLING

**17–48** Early clues about the regulation of S phase came from studies in which human cells at various cell-cycle stages were fused to form single cells with two nuclei. Figure 17–11 shows the outcome of pairwise fusions between $G_1$, S, and $G_2$ cells. Given what we now know about the roles of cyclin–Cdk complexes in progression of the cell cycle, how would you interpret the outcomes of each of these experiments? Do these experiments suggest that there may be a block to re-replication in the cell cycle?

**Problems 5–59** and **5–60** Analyze the block to re-replication in eucaryotic cells.

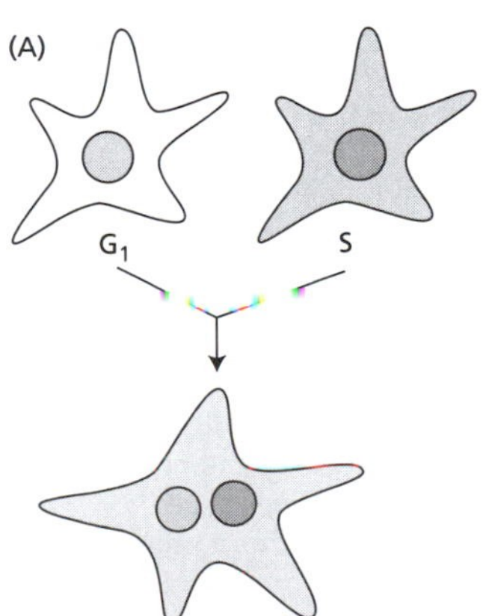


$G_1$-phase nucleus immediately enters S phase; S-phase nucleus continues DNA replication

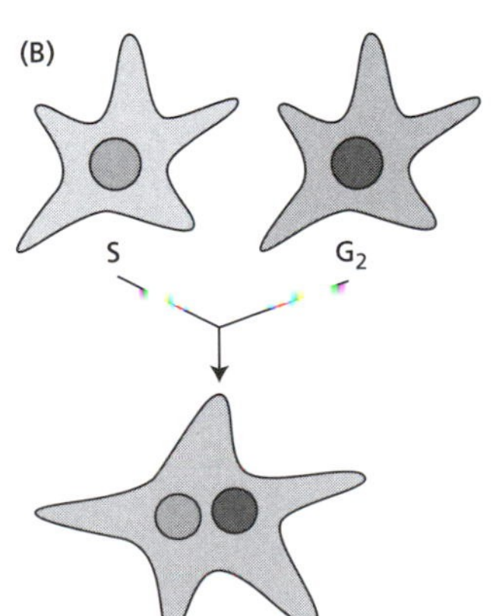


$G_2$-phase nucleus stays in $G_2$ phase; S-phase nucleus continues DNA replication

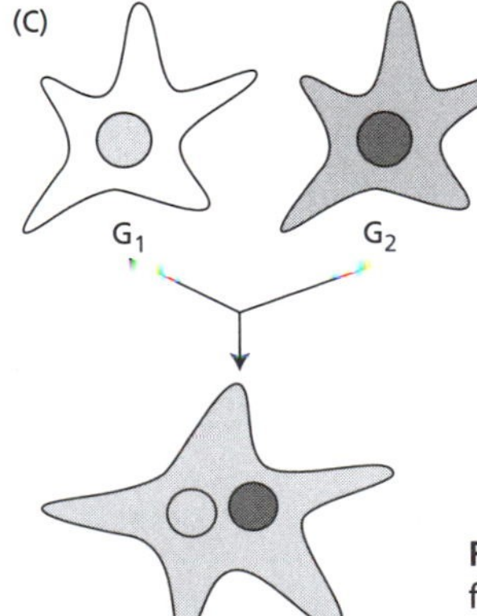


$G_2$-phase nucleus stays in $G_2$ phase; $G_1$-phase nucleus enters S-phase according to its own timetable

**Figure 17–11** Results of cell-fusion experiments using mammalian cells at different stages of the cell cycle (Problem 17–48). (A) Fusion of S and $G_1$ cells. (B) Fusion of S and $G_2$ cells. (C) Fusion of $G_1$ and $G_2$ cells.

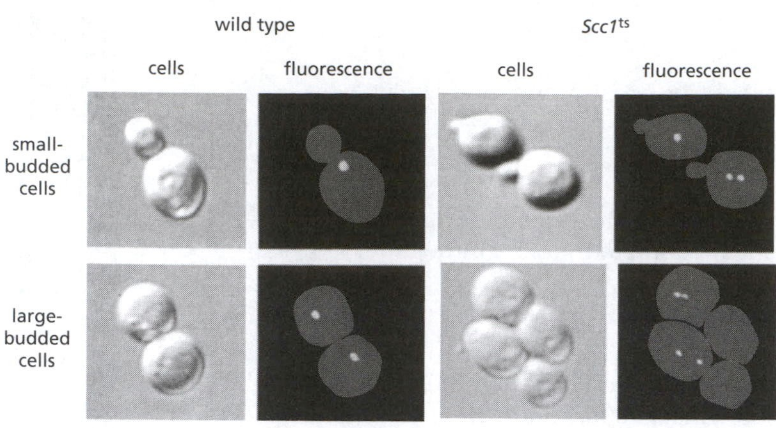


**Figure 17–12** Small- and large-budded cells from wild type and *scc1*$^{ts}$ grown at 37°C (Problem 17–49). For each strain a matched set of pictures shows the appearance of the cells and the corresponding sites of fluorescence.

**17–49** Using a clever genetic screen, you have identified a temperature-sensitive (ts) mutant in a yeast gene (*Scc1*) that appears to be required for sister chromatid cohesion. To assay directly for sister chromatid cohesion, you insert a tandem array of 336 short DNA sequences, to which a bacterial protein can bind tightly, adjacent to the centromere of chromosome V. You then express a fusion of the bacterial protein with GFP in the same cells. When the GFP fusion protein binds to its recognition sequences, it creates a bright dot of green fluorescence on the chromosome. To test for the effects of mutant Scc1 on sister chromatid cohesion, you isolate unbudded cells from wild-type and *Scc1*$^{ts}$ cells that were grown at 25°C, and grow them at 37°C for various times. Representative examples of small-budded cells in S phase and large-budded cells that have passed the metaphase–anaphase transition are shown for both strains in Figure 17–12.

A. Do sister chromatids in wild-type cells adhere to each other normally during S phase, and separate normally during mitosis? How can you tell?

B. Do sister chromatids in *Scc1*$^{ts}$ cells adhere normally in S phase, and separate normally during mitosis? How can you tell?

C. In the large-budded cells from the *Scc1*$^{ts}$ strain, why do both sister chromatids remain in one cell?

**17–50** Scc1, Scc3, Smc1, and Smc3 are subunits of the yeast cohesin complex, which is required to hold sister chromatids together until their separation at mitosis. A fifth protein, Eco1, is also required for sister-chromatid cohesion, but it is not a part of the cohesion complex. To investigate the role of Eco1, you compare the viability of temperature-sensitive (ts) mutants of *Scc1* and *Eco1*. You synchronize cultures of *Scc1*$^{ts}$ and *Eco1*$^{ts}$ in $G_1$ and divide them into three aliquots. You incubate one aliquot at 25°C (the permissive temperature), one at 37°C (the restrictive temperature), and one you incubate at 25°C until the end of S phase and then shift it to 37°C. At different times you measure the number of viable cells in each aliquot by counting colonies that grow at 25°C (Figure 17–13).

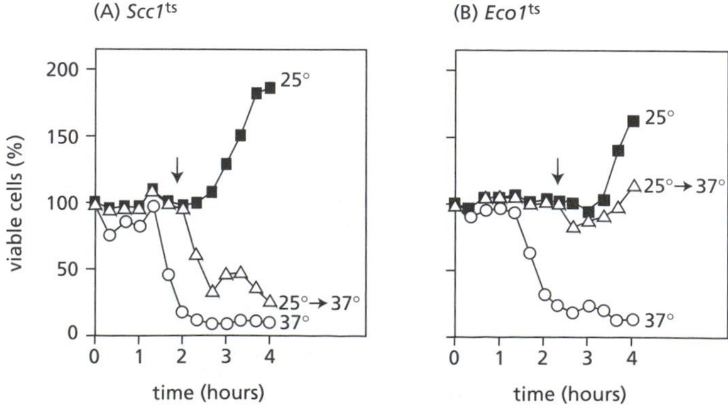


**Figure 17–13** Analysis of the requirement for proteins involved in sister-chromatid cohesion during the cell cycle (Problem 17–50). (A) Viable cells in cultures of *Scc1*$^{ts}$ mutants grown at various temperatures. (B) Viable cells in cultures of *Eco1*$^{ts}$ mutants grown at various temperatures. *Arrows* indicate the time at which an aliquot was shifted from 25°C to 37°C. Viable cells were assayed by their ability to form colonies after plating at 25°C. Number of colonies at time zero was defined as 100%.

A. At 37°C, replication begins at 1.5 hours in the *Scc1*$^{ts}$ and *Eco1*$^{ts}$ mutants. Are Scc1 and Eco1 required during replication? How can you tell from these experiments?
B. At 25°C, S phase ends at about 2 hours. Are Scc1 and Eco1 required after S phase? How can you tell?
C. Propose an explanation for the behavior of *Scc1*$^{ts}$ mutants in these experiments.
D. Propose an explanation for the behavior of *Eco1*$^{ts}$ mutants in these experiments.

# MITOSIS

TERMS TO LEARN

| | | |
|---|---|---|
| anaphase A | condensin | mitotic spindle |
| anaphase B | interpolar microtubule | securin |
| astral microtubule | kinetochore microtubule | separase |
| bi-orientation | metaphase plate | spindle assembly checkpoint |
| catastrophe factor | microtubule-associated protein (MAP) | telophase |
| centrosome | microtubule flux | |

## DEFINITIONS

Match the definition below with its term from the list above.

**17–51** Movement of tubulin subunits toward the spindle poles as a result of addition of new subunits at the plus ends of microtubules and their disassembly at minus ends.

**17–52** Stage of mitosis in which the spindle poles move apart.

**17–53** Mechanism ensuring that cells do not enter anaphase until all chromosomes are correctly bi-oriented on the mitotic spindle.

**17–54** Centrally located organelle of animal cells that after duplication organizes each spindle pole during mitosis.

**17–55** Stage of mitosis in which the chromosomes begin to move toward the two spindle poles.

**17–56** Imaginary plane midway between the spindle poles in which chromosomes are positioned at metaphase.

**17–57** Microtubules that overlap in the spindle midzone and interact via their plus ends, generating an antiparallel array.

**17–58** Final stage of mitosis in which the two sets of separated chromosomes decondense and become enclosed by nuclear envelopes.

**17–59** Complex of proteins that uses the energy of ATP hydrolysis to promote the compaction and resolution of sister chromatids.

**17–60** Protease whose activation at the end of metaphase results in the cleavage of cohesin and the separation of sister chromatids.

**17–61** Microtubule that radiates outward from the spindle pole and contacts the cell cortex, helping to position the spindle in the cell.

## TRUE/FALSE

Decide whether each of these statements is true or false, and then explain why.

**17–62** After the nuclear envelope breaks down, microtubules gain access to the chromosomes and, every so often, a randomly probing microtubule connects with a kinetochore and captures the chromosome.

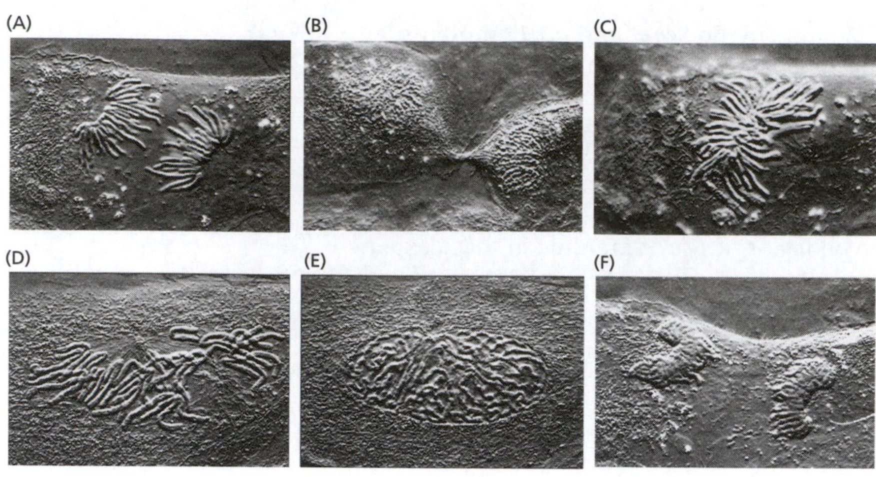


**Figure 17–14** Light micrographs of a single cell at different stages of M phase (Problem 17–66).

**17–63** Chromosomes are positioned on the metaphase plate by equal and opposite forces that pull them toward the two poles of the spindle.

**17–64** Once formed, kinetochore microtubules depolymerize at their plus ends (the ends attached to the kinetochores) throughout mitosis.

**17–65** The six stages of M phase—prophase, prometaphase, metaphase, anaphase, telophase, and cytokinesis—occur in strict sequential order.

## THOUGHT PROBLEMS

**17–66** A living cell from the lung epithelium of a newt is shown at different stages in M phase in Figure 17–14. Order these light micrographs into the correct sequence, and identify the stage in M phase that each represents.

**17–67** It is remarkable that the concentration of cyclin B in the cleaving clam egg rises very slowly and steadily throughout the cell cycle, whereas M-Cdk activity increases suddenly at mitosis (Figure 17–15). How is the activity of M-Cdk so sharply regulated in the presence of a gradual increase in cyclin B?

**17–68** When activated, the DNA replication checkpoint blocks activation of M-Cdk and thereby prevents entry into mitosis. The target of this checkpoint mechanism is the inhibitory phosphate on M-Cdk, which is added by Wee1 tyrosine kinase and removed by Cdc25 tyrosine phosphatase. The level of phosphorylation on M-Cdk is a balance between the activities of Wee1 kinase and Cdc25 phosphatase. The checkpoint mechanism inhibits Cdc25 phosphatase. Why do you suppose it inhibits the phosphatase instead of activating Wee1 kinase?

**17–69** Describe the three main classes of spindle microtubule in animal cells and their functions during mitosis.

**17–70** The lengths of microtubules in various stages of mitosis depend on the balance between the activities of catastrope factors, which destabilize microtubules, and microtubule-associated proteins (MAPs), which stabilize them. If you overexpressed catastrope factors, would you expect the length of the

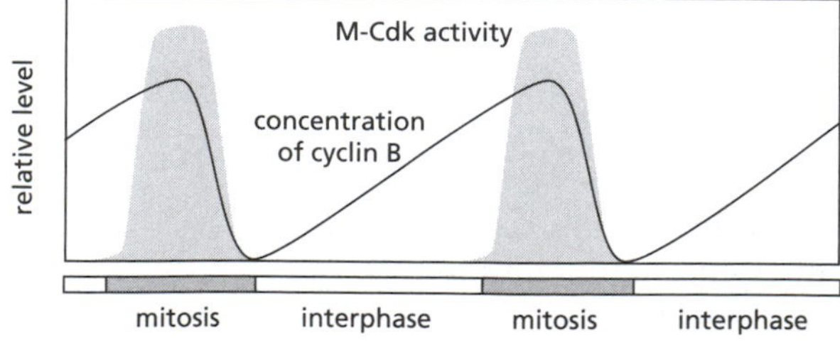


**Figure 17–15** The rise and fall of M-Cdk activity and cyclin B concentration during the cell cycle in a cleaving clam egg (Problem 17–67).

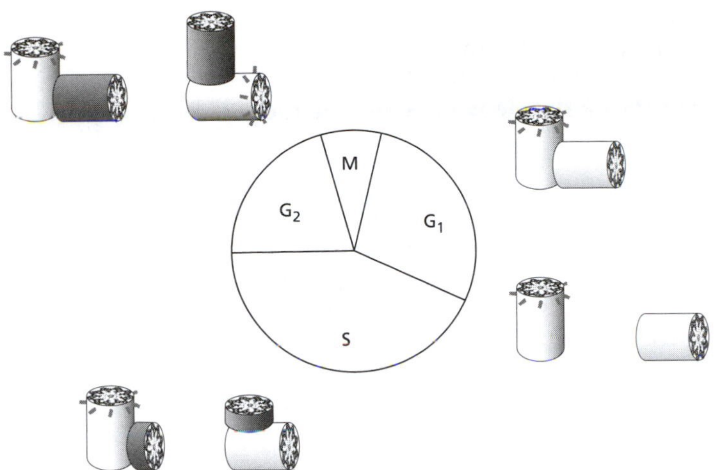


**Figure 17–16** The centrosome duplication cycle (Problem 17–73). The individual centrioles that make up the centrosome are shown as *cylinders*.

mitotic spindle to be longer, shorter, or unchanged relative to the corresponding stage of mitosis in wild-type cells? What would you expect if you overexpressed MAPs? Explain your reasoning.

**17–71** The balance between plus-end-directed and minus-end-directed motors determines spindle length. Where in the spindle are these motors located, and how do they operate to control spindle length?

**17–72** When kinesin-5 motor proteins, which contain two plus-end-directed motor domains, are incubated with microtubules, they will organize the microtubules into an astral array. How do you suppose such an array is generated? Will the plus ends of the microtubules be located in the center of the array or at the periphery? Or will some plus ends be at the center and others at the periphery?

**17–73** Examine the schematic representation of centrosome duplication in Figure 17–16. By analogy with DNA replication, would you classify centrosome duplication as conservative or semi-conservative? Explain your reasoning.

**17–74** Imagine that you could color one sister chromatid and follow it through mitosis and cytokinesis. What event commits this chromatid to a particular daughter cell? Once initially committed, can its fate be reversed? What is the main influence on its initial commitment?

**17–75** How many kinetochores are there in a human cell at mitosis?

**17–76** Both sister chromatids of a chromosome occasionally end up in one daughter cell. Suggest some possible causes for such an event. What would be the consequences for the daughter cells if this event occurred in mitosis?

**17–77** How can there be a constant poleward flux of tubulin subunits in the absence of any visible change in the appearance of the spindle?

**17–78** It is estimated that as many as 25% of kinetochore microtubules and 75% of interpolar microtubules are not anchored to the centrosome. In spite of that, all the microtubules are focused tightly at the spindle pole. Why do you suppose the microtubule ends that weren't nucleated by centrosomes don't splay out away from the poles?

**17–79** Nocodazole reversibly inhibits microtubule polymerization, which is essential for formation of the mitotic spindle. By treating a population of mammalian cells with nocodazole for a time and then washing it out of the medium, it is possible to synchronize the cell population. In the presence of nocodazole, where in the cell cycle would you expect the cells to accumulate? What mechanism do you suppose is responsible for stopping cell-cycle progression in the presence of nocodazole?

**17–80** Budding yeast cells that are deficient for Mad2, a component of the spindle-attachment checkpoint, are killed by treatment with benomyl, which causes microtubules to depolymerize. In the absence of benomyl, however, the cells are perfectly viable. Explain why Mad2-deficient cells live in the absence of benomyl but die in its presence.

**17–81** If a fine glass needle is used to manipulate a chromosome inside a living cell during early M phase, it is possible to trick the kinetochores on the two sister chromatids into attaching to the same spindle pole. This arrangement is normally unstable and is rapidly converted to the standard arrangement with sister chromatids attached to opposite poles. The abnormal attachment can be stabilized, if the needle is used to gently pull the chromosome so that the microtubules that attach it to the same pole are under tension. What does this suggest to you about the mechanism by which kinetochores normally become attached and stay attached to microtubules from opposite spindle poles during M phase? Explain your answer.

**17–82** Compare and contrast the movements of chromosomes and spindle poles, and their underlying mechanisms, during anaphase A and anaphase B.

**17–83** Discuss the following analogy: "Chromosomes are pulled to the spindle pole like fish on a line."

**17–84** Consider the events that lead to formation of the new nucleus at telophase. How do nuclear and cytosolic proteins become properly sorted so that the new nucleus contains nuclear proteins but not cytosolic proteins?

**17–85** Order the following events in animal cell division.

A. Alignment of chromosomes at the spindle equator.
B. Attachment of microtubules to chromosomes.
C. Breakdown of nuclear envelope.
D. Condensation of chromosomes.
E. Decondensation of chromosomes.
F. Duplication of centrosome.
G. Elongation of the spindle.
H. Pinching of cell in two.
I. Re-formation of nuclear envelope.
J. Separation of centrosomes.
K. Separation of sister chromatids.

## CALCULATIONS

**17–86** High doses of caffeine (Figure 17–17) interfere with the DNA replication checkpoint mechanism in mammalian cells. Why then do you suppose the Surgeon General has not yet issued an appropriate warning to heavy coffee and cola drinkers? A typical cup of coffee (150 mL) contains 100 mg of caffeine (196 g/mole). How many cups of coffee would you have to drink to reach the dose (10 mM) required to interfere with the DNA replication checkpoint mechanism? (A typical adult contains about 40 liters of water.)

**17–87** How much DNA does a single microtubule carry in mitosis? From the information in Table 17–2, calculate the average length of chromosomes in each organism in base pairs and in millimeters (1 bp = 0.34 nm), and then calculate how much DNA (in base pairs) each microtubule carries on average in mitosis. Do microtubules carry about the same amount of DNA or does it vary widely in different organisms?

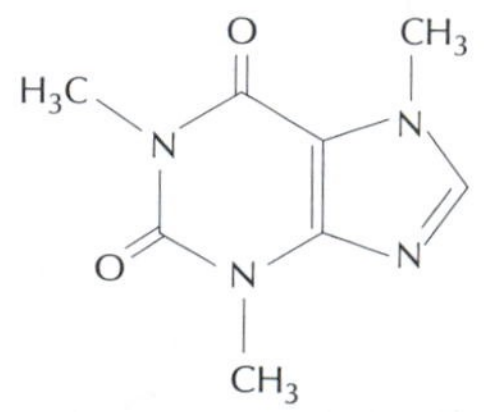

**Figure 17–17** Structure of caffeine (Problem 17–86).

## DATA HANDLING

**17–88** The activities of Wee1 kinase and Cdc25 phosphatase determine the state of phosphorylation of tyrosine 15 in the Cdk1 component of M-Cdk. When tyrosine 15 is phosphorylated, M-Cdk is inactive; when tyrosine 15 is not

**Table 17–2 DNA content, haploid number of chromosomes, and microtubules per chromosome in a variety of organisms (Problem 17–87).**

| TYPE OF ORGANISM | SPECIES | DNA CONTENT (bp) | NUMBER OF CHROMOSOMES | MICROTUBULES/ CHROMOSOME |
|---|---|---|---|---|
| Yeast | *S. cerevisiae* | $1.4 \times 10^7$ | 16 | 1 |
| Yeast | *S. pombe* | $1.4 \times 10^7$ | 3 | 3 |
| Protozoan | *Chlamydomonas* | $1.1 \times 10^8$ | 19 | 1 |
| Fly | *Drosophila* | $1.7 \times 10^8$ | 4 | 10 |
| Human | *Homo sapiens* | $3.2 \times 10^9$ | 23 | 25 |
| Plant | *Haemanthus* | $1.1 \times 10^{11}$ | 18 | 120 |

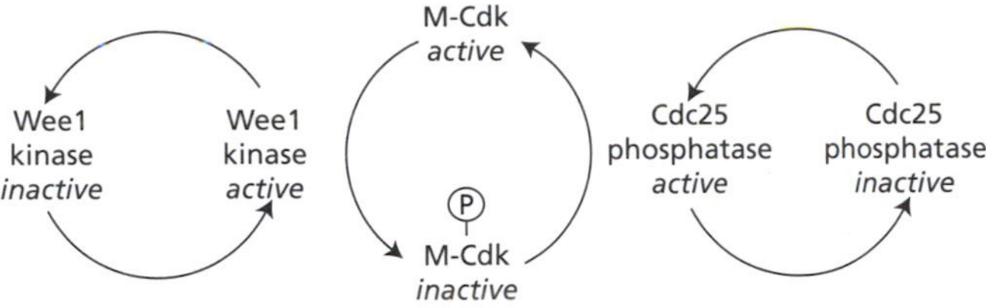


**Figure 17–18** Control of M-Cdk activity by Wee1 tyrosine kinase and Cdc25 tyrosine phosphatase (Problem 17–88).

phosphorylated, M-Cdk is active (Figure 17–18). Just as the activity of M-Cdk itself is controlled by phosphorylation, so too are the activities of Wee1 kinase and Cdc25 phosphatase.

The regulation of these various activities can be studied in extracts of frog oocytes. In such extracts, Wee1 kinase is active and Cdc25 phosphatase is inactive. As a result, M-Cdk is inactive because its Cdk1 component is phosphorylated on tyrosine 15. M-Cdk in these extracts can be rapidly activated by addition of okadaic acid, which is a potent inhibitor of serine/threonine protein phosphatases. Using antibodies specific for Cdk1, Wee1, and Cdc25, it is possible to examine their phosphorylation states by changes in mobility upon gel electrophoresis (Figure 17–19). (Phosphorylated proteins generally migrate more slowly than their nonphosphorylated counterparts.)

A. Based on the results with okadaic acid, decide whether the active forms of Wee1 kinase and Cdc25 phosphatase are phosphorylated or nonphosphorylated. In Figure 17–18, indicate the phosphorylated forms of Wee1 and Cdc25. Also, label the arrows connecting their active and inactive forms to show which transitions are controlled by protein kinases and which are controlled by protein phosphatases.
B. Are the protein kinases and phosphatases that control Wee1 and Cdc25 specific for serine/threonine side chains or for tyrosine side chains? How do you know?
C. How does addition of okadaic acid cause an increase in phosphorylation of Wee1 and Cdc25, but a decrease in phosphorylation of Cdk1?
D. If you assume that Cdc25 and Wee1 are targets for phosphorylation by active M-Cdk, can you explain how the appearance of a small amount of active M-Cdk would lead to its rapid and complete activation?

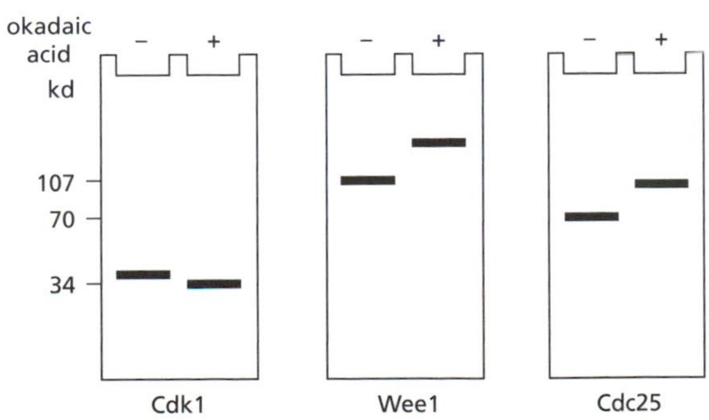


**Figure 17–19** Effects of okadaic acid on the phosphorylation states of Cdk1, Wee1, and Cdc25 (Problem 17–88). Molecular mass markers are shown in kilodaltons on the *left*.

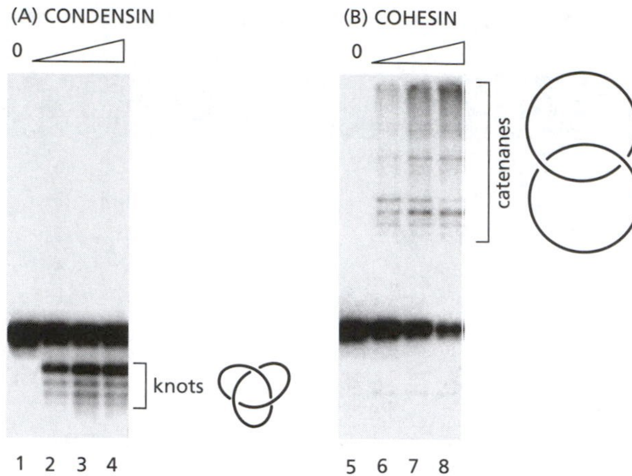


**Figure 17–20** Topological analysis of the functions of condensin and cohesin (Problem 17–89). (A) Electrophoretic analysis of migration of circular DNA incubated with condensin and topoisomerase II. The knots formed by incubation are all like the one illustrated. (B) Electrophoretic analysis of migration of circular DNA incubated with cohesin and topoisomerase II. The products of incubation are catenanes; a dimeric catenane is shown. In both gels the heavy band of material corresponds to the nicked circles that were added to the incubation mixture.

**17–89** Cohesins and condensins are very similar in structure yet carry out quite different biochemical tasks: cohesion of sister chromatids and condensation of chromosomes, respectively. You are skeptical that such similar molecules can perform such distinct functions, and set out to determine if they are truly different using purified components. You incubate pure cohesin or condensin with nicked circular DNA and ATP. You then add topoisomerase II to link duplexes that have been juxtaposed. (Topoisomerase II binds to one duplex and breaks both strands, attaching itself covalently to the ends and holding them together. The topoisomerase II complex can gate the passage of a second duplex through the break and then reseal the original duplex. By linking—or unlinking—duplexes, topoisomerase II can alter their topology in informative ways.)

As shown in Figure 17–20, condensin and cohesin yield very different results upon incubation with topoisomerase II and analysis by gel electrophoresis. Incubation with condensin followed by topoisomerase generates a particular kind of trefoil knot (Figure 17–20A), whereas incubation with cohesin followed by topoisomerase generates a series of catenanes (circles joined like links of a chain, Figure 17–20B).

A. Do these results support the proposed roles of cohesin and condensin? How so?

B. Suggest a plausible mechanism by which binding of cohesin might allow topoisomerase II to link molecules into catenanes.

C. Knots are much more difficult to think about, but often are very revealing of mechanistic details. See if you can figure out a way to use the binding of condensin molecules, coupled with one duplex crossing event catalyzed by topoisomerase II, to tie a circular molecule into any kind of a knot.

**Problem 4–87** probes the function of condensin by analyzing its effects on DNA supercoiling.

**17–90** If the centrosome cycle is disrupted so that centrosomes are not duplicated, continued progression through the cell cycle will ultimately lead to formation of a spindle with a single pole (monopolar spindle), rather than the standard bipolar spindle. The pole of a monopolar spindle could, in principle, contain a single centriole, a centrosome with the usual pair of centrioles, or multiple centrosomes, depending on where the centrosome cycle was interrupted (see Figure 17–16).

A. When defective, the Zyg1 protein kinase of *C. elegans* leads to monopolar spindles that have a single centriole at the pole. At what point is the centrosome cycle interrupted in the absence of functional Zyg1? Explain your reasoning.

B. When a mutant, nonphosphorylatable form of nucleophosmin, a component of centrosomes and a target of $G_1$/S-Cdk, is expressed in cells, monopolar spindles form with a single centrosome (with two centrioles) at the pole. At what point does the mutant form of nucleophosmin interrupt the centrosome cycle? Explain your reasoning.

**17–91** One of the least well-understood aspects of the cell cycle is the duplication of the spindle poles. As illustrated in Figure 17–21, the centrosome normally splits at the beginning of mitosis to form the two spindle poles, which then

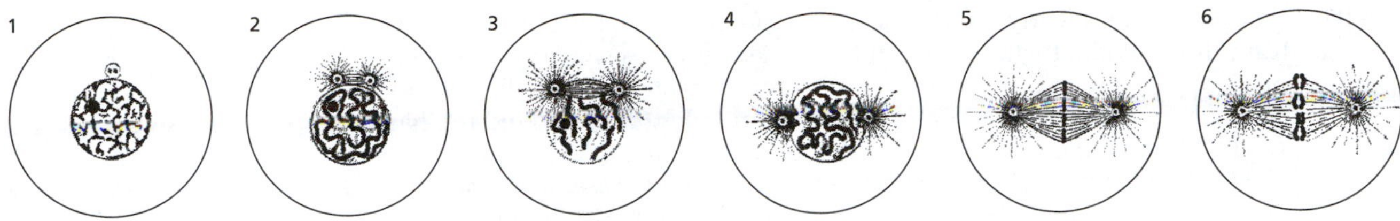


Figure 17–21 Normal process of centrosome splitting during mitosis to form a bipolar spindle (Problem 17–91).

orchestrate chromosome segregation. During the next interphase, the centriole pair within the centrosome is duplicated so that the centrosome can split at the next mitosis. The cycles of centrosome duplication and splitting normally keep step with cell division so that all cells have the capacity to produce bipolar spindles (see Figure 17–16). It is possible, however, to throw the two cycles out of phase as indicated by experiments first performed in the late 1950s.

When a fertilized sea urchin egg at the metaphase stage of the first mitotic division is exposed to mercaptoethanol (MSH), the mitotic spindle disassembles (Figure 17–22A). (It is not known how mercaptoethanol causes this, but the effect is reversible.) While the eggs are kept in mercaptoethanol, the nucleus does not re-form, no DNA synthesis occurs, and the chromosomes stay condensed. When mercaptoethanol is washed out, the spindle re-forms and cell division takes place. Although some eggs re-form a bipolar spindle and divide normally, most eggs form a tetrapolar spindle and divide into four daughter cells (Figure 17–22A). No matter how long the eggs are arrested in mercaptoethanol, they never divide into more than four cells.

The daughter cells from a four-way division re-form the nucleus and traverse the next cell cycle; however, at mitosis they form a monopolar spindle. In a majority of cases, these cells stay in mitosis a little longer than usual and then decondense their chromosomes, disassemble the spindle, and re-form a nucleus (Figure 17–22B). At the next mitosis, these cells form a bipolar spindle and divide normally (Figure 17–22B). More rarely, the cells with monopolar spindles stay in mitosis much longer, the monopole splits to form a bipolar spindle, and the cell divides normally (Figure 17–22B). The daughter cells from such a division once again form a monopolar spindle at the next mitosis (Figure 17–22B).

Describe patterns of centriole duplication and splitting that can account for the observations shown in Figure 17–22.

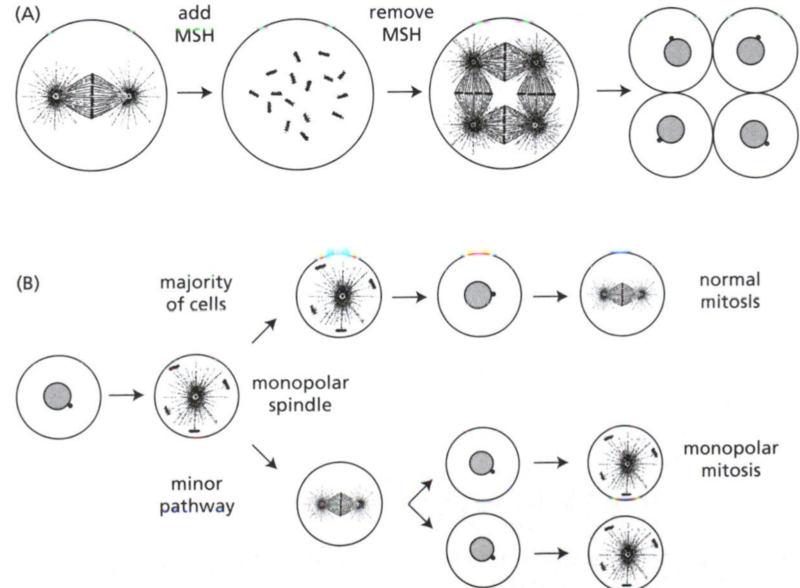


Figure 17–22 Abnormal spindles and cell divisions induced by mercaptoethanol (MSH) (Problem 17–91). (A) Formation of tetrapolar spindles upon treatment of fertilized sea urchin eggs with mercaptoethanol, followed by four-way division. With the breakdown of the spindle, chromosomes disperse while remaining condensed. (B) Major and minor pathways for spindle formation and cell division among the daughter cells from a four-way division. Note that the condensed chromosomes are located at the periphery of a monopolar spindle.

**17–92** A classic paper clearly distinguished the properties of astral microtubules from those of kinetochore microtubules. Centrosomes were used to initiate microtubule growth, and then chromosomes were added. The chromosomes bound to the free ends of the microtubules, as illustrated in Figure 17–23. The complexes were then diluted to very low tubulin concentration (well below the critical concentration for microtubule assembly) and examined again (Figure 17–25). As is evident, only the kinetochore microtubules were stable to dilution.

A. Why do you think the kinetochore microtubules are stable?

B. How would you explain the disappearance of the astral microtubules after dilution? Do they detach from the centrosome, depolymerize from an end, or disintegrate along their length at random?

C. How would a time course after dilution help to distinguish among these possible mechanisms for disappearance of the astral microtubules?

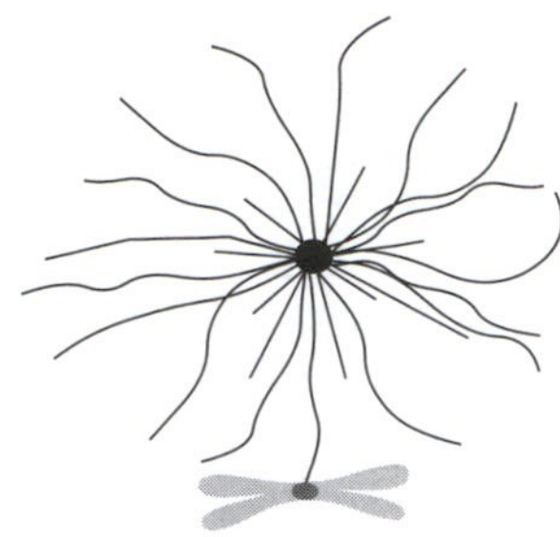

**Figure 17–23** Arrangements of centrosomes, chromosomes, and microtubules before and after dilution to low tubulin concentration (Problem 17–92).

**17–93** In higher eucaryotes, rare chromosomes containing two centromeres at different locations are highly unstable: they are literally torn apart at anaphase when the chromosomes separate. You wonder whether the same phenomenon occurs in yeast, whose chromosomes are too small to analyze microscopically.

You construct a plasmid with two centromeres, as shown in Figure 17–24. Growth of this plasmid in bacteria requires the bacterial origin of replication (*Ori*) and a selectable marker (*Amp*$^R$); its growth in yeast requires the yeast origin of replication (*Ars1*) and a selectable marker (*Trp1*). You prepare a plasmid stock by growth in *E. coli.* This dicentric plasmid transforms yeast with about the same efficiency as a plasmid that contains a single centromere. Individual colonies, however, were found to contain plasmids with a single centromere or no centromere, but never a plasmid with two centromeres. By contrast, colonies that arose after transformation with a monocentric plasmid invariably contained intact plasmids.

A. Considering their extreme instability in yeasts, why are dicentric plasmids stable in bacteria?

B. Why do you suppose the dicentric plasmid is unstable in yeasts?

C. Suggest a mechanism for deletion of one of the centromeres from a dicentric plasmid grown in yeast. Can this mechanism account for loss of both centromeric sequences from some of the plasmids?

**17–94** Circular yeast plasmids that lack a centromere are distributed among individual cells in a peculiar way. Only 5% to 25% of the cells harbor the plasmids, yet the plasmid-bearing cells contain 20 to 50 copies of the plasmid. To investigate the apparent paradox of a high copy number in a small fraction of cells, you perform a pedigree analysis to determine the pattern of plasmid segregation during mitosis. You use a yeast strain that requires histidine for growth and a plasmid that provides the missing histidine gene. The strain carrying the plasmid grows well under selective conditions; that is, when histidine is absent from the medium. By micromanipulation you separate mother and daughter cells for five divisions in the absence of histidine, which yields 32 cells whose pedigree you know (Figure 17–25). You then score each of those cells for their ability to form a colony in the absence of histidine. In Figure 17–25, cells in the last generation that formed colonies are indicated with heavy lines, and those that failed to form colonies are shown with light lines. From this information, the inheritance of the plasmid at each cell division can be inferred, as shown by the black lines in the rest of the pedigree (Figure 17–25).

A. From the pedigree analysis, it is apparent that cells lacking the plasmid can grow for several divisions in the absence of histidine. How can this be?

B. Does this plasmid segregate equally to mother and daughter cells?

C. Assuming that plasmids in yeast cells replicate only once per cell cycle, as the chromosomes do, how can there be 20 to 50 molecules of the plasmid per plasmid-bearing cell?

D. When grown under selective conditions, cells containing plasmids with one centromere (1–2 plasmids per cell) form large colonies, whereas cells

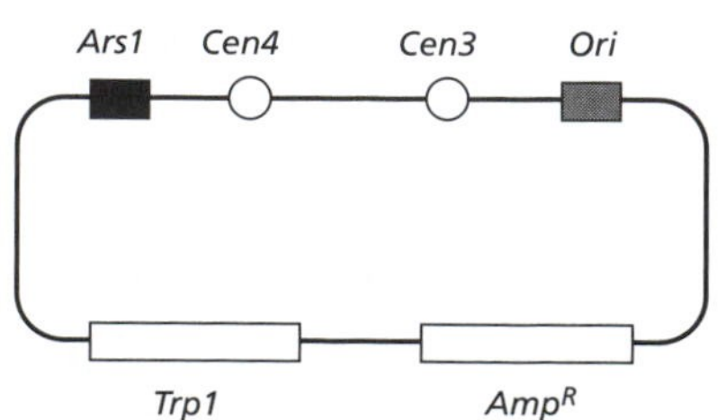


**Figure 17–24** Structure of a dicentric plasmid (Problem 17–93). *Cen3* and *Cen4* refer to centromeres from yeast chromosomes 3 and 4, respectively.

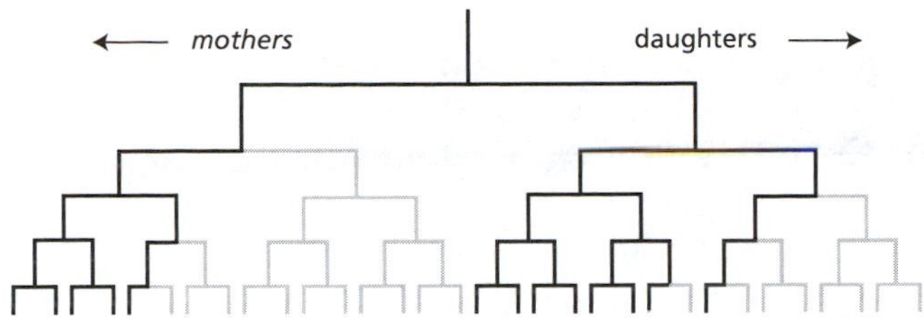


**Figure 17–25** Pedigree analysis showing the inheritance of a plasmid that contains an origin of replication and a selectable histidine marker (Problem 17–94). The *heavy lines* show cells containing the plasmid, and the *light lines* show cells lacking the plasmid. At each division mother cells are shown to the *left* and daughter cells are shown to the *right*.

containing plasmids with no centromere (20–50 plasmids per cell) form small colonies. Does the pedigree analysis help to explain the difference?

**17–95** A strong promoter that directs transcription across a centromere can inactivate the kinetochore. Likewise, kinetochores can provide a strong block to transcription. Evidently, centromere function and transcription are mutually interfering processes. Based on this mutual interference, your advisor has devised a clever scheme that exploits the transcriptional block to measure the strength of DNA–protein interactions in the kinetochore.

To carry out this scheme, you construct a test system consisting of the yeast actin gene fused to the *E. coli LacZ* gene (which encodes β-galactosidase), with the hybrid gene under control of the strong galactose-inducible *Gal10* promoter (Figure 17–26). Into the intron in the actin gene you insert a functional yeast centromere ($Cen6^{wt}$), or a nonfunctional version with an inactivating mutation ($Cen6^{CDEIII}$). When you introduce these plasmids into cells growing on galactose and measure β-galactosidase activity, you are encouraged to find that the wild-type centromere blocks transcription and that the mutant centromere partially relieves the transcription block (Figure 17–26).

You next test a collection of chromosome-transmission-fidelity (*Ctf*) mutants that show elevated rates of chromosome loss. You expect that *Ctf* mutants defective in kinetochore assembly or function will show enhanced expression of β-galactosidase, like the mutation that inactivates the centromere. As shown in Table 17–3, some *Ctf* mutants give substantially elevated levels of β-galactosidase.

A. The test plasmid used in these experiments already contained its own centromere. Why does the introduction of a second centromere in the intron of the hybrid gene not lead to breakage of the plasmid and extreme plasmid instability, as described in Problem 17–93?

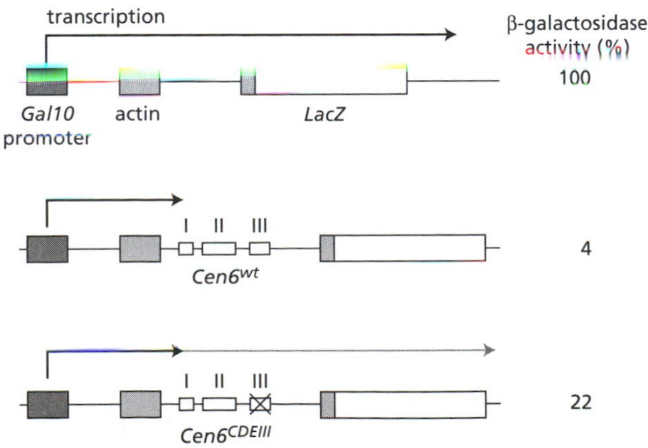


**Figure 17–26** Constructs for testing centromere function (Problem 17–95). β-galactosidase activity observed when these constructs were introduced into cells is shown on the *right*.

**Table 17–3 Transcription through centromeres in *Ctf* mutants** (Problem 17–95).

| HOST STRAIN | TRANSCRIPTION BLOCK | β-GALACTOSIDASE ACTIVITY (nmol/min/mg protein) |
|---|---|---|
| Wild type | *Cen6*$^{wt}$ | 25 |
| Wild type | *Cen6*$^{CDEIII}$ | 135 |
| *Ctf7* | *Cen6*$^{wt}$ | 86 |
| *Ctf8* | *Cen6*$^{wt}$ | 50 |
| *Ctf9* | *Cen6*$^{wt}$ | 19 |
| *Ctf13* | *Cen6*$^{wt}$ | 110 |
| *Ctf17* | *Cen6*$^{wt}$ | 160 |

B. How do you suppose a functional kinetochore might interfere with transcription?

C. Some of the *Ctf* mutants (*Ctf9*, for example) do not show enhanced β-galactosidase activity, yet they do show elevated rates of chromosome loss. What other mechanisms, apart from failure to assemble a functional kinetochore, might lead to the *Ctf* phenotype?

**17–96** The results obtained from the transcription assay described in the previous problem suggest that *Ctf13* may encode a kinetochore protein. As a second and independent check on kinetochore function, you set up a functional assay for kinetochores using dicentric minichromosomes that are not essential for cell survival. One of the two centromeres on the minichromosome is perfectly normal. The other is conditionally active: it is inactive when transcription is directed across it by the galactose-inducible *Gal10* promoter, but fully active when transcription is turned off by addition of glucose (Figure 17–27). As controls, you make two other minichromosomes: one with a weak second centromere due to a small deletion in the centromere sequence (minichromosome B), and the other with a single centromere (minichromosome C) (Figure 17–27).

If both centromeres are functional, they will cause the minichromosomes to break. The loss of an intact minichromosome can be followed visually due to a suppressor tRNA gene (*Sup11*) located next to the normal centromere (Figure 17–27). When the minichromosome breaks, the daughter cell that does not inherit the suppressor tRNA gene will turn red, owing to a suppressible mutation (that is no longer suppressed) in a gene required for adenine biosynthesis. Minichromosome loss gives rise to white colonies containing red sectors. Examining the number and size of such sectors is commonly referred to as a sectoring assay.

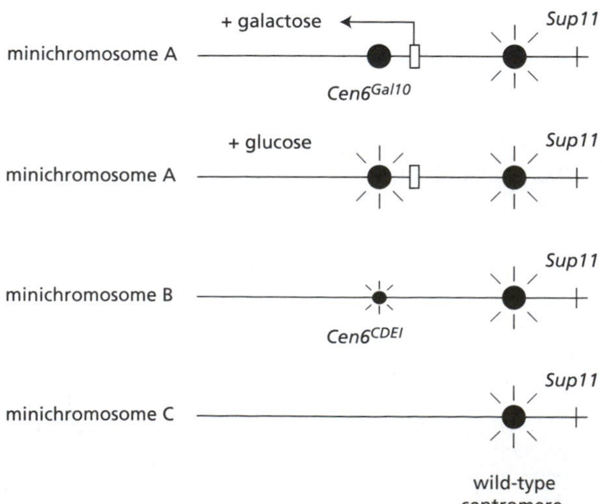


**Figure 17–27** Minichromosomes used to test kinetochore function (Problem 17–96).

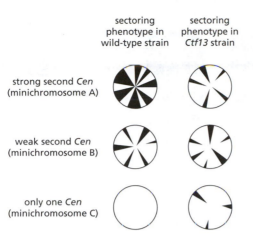

sectoring phenotype in wild-type strain | sectoring phenotype in *Ctf13* strain

strong second *Cen* (minichromosome A)

weak second *Cen* (minichromosome B)

only one *Cen* (minichromosome C)

**Figure 17–28** Sectoring patterns in colonies of wild-type and *Ctf13* yeast carrying monocentric and dicentric minichromosomes (Problem 17–96).

To test the effects of the *Ctf13* mutation on the stability of dicentric minichromosomes, you introduce the three minichromosomes into yeast carrying the *Ctf13* mutation, and into yeast with the normal wild-type gene. In all cases, you grow the cells in the presence of galactose and then test for sectoring on agar plates in the presence of glucose (Figure 17–28).

A. Explain why you placed the second centromere in minichromosome A under control of the *Gal10* promoter, and why you grew the yeasts in galactose before switching to glucose for the sectoring assay.

B. Why do you suppose minichromosome B, which has a weak second centromere, gives less sectoring in the wild-type strain than does minichromosome A, which has two fully functional centromeres?

C. Do your results support the notion that the *Ctf13* gene encodes a kinetochore protein? In your answer, account for the difference in sectoring observed when minichromosome A is introduced into the *Ctf13* strain versus the wild-type strain.

D. The *Ctf13* mutant strain was isolated on the basis of an elevated rate of chromosome loss, as illustrated by the results with minichromosome C (Figure 17–28). Yet in your experiments, it apparently lowers the rate of loss of minichromosome A. How can it be that a mutation that enhances the loss of normal chromosomes actually stabilizes dicentric chromosomes?

**17–97** Among the variety of microtubule-dependent motors associated with mitotic spindles are ones that bind to chromosomes arms. The role of one such motor protein, Xkid, during spindle assembly was investigated by removing the protein (by immunodepletion) from frog egg extracts, which will form spindles under defined conditions. Extracts that have Xkid and immunodepleted extracts, which are missing Xkid, both assemble normal looking spindles, as assessed by tubulin staining (Figure 17–29). In the presence of Xkid the chromosomes are aligned on the metaphase plate (Figure 17–29A), whereas in its absence the chromosomes are dispersed throughout the spindle (Figure 17–29B).

A. Suggest a mechanism by which Xkid might function to align chromosomes on the metaphase plate. Include in your description whether you think Xkid is a plus-end- or a minus-end-directed motor, and why.

B. Is Xkid a plausible candidate for the mediator of the astral ejection force that pushes chromosomes away from the poles?

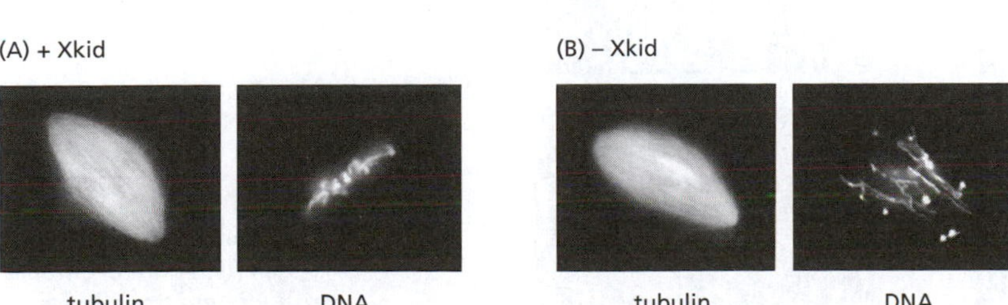


**Figure 17–29** Assembly of mitotic spindles in extracts (Problem 17–97). (A) In the presence of Xkid. (B) In the absence of Xkid. The spindle microtubules were made visible with fluorescent tubulin and the DNA, with a fluorescent stain.

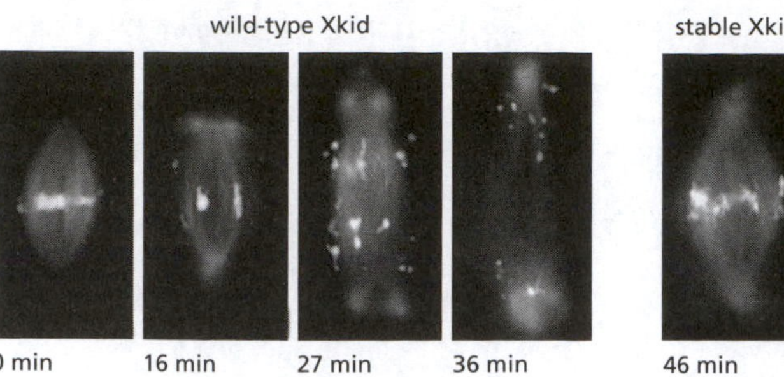


**Figure 17–30** Anaphase in the presence of wild-type and stable Xkid (Problem 17–98). Metaphase spindles were assembled and then anaphase was initiated by addition of $Ca^{2+}$ at time zero. *Faint fluorescence* marks the position of microtubules and *bright spots* mark the position of the chromosomes.

**17–98** At the transition from metaphase to anaphase Xkid is normally degraded. Is the removal of Xkid critical for anaphase to proceed? To answer this question, you assemble metaphase spindles in the presence of wild-type Xkid or a functional, but nondegradable (stable) form of Xkid. You then initiate anaphase and examine the spindles at various times thereafter. Spindles assembled in the presence of wild-type Xkid progress through anaphase normally with the chromosomes reaching the poles within 36 minutes (Figure 17–30). By contrast, the chromosomes on spindles assembled in the presence of stable Xkid remain at the equator even after 46 minutes (Figure 17–30). Examination of the chromosomes shows that in all cases the sister chromatids have separated. Why do you suppose anaphase does not proceed in the absence of Xkid degradation?

**17–99** Bipolar spindles assemble in the absence of centrosomes in *Sciara* when development occurs parthenogenetically. (Normally, the sperm delivers a centrosome to the egg along with a haploid genome.) These spindles look normal except that they lack astral microtubules (Figure 17–31). They can also support the rapid, synchronous series of early nuclear divisions that occur in a common cytoplasm in *Sciara* (similar to the early nuclear divisions in *Drosophila*). The products of these early mitotic events, however, are clearly different in normal embryos and parthenogenetic ones. The nuclei in normal embryos are well distributed in the common cytoplasm (Figure 17–31A), but those in parthenogenetic embryos are clustered together (Figure 17–31B). Can you suggest a way in which astral microtubules might function to keep nuclei well distributed in the common cytoplasm?

**17–100** At the transition from metaphase to anaphase, M-Cdk is inactivated and chromosomes begin to separate into sister chromatids. M-Cdk is inactivated by the anaphase-promoting complex (APC/C), which destroys the cyclin-B

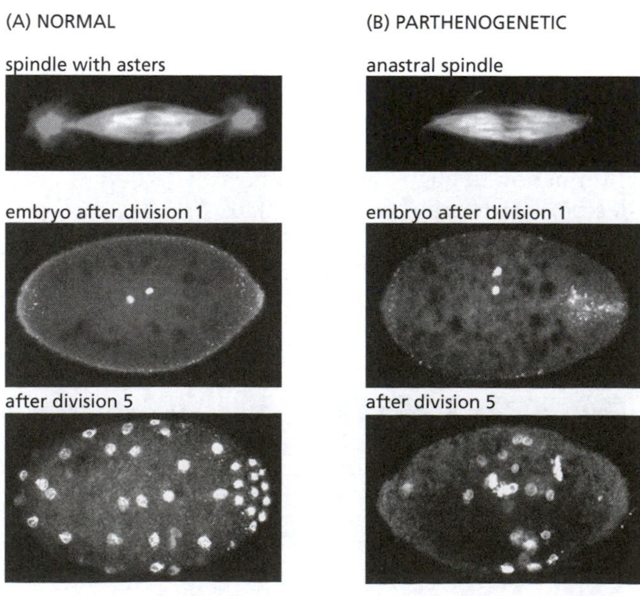


**Figure 17–31** Bipolar spindles and nuclear divisions in *Sciara* (Problem 17–99). (A) In normal embryos. (B) In parthenogenetic embryos.

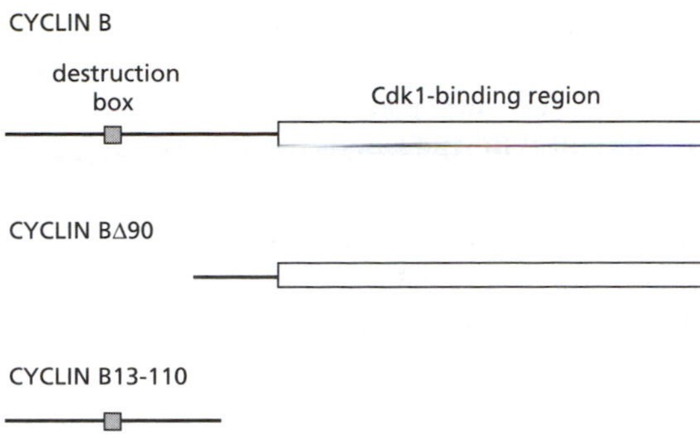


**Figure 17–32** Cyclin B and two mutants (Problem 17–100).

component of M-Cdk, eliminating its kinase activity. You want to know how the separation of sister chromatids is related to M-Cdk inactivation. To answer this question, you make cell-free extracts from unfertilized frog eggs. When nuclei are added to the extract, they spontaneously form spindles with condensed chromosomes aligned on the metaphase plate. Anaphase and the separation of sister chromatids can be triggered by addition of $Ca^{2+}$, which activates APC/C and turns off M-Cdk.

To investigate the control of sister chromatid separation, you make use of two mutant forms of cyclin B (Figure 17–32). Cyclin BΔ90 is missing the destruction box, a sequence of amino acids required for inactivation by APC/C, but it retains its ability to bind to Cdk1 and make functional M-Cdk. Cyclin B13-110 retains the destruction box, but cannot bind to Cdk1. When either protein is added in excess to the extract, M-Cdk activity remains high after addition of $Ca^{2+}$. The two proteins differ, however, in their effects on chromatid separation. In the presence of cyclin BΔ90, sister chromatids separate normally; in the presence of cyclin B13-110, sister chromatids remain linked.

A. Why does M-Cdk remain active in the presence of $Ca^{2+}$ when cyclin BΔ90 is added to the extract?
B. Why does M-Cdk remain active when cyclin B13-110 is added to the extract?
C. How is the separation of sister chromatids related to M-Cdk inactivation? Do sister chromatids separate because a linker protein must be phosphorylated by M-Cdk in order for it to hold the chromatids together? Or do chromatids separate because APC/C degrades the linker protein?

**17–101** Scc1 protein is cleaved by the protease separase. By altering the sites in yeast Scc1 that are recognized and cleaved by separase, you have created an uncleavable form. To analyze the cellular consequences of uncleavable Scc1, you release cells from a $G_1$ cell-cycle block and at the same time turn on expression of uncleavable Scc1 from a regulatable promoter. As expected, sister chromatids do not separate (as shown by the technique described in Problem 17–49). You are puzzled, however, by their progress through the cell cycle, as analyzed by flow cytometry (Figure 17–33). Up to about 3 hours, cells with uncleavable Scc1 are indistinguishable from wild-type cells, but after that they are very different.

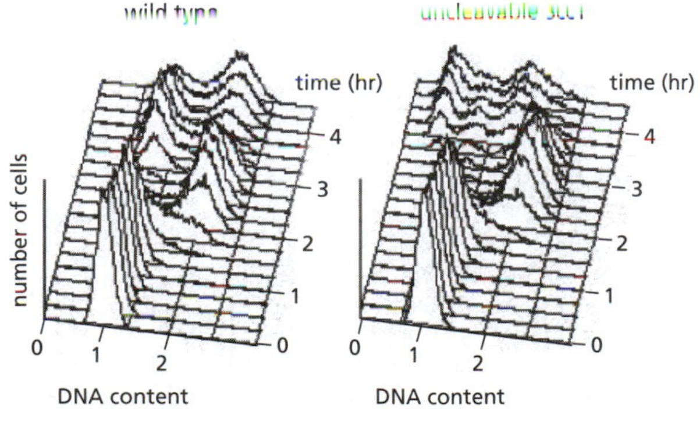


**Figure 17–33** Analysis of cell-cycle progression in wild-type cells and cells expressing an uncleavable form of Scc1 (Problem 17–101). Samples were taken 15-minute intervals after release from $G_1$ block. Successive samples have 'stacked' for clarity.

A. Did you expect cells with uncleavable Scc1 would be the same as wild-type cells up the 3-hour time point? Why or why not?
B. How do cells that express uncleavable Scc1 differ from wild-type cells after the 3-hour time point? Can you suggest an explanation for their behavior?

**17–102** By the turn of the 20th century, it was clear that chromosomes were the carriers of hereditary information. But a fundamental question remained: Does each chromosome carry the total hereditary information or does each one carry a different portion of the hereditary information? According to the first view, multiple chromosomes were required to raise the total quantity of hereditary material above the threshold value needed for proper development. According to the second view, multiple chromosomes were needed so that all portions of the hereditary information would be represented. This question was answered definitively in a classic series of experiments carried out by Theodor Boveri from 1901 to 1905.

As an experimental system, Boveri followed the development of sea urchin eggs that had been fertilized by two sperm, a frequent occurrence during artificial fertilization when large quantities of sperm are added. In contrast to the normal bipolar mitotic spindle and division into two cells, eggs fertilized by two sperm form a tetrapolar mitotic spindle and then divide into four cells. The three sets of chromosomes—one from the egg and two from the sperm—are distributed randomly among the four spindles as shown for four chromosomes in Figure 17–34A. If the dispermic eggs are gently shaken immediately after fertilization, one of the spindle poles often fails to form, resulting in a tripolar mitotic spindle followed by division into three cells (Figure 17–34B).

The species of sea urchin that Boveri studied, *Echinus microtuberculatus*, has a diploid chromosome number of 18, but will develop normally to a pluteus larva—a free-swimming stage in sea urchin development—with a haploid number of 9. Boveri reasoned that for a tripolar or tetrapolar egg to develop to a normal pluteus, each cell resulting from the initial three-way or four-way division would need to have either 9 total chromosomes or 9 different chromosomes—depending on which view of chromosome inheritance was correct. Boveri followed the development of 695 tripolar eggs and found that 58 developed into a normal pluteus. Among 1170 tetrapolar eggs, none formed a normal pluteus.

A. To set up the expectations for this experiment, it is instructive to consider first a hypothetical case in which the egg and two sperm each contribute a single chromosome. For tripolar spindles, there are 10 different arrangements of three chromosomes on three spindles. Sketch these 10 arrangements. Upon separation of sister chromatids and cell division, how many of these arrangements would be expected to produce three cells that each carry at least one chromosome? One arrangement and its division into three cells is shown in Figure 17–34C. (If you want to try your hand at tetrapolar spindles, there are 20 arrangements.)
B. If the total number of chromosomes were the critical factor, the number of tripolar eggs in which each cell gets the minimum number of chromosomes would be the same as that calculated in part A, regardless of the number of

(A) TETRAPOLAR MITOSIS

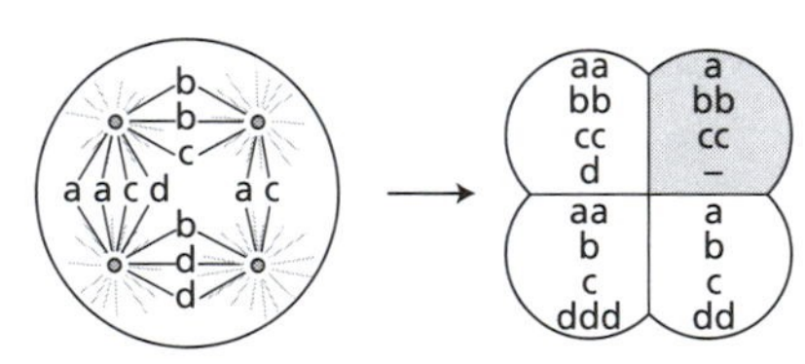


(B) TRIPOLAR MITOSIS: 4 CHROMOSOMES

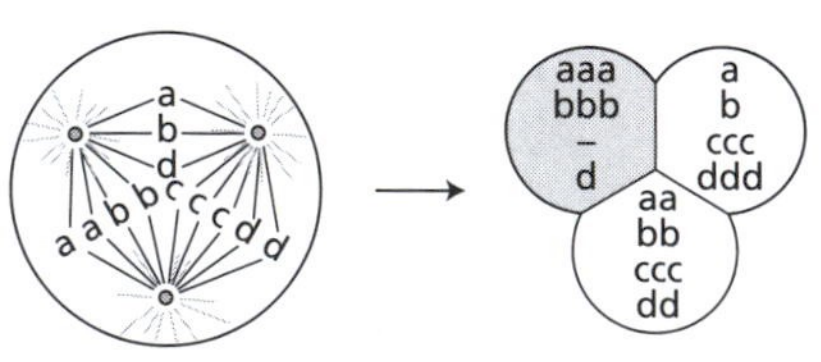


(C) TRIPOLAR MITOSIS: 1 CHROMOSOME

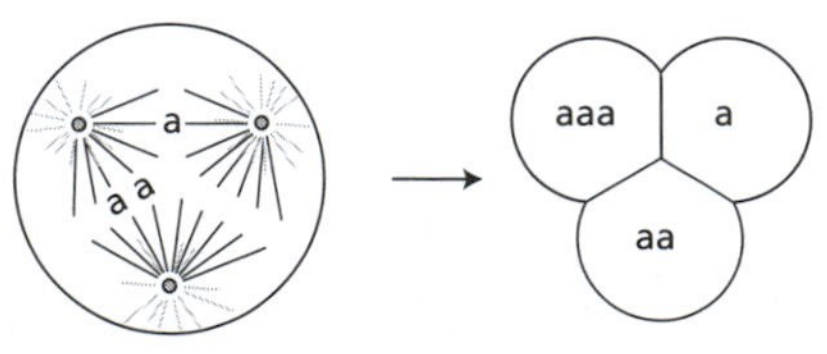


**Figure 17–34** Distributions of chromosomes among multiple spindles (Problem 17–102). (A) Example of a random arrangement of three sets of four chromosomes among the four mitotic spindles in a tetrapolar egg. Note that in this example, all four cells have at least four chromosomes. If total number were the critical aspect of chromosomes in heredity, these cells should develop into a pluteus. On the other hand, if each of the four cells have to have at least one copy of each different chromosome, these cells would not be expected to form a pluteus because one cell (*shaded*) is missing a chromosome. (B) Example of a random arrangement of three sets of four chromosomes among the three mitotic spindles in a tripolar egg. Once again, if total number of chromosomes were critical, these cells should develop into a pluteus; but if chromosome type is critical, they will not since the *shaded* cell is missing one chromosome. (C) Example of one arrangement—out of 10 possible—of three chromosomes on a tripolar spindle.

chromosomes. By contrast, if the distribution of chromosomes were the critical factor, the number of tripolar eggs that generate three cells, each with at least one copy of each different chromosome, would be expected to decrease with increasing numbers of chromosomes. The number of plutei should decrease according to the fraction calculated in part A raised to the power of 9 (the number of different chromosomes). Which hypothesis—total number of chromosomes or distribution of chromosomes—do Boveri's observations support?

# CYTOKINESIS

TERMS TO LEARN

| | | |
|---|---|---|
| cell plate | midbody | preprophase band |
| contractile ring | phragmoplast | syncytium |
| cytokinesis | | |

## DEFINITIONS

Match the definition below with its term from the list above.

**17–103** Cytoplasm containing many nuclei enclosed by a single plasma membrane.

**17–104** Structure formed at the end of cleavage that can persist for some time as a tether between the two daughter cells.

**17–105** Division of the cytoplasm of a plant or animal cell into two.

**17–106** Structure made of microtubules and actin filaments that forms in the prospective plane of division of a plant cell and guides formation of the cell plate.

**17–107** Circular band containing actin and myosin that forms under the surface of animal cells undergoing cell division and contracts to pinch the two daughter cells apart.

## TRUE/FALSE

Decide whether each of these statements is true or false, and then explain why.

**17–108** Cytokinesis follows mitosis as inevitably as night follows day.

**17–109** Whether cells divide symmetrically or asymmetrically, the mitotic spindle positions itself centrally in the cytoplasm.

## THOUGHT PROBLEMS

**17–110** What are the two distinct cytoskeletal machines that are assembled to carry out the mechanical processes of mitosis and cytokinesis in animal cells?

**17–111** You have obtained an antibody to myosin that prevents the movement of myosin molecules along actin filaments. If this antibody were injected into cells, would you expect the movement of chromosomes at anaphase to be affected? How would you expect antibody injection to affect cytokinesis? Explain your answers.

**17–112** If a cell just entering mitosis is treated with nocodazole, which destabilizes microtubules, the nuclear envelope breaks down and chromosomes condense, but no spindle forms and mitosis arrests. In contrast, if such a cell is treated with cytochalasin D, which destabilizes actin filaments, mitosis proceeds normally, but a binucleate cell is generated and proceeds into $G_1$. Explain the basis for the different outcomes of these treatments with cytoskeleton inhibitors. What do these results tell you about cell-cycle checkpoints in M phase?

**17–113** Sketch the way a new cell wall forms between the two daughter cells when a plant cell divides. In particular, show where the membrane proteins of the Golgi-derived vesicles end up, indicating the final location of the part of a protein in the Golgi vesicle membrane that is exposed to the interior of the vesicle.

## CALCULATIONS

**17–114** When cells divide after mitosis, their surface area increases—a natural consequence of dividing a constant volume into two compartments. The increase in surface requires an increase in the amount of plasma membrane. One can estimate this increase by making certain assumptions about the geometry of cell division. Assuming that the parent cell and the two progeny cells are spherical, one can apply the familiar equations for the volume $[(4/3)\pi r^3]$ and surface area $(4\pi r^2)$ of a sphere.

A. Assuming that the progeny cells are equal in size, calculate the increase in plasma membrane that accompanies cell division. (Although this problem can be solved algebraically, you may find it easier to substitute real numbers. For example, let the volume of the parent cell equal 1.) Do you think that the magnitude of this increase is likely to cause a problem for the cell? Explain your answer.

B. During early development, many fertilized eggs undergo several rounds of cell division without any overall increase in total volume. For example, *Xenopus* eggs undergo 12 rounds of division before growth commences and the total cell volume increases. Assuming once again that all cells are spherical and equal in size, calculate the increase in plasma membrane that accompanies development of the early embryo, going from one large cell (the egg) to 4096 small cells (12 divisions).

## DATA HANDLING

**17–115** Cytokinesis—the actual process of cell division—has attracted theorists for well over 100 years. It has been said that all possible explanations of cytokinesis have been proposed; the problem is to decide which one is correct. Consider the following three hypotheses:

1. *Chromosome signaling:* When chromosomes split at anaphase, they emit a signal to the nearby cell surface to initiate furrowing.
2. *Polar relaxation:* Asters relax the tension in the region of the cell surface nearest the spindle poles (the polar region), allowing the region of the membrane farthest from the poles (the equatorial plane) to contract and initiate furrowing.
3. *Aster stimulation:* The asters stimulate contraction in the region of the cell surface where oppositely oriented spindle fibers overlap (that is, the equatorial region), thereby initiating furrowing.

These hypotheses have been tested in a number of ways. One particularly informative experiment involved pressing a glass ball onto the center of a dividing sand dollar egg so as to deform it into a torus (donut shape). As illustrated in Figure 17–35, the egg divided into a single sausage-shaped cell at the first division; at the second division, it divided into four cells.

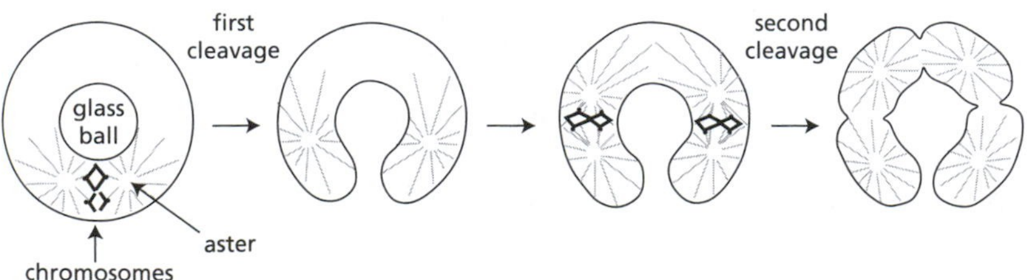


**Figure 17–35** First and second divisions in a torus-shaped sand dollar egg (Problem 17–115).

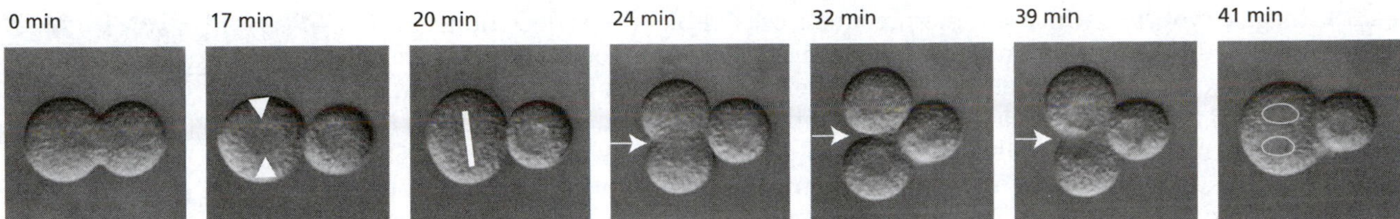


**Figure 17–36** Video recording of a two-cell *C. elegans* embryo treated with brefeldin A (Problem 17–116). *Triangles* mark spindle poles, the *vertical line* indicates spindle orientation, *arrows* show the cleavage furrow forming and regressing. The two nuclei in the final binucleate cell are *circled*.

A. What does the chromosome-signaling hypothesis predict for the result of this experiment? Do the predictions match the experimental observations?
B. What does the polar-relaxation hypothesis predict for the experimental outcome? Do the predictions match the experimental observations?
C. What does the aster-stimulation hypothesis predict? Do the predictions match the experimental observations?

**17–116** At telophase, the cleavage furrow narrows to form the midbody, which contains the remains of the central spindle. This fine intercellular bridge persists for a time before the connection is broken and the membranes reseal to complete the separation of the daughter cells. Little is known about this final phase of cytokinesis in animal cells beyond a requirement for microtubules and possibly centrosomes. Curiously, if exocytosis is blocked by treatment with brefeldin A, cleavage furrows form normally and seem complete, but then they regress to give a binucleate cell (Figure 17–36). Propose a mechanism that links vesicle secretion and microtubules to the final step in cytokinesis.

**17–117** Globoid cell leukodystrophy (GLD, also known as Krabbe's disease) is a hereditary metabolic disorder, characterized morphologically by distinctive multinucleated globoid cells in the white matter of the brain. Deficiency of an enzyme of sphingolipid catabolism leads to accumulation of psychosine in the brain (Figure 17–37A). Psychosine binds to a G protein-coupled receptor that is expressed in only a few cell types. To test whether there might be a relationship between psychosine, its receptor, and multinucleate cells, you express the psychosine receptor in cells that normally lack it, and measure the effects of psychosine treatment by FACS (fluorescence-activated cell sorting) analysis (Figure 17–37B).

Do these results support the idea that psychosine acts through its receptor to inhibit cytokinesis? Explain your reasoning.

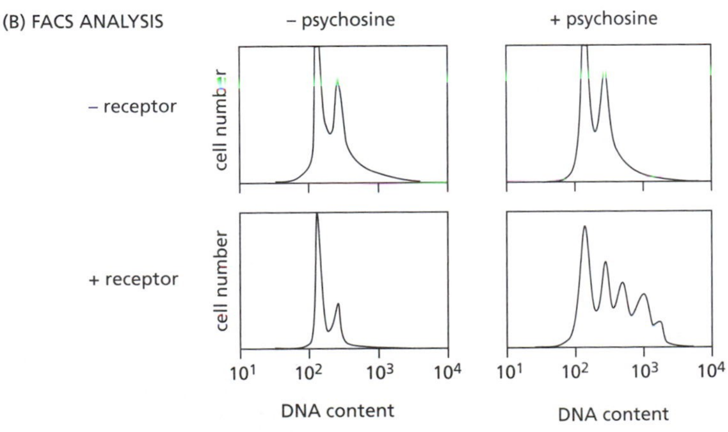

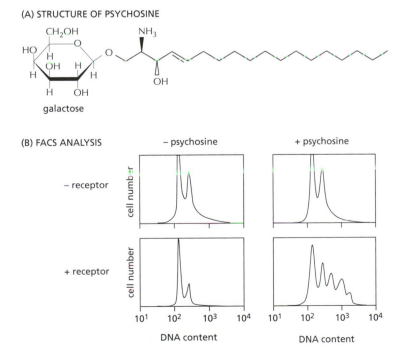

**Figure 17–37** Analysis of role of psychosine in generation of multinucleate cells (Problem 17–117). (A) Structure of psychosine. (B) Effects of psychosine on cells that do or do not express the psychosine receptor. FACS analysis measures the DNA content of individual cells using a fluorescent DNA dye.

**17–118** Megakaryocytes, which are the precursor cells of blood platelets, undergo a unique differentiation program, becoming polyploid through repeated cycles of DNA synthesis without concomitant cell division. Such cells contain some 4 to 128 times the normal DNA content in a single large nucleus. Ultimately, mature megakaryocytes begin to bud off platelets as shown in Figure 17–38. Careful observations of individual precursor cells that were stimulated to undergo polyploidization show the sequence of events in Figure 17–39. How do these events differ from the normal sequence in cell division? At what stage in M phase do these cells deviate from normal cells? What sorts of molecular differences might you expect to find among the components involved in M phase in these cells versus normal cells?

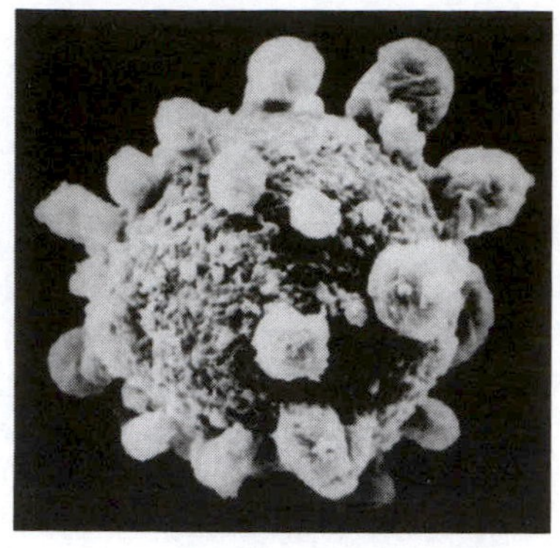

**Figure 17–38** A megakaryocyte budding off platelets (Problem 17–118).

## CONTROL OF CELL DIVISION AND CELL GROWTH

| TERMS TO LEARN | | |
|---|---|---|
| ATM | mitogen | replicative cell senescence |
| ATR | Myc | retinoblastoma protein (Rb) |
| E2F protein | p53 | telomerase |
| $G_0$ | Ras | telomere |
| growth factor | | |

### DEFINITIONS

Match the definition below with its term from the list above.

**17–119** A specialized, nondividing state that cells enter by partly disassembling their cell-cycle control system and exiting from the cell cycle.

**17–120** Extracellular substance that stimulates cell growth.

**17–121** End of a chromosome, associated with a characteristic DNA sequence that is replicated in a special way.

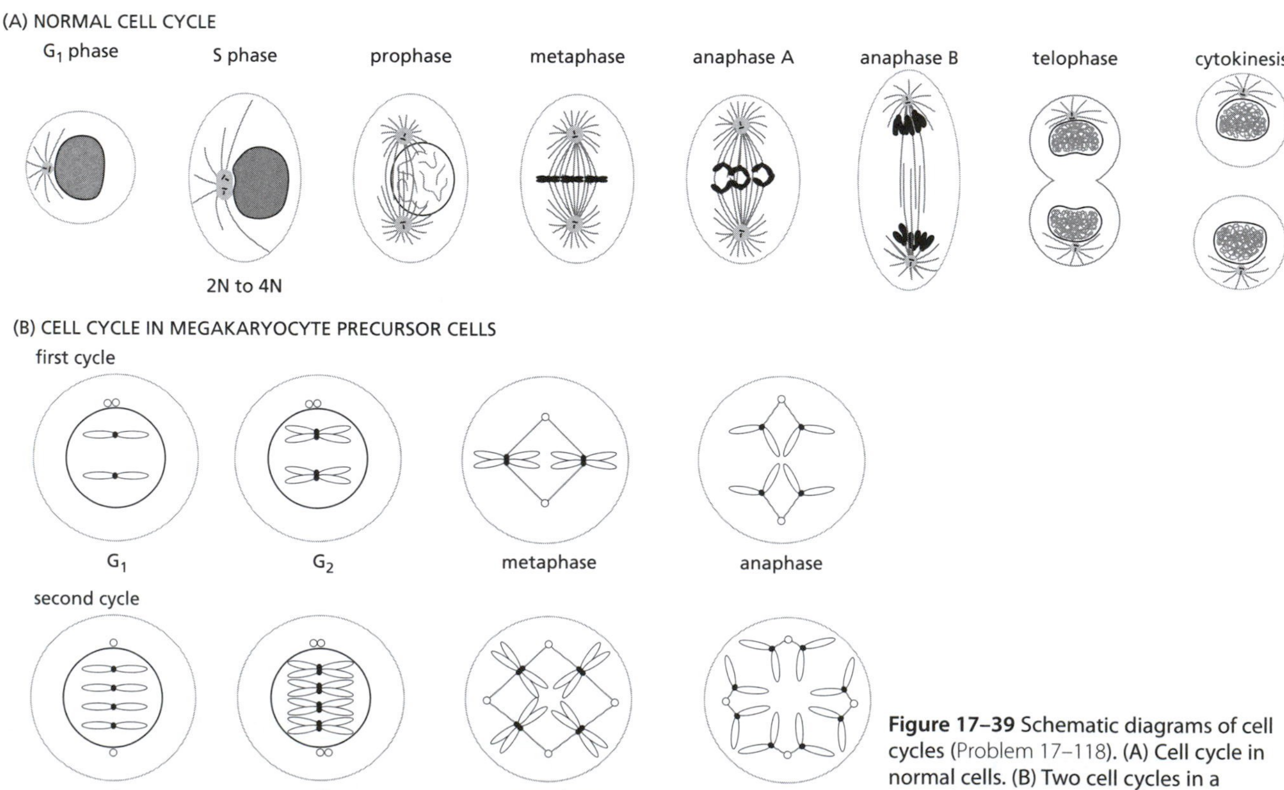


**Figure 17–39** Schematic diagrams of cell cycles (Problem 17–118). (A) Cell cycle in normal cells. (B) Two cell cycles in a megakaryocyte precursor cell.

**17–122** Phenomenon observed in primary cell cultures as they age, in which cell proliferation slows down and finally halts.

**17–123** Extracellular substance that stimulates cell division.

**17–124** Gene regulatory factor that is activated by $G_1$-Cdk complexes in animal cells and binds to specific DNA sequences in the promoters of gene that encode proteins required for S-phase entry.

**17–125** The protein kinase that is initially activated in response to x-ray induced DNA damage and is defective in patients with ataxia telangiectasia.

## TRUE/FALSE

Decide whether each of these statements is true or false, and then explain why.

**17–126** Serum deprivation causes proliferating cells to stop where they are in the cell cycle and enter $G_0$.

**17–127** Budding yeast and mammalian cells respond to DNA damage in the same way: they transiently arrest their cell cycles to repair the damage and if repair cannot be completed, they resume their cycles despite the damage.

**17–128** If we could turn on telomerase activity in all our cells, we could prevent aging.

## THOUGHT PROBLEMS

**17–129** How do mitogens, growth factors, and survival factors differ from one another?

**17–130** For each of the following, decide whether such cells exist in humans and, if they do, give examples.

A. Cells that do not grow and do not divide.
B. Cells that grow, but do not divide.
C. Cells that divide, but do not grow.
D. Cells that grow and divide.

**17–131** Why do you suppose cells have evolved a special $G_0$ state to exit the cell cycle, rather than just stopping in $G_1$ at a $G_1$ checkpoint?

**17–132** PDGF is encoded by a gene that can cause cancer when expressed inappropriately. Why then do cancers not arise at wounds when PDGF is released from platelets?

**17–133** One important biological effect of a large dose of ionizing radiation is to halt cell division.

A. How does a large dose of ionizing radiation stop cell division?
B. What happens if a cell has a mutation that prevents it from halting cell division after being irradiated?
C. What might be the effects of such a mutation if the cell was not irradiated?
D. An adult human who has reached maturity will die within a few days of receiving a radiation dose large enough to stop cell division. What does this tell you (other than that one should avoid large doses of radiation)?

**17–134** What do you suppose happens in mutant cells with the following defects?

A. Cannot degrade M-phase cyclins.
B. Always express high levels of p21.
C. Cannot phosphorylate Rb.

**17–135** Replicative cell senescence occurs at a characteristic number of population doublings, typically about 40 for cells taken from normal human tissue. This observation suggests that in some way individual cells can 'count' the number of times they have divided. How does the structure of telomeres figure into a cell's calculations?

**Table 17–4 Effects of extracellular factors on the timing of entry into S phase (Problem 17–138).**

| EXPERIMENT | ORDER OF ADDITION | | | S PHASE ENTRY |
|---|---|---|---|---|
| | 1 | 2 | 3 | |
| 1 | EGF | PDGF | IGF1 | 12 hours |
| 2 | EGF | IGF1 | PDGF | 12 hours |
| 3 | PDGF | EGF | IGF1 | 1 hour |
| 4 | PDGF | IGF1 | EGF | 6 hours |
| 5 | IGF1 | EGF | PDGF | 12 hours |
| 6 | IGF1 | PDGF | EGF | 6 hours |

Cells were treated for 6 hours with the indicated factors in the order listed. They were thoroughly washed to remove one factor before the next one was added. After the regimen of factor pretreatment, complete medium with $^3$H-thymidine was added and the time when labeled nuclei appeared was determined.

**17–136** Active telomerase comprises two components—a protein and an RNA—that are encoded by distinct genes. Most human somatic cells do not have telomerase activity. If the gene for the protein component of telomerase, hTRT, is expressed in such cells, however, they regain telomerase activity. How is it that expression of just one component can give telomerase activity?

**17–137** Liver cells proliferate both in patients with alcoholism and in patients with liver tumors. What do you suppose are the differences in the mechanisms by which cell proliferation is induced in these diseases?

## DATA HANDLING

**17–138** Vertebrate cells pause in the $G_1$ phase of the cell cycle until conditions are appropriate for their entry into S phase. Some of the requirements for the passage through $G_1$ have been defined using mouse 3T3 cells. In the absence of serum, these cells do not enter S phase. When serum is added to a culture of such arrested cells, they progress through $G_1$ and begin to enter S phase 12 hours later. The serum requirement can be met by supplying three extracellular factors: PDGF, EGF, and IGF1. When these factors are mixed with appropriate nutrients and added to quiescent cells, the cells begin to enter S phase 12 hours later. If any one of the factors is left out, the cells do not enter S phase.

Do all three factors have to be present at the same time? Is their stimulation of cells independent of one another? Or do they stimulate cells in an ordered sequence? To address these questions, you pretreat cells with the factors in a defined order and then add complete medium (containing serum and nutrients) in the presence of $^3$H-thymidine. At various times thereafter, you fix cells and subject them to autoradiography. You define the time of appearance of the first labeled nuclei as the time of entry into S phase. The results of these experiments are given in Table 17–4.

Do the cells require these factors simultaneously, independently, or in an ordered sequence? Explain your answer.

**Problem 15–127** examines the cell signaling pathways that are triggered by the binding of PDGF to the PDGF receptor.

**17–139** A classic paper that first characterized the EGF receptor made use of the mouse fibroblast A-431 cell line, which fortuitously carries enormously increased numbers of EGF receptors. Key observations were made in the following experiments.

1. A plasma membrane preparation contained many proteins as shown on the SDS gel in Figure 17–40A. When $^{125}$I-EGF was added in the presence of a protein cross-linking agent, two proteins were labeled (Figure 17–40B, lane 1). When excess unlabeled EGF was included in the incubation mixture, the labeled band at 170 kd disappeared (lane 2).
2. If the membrane preparation was incubated with $\gamma$-$^{32}$P-ATP, several proteins, including the 170-kd protein, became phosphorylated. Inclusion

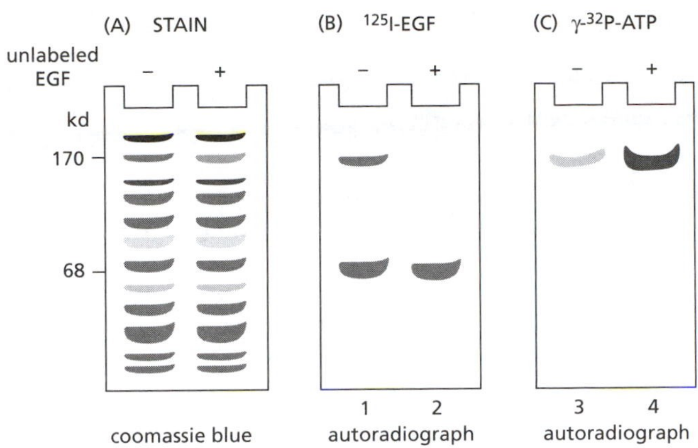


**Figure 17–40** Analysis of the EGF receptor (Problem 17–139). (A) SDS gel of a membrane preparation from A-431 cells. (B) Autoradiograph of an SDS gel of a membrane preparation from A-431 cells incubated with $^{125}$I-EGF and a protein cross-linking agent in the presence and absence of excess unlabeled EGF. (C) Autoradiograph of an SDS gel of antibody-precipitated EGF-receptor incubated with $\gamma$-$^{32}$P-ATP in the presence and absence of EGF.

of EGF in the incubation mixture significantly stimulated phosphorylation.

3. When antibodies specific for the 170-kd protein were used to precipitate the protein, and the incubation with $\gamma$-$^{32}$P-ATP was repeated with the precipitate, the 170-kd protein was phosphorylated in an EGF-stimulated reaction (Figure 17–40C, lanes 3 and 4).
4. When the antibody-precipitated protein was first run on the SDS gel and then renatured in the gel, subsequent incubation with $\gamma$-$^{32}$P-ATP in the presence and absence of EGF yielded the same pattern shown in Figure 17–40C.

A. Which of these experiments demonstrate most clearly that the 170-kd protein is the EGF receptor?
B. Is the EGF receptor a substrate for an EGF-stimulated protein kinase? How do you know?
C. Which experiments show that the EGF receptor is a protein kinase?
D. Is it clear that the EGF receptor is a substrate for its own protein kinase activity?

**17–140** When cells in $G_0$ are exposed to mitogenic growth factors, they enter S phase about 20 hours after stimulation, as can be detected by incorporation of the nucleoside analog BrdU. If antibodies to cyclin D are microinjected into cells up to 12 hours after adding mitogenic growth factors, very few of the injected cells incorporate BrdU. By contrast, cyclin D-specific antibodies have little effect on BrdU incorporation, when injected more than 14 hours after exposure to factors (Figure 17–41). What critical event in $G_1$ do antibodies against cyclin D block? Why do antibodies against cyclin D have no effect after 14 hours?

**17–141** You have found a new way to study radiation-sensitive yeast mutants. By culturing cells on a thin layer of agar on a microscope slide, you can record the fate of individual cells. When you irradiate wild-type cells with x-rays, which cause chromosome breaks, and follow their growth for the next 10 hours, you find that most of the cells arrest temporarily at the large-bud (dumbbell) stage, but about half the cells eventually recover and form small viable colonies after 10 hours (Figure 17–42). The fraction that is still arrested at the dumbbell stage after 10 hours is equal to the fraction of nonviable cells (Table 17–5).

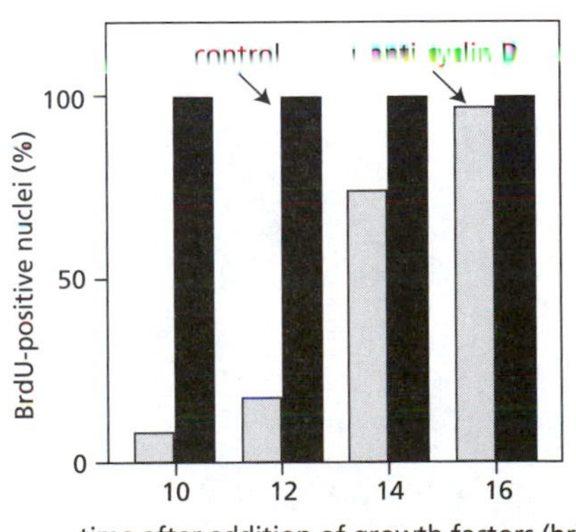


**Figure 17–41** Percentage of control cells and cells injected with cyclin D antibodies that had entered S phase as detected by incorporation of BrdU (Problem 17–140). After addition of mitogenic growth factors, cells were injected with cyclin D antibodies (or not) at the times indicated, and then were assayed for BrdU incorporation 26 hours after addition of the mitogen, a time when uninhibited cells should have been well into S phase. *Black bars* indicate results with control cells that were not injected with cyclin D antibodies. *Gray bars* indicate results with cells that were injected with cyclin D antibodies.

**Table 17–5 Fractions of arrested and nonviable cells after x-ray treatment** (Problem 17–141).

| STRAIN | ARRESTED AT 10 HOURS (%) | NONVIABLE (%) |
|---|---|---|
| Wild type | 50 | 50 |
| *Rad52* | 90 | 95 |
| *Rad9* | 20 | 70 |

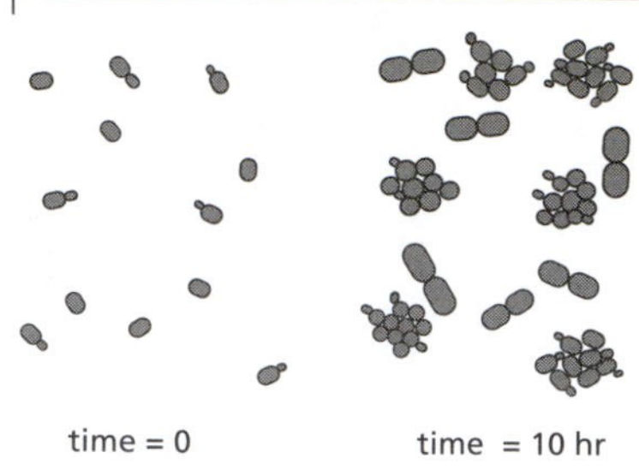


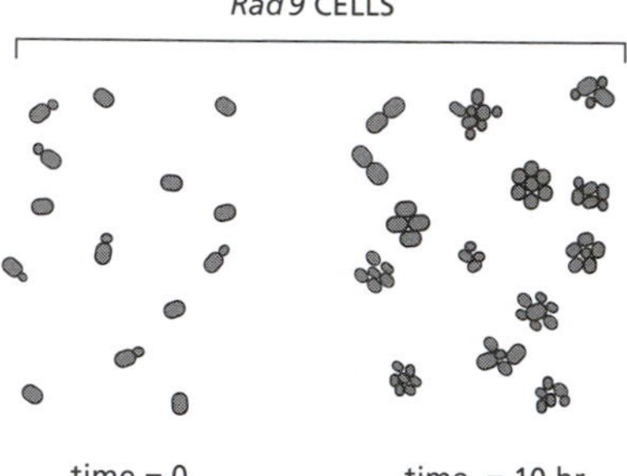


**Figure 17–42** Time-lapse pictures of wild-type and *Rad9* cells at zero and 10 hours after x-irradiation (Problem 17–141).

You repeat these experiments with seven different radiation-sensitive (*Rad*) mutants. For six of the mutants, you observe a similar equality of 10-hour arrested cells and nonviable cells, as shown in Table 17–5 for the *Rad52* mutant. Relative to wild-type cells, however, a higher fraction of *Rad* cells are nonviable (Table 17–5).

One of the seven *Rad* mutants has a strikingly different phenotype. Many fewer *Rad9* cells arrest even temporarily at the dumbbell stage, and after 10 hours only 20% are still arrested (Table 17–5). Although many of the cells divide once or twice, they mostly form nonviable microcolonies (Figure 17–42, Table 17–5).

A. For the wild-type cells, decide which cells in the population appear most likely to remain arrested at the dumbbell stage after 10 hours (Figure 17–42). Given that the cells used in these experiments are haploid and that x-ray-induced breaks are repaired by homologous recombination, decide which stage of the cell cycle the sensitive cells are in.

B. By staining with DNA-binding reagents and tubulin-specific antibodies, you show that cells in the dumbbell stage have a single nucleus and no visible spindle. Given what you know about cell-cycle checkpoints, in what stage of the cell cycle do you think the cells are arrested?

C. Why do half of the wild-type cells temporarily arrest at the dumbbell stage but then go on to form viable colonies after 10 hours?

D. Why do you suppose that so many more *Rad52* mutant cells (relative to wild-type cells) are arrested at the dumbbell stage after 10 hours?

E. Why do you suppose that so few *Rad9* mutant cells arrest even temporarily at the dumbbell stage? Why are so many of the *Rad9* cells nonviable?

F. Would you expect the fraction of nonviable *Rad9* cells to increase, decrease, or stay the same, if the cells were artificially delayed for a couple of hours, using a microtubule inhibitor that reversibly prevents spindle formation? Explain your reasoning?

**17–142** Fission yeast respond to damaged DNA and unreplicated DNA by delaying entry into mitosis. You want to know how the signals from damaged and nonreplicated DNA interact with the mitotic entry checkpoint. You screen a large number of mutant yeast strains and find six that do not delay mitosis in response to DNA damage, unreplicated DNA, or both, as shown in Table 17–6. Which of the signaling pathways shown in Figure 17–43 is supported by your data? On the pathway you choose, indicate where each of the mutant genes acts.

**Table 17–6 Mutant strains that affect the mitotic entry checkpoint** (Problem 17–142).

| | MITOTIC DELAY IN RESPONSE TO | |
|---|---|---|
| MUTANT STRAINS | DAMAGED DNA | UNREPLICATED DNA |
| *Rad24* | No | Yes |
| *Cdc2-3w* | Yes | No |
| *Hus1* | No | No |
| *Hus2* | No | No |
| *Rad1* | No | No |
| *Cdc2-F15* | Yes | No |

**17–143** The extracellular protein factor Decapentaplegic (Dpp) is critical for proper wing development in *Drosophila* (Figure 17–44A). It is normally expressed in a narrow stripe in the middle of the wing, along the anterior-posterior boundary. Flies that are defective for Dpp form stunted 'wings' (Figure 17–44B). If an additional copy of the gene is placed under control of a promoter that is active in the anterior part of the wing, or in the posterior part of the wing, a large mass of wing tissue composed of normal-looking cells is produced at the site of Dpp expression (Figure 17–44C and D). Does Dpp stimulate cell division, cell growth, or both? How can you tell? Give some examples of proteins that you might expect to be activated in response to Dpp stimulation.

**17–144** You have been led by a bizarre accident to a productive line of experimentation. You lost one of your contact lenses while transferring a line of tissue culture cells; a few days later you found the lens on the bottom of a petri dish. Interestingly, the cells attached to the lens were rounded up and very sparsely distributed, whereas those on the rest of the dish were flat and had grown to near confluency, with most cells touching their neighbors. Ah ha! you thought, perhaps this observation can be used to investigate the relationship between cell shape and growth control. As someone once said, "Chance favors the prepared mind."

The manufacturer graciously sends you a supply of the plastic—poly(HEMA)—from which your soft contact lenses were made. When an alcoholic mixture of poly(HEMA) is pipetted into a plastic culture dish, a thin, hard, sterile film of optically clear polymer remains bound to the plastic surface after the alcohol evaporates. Serial dilutions of the alcohol-polymer solution, introduced into each dish at a constant volume, result in decreasing thicknesses of the polymer film. The thinner the film, the more strongly cells adhere to the dish. Moreover, there is a gradual change in cell shape from round to flat with decreasing thickness of the film. Using the height of the cells as an indicator of cell shape (no mean technical feat), you demonstrate that there is a smooth relationship between cell shape and growth potential: the flatter the cells, the better they incorporate $^3$H-thymidine.

Now for the big question: Is density-dependent inhibition of cell growth mediated by changes in cell shape? You grow cells to different densities (different degrees of confluency) on normal plastic dishes and measure the height of the cells and their ability to incorporate $^3$H-thymidine. As shown in Table 17–7, the more confluent the cells, the greater their height and the lower their incorporation of $^3$H-thymidine. Is the decrease in growth due to the increase in cell crowding, or to the change in cell shape? To answer this question, you distribute cells at a low density on plates with poly(HEMA) films of different thickness, such that the height of the cells in the sparse cultures matches the height of cells in the various confluent cultures. Your measurements of $^3$H-thymidine incorporation in these sparse cultures are shown in Table 17–7.

Based on the results in Table 17–7, would you conclude that density-dependent inhibition of cell growth correlates completely, partially, or not at all with changes in cell shape? Explain your reasoning.

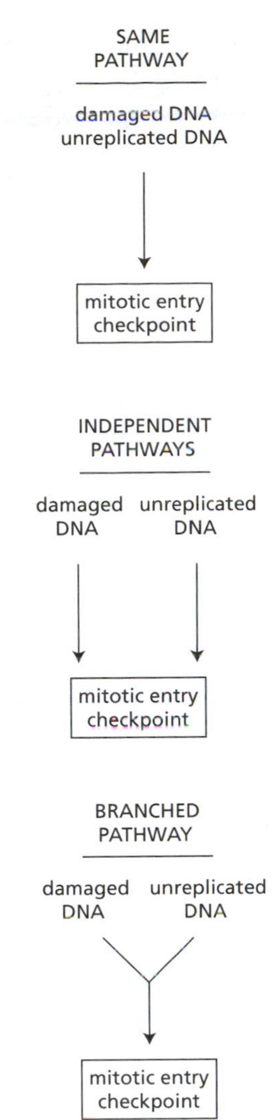


**Figure 17–43** Possible pathways by which signals from damaged DNA and unreplicated DNA might interact with the mitotic entry checkpoint (Problem 17–142).

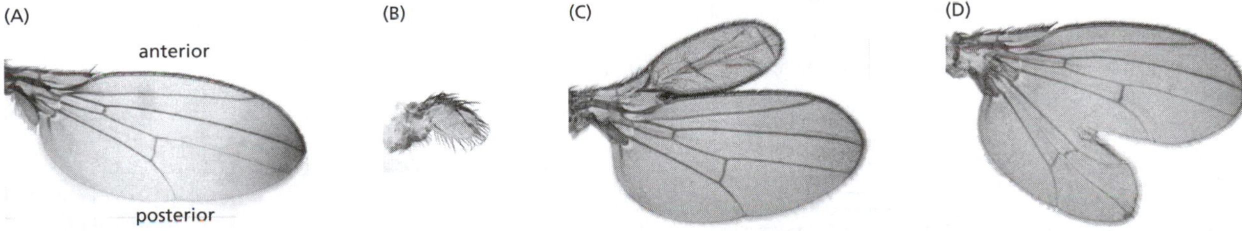

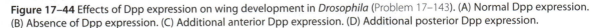

**Figure 17–44** Effects of Dpp expression on wing development in *Drosophila* (Problem 17–143). (A) Normal Dpp expression. (B) Absence of Dpp expression. (C) Additional anterior Dpp expression. (D) Additional posterior Dpp expression.

**Table 17–7 Incorporation of $^3$H-thymidine by cells grown on normal dishes and on poly(HEMA)-treated dishes** (Problem 17–144).

| TYPE OF DISH | CELL DENSITY (cells/dish) | CONFLUENCY | CELL HEIGHT (μm) | $^3$H INCORPORATION (cpm/dish) |
|---|---|---|---|---|
| Normal | 60,000 | subconfluent | 6 | 15,200 |
| Normal | 200,000 | confluent | 15 | 11,000 |
| Normal | 500,000 | confluent | 22 | 3500 |
| Poly(HEMA) | 30,000 | sparse | 6 | 7500 |
| Poly(HEMA) | 30,000 | sparse | 15 | 1500 |
| Poly(HEMA) | 30,000 | sparse | 22 | 210 |

**17–145** In the cell cycle of the fission yeast *S. pombe*, $G_1$ is very short and $G_2$ is very long (just the opposite of the situation in budding yeast). When a temperatire-sensitive (ts) *Wee1* mutant is grown at 25°C, it has a normal cell cycle and a normal size (Figure 17–45A). When shifted to 37°C, the mutant cells undergo a shortened first cell cycle because inactivation of Wee1 kinase reduces the threshold size for mitosis, generating smaller than normal cells (Figure 17–45B). Surprisingly, when these cells are grown at 37°C, the cell cycle is of normal length, but now with a long $G_1$ and a short $G_2$; nevertheless, small cells are still generated. What do you suppose would happen to ts *Wee1* cells if the cell cycle did not increase in length? Propose an explanation for how $G_1$ might be lengthened in ts *Wee1* cells grown at high temperature.

**17–146** Unlike mice and humans, *Drosophila* have a single gene for cyclin D and for its binding partner Cdk4, which greatly simplifies analysis and interpretation. To investigate the role of the cyclin D–Cdk4 complex in the cell cycle, you overexpress both genes using a genetic trick. You deposit the genes into the genome using a transposon and then activate them at various times in development by site-specific recombination. You have arranged it so that activation of these two genes is accompanied by activation of GFP, so that cells expressing cyclin D and Cdk4 also fluoresce bright green. When you examine the pattern of fluorescent cells in developing wings in a large number of flies, you find that the clones of marked cells average 16 cells in size, whereas clones of cells in control flies (expressing GFP alone) average just 12

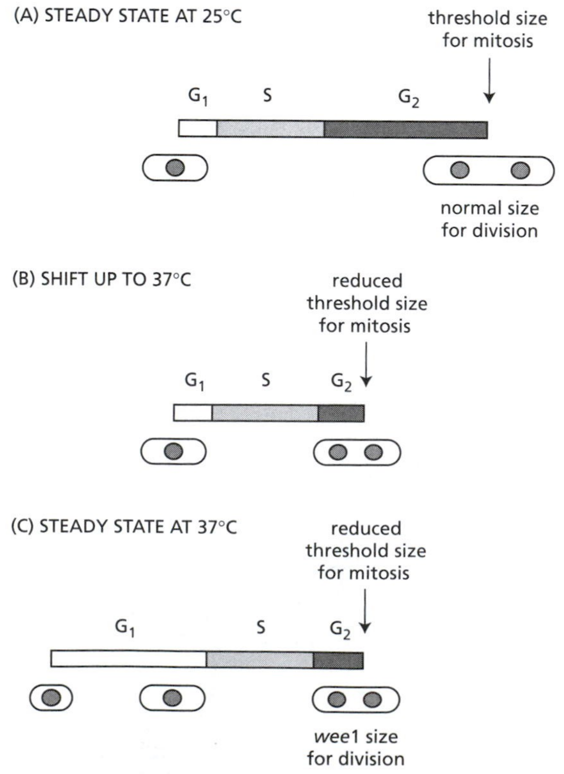


**Figure 17–45** Lengths of the cell cycle and sizes of ts *Wee1* cells grown at 25°C and at 37°C (Problem 17–145). (A) The cell cycle in ts *Wee1* cells grown continuously at 25°C. (B) The cell cycle in ts *Wee1* cells immediately after the shift from 25°C to 37°C. (C) The cell cycle in ts *Wee1* cells grown continuously at 37°C.

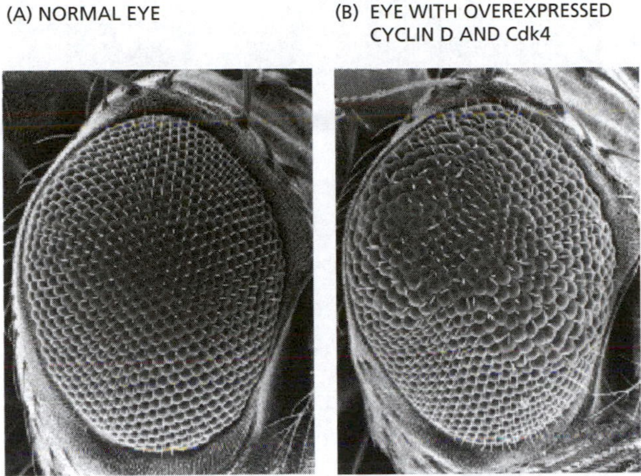


**Figure 17–46** Effects of overexpression of cyclin D and Cdk4 (Problem 17–147). (A) A normal eye. (B) An eye with a patch of cells overexpressing cyclin D and Cdk4.

cells in size. The cells that are overexpressing cyclin D and Cdk4 are exactly the same size as normal cells. (Parallel experiments in which one or the other gene was overexpressed gave results that matched the controls.)

From these results, would you conclude that overexpression of cyclin D and Cdk4 altered the growth rate of the cells, the duration of their cell cycles, or both? Explain your reasoning.

**17–147** To try to decide whether overexpression of cyclin D and Cdk4 in flies results primarily from an effect on growth rate or, alternatively, primarily from an effect on cell cycle, you examine terminally differentiated cells that are no longer dividing. You find that in *Drosophila* eyes, overexpression of cyclin D and Cdk4 causes cell enlargement in post-mitotic (nondividing) cells (Figure 17–46).

Do these results support a primary role of cyclin D and Cdk4 on growth, or do they support a primary effect on cell cycle? Explain your reasoning.

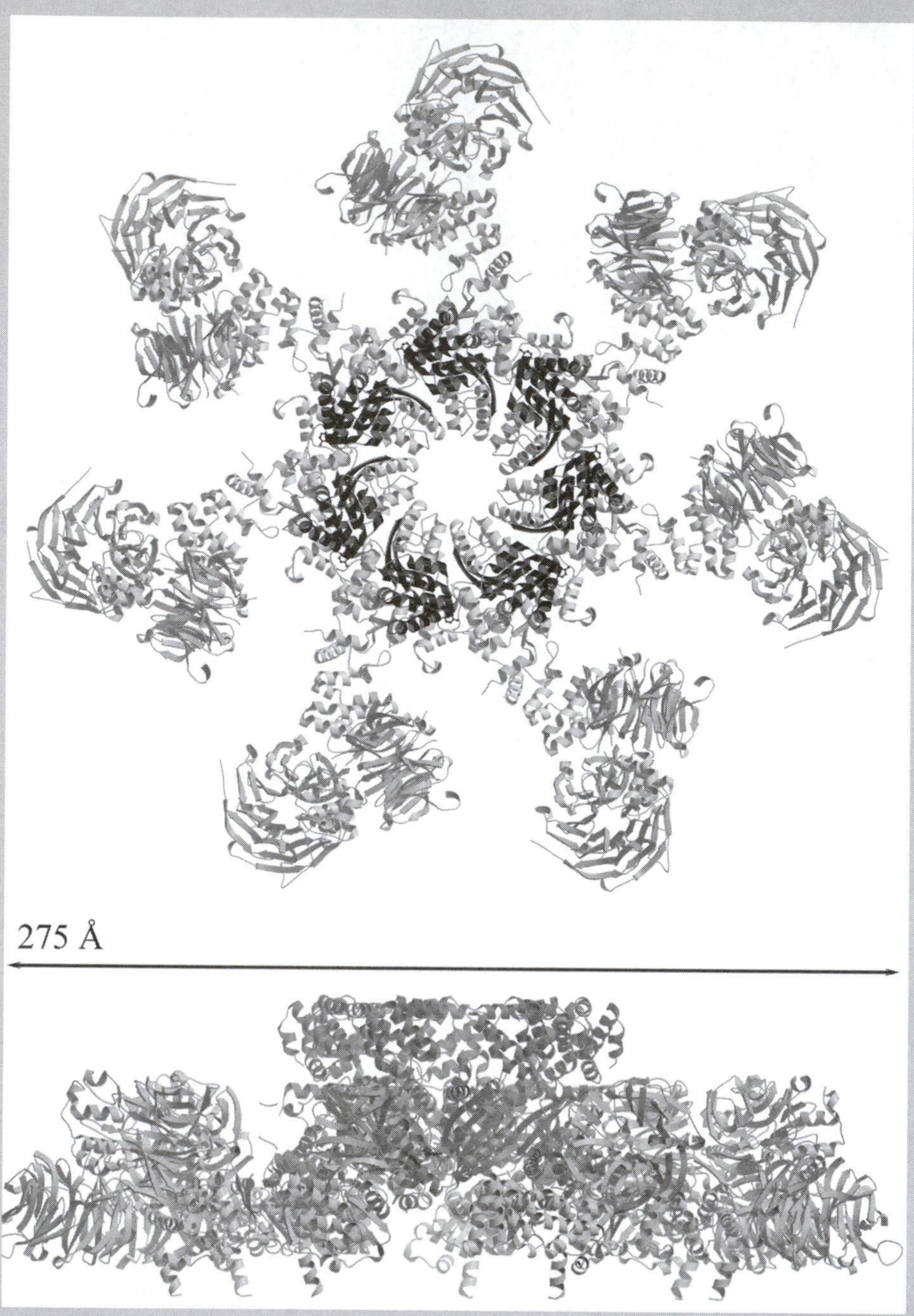


**Two views of a molecular model of the heptameric apoptosome (based on a combination of cryoelectron micrographs and X-ray crystallography).** See AV Diemand and AN Lupas, *J. Struct. Biol.* 156:230–243, 2006. Image provided by Alexander Diemand.

# Apoptosis

18

TERMS TO LEARN

anti-apoptotic Bcl2 protein
anti-IAP
Apaf1
apoptosis
apoptosome
Bak
Bax
Bcl2
Bcl-$X_L$
BH123 protein
BH3-only protein
caspase
cytochrome *c*
death-inducing signaling complex (DISC)
death receptor
executioner procaspase
extrinsic pathway
Fas
Fas ligand
IAP (inhibitor of apoptosis)
initiator procaspase
intrinsic pathway
procaspase
programmed cell death
survival factor

## DEFINITIONS

Match the definition below with its term from the list above.

**18–1** Protease that has a cysteine at its active site and cleaves its target proteins at specific aspartic acids.

**18–2** Wheel-like assembly composed of seven copies of the Apaf-1/cytochrome *c* complex.

**18–3** Form of cell death that leads to fragmentation of the DNA, shrinkage of the cytoplasm, membrane changes and cell death, without lysis or damage to neighboring cells.

**18–4** An assembly of several proteins, including intiator procaspases, on the cytosolic portion of the Fas death receptor.

**18–5** Apoptosis program that is triggered by the binding of an extracellular signal protein.

**18–6** Extracelluar signal molecule that inhibits apoptosis.

**18–7** Apoptosis program triggered by intracellular signals that cause release into the cytosol of proteins from the mitochondrial intermembrane space.

**18–8** Intracellular 'suicide' program that allows cells to kill themselves in a controlled way.

**18–9** Cell-surface molecule that triggers apoptosis when bound by an extracellular signal protein.

**18–10** Inactive form of protease that operates in apoptosis at the start of the proteolytic cascade.

## TRUE/FALSE

Decide whether each of these statements is true or false, and then explain why.

**18–11** In normal adult tissues, cell death usually balances cell division.

**18–12** Mammalian cells that do not have cytochrome *c* should be resistant to apoptosis induced by UV light.

## THOUGHT PROBLEMS

**18–13** In apoptosis the cell destroys itself from within and avoids leakage of the cell contents into the extracellular space. What might be the consequences if programmed cell death were not achieved in so neat and orderly a fashion?

**18–14** Compare the rules of cell proliferation and apoptosis in an animal to the rules that govern human behavior in society. What would happen to an animal if its cells behaved like people normally behave in our society? Could the rules that govern cell proliferation be applied to human societies?

**18–15** Look carefully at the electron micrographs in Figure 18–1. Describe the differences between the cell that died by necrosis and the one that died by apoptosis. How do the pictures confirm the differences between the two processes? Explain your answer.

(A)

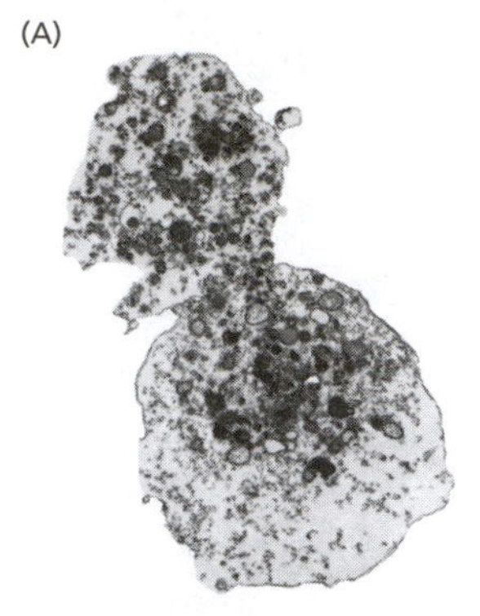

(B)

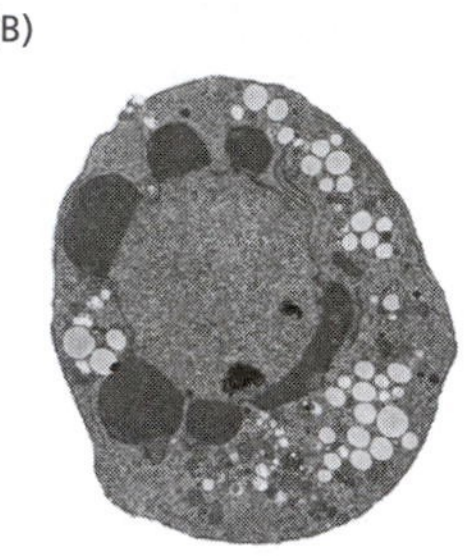

**Figure 18–1** Cell death (Problem 18–15). (A) By necrosis. (B) By apoptosis.

**18–16** One important role of Fas and Fas ligand is to mediate the elimination of tumor cells by killer lymphocytes. In a study of 35 primary lung and colon tumors, half the tumors were found to have amplified and overexpressed a gene for a secreted protein that binds to Fas ligand. How do you suppose that overexpression of this protein might contribute to the survival of these tumor cells? Explain your reasoning.

**18–17** An early argument against the involvement of mitochondria in apoptosis was based on the observation that mammalian cells lacking mitochondrial DNA undergo apoptosis normally. They can also be protected from apoptosis by overexpression of Bcl-2. These cells contain mitochondria that lack all proteins encoded by the mitochondrial genome. As a result, they cannot make ATP by oxidative phosphorylation and rely exclusively on glycolysis to meet their energy needs. Yet these cells carry out cytochrome *c*-mediated apoptosis. How do you suppose that they do it?

**18–18** Development of the nematode *Caenorhabditis elegans* generates exactly 959 somatic cells; it also produces an additional 131 cells that are later eliminated by programmed cell death. Classical genetic experiments in *C. elegans* isolated mutants that identified the first genes involved in apoptosis. Of the many mutant genes affecting apoptosis in the nematode, none have ever been found in the gene for cytochrome *c*. Why do you suppose that such a central effector molecule in apoptosis was not found in the many genetic screens for 'death' genes that have been carried out in *C. elegans*?

**18–19** Imagine that you could microinject cytochrome *c* into the cytosol of wild-type cells and of cells that were doubly defective for Bax and Bak. Would you expect one, both, or neither type of cell to undergo apoptosis? Explain your reasoning.

**18–20** Mice that are defective for Apaf1 ($Apaf1^{-/-}$) or for caspase-9 ($Casp9^{-/-}$) die around the time of birth and exhibit a characteristic set of abnormalities, including brain overgrowth and cranial protrusions. Why do you suppose such abnormalities arise in these deficient mice?

**18–21** In contrast to their similar brain abnormalities, newborn mice deficient in Apaf1 or caspase-9 have distinctive abnormalities in their paws. Apaf1-deficient mice fail to eliminate the webs between their developing digits, whereas caspase-9-deficient mice have normally formed digits (Figure

18–2). If Apaf1 and caspase-9 function in the same apoptotic pathway, how is it possible for these deficient mice to differ in web-cell apoptosis?

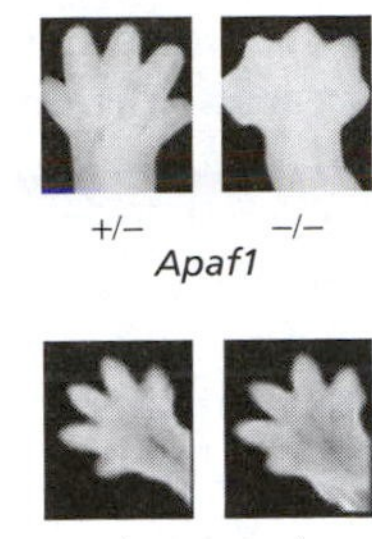


**Figure 18–2** Appearance of paws in *Apaf1*$^{-/-}$ and *Casp9*$^{-/-}$ newborn mice relative to normal newborn mice (Problem 18–21).

## CALCULATIONS

**18–22** Fas ligand is a trimeric, extracellular protein that binds to its receptor, Fas, which is composed of three identical transmembrane subunits (Figure 18–3). The binding of Fas ligand alters the conformation of Fas so that it binds an adaptor protein, which then recruits and activates procaspase-8, triggering a caspase cascade that leads to cell death. In humans the autoimmune lymphoproliferative syndrome (ALPS) is associated with dominant mutations in Fas that include point mutations and C-terminal truncations. In individuals that are heterozygous for such mutations, lymphocytes do not die at their normal rate and accumulate in abnormally large numbers, causing a variety of clinical problems. In contrast to these patients, individuals that are heterozygous for mutations that eliminate Fas expression entirely have no clinical symptoms.

A. Assuming that the normal and dominant forms of Fas are expressed to the same level and bind Fas ligand equally, what fraction of Fas–Fas ligand complexes on a lymphocyte from a heterozygous ALPS patient would be expected to be composed entirely of normal Fas subunits?

B. In an individual heterozygous for a mutation that eliminates Fas expression, what fraction of Fas–Fas ligand complexes would be expected to be composed entirely of normal Fas subunits?

C. Why are the Fas mutations that are associated with ALPS dominant, while those that eliminate expression of Fas are recessive?

## DATA HANDLING

**18–23** When human cancer (HeLa) cells are exposed to UV light at 90 mJ/cm$^2$, most of the cells undergo apoptosis within 24 hours. Release of cytochrome *c* from mitochondria can be detected as early as 6 hours after exposure of a population of cells to UV light, and it continues to increase for more than 10 hours thereafter. Does this mean that individual cells slowly release their cytochrome *c* over this time period? Or, alternatively, do individual cells release their cytochrome *c* rapidly, but with different cells being triggered over the longer time period?

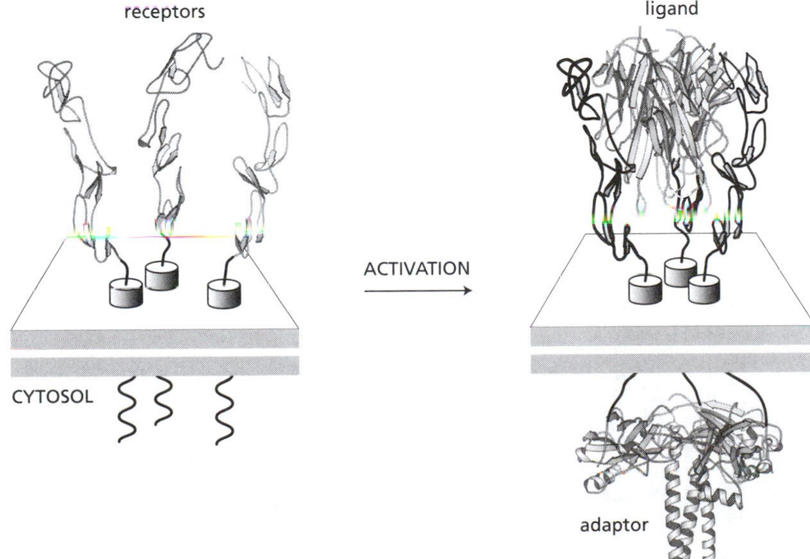


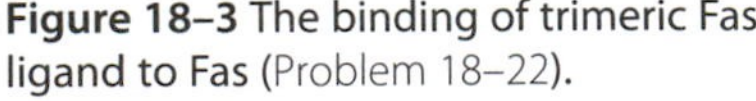

**Figure 18–3** The binding of trimeric Fas ligand to Fas (Problem 18–22).

To answer this fundamental question, you have fused the gene for green fluorescent protein (GFP) to the gene for cytochrome *c*, so that you can observe the behavior of individual cells by confocal fluorescence microscopy. In cells that are expressing the cytochrome *c*–GFP fusion, fluorescence shows the punctate pattern typical of mitochondrial proteins. You then irradiate these cells with UV light and observe individual cells for changes in the punctate pattern. Two such cells (outlined in white) are shown in Figure 18–4A and B. Release of cytochrome *c*–GFP is detected as a change from a punctate to a diffuse pattern of fluorescence. Times after UV exposure are indicated as hours:minutes below the individual panels.

Which model for cytochrome *c* release do these observations support? Explain your reasoning.

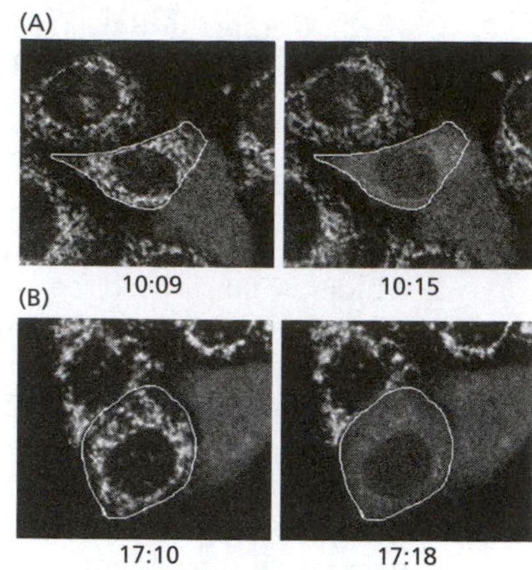


**Figure 18–4** Time-lapse video, fluorescence microscopic analysis of cytochrome *c*–GFP release from mitochondria of individual cells (Problem 18–23). (A) Cells observed for 8 minutes, 10 hours after UV irradiation. (B) Cells observed for 6 minutes, 17 hours after UV irradiation. One cell in (A) and one in (B), each *outlined in white*, have released their cytochrome *c*–GFP during the time frame of the observation, which is shown as hours:minutes below each panel.

**18–24** One common cell strategy for generating a rapid response is to use a positive-feedback loop. The rapid release of cytochrome *c* from mitochondria in response to cell damage could result from such a positive-feedback loop. For example, the caspases activated by cytochrome *c* could act on mitochondria to release additional cytochrome *c*. To test this possibility, cells expressing a cytochrome *c*–GFP fusion were incubated in the presence or absence of the broad-spectrum caspase inhibitor zVAD prior to exposure to a variety of apoptosis-inducing agents. The average time for release of cytochrome *c*–GFP from the mitochondria in individual cells was then compared, as shown in Figure 18–5.

If a caspase-mediated positive-feedback loop were involved in the rapid release of cytochrome *c*–GFP from mitochondria, what results would you have expected for cells that were incubated in the presence of the caspase inhibitor? Do your expectations match the observed results? Is a caspase-mediated positive-feedback loop involved in cytochrome *c* release?

**18–25** Activation of Fas activates caspase-8, which triggers the extrinsic pathway of apoptosis. In some cellls the activation of Fas also engages the intrinsic pathway of apoptosis. In these cells, caspase-8 cleaves the protein Bid to produce an active fragment, tBid, that binds to the mitochondrial membrane. tBid promotes oligomerization of Bax and of Bak, which stimulates the release of cytochrome *c* into the cytosol to trigger events leading to apoptosis (Figure 18–6). To study this pathway in more detail, you've generated mouse embryo fibroblasts (MEFs) that are $Bax^{-/-}$, $Bak^{-/-}$, or $Bax^{-/-}Bak^{-/-}$. You have also constructed a vector that expresses tBid so that you can study the process independent of Fas.

In untreated MEFs and in MEFs treated with the empty vector, very few cells undergo apoptosis (Figure 18–7). By contrast, when MEFs are treated with the vector that expresses tBid, the wild-type cells and the cells individually defective for Bax or Bak show a dramatic increase in apoptotic cells. MEFs that are defective for both Bax and Bak, however, show no increase in the number of apoptotic cells (Figure 18–7). Among the various mutant cells,

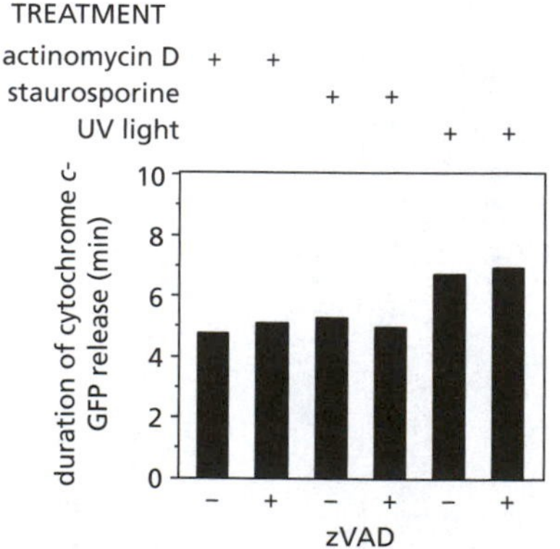


**Figure 18–5** Duration of cytochrome *c*–GFP release from mitochondria in cells treated with various apoptosis-inducing agents in the presence and absence of a caspase inhibitor (Problem 18–24). *Bars* represent the average of measurements on multiple individual cells.

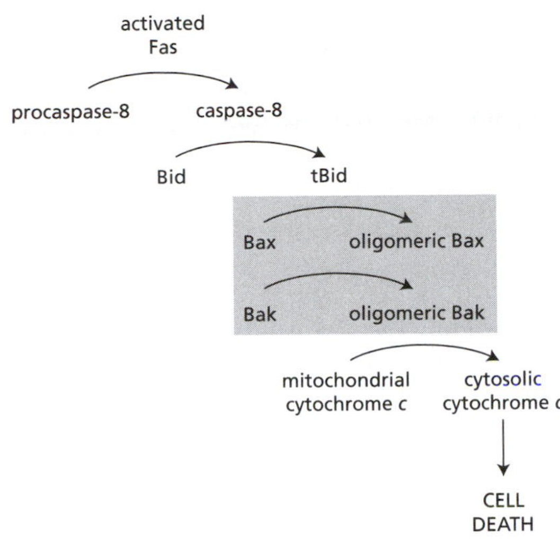


**Figure 18–6** Fas-triggered activation of the intrinsic pathway of apoptosis in MEFs (Problem 18–25). Bax and Bak are *shaded* to indicate that oligomerization of both depends on tBid.

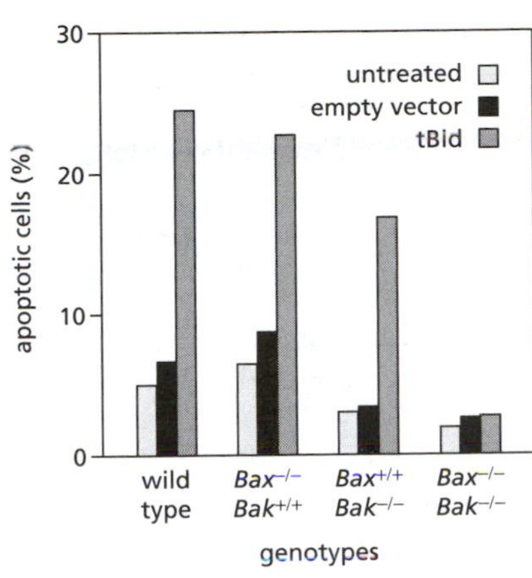


**Figure 18–7** Apoptosis induced by tBid expression in wild-type, Bax-deficient, Bak-deficient, and doubly deficient MEFs (Problem 18–25).

cytochrome *c* is retained in the mitochondria only in the $Bax^{-/-}Bak^{-/-}$ MEFs treated with tBid. What do these results tell you about the requirements for Bax and Bak in cytochrome *c*-induced apoptosis? Explain your reasoning.

**18–26** A variety of treatments can cause cells to undergo apoptosis. You wish to know which of these treatments induce signals that are processed through Bid, Bax, and Bak (see Figure 18–6). You generate $Bid^{-/-}$ and $Bax^{-/-}Bak^{-/-}$ mouse embryo fibroblasts (MEFs) and test them for apoptosis in response to several treatments, with the results shown in Table 18–1.

A. Based on these results, indicate where each of the signals enters the apoptotic pathway relative to Bid, Bax, and Bak.

B. How do you suppose that Fas ligand, which binds to Fas, manages to cause apoptosis in Bid-deficient cells and in Bax- and Bak-deficient cells?

**18–27** The JNK subfamily of MAP kinases is essential for neuronal apoptosis in response to certain stimuli. To test whether JNK family members are required for apoptosis in mouse embryo fibroblasts (MEFs), you knock out the genes for the two JNK proteins—JNK1 and JNK2—that are expressed in MEFs. As shown in Figure 18–8A, $Jnk1^{-/-}$ and $Jnk2^{-/-}$ cells are about as sensitive to UV light-induced apoptosis as wild-type cells, whereas $Jnk1^{-/-}Jnk2^{-/-}$ cells are totally resistant. Apoptosis is accompanied by release of cytochrome *c* into the cytosol, which does not occur in $Jnk1^{-/-}Jnk2^{-/-}$ cells. As measured by DNA fragmentation, $Jnk1^{-/-}Jnk2^{-/-}$

**Table 18–1 Results in $Bid^{-/-}$ and $Bax^{-/-}Bak^{-/-}$ MEFs of various treatments that cause apoptosis in wild-type cells** (Problem 18–26).

| TREATMENT | EFFECT | APOPTOSIS | |
|---|---|---|---|
| | | $Bid^{-/-}$ | $Bax^{-/-}Bak^{-/-}$ |
| Fas ligand | activates Fas | yes | yes |
| Staurosporine | inhibits protein kinases | yes | no |
| UV light | damages DNA | yes | no |
| Etoposide | inhibits topoisomerase II | yes | no |
| Tunicamycin | blocks *N*-linked glycosylation | yes | no |
| Thapsigargin | inhibits $Ca^{2+}$ pump in ER | yes | no |

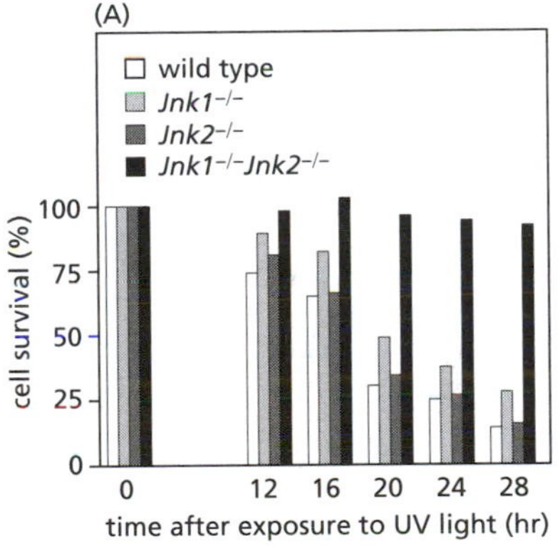


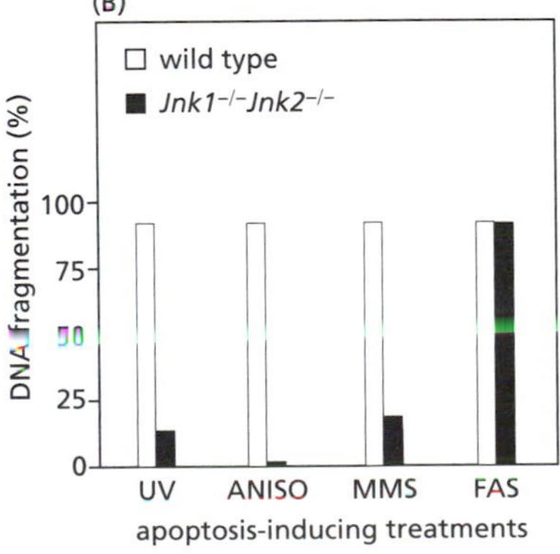


**Figure 18–8** Roles of JNK1and JNK2 in apoptosis in MEFs (Problem 18–27). (A) Cell survival with time after UV irradiation. (B) DNA fragmentation in response to various apoptosis-inducing treatments. ANISO stands for anisomycin; FAS stands for Fas ligand.

cells are protected not only against apoptosis triggered by treatment with UV light, but also against apoptosis induced by treatment with anisomycin (which blocks protein synthesis) and methyl methanesulfonate (MMS, which methylates DNA). By contrast, these cells are not protected from apoptosis induced by activation of Fas by Fas ligand (Figure 18–8B).

Based on these results, decide whether the JNK proteins participate in the intrinsic or in the extrinsic pathway of apoptosis.

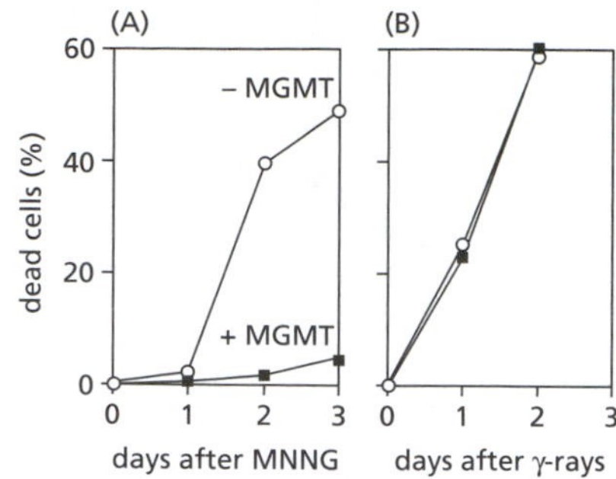


**Figure 18–9** Apoptosis induced in MGMT-deficient and MGMT-overexpressing cells (Problem 18–28). (A) By MNNG. (B) By γ-irradiation. *White circles* indicate cells deficient for MGMT; *black squares* indicate cells that overexpress MGMT.

**18–28** Alkylating agents are commonly used for cancer chemotherapy because they are highly cytotoxic, inducing death by apoptosis. Agents such as *N*-methyl-*N′*-nitro-*N*-nitrosoguanidine (MNNG) alkylate a variety of cellular targets including DNA, RNA, proteins, and lipids. Which of these targets generates the signal for apoptosis? The most common mutagenic lesion to DNA, $O^6$-methylguanine, can be removed by the enzyme $O^6$-methylguanine methyltransferase (MGMT). To test the possibility that alkylation of DNA is responsible for the apoptotic signal, you compare MNNG-induced apoptosis in cells that are deficient for MGMT with apoptosis in cells that overexpress MGMT (Figure 18–9A). As a control, you compare apoptosis mediated by γ-irradiation (Figure 18–9B). Do these results support the idea that alkylation of DNA leads to the apoptotic signal? Why or why not?

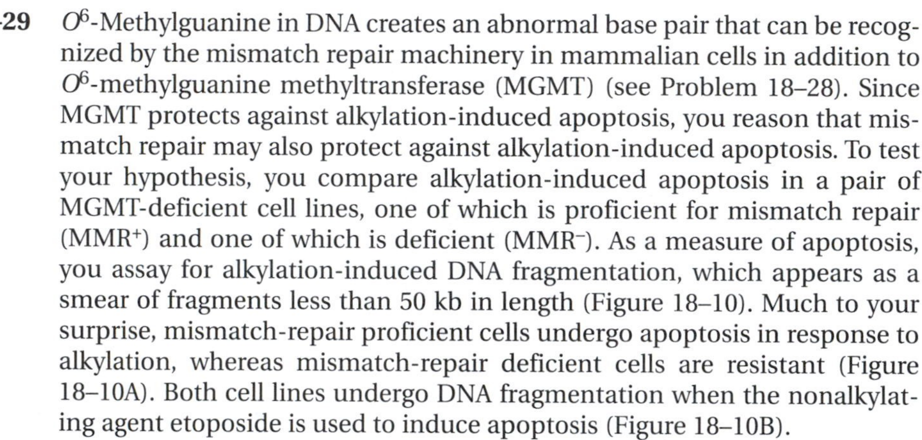

**18–29** $O^6$-Methylguanine in DNA creates an abnormal base pair that can be recognized by the mismatch repair machinery in mammalian cells in addition to $O^6$-methylguanine methyltransferase (MGMT) (see Problem 18–28). Since MGMT protects against alkylation-induced apoptosis, you reason that mismatch repair may also protect against alkylation-induced apoptosis. To test your hypothesis, you compare alkylation-induced apoptosis in a pair of MGMT-deficient cell lines, one of which is proficient for mismatch repair ($MMR^+$) and one of which is deficient ($MMR^-$). As a measure of apoptosis, you assay for alkylation-induced DNA fragmentation, which appears as a smear of fragments less than 50 kb in length (Figure 18–10). Much to your surprise, mismatch-repair proficient cells undergo apoptosis in response to alkylation, whereas mismatch-repair deficient cells are resistant (Figure 18–10A). Both cell lines undergo DNA fragmentation when the nonalkylating agent etoposide is used to induce apoptosis (Figure 18–10B).

Propose an explanation for these results. What do these results suggest about the nature of the apoptosis-inducing lesion?

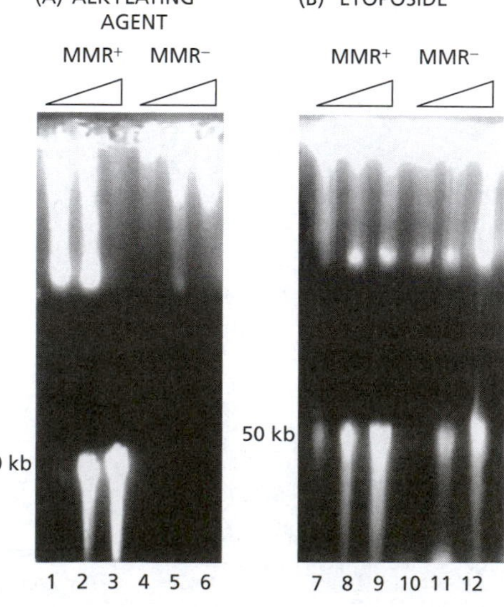


**Figure 18–10** Apoptosis-induced DNA fragmentation in $MMR^+$ and $MMR^-$ cells (Problem 18–29). (A) In response to an alkylating agent. (B) In response to etoposide. *Triangles* above the lanes indicate increasing concentrations of the agents. DNA fragmentation was assayed 72 hours after exposure to the agents.

# Cell Junctions, Cell Adhesion, and the Extracellular Matrix

# 19

## In This Chapter


CADHERINS AND CELL–CELL ADHESION 445

TIGHT JUNCTIONS AND THE ORGANIZATION OF EPITHELIA 449

PASSAGEWAYS FROM CELL TO CELL: GAP JUNCTIONS AND PLASMODESMATA 455

THE BASAL LAMINA 457

INTEGRINS AND CELL–MATRIX ADHESION 459

THE EXTRACELLULAR MATRIX OF ANIMAL CONNECTIVE TISSUES 461

THE PLANT CELL WALL 465


## CADHERINS AND CELL–CELL ADHESION

TERMS TO LEARN

adherens junction
adhesion belt
anchoring junction
cadherin
cadherin superfamily
cell–cell adhesion
channel-forming junction
classical cadherin
connective tissue
desmosome junction
epithelia
epithelial tissue
homophilic
immunoglobulin (Ig) superfamily
integrin
nonclassical cadherin
occluding junction
PDZ domain
scaffold protein
selectin
signal-relaying junction
transmembrane adhesion protein

### DEFINITIONS

Match each definition below with its term from the list above.

**19–1** Member of a family of cell-surface carbohydrate-binding proteins that mediate transient, $Ca^{2+}$-dependent cell–cell adhesion in the bloodstream, for example, between white blood cells and the endothelium of the blood vessel wall.

**19–2** Like-to-like protein interactions in which a molecule on one cell binds to an identical, or closely related, molecule on an adjacent cell.

**19–3** Type of anchoring junction, usually formed between two epithelial cells, characterized by dense plaques of protein into which intermediate filaments in the two adjoining cell insert.

**19–4** Member of a large family of transmembrane proteins involved in the adhesion of cells to the extracellular matrix and to each other.

**19–5** Anchoring junction that connects actin filaments in one cell to those in the next cell.

**19–6** A member of a family of proteins that mediate $Ca^{2+}$-dependent cell–cell adhesion in animal tissues.

**19–7** Beltlike anchoring junction that encircles the apical end of an epithelial cell and attaches it to the adjoining cell.

### TRUE/FALSE

Decide whether each of these statements is true or false, and then explain why.

**19–8** Given the numerous processes inside cells that are regulated by changes in $Ca^{2+}$ concentration, it seems likely that $Ca^{2+}$-dependent cell–cell adhesions are also regulated by changes in $Ca^{2+}$ concentration.

**19–9** Cadherins promote cell–cell interactions by binding to cadherin molecules of the same or closely related subtype on adjacent cells.

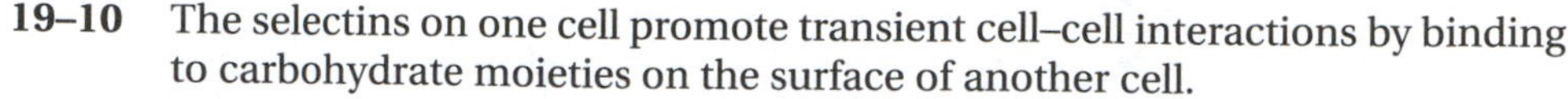

**19–10** The selectins on one cell promote transient cell–cell interactions by binding to carbohydrate moieties on the surface of another cell.

**19–11** Although cadherins and Ig family members are frequently expressed on the same cells, the adhesions mediated by Ig molecules are much stronger and, thus, are largely responsible for holding cells together.

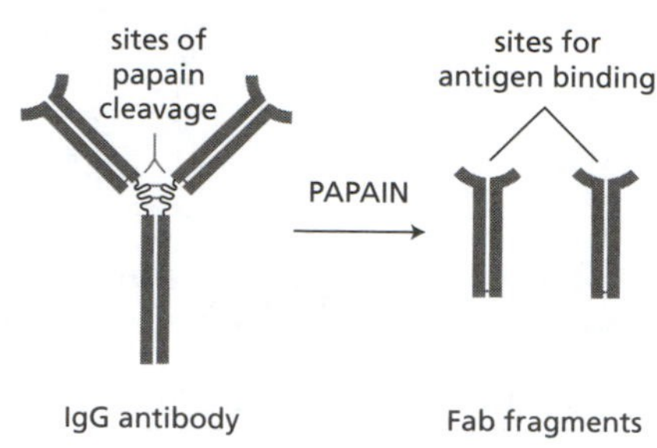


**Figure 19–1** Production of Fab fragments from IgG antibodies by digestion with papain (Problem 19–13).

## THOUGHT PROBLEMS

**19–12** Comment on the following (1922) quote from Warren Lewis, who was one of the pioneers of cell biology. "Were the various types of cells to lose their stickiness for one another and for the supporting extracellular matrix, our bodies would at once disintegrate and flow off into the ground in a mixed stream of cells."

**19–13** Cell adhesion molecules were originally identified using antibodies raised against cell-surface components to block cell aggregation. In the adhesion-blocking assays, the researchers found it necessary to use antibody fragments, each with a single binding site (so-called Fab fragments), rather than intact IgG antibodies, which are Y-shaped molecules with two identical binding sites. The Fab fragments were generated by digesting the IgG antibodies with papain, a protease, to separate the two binding sites (Figure 19–1). Why do you suppose it was necessary to use Fab fragments to block cell aggregation?

**19–14** You have raised an antibody against what you think is E-cadherin, and it blocks the aggregation (cell–cell adhesion) of a mouse embryonic stem cell line. Explain in outline how you could use this reagent to identify a cDNA clone for the protein. How might you check to see whether the cDNA clone encodes a cell adhesion molecule?

**19–15** Mouse L-cells have proven to be an extremely useful model system for investigating the properties of cadherins because they do not normally express any cadherins. If different populations of L-cells are transfected with vectors expressing one or the other of two different cadherins, and are then dissociated and mixed together, they segregate into two separate balls of cells, each held together by a different cadherin. If two populations of cells expressing different levels of the same cadherin are mixed, they segregate into a single ball of cells, with the low-expressing population on the outside (Figure 19–2).

A. Why do you suppose populations of cells expressing different levels of the same cadherin segregate with this characteristic layered structure? Why don't they segregate into two separate balls? Or a mixed ball? Or a ball with the low-expressing population on the inside?

B. What sort of final architecture might you expect if you were to mix together two populations of cells that expressed P-cadherin in common, but, in

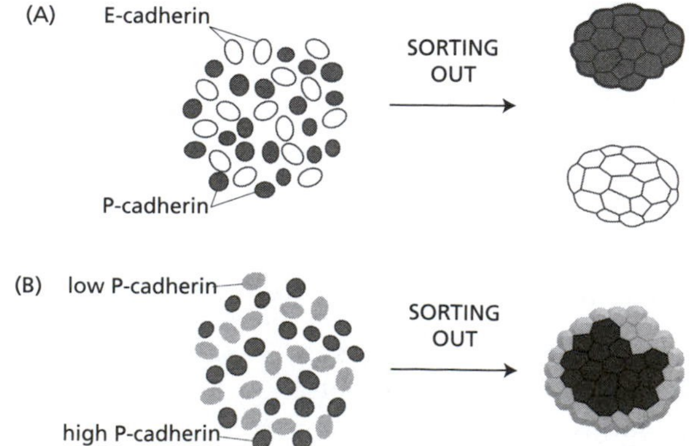


**Figure 19–2** Cadherin-dependent cell sorting (Problem 19–15). (A) Sorting of a mixture of cells that express two different cadherins. (B) Sorting of a mixture of cells that express two different levels of the same cadherin.

addition, one population expressed E-cadherin and the other expressed N-cadherin?

**19–16** A member of the Fat family of cadherins, called Cdh3, is found in the nematode worm. It is widely expressed, and its structure suggests that it plays an important role in cell adhesion, cell recognition, or both, in several morphogenetic processes. You make a null mutation in the *Cdh3* gene, hoping to gain insight into its function. The only phenotype you observe, however, is the presence of malformed tails with kinks and bifurcations that stem from the failure of a single cell, called hyp10, to elongate at the normal time during development. Why do you suppose that no problems arise in the many other cells in the worm that express this protein?

**19–17** It requires a force of about 20 piconewtons (pN) to pull a transmembrane protein like P-selectin out of a pure lipid bilayer. To pull a P-selectin molecule out of the plasma membrane requires more than 110 pN. Why do you suppose it takes so much more force to pull P-selectin out of the plasma membrane than out of a lipid bilayer?

## CALCULATIONS

**19–18** Atomic force microscopy can be used to measure the strength of individual interactions. This technique showed that the binding strength between individual adhesion molecules in a marine sponge averaged 125 piconewtons (pN) under physiological conditions. Just how strong is such an interaction? Since the purpose of these bonds is to hold cells together, this question can be rephrased in terms of the number of cells one interaction could support against the force of gravity in seawater; that is, what number of cells weighs 125 pN?

The weight ($W$) of one cell in seawater is given by the Archimedes equation

$$W = gV(\rho_{\text{cell}} - \rho_{\text{seawater}})$$

where

$g$ is the gravitational constant (9.81 N/kg)
$V$ is the volume of the cell $[(4/3)\pi r^3]$
$\rho_{\text{cell}}$ is the density of the cell (1100 kg/m$^3$)
$\rho_{\text{seawater}}$ is the density of seawater (1018 kg/m$^3$)

Assuming that the cells are spheres with a radius of 5 μm, calculate the weight of one cell in seawater. How many cells could be supported by a bond with a strength of 125 pN?

**19–19** The attachment of bacteriophage T4 to *E. coli* K illustrates the value of multiple weak interactions, which allow relative motion until fixed connections are made (Figure 19–3). During infection, T4 first attaches to the surface of

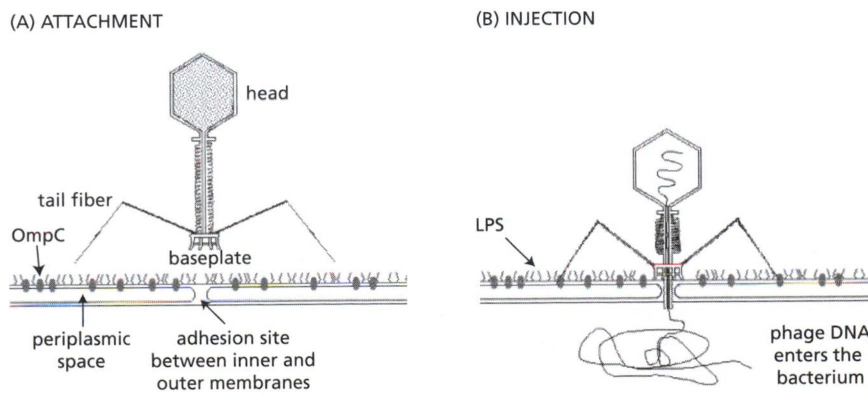


**Figure 19–3** Infection by bacteriophage T4 (Problem 19–19). (A) Attachment to bacterial surface. (B) Injection of its DNA.

**Table19–1 Infectivity of phage T4 on various bacterial mutants** (Problem 19–19).

| BACTERIAL STRAIN | PHAGE T4 INFECTIVITY RELATIVE TO NONMUTANT BACTERIA |
|---|---|
| *ompC*$^+$ LPS$^+$ | 1 |
| *ompC*$^-$ LPS$^+$ | $10^{-3}$ |
| *ompC*$^+$ LPS$^-$ | $10^{-3}$ |
| *ompC*$^-$ LPS$^-$ | $10^{-7}$ |

*E. coli* by the tips of its six tail fibers. It then wanders around the surface until it finds an appropriate place for the attachment of its baseplate. When the baseplate is securely fastened, the tail sheath contracts, injecting the phage DNA into the bacterium (Figure 19–3). The initial tail-fiber cell-surface interaction is critical for infection: phages that lack tail fibers are totally noninfectious.

Analysis of T4 attachment is greatly simplified by the ease with which resistant bacteria and defective phages can be obtained. Bacterial mutants resistant to T4 infection fall into two classes: one lacks a major outer membrane protein called OmpC (outer membrane protein C); the other contains alterations in the long polysaccharide chain normally associated with bacterial lipopolysaccharide (LPS). The infectivity of T4 on wild-type and mutant cells is indicated in Table 19–1. These results suggest that each T4 tail fiber has two binding sites: one for LPS and one for OmpC. Electron micrographs showing the interaction between isolated tail fibers and LPS suggest that individual associations are not very strong, since only about 50% of the fibers appear to be bound to LPS.

A. Assume that at any instant each of the six tail fibers has a 0.5 probability of being bound to LPS and the same probability of being bound to OmpC. With this assumption, the fraction of the phage population on the bacterial surface that will have none of its six tail fibers attached in a given instant is $(0.5)^{12}$ (which is the probability of a given binding site being unbound, 0.5, raised to the number of binding sites, two on each of six tail fibers). In light of these considerations, what fraction of the phage population will be attached at any one instant by at least one tail fiber? (The attached fraction is equal to one minus the unattached fraction.) Suppose that the bacteria were missing OmpC. What fraction of the phage population would now be attached by at least one tail fiber at any one instant?

B. Surprisingly, the above comparison of wild-type and *ompC*$^-$ bacteria suggests only a very small difference in the attached fraction of the phage population at any one instant. As shown in Table 19–1, phage infectivities on these two strains differ by a factor of 1000. Can you suggest an explanation that might resolve this apparent paradox?

## DATA HANDLING

**19–20** You suspect that the cadherin you've cloned is attached to the cytoskeleton, presumably by its cytoplasmic tail. To identify the proteins to which it binds, you express your cadherin in a mouse cell line that doesn't express any other cadherins. You label the cells with $^{35}$S methionine for 16 hours, homogenize them in detergent, and precipitate the cadherin and any associated proteins using antibodies against the extracellular domain of the cadherin. You then analyze the immune precipitates by electrophoresis on SDS-polyacrylamide gels, followed by autoradiography to make the labeled proteins visible. As shown in lane 1 in Figure 19–4A, the antibody precipitates three major proteins in cells transfected with a plasmid expressing the full-length intact cadherin. These proteins are not precipitated from untreated control cells (Figure 19–4A, lane 7), verifying the specificity of your antibody.

To identify the portion of the cadherin molecule that is required for binding the other two proteins, you make a series of constructs with deletions in the cytoplasmic tail, and carry out the same sort of analysis (Figure 19–4A and B).

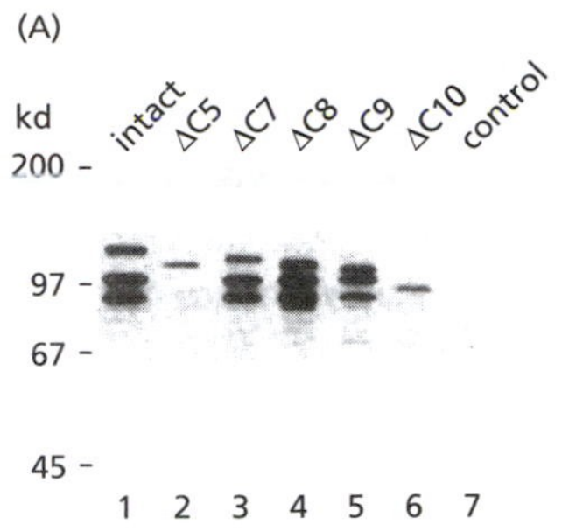


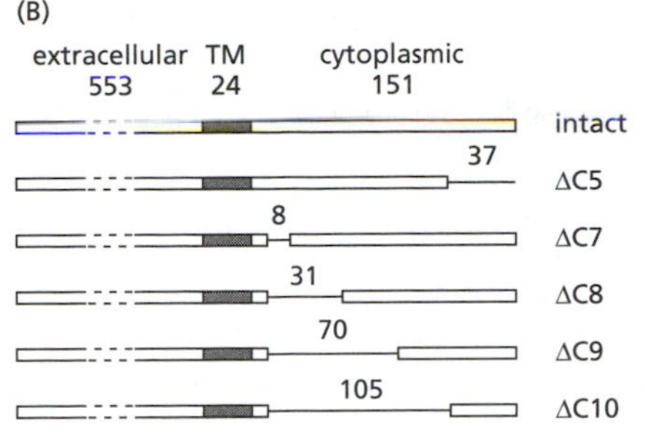


**Figure 19–4** Analysis of cadherin interactions with cellular proteins (Problem 19–20). (A) Immunoprecipitation experiments using anti-cadherin antibodies. (B) Schematic illustration of the various cDNA constructs transfected into cells. The numbers of amino acids in the extracellular, transmembrane (TM), and cytoplasmic domains are indicated. The deleted segments are shown as single lines with the number of deleted amino acids indicated. The left end of the deletion in ΔC5 and the right end of the deletion in ΔC10 are at the same site in the protein.

A. In the autoradiograph in Figure 19–4A, identify the band corresponding to your cadherin and the bands corresponding to proteins that bind to your cadherin. Give your reasoning.
B. Which segment of the cadherin is required for binding to the other proteins?
C. How might you check to see whether the segment you've identified is all that is needed to bind to the other proteins?

**19–21** You want to understand how the density of P-selectin in blood vessel walls affects the rolling interactions of neutrophils when they are subjected to hydrodynamic drag forces in the blood. You introduce P-selectin into a synthetic lipid bilayer and attach it to a glass slide mounted in flow chamber. This arrangement allows you to measure neutrophil attachment at different densities of P-selectin and at different flow rates.

At high densities of P-selectin, from 40 to 400 molecules per μm$^2$, the neutrophils attached to the membrane and rolled jerkily in the direction of the flow. At densities from 1 to 15 molecules per μm$^2$, the cells behaved differently: they either moved at the flow rate of the medium or were transiently tethered before moving again. When the bilayer surface was treated with an antibody against P-selectin, or when EDTA was added to the medium, no tethering occurred.

At the lower densities of P-selectin, the number of tethering events was directly proportional to the density of P-selectin. By recording the results under a videomicroscope, you measured how long each cell remained attached during a tethering event. A plot of the log of the number of cells remaining bound at increasing times shows that the rate of dissociation of the cells followed a simple exponential decay curve ($e^{-kt}$) with a 'cellular off-rate' ($k_{off}$) of about 1 per second. Moreover, the off-rate was unaffected by the density of P-selectin in the range of 1 to 15 per μm$^2$.

A. What is the point of doing the experiments with antibody or EDTA?
B. Is a single interaction between P-selectin and the glycoprotein ligand on the surface of the neutrophil sufficient to tether the cell to the surface transiently? Explain your reasoning.
C. If you increase the flow rate (hence, the shear force exerted on the cells) by a factor of three—equivalent to about 112 piconewtons—the cellular off rate increases to about 3.5 per second. How do you imagine that force might alter a dissociation rate?

## TIGHT JUNCTIONS AND THE ORGANIZATION OF EPITHELIA

TERMS TO LEARN

| | |
|---|---|
| apical | Par6 |
| atypical protein kinase C (aPKC) | planar cell polarity |
| basal | polarized |
| occluding junction | septate junction |
| Par3 | tight junction |

## DEFINITIONS

Match each definition below with its term from the list above.

**19–22** Describes a structural property of epithelial sheets, and their individual cells, which have one surface attached to the basal lamina below and the opposite surface exposed to the medium above.

**19–23** Main type of occluding junction in invertebrates; it seals adjacent epithelial cells together, preventing the passage of most dissolved molecules from one side of the epithelial sheet to the other.

**19–24** Main type of occluding junction in vertebrates; it seals adjacent epithelial cells together, preventing the passage of most dissolved molecules from one side of the epithelial sheet to the other.

**19–25** Describes the tip of a cell. For an epithelial cell, it is the exposed free surface, opposite to the surface attached to the basal lamina.

## TRUE/FALSE

Decide whether each of these statements is true or false, and then explain why.

**19–26** Virtually all epithelia are anchored to other tissues on their basal side and free of such attachment on their apical side.

**19–27** Tight junctions perform two distinct functions: they seal the space between cells to restrict paracellular flow and they fence off membrane domains to prevent the mixing of apical and basolateral proteins.

**19–28** Like the corresponding transport processes through the plasma membrane, paracellular transport can be either active or passive.

**19–29** Septate junctions are found only in insects.

## THOUGHT PROBLEMS

**19–30** Give three examples of epithelial sheets found in the human body.

**19–31** Although we have accumulated a great deal of information about the components and morphological appearance of tight junctions, we still know relatively little about how molecules pass through them. One elegant study used a graded series of polyethylene glycol (PEG) molecules with radii ranging from 0.35 nm to 0.74 nm. By placing the PEG molecules on one side of an epithelial sheet of intestinal cells, and measuring their rate of appearance on the other side, the researchers determined the permeability of the tight junction as a function of PEG radius (Figure 19–5). They interpret these results in terms of a common restrictive pore with a radius of about 0.43 nm and a much rarer, nonrestrictive passageway.

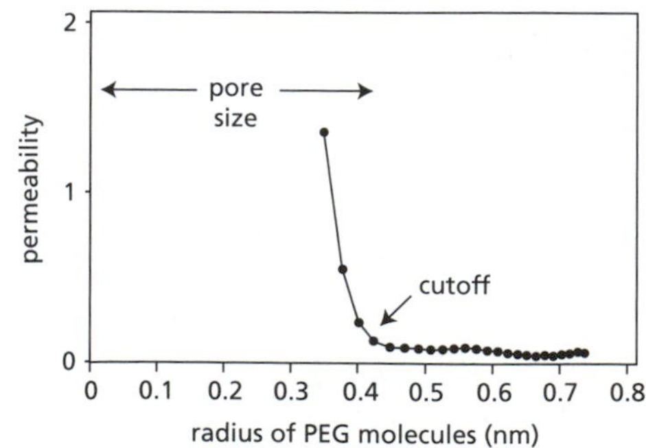


**Figure 19–5** Paracellular permeability of a graded series of PEG molecules (Problem 19–31).

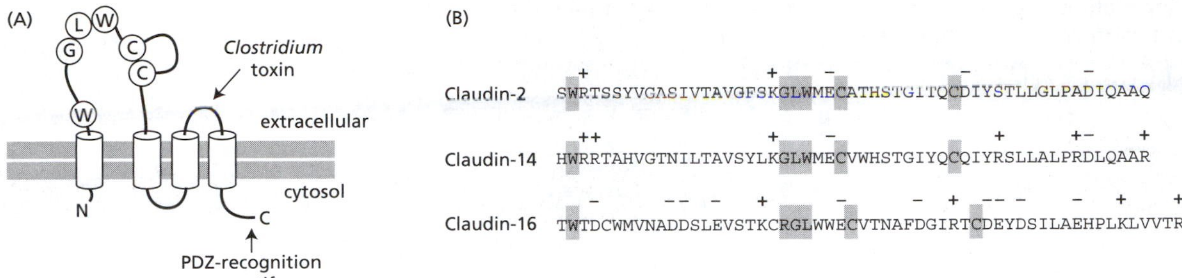


**Figure 19–6** Claudin structure (Problem 19–32). (A) Model depicting the conserved structural and functional features of claudins. *Circled* amino acids in the first extracellular loop indicate the signature residues characteristic of all claudins. Although there has been no direct demonstration of a disulfide bond between the conserved cysteines, it seems likely given the oxidizing extracellular environment and their total conservation between claudins. PDZ indicates the binding sites for proteins that contain PDZ domains, which can bind the C-terminal tails of specific transmembrane molecules. The binding site for the *Clostridium* toxin is also indicated (see Problem 19–34). (B) Amino acid sequences of the first extracellular loop in three different claudins. *Shaded* residues indicate the signature amino acids. Charged residues are indicated with + and – signs.

A. The size of the restrictive pore is defined by the sharp cutoff of the initial part of the curve. Why do you suppose that the permeability of PEG molecules that are smaller that the pore—the four smallest molecules—declines markedly with increasing size? Why don't all of the PEG molecules that can fit through the pore permeate it at the same rate?

B. The identity of the larger, but much rarer, paracellular passageway is not known. Can you suggest some possibilities?

**19–32** Claudin molecules in one cell bind to those in an adjacent cell to form a tight junction. Interactions between the extracellular domains in the paired claudin molecules form the pores that restrict the paracellular transport of small molecules and ions. The first extracellular loop of claudin molecules (Figure 19–6A) is thought to form the pore itself. Figure 19–6B shows the sequences of the first extracellular loop in three claudin molecules. Based on these sequences, which one of the claudins do you suppose might form a cation pore? Explain your reasoning.

## CALCULATIONS

**19–33** You and your advisor noted some time ago that there is a rough correlation between the electrical resistance of an epithelium and the number of sealing strands. This makes intuitive sense because current across an epithelium is carried by small ions that must penetrate the tight junction (Figure 19–7). You can imagine two ways that resistance might depend on the number of sealing strands. If each sealing strand provided a given resistance, then the overall resistance of a tight junction would be linearly related to the number of sealing strands (like electrical resistors in series). On the other hand, if each sealing strand could exist in two states—a closed, high-resistance state and an open, low-resistance state—the resistance of the tight junction would be related to the probability that all strands in a given pathway through the junction would be open at the same time. In that case, the overall resistance would be logarithmically related to the number of sealing strands.

To put the idea on a quantitative basis so that you can distinguish between these two possibilities, you measure the resistance of four different epithelia from a rabbit: the very leaky proximal tubule of the kidney, the less leaky gall bladder, the tight distal tubule of the kidney, and the very tight bladder

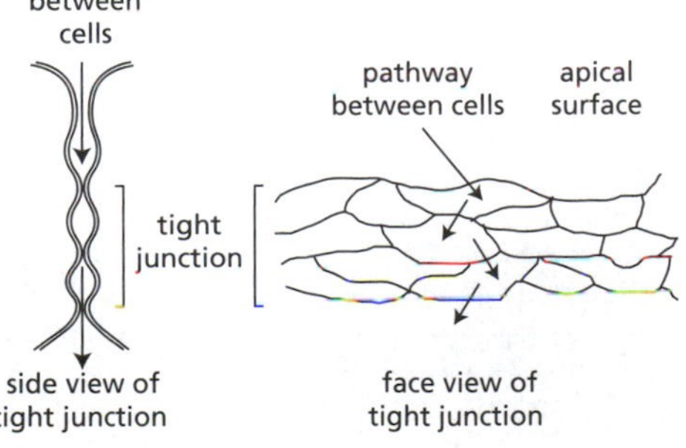


**Figure 19–7** Two views of a tight junction that links cells in an epithelium (Problem 19–33).

**Table 19–2 Correlation between electrical resistance and the number of strands in the tight junctions from various epithelia** (Problem 19–33).

| RABBIT EPITHELIUM | STRANDS IN TIGHT JUNCTION (mean number) | ELECTRICAL RESISTANCE (relative) |
|---|---|---|
| Proximal tubule | 1.2 | 1.0 |
| Gall bladder | 3.3 | 4.7 |
| Distal tubule | 5.3 | 52 |
| Urinary bladder | 8.0 | 470 |

epithelium. In addition, you prepare freeze-fracture electron micrographs, from which you determine the average number of sealing strands in the tight junctions that surround each cell in these epithelia. The results are shown in Table 19–2. Which of the two proposed interpretations of the correlation between electrical resistance of the epithelium and the number of sealing strands in a tight junction is supported by your measurements?

## DATA HANDLING

**19–34** The food poisoning bacterium, *Clostridium perfringens*, makes a toxin that binds to various claudins. When the C-terminus of the toxin is bound to a claudin, the N-terminus can insert into the adjacent cell membrane, forming holes that kill the cell. The portion of the toxin that binds to the claudins has proven to be a valuable reagent for investigating the properties of tight junctions. MDCK cells are a common choice for studies of tight junctions because they can form an intact epithelial sheet with high transepi-thelial resistance. MDCK cells express two claudins: claudin-1, which is not bound by the toxin, and claudin-4, which is.

When an intact MDCK epithelial sheet is incubated with the C-terminal toxin fragment, claudin-4 disappears, becoming undetectable within 24 hours. In the absence of claudin-4, the cells remain healthy and the epithelial sheet appears intact. The mean number of strands in the tight junctions that link the cells also decreases over 24 hours from about 4 to about 2, and they are less highly branched. A functional assay for the integrity of the tight junctions shows that transepithelial resistance decreases dramatically in the presence of the toxin, but the resistance can be restored by washing out the toxin (Figure 19–8A). Curiously, the toxin produces these effects only when it is added to the basolateral side of the sheet; it has no effect when added to the apical surface (Figure 19–8B).

A. How can it be that two tight junction strands remain, even though all of the claudin-4 has disappeared?

B. How do you suppose the toxin fragment causes the tight junction strands to disintegrate?

C. Why do you suppose the toxin works when it is added to the basolateral side of the epithelial sheet, but not when added to the apical side?

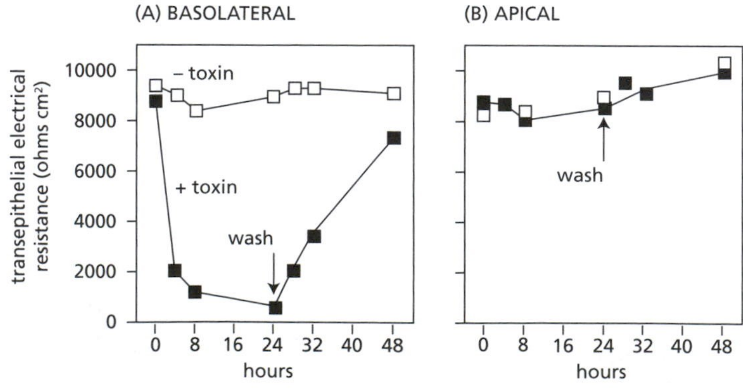


**Figure 19–8** Effects of *Clostridium* toxin on the barrier function of MDCK cells (Problem 19–34). (A) Addition of toxin from the basolateral side of the epithelial sheet. (B) Addition of toxin from the apical side of the epithelial sheet. The higher the resistance (ohms cm$^2$), the less the paracellular current for a given voltage.

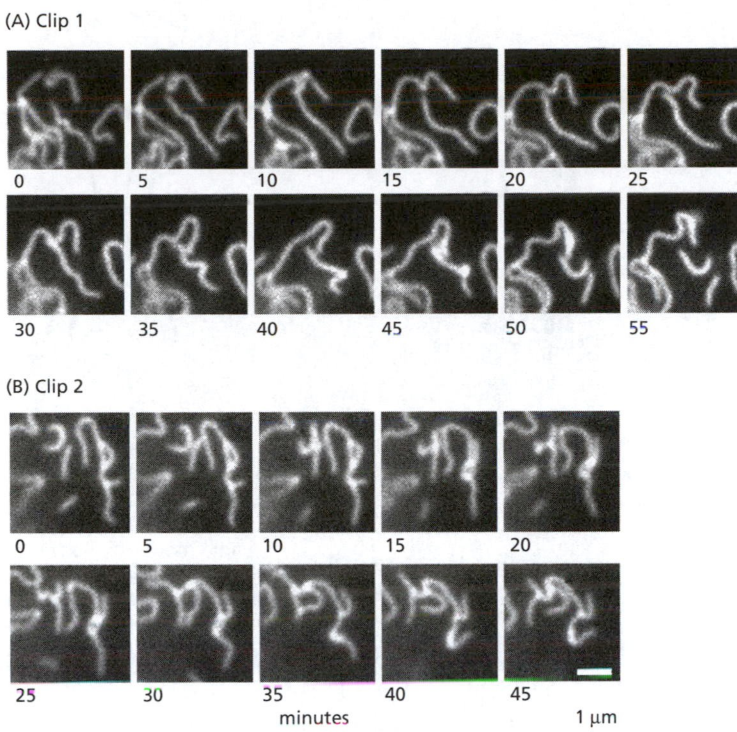


**Figure 19–9** Dynamic behavior of paired claudin strands within apposing plasma membranes (Problem 19–35). (A) Frames from movie clip 1. (B) Frames from movie clip 2.

**19–35** Freeze-fracture electron micrographs provide beautiful images of tight junctions. You are curious to see whether the junctional strands are really as static as those images suggest. By fusing the gene for green fluorescent protein (GFP) to the *claudin-1* gene, you can express a fusion protein in which the cytoplasmic tail of claudin-1 is tagged with GFP. When you transfect this fusion gene into mouse L cells, which express none of their own claudins, fluorescent strands appear on the surface of the cells where they are in contact with one another. It is difficult to maintain focus on these moving cells for extended times, but occasionally you can get 40–50 minutes of clear observation (Figure 19–9).

A. From what you can see in Figure 19–9, does it look like individual strands rapidly elongate or shorten? If your answer is yes, show examples on the figure.

B. Do individual strands join together end-to-end or break apart? If your answer is yes, indicate such areas on the figure.

C. Does the end of an individual strand join to the side of another strand, forming a T-junction? If so, indicate such sites on the figure.

D. A reviewer of your manuscript objects that the plasticity of the strands you observe may be artificial. The reviewer suggests that the presence of GFP at the C-terminus prevents claudin-1 from binding to scaffolding proteins through their PDZ domains. How would you respond?

**19–36** Two fundamentally different structures for tight junctions have been proposed. In addition to the standard protein model, another ingenious model, based on observations from freeze-fracture electron microscopy, proposed that each sealing strand in a tight junction resulted from a membrane fusion that formed cylinders of lipid at the points of fusion (Figure 19–10).

In comparing these lipid and protein models for tight junctions, you realize that they might be distinguishable on the basis of lipid diffusion between the apical and basolateral surfaces. Both models predict that lipids in the cytoplasmic monolayer will be able to diffuse freely between the apical and the basolateral surfaces. However, the models suggest potentially different fates for lipids in the outer monolayer. In the lipid model, lipids in the outer monolayer will be confined to either the apical surface or the basolateral surface, since the cylinder of lipids interrupts the outer monolayer,

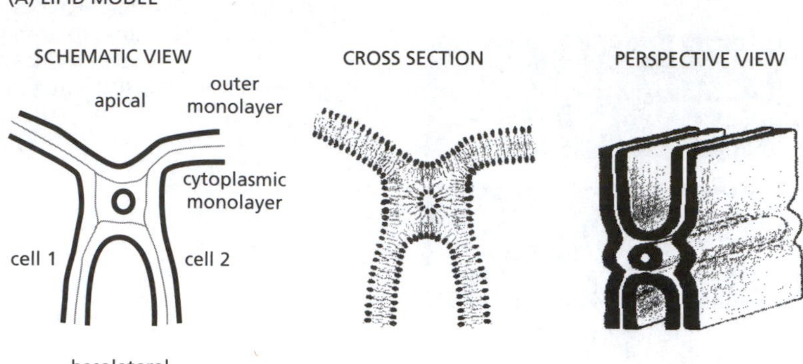


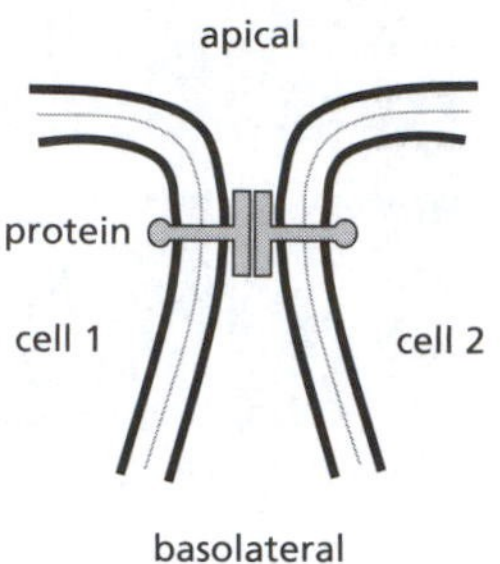


**Figure 19–10** Models for tight-junction structure (Problem 19–36). (A) Three views of the cylinder of lipids that is proposed to form a tight junction according to the lipid model. (B) Schematic representation of the protein model for tight-junction structure.

preventing diffusion through it. By contrast, in the protein model the apical and basolateral surfaces appear to be connected by a continuous outer monolayer, suggesting that lipids in the outer monolayer may be able to diffuse freely between the two surfaces.

You have exactly the experimental tools to resolve this issue! You have been working with the MDCK line of dog kidney cells, which forms an exceptionally tight epithelium with well-defined apical and basolateral surfaces. In addition, after infection with influenza virus, the cells express a fusogenic protein only on their apical surface. This feature allows you to fuse liposomes specifically to the apical surface of infected cells very efficiently by brief exposure to low pH, which activates the fusogenic protein. Thus, you can add fluorescently labeled lipids to the apical surface and detect their migration to the basolateral surface using fluorescence microscopy.

For the experiment, you prepare two sets of labeled liposomes: one with a fluorescent lipid only in the outer monolayer, the other with the fluorescent lipid equally distributed between the cytoplasmic and outer monolayers. You fuse these two sets of liposomes to epithelia in which about half the cells were infected with virus. By adjusting the focal plane of the microscope, you examine the apical and basolateral surfaces for fluorescence. As a control, you remove $Ca^{2+}$ from the medium—a treatment that disrupts tight junctions—and reexamine the basolateral surface. The results are shown in Figure 19–11.

You are delighted! These results show clearly that the lipids in the outer monolayer are confined to the apical surface, whereas lipids in the cytoplasmic monolayer diffuse freely between the apical and basolateral surfaces.

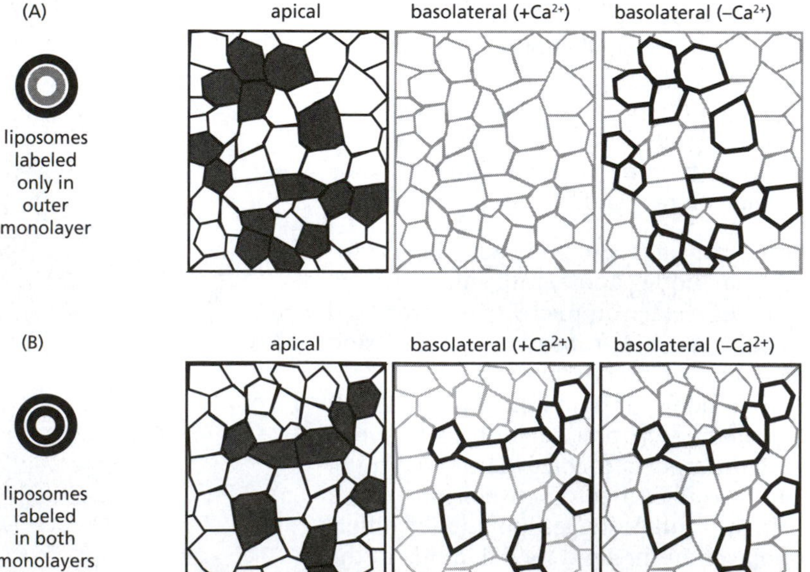


**Figure 19–11** Experimental test of the lipid and protein models for tight-junction structure (Problem 19–36). (A) Liposomes labeled in outer monolayer. (B) Liposomes labeled in both monolayers. Only about half the cells in the epithelium in each experiment were infected with virus. Only infected cells are competent to fuse with the labeled liposomes under the conditions of the experimental protocol.

You show these results to your advisor as proof that the lipid model for tight-junction structure is correct. He examines your results carefully, shakes his head knowingly, gives you that penetrating look of his, and tells you that, although the experiments are exquisitely well done, you have drawn exactly the wrong conclusion. These results prove that the lipid model is incorrect.

What has your advisor seen in the data that you have overlooked? How do your results disprove the lipid model? If the protein model is correct, why do you think it is that the fluorescent lipids in the outer monolayer are confined to the apical surface?

# PASSAGEWAYS FROM CELL TO CELL: GAP JUNCTIONS AND PLASMODESMATA

TERMS TO LEARN

| connexin | connexon | plasmodesmata |
|---|---|---|

## DEFINITIONS

Match each definition below with its term from the list above.

**19–37** Communicating cell–cell junctions in plants in which a channel of cytoplasm lined by plasma membrane connects two adjacent cells through a small pore in their cell walls.

**19–38** Water-filled pore in the plasma membrane formed by a ring of six protein subunits, which link to an identical assembly in an adjoining cell to form a continuous channel between the two cells.

## TRUE/FALSE

Decide whether each of these statements is true or false, and then explain why.

**19–39** Unlike conventional ion channels, individual gap-junction channels remain open continuously once they are formed.

**19–40** The cells in a plant can be viewed as forming a syncytium, in which many cell nuclei share a common cytoplasm.

## THOUGHT PROBLEMS

**19–41** The permeability of gap junctions is regulated by $Ca^{2+}$. Would you expect gap junctions to open or to close when the intracellular concentration of $Ca^{2+}$ rises? Why is this response advantageous?

**19–42** If the cells in a plant are all connected to one another by plasmodesmata, why doesn't the cytoplasm from the entire plant leak out onto the ground when you cut the stem?

## CALCULATIONS

**19–43** Plasmodesmata perform critical functions in plant cells, yet the protein components that presumably confer these functions have proven very difficult to characterize. Part of the problem arises because plasmodesmata are only a small fraction of the cell's total cytoplasm; thus, they are difficult to purify. What fraction of a plant cell's cytoplasm is contained within its plasmodesmata? As the basis for a rough estimate, assume that a plant cell with 1000 $\mu m^3$ of cytoplasm has 100 plasmodesmata, each of which is a cylinder 30 nm in diameter and 100 nm in length.

## DATA HANDLING

COMPACTION

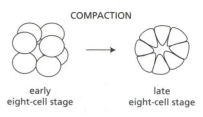

early eight-cell stage

late eight-cell stage

**Figure 19–12** Compaction of the eight-cell mouse embryo (Problem 19–44).

**19–44** Cells in a developing embryo make and break gap-junction connections in specific and interesting patterns, suggesting that gap junctions play an important role in the signaling processes that occur between these cells. At the eight-cell stage, mouse embryos undergo compaction, as illustrated in Figure 19–12. Although the mechanism is not clear, the cells adhere to one another more strongly, changing from a clump of loosely associated cells to a tightly sealed ball. You wish to know whether gap junctions are present before or after this change in adhesion.

Using very fine glass micropipettes, you can measure electrical events and at the same time microinject the enzyme horseradish peroxidase (HRP), 40,000 daltons, or the fluorescent dye fluorescein, 330 daltons. Fluorescein glows bright green under UV illumination, and HRP can be detected by fixing the cells and incubating them with appropriate substrates.

You inject embryos at various stages of development. At both the two-cell and eight-cell stages, different results are obtained, depending on whether the injections are made immediately after cell division or later (Figure 19–13). Immediately after cell division, cytoplasmic bridges linger for a while before cytokinesis is completed.

A. Why do both HRP and fluorescein enter neighboring cells early, but not late, at the two-cell stage?

B. At what stage of embryo development do gap junctions form? Explain your reasoning.

C. In which of the four stages of development diagrammed in Figure 19–13 would you detect electrical coupling if you injected current from the HRP injection electrode and recorded voltage changes in the fluorescein electrode?

**19–45** Tobacco mosaic virus (TMV), like many other plant viruses, spreads through its host by moving from cell to cell via plasmodesmata. For many years this observation presented a puzzle. The normal size-exclusion limit for diffusion through plasmodesmata is around 800 daltons—sufficient for small molecules—yet the TMV particle, by comparison, is huge: a rigid rod 18 nm in diameter and 300 nm in length, formed by many copies of a coat protein arranged helically around a central, single-stranded RNA genome. Two early clues to this puzzle were: (1) The coat protein is not required for the cell-to-cell spread of infection; and (2) TMV encodes a 30-kd 'movement protein' (MP) that is essential for the spread of infection.

To investigate the effects of the MP, the properties of plasmodesmata in normal tobacco cells were compared with those in transgenic cells that expressed the MP. Fluorescence probes of known size were injected into one cell and the appearance of fluorescence was monitored in an adjacent cell, as shown in Figure 19–14. The results for all probes are summarized in Table 19–3. Based on these results, can you suggest a mechanism for how a TMV infection spreads from cell to cell?

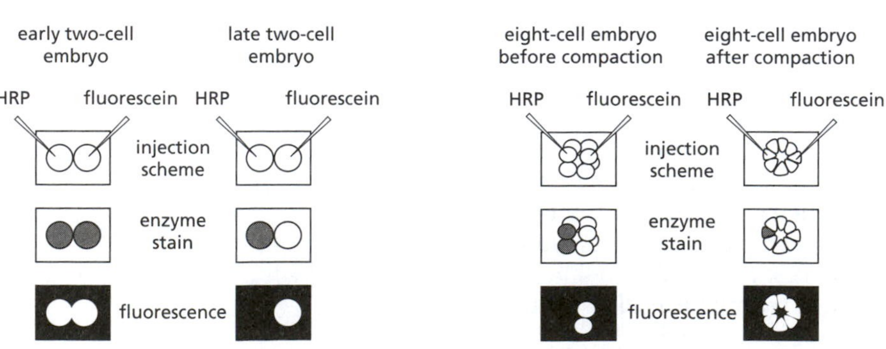


**Figure 19–13** Microinjection of HRP and fluorescein in two-cell and eight-cell mouse embryos (Problem 19–44).

**Table 19–3 Movement of fluorescent probes from one cell to another** (Problem 19–45).

| FLUORESCENT PROBE | | TOBACCO CELLS[a] | |
|---|---|---|---|
| SIZE (d) | RADIUS (nm) | MP⁻ | MP⁺ |
| 457 | 0.6 | 100% | 100% |
| 749 | 0.7 | 50% | 100% |
| 3900 | 1.6 | 14% | 100% |
| 9400 | 2.4 | 0% | 93% |
| 17,200 | 3.1 | 0% | 0% |

[a]The percentage of injections in which the probe was observed to move from the injected cell to the adjacent cell.

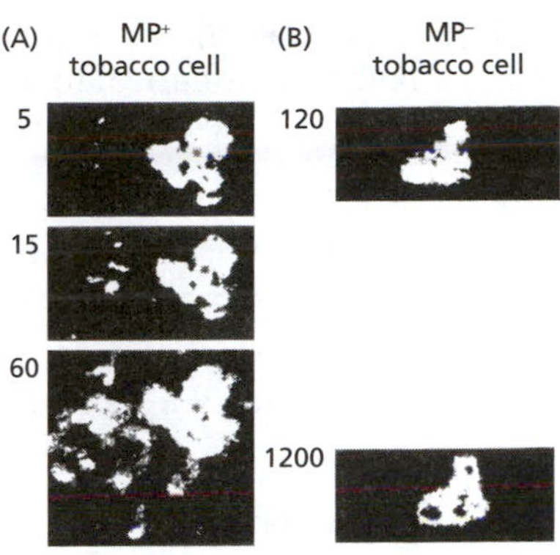


**Figure 19–14** Movement of fluorescent probe after injection into a tobacco cell (Problem 19–45). (A) Injection of the 9400-dalton probe into an $MP^+$ tobacco cell. Fluorescence appears in the adjacent cell over the period of 60 seconds. (B) Injection of the 9400-dalton probe into an $MP^-$ tobacco cell. No fluorescence appears in the adjacent cell within 20 minutes (1200 seconds).

# THE BASAL LAMINA

TERMS TO LEARN

basal lamina (basement membrane)
laminin-1
type IV collagen

## DEFINITIONS

Match each definition below with its term from the list above.

**19–46** Extracellular matrix protein found in basal laminae, where it forms a sheet-like network.

**19–47** Thin mat of extracellular matrix that separates epithelial sheets, and many other types of cells such as muscle or fat cells, from connective tissue.

## TRUE/FALSE

Decide whether each of these statements is true or false, and then explain why.

**19–48** A sheet of basal lamina underlies all epithelia.

**19–49** The basal lamina is constructed from components supplied by different cell types, commonly by epithelial cells on one side and stromal cells on the other.

**19–50** A proteoglycan in the basal lamina of the kidney glomerulus plays a critical role in filtering the molecules that pass from the bloodsteam into the urine.

**19–51** Most muscular dystrophy diseases (muscle-wasting diseases) arise from defective components in the specialized basal lamina (sarcolemma) that surrounds muscle fibers.

## THOUGHT PROBLEMS

**19–52** The glycosaminoglycan polysaccharide chains that are linked to specific core proteins to form the proteoglycan components of the basal lamina are highly negatively charged. How do you suppose these negatively charged polysaccharide chains help to establish a hydrated gel-like environment around the cell? How would the properties of these molecules differ if the polysaccharide chains were uncharged?

**19–53** The unusually thick basal lamina of the kidney glomerulus is a key component of the complex molecular filter that controls passage of solutes into the urine. Typically, 180 L of fluid are filtered through the kidney each day, but most is reabsorbed, with only about 1.5 L being released as urine. The initial filtration is size, shape, and charge selective.

A. The effective pore size is smaller for negatively charged solutes than it is for positively charged ones of the same size. What features of the basal lamina do you suppose might contribute to this difference in filtration of charged molecules?

B. For neutral solutes of the same molecular weight, the effective pore size is smaller for spherical molecules than for elongated molecules. What do you suppose is the basis for this shape selectivity?

**19–54** Discuss the following statement: "The basal lamina of muscle fibers serves as a molecular bulletin board, in which adjoining cells can post messages that direct the differentiation and function of the underlying cells."

**19–55** Certain bacteria secrete enzymes that can digest protein or carbohydrate components of the basal lamina. Why do you suppose they do so?

## DATA HANDLING

**19–56** It is not an easy matter to assign particular functions to specific components of the basal lamina, since the overall structure is a complicated composite material with both mechanical and signaling properties. Nidogen, for example, cross-links two central components of the basal lamina by binding to the laminin-γ1 chain and to type IV collagen. Given such a key role, it was surprising that mice with a homozygous knockout of the gene for nidogen-1 were entirely healthy, with no abnormal phenotype. Similarly, mice homozygous for a knockout of the gene for nidogen-2 also appeared completely normal. By contrast, mice that were homozygous for a defined mutation in the gene for laminin-γ1, which eliminated just the binding site for nidogen, died at birth with severe defects in lung and kidney formation. The mutant portion of the laminin-γ1 chain is thought to have no other function than to bind nidogen, and does not affect laminin structure or its ability to assemble into basal lamina. How would you explain these genetic observations, which are summarized in Table 19–4? What would you predict would be the phenotype of a mouse that was homozygous for knockouts of both nidogen genes?

**19–57** The basal lamina normally provides an impenetrable barrier to cells, but cells such as lymphocytes and macrophages are able to cross the barrier by digesting the components of the lamina using matrix metalloproteases (MMPs). This family of proteases is implicated in normal bodily functions, as well as many diseases. For example, in order for cancer cells to metastasize they have to penetrate the basal lamina. Because of their fundamental importance to basic science and clinical medicine, MMPs have been studied extensively, and many mouse knockouts have been made.

Thus far, these studies have mainly highlighted the complexity of MMP functions. Take the case of MT1-MMP, which is anchored to the cell membrane. MT1-MMP null mice show skeletal abnormalities, grow slowly at birth, and usually die within a few weeks. Cells derived from these animals cannot penetrate collagen gels (unlike their normal counterparts). One consequence is a complete lack of white adipose tissue (Figure 19–15). The adipocytes in the mutant mice are very small compared to those in wild-type mice, and they appear to be trapped in a tangle of collagen fibers. In addition, DNA microarray analysis of their mRNAs shows that they have not fully differentiated. The link between MT1-MMP deficiency and failure of adipocytes to differentiate is not understood. Suggest some possible explanations for how a lack of MT1-MMP might block adipocyte differentiation.

**Table 19–4 Phenotypes of mice with genetic defects in components of the basal lamina** (Problem 19–56).

| PROTEIN | GENETIC DEFECT | PHENOTYPE |
|---|---|---|
| nidogen-1 | gene knockout (–/–) | none |
| nidogen-2 | gene knockout (–/–) | none |
| laminin-γ1 | nidogen binding-site deletion (+/–) | none |
| laminin-γ1 | nidogen binding-site deletion (–/–) | dead at birth |

+/– stands for heterozygous, –/– stands for homozygous.

(A) wild-type mice

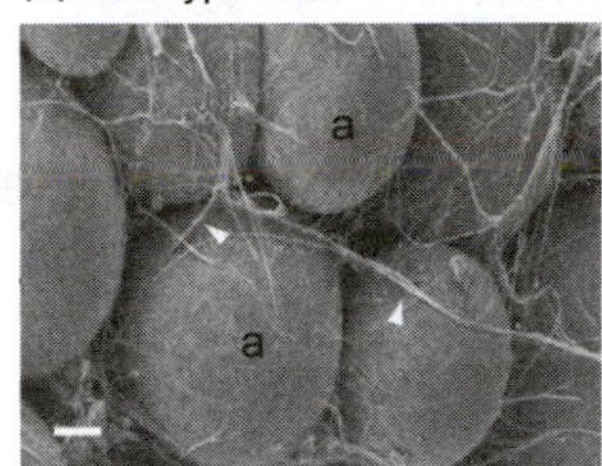

(B) MT1-MMP knockout mice

**Figure 19–15** Adipocytes in normal and MT1-MMP knockout mice (Problem 19–57). (A) Wild-type mice. (B) MT1-MMP knockout mice. A few adipocytes are labeled with 'a.' *White arrowheads* point to collagen fibers. Scale bar is 10 micrometers. Both micrographs are shown at the same magnification.

# INTEGRINS AND CELL–MATRIX ADHESION

TERMS TO LEARN

| | |
|---|---|
| anchorage dependence | integrin |
| focal adhesion kinase (FAK) | |

## DEFINITIONS

Match each definition below with its term from the list above.

**19–58** Cytoplasmic tyrosine kinase present at cell–matrix junctions in association with the cytoplasmic tails of integrins.

**19–59** Principal receptor on animal cells for binding most extracellular matrix proteins, including collagens, fibronectin, and laminins.

**19–60** Dependence of cell growth on attachment to a substratum.

## TRUE/FALSE

Decide whether each of these statements is true or false, and then explain why.

**19–61** Integrins can convert mechanical signals into molecular signals.

**19–62** Various types of integrins connect extracellular binding sites to all the different kinds of cytoskeletal elements, including actin, microtubules, and intermediate filaments.

**19–63** Integrins are thought to be rigid rods that span the membrane and link binding sites outside the cell to those inside the cell.

## THOUGHT PROBLEMS

**19–64** The affinity of integrins for matrix components can be modulated by changes to their cytoplasmic domains: a process known as inside-out signaling. You have identified a key region in the cytoplasmic domains of αIIbβ3 integrin that seems to be required for inside-out signaling (Figure 19–16). Substitution of alanine for either D723 in the β chain or R995 in the α chain leads to a high level of spontaneous activation, under conditions where the wild-type chains are inactive. Your advisor suggests that you convert the aspartate in the β chain to an arginine (D723R) and the arginine in the α chain to an aspartate (R995D). You compare all three α chains (R995, R995A, and R995D) against all three β chains (D723, D723A, and D723R). You find that all pairs have a high level of spontaneous activation,

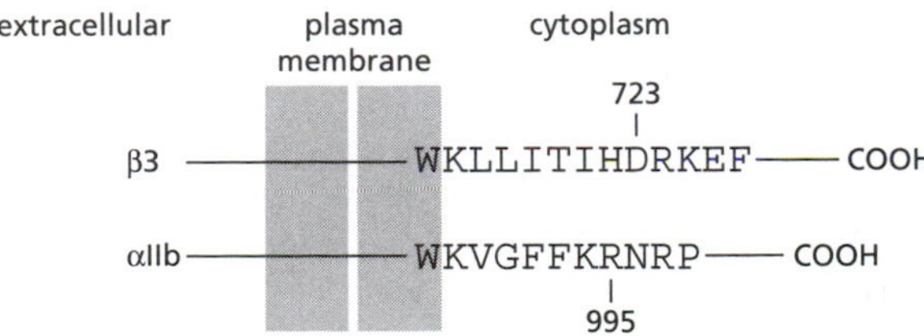


**Figure 19–16** Schematic representation of αIIbβ3 integrin (Problem 19–64). The D723 and R995 residues are indicated.

except D723 vs R995 (the wild type) and D723R vs R995D, which have low levels. Based on these results, how do you think the αIIbβ3 integrin is held in its inactive state?

## CALCULATIONS

**19–65** Platelets are flat, disclike cells about 2 μm in diameter. Estimates of the number of integrin molecules on their surface vary around a mean of about 80,000. If the integrins themselves are about 10 nm in diameter, how tightly packed are they? (Assume that the total membrane area is $2\pi r^2$.)

## DATA HANDLING

**19–66** Integrins are important transmembrane receptors that bind and respond to the extracellular matrix. K562 cells are a line of mouse erythroleukemia cells that are capable of differentiating into red blood cells in response to certain stimuli. These cells express a single type of integrin, α5β1, on their surface, which endows the cells with the ability to bind the RGD motifs of the extracellular matrix protein fibronectin.

Most high-affinity monoclonal antibodies against the separated α5 and β1 chains reduce the amount of fibronectin that binds to the cells; however, one rare antibody specific for the β1 chain gives a 20-fold increase in binding (Table 19–5). Experiments using mixtures of anti-integrin antibodies, anti-fibronectin antibodies, and peptides with and without the RGD motif confirmed that the observed increase in fibronectin binding was due to interactions of fibronectin with the integrin (Table 19–5). In the presence of the anti-β1 antibody, the integrin has an increased affinity for fibronectin (Figure 19–17). Any intracellular influences on the stimulated binding in the presence of the anti-β1 antibody were ruled out by showing that the same results were obtained in the presence of various metabolic poisons such as azide.

A. How do the experiments in Table 19–5 confirm that anti-β1 stimulates fibronectin binding due to its interaction with α5β1 integrin?

B. Which experiments rule out the possibility that the apparent increase in binding affinity in Figure 19–17 is due to increased numbers of integrin molecules on the cell surface: either newly synthesized or newly transferred from some internal compartment?

C. How do you suppose this particular anti-β1 antibody increases the affinity of the α5β1 integrin for fibronectin?

**19–67** The ability of a cell to control integrin–ligand interactions from within is termed inside-out signaling. The major surface protein of blood platelets, αIIbβ3 integrin, binds to fibrinogen when platelets are stimulated with clotting factors such as thrombin. By binding to a receptor on the cell surface, thrombin triggers an intracellular signaling pathway that activates αIIbβ3

**Table 19–5 Effects of antibodies on the binding of fibronectin to cells** (Problem 19–66).

| ADDITIONS | FIBRONECTIN BOUND (cpm) |
|---|---|
| None | 2000 |
| Anti-α5 antibody | 0 |
| Anti-α4 antibody | 2000 |
| Anti-β1 antibody | 40,000 |
| Anti-β1 antibody + anti-α5 antibody | 0 |
| Anti-β1 antibody + anti-α4 antibody | 40,000 |
| Anti-β1 antibody + anti-fibronectin antibody | 500 |
| Anti-β1 antibody + GRGDSP peptide | 3000 |
| Anti-β1 antibody + GRGESP peptide | 40,000 |

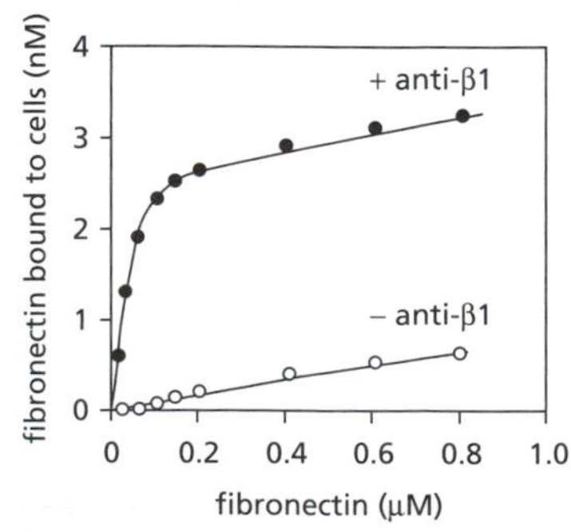


**Figure 19–17** Detailed studies of fibronectin binding to K562 cells (Problem 19–66).

**Table 19–6 Fibrinogen-dependent aggregation of CHO cells expressing various wild-type and mutant αIIb and β3 subunits** (Problem 19–67).

| αIIb CHAIN | β3 CHAIN | AGGREGATION WITHOUT MAb 62 | AGGREGATION WITH MAb 62 |
|---|---|---|---|
| Normal | normal | – | +++ |
| Truncated | normal | +++ | +++ |
| Normal | truncated | – | +++ |
| Truncated | truncated | +++ | +++ |

integrin, allowing platelets to aggregate to form blood clots. Platelets do not bind fibrinogen or aggregate until stimulated, although αIIbβ3 integrin is always present on their surface. What regulates the activity of this all-important integrin?

If the genes for the subunits of αIIbβ3 are expressed in Chinese hamster ovary (CHO) cells, the cells fail to aggregate when incubated with fibrinogen in the presence or absence of thrombin. If the cells are first incubated with MAb 62 antibodies, which bind to αIIbβ3 integrin and activate it (analogous to the anti-β1 antibody described in Problem 19–66), the cells aggregate within minutes of adding fibrinogen. CHO cells without αIIbβ3 do not aggregate when treated this way.

By deleting the short cytoplasmic domains of αIIb and β3, various combinations of truncated and wild-type αIIb and β3 can be tested in CHO cells. All combinations of the αIIb and β3 chains allow cells to aggregate in the presence of fibrinogen and MAb 62; however, the truncated αIIb chain allows aggregation even in the absence of MAb 62 (Table 19–6).

A. Why do you suppose that truncating the cytoplasmic domain of the αIIb subunit increases the affinity of the integrin for fibrinogen and allows the cells to aggregate?

B. The αIIbβ3 integrin is accessible on the surface of the CHO cells, as revealed by the various aggregation studies. Why, then, does thrombin not stimulate the cells to aggregate?

C. There are two genes for αIIb in diploid human cells. If one of the two genes suffered a truncation of the kind described in this problem, do you think the individual would show any blood-clotting problems?

# THE EXTRACELLULAR MATRIX OF ANIMAL CONNECTIVE TISSUES

TERMS TO LEARN

| | |
|---|---|
| collagen | glycosaminoglycan (GAG) |
| collagen fibril | hyaluronan |
| elastin | matrix metalloprotease |
| elastic fiber | proteoglycan |
| fibril-associated collagen | RGD sequence |
| fibrillar collagen | serine protease |
| fibroblast | type III fibronectin repeat |
| fibronectin | |

## DEFINITIONS

Match each definition below with its term from the list above.

**19–68** Fibrous protein rich in glycine and proline that, in its many forms, is a major component of the extracellular matrix and connective tissues.

**19–69** Complex network of polysaccharides (such as glycosaminoglycans or cellulose) and proteins (such as collagens) secreted by cells that serves as a structural element in tissues and also influences tissue development and physiology.

**19–70** General name for long, linear, highly charged polysaccharides composed of a repeating pair of sugars, one of which is always an amino sugar, that is found covalently linked to a protein core in the extracellular matrix.

**19–71** Type of collagen molecule that assembles into ropelike structures and larger, cablelike bundles.

**19–72** Extracellular matrix protein that binds to cell-surface integrins to promote adhesion of cells to the matrix and to provide guidance to migrating cells during embryogenesis.

**19–73** Hydrophobic protein that forms extracellular extensible fibers that give tissues their stretchability and resilience.

**19–74** Common cell type in connective tissue that secretes an extracellular matrix rich in collagen and other extracellular matrix macromolecules.

## TRUE/FALSE

Decide whether each of these statements is true or false, and then explain why.

**19–75** The extracellular matrix is a relatively inert scaffolding that stabilizes the structure of tissues.

**19–76** One of the main chemical differences between proteoglycans and other glycoproteins lies in the structure of their carbohydrate side chains: proteoglycans mostly contain long, unbranched polysaccharide side chains, whereas other glycoproteins contain much shorter, highly branched oligosaccharides.

**19–77** Breakdown and resynthesis of collagen must be important in maintaining the extracellular matrix; otherwise, vitamin C deficiency in adults would not cause scurvy, which is characterized by a progressive weakening of connective tissue due to inadequate hydroxylation of collagen.

**19–78** The elasticity of elastin derives from its high content of $\alpha$ helices, which act as molecular springs.

**19–79** In humans, all forms of fibronectin are produced from one large gene by alternative splicing.

## THOUGHT PROBLEMS

**19–80** Carboxymethyl Sephadex is a negatively charged cross-linked dextran that is commonly used for purifying proteins. It comes in the form of dry beads that swell tremendously when added to water. You packed a chromatography column with the swollen gel. When you start equilibrating the column with a buffer that contains 50 mM NaCl at neutral pH, you are alarmed to see a massive shrinkage in gel volume. Why does the dry Sephadex swell so dramatically when it is placed in water? Why does the swollen gel shrink so much when a salt solution is added?

**19–81** At body temperature, L-aspartate in proteins racemizes to D-aspartate at an appreciable rate. Most proteins in the body have a very low level of D-aspartate, if it can be detected at all. Elastin, however, has a fairly high level of D-aspartate. Moreover, the amount of D-aspartate increases in direct proportion to the age of the person from whom the sample was taken. Why do you suppose that most proteins have little if any D-aspartate, while elastin has high, age-dependent levels?

**19–82** Some have speculated that Abraham Lincoln had Marfan's syndrome, mainly due to his height (6'4" when the average was 5'6") and his thin physique. What protein is mutated in those that suffer from Marfan's syndrome?

## CALCULATIONS

**19–83** Defects in collagen genes are responsible for several inherited diseases, including osteogenesis imperfecta, a disease characterized by brittle bones, and Ehlers–Danlos syndrome, which can lead to sudden death due to ruptured internal organs or blood vessels. In both diseases, the medical problems arise because the defective gene in some way compromises the function of collagen fibrils. For example, homozygous deletions of the type *Iα1* gene eliminates α1(I) collagen entirely, thereby preventing formation of any type I collagen fibrils. Such homozygous mutations are usually lethal in early development. The more common situation is for an individual to be heterozygous for the mutant gene, having one normal gene and one defective gene. Here the consequences are less severe.

A. Type I collagen molecules are composed of two copies of the α1(I) chains and one copy of the α2(I) chain. Calculate the fraction of type I collagen molecules, $[\alpha1(I)]_2\alpha2(I)$, that will be normal in an individual who is heterozygous for a deletion of the entire *α1(I)* gene. Repeat the calculation for an individual who is heterozygous for a point mutation in the *α1(I)* gene.

B. Type III collagen molecules are composed of three copies of the α1(III) chain. Calculate the fraction of type III collagen molecules, $[\alpha1(III)]_3$, that will be normal in an individual who is heterozygous for a deletion of the entire *α1(III)* gene. Repeat the calculation for an individual who is heterozygous for a point mutation in the *α1(III)* gene.

C. Which kind of collagen gene defect—deletion or point mutation—is more likely to be dominant (that is, to cause the heterozygote to display a mutant phenotype)?

## DATA HANDLING

**19–84** Hyaluronan, a polysaccharide composed of strictly alternating *N*-acetylglucosamine (GlcNAc) and glucuronic acid (GlcUA) residues, is spun out directly from the cell surface by hyaluronan synthase, an enzyme embedded in the plasma membrane. Studies with different hyaluronan synthases have failed to agree on such basic questions as whether new sugars are added one at a time or as disaccharide units, and whether they are added to the reducing or nonreducing end of the polysaccharide chain.

You are studying the enzyme from *Pasteurella multocida*. When this enzyme is expressed in *E. coli*, the bacteria spin out a capsule of hyaluronan. To study the reaction mechanism, you use purified membrane fragments containing the recombinant hyaluronan synthase and add a radiolabeled tetrasaccharide acceptor (Figure 19–18A), along with the activated forms of

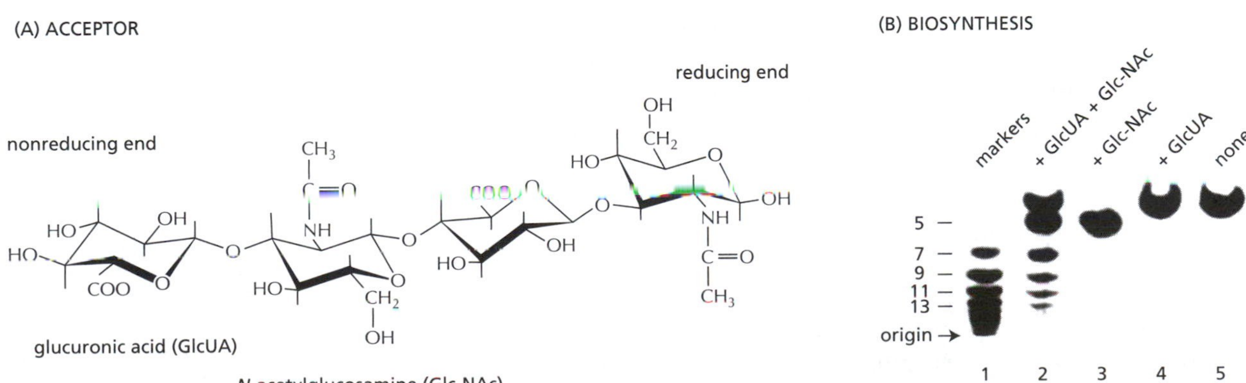


**Figure 19–18** Synthesis of hyaluronan by hyaluronan synthase (Problem 19–84). (A) Tetrasaccharide acceptor. GlcNAc is *N*-acetylglucosamine, GlcUA is glucuronic acid. The reducing and nonreducing ends of the oligosaccharide are indicated. (B) Synthesis of polysaccharide chains by hyaluronan synthase. After incubation of the $^3$H-labeled tetrasaccharide primer with hyaluronan synthase and one or both of the activated sugar monomers, the different-length products were separated by thin-layer chromatography and visualized by fluorography. *Numbers* on the left indicate the length, in sugar residues, of the polysaccharides in the marker lane.

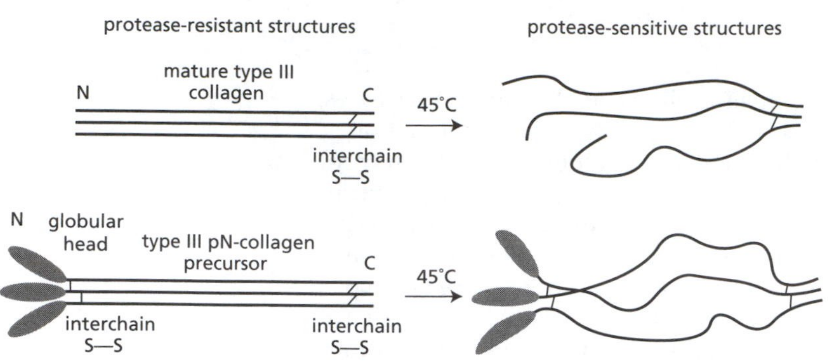


**Figure 19–19** Native and denatured forms of type III collagen and type pN-collagen (Problem 19–85).

the sugar subunits: UDP-*N*-acetylglucosamine and UDP-glucuronic acid. In short incubations with the complete mixture, the tetrasaccharide primer elongates by several units, although it appears that only odd-numbered chains are produced (Figure 19–18B, lane 2). If you leave out one or both activated sugars, you get the results shown in Figure 19–18B, lanes 3–5.

A. Based on these data, would you say that the enzyme normally adds residues one at a time or as disaccharide units? Explain your reasoning.
B. Why do you suppose that only odd-number chains are visible in your assay?
C. Does the enzyme add sugar units to the reducing end or to the nonreducing end of the tetrasaccharide acceptor?

**19–85** The formation of a mature collagen molecule is a complex process. Procollagen is formed first from three collagen chains, which have extensions at their N- and C-termini. Once the chains have been properly wound together, these terminal propeptides are removed. A principal function of the terminal propeptides may be to align the chains correctly to facilitate their assembly.

Two forms of type III collagen molecules have been used to study collagen assembly: the mature type III collagen molecule and type III pN-collagen, which is a precursor with the N-terminal propeptides still attached (Figure 19–19). In both types of collagen molecule the individual chains are held together by disulfide bonds, though the numbers of these bonds differ in the two types. These bonds hold the chains in register, even after they have been completely unwound, so that they reassemble collagen molecules correctly (Figure 19–19).

Reassembly of these two collagen molecules was studied after denaturation at 45°C under conditions that leave the disulfide bonds intact. The temperature was then reduced to 25°C to allow reassembly, and samples were removed at intervals. Each sample was immediately digested with trypsin, which rapidly cleaves denatured chains but does not attack the collagen helix. When the samples were analyzed by SDS-gel electrophoresis in the presence of mercaptoethanol, which breaks disulfide bonds, both collagens yielded identical patterns, as shown for mature collagen in Figure 19–20. Significantly, if the disulfide bonds were left intact during SDS-gel electrophoresis, all bands were replaced by bands with three times the molecular weight.

A. Why did all the resistant peptides increase threefold in molecular weight when the disulfide bonds were left intact?
B. Which set of disulfide bonds in type III pN-collagen (N-terminal or C-terminal) is responsible for the increase in molecular weight of its resistant peptides?
C. Does reassembly of these collagen molecules begin at a specific site or at random sites? If they reassemble from a specific site, deduce its location.
D. Does reassembly proceed zipperlike from an end, or does it occur throughout in an all-or-none mechanism? How do these results distinguish between these possibilities?

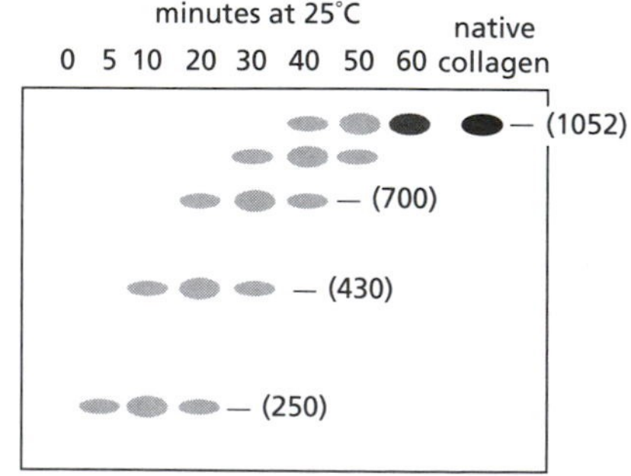


**Figure 19–20** Results of reassembly analysis of type III collagen (Problem 19–85). After various times of reannealing at 25°C, samples were digested with trypsin and subjected to electrophoresis. *Shaded areas* indicate the positions of the trypsin-resistant peptides. *Numbers in parentheses* indicate the approximate number of amino acids in the trypsin-resistant peptides. Reassembly analysis of type III pN-collagen is indistinguishable from that of mature type III collagen.

**19–86** Binding of fragments and competition for binding can be used to identify the portion of a larger ligand that is critical for binding. Fibronectin, which is a large glycoprotein component of the extracellular matrix, binds to

**Table 19–7 Fibronectin-related peptides tested for their ability to promote cell sticking** (Problem 19–86).

| PEPTIDE | SEQUENCE | CONCENTRATION REQUIRED FOR 50% CELL ATTACHMENT (nM) |
|---|---|---|
| Fibronectin | | 0.10 |
| Peptide 1 | YAVTGRGDSPASSKPISINYRTEIDKPSQM(C)* | 0.25 |
| Peptide 2 | VTGRGDSPASSKPI(C) | 1.6 |
| Peptide 3 | SINYRTEIDKPSQM(C) | >100 |
| Peptide 4 | VTGRGDSPA(C) | 2.5 |
| Peptide 5 | SPASSKPIS(C) | >100 |
| Peptide 6 | VTGRGD(C) | 10 |
| Peptide 7 | GRGDS(C) | 3.0 |
| Peptide 8 | RGDSPA(C) | 6.0 |
| Peptide 9 | RVDSPA(C) | >100 |

*The (C) at the C-terminus indicates the cysteine linkage to the carrier protein.

**Table 19–8 Fibronectin-related peptides tested for their ability to block cell sticking** (Problem 19–86).

| PEPTIDE | PERCENT OF INPUT CELLS STICKING |
|---|---|
| GRGDSPC | 2.0 |
| GRGDAPC | 1.9 |
| GKGDSPC | 48 |
| GRADSPC | 49 |
| GRGESPC | 44 |
| None | 47 |

fibronectin receptors on cell surfaces. Fibronectin can stick cells to the surface of a plastic dish, to which they would otherwise not bind, forming the basis of a simple binding assay. By attaching small fragments of fibronectin to dishes, researchers identified the cell-binding domain as a 108-amino acid segment about three-quarters of the way from the N-terminus.

Synthetic peptides corresponding to different portions of the 108-amino acid segment were then tested in the cell-binding assay to localize the active region precisely. Two experiments were conducted. In the first, peptides were linked covalently to plastic dishes via a disulfide bond to an attached carrier protein, and then tested for their ability to promote cell sticking (Table 19–7). In the second experiment, plastic dishes were coated with native fibronectin, and cells that stuck to the dishes in the presence of the synthetic peptides were counted (Table 19–8).

A. The two experiments used different assays to detect the cell-binding segment of fibronectin. Does the sticking of cells to the dishes mean the same thing in both assays? Explain the difference between the assays.

B. From the results in Tables 19–7 and 19–8, deduce the amino acid sequence in fibronectin that is recognized by the fibronectin receptor.

C. How might you make use of these results to design a method for isolating the fibronectin receptor?

# THE PLANT CELL WALL

TERMS TO LEARN

| | |
|---|---|
| cellulose microfibril | primary cell wall |
| cross-linking glycan | secondary cell wall |
| lignin | turgor pressure |
| pectin | |

## DEFINITIONS

Match each definition below with its term from the list above.

**19–87** Thin and extensible cell covering on new plant cells that can accommodate their growth.

**19–88** Bundle of about 40, long, linear chains of covalently linked glucose residues, all with the same polarity, organized in an overlapping parallel array.

**19–89** The large internal hydrostatic pressure that develops in plant cells due to the osmotic imbalance between the cell interior and the fluid in the plant cell wall.

**19–90** A complex network of phenolic compounds that is an abundant polymer in secondary cell walls.

**19–91** Rigid cell covering laid down in layers inside the initial covering once cell growth has stopped.

## TRUE/FALSE

Decide whether each of these statements is true or false, and then explain why.

**19–92** Each cell wall consists of a thin, semirigid primary cell wall adjacent to the cell membrane and a thicker, more rigid secondary cell wall outside the primary wall.

**19–93** Turgor pressure is the main driving force for cell expansion during growth, and it provides much of the mechanical rigidity of living plant tissues.

**19–94** Unlike the extracellular matrix of animal cells, which contains a large amount of protein, plant cell walls are composed entirely of polysaccharides.

**19–95** If the entire cortical array of microtubules were disassembled by drug treatment, new cellulose microfibrils would be laid down in random orientations.

## THOUGHT PROBLEMS

**19–96** Your boss is coming to dinner! All you have for a salad is some wilted, day-old lettuce. You vaguely recall that there is a trick to rejuvenating wilted lettuce, but you can't remember what it is. Should you soak the lettuce in salt water, soak it in tap water, or soak it in sugar water, or maybe just shine a bright light on it and hope that photosynthesis will perk it up.

**19–97** In plant cells the cortical array of microtubules determines the orientation of cellulose microfibrils, which in turn fixes the direction of cell expansion. Cells elongate perpendicular to the orientation of the cellulose microfibrils. The plant growth factors ethylene and gibberellic acid have opposite effects on the orientation of microtubule arrays in epidermal cells of young pea shoots. Gibberellic acid promotes an orientation of the cortical microtubule array that is perpendicular to the long axis of the cell, whereas ethylene treatment causes the microtubule arrays to orient parallel to the long axis of the cell (Figure 19–21).

Which treatment do you think would produce short, fat shoots, and which would produce long, thin shoots?

## CALCULATIONS

**19–98** The hydraulic conductivity of a single water channel is $4.4 \times 10^{-22}$ $m^3$ per second per MPa (megapascal) of pressure. What does this correspond to in terms of water molecules per second at atmospheric pressure? [Atmospheric pressure is 0.1 MPa (1 bar) and the concentration of water is 55.5 M.]

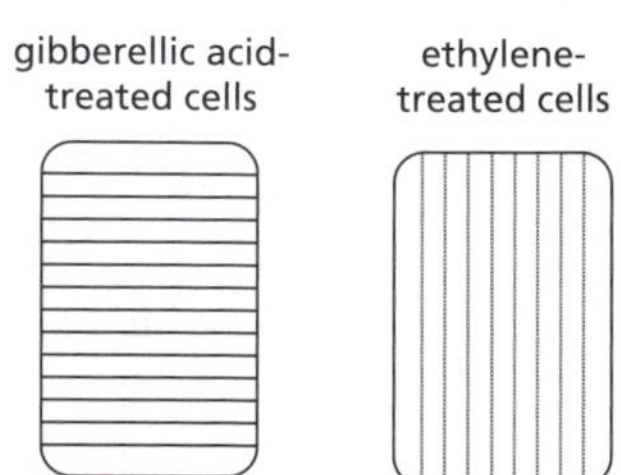


**Figure 19–21** Effects of gibberellic acid and ethylene on the orientation of cortical arrays of microtubules (Problem 19–97).

## DATA HANDLING

**19–99** The synthesis of cellulose is simple from a chemical standpoint: UDP-glucose polymerizes to form cellulose, with release of UDP, which is recycled to form more UDP-glucose. Although cellulose is the most abundant macromolecule on Earth, purifying cellulose synthase from plants proved impossible for many years. Success came from studying bacteria such as *Acetobacter xylinum* that make large amounts of pure cellulose under the right conditions. Curiously, this bacterium requires the signaling dinucleotide,

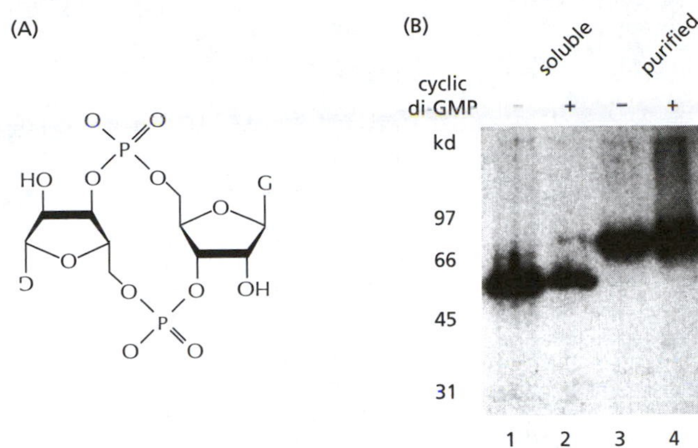


**Figure 19–22** Cellulose synthase (Problem 19–99). (A) The structure of cyclic di-GMP, the activator of cellulose synthase. (B) Photoaffinity labeling of cellulose synthase using $^{32}$P-azido-UDP-glucose activated by UV light. Lanes 1 and 2 are detergent-solubilized cell walls; lanes 3 and 4 are a fraction purified from the cellulose product. Positions of markers of known molecular mass are indicated on the left in kilodaltons. The presence or absence of added cyclic di-GMP is indicated at the top by plusses and minuses.

cyclic di-GMP, for full activation of cellulose synthesis (Figure 19–22A). Active cellulose synthase can be isolated from detergent-extracted bacterial cell walls and readily purified since it gets trapped in its own insoluble cellulose product, rather like a silkworm in its cocoon. To identify the cellulose synthase, you use the affinity label $^{32}$P-azido-UDP-glucose. When exposed to UV light, the azido moiety forms a cross-link to any protein that binds the affinity label.

You test the affinity label with the detergent-solubilized cell walls and with the purified enzyme, with the results shown in Figure 19–23B. Two major bands show labeling: a 57-kd band in the soluble fraction and an 83-kd band in the purified fraction. Both correspond to visible bands on the stained gel. If you omit UV light, the 57-kd band still gets labeled, but the 83-kd band does not. Addition of large amounts of the unlabeled natural substrate, UDP-glucose, blocked labeling of the 83-kd band but not labeling of the 57-kd band.

Which band most likely corresponds to the cellulose synthase? In the explanation of your answer, include the response to cyclic di-GMP, the results in the presence and absence of UV light, and the effects of excess UDP-glucose.

**A pirate with a squamous cell carcinoma in his mouth, from the St Barthomew's Medical Museum in London, photographed by Tim Hunt.** The original painting is in color, which shows that the tumor is well-supplied with blood vessels.

# Cancer

# 20

## In This Chapter


| | |
|---|---|
| CANCER AS A MICROEVOLUTIONARY PROCESS | 469 |
| THE PREVENTABLE CAUSES OF CANCER | 472 |
| FINDING THE CANCER-CRITICAL GENES | 473 |
| THE MOLECULAR BASIS OF CANCER-CELL BEHAVIOR | 477 |
| CANCER TREATMENT: PRESENT AND FUTURE | 481 |


## CANCER AS A MICROEVOLUTIONARY PROCESS

TERMS TO LEARN

| | | |
|---|---|---|
| benign | leukemia | replicative cell senescence |
| cancer stem cells | lymphoma | sarcoma |
| carcinogenesis | malignant | stroma |
| carcinoma | metastases | telomerase |
| chemical carcinogen | primary tumor | tumor progression |
| genetically unstable | | |

### DEFINITIONS

Match each definition below with its term from the list above.

**20–1** The generation of a cancer.

**20–2** Describes a tumor or tumor cell that can invade surrounding tissue or form tumors at other sites in the body.

**20–3** The rare cells within a cancer that are capable of indefinite self-renewal and are responsible for maintaining the cancer.

**20–4** A cancer arising from connective tissue or muscle cells.

**20–5** The neoplasm from which metastases were originally derived.

**20–6** The process by which an initial mildly disordered cell behavior gradually evolves into a full-blown cancer.

**20–7** Describes cells that accumulate genetic and epigenetic changes at an abnormally rapid rate.

**20–8** Describes a tumor that is self-limiting in its growth and noninvasive.

**20–9** Phenomenon observed in primary cell cultures as they age, in which cell proliferation slows down and finally halts.

**20–10** A cancer arising from epithelial cells.

### TRUE/FALSE

Decide whether each of these statements is true or false, and then explain why.

**20–11** All the various cell types in a typical carcinoma, including fibroblasts, inflammatory cells, and blood vessels, evolve from the cancer cell population.

**20–12** Genes can be turned off in an inherited manner without any change in DNA sequence.

**20–13** Genetic instability in the form of point mutations, chromosome rearrangements, and epigenetic changes needs to be maximal to allow the development of cancer.

**20–14** Cancer therapies directed solely at killing the rapidly dividing cells that make up the bulk of a tumor are unlikely to eliminate the cancer from many patients.

## THOUGHT PROBLEMS

**20–15** The incidence of colon cancer increases with age, as shown in Figure 20–1, where the number of newly diagnosed cases in women in 1 year is plotted as a function of age at diagnosis. Studies of many other types of cancer show the same sort of age dependence. Assuming that the rate of mutation is constant throughout life, why do you suppose the incidence of cancer increases so dramatically with age?

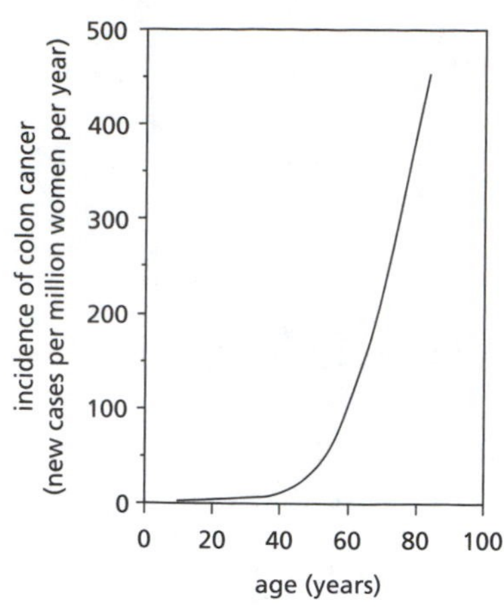


**Figure 20–1** Colon cancer incidence as a function of age (Problem 20–15).

**20–16** In contrast to colon cancer, osteosarcoma, a tumor that occurs most commonly in the long bones, peaks during adolescence. Osteosarcomas are relatively rare in young children (up to age 9) and in adults (over 20). Why do you suppose that the incidence of osteosarcoma does not show the same sort of age-dependence as colon cancer?

**20–17** As shown in Figure 20–2, plots of deaths due to breast cancer and cervical cancer in women differ dramatically from the same plot for colon cancer. At around age 50 the age-dependent increase in death rates for breast and cervical cancer slows markedly, whereas death rates due to colon cancer (and most other cancers) continue to increase. Why do you suppose that the age-dependent increase in death rates for breast and cervical cancer slows after age 50?

## CALCULATIONS

**20–18** Telomere erosion in human somatic cells limits the number of cell divisions to about 50. It has been suggested that this limitation restricts the maximum size of tumors, thus affording some protection against cancer. Assuming that $10^8$ cells have a mass of 1 gram, calculate the mass of a tumor that originated from 50 doublings of a single cancerous cell.

**20–19** Tumor progression—the gradual accumulation of mutations in five or six different genes—provides a natural explanation for the rapid rise in cancer incidence with increasing age. Although this idea is well accepted, it is not the only possible explanation. More than 50 years ago, an entirely different idea was proposed. The central hypothesis was that five or six cancer cells had to be in contact with one another before they could begin to proliferate. (Framed in modern terms, you might imagine that an autocrine growth factor produced by the tumor cells was needed for their growth. Individual cells produced too little to be effective, whereas a small clump of cells secreted enough to trigger their own proliferation.)

If the probability of any cell mutating to a cancer cell is $x$, the probability of it being surrounded by $n$ similarly mutated cells is $x^n$. In a tissue of $N$ cells, the probability of having a critical cluster of cancer cells is $Nx^n$. Suppose, for

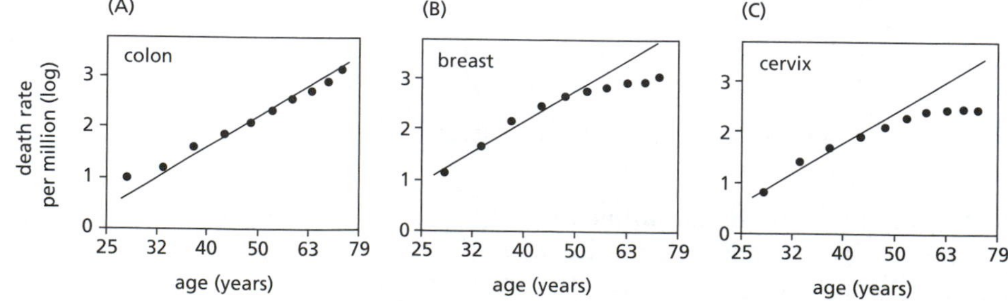


**Figure 20–2** Cancer death rates as a function of age (Problem 20–17). (A) Death rates for colon cancer in females. (B) Death rates for breast cancer in females. (C) Death rates for cervical cancer. The data in all cases are plotted as log of the death rate versus the patient age (on a log scale) at death. The data for colon cancer, displayed on a linear–linear plot, would give the same shape of curve as shown in Figure 20–1. The straight lines in B and C are fit to the data for the earlier age groups, whereas the line in A is fit to all the data points.

the sake of argument, that 1% of the cells in a tissue are precancerous, that five of these cells must be in contact to initiate a cancer, and that there are $10^9$ cells in the tissue. Given these parameters, there would be 0.1 critically sized cancer colonies [$10^9 \times (10^{-2})^5 = 0.1$]. After a doubling in age (hence doubling the number of mutations) there would be 3.2 cancer colonies [$10^9 \times (2 \times 10^{-2})^5 = 3.2$]. This equation predicts that the incidence of a cancer will increase rapidly with age, in much the same way as it would in the tumor-progression model.

In its simplest form, this hypothesis is ruled out by the following experimental observation: an applied carcinogen induces cancers in direct proportion to its concentration. This means that increasing the concentration of carcinogen by a factor of two doubles the number of cancers; increasing it by a factor of four quadruples the number of cancers.

A. How does the linear dependence of cancer on carcinogen concentration rule out the cell-cluster model for cancer formation?

B. How is the linear relationship between cancer and carcinogen concentration explained in the tumor-progression model?

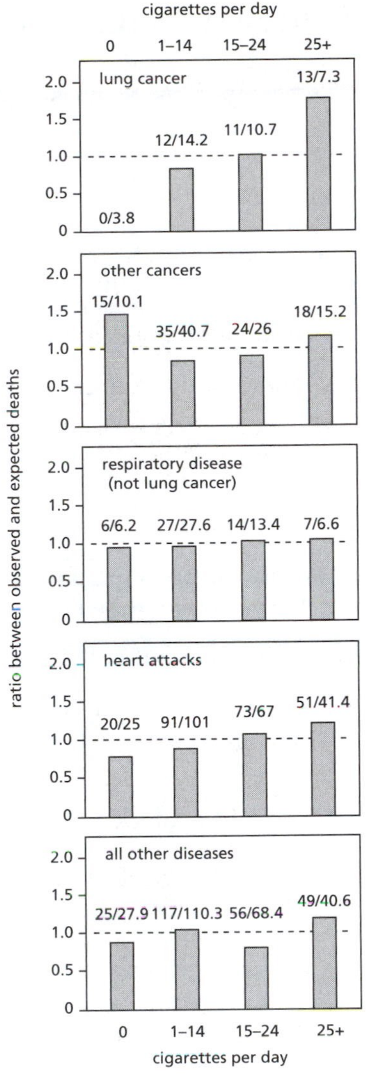


**Figure 20–3** Variation in mortality with amount smoked (Problem 20–20). *Numbers* above the bars give the *observed* deaths/*expected* deaths.

## DATA HANDLING

**20–20** By 1950 it was clear that patients with lung cancer included more heavy smokers than nonsmokers, an association that was not apparent in other diseases. At the time, some considered that the only reasonable interpretation was that smoking is a factor in the disease; others were not prepared to deduce causation from the association. To resolve the issue of causation, a prospective study was carried out to determine the frequency with which lung cancer appeared, in the future, among people whose smoking habits were already known.

A simple questionnaire was sent to about 60,000 doctors in the United Kingdom; about 40,000 responded. Roughly 16,000 were not used because they were from women or men under 35 years old, who were only rarely affected by lung cancer. A preliminary report was published in 1954, 29 months after the questionnaire was sent out. During that time 789 deaths had occurred among the test group, with 36 deaths attributable to a certified diagnosis of lung cancer. Deaths in each of several diseases were analyzed for four groups of smokers: (1) nonsmokers; (2) smokers of 1–14 cigarettes per day; (3) smokers of 15–24 cigarettes per day; and (4) smokers of more than 25 cigarettes per day. The number of deaths in each group was compared with the expected number based on the percentage of all respondents in that group (Figure 20–3).

Among the diseases examined in this preliminary study, which one(s) appear to correlate with amount of tobacco smoked?

**20–21** Mortality due to lung cancer was followed in groups of males in the United Kingdom for 50 years. Figure 20–4 shows the cumulative risk of dying from lung cancer as a function of age and smoking habits for four groups of males: those who never smoked, those who stopped at age 30, those who stopped at age 50, and those who continued to smoke. These data show clearly that an individual can substantially reduce his cumulative risk of dying from lung cancer by stopping smoking. What do you suppose is the biological basis for this observation?

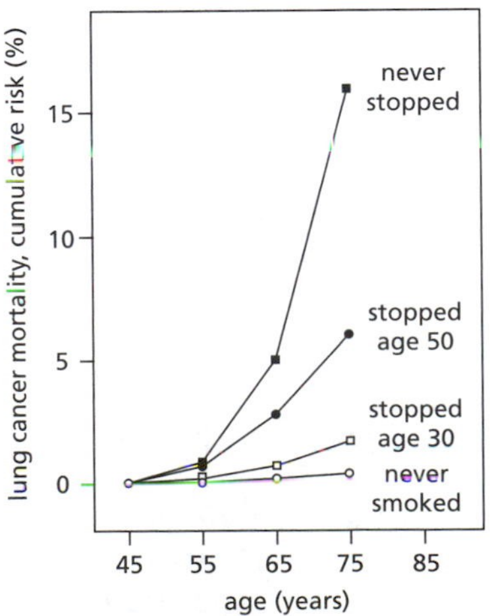


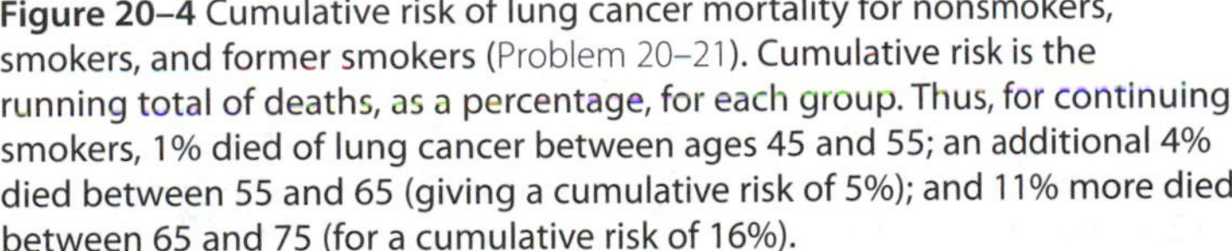
**Figure 20–4** Cumulative risk of lung cancer mortality for nonsmokers, smokers, and former smokers (Problem 20–21). Cumulative risk is the running total of deaths, as a percentage, for each group. Thus, for continuing smokers, 1% died of lung cancer between ages 45 and 55; an additional 4% died between 55 and 65 (giving a cumulative risk of 5%); and 11% more died between 65 and 75 (for a cumulative risk of 16%).

# THE PREVENTABLE CAUSES OF CANCER

TERMS TO LEARN

carcinogen | tumor initiator | tumor promoter

## DEFINITIONS

Match each definition below with its term from the list above.

**20–22** A chemical that is itself not mutagenic but whose repeated application can stimulate tumor formation by mutant cells.

**20–23** An agent that can cause cancer.

## TRUE/FALSE

Decide whether each of these statements is true or false, and then explain why.

**20–24** Many of the most potent carcinogens are chemically inert until after they have been modified by cytochrome P-450 oxidases in the liver.

**20–25** Viruses and other infectious agents play no role in human cancers.

**20–26** The main environmental causes of cancer are the products of our highly industrialized way of life such as pollution and food additives.

## THOUGHT PROBLEMS

**20–27** Epidemiological studies can provide suggestive links between environmental factors and cancer. For example, as shown in Figure 20–5 the curve for deaths due to lung cancer in the US parallels the curve for per capita cigarette consumption. However, the curve for lung cancer is displaced by some 25 years from that for cigarette smoking. What do you suppose is the basis for this delay? What would you say to your uncle, who insists that people who smoke are inherently more cancer prone and that lung cancer really has nothing to do with cigarettes?

## DATA HANDLING

**20–28** The Tasmanian devil, a carnivorous Australian marsupial, is threatened with extinction by the spread of a fatal disease in which a malignant oral–facial tumor interferes with the animal's ability to feed. You have been called in to analyze the source of this unusual cancer. It seems clear to you that the cancer is somehow spread from devil to devil, very likely by their frequent fighting, which is accompanied by biting around the face and mouth. To uncover the source of the cancer, you isolate tumors from 11 devils captured in widely separated regions and examine them. As might be expected, the karyotypes of the tumor cells are highly rearranged relative to that of the wild-type devil (Figure 20–6). Surprisingly, you find that the karyotypes from all 11 tumor samples are very similar. Moreover, one of the Tasmanian devils has an inversion on chromosome 5 that is not present in its facial tumor. How do you suppose this cancer is transmitted from devil to devil? Is it likely to arise as a consequence of an infection by a virus or microorganism? Explain your reasoning.

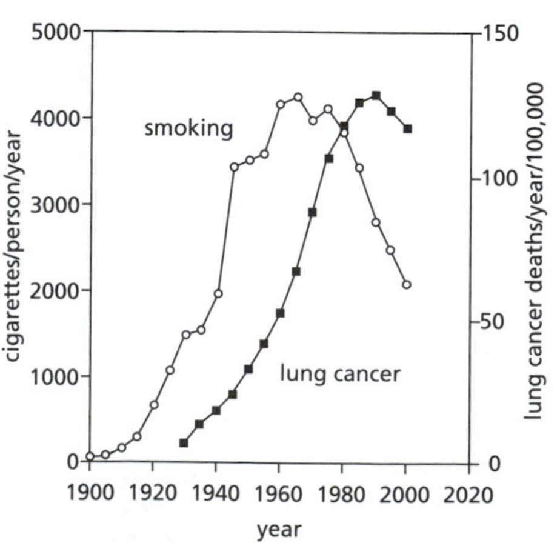


**Figure 20–5** Lung cancer deaths and per capita cigarette consumption in the US from 1930 to 2000 (Problem 20–27).

**20–29** Certain inbred strains of mice suffer tumors of the breast at a relatively high frequency, whereas other inbred strains form breast tumors rarely, or not at all. To investigate the basis for this hereditary difference, you set up a series of genetic crosses between the 'high' and 'low' tumor-forming strains of mice, as shown in Table 20–1. You are amazed to find that high frequencies of tumors appear in F1 female mice only when their mothers were from the

(A)

Tasmanian devil (*Sarcophilus harrisii*)

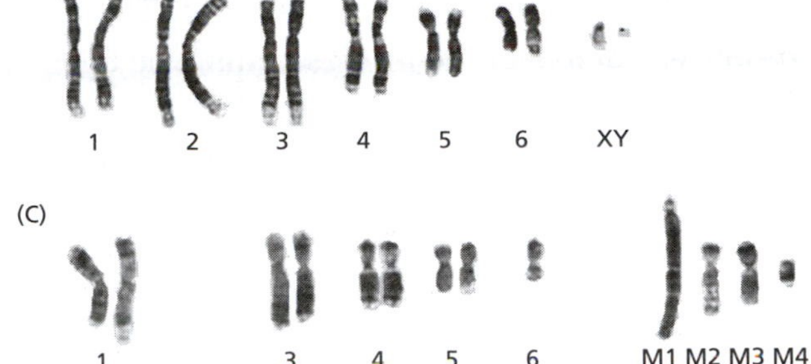


**Figure 20–6** Karyotypes of cells from Tasmanian devils (Problem 20–28). (A) A Tasmanian devil. (B) Normal karyotype for a male Tasmanian devil. The karyotype has 14 chromosomes, including XY. (C) Karyotype of cancer cells found in each of the 11 facial tumors studied. The karyotype has 13 chromosomes, no sex chromosomes, no chromosome-2 pair, one chromosome 6, two chromosomes 1 with deleted long arms, and four highly rearranged chromosomes (M1–M4).

'high'-frequency strains. When you cross the F1 progeny generated in an experiment to produce F2 mice, you find the same result: high frequencies of tumors appear in F2 female mice only when their grandmothers were from the 'high'-frequency strain.

A. Can you explain these results on the basis of inheritance of a chromosomal mutation: recessive, dominant, or X-linked?

B. In Experiment 4 one of your CBA (low) mothers died and you put her pups with an A (high) mother for foster care. Much to your surprise, the fostered female pups developed breast tumors. Moreover, pups from these fostered females passed on the tendency to form breast tumors to their daughters. What do you suppose might be the basis for these results?

# FINDING THE CANCER-CRITICAL GENES

TERMS TO LEARN

| | | |
|---|---|---|
| cancer-critical gene | *Rb* gene | Src |
| oncogene | Rb protein | transformation |
| proto-oncogene | retinoblastoma | tumor suppressor gene |
| *Ras* | retrovirus | tumor virus |

## DEFINITIONS

Match each definition below with its term from the list above.

**20–30** General term for a mutant gene whose overactive form causes cancer.

**20–31** The conversion of normal cells to abnormally proliferating cells with cancerous properties.

**20–32** General term for a normal gene in which a gain-of-function mutation can drive a cell toward cancer.

**Table 20–1 Crosses between high-incidence and low-incidence tumor strains of mice** (Problem 20–29).

| EXPERIMENT | FEMALE PARENT | MALE PARENT | TUMORS IN F1 FEMALES |
|---|---|---|---|
| 1. | D (high) | C57 (low) | 36.1% |
| 2. | C57 (low) | D (high) | 5.5% |
| 3. | A (high) | CBA (low) | 86.3% |
| 4. | CBA (low) | A (high) | 0.0% |
| 5. | Z (high) | I (low) | 90.0% |
| 6. | I (low) | Z (high) | 0.0% |

**20–33** Rare type of human cancer in which cells of the retina are converted to a cancerous state by an unusually small number of mutations.

**20–34** Any one of a number of genes whose mutation frequently contributes to the causation of cancer.

**20–35** General term for a normal gene in which a loss-of-function mutation can contribute to cancer.

## TRUE/FALSE

Decide whether each of these statements is true or false, and then explain why.

**20–36** Oncogenes and tumor suppressor genes can both be detected by introducing fragmented DNA from cancer cells into suitable cell lines and isolating colonies that display cancerous properties.

**20–37** Dimethylbenz[a]anthracene (DMBA) must be an extraordinarily specific mutagen since 90% of the skin tumors it causes have an A-to-T alteration at exactly the same site in the mutant *Ras* gene.

**20–38** Individuals who inherit one inactive copy of a tumor suppressor gene are more likely to develop cancer than individuals with two nonmutant copies.

## THOUGHT PROBLEMS

**20–39** By analogy with automobiles, defects in cancer-critical genes have been likened to broken brakes and stuck accelerators, which are caused in some cases through faulty service by bad mechanics. Using this analogy, decide how oncogenes, tumor suppressor genes, and DNA maintenance genes relate to broken brakes, stuck accelerators, and bad mechanics. Explain the basis for each of your choices.

**20–40** *Rb* is one example of a category of antiproliferative genes in humans. Typically, when both copies of such genes are lost, cancers develop. Do you suppose that cancer could be eradicated if tumor suppressor genes such as *Rb* could be expressed at abnormally high levels in all human cells? What would be the effect on the human? Explain your answers.

**20–41** Overexpression of the Myc protein is a common feature of many types of cancer cells, contributing to their excessive cell growth and proliferation. By contrast, when Myc is overexpressed in most normal cells, the result is not excessive proliferation, but cell-cycle arrest or apoptosis. How do you suppose that overexpression of Myc can have such different outcomes in normal cells and in cancer cells?

## CALCULATIONS

**20–42** The clinical trial of a gene therapy protocol to cure the human genetic disease, severe combined immunodeficiency syndrome (SCID), used retroviruses to carry in the missing *Il2rg* gene. The trial ended in disaster. Nearly 3 years after retroviral gene therapy was completed, two of the treated children developed T-cell leukemia. In both cases, the therapeutic retrovirus had integrated near the *Lmo2* gene, a known human T cell proto-oncogene, causing it to be aberrantly expressed. It had generally been assumed that insertional mutagenesis by replication-defective retroviruses (the kind used in the trial) would be so rare as to be of negligible consequence. Finding two such insertions among 10 treated infants raised serious concerns about the future of retroviral gene therapy.

To gather information on the basis for this affect, another group examined their collection of retrovirally induced blood cell tumors in mice. In a survey of 600 tumors, they found two leukemias with integrations at the *Lmo2* gene and two with integrations at the *Il2rg* gene. Surprisingly, one of these leukemias had one retrovirus integrated at *Lmo2* and a second integrated at *Il2rg*. This observation raised the possibility that the two integrations were co-selected because they cooperate to induce leukemia. The implication is that the gene therapy trial led to cancers because of the leukemia-promoting *combination* of retroviral expression of *Il2rg* and retroviral integration near *Lmo2*. Retroviral expression of other kinds of genes might not cause any problems.

This intriguing explanation for the gene-therapy results rests on the assumption that finding a leukemia with integrations at *Lmo2* and *Il2rg* by random chance is exceedingly small. Just what is the probability of finding such a dual integration by random chance in a survey of 600 tumors? One way to approach this question is to begin by calculating the chance of finding a random integration at *Il2rg* in 600 tumors that all have a retroviral integration at *Lmo2*. Assume that integration in a 100 kb target around the *Il2rg* would be necessary to alter expression of the *Il2rg* gene. Also, assume that there are exactly 2 integration events in each of the 600 tumors: one at *Lmo2* and one that is random.

A. Given that the mouse genome is $2 \times 10^6$ kb, what fraction ($f_i$) of random integration events will be inside the 100-kb target? What fraction ($f_o$) will be outside the target?

B. What is the probability ($P_N$) that in 600 tumors you will not see a second integration in the target? [$P_N = (f_o)^{600}$]

C. What is the probability ($P_Y$) that in 600 tumors you will find a second integration event in the target? ($P_Y = 1 - P_N$)

D. Given that only 2 out of 600 tumors actually had a retroviral integration at *Lmo2*, what is the chance of getting a dual retroviral integration at *Lmo2* and *Il2rg*?

E. Indicate in a general way how each specific assumption affects your calculations. If the *Il2rg* target were 10 kb instead of 100 kb, would the probability calculated in part D be increased or decreased? If there were, on average, fewer than two retroviral integration events per tumor, would the probability in part D be increased or decreased? If there were more than two integrations events per tumor, how would the probability in part D be affected? If retroviral integration were not random, how would the calculation in part D be affected?

## DATA HANDLING

**20–43** Retinoblastoma is an extremely rare cancer of the retina in the eye. The disease mainly affects children up to the age of 5 years because it can only occur while the nerve precursor cells are still dividing. In some cases tumors occur in only one eye, but in other cases tumors develop in both eyes. The bilateral cases all show a familial history of the disease; most of the cases affecting only one eye arise in families with no previous disease history.

An informative difference between unilateral and bilateral cases becomes apparent when the fraction of still undiagnosed cases is plotted against the age at which diagnosis is made (Figure 20–7). The regular decrease with time shown by the bilateral cases suggests that a single chance event is sufficient to trigger the onset of bilateral retinoblastoma. By contrast, the presence of a 'shoulder' on the unilateral curve suggests that multiple events in one cell are required to trigger unilateral retinoblastoma. (A shoulder arises because the events accumulate over time. For example, if two events are required, most affected cells at early times will have suffered only a single event and will not generate a tumor. With time the probability increases that a second event will occur in an already affected cell and therefore cause a tumor.)

One possible explanation for these observations is that tumors develop when both copies of the critical gene (the retinoblastoma, *Rb*, gene) are lost or mutated. In the inherited (bilateral) form of the disease, a child receives a defective *Rb* gene from one parent: tumors develop in an eye when the other

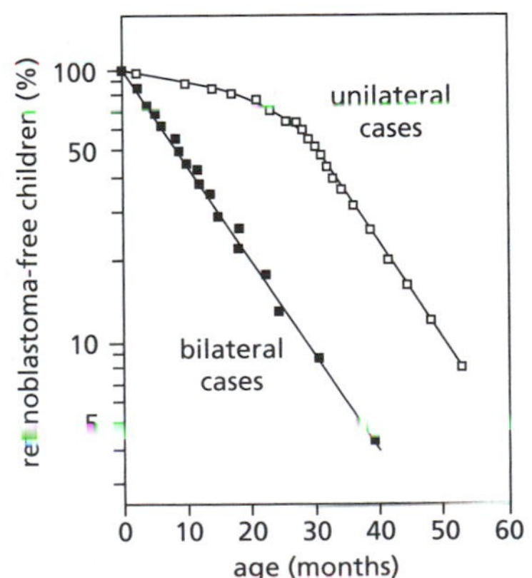


**Figure 20–7** Time of onset of unilateral and bilateral cases of retinoblastoma (Problem 20–43). A population of children, all of whom ultimately developed retinoblastoma, is represented in this graph. The fraction of the population that is still tumor free is plotted against the time after birth.

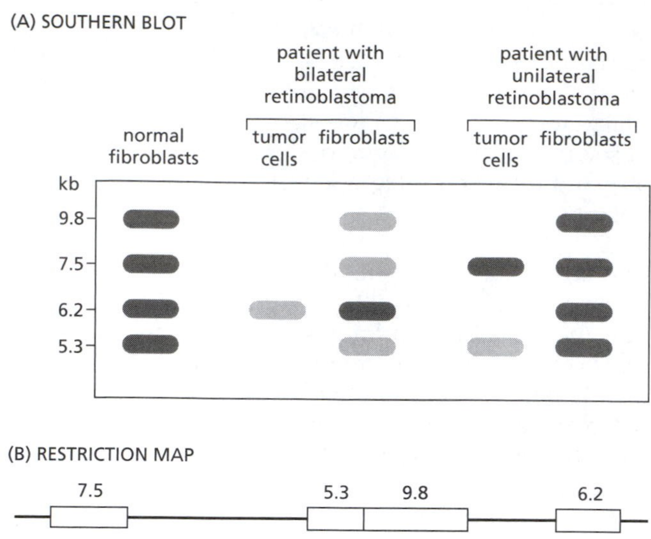


**Figure 20–8** Patterns of blot hybridization of restriction fragments from the retinoblastoma gene (Problem 20–43). (A) Southern blot for normal individuals and for patients with unilateral and bilateral retinoblastoma. *Lighter shading* of some bands indicates half the normal number of copies. (B) The order of the restriction fragments in the *Rb* gene. Fragments that contain exons (*rectangles*) hybridize to the cDNA clone that was used as a probe in these experiments.

copy of the gene is lost through somatic mutation. In fact, the loss of a copy of the gene is frequent enough that tumors usually occur in both eyes. If a person starts with two good copies of the *Rb* gene, tumors arise in an eye only if both copies are lost *in the same cell.* Since such double loss is very rare, it is usually confined to one eye.

To test this hypothesis, you use a cDNA clone of the *Rb* gene to probe the structure of the gene in cells from normal individuals and from patients with unilateral or bilateral retinoblastoma. As illustrated in Figure 20–8, normal individuals have four restriction fragments that hybridize to the cDNA probe (which means each of these restriction fragments contains at least one exon). Fibroblasts (nontumor cells) from the two patients also show the same four fragments, although three of the fragments from the child with bilateral retinoblastoma are present in only half the normal amount. Tumor cells from the two patients are missing some of the restriction fragments.

A. Explain why fibroblasts and tumor cells from the same patient show different band patterns.
B. What are the structures of the *Rb* genes in the fibroblasts from the two patients? What are their structures in the tumor cells from the two patients?
C. Are these results consistent with the hypothesis that retinoblastoma is due to the loss of the *Rb* gene?

**20–44** Now that DNA sequencing is so inexpensive, reliable, and fast, your mentor has set up a consortium of investigators to pursue the ambitious goal of tracking down *all* the mutations in a set of human tumors. He has decided to focus on breast cancer and colorectal cancer because they cause 14% of all cancer deaths. For each of 11 breast cancers and 11 colorectal cancers, you design primers to amplify 120,839 exons in 14,661 transcripts from 13,023 genes. As controls, you amplify the same regions from DNA samples taken from two normal individuals. You sequence the PCR products and use analytical software to compare the 456 Mb of tumor sequence with the published human genome sequence. You are astounded to find 816,986 putative mutations. This represents more than 37,000 mutations per tumor! Surely that can't be right.

Once you think about it for a while, you realize the computer sometimes makes mistakes in calling bases. To test for that source of error, you visually inspect every sequencing read and find that you can exclude 353,738 changes, leaving you with 463,248, or about 21,000 mutations per tumor. Still a lot!

A. Can you suggest at least three other sources of apparent mutations that do not actually contribute to the tumor?
B. After applying a number of criteria to filter out irrelevant sequence changes, you find a total of 1307 mutations in the 22 breast and colorectal cancers, or about 59 mutations per tumor. How might you go about deciding which of

these sequence changes are likely to be cancer-related mutations and which are probably 'passenger' mutations that occurred in genes with nothing to do with cancer (but were found in the tumors because they happened to occur in the same cells with true cancer mutations)?

C. Will your comprehensive sequencing strategy detect all possible genetic changes that affect the targeted genes in the cancer cells?

# THE MOLECULAR BASIS OF CANCER-CELL BEHAVIOR

TERMS TO LEARN

| | | |
|---|---|---|
| colorectal cancer | *p53* | papillomavirus |
| DNA tumor virus | | |

## DEFINITIONS

Match each definition below with its term from the list above.

**20–45** The cause of human warts and a causative factor in carcinomas of the uterine cervix.

**20–46** Tumor suppressor gene—found mutated in about half of human cancers—that encodes a gene regulatory protein that is activated by DNA damage.

**20–47** Common carcinoma of the epithelium lining the colon and rectum.

## TRUE/FALSE

Decide whether each of these statements is true or false, and then explain why.

**20–48** In the cellular regulatory pathways that control cell growth and proliferation, the products of oncogenes are stimulatory components and the products of tumor suppressor genes are inhibitory components.

**20–49** It is clear from studies in mice that mutagenic activation of a single oncogene is sufficient to convert a normal cell into a cancer cell.

**20–50** When analyzed, cancer cells are often found to have mutations in multiple component of a regulatory pathway that controls cell proliferation.

**20–51** The loss of p53 protein makes some cancer cells much less sensitive to irradiation and to many anticancer drugs, which would otherwise destroy tumors by inducing proliferating cells to either stop dividing or undergo apoptosis.

**20–52** Many carcinomas show both chromosomal instability and defective mismatch repair.

## THOUGHT PROBLEMS

**20–53** About 20% of colorectal cancers have mutations in the *B-Raf* gene. B-Raf is a serine/threonine protein kinase that functions in the Ras-Raf-Mek-Erk-MAP kinase cascade, which mediates cellular responses to growth signals. When the pathway is stimulated, Ras activates B-Raf by causing a protein kinase to add phosphates to threonine 598 and serine 601. Activated B-Raf then adds phosphates to key residues in Mek to trigger the rest of the pathway and stimulate cell growth. The mutations of *B-Raf* found in cancer cells give rise to a constantly active form of B-Raf that does not need to be phosphorylated by Ras. In one sample of colorectal cancers, 95% of the mutant *B-Raf* genes had glutamate in place of valine at position 599. Why do you suppose that B-Raf with glutamate at position 599 is active?

**20–54** Mouse mammary tumor virus (MMTV) is an oncogenic retrovirus that causes breast cancer in mice when it integrates into the genome. You want to know whether it carries its own oncogene or generates an oncogene upon integration. You isolate 26 different breast cancers from mice that were exposed to MMTV and determine the sites at which the retroviruses are integrated. In 18 of 26 tumors the viruses are found at a variety of sites that are all located within a 20-kb segment of the mouse genome. Upon closer examination of these 18 tumors, you find that an RNA is expressed from the region of the mouse genome near the integrated virus, but not from the corresponding region in normal mouse breast cells. Do these observations argue for MMTV carrying an oncogene or for it generating an oncogene upon integration? Explain your reasoning.

## CALCULATIONS

**20–55** The *p53* gene encodes a key regulatory protein that can arrest cell growth, induce cell death, or promote cell senescence, in response to DNA damage or other types of cell stress. Its central role in governing a cell's response to stress is highlighted by the finding that it is inactivated by mutation in half of all human cancers. Somewhat surprisingly, mice that lack p53 are fine in all respects—except that they develop tumors by 10 months of age. The product of a second gene, *Mdm2*, negatively regulates p53, targeting it for destruction by attaching ubiquitin to it. Your lab is investigating these genes using mouse knockouts. You can generate *Mdm2*$^{+/-}$ mice perfectly well, but when these mice are mated together, no viable *Mdm2*$^{-/-}$ offspring are born. To investigate the genetic interactions between *p53* and *Mdm2*, you generate doubly heterozygous *p53*$^{+/-}$ *Mdm2*$^{+/-}$ mice and mate them together. The genotypes of the progeny mice are shown in Table 20–2.

A. The *p53* and *Mdm2* genes are on different chromosomes and thus assort independently during meiosis. Assuming that *p53*$^{+}$*Mdm2*$^{+}$, *p53*$^{+}$*Mdm2*$^{-}$, *p53*$^{-}$*Mdm2*$^{+}$, and *p53*$^{-}$*Mdm2*$^{-}$ haploid gametes are produced at equal frequencies by the male and female parents, calculate how frequently each of the progeny genotypes would be generated by random assortment. Which, if any, of the genotypes appear to be significantly under-represented among the progeny?

B. How would you interpret the differences in number of progeny expected and actually generated for *p53*$^{+/+}$ *Mdm2*$^{-/-}$, *p53*$^{+/-}$ *Mdm2*$^{-/-}$, and *p53*$^{-/-}$ *Mdm2*$^{-/-}$ mice?

## DATA HANDLING

**20–56** Remarkably, the *Ink4A-Arf* locus encodes two different tumor suppressor proteins, Ink4A and Arf, which share a common exon but are translated in

**Table 20–2 Genotypes of progeny mice from crosses between doubly heterozygous *p53*$^{+/-}$ *Mdm2*$^{+/-}$ mice** (Problem 20–55).

| GENOTYPE | PROGENY MICE (NUMBER) | PROGENY MICE (EXPECTED) |
|---|---|---|
| $p53^{+/+}$ $Mdm2^{+/+}$ | 3 | |
| $p53^{+/+}$ $Mdm2^{+/-}$ | 5 | |
| $p53^{+/+}$ $Mdm2^{-/-}$ | 0 | |
| $p53^{+/-}$ $Mdm2^{+/+}$ | 7 | |
| $p53^{+/-}$ $Mdm2^{+/-}$ | 11 | |
| $p53^{+/-}$ $Mdm2^{-/-}$ | 0 | |
| $p53^{-/-}$ $Mdm2^{+/+}$ | 1 | |
| $p53^{-/-}$ $Mdm2^{+/-}$ | 7 | |
| $p53^{-/-}$ $Mdm2^{-/-}$ | 2 | |

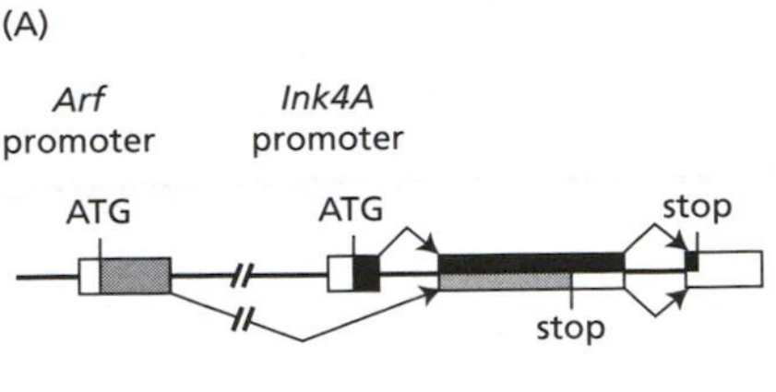


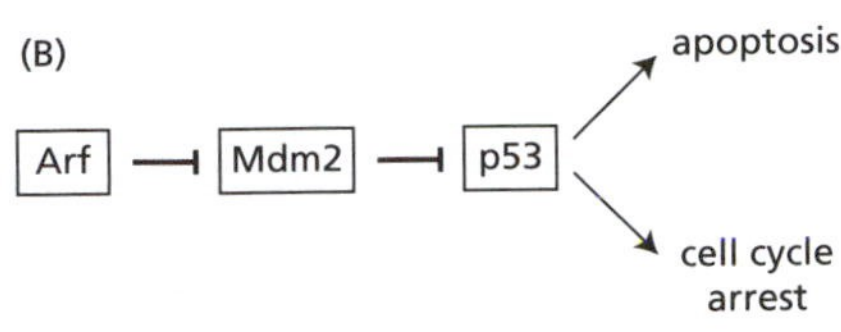


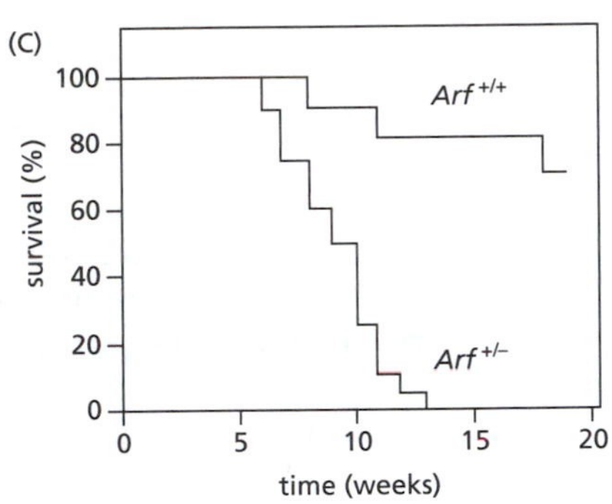


**Figure 20–9** Gene structure and function of Arf (Problem 20–56). (A) Gene structures of *Arf* and *Ink4A*. The ATG in the initial exon of each gene indicates the start site of translation, which corresponds to AUG in the RNA transcript. The reading frame for Ink4A is shown in *black*, and the one for Arf is shown in *gray*. Note that the amino acid sequences of Ink4A and Arf are completely different. (B) Functional relationship between Arf, Mdm2, and p53. The *on-side T* symbol indicates inhibition; for example, Arf inhibits Mdm2. (C) Survival of mice expressing the Myc oncogene on a genetic background that is either $Arf^{+/+}$ or $Arf^{+/-}$. Mice that have survived for a given length of time are expressed as a percentage of the initial population. All the dead mice died of cancer.

different reading frames (Figure 20–9A). Mutations in human tumors that were initially thought to affect Ink4A were later shown to affect a novel protein encoded in a different reading frame; hence, the name of the gene: *Arf*, for alternative reading frame. One of the principal functions of Arf is to inhibit Mdm2, which in turn inhibits p53 (see Problem 20–55). The relationship among Arf, Mdm2, and p53 is commonly represented as shown in Figure 20–9B. The sort of 'double-negative' implied in this relationship can be confusing: Arf is an inhibitor of an inhibitor of p53.

A. Would you expect *Arf*-knockout mice to be more prone, or less prone, to getting tumors than a wild-type mouse? Explain your reasoning.

B. Do you suppose that a $p53^{+/+}$ $Mdm2^{-/-}$ mouse, which will die in early embryogenesis, would be rescued by knockout of the *Arf* gene? That is, would you expect a $p53^{+/+}$ $Mdm2^{-/-}$ $Arf^{-/-}$ mouse to be viable or dead? Explain your reasoning.

C. The *Myc* oncogene, in addition to stimulating cell-proliferation pathways, also activates Arf, thereby indirectly influencing the activity of p53. How would you account for the observation that mice expressing the *Myc* oncogene get tumors more quickly in $Arf^{+/-}$ mice than in $Arf^{+/+}$ mice (Figure 20–9C)?

**20–57** The formation of tumors is a multistep process that may involve the successive activation of several oncogenes. This notion is supported by the discovery that certain pairs of oncogenes, of which *Ras* and *Myc* are among the best-studied examples, transform cultured cells more efficiently than either alone. Similar experiments have been carried out in transgenic mice. In one experiment, the *Ras* oncogene was placed under the control of the MMTV (mouse mammary tumor virus) promoter and incorporated into the germ lines of transgenic mice. In a second experiment, the *Myc* oncogene under control of the same MMTV promoter was introduced into the mouse germ lines. In a third experiment, mice containing the individual oncogenes were mated to produce mice with both oncogenes.

All three kinds of mice developed tumors at a higher frequency than normal animals. Female mice were most rapidly affected because the MMTV promoter, which is responsive to steroid hormones, turns on the transferred oncogenes in response to the hormonal changes at puberty. In Figure 20–10 the rate of appearance of tumors is plotted as the percentage of tumor-free females at different times after puberty.

A. Assume that the lines drawn through the data points are an accurate representation of the data. How many events in addition to expression of the oncogenes are required to generate a tumor in each of the three kinds of mice? (You may wish to reread Problem 20–43.)

B. Is activation of the cellular *Ras* gene the event required to trigger tumor formation in mice that are already expressing the MMTV-regulated *Myc* gene (or vice versa)?

C. Why do you think the rate of tumor production is so high in the mice containing both oncogenes?

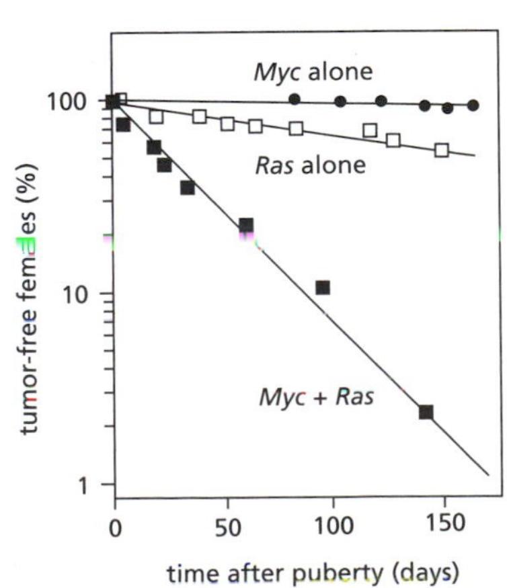


**Figure 20–10** Fraction of tumor-free female mice as a function of time after puberty (Problem 20–57).

**20–58** In a classic experiment, Barbara McClintock studied the genetics of color variegation in corn. She generated speckled kernels by crossing a strain that

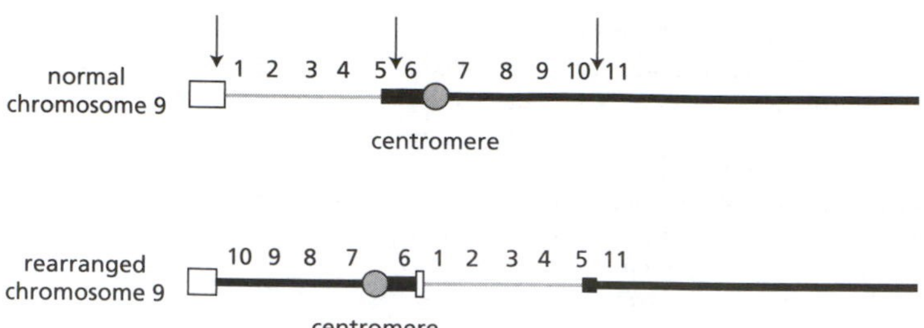


**Figure 20–11** Normal and rearranged chromosomes 9 in corn (Problem 20–58). *Arrows* above the normal chromosome indicate sites of breakage that gave rise to the rearranged chromosome.

carries an x-ray-induced rearrangement of chromosome 9 (Figure 20–11). This chromosome carries a color marker (*C*, colored; recessive form *c*, colorless), which allowed her to follow its inheritance in individual kernels. When strains carrying the rearranged chromosome 9 bearing the dominant *C* allele were crossed with wild-type corn bearing the recessive *c* allele, a small number of kernels in the progeny ears of corn had a speckled appearance.

This color variegation arises by a complex mechanism. In meiosis, recombination within the rearranged segment generates a chromosome with two centromeres as shown in Figure 20–12. In a fraction of these meioses the recombined chromosome with two centromeres gets strung out between the two poles at the first meiotic anaphase, forming a bridge between the two meiotic poles. Some time in anaphase to telophase the strained chromosome breaks at a random position between the duplicated centromeres. The broken ends of the chromosome tend to fuse together after the next S phase, which generates a new dicentric chromosome whose structure depends on where the previous break occurred (Figure 20–12). The forces acting during the subsequent mitosis in turn will break this chromosome, and the fusion–bridge–breakage cycle will repeat itself in the next cell cycle, unless a repair mechanism adds a telomere to the broken end.

Figure 20–12 shows the chromosomal location of another genetic marker on chromosome 9, which can cause a 'waxy' alteration to the starch deposited in the kernels. The waxy allele can be detected by staining with iodine. Waxy (*wx*) is recessive to the normal, nonwaxy (*Wx*) allele. By following the inheritance of the *C* and *Wx* markers in the kernels, McClintock gained an understanding of the behavior of broken chromosomes. She observed three types of patches within the otherwise colored, nonwaxy (*C-Wx*) kernels: colorless, nonwaxy (*c-Wx*) patches; colorless, nonwaxy (*c-Wx*) patches containing one or more colorless, waxy (*c-wx*) spots; and intensely colored, nonwaxy (*?-Wx*) patches (Figure 20–13).

A. Patches arise because cells with a different genetic constitution divide to give identical neighbors that remain together in a cluster. Starting with the

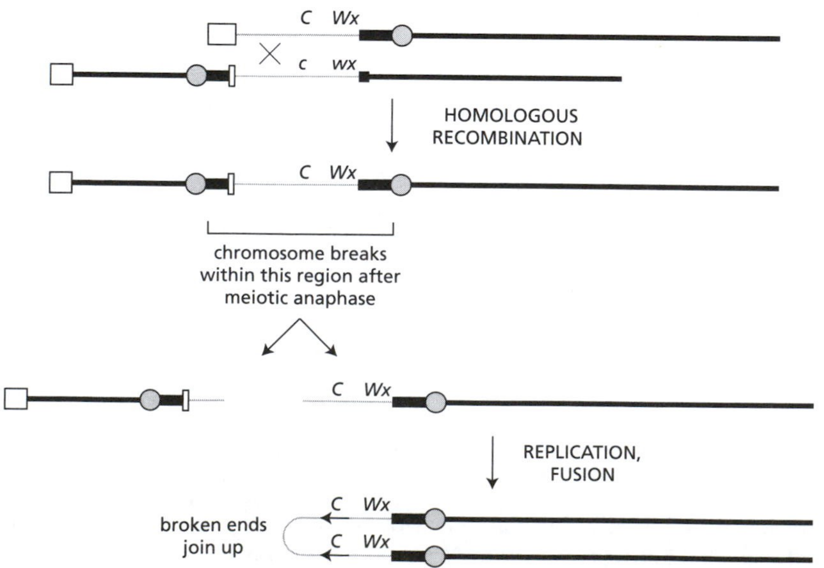


**Figure 20–12** Recombination between a normal and a rearranged chromosome 9 to give a dicentric chromosome (Problem 20–58). Homologous recombination occurs at the X. Only the product with two centromeres is shown; the other product, which has no centromere, will be lost because it cannot attach to the spindle. Breakage of the original dicentric chromosome followed by replication and fusion of the ends gives rise to a second dicentric chromosome.

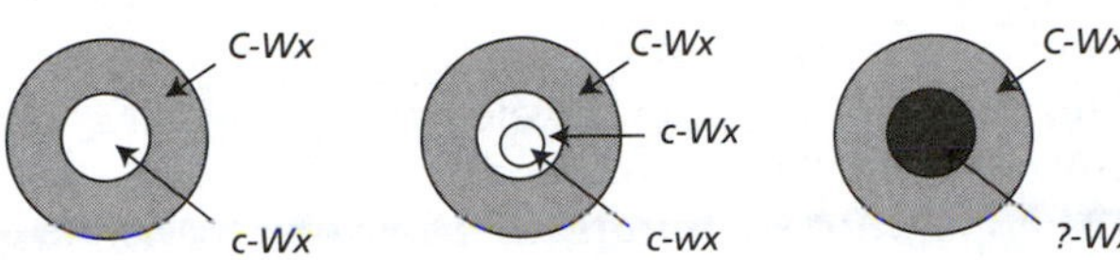


**Figure 20–13** Three types of patches observed in speckled kernels (Problem 20–58).

dicentric chromosome shown at the bottom of Figure 20–12, show how fusion–bridge–breakage cycles might account for the three types of patches shown in Figure 20–13. What is the genetic constitution of the intensely colored patches? (In these crosses, the dominant alleles—*C* and *Wx*—are carried on the rearranged dicentric chromosome at the bottom of Figure 20–12, and the recessive alleles—*c* and *wx*—are carried on the normal, unrearranged chromosome 9.)

B. Would you ever expect to see colored spots within colorless patches? Why or why not?

C. Would you ever expect to see colorless spots within the intensely colored patches? Why or why not?

# CANCER TREATMENT: PRESENT AND FUTURE

| TERMS TO LEARN | | |
|---|---|---|
| gene expression profile | Gleevec | multidrug resistance |

## DEFINITIONS

Match each definition below with its term from the list above.

**20–59** The synthetic drug molecule that can inhibit the chimeric tyrosine kinase encoded by a fusion between the *Bcr* and *Abl* genes.

**20–60** Phenomenon in which a cell becomes insensitive not only to a drug it has been treated with, but also to others to which it has never been exposed.

**20–61** A pattern of expression obtained by simultaneously comparing thousands of genes by DNA microarray analysis of a single sample of normal or abnormal cells.

## TRUE/FALSE

Decide whether each of these statements is true or false, and then explain why.

**20–62** Anticancer therapies take advantage of some molecular abnormality of cancer cells that distinguishes them from normal cells.

**20–63** Tamoxifen kills breast cancer cells by binding to the estrogen receptor and keeping it in its inactive form, even when estrogen is present.

**20–64** The hypermutable nature of tumor cells means that treatments with single anticancer drugs or even combinations of such drugs are unlikely to eradicate all the cancer cells.

## THOUGHT PROBLEMS

**20–65** Progress in cancer therapy is often measured in terms of the fraction of patients that are alive 5 years after their initial diagnosis. For example, in the United States in 1970, 7% of lung cancer patients were alive after 5 years, whereas in 2000, 14% survived for 5 years. Although this modest improvement might suggest a corresponding improvement in lung cancer therapy, many oncologists don't think therapy for this form of cancer has improved at all. In the absence of a significant change in treatment, how can it be that a higher percentage of lung cancer patients now live 5 years after their initial diagnosis?

**20–66** Virtually all cancer treatments are designed to kill cancer cells, usually by inducing apoptosis. However, one particular cancer—acute promyelocytic leukemia (APL)—has been successfully treated with all-*trans*-retinoic acid, which causes the promyelocytes to differentiate into neutrophils. How might a change in the state of differentiation of APL cancer cells help the patient?

**20–67** One major goal of modern cancer therapy is to identify small molecules—anticancer drugs—that can be used to inhibit the products of specific cancer-critical genes. If you were searching for such molecules, would you design inhibitors for the products of oncogenes, tumor suppressor genes, or DNA maintenance genes? Explain why you would (or would not) select each type of gene.

**20–68** You've just read about this really cool technique for high-throughput screens of protein kinase inhibitors. The trouble is, you don't understand it. It is clearly important since it allows one to rapidly screen a large number of potential kinase inhibitors against a large number of protein kinases. There are roughly 500 protein kinases, including about 100 tyrosine kinases, encoded in the human genome. Many of them are critical components in the signal transduction pathways that become misregulated in cancer. Chemicals that inhibit individual protein kinases could serve as important lead compounds for development of drugs that are useful in the fight against cancer (and other diseases). So you want to understand how this technique works.

There are several elements (Figure 20–14). First, individual protein kinase genes are fused to the major capsid (head) protein of T7 phage. When expressed in bacteria, the fusion proteins are assembled into the phage capsid, with the kinases displayed on the outer surface. Second, an analog of ATP, which can bind to the ATP-binding pocket of the kinases, is attached to magnetic beads. Third, a bank of test compounds is prepared.

To measure the ability of a test compound to inhibit the kinase, phage displaying a specific kinase are mixed with the magnetic beads in several wells of a 96-well plate. Then the test compound is added to individual wells over a range of different concentrations. The mixtures are incubated with gentle shaking for 1 hour at 25°C, the beads are pulled to the bottom with a strong magnet, and all the free (unbound) components are washed away. Finally, the remaining, attached phage are dissociated from the beads using an excess of the same ATP analog that is attached to the beads, and counted by measuring the number of plaques they form on a bacterial lawn on a Petri dish (Figure 20–14).

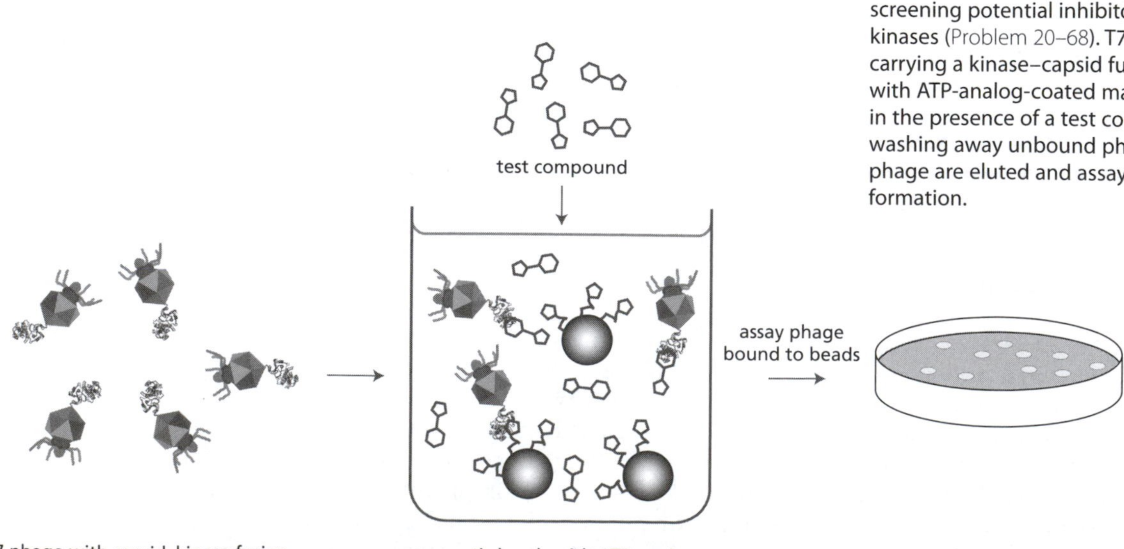


**Figure 20–14** Quantitative assay for screening potential inhibitors of protein kinases (Problem 20–68). T7 phage carrying a kinase–capsid fusion are mixed with ATP-analog-coated magnetic beads in the presence of a test compound. After washing away unbound phage, attached phage are eluted and assayed for plaque formation.

Although the assay is well described and the figure is clear, there are several things you just don't get. For example:

A. What is the point of the one-hour incubation?
B. How does the plaque count relate to the binding efficiency of the test compound? Will a test compound that binds a protein kinase strongly give more plaques or fewer plaques than one that binds the kinase weakly?
C. Do the test compounds compete for binding by the ATP analog? Or will a test compound that binds the kinase tightly someplace else also register in this assay?
D. Assuming that the test compounds bind to the ATP-binding site, how is it possible for them to bind one protein kinase, but not another? After all, every protein kinase has an ATP-binding site; that's how they bind to the ATP analog on the magnetic beads.

## CALCULATIONS

**20–69** In the high-throughput kinase-inhibitor assay described in Problem 20–68, the dissociation constant, $K_d$, for a test compound can be calculated readily from the curve of plaque count versus concentration of the test compound (Figure 20–15). The expression for $K_d$ for the test compound is

$$K_{d(test)} = \frac{K_{d(analog)}}{K_{d(analog)} + [analog]} \times [test]_{½}$$

where $K_{d(analog)}$ is the dissociation constant for binding of the ATP analog to the protein kinase ATP-binding site, [analog] is the concentration of the analog bound to the magnetic beads, and $[test]_{½}$ is the concentration of free test compound that produces a half-maximal response.

If the analog concentration is kept well below $K_{d(analog)}$, then the expression becomes

$$K_{d(test)} \approx [test]_{½}$$

A. Using this approximation, determine the $K_d$ values for the test compounds from the data in Figure 20–15.
B. The above equations assume that the phage concentration is well below $K_{d(test)}$. In these experiments the phage were incubated with the magnetic beads and test compound at $10^{10}$ phage/mL. Is this a low enough concentration that the values calculated in part A are valid?

## DATA HANDLING

**20–70** A major challenge in drug development is to predict clinical responses from research in the laboratory or in animals. Drug development for cancer therapy, as for other diseases, depends on two intertwined objectives. First, a

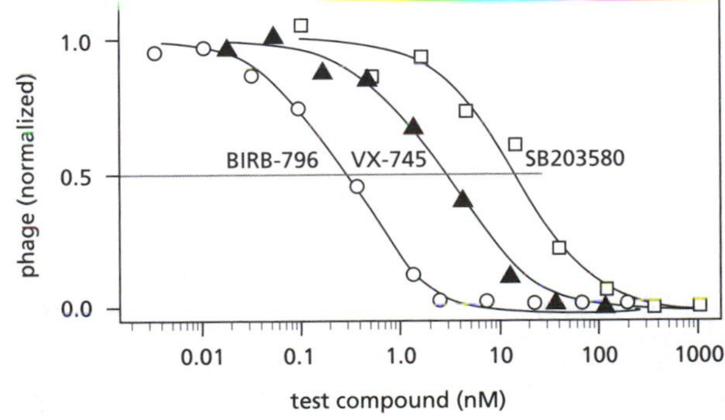


**Figure 20–15** Plot of phage bound versus concentration of test compound (Problem 20–69). All plots have been normalized so that the maximum number of phage bound (highest plaque count) is 1.0 and the minimum is 0.0. For the normalized curve the half-maximal response occurs at 0.5, which is shown by the *gray* line. BIRB-796, VX-745, and SB203580 are the names of the test compounds.

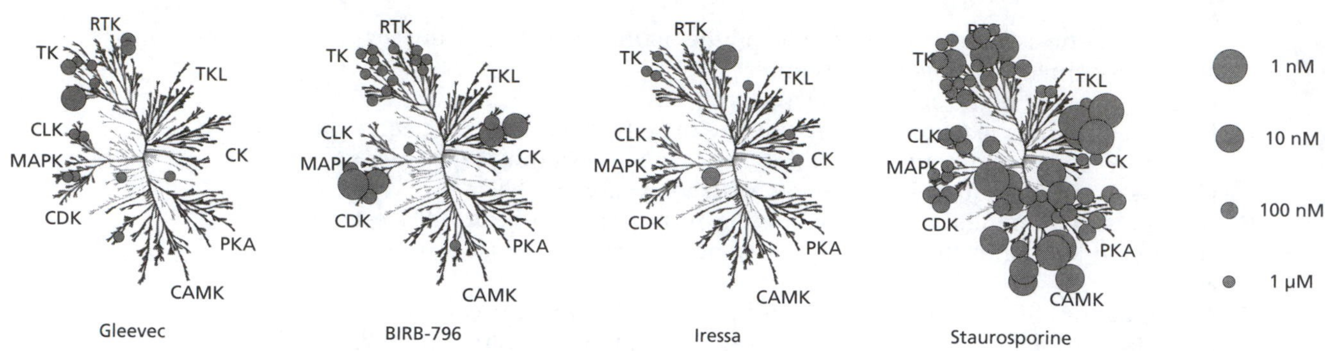


**Figure 20–16** Specificity profiles for protein kinase inhibitors (Problem 20–70). The highly branched structure is the evolutionary tree for the human kinome. Circles of different sizes represent approximate binding constants, as indicated on the *right*. TK, nonreceptor tyrosine kinases; RTK, receptor tyrosine kinases; TKL, tyrosine kinase-like kinases; CK, casein kinase family; PKA, protein kinase A family; CAMK, calcium/calmodulin-dependent kinases; CDK, cyclin-dependent kinases; MAPK, mitogen-activated protein kinases; CLK, CDK-like kinases.

drug must bind its target protein with a low $K_d$ (in the nM range) so that the amount of drug that must be administered to the patient is kept in a reasonable range. Second, the concentration at which a drug affects its intended target protein should be 10-100 fold lower than the concentration at which it affects other (off-target) proteins.

Because protein kinases are key components in the signaling pathways that control cell behavior, they have been intense targets for anticancer drug development. The high-throughput screen described in Problem 20–68 has the potential for measuring $K_d$ values and determining off-target effects in the same assay. The results of the screen are shown for four kinase inhibitors in Figure 20–16. The results are presented schematically on an evolutionary tree of the human kinases (the so-called human kinome). Only 113 of the 500 or so kinases represented on the kinome were tested in the high-throughput screen. Binding affinities are represented by circles overlaid on the position of target kinase on the kinome, with larger circles indicating tighter binding (lower $K_d$ values).

A. Assuming that the largest circle represents the main target of an inhibitor, rank order the inhibitors from the most specific to the least specific.
B. As is true for many of the inhibitors tested in this assay, binding by BIRB-796 appears to be clustered in a few regions of the kinome. Why do you suppose that is?
C. Gleevec, which inhibits the Abl protein kinase (its main target in the kinome), is being used with great success in the treatment of chronic myelogenous leukemia. Many patients, however, ultimately develop cancers that express a mutant form of Abl that is resistant to Gleevec. One of the most common mutant forms of Abl carries an isoleucine in place of threonine at position 315. Abl(T315I) is a poor target for Gleevec but is inhibited by BIRB-796. Do you suppose it would be possible to predict that BIRB-796 would inhibit Abl(T315I) from the data in Figure 20–16?
D. How might you adapt this high-throughput screen to finding inhibitors of clinically important resistant versions of protein kinases?

# Answers to Problems in *Molecular Biology of the Cell,* Fifth Edition

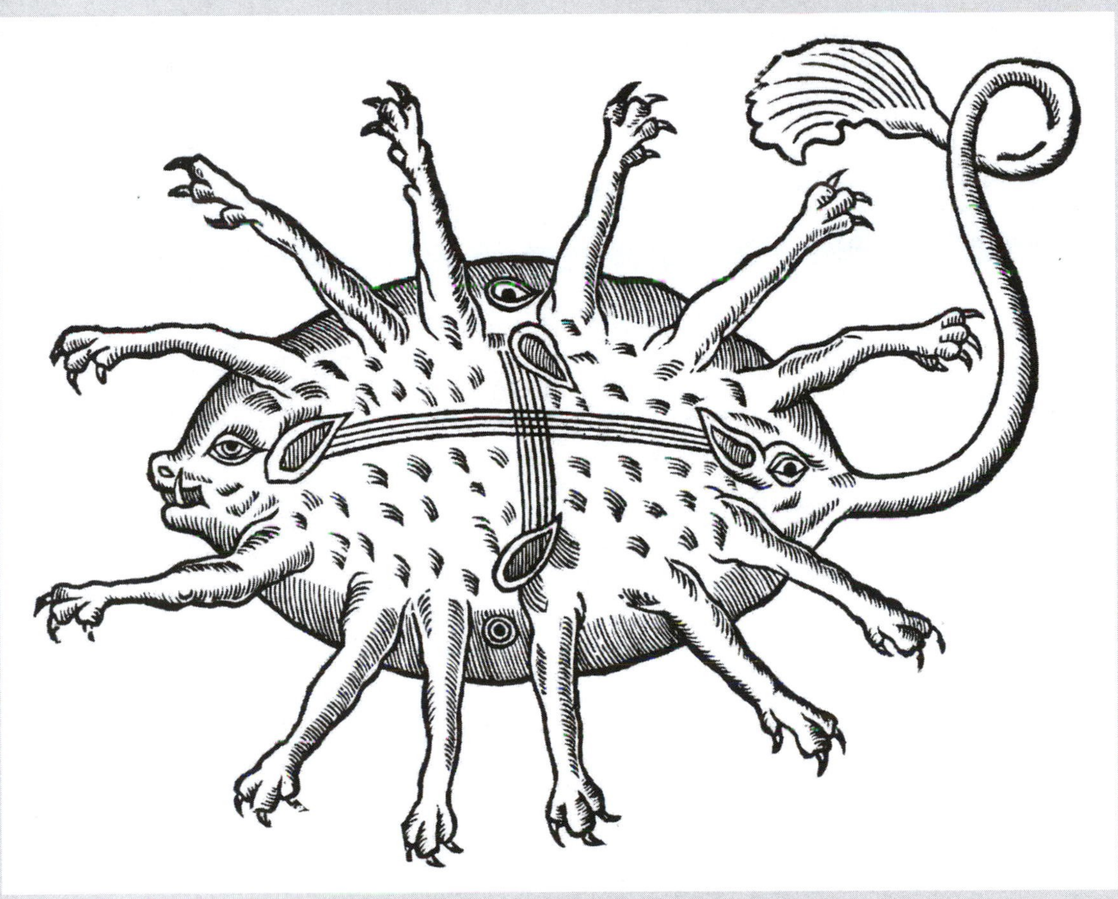

**A sea creature sighted between Antibes and Nice in 1562.** We shall never know exactly what the person who drew this picture actually saw, but it is doubtful if he (or she) made it up completely. Anyone who has taken a histology course will know how very difficult it is to learn to see and to abstract salient, accurate details from a completely unfamiliar scene. This is important for cell biologists; it is all too easy to interpret the unfamiliar in terms of what you already know, blinding you to the more truthful new.

## Chapter 1 Cells and Genomes

**1–1** False. The clusters of human hemoglobin genes arose during evolution by duplication from an ancient ancestral globin gene; thus, they are examples of parologous genes. The human hemoglobin α gene is orthologous to the chimpanzee hemoglobin α gene, as are the human and chimpanzee hemoglobin β genes, etc. All the globin genes, including the more distantly related gene for myoglobin, are homologous to one another.

**1–2** True. In single-celled organisms the genome is the germline and any modification is passed on to the next generation. In multicellular organisms most of the cells are somatic cells and make no contribution to the next generation; thus, modification of those cells by horizontal gene transfer would have no consequence for the next generation. The germline cells are usually sequestered in the interior of multicellular organisms, minimizing their contact with foreign cells, viruses, and DNA, thereby insulating the species from the effects of horizontal gene transfer.

**1–3** True. Bacterial genomes seem to be pared down to the essentials: most of the DNA sequences encode proteins, a small amount of DNA is devoted to regulating gene expression, and there are very few extraneous, nonfunctional sequences. By contrast, only about 1.5% of the DNA sequences in the human genome is thought to code for proteins. Even allowing for large amounts of regulatory DNA, much of the human genome is composed of DNA with no apparent function.

**1–4** On the surface, the extraordinary mutation resistance of the genetic code argues that it was subjected to the forces of natural selection. An underlying assumption, which seems reasonable, is that resistance to mutation is a valuable feature of a genetic code, one that would allow organisms to maintain sufficient information to specify complex phenotypes. This reasoning suggests that it would have been a lucky accident indeed—roughly a one-in-a-million chance—to stumble on a code as error proof as our own.

But all is not so simple. If resistance to mutation is an essential feature of any code that can support the complexity of organisms such as humans, then the only codes we *could* observe are ones that are error resistant. A less favorable frozen accident, giving rise to a more error-prone code, might limit the complexity of life to organisms that would never be able to contemplate their genetic code. This is akin to the anthropic principle of cosmology: many universes may be possible, but few are compatible with life that can ponder the nature of the universe.

Beyond these considerations, there is ample evidence that the code is not static, and thus could respond to the forces of natural selection. Deviant versions of the standard genetic code have been identified in the mitochondrial and nuclear genomes of several organisms. In each case one or a few codons have taken on a new meaning.

**Reference:** Freeland SJ & Hurst LD (1998) The genetic code is one in a million. *J. Mol. Evol.* 47, 238–248.

**1–5** There are several approaches you might try.

1. Analysis of the amino acids in the proteins would indicate whether the set of amino acids used in your organism differs from the set used in Earth organisms. But even Earthly organisms contain more amino acids than the standard set of 20, for example, hydroxyproline, phosphoserine, and phosphotyrosine, which all result from modifications after a protein has been synthesized. Absence of one or more of the common set might be a more significant result.
2. Sequencing DNA from the "Europan" organism would allow a direct comparison with the database of sequences that are already known for Earth organisms. Matches to the database would argue for contamination. Absences of matches would constitute a less strong argument for a novel organism; it is a typical observation that about 15% to 20% of the genes identified in complete genome sequences of microorganisms do not appear to be homologous to genes in the database. Sufficiently extensive sequence comparison should resolve the issue.
3. Another approach might be to analyze the organism's genetic code. We have no reason to expect that a novel organism based on DNA, RNA, and protein would have a genetic code identical to Earth's universal genetic code.

**1–6** Whether it is sunlight or inorganic chemicals, "to feed" means "to obtain free energy and building materials from." In the case of photosynthesis, photons in sunlight are used to raise electrons of certain molecules to a high-energy, unstable state. When they return to their normal, ground state, the released energy is captured by mechanisms that use it to drive the synthesis of ATP. Similarly, lithotrophs at a hydrothermal vent obtain free energy by oxidizing one or more of the reduced components from the vent (for example, $H_2S \rightarrow S + 2\ H^+$), using some common molecule in the environment to accept the electrons (for example, $(2\ H^+ + \frac{1}{2}\ O_2 \rightarrow H_2O)$). Lithotrophs harvest the energy released in such oxidation–reduction (electron-transfer) reactions to drive the synthesis of ATP. For both lithotrophs and phototrophs, the key to success is the evolution of a molecular mechanism to capture the available energy and couple it to ATP synthesis.

For all organisms, be they phototrophs, organotrophs, or lithotrophs, their ability to obtain the free energy needed to support life depends on the exploitation of some nonequilibrium condition. Phototrophs depend on the continual flux of radiation from the sun; organotrophs depend on a supply of organic molecules, provided ultimately by phototrophs, that can be oxidized for energy; and lithotrophs depend on a supply of reduced inorganic molecules, provided, for example, by hydrothermal vents, that can be oxidized to produce free energy.

**1–7** Four (Figure A1–1). All could have split from the common ancestor at the same time. Eubacteria–archaea could have split from eucaryotes, followed by the separation of eubacteria from archaea. Eubacteria–eucaryotes could have split from archaea, followed by the separation of eubacteria from eucaryotes. Archaea–eucaryotes could have split from eubacteria, followed by the separation of archaea from eucaryotes. Although horizontal transfers across these divisions make interpretations problematic, it is thought that archaea–eucaryotes first split from eubacteria, and then archaea split from eucaryotes.

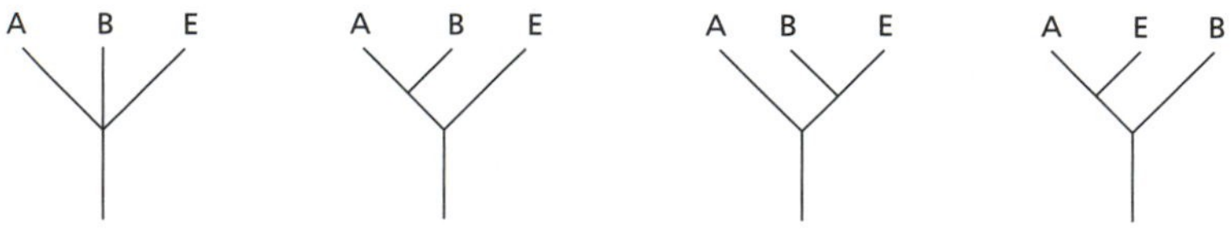


**Figure A1–1** The four possible relationships for the evolution of archaea (A), eubacteria (B), and eucaryotes (E) (Answer 1–7).

**1–8** It is unlikely that any gene came into existence perfectly optimized for its function. It is thought that highly conserved genes such as ribosomal RNA genes were optimized by more rapid evolutionary change during the evolution of the common ancestor to archaea, eubacteria, and eucaryotes. Since ribosomal RNAs (and the products of most highly conserved genes) participate in fundamental processes that were optimized early, there has been no evolutionary pressure (and little leeway) for change. By contrast, less conserved—more rapidly evolving—genes have been continually presented with opportunities to fill new functional niches. Consider, for example, the evolution of distinct globin genes that are optimized for oxygen delivery to embryos, fetuses, and adult tissues in placental mammals.

**1–9**

A. Since it appears that genes involved in informational processes are less subject to horizontal transfer, evolutionary trees derived from such genes should provide a more reliable estimate of evolutionary relationships. Thus, archaea most probably separated from eucaryotes after the archaea–eucaryote lineage separated from eubacteria.

B. Complexity is a logical explanation for the difference in rates of horizontal gene transfer (and it may even be right, although there are other possibilities). Successful transfer of an "informational" gene would require that the new gene product fit into a preexisting, functional complex, perhaps supplanting the original related protein. For a new protein to fit into a complex with other proteins, it would need to have binding surfaces that would allow it to interact with the right proteins in the appropriate geometry. If a new protein had one good binding surface, but not others, it would most probably disrupt the complex and put the recipient at a selective disadvantage. By contrast, a gene product that carries out a metabolic reaction on its own would be able to function in any organism. So long as the metabolic reaction conferred some advantage on the recipient (or at least no disadvantage), the gene transfer could be accommodated.

**Reference:** Jain R, Rivera MC & Lake JA (1999) Horizontal gene transfer among genomes: The complexity hypothesis. *Proc. Natl Acad. Sci. U.S.A.* 96, 3801–3806.

**1–10**

A. The simplest hypothesis is that gene transfer occurred at the point indicated in Figure A1–2. Genera in many of the lineages beyond this point have a nuclear *Cox2* gene, whereas lineages that branched off prior to this point do not.

B. Five genera (*Lespedeza, Dumasia, Pseudeminia, Neonotonia,* and *Amphicarpa*) apparently have functional copies of both the mitochondrial and the nuclear genes, as indicated by shaded boxes in Figure A1–2.

C. Ten genera (*Eriosema, Atylosia, Erythrina, Ramirezella, Vigna, Phaseolus, Ortholobium, Psoralea, Cullen,* and *Glycine*) no longer have a functional mitochondrial gene. The minimum number of inactivation events that could account for the observed data is four, as shown by squares on the tree in Figure A1–2.

D. Six genera no longer have a functional nuclear gene. The minimum number of inactivation events that could account for this is five, as shown by circles on the tree in Figure A1–2.

E. These data argue strongly that transfer of genes from mitochondria to the nucleus is not a one-step process; that is, simultaneous loss of the gene from mitochondria and its appearance in the nucleus. This is an unlikely scenario *a priori* since nuclear versions of mitochondrial genes must acquire a special targeting sequence that allows the encoded proteins to be delivered to mitochondria (see Chapter 12). The data in Figure A1–2 argue that the transfer process begins with the appearance of the gene in the nucleus (presumably followed at some point by its activation via acquisition of a targeting sequence). This first step is not accompanied by loss of the gene from the mitochondria. Once the nuclear gene is activated, there appears to be an

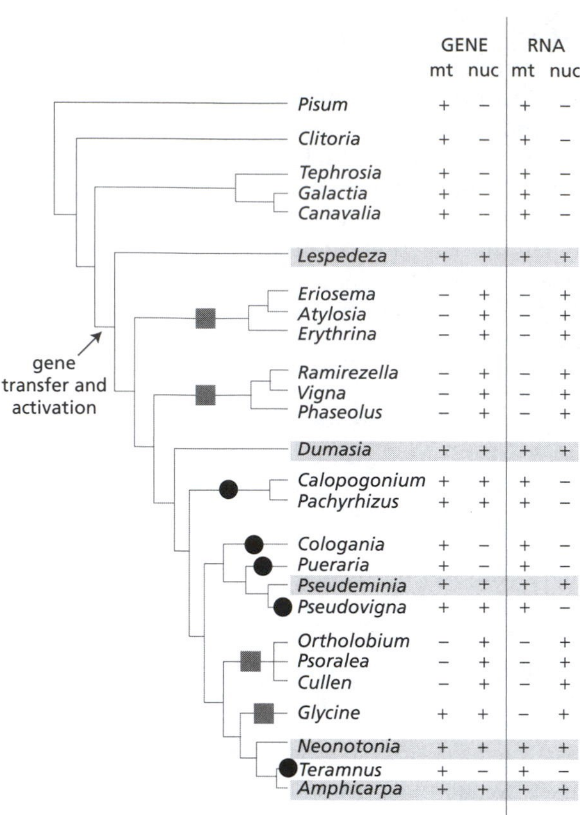

GENE RNA
mt nuc mt nuc
Pisum + − + −
Clitoria + − + −
Tephrosia + − + −
Galactia + − + −
Canavalia + − + −
Lespedeza + + + +
Eriosema − + − +
Atylosia − + − +
Erythrina − + − +
gene transfer and activation
Ramirezella − + − +
Vigna − + − +
Phaseolus − + − +
Dumasia + + + +
Calopogonium + + + −
Pachyrhizus + + + −
Cologania + − + −
Pueraria + − + −
Pseudeminia + + + +
Pseudovigna + + + −
Ortholobium − + − +
Psoralea − + − +
Cullen − + − +
Glycine + + − +
Neonotonia + + + +
Teramnus + − + −
Amphicarpa + + + +


**Figure A1–2** Summary of *Cox2* gene distribution and transcript data in a phylogenetic context, showing the most likely point of gene transfer and the minimal number of points for mitochondrial (*squares*) and nuclear (*circles*) gene inactivation (Answer 1–10). *Boxes* indicate genera with apparently functional copies of both the mitochondrial and nuclear genes.

intermediate stage in which both genes function. Subsequently, one or the other gene is inactivated. If the nuclear gene is inactivated, the transfer process is effectively aborted. If the mitochondrial gene is inactivated (often initially by point mutations), then the transfer can proceed. The final stage of transfer is deletion of the defective mitochondrial gene, a process favored by the economics of genome replication.

**Reference:** Adams KL, Song K, Roessler PG, Nugent JM, Doyle JL, Doyle JJ & Palmer JD (1999) Intracellular gene transfer in action: Dual transcription and multiple silencings of nuclear and mitochondrial *cox2* genes in legumes. *Proc. Natl Acad. Sci. U.S.A.* 96, 13863–13868.

**1–11**

A. The data in the phylogenetic tree (see Figure Q1–3) refute the hypothesis that plant hemoglobin genes arose by horizontal transfer. Looking at the more familiar parts of the tree, we see that the vertebrates (fish to human) cluster together as a closely related set of species. Moreover, the relationships in the unrooted tree shown in Figure Q1–3 are compatible with the order of branching we know from the evolutionary relationships among these species: fish split off before amphibians, reptiles before birds, and mammals last of all in a tightly knit group. Plants also form a distinct group that displays accepted evolutionary relationships, with barley, a monocot, diverging before bean, alfalfa, and lotus, which are all dicots (and legumes). The sequences of the plant hemoglobins appear to have diverged long ago in evolution, at or before the time that mollusks, insects, and nematodes arose. The relationships in the tree indicate that the hemoglobin genes arose by descent from some common ancestor.

B. Had the plant hemoglobin genes arisen by horizontal transfer from a parasitic nematode, then the plant sequences would have clustered with the nematode sequences in the phylogenetic tree in Figure Q1–3.

**1–12** Three general hypotheses have been proposed to account for the differences in rate of evolutionary change in different lineages. The individual hypotheses discussed below are not mutually exclusive and may all contribute to some extent.

The *generation-time* hypothesis proposes that rate differences are a consequence of different generation times. Species such as rat with short generation times will go through more generations and more rounds of germ-cell division, hence more rounds of DNA replication. This hypothesis assumes that errors during DNA replication are the major source of mutations. Tests of this hypothesis in rat versus human tend to support its validity.

The *metabolic-rate* hypothesis postulates a higher rate of evolution for species with a higher metabolic rate. Species with high metabolic rates use more oxygen; hence, they generate more oxygen free radicals, a major source of damage to DNA. This is especially relevant for mitochondrial genomes, because mitochondria are the major cellular site for oxygen utilization and free-radical production.

The *efficiency-of-repair* hypothesis proposes that the efficiency of repair of DNA damage differs in different lineages. Species with highly efficient repair of DNA damage would reduce the fraction of damage events that lead to mutation. There is evidence in cultured human and rat cells that such differences in repair exist, in the expected direction, but it is unclear whether such differences exist in the germlines of these organisms.

**Reference:** Li WH (1997) Molecular Evolution, pp 228–230. Sinauer Associates, Inc.: Sunderland MA.

## Chapter 2 Cell Chemistry and Biosynthesis

**2–1** True. With each half-life, half the remaining radioactivity decays. After 10 half-lives $(1/2)^{10}$, about 1/1000 (exactly 1/1024) of the original radioactivity will remain. (It is useful to remember that $2^{10}$ is about 1000.)

**2–2** False. The pH of the solution will be very nearly neutral, essentially pH 7, because the few $H^+$ ions contributed by HCl will be outnumbered by the $H^+$ ions from dissociation of water. No matter how much a strong acid is diluted it can never give rise to a basic solution. In fact, calculations that take into account both sources of $H^+$ ions and also the effects on the dissociation of water give a pH of 6.98 for a $10^{-8}$ M solution of HCl.

**2–3** False. Many of the functions that macromolecules perform rely on their ability to associate and dissociate readily, which would not be possible if they were linked by covalent bonds. By linking their macromolecules noncovalently, cells can, for example, quickly remodel their interior when they move or divide, and easily transport components from one organelle to another. It should be noted that some macromolecules are linked by covalent bonds. This occurs primarily in situations where extreme structural stability is required, such as in the cell walls of many bacteria, fungi, and plants, and in the extracellular matrix that provides the structural support for most animal cells.

**2–4** True. The difference between plants and animals is in how they obtain their food molecules. Plants make their own using the energy of sunlight, plus $CO_2$ and $H_2O$, whereas animals must forage for their food.

**2–5** True. Oxidation–reduction reactions refer to those in which electrons are removed from one atom and transferred to another. Since the number of electrons is conserved (no loss or gain) in a chemical reaction, oxidation—removal of electrons—must be accompanied by reduction—addition of electrons.

**2–6** False. The equilibrium constant for the reaction $A \rightleftharpoons B$ remains unchanged; it is a constant. Linking reactions together can convert an unfavorable reaction into a favorable one, but it does so not by altering the equilibrium constant, but rather by changing the concentration ratio of products to reactants.

**2–7** True. A reaction with a negative $\Delta G°$, for example, would not proceed spontaneously under conditions where there is already an excess of products over those that would be present at equilibrium. Conversely, a reaction with a positive $\Delta G°$ would proceed spontaneously under conditions where there is an excess of substrates compared to those present at equilibrium.

**2–8** False. Glycolysis is the *only* metabolic pathway that can generate ATP in the absence of oxygen. There are many circumstances in which cells are temporarily exposed to anoxic conditions, during which time they survive by glycolysis. For example, in an all-out sprint the circulation cannot deliver adequate oxygen to leg muscles, which continue to power muscle contraction by passing large amounts of glucose (from glycogen) down the glycolytic pathway. Similarly, there are several human cell types that do not carry out oxidative metabolism; for example, red blood cells, which have no mitochondria, make ATP via glycolysis. Thus, glycolysis is critically important, but it is sort of like insurance: it is not so important until you need it, and then it is hard to do without.

**2–9** False. The oxygen atoms that are part of $CO_2$ do not come from the oxygen atoms that are consumed as part of the oxidation of glucose (or of any other food molecule). The electrons that are abstracted from glucose at various stages in its oxidation are finally transferred to oxygen to produce water during oxidative phosphorylation. Thus, the oxygen used during oxidation of food in animals ends up as oxygen atoms in $H_2O$.

One can show this directly by incubating living cells in an atmosphere that contains molecular oxygen enriched for the isotope, $^{18}O$, instead of the naturally occurring isotope, $^{16}O$. In such an experiment one finds that all the $CO_2$ released from cells contains only $^{16}O$. Therefore, the oxygen atoms in the released $CO_2$ molecules do not come directly from the atmosphere, but rather from the organic molecules themselves and from $H_2O$.

**2–10** Organic chemistry in laboratories—even the very best—is rarely carried out in a water environment because of low solubility of some components and because water is reactive and usually competes with the intended reaction. The most dramatic difference, however, is the complexity. It is critical in laboratory organic chemistry to use pure components to ensure a high yield of the intended product. By contrast, living cells carry out thousands of different reactions simultaneously with good yield and virtually no interference between reactions. The key, of course, is that cells use enzyme catalysts, which bind substrate molecules in an active site, where they are isolated from the rest of the environment. There the reactivity of individual atoms is manipulated to encourage the correct reaction. It is the ability of enzymes to provide such special environments—miniature reaction chambers—that allows the cell to carry out an enormous number of reactions simultaneously without cross talk between them.

**2–11**

A. Ethanol in 5% beer is 0.86 M. Pure ethanol is 17.2 M [(789 g/L) × (mole/46 g)], and thus 5% beer would be 0.86 M ethanol (17.2 M × 0.05).

B. At a legal limit of 80 mg/100 mL, ethanol will be 17.4 mM in the blood [(80 mg/0.1 L) × (mmol/46 mg)].

C. At the legal limit (17.4 mM), ethanol in 5% beer (0.86 M) has been diluted 49.4-fold (860 mM/17.4 mM). This dilution represents 809 mL in 40 L of body water (40 L/49.4). At 355 mL per beer, this equals 2.3 beers (809 mL/355 mL).

D. It would take nearly 4 hours. At twice the legal limit the person would contain 64 g of ethanol [(0.16 g/0.1 L) × (40 L)]. The person would metabolize 8.4 g/hr [(0.12 g/hr kg) × (70 kg)]. Thus, to metabolize 32 g of ethanol (the amount in excess of the legal limit) would require 3.8 hours [(32 g) × (hr/8.4 g)].

**2–12** There is a good reason for the inverse relationship between half-life and maximum specific activity: the shorter the half-life, the greater the number of atoms that will decay per unit time, hence the greater the number of dpm or curies. If the radioactive atoms are present on an equimolar basis, then those with shorter half-lives will give more dpm; that is, they will have more Ci/mmol—a higher specific activity.

**2–13**

A. A solution is said to be neutral when the concentrations of $H^+$ and $OH^-$ are exactly equal. This occurs when the concentration of each ion is $10^{-7}$ M, so that their product is $10^{-14}$ $M^2$.

B. In a 1 mM solution of NaOH, the concentration of $OH^-$ is $10^{-3}$ M. Thus, the concentration of $H^+$ is $10^{-11}$ M, which is pH 11.

$$[H^+] = \frac{K_w}{[OH^-]}$$
$$= \frac{10^{-14}\,M^2}{10^{-3}\,M} = 10^{-11}\,M$$

C. A pH of 5.0 corresponds to an $H^+$ concentration of $10^{-5}$ M. Thus, the $OH^-$ concentration is $10^{-9}$ M ($10^{-14}$ $M^2/10^{-5}$ M).

**2–14** The rank order for p$K$ values is expected to be 4, 1, 2, 3. It is convenient to discuss the rank order starting with the aspartate side chain carboxyl group on the surface of a protein with no other ionizable groups nearby (1). The side chain would be expected to have a p$K$ around 4.5, somewhat higher than observed in the free amino acid because of the absence of the influence of the positively charged amino group. If the side chain were buried in a hydrophobic pocket on the protein (2), its p$K$ would be higher because the presence of a charge in a hydrophobic environment (without the easy bonding to water) would be disfavored. If there were another negative charge in the same hydrophobic environment (3), the p$K$ of the aspartate side chain would be elevated even further because of electrostatic repulsion. (It would be even more difficult to give up a proton and become charged.) If there were a positively charged group in the same environment (4), then the favorable electrostatic attraction would make it very easy for the proton to come off, lowering the p$K$ even below that of the side chain on the surface (1).

**2–15** Assuming that the change in enzyme activity is due to the change in protonation state of histidine, the enzyme must require histidine in the protonated, charged state. The enzyme is active only below the p$K$ of histidine (which is typically around 6.5 to 7.0 in proteins), where the histidine is expected to be protonated.

**2–16** You should advise the runner to breathe rapidly just before the race. Since a sprint will cause a lowering of blood and cell pH, the object of the pre-race routine would be to raise the pH with the idea that the runner could then sprint longer before feeling fatigue. Holding your breath or breathing rapidly both temporarily affect the amount of dissolved $CO_2$ in the bloodstream. Holding your breath will increase the amount of $CO_2$ and push the equilibrium to the right, leading to an increase in [$H^+$] and a lower pH. By contrast, breathing rapidly will reduce the concentration of $CO_2$ and pull the equilibrium to the left, leading to a decrease in [$H^+$] and a higher pH.

**2–17** The functional groups on the three molecules are indicated and named in Figure A2–1.

**2–18** The instantaneous velocities are $H_2O = 3.8 \times 10^4$ cm/sec, glucose = $1.2 \times 10^4$ cm/sec, and myoglobin = $1.3 \times 10^3$ cm/sec. The calculation for a water molecule, which has a mass of $3 \times 10^{-23}$ g [(18 g/mole) × (mole/$6 \times 10^{23}$ molecules)], is shown below.

$$v = (kT/m)^{1/2}$$
$$v = \left(\frac{1.38 \times 10^{-16}\,\text{g cm}^2}{\text{K sec}^2} \times 310\,\text{K} \times \frac{1}{3 \times 10^{-23}\,\text{g}}\right)^{1/2}$$
$$v = 3.78 \times 10^4\ \text{cm/sec}$$

When these numbers are converted to km/hr the results are fairly astounding. Water moves at 1360 km/hr, glucose at 428 km/hr, and myoglobin at

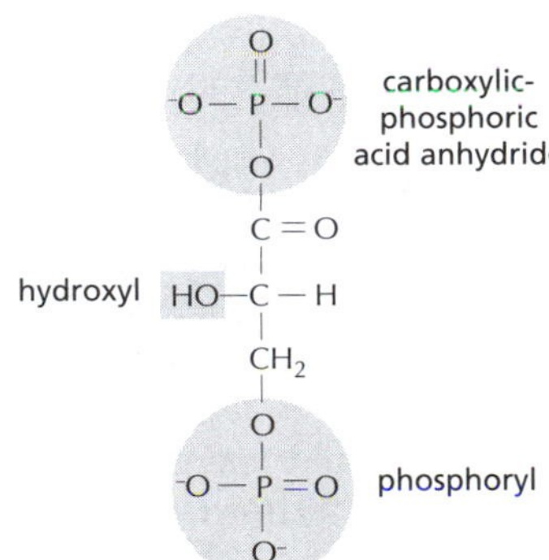


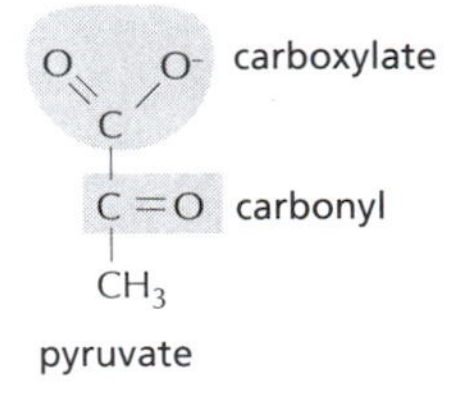


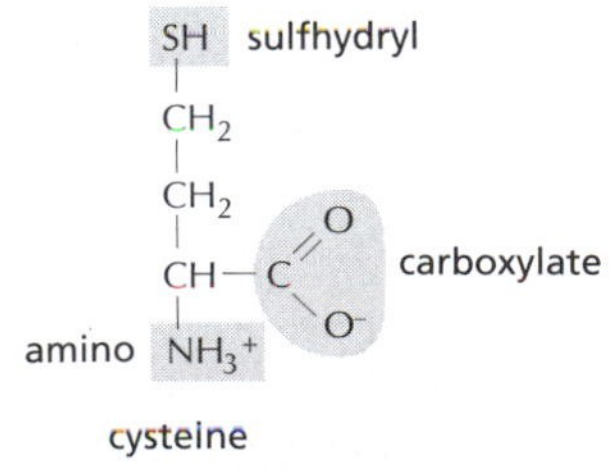


**Figure A2–1** The functional groups in 1,3-bisphosphoglycerate, pyruvate, and cysteine (Answer 2–17).

47 km/hr. Thus, even the largest (slowest) of these molecules is moving faster than the swiftest human sprinter! And water molecules are traveling at Mach 1.1! Unlike a human sprinter, or a jet airplane, these molecules make forward progress only slowly because they are constantly colliding with other molecules in solution.

**Reference:** Berg HC (1993) Random Walks in Biology, Expanded Edition, pp 5–6. Princeton, NJ: Princeton University Press.

**2–19** It seems counterintuitive that polymerization of free tubulin subunits into highly ordered microtubules should occur with an overall increase in entropy (decrease in order). But it is counterintuitive only if one considers the subunits in isolation. Remember that thermodynamics refers to the whole system, which includes the water molecules. The increase in entropy is due largely to the effects of polymerization on water molecules. The surfaces of the tubulin subunits that bind together to form microtubules are fairly hydrophobic, and constrain (order) the water molecules in their immediate vicinity. Upon polymerization, these constrained water molecules are freed up to interact with other water molecules. Their new-found disorder much exceeds the increased order of the protein subunits, and thus the net increase in entropy (disorder) favors polymerization.

**2–20** The whole population of ATP molecules in the body would turn over (cycle) 1800 times per day, or a little more than once a minute. Conversion of 3 moles of glucose to $CO_2$ would generate 90 moles of ATP [(3 moles glucose) × (30 moles ATP/mole glucose)]. The whole body contains $5 \times 10^{-2}$ mole ATP [$(2 \times 10^{-3}$ mole/L) × 25 L]. Since the concentration of ATP does not change, each ATP must cycle 1800 times per day [(90 moles ATP/day)/($5 \times 10^{-2}$ mole ATP)].

**2–21** The human body operates at about 70 watts—about the same as a light bulb.

$$\frac{\text{watts}}{\text{body}} = \frac{10^9 \text{ ATP}}{60 \text{ sec cell}} \times \frac{5 \times 10^{13} \text{ cells}}{\text{body}} \times \frac{\text{mole}}{6 \times 10^{23} \text{ ATP}} \times \frac{12 \text{ kcal}}{\text{mole}} \times \frac{4.18 \times 10^3 \text{ J}}{\text{kcal}}$$
$$= \frac{69.7 \text{ J/sec}}{\text{body}} = \frac{69.7 \text{ watts}}{\text{body}}$$

**2–22** You would need to expend 496 kcal in climbing from Zermatt to the top of the Matterhorn, a vertical distance of 2818 m. Substituting into the equation for work

$$\text{work} = 75 \text{ kg} \times \frac{9.8 \text{ m}}{\text{sec}^2} \times 2818 \text{ m} \times \frac{\text{J}}{\text{kg m}^2/\text{sec}^2} \times \frac{\text{kcal}}{4.18 \times 10^3}$$
$$= 495.5 \text{ kcal}$$

This is equal to about 1.5 Snickers™ (496 kcal/325 kcal), so you would be well advised to plan a stop at Hörnli Hut to eat another one.

In reality the human body does not convert chemical energy into external work at 100% efficiency, as assumed in this answer, but rather at an efficiency of around 25%. Moreover, you will be walking laterally as well as uphill. Thus, you would need more than 6 Snickers™ to make it all the way.

**Reference:** Frayn KN (1996) Metabolic Regulation: A Human Perspective, p 179. London: Portland Press.

**2–23** Under anaerobic conditions, cells are unable to make use of pyruvate—the end product of the glycolytic pathway—and NADH. The electrons carried in NADH are normally delivered to the electron transport chain for oxidative phosphorylation, but in the absence of oxygen the carried electrons are a waste product, just like pyruvate. Thus, in the absence of oxygen, pyruvate and NADH accumulate. Fermentation combines these waste products into a single molecule, either lactate or ethanol, which is shipped out of the cell.

The flow of material through the glycolytic pathway could not continue in the absence of oxygen in cells that cannot carry out fermentation. Because $NAD^+$ + NADH is present in cells in limited quantities, anaerobic glycolysis in the absence of fermentation would quickly convert the pool largely to NADH. The change in the ratio $NAD^+$/NADH would stop glycolysis at the step in which glyceraldehyde 3-phosphate is converted to 1,3-bisphosphoglycerate, a step with only a small negative $\Delta G$ normally. The purpose of fermentation is to regenerate $NAD^+$ by transferring the pair of carried electrons in NADH to pyruvate and excreting the product. Thus, fermentation allows glycolysis to continue.

**2–24** In the absence of oxygen the energy needs of the cell must be met by fermentation to lactate, which requires a high rate of flow through glycolysis to generate sufficient ATP. When oxygen is added, the cell can generate ATP by oxidative phosphorylation, which generates ATP much more efficiently than glycolysis. Thus, less glucose is needed to supply ATP at the same rate.

**2–25** The reverse of the forward reaction is simply not a possibility under physiological conditions. The flow of material through a pathway requires that the $\Delta G$ values for *every* step must be negative. Thus, for a flow from liver glycogen to serum glucose the step from glucose 6-phosphate to glucose must have a negative $\Delta G$. To simply reverse the forward reaction (that is, G6P + ADP → glucose + ATP, $\Delta G° = 4.0$ kcal/mole) would require that the concentration ratio [glucose][ATP]/[G6P][ADP] be less than $10^{-2.84}$ (0.0015) in order to bring the reaction to equilibrium ($\Delta G = 0$).

$$\Delta G = \Delta G° + 1.41 \text{ kcal/mole} \log \frac{[\text{glucose}][\text{ATP}]}{[\text{G6P}][\text{ADP}]}$$

$$0 = 4.0 \text{ kcal/mole} + 1.41 \text{ kcal/mole} \log \frac{[\text{glucose}][\text{ATP}]}{[\text{G6P}][\text{ADP}]}$$

$$\log \frac{[\text{glucose}][\text{ATP}]}{[\text{G6P}][\text{ADP}]} = \frac{-4.0 \text{ kcal/mole}}{1.41 \text{ kcal/mole}} = -2.44$$

$$\frac{[\text{glucose}][\text{ATP}]}{[\text{G6P}][\text{ADP}]} = 0.0015$$

Inside a functioning cell, such as a liver cell exporting glucose, the concentration of ATP always exceeds that of ADP, but just for illustrative purposes let us assume the ratio is 1. Under these conditions the ratio [glucose]/[G6P] must be 0.00145; that is, the concentration of G6P must be nearly 700 times that of glucose. Since the circulating concentration of glucose is maintained at between 4 and 5 mM, this corresponds to about 3 M G6P, an impossible concentration given that the total concentration of cellular phosphate is less than about 25 mM.

**2–26** Knoop's result was surprising at the time. One might have imagined that the obvious way to metabolize fatty acids to $CO_2$ would be to remove the carboxylate group from the end as $CO_2$ and then oxidize the newly exposed carbon atom until it too could be removed as $CO_2$. However, removal of carbon atoms one at a time is not consistent with Knoop's results, since it predicts that odd- and even-number fatty acids would generate the same final product. Similar inconsistencies crop up with removal of fragments containing more than two carbon atoms. For example, removal of three-carbon fragments would work for the eight-carbon and seven-carbon fatty acids shown in Figure Q2–5, but would not work for six-carbon and five-carbon fatty acids.

Removal of two-carbon fragments from the carboxylic acid end is the only scheme that accounts for the consistent difference in the metabolism of odd- and even-number fatty acids. One might ask why the last two-carbon fragment is not removed from phenylacetic acid. It turns out that the benzene ring interferes with the fragmentation process, which involves modification of the third carbon from the carboxylic acid end. Since that carbon is part of the benzene ring in phenylacetic acid, it is protected from modification and further metabolism of phenylacetic acid is blocked.

Knoop's results also specify the direction of degradation. If the nonacidic end of the chain were attacked first, either the benzene ring would make the fatty acids resistant to metabolism, or the same benzene compound would always be excreted, independent of the length of the fatty acid fed to the dogs.

**Reference:** Knoop F (1905) Der Abbau aromatischer Fettsäuren im Tierkörper. *Beitr. Chem. Physiol.* 6, 150–162.

**2–27** The cross-feeding experiments indicate that the three steps controlled by the products of the *TrpB*, *TrpD*, and *TrpE* genes are arranged in the order

$$\text{X} \xrightarrow{\text{TrpE}} \text{Y} \xrightarrow{\text{TrpD}} \text{Z} \xrightarrow{\text{TrpB}} \text{tryptophan}$$

where X, Y, and Z are undefined intermediates in the pathway.

The ability of the *TrpE*⁻ strain to be cross-fed by the other two strains indicates that the *TrpD*⁻ and *TrpB*⁻ strains accumulate intermediates that are farther along the pathway than the step controlled by the *TrpE* gene. The ability of the *TrpD*⁻ strain to be cross-fed by the *TrpB*⁻ strain but not the *TrpE*⁻ strain places it in the middle. The inability of the *TrpB*⁻ strain to be cross-fed by either of the other strains is consistent with its controlling the step closest to tryptophan.

**Reference:** Yanofsky C (2001) Advancing our knowledge in biochemistry, genetics, and microbiology through studies on tryptophan metabolism. *Annu. Rev. Biochem.* 70, 1–37.

## Chapter 3 Proteins

**3–1** True. In a β sheet the amino acid side chains in each strand are alternately positioned above and below the sheet, and the carbonyl oxygens alternate from one side of the strand to the other. Thus, each strand in a β sheet can be viewed as a helix in which each successive amino acid is rotated 180°.

**3–2** True. Chemical groups on such protruding loops can often surround a molecule, allowing the protein to bind to it with many weak bonds.

**3–3** True. Because an enzyme has a fixed number of active sites, the rate of the reaction cannot be further increased once the substrate concentration is sufficient to bind to all the sites. It is the saturation of binding sites that leads to an enzyme's saturation behavior.

**3–4** False. The turnover number is constant since it is $V_{max}$ divided by enzyme concentration. For example, a twofold increase in enzyme concentration would give a twofold higher $V_{max}$, but it would give the same turnover number: $2\ V_{max}/2\ [E] = k_3$.

**3–5** True. The term cooperativity embodies the idea that changes in the conformation of one subunit are communicated to the other identical subunits in any given molecule, so that all of these subunits are in the same conformation.

**3–6** True. Each cycle of phosphorylation–dephosphorylation hydrolyzes one molecule of ATP; however, it is not wasteful in the sense of having no benefit. Constant cycling allows the regulated protein to switch quickly from one state to another in response to stimuli that require rapid adjustments of cellular metabolism or function. This is the essence of effective regulation.

**3–7** Since there are 20 possible amino acids at each position in a protein 300 amino acids long, there are $20^{300}$ (which is $10^{390}$) possible proteins. The mass of one copy of each possible protein would be

$$\text{mass} = \frac{110\ \text{d}}{\text{aa}} \times \frac{300\ \text{aa}}{\text{protein}} \times 10^{390}\ \text{proteins} \times \frac{\text{g}}{6 \times 10^{23}\ \text{d}}$$

$$\text{mass} = 5.5 \times 10^{370}\ \text{g}$$

Thus, the mass of protein would exceed the mass of the observable universe ($10^{80}$ g) by a factor of about $10^{290}$!

**3–8** Generally speaking, an identity of at least 30% is needed to be certain that a match has been found. Matches of 20% to 30% are problematical and difficult to distinguish from background 'noise.' Searching for distant relatives with the whole sequence usually drops the overall identity below 30% because the less conserved portions of the sequence dominate the comparison. Thus, searching with shorter, conserved portions of the sequence gives the best chance for finding distant relatives.

**3–9** The close juxtaposition of the N- and C-termini of this kelch domain identifies it as a 'plug-in' type domain. 'In-line' type domains have their N- and C-termini on opposite sides of the domain.

**3–10**

A. These data are consistent with the hypothesis that the springlike behavior of titin is due to the sequential unfolding of Ig domains. First, the fragment contained seven Ig domains and there are seven peaks in the force-versus-extension curve. In addition, the peaks themselves are what you might expect for sequential unfolding. Second, in the presence of a protein denaturant, conditions under which the domains will already be unfolded, the peaks disappear and the extension per unit force increases. Third, when the domains are cross-linked, and therefore unable to unfold, the peaks disappear and extension per unit force decreases.

B. The spacing between peaks, about 25 nm, is almost exactly what you would calculate for the sequential unfolding of Ig domains. The folded domain occupies 4 nm, but when unfolded, its 89 amino acids would stretch to about 30 nm ($89 \times 0.34$ nm), a change of 26 nm.

C. The existence of separate, discrete peaks means that each domain unfolds when a characteristic force is applied, implying that each domain has a defined stability. The collection of domains unfolds in order from least stable to most stable. Thus, it takes a little more force each time to unfold the next domain.

D. The sudden collapse of the force at each unfolding event reflects an important principle of protein unfolding; namely, its cooperativity. Proteins tend to unfold in an all-or-none fashion. A small number of hydrogen bonds are crucial for holding the folded domain together (Figure A3–1). The breaking of these bonds triggers cooperative unfolding.

**Reference:** Rief M, Gutel M, Oesterhelt F, Fernandez JM & Gaub HE (1997) Reversible folding of individual titin immunoglobulin domains by AFM. *Science* 276, 1109–1112.

**3–11**

A. Synthesis of one 10,000 amino acid protein would be expected to occur correctly 37% of the time.

$$P_C = (f_C)^n = (0.9999)^{10000}$$
$$= 0.37$$

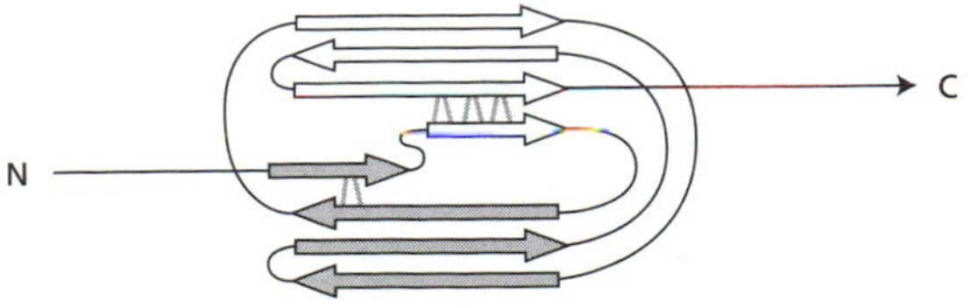


**Figure A3–1** Hydrogen bonds that lock the domain into its folded conformation (Answer 3–10). The indicated hydrogen bonds (*gray lines*), when broken, trigger unfolding of the domain. If you compare this topological diagram with the three-dimensional structure in Figure Q3–2A, you can pick out the two short β strands that are involved in forming these hydrogen bonds.

By contrast, synthesis of each 200 amino acid subunit would be expected to occur correctly 98% of the time [$P_C = (f_C)^n = (0.9999)^{200} = 0.98$]. Assembly of the mixture of subunits into correct ribosomes follows the same equation [$P_C = (f_C)^n = (0.98)^{50} = 0.37$]. Thus, making a ribosome from subunits gives the same final fraction of correct ribosomes, 37%, as making them from one long protein.

B. The assumption in part A that correct and incorrect subunits are assembled into ribosomes with equal likelihood is not true. Any mistake that interferes with the correct folding of a subunit, or that interferes with the ability of the subunit to bind to other subunits, would eliminate the subunit from assembly into a ribosome. As a result, the fraction of correctly assembled ribosomes would be higher than calculated in part A. Thus, the value of subunit synthesis lies not in more accurate synthesis, but rather in permitting quality control mechanisms to reject incorrect subunits efficiently.

**3–12**

A. The relative concentrations of the normal and mutant Src proteins are inversely proportional to the volumes in which they are distributed. The mutant Src is distributed throughout the volume of the cell, which is

$$V_{cell} = (4/3)\pi r^3 = 4\pi(10 \times 10^{-6}\text{ m})^3/3 = 4.1888 \times 10^{-15}\text{ m}^3$$

Normal Src is confined to the 4-nm layer beneath the membrane, which has a volume equal to the volume of the cell minus the volume of a sphere with a radius 4 nm less than that of the cell:

$$\begin{aligned} V_{layer} &= V_{cell} - 4\pi(r - 4\text{ nm})^3/3 \\ &= V_{cell} - 4\pi[(10 \times 10^{-6}\text{ m}) - (4 \times 10^{-9}\text{ m})]^3/3 \\ &= (4.1888 \times 10^{-15}\text{ m}^3) - (4.1838 \times 10^{-15}\text{ m}^3) \\ V_{layer} &= 0.0050 \times 10^{-15}\text{ m}^3 \end{aligned}$$

Thus, the volume of the cell is 838 times greater than the volume of a 4-nm-thick layer beneath the membrane ($4.1888 \times 10^{-15}\text{ m}^3/0.0050 \times 10^{-15}\text{ m}^3$).

Even allowing for the interior regions of the cell from which it would be excluded (nucleus and organelles), the mutant Src would still be a couple of orders of magnitude less concentrated in the neighborhood of the membrane than the normal Src.

B. Its lower concentration in the region of its target X at the membrane is the reason that mutant Src does not cause cell proliferation. This notion can be quantified by a consideration of the binding equilibrium for Src and X:

$$\text{Src} + \text{X} \rightarrow \text{Src} - \text{X}$$

$$K = \frac{[\text{Src} - \text{X}]}{[\text{Src}]\,[\text{X}]}$$

The lower concentration of the mutant Src in the region of the membrane will shift the equilibrium toward the free components, reducing the amount of complex. If the concentration is on the order of 100-fold lower, the amount of complex will be reduced up to 100-fold. Such a large decrease in complex formation could readily account for the lack of effect of the mutant Src on cell proliferation.

**3–13** The antibody binds to the second protein with an equilibrium constant, $K$, of $5 \times 10^7\text{ M}^{-1}$.

A useful shortcut to problems of this sort recognizes that $\Delta G°$ is related to log $K$ by the factor $-2.3\,RT$, which equals $-1.4$ kcal/mole at 37°C. Thus, a factor of ten increase in the equilibrium constant (an increase in log $K$ of 1) corresponds to a decrease in $\Delta G°$ of $-1.4$ kcal/mole. A 100-fold increase in $K$ corresponds to a decrease in $\Delta G°$ of $-2.8$ kcal/mole, and so on. For each factor of ten increase in $K$, $\Delta G°$ decreases by $-1.4$ kcal/mole; for each factor of ten decrease in $K$, $\Delta G°$ increases by 1.4 kcal/mole. This relationship allows a quick estimate of changes in equilibrium constant from free-energy changes and vice versa. In this problem you are told that $\Delta G°$ increased by 2.8 kcal/mole (a weaker binding gives a less negative $\Delta G°$). According to the relationship developed above, this increase in $\Delta G°$ requires that $K$ decrease by a factor of 100 (a decrease by 2 in log $K$); thus, the equilibrium constant for binding to the second protein is $5 \times 10^7$ M$^{-1}$.

The solution to the problem can be calculated by first determining the free-energy change represented by the binding to the first protein:

$$\Delta G° = -2.3\ RT \log K$$

Substituting for $K$,

$$\Delta G° = -2.3\ (1.98 \times 10^{-3}\ \text{kcal/K mole})\ (310\ \text{K})\ \log\ (5 \times 10^9)$$
$$\Delta G° = -1.41\ \text{kcal/mole} \times 9.7$$
$$\Delta G° = -13.68\ \text{kcal/mole}$$

The free-energy change associated with binding to the second protein is obtained by adding 2.8 kcal/mole to the free-energy change for binding to the first protein, giving a value of –10.88 kcal/mole. Thus, the equilibrium constant for binding to the second protein is

$$\log K = (-10.88\ \text{kcal/mole})/(-1.41\ \text{kcal/mole})$$
$$= 7.7$$
$$K = 5 \times 10^7\ \text{M}^{-1}$$

**3–14** The calculated values of fraction of tmRNA bound versus SmpB concentration are shown in Table A3–1. Also shown are rule-of-thumb values, which are easier to remember.

These relationships are useful not only for thinking about $K_d$, but also for enzyme kinetics. The rate of a reaction expressed as a fraction of the maximum rate is

$$\text{rate}/V_{max} = [\text{S}]/([\text{S}] + K_m)$$

which has the same form as the equation for fraction bound. Thus, for example, when the concentration of substrate, [S], is 10-fold above the Michaelis constant, $K_m$, the rate is 90% of the maximum, $V_{max}$. When [S] is 100-fold below $K_m$, the rate is 1% of $V_{max}$.

The relationship also works for the fractional dissociation of an acidic group, HA, as a function of pH. When the pH is 2 units above p$K$, 99% of the acidic group is ionized. When the pH is 1 unit less than p$K$, 10% is ionized.

**3–15** At [S] = zero, the rate equals $0/K_m$ and the rate is therefore zero. At [S] = $K_m$, the ratio of [S]/([S] + $K_m$) equals 1/2 and the rate is 1/2 $V_{max}$. At infinite [S] the ratio of [S]/([S] + $K_m$) equals 1 and the rate is equal to $V_{max}$.

**3–16**

A. An enzyme composed entirely of mirror-image amino acids would be expected to fold stably into a mirror-image conformation; that is, it would look like the normal enzyme when viewed in a mirror.

B. A mirror-image enzyme would be expected to recognize the mirror image of its normal substrate. Thus, 'D' hexokinase would be expected to add a phosphate to L-glucose and to ignore D-glucose.

This experiment has actually been done for HIV protease. The mirror-image protease recognizes and cleaves a mirror-image substrate.

**Reference:** Milton RC, Milton SC & Kent SB (1992) Total chemical synthesis of a D-enzyme: the enantiomers of HIV-1 protease show reciprocal chiral substrate specificity. *Science* 256, 1445–1448.

**3–17** This simple question required decades of research to provide a complete and satisfying answer. At the simplest level, hemoglobin binds oxygen efficiently

**Table A3–1** Calculated values for fraction bound versus protein concentration (Answer 3–14).

| [PROTEIN] | FRACTION BOUND (%) | RULE-OF-THUMB |
|---|---|---|
| $10^4\ K_d$ | 99.99 | 99.99 |
| $10^3\ K_d$ | 99.9 | 99.9 |
| $10^2\ K_d$ | 99 | 99 |
| $10^1\ K_d$ | 91 | 90 |
| $K_d$ | 50 | 50 |
| $10^{-1}\ K_d$ | 9.1 | 10 |
| $10^{-2}\ K_d$ | 0.99 | 1 |
| $10^{-3}\ K_d$ | 0.099 | 0.1 |
| $10^{-4}\ K_d$ | 0.0099 | 0.01 |

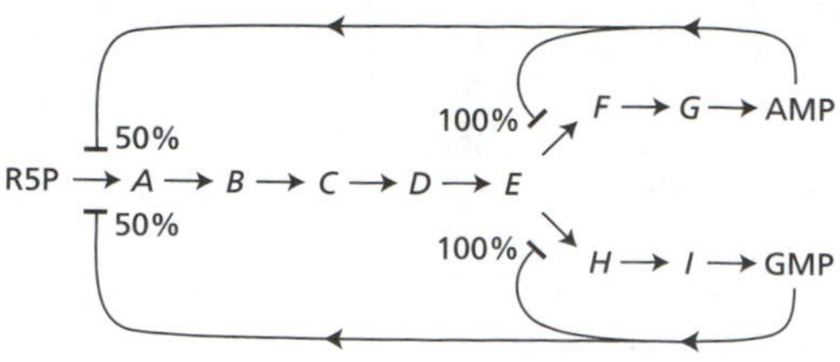


**Figure A3–2** Pattern of inhibition in the metabolic pathway for purine nucleotide synthesis (Answer 3–18).

in the lungs because the concentration (partial pressure) of oxygen is highest there. In the tissues the concentration of oxygen is lower because it is constantly being consumed in metabolism. Thus, hemoglobin will tend to release (bind less) oxygen in the tissues. This natural tendency—an effect on the binding equilibrium—is enhanced by allosteric interactions among the four subunits of the hemoglobin molecule. As a consequence, much more oxygen is released in the tissues than would be predicted by a simple binding equilibrium.

**3–18** One reasonable proposal would be for excess AMP to feedback inhibit the enzyme for converting *E* to *F*, and excess GMP to feedback inhibit the step from *E* to *H*. Intermediate *E*, which would then accumulate, would feedback inhibit the step from R5P to *A*. Some branched pathways are regulated in just this way. Purine nucleotide synthesis is regulated somewhat differently, however (Figure A3–2). AMP and GMP regulate the steps from *E* to *F* and from *E* to *H*, as above, but they also regulate the step from R5P to *A*. Regulation by AMP and GMP at this step might seem problematical since it suggests that a rise in AMP, for example, could shut off the entire pathway even in the absence of GMP. The cell uses a very clever trick to avoid this problem. Individually, excess AMP or GMP can inhibit the enzyme to about 50% of its normal activity; together they can completely inhibit it.

## Chapter 4 DNA, Chromosomes, and Genomes

**4–1** True. The human karyotype comprises 22 autosomes and the two sex chromosomes, X and Y. Females have 22 autosomes and two X chromosomes for a total of 23 different chromosomes. Males also have 22 autosomes, but have an X and a Y chromosome for a total of 24 different chromosomes.

**4–2** True. Humans and mice diverged from a common ancestor long enough ago for roughly 2 out of 3 nucleotides to have been changed by random mutation. The regions that have been conserved are those with important functions, where mutations with deleterious effects were eliminated by natural selection. Other regions have not been conserved because natural selection cannot operate to eliminate changes in nonfunctional DNA.

**4–3** True. All the core histones are rich in lysine and arginine, which have basic—positively charged—side chains that can neutralize the negatively charged DNA backbone.

**4–4** False. By using the energy of ATP hydrolysis, chromatin remodeling complexes can catalyze the movement of nucleosomes along DNA, or even between segments of DNA.

**4–5** True. Duplication of chromosomal segments, which may include one or more genes, allows one of the two genes to diverge over time to acquire different, but related functions. The process of gene duplication and divergence is thought to have played a major role in the evolution of biological complexity.

**4–6** In *all* samples of double-stranded DNA, the mole percents of A and T are equal, as are the mole percents of G and C. Results such as this one stood out as odd in the days before the structure of DNA was known. Now it is clear that, while all cellular DNA is double stranded, certain viruses contain

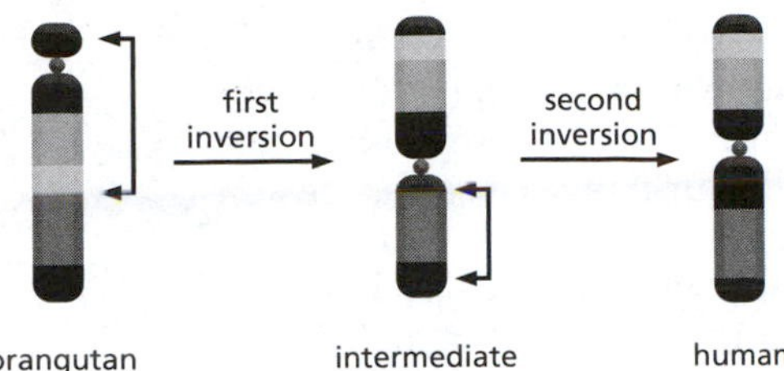


**Figure A4–1** Inversions and intermediate chromosome in the evolution of chromosome 3 in orangutan and humans (Answer 4–9).

single-stranded DNA. The genomic DNA of the M13 virus, for example, is single stranded. In single-stranded DNA, A is not paired with T, nor G with C, and so the A = T and C = G rules do not apply.

**4–7** The segment of DNA in Figure Q4–1 reads from top to bottom 5′-ACT-3′. The carbons in the ribose sugar are numbered clockwise around the ring, starting with C1′, the carbon to which the base is attached, and ending with C5′, the carbon that lies outside the ribose ring.

**4–8** Because C always pairs with G in duplex DNA, their mole percents must be equal. Thus, the mole percent of G, like C, is 20%. The mole percents of A and T account for the remaining 60%. Since A and T always pair, their mole percents are equal to half this value: 30%.

**4–9** The intermediate chromosome and the sites of the inversions are indicated in Figure A4–1.

**4–10** With these assumptions, the DNA is compacted 27-fold in 30-nm fibers relative to the extended DNA. The total length of duplex DNA in 50 nm of the fiber is 1360 nm [(20 nucleosome) × (200 bp/nucleosome) × (0.34 nm/bp) = 1360 nm]. 1360 nm of duplex DNA reduced to 50 nm of chromatin fiber represents a 27-fold condensation [(1360 nm/50 nm) = 27.2]. This level of packing represents 0.27% (27/10,000) of the total condensation that occurs at mitosis, still a long way off what is needed.

**4–11** The biological outcome associated with histone methylation depends on the site that is modified. Each site of methylation has different surrounding amino acid context, which allows the binding of distinct code-reader proteins. It is the binding of different downstream effector proteins that gives rise to different biological outcomes.

**4–12** A dicentric chromosome is unstable because the two kinetochores have the potential to interfere with one another. Normally, microtubules from the two poles of the spindle apparatus attach to opposite faces of a single kinetochore in order to separate the individual chromatids at mitosis. If a chromosome contains two centromeres, half of the time the microtubules from one of the poles will attach to the two kinetochores associated with one chromatid, while the microtubules from the other pole will attach to the two kinetochores associated with the other chromatid. Division can then occur satisfactorily. The other half of the time, the microtubules from each pole will attach to kinetochores that are associated with different chromatids. When that happens, each chromatid will be pulled to opposite spindle poles with enough force to snap it in two. Thus, two centromeres are bad for a chromosome, causing chromosome breaks—rendering it unstable.

**4–13** The control proteins either do not bind to any of the histone N-terminal peptides (Pax5) or bind to all of them (Pc1 and Suv39h1). By contrast, the HP1 proteins all bind specifically to the Lys-9-dimethylated form of the H3 N-terminal peptide. The strong association of HP1 proteins with heterochromatin suggests that the Lys-9-dimethylated form of H3 will be found in heterochromatin.

**Reference:** Lachner M, O'Carroll D, Rea S, Mechtier K & Jenuwein T (2001) Methylation of H3 lysine 9 creates a binding site for HP1 proteins. *Nature* 410, 116–120.

**4–14** The *Hox* gene clusters are packed with complex and extensive regulatory sequences that ensure proper expression of individual *Hox* genes at the correct time and place during development. Insertions of transposable elements into the *Hox* clusters are eliminated by purifying selection, presumably because they disrupt proper regulation of the *Hox* genes. Comparison of the *Hox* cluster sequences in mouse, rat, and baboon reveals a high density of conserved noncoding segments, supporting the idea of a high density of regulatory elements.

**Reference:** Lander ES et al. (2001) Initial sequencing and analysis of the human genome. *Nature* 409, 860–921.

## Chapter 5 DNA Replication, Repair, and Recombination

**5–1** This statement may be true. Each time the genome is copied in preparation for cell division, there is a chance that mistakes (mutations) will be introduced. If a few mutations are made during each replication, then no two daughter cells will be the same. The rate of mutation for humans is estimated to be 1 nucleotide change per $10^9$ nucleotides each time the DNA is replicated. Since there are $6.4 \times 10^9$ nucleotides in each diploid cell, an average of 6.4 random mutations will be introduced into the genome each time it is copied. Thus, the two daughter cells from a cell division will usually differ from one another and from the parent cell that gave rise to them. Occasionally, genomes will be copied perfectly, giving rise to identical daughter cells. The true proportions of identical cells depend on the exact mutation rate. If the true mutation rate in humans were 10-fold less (1 change per $10^{10}$ nucleotides), most of the cells in the body would be identical.

**5–2** True. If the replication fork moves forward at 500 nucleotide pairs per second, the DNA ahead of it must rotate at 48 revolutions per second (500 nucleotides per second/10.5 nucleotides per helical turn) or 2880 revolutions per minute. The havoc this would wreak on the chromosome is prevented by a DNA topoisomerase that introduces transient nicks just in front of the replication fork. This action confines the rotation to a short single-strand segment of DNA.

**5–3** True. Consider a single template strand, with its 5′ end on the left and its 3′ end on the right. No matter where the origin is, synthesis to the left on this strand will be continuous (leading strand), and synthesis to the right will be discontinuous (lagging strand). Thus, when replication forks from adjacent origins collide, a rightward-moving (lagging) strand will always meet a leftward-moving (leading) strand.

**5–4** False. Repair of damage to a single strand by base excision repair or nucleotide excision repair, for example, depends on just the two copies of genetic information contained in the two strands of the DNA double helix. By contrast, precise repair of damage to both strands of a duplex—a double-strand break, for example—requires information from a second duplex, either a sister chromatid or a homolog.

**5–5** The variation in frequency of mutants in different cultures exists because of variations in the time at which the mutations arose. For example, cultures with one mutant acquired the mutation in the last generation; cultures with two mutants likely acquired a mutation in the next to last generation and the mutant cell divided once; cultures with four mutants likely acquired the mutation in the third to last generation and the mutant cell divided twice. Cultures with large numbers of mutant cells acquired a mutation early in growth and those cells divided many times. To understand this variability, it is best to think of the mutation rate (1 mutation per $10^9$ bp per generation) as a probability: a $10^{-9}$ chance of making a mutation each time a nucleotide pair is copied. Thus, sometimes a mutation will occur before $10^9$ nucleotides have been copied and sometimes after.

Analysis of the variation in frequencies among cultures grown in this way (which is known as fluctuation analysis) is a common method for determining rates of mutation. Luria and Delbruck originally devised the method to show that mutations preexist in populations of bacteria; that is, they do not arise as a result of the selective methods used to reveal their presence.

**Reference:** Luria SE & Delbruck M (1943) Mutations of bacteria from virus sensitivity to virus resistance. *Genetics* 28, 491–511.

**5–6** Mismatch repair normally corrects a mistake in the new strand, using information in the old, parental strand. If the old strand were "repaired" using the new strand that contains a replication error as the template, then the error would become a permanent mutation in the genome. The old "correct" information would be erased in the process. Therefore, if repair enzymes did not distinguish between the two strands, there would be only a 50% chance that any given replication error would be corrected.

Overall, such indiscriminate repair would introduce the same number of mutations as would be introduced if mismatch repair did not exist. In the absence of repair a mismatch would persist until the next replication. When the replication fork passed the mismatch, and the strands were separated, properly paired nucleotides would be inserted opposite each of the nucleotides involved in the mismatch. A normal, nonmutant duplex would be made from the strand containing the original information; a mutant duplex would be made from the strand that carried the misincorporated nucleotide. Thus, the original misincorporation event would lead to 50% mutants and 50% nonmutants in the progeny. This outcome is equivalent to that of indiscriminate repair: averaged over all misincorporation events, indiscriminate repair would also yield 50% mutants and 50% nonmutants among the progeny.

**5–7** Clearly, DNA polymerases must be able to extend a mismatched primer occasionally; otherwise no mismatches would be present in the newly synthesized DNA. Most mismatches are removed by the 3′-to-5′ proofreading exonuclease associated with the DNA polymerase. When the exonuclease does not remove the mismatch, the polymerase can extend the growing chain. In reality, DNA polymerase and the proofreading exonuclease are in competition with each other. In the case of bacteriophage T7 DNA polymerase, numbers are available that illustrate this competition. Normally, T7 DNA polymerase synthesizes DNA at 300 nucleotides per second, while the exonuclease removes terminal nucleo-tides at 0.2 nucleotides per second, suggesting that 1 in 1500 (0.2/300) correctly added nucleotides are removed by the exonuclease. When an incorrect nucleotide has been incorporated, the rate of removal increases 10-fold to 2.3 nucleotides per second and the rate of polymerization decreases $10^5$-fold to about 0.01 nucleotide per second. Comparison of these rates for a mismatched primer suggests that about 1 in 200 (0.01/2.3) mismatched primers will be extended by T7 DNA polymerase.

**Reference:** Johnson KA (1993) Conformational coupling in DNA polymerase fidelity. *Annu. Rev. Biochem.* 62, 685–713.

**5–8** As always, you come through with flying colors. Although you were initially bewildered by the variety of structures, you quickly realized that H forms were just like the bubbles except that cleavage occurred within the bubble instead of outside it. Next you realized that by reordering the molecules according to the increasing size of the bubble (and flipping some structures end-for-end), you could present a convincing visual case for bidirectional replication away from a unique origin of replication (Figure A5–1). The case for bidirectional replication is clear since unidirectional replication would give a set of bubbles with one end in common. Replication from a unique origin is likely, but not certain, because you cannot rule out the possibility

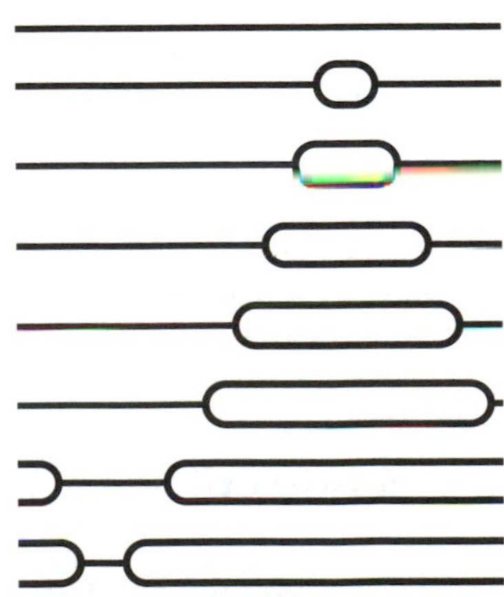

**Figure A5–1** Bidirectional replication from a unique origin (Answer 5–8).

that there are two origins on either side of and equidistant from the restriction site used to linearize the DNA. Repeating the experiment using a different restriction nuclease will resolve this issue and define the exact position of the origin(s) on the viral DNA. Your advisor is pleased.

**5–9** At many sites in vertebrate cells, the sequence 5′-CG-3′ is selectively methylated on the cytosine ring. Spontaneous deamination of methyl-C gives T. A special DNA glycosylase recognizes a mismatched base pair involving T in the sequence TG and removes the T. This DNA repair mechanism is not 100% effective, as methylated C nucleotides are common sites for mutation in vertebrate DNA. Over time, the enhanced mutation of CG dinucleotides has led to their preferential loss, accounting for their underrepresentation in the human genome.

**5–10** If the inaccurately repaired breaks were randomly distributed around the genome, then 2% of them would be expected to alter crucial coding or regulatory information. Thus, the functions of about 40 genes ($0.02 \times 2000$) would be compromised in each cell, although the specific genes would vary from cell to cell. Because the human genome is diploid, the effect of such mutations on cell function would be mitigated by the remaining allele. For most loci, one functional allele (50% of normal protein) is adequate for normal cell function; however, for some loci 50% is not adequate. Thus, the mutations would be expected to compromise the functions of some cells.

**Reference:** Lieber MR, Ma Y, Pannicke U & Schwarz K (2003) Mechanism and regulation of human non-homologous DNA end-joining. *Nat. Rev. Mol. Cell Biol.* 4, 712–720.

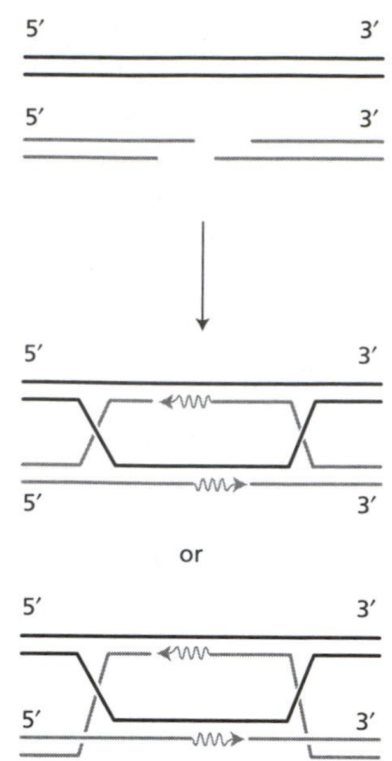


**Figure A5–2** Double Holliday junction (Answer 5–11). New DNA synthesis is indicated by *wavy lines*.

**5–11** The double Holliday junction that would result from strand invasion is shown in Figure A5–2. Two versions are shown, both equally correct. The upper one looks simpler because the invading duplex has been rotated so that the marked 5′ end is on the bottom. This arrangement minimizes the number of lines that must cross, which is why most recombination diagrams are shown in this way. The lower representation is perfectly correct, but it looks more complicated. DNA synthesis uses the 3′ end of the invading duplex as a primer and fills the single-strand gap by 5′-to-3′ synthesis, as indicated.

**5–12** A large percentage of the human genome is made up of repetitive elements such as *Alu* sequences, which are scattered among the chromosomes. If recombination were to occur between two such sequences that were on different chromosomes, for example, a translocation would be generated. Unrestricted recombination between such repeated elements would quickly rearrange the genome beyond recognition. Different rearrangements in different individuals would lead to large numbers of nonviable progeny, putting the species at risk.

This calamity is avoided through the action of the mismatch-repair system. Repeated sequences around the genome differ by a few percent of their sequence. When recombination intermediates form between them, many mismatches are present in the heteroduplex regions. When the mismatch-repair system detects too high a frequency of mismatches, it aborts the recombination process in some way. This surveillance mechanism ensures that recombining sequences are nearly identical, as expected for sequences at the same locus on homologous chromosomes.

## Chapter 6 How Cells Read the Genome: From DNA to Protein

**6–1** True. Errors in DNA replication have the potential to affect future generations of cells, while errors in transcription have no genetic consequence. Errors in transcription lead to mistakes in a small fraction of RNAs, whose functions are further monitored by downstream quality control mechanisms. The essential feature is that errors in DNA replication change the gene and, thereby, affect all the copies of RNA (and protein) made in the original cell and all its progeny cells. By contrast, errors in transcription are limited to a small number of defective RNAs (and proteins) and are not passed on to progeny cells.

These considerations are reflected in the intrinsic error rates for RNA and DNA polymerases: RNA polymerases typically make 1 mistake in copying $10^4$ nucleotides, while DNA polymerases make about 1 error per $10^7$ nucleotides. Such significant differences in error rates suggest that natural selection is stronger against errors in replication than against errors in transcription.

**6–2** False. Although intron sequences are mostly dispensable, they must be removed precisely. An error of even one nucleotide during removal would shift the reading frame in the spliced mRNA molecule and produce an aberrant protein.

**6–3** False. Wobble pairing occurs between the third position in the codon and the first position in the anticodon.

**6–4** False. Although only a few types of reactions are represented among the ribozymes in present-day cells, ribozymes that have been selected in the laboratory can catalyze a wide variety of biochemical reactions, with reaction rates similar to those of proteins. In light of these results, it is unclear why ribozymes are so underrepresented in modern cells. It seems likely that the availability of 20 amino acids versus four bases affords proteins with a greater number of catalytic strategies than ribozymes, as well as endowing them with the ability to bind productively to a wider range of substrates (for example, hydrophobic substrates, which ribozymes have difficulty with).

**6–5** The RNA polymerase must be moving from right to left in Figure Q6–1. If the RNA polymerase does not rotate around the template as it moves, it will overwind the DNA ahead of it, causing positive supercoils, and underwind the DNA behind it, causing negative supercoils. If the RNA polymerase were free to rotate about the template as it moved along the DNA, it would not overwind or underwind the DNA, and no supercoils would be generated.

**Reference:** Liu LF & Wang JC (1987) Supercoiling of the DNA template during transcription. *Proc. Natl Acad. Sci. U.S.A.* 84, 7024–7027.

**6–6** Phosphorylation of the CTD is the event that permits release of RNA polymerase from the other proteins present at the start point of transcription. Phosphorylation also allows association of a new set of proteins that are involved in processing the nascent RNA transcript. These proteins include components required for capping, splicing, and polyadenylation.

**6–7** Statement C is the only one that is necessarily true for exons 2 and 3. It is also true for exons 7 and 8. While statements A and B could be true, they do not have to be. Because the protein sequence is the same in segments of the mRNA that correspond to exons 1 and 10, neither choice of alternative exons (2 versus 3, or 7 versus 8) can alter the reading frame. To maintain the normal reading frame—whatever that is—the alternative exons must have a number of nucleotides that when divided by 3 (the number of nucleotides in a codon) give the same remainder.

Since the sequence of the α-tropomyosin gene is known, it is possible to check to see the actual state of affairs. Exons 2 and 3 both contain the same number of nucleotides, 126, which is divisible by 3 with no remainder. Exons 7 and 8 also contain the same number of nucleotides, 76, which is divisible by 3 with a remainder of 1.

**6–8** The only codon assignments consistent with the observed changes, and with the assumption that single-nucleotide changes were involved, are GUG for valine, GCG for alanine, AUG for methionine, and ACG for threonine. It is unlikely that you would be able to isolate a valine-to-threonine mutant in one step because that would require two nucleotide changes. Typically, two changes would be expected to occur at a frequency equal to the product of the frequencies for each of the single changes; hence, the double mutant would be very rare.

**6–9** When EF-Tu has positioned an aminoacyl-tRNA in the acceptor site on the ribosome, it hydrolyzes its bound GTP and exits, leaving the aminoacyl-tRNA behind. The first delay comes about because the rate of GTP hydrolysis is faster for a correct codon–anticodon pair than for an incorrect one. As a result, an incorrectly bound tRNA has more time to dissociate from the ribosome. The second delay occurs between dissociation of EF-Tu and full accommodation of the tRNA into the A-site. This time delay, which is also shorter for correct than incorrect tRNAs, allows a second opportunity for the incorrect tRNA to dissociate from the ribosome. An incorrect tRNA will, on average, dissociate more rapidly than a correct one because its codon–anticodon match contains fewer hydrogen bonds.

**6–10** In a well-folded protein the majority of hydrophobic amino acids will be sequestered in the interior away from water. Exposed hydrophobic patches thus indicate that a protein is abnormal in some way. Some proteins initially fold with exposed hydrophobic patches that are used in binding to other proteins, ultimately burying those hydrophobic amino acids as well. As a result, hydrophobic amino acids are usually not exposed on the surface of a protein, and any significant patch is a good indicator that something has gone awry. The protein may have failed to fold properly after leaving the ribosome, it may have suffered an accident that partly unfolded it at a later time, or it may have failed to find its normal partner subunit in a larger protein complex.

**6–11** Molecular chaperones fold like any other protein. Molecules in the act of synthesis on ribosomes are bound by Hsp70 chaperones. And incorrectly folded molecules are helped by Hsp60-like chaperones. That they function as chaperones when they have folded correctly makes no difference to the way they are treated before they reach their final, functional conformation. Of course, properly folded Hsp60-like and Hsp70 chaperones must already be present to help fold the newly made chaperones. At cell division, each daughter cell inherits a starter set of such chaperones from the parental cell.

**6–12** RNA has the ability to store genetic information like DNA and the ability to catalyze chemical reactions like proteins. Having both of these essential features of "life" in a single type of molecule makes it easier to understand how life might have arisen from nonliving matter. The use of RNA molecules as catalysts in several fundamental reactions in modern-day cells supports this idea. Nevertheless, it is not yet possible to specify a plausible pathway from the "primordial" soup to an RNA world, and many have speculated that there may have been a precursor molecule to RNA—one that also had catalytic and informational properties.

The deoxyribose sugar of DNA makes the molecule much less susceptible to breakage. The hydroxyl group on carbon 2 of the ribose sugar is an agent for catalysis of the adjacent 3′-5′ phosphodiester bond that links nucleotides together in RNA. Its absence from DNA eliminates that mechanism of chain breakage. In addition, the double-helical structure of DNA provides two complementary strands, which allows damage in one strand to be repaired accurately by reference to the sequence of the second strand. Finally, the use of T in DNA instead of U, as in RNA, builds in a protection against the effects of deamination—a common form of damage. Deamination of T produces an aberrant base (methyl C), whereas deamination of U generates C, a normal base. The cell's job of recognizing damaged bases is much easier when the damage produces an abnormal base.

**6–13** The complement of this hairpin RNA could also form a similar hairpin, as shown in Figure A6–1. The two structures would be identical in the double-

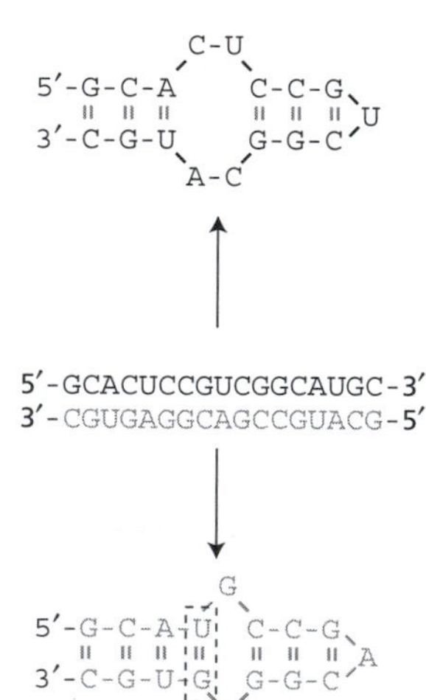


**Figure A6–1** Hairpins formed by an RNA strand and by its complement (Answer 6–13). An RNA and its complement are shown as double-stranded RNA in the middle. The structures formed by each strand are shown *above* and *below* the duplex. The nonstandard GU base pair in the lower hairpin is highlighted with a *dashed box.*

stranded regions that involved standard GC and AU base pairs. They would differ in the sequence of the single-stranded regions. Because GU base pairs are stable in RNA, whereas CA base pairs are not, one hairpin would be predicted to contain an additional base pair, as shown.

## Chapter 7 Control of Gene Expression

**7–1** True. Both the helix-loop-helix motif and the leucine zipper motif are structural motifs that allow gene regulatory proteins to dimerize, so that each member of the pair can position an α helix in the major groove of the DNA.

**7–2** False. Although there are numerous examples of reversible gene rearrangements as gene regulatory mechanisms in procaryotes, there are no known examples of regulation by reversible rearrangements in mammalian cells.

**7–3** True. In unmethylated regions of the genome, spontaneous deamination of Cs (a very common event) gives rise to the novel DNA base, uracil, which can be accurately recognized and repaired. By contrast, deamination of 5-methyl C gives rise to a T, a normal DNA base, which is more difficult for the cell's repair machinery to recognize as incorrect. As a consequence, methylated CG dinucleotides in the germline have tended to be lost during evolution, leaving the CG islands that are associated with active promoters.

**7–4** Each added phosphate alters the charge by one unit, but has relatively little effect on the molecular mass. As a consequence, proteins that differ only in the number of attached phosphates will appear at the same molecular mass, but at different isoelectric points, forming a set of horizontal spots, as shown for a few proteins in Figure A7–1. It is important to keep in mind that a horizontal array of spots does not prove that the proteins are related by phosphorylation; they could be different proteins with the same molecular mass and slightly different isoelectric points. Treatment of the proteins with a protein phosphatase before separation by gel electrophoresis could be used to resolve the issue.

**7–5** Although it is true that cancer cells differ from their normal precursors, they typically differ in their expression of only relatively few genes (oncogenes and tumor suppressor genes). When the abundances of hundreds to thousands of mRNAs are compared, as they are in DNA microarray analysis, the patterns of mRNAs from the unaffected genes (the vast majority) allow a tumor to be definitively assigned to a particular tissue type.

**7–6** Under the specified conditions (equal concentrations of DNA and gene regulatory protein), the protein would find its recognition site equally well in the eucaryotic nucleus and in the bacterium. A nonmathematical way of thinking about this is to imagine a small volume of eucaryotic nucleus, equal in size to that of the bacterium and containing the binding site. That small volume in the eucaryotic nucleus is directly comparable to the interior of the bacterium. In those equal volumes, the ability of the gene regulatory protein

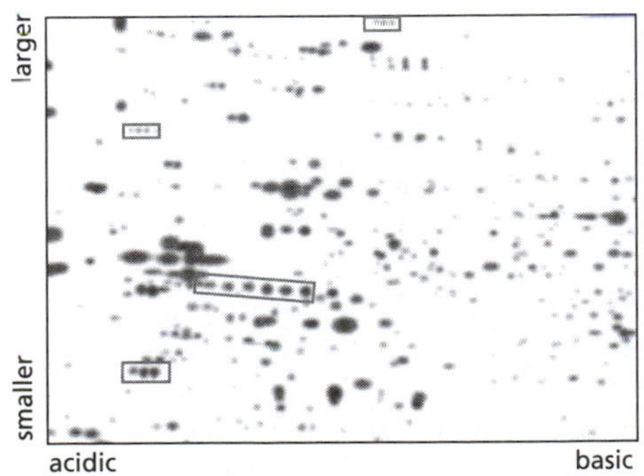


**Figure A7–1** Protein spots in a two-dimensional gel that might differ by the number of attached phosphates (Answer 7–4). A few sets of horizontal spots that could be related by phosphorylation are *boxed*. Not all such sets of proteins are indicated.

to find its binding site is equivalent. So long as the concentrations of the DNA and gene regulatory protein are the same, the total volume will make no difference.

**Reference:** Ptashne M (1986) A Genetic Switch: Gene Control and Phage λ, p 114. Oxford, UK: Blackwell Scientific Press.

**7–7**

A. Some of the individual RNA polymerase molecules deviate from the bulk flow because they bind to the DNA and slide along it for some period of time before they fall off and rejoin the flow. The binding must be nonspecific since the RNA polymerase molecules slide for long distances. Yet some of the same portions of the RNA polymerase molecule normally involved in promoter binding must be involved in sliding, because the nonspecific binding is eliminated when the RNA polymerase molecules have already bound a DNA fragment containing a strong promoter.

B. Sliding along the DNA would allow site-specific DNA-binding proteins to find their targets faster than expected by three-dimensional diffusion because it reduces the search to one dimension. It is thought that most site-specific DNA-binding proteins accelerate their search of DNA by some combination of sliding and intersegment transfer (hopping from one segment to another).

C. If the target sites were present in short DNA molecules, then the search would be expected to approximate a three-dimensional search and, hence, be slow. On the other hand, if the target sites were in long DNA molecules, then the search would be accelerated by sliding. Thus, a site-specific DNA-binding protein would be expected to find its target faster in a population of long DNA molecules than in a population of short ones.

**References:** Kabata H, Kurosawa O, Arai I, Washizu M, Margarson SA, Glass RE & Shimamoto N (1993) Visualization of single molecules of RNA polymerase sliding along DNA. *Science* 262, 1561–1563.

Shimamoto N (1999) One-dimensional diffusion of proteins along DNA. *J. Biol. Chem.* 274, 15293–15296.

**7–8** A specialized group of cells in the hypothalamus—the cells of the suprachiasmatic nucleus (SCN)—regulates our circadian rhythm. These cells receive neural cues from the retina, although not from rods and cones—the light receptors for vision—but from a subset of retinal ganglion cells that responds to light. These retinal signals entrain the cells of the SCN to the daily cycle of light and dark. In totally blind people, information about the light and dark cycle does not reach the SCN cells. As a consequence, they operate on their own inherent rhythm, which is slightly longer than 24 hours. Blind people typically report recurrent periods of insomnia and daytime sleepiness, as their circadian rhythms drift in and out of phase with the normal 24-hour cycle.

**Reference:** Sack RL, Brandes RW, Kendall AR & Lewy AJ (2000) Entrainment of free-running circadian rhythms by melatonin in blind people. *N. Engl. J. Med.* 343, 1114–1116.

**7–9** These results indicate that the tissue-specific synthesis of ApoB100 in liver cells and ApoB48 in intestinal cells results from a difference in the way the RNA transcripts are processed. The hybridization results in Table Q7–1 show that the DNA in both liver and intestine matches the oligo-Q sequence, but not the oligo-STOP sequence. Therefore, the tissue-specific differences cannot be due to separate genes that are transcriptionally regulated. Since the ApoB mRNA in the intestine contains a termination codon at a point where the ApoB mRNA in the liver contains a glutamine codon, the two mRNAs cannot encode the same protein. Thus, ApoB48 and ApoB100 cannot be related to one another by tissue-specific protein cleavage.

The identity of the DNA sequences from liver and intestine and the difference in ApoB mRNA sequences from the same tissues indicate that one tissue must alter at least one specific nucleotide during expression of the *ApoB* gene. The results in Table Q7–1 show that sequences complementary to the oligonucleotide with the termination codon (oligo-STOP) are present only in intestinal RNA. Thus, the intestine specifically alters a nucleotide in the transcript. This work marks one of the earliest discoveries of RNA editing.

**References:** Powell LM, Wallis SC, Pease RJ, Edwards YH, Knott TJ & Scott J (1987) A novel form of tissue-specific RNA processing produces apolipoprotein-B48 in intestine. *Cell* 50, 831–840.

Chen S-H, Habib G, Yang CY, Gu ZW, Lee BR, Weng S-A, Silbermann SR, Cai S-J, Deslypere JP, Rosseneu M, Gotto AM, Li W-H & Chan L (1987) Apolipoprotein B-48 is the product of a messenger RNA with an organ-specific in-frame stop codon. *Science* 238, 363–366.

## Chapter 8 Manipulating Proteins, DNA, and RNA

**8–1** False. A monoclonal antibody recognizes a specific antigenic site, but this does not necessarily mean that it will bind only to one specific protein. There are two complicating factors. First, antigenic sites that are similar but not identical can bind to the same antibody with different affinities. If too much antibody is used in an assay, the antibody may bind to one protein with high affinity and to others with low affinity. Second, it is not uncommon for different proteins to have the same antigenic site; that is, the same cluster of five or six amino acid side chains on their surfaces. This is especially true of members of protein families, which have similar amino acid sequences, and are often identical in functionally conserved regions.

**8–2** False. There are $6 \times 10^{23}$ molecules per mole; hence, only 0.6 molecules in a yoctomole. The limit of detection is one molecule, or 1.7 yoctomole. No instrument can detect less than one molecule (it is either present in the instrument or it is not).

**Reference:** Castagnola M (1998) Sensitive to the yoctomole limit. *Trends Biochem. Sci.* 23, 283.

**8–3** False. By measuring the association and dissociation rates, SPR allows the binding constant to be calculated as $K = k_{off}/k_{on}$. Because the data necessary for determining $K$ are present in SPR measurements, the statement is not correct.

**8–4** True. If each cycle doubles the amount of DNA, then 10 cycles equal a $2^{10}$-fold amplification (which is 1024), 20 cycles equal a $2^{20}$-fold amplification (which is $1.05 \times 10^6$), and 30 cycles equal a $2^{30}$-amplification (which is $1.07 \times 10^9$). (It is useful to remember that $2^{10}$ is roughly equal to $10^3$ or 1000. This simple relationship allows you to estimate the answer to this problem rapidly without resorting to your calculator. It comes in handy in a variety of contexts.)

**8–5** Cells in a tissue are bound together by protein-mediated attachments to one another and to an extracellular matrix containing collagen. Treatment with trypsin, collagenase, and EDTA disrupts these attachments. Trypsin is a protease that will cleave most proteins, but generally only those portions of a native protein that are unstructured. The triple helical structure of collagen, for example, is a poor substrate for trypsin. Collagenase, which is a protease specific for collagen, digests a principal component of the extracellular matrix. EDTA chelates $Ca^{2+}$, which is required for the cell-surface proteins known as cadherins to bind to one another to link cells together. Removal of $Ca^{2+}$ prevents this binding and thereby loosens cell–cell attachments.

The treatment does not kill the cells because all the damage occurs to extracellular components, which the cells can replace. So long as the plasma membrane is not breached, the cells will survive.

**8–6** Yes. In fact, it is common to raise antibodies against other antibodies. Usually, this is done by introducing antibodies from one species into a second species, for example, by injecting mouse antibodies into goats. In this example, the mouse antibodies are recognized as foreign proteins by the goat, which mounts an immune response and generates goat antibodies that bind to the mouse antibodies. It is also possible to raise antibodies against antibodies from the same species. In the same species, most parts of the injected antibody molecules will be indistinguishable from the host antibodies and thus will be treated as 'self' and not elicit an immune response. But the portion of the antibody molecule that binds to an antigen (the so-called idiotype) can be recognized as foreign and elicit an immune response, generating antibodies directed against the antigen-combining site of the injected antibody. This is because an individual can make millions of different antigen-binding sites, so that any one is present in too low a concentration to be recognized as self.

**8–7** Velocity sedimentation is used to separate components that differ in size or shape, or both. It is carried out by layering a solution containing the components to be separated on top of a shallow density gradient formed by increasing concentrations—from top to bottom—of a small molecule such as sucrose. Upon centrifugation, individual components will move through the gradient according to their size and shape. Because identical components have the same properties, they move as a defined band, which can be collected.

Equilibrium sedimentation is used to separate components that differ in their buoyant density. Components to be separated are most often layered on top of a steep sucrose gradient and centrifuged until the components move to their equilibrium density. (The components can also be mixed into the gradient to start with, but for sucrose density gradients it is more common to establish the gradient and then layer the components on top.) Molecules with the same density will form defined bands, which can be collected.

Because most proteins have about the same density, velocity sedimentation would be preferred over equilibrium sedimentation for the separation of two proteins of different size.

**8–8** The rate of sedimentation of a protein is based on size *and* shape. The nearly spherical hemoglobin will sediment faster than the more rod-shaped tropomyosin, even though tropomyosin is the larger protein. Shape comes into play because molecules that are driven through a solution by centrifugal force experience the equivalent of frictional drag. A spherical protein, with its smaller surface-to-volume ratio, will experience less drag than a rod, and therefore will sediment faster. You can demonstrate this difference using two sheets of paper. Crumple one into a sphere and roll the other into a tube. Now drop them. The ball will hit the ground faster than the tube. In this demonstration, the centrifugal force is replaced by gravity and the friction with molecules in solution is replaced by friction with air. The underlying principles are the same.

**8–9**

A. In the beginning, the solution of CsCl is a uniform density throughout, 1.71 g/mL in these experiments, and the DNA is distributed evenly throughout (Figure A8–1A). Under the influence of the centrifugal force (70,000 times gravity), the CsCl is pushed toward the bottom of the tube. This downward force is counterbalanced by random diffusion of the ions. At equilibrium, a linear gradient forms that is typically about 7% denser at the bottom of the tube than at the top (Figure A8–1B). As the gradient of CsCl forms, the DNA floats to its density. The narrowing of the band with time shows how the gradient forms over time. The solution of CsCl becomes less dense at the top and denser at the bottom of the centrifuge tube, as is evident even at 4 hours in Figure Q8–2.

B. The buoyant density of DNA—actually the cesium salt of DNA—is about 1.71 g/mL. At the center of the centrifuge tube, the density of the CsCl solution

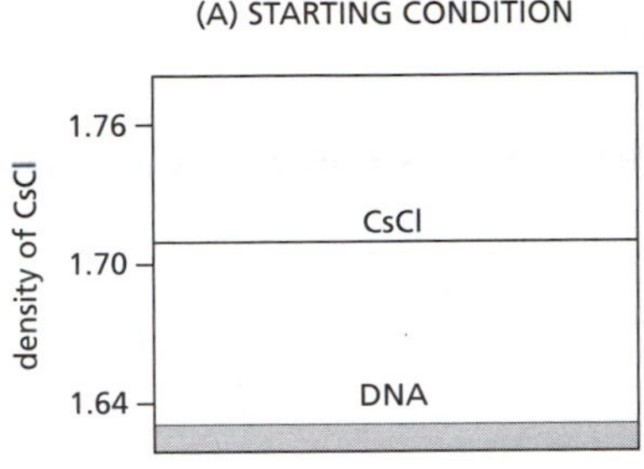


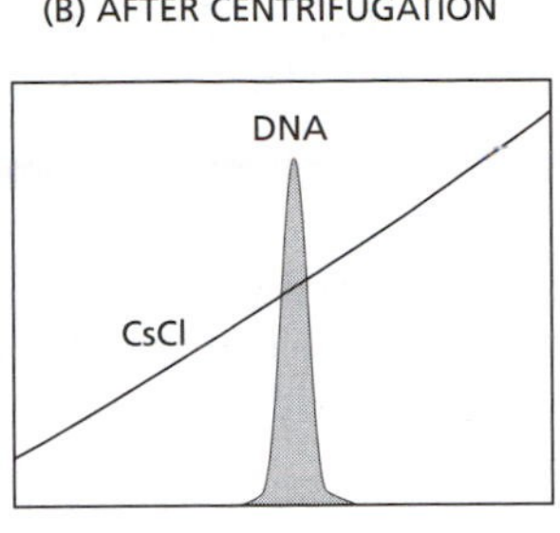


**Figure A8–1** Equilibrium sedimentation of DNA in a solution of CsCl (Answer 8–9). (A) Density of CsCl and distribution of DNA at the beginning of the experiment. (B) Density distribution of CsCl and DNA at equilibrium.

equals the average density, which is the starting density. Since the DNA forms a band near the center, it must have a density near the average. Using a refractometer (or other means), it is possible to determine the density of the gradient at any point, so that a material need not band at the center of the tube to deduce its density.

C. Three major factors contribute to the width of the band of DNA. One is diffusion of the DNA, which opposes the tendency of the DNA to focus at its density. Indeed, the diffusion coefficient of DNA (and other molecules) can be calculated from the width of the band in equilibrium centrifugation. The second factor is that the density of DNA depends on the nucleotide composition. Because the DNA in this experiment was derived by random fragmentation of the *E. coli* genome, the fragments have a range of nucleotide compositions, each with a slightly different density, which also contributes to the width of the bands observed in Figure Q8–2. The third factor is the speed of centrifugation; the higher the speed, the greater the force times gravity. Because the diffusion coefficient does not change with speed, the higher the centrifugal force, the tighter the band.

**References:** Meselson M & Stahl FW (1958) The replication of DNA in *Escherichia coli. Proc. Natl Acad. Sci. U.S.A.* 44, 671–682.

Cantor CR & Schimmel PR (1980) Biophysical Chemistry, pp 632–634. New York: WH Freeman and Company.

**8–10** Although it is invaluable, hybridoma technology is labor intensive and time consuming, requiring several months to isolate a hybridoma cell line that produces a monoclonal antibody of interest. Also, there is no guarantee that the cell line will produce a monoclonal antibody with the specific properties you are after. It is much simpler—a few days work—to add an epitope tag to your protein and then use a commercially available antibody to that epitope. The possibility that the tag may alter the function of the protein is a critical concern, but you can add the epitope easily to the N- or C-terminus and test for the effect on the protein's function. In most cases, a tag at one or the other end of the molecule will be compatible with its function.

**8–11** You would need $10^5$ copies of a 120-kd protein in a mammalian cell (and 100 copies in a bacterial cell) in order to be able to detect it on a gel. The calculation comes in two parts: how many cell-equivalents can be loaded onto the gel, and how many copies of a protein can be detected in the band. As shown below for mammalian cells, 100 μg/mL corresponds to $5 \times 10^5$ mammalian cells and to $5 \times 10^8$ bacterial cells.

$$\frac{\text{cells}}{\text{gel}} = \frac{100\ \mu\text{g}}{\text{gel}} \times \frac{\text{mL}}{200\ \text{mg}} \times \frac{\text{cell}}{1000\ \mu\text{m}^3} \times \frac{\text{mg}}{1000\ \mu\text{g}} \times \frac{(10^4\ \mu\text{m})^3}{(\text{cm})^3} \times \frac{\text{cm}^3}{\text{mL}}$$
$$= 5 \times 10^5\ \text{cells/gel}$$

There are $5 \times 10^{10}$ 120-kd proteins in a 10-ng band.

$$\frac{\text{molecules}}{\text{band}} = \frac{10\ \text{ng}}{\text{band}} \times \frac{\text{nmol}}{120{,}000\ \text{ng}} \times \frac{6 \times 10^{14}\ \text{molecules}}{\text{nmol}}$$
$$= 5 \times 10^{10}\ \text{molecules/band}$$

Thus, if you can detect $5 \times 10^{10}$ proteins in a band and can load the equivalent of $5 \times 10^5$ mammalian cells per gel, there must be $10^5$ copies of the protein per cell ($5 \times 10^{10}/5 \times 10^5$) in order for it to be detectable as a silver-stained band on a gel. For a bacterial cell, there need to be 100 copies of the protein per cell ($5 \times 10^{10}/5 \times 10^8$).

**8–12** The appropriate PCR primers are primer 1 (5′-GACCTGTGGAAGC) and primer 8 (5′-TCAATCCCGTATG). The first primer will hybridize to the bottom strand and prime synthesis in the rightward direction. The second primer will hybridize to the top strand and prime synthesis in the leftward direction. (Remember that strands pair antiparallel.)

The middle two primers in each list (primers 2, 3, 6, and 7) would not hybridize to either strand. The remaining pair of primers (4 and 5) would hybridize, but would prime synthesis in the wrong direction—that is, outward, away from the central segment of DNA. Each of these wrong choices has been made at one time or another in most laboratories that use PCR. In most cases, the confusion arises because the conventions for writing nucleotide sequences have been ignored. By convention, nucleotide sequences are written 5′ to 3′ with the 5′ end on the left. For double-stranded DNA the 5′ end of the top strand is on the left.

**8–13** In PCR amplification, a double-stranded fragment of the correct size is first generated in the third cycle (Figure A8–2).

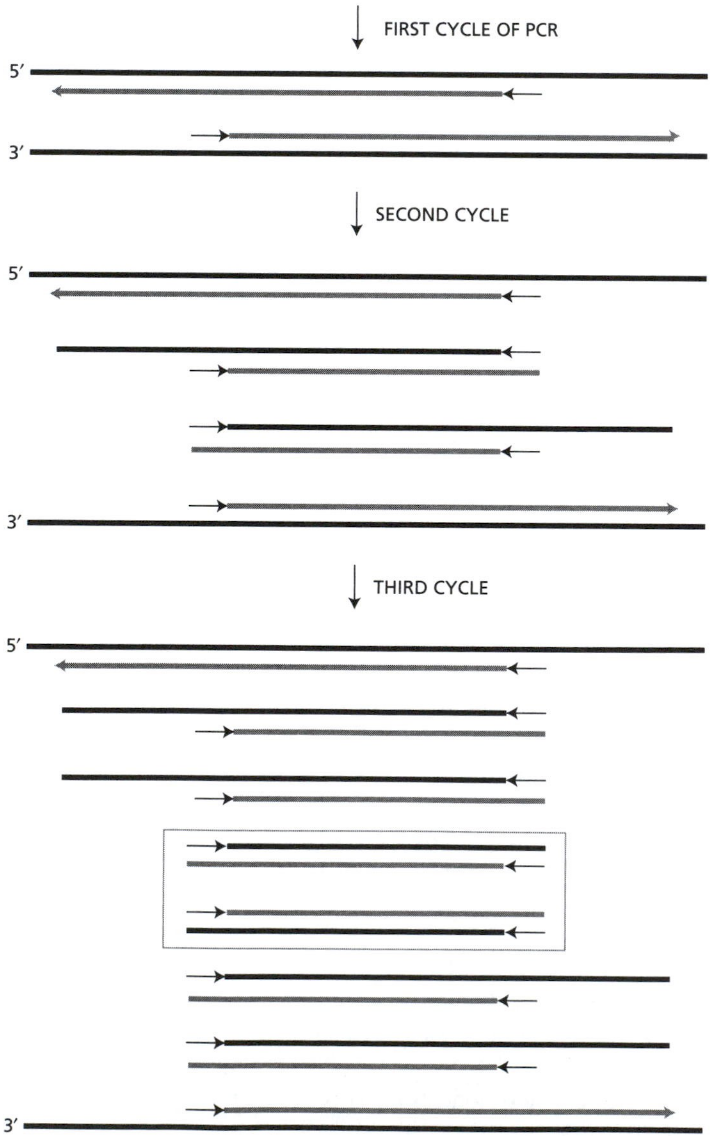


**Figure A8–2** The PCR products generated during the first three cycles (Answer 8–13). DNA that has been synthesized during a cycle is shown as a *gray line*. The positions of the primers in new and old products are clearly indicated. The first products of the correct size are *boxed* in cycle 3.

**8–14** A gain-of-function mutation increases the activity of the protein product of the gene, makes it active in inappropriate circumstances, or gives it a novel activity. The change in activity often has a phenotypic consequence even when the protein is present at only half the concentration of the wild-type protein in normal cells, which is why such mutations are usually dominant. A dominant–negative mutation gives rise to a mutant gene product that interferes with the function of the normal gene product, causing a loss-of-function phenotype even in the presence of a normal copy of the gene. This ability of a single defective allele to determine the phenotype is the reason why such an allele is dominant.

**8–15** This statement is largely true. Diabetes is one of the oldest diseases described by humans, dating back at least to the time of the ancient Greeks. Diabetes itself comes from the Greek word for siphon, which was used to describe the main symptoms—increased production of urine: "The disease was called diabetes, as though it were a siphon, because it converts the human body into a pipe for the transflux of liquid humors." If there were no human disease, the role of insulin would not have come to our attention as soon as it did. It is difficult to overstate the case for the role of disease in focusing our efforts toward a molecular understanding. Even today, the quest to understand and alleviate human disease is a principal driving force in biomedical research.

## Chapter 9 Visualizing Cells

**9–1** False. Although it is not possible to see DNA by light microscopy in the absence of a stain, chromosomes are clearly visible under phase-contrast or Nomarski differential-interference-contrast microscopy when they condense during mitosis. Condensed human chromosomes are more than 1 μm in width—well above the resolution limit of 0.2 μm.

**9–2** True. Longer wavelengths correspond to lower energies. Because some energy is lost during absorption and re-emission, the emitted photon is always of a lower energy (longer wavelength) than the absorbed photon.

**9–3** True. Caged molecules are photosensitive precursors of biologically active substances such as $Ca^{2+}$, cyclic AMP, and inositol trisphosphate. They are designed to carry an inactivating moiety attached by a photosensitive linkage. When exposed to intense light of the correct wavelength, the inactivating group is split off and the active small molecule is released. Because laser beams can be tightly focused, caged molecules can be activated at defined locations in a cell. Thus, the time and location of activation are under the experimenter's control.

**9–4** In a dry lens a portion of the illuminating light is internally reflected at the interface between the coverslip and the air. By contrast, because glass and immersion oil have the same refractive index, there is no interface; hence, no light is lost to internal reflection. In essence the oil-immersion lens increases the width of the cone of light that reaches the objective, which is a key limitation on resolution.

**9–5** The main refraction in the human eye occurs at the interface between air (refractive index 1.00) and the cornea (refractive index 1.38). Because of the small differences in refractive index between the cornea and the lens and between the lens and the vitreous humor, the lens serves to fine-tune the focus in the human eye.

**9–6** Humans see poorly underwater because the refractive index of water (1.33) is very close to that of the cornea (1.38), thus eliminating the main refractive power of the cornea. Goggles improve underwater vision by placing air in front of the cornea, which restores the normal difference in refractive

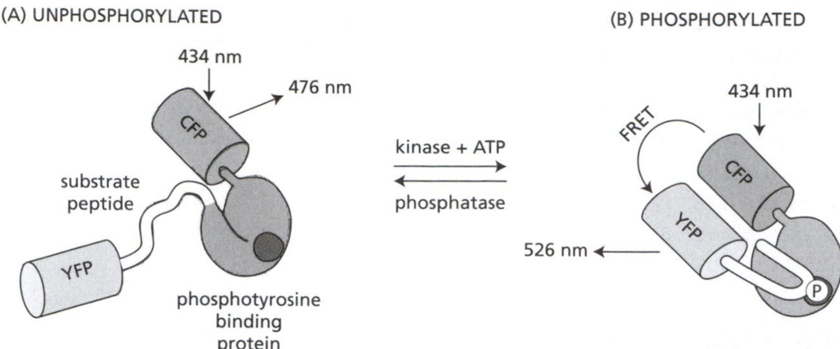


**Figure A9–1** Conformational change in FRET reporter protein upon tyrosine phosphorylation (Answer 9–10).

indices at this interface. The image is still distorted by the refractive index changes at the water-glass and glass-air interfaces of the goggles, but the distortion is small enough that the image can still be focused onto the retina, allowing us to see clearly.

**9–7** Resolution refers to the ability to see two small objects as separate entities, which is limited ultimately by the wavelength of light used to view the objects. Magnification refers to the size of the image relative to the size of the object. It is possible to magnify an image to an arbitrarily large size. It is important to remember that magnification does not change the limit of resolution.

**9–8** Fluorescently tagged antibodies and enzyme-tagged antibodies each have the advantage of amplifying the initial signal provided by the binding of the primary antibody. For fluorescently tagged secondary antibodies, the amplification is usually several fold; for enzyme-linked antibodies, amplification can be more than 1000-fold. Although the extensive amplification makes enzyme-linked methods very sensitive, diffusion of the reaction product (often a colored precipitate) away from the enzyme limits the spatial resolution.

**9–9** The wavelengths at which the chromophore is excited and at which it emits fluorescent light depend critically on its molecular environment. Using a variety of mutagenic and selective procedures, investigators have generated mutant GFPs that fluoresce throughout the visible range. These modified GFPs have a variety of different amino acids around the chromophore, which subtly influence its ability to interact with light.

**Reference:** Service RF (2004) Immune cells speed the evolution of novel proteins. *Science* 306, 1457.

**9–10** The increase in FRET depends on phosphorylation of the protein, since no increase occurs in the absence of Abl protein or ATP, or when the phosphate is removed by a tyrosine phosphatase (see Figure Q9–4B). Thus, phosphorylation must cause CFP and YFP to be brought closer together. A reasonable explanation is that addition of phosphate to the tyrosine in the substrate peptide allows that segment of the protein to fold back to bind to the adjacent phosphotyrosine-binding domain, thereby decreasing the separation of the CFP and YFP domains (Figure A9–1).

**9–11** Substituting numbers into the equation gives a value for θ of 1.4°, which is about 43 times (60°/1.4°) smaller than θ for a typical light microscope.

$$\text{resolution} = \frac{0.61\,\lambda}{n \sin\theta}$$

$$\sin\theta = \frac{0.61\ (0.004\ \text{nm})}{(0.1\ \text{nm})}$$

$$\theta = \arcsin 0.0244 = 1.4°$$

**9–12** Electron microscopists can be sure whether a structure is a pit or a bump. Shadowed structures are unlike shaded circles in a key way: structures that

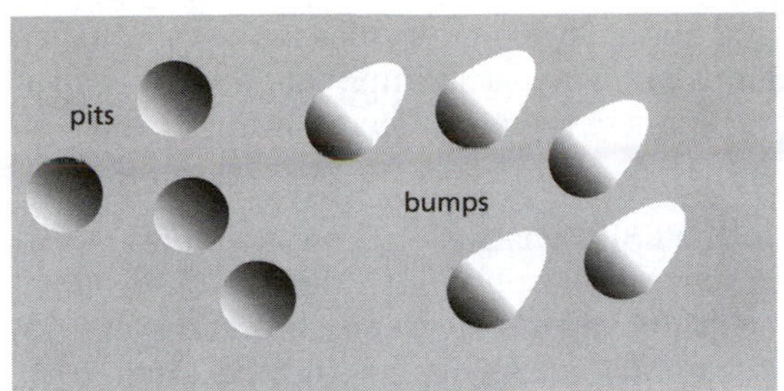


**Figure A9–2** Bumps and pits (Answer 9–12). Bumps and pits can be distinguished because bumps cast shadows.

stand above the surface cast a shadow beyond themselves, whereas pits do not. In everyday experience shadows cast by the sun are dark, but in the world of microscopy, where platinum atoms replace sunlight, the shadow is the absence of metal, hence bright, as shown in Figure A9–2. If you examine the micrographs in Figure Q9–5, especially in the orientation that looks like pits (look at the lower left hand corner in D), you can see that most of the pits are elongated toward the lower left of the micrograph. Thus, these structures are casting shadows; hence, they are bumps. To have the structures perceived as bumps, microscopists arrange the micrographs as shown in Figure Q9–5C, so that the dark portion of each bump is at the bottom. Evidently, this arrangement fits with out hardwired perceptions of the world around us, with the sun coming from above. Note also that we naturally interpret the light areas of such images as structures that reflect light; in reality they are simply the absence of electron dense material and give us no more information about the structure than does a shadow cast by sunlight.

## Chapter 10 Membrane Structure

**10–1** True. The hydrophobic interior of the lipid bilayer acts as a barrier to the passage of the hydrophilic lipid head groups that must occur during flip-flop. The energetic cost of this movement effectively prevents spontaneous flip-flop of lipids, so that it occurs extremely rarely in the absence of specific catalysts, known as phospholipid translocators.

**10–2** False. The carbohydrate on internal membranes is directed away from the cytosol toward the lumen of an internal membrane-bounded compartment. Remember that the lumen of an internal compartment is topologically equivalent to the outside of the cell.

**10–3** False. In addition to lipid rafts, which are microdomains with distinct lipid compositions, the apical and basolateral surfaces of epithelial cells, which are separated by intercellular tight junctions, also have different lipid compositions.

**10–4** The same forces that dictate that certain lipids will form a bilayer, as opposed to micelles, operate in the repair of a tear in the bilayer. The tear will heal spontaneously because a bilayer is the most energetically favorable arrangement. The lipids that make up a bilayer are cylindrical in shape and therefore do not readily form a micelle (or a hemi-micelle), which would require cone-shaped lipids.

**10–5** Vegetable oil is converted to margarine by reduction of double bonds (by hydrogenation), which converts unsaturated fatty acids to saturated ones. This change allows the fatty acids chains in the lipid molecules to pack more tightly against one another, increasing the viscosity, turning oil into margarine.

**10–6** A raft 70 nm in diameter would have an area of $3.8 \times 10^3$ nm$^2$ ($3.14 \times 35^2$), and a lipid molecule 0.5 nm in diameter would have an area of 0.20 nm$^2$ ($3.14 \times 0.25^2$). Thus, there would be about 19,000 lipid molecules per monolayer of raft ($3.8 \times 10^3/0.20 = 19{,}000$), and about 38,000 molecules in the raft bilayer. At a ratio of 50 lipids per protein, a raft would accommodate about 760 protein molecules. The true ratio of lipids to proteins in a raft is unknown.

**10–7**

A. The difference in rate of loss of the ESR signals is due to the different locations of the nitroxide radical on the two phospholipids. The nitroxide radical in phospholipid 1 is on the head group and therefore is in direct contact with the external medium. Thus, it can react quickly with ascorbate. The nitroxide radical in phospholipid 2 is attached to a fatty acid chain and therefore is partially buried in the interior of the membrane. As a consequence, it is less accessible to ascorbate and is reduced more slowly.

B. The key observation is that the extent of loss of ESR signal in the presence and absence of ascorbate is the same in resealed red cell ghosts but different in red cells. These results suggest that there is an undefined reducing agent in the cytoplasm of red cells (which is absent from red cell ghosts). Like ascorbate, this cytoplasmic agent can reduce the more exposed phospholipid 1 but not the less exposed phospholipid 2. Thus, in red cells, phospholipid 2 is stable in the absence of ascorbate; in the presence of ascorbate, the spin-labeled phospholipids in the outer monolayer are reduced, causing loss of half the ESR signal. Phospholipid 1, on the other hand, is not stable in red cells in the absence of ascorbate because the phospholipids in the cytoplasmic monolayer are exposed to the cytoplasmic reducing agent, which destroys half the ESR signal. When ascorbate is added, labeled phospholipids in the outer monolayer are also reduced, causing loss of the remaining ESR signal.

C. The results in Figure Q10–3 indicate that the labeled phospholipids were introduced equally into the two monolayers of the red cell plasma membrane. Phospholipid 2 was 50% sensitive to ascorbate, indicating that half the label was present in the outer monolayer, and 50% insensitive to ascorbate, indicating that half was present in the cytoplasmic monolayer. Similarly, phospholipid 1 was 50% sensitive to the cytoplasmic reducing agent and 50% sensitive to ascorbate, indicating an even distribution between the cytoplasmic and outer monolayers.

**Reference:** Rousselet A, Guthmann C, Matricon J, Bienvenue A & Devaux PF (1976) Study of the transverse diffusion of spin labeled phospholipids in biological membranes: 1. Human red blood cells. *Biochim. Biophys. Acta* 426, 357–371.

**10–8**

A. Sequence A is the actual membrane-spanning α-helical segment of glycophorin, a transmembrane protein from red blood cells. It is composed predominantly of hydrophobic amino acids, although it does contain the uncharged polar amino acids threonine (T) and serine (S), which are not uncommon in membrane-spanning α helices.

Sequence B is unlikely to be a membrane-spanning segment because it contains three prolines (P), which would disrupt an α helix and thereby expose polar groups to the hydrophobic environment of the lipid bilayer.

Sequence C is also unlikely to be a transmembrane segment because it contains three charged amino acids, glutamic acid (E), arginine (R), and aspartic acid (D), whose presence in the hydrophobic lipid bilayer would be energetically unfavorable.

**10–9** Your friend's suggestion is based on an important difference between inside-out and right-side-out vesicles. The contaminating right-side-out vesicles will carry carbohydrate on their exposed surface and, therefore, should be retained on a lectin affinity column. Inside-out vesicles, by contrast, will lack carbohydrate on their exposed surface and, therefore, should pass through the column.

**10–10** Transmembrane domains that are composed entirely of hydrophobic amino acid side chains obviously cannot interact with one another via hydrogen bonds or electrostatic attractions, two of the more important ways to link proteins together noncovalently. Nevertheless, they can interact specifically via van der Waals bonds. If their surfaces are complementary, they can fit together well enough to make a large number of van der Waals contacts, which can hold them together. It should be noted, however, that the transmembrane segment of glycophorin contains a few polar amino acids that may participate in the dimerization process.

## Chapter 11 Membrane Transport of Small Molecules and the Electrical Properties of Membranes

**11–1** True. Transporters bind specific molecules and undergo a series of conformational changes to move the bound molecule across a membrane. They can transport passively down the electrochemical gradient, or the transporters can link the conformational changes to a source of metabolic energy such as ATP hydrolysis to drive active transport. By contrast, channels form aqueous pores that can be open or shut, but always transport downhill; that is, passively. Channels interact much more weakly with the solute to be transported, and they do not undergo conformational changes to accomplish transport. As a consequence, transport through channels cannot be linked to an energy source and is always passive.

**11–2** False. Transporters *and* channels saturate. It is thought that permeating ions have to shed most of their associated water molecules in order to pass, in single file, through the narrowest part—the selectivity filter—of the channel. This requirement limits their rate of passage. Thus, as ion concentrations increase, the flux of ions through a channel increases proportionally, but then levels off (saturates) at a maximum rate.

**11–3** True. It takes a difference of only a minute number of ions to set up the membrane potential.

**11–4** The order is $CO_2$ (small and nonpolar) > ethanol (small and slightly polar) > $H_2O$ (small and polar) > glucose (large and polar) > $Ca^{2+}$ (small and charged) > RNA (very large and highly charged). This list nicely illustrates the two basic properties that govern the capacity of molecules to diffuse through a lipid bilayer: size (small > large) and polarity (nonpolar > polar > charged).

**11–5** The equilibrium distribution of a molecule across a membrane depends on the chemical gradient (concentration) and on the electrical gradient (membrane potential). An uncharged molecule does not experience the electrical gradient and, thus, will be at equilibrium when it is at the same concentration on both sides of the membrane. A charged molecule responds to both components of the electrochemical gradient and will distribute accordingly. $K^+$ ions, for example, are nearly at their equilibrium distribution across the plasma membrane even though they are about 30-fold more concentrated inside the cell. The difference in concentration is balanced by the membrane potential (negative inside), which opposes the movement of cations to the outside of the cell.

**11–6**

A. Yes, they can normalize both the $H^+$ and $Na^+$ concentrations. For every three cycles of the $Na^+$-$H^+$ antiporter, which imports one $Na^+$ and exports one $H^+$, the $Na^+$-$K^+$ pump cycles once, exporting three $Na^+$ ions with each operation.

(You may have wondered how to deal with the hydrolysis of ATP that occurs with each cycle of the $Na^+$-$K^+$ pump. When ATP is hydrolyzed by $H_2O$, the products are ADP and $H_2PO_4^-$. $H_2PO_4^-$ has a $pK$ of 6.86, which means that it is about 70% ionized into $HPO_4^{2-}$ and $H^+$ at the intracellular pH of 7.2. It turns out that you do not need to worry about this $H^+$ because elsewhere in the cell other processes reconvert the products of hydrolysis back into ATP to maintain a steady-state concentration.)

B. The linked action of these two pumps moves $3H^+$ out for every $2K^+$ that are brought into the cell, thereby increasing both the internal $K^+$ concentration and the membrane potential.

**11–7** Based on the scale bars in Figure Q11–1, each microvillus approximates a cylinder 0.1 μm in diameter and 1.0 μm in height. The ratio of the area of the sides of a cylinder, which represent new membrane (new surface area), to the top of a cylinder (which is equivalent to the plasma membrane that would have been present anyway, had the microvillus not been extruded) gives the increase in surface area due to an individual microvillus. The area of the sides of a cylinder ($2\pi rh$, where $r$ is the radius and $h$ is the height) is 0.31 $\mu m^2$ (2 × 3.14 × 0.05 μm × 1.0 μm); the area of the top of the cylinder ($\pi r^2$) is 0.0079 $\mu m^2$ [3.14 × $(0.05)^2$]. Thus, the increase in surface area for one microvillus is 0.31 $\mu m^2$/0.0079 $\mu m^2$ or 40. This value overestimates the increase for the entire plasma membrane, since the microvilli occupy only a portion of the surface. The fraction of plasma membrane occupied by microvilli can be estimated from the cross section in Figure Q11–1. A conservative estimate is that about half the plasma membrane is covered by microvilli. Thus, microvilli increase the surface area in contact with the lumen of the gut by approximately 40/2 or 20-fold.

**Reference:** Adapted from Kristic RV (1997) Ultrastructure of the Mammalian Cell, p 207. Berlin, Germany: Springer-Verlag.

**11–8** Just as a falling body in air reaches a terminal velocity due to friction, an ion in water also reaches a terminal velocity due to friction with water molecules. An ion in water will accelerate for less than 10 nanoseconds before it reaches terminal velocity.

**11–9** The volume of the hemisphere explored by the ball is $2.05 \times 10^4$ $nm^3$ [(2/3) × 3.14 × $(21.4 \text{ nm})^3$]. This volume corresponds to $2.05 \times 10^{-20}$ liters [($2.05 \times 10^4$ $nm^3$) × $(cm/10^7 \text{ nm})^3$ × (liter/1000 $cm^3$)]. One ball in this volume corresponds to $8.13 \times 10^{-5}$ M or 81.3 μM [(1 molecule/$2.05 \times 10^{-20}$ liters) × (mole/$6 \times 10^{23}$ molecules)]. Thus, the local concentration of a tethered ball is about the same as the concentration of free peptide needed to inactivate the channel.

**Reference:** Zagotta WN, Hoshi T & Aldrich RW (1990) Restoration of inactivation in mutants of *shaker* potassium channels by a peptide derived from ShB. *Science* 250, 568–570.

**11–10** The expected membrane potential due to differences in $K^+$ concentration across the resting membrane is

$$V = 58 \text{ mV} \times \log \frac{C_o}{C_i}$$

$$V = 58 \text{ mV} \times \log \frac{9 \text{ mM}}{344 \text{ mM}}$$

$$V = -92 \text{ mV}$$

For $Na^+$, the equivalent calculation gives a value of +48 mV.

The assumption that the membrane potential is due solely to $K^+$ leads to a value near that of the resting potential. The assumption that the membrane potential is due solely to $Na^+$ leads to a value near that of the action potential.

These assumptions approximate the resting potential and action potential because $K^+$ *is* primarily responsible for the resting potential and $Na^+$ *is* responsible for the action potential. A resting membrane is 100-fold more permeable to $K^+$ than it is to $Na^+$ because of the presence of $K^+$ leak channels. The leak channel allows $K^+$ to leave the cell until the membrane potential rises sufficiently to oppose the $K^+$ concentration gradient. The theoretical maximum gradient (based on calculations like those above) is lowered somewhat by the entrance of $Na^+$, which carries positive charge into the cell (compensating for the positive charges on the exiting $K^+$). Were it not for the

$Na^+$-$K^+$ pump, which continually removes $Na^+$, the resting membrane potential would be dissipated completely.

The action potential is due to a different channel, a voltage-gated $Na^+$ channel. These channels open when the membrane is stimulated, allowing $Na^+$ ions to enter the cell. The magnitude of the resulting membrane potential is limited by the difference in the $Na^+$ concentrations across the membrane. The influx of $Na^+$ reverses the membrane potential locally, which opens adjacent $Na^+$ channels and ultimately causes an action potential to propagate away from the site of the original stimulation.

**Reference:** Hille B (1992) Ionic Channels of Excitable Membranes, 2nd ed, pp 23–58. Sunderland, MA: Sinauer.

## Chapter 12 Intracellular Compartments and Protein Sorting

**12–1** False. The interior of the nucleus and the cytosol communicate through the nuclear pore complexes, which allow free passage of ions and small molecules. The cytoplasm and the nucleus are said to be topologically equivalent because the outer and inner nuclear membranes are continuous with one another, so that the flow of material between the nucleus and cytosol occurs without crossing a lipid bilayer. By contrast, the lumen of the ER and the outside of the cell are each separated from the cytosol by a layer of membrane. Thus they are topologically distinct from the cytosol; they are topologically equivalent to each other.

**12–2** True. Ribosomes all begin translating mRNAs in the cytosol. The mRNAs for certain proteins encode a signal sequence for the ER membrane. After this sequence has been synthesized, it directs the nascent protein, along with the ribosome and the mRNA, to the ER membrane. Ribosomes translating mRNAs that do not encode such a sequence remain free in the cytosol.

**12–3** False. Individual nuclear pores mediate transport in both directions. It is unclear how pores coordinate this two-way traffic so as to avoid head-on collisions and congestion.

**12–4** False. All eucaryotic cells contain peroxisomes.

**12–5** False. The first (most N-terminal) transmembrane segment that exits from the ribosome initiates translocation (acts as a start-transfer signal). Its orientation in the ER membrane fixes the reading frame for the insertion of subsequent transmembrane segments. If the first transmembrane segment is oriented with its N-terminus in the cytosol, even-numbered segments will act as stop-transfer signals, and odd-numbered segments will act as start-transfer signals. If the first segment is oriented with its N-terminus in the lumen, then the second segment and subsequent even-numbered segments will act as start-transfer signals. Subsequent odd-numbered segments will act as stop-transfer signals.

**12–6** In the absence of a sorting signal, a protein will remain in the cytosol.

**12–7**

A. The protein would enter the ER. The signal for import into the ER is located at the N-terminus of the protein and functions before the internal signal for nuclear import is synthesized. Once the protein entered the ER the signal sequence for nuclear import could not function because it would be prevented from interacting with cytosolic nuclear import receptors.

B. The protein would enter the ER. Once again, the N-terminal signal for ER import would function before the internal signal for peroxisome import is synthesized. The peroxisome import signal could not function once the protein was sequestered in the ER.

C. The protein would enter the mitochondria. In order to be retained in the ER the protein must first be imported into the ER. Without a signal for ER import, the ER retention signal could not function.

D. A protein with signals for both nuclear import and nuclear export would shuttle between the cytosol and the nucleus. Unlike the other pairs of signals, these signals are not necessarily in conflict. A number of cellular proteins, whose function requires shuttling in and out of the nucleus, are designed in just this way.

**12–8** If the equivalent of one plasma membrane transits the ER every 24 hours and individual membrane proteins remain in the ER for 30 minutes (0.5 hr), then at any one time, 0.021 (0.5 hr/24 hr) plasma membrane equivalents are present in the ER. Since the area of the ER membrane is 20 times greater than the area of the plasma membrane, the fraction of plasma membrane proteins in the ER is 0.021/20 = 0.001. Thus, the ratio of plasma membrane proteins to other membrane proteins in the ER is 1 to 1000. Out of every 1000 proteins in the ER membrane only 1 is in transit to the plasma membrane.

As this calculation illustrates, the sorting of proteins to the plasma membrane represents a substantial purification from the mix of proteins in the ER.

**12–9**

A. The portion of nucleoplasmin responsible for localization in the nucleus must reside in the tail. The nucleoplasmin head does not localize to the nucleus when injected into the cytoplasm, and it is the only injected component that is missing a tail.

B. These experiments suggest that the nucleoplasmin tail carries a nuclear localization signal and that accumulation in the nucleus is not the result of passive diffusion. The observations involving complete nucleoplasmin or fragments that retain the tail do not distinguish between passive diffusion and active transport; they say only that the tail carries the important part of nucleoplasmin—be it a localization signal or a binding site. The key observations that argue against passive diffusion are the results with the nucleoplasmin heads. They do not diffuse into the cytoplasm when they are injected into the nucleus, nor do they diffuse into the nucleus when injected into the cytoplasm, suggesting that the heads are too large to pass through the nuclear pores. Since the more massive forms of nucleoplasmin with tails do pass through the nuclear pores, passive diffusion of nucleoplasmin is ruled out.

**Reference:** Dingwall C, Sharnick SV & Laskey RA (1982) A polypeptide domain that specifies migration of nucleoplasmin into the nucleus. *Cell* 30, 449–458.

**12–10** Each nuclear pore complex must transport about 1 histone molecule per second, on average, throughout a day:

$$\text{transport} = \frac{32 \times 10^6 \text{ octamers}}{\text{day}} \times \frac{8 \text{ histones}}{\text{octamer}} \times \frac{\text{day}}{8.64 \times 10^4 \text{ sec}} \times \frac{1}{3000 \text{ pores}}$$

$$= 0.99 \text{ histones/second/pore}$$

Because histones are synthesized and imported into nuclei only during S phase, which is typically about 8 hours long, the transport rate is about 3 histones per second during S phase (and none during the rest of the cell cycle).

**12–11**

A. Ran is a GTPase and will slowly convert GTP to GDP. Thus, if you had prepared Ran-GTP to start with, by the time you did the experiments you would have had an undefined mixture of Ran-GTP and Ran-GDP, which would have confused the results. By using a form of Ran that cannot hydrolyze GTP, you guaranteed that Ran was in its Ran-GTP conformation. Either RanQ69L-GTP, as was used here, or Ran-GppNp would have served equally well in the experiments described in this problem.

B. Since the Ran-GDP column removed the nuclear import factor, whereas the RanQ69L-GTP column did not, you are looking for a protein that is present in lane 1 but not in lane 2 (see Figure Q12–3). One such protein is present, between the 7-kd and 14-kd markers. Note that the RanQ69L-GTP column binds a set of proteins between the 97-kd and 116-kd markers that the Ran-GDP column does not. They are members of the importin family of nuclear import receptors, which are evidently not required for the nuclear uptake of Ran-GDP.

C. The small protein that binds to Ran-GDP is known as NTF2. In addition to binding tightly to Ran-GDP, NTF2 binds to the FG-repeats present in the nucleoporins of the nuclear pore complex. It is the progressive movement of the NTF2–Ran-GDP complex along the FG tracks in the nucleoporins that allows Ran-GDP to be delivered to the nucleus. In the nucleus, the Ran-GEF converts Ran-GDP to Ran-GTP, causing it to dissociate from NTF2. NTF2 then recycles to the cytoplasm to bring in another Ran-GDP.

D. The information in the problem says only that cytoplasm passed over a Ran-GDP column is depleted of some factor that is essential for nuclear uptake, and the experimental results shown in Figure Q12–3 indicate that NTF2 binds to Ran-GDP. The inference is that NTF2 is the critical factor necessary for nuclear uptake of Ran-GDP. To prove that NTF2 is the import factor, you would need to show that purified or recombinant NTF2 can promote the uptake of Ran-GDP into nuclei. The authors of this study went even further. Using information from the crystal structure of the NTF2–Ran-GDP complex, they mutated the glutamate at position 42 in NTF2 to lysine, thereby disrupting a key salt bridge between the two proteins. This NTF2E42K mutant no longer promoted the nuclear uptake of Ran-GDP. These additional experiments demonstrate that NTF2 is necessary for nuclear uptake of Ran-GDP.

**Reference:** Ribbeck K, Lipowsky G, Kent HM, Stewart M & Görlich D (1998) NTF2 mediates nuclear import of Ran. *EMBO J.* 17, 6587–6598.

**12–12** Normal cells that carry the modified *Ura3* gene make Ura3 that gets imported into mitochondria. It is therefore unavailable to carry out an essential reaction in the metabolic pathway for uracil synthesis. These cells might as well not have the enzyme at all, and they will grow only when uracil is supplied in the medium. By contrast, in cells that are defective for mitochondrial import, Ura3 is prevented from entering mitochondria and remains in the cytosol, where it can function normally in the pathway for uracil synthesis. Thus, cells with defects in mitochondrial import can grow in the absence of added uracil because they can make their own.

**Reference:** Maarse AC, Blom J, Grivell LA & Meijer M (1992) MPI1, an essential gene encoding a mitochondrial membrane protein, is possibly involved in protein import into yeast mitochondria. *EMBO J.* 11, 3619–3628.

**12–13** The binding of methotrexate to the active site prevents the enzyme from unfolding, which is necessary for import into mitochondria. Evidently, methotrexate binds so tightly that it locks the enzyme into its folded conformation and prevents chaperone proteins from unfolding it.

**Reference:** Eilers M & Schatz G (1986) Binding of a specific ligand inhibits import of a purified precursor protein into mitochondria. *Nature* 322, 228–232.

**12–14** The pores formed by porins are large enough for all ions and metabolic intermediates, but not large enough for most proteins. The size cutoff for free passage through the pores of mitochondrial porins is roughly 10 kilodaltons.

**12–15** Catalase is located in the cytosol of peroxisome-deficient cells, as shown by the uniform staining outside the nuclei in Figure Q12–4B. Catalase appears as small dots (punctate staining) in normal cells because it is located in peroxisomes, which are small and distributed throughout the cytosol.

**Reference:** Kinoshita N, Ghaedi K, Shimozawa N, Wanders RJA, Matsuzono Y, Imanaka T, Okumoto K, Suzuki Y, Kondo N & Fujiki Y (1998) Newly identified Chinese hamster ovary cell mutants are defective in biogenesis of peroxisomal membrane vesicles (peroxisome ghosts), representing a novel complementation group in mammals. *J. Biol. Chem.* 273, 24122–24130.

**12–16** As shown in Figure A12–1, elimination of the first transmembrane segment (by making it hydrophilic) would be expected to give rise to a protein with the N-terminal segment in the cytosol (unglycosylated), but with all other membrane-spanning segments in their original orientation. In the unmodified protein, the first transmembrane segment served as a start-transfer signal, oriented so that it caused the N-terminal segment to pass across the ER membrane. The next transmembrane segment is also a start-transfer signal, but oriented so that it passes C-terminal protein across the membrane until it reaches the next transmembrane segment, which serves as a stop-transfer signal. Two more pairs of similarly oriented start- and stop-transfer signals give rise to the final arrangement.

Elimination of the first start-transfer signal would not affect the function of the second transfer signal, which would initiate transfer of C-terminal segments just as it did in the unmodified, original protein.

**12–17** Symmetry of phospholipids in the two leaflets of the ER membrane is generated by a phospholipid translocator, called a scramblase, that rapidly flips phospholipids of all types back and forth between the monolayers of the bilayer. Because it flips phospholipids indiscriminately, the different types of phospholipid become equally represented in the inner and outer leaflets of the bilayer; that is, they become symmetrically distributed. The plasma membrane contains a different kind of phospholipid translocator, which is specific for phospholipids containing free amino groups (phosphatidylserine and phosphatidylethanolamine). These flippases remove these specific phospholipids from the external leaflet and transfer them to the internal leaflet of the plasma membrane, thereby generating an asymmetrical distribution.

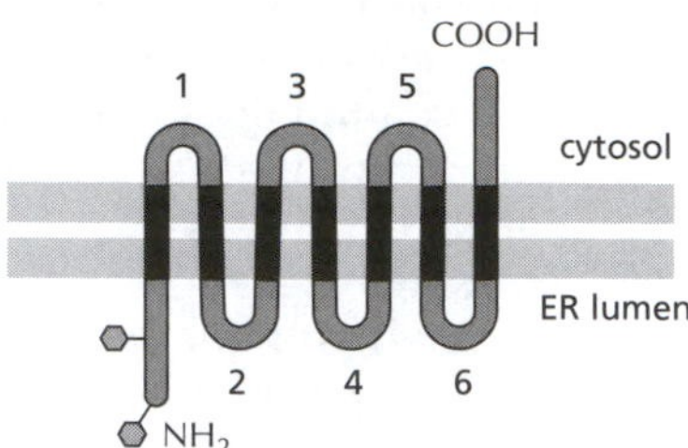


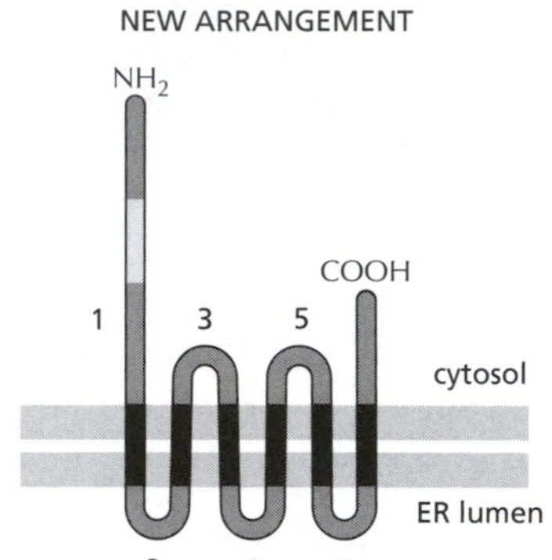


**Figure A12–1** Arrangement of the original multipass transmembrane protein and of the new protein after the first hydrophobic segment was converted to a hydrophilic segment (Answer 12–16).

## Chapter 13 Intracellular Vesicular Traffic

**13–1** True. The cytosolic leaflets of the two membrane bilayers are the first to come into contact and fuse, followed by the noncytosolic leaflets. It is this pattern of leaflet fusion that maintains the topology of membrane proteins, so that protein domains that face the cytosol always do so, regardless of what compartment they occupy.

**13–2** True. A misfolded protein is selectively retained in the ER by binding to chaperone proteins such as BiP and calnexin. Only after it has been released from such a chaperone protein—and thus approved as properly folded—does a protein become a substrate for exit from the ER.

**13–3** True. The oligosaccharide chains are added in the lumens of the ER and Golgi apparatus, which are topologically equivalent to the outside of the cell. This basic topology is conserved in all membrane budding and fusion events. Thus, oligosaccharide chains are always topologically outside the cell, whether they are in a lumen or on the cell surface.

**13–4** False. During transcytosis, vesicles that form from either the apical or basolateral surface first fuse with early endosomes, then move to recycling endosomes, where they are sorted into transport vesicles bound for the opposite surface.

**13–5** If the flow of membrane between cellular compartments were not balanced in a nondividing liver cell, some compartments would grow in size and others would shrink (in the absence of new membrane synthesis.) Keeping all the membrane compartments the same relative size is essential for proper functioning of a liver cell. The situation is different in a growing cell such as a gut epithelial cell. Over the course of a single cell cycle, all of the compartments must double in size to generate two daughter cells. Thus, there will be an imbalance in favor of the outward flow, which will be supported by new membrane synthesis equal to the sum total of all the cell's membrane.

**13–6** The specificity for both the transport pathway and the transported cargo come not from the clathrin coat but from the adaptor proteins that link the clathrin to the transmembrane receptors for specific cargo proteins. The several varieties of adaptor proteins allow different cargo receptors, hence different cargo proteins, to be transported along specific transport pathways.

Incidentally, humans are different from most other organisms in that they have two heavy-chain genes. Like other mammals, they also have two light-chain genes. In addition, in the neurons of mammals the light-chain transcripts are alternatively spliced. Thus, there exists the potential in humans for additional complexity of clathrin coats; the functional consequences of this potential variability are not clear.

**References**: Kirchhausen T (2000) Clathrin. *Annu. Rev. Biochem.* 69, 699–727.

Pearse BMF, Smith CJ & Owen DJ (2000) Clathrin coat construction in endocytosis. *Curr. Opin. Struct. Biol.* 10, 220–228.

**13–7** There will always be some v-SNAREs in the target membrane. Immediately after fusion, the v-SNAREs will be in inactive complexes with t-SNAREs. Once NSF pries the complexes apart, v-SNAREs may be kept inactive by binding to inhibitory proteins. Accumulation of v-SNAREs in the target membrane beyond some minimal population is thought to be prevented by active retrieval pathways that incorporate v-SNAREs into vesicles for redelivery to the original donor membrane.

**13–8** The cell's SNAREs are all bound to the cytosolic surface of whatever membrane they are in. They function by juxtaposing the cytosolic surfaces of the two membranes to be fused. By contrast, enveloped viruses must fuse with a cell membrane by bringing together its external surface with an external surface of a cell membrane. Thus, enveloped viruses cannot make use of a cell's SNAREs because they are located on the wrong side of the membrane. It is for this reason that enveloped viruses make their own fusion proteins, which are properly situated on their external surface.

**13–9** The volume of a cylinder 1.5 nm in diameter and 1.5 nm in height is 2.65 $nm^3$ [$3.14 \times (0.75\ nm)^2 \times 1.5\ nm$], which equals $2.65 \times 10^{-24}$ L [$2.65\ nm^3 \times (cm/10^7\ nm)^3 \times (L/10^3\ cm^3)$]. There are about 88 water molecules in this volume.

$$\frac{\text{water molecules}}{\text{cylinder}} = \frac{2.65 \times 10^{-24}\ \text{L}}{\text{cylinder}} \times \frac{55.5\ \text{mole}}{\text{L}} \times \frac{6 \times 10^{23}\ \text{molecules}}{\text{mole}}$$
$$= 88.2$$

In each monolayer in a circle of membrane 1.5 nm in diameter, there are about 9 phospholipids [$3.14 \times (0.75\ nm)^2 \times (PL/0.2\ nm^2) = 8.8\ PL$]. Thus, there are about 5 water molecules per phospholipid in the area of close approach of the two membranes. This number is slightly less than half the number (10–12) estimated to be associated with phospholipid head groups under normal circumstances. This means that when a vesicle and its target membrane are drawn together in preparation for fusion, somewhat more than half of the water molecules that would normally bind to the membranes must be squeezed out.

**Reference:** Meuse CW, Krueger S, Majkrzak CF, Dura JA, Fu J, Connor JT & Plant AL (1998) Hybrid bilayer membranes in air and water: infrared spectroscopy and neutron reflectivity studies. *Biophys. J.* 74, 1388–1398.

**13–10** To generate maximal alkaline phosphatase activity, vesicles from each strain must carry both v-SNAREs and t-SNAREs (see Figure Q13–2B, experiment 1). If either vesicle is lacking v-SNAREs or t-SNAREs, phosphatase activity is reduced to 30–60% of the maximum (see experiments 3, 4, 6, 7, 8, and 9). If both vesicles are missing either v-SNAREs (see experiment 2) or t-SNAREs (see experiment 5), phosphatase activity is very low, as it is if one vesicle is missing both SNAREs (see experiments 10 and 11). For a reasonable level of fusion, complementary SNAREs must be present on the vesicles. It does not matter which kind of SNARE is on vesicles from strain A so long as vesicles from strain B carry a complementary SNARE (compare experiments 3 and 4, experiments 6 and 7, and experiments 8 and 9).

You might have wondered why there is a low background of phosphatase activity, even where no fusion is expected (see experiments 2, 5, 10, and 11). If a few vesicles were to break, releasing small amounts of pro-Pase and protease, then a small amount of active alkaline phosphatase could be generated in the absence of vesicle fusion.

**Reference:** Nichols BJ, Undermann C, Pelham HRB, Wickner WT & Haas A (1997) Homotypic vacuolar fusion mediated by t- and v-SNAREs. *Nature* 387, 199–202.

**13–11** The modified PDI would be located outside the cell. If PDI were missing the ER retrieval signal, its gradual flow out of the ER to the Golgi apparatus would not be countered by its capture and return to the ER, as normally occurs. Similarly, it would be expected to leave the Golgi apparatus by the default pathway, mixed with the other proteins that the cell is secreting. It would not be expected to be retained anywhere else along the secretory pathway because it presumably has no signals to promote such localization.

**Reference:** Munro S & Pelham HR (1987) A C-terminal signal prevents secretion of luminal ER proteins. *Cell* 48, 899–907.

**13–12** The KDEL receptor binds its ligands more tightly in the Golgi apparatus, where it captures proteins that have escaped the ER, so that it can return them. The receptor binds its ligands more weakly in the ER, so that those proteins that have been captured in the Golgi apparatus can be released upon their return to the ER. The basis for the different binding affinities is thought to be the slight difference in pH; the lumen of the Golgi apparatus is slightly more acidic than that of the ER, which is neutral.

Since the primary job of the KDEL receptor is to capture proteins that have escaped from the ER, it would be reasonable to design the system so that the receptors are found in the highest concentration in the Golgi apparatus. This is, in fact, the way it is in the cell. You would be correct if you predicted that the KDEL receptor does not have a classic ER retrieval signal; after all, the receptor is designed to spend most of its time in the Golgi apparatus, and a classic signal would ensure its efficient return to the ER. It does, however, have a 'conditional' retrieval signal; upon binding to an ER protein in the Golgi apparatus, its conformation is altered so that a binding site for COPI subunits is exposed. That signal allows it to be incorporated into COPI-coated vesicles, which are destined to return to the ER.

**Reference:** Teasdale RD & Jackson MR (1996) Signal-mediated sorting of membrane proteins between the endoplasmic reticulum and the Golgi apparatus. *Annu. Rev. Cell Dev. Biol.* 12, 27–54.

**13–13** The lysosomal enzymes are all acid hydrolases, which have optimal activity at the low pH (about 5.0) in the interior of lysosomes. If a lysosome were to break, the acid hydrolases would find themselves at pH 7.2, the pH of the cytosol, and would therefore do little damage to cellular constituents.

**13–14** Adaptor proteins in general mediate the incorporation of specific cargo proteins into clathrin-coated vesicles by linking the clathrin coat to specific cargo receptors. Because melanosomes are specialized lysosomes, it would seem reasonable that the defect in AP3 affects the pathway from the *trans* Golgi network, which involves clathrin-coated vesicles. AP3 localizes to the *trans* Golgi network, which is consistent with a function in transport from the Golgi to lysosomes. Interestingly, humans with the genetic disorder

Hermansky–Pudlak syndrome have similar pigmentation changes, and they also have bleeding problems and pulmonary fibrosis. These symptoms are all thought to reflect deficiencies in the production of specialized lysosomes, which result from just a single biochemical defect.

**References:** Kantheti P, Qiao X, Diaz ME, Peden AA, Meyer GE, Carskadon SI, Kapfhamer D, Sufalko D, Robinson MS, Noebels JL & Burmeister M (1998) Mutation in AP-3 delta in the *mocha* mouse links endosomal transport to storage deficiency in platelets, melanosomes, and synaptic vesicles. *Neuron* 21, 111–122.

Zhen L, Jiang S, Feng L, Bright NA, Peden AA, Seymour AB, Novak EK, Elliott R, Gorin MB, Robinson MS & Swank RT (1999) Abnormal expression and subcellular distribution of subunit proteins of the AP-3 adaptor complex lead to platelet storage pool deficiency in the *pearl* mouse. *Blood* 94, 146–155.

**13–15**

A. The corrective factors are the lysosomal enzymes themselves. Hurler's cells supply the enzyme missing from Hunter's cells, and Hunter's cells supply the enzyme missing from Hurler's cells. These enzymes are present in the medium because of inefficiency in the sorting process. Since they carry M6P, which normally should direct them to lysosomes, they presumably escaped capture by the lysosomal pathway and were secreted. They are taken into cells and delivered to lysosomes by receptor-mediated endocytosis, which operates due to a small number of M6P receptors on the cell surface. The degradative enzymes, bound to receptors, are taken up through coated pits into endosomes and are eventually delivered to lysosomes. Since lysosomes are the normal site of action for these degradative enzymes, the defect is thereby corrected.

B. Protease treatment destroys the lysosomal enzymes themselves. Periodate treatment and alkaline phosphatase treatment both remove the M6P signal that is required for binding to the receptor, thus preventing the enzymes (which are still active) from entering the cell.

C. Such a scheme is unlikely to work for defects in cytosolic enzymes. External proteins normally do not cross membranes; thus, even when they are taken into cells, they remain in the lumen of a membrane-bounded compartment. In addition, foreign proteins are usually delivered to lysosomes and degraded.

**Reference:** Kaplan A, Achord DT & Sly WS (1977) Phosphohexosyl components of a lysosomal enzyme are recognized by pinocytosis receptors on human fibroblasts. *Proc. Natl Acad. Sci. U.S.A.* 74, 2026–2030.

**13–16** Since the surface area and volume of a macrophage do not change significantly over this time, the rate of exocytosis must also equal 100% of the plasma membrane each half hour.

**13–17**

A. HRP does not bind to a specific cellular receptor and is taken up only by fluid-phase endocytosis. Since endocytosis is a continuous process, HRP gets taken up steadily at a rate that depends only on its concentration in the medium; thus, its uptake rate does not saturate. By contrast, EGF binds to a specific EGF receptor and is internalized by receptor-mediated endocytosis. The limit to the amount of EGF that gets taken up is set by the number of EGF receptors on the cells; when the receptors are saturated, no further increase in uptake occurs (except at enormously high concentrations, where fluid-phase endocytosis becomes significant).

B. At 40 nM, EGF is taken up at a rate of 16 pmol/hr, while at a 1000-fold higher concentration (40 μM), HRP is taken up at 2 pmol/hr. Since the uptake of HRP is linear, the rate at a 1000-fold lower concentration is expected to be $2 \times 10^{-3}$ pmol/hr. Thus, at equal concentrations of 40 nM, EGF should be taken up 8000 times faster than HRP [(16 pmol/hr)/($2 \times 10^{-3}$ pmol/hr)].

If EGF and HRP were present at 40 μM, both would be taken up by pinocytosis at the same rate (2 pmol/hr). EGF, however, would also be taken up by receptor-mediated endocytosis at the saturation rate of 16 pmol/hr. Thus, EGF would be taken up 9 times faster than HRP [(2 pmol/hr + 16 pmol/hr)/(2 pmole/hr)].

C. An endocytic vesicle 20 nm ($2 \times 10^{-6}$ cm) in radius contains $3.4 \times 10^{-17}$ mL of fluid.

$$\text{vesicle volume} = \frac{4\pi r^3}{3}$$
$$= (4/3) \times 3.14 \times (2 \times 10^{-6}\ \text{cm})^3$$
$$= 3.4 \times 10^{-17}\ \text{cm}^3 = 3.4 \times 10^{-17}\ \text{mL}$$

A solution of 40 μM HRP contains $2.4 \times 10^{16}$ molecules/mL of HRP.

$$\text{HRP} = \frac{40\ \mu\text{mol HRP}}{\text{L}} \times \frac{\text{L}}{1000\ \text{mL}} \times \frac{6 \times 10^{17}\ \text{molecules}}{\mu\text{mol}}$$
$$= 2.4 \times 10^{16}\ \text{molecules/mL}$$

Hence each vesicle contains, on average, 0.8 molecule of HRP [($2.4 \times 10^{16}$ molecules/mL)($3.4 \times 10^{-17}$ mL/vesicle)].

D. These calculations, as alluded to by the authors, make the point that by having specific tight-binding receptors on the cell surface, cells can take up molecules from their surroundings at much higher rates—several orders of magnitude higher—than they could simply by taking in fluid, especially at the low concentrations that are typical in biology. Fishing provides an analogy. You could fish by taking random net-fulls from a stream, and occasionally you might catch a fish. But if you put bait where you cast your net, you increase your chances of success enormously. Each time a molecule of EGF hits a receptor, it sticks and subsequently makes its way to a coated pit to be internalized. If the EGF were simply trapped like HRP, its rate of uptake would be infinitesimal at the usual *in vivo* concentrations.

**Reference:** Haigler HT, McKanna JA & Cohen S (1979) Rapid stimulation of pinocytosis in human A-431 carcinoma cells by epidermal growth factor. *J. Cell Biol.* 83, 82–90.

## Chapter 14 Energy Generation: Mitochondria and Chloroplasts

**14–1** False. The three respiratory enzyme complexes exist as independent entities in the mitochondrial inner membrane. The ordered transfers of electrons between complexes are mediated by random collisions between the complexes and two mobile carriers—ubiquinone and cytochrome *c*—with electron transfers occurring when the appropriate components meet.

**14–2** False. Lipophilic weak acids act as uncoupling agents that dissipate the proton-motive force and stop ATP synthesis; however, they increase the flow of electrons through the respiratory chain by eliminating the respiratory control imposed by the electrochemical proton gradient. Normally, the electrochemical proton gradient exerts a backpressure that restricts the flow of electrons down the electron-transport chain. In the absence of the gradient electron transport runs unchecked at the maximum rate.

**14–3** True. Inheritance of organellar genomes is very different from the inheritance of nuclear genes, which is governed by Mendelian rules. A pattern of inheritance that does not obey Mendelian rules is unlikely to be due to a nuclear gene, which leaves the organellar genomes—the only other genomes in a cell.

**14–4** In the presence of oxygen, yeast can generate about 15 times more ATP from each glucose molecule than they can in the absence of oxygen. Thus, to meet their energy needs they need to process about 15-fold fewer glucose molecules; hence the dramatic drop in glucose consumption when $O_2$ is introduced.

**14–5** The number of protons in an actively respiring liver mitochondrion at pH 7.5 ($3.16 \times 10^{-8}$ M $H^+$) is about 10.

$$\frac{H^+}{\text{mitochondrion}} = \frac{3.16 \times 10^{-8}\ \text{mole } H^+}{L} \times \frac{6 \times 10^{23}\ H^+}{\text{mole } H^+} \times \frac{(4/3)(3.14)(0.5\ \mu m)^3}{\text{mitochondrion}} \times \frac{L}{10^{15}\ \mu m^3}$$
$$= 9.9$$

If the matrix of the mitochondrion started at pH 7 ($10^{-7}$ M $H^+$), it originally held about 31 protons (31.4). Thus, to reach pH 7.5, about 21 protons would need to be pumped out. These are remarkable results. Regardless of the particulars of mitochondrial size and exact pHs, it is clear that only a few tens of protons are normally involved in establishing the proton-motive force. More than anything, these results emphasize the dynamic nature of proton pumping and ATP synthesis.

**14–6** It would take the heart 6 seconds to consume its steady-state levels of ATP. Because each pair of electrons reduces one atom of oxygen, the 12 pairs of electrons generated by oxidation of one glucose molecule would reduce 6 $O_2$. Thus, 30 ATP are generated per 6 $O_2$ consumed. At steady state, the rate of ATP production equals its rate of consumption. The time in seconds required to consume the steady-state level of ATP is

$$\text{time} = \frac{5\ \mu\text{mol ATP}}{g} \times \frac{6\ O_2}{30\ \text{ATP}} \times \frac{\text{min g}}{10\ \mu\text{mol } O_2} \times \frac{60\ \text{sec}}{\text{min}}$$
$$= 6\ \text{sec}$$

**14–7** The rates of oxidation of the electron carriers, if measured rapidly enough, reveal their order in the respiratory chain. The carriers closest to oxygen will be oxidized first, and those farthest from oxygen will be oxidized last. This rationale allows you to deduce the order of electron flow through the carriers.

$$\underset{b}{\text{cytochrome}} \rightarrow \underset{c_1}{\text{cytochrome}} \rightarrow \underset{c}{\text{cytochrome}} \rightarrow \underset{(a + a_3)}{\text{cytochrome}} \rightarrow O_2$$

**14–8** An uncoupler promotes weight loss by decreasing the efficiency of oxidative phosphorylation. For example, if sufficient uncoupler were ingested to reduce the efficiency of oxidative phosphorylation to 50%, twice as many calories (from food or internal stores, mainly fat) would have to be burned to generate the same amount of ATP. Dinitrophenol is no longer prescribed because its usage led to several deaths; if oxidative phosphorylation is too efficiently compromised, not enough ATP will be generated to support essential cell functions and death is the result.

**14–9**

A. The energy of a mole of photons at any particular wavelength is the energy of one photon times Avogadro's number ($N$). Therefore, the energy of a mole of photons at a wavelength of 400 nm is

$$E = Nhc/\lambda$$
$$= \frac{6 \times 10^{23}\ \text{photons}}{\text{mole}} \times \frac{1.58 \times 10^{-37}\ \text{kcal/sec}}{\text{photon}} \times \frac{3 \times 10^{17}\ \text{nm}}{\text{sec}} \times \frac{1}{400\ \text{nm}}$$
$$E = 71\ \text{kcal/mole for 400-nm light}$$

This calculation for 680-nm and 800-nm light gives

$E$ = 42 kcal/mole for 680-nm light
$E$ = 36 kcal/mole for 800-nm light

B. If a square meter receives 0.3 kcal/sec of 680-nm light, which is worth 42 kcal/mole of photons, then the time it will take for a square meter to receive one mole of photons is

$$\text{time} = \frac{\text{sec}}{0.3 \text{ kcal}} \times \frac{42 \text{ kcal}}{\text{mole}}$$
$$\text{time} = 140 \text{ sec/mole}$$

C. If it takes 140 seconds for a square meter of tomato leaf to receive a mole of photons and eight photons are required to fix a molecule of $CO_2$, then it will take just under 2 hours to synthesize a mole of glucose.

$$\text{time} = \frac{140 \text{ sec}}{\text{mole photons}} \times \frac{8 \text{ mole photons}}{\text{mole } CO_2} \times \frac{6 \text{ mole } CO_2}{\text{mole glucose}}$$
$$\text{time} = 6720 \text{ seconds or } 112 \text{ minutes}$$

The actual efficiency of photon capture is considerably less than 100%. Under optimal conditions for some rapidly growing plants, the efficiency of utilization of photons that strike a leaf is about 5%. However, even this value greatly exaggerates the true efficiency of utilization of the energy in sunlight. For example, a field of sugar beets converts only about 0.02% of the energy that falls on it during the growing season. Several factors limit the overall efficiency, including saturation of the photosystems far below maximum sunlight, availability of water, and low temperatures.

D. In contrast to the very low overall efficiency of light utilization, the efficiency of conversion of light energy to chemical energy *after photon capture* is 33%.

$$\text{efficiency} = \frac{\text{mole } CO_2}{8 \text{ mole photons}} \times \frac{\text{mole photons}}{42 \text{ kcal}} \times \frac{112 \text{ kcal}}{\text{mole } CO_2}$$
$$= 0.33 \text{ or } 33\%$$

**14–10** The corn plant ($C_4$) eventually will kill the geranium ($C_3$). Because both plants fix $CO_2$, the concentration of $CO_2$ in the chamber will fall. At low $CO_2$ concentrations, the corn plant has a distinct advantage since its enzyme for carbon fixation has a high affinity for $CO_2$. By contrast, the geranium depends on ribulose bisphosphate carboxylase, which has a lower affinity for $CO_2$. At low $CO_2$ concentrations, $O_2$ also competes with $CO_2$ for addition to ribulose 1,5-bisphosphate, ultimately liberating $CO_2$ in the process known as photorespiration. Not only will the geranium give up $CO_2$ in an abortive attempt at photosynthesis, it will continue to respire (using its mitochondria), thereby providing even more $CO_2$ for the corn plant. The corn plant will continue to fix $CO_2$ until the geranium wastes away and dies.

**14–11** Variegation occurs because the plants have a mixture of normal and defective chloroplasts. These sort out by mitotic segregation to give patches of green and yellow in leaves. Many of the green patches have cells that still retain defective chloroplasts in addition to the normal ones. As such patches grow, they can segregate additional cells that have only defective chloroplasts, giving rise upon cell division to an island of yellow cells in a sea of green ones. By contrast, yellow patches are due to cells that retain only defective chloroplasts. Thus, yellow cells cannot give rise to green cells by mitotic segregation; hence, there are no green islands surrounded by yellow.

## Chapter 15 Mechanisms of Cell Communication

**15–1** False. The concentration of a neurotransmitter in the synaptic cleft is much higher than, for example, the concentration of a circulating hormone in the blood or of a local mediator in the neighborhood of the signaling cell. The characteristic differences in concentration of signaling molecules is reflected in differences in ligand affinity of the corresponding receptors. Neurotransmitter receptors typically have a lower affinity for their ligands than do hormone receptors and receptors for local mediators. The combination of high neurotransmitter concentration and low affinity receptors allows the neurotransmitter to dissociate rapidly from the receptor to terminate a response. As a rule of thumb, the $K_d$ for binding to a receptor is about equal to the concentration of signaling molecule to which the receptor is exposed to produce a physiological response.

**15–2** False. Most second messengers, including cyclic AMP, $Ca^{2+}$, and $IP_3$, are water soluble and diffuse freely through the cytosol; however, second messengers such as diacylglycerol are lipid soluble and diffuse in the plane of the membrane.

**15–3** False. GTP-binding proteins are uniformly on when GTP is bound and off when GDP is bound; thus, GEFs turn GTP-binding proteins on and GAPs turn them off. The same is not true for protein kinases and phosphatases. Attachment of a phosphate will turn some target proteins on and others off. Indeed, attachment of a phosphate at one location in a protein can turn it on, while phosphorylation at a different location can turn the same protein off. Thus, while protein kinases throw the molecular switch, it is not always in the same direction.

**15–4** True. Nuclear receptors with a bound ligand bind to DNA sequences in the genome to activate (or inhibit) a specific gene; thus, there is a one-to-one correspondence between the signal and the response. By contrast, signaling pathways that involve enzymes or ion channels can significantly amplify a signal. One activated protein kinase, for example, can phosphorylate many molecules of its target protein.

**15–5** False. Ligand binding usually causes a receptor tyrosine kinase to assemble into dimers, which, because of their proximity, activates the kinase domains. The receptors then phosphorylate themselves to initiate the intracellular signaling cascade. In some cases, the insulin receptor, for example, the receptor exists as a dimer and ligand binding is thought to rearrange their receptor chains, causing the kinase domains to come together.

**15–6** True. Protein tyrosine phosphatases, unlike serine/threonine protein phosphatases, remove phosphate groups only from selected phosphotyrosines on a subset of tyrosine-phosphorylated proteins.

**15–7** False. Although there is some overlap in the cell–cell communication molecules used in plants and animals, there are many significant differences. For example, plants do not use the nuclear receptor family, Ras, JAK, STAT, TGFβ, Notch, Wnt, or Hedgehog proteins.

**15–8** At a circulating concentration of hormone equal to $10^{-10}$ M, about 1% of the receptors will have a bound hormone molecule {$[R–H]/[R]_{TOT} = 10^{-10}$ M/ ($10^{-10}$ M + $10^{-8}$ M) = 0.0099}. Half of the receptors will have a bound hormone molecule when the concentration of hormone equals the $K_d$; that is, at $10^{-8}$ M {$[R–H]/[R]_{TOT} = 10^{-8}$ M/($10^{-8}$ M + $10^{-8}$ M) = 0.5}. The relationships between concentration of ligand (hormone, in this case), $K_d$, and fraction bound are developed in Answer 3–14, p. 499.

**15–9**

A. A telephone conversation is analogous to synaptic signaling in the sense that it is a private communication from one person to another, usually some distance away and sometimes very far away. It differs from synaptic signaling because it is (usually) a two-way exchange, whereas synaptic signaling is a one-way communication.

B. Talking to people at a cocktail party is analogous to paracrine signaling, which occurs between different cells (individuals) and is locally confined.

C. A radio announcement is analogous to an endocrine signal, which is sent out to the whole body (the audience) with only target cells (individuals tuned to the specific radio station) affected by it.
D. Talking to yourself is analogous to an autocrine signal, which is a signal that is sent and received by the same cell.

**15–10** In both cases the signaling pathways themselves are rapid. If the pathway modifies a protein that is already present in the cell, its activity is changed immediately, leading to a rapid response. If the pathway modifies gene expression, there will be a delay corresponding to the time it takes for the mRNA and protein to be made and for the cellular levels of the protein to be altered sufficiently to invoke a response, which would usually take an hour or more.

**15–11** Cells with identical receptors can respond differently to the same signal molecule because of differences in the internal machinery to which the receptors are coupled. Even when the entire signaling pathway is the same, cells can respond differently if they express different effector proteins at the ends of the pathways.

**15–12** Phosphorylation/dephosphorylation offers a simple, universal solution to the problem of controlling protein activity. In a signaling pathway, the activities of several proteins must be rapidly switched from the off state to the on state, or vice versa. Attaching a negatively charged phosphate to a protein is an effective way to alter its conformation and activity. And it is an easy modification to reverse. It is a universal solution in the sense that one activity—that of a protein kinase—can be used to attach a phosphate, and a second activity—a protein phosphatase—can be used to remove it. About 2% of the genes in the human genome encode protein kinases, which presumably arose by gene duplication and modification to create appropriate specificity. Because serines, threonines, and tyrosines are common amino acids on the surfaces of proteins, target proteins can evolve to have appropriate phosphorylation sites at places that will alter their conformations. Finally, phosphorylation/dephosphorylation provides a flexible response that can be adjusted to give rapid on/off switches or more long lasting changes.

All of these attributes of phosphorylation/dephosphorylation are missing with allosteric regulators. While it is possible, in principle, for small molecules to turn proteins on or off, it is not a universal solution. Specific molecules would have to be 'designed' for each target protein, which would require the evolution of a metabolic pathway for the synthesis and degradation of each regulatory molecule. Even if such a system evolved for one target protein, that specific solution would not help with the evolution of a system for any other target protein. In addition, regulation by binding of small molecules is very sensitive to the concentration of the regulator. For a monomeric target protein, the concentration of a small molecule would have to change by 100-fold to go from 9% bound to 91% bound—a minimal molecular switch (see Answer 3–14, p. 499). Few metabolites in cells vary by such large amounts.

**15–13** The use of a scaffolding protein to hold the three kinases into a signaling complex increases the speed of signal transmission and eliminates cross-talk between pathways; however, there is relatively little opportunity for amplification of the signal from the receptor to the third kinase. Freely diffusing kinases offer the possibility for greater signal amplification since the first kinase can phosphorylate many molecules of the second kinase, which in turn can phosphorylate many molecules of the third kinase. The speed of signal transmission is likely to be slower, unless the concentration of kinases (and the potential for amplification) is high enough to compensate for their separateness. Finally, free kinases offer the potential for spreading the signal to other signaling pathways and to other parts of the cell. The organization that a cell uses for a particular signaling pathway depends on what the pathway is intended to accomplish.

**15–14**

1. If more than one effector molecule must bind to activate the target molecule, the response will be sharpened in a way that depends on the number of required effector molecules. At low concentrations of the effector, most target proteins will have a single effector bound (and therefore be inactive). At increasing concentrations of effector the target proteins with the requisite number of bound effectors will rise sharply, giving a correspondingly sharp increase in the cellular response.
2. If the effector activates one enzyme and inhibits another enzyme that catalyzes the reverse reaction, the forward reaction will respond sharply to a gradual increase in effector concentration. This is a common strategy employed in metabolic pathways involved in energy production and consumption.
3. The above mechanisms give sharp responses, but a true all-or-none response can be generated if the effector molecule triggers a positive feedback loop so that an activated target molecule contributes to its own further activation. If the product of an activated enzyme, for example, binds to the enzyme to activate it, a self-accelerating, all-or-none response will be produced.

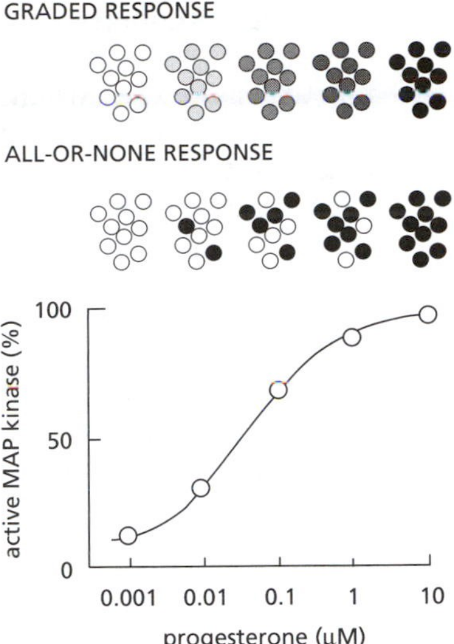


**Figure A15–1** Graded or all-or-none responses in individual oocytes that give rise to a graded response in the population (Answer 15–15).

**15–15** The analysis of individual frog oocytes shows clearly that the response to progesterone is all-or-none, with no oocytes having a partially activated MAP kinase. Thus, the graded response in the population results from an all-or-none response in individual oocytes, with different mixtures of fully mature or immature oocytes giving rise to intermediate levels of MAP kinase activation (Figure A15–1). It is not so clear why individual oocytes respond differently to different concentrations of progesterone, although there is significant variability among oocytes in terms of age and size (and presumably in the number of progesterone receptors and the concentrations of components of the MAP kinase signaling module and downstream targets).

Whether a graded response in a population of cells indicates a graded response in each cell or a mixture of all-or-none responses is a question that arises in many contexts in biology.

**Reference:** Ferrell JE & Machleder EM (1998) The biochemical basis of an all-or-none cell fate switch in *Xenopus* oocytes. *Science* 280, 895–898.

**15–16** Any mutation that generated a regulatory subunit incapable of binding to the catalytic subunit would produce a permanently active PKA. When the catalytic subunit is not bound to the regulatory subunit, it is active.

Two general types of mutation in the regulatory subunit could produce a permanently inactive PKA. A regulatory subunit that was altered so that it could bind the catalytic subunit, but not bind cyclic AMP, would not release the catalytic subunit, rendering PKA permanently inactive. Similarly, a mutant regulatory subunit that could bind cyclic AMP, but not undergo the conformational change needed to release the catalytic subunit, would permanently inactivate PKA.

**15–17** The time that the catalytic kinase subunit spends in its active conformation depends on the extent to which its regulatory subunits are modified. Each modification by phosphorylation or by $Ca^{2+}$ binding nudges the equilibrium toward the active conformation of the kinase subunit; that is, each modification increases the time spent in the active state. By summing the inputs from multiple pathways in this way, phosphorylase kinase integrates the signals that control glycogen breakdown.

**15–18** In order for activation of Ras to depend on inactivation of a GAP, both the GAP and the GEF would need to be active in the absence of the signal. In this way the GEF would constantly load GTP onto Ras and the GAP would keep the concentration of Ras–GTP low by constantly inducing GTP hydrolysis to return Ras to its GDP-bound state. Under these conditions, inactivation of the GAP would result in a rapid increase in the Ras–GTP level, allowing rapid signaling. Although this would be a perfectly effective way to regulate the level of active Ras, it would be wasteful of energy. In order to keep Ras in its inactive state, GTP would be constantly hydrolyzed to GDP, which would then need to be reconverted to GTP (by ATP)—a drain on cellular energy metabolism. Regulation by activation of a GEF avoids this problem.

Although avoiding constant GTP hydrolysis is a rational explanation, eucaryotic cells are notoriously profligate in their energy expenditures. At several points in energy metabolism, for example, they operate so-called 'futile' cycles that hydrolyze ATP as a means for rapid regulation of the flux through metabolic pathways. Thus, it could be that constant hydrolysis of GTP by a Ras GEF and GAP would not unduly tax the cell's energy budget. Perhaps the cell's method of regulating Ras by controlling the activity of a GEF is simply an evolutionary happenstance.

**15–19** Cells of flies with the heterozygous $Dsh^{\Delta}$/+ genotype probably make just half the normal amount of Dishevelled. Thus, underexpression of Dishevelled corrects the multi-hair phenotype generated by the overexpression of Frizzled. This relationship suggests that Frizzled acts upstream of Dishevelled; it is easy to imagine how underexpression of a downstream component could correct the overexpression of an upstream component. All this makes sense, as Frizzled is a Wnt receptor and Dishevelled is an intracellular signaling protein. However, if you knew nothing of the functions of Dishevelled and Frizzled, with only the genetic interactions as a guide, it would be possible to imagine more complex relationships (involving other unknown components) with Dishevelled acting upstream of Frizzled that could account for the phenotypes given in this problem. See if you can design such a pathway.

**Reference:** Winter CG, Wang B, Ballew A, Royou A, Karess R, Axelrod JD & Luo L (2001) *Drosophila* Rho-associated kinase (Drok) links Frizzled-mediated planar cell polarity signaling to the actin cytoskeleton. *Cell* 105, 81–91.

## Chapter 16 The Cytoskeleton

**16–1** True. When ATP in actin filaments (or GTP in microtubules) is hydrolyzed, much of the free energy released by cleavage of the high-energy bond is stored in the polymer lattice, making the free energy of the ADP-containing polymer higher than that of the ATP-containing polymer. This shifts the equilibrium toward depolymerization so that ADP-containing actin filaments disassemble more readily than ATP-containing actin filaments.

**16–2** False. The centrosome, which establishes the principal array of microtubules in most animal cells, nucleates microtubule growth at the minus end. Thus, the plus ends of the microtubules are near the plasma membrane, and the minus ends are buried in the centrosome at the center of the cell. This orientation of the array requires that plus end-directed motors be used to transport cargo to the cell periphery and that minus end-directed motors be used for cargo delivery to the center of the cell.

**16–3** False. The entry of $Ca^{2+}$ through the voltage-sensitive $Ca^{2+}$ channels in T-tubules is not sufficient, by itself, to trigger rapid muscle contraction. Instead, this initial burst of $Ca^{2+}$ opens $Ca^{2+}$-release channels in the sarcoplasmic reticulum, which flood the cytoplasm with $Ca^{2+}$, initiating rapid muscle contraction by binding to troponin C.

**16–4** A growth rate of 2 μm/min (2000 nm/60 sec = 33 nm/sec) corresponds to the addition of 4.2 αβ-tubulin dimers [(33 nm/sec) × (αβ-tubulin/8 nm) = 4.17 dimers/sec] to each of 13 protofilaments, or about 54 αβ-tubulin dimers/sec to the ends of a microtubule.

**Reference:** Detrich WH, Parker SK, Williams RC, Nogales E & Downing KH (2000) Cold adaptation of microtubule assembly and dynamics. *J. Biol. Chem.* 275, 37038–37047.

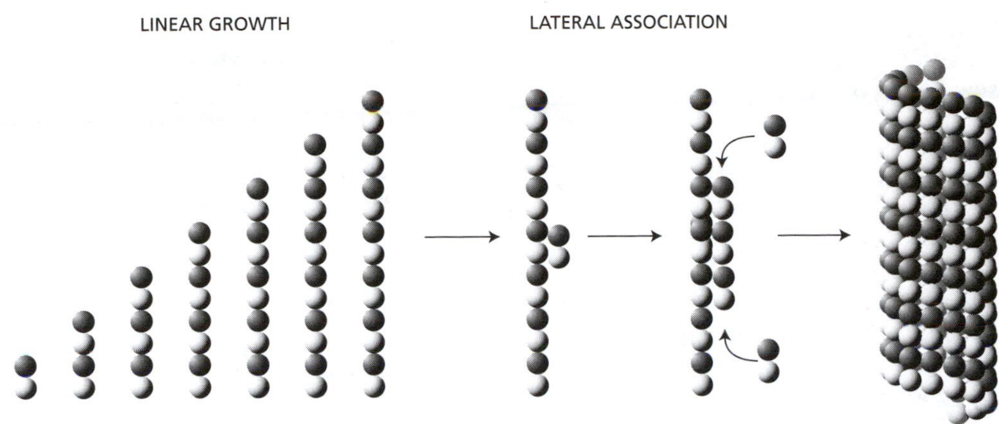


**Figure A16–1** Rapid addition of αβ-tubulin dimers to nucleation structure (Answer 16–5).

**16–5** Once the first lateral association has occurred, the next αβ-dimer can bind much more readily because it is stabilized by both lateral and longitudinal contacts (Figure A16–1). The formation of a second protofilament stabilizes both protofilaments, allowing the rapid addition of new αβ-tubulin dimers to form adjacent protofilaments and to extend existing ones. At some point the initial sheet of tubulin curls into a tube to form the microtubule.

**Reference:** Leguy R, Melki R, Pantaloni D & Carlier M-F (2000) Monomeric γ-tubulin nucleates microtubules. *J. Biol. Chem.* 275, 21975–21980.

**16–6** The centrosome nucleates a three-dimensional, star-burst array of microtubules that grow until they encounter an obstacle, ultimately the plasma membrane. Dynamic instability of the microtubules, coupled to the requirement for equal pushing of oppositely directed microtubules, eventually positions the centrosome in the middle of the cell. One way to think about the notion of equal and opposite forces is to realize that the microtubules are not absolutely rigid structures. Imagine pushing an object with a short steel rod versus a very long one; the short rod transmits force effectively, but the long rod will bend, delivering less force. The same principle may operate inside the cell, with microtubules of equal length delivering the same force. When all the oppositely directed microtubules emanating from a centrosome are the same length, the centrosome will be in the center of the cell.

**16–7** In cells, most of the actin subunits are bound to thymosin, which locks actin into a form that cannot hydrolyze its bound ATP and cannot be added to either end of a filament. Thymosin reduces the concentration of free actin subunits to around the critical concentration. Actin subunits are recruited from this inactive pool by profilin, whose activity is regulated so that actin polymerization occurs when and where it is needed. The advantage of such an arrangement is that the cell can maintain a large pool of subunits for explosive growth at the sites and times of its choosing.

**16–8**

A. The unidirectional movement of kinesin along a microtubule is driven by the free energy of ATP hydrolysis. ATP binding and hydrolysis are coupled to a series of conformational changes in the kinesin head that bring about the unidirectional stepping of the kinesin motor domains along the microtubule.
B. In the trace, the kinesin moves 80 nm in 9 sec, an average rate of about 9 nm/sec. This rate is about 100-fold slower than the *in vivo* rate because the experimental conditions (ATP concentration and force exerted by the interference pattern) were adjusted to slow the movements of kinesin so that individual steps could be observed.
C. As can be seen in Figure A16–31B, the kinesin molecule took 10 steps to move 80 nm, indicating that the length of an individual step is about 8 nm.
D. Since the step length and the interval between β-tubulin subunits along a microtubule protofilament are both 8 nm, a kinesin appears to move by

stepping from one β-tubulin to the next along a protofilament. Because kinesin has two domains that can bind to β-tubulin, it presumably keeps one domain anchored as it swings the other domain to the next β-tubulin-binding site—much like a person walking along a path of stepping-stones.

E. The data in Figure A16–31B contain no information about the number of ATP hydrolyzed per step. Other experiments by these same investigators suggest that hydrolysis of one ATP does not cause multiple steps. By lowering ATP concentrations to slow movement along the microtubule, the investigators showed the same sort of stepping patterns as in Figure A16–31B, although on a longer time scale. If hydrolysis of a single ATP could cause multiple steps, a clustering of steps might have been expected under these experimental conditions. None of these experiments rule out the possibility that more than one ATP might need to be hydrolyzed for each step.

**Reference:** Svoboda K, Schmidt CF, Schnapp BJ & Block SM (1993) Direct observation of kinesin stepping by optical trapping interferometry. *Nature* 365, 721–727.

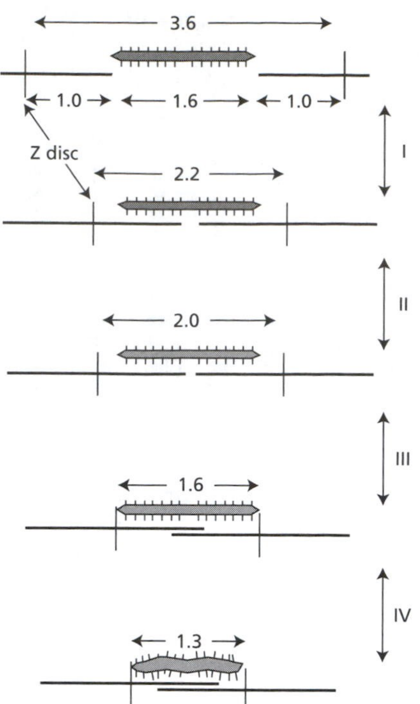


**Figure A16–2** Schematic diagrams of sarcomeres at the points indicated by *arrows* in Figure Q16–3 (Answer 16–10). Numbers refer to lengths in micrometers.

**16–9** The unidirectional motion of a lamellipodium results from the nucleation and growth of actin filaments at the leading edge of the cell and depolymerization of the older actin meshwork more distally. Cofilin plays a key role in differentiating the new actin filaments from the older ones. Because cofilin binds cooperatively and preferentially to actin filaments containing ADP-actin, the newer filaments at the leading edge, which contain ATP-actin, are resistant to depolymerization by cofilin. As the filaments age and ATP hydrolysis proceeds, cofilin can efficiently disassemble the older filaments. Thus, the delayed ATP hydrolysis by filamentous actin is thought to provide the basis for a mechanism that maintains an efficient, unidirectional treadmilling process in the lamellipodium.

**16–10** Sketches representing sarcomeres at each of the arrows in Figure Q16–3 are shown in Figure A16–2. As illustrated in these pictures, the increase in tension with decreasing sarcomere length in segment I is due to increasing numbers of interactions between myosin heads and actin. In segment II, actin begins to overlap with the bare zone of myosin, yielding a plateau at which the number of interacting myosin heads remains constant. In segment III, the actin filaments begin to overlap with each other, thereby interfering with the optimal interaction of actin and myosin and producing a decrease in tension. In segment IV, the spacing between the Z discs is less than the length of the myosin thick filaments, causing their deformation and a precipitous drop in muscle tension.

**Reference:** Gordon AM, Huxley AF & Julian FJ (1966) The variation in isometric tension with sarcomere length in vertebrate muscle fibres. *J. Physiol.* 184, 170–192.

## Chapter 17 The Cell Cycle

**17–1** False. Although a number of cells equivalent to an adult human is replaced about every three years, not all cells are replaced at the same rate. Blood cells and cells that line the gut are replaced at a high rate, whereas cells in most organs are replaced more slowly, and neurons are rarely replaced.

**17–2** False. Although cyclin–Cdk complexes are indeed regulated by phosphorylation and dephosphorylation, they can also be regulated by the binding of Cdk inhibitor (CKI) proteins. Moreover, the rates of synthesis and proteolysis of the cyclin subunits are extremely important for regulating Cdk activity.

**17–3** True. If the length of the cell cycle were shorter than it takes for the cell to double in size, the cell would get progressively smaller with each division; if it were longer, the cells would get bigger and bigger.

**17–4** True. The origin recognition complex serves as a scaffold at origins of replication in eucaryotic cells around which other proteins are assembled and activated to initiate DNA replication.

**17–5** False. Equal and opposite forces that tug the chromosomes toward the two spindle poles would tend to position them at random locations between the poles. The poleward force on each chromosome is opposed by a polar ejection force that pushes the chromosomes away from the pole. As a chromosome approaches a spindle pole, the ejection force is thought to increase, thereby raising the tension on the kinetochore, temporarily shutting off the poleward force, allowing the chromosome to be pushed away from the pole. This balance of forces at each pole tends to position the chromosomes at the midpoint between the poles—the metaphase plate.

**17–6** False. Organism senescence (aging) is distinct from replicative cell senescence, which occurs in the absence of telomerase. Aging is thought to depend largely on progressive oxidative damage to macromolecules. Strategies that reduce metabolism—for example, restricted caloric intake—decrease the production of reactive oxygen species, and can extend the lifespan of experimental animals.

**17–7** Enzymes for most metabolic reactions function in isolation; that is, their enzymatic competence does not depend on critical interactions with other proteins. So long as the enzyme folds properly and its small molecule substrate is present, the reaction will proceed. By contrast, cell-cycle proteins must interact with many other proteins to form the complexes that are critical for coordinated progression through the cell cycle. The ability of many human cell-cycle proteins to interact with yeast components implies that the binding surfaces responsible for these interactions have been preserved through more than a billion years of evolution. That is remarkable.

**17–8** To isolate the wild-type gene, you would first grow a population of the *Cdc* mutant cells at 25°C. These cells would then be transfected with the DNA library and grown at 37°C. Cells that receive a plasmid expressing a DNA that does not correspond to the mutant gene will not be able to form a colony at the restrictive temperature. However, cells that receive the wild-type version of the mutant gene will be able to progress through the cell cycle normally at the restrictive temperature and thus will form a colony. Plasmid DNA isolated from such colonies can be sequenced to determine the identity of the gene.

Among the DNAs isolated in experiments like this one, it is not uncommon to find unexpected genes in addition to the one corresponding to the mutant gene. Occasionally, expression of a second gene at higher than normal levels, as often occurs from plasmids, can compensate for the decreased function of the mutant gene. Characterization of such suppressor genes can provide insight into the biological functions of both genes.

**17–9**

A. The execution point for your temperature-sensitive mutant is marked on Figure A17–1. Cells that had not reached the execution point in the cell cycle when the temperature was raised grew to the characteristic landmark morphology (large buds) but did not divide. Cells that were beyond the execution point when the temperature was raised divided and then stopped at the landmark morphology during the next cell cycle.

B. The characteristic landmark morphology defines the time at which the cell stops its progress through the cell cycle, as indicated for your mutant in Figure A17–1. The landmark morphology is clearly different from the morphology at the execution point. Therefore, the execution point and the point of cell-cycle arrest do not correspond in your mutant.

At first glance, it may seem odd that the execution point and the point of arrest do not coincide. An analogy may make the situation clearer. The addition of engine mounts to the chassis is an early step in the assembly of an automobile. Without engine mounts the engine cannot be added and a

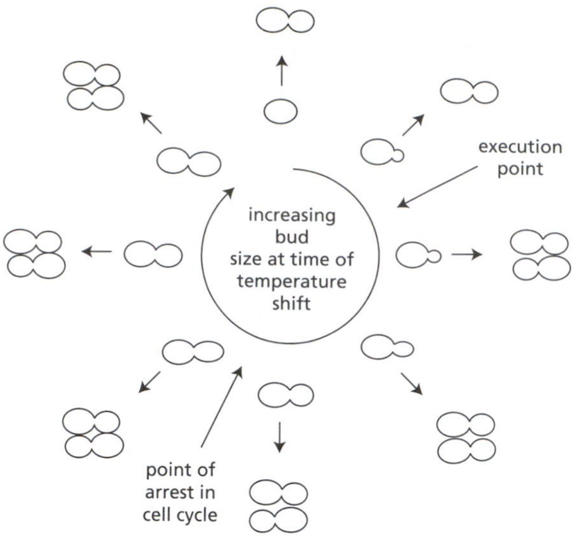

**Figure A17–1** The execution point and the point of arrest for the gene product affected in your mutant (Answer 17–9).

complete car cannot be built. In the absence of engine mounts, assembly of other parts of the automobile can continue until a point is reached at which all further assembly depends on the engine mounts. In this case the normal execution point for engine-mount addition is early, but the arrest point for assembly is relatively late, with a characteristic landmark morphology, which resembles a complete automobile (until one looks under the hood).

**Reference:** Hartwell LH (1978) Cell division from a genetic perspective. *J. Cell Biol.* 77, 627–637.

**17–10** Cohesins must be present during S phase because it is only while DNA is being replicated that sister chromatids can be reliably identified by the cellular machinery that links them together. Once sister chromatids have separated, it is impossible for a nonspecific DNA-binding protein like cohesin to tell which chromosomes are sisters. And it would be virtually impossible for any protein to distinguish sister chromatids from homologous chromosomes. If sister chromatids are not kept together after their formation, they cannot be accurately segregated to the two daughter cells during mitosis.

**Reference:** Uhlmann F & Nasmyth K (1998) Cohesion between sister chromatids must be established during DNA replication. *Curr. Biol.* 8, 1095–1101.

**17–11** In cells that fully condense their chromosomes at mitosis, as vertebrate cells do, much of the cohesin is released from the chromosome arms at the start of mitosis, when condensins begin to drive condensation. With most of the cohesin out of the way, the condensins can coil individual sister chromatids into separate domains. The small amount of cohesin that remains is sufficient to hold sister chromatids together until anaphase, when the residual cohesins are degraded.

**17–12** The dose of caffeine required to interfere with the DNA replication checkpoint mechanism is much higher than the amount imbibed by even the most excessive drinkers of coffee and colas. The concentration of caffeine in a cup of coffee is about 3.4 mM.

$$[\text{caffeine}] = \frac{100\text{ mg}}{150\text{ mL}} \times \frac{\text{g}}{1000\text{ mg}} \times \frac{\text{mole}}{196\text{ g}} \times \frac{1000\text{ mL}}{\text{L}}$$
$$= 3.4 \times 10^{-3}\text{ M} = 3.4\text{ mM}$$

Since the concentration in a cup is less than the 10 mM required to interfere with the DNA replication checkpoint mechanism, you cannot get a higher concentration by drinking it and diluting it in the water volume of the body. If you assume for the purposes of calculation that the caffeine is not metabolized or excreted (but that all the liquid is), then you can ask how many cups of coffee would you need to drink (at 100 mg of caffeine per cup) to reach a concentration of 10 mM in 40 L of body water. The answer is: you would need to drink 784 cups of coffee!

**17–13** Prophase (see Figure Q17–3E), prometaphase (see Figure Q17–3D), metaphase (see Figure Q17–3C), anaphase (see Figure Q17–3A), telophase (see Figure Q17–3F), and cytokinesis (see Figure Q17–3B).

**17–14** There are 46 human chromosomes, each with two kinetochores—one for each sister chromatid—thus, there are 92 kinetochores in a human cell at mitosis.

**17–15**

A. A kinetochore microtubule is relatively stable because both of its ends are protected from disassembly: one by attachment to the centrosome, and the other by attachment to the kinetochore. This stabilization suggests that kinetochores cap the plus ends of the microtubules, thereby altering the equilibrium for subunit dissociation.

B. Astral microtubules disassemble when tubulin is below the critical concentration for microtubule assembly, as discussed in Chapter 16. Under these conditions, the rate of addition does not balance the rate of dissociation, and microtubules get progressively shorter. The presence of kinetochore microtubules strengthens this interpretation relative to the other two possibilities mentioned. Neither detachment from the centrosome nor random breakage would explain the stability of kinetochore microtubules.

C. The possible mechanisms of disappearance would be readily distinguished by a time course. At intermediate time points, the number and length of microtubules will be sensitive indicators of which mechanism operates. If microtubules detached, the number of microtubules per centrosome would decrease, but the length would remain the same. If microtubules depolymerized from the end, the number would remain constant and their lengths would decrease relatively uniformly. If the microtubules broke at random, the number would remain relatively constant, but the distribution of lengths would be very broad.

**Reference:** Mitchison TJ & Kirschner MW (1985) Properties of the kinetochore *in vitro*. II. Microtubule capture and ATP-dependent translocation. *J. Cell. Biol.* 101, 766–777.

**17–16** The two cytoskeletal machines are the mitotic spindle and the contractile ring. The segregation of chromosomes and their distribution to daughter cells is accomplished by the bipolar mitotic spindle, which is composed of microtubules and a variety of microtubule-dependent motors and other proteins. The division of an animal cell into daughter cells by cytokinesis is accomplished by the contractile ring, which is composed of actin and myosin filaments and is located just under the plasma membrane. As the ring constricts, it pulls the membrane inward, ultimately dividing the cell in two.

**17–17** Mitogens stimulate cell division, primarily by relieving intracellular negative controls that otherwise block progress through the cell cycle. Growth factors stimulate cell growth (an increase in cell mass) by promoting the synthesis of proteins and other macromolecules and by inhibiting their degradation. Survival factors promote cell survival by suppressing apoptosis.

## Chapter 18 Apoptosis

**18–1** True. Adult tissues are maintained at a constant size, so that there must be a balance between cell death and cell division. If this were not so, the tissue would grow or shrink.

**18–2** True. Cytochrome *c* mediates apoptosis from signals within a mammalian cell—the intrinsic pathway of apoptosis. This has been confirmed directly by generating cytochrome *c*-deficient mouse embryo fibroblasts (MEFs) by reverse genetics. Although mice with knockouts of their cytochrome *c* genes die about midway through gestation, fibroblasts from such embryos can be cultured under special conditions and tested for sensitivity to various apoptotic signals. They are resistant to a variety of agents that induce the intrinsic pathway of apoptosis.

**Reference:** Li K, Li Y, Shelton JM, Richardson JA, Spencer E, Chen ZJ, Wang X & Williams RS (2000) Cytochrome *c* deficiency causes embryonic lethality and attenuates stress-induced apoptosis. *Cell* 101, 389–399.

**18–3** Overexpression of a secreted protein that binds to Fas ligand would protect tumor cells from attack by killer lymphocytes. By binding to the Fas ligand on the surface of killer lymphocytes, the secreted protein would prevent the Fas ligand from binding to Fas on the surface of tumor cells, thereby insulating them from death-inducing interactions with killer lymphocytes. Secreted proteins that bind to Fas ligand are commonly known as decoy receptors. They play a normal role in modulating the killing induced by interactions between Fas ligand and Fas. When tumor cells overproduce such decoy receptors, they subvert this normal mechanism into a cellular defense against Fas-mediated killing.

**Reference:** Pitti RM, Marsters SA, Lawrence DA, Roy M, Kischkel FC, Dowd P, Huang A, Donahue CJ, Sherwood SW, Baldwin DT, Godowski PJ, Wood WI, Gurney AL, Hillan KJ, Cohen RL, Goddard AD, Botstein D & Ashkenazi A (1998) Genomic amplification of a decoy receptor for Fas ligand in lung and colon cancer. *Nature* 396, 699–703.

**18–4** Somewhat surprisingly, cytochrome c seems not to be required for apoptosis in *C. elegans.* However, even if it were required, *C. elegans* mutants that were defective for cytochrome *c* would not have been isolated because they would not be viable. Cytochrome *c* is an essential component of the electron transport chain in mitochondria. Without it, no production of ATP by oxidative phosphorylation would be possible, and such a mutant organism could not survive.

**Reference:** Ellis HM & Horvitz RH (1986) Genetic control of programmed cell death in the nematode *C. elegans. Cell* 44, 817–829.

**18–5** Upon microinjection of cytochrome *c,* both types of cell undergo apoptosis. The presence of cytochrome *c* in the cytosol is a signal for the assembly of apoptosomes and the downstream events that lead to apoptosis. Cells that are defective for both Bax and Bak cannot release cytochrome *c* from mitochondria in response to upstream signals, but there is no defect in the downstream part of the pathway that is triggered by cytosolic cytochrome *c.* Thus, microinjection bypasses the defects in the doubly defective cells, triggering apoptosis.

**Reference:** Wei MC, Zong W-X, Cheng EH-Y, Lindsten T, Panoutsakopoulou V, Ross AJ, Roth KA, MacGregor GR, Thompson CB & Korsmeyer SJ (2001) Proapoptotic BAX and BAK: A requisite gateway to mitochondrial dysfunction and death. *Science* 292, 727–730.

**18–6** The retention of the web cells in *Apaf1*$^{-/-}$ mice indicates that Apaf1 is essential for web-cell apoptosis, presumably in conjunction with cytochrome *c.* The absence of web cells in *Casp9*$^{-/-}$ mice indicates that caspase-9 is not required for web-cell apoptosis. These observations suggest that Apaf1 may activate a different caspase in web cells, in addition to or instead of caspase-9.

**Reference:** Earnshaw WC, Martins LM & Kaufmann SH (1999) Mammalian caspases. *Annu. Rev. Biochem.* 68, 383–424.

**18–7** The two cells in Figure Q18–2 have released cytochrome *c*–GFP from all their mitochondria within a few minutes: within 6 minutes for the cell in Figure Q18–2A and within 8 minutes for the cell in Figure Q18–2B. The time after exposure to UV at which the release occurred varied dramatically for the two cells: after 10 hours for the cell in Figure Q18–2A and after 17 hours for the cell in Figure Q18–2B. These observations indicate that individual cells release cytochrome *c* from all their mitochondria very rapidly, but that release is triggered in different cells at widely varying times after exposure to apoptosis-inducing levels of UV light.

**Reference:** Goldstein JC, Waterhouse NJ, Juin P, Evan GI & Green DR (2000) The coordinate release of cytochrome *c* during apoptosis is rapid, complete and kinetically invariant. *Nat. Cell Biol.* 2, 156–162.

## Chapter 19 Cell Junctions, Cell Adhesion, and the Extracellular Matrix

**19–1** False. Although cells can be readily dissociated by removing $Ca^{2+}$ from the external medium, it is unlikely that $Ca^{2+}$-dependent cell–cell adhesions are regulated by changes in $Ca^{2+}$ concentration. Cells have no way to control the $Ca^{2+}$ concentration in their environment.

**19–2** True. The barriers formed by tight junction proteins restrict the flow of molecules between cells and the diffusion of proteins (and lipids) from the apical to the basolateral domain and vice versa.

**19–3** True. Tension—a mechanical signal—applied to an integrin can cause it to tighten its grip on intracellular and extracellular structures, including not only cytoskeletal and matrix components, but also molecular signaling complexes. Similarly, loss of tension can loosen its hold, so that molecular signaling complexes fall apart on either side of the membrane. Thus, the tension on the integrin can trigger or inhibit molecular signaling.

**19–4** False. The elasticity of elastin fibers derives from their lack of secondary structure: elastin forms random coils that are easily stretched. The set of hydrogen bonds that stabilizes an $\alpha$ helix is too strong, in aggregate, to be disrupted by the kinds of forces that deform elastin.

**19–5** This quote is correct in spirit, though incorrect in detail. Warren Lewis was trying to draw attention to the importance of the adhesive properties of cells in tissues at a time when the problem had been largely ignored by the biologists of the day. The quote is incorrect because a large fraction of our bodies is made up of connective tissue such as bone and tendon, whose integrity depends on the quality of the matrix rather than on the cells that inhabit it. It is not at all easy to dissociate cells from tissues, as anyone who has eaten a tough piece of steak can testify.

**19–6** IgG antibodies contain two identical binding sites; thus, they are able to cross-link the molecules they recognize (this is the basis for immune precipitation). If whole antibodies were used to block aggregation, they might cross-link the cells rather than inhibit their aggregation. By contrast, monovalent Fab fragments cannot cross-link cells. They bind to the cell adhesion molecules and prevent them from binding to their partners, thus preventing cell aggregation (Figure A19–1).

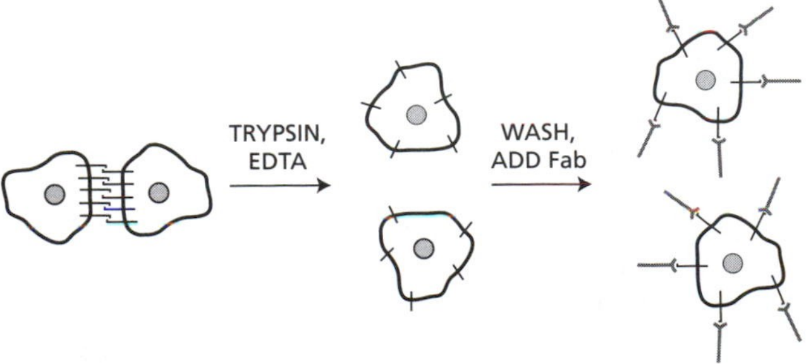


**Figure A19–1** Fab antibody fragments block cell adhesion (Answer 19–8).

**Reference:** Beug H, Katz FE & Gerisch G (1973) Dynamics of antigenic membrane sites relating to cell aggregation in *Dictyostelium discoideum. J. Cell Biol.* 56, 647–658.

**19–7**

A. Even though all of the claudin-4 has disappeared, the cells still express claudin-1, which is not affected by the toxin. Using antibodies specific for claudin-1, the authors showed that it remained intact at the sites of the tight junctions in the presence of the toxin.

B. There are several possible ways that binding of the toxin might cause the claudin-4 strands to disintegrate. Bound toxin could alter the conformation of claudin-4, weakening its affinity for other claudin-4 molecules in the same strand. Alternatively, toxin binding might alter the lateral affinity between claudin-4 molecules in the paired strands from different cells. Finally, the toxin might bind preferentially to claudin-4 monomers, shifting the equilibrium toward the dissociated form. Since claudin-4 actually disappears when the cells are treated with toxin, it must also be true—regardless of which possibility is correct—that the claudin-4 molecules outside of paired strands are degraded fairly quickly.

C. Because the tight junction prevents molecules from penetrating the junction, added toxin will have access to only one side of the junction. Its inability to work from the apical side suggests that its binding sites on the claudin-4 molecules are accessible only from the basolateral side. If the toxin binds to monomers, as suggested above, then it could be that the monomers are delivered to the basolateral membrane domain, and therefore accessible only from that side of the epithelial sheet. Alternatively, if the strands of claudin molecules are all oriented in the same way—that is, with their 'top' surfaces all facing the apical side and their 'bottom' surfaces all facing the basolateral side, as would be expected from symmetry principles—then a toxin binding site on the 'bottom' surface would only be accessible from the basolateral side.

**Reference:** Sonoda N, Furuse M, Sasaki H, Yonemura S, Katahira J, Horiguchi Y & Tsukita S (1999) *Clostridium perfringens* enterotoxin fragment removes specific claudins from tight junction strands: evidence for direct involvement of claudins in tight junction barrier. *J. Cell Biol.* 147, 195–204.

**19–8** Mice that are homozygous for knockout of the gene for either nidogen-1 or nidogen-2 presumably have no phenotype because the two forms of nidogen can substitute for one another. Mice that are homozygous for the mutant form of laminin-$\gamma$1, which does not bind nidogen, have a much more severe phenotype than either of the individual nidogen gene knockouts because they eliminate the ability of both nidogens to bind to laminin. As a result, these mice do not form proper basal lamina and die at birth with severe defects in kidney and lung. If this is the correct explanation for the genetic observations, then you would predict that mice that are homozygous for knockouts of both nidogen genes would have a very severe phenotype, comparable to that of the laminin-$\gamma$1 mutant. Such mice have been made and they do have a severe phenotype.

**Reference:** Sasaki T, Fassler R & Hohenester E (2004) Laminin: the crux of basement membrane assembly. *J. Cell Biol.* 164, 959–963.

**19–9** This statement encapsulates our growing recognition of the diverse roles the basal lamina plays. Although it provides structural support for the cells that rest upon it, mechanical stability is only one of the several functions the basal lamina supplies. For example, during the regeneration of muscles or motor neurons, the neuromuscular junction is re-established based on information contained in the basal lamina. Special molecules stuck in the basal lamina—like messages on a bulletin board—mark the site of the junction and allow it to be reconstituted exactly. Growing evidence indicates that similar processes occur during the original development of muscle and neuromuscular junctions.

**Reference:** Sanes JR (2003) The basement membrane/basal lamina of skeletal muscle. *J. Biol. Chem.* 278, 12601–12604.

**19–10** The high level of activation when alanine was substituted for D723 in the β chain, or for R995 in the α chain, indicates that those residues are somehow important for holding the αIIbβ3 integrin in an inactive state. The 'charge-swap' experiment, which showed that D723R paired with R995D was as inactive as the wild type, suggests strongly that these two residues form an electrostatic attraction—a salt bridge—that helps to hold αIIbβ3 integrin in its inactive configuration. It follows that inside-out signaling is probably triggered by breaking this salt bridge.

**Reference:** Hughes PE, Diaz-Gonzalez F, Leong L, Wu C, McDonald JA, Shattil SJ & Ginsberg MH (1996) Breaking the integrin hinge: A defined structural constraint regulates integrin signaling. *J. Biol. Chem.* 271, 6571–6574.

**19–11** Individual beads of carboxymethyl Sephadex swell because of osmotic effects. The negative charges on the cross-linked polymer trap an equal number of cations (which are present in the dry gel) to maintain electrical neutrality. These charges—both the fixed negative charges on the polymer and the mobile cations—are confined by electrostatic forces to the volume occupied by the cross-linked polymer. In a solution of pure water, the concentration of particles in the volume of the gel is higher than in the water; hence, the water flows into the gel to try to equalize the concentrations inside and outside the gel bead. The beads swell as the water enters until they reach a maximum size dictated by their cross-linked structure. At that point, further entry of water molecules is opposed by the pressure exerted by the cross-linked structure of the polymer. It is more common to think of osmotic effects across a semi-permeable membrane, but the effects are exactly analogous in a cross-linked gel.

When a salt solution is added to the swollen gel, the osmotic effects are reversed. The concentration of particles in 50 mM NaCl is much higher than in the swollen gel, so water flows out of the gel (and $Na^+$ and $Cl^-$ ions flow in) to equalize the concentration of ions. In the absence of the osmotic water pressure, the gel shrinks.

**19–12** Because the racemization of L-aspartate to D-aspartate occurs slowly, proteins that turn over rapidly will have very low levels of D-aspartate, if it can be detected at all. Proteins that are degraded and replaced more slowly will be expected to have a higher percentage of D-aspartate, with the absolute level depending on the rate of turnover. What makes the observations on elastin remarkable is the age dependence of the D-aspartate levels. These observations have been interpreted to mean that our lifetime supply of elastin is made early on and never degraded. In addition, studies in humans that made use of the inadvertent metabolic $^{14}C$-labeling due to atmospheric testing of nuclear weapons, led to the conclusion that elastin synthesis occurs almost exclusively during the fetal and postnatal periods of development. Experiments in mice support this idea.

**Reference:** Shapiro SD, Endicott SK, Province MA, Pierce JA & Campbell EJ (1991) Marked longevity of human lung parenchymal elastic fibers deduced from prevalence of D-aspartate and nuclear weapons-related radiocarbon. *J. Clin. Invest.* 87, 1828–1834.

**19–13** If you soak the lettuce in tap water, it will take up water due to osmosis and become crisper. Soaking the lettuce in salt water or sugar water will have the opposite effect, sucking even more water out of the lettuce, making it even limper. Your day-old lettuce is long past the point at which photosynthesis can do it any good, and the bright light will dry it out even more.

**19–14** At 0.1 MPa the hydraulic conductivity of a single water channel is $4.4 \times 10^{-23}$ $m^3/s$ [$(4.4 \times 10^{-22}$ $m^3/s$ MPa$) \times 0.1$ MPa $= 4.4 \times 10^{-23}$ $m^3/s$]. Thus, the question becomes how many water molecules are in $4.4 \times 10^{-23}$ $m^3$. There are 3.33 $\times 10^{28}$ water molecules/$m^3$ [(55.5 moles/L)($10^3$ L/$m^3$)($6 \times 10^{23}$ water molecules/mole)]. Therefore, $1.5 \times 10^6$ water molecules flow through a water channel each second at 1 atmosphere of pressure [$(4.4 \times 10^{-23}$ $m^3/s)(3.33 \times 10^{28}$ water molecules/$m^3$)].

**Reference:** Tyerman SD, Bohnert HJ, Maurel C, Steudle S & Smith JAC (1999) Plant aquaporins: their molecular biology, biophysics and significance for plant water relations. *J. Exp. Bot.* 50, 1055–1071.

## Chapter 20 Cancer

**20–1** False. There is an optimum level of genetic instability for the development of cancer. A cell must be mutable enough to evolve rapidly, but not so mutable that it accumulates too many harmful changes and dies.

**20–2** True. Many cancers appear to be maintained by a small population of stem cells. These cancer stem cells usually divide more slowly than the cells in the bulk of the tumor, and they are less sensitive to treatments aimed at rapidly dividing cells. If the stem cells are not killed, the cancer is likely to return.

**20–3** False. Although it is popular to think so, there is scant evidence to support those ideas, except in very specific instances like 2-naphthylamine and asbestos.

**20–4** False. It is not that DMBA is a specific mutagen, but rather that the *Ras* gene is converted to its activated, cancer-causing form by a particular A-to-T alteration that leads to a very specific amino acid change. DMBA causes mutations throughout the genome, but only those at the specific site in the *Ras* gene give rise to cells that have cancerous properties and thus are identified in the assay.

**20–5** True. That is why oncogenes in their overactive, mutant form tend to drive cell growth and proliferation, and why the loss of tumor suppressor genes removes natural blocks in these pathways, which also promotes cell growth and proliferation.

**20–6** The key difference in the incidences of colon cancer and osteosarcomas is the size of the population of cells at risk for the disease. Colon cancer arises from the population of proliferating cells in the colon, which are present in roughly the same number throughout life. This population can accumulate mutations over time, giving rise to the age-dependent increase in cancer incidence. By contrast, the cells responsible for osteosarcomas are present in much greater numbers during adolescence, when their proliferation is required to increase the size of the skeleton, than they are in young children or adults. In this case, it is the number of cells at risk that is the most important determinant of the frequency of cancer.

**Reference:** Knudson AG (2001) Two genetic hits (more or less) to cancer. *Nat. Rev. Cancer* 1, 157–162.

**20–7** The time at which the death rates due to breast and cervical cancer slow corresponds to the menopause, at which time the production of estrogen declines. Estrogen normally promotes the proliferation of cells in the breast and uterus. Thus, the decline in estrogen would reduce the population of proliferating cells, thereby reducing the risk of cancer in these tissues.

**Reference:** Armitage R & Doll R (1954) The age distribution of cancer and a multi-stage theory of carcinogenesis. *Br. J. Cancer* 8, 1–12. Reprinted in (2004) *Br. J. Cancer* 91, 1983–1989.

**20–8** Oncogenes correspond to stuck accelerators. In their mutated, overactive form they drive a cell to proliferate in a way that is not responsive to normal controls. Defective tumor suppressor genes correspond to broken brakes.

They normally function to inhibit steps in signaling pathways; that is, to act as brakes. When tumor suppressor genes are defective, signaling pathways are unrestrained. Defective DNA maintenance genes correspond to bad mechanics. These genes normally operate to maintain the genome during its propagation and in the face of DNA damage. Defective DNA maintenance genes lead to genetic rearrangements or increased point mutations, either of which can convert a proto-oncogene to an oncogene, or eliminate a tumor suppressor gene.

**Reference:** Vogelstein B & Kinzler KW (2004) Cancer genes and the pathways they control. *Nat. Med.* 10, 789–798.

**20–9** Development of most cancers seems to require a gradual accumulation of mutations in a number of different genes—at least five or six. In the ongoing presence of cigarette smoke these mutations evidently accumulate at an increased rate (over their accumulation in the absence of cigarette smoke). By stopping smoking, an individual returns to the normal, slower rate of mutation accumulation. Thus, whatever mutations remain to be generated in a reformed smoker are generated at a slower rate than in a continuing smoker. The slower rate of accumulation of mutations translates into a lower cumulative risk.

**Reference:** Peto R, Darby S, Deo H, Silcocks P, Whitely E & Doll R (2000) Smoking, smoking cessation, and lung cancer in the UK since 1950: combination of national statistics with two case-control studies. *Br. Med. J.* 321, 323–329.

**20–10** The highly rearranged karyotypes and their similarity from tumor to tumor suggest that the cancer cells, themselves, are being transmitted from devil to devil. It is extremely unlikely that an infectious agent such as a virus or a microorganism could induce the same set of complicated rearrangements in different animals. Most importantly, the existence of a chromosome-5 inversion in one Tasmanian devil, which is not present in chromosome 5 of its tumor cells, argues strongly that the tumors are not generated from the host devil's own cells. It appears that this cancer has arisen from a rogue line of cancer cells, from a tumor of unknown origin, that has acquired the capability for parasitic existence. This is one of just two examples of natural transmission of cancer by tumor cells, the other being a venereal disease in dogs. A special case of such transmission occurs occasionally during organ transplantation in humans. But the requirements for organ transplantation—matching tissue and immune suppression—highlight just how unusual natural transmission is. The cancer cells responsible for facial tumors in Tasmanian devils must somehow evade the new host's immune defenses.

**Reference:** Pearse A-M & Swift K (2006) Transmission of devil facial-tumour disease. *Nature* 439, 549.

**20–11**

A. Most of the 463,248 sequence changes that remained after the reading errors were removed have nothing to do with cancer. The authors of this study applied six additional filters to eliminate sequence changes that are unlikely to contribute to the functional differences between the normal cells and the tumor cells. See if your suggestions are included in this list.

1. Filter out changes that do not alter the encoded amino acid sequence; for example, mutations to synonymous codons. (259,957 changes were eliminated by this criterion.)
2. Filter out changes that are also present in the DNA from the two normal individuals that were included in the analysis. (163,006 changes were eliminated by this criterion.)
3. Filter out changes that correspond to known sequence polymorphisms in the human population. (11,004 changes were eliminated by this criterion.)

4. Filter out changes that cannot be confirmed upon reamplifying and resequencing the sample. (Of the 29,281 sequence differences that remained after applying the above filters, 9295 were not confirmed and therefore eliminated.)
5. Filter out changes that are also present in normal tissue from the same individual that had the tumor. (18,414 of the 19,986 sequence differences that remained after applying filter 4 were eliminated were eliminated by this criterion.)
6. Filter out changes in sequences that have closely related sequences elsewhere in the genome. There can be problems deciding which genomic location is the true source of the sequence read. (265 of the remaining 1572 sequence differences were eliminated by this criterion, leaving 1307 potential cancer-relevant mutations.)

B. Deciding which of these 1307 mutations are likely to contribute to the cancers is not an easy task. One approach is to look for mutant genes that are found in multiple breast tumors or in multiple colorectal tumors. The underlying assumption is that similar cancers should have similar sets of causative mutations. The authors used this sort of analysis to identify roughly 12 cancer-related mutations in breast tumors, and about 9 in colorectal tumors. (The rest of the mutations are likely to be passenger mutations.) Because only about half the genes in the genome were analyzed (13,023/25,000), the real number of cancer-relevant mutations in these tumors may be closer to 20.

C. The sequencing strategy used here—amplifying and sequencing exons—was designed to detect small changes in sequence: point mutations and short deletions. Larger deletions and gene rearrangements would not be detected because they would not give an informative PCR product.

**Reference:** Sjoblom T et al (2006) The consensus coding sequences of human breast and colorectal cancers. *Science* 314, 268–274.

**20–12** The promyelocytes of APL are blocked at an intermediate stage in their development, at a point where they still divide and increase in number. It is this unchecked increase in number that causes problems for the cancer patient. Normally, such precursor cells divide only a few times before they terminally differentiate into a nondividing blood cell. By triggering the differentiation of promyelocytes into terminally differentiated neutrophils, which no longer divide, treatment with all-*trans*-retinoic acid eliminates the problems caused by unchecked proliferation.

APL arises by one of a few types of translocation that fuses the retinoic acid receptor (RAR) gene on chromosome 17 with a gene on another chromosome to make a hybrid protein that interferes with the normal developmental program. It is not yet clear how the fusion protein blocks development, although it likely does so by interfering with the function of the normal RAR receptor. In some way, treatment with all-*trans*-retinoic acid allows APL cells to move through the block.

**Reference:** Warrell RP Jr, de The H, Wang Z-Y & Degos L (1993) Acute promyelocytic leukemia. *N. Engl. J. Med.* 329, 177–189.

**20–13** The products of oncogenes are the only feasible targets for such small molecules. The product of an oncogene has a dominant, growth-promoting effect on the cell. Thus, if the growth-promoting oncogene product were inhibited, the cell might return to a more normal state. This is the underlying rationale for searching for drugs that inhibit oncoproteins.

By contrast, the products of tumor suppressor genes and DNA maintenance genes are not targets for anticancer drug development. These two classes of genes cause cancer by *not* making their product. Thus, there is no abnormal product to be inhibited in cancer cells that arise by mutation of tumor suppressor or DNA maintenance genes.

# Credits

## Chapter 1

**Figure 1–1** Data courtesy of Steve Freedland.

## Chapter 2

**Figure 2–25** From HC Berg, Random Walks in Biology. Princeton: Princeton University Press, 1993.
**Figure 2–31** Courtesy of Zermatt Tourism.
**Figure 2–39** Adapted from AL Lehninger, DL Nelson and MM Cox, Principles of Biochemistry, 3rd ed. New York: Worth Publishers, 2000.

## Chapter 3

**Figure 3–14** From FA Smit, *Svenska Vetensk. Acad. Handl.* 24:1–40, 1898.
**Figure 3–14 (top)** From GA Boulenger, in *Report on the collections of natural history made in the Antartic regions during the voyage of the "Southern Cross"* 174–189. London: British Museum (Natural History), 1902.
**Figure 3–25 (bottom)** Courtesy of Eli Lilly and Company. Reprinted from JE Thomas, M Smith, B Rubinfeld, M Gutowski, RP Beckmann and P Polakis, *J. Biol. Chem.* 271:28630–28635, 1996. With permission from American Society for Biochemistry and Molecular Biology.
**Figure 3–27** From AW Karzai, MM Susskind and RT Sauer, *EMBO J.* 18:3793–3799, copyright 1999. With permission from Macmillan Publishers Ltd.
**Figure 3–32** From NR Brown, ME Noble, AM Lawrie, MC Morris, P Tunnah, G Divita, LN Johnson and JA Endicott, *J. Biol. Chem.* 274:8746–8756, 1999. With permission from American Society for Biochemistry and Molecular Biology.
**Figure 3–33** From AJ Flint, T Tiganis, D Barford, and NK Tonks, *Proc. Natl Acad. Sci. U.S.A.* 94:1680–1685, 1997. With permission from National Academy of Sciences.

## Chapter 4

**Figure 4–5** Courtesy of Laurent J Beauregard.
**Figure 4–7** From D Vollrath and RW Davis, *Nucleic Acids Res.* 15:7865–7876, 1987. With permission from Oxford University Press.
**Figure 4–16** Adapted from EPM Candido, R Reeves and JR Davie, *Cell* 14:105–113, 1978. With permission from Elsevier.
**Figure 4–17** Adapted from M Lachner, D O'Carroll, S Rea, K Mechtier and T Jenuwein, *Nature* 410:116–120, 2001. With permission from Macmillan Publishers Ltd.
**Figure 4–27** From K Shibahara and B Stillman, *Cell* 96:575–585, 1999. With permission from Elsevier.
**Figure 4–30** From K Kimura and T Hirano, *Cell* 90:625–634, 1997. With permission from Elsevier.

## Chapter 5

**Figure 5–14** Courtesy of Victoria Foe.
**Figure 5–22 and 5–62** From RD Klemm, RJ Austin and SP Bell, *Cell* 88:493–502, 1997. With permission from Elsevier.
**Figure 5–23** From K Shibahara and B Stillman, *Cell* 96:575–585, 1999. With permission from Elsevier.
**Figure 5–24** From TM Nakamura, GB Morin, KB Chapman, SL Weinrich, WH Andrews, J Lingner, CB Harley and TR Cech, *Science* 277:955–959, 1997. With permission from AAAS.
**Figure 5–27** From C Masutano, M Araki, A Yamadam, R Kusumoto, T Nogimori, T Maekawa, S Iwai and F Hanoaka, *EMBO J.* 18: 3491–3501, 1999. With permission from Macmillan Publishers Ltd.
**Figure 5–34** From E Van Dyck, AZ Stasiak, A Stasiak and SC West, *Nature* 398:728–731, 1993. With permission from Macmillan Publishers Ltd.
**Figure 5–42** From JR Shah, R Cosstick and SC West, *EMBO J.* 16:1464–1472, 1997. With permission from Macmillan Publishers Ltd.

## Chapter 6

**Figure 6–1** Courtesy of Ulrich Sheer.
**Figure 6–10** From RD MJ Thomas, J Platas and DK Hawley, *Cell* 93:627–637, 1998. With permission from Elsevier.
**Figure 6–12** From SL McKnight and R Kingsbury, *Science* 217:316–324, 1982. With permission from AAAS.
**Figure 6–14** Courtesy of Fred Stevens.
**Figure 6–16 C** From E Nudler, A Mustaev, E Lukhtanov and A Goldfarb, *Cell* 89:33–41, 1997. With permission from Elsevier.
**Figure 6–17** From A Dugaiczyk, SL Woo, EC Lai, ML Mace Jr, L McReynolds and BW O'Malley, *Nature* 274:328–33, 1978. With permission from Macmillan Publishers Ltd.
**Figure 6–25** From BE Jády, T Kiss, *EMBO J.* 20:541–551, 2001. With permission from Macmillan Publishers Ltd.
**Figure 6–35** From AW Karzai, MM Susskind and RT Sauer, *EMBO J.* 18:3793–3799, 1999. With permission from Macmillan Publishers Ltd.
**Figure 6–36** From S Horowitz and MA Gorovsky, *Proc. Natl Acad. Sci. U.S.A.* 82:2452–2455, 1985. With permission from National Academy of Sciences.

**Figure 6–37** From SA Teter, WA Houry, D Ang, T Tradler, D Rockabrand, G Fischer, P Blum, C Georgopoulos and FU Hartl, *Cell* 97:755–765, 1999. With permission from Elsevier.
**Figure 6–38** From E Deuerling, A Schulze-Specking, T Tomoyasu, A Mogk and B Bukau, *Nature* 400:693–696, 1999. With permission from Macmillan Publishers Ltd.
**Figure 6–41 A** From U Schubert, LC Anton, J Gibbs, CC Norbury, JW Yewdell and JR Bennink, *Nature* 404:770–774, 2000. With permission from Macmillan Publishers Ltd.
**Figure 6–42** From H Yamano, C Tsurumi, J Gannon and T Hunt, *EMBO J.* 17:5670–5678, 1998. With permission from Macmillan Publishers Ltd.
**Figure 6–44 and 6–45** From A Bachmair, D Finley and A Varshavsky, *Science* 234:179–186, 1986. With permission from AAAS.

## Chapter 7

**Figure 7–1** Courtesy of Tim Myers and Leigh Anderson, Large Scale Biology Corporation
**Figure 7–3** From I Wilmut, AE Schnieke, J McWhir, AJ Kind and KHS Campbell, *Nature* 385:810–813, 1997. With permission from Macmillan Publishers Ltd.
**Figure 7–7** Adapted from MS Lee, GP Gippert, KV Soman, DA Case and PE Wright, *Science* 245:635–637, 1989. With permission from AAAS.
**Figure 7–11 B** From H Kabata, O Kurosawa, I Arai, M Washizu, SA Margarson, RE Glass and N Shimamoto, *Science* 262: 1561–1563, 1993. With permission from AAAS.
**Figure 7–13 B** From LR Patel, T Curran and TK Kerppola, *Proc. Natl Acad. Sci. U.S.A.* 91:7360–7364, 1994. With permission from National Academy of Sciences.
**Figure 7–16** Courtesy of Michele Sawadogo and Robert Roeder. From M Sawadogo and RG Roeder, *Cell* 43:165–175, 1985. With permission from Elsevier.
**Figure 7–19** Adapted from M Ptashne, A Genetic Switch: Phage λ and Higher Organisms. Cambridge, MA: Cell Press, 1992.
**Figure 7–27 B** From E Larshan and F Winston, *Genes Dev.* 15:1946–1956, 2001. With permission from Cold Spring Harbor Press.
**Figure 7–31 B** From X Bi, M Braunstein, GJ Shei and JR Broach, *Proc. Natl Acad. Sci. U.S.A.* 96:11934–11939, 1999. With permission from National Academy of Sciences.
**Figure 7–36** From J Zieg, M Silverman, M Hilmen and M Simon, *Science* 196:170–172, 1977. With permission from AAAS.
**Figure 7–38** From D Whitmore, NS Foulkes and P Sassone-Corsi, *Nature* 404:87–91, 2000. With permission from Macmillan Publishers Ltd.
**Figure 7–43** From GD Penny, GF Kay, SA Sheardown, S Rastan and N Brockdorff, *Nature* 379:131–137, 1996. With permission from Macmillan Publishers Ltd.
**Figure 7–54** From SR Thompson, EB Goodwin and M Wickens, *Mol. Cell Biol.* 20:2129–2137, 2000. With permission from American Society for Microbiology.
**Figure 7–56** From CY Chen and P Sarnow, *Science* 268:415–417, 1995. With permission from AAAS.
**Figure 7–57** Courtesy of Tim Myers and Leigh Anderson, Large Scale Biology Corporation

## Chapter 8

**Figure 8–3** From M Meselson and FW Stahl, *Proc. Natl Acad. Sci. U.S.A.* 44:671–682, 1958. With permission from National Academy of Sciences.
**Figure 8–7** All attempts have been made to contact the copyright holder.
**Figure 8–10** Reprinted from T Tugal, XH Zou-Yang, K Gavin, D Pappin, B Canas, R Kobayashi, T Hunt and B Stillman, *J. Biol. Chem.* 273:32421–32429, 1996. With permission from American Society for Biochemistry and Molecular Biology.
**Figure 8–14** Courtesy of JC Revy/Science Photo Library.
**Figure 8–26** Courtesy of Leander Lauffer and Peter Walter.
**Figure 8–42** From AM MacNicol, AJ Muslin and LT Williams, *Cell* 73:571–583, 1993. With permission from Elsevier.

## Chapter 9

**Figure 9–8** From RF Service, *Science* 306:1457, 2004. With permission from AAAS.
**Figure 9–8** Picture from Roger Tsien and colleagues, UCSD. From L Wang, WC Jackson, PA Steinbach, and RY Tsien, *Proc. Natl Acad. Sci. U.S.A.* 101:16745–16749, 2004. With permission from National Academy of Sciences.
**Figure 9–10** Adapted from BD Biosciences website. Courtesy and © Becton, Dickinson and Company.
**Figure 9–11** From AY Ting, KH Kain, RL Klemke and RY Tsien, *Proc. Natl Acad. Sci. U.S.A.* 98:15003–15008, 2001. With permission from National Academy of Sciences.
**Figure 9–12** From AY Ting, KH Kain, RL Klemke and RY Tsien, *Proc. Natl Acad. Sci. U.S.A.* 98:15003–15008, 2001. With permission from National Academy of Sciences.
**Figure 9–13** Courtesy of LA Staehelin.
**Figure 9–14** From KJ Fujimoto, *Cell Sci.* 108, 3443–3449. With permission from The Company of Biologists.
**Figure 9–15** From JE Rash, T Ysumura, CS Huson, P Agre, and S Nielsen, *Proc. Natl Acad. Sci. U.S.A.* 195:11981–11986, 1998. With permission from National Academy of Sciences.
**Figure 9–19** Adapted from R Heim, DC Prasher, and RY Tsien, *Proc. Natl Acad. Sci. U.S.A.* 91:12501–12504, 1994. With permission from National Academy of Sciences.

## Chapter 11

**Figure 11–6 (top)** From Rippel Electron Microscope Facility, Dartmought College. **Figure 11–6 (bottom)** From David Burgess.
**Figure 11–11 A** From T Hoshi, WN Zagotta and RW Aldrich, *Science* 250:568–571, 1990. With permission from AAAS.

## Chapter 12

**Figure 12–8** From K Ribbeck, G Lipowsky, HM Kent, M Stewart and D Gorlich, *EMBO J.* 17:6587–6598, 1998. With permission from Macmillan Publishers Ltd.
**Figure 12–9** From B Wolff, JJ Sanglier and Y Wang, *Chem. Biol.* 4:139–147, 1997. With permission from Elsevier.
**Figure 12–10 and 12–11** From N Kudo, N Matsumori, H Taoka, D Fujiwara, EP Schreiner, B Wolff, M Yoshida and S Horinouchi, *Proc. Natl Acad. Sci. U.S.A.* 96:9112–9117, 1999. With permission from National Academy of Sciences.

**Figure 12–12** From M Fornerod, M Ohno, M Yoshida and IW Mattaj, *Cell* 90:1051–1060, 1997. With permission from Elsevier.
**Figure 12–13** From MV Nachury and K Weis, *Proc. Natl Acad. Sci. U.S.A.* 96:9622–9627, 1999. With permission from National Academy of Sciences.
**Figure 12–16B** From M Donzeau, K Kaldi, A Adam, S Paschen, G Wanner, B Guiard, MF Bauer, W Neupert and M Brunner, *Cell* 101:401–412, 2000. With permission from Elsevier.
**Figure 12–19** From N Kinoshita, K Ghaedi, N Shimozawa, RJ Wanders, Y Matsuzono, T Imanaka, K Okumoto, Y Suzuki, N Kondo and Y Fujiki, *J. Biol. Chem.* 273:24122–24130, 1998. With permission from American Society for Biochemistry and Molecular Biology.
**Figure 12–22** Adapted from V Dammai and S Subramani, *Cell* 105:187–196, 2001. With permission from Elsevier.

## Chapter 13

**Figure 13–2B** Courtesy of BMF Pearse, adapted from CJ Smith, N Grigorieff and BM Pearse, *EMBO J.* 17:4943–4953, 1998. With permission from Macmillan Publishers Ltd.
**Figure 13–2C** Courtesy of Wyn Locke.
**Figure 13–7 and 13–8** From H Tokumaru, K Umayahara, LL Pellegrini, T Ishizuka, H Saisu, H Betz, GJ Augustine and T Abe, *Cell* 104:421–432, 2001. With permission from Elsevier.
**Figure 13–13** Courtesy of Margit Burmeister.
**Figure 13–15** From SM Wilson, R Yip, DA Swing, TN O'Sullivan, Y Zhang, EK Novak, RT Swank, LB Russell, NG Copeland and NA Jenkins, *Proc. Natl Acad. Sci. U.S.A.* 97:7933–7938, 2000. With permission from National Academy of Sciences.
**Figure 13–16 A** From MM Perry and AB Gilbert, *J.Cell Sci.* 39:257–272, 1979. With permission from The Company of Biologists.
**Figure 13–20** From JH Koenig amd K Ikeda, *J. Neuroscience* 9:3844–3960, 1989. With permission from The Society of Neuroscience.
**Figure 13–23** Reproduced from M Matteoli, K Takei, MS Perin, TC Sudhof and P De Camilli, *J. Cell Biol.* 117:849–861, 1992. With permission from The Rockefeller University Press.
**Figure 13–24B and C** From T Xu, U Ashery, RD Burgoyne and E Neher, *EMBO J.* 18:3293–304, 1999. With permission from Macmillan Publishers Ltd.

## Chapter 14

**Figure 14–4B** From R Yasuda, H Noji, K Kinosita Jr, and M Yoshida, *Cell* 93:1117–1124, 1998. With permission from Elsevier.
**Figure 14–8 and 14–9** Adapted from David Keilin, The History of Cell Respiration and Cytochrome. Cambridge: Cambridge University Press, 1966.

## Chapter 15

**Figure 15–10** Courtesy of Helfrid Hochegger.
**Figure 15–11** From JE Ferrell Jr and EM Machleder, *Science* 280:895–898, 1998. With permission from AAAS.
**Figure 15–29** From A Toker and A Newton, *J. Biol. Chem.* 275:8271–8274, 2000. With permission from American Society for Biochemistry and Molecular Biology.
**Figure 15–32** From CG Winter, B Wang, A Ballew, A Royou, R Karess, JD Axelrod and L Luo, *Cell* 105:81–91, 2001. With permission from Elsevier.
**Figure 15–33** From H Aberle, A Bauer, J Stappert, A Kispert and R Kemler, *EMBO J.* 16:3797–3804, 1997. With permission from Macmillan Publishers Ltd.
**Figure 15–34** From JA Porter, DP von Kessler, SC Ekker, KE Young, JJ Lee, K Moses and PA Beachy, *Nature* 374:363–366, 1995. With permission from Macmillan Publishers Ltd.
**Figure 15–35B and C** From JA Porter, DP von Kessler, SC Ekker, KE Young, JJ Lee, K Moses and PA Beachy, *Nature* 374:363–366, 1995. With permission from Macmillan Publishers Ltd.
**Figure 15–36** From JA Porter, DP von Kessler, SC Ekker, KE Young, JJ Lee, K Moses and PA Beachy, *Nature* 374:363–366, 1995. With permission from Macmillan Publishers Ltd.

## Chapter 16

**Figure 16–5** Reproduced from EM Mandelkow, E Mandelkow and RA Milligan, *J. Cell Biol.* 114:977–991, 1991. With permission from The Rockefeller University Press.
**Figure 16–6** Electron micrograph courtesy of M Runge and TD Pollard.
**Figure 16–10** Reproduced from M Yoon, RD Moir, V Prahlad and RD Goldman, *J. Cell Biol.* 143:147–157, 1998. With permission from The Rockefeller University Press.
**Figure 16–14A** From MR Bubb, I Spector, AD Bershadsky and ED Korn, *J. Biol. Chem.* 275:5163–70, 2000. With permission from American Society for Biochemistry and Molecular Biology.
**Figure 16–14B and C** From MO Steinmetz, D Stoffler, A Hoenger, A Bremer and U Aebi, *J. Struct. Biol.* 119:295–320, 1997. With permission from Elsevier.
**Figure 16–21** From R Leguy, R Melki, D Pantaloni and MF Carlier, *J. Biol.Chem.* 275:21975–21980, 2000. With permission from American Society for Biochemistry and Molecular Biology.
**Figure 16–22** From Y Zheng, ML Wong, B Alberts and T Mitchison, *Nature* 378:578–583, 1995. With permission from Macmillan Publishers Ltd.
**Figure 16–23A** From JA Theriot and DC Fung, *Methods Enzymol.* 298:114–122, 1998. With permission from Elsevier.
**Figure 16–26 and 16–27** From J Fan, AD Griffiths, A Lockhart, RA Cross and LA Amos, *J. Mol. Biol.* 259:325–330, 1996. With permission from Elsevier.
**Figure 16–32** Courtesy of Lewis Tilney.
**Figure 16–33** Courtesy of Kate Nobes.
**Figure 16–38** Courtesy of Keith Roberts.
**Figure 16–41A** From B Huang, G Piperno, Z Ramanis, and DJ Luck, *J. Cell Biol.* 88:80–88, 1981. With permission from The Rockefeller University Press.
**Figure 16–41B** From G Piperno, B Huang, Z Ramanis and DJ Luck, *J. Cell Biol.* 88:73–79, 1981. With permission from The Rockefeller University Press.
**Figure 16–43A** From A Mallavarapu and T Mitchison, *J. Cell Biol.* 146:1097–1106, 1999. With permission from The Rockefeller University Press.
**Figure 16–50** From CL Ho, JL Martys, A Mikhailov, GG Gundersen, RK Liem, *J. Cell Sci.* 111:1767–1778, 1998. Reproduced with permission of The Company of Biologists Ltd.
**Figure 16–52** From MO Steinmetz, D Stoffler, A Hoenger, A Bremer and U Aebi, *J. Struct. Biol.* 119:295–320, 1997. With permission from Elsevier.

## Chapter 17

**Figure 17–10** Courtesy of Terry D Allen.
**Figure 17–12** From C Michaelis, R Ciosk and K Nasmyth, *Cell* 91:35–41, 1997. With permission from Elsevier.
**Figure 17–14** Courtesy of Conly L Rieder, Wadsworth Center.
**Figure 17–20** From A Losada and T Hirano, *Curr. Biol.* 11:268–272, 2001. With permission from Elsevier.
**Figure 17–29** From C Antonio, I Ferby, H Wilhelm, M Jones, E Karsenti, AR Nebreda and I Vernos, *Cell* 102:425–435, 2000. With permission from Elsevier.
**Figure 17–30** From H Funabiki and A Murray, *Cell* 102:411–424, 2000. With permission from Elsevier.
**Figure 17–31** From B de Saint Phalle and W Sullivan, *J. Cell Biol.* 1141:1383–139, 1998. With permission from The Rockefeller University Press.
**Figure 17–33** Reprinted with permission from F Uhlmann, F Lottspeich and K Nasmyth, *Nature* 400:37–42, 1999. With permission from Macmillan Publishers Ltd.
**Figure 17–36** From AR Skop and JG White, *Curr. Biol.* 11:735–746, 2001. With permission from Elsevier.
**Figure 17–38** From A Bassini, S Pierpaoli, E Falcieri, M Vitale, L Guidotti, S Capitani and G Zauli, *Br. J. Haematol.* 104:820–828, 1999. With permission from Blackwell Science.
**Figure 17–44** From M Zecca, K Basler and G Struhl, *Development* 121:2265–2278, 1995. With permission from The Company of Biologists.
**Figure 17–46** From SA Datar, HW Jacobs, AF de la Cruz, CF Lehner and BA Edgar, *EMBO J.* 19:4543–4554, 2000. With permission from Macmillan Publishers Ltd.

## Chapter 18

**Figure 18–1** Courtesy of Julia Burne.
**Figure 18–2** From H Yoshida, YY Kong, R Yoshida, AJ Elia, A Hakem, R Hakem, JM Penninger and TW Mak, *Cell* 94:739–750, 1998. With permission from Elsevier.
**Figure 18–4** From JC Goldstein, NJ Waterhouse, P Juin, G Evanl and DR Green, *Nat. Cell Biol.* 2:156–162, 2000. With permission from Macmillan Publishers Ltd.
**Figure 18–10** From S D'Atri, L Tentori, PM Lacal, G Graziani, E Pagani, E Benincasa, G Zambruno, E Bonmassar and J Jiricny, *Mol. Pharmacol.* 54:334–341, 1998. With permission from The American Society for Pharmacology and Experimental Therapeutics.

## Chapter 19

**Figure 19–2** From A Nagafuchi, and M Takeichi, *Cell* 54:993–1001,1988, with permission from Elsevier, and from M Steinberg and M Takeichi, *Proc. Natl Acad. Sci. U.S.A.* 91:206–209, 1994, with permission from National Academy of Sciences.
**Figure 19–4** Adapted from M Ozawa, M Ringwald and R Kemler, *Proc. Natl Acad. Sci. U.S.A.* 87:4246–4250, 1990. With permission from National Academy of Sciences.
**Figure 19–5** Modified from CJ Watson, M Rowland, and G Warhurst, *Am. J. Physiol. Cell Physiol.* 281:C388–C397, 2001. With permission from American Physiological Society.
**Figure 19–6** From CM Van Itallie and JM Anderson, *Annu. Rev. Physiol.* 68:403–429, 2006. With permission from Annual Reviews
**Figure 19–9** From Sasaki et al, *Proc. Natl Acad. Sci. U.S.A.* 100:3971–3976, 2003. With permission from National Academy of Sciences.
**Figure 19–14** From S Wolf, CM Deom, RN Beachy, and WJ Lucas, *Science* 246:377–379, 1989. With permission from AAAS.
**Figure 19–15** From TH Chun, et al, *Cell* 125:577–591, 2006. With permission from Elsevier.
**Figure 19–16** From RW Orkin, P Gehron, EB McGoodwin, GR Martin, T Valentine, and RJ Swarm, *Exp. Med.* 145:204–220, 1977. With permission from The Rockefeller University Press.
**Figure 19–17** From PE Hughes, F Diaz-Gonzalez, L Leong, C Wu, JA McDonald, SJ Shattil, MH Ginsberg, *J. Biol. Chem.* 271:6571–4, 1996. With permission from American Society for Biochemistry and Molecular Biology.
**Figure 19–19** From PL DeAngelis, *J. Biol. Chem.* 274:26557–26562, 1999. With permission from American Society for Biochemistry and Molecular Biology.
**Figure 19–23** Adapted from FC Lin, RM Brown, RR Drake and BE Haley, *J. Biol. Chem.* 265:4782–4784, 1990. With permission from American Society for Biochemistry and Molecular Biology.
**Figure 19–28** From Pavenstädt, Kriz and Kretzle, *Physiol. Rev.* 83:253–307, 2003. With permission from American Physiological Society.

## Chapter 20

**Figure 20–2** Data from P Armitage, and R Doll, *Br. J. Cancer* 91:1983–1989, 2004. With permission from Macmillan Publishers Ltd.
**Figure 20–3** From AM Pearse, and K Swift, *Nature* 439:549, 2006. With permission from Macmillan Publishers Ltd.
**Figure 20–6** Reproduced courtesy of Museum Victoria.

# References


Aberle H, Bauer A, Stappert J, Kispert A & Kemler R (1997) β-Catenin is a target for the ubiquitin-proteasome pathway. *EMBO J.* 16, 3797–3804. (Problem 15–148)

Abrahams JP, Leslie AG, Lutter AGW & Walker JE (1994) Structure at 2.8 Å resolution of F1-ATPase from bovine heart mitochondria. *Nature* 370, 621–628. (Problem 14–32)

Abrami L, Simon M, Rousselet G, Berthonaud V, Buhler JM & Ripoche P (1994) Sequence and functional expression of an amphibian water channel, FA-CHIP: a new member of the MIP family. *Biochim. Biophys. Acta* 1192, 147–151. (Problem 11–72)

Adams GA & Rose JK (1985) Structural requirements of a membrane-spanning domain for protein anchoring and cell surface transport. *Cell* 41, 1007–1015. (Problem 13–56)

Adams KL, Song K, Roessler PG, Nugent JM, Doyle JL, Doyle JJ & Palmer JD (1999) Intracellular gene transfer in action: Dual transcription and multiple silencings of nuclear and mitochondrial *cox2* genes in legumes. *Proc. Natl Acad. Sci. U.S.A.* 96, 13863–13868. (Problem 1–47)

Adelstein RS & Klee CB (1980) Smooth muscle myosin light-chain kinase. In Calcium and Cell Function (WY Cheung ed), Vol. 1, pp 167–182. New York: Academic Press. (Problem 15–96)

Alberts BM & Frey L (1970) T4 bacteriophage gene 32: a structural protein in the replication and recombination of DNA. *Nature* 227, 1313–1318. (Problem 5–37)

Allard WJ & Lienhard GE (1985) Monoclonal antibodies to the glucose transporter from human erythrocytes: identification of the transporter as a $M_r$ = 55,000 protein. *J. Biol. Chem.* 260, 8668–8675. (Problem 11–17)

Alon R, Hammer DA & Springer TA (1995) Lifetime of the P-selectin-carbohydrate bond and its response to tensile force in hydrodynamic flow. *Nature* 374, 539–542. (Problem 19–21)

Amann KJ & Pollard TD (2001) The ARP2/3 complex nucleates actin filament branches from the sides of preexisting filaments. *Nat. Cell Biol.* 3, 306–310. (Problem 16–64)

Anderson JE (1993) Restriction endonucleases and modification methylases. *Curr. Opin. Struct. Biol.* 3, 24–30. (Problem 8–70)

Andreasen PH, Dreisig H & Kristiansen K (1987) Unusual ciliate-specific codons in *Tetrahymena* mRNAs are translated correctly in a rabbit reticulocyte lysate supplemented with a subcellular fraction from *Tetrahymena*. *Biochem. J.* 244, 331–335. (Problem 6–93)

Antonio C, Ferby I, Wilhelm H, Jones MJ, Karsenti E, Nebreda AR & Vernos I (2000) Xkid, a chromokinesin required for chromosome alignment on the metaphase plate. *Cell* 102, 425–435. (Problem 17–97)

Armitage P & Doll R (1954) The age distribution of cancer and a multi-stage theory of carcinogenesis. Reprinted in 2004 in *Br. J. Cancer* 91, 1983–1989. (Problems 20–17 and 20–19)

Artandi SE, Chang S, Lee S-L, Alson S, Gottlieb GJ, Chin L & DePinho RA (2000) Telomere dysfunction promotes non-reciprocal translocations and epithelial cancers in mice. *Nature* 406, 641–645. (Problem 4–41)

Ashburner M, Chihara C, Meltzer P & Richards G (1973) Temporal control of puffing activity in polytene chromosomes. *Cold Spring Harbor Symp. Quant. Biol.* 38, 655–662. (Problem 15–49)

Bachinger HP, Bruckner P, Timpl R, Prockop DJ & Engel J (1980) Folding mechanism of the triple helix in type III collagen and type III pN-collagen. *Eur. J. Biochem.* 106, 619–632. (Problem 19–86)

Bachmair A, Finley D & Varshavsky A (1986) *In vivo* half-life of a protein is a function of its amino-terminal residue. *Science* 234, 179–186. (Problems 6–99 and 6–100)

Bachurski CJ, Theodorakis NG, Coulson RM & Cleveland DW (1994) An amino-terminal tetrapeptide specifies cotranslational degradation of β-tubulin but not α-tubulin mRNAs. *Mol. Cell Biol.* 14, 4076–4086. (Problem 7–114)

Baeuerle PA & Baltimore D (1988) Activation of DNA-binding activity in an apparently cytoplasmic precursor of the NF-kB transcription factor. *Cell* 53, 211–217. (Problem 12–61)

Baldin V, Lukas J, Marcote MJ, Pagano M & Draetta G (1993) Cyclin D1 is a nuclear protein required for cell cycle progression in $G_1$. *Genes Dev.* 7, 812–821. (Problem 17–140)

Baltzer F (1967) Theodor Boveri: Life and Work of a Great Biologist. Berkeley: University of California Press. (Problem 17–102)

Bamberg E & Lauger P (1974) Temperature-dependent properties of gramicidin A channels. *Biochim. Biophys. Acta* 367, 127–133. (Problem 11–19)

Barnes G & Rine J (1985) Regulated expression of endonuclease EcoRI in *Saccharomyces cerevisiae*: nuclear entry and biological consequences. *Proc. Natl Acad. Sci. U.S.A.* 82, 1354–1358. (Problems 12–52 and 12–53)

Barron JT, Gu L & Parrillo JE (2000) NADH/NAD redox state of cytoplasmic glycolytic compartments in vascular smooth muscle. *Am. J. Physiol. Heart Circ. Physiol.* 279, H2872–H2878. (Problem 14–55)

Bartel DP & Szostak JW (1993) Isolation of new ribozymes from a large pool of random sequences. *Science* 261, 1411–1418. (Problems 6–105, 6–109, 6–110, and 6–111)

Baserga R & Wiebel F (1969) The cell cycle of mammalian cells. *Intern. Rev. Exp. Pathol.* 7, 1–30. (Problem 17–19)

Beck M, Forster F, Ecke M, Plitzko JM, Melchior F, Gerisch G, Baumeister W & Medalia O (2004) Nuclear pore complex structure and dynamics revealed by cryoelectron tomography. *Science* 306, 1387–1390. (Problem 9–45)

Bell SP & Dutta A (2002) DNA replication in eukaryotic cells. *Annu. Rev. Biochem.* 71, 333–374. (Problem 17–45)

Bender J & Kleckner N (1986) Genetic evidence that Tn10 transposes by a nonreplicative mechanism. *Cell* 45, 801–815. (Problem 5–110)

Benezra R, Davis RL, Lockshon D, Turner DL & Weintraub H (1990) The protein Id: a negative regulator of helix–loop–helix DNA binding proteins. *Cell* 61, 49–59. (Problem 7–29)

Bennett V & Stenbuck PJ (1979) The membrane attachment protein for spectrin is associated with band 3 in human erythrocyte membranes. *Nature* 280, 468–473. (Problem 10–66)

Bennett V & Stenbuck PJ (1980) Association between ankyrin and the cytoplasmic domain of band 3 isolated from the human erythrocyte membrane. *J. Biol. Chem.* 255, 6424–6432. (Problem 10–66)

Berg HC (1993) Random Walks in Biology, Expanded Edition, pp 5–6. Princeton, NJ: Princeton University Press. (Problems 2–91 and 2–92)

Berg JM, Tymoczko JL & Stryer L (2002) Biochemistry, Fifth Edition, pp 436–437. New York: WH Freeman and Co. (Problem 2–97)

Berget SM, Berk AJ, Harrison T & Sharp PA (1977) Spliced segments at the 5′ termini of adenovirus-2 late mRNA: a role for heterogeneous nuclear RNA in mammalian cells. *Cold Spring Harbor Symp. Quant. Biol.* 42, 523–529. (Problem 6–26)

Bestvater F, Spiess E, Stobrawa G, Hacker M, Feurer T, Porwol T, Berchner-Phannschmidt U, Wotzlaw C & Acker H (2002) Two-photon fluorescence absorption and emission spectra of dyes relevant for cell imaging. *J. Microscopy* 208, 108–115. (Problem 9–30)

Beug H, Katz FE & Gerisch G (1973) Dynamics of antigenic membrane sites relating to cell aggregation in *Dictyostellium discoideum. J. Cell Biol.* 56, 647–658. (Problem 19–13)

Bi X, Braunstein M, Shei G-J & Broach J (1999) The yeast *HML* I silencer defines a heterochromatin domain boundary by directional establishment of silencing. *Proc. Natl Acad. Sci. U.S.A.* 96, 11934–11939. (Problem 7–68)

Bittner JJ (1936) Some possible effects of nursing on the mammary gland tumor incidence in mice. *Science* 84, 2172. (Problem 20–29)

Bleil JD & Bretscher MS (1982) Transferrin receptor and its recycling in HeLa cells. *EMBO J.* 1, 351–355. (Problems 13–98 and 13–99)

Bloom K (1993) The centromere frontier: kinetochore components, microtubule-based motility, and the CEN-value paradox. *Cell* 73, 621–624. (Problem 17–87)

Bloom KS & Carbon J (1982) Yeast centromere DNA is in a unique and highly ordered structure in chromosomes and small circular minichromosomes. *Cell* 29, 305–317. (Problem 4–73)

Boeke JD, Garfinkel DJ, Styles CA & Fink GR (1985) Ty elements transpose through an RNA intermediate. *Cell* 40, 491–500. (Problem 5–111)

Bogenhagen D & Clayton DA (1977) Mouse L cell mitochondrial DNA molecules are selected randomly for replication throughout the cell cycle. *Cell* 11, 719–727. (Problem 14–112)

Bonfanti L, Mironov Jr. AA, Martinez-Menarguez JA, Martella O, Fusella A, Baldassarre M, Buccione R, Geuze HJ, Mironov AA & Luini A (1998) Procollagen traverses the Golgi stack without leaving the lumen of cisternae: evidence for cisternal maturation. *Cell* 95, 993–1003. (Problem 13–60)

Bonner JT & Savage LJ (1947) Evidence for the formation of cell aggregates by chemotaxis in the development of the slime mold *Dictyostelium discoideum. J. Exp. Zool.* 106, 1–26. (Problem 15–46)

Bootsma D, Budke L & Vos O (1964) Studies on synchronous division of tissue culture cells initiated by excess thymidine. *Exp. Cell Res.* 33, 301–309. (Problem 17–14)

Bostock CJ, Prescott DM & Kirkpatrick JB (1971) An evaluation of the double thymidine block for synchronizing mammalian cells at the $G_1$-S border. *Exp. Cell Res.* 68, 163–168. (Problem 17–14)

Boveri T (1902) Über mehrpolige Mitosen als mittel zur Analyse des Zellkerns. *Verh. d. phys.-med. Ges. Würzburg, N.F.* 35, 67–90. (Available in English translation, Foundations of Experimental Embryology, B Willier and J Oppenheimer eds. Englewood Cliffs, NJ: Prentice-Hall, 1964.) (Problem 17–102)

Boveri T (1907) Zellenstudien VI: Die Entwicklung dispermer Seeigeleier. Ein Beitrag zur Befruchtungslehre und zur Theorie des Kernes. *Jenaische Zeitschr. Naturwissen.* 43, 1–292. (Problem 17–102)

Bretscher M (1972) Asymmetrical lipid bilayer structure for biological membranes. *Nat. New Biol.* 236, 11–12. (Problem 10–29)

Brewer BJ & Fangman WL (1987) The localization of replication origins on ARS plasmids in *S. cerevisiae. Cell* 51, 463–471. (Problem 5–61)

Brinker A, Pfeifer G, Kerner MJ, Naylor DJ, Hartl FU & Hayer-Hartl M (2001) Dual function of protein confinement in chaperonin-assisted protein folding. *Cell* 107, 223–233. (Problem 6–96)

Brokaw CJ, Luck DJL & Huang B (1982) Analysis of the movement of *Chlamydomonas* flagella: the function of the radial-spoke system is revealed by comparison of wild-type and mutant flagella. *J. Cell Biol.* 92, 722–732. (Problem 16–107)

Brown CJ, Ballabio A, Rupert JL, Lafreniere RG, Grompe M, Tonlorenzi R & Willard HF (1993) A gene from the region of the human X inactivation center is expressed exclusively from the inactive X chromosome. *Nature* 349, 38–44. (Problem 7–93)

Brown DA & Rose JK (1992) Sorting of GPI-anchored proteins to glycolipid-enriched membrane subdomains during transport to the apical cell surface. *Cell* 68, 533–544. (Problem 10–30)

Brown MS & Goldstein JL (1979) Receptor-mediated endocytosis: insights from the lipoprotein receptor system. *Proc. Natl Acad. Sci. U.S.A.* 76, 3330–3337. (Problems 13–100 and 13–101)

Brown NR, Noble MEM, Lawrie AM, Morris MC, Tunnah P, Divita G, Johnson LN & Endicott JA (1999) Effects of phosphorylation of threonine 160 on cyclin-dependent kinase 2 structure and activity. *J. Biol. Chem.* 274, 8746–8756. (Problem 3–109)

Brutlag D & Kornberg A (1972) Enzymatic synthesis of deoxyribonucleic acid. 36. A proofreading function for the 3' to 5' exonuclease activity in deoxyribonucleic acid polymerases. *J. Biol. Chem.* 247, 241–248. (Problem 5–41)

Bubb MR, Spector I, Bershadsky AD & Korn ED (1995) Swinholide A is a microfilament disrupting marine toxin that stabilizes actin dimers and severs actin filaments. *J. Biol. Chem.* 270, 3463–3466. (Problem 16–43)

Burch RM., Luini A & Axelrod J (1986) Phospholipase $A_2$ and phospholipase C are activated by distinct GTP-binding proteins in response to $\alpha_1$-adrenergic stimulation in FRTL5 thyroid cells. *Proc. Natl Acad. Sci. U.S.A.* 83, 7201–7205. (Problem 15–94)

Byers D, Davis RL & Kiger JA (1981) Defect in cyclic AMP phosphodiesterase due to the *dunce* mutation of learning in *Drosophila melanogaster. Nature* 289, 79–81. (Problem 15–93)

Callan HG (1963) The nature of lampbrush chromosomes. *Int. Rev. Cytol.* 15, 1–34. (Problem 4–82)

Candido EPM, Reeves R & Davie JR (1978) Sodium butyrate inhibits histone deacetylation in cultured cells. *Cell* 14, 105–113. (Problem 4–69)

Cantor CR & Schimmel PR (1980) Biophysical Chemistry. New York: WH Freeman and Company. (Problems 2–59, 3–107, 3–108, 8–17, 8–23, and 8–24)

Capaldi RA, Darley-Usmar V, Fuller S & Millet F (1982) Structural and functional features of the interaction of cytochrome *c* with complex III and cytochrome *c* oxidase. *FEBS Lett.* 138, 1–7. (Problem 14–62)

Carlier M-F, Pantaloni D & Korn ED (1984) Evidence for an ATP cap at the ends of actin filaments and its regulation of the F-actin steady state. *J. Biol. Chem.* 259, 9983–9986. (Problem 16–37)

Carlier M-F, Pantaloni D & Korn ED (1985) Polymerization of ADP- and ATP-actin under sonication and characteristics of the ATP-actin equilibrium polymer. *J. Biol. Chem.* 260, 6565–6571. (Problem 16–12)

Castagnola M (1998) Sensitive to the yoctomole limit. *Trends Biochem. Sci.* 23, 283. (Problem 8–32)

Cecconi F, Alvarez-Bolado G, Meyer BI, Roth KA & Gruss P (1998) Apaf1 (CED-4 homolog) regulates programmed cell death in mammalian development. *Cell* 94, 727–737. (Problem 18–20)

Cha TA & Alberts BM (1986) Studies of the DNA helicase–RNA primase unit from bacteriophage T4. A trinucleotide sequence on the DNA template starts RNA primer synthesis. *J. Biol. Chem.* 261, 7001–7010. (Problem 5–42)

Chalfie M, Tu Y, Euskirchen G, Ward WW & Prasher DC (1994) Green fluorescent protein as a marker for gene expression. *Science* 263, 802–805. (Problem 9–25)

Chamberlain JS, Gibbs RA, Ranier JE & Caskey CT (1990) Multiplex PCR for the diagnosis of Duchenne muscular dystrophy. In PCR Protocols (MA Innis, DH Gelfand, JJ Sninsky & TJ White eds), pp 272–281. San Diego, CA: Academic Press. (Problem 8–96)

Chant J & Herskowitz I (1991) Genetic control of bud site selection in yeast by a set of gene products that constitute a morphogenetic pathway. *Cell* 65, 1203–1212. (Problem 16–112)

Chant J, Corrado K, Pringle JR & Herskowitz I (1991) Yeast BUD5, encoding a putative GDP-GTP exchange factor, is necessary for bud site selection and interacts with bud formation gene *BEM1*. *Cell* 65, 1213–1224. (Problem 16–112)

Chao S-H, Fujinaga K, Marion JE, Taube R, Sausville EA, Senderowicz AM, Peterlin BM & Price DH (2000) Flavopiridol inhibits P-TENb and blocks HIV-1 replication. *J. Biol. Chem.* 275, 28345–28348. (Problem 7–104)

Chapeville F, Lipmann F, von Ehrenstein G, Weisblum B, Ray WJ & Benzer S (1962) On the role of soluble ribonucleic acid in coding for amino acids. *Proc. Natl Acad. Sci. U.S.A.* 48, 1086–1092. (Problem 6–72)

Chardin P & McCormick F (1999) Brefeldin A: the advantage of being uncompetitive. *Cell* 97, 153–155. (Problem 13–29)

Chen C-N, Denome S & Davis RL (1986) Molecular analysis of cDNA clones and the corresponding genomic coding sequences of the *Drosophila dunce*$^+$ gene, the structural gene for cyclic AMP phosphodiesterase. *Proc. Natl Acad. Sci. U.S.A.* 83, 9313–9317. (Problem 15–93)

Chen C-Y & Sarnow P (1995) Initiation of protein synthesis by the eucaryotic translation apparatus on circular RNAs. *Science* 268, 415–417. (Problem 7–119)

Chen S-H, Habib G, Yang CY, Gu ZW, Lee BR, Weng S-A, Silbermann SR, Cai S-J, Deslypere JP, Rosseneu M, Gotto AM, Li W-H & Chan L (1987) Apolipoprotein B-48 is the product of a messenger RNA with an organ-specific in-frame stop codon. *Science* 238, 363–366. (Problem 7–113)

Cheng CHC & Chen L (1999) Evolution of an antifreeze protein. *Nature* 401, 443–444. (Problem 3–64)

Chrétien D, Fuller SD & Karsenti E (1995) Structure of growing microtubule ends: Two-dimensional sheets close into tubes at variable rates. *J. Cell Biol.* 117, 1311–1328. (Problem 16–17)

Chun T-H, Hotary KB, Sabeh F, Saltiel AR, Allen ED, Weiss SJ (2006) A pericellular collagenase directs the three-dimensional development of white adipose tissue. *Cell* 125, 577–591. (Problem 19–57)

Claude P (1978) Morphological factors influencing transepithelial permeability: a model for the resitance of the zonula occludens. *J. Memb. Biol.* 39, 219–232. (Problem 19–33)

Cohen S, Ushiro H, Stocheck C & Chinkers M (1982) A native 170,000 epidermal growth factor receptor-kinase complex from shed plasma membrane vesicles. *J. Biol. Chem.* 257, 1523–1531. (Problem 17–139)

Coluccio LM & Tilney LG (1984) Phalloidin enhances actin assembly by preventing monomer dissociation. *J. Cell Biol.* 99, 529–535. (Problem 16–42)

Conway L & Wickens M (1987) Analysis of mRNA 3′-end formation by modification interference: the only modifications which prevent processing lie in AAUAAA and the poly(A) site. *EMBO J.* 6, 4177–4184. (Problem 6–46)

Cox M & Lehman IR (1981) The polarity of the recA protein-mediated branch migration. *Proc. Natl Acad. Sci. U.S.A.* 78, 6023–6027. (Problem 5–96)

Craigen WJ, Cook RG, Tate WP & Caskey CT (1985) Bacterial peptide chain release factors: conserved primary structure and possible frameshift regulation of release factor 2. *Proc. Natl Acad. Sci. U.S.A.* 82, 3616–3620. (Problem 6–91)

Creighton TE (1993) Proteins, 2nd ed. New York: WH Freeman. (Problems 3–13, 3–37, and 3–41)

Crick F (1988) What Mad Pursuit: A Personal View of Scientific Discovery, p 109. New York: Basic Books, Inc. (Problem 6–17)

D'Atri S, Tentori L, Lacal PM, Graziani G, Pagani E, Benincasa E, Zambruno G, Bonmassar E & Jiricny J (1998) Involvement of the mismatch repair system in temozolomide-induced apoptosis. *Mol. Pharmacol.* 54, 334–341. (Problem 18–29)

Dammai V & Subramani S (2001) The human peroxisomal targeting signal receptor, Pex5p, is translocated into the peroxisomal matrix and recycled to the cytosol. *Cell* 105, 187–196. (Problem 12–98)

Dammer U, Popescu O, Wagner P, Anselmetti D, Guntherodt H-J & Misevic GN (1995) Binding strength between cell adhesion proteoglycans measured by atomic force microscopy. *Science* 267, 1173–1175. (Problem 19–18)

Datar SA, Jacobs HW, Flor A, de la Cruz A, Lehner CF & Edgar BA (2000) The *Drosophila* cyclin D-Cdk4 complex promotes cellular growth. *EMBO J.* 19, 4543–4554. (Problem 17–146, Problem 17–147)

Dave UP, Jenkins NA & Copeland NG (2004) Gene therapy insertional mutagenesis insights. *Science* 303, 333. (Problem 20–42)

Davis RL, Weintraub H & Lassar AB (1987) Expression of a single transfected cDNA converts fibroblasts to myoblasts. *Cell* 51, 987–1000. (Problem 7–92)

de Oca Luna RM, Wagner DS & Lozano G (1995) Rescue of early embryonic lethality in *mdm2*-deficient mice by deletion of *p53*. *Nature* 378, 203–206. (Problem 20–55)

de Saint Phalle B & Sullivan W (1998) Spindle assembly and mitosis without centrosomes in parthenogenetic *Sciara* embryos. *J. Cell Biol.* 141, 1383–1391. (Problem 17–99)

DeAngelis PL (1999) Molecular directionality of polysaccharide polymerization by the *Pasteurella multocida* hyaluronan synthase. *J. Biol. Chem.* 274, 26557–26562. (Problem 19–85)

Deenen LLM & DeGier J (1974) Lipids of the red cell membrane. In The Red Blood Cell (D MacN Surgenor, ed), pp 147–211. New York: Academic Press. (Problem 10–29)

DeFranco D & Yamamoto KR (1986) Two different factors act separately or together to specify functionally distinct activities at a single transcriptional enhancer. *Mol. Cell. Biol.* 6, 993–1001. (Problem 15–47)

Delagado-Partin VM & Dalbey RE (1998) The proton motive force, acting on acidic residues, promotes translocation of amino-terminal domains of membrane proteins when the hydrophobicity of the translocation signal is low. *J. Biol. Chem.* 273, 9927–9934. (Problem 12–27)

Desai A & Mitchison TJ (1997) Microtubule polymerization dynamics. *Annu. Rev. Cell Dev. Biol.* 13, 83–117. (Problem 16–15)

Deshaies RJ & Schekman R (1987) A yeast mutant defective at an early stage in import of secretory protein precursors into the endoplasmic reticulum. *J. Cell Biol.* 105, 633–645. (Problem 12–131)

Detrich WH, Parker SK, Williams RC, Nogales E & Downing KH (2000) Cold adaptation of microtubule assembly and dynamics. *J. Biol. Chem.* 275, 37038–37047. (Problems 16–29 and 16–35)

Deuerling E, Schulze-Specking A, Tomoyasu T, Mogk A & Bukau B (1999) Trigger factor and DnaK cooperate in folding of newly synthesized proteins. *Nature* 400, 693–696. (Problems 6–94 and 6–95)

Dingwall C, Sharnick SV & Laskey RA (1982) A polypeptide domain that specifies migration of nucleoplasmin into the nucleus. *Cell* 30, 449–458. (Problem 12–51)

Dittmer F, Ulbrich EJ, Hafner A, Schmahl W, Meister T, Pohlmann R & von Figura K (1999) Alternative mechanisms for trafficking of lysosomal enzymes in mannose 6-phosphate receptor-deficient mice are cell type specific. *J. Cell Sci.* 112, 1591–1597. (Problem 13–70)

Dodson M, Dean FB, Bullock P, Echols H & Hurwitz J (1987) Unwinding of duplex DNA from the SV40 origin of replication by T-antigen. *Science* 238, 964–967. (Problem 5–59)

Doheny KF, Sorger PK, Hyman AA, Tugendreich S, Spencer F & Hieter P (1993) Identification of essential components of the *S. cerevisiae* kinetochore. *Cell* 73, 761–774. (Problems 17–95 and 17–96)

Doll R & Hill AB (1954) The mortality of doctors in relation to their smoking habits: A preliminary report. Reprinted in 2004 in *Br. Med. J.* 328, 1529–1533. (Problem 20–20)

Doll R, Peto R, Boreham J & Sutherland I (2004) Mortality in relation to smoking: 50 years' observations on male British doctors. *Br. Med. J.* 328, 1519–1528. (Problem 20–20)

Dolznig H, Bartunek P, Nasmyth K, Mullner EW & Beug H (1995) Terminal differentiation of normal chicken erythroid progenitors: shortening of $G_1$ correlates with loss of D-cyclin/cdk4 expression and altered cell size control. *Cell Growth Differ.* 6, 1341–1352. (Problem 17–130)

Donaldson JG, Finazzi D & Klausner RD (1992) Brefeldin A inhibits Golgi membrane-catalyzed exchange of guanine nucleotide into Arf protein. *Nature* 360, 350–352. (Problem 13–29)

Donzeau M, Káldi K, Adam A, Paschen S, Wanner G, Guiard B, Bauer MF, Neupert W & Brunner M (2000) Tim23 links the inner and outer mitochondrial membranes. *Cell* 101, 401–412. (Problem 12–86)

Dramsi S & Cossart P (1998) Intracellular pathogens and the actin cytoskeleton. *Annu. Rev. Cell Dev. Biol.* 14, 137–166. (Problem 16–65)

Dugaiczyk A, Woo SL, Lai EC, Mace Jr ML, McReynolds L & O'Malley BW (1978) The natural ovalbumin gene contains seven intervening sequences. *Nature* 274, 328–333. (Problem 6–43)

Dunham I et al. (1999) The DNA sequence of human chromosome 22. *Nature* 402, 489–495. (Problem 4–45, Problem 4–46)

Dunn TM, Hahn S, Ogden S & Schleif RF (1984) An operator at –280 base pairs that is required for repression of *araBAD* operon promoter: addition of DNA helical turns between the operator and promoter cyclically hinders repression. *Proc. Natl Acad. Sci. U.S.A.* 81, 5017–5020. (Problem 7–62)

Duysens LNM, Amesz J & Kamp BM (1961) Two photochemical systems in photosynthesis. *Nature* 190, 510–511. (Problem 14–94)

Earnshaw WC, Martins LM & Kaufmann SH (1999) Mammalian caspases. *Annu. Rev. Biochem.* 68, 383–424. (Problem 18–21)

Echols H (2001) Operators and Promoters: The Story of Molecular Biology and Its Creators, pp 46–47. Berkeley: University of California Press. (Problem 7–31)

Edelman P & Gallant J (1977) Mistranslation in *E. coli*. *Cell* 10, 131–137. (Problem 6–88)

Edgar BA & Lehner CF (1996) Developmental control of cell cycle regulators: A fly's perspective. *Science* 274, 1646–1652. (Problem 17–143)

Eilers M & Schatz G (1986) Binding of a specific ligand inhibits import of a purified precursor protein into mitochondria. *Nature* 322, 228–232. (Problem 12–81)

Eisenberg E & Hill TL (1985) Muscle contraction and free energy transduction in biological systems. *Science* 227, 999–1006. (Problem 16–76)

Eliasson C, Sahlgren C, Berthold CH, Stakeberg J, Celis JE, Betsholtz C, Eriksson JE & Pekny M (1999) Intermediate filament protein partnership in astrocytes. *J. Biol. Chem.* 274, 23996–24006. (Problem 16–26)

Elledge SJ, Mulligan JT, Ramer SW, Spottswood M & Davis RW (1991) λYES: a multifunctional cDNA expression vector for the isolation of genes by complementation of yeast and *Escherichia coli* mutations. *Proc. Natl Acad. Sci. U.S.A.* 88, 1731–1735. (Problem 8–88)

Ellis HM & Horvitz RH (1986) Genetic control of programmed cell death in the nematode *C. elegans*. *Cell* 44, 817–829. (Problem 18–18)

Emerson R (1958) The quantum yield of photosynthesis. *Annu. Rev. Plant Physiol.* 9, 1–24. (Problem 14–93)

Fabian MA et al. (2005) A small molecule-kinase interaction map for clinical kinase inhibitors. *Nat. Biotechnol.* 23, 329–336. (Problems 20–68, 20–69, and 20–70)

Fan J, Griffiths AD, Lockhart A, Cross RA & Amos LA (1996) Microtubule minus ends can be labeled with a phage display antibody specific to α-tubulin. *J. Mol. Biol.* 259, 325–330. (Problems 16–32, 16–74, and 16–75)

Farrell PJ, Balkow K, Hunt T, Jackson RJ & Trachsel H (1977) Phosphorylation of initiation factor eIF-2 and the control of reticulocyte protein synthesis. *Cell* 11, 187–200. (Problem 7–112)

Faull RJ, Kovach NL, Harlan JM & Ginsberg MH (1993) Affinity modulation of integrin $\alpha_5\beta_1$: regulation of the functional response by soluble fibronectin. *J. Cell Biol.* 121, 155–162. (Problem 19–66)

Federman AD, Conklin BR, Schrader KA, Reed RR & Bourne HR (1992) Hormonal stimulation of adenylyl cyclase through $G_i$ protein βγ subunits. *Nature* 356, 159–161. (Problem 15–91)

Ferrell JE & Machleder EM (1998) The biochemical basis of an all-or-none cell fate switch in *Xenopus* oocytes. *Science* 280, 895–898. (Problem 15–50)

Fersht A (1999) Structure and Mechanism in Protein Science. New York: WH Freeman. (Problems 3–23 and 3–96)

Fields S & Song O (1989) A novel genetic system to detect protein–protein interactions. *Nature* 340, 245–246. (Problem 8–47)

Fijalkowska IJ, Jonczyk P, Tkaczyk MM, Bialoskorska M & Schaaper RM (1998) Unequal fidelity of leading strand and lagging strand DNA replication on the *Escherichia coli* chromosome. *Proc. Natl Acad. Sci. U.S.A.* 95, 10020–10025. (Problem 5–44)

Fisher JC & Hollomon JH (1951) A hypothesis for the origin of cancer foci. *Cancer* 4, 916–918. (Problem 20–19)

Flint AJ, Tiganis T, Barford D & Tonks NK (1997) Development of substrate-trapping mutants to identify physiological substrates of protein tyrosine phosphatases. *Proc. Natl Acad. Sci. U.S.A.* 94, 1680–1685. (Problem 3–110)

Folkman J & Moscona A (1978) Role of cell shape in growth control. *Nature* 273, 345–349. (Problem 17–144)

Forbush B, Kok B & McGloin M (1971) Cooperation of charges in photosynthetic oxygen evolution II. Damping of flash yield, oscillation and deactivation. *Photochem. Photobiol.* 14, 307–321. (Problem 14–95)

Fornerod M, Ohno M, Yoshida M & Mattaj IW (1997) CRM1 is an export receptor for leucine-rich nuclear export signals. *Cell* 90, 1051–1060. (Problem 12–60)

Fox TD (1986) Nuclear gene products required for translation of specific mitochondrially coded mRNAs in yeast. *Trends Genet.* 2, 97–100. (Problem 14–116)

Foyer CH (1984) Photosynthesis, pp 176–195. New York: Wiley. (Problem 14–91)

Frayn KN (1996) Metabolic Regulation: A Human Perspective, p 179. London: Portland Press. (Problem 2–120)

Freeland FJ & Hurst LD (1998) The genetic code is one in a million. *J. Mol. Evol.* 47, 238–248. (Problem 1–14)

Friedberg EC, Walker GC & Siede W (1995) DNA Repair and Mutagenesis, pp 92–103. New York: WH Freeman. (Problem 5–76)

Fries E, Gustafsson L & Peterson PA (1984) Four secretory proteins synthesized by hepatocytes are transported from the endoplasmic reticulum to Golgi complex at different rates. *EMBO J.* 3, 147–152. (Problem 13–114)

Fujiki Y (2000) Peroxisome biogenesis and peroxisome biogenesis disorders. *FEBS Let.* 476, 42–46. (Problem 12–94)

Fujimoto K (1995) Freeze-fracture replica electron microscopy combined with SDS digestion for cytochemical labeling of integral membrane proteins. *J. Cell Sci.* 108, 3443–3449. (Problem 9–47)

Funabiki H & Murray AW (2000) The *Xenopus* chromokinesin Xkid is essential for metaphase chromosome alignment and must be degraded to allow anaphase chromosome movement. *Cell* 102, 411–424. (Problem 17–98)

Fung BK-K & Stryer L (1980) Photolyzed rhodopsin catalyzes the exchange of GTP for bound GDP in retinal rod outer segments. *Proc. Natl Acad. Sci. U.S.A.* 77, 2500–2504. (Problem 15–89)

Fung BK-K, Hurley JB & Stryer L (1981) Flow of information in the light-triggered cyclic nucleotide cascade of vision. *Proc. Natl Acad. Sci. U.S.A.* 78, 152–156. (Problem 15–89)

Fung Y-KT, Murphree AL, T'Ang A, Qian J, Hinrichs SH & Benedict WF (1987) Structural evidence for the authenticity of the human retinoblastoma gene. *Science* 236, 1657–1661. (Problem 20–43)

Gall JG & Murphy C (1998) Assembly of lampbrush chromosomes from sperm chromatin. *Mol. Biol. Cell* 9, 733–747. (Problem 4–80)

Garapin AC, Cami B, Roskam W, Kourilsky P, Le Pennec JP, Perrin F, Gerlinger P, Cochet M & Chambon P (1978) Electron microscopy and restriction enzyme mapping reveal additional intervening sequences in the chicken ovalbumin split gene. *Cell* 14, 629–639. (Problem 6–43)

Gartenberg MR & Crothers DM (1988) DNA sequence determinants of CAP-induced bending and protein binding affinity. *Nature* 333, 824–829. (Problem 7–33)

Gay DA, Yen TJ, Lau JTY & Cleveland DW (1987) Sequences that confer β-tubulin autoregulation through modulated mRNA stability reside within exon 1 of a β-tubulin mRNA. *Cell* 50, 671–679. (Problem 7–114)

Gelehrter TD & Collins FS (1990) Principles of Medical Genetics, p 80. Baltimore: Williams & Wilkins. (Problem 4–102)

Gerhart JH & Pardee AB (1963) The effect of the feedback inhibitor, CTP, on subunit interactions in aspartate transcarbamylase. *Cold Spring Harbor Symp. Quant. Biol.* 28, 491–496. (Problem 3–107)

Gibbons IR (1981) Cilia and flagella of eucaryotes. *J. Cell Biol.* 91, 107s–124s. (Problem 16–107)

Gilbert W & Müller-Hill B (1966) Isolation of the Lac repressor. *Proc. Natl Acad. Sci. U.S.A.* 56, 1891–1898. (Problem 3–101)

Glandsdorff N (1987) Biosynthesis of arginine and polyamines. In *Eschericia coli* and *Salmonella typhimurium*: Cellular and Molecular Biology (FC Neidhardt ed), pp 321–344. Washington DC: American Society for Microbiology. (Problem 7–48)

Godowski PJ, Rusconi S, Miesfeld R & Yamamoto KR (1987) Glucocorticoid receptor mutants that are constitutive activators of transcriptional enhancement. *Nature* 325, 365–368. (Problem 7–64)

Goldberg E (1983) Recognition, attachment, injection. In Bacteriophage T4 (CK Mathews, EM Kutter, G Mosig, PB Berget, eds), pp 32–39. Washington, DC: American Society for Microbiology. (Problem 19–19)

Goldberg J (1999) Structural and functional analysis of the Arf1–ArfGAP complex reveals a role for coatomer in GTP hydrolysis. *Cell* 96, 893–902. (Problem 13–30)

Goldman YE, Hibberd MG, McCray JA & Trentham DR (1980) Relaxation of muscle fibers by photolysis of caged ATP. *Nature* 300, 701–705. (Problem 16–104)

Goldstein JC, Waterhouse NJ, Juin P, Evan GI & Green DR (2000) The coordinate release of cytochrome *c* during apoptosis is rapid, complete and kinetically invariant. *Nat. Cell Biol.* 2, 156–162. (Problem 18–23, Problem 18–24)

Gordon AM, Huxley AF & Julian FJ (1966) The variation in isometric tension with sarcomere length in vertebrate muscle fibres. *J. Physiol.* 184, 170–192. (Problem 16–105)

Görlich D, Prehn S, Laskey RA & Hartmann E (1994) Isolation of a protein that is essential for the first step of nuclear protein import. *Cell* 79, 767–776. (Problem 12–54)

Gottesman MM (1985) Genetics of cyclic-AMP-dependent protein kinases. In Molecular Cell Genetics (MM Gottesman ed), pp 711–743. New York: Wiley. (Problem 15–92)

Gottlieb TA, Beaudry G, Rizzolo L, Colman A, Rindler MJ, Adesnik M & Sabatini DD (1986) Secretion of endogenous and exogenous proteins from polarized MDCK monolayers. *Proc. Natl Acad. Sci. U.S.A.* 83, 2100–2104. (Problem 13–117)

Goutte C & Johnson AD (1988) a1 protein alters the DNA-binding specificity of α2 repressor. *Cell* 52, 875–882. (Problem 7–87)

Griffin JD (2005) Interaction maps for kinase inhibitors. *Nat. Biotechnol.* 23, 308–309. (Problem 20–68)

Guo W, Roth D, Walch-Solimena C & Novick P (1999) The exocyst is an effector for Sec4p, targeting secretory vesicles to sites of exocytosis. *EMBO J.* 18, 1071–1080. (Problem 13–35)

Hacein-Bey-Abina S, von Kalle C, Schmidt M, Le Deist F, Wulffraat N, McIntyre E, Radford I, Villeval J-L, Fraser CC, Cavazzana-Calvo M & Fischer A (2003) A serious adverse event after successful gene therapy for severe combined immunodeficiency syndrome. *N. Engl. J. Med.* 348, 255–256. (Problem 20–42)

Hadwiger JA, Wittenberg C, Richardson HE, de Barros Lopes M & Reed SI (1989) A family of cyclin homologs that control the $G_1$ phase in yeast. *Proc. Natl Acad. Sci. U.S.A.* 86, 6255–6259. (Problem 17–36)

Haigler HT, McKenna JA & Cohen S (1979) Rapid stimulation of pinocytosis in human A-431 carcinoma cells by epidermal growth factor. *J. Cell Biol.* 83, 82–90. (Problem 13–96)

Hakem R, Hakem A, Duncan GS, Henderson JT, Woo M, Soengas MS, Elia A, de la Pompa JL, Kagi D, Khoo W, Potter J, Yoshida R, Kaufman SA, Lowe SW, Penninger JM & Mak TW (1998) Differential requirement for caspase 9 in apoptotic pathways *in vivo*. *Cell* 94, 339–352. (Problem 18–20)

Hale SP, Auld DS, Schmidt E & Schimmel P (1997) Discrete determinants in transfer RNA for editing and aminoacylation. *Science* 276, 1250–1252. (Problem 6–89)

Hall A (1998) Rho GTPases and the actin cytoskeleton. *Science* 279, 509–514. (Problem 16–98)

Hancock WO & Howard J (1998) Processivity of the motor protein kinesin requires two heads. *J. Cell Biol.* 140, 1395–1405. (Problem 16–77)

Harada Y, Ohara O, Takatsuki A, Itoh H, Shimamoto N & Kinosita K (2001) Direct observation of DNA rotation during transcription by *Escherichia coli* RNA polymerase. *Nature* 409, 113–115. (Problem 6–23)

Harder T & Simons K (1997) Caveolae, DIGs, and the dynamics of sphingolipid-cholesterol microdomains. *Curr. Opin. Cell Biol.* 9, 534–542. (Problem 10–20)

Harland RM & Laskey RA (1980) Regulated replication of DNA microinjected into eggs of *Xenopus laevis*. *Cell* 21, 761–771. (Problem 5–60)

Hartwell LH (1978) Cell division from a genetic perspective. *J. Cell Biol.* 77, 627–637. (Problems 17–17 and 17–18)

Hartwell LH, Hood L, Goldberg ML, Reynolds AE, Silver LM & Veres RC (2000) Genetics: From Genes to Genomes. New York: McGraw Hill. (Problems 5–5, 7–84, and 8–121)

Hartwell LH & Weinert TA (1989) Checkpoints: controls that ensure the order of cell cycle events. *Science* 246, 629–634. (Problem 17–141)

Heald R & McKeon F (1990) Mutations of phosphorylation sites in lamin A that prevent nuclear lamina disassembly in mitosis. *Cell* 61, 579–589. (Problem 16–39)

Hedjran F, Yeakley JM, Huh GS, Hynes RO & Rosenfeld MG (1997) Control of alternative pre-mRNA splicing by distributed repeats. *Proc. Natl Acad. Sci. U.S.A.* 94, 12343–12347. (Problem 7–112)

Heim R, Prasher DC & Tsien RY (1994) Wavelength mutations and posttranslational autoxidation of green fluorescent protein. *Proc. Natl Acad. Sci. U.S.A.* 91, 12501–12504. (Problem 9–25)

Helms JB & Rothman JE (1992) Inhibition by brefeldin A of a Golgi membrane enzyme that catalyzes exchange of guanine nucleotide bound to ARF. *Nature* 360, 352–354. (Problem 13–29)

Hentze MW & Kühn LC (1996) Molecular control of vertebrate iron metabolism: mRNA-based regulatory circuits operated by iron, nitric oxide, and oxidative stress. *Proc. Natl Acad. Sci. U.S.A.* 93, 8175–8182. (Problems 7–107 and 7–108)

Herrmann H & Aebi U (2000) Intermediate filaments and their associates: multi-talented structural elements specifying cytoarchitecture and cytodynamics. *Curr. Opin. Cell Biol.* 12, 79–90. (Problem 16–26)

Hershey AD & Chase M (1952) Independent functions of viral protein and nucleic acid in growth of bacteriophage. *J. Gen. Physiol.* 36, 39–56. (Problem 4–17)

Hertel C, Muller P, Portenier M & Staehelin M (1983) Determination of the desensitization of β-adrenergic receptors by [$^3$H]CGP-12177. *Biochem. J.* 216, 669–674. (Problem 15–51)

Higgs HN & Pollard TD (2001) Regulation of actin filament network formation through ARP2/3 complex: Activation by a diverse array of proteins. *Annu. Rev. Biochem.* 70, 649–676. (Problem 16–64)

Higuchi R (1990) Recombinant PCR. In PCR Protocols (MA Innis, DH Gelfand, JJ Sninsky & TJ White, eds), pp 177–183. San Diego, CA: Academic Press. (Problem 8–84)

Hille B (1992) Ionic Channels of Excitable Membranes, 2nd ed. Sunderland, MA: Sinauer. (Problems 11–84, 11–86, and 11–91)

Holloway SL, Glotzer M, King RW & Murray AW (1993) Anaphase is initiated by proteolysis rather than by the inactivation of maturation-promoting factor. *Cell* 73, 1393–1402. (Problem 17–100)

Honegger AM, Kris RM, Ullrich A & Schlessinger J (1989) Evidence that autophosphorylation of solubilized receptors for epidermal growth factor is mediated by intermolecular cross-phosphorylation. *Proc. Natl Acad. Sci. U.S.A.* 86, 925–929. (Problem 15–126)

Honegger AM, Schmidt A, Ullrich A & Schlessinger J (1990) Evidence for epidermal growth factor (EGF)-induced intermolecular autophosphorylation of the EGF receptors in living cells. *Mol. Cell. Biol.* 10, 4035–4044. (Problem 15–126)

Horio T & Hotani H (1986) Visualization of the dynamic instability of individual microtubules by dark-field microscopy. *Nature* 321, 605–607. (Problems 16–33 and 16–55)

Horowitz S & Gorovsky MA (1985) An unusual genetic code in nuclear genes of *Tetrahymena*. *Proc. Natl Acad. Sci. U.S.A.* 82, 2452–2455. (Problem 6–93)

Howard J (2001) Mechanics of Motor Proteins and the Cytoskeleton, pp 151–163. Sunderland, MA: Sinauer Associates, Inc. (Problem 2–96)

Huang C-Y F & Ferrell JE (1996) Ultrasensitivity in the mitogen-activated protein kinase cascade. *Proc. Natl Acad. Sci. U.S.A.* 93, 10078–10083. (Problem 15–128)

Huang S, Ratliff KS, Schwartz MP, Spenner JM & Matouschek A (1999) Mitochondria unfold precursor proteins by unraveling them from their N-termini. *Nat. Struct. Biol.* 6, 1132–1138. (Problems 12–84 and 12–85)

Huberman JA & Riggs AD (1968) On the mechanism of DNA replication in mammalian chromosomes. *J. Mol. Biol.* 32, 327–341. (Problem 5–58)

Huganir RL, Delcour AH, Greengard P & Hess GP (1986) Phosphorylation of the nicotine acetycholine receptor regulates its rate of desensitization. *Nature* 321, 774–776. (Problem 15–52)

Hughes PE, Diaz-Gonzalez F, Leong L, Wu C, McDonald JA, Shattil SJ & Ginsberg MH (1996) Breaking the integrin hinge: A defined structural constraint regulates integrin signaling. *J. Biol. Chem.* 271, 6571–6574. (Problem 19–65)

Hume AN, Tarafder AK, Ramalho JS, Sviderskaya EV & Seabra MC (2006) A coiled-coil domain of melanophilin is essential for myosin Va recruitment and melanosome transport in melanocytes. *Mol. Biol. Cell* 17, 4720–4735. (Problem 13–75)

Hunt T, Luca FC & Ruderman JV (1992) The requirement for protein synthesis and degradation, and the control of destruction of cyclins A and B in the meiotic and mitotic cell cycles of the clam embryo. *J. Cell Biol.* 116, 707–724. (Problem 17–35)

Hyman AA & Karsenti E (1996) Morphogenetic properties of microtubules and mitotic spindle assembly. *Cell* 84, 401–410. (Problem 17–72)

Im D-S, Heise CE, Nguyen T, O'Dowd BF & Lynch KR (2001) Identification of a molecular target of psychosine and its role in globoid cell formation. *J. Cell Biol.* 153, 429–434. (Problem 17–117)

Ingledew JW (1982) *Thiobacillus ferrooxidans*: the bioenergetics of an acidophilic chemolithotroph. *Biochim. Biophys. Acta* 683, 89–117. (Problems 14–26 and 14–57)

Inman RB & Schnos M (1971) Structure of branch points in replicating DNA: presence of single-stranded connections in lambda DNA branch points. *J. Mol. Biol.* 56, 319–325. (Problem 5–39)

Jacks T & Varmus H (1985) Expression of the Rous sarcoma virus *pol* gene by ribosomal frameshifting. *Science* 230, 1237–1242. (Problem 6–91)

Jacobson MD, Burne JF, King MP, Miyashita T, Reed JC & Raff MC (1993) Bcl-2 blocks apoptosis in cells lacking mitochondrial DNA. *Nature* 361, 365–369. (Problem 18–17)

Jady B & Kiss T (2001) A small nucleolar guide RNA functions both in 2′-*O*-ribose methylation and pseudouridylation of the U5 spliceosomal RNA. *EMBO J.* 20, 541–551. (Problems 6–48 and 6–49)

Jagendorf AT & Uribe E (1966) ATP formation caused by acid–base transition of spinach chloroplasts. *Proc. Natl Acad. Sci. U.S.A.* 55, 170–177. (Problem 14–97)

Jain R, Rivera MC & Lake JA (1999) Horizontal gene transfer among genomes: The complexity hypothesis. *Proc. Natl Acad. Sci. U.S.A.* 96, 3801–3806. (Problem 1–37)

Jensen RA, Thompson ME, Jetton TL, Szabo CI, van der Meer R, Helou B, Tronick SR, Page DL, King MC & Holt JT (1996) BRCA1 is secreted and exhibits properties of a granin. *Nat. Genet.* 12, 303–308. (Problem 3–100)

Jia Z & Davies PL (2002) Antifreeze proteins: an unusual receptor–ligand interaction. *Trends Biochem. Sci.* 27, 101–106. (Problem 3–64)

Johnson KA (1993) Conformational coupling in DNA polymerase fidelity. *Annu. Rev. Biochem.* 62, 685–713. (Problem 5–34)

Johnson PJ, Kooter JM & Borst P (1987) Inactivation of transcription by UV irradiation of *T. brucei* provides evidence for a multicistronic transcription unit including a VSG gene. *Cell* 51, 273–281. (Problem 6–34)

Johnson RT & Rao PN (1971) Nucleo-cytoplasmic interactions in the achievement of nuclear synchrony in DNA synthesis and mitosis in multinucleate cells *Biol. Rev. Camb. Philos. Soc.* 46, 97–155. (Problem 17–48)

Jones KA (1997) Taking a new TAK on Tat transactivation. *Genes Dev.* 11, 2593–2599. (Problem 7–104)

Jones SN, Roe AE, Donehower LA & Bradley A (1995) Rescue of embryonic lethality in Mdm2-deficient mice by absence of p53. *Nature* 378, 206–208. (Problem 20–55)

Kabata H, Kurosawa O, Arai I, Washizu M, Margarson SA, Glass RE & Shimamoto N (1993) Visualization of single molecules of RNA polymerase sliding along DNA. *Science* 262, 1561–1563. (Problem 7–32)

Kantheti P, Qiao X, Diaz ME, Peden AA, Meyer GE, Carskadon SI, Kapfhamer D, Sufalko D, Robinson MS, Noebels JL & Burmeister M (1998) Mutation in AP-3 delta in the *mocha* mouse links endosomal transport to storage deficiency in platelets, melanosomes, and synaptic vesicles. *Neuron* 21, 111–122. (Problem 13–71)

Kaplan A, Achord DT & Sly WS (1977) Phosphohexosyl components of a lysosomal enzyme are recognized by pinocytosis receptors on human fibroblasts. *Proc. Natl Acad. Sci. U.S.A.* 74, 2026–2030. (Problem 13–73)

Karaiskou A, Jessus C, Brassac T & Ozon R (1999) Phosphatase 2A and Polo kinase, two antagonistic regulators of Cdc25 activation and MPF auto-amplification. *J. Cell Sci.* 112, 3747–3756. (Problem 17–34)

Karrer KM & Gall JG (1976) The macronuclear ribosomal DNA of *Tetrahymena pyriformis* is a palindrome. *J. Mol. Biol.* 104, 421–453. (Problem 8–92)

Karzai AW, Susskind MM & Sauer RT (1999) SmpB, a unique RNA-binding protein essential for the peptide-tagging activity of SsrA (tmRNA). *EMBO J.* 18, 3793–3799. (Problems 3–102 and 6–92)

Keilin D (1966) The History of Cell Respiration and Cytochrome. Cambridge, UK: Cambridge University Press. (Problem 14–59)

Kellems RE, Allison VF & Butow RA (1975) Cytoplasmic type 80S ribosomes associated with yeast mitochondria. IV. Attachment of ribosomes to the outer membrane of isolated mitochondria. *J. Cell Biol.* 65, 1–14. (Problem 12–74)

Kimura K & Hirano T (1997) ATP-dependent positive supercoiling of DNA by 13S condensin: A biochemical implication for chromosome condensation. *Cell* 90, 625–634. (Problem 4–87)

Kimura K, Rybenkov VV, Crisona NJ, Hirano T, & Cozzarelli NR (1999) 13S condensin actively reconfigures DNA by introducing global positive writhe: Implications for chromosome condensation. *Cell* 98, 239–248. (Problem 17–89)

Kinoshita N, Ghaedi K, Shimozawa N, Wanders RJA, Matsuzono Y, Imanaka T, Okumoto K, Suzuki Y, Kondo N & Fujiki Y (1998) Newly identified Chinese hamster ovary cell mutants are defective in biogenesis of peroxisomal membrane vesicles (peroxisome ghosts), representing a novel complementation group in mammals. *J. Biol. Chem.* 273, 24122–24130. (Problems 12–92 and 12–97)

Kirchhausen T (2000) Clathrin. *Annu. Rev. Biochem.* 69, 699–727. (Problems 13–19 and 13–20)

Kirschner MW & Schachman HK (1973) Local and gross conformational changes in aspartate transcarbamylase. *Biochemistry* 12, 2997–3004. (Problem 3–108)

Klemm RD, Austin RJ & Bell SP (1997) Coordinate binding of ATP and origin DNA regulates the ATPase activity of the origin recognition complex. *Cell* 88, 493–502. (Problem 5–63)

Knoop F (1905) Der Abbau aromatischer Fettsäuren im Tierkörper. *Beitr. Chem. Physiol.* 6, 150–162. (Problem 2–125)

Knudson AG (1971) Mutation and cancer: statistical study of retinoblastoma. *Proc. Natl Acad. Sci. U.S.A.* 68, 820–823. (Problem 20–43)

Knudson AG (2001) Two genetic hits (more or less) to cancer. *Nat. Rev. Cancer* 1, 157–162. (Problem 20–16)

Kobayashi T, Rein T & DePamphilis ML (1998) Identification of primary initiation sites for DNA replication in the hamster dihydrofolate reductase gene initiation zone. *Mol. Cell. Biol.* 18, 3266–3277. (Problem 5–64)

Koenig JH & Ikeda K (1999) Contribution of active zone subpopulation of vesicles to evoked and spontaneous release. *J. Neurophysiol.* 81, 1495–1505. (Problem 13–113)

Koleske AJ, Buratowski S, Nonet M & Young RA (1992) A novel transcription factor reveals a functional link between the RNA polymerase II CTD and TFIID. *Cell* 69, 883–894. (Problem 6–41)

Kondor-Koch C, Bravo R, Fuller SD, Cutler D & Garoff H (1985) Protein secretion in the polarized epithelial cell line MDCK. *Cell* 43, 297–306. (Problem 13–117)

Konkel DA, Maizel JV & Leder P (1979) The evolution and sequence comparison of two recently diverged mouse chromosomal β-globin genes. *Cell* 18, 865–873. (Problem 4–99)

Krause M & Hirsh D (1987) A trans-spliced leader sequence on actin mRNA in *C. elegans*. *Cell* 49, 753–761. (Problem 6–45)

Krebs HA & Johnson WA (1973) The role of citric acid in intermediate metabolism in animal tissues. *Enzymologia* 4, 148–156. (Problem 2–126)

Krings M, Stone A, Schmitz RW, Krainitzki H, Stoneking M & Pääbo S (1997) Neanderthal DNA sequences and the origin of modern humans. *Cell* 90, 19–30. (Problem 4–100)

Kristic RV (1997) Ultrastructure of the Mammalian Cell, p 207. Berlin, Germany: Springer-Verlag. (Problem 11–40)

Kudo N, Matsumori N, Taoka H, Fujiwara D, Schreiner EP, Wolff B, Yoshida M & Horinouchi S (1999) Leptomycin B inactivates CRM1/exportin 1 by covalent modification at a cysteine residue in the central conserved region. *Proc. Natl Acad. Sci. U.S.A.* 96, 9112–9117. (Problems 12–58 and 12–59)

Kuhne T, Wieringa B, Reiser J & Weissmann C (1983) Evidence against a scanning model for RNA splicing. *EMBO J.* 2, 727–733. (Problem 6–30)

Kuida K, Haydar TF, Kuan C-Y, Gu Y, Taya C, Karasuyama H, Su MS-S, Rakic P & Flavell RA (1998) Reduced apoptosis and cytochrome *c*-mediated caspase activation in mice lacking caspase 9. *Cell* 94, 325–337. (Problem 18–20)

Kumagai A & Dunphy WG (1992) Regulation of the Cdc25 protein during the cell cycle in *Xenopus* extracts. *Cell* 70, 139–151. (Problem 17–88)

Kurjan I (1992) Pheromone response in yeast. *Annu. Rev. Biochem.* 61, 1097–1129. (Problem 15–90)

Kyte J (1995) Mechanism in Protein Chemistry. New York: Garland Publishing. (Problems 3–73 and 3–101)

Lachner M, O'Carroll D, Rea S, Mechtier K & Jenuwein T (2001) Methylation of H3 lysine 9 creates a binding site for HP1 proteins. *Nature* 410, 116–120. (Problem 4–70)

Landegren U, Kaiser R, Sanders J & Hood L (1988) A ligase-mediated gene detection technique. *Science* 241, 1077–1080. (Problem 8–94)

Lander ES et al. (2001) Initial sequencing and analysis of the human genome. *Nature* 409, 860–921. (Problems 4–35, 4–45, 4–51, 4–52, 4–96, and 6–69)

Larschan E & Winston F (2001) The *S. cerevisiae* SAGA complex functions *in vivo* as a coactivator for transcriptional activation by Gal4. *Genes Dev.* 15, 1946–1956. (Problem 7–66)

Lawlor DW (1987) Photosynthesis: Metabolism, Control, and Physiology. New York: Wiley. (Problem 14–93)

LeBowitz JH & McMacken R (1986) The *Escherichia coli* DnaB replication protein is a DNA helicase. *J. Biol. Chem.* 261, 4738–4748. (Problem 5–43)

Lee JT& Lu N (1999) Targeted mutagenesis of *Tsix* leads to nonrandom X inactivation. *Cell* 99, 47–57. (Problem 7–95)

Lee M-H & Schedl T (2001) Identification of *in vivo* mRNA targets of GLD-1, a maxi-KH motif containing protein required for *C. elegans* germ cell development. *Genes Dev.* 15, 2408–2420. (Problem 7–118)

Leff SE, Evans RM & Rosenfeld MG (1987) Splice commitment dictates neuron-specific alternative RNA processing in calcitonin/CGRP gene expression. *Cell* 48, 517–524. (Problem 7–112)

Lefkowitz RJ, Limbird LE, Mukherjee C & Caron MG (1976) The β-adrenergic receptor and adenylate cyclase. *Biochim. Biophys. Acta* 457, 1–39. (Problem 15–88)

Leguy R, Melki R, Pantaloni D & Carlier M-F (2000) Monomeric γ-tubulin nucleates microtubules. *J. Biol. Chem.* 275, 21975–21980. (Problems 16–50, 16–60, and 16–62)

Li K, Li Y, Shelton JM, Richardson JA, Spencer E, Chen ZJ, Wang X & Williams RS (2000) Cytochrome *c* deficiency causes embryonic lethality and attenuates stress-induced apoptosis. *Cell* 101, 389–399. (Problem 18–12)

Li L, Zhou J, James G, Heller-Harrison R, Czech MP & Olsen EN (1992) FGF inactivates myogenic helix–loop–helix proteins through phosphorylation of a conserved protein kinase C site in their DNA-binding domains. *Cell* 71, 1181–1194. (Problem 7–29)

Li R & Murray AW (1991) Feedback control of mitosis in budding yeast. *Cell* 66, 519–531. (Problem 17–80)

Li WH (1997) Molecular Evolution. Sinauer Associates, Inc.: Sunderland MA. (Problems 1–46 and 1–49)

Lieber MR, Ma Y, Pannicke U & Schwarz K (2003) Mechanism and regulation of human non-homologous DNA end-joining. *Nat. Rev. Mol. Cell Biol.* 4, 712–720. (Problems 5–77 and 5–78)

Liebold EA & Munro HN (1988) Cytoplasmic protein binds *in vitro* to a highly conserved sequence in the 5′ untranslated region of ferritin heavy- and light-subunit mRNAs. *Proc. Natl Acad. Sci. U.S.A.* 85, 2171–2175. (Problem 7–111)

Lin FC, Brown RM, Drake RR & Haley BE (1990) Identification of the uridine 5′-diphosphoglucose (UDP-Glc) binding subunit of cellulose synthase in *Acetobacter xylinum* using the photoaffinity probe 5-azido-UDP-Glc. *J. Biol. Chem.* 265, 4782–4784. (Problem 19–100)

Lindahl T, Demple B & Robins P (1982) Suicide inactivation of the *E. coli* $O^6$-methylguanine-DNA methyltransferase. *EMBO J.* 1, 1359–1363. (Problem 5–84)

Lipmann F (1941) Metabolic generation and utilization of phosphate bond energy. *Adv. Enzymol.* 1, 99–162. (Problem 2–122)

Little CC (1933) The existence of non-chromosomal influence in the incidence of mammary tumors in mice. *Science* 78, 465–466. (Problem 20–29)

Liu LF & Wang JC (1987) Supercoiling of the DNA template during transcription. *Proc. Natl Acad. Sci. U.S.A.* 84, 7024–7027. (Problem 6–22)

Liu X, Kim CN, Yang J, Jemmerson R & Wang X (1996) Induction of apoptotic program in cell-free extracts: Requirement for dATP and cytochrome *c*. *Cell* 86, 147–157. (Problem 18–18)

Lo CW & Gilula NB (1979) Gap junctional communication in the preimplantation mouse embryo. *Cell* 18, 399–409. (Problem 19–44)

Lodish HF, Kong N, Snider M & Strous GJAM (1983) Hepatoma secretory proteins migrate from rough endoplasmic reticulum to Golgi at characteristic rates. *Nature* 304, 80–83. (Problem 13–114)

Logothetis DE, Kurachi Y, Galper J, Neer EJ & Clapham DE (1987) The βγ subunits of GTP-binding proteins activate the muscarinic $K^+$ channel in heart. *Nature* 325, 321–326. (Problem 15–97)

Losada A & Hirano T (2001) Intermolecular DNA interactions stimulated by the cohesin complex *in vitro*: Implications for sister chromatid cohesion. *Curr. Biol.* 11, 268–272. (Problem 17–89)

Lucas WJ (2006) Plant viral movement proteins: agents for cell-to-cell trafficking of viral genomes. *Virology* 344, 169–184. (Problem 19–45)

Luck DJL (1984) Genetic and biochemical dissection of the eucaryotic flagellum. *J. Cell Biol.* 98, 789–794. (Problem 16–108)

Luria SE & Delbrück M (1943) Mutations of bacteria from virus sensitivity to virus resistance. *Genetics* 28, 491–511. (Problem 5–10)

Maarse AC, Blom J, Grivell LA & Meijer M (1992) MPI1, an essential gene encoding a mitochondrial membrane protein, is possibly involved in protein import into yeast mitochondria. *EMBO J.* 11, 3619–3628. (Problem 12–77)

MacKinnon R, Aldrich RW & Lee AW (1993) Functional stoichiometry of *shaker* potassium channel inactivation. *Science* 262, 757–759. (Problem 11–89)

MacLean-Fletcher S & Pollard TD (1980) Mechanism of action of cytochalasin B on actin. *Cell* 20, 329–341. (Problem 16–40)

MacNicol A, Muslin AJ & Williams LT (1993) Raf-1 kinase is essential for early *Xenopus* development and mediates the induction of mesoderm by FGF. *Cell* 73, 571–583. (Problem 8–124)

Madhuri HD & Fink GR (1998) The control of filamentous differentiation and virulence in fungi. *Trends Cell Biol.* 8, 348–353. (Problem 15–162)

Madison-Antenucci S, Grams J & Hajduk SL (2002) Editing machines: the complexities of *Trypanosome* RNA editing. *Cell* 108, 435–438. (Problem 7–106)

Maizia D, Harris PJ & Bibring T (1960) The multiplicity of mitotic centers and the time-course of their duplication and separation. *J. Biophys. Biochem. Cytol.* 7, 1–20. (Problem 17–91)

Mallavarapu A & Mitchison TJ (1999) Regulated actin cytoskeleton assembly at filopodium tips controls their extension and retraction. *J. Cell Biol.* 146, 1097–1106. (Problem 16–111)

Mann C & Davis RW (1983) Instability of dicentric plasmids in yeast. *Proc. Natl Acad. Sci. U.S.A.* 80, 228–232. (Problem 17–93)

Manoil C & Beckwith J (1986) A genetic approach to analyzing membrane protein topology. *Science* 233, 1403–1408. (Problem 12–132)

Manson MD, Blank V, Brade G & Higgins CF (1986) Peptide chemotaxis in *E. coli* involves the Tap signal transducer and the dipeptide permease. *Nature* 321, 253–258. (Problem 15–132)

Manson MD, Tedesco P, Berg HC, Harold FM & van der Drift C (1977) A protonmotive force drives bacteria flagella. *Proc. Natl Acad. Sci. U.S.A.* 74, 3060–3064. (Problem 14–64)

Marks B, Stowell MHB, Vallis Y, Mills IG, Gibson A, Hopkins CR & McMahon HT (2001) GTPase activity of dynamin and resulting conformation changes are essential for endocytosis. *Nature* 410, 231–235. (Problem 13–22)

Masaike T, Mitome N, Noji H, Muneyuki E, Yasuda R, Kinosita K & Yoshida M (2000) Rotation of $F_1$-ATPase and the hinge residues of the β subunit. *J. Exp. Biol.* 203, 1–8. (Problem 14–31)

Masutani C, Araki M, Yamada A, Kusumoto R, Nogimori T, Maekawa T, Iwai S & Hanaoka F (1999) Xeroderma pigmentosum variant (XP-V) correcting protein from HeLa cells has a thymine dimer bypass DNA polymerase activity. *EMBO J.* 18, 3491–3501. (Problem 5–82)

Matteoli M, Takei K, Perin MS, Südhof TC & DeCamilli P (1992) Exo-endocytotic recycling of synaptic vesicles in developing processes of cultured hippocampal neurons. *J. Cell Biol.* 117, 849–861. (Problem 13–118)

McClintock B (1939) The behavior of successive nuclear divisions of a chromosome broken at meiosis. *Proc. Natl Acad. Sci. U.S.A.* 25, 405–416. (Problem 20–58)

McKeon FD, Kirschner MW & Caput D (1986) Homologies in both primary and secondary structure between nuclear envelope and intermediate filament proteins. *Nature* 31, 463–468. (Problem 3–34)

McKnight SL & Kingsbury R (1982) Transcriptional control signals of a eukaryotic protein-coding gene. *Science* 217, 316–324. (Problem 6–38)

Meikrantz W, Bergom MA, Memisoglu A & Samson L (1998) $O^6$-Alkylguanine DNA lesions trigger apoptosis. *Carcinogenesis* 19, 369–372. (Problems 5–85 and 18–28)

Meselson M & Stahl FW (1958) The replication of DNA in *Escherichia coli*. *Proc. Natl Acad. Sci. U.S.A.* 44, 671–682. (Problem 8–23)

Meuse CW, Krueger S, Majkrzak CF, Dura JA, Fu J, Connor JT & Plant AL (1998) Hybrid bilayer membranes in air and water: infrared spectroscopy and neutron reflectivity studies. *Biophys. J.* 74, 1388–1398. (Problem 13–28)

Meyer CA, Jacobs HW, Datar SA, Du W, Edgar BA & Lehner CF (2000) *Drosophila* Cdk4 is required for normal growth and is dispensable for cell cycle progression. *EMBO J.* 19, 4533–4542. (Problems 17–146 and 17–147)

Michaelis C, Ciosk R & Nasmyth K (1997) Cohesins: chromosomal proteins that prevent premature separation of sister chromatids. *Cell* 91, 35–45. (Problem 17–49)

Miller JH (1985) Mutagenic specificity of ultraviolet light. *J. Mol. Biol.* 182, 45–65. (Problem 5–81)

Milton RC, Milton SC & Kent SB (1992) Total chemical synthesis of a D-enzyme: the enantiomers of HIV-1 protease show reciprocal chiral substrate specificity. *Science* 256, 1445–1448. (Problem 3–72)

Minakami S, Suzuki C, Saito T & Yoshikawa H (1965) Studies on erythrocyte glycolysis. I. Determination of the glycolytic intermediates in human erythrocytes. *J. Biochem.* (*Tokyo*) 58, 543–550. (Problem 2–97)

Mitchell MB & Mitchell HK (1952) A case of "maternal" inheritance in *Neurospora crassa*. *Proc. Natl Acad. Sci. U.S.A.* 38, 442–449. (Problem 14–115)

Mitchell MB, Mitchell HK & Tissieres A (1953) Mendelian and non-Mendelian factors affecting the cytochrome system in *Neurospora crassa*. *Proc. Natl Acad. Sci. U.S.A.* 39, 605–613. (Problem 14–115)

Mitchison TJ (1993) Localization of an exchangeable GTP binding site at the plus end of microtubules. *Science* 261, 1044–1047. (Problem 16–32)

Mitchison TJ (2001) Psychosine, cytokinesis, and orphan receptors: Unexpected connections. *J. Cell Biol.* 153, F1–F3. (Problem 17–117)

Mitchison TJ & Kirschner MW (1984) Microtubule assembly nucleated by isolated centrosomes. *Nature* 312, 232–237. (Problem 16–59)

Mitchison TJ & Kirschner MW (1985) Properties of the kinetochore *in vitro*. I. Microtubule nucleation and tubulin binding. *J. Cell Biol.* 101, 755–765. (Problem 16–61)

Mitchison TJ & Kirschner MW (1985) Properties of the kinetochore *in vitro*. II. Microtubule capture and ATP-dependent translocation. *J. Cell. Biol.* 101, 766–777. (Problem 17–92)

Miyawaki A, Llopis J, Heim R, McCafferty JM, Adams JA, Ikura M & Tsien RY (1997) Fluorescent indicators for $Ca^{2+}$ based on green fluorescent protein and calmodulin. *Nature* 388, 882–887. (Problem 9–34)

Mlynarczyk SK & Panning B (2000) X inactivation: *Tsix* and *Xist* as yin and yang. *Curr. Biol.* 10, R899–R903. (Problem 7–96)

Molecular Probes Handbook, 1999 (www.probes.com/handbook). (Problem 11–42)

Monod J (1947) The phenomenon of enzymatic adaptation. Growth Symposium XI:223–289. [Reprinted in Selected Papers in Molecular Biology by Jacques Monod (A Lwoff, A Ullmann, eds), pp 68–134. New York: Academic Press, 1947.] (Problem 7–60)

Montoya J, Ojala D & Attardi G (1981) Distinctive features of the 5′-terminal sequences of the human mitochondrial mRNAs. *Nature* 290, 465–470. (Problem 14–114)

Moore MS & Blobel G (1993) The GTP-binding protein Ran/TC4 is required for protein import into the nucleus. *Nature* 365, 661–663. (Problem 12–54)

Moore MS & Blobel G (1994) Purification of a Ran-interacting protein that is required for protein import into the nucleus. *Proc. Natl Acad. Sci. U.S.A.* 91, 10212–10216. (Problem 12–54)

Moore T & Haig D (1991) Genomic imprinting in mammalian development: a parental tug of war. *Trends Genet.* 7, 45–49. (Problem 7–84)

Morand OH, Allen LA, Zoeller RA & Raetz CR (1990) A rapid selection for animal cell mutants with defective peroxisomes. *Biochim. Biophys. Acta* 1034, 132–141. (Problem 12–93)

Moriyoshi K, Masu M, Ishii T, Shigemoto R, Mizuno N & Nakanishi S (1991) Molecular cloning and characterization of the rat NMDA receptor. *Nature* 354, 31–37. (Problem 11–93)

Mowry KL & Steitz JA (1987) Identification of human U7 snRNP as one of several factors involved in the 3′-end maturation of histone premessenger RNAs. *Science* 238, 1682–1687. (Problem 6–47)

Mueller-Storm HP, Sogo JM & Schaffner W (1989) An enhancer stimulates transcription in trans when attached to the promoter via a protein bridge. *Cell* 58, 767–777. (Problem 7–52)

Munro S & Pelham HR (1987) A C-terminal signal prevents secretion of luminal ER proteins. *Cell* 48, 899–907. (Problem 13–50)

Murray A & Hunt T (1993) The Cell Cycle: An Introduction, pp 143–144. New York: WH Freeman. (Problem 17–142)

Murray A & Szostak JW (1982) Pedigree analysis of plasmid segregation in yeast. *Cell* 34, 961–970. (Problem 17–94)

Murray EJ & Grosveld F (1987) Site-specific demethylation in the promoter of human γ-globin gene does not alleviate methylation mediated suppression. *EMBO J.* 6, 2329–2335. (Problem 7–91)

Nachury MV & Weis K (1999) The direction of transport through the nuclear pore can be inverted. *Proc. Natl Acad. Sci. U.S.A.* 96, 9622–9627. (Problems 12–48, 12–49, and 12–50)

Nagafuchi A, Shirayoshi Y, Okazaki K, Yasuda K & Takeichi M (1987) Transformation of cell adhesion properties by exogenously introduced E-cadherin cDNA. *Nature* 329, 341–343. (Problem 19–14)

Nagai T, Yamada S, Tominaga T, Ichikkawa M & Miyawaki A (2004) Expanded dynamic range of fluorescent indicators for $Ca^{2+}$ by circularly permuted yellow fluorescent proteins. *Proc. Natl Acad. Sci U.S.A.* 101 10554–10559. (Problem 9–34)

Nagata K & Handa H (2000) Real-Time Analysis of Biomolecular Interactions: Applications of BIACORE. Tokyo, Japan: Springer. (Problem 8–51)

Nagata Y, Muro Y & Todokoro K (1997) Thrombopoietin-induced polyploidization of bone marrow megakaryocytes is due to a unique regulatory mechanism in late mitosis. *J. Cell Biol.* 139, 449–457. (Problem 17–118)

Nakamura TM, Morin GB, Chapman KB, Weinrich SL, Andrews WH, Lingner J, Harley CB & Cech TR (1997) Telomerase catalytic subunit homologs from fission yeast and human. *Science* 277, 955–959. (Problem 5–66)

Nathans J, Piantanida TP, Eddy RL, Shows TB & Hogness DS (1986) Molecular genetics of inherited variation in human color vision. *Science* 232, 203–210. (Problem 4–52)

Nathans J, Thomas D & Hogness DS (1986) Molecular genetics of human color vision: the genes encoding blue, green, and red pigments. *Science* 232, 193–202. (Problem 4–52)

Neupert W (1997) Protein import into mitochondria. *Annu. Rev. Biochem.* 66, 863–917. (Problem 12–83)

Newman M, Lunnen K, Wilson G, Greci J, Schildkraut I & Phillips SEV (1998) Crystal structure of restriction endonuclease BglI bound to its interrupted DNA recognition sequence. *EMBO J.* 17, 5466–5476. (Problem 8–70)

Nicholls DG & Ferguson SJ (1992) Bioenergetics 2. London: Academic Press. (Problems 14–34, 14–35, 14–61, and 14–63)

Nichols BJ, Undermann C, Pelham HRB, Wickner WT & Haas A (1997) Homotypic vacuolar fusion mediated by t- and v-SNAREs. *Nature* 387, 199–202. (Problems 13–31 and 13–48)

Ninfa AJ, Reitzer LJ & Magasanik B (1987) Initiation of transcription at the bacterial *glnAp2* promoter by purified *E. coli* components is facilitated by enhancers. *Cell* 50, 1039–1046. (Problem 7–61)

Nishizuka Y (1983) Calcium, phospholipid turnover and transmembrane signaling. *Phil. Trans. R. Soc. Lond. Biol.* 302, 101–112. (Problem 15–95)

Nogales E, Whittaker M, Milligan RA & Downing KH (1999) High-resolution model of the microtubule. *Cell* 96, 79–88. (Problem 16–32)

Noji H, Yasuda R, Yoshida M & Kinosita K (1997) Direct observation of the rotation of $F_1$-ATPase. *Nature* 386, 299–302. (Problem 14–32)

Norby JG (2000) The origin and the meaning of the little p in pH. *Trends Biochem. Sci.* 25, [illegible]. (Problem [illegible])

Norman C, Runswick M, Pollock R & Treisman R (1988) Isolation and properties of cDNA clones encoding SRF, a transcription factor that binds to the *c-fos* serum response element. *Cell* 55, 989–1003. (Problem 7–35)

Nudler E, Mustaev A, Lukhtanov E & Goldfarb A (1997) The RNA–DNA hybrid maintains the register of transcription by preventing backtracking of RNA polymerase. *Cell* 89, 33–41. (Problem 6–42)

Nugent JM & Palmer JD (1991) RNA-mediated transfer of the gene *coxII* from the mitochondrion to the nucleus during flowering plant evolution. *Cell* 66, 473–481. (Problem 1–48)

Nurse P (1975) Genetic control of cell size at cell division in yeast. *Nature* 256, 547–551. (Problems 17–38 and 17–145)

Nurse P & Thuriaux P (1980) Regulatory genes controlling mitosis in the fission yeast *Schizosaccharomyces pombe*. *Genetics* 96, 627–637. (Problem 17–38)

Nusse R & Varmus HE (1982) Many tumors induced by the mouse mammary tumor virus contain a provirus integrated in the same region of the host genome. *Cell* 31, 99–109. (Problem 20–54)

O'Connell KF, Caron C, Kopish KR, Hurd DD, Kemphues KJ, Li Y & White JG (2001) The *C. elegans zyg-1* gene encodes a regulator of centrosome duplication with distinct maternal and paternal roles in the embryo. *Cell* 105, 547–558. (Problem 17–90)

O'Keefe EJ & Pledger WJ (1983) A model of cell-cycle control: sequential events regulated by growth factors. *Mol. Cell. Endocrinol.* 31, 167–186. (Problem 17–138)

O'Toole TE, Mandelman D, Forsyth J, Shattil SJ, Plow EF & Ginsberg MH (1991) Modulation of the affinity of integrin $\alpha_{IIb}\beta_3$ (GPIIb-IIIa) by the cytoplasmic domain of $\alpha_{IIb}$. *Science* 254, 845–847. (Problem 19–67)

Oeller PW, Min-Wong L, Taylor LP, Pike DA, Theologis A (1991) Reversible inhibition of tomato fruit senescence by antisense RNA. *Science* 254, 437–439. (Problem 15–164)

Ojala D, Montoya J & Attardi G (1981) tRNA punctuation model of RNA processing in human mitochondria. *Nature* 290, 470–474. (Problem 14–114)

Oka Y & Czech MP (1984) Photoaffinity labeling of insulin-sensitive hexose transporters in intact rat adipocytes: direct evidence that latent transporters become exposed to the extracellular space in response to insulin. *J. Biol. Chem.* 259, 8125–8133. (Problem 11–18)

Oosawa F (2001) A historical perspective of actin assembly and its interactions. *Res. Prob. Cell Diff.* 32, 9–21. (Problem 16–31)

Orci L, Glick BS & Rothman JE (1986) A new type of coated vesicular carrier that appears not to contain clathrin: its possible role in protein transport within the Golgi stack. *Cell* 46, 171–184. (Problem 13–115)

Orkin RW, Gehron P, McGoodwin EB, Martin GR, Valentine T & Swarm R (1977) A murine tumor producing a matrix of basement membrane. *J. Exp. Med.* 145, 204–220. (Problem 19–58)

Orr-Weaver TL & Spradling AD (1986) *Drosophila* chorion gene amplification requires an upstream region regulating s18 transcription. *Mol. Cell. Biol.* 6, 4624–4633. (Problem 5–62)

Osinga KA, Swinkels BW, Gibson WC, Borst P, Veeneman GH, Van Boom JH, Michels PAM & Opperdoes FR (1985) Topogenesis of microbody enzymes: sequence comparison of the genes for the glycosomal (microbody) and cytosolic phosphoglycerate kinases of *Trypanosoma brucei*. *EMBO J.* 4, 3811–3817. (Problem 12–95)

Otero LJ, Devaux A & Standart N (2001) A 250-nucleotide UA-rich element in the 3′ untranslated region of *Xenopus laevis* Vg1 mRNA represses translation both *in vivo* and *in vitro*. *RNA* 7, 1753–1767. (Problem 7–109)

Ottaviano Y & Gerace L (1985) Phosphorylation of the nuclear lamins during interphase and mitosis. *J. Biol. Chem.* 260, 624–632. (Problem 16–39)

Ovchinnikov IV, Gotherstrom A, Romanova GP, Kharitonov VM, Liden K & Goodwin W (2000) Molecular analysis of Neanderthal DNA from the northern Caucasus. *Nature* 404, 490–493. (Problem 4–100)

Ozawa M, Ringwald M & Kemler R (1990) Uvomorulin-catenin complex formation is regulated by a specific domain in the cytoplasmic region of the cell adhesion molecule. *Proc. Natl Acad. Sci. U.S.A.* 87, 4246–4250. (Problem 19–20)

Pace NR (2001) The universal nature of biochemistry. *Proc. Natl Acad. Sci. U.S.A.* 98, 805–808. (Problem 1–11)

Paigen K (2003) One hundred years of mouse genetics: An intellectual history. I. The classical period (1902–1980). *Genetics* 163, 1–7. (Problem 20–29)

Park H-O, Chant J & Herskowitz I (1993) *BUD2* encodes a GTPase-activating protein for Bud1/Rsr1 necessary for proper bud-site selection in yeast. *Nature* 365, 269–274. (Problem 16–112)

Paschen SA, Rothbauer U, Kaldi K, Bauer MF, Neupert W & Brunner M (2000) The role of the TIM8-13 complex in the import of Tim23 into mitochondria. *EMBO J.* 19, 6392–6400. (Problem 12–86)

Patel LR, Curran T & Kerppola TK (1994) Energy transfer analysis of Fos–Jun dimerization and DNA binding. *Proc. Natl Acad. Sci. U.S.A.* 91, 7360–7364. (Problem 7–34)

Pauling L (1948) Chemical achievement and hope for the future. *Am. Sci.* 36, 50–58. (Problem 3–73)

Payvar F, DeFranco D, Firestone GL, Edgar B, Wrange O, Okret S, Gustafsson J-A & Yamomoto KR (1983) Sequence-specific binding of glucocorticoid receptor to MMTV DNA at sites within and upstream of the transcribed region. *Cell* 35, 381–392. (Problem 15–48)

Pearse A-M & Swift K (2006) Transmission of devil facial-tumour disease. *Nature* 439, 549. (Problem 20–28)

Pearse BMF, Smith CJ & Owen DJ (2000) Clathrin coat construction in endocytosis. *Curr. Opin. Struct. Biol.* 10, 220–228. (Problems 13–18 and13–20)

Pelham HRB & Rothman JE (2000) The debate about transport in the Golgi—two sides of the same coin? *Cell* 102, 713–719. (Problem 13–55)

Penny GD, Kay GF, Sheardown SA, Rastan S & Brockdorff N (1996) Requirement for *Xist* in X chromosome inactivation. *Nature* 379, 131–137. (Problem 7–94)

Perara E, Rothman RE & Lingappa VR (1986) Uncoupling translocation from translation: implications for transport of proteins across membranes. *Science* 232, 348–352. (Problem 12–130)

Peter M, Nakagawa J, Dorée M, Labbé JC & Nigg EA (1990) *In vitro* disassembly of the nuclear lamina and M phase-specific phosphorylation of lamins by cdc2 kinase. *Cell* 61, 591–602. (Problem 16–39)

Peto R, Darby S, Deo H, Silcocks P, Whitely E & Doll R (2000) Smoking, smoking cessation, and lung cancer in the UK since 1950: combination of national statistics with two case-control studies. *Br. Med. J.* 321, 323–329. (Problem 20–21)

Pettit J, Wood WB & Plasterk RHA (1996) *cdh-3*, a gene encoding a member of the cadherin superfamily, functions in epithelial cell morphogenesis in *Caenorhabditis elegans. Development* 122, 4149–4157. (Problem 19–16)

Pierschbacher MD & Ruoslahti E (1984) Cell attachment activity of fibronectin can be duplicated by small synthetic fragments of the molecule. *Nature* 309, 30–33. (Problem 19–86)

Pitti RM, Marsters SA, Lawrence DA, Roy M, Kischkel FC, Dowd P, Huang A, Donahue CJ, Sherwood SW, Baldwin DT, Godowski PJ, Wood WI, Gurney AL, Hillan KJ, Cohen RL, Goddard AD, Botstein D & Ashkenazi A (1998) Genomic amplification of a decoy receptor for Fas ligand in lung and colon cancer. *Nature* 396, 699–703. (Problem 18–16)

Pollard TD (1986) Rate constants for the reactions of ATP- and ADP-actin with the ends of actin filaments. *J. Cell Biol.* 103, 2747–2754. (Problem 16–34)

Pollard TD, Blanchoin L & Mullins RD (2000) Molecular mechanisms controlling actin filament dynamics in nonmuscle cells. *Annu. Rev. Biophys. Biomol. Struct.* 29, 545–576. (Problems 16–31 and 16–34)

Pollock R & Treisman R (1990) A sensitive method for the determination of protein–DNA binding specificities. *Nucleic Acids Res.* 18, 6197–6204. (Problems 7–30 and 7–36)

Ponticelli AS, Schultz DW, Taylor AF & Smith GR (1985) Chi-dependent DNA strand cleavage by RecBCD enzyme. *Cell* 41, 145–151. (Problem 5–100)

Porter JA, von Kessler DP, Ekker SC, Young KE, Lee JJ, Moses K & Beachy PA (1995) The product of *hedgehog* autoproteolytic cleavage active in local and long-range signaling. *Nature* 374, 363–366. (Problems 15–150, 15–151, and 15–152)

Potter H & Dressler D (1976) On the mechanism of genetic recombination: electron microscopic observation of recombination intermediates. *Proc. Natl Acad. Sci. U.S.A.* 73, 3000–3004. (Problem 5–95)

Powell LM, Wallis SC, Pease RJ, Edwards YH, Knott TJ & Scott J (1987) A novel form of tissue-specific RNA processing produces apolipoprotein-B48 in intestine. *Cell* 50, 831–840. (Problem 7–115)

Prescott DM (1975) Reproduction of Eucaryotic Cells, pp 85–86. New York: Academic Press. (Problem 17–10)

Prober DA & Edgar BA (2001) Growth regulation by oncogenes—new insights from model organisms. *Curr. Opin. Genet. Dev.* 11, 19–26. (Problem 17–143)

Prunell A, Kornberg RD, Lutter L, Klug A, Levitt M & Crick FH (1979) Periodicity of deoxyribonuclease I digestion of chromatin. *Science* 204, 855–858. (Problem 4–54)

Ptashne M & Gann A (2002) Genes & Signals, pp 75–76. Cold Spring Harbor, New York: Cold Spring Harbor Laboratory Press. (Problem 7–54)

Ptashne M (1986) A Genetic Switch: Gene Control and Phage λ, p 114. Oxford, UK: Blackwell Scientific Press. (Problem 7–25)

Purdue PE, Allsop J, Isaya G, Rosenberg LE & Danpure CJ (1991) Mistargeting of peroxisomal L-alanine:glyoxylate aminotransferase to mitochondria in primary hyperoxaluria patients depends upon activation of a cryptic mitochondrial targeting sequence caused by a point mutation. *Proc. Natl Acad. Sci. U.S.A.* 88, 10900–10904. (Problem 12–91)

Pytela R, Pierschbacher MD & Ruoslahti E (1985) Identification and isolation of a 140 kd cell surface glycoprotein with properties expected of a fibronectin receptor. *Cell* 40, 191–198. (Problem 19–86)

Quelle DE, Zindy F, Ashmun RA & Sherr CJ (1995) Alternative reading frames of the *INK4a* tumor suppressor gene encode two unrelated proteins capable of inducing cell cycle arrest. *Cell* 83, 993–1000. (Problem 20–56)

Racker E & Stoeckenius W (1974) Reconstitution of purple membrane vesicles catalyzing light-driven proton uptake and adenosine triphosphate formation. *J. Biol. Chem.* 249, 662–663. (Problem 14–21)

Rao PN & Johnson RT (1970) Mammalian cell fusion: I. Studies on the regulation of DNA synthesis and mitosis. *Nature* 225, 159–164. (Problem 17–14)

Rappaport R (1986) Establishment of the mechanism of cytokinesis in animal cells. *Int. Rev. Cytol.* 105, 245–281. (Problem 17–115)

Rash JE, Yasumura T, Hudson CS, Agre P & Nielsen S (1998) Direct immunogold labeling of aquaporin-4 in square arrays of astrocyte and ependymocyte plasma membranes in rat brain and spinal cord. *Proc. Natl Acad. Sci. U.S.A.* 95, 11981–11986. (Problem 9–48)

Rasko JEJ, Battini J-L, Kruglyak L, Cox DR & Miller AD (2000) Precise gene localization by phenotypic assay of radiation hybrid cells. *Proc. Natl Acad. Sci. U.S.A.* 97, 7388–7392. (Problem 8–12)

Ren R, Mayer BJ, Cicchetti P & Baltimore D (1993) Identification of a ten-amino-acid proline-rich SH3 binding site. *Science* 259, 1157–1161. (Problem 15–119)

Reuveny E, Slesinger PA, Inglese J, Morales JM, Iniguez-Lluhi JA, Lefkowitz RJ, Bourne HR, Jan YN & Jan LY (1994) Activation of the cloned muscarinic potassium channel by G protein βγ subunits. *Nature* 370, 143–146. (Problem 15–97)

Ribbeck K, Lipowsky G, Kent HM, Stewart M & Görlich D (1998) NTF2 mediates nuclear import of Ran. *EMBO J.* 17, 6587–6598. (Problem 12–55)

Richardson HE, Wittenberg C, Cross F & Reed SI (1989) An essential $G_1$ function for cyclin-like proteins in yeast. *Cell* 59, 1127–1133. (Problem 17–37)

Rief M, Gutel M, Oesterhelt F, Fernandez JM & Gaub HE (1997) Reversible folding of individual titin immunoglobulin domains by AFM. *Science* 276, 1109–1112. (Problem 3–43)

Roeder RG (1974) Multiple forms of deoxyribonucleic acid-dependent ribonucleic acid polymerase in *Xenopus laevis*. Isolation and partial characterization. *J. Biol. Chem.* 249, 241–248. (Problem 6–40)

Roger AJ, Svard SG, Tovar J, Clark CG, Smith MW, Gillin FD & Sogin ML (1998) A mitochondrial-like chaperonin 60 gene in *Giardia lamblia*: Evidence that diplomonads once harbored an endosymbiont related to the progenitor of mitochondria. *Proc. Natl Acad. Sci. U.S.A.* 95, 229–234. (Problem 1–45)

Rohatgi R, Ma L, Miki H, Lopez M, Kirchhausen T, Takenawa T & Kirschner MW (1999) The interaction between N-WASp and the Arp2/3 complex links Cdc42-dependent signals to actin assembly. *Cell* 97, 221–231. (Problems 16–113 and 16–114)

Roth DB, Porter TN & Wilson JH (1985) Mechanisms of nonhomologous recombination in mammalian cells. *Mol. Cell. Biol.* 5, 2599–2607. (Problem 5–86)

Rothman JE & Dawidowicz EA (1975) Asymmetric exchange of vesicle phospholipids catalyzed by the phosphatidylcholine exchange protein. Measurement of inside–outside transitions. *Biochemistry* 14, 2809–2816. (Problem 12–133)

Rothman JE, Miller RL & Urbani LJ (1984) Intercompartmental transport in the Golgi complex is a dissociative process: facile transfer of membrane protein between two Golgi populations. *J. Cell Biol.* 99, 260–271. (Problem 13–59)

Rousselet A, Guthmann C, Matricon J, Bienvenue A & Devaux PF (1976) Study of the transverse diffusion of spin labeled phospholipids in biological membranes: 1. Human red blood cells. *Biochim. Biophys. Acta* 426, 357–371. (Problem 10–28, Problem 10–32)

Royzman I, Austin RJ, Bosco G, Bell SP & Orr-Weaver TL (1999) ORC localization in *Drosophila* follicle cells and the effects of mutations in *dE2F* and *dDP*. *Genes Dev.* 13, 827–840. (Problem 5–62)

Ruffner DE, Sprung CN, Minghetti PP, Gibbs PEM & Dugaiczyk A (1987) Invasion of the human albumin-α-fetoprotein gene family by *Alu*, *Kpn*, and two novel repetitive DNA elements. *Mol. Biol. Evol.* 4, 1–9. (Problem 4–101)

Ruoslahti E & Pierschbacher MD (1986) Arg-Gly-Asp: a versatile cell recognition signal. *Cell* 44, 517–518. (Problem 19–86)

Russo AF & Koshland DE (1983) Separation of signal transduction and adaptation functions of the aspartate receptor in bacterial sensing. *Science* 220, 1016–1020. (Problem 15–133)

Sack RL, Brandes RW, Kendall AR & Lewy AJ (2000) Entrainment of free-running circadian rhythms by melatonin in blind people. *N. Engl. J. Med.* 343, 1114–1116. (Problem 7–80)

Sadowski HB, Shuai K, Darnell JE & Gilman MZ (1993) A common nuclear signal transduction pathway activated by growth factor and cytokine receptors. *Science* 261, 1739–1744. (Problem 15–131)

Safer B, Kemper W & Jagus R (1978) Identification of a 48S preinitiation complex in reticulocyte lysate. *J. Biol. Chem.* 253, 3384–3386. (Problem 6–80)

Sakmann B (1992) Elementary steps in synaptic transmission revealed by currents through single ion channels. *Science* 256, 503–512. (Problems 11–88 and 11–92)

Sanes JR (2003) The basement membrane/basal lamina of skeletal muscle. *J. Biol. Chem.* 278, 12601–12604. (Problems 19–51 and 19–54)

Sasaki H, Matsui C, Furuse K, Mimori-Kiyosue Y, Furuse M, & Tsukita S (2003) Dynamic behavior of paired claudin strands within apposing plasma membranes. *Proc. Natl Acad. Sci. U.S.A.* 100, 3971–3976. (Problem 19–35)

Sasaki T, Fassler R & Hohenester E (2004) Laminin: the crux of basement membrane assembly. *J. Cell Biol.* 164, 959–963. (Problem 19–56)

Satir P & Matsuoka T (1989) Splitting the ciliary axoneme: implications for a "switch-point" model of dynein arm activity in ciliary motion. *Cell Motil. Cytoskel.* 14, 345–358. (Problem 16–94)

Sawadogo M & Roeder RG (1985) Factors involved in specific transcription by human RNA polymerase II: analysis by a rapid and quantitative *in vitro* assay. *Proc. Natl Acad. Sci. U.S.A.* 82, 4394–4398. (Problems 6–35 and Problem 6–39)

Sawadogo M & Roeder RG (1985) Interaction of a gene-specific transcription factor with the adenovirus major late promoter upstream of the TATA box region. *Cell* 43, 165–175. (Problem 7–63)

Scheidereit C, Geisse S, Westphal HM & Beato M (1983) The glucocorticoid receptor binds to defined nucleotide sequences near the promoter of mouse mammary tumor virus. *Nature* 304, 749–752. (Problem 15–48)

Schibler U (2000) Heartfelt enlightenment. *Nature* 404, 25–28. (Problem 7–90)

Schleif RF (1986) Genetics and Molecular Biology, Chap. 13. Reading, MA: Addison Wesley. (Problem 7–59)

Schubert U, Antón LC, Gibbs J, Norbury CC, Yewdell JW & Bennink JR (2000) Rapid degradation of a large fraction of newly synthesized proteins by proteasomes. *Nature* 404, 770–774. (Problem 6–97)

Schwoebel ED, Talcott B, Cushman I & Moore MS (1998) Ran-dependent signal-mediated nuclear import does not require GTP hydrolysis by Ran. *J. Biol. Chem.* 273, 35170–35175. (Problem 12–56)

Service RF (2004) Immune cells speed the evolution of novel proteins. *Science* 306, 1457. (Problem 9–26)

Shah R, Cosstick R & West SC (1997) The RuvC protein dimer resolves Holliday junctions by a dual incision mechanism that involves base-specific contacts. *EMBO J.* 16, 1464–1472. (Problem 5–101)

Shapiro SD, Endicott SK, Province MA, Pierce JA & Campbell EJ (1991) Marked longevity of human lung parenchymal elastic fibers deduced from prevalence of D-aspartate and nuclear weapons-related radiocarbon. *J. Clin. Invest.* 87, 1828–1834. (Problem 19–82)

Shea TB & Yabe J (2000) Occam's razor slices through the mysteries of neurofilament axonal transport: Can it really be so simple? *Traffic* 1, 522–523. (Problem 16–110)

Shibahara K & Stillman B (1999) Replication-dependent marking of DNA by PCNA facilitates CAF-1-coupled inheritance of chromatin. *Cell* 96, 575–585. (Problem 5–65)

Shimamoto N (1999) One-dimensional diffusion of proteins along DNA. *J. Biol. Chem.* 274, 15293–15296. (Problem 7–32)

Shtilerman M, Lorimer GH & Englander SW (1999) Chaperonin function: Folding by forced unfolding. *Science* 284, 822–825. (Problem 6–96)

Sickles DW, Pearson JK, Beall A & Testino A (1994) Toxic axonal degeneration occurs independent of neurofilament accumulation. *J. Neurosci. Res.* 39, 347–354. (Problem 16–28)

Siegel RM, Chan FK-M, Chun HJ & Lenardo MJ (2000) The multifaceted role of Fas signaling in immune cell homeostasis and autoimmunity. *Nat. Immunol.* 1, 469–474. (Problem 18–22)

Sinn E, Muller W, Pattengale P, Tepler I, Wallace R & Leder P (1987) Coexpression of MMTV/v-Ha-*ras* and MMTV/c-*myc* genes in transgenic mice: synergistic action of oncogenes *in vivo*. *Cell* 49, 465–475. (Problem 20–57)

Sjoblom T et al. (2006) The consensus coding sequences of human breast and colorectal cancers. *Science* 314, 268–274. (Problem 20–44)

Skibbens RV, Corson LB, Koshland D & Hieter P (1999) Ctf7 is essential for sister chromatid cohesion and links mitotic chromosome structure to the DNA replication machinery. *Genes Dev.* 13, 307–319. (Problem 17–50)

Skop AR, Bergmann D, Mohler WA & White JG (2001) Completion of cytokinesis in *C. elegans* requires a brefeldin A-sensitive membrane accumulation at the cleavage furrow apex. *Curr. Biol.* 11, 735–746. (Problem 17–116)

Sloboda RD, Dentler WL & Rosenbaum JL (1976) Microtubule-associated proteins and the stimulation of tubulin assembly *in vitro*. *Biochemistry* 15, 4497–4505. (Problem 16–36)

Sluder G & Rieder CL (1985) Centriole number and the reproductive capacity of spindle poles. *J. Cell Biol.* 100, 887–896. (Problem 17–91)

Smeekens S, Bauerle C, Hageman J, Keegstra K & Weisbeek P (1986) The role of the transit peptide in the routing of precursors toward different chloroplast compartments. *Cell* 46, 365–375. (Problem 12–87)

Smith HT, Ahmed AJ & Millet F (1981) Electrostatic interaction of cytochrome *c* with cytochrome $c_1$ and cytochrome oxidase. *J. Biol. Chem.* 256, 4984–4990. (Problem 14–62)

Söllner T, Whiteheart SW, Brunner M, Erdjument-Bromage H, Geromanos S, Tompst P & Rothman JE (1993) SNAP receptors implicated in vesicle targeting and fusion. *Nature* 362, 318–324. (Problem 13–34)

Sonoda N, Furuse M, Sasaki H, Yonemura S, Katahira J, Horiguchi Y & Tsukita S (1999) *Clostridium perfringens* enterotoxin fragment removes specific claudins from tight junction strands: evidence for direct involvement of claudins in tight junction barrier. *J. Cell Biol.* 147, 195–204. (Problem 19–34)

Spierer A & Spierer P (1984) Similar levels of polyteny in bands and interbands of *Drosophila* giant chromosomes. *Nature* 307, 176–178. (Problem 4–83)

Stanners CP & Till JE (1960) DNA synthesis in individual L-strain mouse cells. *Biochim. Biophys. Acta* 37, 406–419. (Problem 17–20)

Stanojevic D, Small S & Levine M (1991) Regulation of a segmentation stripe by overlapping activators and repressors in the *Drosophila* embryo. *Science* 254, 1385–1387. (Problem 7–67)

Stare FJ & Baumann CA (1936) The effect of fumarate on respiration. *Proc. R. Soc. Lond.* B121, 338–357. (Problem 2–126)

Stearns T (2001) Centrosome duplication: A centriolar pas de deux. *Cell* 105, 417–420. (Problem 17–73)

Steinmetz MO, Stoffler D, Hoenger A, Bremer A & Aebi U (1997) Actin: From cell biology to atomic detail. *J. Struct. Biol.* 119, 295–320. (Problem 16–41)

Stern DB & Palmer JD (1984) Extensive and widespread homologies between mitochondrial DNA and chloroplast DNA in plants. *Proc. Natl Acad. Sci. U.S.A.* 81, 1946–1950. (Problem 14–113)

Stone JD, Peterson AP, Eyer J & Sickles DW (2000) Neurofilaments are nonessential elements of toxicant-induced reductions in fast axonal transport: pulse labeling in CNS neurons. *Neurotoxicology* 21, 447–457. (Problem 16–28)

Svejstrup JQ, Li Yang, Fellows J, Gnatt A, Bjorklund S & Kornberg RD (1997) Evidence for a mediator cycle at the initiation of transcription. *Proc. Natl Acad. Sci. U.S.A.* 94, 6075–6078. (Problem 6–41)

Svoboda K, Schmidt CF, Schnapp BJ & Block SM (1993) Direct observation of kinesin stepping by optical trapping interferometry. *Nature* 365, 721–727. (Problem 16–78)

Szent-Györgyi AV (1924) Über den mechanismus de Succin- und Paraphenylendiaminoxydation. Ein Betrag der Zellatmung. *Biochem Z.* 150, 141–149. (Problem 2–126)

Szostak JW & Blackburn EH (1982) Cloning yeast telomeres on linear plasmid vectors. *Cell* 19, 245–255. (Problem 4–53)

Szostak JW, Bartel DP & Luisi L (2001) Synthesizing life. *Nature* 409, 387–390. (Problem 6–107)

Takeuchi Y, Porter CD, Strahan KM, Preece AF, Gustafsson K, Cosset FL, Weiss RA & Collins MK (1996) Sensitization of cells and retroviruses to human serum by (α1-3) galactosyltransferase. *Nature* 379, 85–88. (Problem 13–53)

Tang W-J & Gilman AG (1991) Type-specific regulation of adenylyl cyclase by G protein βγ subunits. *Science* 254, 1500–1503. (Problem 15–91)

Teasdale RD & Jackson MR (1996) Signal-mediated sorting of membrane proteins between the endoplasmic reticulum and the Golgi apparatus. *Annu. Rev. Cell Dev. Biol.* 12, 27–54. (Problem 13–51)

Teo I, Sedgwick B, Kilpatrick MW, McCarthy TV & Lindahl T (1986) The intracellular signal for induction of resistance to alkylating agents in *E. coli.* *Cell* 45, 315–324. (Problem 5–83)

Teter SA, Houry WA, Ang D, Tradler T, Rockabrand D, Fischer G, Blum P, Georgopoulos C & Hartl FU (1999) Polypeptide flux through bacterial Hsp70: DnaK cooperates with trigger factor in chaperoning nascent chains. *Cell* 97, 755–765. (Problems 6–94 and 6–95)

Thomas JE, Smith M, Rubinfeld B, Gutowski M, Beckmann RP & Polakis P (1996) Subcellular localization and analysis of apparent 180-kDa and 220-kDa proteins of the breast cancer susceptibility gene, BRCA1. *J. Biol. Chem.* 271, 28630–28635. (Problem 3–100)

Thomas MJ, Platas AA & Hawley DK (1998) Transcriptional fidelity and proofreading by RNA polymerase II. *Cell* 93, 627–637. (Problem 6–36)

Thompson CM, Koleske AJ, Chao DM & Young RA (1993) A multisubunit complex associated with the RNA polymerase II CTD and TATA-binding protein in yeast. *Cell* 73, 1361–1375. (Problem 6–41)

Thompson NE & Burgess RR (1996) Immunoaffinity purification of RNA polymerase II and transcription factors using polyol-responsive monoclonal antibodies. *Methods Enzymol.* 274, 513–526. (Problem 3–67)

Thompson SR, Goodwin EB & Wickens M (2000) Rapid deadenylation and poly(A)-dependent translational repression mediated by the *Caenorhabditis elegans tra-2 3′* untranslated region in *Xenopus* embryos. *Mol. Cell. Biol.* 20, 2129–2137. (Problem 7–119)

Tilney LG & Inoue S (1982) Acrosomal reaction of *Thyone* sperm. II. The kinetics and possible mechanism of acrosomal process elongation. *J. Cell Biol.* 93, 820–827. (Problem 16–109)

Timpl R & Heilwig R (1979) Laminin—A glycoprotein from basement membranes. *J. Biol. Chem.* 254, 9933–9937. (Problem 19–58)

Ting AY, Kain KH, Klemke RL & Tsien RY (2001) Genetically encoded fluorescent reporters of protein tyrosine kinase activities in living cells. *Proc. Natl Acad. Sci. U.S.A.* 98, 15003–15008. (Problem 9–33)

Toker A & Newton AC (2000) Akt/protein kinase B is regulated by autophosphorylation at the hypothetical PDK-2 site. *J. Biol. Chem.* 275, 8271–8274. (Problem 15–130)

Tokumaru H, Umayahara K, Pellegrini LL, Ishizuka T, Saisu H, Betz H, Augustine GJ & Abe T (2001) SNARE complex oligomerization by synaphin/complexin is essential for synaptic vesicle exocytosis. *Cell* 104, 421–432. (Problems 13–32 and 13–33)

Tokuyama Y, Horn HF, Kawamura K, Tarapore P & Fukasawa K (2001) Specific phosphorylation of nucleophosmin on Thr199 by cyclin-dependent kinase 2-cyclin E and its role in centrosome duplication. *J. Biol. Chem.* 276, 21529–21537. (Problem 17–90)

Tonegawa, S. (1983) Somatic generation of antibody diversity. *Nature* 302, 575–581. (Problem 7–12)

Toth A, Ciosk R, Uhlmann F, Galova M, Schleiffer A & Nasmyth K (1999) Yeast cohesin complex requires a conserved protein, Eco1p(Ctf7), to establish cohesion between sister chromatids during DNA replication. *Genes Dev.* 13, 320–333. (Problem 17–50)

Tournier C, Hess P, Yang DD, Xu J, Turner TK, Nimnual A, Bar-Sagi D, Jones SN, Flavell RA & Davis RJ (2000) Requirement of JNK for stress-induced activation of the cytochrome *c*-mediated death pathway. *Science* 288, 870–874. (Problem 18–27)

Treisman R (1985) Transient accumulation of *c-fos* RNA following serum stimulation requires a conserved 5′ element and *c-fos* 3′ sequences. *Cell* 42, 889–902. (Problem 7–117)

Treisman R (1987) Identification and purification of a polypeptide that binds to the c-fos serum response element. *EMBO J.* 6, 2711–2717. (Problem 7–35)

Tsukamoto T, Yokota S & Fujiki Y (1990) Isolation of Chinese hamster ovary cell mutants defective in assembly of peroxisomes. *J. Cell. Biol.* 110, 651–660. (Problem 12–96)

Tugal T, Zou-Yang XH, Gavin K, Pappin D, Canas B, Kobayashi R, Hunt T & Stillman B (1998) The Orc4p and Orc5p subunits of the *Xenopus* and human origin recognition complex are related to Orc1p and Cdc6p. *J. Biol. Chem.* 273, 32421–32429. (Problem 8–46)

Tyerman SD, Bohnert HJ, Maurel C, Steudle S & Smith JAC (1999) Plant aquaporins: their molecular biology, biophysics and significance for plant water relations. *J. Exp. Bot.* 50, 1055–1071. (Problem 19–99)

Tzagoloff A (1982) Mitochondria, pp 212–213. New York: Plenum Press. (Problem 14–34)

Uhlmann F & Nasmyth K (1998) Cohesion between sister chromatids must be established during DNA replication. *Curr. Biol.* 8, 1095–1101. (Problem 17–46)

Uhlmann F, Lottspeich F & Nasmyth K (1999) Sister-chromatid separation at anaphase onset is promoted by cleavage of the cohesin subunit of Scc1. *Nature* 400, 37–42. (Problem 17–101)

Valius M & Kazlauskas A (1993) Phospholipase C-γ1 and phosphatidylinositol 3 kinase are the downstream mediators of the PDGF receptor's mitogenic signal. *Cell* 73, 321–334. (Problem 15–127)

Valius M, Secrist JP & Kazlauskas A. (1995) The GTPase-activating protein of Ras suppresses platelet-derived growth factor β receptor signaling by silencing phospholipase C-γ1. *Mol. Cell. Biol.* 15, 3058–3071. (Problem 15–127)

van Arsdell SW & Weiner AM (1984) Human genes for U2 small nuclear RNA are tandemly repeated. *Mol. Cell. Biol.* 4, 492–499. (Problem 4–51)

Van Dyck E, Stasiak AJ, Stasiak A & West SC (1999) Binding of double-strand breaks in DNA by human Rad52 protein. *Nature* 398, 728–731. (Problem 5–87)

Van Itallie CM & Anderson JM (2006) Claudins and epithelial paracellular transport. *Annu. Rev. Physiol.* 68, 403–429. (Problems 19–32 and 19–33)

Van Meer G & Simons K (1986) The function of tight junctions in maintaining differences in lipid composition between the apical and the basolateral cell surface domains of MDCK cells. *EMBO J.* 5, 1455–1464. (Problem 19–36)

Varma R & Mayor S (1998) GPI-anchored proteins are organized in submicron domains at the cell surface. *Nature* 394, 798–801. (Problem 10–31)

Vogelstein B & Kinzler KW (2004) Cancer genes and the pathways they control. *Nat. Med.* 10, 789–798. (Problem 20–39)

Vojtek AB, Hollenberg SM & Cooper JA (1993) Mammalian Ras interacts directly with the serine/threonine kinase Raf. *Cell* 74, 205–214. (Problems 8–47, 8–48, and 8–49)

Vollrath D, Nathans J & Davis RW (1988) Tandem array of human visual pigment genes at Xq28. *Science* 240, 1669–1671. (Problem 4–52)

Wahl GM, Padgett RA & Stark GR (1979) Gene amplification causes overproduction of the first three enzymes of UMP synthesis in *N*-(phosphonacetyl)-L-aspartate (PALA)–resistant hamster cells. *J. Biol. Chem.* 254, 8679–8689. (Problem 3–97)

Walker RA, Inoué S & Salmon ED (1989) Asymmetric behavior of severed microtubule ends after ultraviolet-microbeam irradiation of individual microtubules *in vitro*. *J. Cell Biol.* 108, 931–937. (Problem 16–19)

Walworth NC, Goud B, Kabcenell AK & Novick PJ (1989) Mutational analysis of *SEC4* suggests a cyclical mechanism for the regulation of vesicular traffic. *EMBO J.* 8, 1685–1693. (Problem 13–35)

Wang L, Cunningham JM, Winters JL, Guenther JC, French AJ, Boardman LA, Burgart LJ, McDonnell SK, Schaid DJ & Thibodeau SN (2003) *B-Raf* mutations in colon cancer are not likely attributable to defective DNA mismatch repair. *Cancer Res.* 63, 5209–5212. (Problem 20–53)

Wang YH & Griffith J (1995) Expanded CTG triplet repeat blocks from the myotonic dystrophy gene create the strongest known natural nucleosome positioning elements. *Genomics* 25, 570–573. (Problem 4–43)

Ward GE & Kirschner MW (1990) Identification of cell cycle-related phosphorylation sites on nuclear lamin C. *Cell* 61, 561–577. (Problem 16–39)

Warrell RP Jr, de The H, Wand ZY & Degos (1993) Acute promyelocytic leukemia. *N. Engl. J. Med.* 329, 177–189. (Problem 20–66)

Wasserman WJ & Masui Y (1975) Effects of cycloheximide on a cytoplasmic factor initiating meiotic maturation in *Xenopus* oocytes. *Exp. Cell Res.* 91, 381–388. (Problem 17–34)

Watson CJ, Rowland M & Warhurst G (2001) Functional modeling of tight junctions in intestinal cell monolayers using polyethylene glycol oligomers. *Am. J. Physiol. Cell Physiol.* 281, C388–C397. (Problem 19–31)

Webb MR, Grubmeyer C, Penefsky HS & Trentham DR (1980) The stereochemical course of phosphoric residue transfer catalyzed by beef heart mitochondrial ATPase. *J. Biol. Chem.* 255, 11637–11639. (Problem 14–30)

Wei MC, Zong W-X, Cheng EH-Y, Lindsten T, Panoutsakopoulou V, Ross AJ, Roth KA, MacGregor GR, Thompson CB & Korsmeyer SJ (2001) Proapoptotic BAX and BAK: A requisite gateway to mitochondrial dysfunction and death. *Science* 292, 727–730. (Problems 18–19, 18–25, and 18–26)

Weinberg RA (2006) The Biology of Cancer, p726. New York: Garland Science. (Problem 20–65)

Weinert TA & Hartwell LH (1988) The *rad9* gene controls the cell cycle response to DNA damage in *Saccharomyces cerevisiae*. *Science* 241, 317–322. (Problem 17–141)

Weinrich SL, Pruzan R, Ma L, Ouellette M, Tesmer VM, Holt SE, Bodnar AG, Lichtsteiner S, Kim NW, Trager JB, Taylor RD, Carlos R, Andrews WH, Wright WE, Shay JW, Harley CB & Morin GB (1997) Reconstitution of human telomerase with the template RNA component hTR and the catalytic protein subunit hTRT. *Nat. Genet.* 17, 498–502. (Problem 17–136)

Weintraub H & Groudine M (1976) Chromosomal subunits in active genes have an altered conformation. *Science* 193, 848–856. (Problem 4–68)

Welch MD, Rosenblatt J, Skoble J, Portnoy DA & Mitchison TJ (1998) Interaction of human Arp2/3 complex and the *Listeria monocytogenes* ActA protein in actin filament nucleation. *Science* 281, 105–108. (Problem 16–65)

Whitehouse I, Flaus A, Cairns BR, White MF, Workman JL & Owen-Hughes T (1999) Nucleosome mobilization catalysed by the yeast SWI/SNF complex. *Nature* 400, 784–787. (Problem 4–57)

Whitmore D, Foulkes NS & Sassone-Corsi P (2000) Light acts directly on organs and cells in culture to set the vertebrate circadian clock. *Nature* 404, 87–91. (Problem 7–90)

Wickens M, Goodwin EB, Kimble J, Strickland S & Hentze M (2000) Translational control of developmental decisions. In Translational Control of Gene Expression (N Sonenberg, JWB Hershey, MB Mathews eds), pp 295–370. Cold Spring Harbor, NY: Cold Spring Harbor Laboratory Press. (Problem 7–118)

Wilmut I, Schnieke AE, McWhir J, Kind AJ & Campbell KHS (1997) Viable offspring derived from fetal and adult mammalian cells. *Nature* 385, 810–813. (Problem 7–11)

Wilson SM, Yip R, Swing DA, O'Sullivan TN, Zhang Y, Novak EK, Swank RT, Russell LB, Copeland NG & Jenkins NA (2000) A mutation in *Rab27* causes the vesicle transport defects observed in *ashen* mice. *Proc. Natl Acad. Sci. U.S.A.* 97, 7933–7938. (Problem 13–75)

Wilson T & Treisman R (1988) Removal of poly(A) and consequent degradation of c-*fos* mRNA facilitated by 3′ AU-rich sequences. *Nature* 336, 396–399. (Problem 7–117)

Winter CG, Wang B, Ballew A, Royou A, Karess R, Axelrod JD & Luo L (2001) *Drosophila* Rho-associated kinase (Drok) links Frizzled-mediated planar cell polarity signaling to the actin cytoskeleton. *Cell* 105, 81–91. (Problem 15–145)

Witke W, Schleicher M & Noegel AA (1992) Redundancy in the microfilament system: abnormal development of *Dictyostelium* cells lacking two F-actin cross-linking proteins. *Cell* 68, 53–62. (Problem 16–67)

Wolf S, Deom CM, Beachy RN & Lucas WJ (1989) Movement protein of tobacco mosaic virus modifies plamodesmatal size exclusion limit. *Science* 246, 377–379. (Problem 19–45)

Wolff B, Sanglier J-J & Wang Y (1997) Leptomycin B is an inhibitor of nuclear export: inhibition of nucleo-cytoplasmic translocation of the human immunodeficiency virus type 1 (HIV-1) Rev protein and Rev-dependent mRNA. *Chem. Biol.* 4, 139–147. (Problem 12–57)

Wolffe AP & Brown DD (1986) DNA replication *in vitro* erases a *Xenopus* 5S RNA gene transcription complex. *Cell* 47, 217–227. (Problem 7–91)

Workman JL & Roeder RG (1987) Binding of transcription factor TFIID to the major late promoter during *in vitro* nucleosome assembly potentiates subsequent initiation by RNA polymerase II. *Cell* 51, 613–622. (Problem 7–65)

Wyrick JJ, Holstege FC, Jennings EG, Causton HC, Shore D, Grunstein M, Lander ES & Young RA (1999) Chromosomal landscape of nucleosome-dependent gene expression and silencing in yeast. *Nature* 402, 418–421. (Problem 4–72)

Xeros N (1962) Deoxyriboside control and synchronization of mitosis. *Nature* 194, 682–683. (Problem 17–14)

Xu T, Ashery U, Burgoyne RD & Neher E (1999) Early requirement for a-SNAP and NSF in the secretory cascade in chromaffin cells. *EMBO J.* 18, 3293–3304. (Problem 13–119)

Yabe JT, Pimenta A & Shea TB (1999) Kinesin-mediated transport of neurofilament protein oligomers in growing axons. *J. Cell Sci.* 112, 3799–3814. (Problem 16–110)

Yalow RS (1978) Radioimmunoassay: a probe for the fine structure of biologic systems. *Science* 200, 1236–1245. (Problem 15–42)

Yamano H, Tsurumi C, Gannon J & Hunt T (1998) The role of the destruction box and its neighbouring lysine residues in cyclin B for anaphase ubiquitin-dependent proteolysis in fission yeast: defining the D-box receptor. *EMBO J.* 17, 5670–5678. (Problem 6–98)

Yanofsky C (2001) Advancing our knowledge in biochemistry, genetics, and microbiology through studies on tryptophan metabolism. *Annu. Rev. Biochem.* 70, 1–37. (Problem 2–127)

Yao M-C, Zhu S-G & Yao C-H (1985) Gene amplification in *Tetrahymena thermophila*: formation of extrachromosomal palindromic gene coding for rRNA. *Mol. Cell. Biol.* 5, 1260–1267. (Problem 8–92)

Ybe JA, Brodsky FM, Hofmann K, Lin K, Liu SH, Chen L, Earnest TN, Fletterick RJ & Hwang RK (1999) Clathrin self-assembly is mediated by a tandemly repeated superhelix. *Nature* 399, 371–375. (Problem 13–19)

Yen TJ, Machlin PS & Cleveland DW (1988) Autoregulated instability of β-tubulin mRNAs by recognition of the nascent amino terminus of β-tubulin. *Nature* 334, 580–585. (Problem 7–116)

Yoon M, Moir RD, Prahlad V & Goldman RD (1998) Motile properties of vimentin intermediate filament networks in living cells. *J. Cell Biol.* 143, 147–157. (Problem 16–38)

Yoshida H, Kong Y-Y, Yoshida R, Elia AJ, Hakem A, Hakem R, Penninger JM & Mak TW (1998) Apaf1 is required for mitochondrial pathways of apoptosis and brain development. *Cell* 94, 739–750. (Problem 18–20)

Zagotta WN, Hoshi T & Aldrich RW (1990) Restoration of inactivation in mutants of *shaker* potassium channels by a peptide derived from ShB. *Science* 250, 568–570. (Problem 11–82)

Zagouras P & Rose JK (1989) Carboxy-terminal SEKDEL sequences retard but do not retain two secretory proteins in the endoplasmic reticulum. *J. Cell Biol.* 109, 2633–2640. (Problem 13–52)

Zahringer J, Baliga BS & Munro HN (1976) Novel mechanism for translational control in regulation of ferritin synthesis by iron. *Proc. Natl Acad. Sci. U.S.A.* 73, 857–861. (Problem 7–111)

Zecca M, Basler K & Struhl G (1995) Sequential organizing activities of engrailed, hedgehog and decapentaplegic in the *Drosophila* wing. *Development* 121, 2265–2278. (Problem 17–143)

Zerangue N, Malan MJ, Fried SR, Dazin PF, Jan YN, Jan LY & Schwappach B (2001) Analysis of endoplasmic reticulum trafficking signals by combinatorial screening in mammalian cells. *Proc. Natl Acad. Sci. U.S.A.* 98, 2431–2436. (Problem 13–49)

Zhang X-F, Settleman J, Kyriakis JM, Takenchi-Suzuki E, Elledge SJ, Marshall MS, Bruder JT, Rapp UR & Avruch J (1993) Normal and oncogenic $p21^{ras}$ proteins bind to the amino-terminal domain of c-Raf-1. *Nature* 364, 308–313. (Problems 8–48 and 8–49)

Zhao C, Takita J, Tanaka Y, Setou M, Nakagawa T, Takeda S, Yang HW, Terada S, Nakata T, Takei Y, Saito M, Tsuji S, Hayashi Y & Hirokawa N (2001) Charcot-Marie-Tooth disease type 2A caused by mutation in a microtubule motor KIF1Bβ. *Cell* 105, 587–597. (Problem 16–101)

Zhen L, Jiang S, Feng L, Bright NA, Peden AA, Seymour AB, Novak EK, Elliott R, Gorin MB, Robinson MS & Swank RT (1999) Abnormal expression and subcellular distribution of subunit proteins of the AP-3 adaptor complex lead to platelet storage pool deficiency in the *pearl* mouse. *Blood* 94, 146–155. (Problem 13–71)

Zheng Y, Wong ML, Alberts B & Mitchison TJ (1995) Nucleation of microtubule assembly by a γ-tubulin-containing ring complex. *Nature* 378, 578–583. (Problem 16–63)

Zhuang Y & Weiner AM (1986) A compensatory base change in U1 snRNA suppresses a 5′ splice site mutation. *Cell* 46, 827–835. (Problem 6–44)

Zieg J, Silverman M, Hilmen M & Simon M (1977) Recombinational switch for gene expression. *Science* 196, 170–172. (Problem 7–87)

Zimmet J & Ravid K (2000) Polyploidy: occurrence in nature, mechanisms, and significance for the megakaryocyte-platelet system. *Exp. Hematol.* 28, 3–16. (Problem 17–118)

Zindy F, Williams RT, Baudino TA, Rehg JE, Skapek SX, Cleveland JL, Roussel MF & Sherr CJ (2003) *Arf* tumor suppressor promoter monitors latent oncogenic signals *in vivo*. *Proc. Natl Acad. Sci. U.S.A.* 100, 15930–15935. (Problem 20–56)

Zinkel SS & Crothers DM (1987) DNA bend direction by phase sensitive detection. *Nature* 328, 178–181. (Problem 7–33)

# Index

Page numbers refer to a major text discussion of the entry; page numbers with an F refer to a figure, with a T refer to a table.

## A


- Abl protein tyrosine kinase, 227
  - inhibitors, 484
- ACC synthase, 372
- Acentric chromosomes, 420
- Acetate, bacterial membrane transport, 245
- Acetylation of histone tails, 72–73, 74F, 76–77
- Acetylcholine, 258, 259
  - actions, 344
  - potassium channel activation, 360
  - receptors *see* Acetylcholine receptors
  - synaptic vesicles, 254
- Acetylcholine receptors, 254, 259F
  - conduction differences, 259–260
  - desensitization, 352
  - ionic permeability, 254
  - muscarinic, 360
    - *see also* G-protein-linked (coupled) receptors (GPCRs)
  - myasthenia gravis autoantibodies, 254–255
  - neuromuscular junction, 254, 257
  - nicotinic, 352
    - *see also* Ion channels
  - pentameric structure, 259–260
  - phosphorylation, 352
  - succinylcholine antagonist, 344
- Acetylcholinesterase, inhibition, 255
- Acetyl CoA, 32
- Acid hydrolases, 302–303
- Acids, 14–16, 21T
  - DNA, 63
  - titration, 14
  - weak lipophilic, 323
- Acrosome reaction, 380–381
- Acrosomes, actin/actin filaments, 369, 370, 380–381
- Acrylamide, 376
- ActA protein, 386–387
- Actin/actin filaments, 389F
  - acrosomes, 369, 370, 380–381
  - ATP caps, 378–379
  - binding proteins, 375
  - cellular concentrations, 383
  - cytochalasin B effects, 245, 369, 380, 413
  - depolymerization, 380–381
    - assay, 381–382
    - Cofilin, 383
  - evolutionary conservation, 375
  - filamin cross linking, 383
  - functions, 373
    - cell motility, 369–370, 386–387
    - cytokinesis, 413, 427
    - *Dictyostelium* aggregation, 387
    - ECM interactions, 393
    - lamellipodia/filopodia, 359–361, 383
  - gene fusion experiments, 421–422, 422T
  - GTPase effects, 394
  - minus end, 373, 374F, 377, 380–381
  - mRNA splicing, 130, 130F, 130FF
  - myosin-decorated, 376, 380
  - nucleation by ARP complex, 360, 383, 386–387
  - phalloidin binding, 380
  - plus end, 373, 374F, 377, 380–381
  - polymerization, 374
    - ActA effects, 386–387
    - assay, 378–379
    - ATP role, 378–379
    - Cdc42/N-WASp roles, 400–401, 400F
    - critical concentration, 374, 377
    - as a function of actin concentration, 377
    - kinetics, 378–379, 386–387
    - mechanism, 376–377
    - time course, 374
  - protein-associations, 242
  - random thermal motion, 393
  - sliding filament, 389F, 400–401
  - swinholide A effects, 381–382
  - thick filaments (skeletal muscle), 391, 392F
  - thin filaments (skeletal muscle), 366, 391, 392F
  - *see also* Myosins
- $\alpha$-Actinin, 92, 392
- Actinomycetes, 381
- Action potential
  - muscle cells, 253, 366
  - Nernst equation, 256
  - propagation, 254
  - sodium ion stoichiometry, 256–257
  - squid giant axons, 255–256
  - voltage-gated sodium channels, 254, 256–257
  - *see also* Membrane potential
- Action spectra, 334, 334F
- Activated carriers, 26
  - *see also* ATP (adenosine triphosphate)
- Active transport
  - antiporters, 247, 249
  - ATP-driven pumps, 247, 250–251
  - coupled transport, 247
  - energetics, 250–251, 389
  - light-driven pumps, 247
  - passive *versus,* 243
  - symporters, 247
  - uniporters, 248
  - *see also* Carrier proteins
- Adaptins, 279, 290, 291
- Adaptive response, 108
- Adenine, base pairing, 63, 63F
- Adenosine diphosphate *see* ADP (adenosine diphosphate)
- Adenosine monophosphate, synthesis, 51, 51F
- Adenosine triphosphate *see* ATP (adenosine triphosphate)
- Adenovirus, DNA–RNA hybrids, 121
- Adenylate cyclase *see* Adenylyl cyclase
- Adenylyl cyclase
  - $\beta$-adrenergic receptor-activation, 325, 351, 357
  - cholera toxin, 358
  - GPCR-coupling, 355–356, 357
  - type II, 357
- Adipocytes, 458, 459F
- ADP (adenosine diphosphate)
  - antiporter, 321
  - mitochondrial membrane transport, 321
  - structure, 321F
- ADP–ATP antiporter, 321
- Adrenaline, actions, 332, 333, 355–356
- Adrenergic receptors (adrenoceptors)
  - $\alpha_1$-adrenergic, signaling, 359
  - $\beta$-adrenergic
    - adenylyl cyclase, 351, 356, 357
    - competitive inhibitor, 355–356
    - desensitization, 351
  - muscle glycogen metabolism, 332, 333, 353
- Affinity chromatography, 48–49, 154, 159–160, 183
  - antibody columns, 48–49
  - biotin–streptavidin, 272
  - DNA, 159–160
  - lectin columns, 240
  - tagged proteins, 207–208, 248–249
- Agarose gel electrophoresis, DNA analysis, 71, 71F, 85–86, 210, 212, 321
- Aggregation centers, *Dictyostelium,* 348
- Aging, 431
  - cancer and, 470
- Alanine, 23, 23F
- Alanine:glyoxylate aminotransferase (AGT) deficiency, 279
- Albumin
  - gene family, human *Alu* sequences, 85
  - secretion, 310–311
- Alcohol *see* Ethanol
- Alcoholism, 432
- Algae, photosynthesis, 333, 333F
- Alkaline phosphatase, 278, 287
  - cloning vector treatment, 213–214
  - SNARE assay, 294
- Alkylating agents, 107–109
  - apoptosis, 444
- Alleles, 215
  - conditional, 219
  - disease, 217–218
  - null, 219
- Allele-specific oligonucleotides (ASOs), 217–218
- Allosteric regulation, 52–53
  - MWC postulates, 52–53
  - *see also* Protein–protein interactions
- $\alpha$-helices, 40
  - amphipathic, 40, 253, 257
  - hydropathy analysis, 239, 255, 287
  - membrane proteins, 239–240
    - *see also* Membrane proteins
  - net dipole, 40
- Alprenolol, 355–356
- ALPS (autoimmune lymphoproliferative syndrome), 441
- Alternative splicing, 121–122, 122F
  - gene expression regulation, 185–186
- AluI restriction enzyme, 204
- *Alu* sequences, 85, 85F
- Alzheimer's disease, amyloid plaques, 368
- *Amanita phalloides*
  - $\alpha$-amanitin, 127, 127F
  - phalloidin, 380
- $\alpha$-Amanitin, RNA polymerase inhibition, 127, 127F
- Amine groups, 14, 18, 20
- Amino acid(s)
  - chirality, 17
  - codon co-linearity, 1
  - effect of substitutions, 8, 8T
  - electrostatic interactions, 14
  - hydrophobicity, 39, 40, 240
  - pH, 14, 14F, 23, 23F, 45
  - sequence analysis, frameshift mutation, 135

side chains, 40, 45, 240
synthesis, 36–37
*see also individual amino acids*
Aminoacyl-tRNA synthetases, 48
aminoacylation reaction, 136
proofreading, 136
tmRNA and, 142
*see also* Amino acid(s); Transfer RNA
Ammonia, 283, 302
Amniocentesis, 218
Amoebae, *Dictyostelium,* 348, 387
AMP (adenosine monophosphate)
cyclic *see* Cyclic AMP (cAMP)
synthesis, 51, 51F
Amphipathic molecules, α-helices, 253, 257
Amphiphilic molecules, 17, 17F
β-sheets, 41, 41F
α-helices, 41
Amplification-control element, 101
Amyloid plaques, 368
β-Amyloid precursor protein (APP), 368
Amylose, structure, 16, 16F
Anabolic pathways, 24
Anerobic metabolism, 31
glucose consumption, 31
muscle, 14–15
Anaphase, 412, 414, 415
Anaphase-promoting complex (APC), 145, 381–382
Cdc20–APC complex, 424–425
Hct1–APC complex, 424–425
Ancient DNA, sequence analysis, 83–85
Ankyrin, protein-associations, 242
Antenna complex, 330
Antibiotics
macrolides, 381–382
protein synthesis inhibition, 137
*see also specific drugs*
Antibodies, 362
affinity, 48–49
antigen binding *see* Antigen–antibody interactions
applications
chromatography, 48–49
*see also* Affinity chromatography
fluorescence, 224
immunoblotting *see* Immunoblots
immunoprecipitation *see* Immunoprecipitation
protein-association experiments, 242
protein purification, 48–49
radioimmunoassay, 347
autoantibodies, 255, 347
gene rearrangement, 152–153
HRP conjugation, 198
monoclonal *see* Monoclonal antibodies
NFκB gene regulatory protein, 273
*see also* Immunoglobulin(s)
Anticancer drugs *see* Cancer chemotherapy
Antifreeze protein, notothenoid fish, 48
Antigen–antibody interactions, 46, 49, 191
binding sites, 362
*see also* Protein–ligand interactions
Antiporters, 247, 249
ADP–ATP antiporter, 321
$Ca^{2+}$–$Na^{+}$ antiporter, 249
$Na^{+}$–$H^{+}$ antiporter, 248
Antisense RNA, genetic modification, 372
Antitrypsin, deficiency, 311
AP1 DNA binding site, 158–159
Apaf1 adaptor protein, 440–441, 441F
APC *see* Anaphase-promoting complex (APC)
*Apc* (adenomatous polyposis coli) gene, 368–369
AP endonuclease, 104
Apolipoprotein B, tissue-specific mRNA editing, 186, 186F
Apoptosis, 439–444
activation
Apaf1, 440–441
Fas/Fas ligand, 441–442, 443
ALPS, 441
apoptotic signals, alkylation, 444
cancer role, 482
killer lymphocytes, 440
caspase cascade (proteolysis), 390, 440, 441
cell morphology, 440, 440F
cytochrome *c*-dependent, 390, 391–392, 441–442
death genes, 440
developmental
*C. elegans,* 440
mouse paws, 441
DNA damage response, 431
mitochondria role, 440
necrosis *versus,* 440, 440F
neuronal, 443–444
regulation
Bcl2 family, 440, 443
feedback, 442
JNK kinases, 443–444
AQP4, 230
Aquaporins, 230, 254
*AraBAD* gene cluster, 155, 164, 166F
Arabinose metabolism, 164, 164F, 166–167
Arachidonic acid, $\alpha_1$-adrenoceptor signaling, 359
AraC protein, transcriptional regulation, 164, 166
Archaea, tree of life, 5
ARF GTPase, 293–294
Arginine biosynthesis, genetic switches, 161
ARP complex, 360, 386–387, 400
actin filament nucleation, 383
Arsenate, 31, 34
mechanism of action, 31
ARSs (autonomously replicating sequences), 100–101, 101F
Artificial chromosomes, 70
*Ashen* mice, 305, 305F
ASOs (allele-specific oligonucleotides), 217–218
Aspartate transcarbamoylase, 55, 55F
allosteric regulation, 60, 60F
conformational changes, 61F
Aspirin, 15, 15F
Asters, 424, 428–429
Aster stimulation cytokinesis hypothesis, 428–429
Astral array, 415
Astral microtubules, 412, 415, 420
Astrocytes
intermediate filaments role, 375
visualization, 230, 230F
ATCase *see* Aspartate transcarbamoylase
Atomic force microscopy, 447
Atomic number, 12
Atomic structure, 12–13
electron shells, 13
Atomic weight, 12
ATP (adenosine triphosphate), 52
caged, 394–395
charge, 322
cycling, 29, 32
energy carrier, 26, 29
export, 322
hydrolysis *see* ATP hydrolysis
phosphoanhydride bond, 26
regeneration, 26
structure, 321F
synthesis *see* ATP synthesis
*see also* ADP (adenosine diphosphate)
ATP caps, actin filaments, 378–379
ATP-driven pumps (ATPases)
active transport, 247, 250–251
*see also specific proteins*
ATP hydrolysis, 13, 32
cytoskeletal polymerization, actin filaments, 378–379
DNA unwinding, 95
exercise, 316
membrane transport, 247
energetics, 250–251
mitochondrial import, 275
Sodium-potassium pump, 249, 251
motor proteins, 53, 390
*see also* Motor proteins
muscle contraction, 392
phosphorylation reaction, 48
protein folding, 143–144
stereochemistry, 319
thermodynamics, 26, 29, 250–251, 295
Walker motif, 101
ATP-like signal metabolites, 52
ATP synthase
αβ dimers, 320–321
conformational changes, 320–321, 320F, 320FF
γ subunit, 320–321
mechanism of action, 320–321
proton flow, 316–317, 320–321
*see also* Electrochemical proton gradients
rotary motor, 320–321
ATP synthesis, 32
chloroplasts, 330, 335–336, 336
glycolysis, 26, 30–31, 254, 290
mitochondrial, 275
oxidative phosphorylation, 295, 316
phosphate origin, 319
proton gradients *see* Electrochemical proton gradients
stereochemistry, 319
stoichiometry, 321–322
synthase *see* ATP synthase
thermodynamics, 318–319, 332
*Thiobacillus ferrooxidans,* 318
uncoupling from electron transport, 324
Atractyloside, 327, 327T
Autoantibodies, 255, 347
Autocrine signaling, 344
Autoimmune lymphoproliferative syndrome (ALPS), 441
Autonomously replicating sequences (ARSs), 100–101, 101F
Autophagosomes, 302
Autophosphorylation, 364
Autoproteolytic cleavage, hedgehog precursor protein, 369
Autoradiography
DNA synthesis, 98–99, 98F, 375–376, 393–394
gel electrophoresis, 69, 69F, 78, 78F, 85, 94, 227
protein, 198, 198F, 334
*see also* Electrophoresis
radiolabeled markers, 198
sensitivity, 225
Autoregulatory pathways, tubulin gene expression, 186
Avidin, 51
Axonal transport, 394, 399
Axonemes, 385, 393
5-Azacytidine, DNA methylation effect, 180
Azide, 460


## B


Bacteria
amino acid biosynthesis, 36–37
carbon sources, 245
cell division, 3
cell growth, 4
cell motility, 386–387
flagellar motor, 328–329
*see also* Cell motility; Flagella
cell wall, 59–60
chemotaxis, 366–367
citric acid cycle, 36
environmental specialization, 297–298, 302, 318
enzyme secretion, 458
gene expression, 161
eucaryotic genes *versus,* 209
genetic switches, 160–162
sigma (σ) factors, 120
*see also* Operons
genomes, 6
evolutionary pressure, 161
immune evasion, 175
membrane transport, 245
methanogenic, 328
phase variation, 175
photosynthesis, 4–5, 316–317, 330
plasmids, 79, 79F
*see also* Plasmids
protein synthesis termination, 141

surface-to-volume ratio, 265
tree of life, 5
viruses *see* Bacteriophage
*see also individual bacteria*
Bacteriophage
ϕX174, 3
host resistance, 87
lambda (λ) *see* Lambda (λ) bacteriophage
M13 *see* M13 bacteriophage
replication, 85, 92–93
T4 *see* T4 bacteriophage
Bacteriorhodopsin, 239, 316–317, 317F
Bak, apoptosis regulation, 442–443
Ball-and-chain model, voltage-gated potassium channel inactivation, 255, 255F
BamHI restriction enzyme, 204
Band 3 protein, 241
Barnase, protein folding/unfolding model, 276, 277F
Barr body, 180–181
Basal lamina, 457–458
enzymatic digestion, 458
genetic defects, 458, 458T
nidogen, 458, 458T
Base excision repair, 107
Base pairing
Chargaff's rules, 3
DNA, 3, 63–64, 63F, 64F
*see also* DNA structure
wobble, 136
Bases
amine groups, 14, 14F
ammonia, 283, 302
Baumann, C.A., 35
Bax, apoptosis regulation, 443
β-barrels, 240
β-particles, 12, 12T
β-propeller, 42
β-sheets, 41, 41F
amphiphilic, 41
parallel/antiparallel strands, 41
Bcl2 family, apoptosis regulation, 440, 442–443
Benzer, Seymour, 139
Bicarbonate/$CO_2$ buffering system, 22
Bid, apoptosis regulation, 443, 443T
Bidirectional DNA replication, 96
Binding surfaces, 43, 43F
Biosensors, 202
Biosynthetic pathways, 24
*see also individual pathways*
Biosynthetic-secretory pathway(s), 290, 290F
*see also* Exocytosis; Protein translocation
Biotin–avidin interaction, 163F
Biotin–streptavidin affinity chromatography, 272
BIRB-796, 484F
Birth defects, thalidomide, 16–17
1,3-Bisphosphoglycerate
ATP synthesis, 31, 31F
structure, 15F
Blindness, circadian rhythm alteration, 173
Blood
alcohol content, 20
buffering systems, 22–23
cells *see* Red blood cells
clotting cascade, 460–461
*see also* Platelet(s)
glucose concentration, 19, 19F
Blood clotting, 160–161
Boveri, Theodore, 426
Box C/D snoRNAs, 132
Box elements (snoRNAs), 132, 133
Box H/ACA snoRNAs, 132
*Brac1* gene, 56–57, 57F
B-Raf, colon cancer, 477
Brain, glucose metabolism, 244
Branch migration (DNA), 107, 112
BrdU labeling, 99–100
DNA analysis, 102, 395
mtDNA, 338–339
BrdUTP, 99–100
Breast cancer
familial, *Brac1* gene, 56–57, 57F
incidence, 470, 470F
in mice, 472–473
Brefeldin A, 293, 413
Bright-field microscopy, 223, 223F
Bromodeoxyuridine labeling *see* BrdU labeling
Buffers
cellular, 22
definition, 22
Buoyant density, 194
Butylmalonate, 327, 327T


## C


$C_3$ plants, carbon fixation, 333F
$C_4$ plants, carbon fixation, 333F
$Ca^{2+}$ *see* Calcium ions ($Ca^{2+}$)
Cadherin-dependent cell sorting, mouse L-cells, 446, 446F
Cadherins, cell adhesion, 446–448
analysis, 449, 449F
*Caenorhabditis elegans*
cytokinesis, 429F
developmental apoptosis, 440
CAF1 (chromosome assembly factor-1), 102–103, 103F
Caffeine
cAMP effects, 354
cell cycle effects, 416
structure, 416F
Caged molecules, 222, 313–314, 371, 394–395, 395F
Calcitonin, alternative splicing, 185–186, 185F
Calcitonin gene-related peptide (CGRP), alternative splicing, 185–186
Calcium, abundance, 13F
Calcium–calmodulin-dependent protein kinase II (CaM-kinase II), 355
Calcium channels
kinetics, 257
NMDA receptors, 260–261
voltage-gated, 256
Calcium ions ($Ca^{2+}$)
chelation, 191, 223, 329
intracellular, 249
ionophores, 256, 329
muscle contraction role, 253, 394–395
speed, 323
sperm–egg interactions, 256
Calcium pump ($Ca^{2+}$-ATPase), 240, 249, 366
Calcium signaling, 353–354, 445
calmodulin, 359–360
CaM kinase II, 355
$IP_3$-mediated, 329, 359
phosphorylase kinase signal integration, 359–360
Calcium–sodium ($Ca^{2+}$–$Na^{+}$) antiporter, 249
Calmodulin, calcium signaling role, 227, 359–360
Calnexin, 297
Calreticulin, 297
Calvin cycle, 330, 330T, 332
*see also* Carbon dioxide; Photosynthesis
CaM-kinase II ($Ca^{2+}$–calmodulin-dependent protein kinase II), 355
Cancer, 469–484
cell behavior, 477–481
cell cycle disinhibition, 475
DNA microarray typing, 152
drug development, 482–483
incidence as function of age, 470
metastases, 469
in mice, 472–473, 473T
mutations causing, 368–369
Fas, 441
Myc, 474
proto-oncogenes, 368–369
*see also* Oncogenes
Ras GTPase mutation, 363
tumor suppressor genes, 368–369
*see also* Tumor suppressor genes
*see also* Mutagenesis; Mutation
preventable causes, 472–473
treatment *see* Cancer chemotherapy
*see also* Tumors
Cancer chemotherapy, 470, 481–484
alkylating agents, 107–109, 392
microtubule dynamics, 375–376
Cancer critical genes, 473–477
Cap-binding complex, 123
Caps on eucaryotic mRNA, 120, 133
Carbohydrates, 298–300
general formula, 13
membrane proteins, 239
structures, 16
*see also entries beginning glyco-;* Oligosaccharides; Polysaccharides
Carbon, 32
$^{14}C$ isotope, 12, 12T
abundance, 13F
atomic structure, 12
carbon–carbon double bonds, 13
oxidation states, 24, 27F
sources, bacterial cells, 245
Carbon–carbon double bonds, 13
Carbon dioxide
atmospheric, 19
*see also* Carbon fixation
blood
buffering, 22
transport, 248
excretion, 248–249
intracellular generation, 248
Carbon fixation, 5
$C_3$ *versus* $C_4$ plants, 332–333, 333F
Calvin cycle, 330, 330T, 332, 333F
$CO_2$ pump, 332–333, 333F
energetics, 330
ribulose bisphosphate carboxylase, 333
starch synthesis, 332–333
stoichiometry, 332
*Thiobacillus ferrooxidans*, 332
*see also* Photosynthesis
Carbonic anhydrase
catalytic rate, 25
intracellular pH and, 248
kinetics, 55
Carbon monoxide, poisoning, 323
Carbon sources, bacterial cells, 245
Carbonyl groups, 15–16
Carboxylate groups, 14, 14F, 18, 20
Carboxylic-phosphoric acid anhydride, 15
Cardiac muscle
contraction, 249
energy requirement, 316
muscarinic receptors, 360
oxidative phosphorylation, 318
Carrier proteins
active *versus* passive transport, 243
*see also* Active transport
antiporters *see* Antiporters
ATP-driven, 248
*see also* ATP hydrolysis
mechanism of action, conformation changes, 248
symporters, 247
uniporters, 248
*see also specific proteins*
Caspase(s)
cascade, 441
caspase-8, 442–443
caspase-9, 440–441
control, 442–443
Catabolic pathways, 24, 29
*see also specific pathways*
Catabolite activator protein (CAP)
binding sites, 162
DNA bending, 157–158
domain structure, 42, 42F
Catalase, 279, 279F
Catalysis, 49F
Catalytic RNA
ribozymes, 147–148
self-splicing introns, 124
*in vitro* evolution, 147–148
Catastrophe factors, 414–415
Catenanes, 418
β-Catenin, 368–369
Caveolae, 306–307


Page numbers refer to a major text discussion of the entry; page numbers with an F refer to a figure, with a T refer to a table.

CCCP ionophore, 266, 266F
Cdc6, 411
Cdc13 *see* Cyclin B (cdc13)
Cdc20–APC complex, 424–425
Cdc25 phosphatase, 414
Cdc42, actin polymerization, 394, 400–401
*Cdc* genes, 379, 404–405
Cdh3, 447
Cdk1, 410
Cdk2, 61
Cdk4, 436–437
Cdk9, inhibition, 183
Cdk inhibitors (CKIs), 410
Cdks *see* Cyclin-dependent protein kinases (Cdks)
cDNA cloning, 209, 214T
  cell-surface receptors, 347–348
  controls, 213–214
  efficiency, 209
  expression vector, 213–214
  genomic cloning *versus,* 209
  NMDA receptors, 260–261
  oligonucleotide splints, 205
  subtractive hybridization, 188, 188F
CDNA libraries, 70
cDNA libraries, 210, 334
  cloning *see* cDNA cloning
  construction, 209
    pooling strategy, 260–261, 347–348
  gene isolation, 404
  genomic libraries *versus,* 209
  screening, 180
Cell(s)
  animal *versus* plant, 6, 342
  aqueous environment, 12
  buffering systems, 22
  chemical components, 11–23
  communication *see* Cell communication
  compartmentalization, 147, 243–246
    *see also individual organelles*
  culture *see* Cell culture
  differentiation, 155
    *see also* Development
  electrical neutrality, 247
  energy sources, 30–37
  energy use, 23–30
    measurement, 139
    *see also* Enzyme catalysis; Thermodynamics
  fractionation *see* Cell fractionation
  replacement rate, 393, 403
  senescence, 431
  universal features, 1–3
  visualization, light microscope, 221–228
Cell adhesion, 435, 435T, 445–467
Cell communication, 343–372
  *Dictyostelium,* 348
  general principles, 343–351
  neuronal *see* Neurotransmission
  phosphorylation role, 350–351
    *see also* Phosphorylation; Protein kinase(s)
  plants, 370–372
  receptor-mediated, 352–367
    enzyme-linked *see* Enzyme-linked cell-surface receptors
    G-protein linked *see* G-protein-linked (coupled) receptors (GPCRs)
  regulated proteolysis-linked, 367–370
  *see also* Signaling molecules; Signaling pathways; *specific pathways*
Cell culture, 191–192
  cell lines *see* Cell lines
  confluency, 435, 435T
  density-dependent cell growth, 435, 435T
  DNA synthesis analysis, 98–99
  homogenous cell populations, 191
  primary cultures, 191
  secondary cultures, 191
Cell cycle, 403–437, 415, 430F
  analysis
    $^3$H-thymidine incorporation, 393–394, 406–407
    cell fusions, 411
    flow cytometry, 407, 410, 410F
    gel mobility shift assay, 417F
    microinjection experiments, 408–409
    video microscopy, 386, 405–406, 414
    yeast mutants, 405–406, 410, 412
  blocking, 183
  cancer and, 475–476
  cell size control, 384, 408, 410
  chromosome structures, 67, 405–406
    *see also* Chromosome(s)
  control *see* Cell cycle control
  DNA damage effects, 431
  DNA synthesis regulation, 99–100
  function, 403
  $G_0$, 433
  $G_1$, 384–385, 393, 403, 406–407, 408, 415, 433
  $G_2$, 406–407
  gaps, 393, 403
  genes
    *Cdc* genes, 404–405, 410
    evolutionary conservation, 403–404
    execution point, 405–406
    *Rb* gene, 475–476
  length, 406–407, 436
    mitotic index estimation, 405
    phases, 403–404
  megakaryocytes, 430
  microtubule role, 409
  M phase, 403, 412–427, 424–425
    cytokinesis *see* Cytokinesis
    general principles, 412–427
    mitosis *see* Mitosis
  mtDNA *versus* nuclear DNA replication, 338–339
  nucleolus, 123
  overview, 403–407
  protein degradation, 145, 409
  S phase, 93, 96, 98, 339, 403–404, 415
    extracellular control, 432, 432T
    length *versus* DNA content, 403–404
    measurement, 406–407
    regulation, 410–413
    *see also* DNA synthesis
  synchronization, 96–97
    protocols, 404
  *see also* Cell division; DNA replication
Cell cycle control, 61, 407–410
  checkpoints, 415
    caffeine effect, 416
    cancer and, 474
    cell size ($G_1$), 408, 431
    DNA damage, 433–434
    DNA replication, 416
    mitotic entry, 433–434
    negative signaling, 408
    spindle-attachment, 415
  components, 407–410
    Cdks *see* Cyclin-dependent protein kinases (Cdks)
    colchicine effect, 409–410
    cyclins *see* Cyclin(s)
    E2F gene regulatory protein, 475–476
    emetine effect, 409
    *see also specific cyclins/Cdks*
  extracellular, 432
  intracellular
    phosphorylation/dephosphorylation, 414
    positive feedback loops, 414
    Scc1 protein, 412
  retinoblastoma, 475–476
  S-phase, 410–413, 411F
Cell death
  apoptosis *see* Apoptosis
  necrosis, 440, 440F
Cell division, 3
  bacterial, 3
    eucaryotic *versus,* 420
  doubling time, 405
  extracellular control, 432
  generation time, 6
  human, 3, 393
  ionizing radiation effects, 433–434
  limits, 431
  mechanics
    centrosome cycle, 418–419
    chromatid fate, 415
    cytokinesis *see* Cytokinesis
    cytoskeletal filaments, 379–380, 427
      *see also* Actin/actin filaments; Microtubules
    mitosis *see* Mitosis
    *see also* Chromosome(s); Meiosis
  mercaptoethanol effects, 419
  sequence of events, 416
  uncontrolled *see* Cancer
  *see also* Cell cycle
Cell fractionation
  centrifugation (sedimentation), 194
  ER analysis (microsomes), 285
  peroxisome analysis, 281
  submitochondrial fractions, 291, 316–317
  thylakoid vesicles, 336
  *see also individual techniques*
Cell-free extracts, DNA replication experiments
  replication mutants, 90–91
  SV40, 102–103
Cell fusions, cell cycle analysis, 411
Cell growth, 3
  bacterial, 3
  cell size control, 393, 408
  density-dependent inhibition, 435, 435T
  extracellular control, 435
  human, 3
  surface-to-volume ratio–growth rate correlation, 3
Cell hybrids, 192
  *see also* Monoclonal antibodies
Cell isolation, 191
Cell junctions, 445–467
Cell lines, 182, 191
  10T and a half (10T½), 180
  CHO cells, 461
  embryonic stem cells *see* Embryonic stem cells
  HeLa cells, 441–442
  MDCK cells, 313
  mouse embryo fibroblasts, 442–443
  peroxisome deficient, 281
Cell matrix adhesion, 459–461
Cell membrane, actin filament interaction, 393
Cell memory, 173
Cell motility, 393
  actin-based, 393
  bacterial, 386–387
  cancer cells, 383
  cilia *see* Cilia
  colchicine effects, 393
  crawling movements, 393
  cytochalasin B effects, 380, 393
  fibroblasts, 393
  flagella *see* Flagella
  lamellipodia, 383
  neutrophils, 391
  *see also* Cytoskeleton
Cell senescence, 431
Cell-sorting, 191
Cell-surface receptors, cloning, 347–348
Cellulose
  microfibrils, 466
  structure, 16, 16F, 18F, 21
Cell volume, cytokinesis, 428
Cell wall, bacterial, 59–60
CEN3, 76F
Central dogma, 120
Centrifugation, 60–61
  density-gradient *see* Density-gradient centrifugation
  equilibrium sedimentation, 194, 216
  sedimentation coefficient (S), 193
  velocity sedimentation, 193
Centrioles, 382, 418
Centromeres, 67
  acentric chromosomes, 420
  conditional, 422
  dicentric chromosomes, 73, 420, 480
  functional studies, 420–421
  meiotic recombination, 480
  monocentric chromosomes, 420
  transcription interference, 421–422
    *Ctf* mutants, 421–423

yeast (CEN3), 76
*see also* Kinetochores
Centrosome cycle, 418–419
Centrosomes, 383
arrangement, 420
duplication cycle, 418–419
microtubule nucleation sites, 383, 385, 420
Cesium chloride gradients, 194
CG (CpG) islands, 172, 178–179
*see also* DNA methylation
CGRP (calcitonin gene-related peptide), alternative splicing, 185–186
Chair conformation, glucose, 16, 16F
Channel-forming ionophores, gramicidin A, 246
Chaperones, 143–144, 284
endoplasmic reticulum, 284, 286
Hsp60-like, 144–145
Hsp70, 143–144, 254, 283
Charcot–Marie–Tooth disease, kinesin defect, 394
Chargaff, Edwin, 3
Chargaff's rules, 3
Chase, Martha, 65
Chemical bonds, 13
polarity, 15
*see also individual types*
Chemical elements
abundance, 13, 13F
chemical properties, 13, 15
*see also* Atomic structure; *individual elements*
Chemical gradients, 247
Chemiosmotic model, 335–336
Chemistry, 2–3
Chemotaxis
bacteria, 366–367, 366T
*Dictyostelium discoidium,* 348
neutrophils, 391
Chemotaxis receptors, 366–367
Chiral molecules, 16–17
enzyme specificity, 49–50
*see also* Isomers
Chi sites, 112–113, 113F
*Chlamydomonas,* 330, 366, 368–369
Chloramphenicol, alkylation repair effects, 108, 108F
Chloramphenicol acetyltransferase (CAT), reporter gene, 168, 321
Chloride channels, inhibitory neurotransmission, 254
Chlorophyll(s), 331
action spectra, 334, 334F
P700, 331
photon absorption, 331
Chloroplast(s)
antenna complex, 330
chlorophyll *see* Chlorophyll(s)
compartments, 278F, 278FF
DCMU effects, 334, 336
energy conversion, 329–336
ATP synthesis, 330, 335–336, 336
electron transport, 304–307, 330
*see also* Electron transport chains
*see also* Photosynthesis
evolutionary origins, 337
genome, 336–341
DNA transfer, 339
introns, 336
replication, 336
variegation mechanism, 337–338
inheritance, 337
pH effects, 335F
proton gradients, 332
chemiosmotic model, 335–336
*see also* Electrochemical proton gradients
structure, 277, 278F
mitochondria *versus,* 330
transmembrane transport
ferredoxin, 277–278
mitochondria *versus,* 274
plastocyanin, 277–278
proton pumps, 331
signal peptides, 278
CHO cells, 461
Cholera toxin, adenylyl cyclase stimulation, 358
Cholesterol
hedgehog signaling and, 369–370
hypercholesterolemia, 308–309
lipid raft formation, 233
membrane composition, 234–235
receptor-mediated endocytosis, 308–310
steroid hormone synthesis, 345
structure, 345F
Cholinesterase, inhibition, 255
Chorion proteins, gene amplification, 101, 101F
Chromatids
cohesins, 399–401, 412
condensation, 77, 406
fate during cell division, 415
unequal, consequences, 415
separation, 412, 416
Chromatin, 65–76, 169F
30 nm-fiber, 68
beads-on-a-string, 66
DNase I digestion, 71, 71F, 73
gene expression role, 73, 76, 168–169
higher-order structures, 73
immunoprecipitation, 169–170
loops, 78, 78F
micrococcal nuclease digestion, 71, 73
puffs, 350, 350F
remodeling, 68, 168–169
*see also* Histone(s); Nucleosomes
Chromatin immunoprecipitation, 169–170
Chromatin remodeling complexes, 168–169
Chromatography
affinity *see* Affinity chromatography
gel-filtration, 185, 193–194
globin chains, 141
HPLC, 141
hydrophobic, 193
ion-exchange, 184–185, 193
methylated DNA, 108
protein purification methods, 192–195
RNA polymerases, 127
Chromic transients, 334
Chromosome(s)
artificial, 70
bands, 78, 80
Boveri's experiments, 426
duplications, 69–70
evolution, 67, 67F
hereditary information, 426
inversions, 67F
karyotyping, 66, 66F
mammalian, 77
human, 64–65, 66, 80, 92
*see also* Human genome
McClintock's experiments, 479–480
mitotic, 73, 381–382, 405–406, 412–427, 414
*see also* Mitosis
pulsed-field gel electrophoresis, 68
rearrangement, 479–480
replication, 70
end-replication problem, 97
*see also* Cell cycle; Mitosis
sex, 66
X-inactivation, 180–181
structure, 72–80
centromeres *see* Centromeres
dicentric, 73, 404–407
DNA packaging, 65–72
lampbrush loops, 77–78, 78F
origins of replication, 70
polyteny, 78, 322
puffs, 350, 350F
telomeres *see* Telomeres
*see also* Chromatin; Supercoiling
yeast, 66, 74–76, 75F
X-ray damage, 433–434
*see also* DNA
Chromosome assembly factor-1 (CAF1), 102–103, 103F
Chromosome puffs, 350, 350F
Chromosome signaling cytokinesis hypothesis, 428–429
Chromosome-transmission-fidelity *(Ctf)* mutants, 421–423
Chronic myelogenous leukemia, karyotype, 66, 66F
Cilia, 395–397
axoneme, 397–398
coordinated movement, 395–396, 396F
dynein motor, 397
sliding microtubule mechanism, 393
structure, 395–396
*see also* Flagella; Microtubules
Circadian rhythms, 173, 176, 177F
cI repressor, 172, 172F
Citric acid cycle, 27F, 35–36, 35F
electron carriers, 316
mitochondrial metabolism, 316
oxygen consumption, 36
stoichiometry, 36
Civ1 protein kinase, 61
Clathrin-coated pits, 290, 291, 306–307
Clathrin-coated vesicles, 291, 291F, 306–307
adaptins, 279, 290, 291
cargo particles, 306–307
dynamin, 291
GTP requirement, 291
pinching-off, 291
Clathrin coats, 306–307
heavy/light chains, 291
triskelions, 290–291
Claudin
behavior, 453F
*Clostridium perfirngens* binding, 452
role tight junctions, 451
structure, 451F
Cleavage furrow, cytokinesis, 429
Clock genes, circadian rhythms, 176
Clock protein, 176
Cloning (whole animal), 152
Cloning vectors, 205
alkaline phosphatase treatment, 213–214
shuttle vectors, 209
*Clostridium perfirngens,* claudin binding, 452, 452F
Coat color, 305, 305F
Codons, 3, 127
amino acid co-linearity, 1
stop *see* Stop codons
usage, 336
wobble bases, 136
Codon usage, 336
Cofilin
actin filaments depolymerization, 383
actin filaments polymerization, 400
Cofilin-homology domain, 400F
Cohesins
DNA topology, 418
Eco1 protein, 412–413
Scc1 protein, 411, 412
Scc3 protein, 412
Smc1 protein, 412
Smc3 protein, 412
Colchicine (colcemid)
cyclin B effects, 409–410
microtubule inhibition, 375–376, 393
Collagen, 462–464
defects, 463
Collagenase, cell isolation, 191
Colon cancer
age-related incidence, 470, 470F, 470FF
B-Raf, 477
Color blindness, RFLP analysis, 69–70, 70F
Compartmentalization, 263–266
evolutionary role, 147
maintenance, 289–296
Competitive inhibition, glucose transporter (GLUT1), 245
Complementation analysis, 217, 258–259
Condensation reaction, 17
Condensins, 80, 80F, 400
DNA topology, 418, 418F
Conditional-lethal mutations, 90–91, 209, 247, 262
*Cdc* genes, 405
temperature-sensitive *see* Temperature-sensitive mutants
Connexin, 455
Connexon, 455


Page numbers refer to a major text discussion of the entry; page numbers with an F refer to a figure, with a T refer to a table.

Consensus sequences
  bacterial promoters, 115, 120
  glucocorticoid response element, 349
Constant domain (immunoglobulins), 152–153
Contamination
  ancient DNA analysis, 83–85
  extraterrestrial life, 2–3
  membrane vesicles, 240
  noncatalytic RNA in ribozyme selection experiment, 147–148
  Pasteur's experiment on spontaneous generation, 2–3
Contractile ring, cytokinesis, 427
Convergent evolution, 274
Coomassie blue, 433F
Cooperativity, 180–181
  allosteric transitions, 52–53
  hemoglobin, 51
  single-stranded DNA-binding proteins, 92
  transcriptional regulation, 180–181
COPI vesicle formation, 292–293
COPII vesicle formation, 272, 292
Corn, meiotic recombination, 479–480
Cortisol, structure, 345F
Cosmic radiation, 19
Coupled transport
  active transport, 247
  mitochondria, 317
Covalent bonds, 13, 22
  permanent dipoles, 13
  polar, 13
*Cox2* gene
  evolution, 7, 7F
  gene transfer, 7
Creatine phosphate, 34
Cre–Lox system, 115, 115F
  site-directed mutagenesis, 219
Cre recombinase, 115, 115F, 219
Crick, Francis, central dogma, 120
Cro repressor, 42, 172, 172F
Cross-feeding experiments, 36–37, 36F, 37F
Crossing-over
  general recombination, 111
  homologous recombination, 111, 111F
  unequal, 69–70
C-terminal domain (CTD) RNA polymerase II, 127–128
  hypophosphorylation in HIV gene expression, 183
  phosphorylation, 121
*Ctf* (chromosome-transmission-fidelity) mutants, 421–423, 423F
Curie (Ci), 13
Cyan fluorescent protein (CFP), 226–227
Cyanide, cytochrome oxidase binding, 323, 327, 327T
Cyanobacteria, photosynthesis, 330
Cyclic AMP (cAMP)
  G-protein-linked receptors, 353–354
  phosphorylase kinase signal integration, 359–360
  PKA-mediated effects, 354
  structure, 353
  synthesis *see* Adenylyl cyclase
Cyclic AMP phosphodiesterase
  caffeine inhibition, 354
  *Drosophila* gene location, 358
  theophylline inhibition, 358
Cyclic GMP (cGMP)
  hydrolysis, 345
    rhodopsin-mediated, 356
    *see also* Cyclic GMP phosphodiesterase
  synthesis, 345
Cyclic GMP phosphodiesterase
  Viagra inhibition, 345
  visual transduction, 356
Cyclic photophosphorylation, 332
Cyclin(s)
  Cdk-binding *see* Cyclin–Cdk complexes
  M-phase, 424–425
  overexpression analysis, 410, 437
  yeast cyclins, 384–385, 410
  *see also individual cyclins*
Cyclin A, 61
  Cdk2 binding, 61
  hydrophobic cleft, 201
  phage-display, 201
Cyclin A proteins, 201
Cyclin B (Cdc13)
  colchicine (colcemid) effects, 409–410
  concentration, 414
  degradation, 145, 409–410
  destruction box, 424–425
  mutants, 424–425, 425F
Cyclin–Cdk complexes, 409–410, 411
  cyclin D–Cdk4 complex, 436–437
  $G_1$-Cdk, 410, 433
  M-Cdk, 408, 414, 414F, 424–425
    positive feedback, 414
  regulation by phosphorylation, 410, 414
    Cdc25 phosphatase, 414
    Cdk inhibitors, 410
    Wee1 kinase, 414, 436, 436F
  S-Cdk, 411
Cyclin D, 433, 436–437
Cyclin-dependent protein kinase 1 (Cdk1), 410
Cyclin-dependent protein kinase 2 (Cdk2), 61
Cyclin-dependent protein kinase 4 (Cdk4), 436–437
Cyclin-dependent protein kinase 9 (Cdk9), inhibition, 183
Cyclin-dependent protein kinases (Cdks), 410, 436–437
  cyclin-binding *see* Cyclin–Cdk complexes
  inhibition, 183
  overexpression analysis, 437
  phosphorylation, 61
  yeast, 410
Cyclin T1, 183
Cyclobutane dimer, 107
Cycloheximide, 99–100, 134, 135, 145, 252, 322
  mechanism of action, 274–275
Cysteine misincorporation experiment, 139
Cysteine residues
  disulfide bonds, 46
  structure, 15F
Cystic fibrosis, 272
Cytochalasin B, 245
  cell motility effects, 380, 393
Cytochalasin D, cytokinesis effects, 427
Cytochrome(s), 322–324, 326–328
  heme group, 323
  photosynthesis, 334
  redox potentials, 322
  spectroscopic absorption bands, 326–327
  *see also individual cytochromes*
Cytochrome $a_3$, 323
Cytochrome $b_6$-*f* complex, photosystem II, 334
Cytochrome $b$-$c_1$ complex, 238
  *petite* mutants, 341, 341F
Cytochrome *c*
  apoptosis, 391–392, 441–442, 442
  electron transfer, 238, 323
Cytochrome oxidase complex, poisoning, 323, 327, 327T
Cytokine-mediated signaling, 273
  interferon-γ, 365–366
Cytokinesis, 427–430
  asters, 424, 428–429
  *C. elegans,* 429F
  cell volume effects, 428
  cleavage furrow, 429
  contractile ring, 427
    *see also* Actin/actin filaments
  cytochalasin D effects, 427
  models, 428–429
  plant cells, 428
  sand dollar egg, 428
  symmetrical *versus* asymmetrical, 412
  timing, 415, 427
  vesicles/vesicular transport and, 429
  *see also* Mitosis
Cytosine, base pairing, 63, 63F
Cytoskeleton, 373–401
  actin filaments *see* Actin/actin filaments
  cellular behavior role, 391–401
    axonal transport, 394, 399
    budding in yeast, 399–400
    cell division, 427
    cell movement *see* Cell motility
  evolution, 375
  filament polymerization, 373–382, 376
  integrin role, 459
  intermediate filaments *see* Intermediate filaments
  microtubules *see* Microtubules
  molecular motors *see* Motor proteins
  regulation, 382–387
    nucleation, 382–387
Cytosol
  pH, 22
  as reducing environment, 240, 352
    NADPH role, 240


## D


*Dam* methylase, 177–178
Dark-field microscopy, 223, 223F
Databases
  genomic, 5, 45
    exon identification, 66–67
  protein, 42, 198
  searching, 42, 69
D-box, 145
DCMU, 331, 334, 336
DEAE-sepharose, 195
Deamination, 104
Death genes, 440
Decapentaplegic protein (DPP), 435, 435F
Deletion mutation, 37
Delta G ($\Delta G$), 25–26, 27, 28, 32, 61
  ATP hydrolysis, 318
  electron transport, 324–326, 326
  membrane transport, 251
Delta H ($\Delta H$), 25
Delta S ($\Delta S$), 25
Denaturation, protein, 40, 43, 45F, 185
Density-gradient centrifugation, 137
  CsCl gradients, 100F
  DNA, 93, 93F
    buoyant density, 194
  membrane proteins, 235–236
  sucrose gradients, 235–236
Deoxyribonucleic acid *see* DNA
Dephosphorylation, 61–62
Depurination, 104
Destruction-box, 145
  cyclin B (Cdc13), 424–425
Detergents, 240
  lipid bilayer effects, 235–236
  protein denaturation, 40
Development
  apoptosis, 440–441
  *Caenorhabditis elegans,* 414, 440
  cell volume (early development), 428
  combinatorial gene control, 167, 173
    *Eve (Even-skipped)* gene, 170–171
  DNA rearrangements, 152–153, 163
  *Drosophila see under Drosophila melanogaster*
  egg, 101
  eye, 437
  FGF role, 218
  Hox genes, 81, 81F
  human, 3
  mouse paws, 441
  muscle, 155, 166
  neuronal growth cones, 399
  *Raf1* genes, 218–219
  signaling, gap-junctions, 456
*Dhfr* (dihydrofolate reductase) gene, 102, 102F
Diagon plots, 83, 83F
Dialysis, equilibrium, 57
Dicentric chromosomes, 73, 420F
  artificial (plasmid constructs), 420
  meiotic recombination, 479–480
*Dictyostelium discoidium*
  aggregation
    actin filaments, 387
    chemical signaling, 348, 348F
  nuclear pore complexes, 228
Dideoxy DNA sequencing, 206

Dideoxy nucleotides, 89–90, 90F, 206
Diffusion
  cytoskeletal polymerization, 376
  equation, 376
  rate, 28, 56, 376
  size/polarity effects, 243
  transport *versus*, 244–245, 370
Diffusion coefficient, 28, 376
Digitalis, sodium–potassium pump effects, 249
Dihydrofolate reductase (DHFR)
  *Dhfr* gene, 102, 102F
  mitochondrial import, 276
Dihydroxyacetone phosphate, 50
*dilute* mice, 305, 305F
Dimers
  cyclobutane, 107
  DNA (pyramidine nucleotides), 91, 98–99, 100, 104
  protein
    ATP synthase, 320–321
    Fos–Jun heterodimers, 158–159
    gene regulatory proteins, 154–155
    glycophorin, 240–241
    head-to-tail, 42–43
    HLH proteins, 155
    myogenin, 155
    protein, 43
    receptors, 334, 362
    tail-to-tail, 42–43
    tubulin, 355, 356, 361, 375
    *see also* Protein complexes (assemblies)
Dinitrophenol, 317, 324
Dipoles
  α-helix, 40
  covalent bonds, 13
*Dishevelled* gene, Wnt signaling pathway, 368
Dissociation constants ($K_d$), 61, 61T
Distance-matrix tree construction, 83
Disulfide bonds, 40
  collagen synthesis, 464
  cytosolic reduction of, 240, 375
  endoplasmic reticulum, 283
  keratin filaments, 375
  protein structure, 46
    barnase, 276
  SDS-PAGE and, 196
  test for, 46, 46F
Dithiothreitol (DTT), 46
DNA, 1, 3, 64F, 64FF, 154F, 154FF
  ancient, 83–85
  base composition, 63–64, 98
  branch migration, 107, 112
  charge, 205
  circular, 63
    replication, 96–97
    topology, 79
  cloning *see* Gene cloning
  compaction, 66, 68
    *see also* Chromatin
  damage *see* DNA damage
  density-gradient centrifugation, 93, 93F, 194
  electrophoresis *see under* Electrophoresis
  fingerprinting, 85, 85F
  human, 64
    *see also* Human genome
  hybridization *see* DNA hybridization
  indirect end labeling, 76
  inheritance, 93
  isolation, 66, 98, 194
  linker, 68
  loops, 77–78, 78F
  M13 virus, 64
  manipulation, 203–214
  methylation *see* DNA methylation
  microarrays *see* DNA microarrays
  polarity, 63, 82–83, 120, 137
  rearrangements, 152–153, 163
  repair *see* DNA repair
  repetitive *see* Repetitive DNA
  replication *see* DNA replication
  restriction analysis, 69–70, 69F, 204–205, 209–210
  RNA *versus*, 147
  sequence maintenance, 87–88
  single-stranded, 64, 64F
  stability, 87, 141
  supercoiling *see* Supercoiling
  topology *see* DNA topology
  unwinding, 94–95, 95F
    *see also* DNA helicases
  visualization, 222
  *see also* Chromosome(s); Genome(s)
DNA amplification
  chorion genes, 101, 101F
  gene, 101
  PCR *see* Polymerase chain reaction (PCR)
  techniques *see* Polymerase chain reaction (PCR)
DNA bending
  CAP-mediated, 157–158, 157F
  protein–DNA interactions, 156–157, 157–158
DnaB helicase, 94–95, 94F, 95F
DNA binding, stability, half-life, 167
DNA-binding proteins, 43
  binding sites, 160
    deletion analysis, 167
    footprinting, 101, 160, 160F
    gel mobility shift assays, 159
  DNA bending, 157–158, 163
  domains, 163
  hormone receptors, 168
  kinetics, 156–157
  motifs, 153–160
    HLH domains, 155
    leucine zipper, 158–159
    zinc fingers, 154
  origin recognition complex, 101–102, 102F
  purification, 159
  repressors *see* Repressor proteins
  single-strand binding, 90, 92, 92F
  specificity, 99, 154
  *see also* Gene regulatory proteins; Protein–DNA interactions
DNA damage, 91, 91F
  alkylating agents, 107–109, 444
  apoptosis induction, 431, 444
  cell cycle effects, 431
  deamination, 104
  depurination, 104
  dimer formation, 91, 91F, 98–99, 100, 104
  radiation and, 12, 88–89, 431
  responses to, 431
  *see also* DNA repair; Mutagenesis
DNA fingerprinting, 85, 85F
DNA footprinting, 101, 160, 160F
DNA glycosylase, 104
DNA helicases
  DnaB helicase, 94–95, 94F, 95F
  M13 primosome, 94, 94F
  RecBCD, 113
DNA hybridization, 116
  indirect end labeling, 76
  polytene chromosomes, 78
  reannealing experiments, 175
  replication analysis, 100–101
  *see also* Southern blotting
DnaK, protein folding, 143–144, 143F
DNA libraries, 70, 210
  cDNA *see* cDNA libraries
  construction, 209
  genomic, 209
  overlapping clones, 209
  screening, 209
  *see also* Gene cloning
DNA ligases
  T4 ligase, 204
  temperature-sensitive, 90–91
DNA ligation, 71, 204–205
  cDNA library construction, 209
  ligases *see* DNA ligases
  oligonucleotide-ligation assay, 212
DNA loops, 77–78, 78F
  repression mechanism, 162–163
  transcriptional role, 77–78
    enhancer model, 162–163
DNA melting, 94–95, 95F
DNA methylation
  5-azacytidine effect, 180
  CG (CpG) islands, 178–179
    evolution, 172
  chromatography, 108
  DNA repair role, 89, 98
  gene expression role, 169–172, 177–178
    γ-globin gene, 178–179
  methylation-dependent repair, 91
  methyltransferases *see* DNA methyltransferases
  replication and, 170, 177–178
  restriction–modification system, 203
DNA methyltransferases, 108–109
  *Dam* methylase, 177–178
  *de novo,* 174
  maintenance methyltransferase, 174
DNA microarrays, 75–76, 206–207, 218F
  ASOs, 217
  cancer cell typing, 152–153
DNA polymerase(s)
  DNA pol α *versus* XP-variant polymerase, 107
  DNA polymerase I, 91, 93
  DNA polymerase III, 95
  mismatch extension, 91, 95
  proofreading exonuclease, 90, 93, 93F
  restriction digests, 204
DNA rearrangements, 172
  developmental, 152–153
  phase variation, 175
DNA repair, 89–92, 103–110
  alkylation, 107–108, 444
  base excision repair, 104
  deficiency (xeroderma pigmentosum), 107
  double-strand breaks, 110
    *see also* Homologous recombination
  enzymes, 91, 98, 105–106
    *see also specific enzymes*
  error-prone, 106
    *see also* SOS response
  methylation role, 89, 98
  mismatch repair, 91, 113, 116, 444
  nucleotide excision repair, 105–106
  photoreactivation, 104
  RecA role, 103–104, 105–106
  UvrABC endonuclease, 105–106
  variety, 104
  *see also* DNA damage
DNA replication, 89–96, 92F, 97F
  2D electrophoresis, 100, 100F
  bidirectional, 96
  cell cycle checkpoint, 414
  circular dsDNA, 96–97
  conditional-lethal mutants, 90–91
  end-replication problem, 97
  eucaryotes, 99
  fidelity, 90
    leading *versus* lagging strand, 95, 95F
  gene expression and, 177–178
  mechanism, 88–96
  mitochondrial, 336, 338
  Okazaki fragments, 89
  origin recognition complex, 101–102, 102F, 199, 380
  origins *see* Replication origin(s)
  overreplication, 101
  polymerases *see* DNA polymerase(s)
  polymerization reaction, 89, 90, 93
  primosome, 94
  replication fork, 88, 89, 91, 96
  RNA priming, 94
  semi-conservative nature, Messelson–Stahl experiment, 93, 93F, 194
  strand-labeling experiments, 99–100
  structure effects, 90, 92
  *see also* Cell cycle; DNA synthesis; Mitosis
DNA–RNA hybrids, 121, 121F, 125, 128–129, 128F
DNase
  chromatin digestion, 71, 71F, 73–74
  footprinting, 101, 102F
DNA sequence
  analysis *see* DNA sequence analysis
  genomic diversity, 3
  mutation effect, 107–108


Page numbers refer to a major text discussion of the entry; page numbers with an F refer to a figure, with a T refer to a table.

  - palindromic, 160, 190
  - structure effects, 211
- DNA sequence analysis, 83–85, 206, 206F, 208F
  - ancient DNA, 83–85
  - $\beta$-globin genes, 83, 122
  - diagon plots, 83, 83F
  - eucaryotes, 6, 7–9
  - exon identification, 66–67
  - footprinting, 160
  - human tumors, 476–477
  - intron/exon evolution, 122
  - mitochondrial DNA, 83–85
  - *see also* Genome sequencing; Restriction digests
- DNA structure, 63–65
  - base pairing, 3, 63–64, 63F
  - Chargaff's rules, 3
  - coiling, 67
    - *see also* Supercoiling
  - complementary strands, 64, 82–83
  - double helix, 67, 145–146
    - major/minor grooves, 63–64, 146, 189
    - right-handedness, 64, 64F
  - flexibility, 67
  - hairpin formation, 211
  - nucleotides, 64
    - *see also* Nucleotides
  - packaging, 68
    - *see also* Chromatin; Nucleosomes
  - plectonemic, 79–80, 79F
  - radiation effects, 12
  - replication effects, 90
  - sequence effects, 211
  - single-stranded, 64, 64F, 86
  - solenoidal, 78–80, 79F
  - *see also* Chromosome(s); DNA topology
- DNA synthesis
  - autoradiography, 98–99, 98F, 375–376, 393–394
  - cell culture, 98–99
  - cell-cycle regulation, 99–100, 403–404
  - hydroxyurea blocking, 405
  - PDGF receptor stimulation, 364
  - *see also* Cell cycle; DNA replication
- DNA topoisomerases, 79F
  - energetics, 80, 89
  - enzyme kinetics, 80
  - topoisomerase I, 80, 89, 92
  - topoisomerase II, 92, 400
  - transcription role, 121
  - *see also* DNA topology
- DNA topology, 67
  - catenanes, 418
  - knots, 418
  - nucleosome effects, 79–80
  - supercoiling, 78–80, 79F
  - transcription effects, 121
  - *see also* DNA structure; DNA topoisomerases
- DNA transposition, 81, 108–109
  - replicative *versus* nonreplicative, 116F
  - *see also* Site-specific recombination
- DNA viruses, 121
- Dolichol phosphate, 284
- Dolly the sheep, 152
- Dominant-negative mutation, 208, 215
- Dopamine receptors, 357
- Dosage compensation, 174
- Double-strand breaks (DSBs), repair, 110
- Double thymidine block, 406
- Drift velocity, ion flow, 253–254
- *Drosophila melanogaster*
  - cAMP phosphodiesterase, gene location, 358
  - chromosome puffs, 350, 350F
  - development
    - cyclin D–Cdk4 complex, 436–437
    - DPP and wing development, 435, 435F
    - egg development, 101
    - eye development, 217F, 217T, 437, 437F
    - *Raf1* mutants, 218–219
    - segmentation genes, 170–171
  - evolutionary relationships, 41
  - gene dosage compensation, 174
  - genetics (classical), 217
  - genome mapping, 78
  - mutations
    - *Dunce* mutants, 358
    - eye color, 217
    - *Shibire* mutants, 311
  - polytene chromosomes, 78, 322
  - shaker $K^+$ channels, 257–258, 258F
  - signaling pathways
    - ecdysone signaling, 350
    - Wnt/hedgehog signaling, 368, 370
- Drug development, cancer, 482–483
- DTT (dithiothreitol), 46
- Duchenne's muscular dystrophy (DMD), 213
- *Dunce* mutants, 358
- Dynamic instability, microtubules, 377–378, 406
- Dynamin
  - mechanism of action, 291
  - synaptic transmission, 311
- Dynein, 397


## E


- E2F gene regulatory protein, 475–476
- E-cadherin, 446
- Ecdysone, 350
- Eco1 cohesin protein, 412–413
- EcoRI restriction enzyme, 189, 204, 205, 247
- Edeine, 137
- EDTA, cell isolation, 191
- EF-Tu elongation factor, 137
- Egg
  - activation, 350–351, 377–378
  - development, 101, 187–188
  - dispermic, 426
  - sand dollar, cytokinesis, 428
  - sea urchin, 426
  - sperm interactions, 256
  - tetrapolar, 426
  - *Xenopus*, 404
- Egg–sperm interactions, 256
- Egg white, 51
- Ehlers–Danlos syndrome, 463
- Electrochemical proton gradients
  - ATP synthesis, 319
    - chloroplasts, 330, 331
    - coupling mechanism, 319
    - *Thiobacillus ferrooxidans*, 318
    - uncoupling, 324
    - *see also* ATP synthase
  - bacterial flagella, 328–329
  - chemiosmotic model, 335–336
  - chloroplasts, 330, 335–336
  - coupled transport, 317
  - dinitrophenol effect, 317, 324
  - methanogenic bacteria, 328
  - microscopic reversibility, 328–329
  - mitochondrial, 316–317
  - nigericin effect, 317
  - proton pumps *see* Proton pumps
  - *see also* Membrane potential
- Electrodes
  - DNA electrophoresis, 205
  - protein electrophoresis, 196
- Electron microscopy, 228–230
  - resolving power, 229
  - tight junctions, 453, 453F
  - transmission *vs.* scanning, 228
- Electron shells, 12, 19–20, 19F
- Electron spin-resonance (ESR) spectroscopy, lipid bilayer, 217–218, 234
- Electron transport chains, 322–329
  - bacterial
    - methanogens, 328
    - *Thiobacillus ferrooxidans*, 325–326, 332
  - cofactor recycling, 316, 325
  - inhibitors, 327, 327T, 336
  - microscopic reversibility, 328–329
  - mitochondrial respiratory chain
    - cytochromes *see* Cytochrome(s)
    - electron affinity, 316
    - order assignment, 323
    - redox potentials, 323, 323T
    - *see also* Mitochondria
  - oxygen requirement, 316
  - photosynthetic, 304–307, 331
    - photosystem electron flow, 334–335
    - plastoquinone, 331
    - *see also* Chloroplast(s)
  - redox reactions, 301, 324–326, 332
  - uncoupling, 324, 336
- Electrophoresis
  - agarose, 71, 71F, 85–86, 210, 212, 321
  - DNA analysis, 68, 77, 100, 100F, 101F, 109, 203, 204F
    - DNA fingerprinting, 85, 85F
    - restriction digest analysis, 71, 71F, 106, 188, 204, 210–211
    - RFLP analysis, 69–70, 70F
    - sequencing, 208
    - topology effects, 79, 80, 95, 400, 401
    - *see also* Southern blotting
  - electrodes, 196, 205
  - equilibrium constant determination, 58
  - gel mobility shift assay, 417F
  - polyacrylamide *see* Polyacrylamide gel electrophoresis (PAGE)
  - protein analysis, 196–199
  - proteins, 100, 144, 151F, 198, 467
    - $\beta$-catenin, 369F
    - EGF receptor, 432–433, 433F
    - enzyme purification, 49–50
    - membrane protein analysis, 236, 256
    - post-translational modification effects, 186
    - *see also* Immunoblots
  - pulsed-field, 68, 68F, 203
  - RNA analysis, 116, 126, 179
  - two-dimensional, 100, 144, 196, 351, 368
  - visualization *see* Autoradiography; Gel staining
- Electrostatic interactions, amino acids, 14
- Elongation factors, EF-Tu, 137
- Embryonic stem cells
  - cell lines, 191
  - targeted recombination, 219
  - X-inactivation analysis, 180–181
- Emerson, R., 334
- Emetine, cell cycle effect, 409–410
- Endocrine signaling, 344
  - neurotransmission *versus*, 344
- Endocytosis, 290, 290F, 291, 305–310
  - caveolae, 306–307
  - early endosomes, 291
  - exocytosis *versus*, 310–311
  - fluid-phase (pinocytosis), 307–308
  - late endosomes, 277, 278, 291
  - phagocytosis, 306
  - receptor-mediated *see* Receptor-mediated endocytosis
  - temperature effect, 308
- Endoglycosylase H (endo H), 273, 285
- Endoplasmic reticulum (ER)
  - disulfide bond formation, 283
  - ER chaperones, 284, 286
  - ER to Golgi transport, 291, 292, 293, 297–298
    - COPII vesicle formation, 272, 292
  - glycosylation, 284, 285
    - mannosidase, 275, 299–300, 300F
  - lumen, 283
    - enzymes, 272, 285
    - nucleoside triphosphates, 284
  - microsomes, 283
  - misfolded proteins, 272, 284, 286
  - multipass transmembrane proteins, 284, 284F
  - phospholipids, 285
  - quality control, 272, 284
  - resident proteins, 297
  - rough ER, 265, 283
  - scramblases, 285
  - signal sequences, 264, 284
    - retrieval signals, 297–298
  - smooth, 283
  - transmembrane transport, 259–264, 264
    - cotranslational *versus* post-translational import, 285–286
    - Hsp70 role, 284

nuclear transport *versus,* 284
Sec61 translocator complex, 283, 284, 286
translocation criteria, 285
translocation mutant selection, 286
*see also* Golgi apparatus; Membrane proteins
Endosomes
early, 282, 291
late, 277, 278, 291
*see also* Lysosomes
pH, 307
vesicular transport, 264
Endosymbionts, mitochondria/chloroplast origins, 337
End-replication problem, 97
Energy
cellular processes, 13, 23–30
conversion, 315–341
chemical–mechanical transduction, 389
*see also* ATP (adenosine triphosphate); Chloroplast(s); Mitochondria
light, 331
oxidation reactions, 24
requirements, 30
sources, 30–37
storage, 30
thermodynamics *see* Thermodynamics
Engelmann, T.W., 333
Enhancers, 162–163
DNA looping model, 162–163
Maloney murine sarcoma virus, 348–349
NFκB gene regulatory protein, 273
scanning/entry site model, 162–163
Enolase, 33
Enthalpy *(H),* 25, 44
Entropy *(S),* 25, 44
Enzyme(s)
active site, 49
allosteric, 52–53
catalysis *see* Enzyme catalysis
chiral specificity, 49–50
histidine side chains, 23
kinetics *see* Enzyme kinetics
purification, 49, 183–184
*see also* Protein purification
regulation, 49, 57, 59
*see also* Phosphorylation
restriction enzymes, 69, 188
ribozymes *versus,* 147
signaling pathways, 362–367
specific activity, 193
Enzyme catalysis
equilibrium constants *see* Equilibrium constants *(K)*
equilibrium ratio, 26
histidine side chains, 23
kinetics *see* Enzyme kinetics
multienzyme complexes, 51
nucleophilic, 62
pH effects, 22
product concentration, 25, 34
reaction equation, 244
substrate concentration, 25, 34, 49
thermodynamics, 23–30, 34, 49–50
transition states, 50
uncatalyzed reactions *versus,* 49
Enzyme inhibitors, 50
methotrexate, 276
PALA, 53, 53F
phosphoglycolate, 50
resistance, 55
telomerase inhibition, 67
transition state analogs, 50
*see also* Poisons
Enzyme kinetics, 49–50
Lineweaver–Burke equation, 59
Michaelis constant, 56, 59
Michaelis–Menten kinetics, 59, 244
multienzyme complexes, 51
reaction rate, 26, 28, 49, 56
enhancement, 25, 25F
rate constants, 49
saturation behavior, 49
topoisomerases, 80
turnover number, 55
Enzyme-linked cell-surface receptors, 361–367
autophosphorylation, 364
bidirectional signaling, 362
chemotaxis receptors, 366–367
dimerization, 338, 362, 364
ephrin receptors, 362
JAK–STAT signaling, 365–366
MAP kinases *see* MAP kinase(s)
PDGF receptor *see* Platelet-derived growth factor (PDGF) receptor
PI 3-kinase, 56–57, 56F, 336–337, 364
receptor-like tyrosine phosphatases, 363–364
receptor tyrosine kinases, 364
*see also* GTP-binding proteins (GTPases); Protein kinase(s)
Enzyme purification, 193T
Enzyme–substrate complex, 49–50
Ephrin receptors, 362
Epidermal growth factor (EGF), 307–308
cell cycle control, 432, 432T
receptor characterization, 432–433, 433F
Epigenetic inheritance, 74–75
Epithelial cells
endocytosis experiments, 307–308
microvilli, 250
polarized, 312–313
protein sorting, 312–313
transcytosis, 312–313
Epithelial sheets
electrical resistance, 451, 452T
structural properties, 449–450
types, 450
Epitopes, 191
Epitope tagging, 198
FLAG, 188, 259
hemagglutinin, 169–170
Equilibrium constants *(K),* 26, 27, 46, 49, 57
determination, 57–58
Equilibrium dialysis, 57
Equilibrium ratio, 26
Equilibrium sedimentation, 194, 216
Equivalence point, 14, 14F
Error-prone PCR, 149
Erythrocytes *see* Red blood cells
Erythromycin, 381
*Escherichia coli*
adaptive response to alkylating agents, 108
arabinose metabolism, 164, 166–167
arginine biosynthesis, 161
CAP protein, 42, 157–158
chemotaxis, 366
citric acid cycle, 36
DNA polymerase I, 90, 93
DNA repair, 105–106
mismatch repair, 89
photoreactivation, 104
DNA replication, 85–86, 93
DNA supercoiling, 79–80, 79F
homologous recombination, 111–112
lactose metabolism *see* Lactose *(Lac)* operon
membrane protein organization, 287
mutation rate, 88
plasmids, 79–80, 79F
SOS response, 104
surface-to-volume ratio, 265
T4 infection, 447–448
topoisomerase I, 80
UV-induced mutation, 100, 104–106
Estradiol, structure, 345F
Ethanol
bacterial membrane transport, 245
metabolism, 20
Ethidium bromide, DNA staining, 204
Ethylene, plant signaling, 371–372
Eubacteria *see* Bacteria
Eucaryotes
DNA replication, 99
*see also* Cell cycle; DNA replication; Mitosis
gene expression
bacterial genes *versus,* 213–214
control mechanisms, 152
gene regulatory proteins, 161
genomes, 6–9
phylogeny, 5, 371, 371F
procaryotes *versus,* 137, 350
protein synthesis, 137
amount, 265
Eucaryotic cells
cellular organization, 6
organelles, 263–266
mammalian *see* Mammalian cells
plant *see* Plant cells
surface-to-volume ratio, 265
yeast *see* Yeast
*Eve (Even-skipped)* gene control, 170–171, 170F, 171F
Bicoid/Hunchback activation, 170–171
Krüppel/Giant repression, 170
Evolution, 6–7
CG (CpG) islands, 172
chromosome 3 (apes), 67, 67F
compartmentalization, 147
convergent, 274
cytoskeletal components, 375
eucaryotes, 371, 371F
genetic code, 135–136
genomes
human genome, 66–67, 83–85, 110
imprinting, 174
mitochondrial genome, 7
new genes, 5, 8
mitochondrial origins, 337
multicellular organisms, 371, 371F
natural selection *see* Natural selection
plant hemoglobins, 8–9, 9F
rate, 6–7, 8
efficiency-of-repair hypothesis, 6
generation time hypothesis, 6
metabolic rate hypothesis, 6
variations, 6
*see also* Mutation rate
ribosomal RNA, 5
RNA significance, 147–149
*in vitro,* 147–148
*see also* Phylogenetic trees; Tree of life
Evolutionary conservation
actin/actin filaments, 375
genes, 5
cell cycle, 403–404
glycolysis, 31
histones, 8, 8T, 71
protein folds, 42
tubulin, 375
Evolutionary distance, 6–7
Evolutionary tracing, 47
Excitatory neurotransmitters, sodium channels and, 254
Exercise
ATP hydrolysis, 316
energetics, 34
glucose metabolism, 316
glycogen catabolism, 354
Exocytosis, 310–314
constitutive (regulated) pathway, 311, 312
default pathway (bulk transport), 312
endocytosis *versus,* 310–311
neurotransmitter release, 313
*see also* Synaptic vesicles
rates, 311–312
secretory signals, 313
secretory vesicles *see* Secretory vesicles
sorting signals, 312–313
VSV G protein studies, 312
Exons, 68, 121–123
β-globin gene, 83
computer algorithm for identification, 66–67
evolution, 122
splicing, 122
splice variants, 121–122
*see also* Splicing
Expression vectors, 213–214
Extracellular matrix
animal, 461–465
fibronectin, 460, 464–465

Extraterrestrial life, 2
Eye, human, refractive indices, 223
Eye color, *Drosophila* mutations, 217
Eye development, *Drosophila* mutations, 437, 437F

## F

FACS (fluorescence-activated cell sorter), 191, 414
Factor VIII gene, degenerate probes, 210
FAD/$FADH_2$
  citric acid cycle, 316
  redox reactions, 325
Familial hypercholesterolemia, 308–310
Fas/Fas ligand, 441–442, 441F
  cancer role, 441
  pathway, 442–443
Fast axonal transport, 399
Fat cells, energy storage, 30
Fats *see* Lipid(s)
Fatty acids
  glucose oxidation *versus,* 32
  hydrogenation, 233
  membrane protein anchors, 239
  oxidation, 34–35, 34F
  saturated, 233
  structure, 17, 17F
  synthesis, 215
  unsaturated, 233
Fatty acid synthase, 215
Fava beans, 240
FCCP, 302, 327, 336
Feedback control
  negative
    gene expression control, 161
    purine synthesis, 52
    *see also* Repressor proteins
  positive, 442
    cell cycle control, 414
Feeding/food sources, 4, 24
  energetics, 30–37
  storage, 30
Feline leukemia virus type C (FeLV-C) receptor, gene mapping, 192, 192F
Fermentations, 31
  *see also* Anaerobic metabolism; Glycolysis
Ferredoxin, chloroplast membrane transport, 277–278
Ferritin, 184T, 311
  iron regulation, 183
  synthesis, 183
ϕX174, 3
Fibrillin, 462
Fibrinogen, integrin-binding, 460–461
Fibrinopeptides, mutation rate, 87–88
Fibroblast growth factor (FGF), developmental role, 218
Fibroblasts
  differentiation, 171–172, 172
  locomotion, 393
Fibronectin
  integrin-binding, 460–461
  receptor binding, 464–465
  synthetic peptides, 464–465
Filamin, lamellipodia role, 383
Filopodia, 394, 394F, 399F
  actin role, 394F, 399
  neuronal growth cones, 399
  platelets, 359
  retrograde flow, 399
Filters, light microscopy, 224
First-order processes, radioactive decay, 19
Fischer projection, glucose, 16, 16F
Flagella
  beat cycles, 396, 396F
  *Chlamydomonas,* 393, 393F, 397–398
  mechanism, 396, 396F
  motor, 328–329, 339
  mutation analysis, 397–398
  structure, 393F
    axonemes, 385, 397, 397F
    radial spokes, 397, 397F
  *see also* Cilia; Microtubules
Flagellins, 139, 175
FLAG epitope tagging, 18, 259
Flippases, 285
Flow cytometry, cell cycle analysis, 407, 410
'Floxing,' 219
FLP–FRT system, site-specific recombination, 115, 115F
FLP recombinase, 115, 115F
Fluorescein, 456, 456F
Fluorescence-activated cell sorter (FACS), 191, 414
Fluorescence microscopy, 56, 222, 224, 224F, 257
Fluorescence recovery after photobleaching (FRAP), 242
Fluorescence resonance energy transfer *see* FRET analysis
Fluorescent dyes, 456
Fluorescent proteins, 224, 226–227, 457F, 457T
  *see also* Green fluorescent protein (GFP); Protein tags
Fluorophores, 226T, 312
  pH-sensitive, 251
Folate receptors, GPI-anchor, 236–237
Food *see* Feeding/food sources
*Fos* gene, 187, 187F
Fos–Jun heterodimerization, 158–159, 158F
Frameshifting (ribosomal), 139
Frameshift mutation, 106, 120
  amino acid sequence analysis, 135
  discovery, 139
FRAP (fluorescence recovery after photobleaching), 242
Free energy, 25
  changes *see* Delta G (ΔG)
  transduction principles, 389
Free-running circadian rhythms, 173
FRET analysis, 158–159, 226–227, 227F, 227FF
  lipid rafts, 236–237
*Frizzled* gene, 368
FRT recombination element, 115, 115F
Fructose-6-phosphate, 28
Fruit ripening, 371–372
Fungi
  cell characteristics, 6
  phylogeny, 371, 371F
  signaling pathways, 371
  *see also* Yeast
Fusion protein(s), 146–147, 146F, 199–200, 287
  fluorescent, 226–227
  GST, 201, 269–270
  MBP, 201
Fusogenic protein, tight junction studies, 454

## G

Gain-of-function mutation, 215
Gal4 transcriptional activators, 163
  coactivation, 169–170
Galactokinase streak test, 156, 166, 166F
Galactose metabolism, control, 163
Galactose oxidase, 42
Galactose transferase, 275, 299–300, 300F
β-Galactosidase
  fusion proteins, 146–147
  reporter gene, 170–171, 200–201, 421–422
Gametogenesis, 187–188
Gangliosides, sialic acid residues, 234
Gap junctions, 455–456
  developmental role, 456
  permeability, 455
  visualization, 229, 229F
Gated transport, nuclear–cytoplasmic, 264
GEF *see* Guanine nucleotide-exchange factor (GEF)
Gelation factor, 383
Gel-filtration chromatography, 185, 193
Gel-mobility shift assay(s), 159, 159F, 417, 417F
Gel staining
  Coomassie blue, 433F
  ethidium bromide (DNA), 204
  silver staining (proteins), 197
Gene(s)
  amplification, 101
  chromatin structure, 73–74
  conserved, 5
  conversion, 115
  definition, 67
  developmental rearrangement, 152–153, 163
  dominant, 215
  dosage, 416
    compensation, 174
  evolution of, 5, 8
  expression *see* Gene expression
  function determination, 5
  homologous, 215
  hybrid, 187, 187F, 188–189, 199–201
    *see also* Fusion protein(s); Reporter genes
  identification, 3, 4, 69, 182, 194, 375
    *see also* Gene cloning; Gene mapping
  information process *versus* metabolic, 5
  lethal, recessive, 88
  number, 3
  orthologous, 4
  protein co-linearity, 1
  recessive, 215
  structure *see* Gene structure
Gene cloning, 205, 211
  cDNA cloning *see* cDNA cloning
  cloning vectors, 197, 205
    shuttle vectors, 209
  controls, 213–214
  efficiency, 209
  genomic cloning, 209
  ligation, 204–205
    *see also* DNA ligation
  oligonucleotide splints, 205
  overlapping clones, 209
  probe choice, 210
  restriction digests, 69–70, 69F
    *see also* Restriction digests
  *see also* DNA libraries
Gene conversion, 115
Gene expression
  analysis, 214–220
    DNA microarrays, 75–76, 144, 206–207
    *in situ* hybridization, 217
  chromatin role, 73–74
  control, 152F
  developmental, 170–171, 218–219
  eucaryotic genes in bacteria, 198–199, 209
  eucaryotic *versus* bacterial genes, 209
  imprinting, 174
  position effect, 74–75
  regulation *see* Gene expression control
  time-course, 345
  tissue-specific, 186
  yeast, 75F
  *see also* Messenger RNA; Transcription
Gene expression control, 151–189
  cell memory, 173
  developmental, 166, 170–171
  DNA methylation role, 169–172, 177–178
  eucaryotic, 163, 167–171
    alternative splicing, 185–186
      *see also* Splicing
    AP1 binding site, 158–159
    bacteria *versus,* 154
    coactivators, 169–170
    hormone receptors, 168
    imprinting, 174
    insulators, 163
    LCRs, 163
    mechanisms, 152
    mRNA editing, 183, 186
    X-inactivation *see* X-inactivation
  feedback inhibition, 161
  gene regulatory circuits, 173
  genetic switches *see* Genetic switches
  overview, 151–153
  positive *versus* negative, 161
  post-transcriptional control, 182–189
    *see also* Protein synthesis; Translation
  procaryotic, 152, 160–162, 162
    eucaryotic *versus,* 152

*Lac* operon *see* Lactose *(Lac)* operon
lysis/lysogeny (phage λ), 172
phase switching, 175
promoters *see* Promoter elements
regulatory proteins *see* Gene regulatory proteins
specialization, 172–182
structural
chromatin role, 73–74, 76, 168–169
DNA bending, 157–158, 163
DNA looping, 77–78, 78F, 156, 162–163
nucleosome role, 71–72, 73, 75–76, 168–169
transcriptional control, 151–182
*see also* Transcription
Gene families
evolutionary tracing, 47
Hox gene clusters, 81
human albumin genes, *Alu* sequences, 85
single gene *versus*, 206
Gene knockouts
appotosis genes, 391, 442–443
intermediate filaments, 375
kinesin, 394
Gene locus, 215
Gene mapping
cell hybrid mapping of FeLV-C receptor, 192
linkage analysis, 215
oligonucleotide probes, 210
Gene organization, U2 snRNA, 68–69
General recombination
branch migration, 107
crossing-over, 111
DSB repair, 110
homologous DNA, 111
proteins
Rad52, 110
RecA, 106
Generation time, 6
Gene regulatory circuits, 173
Gene regulatory proteins, 345
developmental, 170–171
dimerization, 154–155
*see also* Protein complexes (assemblies)
DNA-bending, 157–158, 163
DNA-binding, 145–151
*see also* Protein–DNA interactions
E2F and cell cycle control, 475–476
eucaryotic, 161
inducers, 49, 58
mechanisms, 160–172
*see also* Genetic switches
regulation, 169–170, 251, 342
repressors *see* Repressor proteins
*see also* DNA-binding proteins; *specific proteins*
Gene structure
eucaryotic, 68, 120, 122
*see also* Exons; Introns
mammalian *versus* bacterial, 209
Gene targeting, 116–117, 209
Gene therapy, retroviral, 474
Genetically modified organisms (GMOs), 215–216
antisense RNA, 372
Genetic code, 3, 127–128
amino acid sequence deduction, 206
codon–anticodon interaction, 136, 136T
wobble, 136
*see also* Transfer RNA
codons *see* Codons
computer-generated *versus* natural, 2, 2F
deciphering, 135
evolution, 135–136
hypothetical, 135
mutation resistance, 2
natural selection, 2
properties, 135
reading frames, 120, 127, 134, 206
reduced, 197
variations, 142
mtDNA, 339–340, 340
*see also* Translation
Genetic disease
Duchenne's muscular dystrophy, 213
familial hypercholesterolemia, 308–310
G6PD deficiency, 240
inheritance patterns, 338, 340
lysosomal storage diseases, 304
microarray analysis, 217–218
mutation, 212, 213
primary hyperoxaluria type 1, 279
sickle-cell anemia, 206, 212
xeroderma pigmentosum, 107, 107F
Genetic elements
mating-type, 171, 175–176
phase switching, 175
transposable *see* Transposable elements
Genetic engineering, 215–216
gene targeting, 116–117, 209
*in vitro* mutagenesis, 216–217, 219
*see also* Gene cloning; Protein engineering
Genetic instability, cancer development, 469, 477
Genetics (classical)
complementation, 217
*Drosophila melanogaster*, 217
pedigrees, 174, 338, 338F, 407
reverse *versus*, 215
Genetics (reverse), classical *versus*, 215
Genetic switches, 160–172
arabinose metabolism, 164, 166–167
arginine biosynthesis, 161
components, 154
DNA looping, 162–163
glutamine synthetase regulation, 165–166
hormone receptors, 168
lactose metabolism *see* Lactose *(Lac)* operon
negative control, 161, 162–163
repressors *see* Repressor proteins
positive control, 161, 162–163
coactivators, 169–170
enhancers, 162–163
transcriptional activators, 163
*see also* Transcription factors
tryptophan biosynthesis, 154, 161
Genome(s)
databases, 5, 45, 69
DNA *versus* RNA as genetic material, 147
eucaryotic, 6–9
organellar, 336–341
*see also* Chloroplast(s); Mitochondrial genome
evolutionary diversity, 4–6
'privileged sites,' 8
procaryotic, 5
replication *see* DNA replication
sequence diversity, 5
*see also* specific genomes
Genome sequencing, 5
procaryotes, 5
protein tyrosine phosphatases, 61
*see also* DNA sequence analysis
Genomic DNA libraries, 210
Genotype, 215
Genotype–phenotype relationship, 88
Germ cells
DNA stability, 87
gene expression regulation, 187–188
Germinal vesicle breakdown, 408
GFAP (glial fibrillary acid protein), gene knockouts, 375
*Giant* yeast, 410
*Giardia*, lack of mitochondria, 337
*Giardia lamblia*, 6
Giardiasis, 6
Gibberellic acid, plant signaling, 466–467
Gibbs free energy, 25
changes *see* Delta G (ΔG)
Gilbert, Walter, 57
Glass beads, 223, 223F
GlcNAc phosphoglycosidase, 303
GlcNAc phosphotransferase, 303
GlcNAc transferase II, 275, 300F
GLD (globoid cell leukodystrophy; Krabbe's disease), 429
Gleevec, 484, 484F
Glial fibrillary acid protein (GFAP), gene knockouts, 375
Globin chains
β chains, 141
*see also* β-Globin genes
α chains, 8, 141
γ-globin gene, DNA methylation, 78–179
histidine residues, 22
synthesis, 141, 141F
reticulocytes, 185
γ-Globin gene, DNA methylation, 178–179
Globin genes, 73T
methylation, 179F
*see also specific genes*
β-Globin genes, 83
enhancer experiments, 162–163, 163F
evolution, 122
human sequence analysis, 83, 121–122, 206
mouse *versus*, 83
intron–exon boundaries, 122
introns, 122
U85 snoRNA production, 132, 132F
LCR control, 163
mutation, 206
ASO analysis, 217–218
sickle-cell anemia, 206, 212
Globoid cell leukodystrophy (GLD; Krabbe's disease), 429
Glucagon, glycogen breakdown, 354
Glucocorticoid-mediated signaling, 348–350
Glucocorticoid receptors, 348–350
transcription regulation, 168, 168F, 349–350
Glucocorticoid response element, 168, 349
Glucose
blood concentration, 19, 19F
insulin effects, 245–246
isomers, 15–16, 16F
metabolism
aerobic consumption, 316
anaerobic consumption, 31, 316
brain, 244
exercise, 316
liver, 244
muscle, 14–15, 316
oxidation, 19, 28, 34
*see also* Glycolysis
phosphorylation, 25–26
structure, 15–16, 15F, 19
fatty acids *versus*, 32, 32F
transporters, 245
active transport, 248
GLUT1, 245
GLUT2, 244
GLUT3, 228, 244
GLUT4, 246
kinetics, 244
Glucose-6-phosphate, 27, 28
Glucose-6-phosphate dehydrogenase (G6PD), deficiency, 240
Glucosidase I, 275, 300F
Glucosidase II, 275, 300F
Glutamate, excitatory neurotransmitter, 260–261
Glutamate receptors, 260–261
Glutamine synthetase
helix-wheel projection, 275
transcription regulation, 165–166
NtrC-binding sites, 165
Glutathione S-transferase
GST 'pull down' assay, 295
protein tagging, 62, 62F, 201, 269–270, 334
Glyceraldehyde 3-phosphate, 30–31, 50
Glycine, 14, 14F
Glycogen
catabolism, 31, 52F, 57, 332, 333, 354, 355F
adrenergic receptor signaling, 353
exercise, 354
glucagon-mediated, 354
vasopressin-mediated, 354
energy storage, 31, 316
synthesis, 52, 52F
Glycogen phosphorylase, allosteric regulation, 52
Glycogen synthase, 52
Glycogen synthase kinase-3β (GSK-3β), β-catenin phosphorylation, 369

Page numbers refer to a major text discussion of the entry; page numbers with an F refer to a figure, with a T refer to a table.

Glycolipids
membrane composition, 234
membrane distribution, 239
synthesis, 239, 274
Glycolysis, 29F, 30–31
ATP synthesis, 275
energetics, 26, 31, 316–317
muscle cells, 316
evolutionary conservation, 31
glycolytic pathway, 26, 29F
phosphoglycerate kinase, 280
red blood cells, 29T
stoichiometry, 31
trypanosome glycosomes, 280
*see also* Glucose
Glycophorin
homodimerization, 240–241
quantitative analysis, 240–241
Glycoproteins
analysis, 245
fibronectin, 464–465
quality control, 284
synthesis, 274–275, 284
oligosaccharide processing, 298
oligosaccharide transferases, 285
preassembled sugars, 284
Glycosaminoglycans, lysosomal accumulation, 304
Glycosidic bonds, 16
Glycosomes, 280
Glycosylphosphatidylinositol (GPI) anchor
analysis, 235–236, 236F
lipid rafts, 236–237
membrane proteins, 235–236, 283
GMP (guanosine monophosphate)
cyclic *see* Cyclic GMP (cGMP)
synthesis, 52
Golgi apparatus, 300F
cisternae, 277, 296, 300–301
glycosylation, 239, 274–275
mannosidases, 275, 299–300, 300F
membrane protein sorting in RER, 265
reassembly, microtubules, 264
*trans*-Golgi network, 275, 276, 277, 292
protein sorting, 311–312
vesicular transport, 264, 291, 293
brefeldin A effect, 293, 413
to cell surface, 284–287, 292
*see also* Exocytosis
cisternal maturation model, 300–301
COPI vesicle formation, 293
from ER, 293
rate, 298
transport through Golgi, 274–277
transport to lysosomes, 302–305
vesicular transport model, 300–301
VSV G protein experiments, 298–299, 301
*see also* Endoplasmic reticulum (ER)
GPI-anchored proteins *see* Glycosylphosphatidylinositol (GPI) anchor
G-protein-linked (coupled) receptors (GPCRs), 352–360
adenylyl cyclase coupling, β-adrenergic receptors, 355–356, 357
andenylyl cyclase coupling, 357
α-factor pheromone signaling (yeast mating), 356–357
GTPases *versus*, 363
*see also* GTP-binding proteins (GTPases)
muscarinic receptors, 360
pertussis toxin, 329, 357, 358
phosphorylase kinase integration, 354
RGS (regulator of G protein signaling), 353
visual transduction, 356
*see also* G proteins; *specific receptors*
G proteins, 353
βγ complex, 357
inhibitory ($G_i$), 357
transducin ($G_t$), 332, 358
Gramicidin A, 246, 246F
Green fluorescent protein (GFP)
microscopy, 221, 224
modified, 224, 225F
nuclear export experiments, 267–268
vimentin analysis, 379
GroEL chaperones, 144–145, 144F
Group II self-splicing introns, 124
Group I self-splicing introns, 124
Growth cones, neuronal, 399
Growth factors, 431
*see also individual factors*
GTP (guanosine triphosphate)
cAMP generation in GPCRs, 357
hydrolysis, 292, 293–294
energetics, 363
microtubule polymerization, 375, 376–377, 378
*see also* GTP-binding proteins (GTPases)
nonhydrolyzable analogs, 271, 329, 331, 332, 353
GTPase-activating protein (GAP), 53, 293–294
actin effects, 394
*Bud2*, 399–400
ER-to-Golgi transport, 292
phosphorylation control, 345–346
Ran-GAP, 271
Ras-GAP, 363–364
RGS (regulator of G protein signaling), 353
tubulin, 375
GTP-binding proteins (GTPases)
ARF, 293–294
*Bud1*, 399–400
Cdc42, 394, 400–401
dynamin, 291, 311
GPCRs *versus*, 363
kinetics, 293–294
Rab, 271, 290
Rac, 394
Ran *see* Ran GTPase
Ras *see* Ras GTPase
regulation, 53
GAP *see* GTPase-activating protein (GAP)
GEF *see* Guanine nucleotide-exchange factor (GEF)
Rho, 394
Sar1 protein, 292
Sec4, 296
trimeric *see* G proteins
GTP caps, microtubules, 376–377, 378
Guanine, base pairing, 63, 63F
Guanine nucleotide-exchange factor (GEF), 53, 292
*Bud5*, 399–400
ER-to-Golgi transport, 292
phosphorylation control, 345–346
Ran-GEF, 271
Ras-GEF, 363
Guanosine monophosphate
cyclic *see* Cyclic GMP (cGMP)
synthesis, 52
Guanosine triphosphate *see* GTP (guanosine triphosphate)
Guanylyl cyclase, nitric oxide (NO) stimulation, 345
Guide RNA, 183

## H

Hairpins
DNA structure, 90, 211
RNA structure, 147, 147F, 183
Half-life, radioactive, 12, 12T
autoradiography and, 225
*Halobacterium*, light-driven proton pump, 316–317
Haworth projection, glucose, 16, 16F
Head-to-tail dimers, 42–43
Heart muscle *see* Cardiac muscle
'Heavy medium,' 93, 93F
Hedgehog signaling pathway, 368, 369
Helices
DNA double helix, 64, 64F, 78–79
*see also* DNA structure
hairpins, 90
α helix *see* α-helices
helix–helix interactions, 56
helix wheel projection, 40, 40F, 275
Helix–helix interactions, 56
Helix-loop-helix (HLH) proteins, heterodimerization, 155
Helix-wheel projection, 40, 40F, 275, 275F
Hemagglutinin (HA), epitope tag, 169–170
Heme, 307
carbon monoxide binding, 323
cytochromes, 327
protein synthesis role, 185
Heme-controlled repressor (HCR), 185
Hemoglobin, 82T
centrifugation, 193
α chains, 141
conformational changes, 248
cooperativity, 51
gas transport, 248
globin chains *see* Globin chains
heme *see* Heme
human genes, 4
oxygen binding affinity, 51
phylogenetic tree construction, 9, 9F
plant, 8–9, 9F
structure, 193F
tube worms, 4
Hemolysis, snake venom, 234
Hemolytic anemia, G6PD deficiency, 240
Henderson–Hasselbach equation, 19, 21
Herbicides, 331
Herpesvirus, VP16 activation domain, 199–200
Hershey, Alfred, 65
Hershey–Chase experiment, 65
Heterocaryons
nuclear–cytoplasmic transport analysis, 268
peroxisome transport analysis, 281
Heteroduplex formation, DNA transposition, 116, 116F
Heterologous injection experiments, 77
Heterozygote, 88, 215
Hexokinase, chiral specificity, 50
Hexoses, 16
High-performance liquid chromatography (HPLC), globin chains, 141
High-throughput screens, cancer treatment, 482–483
Histidine residues
enzyme catalysis, 23, 45
hemoglobin, 248
Histidine-tagged proteins, 207–208
Histone(s)
acetylation, 74F
charge, 66
conservation, 8, 8T, 71
core, 66
gene expression, 75F
modification, 72–73, 76–77
acetylation, 163, 169
phosphorylation, 61, 61F
mRNAs, 131, 131F
nuclear–cytoplasmic transport, 268
octamers, 68, 71, 268
tails, 72–73
transcription regulation, 169
*see also* Chromatin
Histone acetylases, transcription regulation, 163, 169
Histone H1, 61, 61F
Histone H3, 8, 8T
Histone H4, 75, 75F, 77
HIV (human immunodeficiency virus)
gene expression, 182–183
Rev-mediated nuclear export, 272
HLH (helix-loop-helix) proteins, heterodimerization, 155
HMG CoA reductase, 297
*Hml* mating-type locus, 171
*Hmr* mating-type locus, 171
Hoechst 33342, 223, 224, 407
Holliday junctions, 113–114, 114F
Homeodomain proteins, 41
Homoduplex formation, DNA transposition, 116, 116F
Homologous genes, 215
Homologous recombination, 111–114
branch migration, 112
Chi sites, 112–113, 113F

  crossing-over, 111
  Holliday junctions, 113–114, 114F
  intermediate structures, 111, 111F
  meiotic, 479–480
  proteins
    RecA, 112
    RecBCD, 113
Homozygote, 88, 215
Horizontal gene transfer, 5, 6
  mitochondrial genes, 6–8, 339
Hormone receptors
  receptor–ligand interactions, 346
  transcription regulation, 168, 349–350
    *see also* Glucocorticoid receptors
Hormones, 346
  plant growth regulators *versus,* 372
  receptors *see* Hormone receptors
  steroid, 345, 345F
  *see also individual hormones*
Horseradish peroxidase (HRP), 307–308, 456
  antibody conjugation, 198
*Hox* gene clusters, 81
HPLC, globin chains, 141
Hsp60-like chaperones, protein folding, 144–145
Hsp70 chaperones
  DnaK, 143–144, 143F
  ER import role, 284
  mitochondrial import role, 275, 276, 283
  protein folding, 143–144
  trigger factor, 143–144
Human cells
  human–rodent cell hybrids, 192
  kinetochores, 415
  replacement, 403
  telomerase, 432
Human genome
  *Alu* sequences, 85
  average protein, 265
  coding sequences, 6, 67
  CpG islands, 104
  DNA content, 64
  DNA sequence diversity, 3
  duplications, 69
  evolution, 85
    chromosomes, 67, 67F
    mitochondrial DNA analysis, 83–85
  gene identification, 192
  mitochondrial DNA, 83–85, 339–340
  replication, 97
    *see also* Cell cycle; DNA replication; Mitosis
  U2 snRNA gene organization, 68–69, 69F
Human–rodent cell hybrids, 192
Hunter's disease, 304
Hurler's disease, 304
Hyaluronan, synthesis, 463, 463F
Hybrid cells, 192
Hybrid genes, 187, 187F, 199–201, 210, 210F, 421–422
  *see also* Fusion protein(s); Reporter genes
Hybridomas, 192
  *see also* Monoclonal antibodies
Hydride ion, 323
Hydrogen, 323
  $^3H$ (tritium), 12, 12T
  abundance, 13F
  atomic structure, 12
Hydrogenation, 233
Hydrogen bonds, 13
  membrane protein role, 240
  protein–DNA interactions, 154
Hydrogen sulfide, 13, 13F
  tube worm hemoglobin, 4
Hydronium ions, 20
Hydropathy plots, membrane protein analysis, 239, 255, 287
Hydrophilic molecules, 17
  amino acids, 40
  signaling molecules, 345
Hydrophilic pores, membrane proteins, 239–240
Hydrophobic chromatography, 193
Hydrophobic interactions
  membrane proteins, 239
  protein–DNA interactions, 154
  protein folding, 239
  protein translocation, 266
  water molecules, 232
Hydrophobic molecules, 17, 232
  amino acids, 39, 40
Hydrothermal vents
  microorganism DNA replication, 90
  tube worms, 4
Hydroxyl group, 15–16
Hypercholesterolemia, familial, 308–310

**I**

I-cell disease, 303
Ice protein, 215–216
Id protein, 155
IκB (inhibitor of NFκB), 273
Imidazole, 207–208, 207F
Immersion oil, 222, 223
Immune evasion, phase switching, 175
Immunoaffinity purification, 199
Immunoblots, 56–57, 57F, 142, 142F, 197–198
  detection limit, 197–198
  myelin basic protein, 197–198
  principle, 197–198
  sensitivity, 198F
  syntaxin, 295
Immunofluorescence microscopy, 46
Immunoglobulin(s)
  domain, 56
  Fab fragments, 446, 446F
  gene rearrangement, 152–153
  IgG, 446, 446F
  NFκB gene regulatory protein, 273
  *see also* Antibodies
Immunoprecipitation
  acyl CoA oxidase, 280–281
  chromatin, 169
  GPI-anchored protein, 236
  PDGF receptor, 56–57
  receptor tyrosine kinases, 364
Imprinting, maternal *versus* paternal, 174
Inclusion-cell (I-cell) disease, 303
Indirect end labeling, 76
Influenza viruses, cell entry, 307
Inhibitor of NFκB (IκB), 273
Inhibitory neurotransmitters, 254
In-line protein domains, 42
Inositol triphosphate ($IP_3$)
  $\alpha_1$-adrenergic receptors, 359
  calcium signaling, 329, 359
Inside-out membrane vesicles, 240
Inside-out signaling, 459
*In situ* hybridization, 217
Instantaneous velocity, 28
Insulator elements, 163
Insulin
  glucose uptake, 245–246
  processing, 312
  radioimmunoassay, 347
  secretion, 312
Insulin-like growth factor (IGF), cell cycle control, 432, 432T
Insulin-like growth factor-2 (IGF2), imprinting, 174
Integrins, 459–461
  amino acid substitutions, 459
Interferon-γ (IFNγ), 365–366
Intermediate filaments, 43, 379
  astrocyte role, 375
  cell division, 379–380
  disassembly, 379–380
    acrylamide effects, 376
  distribution, 375
  disulfide bonds, 375
  functions, 373
  keratin, 375
  lack of polarity, 375, 388
  mouse knockouts, 375
  myosin II filaments *versus,* 391
  phosphorylation, 379–380
  *see also* Nuclear lamins; *individual filaments*
Internal ribosome entry sites (IRESs), 189
Interphase chromosomes, 419
  human, 77
  topology, 80
Intracellular compartments, 290F
Intracellular dynamics, 313–314
Intracellular vesicular traffic, 289–314
  pathways, 291
Intron–exon boundary, 122
Introns, 66–67, 122, 123
  *Alu* sequences, 85
  evolution, 122
  organellar, 336
  removal by splicing, 121–122
    *see also* Splicing
  self-splicing, 124
  snoRNA processing, 132–133
  substitution rate, 85
Intron scanning, 122–123
Invertebrates
  phylogeny, 9, 9F
  septate junctions, 449–450
*In vitro* evolution, ribozymes, 147–148
*In vitro* mutagenesis, 186, 216–217, 219
*In vitro* transcription, 157–158, 159
  components, 167, 168
  efficiency, 127
  promoter assays, 116–117
*In vitro* translation, 131, 135, 142, 159
  chloroplast mRNAs, 278
  reticulocyte lysates, 185
Ion channels, 252–261
  all-or-nothing opening, 254
  aqueous pore *versus,* 254
  carrier proteins *versus,* 233
  ligand-gated, 237–241, 254
  mechanically-gated, 254
  membrane potential role, 253
  patch-clamp recording, 235, 255, 257, 258
  pharmaceutical targets, 260–261
  saturation, 253
  signal amplification, 345
  voltage-gated, 254
  *see also specific channels*
Ion-exchange chromatography, 193, 195F
  pH effects, 195
  resins, 194–195
Ion flow
  drift velocity, 253–254
  Nernst equation, 256, 295
    *see also* Action potential
Ionophores, 328–329
  calcium, 256, 329
  CCCP, 266
  channel-forming, gramicidin A, 246
  FCCP, 302, 327, 336
  valinomycin, 328–329, 329T
IRESs (internal ribosome entry sites), 189, 190F
Iressa, 484F
Iron, 307
  endocytic uptake, 284, 307
  gene expression regulation, 183
    iron-response element, 183
  ovotransferrin binding, 51
Iron-response element (IRE), 183
Iron-response protein (IRP), 183
Iron–sulfur centers, redox potentials, 322–323
Isoelectric focusing, 151–152, 186, 351
Isoelectric point, 14, 152
Isomers
  glucose, 15–16, 16F
  thalidomide, 16–17, 16F
Isoproterenol, β-adrenoceptor binding, 351, 357

**J**

JAK-STAT signaling, interferon-γ (IFNγ), 365–366
Jun–Fos heterodimerization, 158–159, 158F
Junk DNA, 66


Page numbers refer to a major text discussion of the entry; page numbers with an F refer to a figure, with a T refer to a table.

## K


Karyotyping
chronic myelogenous leukemia, 66, 66F
oral–facial tumors, 472
Katanin, microtubule severing, 383
KDEL receptor, ER retrieval signal, 297–298
Keilin, David, 326–327
Kelch motif, 42, 42F
Kelner, Albert, 104
Keratin filaments, disulfide bonds, 375
Kidney glomerulus, filtering, 457
Kidney tubules, tight junctions, 451–452, 452T
Kinesin motors, 389–390
axonal transport, 399
Charcot–Marie–Tooth disease, 394
microtubule tracks, 389–390
mouse knockouts, 394
optical tweezer experiments, 390
processivity, 389–390
Kinetochores, 412, 413, 415, 420
dicentric minichormosome assays, 421–423
human cells, 41
microtubule nucleation sites, 385
polymerization/depolymerization, 414
protein–DNA interactions, 421–423
proteins, 421–423
tension effects, 415
transcription block, 421–422
*see also* Centromeres
Knockouts *see* Gene knockouts
Knoop, Franz, 34–35
Knots, 418
Krabbe's disease (GLD; globoid cell leukodystrophy), 429
Krebs, Hans, 34, 35–36
Krebs cycle *see* Citric acid cycle
Krüppel protein, *Eve* gene repression, 170–171


## L


*LacI* gene, 106, 106F
Lac repressor, 57, 59, 106
binding sites, 162
kinetics, 156–157
β-Lactamase enzyme, amino acid sequence, 135
Lactic acid, 14–15, 31
Lactose *(Lac)* operon, 59, 146, 162, 162F, 162FF, 164–165
gene control, 164–165
CAP *see* Catabolite activator protein (CAP)
repression *see* Lac repressor
*LacI* mutations, 106, 106F, 106FF
*LacZ* gene
mutations, 95, 108
reporter gene, 116
*see also* β-Galactosidase
*LacZ* gene, 96T, 200
Lagging strand synthesis
fidelity, 95, 95F
Okazaki fragments, 91, 91F
Lambda (λ) bacteriophage, 69, 69F, 172F
λ repressor, 154–155, 155F, 156
lysis/lysogeny, 156, 172
cI *versus* Cro, 172
prophage, 156, 172
replication, 93, 93F
restriction map, 69F
Lamellipodia, 394, 394F
filamin role, 383
platelets, 383
Lamin C, amino acid sequence, 43F
Laminin-γ1, 458
Lamins, nuclear *see* Nuclear lamins
Lampbrush chromosomes, 77–78
transcription, 77–78
Lariat structures, 124
Laser capture microdissection, 191
LCRs (locus control regions), 163
LDL *see* Low-density lipoprotein(s) (LDLs)
*Leaden* mice, 305, 305F
Leading strand synthesis
fidelity, 95, 95F
template, 91
Lectins, 240
Lens(es), light microscopy, 222, 222F
Leptomycin B, 272, 273F
Leucine zippers, 158–159
Leukemia, 474–475
Levinthal, Cyrus, 41
Levithal paradox, 41
LexA repressor, gene hybrids, 199–201
Life
chemistry, 2–3
definition, 1, 2
diversity of, 4–6
extraterrestrial, 2
origins, 147–149
physics, 3
spontaneous generation, 2–3
Life scores, 1–2
Ligand-gated ion channels, 237–241, 254
*see also specific channels*
Ligation
DNA, 71, 204–205
*see also* DNA ligases
RNA, 149
Light
energetics, 331
rays, microscopy, 222
Light–dark cycles, circadian rhythms, 173, 176
Light-driven pumps
active transport, 247, 316–317
bacteriorhodopsin, 239, 316–317
Light microscope
microscopy, 221–228
refractive indices, 222
resolving power, 225
schematic, 222F
Light scattering, polymerization assay, 378–379
LINE elements, 81
Lineweaver–Burke equation, 59
Linkage analysis, SNPs, 215
Linker DNA, 71
Linker-scanning analysis, promoter elements, 126, 126F
Lipid(s), 232–238
gangliosides, 234
glycolipids, 234, 239
phospholipids *see* Phospholipids
sphingolipids, 233, 234–235
structure, 17, 17F
*see also* Lipid bilayer
Lipid anchors, 235–237, 239
Lipid bilayer, 232–238
composition, 234–235
cholesterol *see* Cholesterol
glycolipids *see* Glycolipids
phospholipids *see* Phospholipids
physical effects, 232
*see also* Membrane proteins
detergent effects, 235–236, 236–237
dynamics, 233
ESR spectroscopy, 234, 237–238
flip–flop, 231, 237–238
flippases, 285
fluidity, 231, 233
temperature effects, 233
lysis, 232
microdomains, 231
rafts *see* Lipid rafts
permeability, 243
properties, 232
scramblases, 285
tear repair, 232, 232F
*see also* Lipid(s)
Lipid diffusion, tight junctions, 453
Lipid rafts, 233
analysis
density-gradient centrifugation, 236
FRET, 236–237
SDS-PAGE, 235–236
caveolae, 306–307
cholesterol, 233
GPI anchors, 235–236
sphingolipids, 233
Lipid vesicles, 235, 249
Lipopolysaccharide (LPS), bacteriophage T4 attachment, 447–448
Lipoproteins, LDL *see* Low-density lipoprotein(s) (LDLs)
Liposomes, 454
*Listeria monocytogenes*, motility, 383
Lithotrophs, 4
Liver cells
energy storage, 31
G6P concentration, 29
glucose metabolism, 244
microsomes, 283
mitochondria, 316
proliferation, 432
protein synthesis, 264
*Lmo2* gene, 474–475
Locus control regions (LCRs), 163
Long-terminal repeats (LTRs)
LTR retrotransposons, 81
MMTV, 349
Loops
DNA *see* DNA loops
protein structure, 41
RNA, 129
Low-density lipoprotein(s) (LDLs)
metabolism, 308–310
receptor-mediated endocytosis, 308–310
Low-density lipoprotein (LDL) receptors, 309–310
Lox recombination element, 115, 209
LTR retrotransposons, 81
Lung cancer, smoking related, 471–472, 471F, 471FF, 472F
Lymphocytes, killer, 440
*see also* Tumors
Lysis/lysogeny, lambda (λ) bacteriophage, 156, 172
Lysosomal storage diseases, 304
Lysosomes, 309
conversion from late endosomes, 302
enzymes (acid hydrolases), 302
M6P-binding, 302–303, 303–304
LDL metabolism, 309
melanosomes, 303
membranes, 302
pH, 278, 302
proton pump, 302
sorting, 302–304
vesicular transport, 264, 290
Golgi-to-lysosome, 302–305
Lysozyme, 59–60
stability, 44
structure, 51, 51F
prediction, 44


## M


M6P *see* Mannose-6-phosphate (M6P)
M13 bacteriophage
phage-display libraries, 201
replicative RNA priming, 94, 94F
single-stranded DNA, 64
M13 primosome, 94, 94F
Macrolides, 381
Macromolecules, 17, 17F
*see also specific molecules*
Macronucleus, *Tetrahymena*, 211
Macrophage, phagocytosis, 306
Magnesium, abundance, 13F
Magnification, *vs.* resolution, 223
Malaria, G6PD deficiency, 240
Malate, 60
MALDI-TOF mass spectrometry, 198, 199F
Malonate, 34, 35
mitochondrial inhibition, 327T
Maloney murine sarcoma virus, glucocorticoid responsiveness, 348–349
Maltose-binding protein, fusion proteins, 201
Mammalian cells
chromosomes, 77

  - DNA damage response, 431
    - *see also* Apoptosis
  - ion transport, 247
  - protein content, 23
  - replication origins, 102, 102F
- Mannose-6-phosphate (M6P)
  - acid hydrolase-binding, 302–303, 303–304
  - lysosomal sorting, 302–304
  - receptors, 302–303
  - synthesis, 304–305
- Mannosidase(s), 299–300, 300F
- Manometer, 35
- MAP kinase(s) (MAPKs), 365
  - activation, 365F
  - conformational changes, 363, 363F
  - JNK subfamily, 443–444
  - scaffolding, 363
  - signaling cascade, 201, 365
    - oocyte activation, 350–351
    - stimulus response curves, 365F
- MAPKK (MAP kinase kinase), 365
- MAPKKK (MAP kinase kinase kinase), 365
- Marfan's syndrome, 462
- Mass spectrometry, MALDI-TOF, 198
- Maternal imprinting, 174
- Mating-type
  - gene control, 171, 183
    - **a**1 gene product, 175–176
    - α2 repressor, 175–176
    - regulatory sequences, 175–176
  - loci, 171
  - mutant analysis, 176
  - silent locus, 171, 171F
  - **a** type, 175–176
  - α type, 175–176
- *Mat* locus, 171, 171F
- Matrix metalloproteases (MMPs), 458
- Maxigenes, 168, 178
- McClintock, Barbara, 479–480
- M-Cdk *see under* Cyclin–Cdk complexes
- Mcm proteins, 411
- MDCK cells
  - protein sorting, 313
  - tight junction studies, 452
- *Mdm2* gene, role in cancer, 478, 478T
- Mechanically-gated ion channels, 254
- Megakaryocytes, 430, 430F
- Meiosis, 187–188
  - meiotic recombination, 479–480
  - meiotic spindle, 408
- Meiotic recombination, 479–480
- Meiotic spindle, 408
- Melanocytes, 305F
- Melanoma cells, motility, 383
- Melanosomes, 303
- Membrane potential, 236, 247, 248
  - resting, 253
  - squid giant axons, 255–256
  - strength, 255–256
  - *see also* Action potential
- Membrane proteins, 238–242, 239F
  - analysis, 235–236, 287
    - FRAP, 242
    - hydropathy plots, 239, 255, 287
    - quantitative, 241
    - SDS PAGE, 226, 227, 236, 242
  - anchors, 235
    - fatty acid chains, 239
    - GPI anchor, 235–236, 239, 283
    - prenyl groups, 239
  - arrangements, 239
  - carrier proteins *see* Carrier proteins
  - conformation changes, 242, 270
  - distribution, 241
  - functions, 219, 238
  - hydrogen bonds, 240
  - hydrophilic pores, 239–240
  - hydrophobic forces, 239
  - insertion, 246, 266, 283
  - OmpC, 448
  - protein–protein interactions, 240
  - structure
    - β barrels, 240
    - α helices, 239–240
    - multipass, 284, 284F
  - synthesis, 265–266
  - tight junctions, 453–455
  - viral attachment, 448
  - VSV G protein, 298–299
  - *see also specific proteins*
- Membranes
  - intracellular, 263
    - flow, 290
    - *see also specific organelles*
  - plasma *see* Plasma membrane
  - potential *see* Membrane potential
  - structure, 231–242
    - carbohydrate localization, 239
    - lipid bilayer *see* Lipid bilayer
    - lipid content (by volume), 265
    - lipid:protein mass ratio, 239, 241
    - protein *see* Membrane proteins
  - transport, 243–263
    - active *versus* passive, 243
      - *see also* Active transport
    - aspirin, 15
    - carrier proteins *see* Carrier proteins
    - diffusion, 28, 56, 243, 244–245
    - energetics, 250–251
    - ion channels *see* Ion channels
    - kinetics, 244–245
    - molecular mechanisms, 289–296
    - principles, 243–246
    - protein numbers, 258–259
    - vesicles *see* Vesicles/vesicular transport
  - visualization, 228, 229F
- Mercaptoethanol, 196
  - cell division effects, 419
- Mesoderm induction, 218
  - transforming growth factor-β, 184
- Messelson–Stahl experiment, 93, 93F, 183
- Messenger RNA, 68
  - broken, translation prevention, 137
  - cellular localization, 183
  - cleavage, 124
  - eucaryotic, 120
    - 3′ end (poly-A tail), 131
    - 5′ end (cap), 131, 133
    - ER attachment, 283
    - nuclear–cytoplasmic transport, role, 264
    - splice variants, 121–122
  - export ready, 123
  - gene expression regulation, 182–189
    - editing, 183, 186
    - hairpin structure, 183
    - polyadenylation, 178–179, 182
    - stability, 186
    - tissue-specific, 186
  - histones, 131
  - IRESs, 189
  - polycistronic, 137
  - primer extension mapping, 130, 130F
  - procaryotic, 137
    - tmRNA, 142
  - S1 mapping, 130
  - splicing *see* Splicing
- Metabolic genes, 5
- Metabolism, 23–30
  - anabolism, 24
  - catabolism, 24, 29
  - energetics *see* Energy; Thermodynamics
  - glucose *see* Glucose
  - intermediates, 26
  - labeling experiments, 34–35
  - mitochondrial, 316
  - pathway genes, 36–37
  - regulation, 51
  - thermodynamics, 26, 28–29
  - *see also specific pathways*
- Metal shadowing, electron microscopy, 228
- Metaphase, 403, 414, 415, 419, 424
- Metaphase plate, 414
- Methanogenesis, 328
- Methotrexate, 276
- Methylation
  - DNA *see* DNA methylation
  - histone tails, 72–73
  - RNA, 133, 133F
- $O^6$-Methylguanine, 108F, 444
- $O^6$-Methylguanine methyltransferase (MGMT), 108F, 444
- *N*-methyl-*N′*-nitro-*N*-nitrosoguanidine (MNNG), 107–108, 108F, 444
- Methyltransferases, 108, 108F, 167, 174
- MGMT ($O^6$-Methylguanine methyltransferase), 108F, 444
- Micelle formation, 232
- Michaelis constant ($K_m$), 56, 59
- Michaelis–Menten kinetics
  - enzymes, 59
  - transport proteins, 244
- Micrococcal nuclease, 71, 71F, 73–74, 76, 76F
- Microdissection, laser capture, 191
- Microhomology, 109F, 110T
- Microinjection experiments, 151, 239, 240
  - cell cycle control, 408–409
  - mouse embryo compaction, 456, 456F
  - nuclear export studies, 272–273
  - nucleoplasmin localization, 269
- Micronucleus, *Tetrahymena*, 211
- Microsatellite DNA, cloned animal analysis, 152–153, 153F
- Microscopy
  - electron, 228–230
  - fluorescence, 56, 257
  - light, 221–228
    - optical density, 222
    - rays, 222
    - schematic, 222F
  - video, 386, 405–406, 414, 449
- Microsomes, 283
- Microspikes, 400
- Microtubule-associated proteins (MAPs), 383, 414–415
- Microtubule-organizing center (MTOC), 382
- Microtubules
  - centromere attachment, 76
  - colchicine inhibition, 375–376, 378–379, 393
  - depolymerization, 264, 378
    - benomyl, 416
  - dynamic instability, 377–378, 414–415
    - catastrophe factors, 414–415
  - dynamin, 311
  - functions, 355, 373
  - Golgi apparatus reassembly, 264
  - GTP caps, 375, 376–377, 378
  - katanin, 383
  - kinetics, 377F
  - length, 414–415
  - microtubule-associated proteins, 383, 414–415
  - mitosis, 413
  - mitotic, 383, 414–415, 423–424, 427
    - *see also* Mitosis
  - motor proteins, 388–390
    - directionality, 370, 389–390
    - landing rates, 390, 390F
    - mitosis, 423–424
    - processivity, 389–390
    - *see also* Motor proteins
  - nocodazole destabilization, 413, 415
  - nucleation reaction, 384, 386
  - nucleation sites, 382, 385
    - axonemes, 385, 396
    - centrosomes, 383, 385, 420
    - γ-TuRC, 384–385, 385–386
    - kinetochores, 385
  - plant cells, 466
  - polymerization, 374, 374F
    - critical concentration, 384, 384F
    - dynamics, 375, 377–378, 414–415
    - energetics, 24–25, 25F
    - hetero-*versus* homotypic interaction, 374
    - kinetics, 376–377, 377F, 384F
    - lateral interactions, 374, 383F
    - stoichiometry, 385
  - spatial organization, 384

species differences, 377–378
structure, 374F
taxol stabilization, 375–376
visualization, 221
*see also* Cilia; Flagella; Tubulin
Microvilli, 250, 250F
Minichromosomes
dicentric, kinetochore function assay, 421–423, 422F
*Tetrahymena*, rRNA gene analysis, 211
Minigenes, 123
Mismatch extension, DNA polymerase(s), 91, 95
Mismatch proofreading
DNA polymerase, 89
RNA polymerase, 125
Mismatch repair, 109, 113
apoptosis role, 444
methylation-dependent, 89, 91
*in vitro* mutagenesis, 216–217
Missense mutation, 106
Mitochondria
apoptosis role, 440
cytochrome-*c* release, 391–392, 440, 442
chloroplasts *versus*, 330
division, 337
energy conversion, 315–322
ATP synthesis, 275
inhibitors, 327, 327T
oxygen consumption, 299–300, 322
respiratory enzyme complexes, 300, 316, 339
*see also* Electron transport chains
*see also* Electrochemical proton gradients
functional organization, 316
genome *see* Mitochondrial genome
inheritance, 84F, 84FF, 340
membranes
inner membrane, 275
intermembrane space, 315
outer membrane, 276
permeability, 290, 316
phospholipids, 285
proton pumps, 316
metabolic roles, 316
organisms lacking, 337
origin, 7, 337
endosymbionts, 337
pH, 318
preparation, 193
submitochondrial fractions, 291, 316
transmembrane transport, 252–255, 264
ADP–ATP antiporter, 321
chloroplasts *versus*, 274
Hsp70 role, 275, 276, 283
import-defective mitochondria, 275
posttranslational, 275
protein folding/unfolding, 276
signal sequences, 274
thermal-ratchet model, 275
TIM complexes, 275, 277
TOM complexes, 276
Mitochondrial genome, 7, 336, 338
copy number, 338–339
deletion mutants, 341
DNA sequence analysis, 83–85
evolution, 7, 337
fungal, 340–341, 341
genetic code variations, 336, 340
gene transfer, 6–8, 339
human, 83–85, 339–340
codon usage, 339–340
introns, 336
non-Mendelian inheritance, 337, 341
plant, 7
protein synthesis, 339–340
replication, 336, 338
RNA editing, 183
transcription maps, 339F
Mitogens, 431, 433
*see also specific factors*
Mitosis, 413–426, 413–427, 414F
analysis
Boveri's experiments, 426
plasmid constructs, 420–421
chromosomes
cohesins, 411–413, 412
cohesion, 411–413
condensation, 77, 406, 418
condensins, 418
kinetochore attachment, 412, 413
movements, 416
positioning, 423–424
separation, 416
*see also* Chromatids; Chromosome(s)
cytoskeleton, 427
DNA amounts, 416
intermediate filaments, 379–380
mitotic spindle, 427, 430
motor proteins, 423–424
*see also* Microtubules
DNA alterations
compaction, 68, 70
*see also* Chromatin
damage effects, 431
topology, 80, 80F, 418
energetics, 24–25
frequency (mitotic index), 405
metaphase arrest, 414
metaphase plate, 414
microtubules, 413
nuclear re-import, 268
nucleolus, 123
phases
anaphase (A/B), 412, 414, 415
metaphase, 410, 411, 414, 415, 419, 424
order of events, 415, 419F
prometaphase, 414
prophase, 414, 415
telophase, 412, 413, 414, 415
spindles
bipolar, 419, 419F, 424
formation, 419, 419F
microtubules, 412, 414, 418–419
multiple, 426, 426F
parthenogenesis, 424
tetrapolar, 426, 426F
transcription blocks, 421–422
tripolar, 426, 426F
*see also* Cytokinesis; Meiosis
Mitosis-promoting factor (MPF)
oocyte maturation, 408–409
protein kinase activity, 408–409
Mitotic index, 405
MMTV (mouse mammary tumor virus), 349–350, 477–479
MNNG (*N*-methyl-*N'*-nitro-*N*-nitrosoguanidine), 107–108, 108F, 109F, 444
*Mocha* mice, 303, 303F
Molarity, 19
Molecular motions, lipid bilayer, 233
Molecules
detection, 198–199
diffusional properties, 243
equilibrium distributions, 244
per mole, 198–199
salts *versus*, 15
Monoclonal antibodies, 199
anti-Orc1, 199
epitope-binding, 191
hybridoma production, 192
immunoaffinity purification, 199
immunoblotting, 197–198
Monod, Wyman and Changeux (MWC) postulates, 52–53
Morgan, Thomas Hunt, 217
Motor proteins, 389–390
cilia motor, 397
dynein, 388, 397
energetics, 53, 389
flagellar motor, 328–329
kinesins *see* Kinesin motors
landing rates, 390, 390F
mechanism of action, 390
conformational changes, 389
mitotic, 423–424
myosins *see* Myosins
processivity, 389–390
techniques for study
optical tweezers, 390
tethering, 368, 388
Xkid, 423–424
*see also* Actin/actin filaments; Microtubules
Mouse coat color, 305
Mouse embryo, compaction, 456
Mouse embryo fibroblasts (MEFs), 442–443
Mouse genome, 5
β-globin genes, 69
Mouse knockouts
intermediate filaments, 375
kinesin, 394
Mouse L-cells, cadherin-dependent cell sorting, 446, 446F
Mouse mammary tumor virus (MMTV), 349–350, 477–479
mRNA *see* Messenger RNA
Muller-Hill, Benno, 57
Multicellular organisms, evolution, 371, 371F
Multinucleate cells, 416, 430
Multiplex polymerase chain reaction (PCR), 213
Muscarinic acetylcholine receptors, 360
Muscle
cardiac *see* Cardiac muscle
contraction, 249, 391–392, 392F
acetylcholine, 257
ATP hydrolysis, 392
calcium role, 253, 394–395
sliding filament model, 395
troponin/tropomyosin role, 391, 392
creatine phosphate, 34
differentiation, 155, 166, 171–172
glucose metabolism, 14–15, 316
glycogen metabolism, 332, 333, 353, 354
myosin, 359–360
skeletal, 359–360, 391–393
smooth *versus* skeletal, 359–360
structure, 391–393, 392F
triglyceride metabolism, 354
Muscle fibers, basal lamina, 458
Muscular dystrophy diseases, basal lamina role, 457
Mushrooms, poisonous, 127, 380
Mutagenesis
genetic code and, 135
transposable elements, 115
UV-mediated *see* Ultraviolet (UV) radiation
*in vitro*, 216–217, 219
site-directed, 186, 209
*see also* DNA damage; Mutagens; Mutation
Mutagens, 134–135, 205
chemical, 107–108, 205, 444
alkylating agents *see* Alkylating agents
nitrosoguanidines, 107–108
radiation, 12, 88–89, 386
*see also* Mutagenesis; Mutation
Mutants
biosynthetic, 36–37
conditional-lethal *see* Conditional-lethal mutations
DNA repair, 104–106, 106T
*see also* Mutation
Mutation
aging, 470
cell cycle, 405–406
conditional *see* Conditional-lethal mutations
consequences, 135, 206, 215
deletion, 37, 128
dominant-negative, 208, 215
elimination, 88
frameshift, 106, 135
frequency
UV-induced, 106
variation, 88
gain-of-function, 215
genetic code resistance, 2
human genetic disease *see* Genetic disease
missense, 106
natural selection, 6, 88
null, 219
organellar *versus* nuclear inheritance, 340

rates *see* Mutation rate
silent, 106–107
substitution, 8, 8T, 128, 192
suppressor, 136, 313
susceptibility, 2F
*see also* Mutants
Mutation rate, 8
frequency *versus,* 88
introns, 85
SOS response effect, 104
underestimation, 88
MWC postulates, allosteric regulation, 52–53
Myasthenia gravis, 254
*Myc* oncogene, 479
Myelin basic protein, immunoblots, 197–198
Myelin sheath, saltatory conduction, 254
Myoblasts, differentiation, 180
MyoD gene regulatory protein, 173–174
Myogenic proteins, 155, 166, 172
Myogenin
gene regulation, 155, 155F, 166
heterodimerization, 155
phosphorylation, 155
Myosin II
intermediate filaments *versus,* 391
motor domains, 388
processivity, 389–390
skeletal muscle, 391–393
sliding, 388, 389–390
Myosin light-chain kinase, 359–360
Myosins, 359–360
actin-binding, 376
phosphorylation, 359–360
*see also* Actin/actin filaments; *individual myosins*

## N

*N*-acetylglucosamine (NAG), 59–60, 60F
*N*-acetylglucosaminoglycan (GlcNAc) *see* GlcNAc
*N*-acetylmuramate (NAM), 59–60, 60F
NADH/NAD$^+$
citric acid cycle, 316
redox reactions, 324
NADPH/NADP$^+$, 240
photosynthesis, 332
NAG, 59–60, 60F
NAM, 59–60, 60F
NANA transferase, 274, 299–300, 300F
Natural selection, 6
genetic code, 2
mutation elimination, 88
Neanderthal DNA, 83–85
Nebulin, 392
Necrosis, 440, 440F
Negative staining, electron microscopy, 228
Negative supercoiling, 79, 79F
NEM (*N*-ethylmaleimide), 46, 185, 287
Nematodes
actin mRNA splicing, 130
dosage compensation, 174
gametogenesis, 187–188
GLD1 mutants, 188–189
*Tra2* regulatory elements, 187–188
genome, 130
mRNA stability, 188–189
*see also Caenorhabditis elegans*
Neomycin, 359
Neostigmine, myasthenia gravis treatment, 254
Nernst equation, 256, 295
*N*-ethylmaleimide (NEM), 46, 185, 287
Neuraminidase, 234
Neurofilaments
acrylamide-mediated depolymerization, 376
transport, 399
Neuromuscular junction, acetylcholine receptors, 254, 257
Neuronal growth cones, 399
Neurons
apoptosis, 443–444
axonal transport, 394
morphology, 313, 313F
Neuropathies, mouse models, 394
*Neurospora,* genetic analysis, 340, 340T
Neurotoxins
acrylamide, 376
pertussis toxin, 329, 357, 358
saxitoxin, 258
Neurotransmission, 344
action potential *see* Action potential
endocrine signaling *versus,* 344
neuromuscular junction, 254, 257
Neurotransmitter-gated channels (receptors), 253–257
affinities, 344
as pharmaceutical targets, 260–261
phosphorylation-mediated desensitization, 352
signaling speed, 344
*see also specific receptors*
Neurotransmitters, 344
inhibitory, 254
receptors *see* Neurotransmitter-gated channels (receptors)
release, 313
*see also* Synaptic vesicles
*see also specific transmitters*
Neutral pH, 20
Neutron, 12
Neutrophils, cell movement, 391
Newton's laws of motion, 253–254
Nexin, 396
NFκB gene regulatory protein, 273
N-formylated peptides, neutrophil chemotaxis, 391
Nicotinic acetylcholine receptors, 352
Nidogen, basal lamina cross-linking, 458
Nigericin, 317
Nightblindness, Oguchi's disease, 355
Nitric oxide (NO), guanylyl cyclase stimulation, 345
Nitrocellulose filters, 198
Nitrogen availability, 165–166
Nitrosoguanidines, 107–108
Nitroxide radical, 234
NMDA receptors
cDNA cloning, 260–261
pharmacology, 260–261
Nocodazole, 427
Nodes of Ranvier, saltatory conduction, 254
Nomarski differential-interference-contrast microscopy, 223, 223F
Noncoding DNA, 66–67
junk DNA, 66
*see also* Repetitive DNA
Noncovalent bonds, 15
molecule aggregation, 15
protein conformation, 39
Noncyclic photophosphorylation, 355
Nonhomologous end-joining, 109–110, 109F
Non-Mendelian inheritance, 337
Nonsynonomous nucleotide substitutions, 8, 8T
Noradrenaline, effects, 359
Notch signaling pathway, 368
Notothenoid fish
antifreeze protein, 48, 48F
microtubule dynamics, 377–378
NSF (NEM-sensitive factor), SNARE protein disassembly, 287, 289, 292, 295, 297
NtrC protein
phosphorylation by NtrB, 165
transcriptional regulation, 165–166, 165F
Nuclear–cytoplasmic transport, 266–274
analysis in yeast, 269–270
directionality, 268
gated transport, 264
heterocaryon analysis, 268
leptomycin B, 272
mitosis and, 268
NPCs *see* Nuclear pore complexes (NPCs)
NTF2, 271
nuclear export, 270–271
nuclear export signal, 267, 268–269, 272–273
oocyte studies, 272–273
Rev protein, 272
stoichiometry, 268–269
nuclear import, 270
ER import *versus,* 284
forced nuclear import experiments, 269
receptors, 270–271
nuclear localization signal, 268, 269
phosphorylation, 267
nucleoplasmin localization experiments, 269
protein shuttling, 267, 272
Ran GTPase, 268, 271
regulation by phosphorylation, 267
uptake studies, 271
Nuclear envelope, 290
Nuclear import receptor, 270
Nuclear lamins, 43
disassembly, 379–380
lamin C
amino acid sequence, 43, 43F
hybrid genes, 199–201
phosphorylation, 379–380
Nuclear pore complexes (NPCs), 269
dilation, 268
two-way traffic, 267
Nuclear pore complexes, *Dictyostelium discoidium,* 228
Nuclease(s)
DNase I, 71, 73–74
micrococcal, 71, 73–74
Nucleic acids
DNA *versus* RNA as genetic material, 147
hybridization
DNA *see* DNA hybridization
*in situ* hybridization, 217
polarity, 18, 23
structure, 18
*see also* DNA; Nucleotides; RNA
Nucleolus, cell cycle variations, 123
Nucleophilic attack, 62
Nucleophosmin, centrosome cycle, 418–419
Nucleoplasmin, 269, 269F
Nucleosides, 18
energy carriers, 26
Nucleosomes, 66, 67, 68
assay, 130F
assembly, 68, 71–72, 95
DNA topology effects, 79–80
gene expression role, 71–72, 73–74, 75–76, 168–169
histones *see* Histone(s)
linker DNA, 68
packing ratio, 68
positioning, 71–72
SW1/SNF complex, 71–72, 72F
yeast chromosomes, 76
*see also* Chromatin
Nucleotide excision repair (NER), 107
deficiency (xeroderma pigmentosum), 107
Nucleotides, 18
dideoxy, 89–90, 90F, 206
energy carriers, 29
*see also* ATP (adenosine triphosphate)
modified, RNA, 128–129
purines *see* Purine nucleotides
pyramidines *see* Pyramidine nucleotides
substitutions, 8, 8T, 127, 192
synthesis, 55, 57
Nucleotide sequence comparison *see* DNA sequence analysis
Nucleotide substitutions, 8, 8T, 127, 192
Nucleotide triphosphates (NTPs), endoplasmic reticulum lumen, 284
Nucleus, 154
envelope, 290
breakdown, 413
inner membrane, 267
outer membrane, 267
functions, 264
human, 64
mitochondrial gene transfer, 6–8, 339
telophase, formation, 416
transport *see* Nuclear–cytoplasmic transport
Nutrients, membrane transport, 250
N-WASp, actin polymerization, 400–401, 400F

## O


Oguchi's disease, 355
Okadaic acid, 417
Okazaki fragments, 91, 93
Oligomycin, 327T
Oligonucleotide-ligation assay, sickle-cell anemia, 212, 212F
Oligonucleotides, 18, 18F, 23
allele-specific, 217–218
degenerate probes, 210
synthesis, 197, 205
Oligonucleotide splints, 205
Oligosaccharides
*N*-linked , 274, 298
*O*-linked , 274–275
lysosomal sorting, 303–304
processing, 274–275, 285, 299T, 300F
endoglycosylase H, 273, 285
species differences, 274
synthesis, energetics, 287
*see also* Carbohydrates; Polysaccharides
Oligosaccharide transferases, 285
OmpC membrane protein, T4 infection, 448, 448T
Oncogenes
cellular homologs, 187
mutation and cancer, 368–369
*Fos*, 158–159, 187
*Jun*, 158–159
*Myc*, 474
*Ras*, 479
*Src*, 55–56
Onion-skin structure (DNA), 101
Oocyte, 350–351
maturation, 408–409
*Xenopus*, 404
*see also* Microinjection experiments
*see also* Egg
Oogenesis, nematode gene control, 187–188
Open reading frame, 206
Operons, 161–162
lactose metabolism *see* Lactose *(Lac)* operon
*Trp* operon, 162
Optical density, light microscopy, 222
Optical tweezers, 390
Organelles
compartmentalization, 263–266
genomes, 336–341
mitochondrial *see* Mitochondrial genome
human, 264
inheritance, 337
numbers, 264
transport mechanisms, 264
*see also specific organelles*
Organic chemistry, 12
Origin of replication *see* Replication origin(s)
Orthologous genes, 4
Osmotic balance, 249, 254
Osteogenesis imperfecta, 463
Ouabain, sodium-potassium pump effects, 249
Ovalbumin, 129F
Overlap microtubules, 415, 420
Ovotransferrin, 51
Oxaloacetate, 35
regeneration, 36
Oxidation reactions, 26–27, 324
palmitic acid, 32
Oxidation–reduction (redox) reactions, 26–27
general principles, 324–326
lithotrophs, 4
standard redox potential ($E_0'$), 324
thermodynamics, 301, 324–326, 332
Oxidation states, 27, 27F
Oxidative phosphorylation
ATP synthesis, 316, 318
uncoupling, 324
electron transport *see* Electron transport chains
Oxygen
abundance, 13F
ATP regeneration, 32
cellular consumption, 139
cardiac muscle, 318
mitochondria, 299–300, 322
chemical properties, 13
citric acid cycle stoichiometry, 36
electron transport, 316, 321–322
hemoglobin binding affinity, 51
measurement, 321–322
photosynthesis, 4–5, 334–335
Oxygen electrodes, 321–322, 327


## P


P9OH incorporation, peroxisomes, 279
p53, role in cancer, 477–478, 478T
PAGE *see* Polyacrylamide gel electrophoresis (PAGE)
Palindromic DNA sequence, 160, 190
Palmitic acid, 32, 32F
*Paracoccus denitrificans,* respiratory chain, 337
Paracrine signaling, 344
Parthenogenesis, mitosis, 424, 424F
Pasteur, Louis, 2–3, 316
Pasteur effect, 316
Pasteur's experiment (spontaneous generation), 2–3
Patch clamp analysis
inside-out, 360
membrane transport, 246F
neuromuscular junction, 257
potassium channels, 255, 255F, 257–258
acetylcholine activation, 360
whole-cell, 313–314
Paternal imprinting, 174
Pauling, Linus, 50
PCR *see* Polymerase chain reaction (PCR)
PDGF *see* Platelet-derived growth factor (PDGF)
PDGF-R (platelet-derived growth factor receptor) *see* Platelet-derived growth factor (PDGF) receptor
Pedigrees, 174, 174F, 338, 338F, 407, 421F
Penile erection, 345
Periodate, 304
Peroxisomes
alanine:glyoxylate aminotransferase (AGT), 279
assay, 280–281
assembly, 281–282
catalase, 279
deficient cell selection, 281
distribution, 279
glycosomes, 280
membrane phospholipids, 287–288
P9OH incorporation, 279
primary hyperoxaluria type 1 (PH1), 279
transmembrane transport, 256–259, 264
analysis, 281
cytosolic receptor, 281–282
folded proteins, 281–282
*Pex2,5,6* genes, 281–282
Pertussis toxin, GPCR effects, 329, 357, 358
Petite *(Pet)* mutants, 341, 341F
*Pex2,5,6* genes, peroxisome membrane transport, 281–282, 282F
pH, 13–15
gradients, 318
*see also* Electrochemical proton gradients
Henderson–Hasselbach equation, 19, 21
intracellular, 22, 248
endosomes, 307
lysosomes, 278, 302–303
measurement, 251
ion-exchange chromatography, 195
isoelectric point, 14
mitochondria, 318
protein denaturation, 45
titration curves, 14, 14F, 14FF
*see also* Acids; Bases; Buffers
Phage-display, 197
cyclin A, 201
Phage-display libraries, 197, 201
panning, 201
reduced genetic code, 197
Phagocytosis, 306
Phalloidin, actin binding, 380, 381F
Phase contrast microscopy, 223, 223F
Phase variation, DNA rearrangement, 175
Phenotype, 215
genotype relationship, 88
Phorbol esters, 273, 359
Phosphate bonds, energetics, 33, 92
Phosphate buffer, 22
Phosphatidylcholine, 233
membrane exchange, 287–288
Phosphatidylcholine exchange protein, 287–288
Phosphatidylethanolamine, 233
Phosphatidylinositol (PI), cell signaling, 233
Phosphatidylinositol 3′-kinase (PI 3-kinase), 56–57, 56F, 336–337, 364
Phosphatidylinositol-dependent protein kinase (PDK1), Akt activation, 365
Phosphatidylserine, 233
membrane redistribution, 238
Phosphoanhydride bond, 26
Phosphodiester bond, 92
Phosphoenolpyruvate, 32
Phosphoglucose isomerase, thermodynamics, 28–29
2-Phosphoglycerate, 33
3-Phosphoglycerate, 31F, 333F
conversion to pyruvate, 31, 32–33, 33F
oxidation state, 247
structure, 33
Phosphoglycerate kinase (PKG), 280
Phosphoglycerate mutase, 33
Phosphoglycolate, enzyme inhibition, 50
Phospholipase C
neomycin inhibition, 359
PDGF receptor, 364
Phospholipases
membrane lysis, 233, 234
phospholipase C, 359, 364
Phospholipid exchange proteins, 287–288
Phospholipids
charge (no net), 231
ER membrane, 284
ESR spectroscopy, 234, 237–238
membrane, 233
distribution, 233, 237–238
mitochondrial, 287–288
peroxisomes, 287–288
red blood cells, 241
phospholipase action, 234
structure, 17, 17F, 234F
vesicle fusion, 292
*see also specific lipids*
Phospholipid vesicles, synthetic, 287–288
*N*-Phosphonacetyl-L-aspartate (PALA), 55, 55F
Phosphorus-32, 12, 12T
Phosphorylase kinase, signal integration, 354
Phosphorylation, 53F, 61
ATP requirement, 61
autophosphorylation, 364
axonal transport role, 399
cell cycle control, 61
histone tails, 61, 72–73
immunoblots, 197–198
intermediate filament disassembly, 379–380
myogenin, 155
nuclear import regulation, 267, 273
protein regulation, 353, 358
coagulation cascade, 359
enzyme regulation, 52, 61
PDGF autophosphorylation, 364
receptor desensitization, 352
signaling pathways, 345–346, 350–351
Trk autophosphorylation, 364
proteosome sensitization, 369
RNA pol CTD, 121
substrate-level, 33
*see also* Protein kinase(s)
Phosphotransfer reactions, stereochemistry, 319
Phosphotyrosine phosphorylase, 364
Photochemical reaction center, 331
Photomicrographs, light microscopy, 223F
Photophosphorylation, 331
6-4 Photoproduct, 107
Photoreactivation, 104
Photoreceptors, rod cells, 355

Photorespiration, 330–331
  *see also* Carbon fixation
Photosynthesis, 4, 23, 329–336
  algal, 333, 333F
  bacterial, 316–317
    cyanobacteria, 330
    purple sulfur bacteria, 5
  cytochromes, 334
  electron transport *see* Electron transport chains
  energetics, 331
  oxygen evolution, 5, 334–335
  photophosphorylation, 331
    cyclic, 332
    noncyclic, 332
  photosystems, 331, 334–335
  stoichiometry, 5
  Z scheme, 334
  *see also* Carbon fixation; Chloroplast(s)
Photosystems, 331, 334–335
  cooperation, 333
  electron flow, 334–335
  photosystem I, 331, 334
  photosystem II, 334–335
Phototransduction, 356
Phototrophs, 4
pH-sensitive fluorophores, SNARF-1, 251
Phylogenetics, eucaryotes, 5, 82F, 371, 371F
Phylogenetic trees, 9
  hemoglobin genes, 8, 9, 9F
  mitochondrial genes, 7
Physics, life, 3
PI 3-kinase (phosphatidylinositol 3′-kinase), 56–57, 56F
Pigmentation, defects, 334
Pinocytosis, 307–308
p$K_a$ values, 14–15, 14T, 21–23, 45
Planck's constant, 331
Plant(s)
  cell signaling, 371–372
  fruit ripening, 371–372
  green appearance, 331
  phylogeny, 9, 9F, 371, 371F
Plant cells
  animal cells *versus,* 6, 371, 371F
  cell wall, 465–467
  characteristics, 6
  chloroplasts *see* Chloroplast(s)
  cytokinesis, 428
  energy conversion, 329–336
    efficiency, 331
    *see also* Photosynthesis
  expansion, 466
  microtubules, 466
  organellar genomes
    chloroplasts, 336–341
    mitochondrial, 7
  variegation, 337, 337F, 404–405
Plant growth factors, 371–372
Plant growth regulators, 372
Plant hemoglobins, evolution, 8–9, 9F
Plaque assay, 482
Plasma membrane
  cholesterol, 308–309
  functions, 243, 250
  membrane protein sorting in ER, 265–266
  permeability, 243
  *see also* Lipid bilayer; Membrane proteins
Plasmids
  bacterial, 79–80
  engineered, 116, 210
    dicentric, 420
  structures, 111–112
  supercoiling, 79–80
  yeast, 70–71, 76, 76F, 420
Plasmodesmata, 455–456
  TMV diffusion, 456, 457F, 457T
Plastocyanin, chloroplast membrane transport, 277–278
Plastoquinone, 331
Platelet(s)
  clotting cascade, 359, 460–461
    actin filaments and, 383
    regulation by phosphorylation, 359
  integrin expression, 460
  megakaryocyte precursors, 430
Platelet-derived growth factor (PDGF), 56–57, 227
  cancer and, 431
  cell cycle control, 432, 432T
Platelet-derived growth factor (PDGF) receptor, 56–57
  autophosphorylation, 364
  DNA synthesis, 364
  signaling complex, 364, 364F
Pleckstrin homology (PH) domain, 345
Plectonemic DNA supercoiling, 79–80
Plug-in protein domains, 42
Poisons
  α-amanitin, 127
  arsenate, 31, 33, 34
  azide, 460
  carbon monoxide, 323
  cyanide, 323, 327, 327T
  cycloheximide *see* Cycloheximide
  digitalis, 249
  herbicides, 331
  malonate, 34, 35
  metabolic, 31, 33, 34, 35, 318
    electron transport inhibition, 327, 327T
  oubain, 249
  protein synthesis inhibition *see under* Protein synthesis
  RNA polymerase inhibition, 127
  snake venom, 234
  urethane, 327, 327T
  *see also* Antibiotics; Enzyme inhibitors; Mutagens; Toxins
Polarity
  chemical bonds, 15
  nucleic acids, 18, 23
Polar relaxation cytokinesis hypothesis, 428
Polyacrylamide gel electrophoresis (PAGE)
  DNA, 112, 147
    *see also* Southern blotting
  proteins, 46, 131, 137, 185–186, 199
    membrane, 226–227, 236, 242
    mercaptoethanol, 196
    sensitivity, 197–198
    silver staining, 197
  radioactive markers, 198
  SDS-PAGE, 137, 138, 196, 199, 236, 242, 336
    autoradiography, 198
    gel-filtration *versus,* 196
  two-dimensional, 100, 151, 196, 351, 368
Polyadenylation signal sequence, 131
Polyadenylation site, 131
Poly-A tails (mRNA 3′ end), 131
  gene expression role, 182
  polyadenylation reaction, 131
Polycistronic mRNA, 137
Polyclonal antibodies, immunoblotting, 197–198
Polyethylene glycol (PEG), tight junctions, 450, 450F
Polymerase chain reaction (PCR), 148F, 213
  amplification, amount, 213
  ancient DNA, 83–85
  DMD gene, 213
  DNA-binding sequence, 159–160
  error-prone, 149
  multiplex analysis, 213
  primers, 198, 206–207
  principles, 213
  quantitative, 169, 169F
  recombinant, 08
  ribozymes, 147–148
Polymerization reactions, 17, 29F, 374, 376–377
  condensation reaction, 17
  DNA, 89, 90, 93
    *see also* DNA synthesis
  temperature dependence, 24–25
  thermodynamics, 24–25, 25F, 25FF
  *see also specific reactions*
Polymorphisms, 181
  RFLPs, 69–70, 70F
  SNPs, 217
  VNTRs, 85
Polypeptides *see* Protein(s)
Polysaccharides
  bacterial cell wall, 59–60
  fatty acid oxidation *versus,* 32
  glycosidic bonds, 16
  lipopolysaccharide, 448
  lysozyme cleavage, 50
  structure, 16, 18, 18F, 21
  *see also* Carbohydrates; Oligosaccharides; *individual sugars*
Polytene chromosomes, 78, 78F, 79F
  bands/interbands, 76, 78
  diploid *versus,* 78
  ecdysone effects, 350
  puffs, 350, 350F
Porins, mitochondrial membranes, 315
Position effect, 74–75
Positive feedback loops, 442
  cell cycle control, 414
Positive supercoiling, 79, 79F, 121
Potassium, abundance, 13F
Potassium channels
  acetylcholine activation, 360
  inhibitory neurotransmission, 254
  patch clamp analysis, 255, 255F, 257–258, 360
  selectivity filter, 255
  voltage-gated, 255, 257–258
Prenatal diagnosis, DNA microarrays, 218
Prenyl groups, 239
Pre-pro-proteins, 312
Priestly, Joseph, 330–331
Primary cell culture, 191
Primary hyperoxaluria type 1 (PH1), 279
Primase, 94
Primer extension mapping, 130, 130F
Primosome, 94
Procaryotes
  eucaryotes *versus*
    cytoskeleton, 375
    protein synthesis, 137
  genome sequence, 5
  transposons, 115–116
  *see also specific types*
Procollagen type I, 301
Progesterone, oocyte activation cascade, 350–351, 408–409
Programmed cell death *see* Apoptosis
Proline, biosynthesis, 37
Proline hydroxylase, 301
Promoter elements, 120, 123, 125–126
  consensus sequences, 116
    bacterial, 114, 124
  inducible, 408, 409
  *Lac* operon, 162
  linker-scanning analysis, 116
  thymidine kinase gene, 116
  trypanosome *Vsg* gene, 123–124
  UV mapping experiments, 124
  *in vitro* transcription assays, 116–117
Proofreading
  aminoacyl-tRNA synthetases, 136
  DNA polymerases, 90, 93, 93F
  RNA polymerases, 125
Prophase, 414, 415
Proteases, 134, 234, 278, 285
  caspases, 441
Proteasomes, 145, 145F
  ALLN inhibition, 369
  β-catenin degradation, 369
  phosphorylation sensitization, 369
  protein degradation, 145
    *see also* Proteolysis
  *see also* Ubiquitination
Protein(s), 2
  analysis, 195–202
    electrophoresis *see under* Electrophoresis
    footprinting, 160, 160F
    FRET, 226–227
    MALDI-TOF, 198
    membrane proteins, 235–236
    phage-display, 197, 201
    SPR, 202

Page numbers refer to a major text discussion of the entry; page numbers with an F refer to a figure, with a T refer to a table.

yeast-two hybrid system, 199–201
*see also specific methods*
average molecular weight, 139, 243
binding surfaces, 42
cellular fate, 264
*see also* Protein sorting
conformational changes, 48
ATP synthase, 293, 320–321
bacteriorhodopsin, 239
hemoglobin, 248
integrins, 460–461
MAP kinases, 363
motor proteins, 389
transport proteins, 248
conformational interactions, 39
C-terminal, 17
databases, 42, 198
degradation *see* Proteolysis
denaturation, 40, 43, 45, 45F, 196
domains, 163
fluorescent, 226–227
folded *versus* unfolded, 44
folding *see* Protein folding
function, 47–62
fusion, 146–147, 199–200
glycosylation, 284, 285
*see also* Glycoproteins
membrane *see* Membrane proteins
near entropy minimum, 39
post-translational modification, 153, 186
purification *see* Protein purification
secretory, 312
side chains, 17, 40
signaling, 345–346
spring-like, 46, 46F
stability, 44, 154
pH effects, 45, 195
structure *see* Protein structure
subunits, 45
*see also* Protein complexes (assemblies)
synthesis *see* Protein synthesis
synthetic peptides, 464–465
tagging *see* Protein tags
transport, 263–288
*see also* Protein sorting
Protein analysis, electrophoresis, 196–199
Protein complexes (assemblies), 215
assembly, 42, 44, 52–53
binding surfaces, 43
dimerization, 42–43, 48, 242
advantages, 154
heterodimerization, 48, 149–150
homodimerization, 154–155
energetics, 24–25
membrane protein-associations, 240, 242
tetramers, 52
*see also* Protein–protein interactions
Protein degradation *see* Proteolysis
Protein disulfide isomerase (PDI), 297
Protein–DNA interactions, 43, 48, 49
DNA bending, 157–158, 163
DNA footprinting, 101, 160, 160F
gel-mobility shift assays, 159
hydrogen bonds, 154
hydrophobic interactions, 154
kinetochore, 421–423
motifs *see* Protein motifs
specificity, 99, 154
strength, 153
*see also* DNA-binding proteins
Protein engineering, 207, 208
synthetic peptides, 464–465
*see also* Genetic engineering
Protein families, 42
Protein folding
ATP requirement, 143–144
barnase, 276
chaperones *see* Chaperones
conservation, 42
hydrophobic forces, 239
Levinthal paradox, 41
misfolded proteins, 144, 264, 272, 286
pathways, 41
problem, 45
purification problems, 208
refolding, 144
stability effect, 44
topological representation, 41, 41F
unfolding, 40, 43, 45F, 185
transport requirement, 276, 281–282
Protein kinase(s)
cell cycle regulation, 61
protein tyrosine kinases, 55–56, 482–483, 483F
serine/threonine kinases, 218–219
signaling pathways, 333–339, 350–351
*see also* Phosphorylation; *individual kinases*
Protein kinase A (PKA), 358
Protein kinase B (PKB/Akt), 365
Protein kinase C (PKC), clotting cascade, 359
Protein–ligand interactions
binding-site determination, 56
dissociation rate, 202
equilibrium dialysis, 57
kinetics, 49, 53–55
protein–DNA *see* Protein–DNA interactions
protein–protein *see* Protein–protein interactions
Scatchard analysis, 57–58
SPR analysis, 202
two-hybrid analysis, 199–201
*see also* Antigen–antibody interactions; Enzyme catalysis
Protein motifs, 42
DNA-binding, 153–160
HLH, 155
leucine zipper, 158–159
zinc finger, 154
RGD motif, 460
Walker motif, 101
*see also* Protein–ligand interactions
Protein–protein interactions, 43
cooperativity *see* Cooperativity
energetics, 53–55
helix–helix interactions, 56
PDGF–PDGF receptor, 56–57, 364
pleckstrin homology domain, 345
pulse–chase experiments, 143
Src homology 2 domain, 346, 364, 366
Src homology 3 domain, 346, 364
strength, 163
string–surface interactions, 56
surface–surface interactions, 56
van der Waals interactions, 242
*see also* Protein complexes (assemblies)
Protein purification, 48, 60, 183–184, 192–195, 462
chromatography, 192–195
DNA-binding proteins, 159
EGF receptor, 432–433, 433F
immunoaffinity purification, 199
tagged proteins, 207–208
*see also* Affinity chromatography; Protein tags
tubulin, 378
Protein sorting
cytosolic proteins, 264
polarized epithelium, 312–313
problem, 265
sorting receptors, 264
sorting signals, 245, 260, 264
nuclear localization signal, 269
secretory proteins, 312–313
signal sequences, 264, 273
start/stop-transfer signals, 284, 284F
*trans*-Golgi network, 311–312
*see also* Golgi apparatus
VSV G protein studies, 298–299
*see also* Membrane proteins; Protein translocation
Protein structure, 39–46
β-sheets, 40, 41F, 41FF
buried side-chains, 45, 45T
C-terminal, 17
databases, 42, 198
determination, 42
disulfide bonds, 46, 46F
domains, 163
folded *versus* unfolded, 44
*see also* Protein folding
α-helices, 40
in-line domains, 42
loops, 39
motifs, 42
N-terminal, 17
plug-in domains, 42
prediction, 44
primary, 17, 45, 198
deduction, 135, 195
*see also* Amino acid(s)
Stokes radius, 194
topology, 41
*see also* Amino acid(s)
Protein synthesis, 134–149
accuracy, 139
cysteine misincorporation, 139
error rate, 45, 129
daily net synthesis, 32
direction, 141
EF-Tu elongation factor, 137
energetics, 139, 284
eucaryotes *versus* procaryotes, 137
folding *see* Protein folding
inhibition, 99–100, 137
cycloheximide *see* Cycloheximide
edeine, 137
emitine, 409
*see also* Antibiotics; Poisons
location, 264
machinery
evolution, 135–136
*see also* Ribosomes; Transfer RNA
measurement, 183
mitochondrial, 339–340
rate, 138–139
termination
bacterial release factors, 141
*see also* Stop codons
*Tetrahymena*, 142
*see also* Protein sorting; Translation
Protein tags, 62
epitope *see* Epitope tagging
fluorescent, 158–159, 226–227, 250–251, 271, 285
glutathione S-transferase, 62F, 201, 269–270, 334
histidine tagging, 207–208
tmRNA-mediated tagging, 142
*see also* Affinity chromatography
Protein translocation
experimental analysis, 266
microsomal membranes, 285–286
protein folding/unfolding, 276
translocators, 264
mitochondrial TIM complex, 275, 277
mitochondrial TOM complex, 276
mutant selection, 286
Sec61 translocator complex, 283, 284, 286
*see also* Protein sorting; *specific systems*
Protein tyrosine kinases
Abl, 227
kinetic parameters, 62T
receptors, 364
Src, 55–56
Protein tyrosine phosphatases (PTPs), 61, 362
Proteoglycan
carbohydrate side chains, 462
synthesis, 297
Proteolysis, 145–147
autoproteolytic cleavage, hedgehog, 369
caspase cascade, 441
cell cycle role, 145, 378–379
D-box, 145
inhibition, 145
kinetics, 146
proteases *see* Proteases
proteasomes *see* Proteasomes
regulated proteolysis-linked signaling, 367–370
tmRNA-mediated tagging, 142
ubiquitination, 134, 139, 145, 339–340
Proton-motive force, 319–320
flagellar motor, 328–329
Proton pumps

light-activated *see* Light-driven pumps
lysosomal membranes, 302
mitochondrial, 316–317
sodium–proton ($Na^+$–$H^+$) antiporter, 248
*see also* Electrochemical proton gradients
Protons, 12, 324
Protozoa
*Giardia lamblia,* 6
phylogeny, 9, 9F
telomerase, 103
P-selectin
blood vessel walls, 449
neutrophil rolling interactions, 449
removal from membranes, 447
Pseudocholinesterase, 344
Pseudouridination, RNA modification, 133, 133F
PstI restriction enzyme, 204
Psychosine accumulation, GLD (Krabbe's disease), 429
PTPs (protein tyrosine phosphatases), 61, 362
Pulse–chase experiments
Hsp70–protein binding, 143
mtDNA replication, 338
peroxisome function, 280–281
proteasome-mediated degradation, 145, 145F
Pulsed-field gel electrophoresis, DNA separation, 68, 68F, 203
Purine nucleotides
DNA structure, 64
synthesis regulation, 52
Purple sulfur bacteria, photosynthesis, 5
Pyramidine nucleotides
dimer formation, 91, 100, 104
UV mapping, 124
DNA structure, 64
synthesis inhibition, 55
Pyruvate
fermentation, 31, 31F
glycolytic pathway, 33
structure, 15F
Pyruvate kinase, 33


## Q


'Quick-stop' mutants, 90–91
Quinpirole, dopamine receptor activation, 357


## R


Rab GTPases, 271, 290
Rac GTPase, 394
Rad52 (and homologs), 110, 110F, 434
Radiation
cosmic, 19
DNA damage, 12, 88–89, 433–434
ionizing, 433–434
*see also* Ultraviolet (UV) radiation
Radiation-sensitive mutants, yeast, 433–434
Radioactive isotopes, 12–13, 12T
autoradiography *see* Autoradiography
DNA experiments, 65, 90–91, 93
iodine, 347
metabolic pathway analysis, 34–35
RIA, 347
Radioactive markers, 198
Radioactivity, 12, 13, 12T
decay constant (λ), 19, 19F
half-life, 12, 12T, 19, 225
specific activity, 225
standard unit (Ci), 13
Radioimmunoassay (RIA), 347
*Rad* mutants (yeast), 433–434
*Raf1* gene, 218–219, 219T
Raf protein, 201
Random thermal motion, 393
Random walks, 28, 28F
Ran-GEF, 272
Ran GTPase
nuclear transport, 251, 271
Ran-GDP structure, 271, 271F
Ran-GTP structure, 271, 271F
Ras GTPase, 53, 293, 363, 375, 395
gene hybrids, 199–201
GPCRs *versus,* 363
molecular switch, 363
mutation in cancers, 363
protein interactions, 199–200
*Ras* oncogene, 479
Rate constants, 49
Rb gene, 474
Reaction coupling, uncoupling, 31
Reactive groups, 15
activated carriers, 26
*see also specific groups*
Reading frame, 120, 127, 134, 142
open reading frames, 206
RecA protein
DNA repair role, 105–106, 112F, 112FF
homologous recombination, 112
strand assimilation assay, 112F
RecBCD, 113, 113F
Receptor-mediated endocytosis, 305–310
clathrin-coated vesicles *see* Clathrin-coated vesicles
kinetics, 308–309
LDL metabolism, 308–310
phagocytosis, 306
pinocytosis *versus,* 307–308
Receptors
cell-surface (transmembrane)
affinities, 460
cloning, 347–348
desensitization, 351
enzyme-linked *see* Enzyme-linked cell-surface receptors
G-protein-linked *see* G-protein-linked (coupled) receptors (GPCRs)
hormone *see* Hormone receptors
neurotransmitter *see* Neurotransmitter-gated channels (receptors)
radioimmunoassay, 347–348
*see also individual receptors*
nuclear, 345
*see also* Gene regulatory proteins
Receptor tyrosine kinases (trks), 364
Recombinant DNA technology, 203–214
*see also* Gene cloning; Genetic engineering; *specific techniques*
Recombinant PCR, 208
Recombination
homologous/general *see* Homologous recombination
meiotic, 479–480
site-specific *see* Site-specific recombination
Red blood cells
$CO_2$ transport, 248
ghosts, 226, 234, 235F, 235T, 242, 242F, 288
glycolysis, 29, 29F
malaria and, 240
membranes, 239
composition, 241
ESR spectroscopy, 234
intracellular, lack of, 239
permeability, 254
phospholipid distribution, 234
proteins, 241
Redox potential, standard ($E_0'$), 324
respiratory chain, 326
Redox reactions *see* Oxidation–reduction (redox) reactions
Reduction reactions, 324
Refraction, light, microscopy, 222, 222F
Refractive indices, 225, 225F
eye, human, 223
light microscopy, 222
Release factors, bacterial, 141
Repetitive DNA
alternative splicing, 185
microsatellites, 152–153
recombination prevention, 115
tandem repeats, 85
telomeres, 67
trinucleotide repeats, 67
VNTRs, 85
Replica plating, 87
Replication bubbles, 91, 92–93, 96, 97F, 98F
Replication fork, 88, 89, 91, 96
Replication origin(s), 70, 94
bacterial *(OriC),* 420
bidirectional, 96
cell cycle control, 411
mammalian, 102, 102F
minimum number, 98
origin recognition complex, 102F, 380
SV40, 99, 99F
yeast, 411, 420
Replicative cell senescence, 431
Reporter genes, 116
chloramphenicol acetyltransferase, 168, 321
β-galactosidase, 170–171, 201–202, 421–422
*LacZ* gene, 116
yeast-two hybrid system, 199–201
Repressor proteins
arginine repressor, 164
cI repressor, 172
Cro repressor, 42–43, 172
dimerization, 154–155
DNA looping and, 162–163
feedback inhibition, 161
heme-controlled repressor, 185
Lac repressor, 49, 59, 100, 156–157
LexA repressor, 199–201
λ repressor, 154–155, 156
tryptophan repressor, 161
yeast α2, 175–176
Resistance, enzyme inhibitors, 55
Resolution, *vs.* magnification, 223
Resonance energy transfer, 331
Respiration, Krebs' experiments, 35–36, 35T
Restriction digests, 69F, 69FF, 204–206
centromeric DNA, 76
double digests, 205
gene cloning, 68–69, 69F
*see also* Gene cloning
helical DNA sites, 204, 204F
incompatible ends, 205
mapping *see* Restriction mapping
mtDNA, 339, 339F
nucleosomal DNA, 71–72
partial digestion, 210
recombination analysis, 112, 112F
*Tetrahymena* rRNA minichromosome, 211
unique sites, 70
Restriction enzymes, 69, 70, 204–205
Restriction fragment length polymorphism (RFLP), 69–70, 70F
Restriction mapping, 69–70, 205
cloned gene, 211
viral DNA, 69–70
Restriction–modification system, 203
Reticulocyte lysates, 185
Retinoblastoma, 475–476, 475F
restriction fragments, 476F
*trans*-Retinoic acid, cancer treatment, 482
Retrieval transport pathways, 291
Retrotransposons, 81
Reverse transcriptase, 116–117
Rev-mediated nuclear export, 272, 272F
RFLP (restriction fragment length polymorphism), 69–70
RGD motif, 460
RGS (regulator of G protein signaling), 353
Rhodamine, caged, 399
Rhodopsin, 356
Rhodopsin-specific kinase, 355
Rho GTPase, 394
Ribonucleases
barnase, 276
RNase H, 132
unfolding, 45, 276
Ribonucleic acid *see* RNA
Ribose 5-phosphate, feedback inhibition, 51
Ribosomal RNA
evolution, 5
genes, 339–340
*Tetrahymena* minichromosome, 211


Page numbers refer to a major text discussion of the entry; page numbers with an F refer to a figure, with a T refer to a table.

transcription, 120, 120F, 124F, 125, 167
Ribosomes
assembly, 45
bacterial, 45
cyclohexamide effect, 274–275
frameshifting, 139
free *versus* membrane-bound, 259, 263
*see also* Endoplasmic reticulum (ER)
IRESs, 189
protein translocation across ER, 285–286
RNA *see* Ribosomal RNA
'roadblocks,' 141
tmRNA association, 142
*see also* Protein synthesis; Translation
Ribozymes, 147–148, 148F, 149T
Ribulose bisphosphate carboxylase, photorespiration, 330
RME-2 protein, 189
RNA, 1–3
amplification, 148
biological functions, 120, 139
catalytic *see* Catalytic RNA
editing *see* RNA editing
evolutionary significance, 147–149
as genetic material, 147
guide RNAs, 183
isolation, 180
ligation, 149
loops, 129
modification, 133
mRNA *see* Messenger RNA
polymerization, 148–149
rRNA *see* Ribosomal RNA
structure *see* RNA structure
transport, 183
tRNA *see* Transfer RNA
*Xist* RNA, 181–182
RNA–DNA hybrids, 121, 125, 128–129, 128F
RNA editing, 8, 183F
guide RNAs, 183
tissue specific, 186
RNA modification, 133
RNA polymerase(s), 120–121, 157F
active site, 128–129, 128F
bacterial, 120
characterization, 127
DNA rotation, 121
*see also* DNA topology
eucaryotic, 127–130
fluorescent, 156
inhibition, 127
kinetics, 125, 125F, 156–157
open *versus* closed complex, 165
proofreading, 125
protein–DNA interactions, 156–157
RNA pol II
C-terminal domain, 127–128, 173
HIV transcription, 182–183
scanning/entry-site model of enhancer function, 157, 162–163
sigma factors, 120, 125, 125F
RNA priming in DNA replication, 94, 94F
RNA–RNA interactions, 130
RNase H, 132
RNase protection assay, 133, 133F
RNA structure
hairpins, 147, 183
snoRNAs, 132
stem–loop, 132F
tRNAs, 140, 140F
RNA transport, 183
RNA world, 147–149
Rod photoreceptors, 355
Rous sarcoma virus, 55–56
rRNA *see* Ribosomal RNA
*Rsa* genes, 69–70, 70F
RuvC, Holliday junction cleavage, 113–114, 114F

## S

S1 mapping, 130
*Saccharomyces cerevisiae*
*Cdc* genes, 404
chromosomal DNA, 68
Ty elements, 116–117, 117F
*see also* Yeast
SAGA complex, yeast gene expression, 169–170
*Salmonella*
chemotaxis receptors, 366–367
phase variation, 175
Saltatory conduction, 254
Salts
glycine, 14
molecules *versus,* 15
Sand dollar egg, cytokinesis, 428, 428F
Sar1 protein, 292
Sarcomeres, 391
Sarcoplasmic reticulum, 260, 391
Sau3A restriction enzyme, 209
Saxitoxin, 258
Scaffolding proteins, 345, 363
Scatchard, George, 57
Scatchard analysis
microtubule assembly, 385, 385F
protein–ligand interactions, 57–58, 57F
Scc1 protein, 412
Scc3 protein, 412
*Schizosaccharomyces pombe* (fission yeast)
*Cdc* genes, 404
DNA damage response pathway, 431, 433–434
Wee1 mutants, 436
Schrödinger, Erwin, 3
*Sciara,* parthenogenesis, 424
Scramblases, 285
SDS (sodium dodecyl sulfate), 240, 240F
SDS-PAGE *see under* Polyacrylamide gel electrophoresis (PAGE)
Sea urchin eggs, 419F, 426
Sec4 GTPase, 296
Sec61 translocator complex, 283, 284
Secondary cell culture, 191
Second messengers, 345
Secretory proteins, 312
Secretory vesicles, 265, 285, 290–291
secretion signal, 313
synaptic *see* Synaptic vesicles
transport, 264
*see also* Exocytosis
Sectoring assays, 422–423, 423F
Sedimentation *see* Centrifugation
Sedimentation coefficient (S), 193
Segmentation genes, 170–171
Self-splicing introns, 123F, 124
Separase, 425
Septate junctions, invertebrates, 449–450
Sequence hypothesis, 120
Serine/threonine protein kinases, 218–219
Serine/threonine protein phosphatases, 362
Serotonin, platelet secretion, 359
Serum response element (SRE), 187
Severe combined immunodeficiency syndrome (SCID), 474–475
Sex chromosomes, 66
DMD gene, 213
X-inactivation, 180–181
*Shibire* mutants, 311, 311F
Shuttle cloning vectors, 209
Sialic acid residues
neuraminidase, 234
sialidase, 242
Sialidase, 242
Sickle-cell anemia, 206, 212
Sigma (σ) factors, bacterial transcription, 120, 125, 125F
Signaling complexes, 345–346
PDGF receptor, 364
Signaling molecules
cascades, 365
cell survival, 345
complexes, 345–346, 364
differential responses, 345
ephrins, 362
extracellular *versus* intracellular, 344
hormones *see* Hormones
hydrophilic nature, 345
second messengers, 345
calcium *see* Calcium signaling
cAMP *see* Cyclic AMP (cAMP)
cGMP *see* Cyclic GMP (cGMP)
turnover, 347
Signaling pathways
amplification, 331, 332, 345
ATP-like signal metabolites, 52
autocrine, 344
cross-talk, 345
cytokine-mediated, 273
defining, 345
endocrine, 344
inside-out signaling, 459
paracrine, 344
phosphorylase kinase, 354
protein kinases *see* Protein kinase(s)
Ras GTPase, 53
synaptic *see* Neurotransmission
termination, 332, 354
time-course, 345, 346
*see also* Cell communication; *specific pathways*
Signal peptidases, 260
Signal peptides, 260, 264
chloroplast transmembrane transport, 278
Signal-recognition particle(s) (SRPs), 283
SINE elements, 81
Single-nucleotide polymorphisms (SNPs), 215
Single-stranded DNA, 64
binding proteins *see* Single-stranded DNA-binding proteins (SSBs)
hairpins, 90
Single-stranded DNA-binding proteins (SSBs), 90, 92
cooperativity, 92
T antigen-binding, 99, 99F
*Sir3* gene, 75–76
Sir protein complex, 75–76
Site-directed mutagenesis, 186, 209, 216F
Site-specific recombination, 114–117, 436–437
Cre–Lox system, 115, 209
efficiency, 115
FLP–FRT system, 115
*see also* DNA transposition; Transposable elements
Skeletal muscle, 359–360, 391–393
Skin cells, energy requirement, 316
Sliding filament model, muscle contraction, 395
Slow axonal transport, 399
Slugs, *Dictyostelium,* 348, 359
Smc1 protein, 412
Smc3 protein, 412
SmpB protein, 58, 58F, 58FF, 142, 142F
Snake venom, membrane lysis, 234
Snap25, 295
SNAREs, 294–295, 294F, 313–314
biosynthesis, 295
disassembly, NSF, 292, 295, 297, 313–314
docking control, 295
GST 'pull down' assay, 295
oligomerization, 295
t-SNAREs, 294–295
Snap25, 295
syntaxin, 295
v-SNAREs, 294–295
synaptobrevin, 295
yeast vacuolar fusion, 294–295, 297
alkaline phosphatase assay, 294
SNARF-1, 251, 251F
snoRNAs, 132–133
box elements, 132, 133
human U85, 132–133, 132F, 132FF, 133F, 133T
RNA modification role, 133
structure, 132F
SNPs (single-nucleotide polymorphisms), 215
snRNAs
cloning, 68–69
splicing role, 132
U1 snRNA, splicing role, 129
U2 snRNA, gene organization, 68–69, 69F
U5 snRNA, modification, 133
snRNPs

  - characterization, 129F
  - splicing role, 129
  - U1 snRNP, 129
  - U7 snRNP, 131
- Sodium, abundance, 13F
- Sodium channels
  - excitatory neurotransmission, 254
  - voltage-gated, 256–257
- Sodium ions
  - action potential role, 253, 256
  - equilibrium potential, 253
  - neuromuscular junction, 257
- Sodium nitrite, 323
- Sodium-potassium pump(s) (ATPase), 248
  - ATP hydrolysis, 249, 249F, 251, 251F
  - efficiency, 251
  - heart contraction, 249
  - oubain/digitalis effects, 249
- Sodium–proton ($Na^+$–$H^+$) antiporter, 248
- Solenoidal DNA supercoiling, 79–80
- Solutions, molecules in, 26, 28
- Somatic cells
  - cloning from, 151
  - DNA stability, 87
- Somites, 155
- SOS response, 106
  - damage-inducible genes, 104
  - LexA, 199–201
    - *see also* Repressor proteins
  - RecA, 105–106
    - *see also* RecA protein
  - *see also* DNA repair
- Southern blotting, 76, 94, 153F, 206, 209, 212
  - DNA rearrangements, 153
- Specific activity, enzymes, 193–194
- Spectrin
  - protein-associations, 241
  - quantitative analysis, 241
- Spectroscopy
  - absorption spectra, 334, 334F
  - action spectra, 334, 334F
  - cytochrome absorption bands, 326–327
- Spermatogenesis, nematode gene control, 187–188, 188F
- Sperm–egg interactions, calcium role, 256
- S phase, cell cycle control, 98, 410–413, 411F
- Sphingolipids
  - GLD (Krabbe's disease), 429
  - lipid rafts, 233
- Sphingomyelin, 233
- Sphingomyelinase, 233
- Spindle-attachment checkpoint, 415
- Spin-labeled lipids, 234, 237–238
- Splicing, 121–123
  - actin mRNA, 130
  - alternative *see* Alternative splicing
  - α-tropomyosin pre mRNA, 121–122
  - calcitonin/CGRP RNA, 185–186
  - frameshift mutations, 121–122
  - gene expression regulation, 185–186
  - intron removal, 121–122
  - intron scanning, 122–123
  - self-splicing, 124
  - snRNP role, 129
  - splice signals, 130
  - splice sites, 133
  - *trans*-splicing, 130
- Spontaneous generation, Pasteur's experiment, 2–3
- SPR (surface plasmon resonance), 202, 202F
- Spring-like protein(s), 46, 46F
- Squid giant axons, 255–256
  - synaptic vesicles, 295
- *Srb2* gene, 127–128
- Srb2 protein, transcription role, 127–128, 128F
- Src homology 2 domain (SH2), 346, 362, 364
- Src homology 3 domain (SH3), 346, 364
- Src oncogene, 56–57
- SRE (serum response element), 187
- Staining, negative, electron microscopy, 228
- Stare, F.J., 35
- Staurosporine, 484F
- Stem cells, embryonic cell lines, 191
- Steroid hormones, 345, 345F
  - ecdysone, 350
  - *see also* Cholesterol
- Stokes radius, 194
- Stomach, pH, 15
- Stop codons, 67, 129, 133
  - mitochondrial genome, 337
  - ORF analysis, 206
  - variations, 142
- Stress fibers, 394F
- String–surface interactions, 56
- Stroma, 277, 277F, 331
- Substrate-level phosphorylation, 33
- Succinylcholine, 344
- Sugar tree, 284
  - *see also* Glycoproteins; Polysaccharides
- Sulfhydryl groups, 15F, 185
- Sulfur, chemical properties, 13
- Supercoiling, 78–80, 121F
  - degrees of, 79
  - electrophoretic analysis, 79, 79F, 80, 95
  - mitotic chromosomes, 80
  - negative, 79, 79F
  - plectonemic DNA, 79–80, 79F
  - positive, 79, 79F, 121
  - RNA polymerases and, 121
  - solenoidal DNA, 79–80, 79F
- Suppressor genes, 404, 409
- Suppressor mutations, 136, 313
- Suprachiasmatic nucleus, 173
- Surface plasmon resonance (SPR), 194, 202
  - protein-ligand dissociation, 202
- Surface–string interactions, 56
- Surface–surface interactions, 56
- Surface-to-volume ratio
  - bacteria *versus* eucaryotic cells, 265
  - cell growth rate correlation, 3
  - microvillli, 250
  - sedimentation coefficient and, 193
  - water permeability, 254
- Survival (trophic) factors, 431
- SV40 virus
  - replication, 91–92, 92, 95
  - T antigen, 99, 99F
- SW1/SNF complex, 71–72, 72F
- Swinholide A
  - actin effects, 381–382
  - structure, 382F
- Symbioses, tube worms, 4
- Symporters, 247
- Synaphin, 295
- Synaptic vesicles, 254, 295
  - acetylcholine, 254
  - axonal transport, 394
  - exocytosis, 313
  - fusion, 295, 313–314
  - pre-synaptic location, 313
  - proteins, 295, 313
    - *see also* SNAREs
- Synaptobrevin, 295
- Synaptotagmin, 313
- Synonymous nucleotide substitutions, 8
- Syntaxin, 295
- Szent–Györgyi, A.V., 35

## T

- T4 bacteriophage
  - attachment, 447–448, 447F
  - DNA injection, 65, 447–448
  - DNA ligase, 204
  - Hershey–Chase experiment, 65
  - infection by, 447–448
  - SSB, 92, 92F
- T4 ligase, 204
- T7 phage, kinase-capsid fusion, 482, 482F
- Tail-to-tail dimers, 42–43
- Tamoxifen, 481
- Tandem repeats, 85
  - VNTRs, 85
- T antigen (SV40 virus), 99, 99F
- TATA-binding factor (TFIID), 167, 168–169
- TATA-binding protein association factors (TAFs), 169
- TATA box, 116
- Tat protein, transcriptional role, 182–183
- Taxol, effect on microtubule organization, 375–376
- TCA cycle *see* Citric acid cycle
- Telomerase, 67, 432
  - inhibitors, 67
  - tumor formation, 67
  - yeast, 103
- Telomeres, 70–71, 130F
  - analysis, 70–71, 70F
  - DNA repeats, 67
  - gene expression effects, 74–75
  - replicative cell senescence, 431
  - shortening, 103, 103F
    - related to cancer, 470
  - Sir protein complex, 75
  - *Tetrahymena*, 70–71
- Telophase, 412, 414, 415, 429
- Temperature-sensitive mutants, 37, 87, 247, 286–287
  - *Cdc* genes, 404–405, 410
  - cohesins, 12, 399–400
- Teratogens, thalidomide, 16–17
- Testosterone, structure, 345F
- *Tetrahymena*
  - cilia, 395–396, 396F
  - genetic code variation, 142
  - minichromosomes, 211
    - restriction analysis, 211
  - nuclei, 211
  - telomeres, 70–71, 70F
- Tetrapolar eggs, 426, 426F
- TFIIA transcription factor, 168
- TFIIB transcription factor, 167, 168
- TFIID (TATA-binding factor), 167
- TFIIE transcription factor, 167, 168
- Thalidomide, 16–17
  - structure, 16F
- Theophylline, cAMP phosphodiesterase inhibition, 358
- Thermal motions, 13
- Thermal-ratchet model of mitochondrial import, 275
- Thermodynamics
  - ATP hydrolysis, 28, 29, 250–251, 295
  - ATP synthesis, 319, 332
  - enthalpy, 25, 38
    - changes (delta H ($\Delta H$)), 25
  - entropy, 25, 38
    - changes (delta S ($\Delta S$)), 25
  - enzyme catalysis, 23–30
  - folded *versus* unfolded proteins, 39
  - Gibbs free energy, 25
    - changes *see* Delta G ($\Delta G$)
    - transduction, 389
  - membrane transport, 250–251
  - phosphoglucose isomerase, 28–29
  - polymerization reactions, 24–25
  - protein–protein interactions, 56
  - redox reactions, 301, 324–326, 332
  - second law, 24
- *Thiobacillus ferrooxidans*
  - ATP synthesis, 318
  - carbon fixation, 332
  - electron transport, 325–326, 332
- Thioredoxin, 41F
  - structure, 41
- Threading, 42
- Threonine, structure, 48, 48F
- Thrombin, 460
- Thylakoid(s)
  - lumen, 277, 277F, 331
  - membrane, 277, 277F, 302
  - vesicles, 336
- $^3$H-Thymidine autoradiography, 98–99, 98F, 375–376, 393–394
- Thymidine kinase (tk), promoter element, 116
- Thymine
  - base pairing, 63, 63F
  - UV-induced dimers, 91, 100
- Thyroxine, 359
- Tight junctions, 449–450

  - claudin, 451
  - electron micrographs, 453F
  - functions, 450
  - membrane proteins, 453–454
  - protein model *versus* lipid model, 453–454, 454F, 454FF
  - schematic, 451
- TIM complexes, 275, 277, 277F
- Timeless protein, 176–177
- *Tiny* yeast, 410
- Titin, 46, 46F, 132
  - skeletal muscle, 392
- Titration, 14, 20F
  - amino acids, 14, 14F, 14FF
  - equivalence point, 14, 14F, 14FF
  - ribonuclease unfolding, 45
- tmRNA, 142
- Tn10 transposons, 115–116, 116F
- Tobacco mosaic virus (TMV)
  - diffusion through plasmodesmata, 456, 457F, 457T
  - mRNA *in vitro* translation, 142
    - protein synthesis rate, 138–139, 138F
- TOM complex, 276
- Topoisomerases *see* DNA topoisomerases
- Topology, DNA *see* DNA topology
- Toxins
  - acrylamide, 376
  - cholera toxin, 358
  - pertussis toxin, 329, 357, 358
  - saxitoxin, 258
  - *see also* Poisons
- Transcription, 120–133
  - assays, 126F
  - bacterial, sigma factors, 115, 120
  - centromere interference, 421–422
  - coding relationships, 3
  - direction, 120
  - elongation, topological considerations, 121
  - energetics, 139
  - errors, 120
  - gene expression control, 151–182
    - chromatin role, 73–74, 76, 168–169
    - complex stability, 177–178
    - developmental regulation, 166, 170–171
    - DNA methylation role, 169–172, 177–178
    - DNA replication and, 177–178
    - nucleosome role, 71–72, 73, 75–76, 168
    - *see also* Gene regulatory proteins; *regulatory elements (below)*
  - initiation, 165
    - preinitiation complex, 127
    - promoters *see* Promoter elements
    - *see also* Transcriptional activators; Transcription factors
  - kinetochore block, 421–422
  - lampbrush chromosomes, 77–78
  - regulatory elements
    - AP1 site, 158–159
    - glucocorticoid response element, 168, 321–322
    - insulators, 163
    - iron-response element, 183
    - LCRs, 163
    - segmentation genes, 170–171
    - serum response element, 187
    - TATA box, 116
    - *see also* Gene regulatory proteins; Transcription factors
  - ribosomal RNA genes, 115, 120F, 125, 177–178
    - *Tetrahymena* minichromosome, 211
  - Srb2 protein role, 127–128
  - viral promoters, 167F
  - *in vitro see In vitro* transcription
  - *see also* Gene expression; Messenger RNA; RNA polymerase(s)
- Transcriptional activators
  - CAP *see* Catabolite activator protein (CAP)
  - coactivators, 169–170
  - Gal4 activator (yeast), 163
  - *see also* Repressor proteins; Transcription factors
- Transcription factors, 116–117
  - binding-site analysis, 159, 167
    - deletion analysis, 167
    - gel mobility shift assays, 159
  - complex formation, 168
  - general, 120
  - hybrid, 199–200
  - interferon-γ response, 365–366
  - modular nature, 199–200
  - purification, 126
  - TFIIA, 168
  - TFIIB, 167, 168
  - TFIID (TATA-binding factor), 167, 168
  - TFIIE, 167, 168
  - *in vitro* assays, 116–117
  - *see also* RNA polymerase(s)
- Transcription maps, human mitochondrial DNA, 339F
- Transcytosis, 312–313
- Transducin ($G_t$), 332, 358
- Transferrin, 308, 311–312
  - 3, 11–312
  - receptor
    - recycling, 308
    - regulation, 183
  - receptor-mediated endocytosis, 308, 311–312
- Transfer RNA, 58
  - aminoacylation, 136
    - accuracy, 136
    - editing, 140, 140F
    - energetics, 136–137
    - *see also* Aminoacyl-tRNA synthetases
  - anticodons, 136
    - loop structure, 140, 140F
    - wobble, 136
    - *see also* Codons; Genetic code
  - genes, 339–340
  - minimum number, 136
  - structure, 140F
  - suppressor genes, 422
  - suppressor mutations, 136
  - valine-specific, 48, 140
- Transforming growth factor-β (TGFβ), Vg1 mRNA, 184
- Transition state, 50
- Transition state analogs, 50
  - *see also* Enzyme inhibitors
- Translation, 134–149
  - assay, 189
  - broken mRNA, 137
  - coding relationships, 3
  - coupling to protein translocation, 285–286
  - elongation factors, 137
  - frameshifting, 139
  - gene expression control, 183
    - control elements, 180, 187
    - IRES effect, 189
    - nematode gametogenesis, 187–188
    - repression, 185
  - rate, 138–139
  - *in vitro see In vitro* translation
  - *see also* Genetic code; Messenger RNA; Ribosomes; Transfer RNA
- Translocations, chromosomal, 81
- Transmembrane kinase, 285
- Transmembrane transport *see* Protein translocation
- Transporter proteins
  - conformational change, 248F
  - mechanism of action, 244
  - transport kinetics, 244
- Transporters *see* Carrier proteins
- Transport proteins, 246–252
  - active *versus* passive transport, 243
  - ATP-driven, 247, 250–251
  - channel proteins *versus,* 254
  - energetics, 250–251
  - saturation, 253
- Transposable elements, 81, 81F, 108–110, 116F
  - gene disruption, 115
  - types, 81
- Transposition, 114–117, 116F
- Transposons
  - DNA, 81, 108
  - LTR retrotransposons, 81
- Tree of life, 4–6
  - possible relationships, 5
- Triacylglycerol
  - energy storage, 30
  - structure, 17, 17F
- Tricarboxylic acid cycle *see* Citric acid cycle
- Trigger factor (TF), protein folding, 143
- Triglyceride metabolism, muscle, 354
- Tri-*N*-acetylglucosamine (tri-NAG), 59–60
- Trinucleotide repeats, structural effects, 67
- Triosephosphate isomerase, 50, 50F
- Triskelions, clathrin coats, 290–291
- Tritium ($^3$H), radioactive decay, 12, 12T
- Triton X-100, 235–236
  - structure, 240F
- Trophic (survival) factors, 431
- Tropomyosin(s)
  - muscle contraction, 391, 395
  - sedimentation, 193
  - structure, 193F
  - α-tropomyosin, alternative splicing, 121–122
- Troponin C, muscle contraction, 391, 395
- Trypanosomes
  - mitochondrial mRNA editing, 183
  - peroxisomes/glycosomes, 280
  - UV mapping, 124F
  - VSG gene transcription, 123–124
- Trypsin, 308
  - cell isolation, 191
- Tryptophan
  - biosynthesis, 36, 152
  - biosynthetic mutants, 35–36, 36T
  - *trp* operon control, 162
- Tube worms, 4
- Tubulin
  - αβ-tubulin dimers, 375
    - cellular concentration, 384
    - lateral association, 383, 383F
    - orientation, 376, 389
  - binding proteins, 375
  - evolutionary conservation, 375
  - gene expression regulation, 186
  - GTP cap, 375, 376–377, 378
  - microtubule nucleation, 384–386
  - mitotic flux, 415
  - mutational effects, 186F
  - polymerization, 24–25, 374
    - heterotypic *versus* homotypic interactions, 374
    - *in vitro,* 384
  - purification scheme, 378
  - *see also* GTP-binding proteins (GTPases); Microtubules
- α-Tubulin, 374, 385–386
- β-Tubulin, 374, 375, 385–386
- γ-Tubulin, 384–386
- γ-Tubulin ring complex (γ-TuRC), 385–386
  - microtubule nucleation sites, 384–385
- Tumors
  - elimination by apoptosis, 440
  - formation, 479
    - *see also* Oncogenes; Tumor suppressor genes
  - liver, 432
  - progression, 470
  - telomerase role, 67
  - *see also* Cancer
- Tumor suppressor genes
  - *Apc* gene, 368–369
  - cancer association, 368–369
  - *Retinoblastoma* gene, 475–476
- Turnover number, enzyme kinetics, 55
- Twins, VNTR analysis, 85
- Ty elements, 116–117, 117F
- Tyrosine kinase receptors (trks), 364
- Tyrosine phosphorylation, 227F, 365–366
  - cell cycle control, 414


## U


- Ubiquitin-activating enzyme, 145
- Ubiquitination, 134, 139, 145
  - β-catenin, 369
  - enzymes, 145
  - *see also* Proteasomes
- Ubiquitin ligase, 145

Ultracentrifugation, 194
Ultraviolet (UV) mapping, 115
Ultraviolet (UV) radiation, 106F, 106T
absorption, DNA bands, 194
cytochrome-*c*-mediated apoptosis, 440
DNA damage, 91, 104–106, 105F
bacterial response, 104
*LacI* gene, 106
mutation frequency, 106
promoter mapping, 115
sensitive-mutants, 105–106, 106T
Uncoupling agents, 31
Unequal crossing-over, 69–70
Uniporters, 248
Uracil, 275
Uridine, reactive analog, 128F
UV *see* Ultraviolet (UV) radiation
*UvrA,B,C* genes, 105–106, 105F

## V

Valine, structure, 48, 48F
Valinomycin, 328–329, 329T
Valyl-tRNA synthetase, 48
van der Waals attractions, 13, 19
membrane protein interactions, 241
van der Waals contact distance, 19
van der Waals radius, 19
Variable domain (immunoglobulins), 152–153
Variable number tandem repeats (VNTRs), DNA fingerprinting, 85
Variable surface glycoprotein (VSG), transcription, 123–124
Variegation, 337, 337F, 404–405
Vasopressin, glycogen catabolism, 354
Velocity sedimentation, 193
Venom, snake, 234
Vernier protein assembly, 44, 44F
Verprolin homology domain, 400F
Vertebrates
phylogeny, 9, 9F
tight junctions, 449–450
Vesicles/vesicular transport, 264, 289–314
clathrin-coated *see* Clathrin-coated vesicles
coats, 290
cytokinesis and, 429
docking, 295
Rab effectors, 295
Rab GTPases, 295
ER to Golgi, 293, 297–298
fusion, 269, 286–287, 290, 292, 297
SNAREs *see* SNAREs
Golgi to cell surface, 284–287, 292
GTP requirement, 293
inside-out membranes, 240
lipid, 249
secretory *see* Secretory vesicles
synaptic *see* Synaptic vesicles
synthetic, 287–288
through Golgi, 274–277
thylakoid, 336
viral, 292
*see also* Vesicular stomatitis virus (VSV)
*see also* Endocytosis; Exocytosis; *specific organelles*
Vesicular stomatitis virus (VSV), G protein
exocytosis experiments, 312
membrane-spanning domain, 298–299
vesicular transport experiments, 298–299, 301
Viagra, 345
Video microscopy, cell cycle analysis, 386, 405–406, 414
Vimentin filaments, 379, 379F
mouse knockouts, 375
Viruses
bacterial *see* Bacteriophage
cell entry, 292
DNA, 121
integration, 212
mapping, 212
enveloped, 292
Influenza viruses, cell entry, 307
Maloney murine sarcoma virus, 348–349
mouse mammary tumor virus, 349–350
Rous sarcoma virus, 55–56
SV40 *see* SV40 virus
VSV *see* Vesicular stomatitis virus (VSV)
Visual transduction, 356
VNTRs, 85
Voltage-gated calcium channels, 256
Voltage-gated ion channels, 252–255
*see also specific channels*
Voltage-gated potassium channel(s)
ball-and-chain inactivation, 257–258, 258F
patch-clamp recording, 255, 257–258, 258
shaker channels, 257–258
Voltage-gated sodium channels
action potential role, 237, 256–257
measurement, 258–259
saxitoxin, 258
VP16 activation domain, 199–200
VSV *see* Vesicular stomatitis virus (VSV)

## W

Walker protein motif, 101
Water
cage formation, 212, 232, 232F
channels (aquaporins), 254, 466
chemical properties, 20
Wee1 tyrosine kinase, 414, 436, 436F
Weight loss, electron transport uncoupling, 324
Western blotting, 197–198
*see also* Immunoblots
Wnt signaling pathway, 368, 370F
Wobble base-pairing, 136, 136T

## X

X-chromosome inactivation, 180–181
*Xenopus laevis*
heterologous injection experiments, 77
oocytes, 404, 412
*see also* Microinjection experiments
Xeroderma pigmentosum (XP), 107, 107F
XIC (X-inactivation center), 181–182
X-inactivation, 180–181
X-inactivation center (XIC), 181–182
*Xist* RNA, 181–182, 182F
gene knockouts, 181
Xkid motor protein, 423–424, 423F, 424F
XP-variant polymerase, 107, 107F

## Y

Yeast
budding, 412F
bud-site selection genes, 399–400
cell polarization, 399–400
cytoskeleton role, 399–400
dumbell stage, 431, 433–434
Mad2 mutants, 416
patterns, 399–400
Sec4 GTPase, 296
vacuolar fusion, 294, 297
X-ray treatment, 433–434
cell cycle
Cdk1, 410
cyclins, 384–385, 410
DNA damage response, 431, 433–434
size control, 410
*see also* Cell cycle
cell cycle analysis, 404–406
*Cdc* mutants, 404–405, 410
execution point, 405–406
landmark morphology, 405
mitosis, 420–422
Scc1 mutants, 412
suppressor genes, 404
chromosomes, 74–76, 433–434
centromeres, 76, 421–423
telomeres, 103, 103F
citric acid cycle, 36
evolutionary relationships, 41
gene expression
*Ade2* gene position effect, 74–75, 75F
analysis, 75–76
Gal4 transcriptional activators, 163, 169–170
SAGA complex, 169–170
mating, 171F, 177F
α-factor pheromone signaling, 356–357
*see also* Mating-type
nuclear localization, 269
nuclear export signal, 272
petite *(Pet)* mutants, 341, 341F
plasmids, 70–71, 76, 420
radiation-sensitive mutants, 433–434
sectoring assays, 422–423, 423F
telomerase, 103
*see also Saccharomyces cerevisiae; Schizosaccharomyces pombe* (fission yeast)
Yeast two-hybrid system, 199–201
bait and prey, 199–200, 200T
hybrid transcription factors, 199–200
modification, 200–201
Yellow fluorescent protein (CFP), 226–227

## Z

Z disc (muscle), 392, 395
Zinc finger proteins, DNA-binding, 154
Z scheme, photosynthesis, 334
Zyg1 protein, 418
Zygotic (erotic) induction, 156

# Prefixes

| SYMBOL | NAME | VALUE | SYMBOL | NAME | VALUE |
|---|---|---|---|---|---|
| d- | deci- | $10^{-1}$ | da- | deca- | $10^{1}$ |
| c- | centi- | $10^{-2}$ | h- | hecto- | $10^{2}$ |
| m- | milli- | $10^{-3}$ | k- | kilo- | $10^{3}$ |
| µ- | micro- | $10^{-6}$ | M- | mega- | $10^{6}$ |
| n- | nano- | $10^{-9}$ | G- | giga- | $10^{9}$ |
| p- | pico- | $10^{-12}$ | T- | tera- | $10^{12}$ |
| f- | femto- | $10^{-15}$ | P- | peta- | $10^{15}$ |
| a- | atto- | $10^{-18}$ | E- | exa- | $10^{18}$ |
| z- | zepto- | $10^{-21}$ | Z- | zetta- | $10^{21}$ |
| y- | yocto- | $10^{-24}$ | Y- | yotta- | $10^{24}$ |

# Geometric Formulas

| FIGURE | AREA | SURFACE AREA | VOLUME |
|---|---|---|---|
| square | $l^2$ | | |
| circle | $\pi r^2$ | | |
| ellipse | $\pi r_1 r_2$ | | |
| cube | | $6\,l^2$ | $l^3$ |
| cylinder | | $2\,\pi r h + 2\,\pi r^2$ | $\pi r^2 h$ |
| sphere | | $4\,\pi r^2$ | $^4/_3\,\pi r^3$ |
| cone | | | $^1/_3\,\pi r^2 h$ |

# Radioactive Isotopes

| ISOTOPE | EMISSION | HALF-LIFE | COUNTING EFFICIENCY[a] | MAXIMUM SPECIFIC ACTIVITY[b] |
|---|---|---|---|---|
| $^{14}C$ | beta | 5730 years | 96% | 0.062 Ci/mmol |
| $^{3}H$ | beta | 12.3 years | 65% | 29 Ci/mmol |
| $^{35}S$ | beta | 87.4 days | 97% | 1490 Ci/mmol |
| $^{125}I$ | gamma, Auger, and conversion electrons | 60.3 days | 78% | 2400 Ci/mmol |
| $^{32}P$ | beta | 14.3 days | 100% | 9120 Ci/mmol |
| $^{131}I$ | beta and gamma | 8.04 days | 100% | 16,100 Ci/mmol |

[a]Maximum efficiency for an unquenched sample in a liquid scintillation counter. Most real samples are quenched to some extent.
[b]This value assumes one atom of radioisotope per molecule. If there are two radioactive atoms per molecule, the specific activity will be twice as great, and so on.